The Geology of North America
Volume J

The Gulf of Mexico Basin

Edited by

Amos Salvador
Department of Geological Sciences
The University of Texas at Austin
P.O. Box 7909
Austin, Texas 78713-7909

1991

Acknowledgment

Publication of this volume, one of the synthesis volumes of *The Decade of North American Geology Project* series, has been made possible by members and friends of the Geological Society of America, corporations, and government agencies through contributions to the Decade of North American Geology fund of the Geological Society of America Foundation.

Following is a list of individuals, corporations, and government agencies giving and/or pledging more than $50,000 in support of the DNAG Project:

Amoco Production Company
ARCO Exploration Company
Chevron Corporation
Cities Service Oil and Gas Company
Diamond Shamrock Exploration
  Corporation
Exxon Production Research Company
Getty Oil Company
Gulf Oil Exploration and Production
  Company
Paul V. Hoovler
Kennecott Minerals Company
Kerr McGee Corporation
Marathon Oil Company
Maxus Energy Corporation
McMoRan Oil and Gas Company
Mobil Oil Corporation
Occidental Petroleum Corporation

Pennzoil Exploration and
  Production Company
Phillips Petroleum Company
Shell Oil Company
Caswell Silver
Standard Oil Production Company
Oryx Energy Company (formerly
  Sun Exploration and Production
  Company)
Superior Oil Company
Tenneco Oil Company
Texaco, Inc.
Union Oil Company of California
Union Pacific Corporation and
  its operating companies:
    Union Pacific Resources Company
    Union Pacific Railroad Company
    Upland Industries Corporation
U.S. Department of Energy

Published by The Geological Society of America, Inc.
3300 Penrose Place, P.O. Box 9140, Boulder, Colorado 80301

Printed in U.S.A.

Library of Congress Cataloging-in-Publication Data
The Gulf of Mexico Basin / edited by Amos Salvador.
    p.   cm.
    Includes bibliographical references and index.
    ISBN 0-8137-5216-7
    1. Geology—Mexico, Gulf of.   2. Marine mineral resources—Mexico, Gulf of.   I. Salvador, Amos.
    QE71.G48   1986   vol. J
    [QE350.22.M48]
    557 s—dc20
    [551.46'08'09364]   91-36132
                                            CIP

Front Cover: Part of a relief map of North America showing the Gulf of Mexico region, produced by Johann Baptist Homann (1664–1724), who ranked next in greatness to Mercator among German cartographers. The map is from an atlas published in the early 1700s in Nuremberg, Germany. Reproduced courtesy of the Nettie Lee Benson Latin American Collection, University of Texas at Austin.

10   9   8   7   6   5   4   3   2   1

Contents

iii

Plates
(in accompanying slipcase)

Preface

The Geology of North America series has been prepared to mark the Centennial of The Geological Society of America. It represents the cooperative efforts of more than 1,000 individuals from academia, state and federal agencies of many countries, and industry to prepare syntheses that are as current and authoritative as possible about the geology of the North American continent and adjacent oceanic regions.

This series is part of the Decade of North American Geology (DNAG) Project, which also includes seven wall maps at a scale of 1:5,000,000 that summarize the geology, tectonics, magnetic and gravity anomaly patterns, regional stress fields, thermal aspects, and seismicity of North America and its surroundings. Together, the synthesis volumes and maps are the first coordinated effort to integrate all available knowledge about the geology and geophysics of a crustal plate on a regional scale.

The products of the DNAG Project present the state of knowledge of the geology and geophysics of North America through the 1980s, and they point the way toward work to be done in the decades ahead.

A. R. Palmer
General Editor for the volumes
published by The Geological
Society of America

J. O. Wheeler
General Editor for the volumes
published by the Geological
Survey of Canada

Foreword

The goal of *The Geology of North America* series of volumes has been to provide a comprehensive summary or synthesis of the current geological and geophysical knowledge of the regions or subjects covered by each of the volumes. The volumes are intended not so much for those most familiar with the region or subject, but for a broader audience of earth scientists—geologists and geophysicists from anywhere in the world who want to inform themselves about the geological and geophysical configuration of the North American Plate and its surrounding regions.

This volume, covering the Gulf of Mexico basin, was planned with these goals in mind. Assuming that a good number of the readers may not be familiar with the geography of the region, much less with the multitude of structural and stratigraphic terms in common use, efforts were made, with varying degrees of success, to keep geographic and geologic terminology as simple as possible. Keeping in mind, for instance, that earth scientists unacquainted with the stratigraphy of the Gulf of Mexico basin may be confused when faced with the countless number of group, formation, and member terms used in the area, the stratigraphy of the basin was described whenever possible in terms of internationally recognized chronostratigraphic terms. As further assistance to the unfamiliar reader in this respect, those chapters of this volume that describe the stratigraphy of the basin contain simplified stratigraphic correlation charts that show the relation of the local lithostratigraphic units discussed in the volume to the global chronostratigraphic scale. A more detailed stratigraphic correlation chart, Plate 5, is also included in the volume. For the same reasons, an effort was also made to show in the figures and plates of the volume all geographic and structural terms used in the text.

Another important goal of this volume, of course, was for it to be useful to the hundreds of geologists and geophysicists working in the Gulf of Mexico basin, in both the United States and Mexico. Even though the space limitations have not permitted discussion of all aspects of the geology of the basin in as much detail as the local geoscientists may have wanted, I am hopeful that by having available in one place a much-needed summary of a wide range of geological and geophysical information on the basin, as well as a review of current thinking about its interpretation, they may find this volume a valuable reference. For this purpose, emphasis was placed during the preparation of the volume on including an extensive list of references on the subject of each chapter, particularly recent, significant, and up-to-date references. This will allow readers interested in particular subjects to enlarge their knowledge and satisfy their need for additional information. The lists of references, in fact, may be one of the most valuable contributions of the volume.

I also see this volume as more than just a valuable reference and a useful summary or synthesis of what is now known about the geology of the Gulf of Mexico basin. I want to think that it will become the foundation, the point of departure for future geological and geophysical studies on the basin. Some such studies have been suggested in the last chapter of the volume, but many others need to be undertaken in the years to come. The field is certainly wide open for more studies, and I am hopeful that the publication of this volume will provide the encouragement for undertaking them.

Future studies no longer need to be based on old, obsolete information or on only partial or limited kinds of information. New and better interpretations and concepts about the geologic history of the basin can now be based on the broader and essentially complete, up-to-date geological and geophysical information that has either been compiled in the chapters of this volume or is contained in the numerous references listed in each of them. This volume should make it possible to readily find the sources for the bulk of the geological and geophysical information on the Gulf of Mexico basin.

For this volume to serve as a foundation for future geological and geophysical studies, a concerted effort was made to separate, during its preparation, facts from interpretations. Most chapters are primarily descriptive of the physiography, stratigraphy, and present structure of the basin and of its natural resources. Interpretation of this information is mainly addressed in the chapter that discusses the origin and development of the basin.

To attain uniformity and consistency was another of my aims in editing this volume. Nearly all maps, those of the large plates as well as those in the text figures, are on a common projection—Transverse Mercator Projection, with 100° west as the central meridian—the projection to be used in all DNAG (Decade of North American Geology) maps. Good uniformity was also achieved in the terminology of structural features and stratigraphic units. This volume could be considered to have made an important contribution if this terminology will become the accepted standard terminology for the Gulf of Mexico basin where it is not uncommon for structural features and stratigraphic units to have a confusing number of different names. More limited success was achieved concerning uniformity in the dating of some stratigraphic units, particularly those for which the evidence of age is poor or controversial. The preferences of the authors were respected in these cases.

No major conflicts were experienced in coordinating the information and ideas contained in the various chapters of the volume, a remarkable and surprising achievement considering the well-known independence of thought of earth scientists. The volume deals with a single basin, a large basin to be sure, but a simple one in both stratigraphy and present structure; and while theories about its origin varied considerably at one time, there has been in the last few years a trend toward closer agreement on the subject. As more and better information has become available, most of the early differences of opinion seem to have been settled. I am not sure, however, if this is good or bad.

The preparation of this volume was not expected to answer all questions or solve all problems concerning the geology of the Gulf of Mexico basin. It did not. I am hopeful, however, that it may serve, as expected, to clarify some problems and answer some questions; support some assumptions, hypotheses, and concepts; and contradict or weaken others.

Many unsolved questions remain. Questions concerning the orogenic belts rimming the Gulf of Mexico basin—the Sierra Madre Oriental in Mexico, the Ouachita orogenic belt, the Ouachita Mountains and the Central Mississippi deformed belt in the United States—as well as those about the pre-Mesozoic geologic history of the region that would become the Gulf of Mexico basin, for instance, are far from answered. The orogenic belts, fortunately, are discussed in other volumes of *The Geology of North America* series; the pre-Mesozoic history is treated briefly in two chapters of this volume, but the limited amount of information about Precambrian and Paleozoic rocks in the Gulf of Mexico basin region allowed only broad generalizations and poorly supported speculations. Debate is very much alive on these subjects. As discussed in the last chapter of the volume, many others require considerable further study.

The completion of a volume like this would not have been possible without the contributions of many individuals and organizations. All, unfortunately, cannot be properly acknowledged here. Special recognition is due to the authors of the chapters who devoted so much of their valuable time to make this volume a success. With my thanks to them also go my apologies to those who submitted their contributions promptly, as the delay in completing the volume may make their chapters somewhat outdated. All chapters were read by at least two knowledgeable reviewers. I am most grateful to them for their prompt, valuable, and constructive comments and suggestions; they contributed substantially to the improvement of the volume.

Outstanding support was received from academic institutions, government agencies, and industrial organizations. To all of them I express my deep gratitude. The Bureau of Economic Geology of the University of Texas deserves special mention—it contributed authors and information to the preparation of several chapters and provided the drafting of Plates 2 to 6, a major undertaking carried out by Margaret Koenig and John Ames under the able direction of Richard Dillon. Petroleos Mexicanos and the Instituto Mexicano del Petroleo kindly furnished valuable information for the compilation of Plates 2 to 5 and made available a number of their earth scientists for the preparation of several chapters. Arco Oil and Gas Company, Chevron U.S.A., Conoco Inc., Exxon Company U.S.A., Standard Oil Production Company, and Texaco U.S.A. helped defray the cost of the drafting for several chapters of the volume. Richard Platt, Patrice Porter, and Maria Saenz did the drafting.

Several individuals also deserve exceptional recognition: William A. Thomas, one of the editors of neighboring Volume F-2 (The Appalachian-Ouachita Orogen in the United States), made available copies of manuscripts and maps of his volume in advance of their publication and helped in the preparation of this volume in many other ways; Thomas M. Scott, of the Florida Geological Survey, provided much-needed information and advice about the geology of a part of the Gulf of Mexico basin with which I was very poorly acquainted. The patience and good humor of "Pete" Palmer will always be remembered gratefully; without his enthusiastic encouragement and support, the task of completing this volume would have been much more difficult. It was a real pleasure working with him. Last, but far from least, I want to thank Betty Kurtz, who typed and retyped innumerable manuscripts and letters, shared with me the ups and downs of the preparation of the volume, and cheerfully saw me through the long ordeal.

Amos Salvador
Volume editor
August 1991

The Geology of North America
Vol. J, The Gulf of Mexico Basin
The Geological Society of America, 1991

Chapter 1

Introduction

Amos Salvador
Department of Geological Sciences, The University of Texas at Austin, P.O. Box 7909, Austin, Texas 78713-7909

THE GULF OF MEXICO—DISCOVERY AND EARLY EXPLORATION

By the year 1517, 25 years after Christopher Columbus discovered the New World, most of the Atlantic coasts of both North and South America had been sighted and reasonably well surveyed. Most of the islands of the Caribbean had been colonized by the Spaniards, and northern South America and the Caribbean coast of Central America, from Panama to Honduras, had been explored and mapped. Four years earlier, in 1513, Vasco Nuñez de Balboa had reached the Pacific Ocean by crossing the isthmus of Panama, but the search for a sea route to the Pacific and to the fabulous kingdoms of the Orient, the original objective of Columbus' trips, had not yet met with success. The great Gulf of Mexico—the "Sinus Mexicanus" of many old maps—also remained unknown, and there was considerable confusion concerning which of the discovered lands were islands and which were parts of the mainland.

Juan Ponce de Leon, in his search for the legendary fountain that would restore youth to old men, had explored in 1513 the east coast of the Florida Peninsula as far north as the present location of St. Augustine, but had only ventured a short distance north along the west coast. He was convinced that he had discovered an immense island. Diego Miruelo, in 1516, explored the east coast of Florida and seems to also have sailed some distance along the west coast of the peninsula. But neither Ponce de Leon nor Miruelo realized they had sailed into the entrance to a vast gulf or inland sea around whose shores advanced civilizations had flourished for many centuries and which would become a major political and economic center of the New World.

The Gulf of Mexico was discovered in 1517. A three-ship expedition headed by Francisco Hernández de Córdoba on its way from Cuba to the island of Guanaja, near the northern coast of Honduras, was blown off course by a violent storm and after sailing for three weeks reached land in the northeastern part of the Yucatan Peninsula, near what is now called Cabo Catoche. There the Spaniards were astonished by the sight of a sizable town with masonry houses and temples, by the advanced cultivation of the land, and by the fine garments and gold ornaments worn by the inhabitants—nothing that had previously been seen in the New World had indicated the existence of such a superior culture! They were met with fierce hostility by the Indians at Cabo Catoche and by those of other settlements as they sailed along the north shore of Yucatan. Hernández de Córdoba followed the coast as far west as the present town of Champotón, in the Bay of Campeche, still uncertain as to whether Yucatan was part of an island or not. The hostility of the Indians, who killed or wounded all but one of the members of the expedition, and the consequent difficulty in putting ashore to obtain fresh water, forced a badly wounded Hernández de Córdoba to return to Cuba by way of Florida where the Indians proved no less dangerous than those of Yucatan. He died soon after his return, but brought back samples of gold ornaments and news of an advanced and rich civilization, and of a previously unknown large body of water that could provide the long-sought sea route to the Pacific.

It did not take long before another expedition sailed from Cuba to confirm the discoveries of Hernández de Córdoba. It was commanded by Juan de Grijalva and left in four vessels in the spring of 1518. Grijalva first sighted land at the island of Cozumel and from there he coasted the Yucatan Peninsula, then followed the shore of east-central Mexico as far north as a large headland, probably Cabo Rojo, just north of the present town of Túxpan, where strong currents and unfavorable winds made it difficult to continue.

Juan de Grijalva experienced the same hostility that Hernández de Córdoba had endured before him, but he was better prepared to handle it and succeeded in having some amicable communication with the natives. Before reaching the farthest point in his voyage, he sent Pedro de Alvarado back to Cuba in one of his caravels, with the gold objects obtained by bartering with the Indians and the news of his discoveries. Grijalva had confirmed the existence of an advanced, powerful and wealthy society. He had found, too, that Yucatan was probably not an island but part of the mainland—a spacious country, as evidenced by the eternal snows in the high peaks he had seen from afar. A passage to the Pacific remained undiscovered.

Salvador, A., 1991, Introduction, *in* Salvador, A., The Gulf of Mexico Basin: Boulder, Colorado, Geological Society of America, The Geology of North America, v. J.

On arriving back in Cuba, Grijalva found out that a new expedition to the newly discovered country was already being fitted out—a much larger and more powerful expedition. It was to be commanded by Hernán Cortés at the head of 553 heavily armed soldiers, 100 seamen, 200 Indians, and 16 horses, in 11 ships. Cortés and his followers sailed from Cuba in February 1519.

The flotilla under the command of Hernán Cortés was not the only expedition to sail into the Gulf of Mexico to explore the newly discovered territories. Impressed by the reports of Hernández de Córdoba and Grijalva, Francisco de Garay, the governor of Jamaica, decided to send his own expedition in order to claim and settle the new land as well as to search for the elusive passage to the Pacific and the Orient. For this task he enlisted the services of Alonso Alvarez de Pineda.

In the spring of 1519, just a few months after the departure of Cortés from Cuba, Alvarez de Pineda sailed from Jamaica with four ships, 107 soldiers, horses and supplies. During a voyage that lasted 9 months, he entered the Gulf of Mexico, sailed north to somewhere along the coast of the Florida panhandle and from there followed the shores of the Gulf of Mexico to the site of the present Veracruz, Mexico. During this trip he mapped every bay and river mouth, including that of the Mississippi River, which he named Río del Espiritu Santo, 22 years before Hernando de Soto was to cross the big river 800 km upstream.

Alvarez de Pineda arrived in Veracruz on August 1, 1519, to find that Hernán Cortés was already there and had claimed the land. Distrustful of Cortés, Alvarez de Pineda did not land and started his return trip to Jamaica a few days later. The expedition went back the same way it came, continuing to explore and map the northwestern and northern coasts of the Gulf of Mexico. According to Bernal Díaz del Castillo, the remarkable chronicler who participated in the voyages of Hernández de Córdoba and Grijalva, and who accompanied Cortés in his campaigns, Alvarez de Pineda died on his return trip of wounds received at the Pánuco River while fighting the Indians. The Alvarez de Pineda expedition proved that the eastern, northern, and western coasts of the Gulf of Mexico were continuous. There was no passage to the Pacific, and the Gulf was indeed a very large inland sea with only one opening: to the south toward the Caribbean Sea.

After Alvarez de Pineda, many navigators sailed, explored and mapped the Gulf of Mexico and its coasts. In 1520, Diego de Camargo and Miguel Díaz de Auz made separate and unsuccessful attempts to settle somewhere near the mouth of the Rio Grande. Juan Ponce de Leon returned to Florida in 1521 to try, also unsuccessfully, to establish a colony. He was wounded in an Indian attack and died soon after in Cuba. In 1523, Francisco de Garay, the governor of Jamaica, personally headed an expedition to establish a settlement in eastern Mexico but ended up joining Hernán Cortés.

In 1528, Pánfilo de Narváez landed near modern Tampa to take possession of the land. He struck inland with 300 men, but became disillusioned at finding no wealth. Suffering from exposure, hunger, and fierce Indian attacks, he returned to the coast with his force greatly reduced in number. Not finding their ships, they built new ones and sailed for Mexico, but a storm destroyed the ships, and only four men, including Alvar Núñez Cabeza de Vaca, reached shore. For eight years the four survivors wandered among the Indians of the Gulf coast, from Florida to northern Mexico, although their exact route has long been debated. They eventually found their way to Spanish settlements in Mexico. The careful journal of Núñez Cabeza de Vaca, published in Spain in 1542 under the title *La Relación,* includes some of the earliest and more detailed descriptions of the regions surrounding the Gulf of Mexico.

The tales of the adventures of Núñez Cabeza de Vaca probably inspired the expedition of Hernando de Soto. De Soto left Cuba in 1539 with nine ships, 500 men, and 200 horses. They sailed along the west coast of Florida, reached the area of present Tallahassee, and travelled from there through Georgia, South Carolina, North Carolina, Alabama, Mississippi, northern Arkansas and Louisiana. He first crossed the Mississippi River below Memphis, Tennessee, and died near Ferriday, Louisiana on April 21, 1542. The rest of the expedition, under Luis de Moscoso, floated down the Mississippi into the Gulf of Mexico and followed the coast westward until they reached Tampico on September 10, 1543.

For still a few more years, other Spanish sailors and explorers travelled through the eastern and northern regions of the Gulf of Mexico and surrounding lands, but not finding the riches in search of which they had come to the New World, they made no effort to settle in the region. Their interest was increasingly directed to the wealthier regions of Mexico and the west coast of North America.

It was not until more than 100 years later that the lands bordering the northern coast of the Gulf of Mexico became again the objective of European explorers, this time reaching the region from the north. In 1682, Rene Robert Cavalier, Sieur de la Salle, sailed down the Mississippi River from the French possessions in Canada. He claimed all the land drained by the river and its tributaries for Louis XIV of France and named it Louisiana. He was followed a few years later by Pierre de Moyne, Sieur d'Iberville, who took up the task of colonizing Louisiana in the name of the French crown.

The early navigators, explorers, and conquistadores who searched the Gulf of Mexico and settled the lands surrounding it during the 16th and 17th centuries were driven by the lofty ideals of claiming the country for their sovereigns and spreading the faith, or by the less altruistic desire to seek adventure and riches. More than 200 years later a new breed of explorers would venture in these parts of the New World—the geologists from the universities, government agencies, and the petroleum industry. They, too, took it upon themselves to discover unknown territories and new wealth, a task no less challenging and difficult than that of the earlier conquistadores.

GEOGRAPHIC LIMITS AND GEOLOGIC FRAMEWORK OF THE GULF OF MEXICO BASIN

The Gulf of Mexico basin, as defined for the purpose of this volume, is a relatively simple, roughly circular structural basin approximately 1,500 km in diameter, filled in its deeper part with 10 to 15 km of sedimentary rocks that range in age from Late Triassic to Holocene. As shown in Figure 1, the northern part of the basin is in the U.S., its southern part in Mexico. The central part of the basin is occupied by the Gulf of Mexico, which covers an area of more than 1,500,000 km^2, and attains water depths of 3,750 m (12,303 ft) in its abyssal plain (Plate 1). The floor of the Gulf of Mexico rises steeply to the east and south along the Florida and Campeche escarpments, which form the gulfward limits of the shallow underwater parts of the Florida and Yucatan platforms. Elsewhere, the floor of the Gulf rises more gently and smoothly toward the coast through generally well-defined rise,

slope, and continental shelf physiographic provinces. The Gulf of Mexico is surrounded to the north and west by a low coastal plain ranging in width from less than 50 km in east-central Mexico to more than 550 km in the central part of the U.S. Gulf Coastal Plain, in the states of Louisiana, Mississippi, and Arkansas. The physiographic and bathymetric provinces of the Gulf of Mexico basin will be described in detail in Chapter 2 of this volume.

The limits of the Gulf of Mexico basin correspond, for the most part, with structural features (Fig. 2). The Florida carbonate platform to the east and the Yucatan carbonate platform to the south are considered to be the very gentle eastern and southern flanks, respectively, of the basin and are included within the limits of the area described in this volume. The western limit of the basin has been placed roughly at the foot of the Chiapas massif and the Sierra Madre Oriental of Mexico, and along the eastern edge of the Coahuila platform. To the north, the structural limits

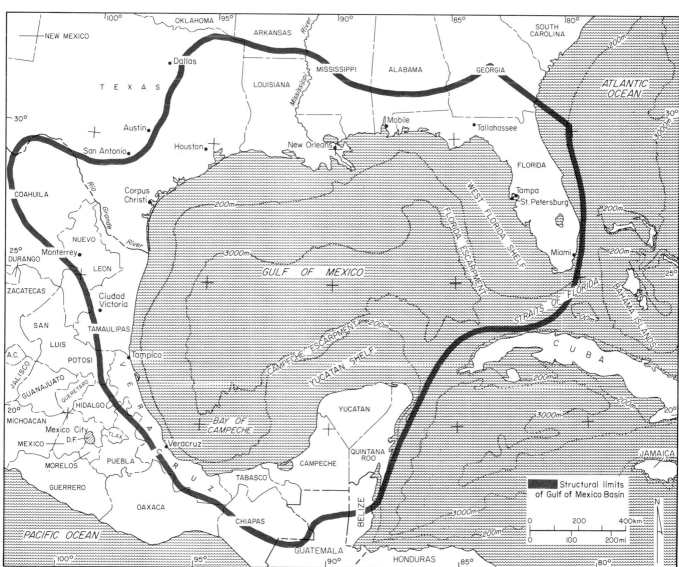

Figure 1. Location of most important geographic features.

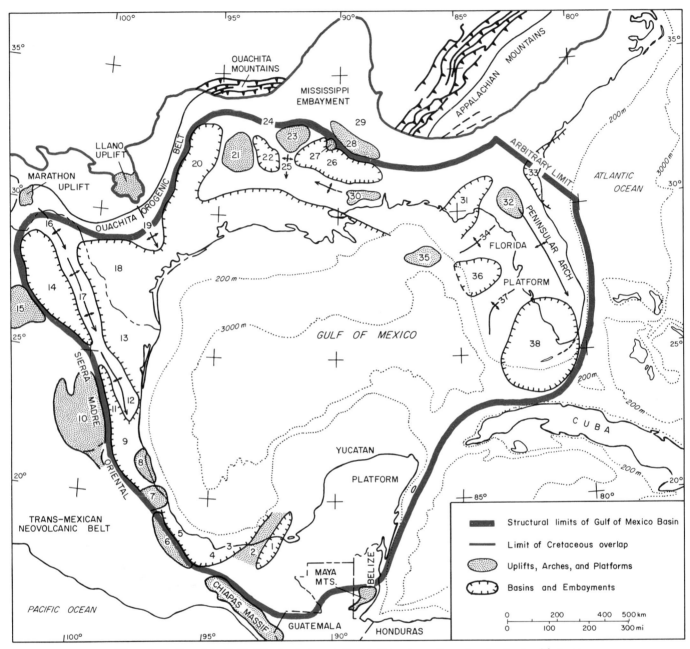

Figure 2. Outline of the Gulf of Mexico basin as defined in this volume. Second-order structural features within the basin: 1, Macuspana basin; 2, Villahermosa uplift; 3, Comalcalco basin; 4, Isthmus Saline basin; 5, Veracruz basin; 6, Córdoba platform; 7, Santa Ana massif; 8, Tuxpan platform; 9, Tampico-Misantla basin; 10, Valles–San Luis Potosí platform; 11, Magiscatzin basin; 12, Tamaulipas arch; 13, Burgos basin; 14, Sabinas basin; 15, Coahuila platform; 16, El Burro uplift; 17, Peyotes-Picachos arches; 18, Rio Grande embayment; 19, San Marcos arch; 20, East Texas basin; 21, Sabine uplift; 22, North Louisiana salt basin; 23, Monroe uplift; 24, Desha basin; 25, La Salle arch; 26, Mississippi salt basin; 27, Jackson dome; 28, Central Mississippi deformed belt; 29, Black Warrior basin; 30, Wiggins uplift; 31, Apalachicola embayment; 32, Ocala uplift; 33, Southeast Georgia embayment; 34, Middle Ground arch; 35, Southern platform; 36, Tampa embayment; 37, Sarasota arch; 38, South Florida basin.

of the basin correspond, from west to east, with the basinward flanks of the Marathon uplift, the Ouachita orogenic belt, the Ouachita Mountains, the Central Mississippi deformed belt, and the southern reaches of the Appalachian Mountains. From the Appalachians to the shores of the Atlantic Ocean, the limit of the basin is arbitrary because no distinct structural feature clearly separates it from the Atlantic Coastal Plain.

This volume includes the description of the Cretaceous and Tertiary rocks to their stratigraphic pinchout beyond the structural limits of the Gulf of Mexico basin along its northwestern and northern flanks. The volumes in this series that cover the adjoining areas treat the pre-Cretaceous section below the Cretaceous/Tertiary overlap (Sloss, 1988; Hatcher and others, 1989).

The broad outline of a simple, roughly circular structural basin is locally modified by a number of significant, second-order structural elements—highs, arches, uplifts, basins, and embayments. Some of the most prominent among these second-order structural elements are shown in Figure 2 and Plate 2. They are described in Chapter 3 of this volume.

The Upper Triassic to Holocene sedimentary section filling the Gulf of Mexico basin overlies unconformably a complex and poorly known pre-Triassic "basement." Precambrian and Paleozoic rocks are known from only a few outcrop areas around the periphery of the basin and from modest penetration by a limited number of wells in the shallow parts of the basin flanks.

The "basement" in the central, deepest part of the basin has been estimated to be at a depth of about 12 to 16 km below sea level (Plate 3). Geophysical surveys indicate that this deep central part of the basin is underlain by an oceanic-type crust, whereas most of the U.S. Gulf Coastal Plain, the Florida and Yucatan platforms, and the remainder of Mexico are underlain by continental or thick "transitional" crust. "Transitional" crust, inferred to represent "rifted" or "stretched" continental crust, occurs also between the continental and oceanic crust under the northern and southern parts of the Gulf of Mexico. Along the western and eastern limits of the deeper central part of the Gulf, the boundary between oceanic and continental crust is represented by a narrower belt of "transitional" crust. The continental crust and the "transitional" crust consist of a variety of igneous, metamorphic, and older sedimentary rocks. Two cross sections, one east-west and one north-south, depict the generalized distribution and nature of the crust under the Gulf of Mexico basin (Fig. 3). A complete discussion of the type and distribution of the crust is included in Chapter 4 of this volume.

As in the case of the Precambrian and Paleozoic, information concerning the Triassic, the Lower Jurassic, and most of the Middle Jurassic rocks of the Gulf of Mexico basin is limited. The earliest stratigraphic interval of which more is known is the uppermost Middle Jurassic, represented over most of the Gulf of Mexico basin by extensive salt deposits. The plastic behavior of this salt interval under the load of the overlying sediments has resulted in the formation of many diapiric salt domes, pillows, and ridges throughout the basin, some of which are still active.

Overlying the salt, or the oceanic "basement" in the central part of the basin, is a very thick sequence of sediments, deposited for the most part in marine environments. This thick sequence of sediments filled the basin as it subsided persistently from Late Jurassic time to the present. The thickest sections are found along the north flank of the basin where up to 15 km of sediments, mainly Tertiary clastics, have prograded into the basin (see Fig. 3 and Plate 3).

The Upper Jurassic, Cretaceous, and Cenozoic sedimentary section of the Gulf of Mexico basin was deposited under remarkably stable tectonic conditions; the persistent subsidence of the basin was modified only by local deformation of the Jurassic salt and by growth-faulting bordering major depocenters. As a result of this tectonic stability, the environments of deposition, the main avenues of terrigenous clastic influx, and consequently, the lithologic composition of the sedimentary sequence persisted, with minor modifications, from the Late Jurassic to the present. In a broad sense, three distinctive "lithofacies provinces," clearly represented in the accompanying Stratigraphic Correlation Chart (Plate 5), can be recognized in the sedimentary sequence of the Gulf of Mexico basin, particularly in the uppermost Jurassic (Tithonian), Cretaceous, and Tertiary part of the section:

1. To the east and southeast, in the area of the Florida and Yucatan platforms, the sedimentary section is predominantly composed of carbonates and lesser amounts of evaporites.

2. To the southwest and west, in the area of the Isthmus of Tehuantepec and in east-central Mexico, carbonates and fine-grained terrigenous clastics predominate. Coarser-grained clastics are much less common in the Jurassic, Cretaceous, and lower part of the Tertiary section. They are, however, the main component of the Miocene and younger sequences, particularly on the Comalcalco and Macuspana basins of southern Mexico.

3. To the northwest and north, from northeastern Mexico to Alabama, coarse-grained terrigenous clastics are much more prevalent, indicating the presence in this area of important streams draining the continental interior and depositing their load along the northwestern and northern shores of the Gulf of Mexico of the time.

HISTORY OF GEOLOGICAL AND GEOPHYSICAL STUDIES

Many references to the mines, minerals, volcanoes, earthquakes, thermal springs, and prolific oil seeps of Mexico can be found in the writings of the Spanish chroniclers and historians of the 16th, 17th, and 18th centuries.

Alexander von Humboldt, the famous German naturalist, visited Mexico for a year, from March 1803 to March 1804, and described many features of the geography and geology of the central part of the country, in particular its mines and volcanoes.

During most of the remaining first half of the 19th century, geological studies of the Gulf of Mexico Basin and surrounding areas, in both the U.S. and Mexico, were restricted to broad reconnaissance trips by individual European and American geologists and paleontologists. Some of the most distinguished

6
A. Salvador

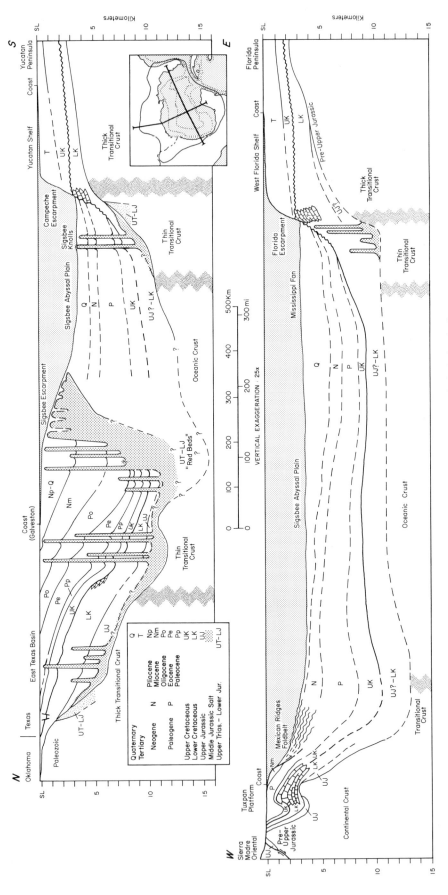

Figure 3. Generalized cross sections, Gulf of Mexico basin.

among these geological pioneers are Ferdinand Römer and Timothy A. Conrad, who established the foundations of the stratigraphy of Texas and Alabama, respectively. Sir Charles Lyell, on his second trip to the United States in 1845, sailed up the Mississippi river from New Orleans to Memphis, Tennessee. The German geologists Johannes Felix and H. Lenk, in their *Beitrage zur Geologie und Paleontologie der Republik Mexico* (1890–1899), compiled the results of much of the early reconnaissance work in Mexico, in addition to original work on the regions they themselves had visited.

Toward the middle part of the 19th century, geological investigations became progressively more systematic. Government geological surveys were established in most of the states surrounding the Gulf of Mexico. Alabama instituted its Geological Survey in 1848 and by 1856 Michael Tuomey, the State Geologist, had produced two geological maps of the state. Mississippi followed in 1850, and Arkansas in 1857. The Mississippi Geological Survey issued several comprehensive reports between 1857 and 1860; the most remarkable was an 1860 report by Eugene W. Hilgard summarizing the geology of the state. Hilgard's important contributions to the geology of the U.S. Gulf coastal area continued until 1912. He laid the foundation on which subsequent geological work would securely rest.

The first official Geological Survey of Texas was established in 1858 under Benjamin F. Shumard, but survived only three years. A second survey, with Samuel B. Buckley as State Geologist, was active between 1867 and 1875. In 1888, the Texas legislature authorized the establishment of the Geological and Mineralogical Survey of Texas, the third official survey, often called the "Dumble Survey" after its able director, E. T. Dumble. But it, too, was disbanded in 1894. Meanwhile, Louisiana had instituted its Geological Survey in 1869. In Georgia and Florida, short-lived geological organizations operated between 1874 and 1879, and between 1886 and 1887, respectively.

In Mexico, the Comisión Geológica de México was created in 1888 and in 1891 developed into the Instituto Geológico de México. The Georgia Geological Survey was reorganized in 1889; the Texas Mineral Survey, as part of the University of Texas, was established in 1901 and changed its name to Bureau of Economic Geology in 1909. Florida established its Geological Survey in 1907.

By the early part of this century, therefore, well-established geological surveys were functioning in all states surrounding the Gulf of Mexico and in Mexico, and the many able geologists and paleontologists serving in these organizations, in addition to those of the U.S. Geological Survey, had adequately described in a general way the geology of the rock sequences cropping out around the periphery of the Gulf of Mexico Basin. Deserving special mention among those working in Mexico are José G. Aguilera, Emil Böse (who also worked for the Texas Geological Survey), C. Burckhardt, Antonio del Castillo, W. H. Dalton, Ezequiel Ordóñez, and Karl Sapper; in Texas, E. T. Dumble, Robert T. Hill, William Kennedy, R.A.F. Penrose, Jr., and B. F. Shumard; in Arkansas, Louisiana, and Mississippi, John C.

Branner, Gilbert D. Harris, E. W. Hilgard, W. J. McGee, and Arthur C. Veatch; in Alabama, Eugene A. Smith and Michael Tuomey.

But even if the efforts of the geologists and paleontologists of the state and national geological surveys had succeeded in putting together a reasonably good picture of the geology along the periphery of the Gulf of Mexico Basin, the flanks and central part of the basin were entirely unknown. It would be up to the petroleum industry to fill this enormous gap and to make by far the most important contribution to the understanding of the geological and geophysical framework of the Gulf of Mexico Basin. The study of the surface geology had reached the limits of its applicability, and the geology of the basin as a whole could be correctly interpreted only with the help of adequate subsurface information.

The vast contribution of the petroleum industry to the understanding of the geology of the Gulf of Mexico basin has been the result of the work of thousands of hard-working and dedicated geologists, paleontologists, and geophysicists, some of them widely recognized and honored, but most of them anonymous and unacknowledged. All of them—recognized and unrecognized, American, Mexican, or from many other countries—deserve great credit for having accumulated a prodigious amount of invaluable information, and for having contributed to the development of the advanced understanding of the Gulf of Mexico basin that we have today.

The first record of a concerted effort to search for oil in the Gulf of Mexico basin dates back to the mid-1860s, when drilling was undertaken in east Texas in the neighborhood of oil seeps. The first discovery was made in 1867 near Nacogdoches, but in spite of active drilling, which soon followed, no important production was developed in the area for many years.

The large and numerous oil seeps in the Gulf Coastal Plain of Mexico (locally called "chapopoteras"), had been known to the Indians of the region for many centuries; they were recorded in Aztec and Mayan histories and reported by the earliest Spanish explorers. The oil seeps first attracted the attention of the oil finders in the 1860s. The first local oil company was formed in 1896, and drilling started soon after. But, as in east Texas, the first results were disappointing, and no major discoveries were made for the remaining years of the 19th century.

With the arrival of the new century, exploratory activities were intensified in both the U.S. and Mexico. Spectacular results were not long in coming. On January 10, 1901, a well at Spindletop, near Beaumont, Texas, blew in from a depth of 347 m (1,139 ft), ejecting some 6 tons of drill pipe 100 m in the air and flowing at an estimated rate of 60,000 to 100,000 barrels of oil per day, more than all the other oil wells in the United States combined. The well was located on a small mount rising above the flat coastal plain and had been drilled on the strength of a few small gas seeps. The productive interval was the limestone caprock of a salt dome. The search for similar features spread rapidly over the coastal plain area of Texas and Louisiana. Several new fields were discovered in 1901, and production in Texas rapidly increased from 2,300 barrels of oil per day in 1900 to 50,000 in 1902. During the first decade of the 20th century, petroleum exploration along the U.S. Gulf Coastal Plain was done primarily by locating salt domes indicated by surface features, principally mounts, oil and gas seeps, salt water springs, and accumulations of "paraffin dirt."

In Mexico, meantime, the birth of the petroleum industry was no less spectacular. After the disappointing results of the early ventures during the last three decades of the 19th century, which succeeded in obtaining only minor production of oil, the first significant oil discovery in Mexico was at last made in 1904: La Pez No. 1, the discovery well of the Ebano Field, about 50 km west of Tampico, came in producing 1,500 barrels of oil per day.

Other, more dramatic, discoveries followed in the next few years. In 1908, San Diego de la Mar No. 2 was completed at the rate of 2,500 barrels of oil per day as the discovery well of the Dos Bocas Field. Later that year, San Diego de la Mar No. 3 blew out and immediately caught fire; production was variously estimated at between 15,000 and 80,000 barrels of oil per day. Two years later, in 1910, Potrero del Llano No. 4, the discovery well of the Pánuco Field, 40 km southwest of Tampico, blew out and flowed wild for two months. After it was brought under control, it was gauged at 100,000 to 110,000 barrels of oil per day. The first field of the prodigious Golden Lane trend was discovered in 1916 when Cerro Azul No. 4 was completed at a rate of 260,000 barrels of oil per day, the most prolific well ever known up to that time! The petroleum industry took good notice of the remarkable Mexican discoveries.

Until about 1920, petroleum exploration around the Gulf of Mexico, in both the U.S. and Mexico, continued to be guided mainly by surface geologic or topographic features, by the occurrence of oil and gas seeps and, to a lesser extent, by subsurface geological studies. In an effort to improve their success, several oil companies created geological departments between 1913 and 1918, and searched for better approaches to petroleum exploration. One such new and better approach was the use of micropaleontology to zone and map the monotonous Tertiary sections of the Gulf of Mexico basin, the first application of this new branch of paleontology to practical stratigraphic work. A course on micropaleontology was initiated at the University of Texas in 1915, and it did not take long for the oil industry to realize the great utility of the procedure in subsurface mapping. The first micropaleontology laboratories were established in 1920, and by 1925 most of the oil companies operating in the Gulf Coastal Plain of the U.S. and Mexico were doing micropaleontological work. Some of the pioneers in applying this field to stratigraphic work in the Gulf of Mexico Basin area were Alva Ellisor, Hedwig Kniker, and Esther Richards (later Mrs. Paul L. Applin) in the U.S., and W. S. Adkins, R. W. Barker, Joseph A. Cushman, W.L.F. Nuttall, and Hans Thalmann in Mexico. Ellisor, Kniker, and Richards (Applin) were among the first women to make their mark in geological work.

The increasing use of well-trained geologists, the recent addition of micropaleontology as a new and reliable tool in subsurface geological work, and improved sampling and coring techniques contributed greatly to the efforts of the petroleum industry to discover new oil fields. By the early 1920s, however, the oil finders found themselves at the limit of their capabilities in attempting to map by geological methods favorable prospective petroleum traps in the featureless coastal plains surrounding the Gulf of Mexico. A new exploration approach, a new tool, was obviously needed. Geophysical prospecting came to the rescue—first, the torsion balance and refraction seismograph; then the reflection seismograph, the principal geophysical exploration tool in the area today.

The torsion balance, an instrument invented around 1880 by Baron Roland von Eötvös, a Hungarian physicist, for the purpose of measuring distortions in the gravitational field, had been used with some success in 1917 to map a salt dome near Hannover, Germany. Oil explorers in the Gulf coast area, having nearly exhausted their capabilities of locating salt domes on the basis of surface indicators, decided to try the new geophysical method. A torsion balance was brought from Hungary in 1922 and tested successfully at the known Spindletop and Pierce Junction salt domes. In the next 7 or 8 years, torsion balances were instrumental in discovering numerous salt domes in the U.S. Gulf Coastal Plain. The Nash dome in Brazoria County, Texas, was mapped in 1924, the first prospective petroleum trap ever identified by geophysical methods. The initial productive well was drilled on the structure in 1926.

The torsion balance was ideally suited to search for salt diapirs but was not effective in locating other types of structures. It is not surprising, therefore, that it was of little use in the Tampico-Misantla basin of east-central Mexico, where salt structures are not present.

Almost simultaneously with the first use of the torsion balance for locating salt domes, a second geophysical method—the refraction seismograph—was first applied in petroleum prospecting in the Gulf of Mexico coastal plains. Experimental refraction shots had been conducted across salt domes in Germany in 1920 and 1921, and the first German refraction crews started to operate in the U.S. and Mexico in 1923. The first discovery of a salt dome with this technique, the Orchard salt dome in Fort Bend County, Texas, was made in 1924. From that time on, the refraction seismic method shared with the torsion balance the credit for the discovery of a great many salt structures on which oil production was found in succeeding years. "Fan shooting," the shooting of refraction fans to detect time-of-arrival differences that indicate the presence of higher-velocity salt domes, was developed and adopted in late 1925 or early 1926 and proved to be particularly successful.

By 1929 or 1930, however, the torsion balance and refraction seismograph, suited primarily for locating shallow or intermediate-depth salt domes, had reached the point of diminishing returns; most of these structures had already been located over the coastal plains of the Gulf of Mexico. A new geophysical method was needed—the stage was set for seismic reflection prospecting.

Unlike the torsion balance and refraction seismograph that were developed in Europe, the seismic reflection method was conceived and perfected in the U.S. Experiments were conducted as early as 1913, but the first tests of its application to prospecting for oil and gas were not carried out until 1921 and 1922, and satisfactory results were not obtained until several years later, in 1926 or 1927, when seismic reflection crews were assigned to the Gulf Coastal Plains of Louisiana and Texas.

The 1920s were the heyday of the torsion balance and refraction seismograph, but after those years they were rapidly replaced by the reflection seismograph, which has become the preeminent, if not the sole, geophysical prospecting method in the search for oil and gas accumulations. In the first five years after the introduction of the reflection seismograph, oil discoveries in the U.S. part of the Gulf of Mexico Coastal Plain area doubled all previous discoveries.

Since the early 1930s, but particularly after World War II, reflection seismic surveys have not only been extremely successful in petroleum exploration but have also contributed immeasurably to the knowledge of the geology of the Gulf of Mexico basin.

To the invaluable information provided by the reflection seismograph, the petroleum industry has added in the last 50 years important subsurface stratigraphic and structural data from hundreds of thousands of wells drilled in search of oil and gas around the perimeter of the Gulf of Mexico basin. The value of these subsurface data was greatly enhanced when wire-line well logging was introduced in 1933. This new information proved to be an unequaled register of stratigraphic details in the monotonous Tertiary sand-shale section, and it provided the means to establish stratigraphic correlations previously possible only by means of microfossils. Acquisition of important new subsurface information was also made possible by the increasingly greater depths to which wells were able to penetrate, the result of extraordinary advances in drilling technology.

The progressive combination of more and better subsurface geological and geophysical data made possible increasingly better interpretations of the geologic and geophysical framework of the basin, and of its geologic history.

Until the mid 1930s, however, geological knowledge of the Gulf of Mexico basin was restricted to the area of its coastal plains. Little was known about the part of the basin below the waters of the Gulf of Mexico. Offshore exploration did not start until 1938, when a well was spudded in 4 m (14 ft) of water, a little more than a kilometer offshore from the coast of southwestern Louisiana, a few kilometers east of the town of Cameron. It was completed the next year as an oil producer. The first offshore seismic survey was conducted in 1944, and the first offshore field out of sight of land was discovered in 1947, in 5.5 m (18 ft) of water and about 20 km from the coast of Terrebone Parish, Louisiana. Since then, exploratory drilling and reflection seismic surveys have progressed to increasingly deeper waters. Exploratory wells have been drilled in water as deep as 2,292 m (7,520 ft), a world record for water-depth petroleum exploration drilling, and oil and gas are being produced in fields discovered in water

depths of as much as 683 m (2,243 ft), also a world record. Petroleum companies now hold leases in water almost 3,000 m deep.

Geological and geophysical information about the central, deeper part of the basin did not become available until the last 20 or 25 years, mostly through the results of the Deep Sea Drilling Project and of regional seismic reflection and refraction surveys conducted by the U.S. Geological Survey and academic institutions, particularly by the Institute for Geophysics of the University of Texas at Austin (see Chapter 13, this volume).

These geophysical surveys have shed much light on the seismic nature of the crust underlying the Gulf of Mexico basin and have contributed, for the first time, information concerning the distribution, thickness, and structural configuration of the extensive salt deposits and of the thick sedimentary section that fills the central part of the basin.

This review of the progress of geological and geophysical studies in the Gulf of Mexico basin may be summarized by recognizing five distinct, though not necessarily consecutive, stages:

1. During the first half of the 19th century, geological work was conducted mostly by individual scientists engaged in broad reconnaissance journeys through the region.

2. The second half of the 19th century witnessed the establishment of geological surveys in the states bordering the Gulf of Mexico in the U.S. as well as the creation of a national geological organization in Mexico. The efforts of the geologists engaged by these geological surveys were limited to the field observation of rock outcrops around the periphery of the basin.

3. The successful discovery of oil under the coastal plains surrounding the Gulf of Mexico during the early years of the 20th century marked the birth of a strong and aggressive petroleum industry in the region, and the beginning of the accumulation of invaluable and previously inaccessible subsurface geological information. The addition of this new information was a major step toward better understanding of the geology of the Gulf of Mexico coastal plains.

4. The beginning of the next stage was marked by the introduction of geophysical methods: first the torsion balance and refraction seismograph in the early 1920s, then the reflection seismograph in the late 1920s. The blending of geophysical information with increasing amounts of subsurface geological information from the hundreds of thousands of wells drilled in search of oil and gas by the petroleum industry, onshore and offshore, has made possible over the last 50 or 55 years a progressively better and more detailed interpretation of the local and regional aspects of the stratigraphy, structure, and geologic history of the Gulf of Mexico basin.

5. The geophysical surveys of the central, deeper part of the Gulf of Mexico basin mark the most recent stage in the study of the basin. The reflection and refraction surveys, conducted for the most part by the U.S. Geological Survey and various academic institutions, have allowed not only a more complete understanding of the sedimentary section of the basin, but also a better

knowledge of the nature of the crust that underlies it. With this important addition, the entire area of the Gulf of Mexico basin has now been studied to a greater or lesser degree.

The study of the geologic constitution of the Gulf of Mexico basin has thus progressed from the periphery to the center, from the investigation of the rock outcrops around its margins to the interpretation of large amounts of geological and geophysical subsurface information, first from onshore areas, then from the shallow-water shelves, and finally from the central part of the basin. Geological knowledge has grown little by little, step by step, a most stimulating process accomplished through the untiring, cooperative efforts of geologists and geophysicists from the petroleum industry, government agencies, and academic institutions.

Geological and geophysical investigations in the Gulf of Mexico basin can claim several "firsts," most notable the first use of micropaleontology in stratigraphic work, the first application of the study of recent sediments to the understanding of older sedimentary sequences, and the first use of geophysics in prospecting for oil and gas—the torsion balance and the seismograph. Even though these geophysical techniques had been tried and had demonstrated their capabilities elsewhere, they achieved their first impressive success in the coastal plains surrounding the Gulf of Mexico.

OBJECTIVE, SCOPE, AND ORGANIZATION OF THIS VOLUME

The large amount of geological and geophysical information on the Gulf of Mexico basin is now scattered in hundreds, if not thousands, of publications, some readily accessible but many in technical and professional journals of limited distribution. Nowhere is there available in a single publication an integrated synthesis and interpretation of this large amount of data. It is the objective of this volume, therefore, to provide such a synthesis of the geological and geophysical information now available on the Gulf of Mexico basin—its physiography, structural framework, and sedimentary and magmatic record—and to interpret, on the basis of this information, the origin and development of the basin.

The volume has been planned as a systematic, cohesive, and we hope, comprehensive description of the geology of the Gulf of Mexico basin, not a collection of independent papers brought together because they deal in general with a common area. It is divided into 18 chapters.

Chapter 1 is a general introduction. It covers the discovery and early exploration of the Gulf of Mexico, the history of geological and geophysical studies in the Gulf of Mexico basin, a general description of the geography and the geologic framework of the basin, the goals and organization of the volume, and the geographic and geological nomenclature. It concludes with a brief summary of the geologic history of the Gulf of Mexico basin.

The physiographic and bathymetric provinces are discussed in Chapter 2, the present structural framework of the basin in Chapter 3, and the nature of the crust underlying the basin in

Chapter 4. Chapter 5 reviews the most important structural processes that have persisted during most of the development of the Gulf of Mexico basin and have contributed to its present configuration, and Chapter 6 summarizes the Mesozoic and Cenozoic igneous activity.

Following are seven chapters (Chapters 7 through 13) describing the stratigraphy of the rock sequences of the Gulf of Mexico basin, from the "basement" to the recent sediments: Chapter 7 covers pre-Triassic rocks—the nature, age, and occurrence of the Precambrian and the Paleozoic rocks; it is followed by chapters on the Triassic-Jurassic (Chapter 8), Lower Cretaceous (Chapter 9), Upper Cretaceous (Chapter 10), and Cenozoic (Chapter 11). Chapter 12 discusses the Quaternary sediments and some of the sedimentary processes responsible for their formation. These same or similar sedimentary processes, it is believed, have been active during most of the history of the basin and have contributed to some of its most impressive sedimentary, physiographic, and bathymetric features (deltas, carbonate platforms, etc.), past as well as present. The last of the "stratigraphic" chapters (Chapter 13), discusses the stratigraphy of the central part of the basin based on seismic information, the only information available on the area.

A unifying interpretation of the geologic history of the Gulf of Mexico basin, its origin and subsequent development, is attempted in Chapter 14.

The next three chapters cover the natural resources of the basin: oil and gas resources (Chapter 15), mineral and geothermal resources (Chapter 16), and the all-important water resources (Chapter 17).

The volume concludes with a summary chapter (Chapter 18) that discusses some of the major controversies and most significant unanswered questions concerning the geologic and geophysical makeup of the Gulf of Mexico basin, and the possible directions and approaches (surveys, studies) for settling controversies and answering questions.

Most of the chapters of the volume are not meant to be complete in themselves; they all complement each other to provide a general synthesis of the present geological and geophysical knowledge of the Gulf of Mexico basin.

Six plates are in the separate pocket:

Plate 1—Bathymetric map of the Gulf of Mexico
Plate 2—Map of the principal structural features
Plate 3—Structural map at the base and subcrop map below the Mesozoic marine section (top of the "basement" and types of "basement")
Plate 4—Map of natural resources
Plate 5—Stratigraphic correlation chart
Plate 6—Cross sections

The synthesis and interpretation of the geological and geophysical information now available on the Gulf of Mexico basin should serve to clarify some problems; answer many questions; support some assumptions, hypotheses and ideas; and contradict or weaken others. But it should not be expected to answer all questions or clarify all problems. It will represent not only a much-needed summary of what we now know about the basin but also a statement of what we do not know. It is hoped that it will also suggest what programs of geological and geophysical research may be needed to answer some of the most important questions concerning the origin and development of the Gulf of Mexico basin and to stimulate the initiation of such programs.

In addition, it is hoped that a better description and understanding of the origin and development of the Gulf of Mexico basin, for which there is extensive geological and geophysical information, will serve as an example that may be used to better interpret the history of other basins and other areas with similar sedimentary and tectonic characteristics but for which much less information is now available.

GEOGRAPHICAL AND GEOLOGICAL TERMINOLOGY

A serious problem inherent to a regional geological synthesis such as the one attempted in this volume is the choice of a properly balanced geographical and geological nomenclature. Too many names, geographic, structural, and stratigraphic, inevitably lead to hopeless confusion; too few names impair the proper description of the geology and the discussion of concepts and ideas. To attain a balance is not always easy.

An equally serious problem has been the need to locate in the figures illustrating the chapters of this volume all geographic and structural features mentioned in the text, as well as to show in correlation charts or columnar sections all stratigraphic terms used. Nothing is more frustating than to read about geographic localities or structural features that cannot be located in the maps accompanying a text, or about stratigraphic terms whose whereabouts in space and time cannot be properly ascertained.

A concerted effort to avoid these pitfalls has been made in the preparation of this volume. Several obstacles, however, have been persistently encountered; the main ones are the complexity of the structural and stratigraphic terminology and the small size of most of the maps.

Names of structural features

An effort has been made to standardize the terminology of the structural features within the Gulf of Mexico basin—uplifts, arches, basins, embayments, faults, etc. This was not easy and the results are not expected to please all. Certain of these structural features are known by several names—Arthur (1988), for example, mentions that what has been called in this volume the Apalachicola embayment has received 14 different names in the literature on the area. The choice of the terms used in the text and figures of this volume was made on the basis of what was per-

ceived to be the most common current use. All structural features recognized for the purposes of this volume are shown in Plate 2, which is located in the pocket.

Stratigraphic terminology

The problem with the stratigraphic terminology of the Gulf of Mexico basin is that it has unfortunately been overburdened with many hundreds of group, formation, and member names, many of them unnecessary, poorly defined, vague, or obsolete. Even in the case of properly defined and valid terms, it is common to find that different terminologies have been applied in different areas to essentially equivalent and lithologically similar sequences. In addition, there are a good number of local chronostratigraphic units of very limited use, and innumerable biostratigraphic units of variable utility.

To simplify this complex stratigraphic terminology, facilitate regional correlations and the synthesis of the stratigraphy of the basin, and to aid those not familiar with the profusion of local stratigraphic terms, the number of such local terms has been reduced to a minimum, and the stratigraphic sequences of the area have been described, whenever possible, in terms of internationally recognized chronostratigraphic units (systems, series, and stages). A simplified stratigraphic correlation chart showing the stratigraphic terms most commonly used in the Gulf of Mexico basin, Plate 5, is included in the pocket of this volume. In it, the dominant lithology of each of the units is indicated by distinctive colors. The numerical ages shown in this chart and used throughout the volume are those of Palmer (1983).

The problems of stratigraphic nomenclature are particularly acute when dealing with the extremely thick Tertiary sandstone-siltstone-shale sections of the U.S. Gulf coastal area and those of southern Mexico, in which a distinctive and useful lithostratigraphic subdivision is all but impossible. Group, formation, or member names in these sequences have little significance and even less utility. The section can best be zoned on the basis of its contained fossils, predominantly foraminifers, into biostratigraphic units, and that is the procedure generally followed in the region. In southern Mexico, some of these biostratigraphic units have been given geographic names as if they were lithostratigraphic units. In the attached stratigraphic correlation chart, these spurious units are shown in quotation marks.

Also peculiar to the stratigraphic nomenclatural procedures in the Gulf of Mexico basin is the extension of lithostratigraphic unit-terms on the basis of correlations based on wire-line logs (electrical, radioactive, sonic) with almost complete disregard for the lithologic composition of the stratigraphic interval in question. Some surface lithostratigraphic units, for example, have been carried into the subsurface in this manner even when the lithology has radically changed.

These and other problems of stratigraphic nomenclature will be discussed in more detail in the chapters dealing with the stratigraphic section, particularly Chapters 9, 10, and 11.

Names of geographic features

To avoid undesirable clutter in the maps, all but the most indispensable geographic features and their names have been excluded. To aid those not familiar with the geography and culture of the area, a map is included with this introductory chapter (Fig. 2) showing the location of the most important geographic features (states, main cities, etc.). Other maps will not include most of these features, and Figure 2 should be referred to when searching for their location.

SUMMARY OF GEOLOGIC HISTORY: ORIGIN, AND DEVELOPMENT OF THE GULF OF MEXICO BASIN

Little is known about the geologic history of the Gulf of Mexico Basin before Late Triassic time. Since pre-Triassic rocks are known from only a few widely separated outcrop areas and wells, much of the geologic history of the basin during Paleozoic time needs to be inferred from the study of neighboring areas. Some authors have postulated the presence of a basin in the area during most of Paleozoic time, but present evidence seems to indicate that Paleozoic rocks do not underlie most of the Gulf of Mexico basin and that the area was, at the end of Paleozoic time, part of the large supercontinent of Pangea, the result of the collision of several continental plates. The present Gulf of Mexico basin, in any case, is believed to have had its origin in Late Triassic time as the result of rifting within the North American Plate at the time it began to crack and drift away from the African and South American plates. Rifting probably continued through Early and Middle Jurassic time with the formation of "stretched" or "transitional" continental crust throughout the central part of the basin. Intermittent advance of the sea into the continental area from the west during late Middle Jurassic time resulted in the formation of the extensive salt deposits known today in the Gulf of Mexico basin. It appears that the main drifting episode, during which the Yucatan block moved southward and separated from the North American Plate and true oceanic crust formed in the central part of the basin, took place during the early Late Jurassic, after the formation of the salt deposits.

Since Late Jurassic time, the basin has been a stable geologic province characterized by the persistent subsidence of its central part, probably due at first to thermal cooling and later to sediment loading as the basin filled with thick prograding clastic wedges along its northwestern and northern margins, particularly during the Cenozoic. To the east, the stable Florida platform was not covered by the sea until the latest Jurassic or the beginning of

Cretaceous time. The Yucatan platform was emergent until the mid-Cretaceous. After both platforms were submerged, the formation of carbonates and evaporites has characterized the geologic history of these two stable areas. Most of the basin was rimmed during the Early Cretaceous by carbonate platforms, and its western flank was involved during the latest Cretaceous and early Tertiary in a compressive deformation episode, the Laramide Orogeny, which created the Sierra Madre Oriental of eastern Mexico.

REFERENCES CITED

The notes on the discovery and early exploration of the Gulf of Mexico are based mainly on the entrancing chronicles of Bernal Díaz del Castillo, *Historia Verdadera de la Conquista de la Nueva España (1632),* and of William H. Prescott, *History of the Conquest of Mexico* (1843). The circumnavigation of the Gulf of Mexico by Alonso Alvarez de Pineda has been described by C. P. Garcia in *Captain Alonso Alvarez de Pineda and the exploration of the Texas coast and the Gulf of Mexico* (1982, Jenkins Publishing Co., 64 p.).

There is extensive literature on the history of geological and geophysical studies on the Gulf of Mexico basin. Most of the information on the subject used in this chapter was obtained from the following references:

Eckhardt, E. A., 1940, A brief history of the gravity method in prospecting for oil: Geophysics, v. 5, p. 231–242.

Ferguson, W. K., 1969, Geology and politics in Frontier Texas, 1845–1909: Austin, University of Texas Press, 233 p.

—— , 1981, History of the Bureau of Economic Geology, 1909–1960: Bureau of Economic Geology, 329 p.

Owen, E. W., 1975, Trek of the oil finders; A history of exploration for petroleum: American Association of Petroleum Geologists Memoir 6, 1647 p. (by far the most comprehensive and encyclopedic publication on its subject).

Smith, E. A., 1914, Pioneers in Gulf Coastal Plain geology: Geological Society of America Bulletin, v. 25, p. 157–178.

Sweet, G. E., 1966, The history of geophysical prospecting: Science Press, 326 p.

Other references cited in this chapter are:

Arthur, J. D., 1988, An overview of Florida basement geology: Florida Geologic Survey Report of Investigations 97, p. 33–39.

Hatcher, R. D., Jr., Thomas, W. A., and Viele, G. W., eds., 1989, The Appalachian-Ouachita Orogen in the United States: Boulder, Colorado, Geological Society of America, The Geology of North America, v. F-2, 748 p.

Palmer, A. R., 1983, The Decade of North American Geology 1983 Geologic Time Scale: Geology, v. 11, p. 503–504.

Sloss, L. L., ed., 1988, Sedimentary Cover—North American Craton; U.S.: Boulder, Colorado, Geological Society of America, The Geology of North America, v. D-2, 506 p.

MANUSCRIPT ACCEPTED BY THE SOCIETY, NOVEMBER 14, 1989

The Geology of North America
Vol. J, The Gulf of Mexico Basin
The Geological Society of America, 1991

Chapter 2

Physiography and bathymetry

William R. Bryant
Department of Oceanography, Texas A&M University, College Station, Texas 77843
José Lugo and Carlos Córdova
Instituto de Geografía, Universidad Nacional Autónoma de México, Apartado Postal 20-850, 01000 Mexico, D.F., Mexico
Amos Salvador
Department of Geological Sciences, The University of Texas at Austin, P.O. Box 7909, Austin, Texas 78713-7909

INTRODUCTION

The topographic relief and bathymetry of the Gulf of Mexico basin area reflect quite closely the geologic structure of the basin (Fig. 1 and Plates 1 and 3). Parts of the structural rims along the northern, northwestern, and western flanks of the basin are marked by mountain ranges and highlands: the southern plunge of the Appalachians and the Ouachita Mountains to the north, the Edwards Plateau and the low ridges of the Marathon area to the northwest, and the Sierra Madre Oriental to the west. From the foothills of these highlands, the coastal plains slope toward the Gulf of Mexico, a small ocean basin that occupies the central and deeper part of the basin. To the north and northwest, the coastal plains and the continental shelf of the Gulf of Mexico are widest and have a gentler slope toward the center of the Gulf, corresponding to the gentle slope of the "basement" in the region. To the west, in eastern Mexico, the coastal plain and the shelf are much narrower and steeper, as is the "basement" surface. To the southeast and east, the floor of the Gulf of Mexico, which in its deepest part reaches depths of a little more than 3,700 m, rises steeply along the Campeche and Florida submarine escarpments to the flat Yucatan and Florida carbonate platforms, under which the "basement" is similarly flat and featureless. Much of these two platforms lies submerged below the waters of the Gulf of Mexico at depths of less than 200 m; the rest is above sea level and forms the low Yucatan and Florida Peninsulas, which rise to only under 100 m above sea level. A less prominent submarine escarpment, the Sigsbee Escarpment, bounds to the north the deep central plain of the Gulf of Mexico.

The physiography of the Gulf of Mexico basin area is also strongly influenced by the appreciable changes in sea level that took place during the Quaternary as a result of the alternating glacial and interglacial episodes that affected the North American continent. Even though the ice sheets did not reach the region, the drastic changes in climate, and the sea-level changes caused by the recurring glaciation events, as well as the periodic influx of meltwater, controlled to a great extent the drainage systems of the region, the morphology of the coastal-plain alluvial systems, and the sediment volumes supplied to the sedimentary basin. During times of low sea level, large areas that are presently submerged were dry land. The effects of this subaerial exposure and erosion, however, were different in different areas depending on the structure, the lithology of the sediments exposed, and the climate.

The structural Gulf of Mexico basin, as defined in this volume, includes an area of 2.7 million km^2 of which 1.5 million km^2 is covered by the Gulf of Mexico (see Chapter 1 of this volume).

ONSHORE PHYSIOGRAPHY

Two sharply contrasting physiographic provinces can be distinguished in the onshore part of the Gulf of Mexico basin. In addition to reflecting the overall shape of the basin, they are indicative of the distinctive lithology of the rocks that underlie them: to the east and southeast, the Florida and Yucatan Peninsulas are underlain by essentially flat carbonates; to the north, northwest and west, terrigenous clastic sediments dipping gently toward the center of the basin are the predominant rocks under the coastal plains surrounding those parts of the Gulf of Mexico.

The Florida and Yucatan Peninsulas

The Florida and Yucatan Peninsulas, to the east and south, respectively, of the Gulf of Mexico are the emergent parts of the two large carbonate platforms of the same name. They are both low, generally less than 100 m above sea level, and show little relief, reflecting the flat attitude of the predominantly carbonate Cretaceous and Cenozoic section that underlies them. The Cretaceous-Cenozoic section attains thicknesses of 3,000 to 4,000 m in the Yucatan Peninsula and as much as 7,000 m in southernmost Florida; it overlies a "basement" of variable age and composition (see Plate 3 and Chapter 7 of this volume).

Bryant, W. R., Lugo, J., Córdova, C., and Salvador, A., 1991, Physiography and bathymetry, *in* Salvador, A., ed., The Gulf of Mexico Basin: Boulder, Colorado, Geological Society of America, The Geology of North America, v. J.

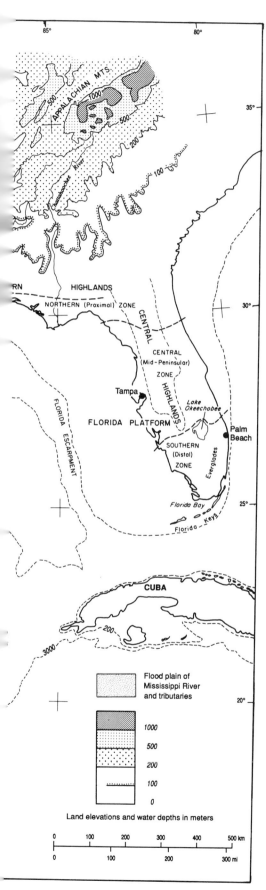

Figure 1. Onshore physiographic provinces and features, Gulf of Mexico basin. 1. Soto la Marina River, 2. Pánuco River, 3. Moctezuma River, 4. Sierra Tantima, 5. Tuxpan River, 6. Tecolutla River, 7. Papaloapan River, 8. Coatzacoalcos River, 9. Grijalva River, 10. Usumacinta River.

Florida Peninsula. Puri and Vernon (1964) and White (1970) have divided the Florida Peninsula into three generalized physiographic zones separated by roughly east-west boundaries (Fig. 1). From north to south, they are the Northern or Proximal zone, the Central or Mid-peninsular zone, and the Southern or Distal zone. Superimposed on this physiographic subdivision, these authors recognize a Central Highlands area trending NNW-SSE through the center of the peninsula as far south as Lake Okeechobee, flanked to the east, south, and west, by plains and coastal lowlands.

The Northern physiographic zone includes a broad upland, the so-called Northern Highlands, that includes northern Florida and southern Georgia, and extends westward to the Florida panhandle and southeastern Alabama. It reaches elevations of 60 to 90 m. The Northern zone is, for the most part, high enough to be above the piezometric surface and is therefore characterized by dry karst sinks, abandoned spring heads, dry steam courses, and dry beds of former shallow lakes that are now prairies. The Northern Highlands are bound on the south and east by a scarp, perhaps the most conspicuous and persistent topographic feature in Florida. It is interrupted only by the valleys of major streams.

The Mid-peninsular zone is characterized by discontinuous, subparallel ridges that rise to about 60 m above sea level, are parallel to the length of the peninsula, and are separated by broad valleys that often contain numerous shallow lakes. The Southern zone, in turn, is characterized by a broad, gently sloping and poorly drained plain for the most part less than 10 m above sea level; it forms the southern third of the peninsula and is most everywhere below the piezometric surface and covered, therefore, by extensive swamps in which the carbonate rocks may be overlain by as much as 3.5 m of peat. This vast area of saltwater marshes and swamps south of Lake Okeechobee is known as the Everglades.

Both the Northern and Central Highlands are believed to be the remnants of a once much larger highland, which has been dissected by erosion and differential dissolution of the underlying carbonate rocks. All major ridges of the Mid-peninsular zone are believed to represent relict coastal features that reflect the many advances and retreats of the shoreline resulting from changes of sea level during the Quaternary.

The coastal lowlands bordering the coastline of Florida are low in elevation, and their topographic features—relict lagoons, coastal marine terraces, dune and beach ridges, relict shorelines, barrier islands, and coral reefs—are generally parallel to the coast. They also reflect sea-level changes during the Quaternary and their formation by marine processes. Inland, the gently sloping plains often contain well-developed karst features—sinkholes and caves. Many isolated dunes or groups of dunes are present along the west coast of the Florida Peninsula north of Tampa, and along the east coast as far south as Palm Beach.

At the southern tip of the Florida Peninsula are the Florida Keys, an alignment of low islands forming a NE-SW–trending arc. They are composed predominantly of fragmental calcareous sediments. Northwest of the keys is Florida Bay, a large triangular expanse of warm and very shallow water where carbonate material is actively being deposited.

Yucatan Peninsula. Four main physiographic elements can be distinguished in the Yucatan Peninsula (Fig. 1): (1) a coastal plain, which ranges in width between 10 and 35 km along the Gulf of Mexico coast and from only a few tens of meters to 12 km along the eastern Caribbean coast; lagoons and bars are common; reefs rise to sea level offshore; (2) a flat, undissected plain along the north coast that reaches maximum elevations of 40 to 50 m and is underlain by near-horizontal carbonate rocks of latest Neogene age; (3) an area of alternating gently rolling plains and low hills with well-developed karst features, which characterize the northeastern part of the peninsula; and (4) low hills underlain by flat-lying Cenozoic rocks, which rise in elevation southward toward Guatemala and Belize and distinguish the southwestern part of the peninsula.

The most prominent topographic feature of the Yucatan Peninsula is the Sierrita de Ticul, a northwest-trending, elongated low ridge 125 km long, with average elevations of 100 m above sea level. It has a gentle southwestern slope, but well-developed scarps to the northeast. This steeper side has been interpreted to reflect a structural step caused by a down-to-the-northeast normal fault (Geological Map of Mexico, scale 1:1,000,000, 1981).

Most of the Yucatan Peninsula lacks a surface integrated fluvial drainage system. In contrast to the Florida Peninsula, which has many rivers, Yucatan has none. Drainage is subterranean, mostly at shallow depths, fed by abundant rainfall that ranges between 1,100 and 1,500 mm/yr in the south and east, between 900 and 1,200 mm/yr in the central part, and between 500 and 1,000 mm/yr along the Gulf of Mexico coast. Rainwater enters the subsurface to a depth of a few meters mainly through fractures in the low-relief carbonate section, resulting in the formation of an extraordinary number of karst features—sinkholes (dolines) with vertical walls and broad bottoms, often covered by water, known in Mexico as "cenotes," a term of Mayan origin (see Chapter 17 of this volume).

Gerstenhauer (1969) believes that the karst features of the Yucatan Peninsula have been generated in two stages: the older one took place during the Paleogene and developed karst cones, which are common for a tropical climate, over an uplifted surface. The karst features of the second stage developed in flat uplands, 25 to 30 m above sea level, and due to the gradual ascent of the ground-water level. The Yucatan Peninsula did not attain its present configuration until it was uplifted during the late Pliocene and the Quaternary, and records a raise in the sea level toward the end of the Pleistocene that can also be detected in the Florida Peninsula. It is one of the youngest geologic and geomorphologic structures in Mexico.

The difference of the submarine relief of the margins of the Yucatan Peninsula is remarkable: on the Gulf of Mexico side the continental shelf is broad and structurally undeformed; the Caribbean side, on the other hand, has a very narrow shelf and the sea floor drops abruptly into the Yucatan Basin to depths of more than 4,000 m just a few tens of kilometers from the coast.

This eastern margin of the Yucatan Peninsula is characterized by a complex zone of fault blocks bounded by normal faults parallel to the coast, and reflected in the sea floor as NNE-trending ridges and troughs.

The northern and western coastal plains

Surrounding the Gulf of Mexico to the north and west is a belt of low coastal plains of varied width and configuration (Fig. 1). Whereas the Florida and Yucatan Peninsulas are underlain by a nearly horizontal, predominantly carbonate section, the Gulf's northern and western coastal plains are underlain by Cretaceous to Quaternary sequences composed for the most part of terrigenous clastics dipping gently toward the Gulf. Only around the periphery of the basin, along the foothills of the surrounding highlands are carbonate rocks of Early Cretaceous age found at the surface. The regional gulfward dip of the sedimentary section underlying the coastal plains reflects the subsidence of the central part of the Gulf of Mexico basin, active recurrently since at least Late Jurassic time. Older sediments are more indurated and have a greater dip toward the center of the basin.

The Gulf of Mexico coastal plains are widest to the north and northwest of the Gulf where they are dominated by the broad valleys of the Mississippi River and the Rio Grande, which run roughly normal to the coast of the Gulf. Both rivers have their headwaters far from the Gulf of Mexico basin area, and their valleys form lengthy inland extensions of the coastal plains. The Gulf coastal plains are narrowest in east-central Mexico between the Sierra Madre Oriental and the shores of the Gulf.

Local relief in the coastal plains tends to increase away from the Gulf of Mexico shore and the Mississippi and Rio Grande valleys.

The northern coastal plains. The northern coastal plains of the Gulf of Mexico, as defined here, extend from the western limit of the Florida Peninsula to the valley of the Rio Grande, which marks the boundary between the United States and Mexico. The broad flood plain of the Mississippi River separates an eastern Gulf coastal plain from a western Gulf coastal plain (Fig. 1).

The physiography of the northern coastal plains has been summarized by Murray (1961) and Thornbury (1965). More recently, Walker and Coleman (1987) discussed the geomorphic systems of the U.S. Gulf Coastal Province. The brief review that follows is based to a great extent on these three publications.

The most prominent physiographic features of the northern coastal plain of the Gulf of Mexico are related to the course of the Mississippi River. The activity of the Mississippi is responsible for a large northward extension of the coastal plains—the so-called Mississippi Embayment—that extends 800 km from southern Louisiana to southern Illinois. As it enters the Gulf of Mexico, the Mississippi forms a great deltaic complex, the most intensely studied delta in the world. The Mississippi Delta is also one of the largest deltas, covering an area of 28,600 km^2, of which at least 4,700 km^2 is now under the waters of the Gulf of Mexico. The

Mississippi Delta will be discussed in Chapter 12 of this volume, and its underwater part—the Mississippi Fan—will be described later in this chapter as well as in Chapter 12.

The Mississippi River is not only a prominent feature of the present physiography of the northern flank of the Gulf of Mexico basin; it is also a dominant element in the geologic history of the basin. The river is known to have flowed into the basin, essentially along its current course, since Late Jurassic time and to have been a major source of terrigenous clastics since then (see Chapters 8, 9, 10, and 11 of this volume).

The physiography of the Lower Mississippi Valley also has received considerable attention. It was discussed most thoroughly in the classic publications of Fisk, particularly in his 1944 report of the Mississippi River Commission. In that report, Fisk described the extensive fluvial terraces present along the course of the Mississippi and some of its tributaries, particularly the Arkansas and Red Rivers, and recognized four major systems: from older to younger, the Williana, Bentley, Montgomery, and Prairie Terraces. The oldest of these terraces are topographically the highest. Fisk's terrace systems were generally accepted during the following decades by most investigators. As additional information became available, however, it became evident that refinements were necessary. Autin and others (1991) have recently provided such refinements in their overview of the geology and physiography of the Lower Mississippi Valley. They summarize alternative concepts and attempt to systematize and standardize the terminology used in previous studies. To avoid the apparent ambiguity concerning the meaning of the term "terrace" and the definition of terrace sequences, Autin and others proposed the term "complex" to apply to "a geomorphologic surface or set of temporally related surfaces with an associated sedimentary sequence that may represent more than one depositional environment." They recognize four complexes in the Lower Mississippi Valley: from older to younger, the Upland Complex, the Intermediate Complex, the Prairie Complex, and the Deweyville Complex.

The Upland Complex, which corresponds to the combined Williana and Bentley terrace systems of Fisk, though discontinuous, is the most widespread Quaternary unit of the Lower Mississippi Valley area. It is more widely distributed along the eastern part of the valley where it occurs as erosional remnants of a once extensive unit. The age of the Upland Complex is controversial—either late Pliocene, or early Pleistocene, or both—as is its origin—whether wholly or partly the result of glacial events.

The Intermediate Complex, which contains the Montgomery terrace system of Fisk, is more restricted in its geographic distribution and less well understood; it apparently represents much of Pleistocene time. The Prairie Complex includes "a widespread sequence of morphostratigraphic and depositional units loosely and controversially tied together by a single designation. As presently mapped it includes fluvial, colluvial, deltaic, estuarine, and marine deposits" (Autin and others, 1991). It is late Pleistocene in age and corresponds to the Prairie terrace system of Fisk.

The Deweyville Complex, finally, is developed along the course of many tributaries of the Mississippi River, but has no expression in the main Mississippi Valley itself. It overlies the Prairie Complex and underlies the Holocene flood-plain sediments. The Deweyville Complex does not seem to have equivalent units along the coast of Louisiana, but has been identified in the inner continental shelf.

Loess deposits of probable late Pleistocene age parallel the course of the Mississippi River from western Kentucky to southern Louisiana. (Russell, 1944; Autin and others, 1991). They are present on both sides of the valley, but are thickest and more continuous east of the river.

Tectonic activity, particularly subsidence, glacioeustatic changes in sea level, and climatic changes as they influence the supply of sediments are probably the main factors controlling the distribution and formation of sedimentary deposits and landforms in the Lower Mississippi Valley.

Fluvial terraces along the courses of other rivers flowing through the northern coastal plains are of lesser importance (Doering, 1956, 1958). They have received different names in different regions and even different names at different times by different authors in the same areas. In Texas, step-like Quaternary fluvial terrace systems have most commonly been named, from older to younger, the Willis, Lissie, and Beaumont Terraces. As in the Mississippi River valley, the gentle differential subsidence of the central part of the Gulf of Mexico basin has resulted in the older terrace systems having a greater elevation above the level of the Gulf, and a steeper Gulfward slope, as well as being more fully dissected.

East and west of the Mississippi River flood plain, the more inland parts of the northern coastal plains of the Gulf of Mexico are characterized by a "belted" topography. Between the valleys of the principal rivers crossing the coastal plains, the presence at the surface of alternating softer and harder beds has resulted in the formation of parallel low ridges with landward-facing escarpments (cuestas) and lowlands or vales forming distinctive topographic belts, which have received numerous local names (Thornbury, 1965). Particularly in northeast Texas, this kind of topography is characterized by prominent inland-facing cuestas developed in the more resistant stratigraphic units. Some of these differentially eroded cuesta systems clearly show the major structural framework of the area by wrapping around such features as the Sabine uplift and the northern rim of the East Texas basin. In south Texas and northeastern Mexico the belted topography is less evident due to a more arid climate and the increasing influence of the Sierra Madre Oriental.

Larger rivers, such as the Brazos River and the Rio Grande, empty directly into the Gulf of Mexico and have broad alluvial valleys and deltaic plains; smaller rivers have narrower valleys and frequently flow into estuaries or lagoons separated from the Gulf by barrier islands or offshore bars.

Locally along the coastal plain regions closer to the shore of the Gulf of Mexico, particularly west of the Mississippi Valley, small rounded mounds and low ridges surrounding circular depressions, commonly filled with water, reflect the presence of salt domes that have nearly reached the surface in very recent times.

Along the northern shore of the Gulf of Mexico, landforms are extremely diverse and variable, indicating that different processes of construction and modification have been active during the Quaternary (Deussen, 1924; Russell, 1940; Weeks, 1945; Carlston, 1950; Shepard, 1956; LeBlanc and Hodgson, 1959; Murray, 1961; Walker and Coleman, 1987). The effects of these processes also depended on the local topography, the type and supply of sediments in the area, the nature of offshore and longshore currents, the climate and wave energy, and the sea-level changes, among other factors. Most common among coastal features are lengthy barrier islands and their associated lagoons and beach ridges, but also common are open sand beaches, dunes, mudflats, marshes, mangrove swamps, and deltas.

Long barrier islands are particularly distinctive on the Texas coast (Fisk, 1959). They are among the world's longest, with few tidal inlets, and extensive adjoining lagoons. Dunes are common in the barrier islands. Padre Island, in south Texas, with a length of 210 km the longest barrier island bordering the Gulf of Mexico, has two dune fields—a system on the Gulf side of the island, which rises to as much as 15 m, and a system developed on the side of the lagoon (Laguna Madre). Smaller, more or less isolated barrier islands are present along other sectors of the northern coastline of the Gulf of Mexico, from the Florida panhandle to Mississippi.

The coast of southwestern Louisiana, on the other hand, is distinguished by a low coastal tidal marsh and mud flats, within which are numerous long, narrow sand and shell relict beach ridges, which rise only about 1 to 3 m above the marsh surface, vary in width between a few meters and 500 m, and may extend for many kilometers, generally parallel to the coast. They have been called "cheniers," from the French word for oak (chêne), a tree that commonly grows over these coastal ridges (Price, 1955; Byrne and others, 1959; Gould and McFarlan, 1959; Autin and others, 1991).

Along the entire northern coast of the Gulf of Mexico, near the river mouths and deltas, riverine processes dominate as the river discharge is high and the wave power generally very low.

The western coastal plains in eastern Mexico. The wide Gulf Coastal Plain of south Texas continues into Mexico where it narrows gradually southward until it disappears against the Quaternary volcanic rocks of the Trans-Mexican Neovolcanic Belt, which nearly reaches the coast of the Gulf of Mexico at about latitude 20°N (Fig. 1). The coastal plain widens again south of this volcanic range and stretches as far as the states of Tabasco and western Campeche. From the Rio Grande, south to the Laguna de Términos, the Mexican part of the Gulf Coastal Plain has a length of 1,320 km, 700 km north of the Neovolcanic Belt and 620 km south of it.

The western Gulf Coastal Plain shows very varied morphology and reflects diverse processes. It corresponds to the western flank of the Gulf of Mexico Basin and borders the Sierra Madre

Oriental orogenic belt. The coastal plain is underlain by a great thickness of Mesozoic and Tertiary rocks, which generally form a homocline sloping gently toward the Gulf of Mexico. Its subsurface is well known from the information supplied by many thousands of wells drilled since the early years of this century in search of oil and gas. The physiography of the Gulf Coastal Plain in Mexico has been discussed by Ordoñez (1936), Alvarez (1961), and Murray (1961), and illustrated by Raisz (1964) in his map of the landforms of Mexico.

Besides the Trans-Mexican Neovolcanic Belt, two other highlands—the Sierra de Tamaulipas to the north, and the Los Tuxtlas volcanic center to the south of the Neovolcanic Belt— break the Gulf Coastal Plain into four separate physiographic "embayments": from north to south, the southern part of the Rio Grande Embayment, the Tampico Embayment, the Veracruz Embayment, and the Isthmus of Tehuantepec Embayment (Fig. 1). These four physiographic embayments correspond closely with known structural basins—the Burgos basin north of the Sierra de Tamaulipas, the Tampico-Misantla basin between the Sierra de Tamaulipas and the Neovolcanic Belt, the Veracruz basin from the Neovolcanic Belt and the Los Tuxtlas volcanic center, and the basins of the Isthmus region between this last highland and the western boundary of the Yucatan platform (see Chapter 3 of this volume). The portion of the Gulf Coastal Plain corresponding to each of these basins is described separately in the following paragraphs.

The coastal plain of the southern Rio Grande Embayment. This segment of the Gulf Coastal Plain stretches from the Rio Grande (known in Mexico as the Rio Bravo) to the Sierra de Tamaulipas, a distance of about 100 km. It is widest, 165 km, in its northern part. From east to west it includes the outcrop of Quaternary alluvial and marine beds (110 km), and rocks of Pliocene (25 km), Miocene (10 km), and Oligocene (20 km) age.

The surface of the coastal plain slopes gently to the east and includes subparallel cuestas and hillocks resulting from the differential erosion of tilted and folded Oligocene and Neogene beds. Elevations reach 200 to 300 m above sea level.

South of latitude 24°N, this gently sloping surface is interrupted by the Sierra de Tamaulipas, which reaches elevations of 1,200 to 1,400 m. The Sierra de Tamaulipas is part of an elongated range that extends in a north-northwest direction for more than 300 km roughly parallel to and west of the Rio Grande Valley. The Upper Jurassic and Cretaceous rocks that crop out in the Sierra de Tamaulipas form a broad anticlinorium underlain by a Permian-Triassic granitic batholith. Tertiary and Quaternary basic volcanic rocks are also common along this elongated range, which separates the Burgos and Tampico-Misantla structural basins. The proximity of the Sierra Madre Oriental is reflected in the Mexican part of the Rio Grande Embayment by the occurrence of extensive gravel deposits and an increase in topographic relief toward the mountains. The Sierra de Cruillas and the Sierra de San Carlos, in the southern part of the Rio Grande Embayment, are domal uplifts related to igneous intrusions.

The Gulf of Mexico coast in the southern Rio Grande Embayment, as that in south Texas, is characterized by long barrier islands in front of extensive lagoons. Of the latter, the most important is Laguna Madre. The barrier islands stretch over a distance of 200 km interrupted only by a few inlets. Estavillo and Aguayo (1985) consider that Laguna Madre is the result of the delta-building processes of the Rio Grande while the smaller Conchos and Soto la Marina Lagoons, to the south of it, were formed due to changes in the hydrologic systems that took place during the Pleistocene.

Several rivers, sourced in the Sierra Madre Oriental, cross the coastal plain south of the Rio Grande; the principal ones are the Conchos and Soto la Marina Rivers, which are fed by rainfall of 600 to 800 mm/yr and have a flow volume that considerably surpasses the loss by evaporation.

The Tertiary section of the Burgos basin reaches, according to Lopez Ramos (1980), a thickness of 10,000 m near the coast of the Gulf of Mexico, of which 3,000 m are Miocene continental and littoral sediments and 15 to 35 m are a predominantly deltaic sequence of Pliocene age. The Quaternary section consists of a fluvial and littoral sequence that reaches up to 300 m near the coast.

The sedimentary characteristics of the Neogene-Quaternary section and the change from subsidence to uplift during the time of their deposition indicate considerable neotectonic activity in spite of an apparent stability. Sedimentation continues in the continental shelf while gradual uplift takes place from west to east.

The coastal plain of the Tampico Embayment. This segment of the western Gulf Coastal Plain extends from the Sierra de Tamaulipas to the Trans-Mexican Neovolcanic Belt, a distance of approximately 415 km. It narrows progressively southward from 140 km at the Pánuco River, to 115 km at the Tamiahua Lagoon, 60 km at the Tecolutla River and only a few kilometers where the Trans-Mexican Neovolcanic Belt almost reaches the shore of the Gulf of Mexico. Its relief is different than that of the Rio Grande Embayment region. The coastal plain rises westward from the coast to an elevation of about 400 m along the foothills of the Sierra Madre Oriental whose structural and physiographic front forms a distinctive inner limit to the plain. The Tampico-Misantla basin, which underlies the Tampico Embayment, is an elongated structural depression trending north-northwest, and containing a sedimentary section 2,000 to 4,000 m thick, which ranges in age from Late Triassic to Miocene. The Miocene section, predominantly composed of sandstones, is 450 to 500 m thick. It crops out along a belt averaging 12 km in width near the coast south from the Timiahua Lagoon. Pliocene sediments are not known in the Tampico-Misantla Basin. The Miocene is overlain along the coast by Quaternary fluvial and littoral deposits.

The coastal plain of the Tampico Embayment includes hillocks, undulated plains sloping gently toward the east, and irregular belts of roughly parallel low hills trending north-northwest, formed by the outcrop of Upper Cretaceous, Paleocene, Eocene, and Oligocene sedimentary rocks. The higher elevations correspond to anticlinal folds. Small, isolated Quaternary volcanic hills

and knobs dot the coastal plain between the Tamiahua Lagoon and the Tecolutla River. Along the coast, the flat Quaternary surfaces are limited and are no more than 14 km wide. A significant topographic feature caused by a Tertiary volcanic body is the Sierra Tantima, which rises above the coastal plain to an elevation of 1,200 m, 40 km inland from the southern end of the Tamiahua Lagoon.

North of Tuxpan, barrier islands and lagoons are well developed. The widest and most prominent are those enclosing the Tamiahua Lagoon. They have a cuspate shape, the tip of which is the prominent Cabo Rojo. The formation of the barrier islands has been related to the sea-level rise at the end of the last Pleistocene glacial episode. Also common along the coast of the Gulf of Mexico are marshes and dune fields, which in some cases have contributed to the growth of the barrier islands.

Numerous rivers flow from the Sierra Madre Oriental across the coastal plain of the Tampico Embayment, fed by a rainfall of 1,000 to 1,500 mm/yr in the plains, and even higher rates in the mountains to the west. The most important are, from north to south, the Moctezuma-Pánuco, Tuxpan, and Tecolutla Rivers. They all form estuaries as they enter the Gulf of Mexico, except the Pánuco River, which forms a small delta.

Malpica Cruz (1986) studied the marine terraces of the Gulf of Mexico coast where the coastal plain narrows in front of the Neovolcanic Belt. He observed three levels of terraces at 5, 8, and 12 m above sea level. He believed this to be one of the few areas in Mexico where these kinds of landforms are well preserved and ascribed their origin to sea-level fluctuations due to glacial episodes and regional uplift during the Pleistocene.

The coastal plain of the Veracruz Embayment. This part of the Mexican Coastal Plain extends for about 180 km from the eastern termination of the Trans-Mexican Neovolcanic Belt on the north to the western boundary of the Los Tuxtlas volcanic center. It is bounded to the west by the Sierra Madre Oriental. The Veracruz structural basin, which underlies it, contains a sedimentary section more than 11,0000 m thick.

The coastal plain of the Veracruz Embayment has a steeper slope than that of neighboring areas, reaching an elevation of 400 to 500 m along the foothills of the Sierra Madre Oriental. Quaternary piedmont deposits predominate over most of the area. The coastal plain is cut by numerous rivers that have their headwaters in the Sierra Madre Oriental. The coast is more irregular than to the north and south, and includes some barrier islands and lagoons, of which the largest is the Laguna de Alvarado at the mouth of the Papaloapan River.

The Los Tuxtlas volcanic uplift is a mountainous area underlain by Tertiary and Quaternary basic volcanic rocks. Four volcanic cones have been recognized, which form a high structure, 75 by 50 km in size, with a lake—Lago Catemaco—in the center. The highest of the volcanoes is Volcán San Martín, which reaches an altitude of 1,700 m and was active in 1664 and 1793.

The coastal plain of the Isthmus of Tehuantepec. The coastal plain in the region of the Isthmus of Tehuantepec extends from the Los Tuxtlas volcanic center to the western boundary of the

Yucatan Platform. To the south it is bounded by the Sierra de Chiapas. It ranges in width between 80 and 100 km.

The coastal plain of the Isthmus region is wide and has a more gentle slope than that of the Veracruz Embayment. It reaches maximum elevations of only about 200 m where it abuts against the Mesozoic sediments of the foothills of the Sierra de Chiapas, but is generally less than 100 m in elevation. The surface of the coastal plain is composed of Quaternary alluvial deposits and Tertiary sediments of Eocene, Oligocene, and most commonly Miocene and Pliocene age. The late Tertiary was a time of rapid deposition in the sedimentary basins of the Isthmus region, as well as in the Veracruz basin; the section is predominantly made up of shales in its lower part, while coarser clastics, sandstones and conglomerates, are the main components of the upper part of the section. It attains thicknesses of 2,500 to 5,000 m.

The most important rivers crossing the coastal plain of the Isthmus area are, from west to east, the Coatzacoalcos, Grijalva, and Usumacinta Rivers. The last two meander through the broad low coastal plain of the state of Tabasco where extensive coastal marshes, swamps, and numerous lakes cover at least one-third of the area. They join before emptying into the Gulf of Mexico where they form a small delta. The rainfall in the region ranges between 2,000 and 2,500 mm/yr.

West and others (1985) divided the coastal plain of the state of Tabasco into three units: (1) farthest inland, Pleistocene fluvial terraces form a belt 50 to 75 km wide parallel to the coast, (2) gulfward from this belt, the fluvial terraces grade into Holocene fluvial plains in part derived from the erosion of the terraces, and (3) closest to the coast are low Holocene plains related to the outer part of the delta of the Grijalva and Usumacinta Rivers. Along this belt, which can reach 42 km in width, numerous beach ridges record the position of former coast lines.

Along the coast, barrier islands separate the Gulf of Mexico from numerous lagoons, the largest of which is the Laguna de Términos.

Occasional mounds are known in the low coastal plain. They are the expression of shallow salt domes. One of these mounds, southeast of Coatzacoalcos, is the site of the famous archeological locality of El Plan.

Northeastern Mexico Highlands

Between the Transversal Ranges of the Sierra Madre Oriental to the south and the Rio Grande Valley on the north, in the northern part of the states of Coahuila and Nuevo Leon, is a large semi-desertic upland, 70 to 150 km wide and 530 km long, trending in a general northwest direction. Elevations range from 1,000 m along the foothills of the Sierra Madre Oriental to 200 m in the valley of the Rio Grande.

The region has the typical basin-and-range topography, characterized by northwest-trending, long and narrow ridges separated by broad, flat, intermontane valleys ("bolsons"). The ridges are commonly faulted, often breached anticlinal structures

formed by Jurassic and Cretaceous rocks. The valleys are covered by Quaternary alluvial and piedmont deposits. The center of the elongated breached anticlinal ridges are often occupied by almost completely closed hidden valleys, locally called "potreros," surrounded in all directions by scarps formed by the flanks of the structures.

Geologically, the region corresponds with the Sabinas basin in which up to 4,500 m of Jurassic and Cretaceous rocks overlie a Paleozoic and early Mesozoic "basement". The region, in a broad sense, has been interpreted to represent the southern boundary of the North American craton.

The area of the Northeastern Mexico Highlands was uplifted during the Oligocene. Since then, fluvial erosion has been mainly responsible for the sculpturing of the present surface. Several tributaries of the Rio Grande cross the area; most are intermittent due to the low rainfall in this arid region, which ranges from 200 to 300 mm/yr in northernmost Coahuila to 500 to 600 mm/yr farther to the south along the Rio Grande valley.

SUBMARINE PHYSIOGRAPHY AND BATHYMETRY OF THE GULF OF MEXICO

The Gulf of Mexico is a semienclosed, small ocean basin (1.5 million km^2) that, along with the Caribbean, forms the American Mediterranean. A bathymetric chart of the Gulf of Mexico at a scale of 1:2,500,000 is enclosed (Plate 1). The most recent bathymetric charts of the Gulf of Mexico were compiled by Sorensen and others (1975) and Bryant and others (1984). However, the bathymetric chart of the Gulf of Mexico accompanying this volume (Plate 1) is a revised edition (with a different projection) of the Bryant and others (1984) chart.

Figure 2 is a chart showing the major bathymetric and physiographic subprovinces. The subprovinces displayed are similar to those displayed by Martin and Bouma (1978), and their terminology has been followed. There are many excellent articles in print on the physiography of the Gulf of Mexico (Bergantino, 1971; Garrison and Martin, 1973; Uchupi, 1967, 1975; Martin and Bouma, 1978; Walker and Coleman, 1987). Little can be added on the physiography of the Gulf that has not been covered by these authors and in Chapter 12 of this volume. Therefore, in order not to be redundant, only a brief description of each subprovince within the Gulf of Mexico will follow.

The physiography of the Gulf of Mexico is a study in contrast. In the north, south, and east portions of the Gulf the continental shelves are very broad, up to 170 km in width. In the western portion the continental shelf off Mexico is less than 13 km wide in places. The extremely flat Sigsbee Abyssal Plain is broken only by the intrusion of seamount-like salt diapirs. The hummocky low-angle continental slope off Texas and Louisiana is in stark contrast to the precipitous West Florida and Campeche Slopes. The erosional valley called the DeSoto Canyon divides the terrigenous western Gulf from the carbonate province of the Florida Platform, while the structure-controlled Campeche Canyon divides the carbonate Yucatan Platform from the terrigenous salt-controlled region called the Campeche Knolls. The East Mexico Slope, consisting of long, ridge-like structures attributed to mass movement processes, is separated from the Campeche Knolls region, an area of intense diapiric activity, by a section of little relief called the Veracruz Tongue. The ciassic submarine canyons of the DeSoto and Mississippi Canyons are contrasted to the Alaminos, Rio Perdido, and Keathley Canyons that were formed or deformed in part by diapiric activity. The Mississippi Fan, the largest structural interbasin feature, is Plio-Pleistocene in age and stands in contrast to the great age of the basin.

The present-day physiography of the Gulf of Mexico reflects events that happened during the birth of the basin. Massive amounts of sediments and substantial subsidence in all areas of the Gulf have obscured the rift that gave birth to the basin and captured the salt now mobilized by the sediment deposited within the basin. The physiography of the Gulf of Mexico basin has been controlled by the following processes: (1) rifting, (2) subsidence, (3) carbonate platform development, (4) eustatic changes in sea level, (5) salt diapirism, (6) gravity slumping, and (7) density flow.

Carbonate platforms

The carbonate platforms of Florida and Yucatan were divided by a deep seaway throughout the Cretaceous and Cenozoic. The structural positions of these platforms relative to one another have remained unchanged since the Cretaceous (Schlager and others, 1984). The pre-Cretaceous history of the area between the platforms is not known for certain, but it is suggested by Schlager and others (1984) that the tilted graben fills on the faulted basement revealed by seismic profiles represents Triassic–Lower Jurassic rift sediments composed of alluvial, littoral, and neritic clastics and bathyal limestones.

Florida Platform. The west Florida continental margin is a massive upbuilding of shallow water carbonate and evaporite accumulation similar in most respects to the Yucatan Platform. The west Florida continental margin consists of a broad, flat carbonate-dominated shelf up to 240 km wide at its maximum width (Howell Hook). Reef structures are common features. A living reef complex called the Florida Middle Ground dominates the area of the northwest West Florida Shelf. The shelf narrows toward the northwest and is bounded to the northwest by the DeSoto Canyon, to the west by the West Florida Terrace, and to the south by the Pourtales Terrace. The Pourtales Terrace contains karst-like topography produced by solution, and collapse caused by seeping waters from the underlying Florida aquifer (Mallory and Hurley, 1970).

The West Florida Terrace occupies the area between the 200- and 1,000-m isobaths. It narrows to the south and achieves its maximum width near the DeSoto Canyon. The western boundary of the Florida Platform is the Florida Escarpment. This escarpment contains some of the steepest submarine slopes (45°) in the world and separates the continental slope from the deeper portions of the Mississippi Fan and the Florida Plain. Seismic

profiles indicate that the western portion of the Florida Platform consists mostly of upbuilding of carbonate and evaporite deposits. The Florida Escarpment is the western edge of exposed Lower Cretaceous and younger limestones. The escarpment is free of outbuilding sediment layers, possibly due to the Gulf Loop Current that appears at times to sweep parallel to the escarpment. Sediment cores obtained along the escarpment between 27 and 28° N contain sections of Mississippi River muds layered between carbonate oozes. The surficial sediments retained on the escarpment face are the muds deposited by the Loop Current during low sea stands and the pelagic rain of carbonate sediments from above when the Loop Current swings more to the west.

The unevenness of the Florida Escarpment south of 26°50′N may be due to more uneven subsidence of this area compared to that farther north, resulting in more irregular linear reefs (Bergantino, 1971). The Florida Escarpment merges with the Mitchell Escarpment after being dissected by the Tortugas and Agassiz Valleys. Some of the most striking features of the West Florida upper continental slope are the irregular morphology associated with mass-movement processes. The mass-movement features range from creep to massive slides, to gravity-induced folds tens of kilometers long (Doyle and Holmes, 1985).

Yucatan Platform. The submerged northern and western part of the Yucatan Platform is called the Banco de Campeche (Campeche Bank). The land portion of the platform is a region of karst topography with little surface drainage and consequently has had little influence on the depositional history of the shelf and slope of the platform. The continental shelf of the Yucatan Platform is broad on the north, reaching 240 km wide, and is narrow on the eastern portion of the platform. The shelf has been the site of limestone and evaporite deposition since the Early Cretaceous. At present, numerous living reefs and biohermal mounds, wave-cut terraces, and small-scale karst features dominate the topography of the area (Logan and others, 1969; Martin and Bouma, 1978). The broad shelf is surrounded on three sites by relatively steep slopes that plunge as much as 3,600 m from the shelf edge to the Gulf floor. The slope is relatively steep (up to 35°) and is broken only by broad terraces, primarily the Campeche Terrace. This terrace, found at a depth of approximately 1,000 m, has a width of 200 km and an average slope of only 5°. The slope is scarred by numerous small canyons and large-scale slumping.

The Campeche Escarpment is the product of Early Cretaceous reef building and upward growth of the Yucatan Platform through the slow accumulation of carbonate sediments and evaporites that have kept pace or exceeded subsidence. Seismic profiles across the shelf-slope area of the northern portion of the platform reveal an upbuilding and outbuilding of carbonate material likened in some cases to that of delta building. The outbuilding portion of the shelf may be associated with sediments deposited from that portion of the Gulf Stream that sweeps northwest across the Banco de Campeche as it divides at the Yucatan Channel. The western portion of the Banco de Campeche forms the eastern flank of the Campeche Canyon. For

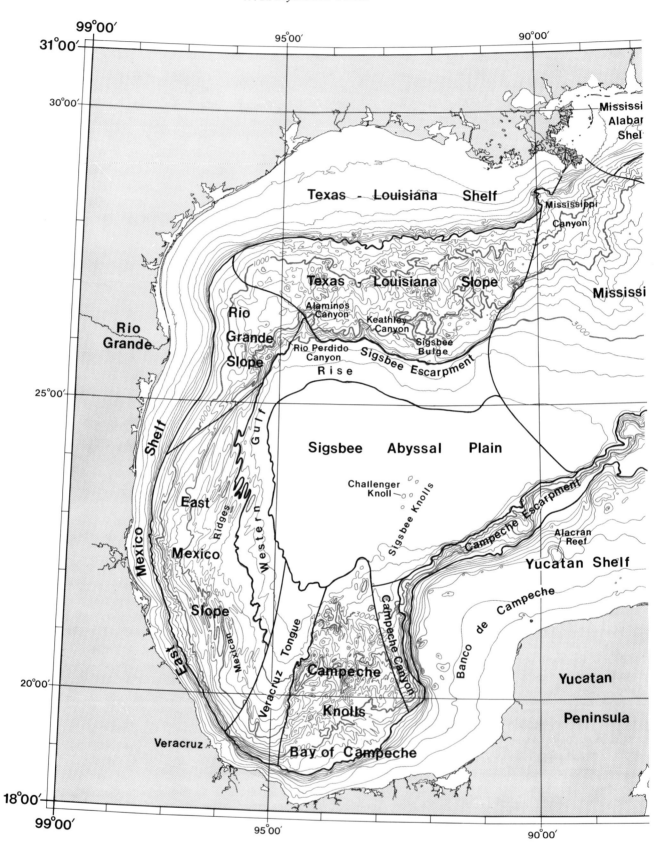

Figure 2. Bathymetric-physiographic map, Gulf of Mexico.

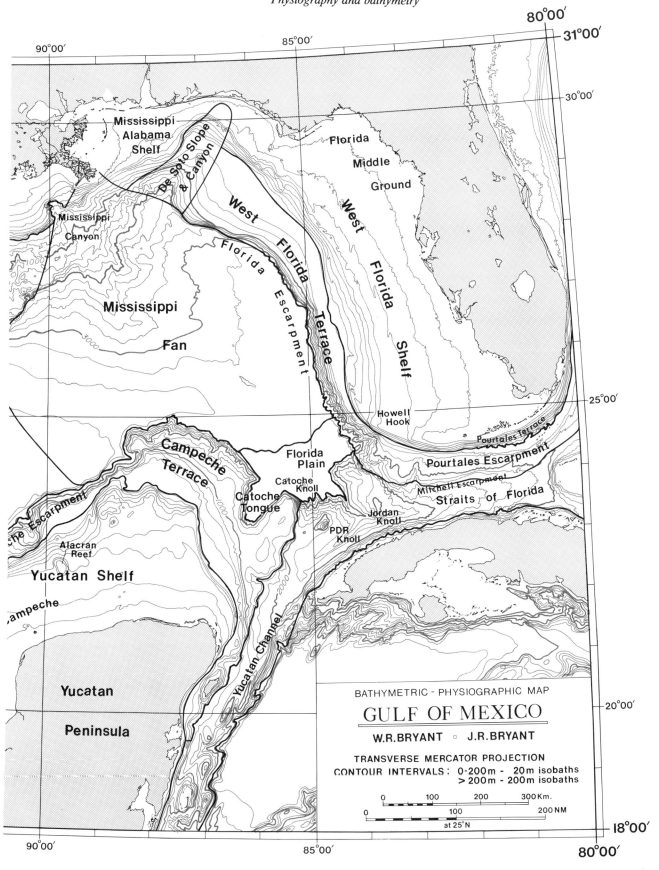

90°00' **85°00'** **80°00'**
80°00'
31°00'
30°00'

Mississippi
Alabama
Shelf

De Soto Slope & Canyon

Florida
Middle
Ground

Mississippi
Canyon

West
Florida
Terrace

West
Florida
Shelf

Florida Escarpment

Mississippi

Fan

25°00'

Howell
Hook

Pourtales Terrace

Campeche

Terrace

Florida
Plain

Pourtales Escarpment

Catoche
Knoll

Mitchell Escarpment

Straits of Florida

Catoche
Tongue

Jordan
Knoll

Escarpment

PDR
Knoll

Alacran
Reef

Yucatan Shelf

Campeche

Yucatan Channel

Yucatan

Peninsula

BATHYMETRIC - PHYSIOGRAPHIC MAP

GULF OF MEXICO

W.R.BRYANT □ J.R.BRYANT

TRANSVERSE MERCATOR PROJECTION

CONTOUR INTERVALS : 0-200m - 20m isobaths
> 200m - 200m isobaths

0   100   200   300 Km.
0   100   200 NM
at 25°N

20°00'

18°00'

90°00' **85°00'** **80°00'**

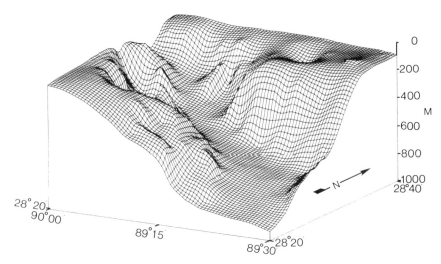

Figure 3. Isometric projection of the Mississippi Canyon area as viewed from the south-southeast at an elevation of zero degrees. Depth in meters.

more information on the Banco de Campeche and its association with the Florida Platform, see Chapter 13 of this volume.

The continental shelf

Texas-Louisiana Shelf. The Texas-Louisiana Shelf is a broad, flat portion of the continental margin that ranges in width from 32 to 90 km. Its surface is highly dissected by Wisconsinan-age channels now filled with sediments. Up to four sets of ancient fluvial systems have been recorded by Berryhill and others (1984) and Suter and Berryhill (1985). Each set represents a previous low stand of sea level during a Pleistocene ice age. This shelf is highly faulted and contains diapiric structures of both shale and salt. For a detailed discussion of the Texas-Louisiana Shelf area, see Chapter 12 this volume.

East Mexico Shelf. The broad continental shelf off Texas narrows progressively in Mexican waters, along the western shores of the Gulf of Mexico. East of the mouth of the Rio Grande the shelf is 72 to 80 km wide, but it narrows to 33 to 37 km at 23°N, and is narrowest, 6 to 16 km, offshore from the Los Tuxtlas volcanic uplift, near Veracruz. Eastward from this area the shelf widens into the Bay of Campeche.

The East Mexico Shelf belongs to three different structural elements. North of 22°45′N latitude the shelf belongs to the Rio Grande Embayment (Burgos basin). From 22°45′N to 20°N, the shelf is part of the Tampico-Misantla structural basin. South of there the area is within the Veracruz and Isthmian structural basins. East of 95°W longitude the shelf is influenced by salt diapirs of the Tabasco-Campeche salt basin (Garrison and Martin, 1973).

The continental slope

Texas-Louisiana Slope. The Texas-Louisiana Slope is bounded to the north by the shelf break and on the south by the Sigsbee Escarpment. Its eastern boundary is the west flank of the Mississippi Canyon. It is an area of extreme complexity rising

from the sculpturing effects of salt and shale diapiric activity. Figure 3 is an isometric projection of the western lower slope illustrating the hummocky nature of the area.

Gealy (1955) originally interpreted the origin of the hummocky topography of the area to be the result of gravity slumping. Extensive seismic surveying and drilling revealed the role that salt plays in controlling the geomorphic and geologic features of the area (Moore and Curray, 1963; Lehner, 1969; Garrison and Martin, 1973; Buffler and Worzel, 1978; Buffler, 1983; Humphries, 1978; Martin, 1980; Shaub and others, 1984; Watkins and others, 1978; Winker and Edwards, 1983).

The lower Texas-Louisiana Continental Slope is a unique geomorphic and geological province. Buffler (1983) divides this portion of the slope into two major provinces—the Sigsbee Bulge area and the central slope/Mississippi Fan area. The Sigsbee Bulge, a terrace-like feature 50 to 75 km wide, has as its southern border the Sigsbee Escarpment The bulge area is characterized by a shallow, irregular high-amplitude reflector overlain by a relatively thin sedimentary section. The reflector is interpreted to be the top of mobilized salt (Buffler and Worzel, 1978). The Sigsbee Escarpment east of the bulge area appears to be formed by vertically symmetrical salt ridges that have been ponded and loaded by thick sedimentary deposits. Buffler and Worzel (1978) state that these salt ridges continue to the east where they are buried by the Mississippi Fan. At that point, the Sigsbee Escarpment loses its identity as a topographic feature.

Rio Grande Slope. The Rio Grande Slope area is a complex region where three major tectonic provinces overlap: the Mexican Ridges foldbelt to the south, the Perdido foldbelt to the east, and the salt province to the north. Little is known of this area; only two multichannel seismic lines have been shot within this area as of 1989. Such basic questions as how the Mexican Ridges foldbelt relates to the Perdido foldbelt and how the salt provinces overlap and interact with the northernmost extension of the Mexican Ridges foldbelt are unknown.

East Mexico Slope. The East Mexico Slope contains some of the most unusual topography in the world's oceans. The northern slope contains salt diapirs that are the southern extension of evaporite diapirs and massifs of the Texas-Louisiana Continental Slope and are contained in the Rio Grande Slope Province (Fig 2). To the south of the East Mexico Slope and extending seaward north of the Los Tuxtlas uplift is the even-bottomed Veracruz Tongue. Its eastern boundary is flanked by the irregular topography of the Campeche Knolls. Between the Veracruz Tongue and the Rio Grande Slope Province are the Mexican Ridges.

First described by Bryant and others (1968), the Mexican Ridges consist of a series of linear bathymetric ridges formed by anticlinal folds that trend parallel to subparallel to the coast. The ridges form a regular regional pattern with a wavelength averaging 10 to 12 km and extend north to south 500 km, including their buried portion. They have a maximum relief of 500 m. They are effective sediment traps; only pelagic sediments are found seaward of the first or second ridge nearest the shore. Individual ridges extend for great distances. Bryant and others (1968) traced one ridge for a distance of over 96 km by following along its crest. They described four zones within the ridge complex. The first zone is associated with the linear salt ridges of the Rio Grande Slope complex. The second zone consists of NNE-SSW–trending linear ridges extending from 22°20′N to 24°30′N. The third zone consists of an area between two major fold systems, which they considered a fault. The fourth zone extends from 20°30′ to 22°N and consists of arcuate ridges trending south-southeast to north-northwest. Zone 2 is wider than zone 4, each exceeding 120 km in width. Based on subsequent analysis of CDP seismic data, Pew (1982) added a fifth zone that consists of SSE- to NNW-trending parallel buried ridges extending from 19° to 20°N.

Bryant and others (1968) hypothesized that the Mexican Ridges can be accounted for by one of the following processes: (1) folds resulting from the mobilization of salt, (2) folding caused by compressional tectonic stress, (3) folding related to faulting, (4) gravity sliding of sedimentary rocks on a decollement surface. Garrison and Martin (1973) suggested that since no salt was present in the area, gravity sliding perhaps offered the best mechanism for the development of the ridges. The impression of large-scale submarine landsliding is heightened, Garrison and Martin (1973) suggested, by the grouping of the fold ridges in two lobate areas whose general trend is consistent within each lobe but differs between lobes. Buffler and others (1979) stipulated that the Mexican Ridges folds were adequately explained by two separate mechanisms: (1) massive gravity sliding, possibly triggered by regional uplift and supplemented by sediment loading at the head of the slide; and (2) compressional tectonic stresses originating within the deeper crust beneath Mexico and transmitted into the fold area through deep thrust zones. Both processes take place above a decollement or deformed zone located within mobile substrata.

Pew (1982) reported that the folds east of Tampico (zone 4 of Bryant and others, 1968) appeared to be massive gravity slides. He interpreted a detachment or decollement in a thick Upper Cretaceous or lower Tertiary pelagic shale sequence. Folded Plio-Pleistocene strata established the youth of the slide. The buried folds northeast of Veracruz were gravity-induced folds active in the middle Miocene. Subsequent loading by thick middle Miocene–Holocene section gradually halted downslope movement and initiated flow of plastic substrata from beneath loaded synclinal troughs into anticlinal cores. This deformation, Pew suggested, has continued to the present in some folds. It appears that the Mexican Ridges are the largest submarine slides in the world.

Veracruz Tongue

The Veracruz Tongue is that portion of the Gulf of Mexico that is positioned between the Mexican Ridges on the west and the diapirically controlled Campeche Knolls to the east. The southwestern portion of the Veracruz Tongue covers the southernmost buried portions of the Mexican Ridges. The continental shelf adjacent to the Veracruz Tongue is the narrowest continental shelf in the Gulf of Mexico. The funnelling effect of the diapiric structures on the east and the ridges on the west has concentrated the sediment supply from the drainage area of mainland Mexico into the trough that makes up the Veracruz Tongue.

Campeche Knolls

The Campeche Knolls province, located in the Bay of Campeche, consists of a hummocky topography similar to the continental slope off Texas and Louisiana. This hummocky topography is attributed to salt-diapir activity. Worzel and others (1968) were the first to survey the region in detail and suggested that salt was the mechanism that controlled the topography. The lineation formed by the Isthmus Saline basin, the Campeche Knolls, and the Sigsbee Knolls suggests that salt underlies the entire continental slope in the Bay of Campeche. Garrison and Martin (1973) presented a fairly comprehensive review of the area. They state that in the central region of the upper slope near 20°N, an area of more than 20,000 km^2, is underlain by salt massifs. The top of the salt mass in this region is thrust upward into a multitude of peaks and spines, generally covered by about 1 km of strongly faulted and folded sedimentary beds. Because of the vertical displacement of these formerly horizontal sediments, the sea floor is hummocky. Probable top-of-salt reflectors can be carried at shallow depths on all seismic lines crossing the area, suggesting that the central region is probably the crest of a single salt mass. North of the region of massive salt is a rather narrow belt of isolated salt domes that, in most cases, penetrate the sea floor and form knolls with relief as great as 1,500 m. South of the central region of massive salt, another grouping of salt massifs appears to be present beneath deposits of the shelf and upper slope. The main body of salt, bounded on the north and separated from the central salt mass by a belt of isolated diapirs, probably extends shoreward beneath the shelf and is associated with the salt domes of the Isthmus Saline basin in southeastern Veracruz and western Tabasco. The eastern flanks of the Campeche Knoll province constitute the western boundary of Campeche Canyon. The west flank of the area is the eastern boundary of the Veracruz Tongue.

Submarine canyons

Both of the carbonate platforms (Florida and Yucatan) have a terminus on their western flank consisting of a submarine canyon. On the western boundary of the Banco de Campeche the eastern flank of the Campeche Canyon consists of an upbuilding carbonate structure while its western flank consists of diapiric structures of the Campeche Knolls. The DeSoto Canyon marks the limits of the carbonate platform of Florida and the mud and clay provinces of the Mississippi and Alabama shelf area.

The DeSoto Canyon has been an area of nondeposition and some sediment erosion since the Late Cretaceous. The apparent westward extension of the Suwannee Strait during the Late Cretaceous and early Tertiary may have been the mechanism responsible for the formation of the DeSoto Canyon (Mitchum, 1978). The offset of the channel axis south of the Pleistocene axis suggests asymmetric Pleistocene sediment filling from the north. Sediment choking in the north-south segment of the canyon and the lack of a well-defined channel in the lower canyon suggest little recent sediment transport (Mitchum, 1978).

The Mississippi Canyon was originally interpreted by Shepard (1937) and later by Gealy (1955) as resulting from massive submarine slumping in combination with diapir activity. Bates (1953), Fisk and McFarlan (1955), and Phleger (1955) stressed subaerial erosion and cutting of the canyon by sand-laden density flows. Ferebee and Bryant (1979) summarized a large amount of seismic data and information from shallow cores obtained in the vicinity of the canyon and added to knowledge of the geometry of the filled parts of the Mississippi Canyon.

Coleman and others (1983) suggest that the present data show definitely that the Mississippi Canyon formed 25,000 to 27,000 yr B.P. and that the lower infilling of the canyon commenced by 20,000 yr B.P. Thus, given even the most generous interpretation, the canyon cannot have had more than 7,000 yr in which to form and remove 1,500 to 2,000 km^3 of material. Density currents are at best infrequent events and are not the most probable erosional mechanism for cutting such large features in such a short time. Subaerial erosion by ancestral rivers is also not a likely mechanism, as the river would have had to erode to more than 1,200 m below present sea level in a short time during a period when sea level could not have been lower than 100 to 150 m.

The most likely explanation for the formation of the Mississippi Canyon is large-scale slumping on an unstable continental margin. The most obvious indication of large-scale instability comes from a lowermost unit of canyon fill, where individual sedimentary masses can be related back to scars along the canyon wall. The first stage of canyon fill could thus be interpreted as the final phase of canyon formation, when the canyon walls themselves were achieving equilibrium. The overall morphology also suggests massive failures. The low slope of the canyon base (0.5°) implies that failure continued until an equilibrium slope was reached, lower than the primary depositional slopes surrounding the canyon. In addition, the floor of the canyon is rather flat (0.5°), and the canyon form is a shallow U-shape (Fig. 3). Thus, it appears that the Mississippi Canyon formed by a series of successive failures, each one creating the upslope and side instability that triggered the next, and thus migrating rather rapidly upslope. Once initiated, the canyon then acted as its own conduit, directing the failed material basinward until equilibrium was reached. The primary instability was probably produced by rapid sedimentation on the upper slope close to the shelf edge during an earlier phase of delta building. The irregular upper slump fill was completed by 20,000 yr B.P. During the interval from 20,000 yr B.P. to 12,000 to 10,000 yr B.P. a series of late Wisconsinan delta lobes deposited north of the present canyon were probably responsible for the major fill of the canyon. Since about 10,000 years ago, only pelagic sediments have been deposited in the area (Coleman and others, 1983).

There are three other secondary-type submarine canyons—Alaminos, Keathly, and Rio Perdido Canyons—all located in the northwestern continental slope of the Gulf of Mexico and all associated in some fashion with salt-related processes (Fig. 4). The largest of the three is Alaminos Canyon, which separates the hill-and-basin sector of the Texas-Louisiana slope from the relatively smooth sea floor on the west and is the product of a complex history of faulting, folding, diapiric uplift, salt-front movement, and submarine erosion (Bouma and others, 1972). More recent studies by Hardin (1989) suggest that the Alaminos Canyon is in part the result of coalescing of salt canopies. The Rio Perdido Canyon may be associated with the Rio Grande River and diapiric processes.

The lower continental slope of the northwestern Gulf of Mexico is underlain by shallow, allochthonous salt. Keathley Canyon, due east of Alaminos Canyon, is said to have formed in a diapir-free area between separate lobes of a salt front (Martin and Bouma, 1978). Keathley Canyon is one of several large canyons that locally define the salt front, extending into the interior of the slope province. Keathley Canyon can be divided into two morphologically contrasting parts: (1) the upper canyon, consisting of a narrow valley, and (2) the lower canyon, consisting of a broad reentrant. Unlike most other canyon systems, it contains no levees, tributary valleys, distributaries, nor the expected sedimentary fan at its mouth. Also, it is too far seaward to be related to onland rivers, which initiate many submarine canyons during sea-level lowstands. Multichannel seismic reflection profiles provide evidence of salt control on the formation of the canyon. In the upper canyon, two large salt lobes and a smaller one in the middle, moving laterally, elevated the continental slope or rise sediments and collided, resulting in a narrow valley with walls of uplifted and deformed strata. Erosion and mass movement of destabilized sediments on the canyon walls caused a

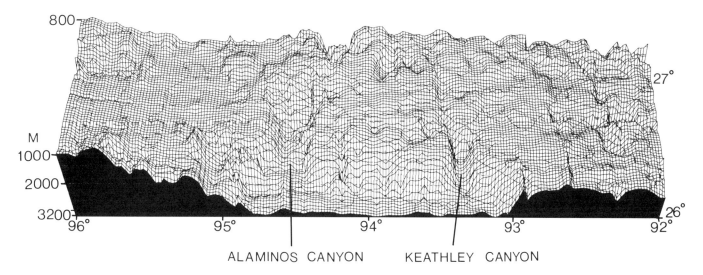

Figure 4. Isometric projection of the lower continental slope in the northwestern Gulf of Mexico as viewed from the south at an azimuth angle of 0° and an elevation of 50°. The Alaminos and Keathley Canyons are the most visible structures in the area. Depth in meters.

minor enlargement and/or steepening of the canyon. On the downslope side, the broad interlobal area of the two large lobes forms the lower canyon (Lee and others, 1989).

Mississippi Fan

The Mississippi Fan is a large, deep-sea fan in the northeast Gulf of Mexico that extends seaward from the continental shelf of Louisiana to abyssal depths. The fan itself consists of a broad arcuate pile of sediments marked by a seaward bowing of the bathymetric contours as shown in Plate 1 and Figure 2. The fan is flanked on the east by the west Florida carbonate platform and by the Texas-Louisiana-Florida continental slope on the north and west. The deeper portions of the fan merge with the Florida Plain to the southeast and the Sigsbee Abyssal Plain on the southwest.

The Mississippi Fan was first recognized by Bates (1953), who called the accumulation of sediments a submarine delta. Two years later, Fisk and McFarlan (1955) presented a detailed study of the late Pleistocene history of the Mississippi River, emphasizing the interaction of the fluviodeltaic and marine processes resulting from lowered sea level. They termed the fan a submarine bulge. Since that time the fan has been studied by numerous early investigators (Ewing and others, 1958; Ewing and others, 1962; Walker and Massingill, 1970; Huang and Goodell, 1970; Garrison and Martin, 1973; Stuart and Caughey, 1976; Martin and Bouma, 1978; Moore and others, 1978). These studies were limited by the vast size of the fan to only defining basic physiography and morphology.

The Mississippi Fan can be subdivided into three major physiographic provinces: upper, middle, and lower fan (Stuart and Caughey, 1976; Moore and others, 1978). The upper fan occurs in water depth from about 1,300 m to 2,500 m, has an average surface gradient of 1° (Stuart and Caughey, 1976), and is characterized by irregular, hummocky topography and an ero-

sional unleveed channel. The channel, the youngest of at least a dozen or more mapped in the subsurface, is dated as late Wisconsinan and connects to the Mississippi Canyon to the north. The structural and topographic complexities of the upper fan region resulted from salt diapirism (Shih and others, 1977; Buffler and Worzel, 1978), slumping (Walker and Massingill, 1970; Stuart and Caughey, 1976), and current scour (Stuart and Caughey, 1976).

The middle fan forms a broad depositional lobe. It is characterized by a low-gradient (<0.25°), moderately smooth surface and is composed of a massive complex of filled channels with natural levees and extensive slumped material (Walker and Massingill, 1970; Stuart and Caughey, 1976). Interbedded units of continuous reflectors are interpreted to be turbidites, hemipelagites, and possibly contourites (Stuart and Caughey, 1976). The lower fan is very smooth, with a gently sloping surface that merges with the abyssal plain to the southeast and southwest. Numerous small distributary channels with natural levees are clustered at the base of the fan, suggesting sedimentation by turbidity currents moving in discrete channels (Moore and others, 1978).

From seismic reflection profiles, Moore and others (1978) subdivided the upper sedimentary section of the fan into three stratigraphic units. The youngest, unit A, is equivalent to the Sigsbee (Pleistocene) seismic unit of Watkins and others (1976). Two distinct facies are apparent within the unit. The proximal facies is defined as containing greater than 10 percent disruptive seismic zones and is composed of channel, slump, and debris-flow deposits. The distal facies of unit A is characterized by parallel to slightly divergent, discontinuous reflectors, which are interpreted to be interbedded turbidite and hemipelagic sediments. The fan is believed to have formed during the time represented by unit A (i.e., mainly Pleistocene). At its apex the fan is over 3 km thick.

It is generally agreed that major fan development occurred

during the Pleistocene (Moore, 1969; Woodbury and others, 1973; Stuart and Caughey, 1976; Worzel and Burk, 1978; Moore and others, 1978). Stages of development are closely linked to eustatic sea-level changes and to the course of the ancestral Mississippi (Osterhoudt, 1946; Fisk and McFarlan, 1955; Woodbury and others, 1973; Bouma and others, 1989; Weimer, 1989, 1990; Feeley and others, 1990). The position of the river at the onset of Nebraskan, Kansan, and Illinoian glaciations is unknown. Shelf progradation and river migration have largely obscured the early history of the fan development on the outer shelf and upper slope. However, extensive mapping, dating, and volume calculations of the late Pleistocene channel on the shelf (Osterhoudt, 1946; Carsey, 1950; Stuart and Caughey, 1976) indicate that the majority of sediment reaching the slope and forming at least the upper fan was deposited during the latter part of the Pleistocene, probably during Wisconsinan glaciation.

In September and October, 1983, the D/V *Glomar Challenger* drilled nine holes on the middle and lower fan during the Deep Sea Drilling Project (DSDP) Leg 96. The results of the DSDP drilling program on the Mississippi Fan are presented in Chapter 12 of this volume.

According to Bouma and others (1989), the bulk of the Mississippi Fan was deposited during the Pleistocene; accumulation occurred in several stages as fanlobes and the time interval required to construct the Mississippi Fan (about 2.1 m.y.) approximates one major eustatic sea-level cycle. Bouma and others (1983/1984) introduced the term "fanlobe" as a depositional unit; a channel-levee-overbank complex. A sea-level-driven depositional model for individual fanlobes on the Mississippi Fan was developed by Bouma and others (1989) in which they divided the fan into four parts: (1) the canyon—the canyon fill is considered to be the upper part of a fanlobe; (2) the upper fan—the continuation of the canyon into the middle fan, the updip portion of which is erosional in nature gradually changing to aggradational near the base of the slope; (3) the middle fan—the constructional part of the fanlobe, lenticular in cross section with a sinuous migratory channel located at its apex; and (4) the lower fan—the area where the sinuosity and size of the fan channel decrease along with the thickness of the channel fill. Channel bifurcation and the deposition of sheet sands are common features.

Feeley and others (1990) and Weimer (1989, 1990) examined extensive multichannel and single-channel data across the Mississippi Fan, which revealed that at least 16 to 17 seismic sequences make up the Pliocene/Pleistocene section. The sequence boundaries were described as basinwide unconformities. In general, each sequence is lens shaped in cross section, thinning laterally from an area of maximum thickness. Seven seismic facies were identified within the fan units, each with one or more facies types. Each facies occupied a unique position in the evolution of each sequence, reflecting a succession of depositional regimes. This succession is related to cycles of sea-level fluctuations.

At the end of the last major period of glaciation, the Wisconsinan stage (11,000 yr B.P.), the course of the Mississippi River moved to the east to essentially its present position. This change cut off the Mississippi Canyon system as a conduit for sediment transport to the fan. Recently, extensive channels have been reported directly downslope from the present delta system (Walker and Massingill, 1970). At the present time, the modern delta is within several kilometers of completely crossing the shelf. Most of the material discharge from the river is currently trapped landward of the 1,000-m isobath (Moore and others, 1978).

Sigsbee Abyssal Plain

The Sigsbee Abyssal Plain is one of the flattest surfaces on the face of the Earth. The plain extends from 90° to 95°W and from 22° to 25°N. It is 450 km long and 290 km wide and covers over 103,600 km^2. The slope of its surface is 1:10,000 in a westerly direction. The plain is underlain by a very thick (up to 9 km) sediment section. The main source of Miocene sediments is the Rio Grande, and the Pliocene and Pleistocene sediments are from the Mississippi River system (Worzel and others, 1973). The very flat floor of the plain is broken by a series of salt diapirs that form the Sigsbee Knolls (Ewing and Antoine, 1966). Cores taken during DSDP Leg 1 from the Challenger Knoll in the Sigsbee Knolls complex recovered caprock, oil, and halite (Ewing and others, 1969). There is little doubt that the Sigsbee Knolls are salt diapirs and may be associated with the Campeche Knolls and the salt domes in the Isthmus Saline Basin.

The Sigsbee Abyssal Plain is an excellent example of turbidity-flow processes. The surficial sediments covering the tops of the Sigsbee Knolls consists of pelagic oozes, mainly foraminiferal ooze and a few very thin turbidite layers. This is in contrast to the thick sequence of silty-muddy turbidites of the Sigsbee Abyssal Plain, and attests to the shallow heights of the turbidity flow in the area during the late Pleistocene. Holocene sediments of the area consist mostly of foraminiferal ooze.

REFERENCES CITED

Alvarez, M., Jr., 1961, Provincias fisiográficas de la República Mexicana: Sociedad Geologica Mexicana Boletín, v. 24, p. 3–20.

Autin, W. J., Burns, S. F., Miller, B. J., Saucier, R. T., and Snead, J. I., 1991, Quaternary geology of the Lower Mississippi Valley, *in* Morrison, R. B., ed., Quaternary nonglacial geology, conterminous U.S.: Boulder, Colorado, Geological Society of America, The Geology of North America, v. K-2 (in press).

Bates, C. C., 1953, Rational theory of delta formation: American Association of Petroleum Geologists Bulletin, v. 37, p. 2119–2162.

Bergantino, R. N., 1971, Submarine regional geomorphology of the Gulf of Mexico: Geological Society of America Bulletin, v. 82, p. 741–752.

Berryhill, H. L., Owen, D. E., and Suter, J. R., 1984, Distribution of ancient fluvial sediments of probable early and late Wisconsinan age: Minerals Management Service, Outer Continental Shelf Map Series MMS 84-0003, sheet 2, scale 1:250,000.

Bouma, A. H., Chancey, O., and Merkel, G., 1972, Alaminos Canyon area, *in* Rezak, R., and Henry, V. J., eds., Contributions on the geological oceanography of the Gulf of Mexico: College Station, Texas A&M University Oceanographic Studies, v. 3, p. 153–179.

Bouma, A. H., Stelting, C. E., and Coleman, J. M., 1983/1984, Mississippi Fan; Internal structure and depositional processes: Geo-Marine Letters, v. 3, p. 147–154.

Bouma, A. H., Coleman, J. M., Stelting, C. E., and Kohl, B., 1989, Influence of relative sea level changes on the construction of the Mississippi Fan: Geo-Marine Letters, v. 9, no. 3, p. 161–170.

Bryant, W. R., Antoine, J. W., Ewing, M., and Jones, B., 1968, Structure of the Mexican continental shelf and slope, Gulf of Mexico: American Association of Petroleum Geologists Bulletin, v. 52, p. 1204–1228.

Bryant, W. R., Ziegler, M. D., Joyce, P. L., Feeley, M. H., and Bryant, J. R., 1984, Gulf of Mexico bathymetric chart: Ocean Drilling Program Regional Atlas Series 6, scale 1:2,000,000 at 24°N.

Buffler, R. T., 1983, Structure of the Sigsbee Scarp, Gulf of Mexico, *in* Bally, A. W., ed., Seismic expression of structural styles; A picture and work atlas: American Association of Petroleum Geologists Studies in Geology 15, v. 2, p. 2.3–2.50.

Buffler, R. T., and Worzel, J. L., 1978, Deformation and origin of the Sigsbee Scarp, lower continental slope, northern Gulf of Mexico: Offshore Technology Conference Proceedings, v. 3, p. 1425–1433.

Buffler, R. T., Shaub, F. J., Watkins, J. S., and Worzel, J. L., 1979, Anatomy of the Mexican Ridges, southwestern Gulf of Mexico, *in* Watkins, J. S., ed., Geological and geophysical investigations of continental margins: American Association of Petroleum Geologists Memoir 29, p. 319–327.

Byrne, J. V., LeRoy, D. O., and Riley, C. M., 1959, The chenier plain and its stratigraphy, southwestern Louisiana: Gulf Coast Association of Geological Societies Transactions, v. 9, p. 237–259.

Carlston, C. W., 1950, Pleistocene history of coastal Alabama: Geological Society of America Bulletin, v. 61, p. 1119–1130.

Carsey, J. B., 1950, Geology of Gulf coastal area and continental shelf: American Association of Petroleum Geologists Bulletin, v. 34, p. 361–385.

Coleman, J. M., Prior, D. B., and Lindsay, J. F., 1983, Deltaic influences on shelf edge instability processes, *in* Stanley, D. J., ed., The shelfbreak; critical interface on continental margins: Society of Economic Paleontologists and Mineralogists Special Publication 33, p. 121–137.

Deussen, A., 1924, Geology of the coastal plain of Texas west of the Brazos River: U.S. Geological Survey Professional Paper 126, 145 p.

Doering, J. A., 1956, Review of Quaternary surface formations of Gulf Coast region: American Association of Petroleum Geologists Bulletin, v. 40, p. 1816–1862.

—— , 1958, Citronelle age problem: American Association of Petroleum Geologists Bulletin, v. 42, p. 764–786.

Doyle, L. J., and Holmes, C. W., 1985, Shallow structure, stratigraphy, and carbonate sedimentary processes of west Florida upper continental slope: American Association of Petroleum Geologists Bulletin, v. 69, p. 1133–1144.

Estavillo, G. C., and Aguayo, C. J., 1985, Ambientes sedimentarios recientes en Laguna Madre: Sociedad Geológica Mexicana Boletín, v. 46, p. 29–64.

Ewing, J., Worzel, J. L., and Ewing, M., 1962, Sediments and oceanic structural history of the Gulf of Mexico: Journal of Geophysical Research, v. 67, p. 2509–2527.

Ewing, M., and Antoine, J., 1966, New seismic data concerning sediments and diapiric structures in Sigsbee Deep and upper continental slope, Gulf of Mexico: American Association of Petroleum Geologists Bulletin, v. 50, p. 479–504.

Ewing, M., Erickson, D. B., and Heezen, B. C., 1958, Sediments and topography of the Gulf of Mexico, *in* Weeks, L. G., ed., Habitat of oil: American Association of Petroleum Geologists, p. 995–1053.

Ewing, M., Worzel, J. L., and Burk, C. A., 1969, Initial reports of the Deep Sea Drilling Project: Washington, D.C., U.S. Government Printing Office, v. 1, 672 p.

Feeley, M. H., Moore, T. C., Jr., Loutit, T. S., and Bryant, W. R., 1990, Sequence stratigraphy of Mississippi Fan related to oxygen isotope sea level index: American Association of Petroleum Geologists Bulletin, v. 74, p. 407–424.

Ferebee, T. W., and Bryant, W. R., 1979, Sedimentation in the Mississippi Trough: College Station, Texas A&M University Department of Oceanography Technical Report 79-4-T, 178 p.

Fisk, H. N., 1944, Geological investigations of the alluvial valley of the Lower Mississippi River: U.S. Army Corps of Engineers Mississippi River Commission, 78 p.

—— , 1959, Padre Island and the Laguna Madre flats, coastal south Texas, *in* Russel, R. G., ed., 2nd Coastal Geography Conference: Baton Rouge, Louisiana State University Coastal Studies Institute, p. 103–151.

Fisk, H. N., and McFarlan, E., Jr., 1955, Late Quaternary deltaic deposits of the Mississippi River, *in* Poldevaard, A., ed., Crust of the Earth: Geological Society of America Special Paper 62, p. 279–302.

Garrison, L. E., and Martin, R. G., Jr., 1973, Geologic structure in the Gulf of Mexico: U.S. Geological Survey Professional Paper 773, 85 p.

Gealy, B. L., 1955, Topography of the continental slope, northwest Gulf of Mexico: Geological Society of America Bulletin, v. 66, p. 203–227.

Geological Map of Mexico, 1981: Secretaria de Programación y Presupuesto, Dirección General de Geografía del Territorio Nacional, scale 1:1,000,000.

Gerstenhauer, A., 1969, Ein Karstmorphologischer vergleich zwischen Florida und Yucatan: Verhandlungen Deutschen Geographentag, v. 36, p. 332–341.

Gould, H. R., and McFarlan, E., Jr., 1959, Geologic history of the chenier plain, southwestern Louisiana: Gulf Coast Association of Geological Societies Transactions, v. 9, p. 261–270.

Hardin, N. S., 1989, Salt distribution and emplacement processes, northwest Gulf of Mexico lower slope; A suture between two provinces: Gulf Coast Section, Society of Economic Paleontologists and Mineralogists Foundation 10th Annual Research Conference Program and Abstracts, p. 55–59.

Huang, T. C., and Goodell, H. G., 1970, Sediments and sedimentary processes of eastern Mississippi Cone, Gulf of Mexico: American Association of Petroleum Geologists Bulletin, v. 54, p. 2070–2100.

Humphris, C. C., Jr., 1978, Salt movement on continental slope, northern Gulf of Mexico, *in* Bouma, A. H., Moore, G. T., and Coleman, J. M., eds., Framework, facies, and oil-trapping characteristics of the upper continental margin: American Association of Petroleum Geologists Studies in Geology 7, p. 69–85.

LeBlanc, R. J., and Hodgson, W. D., 1959, Origin and development of the Texas shoreline, *in* Russel, R. J., ed., 2nd Coastal Geography Conference: Baton Rouge, Louisiana State University Coastal Studies Institute, p. 57–101.

Lee, G. H., Bryant, W. R., and Watkins, J. S., 1989, Salt structures and sedimentary basins in the Keathly Canyon area, northwestern Gulf of Mexico: Gulf Coast Section, Society of Economic Paleontologists and Mineralogists Foundation 10th Annual Research Conference Program and Abstracts, p. 90–93.

Lehner, P., 1969, Salt tectonic and Pleistocene stratigraphy on continental slope of northern Gulf of Mexico: American Association of Petroleum Geologists Bulletin, v. 53, p. 2431–2479.

Logan, B. W., Harding, J. L., Ahr, W. M., Williams, J. D., and Snead, R. G., 1969, Carbonate sediments and reefs, Yucatan Shelf, Mexico: American Association of Petroleum Geologists Memoir 11, 198 p.

Lopez Ramos, E., 1980, Geologia de Mexico, v. 2, 2nd ed.: Printed privately by the author, 454 p.

Mallory, R. J., and Hurley, R. J., 1970, Geomorphology and geologic structure; Straits of Florida: Geological Society of America Bulletin, v. 81, p. 1947–1972.

Malpica Cruz, V. M., 1986, Terrazas marinas pleistocénicas en la costa central del Estado de Veracruz: 8th Convención Geológica Nacional Resúmenes, p. 164.

Martin, R. G., 1980, Distribution of salt structures in the Gulf of Mexico: U.S. Geological Survey Map MF-1213, map and descriptive text, scale 1:2,500,000 at 19°N.

Martin, R. G., and Bouma, A. H., 1978, Physiography of Gulf of Mexico, *in* Bouma, A. H., Moore, G. T., and Coleman, J. M., eds., Framework, facies, and oil-trapping characteristics of the upper continental margin: American Association of Petroleum Geologists Studies in Geology 7, p. 3–19.

Mitchum, R. M., Jr., 1978, Seismic stratigraphy investigation of West Florida

Slope, Gulf of Mexico, *in* Bouma, A. H., Moore, G. T., and Coleman, J. M., eds., Framework, facies, and oil-trapping characteristics of the upper continental margin: American Association of Petroleum Geologists Studies in Geology 7, p. 193–223.

Moore, D. G., and Curray, J. R., 1963, Structural framework of the continental terrace, northwest Gulf of Mexico: Journal of Geophysical Research, v. 68, p. 1725–1747.

Moore, G. T., 1969, Interaction of rivers and ocean; Pleistocene petroleum potential: American Association of Petroleum Geologists Bulletin, v. 53, p. 2421–2430.

Moore, G. T., Starke, G. W., Bonham, L. C., and Woodbury, H. O., 1978, Mississippi Fan, Gulf of Mexico; Physiography, stratigraphy, and sedimentation patterns, *in* Bouma, A. H., Moore, G. T., and Coleman, J. M., eds., Framework, facies, and oil-trapping characteristics of the upper continental margin: American Association of Petroleum Geologists Studies in Geology 7, p. 155–191.

Murray, G. E., 1961, Geology of the Atlantic and Gulf Coast Province of North America: New York, Harper and Brothers, 692 p.

Ordoñez, E., 1936, Principal physiographic provinces of Mexico: American Association of Petroleum Geologists Bulletin, v. 20, p. 1277–1307.

Osterhoudt, W. J., 1946, The seismograph discovery of an ancient Mississippi River channel [abs.]: Geophysics, v. 11, p. 417.

Pew, E., 1982, Seismic structural analysis of deformation in the southern Mexican Ridges [M.A. thesis]: Austin, University of Texas at Austin, 102 p.

Phleger, F. B., 1955, Foraminiferal faunas in cores offshore from the Mississippi Delta; Papers in marine biology and oceanography: Deep Sea Research, v. 3 supplement, p. 45–57.

Price, W. A., 1955, Environment and formation of the chenier plain: Quaternaria, v. 2, p. 75–86.

Puri, H. S., and Vernon, R. O., 1964, Summary of the geology of Florida and a guidebook to the classic exposures: Florida Geological Survey Special Publication 5, 255 p.

Raisz, E., 1964, Landforms of Mexico, 2nd corrected ed.: Prepared for the Office of Naval Research Geography Branch, scale ~1:3,000,000.

Russell, R. J., 1940, Quaternary history of Louisiana: Geological Society of America Bulletin, v. 51, p. 1199–1233.

—— , 1944, Lower Mississippi Valley loess: Geological Society of America Bulletin, v. 55, p. 1–40.

Schlager, W., Buffler, R. T., and Scientific Party, 1984, Deep Sea Drilling Project, Leg 77, southeastern Gulf of Mexico: Geological Society of America Bulletin, v. 95, p. 226–236.

Shaub, F. J., Buffler, R. T., and Parsons, J. G., 1984, Seismic stratigraphic framework of the deep central Gulf of Mexico Basin: American Association of Petroleum Geologists Bulletin, v. 68, p. 1790-1802.

Shepard, F. P., 1937, "Salt" domes related to Mississippi submarine trough: Geological Society of America Bulletin, v. 48, p. 1354–1361.

—— , 1956, Late Pleistocene and Recent history of the central Texas coast: Journal of Geology, v. 64, p. 56–69.

Shih, T., Worzel, J. L., and Watkins, J. S., 1977, Northeastern extension of Sigsbee Scarp, Gulf of Mexico: American Association of Petroleum Geologists Bulletin, v. 61, p. 1962–1978.

Sorensen, F. H., and 5 others, 1975, Preliminary bathymetric map of Gulf of Mexico region: U.S. Geological Survey Open-File Report 75-140, scale 1:2,500,000.

Stuart, C. J., and Caughey, C. A., 1976, Form and composition of the Mississippi Fan: Gulf Coast Association of Geological Societies Transactions, v. 26, p. 333–343.

Suter, J. R., and Berryhill, H. L., Jr., 1985, Late Quaternary shelf margin deltas, northwest Gulf of Mexico: American Association of Petroleum Geologists Bulletin, v. 69, p. 77–91.

Thornbury, W. D., 1965, Regional geomorphology of the United States: New York, John Wiley and Sons, p. 53–69.

Uchupi, E., 1967, Bathymetry of the Gulf of Mexico: Gulf Coast Association of Geological Societies Transactions, v. 17, p. 161–172.

—— , 1975, Physiography of the Gulf of Mexico and Caribbean Sea, *in* Nairn, A.E.M., and Stehli, F. G., eds., Ocean basins and margins; Volume 3, The Gulf of Mexico and the Caribbean: New York, Plenum Press, v. 3, p. 1–64.

Walker, H. J., and Coleman, J. M., 1987, Atlantic and Gulf Coastal Province, *in* Graf, W. L., ed., Geomorphic systems of North America: Boulder, Colorado, Geological Society of America, Centennial Special Volume 2, p. 51–110.

Walker, J. R., and Massingill, J. V., 1970, Slump features on the Mississippi Fan, northeastern Gulf of Mexico: Geological Society of America Bulletin, v. 81, p. 3101–3108.

Watkins, J. S., Worzel, J. L., and Ladd, J. W., 1976, Deep seismic reflection investigation of occurrence of salt in the Gulf of Mexico, *in* Bouma, A. H., Moore, G. T., and Coleman, J. M., eds., Beyond the shelf break: American Association of Petroleum Geologists Marine Geology Committee Short Course, p. G1–G34.

Watkins, J. S., and 5 others, 1978, Occurrence and evolution of salt in the deep Gulf of Mexico, *in* Bouma, A. H., Moore, G. T., and Coleman, J. M., eds., Framework, facies, and oil-trapping characteristics of the upper continental margin: American Association of Petroleum Geologists Studies in Geology 7, p. 43–65.

Weeks, A. W., 1945, Quaternary deposits of Texas Coastal Plain between Brazos River and Rio Grande: American Association of Petroleum Geologists Bulletin, v. 29, p. 1693–1720.

Weimer, P., 1989, Sequence stratigraphy of the Mississippi Fan (Plio-Pleistocene), Gulf of Mexico: Geo-Marine Letters, v. 9, no. 4, p. 185–272.

—— , 1990, Sequence stratigraphy, facies geometries, and depositional history of the Mississippi Fan, Gulf of Mexico: American Association of Petroleum Geologists Bulletin, v. 74, p. 425–453.

West, R. C., Psuty, P. C., and Thom, B. G., 1985, Las tierras bajas de Tabasco: Gobierno del Estado de Tabasco, Biblioteca Basica Tabasqueña 8, 490 p.

White, W. A., 1970, The geomorphology of the Florida Peninsula: Florida Bureau of Geology Geological Bulletin 51, 164 p.

Winker, C. D., and Edwards, M. B., 1983, Unstable progradational clastic shelf margins: Society of Economic Paleontologists and Mineralogists Special Publication 33, p. 139–157.

Woodbury, H. O., Murray, I. B., Pickford, P. J., and Akers, W. H., 1973, Pliocene and Pleistocene depocenters, outer continental shelf, Louisiana and Texas: American Association of Petroleum Geologists Bulletin, v. 57, p. 2428–2439.

Worzel, J. L., and Burk, C. A., 1978, The margins of the Gulf of Mexico, *in* Watkins, J. S., ed., Geological and geophysical investigations of continental margins: American Association of Petroleum Geologists Memoir 29, p. 403–419.

WorzeL, J. L., Leyden, R., and Ewing, M., 1968, Newly discovered diapirs in the Gulf of Mexico: American Association of Petroleum Geologists Bulletin, v. 52, p. 1194–1203.

Worzel, J. L., Bryant, W. R., and Scientific Party, 1973, Initial reports of the Deep Sea Drilling Project: Washington, D.C., U.S. Government Printing Office, v. 10, 748 p.

MANUSCRIPT ACCEPTED BY THE SOCIETY OCTOBER 9, 1990

ACKNOWLEDGMENTS

Arnold Bouma and Richard Rezak reviewed the manuscript and offered useful comments and suggestions for its improvement. Roger Saucier called attention to important recent publications on the lower Mississippi Valley. Thanks to G. H. Lee for the 3D views of the slope and to Jim Coleman for providing the figure of the Mississippi Canyon and supplying information on the canyon.

Chapter 3

Structural framework

Thomas E. Ewing
Frontera Exploration Services, San Antonio, Texas 78209

INTRODUCTION

The Gulf of Mexico basin is a roughly circular structural basin that has been filled with 0 to 15 km of sedimentary rocks ranging in age from Late Triassic to Holocene. The crust beneath the central part of the basin is oceanic in character; this is surrounded by continental crust, which underneath much of the basin has been greatly attenuated by rift-related extension (Worzel and Burke, 1978; Buffler and Sawyer, 1985; and Chapter 4, this volume).

Superimposed on the basin are second-order structural features that modify the overall simple geometry (Plate 2):

1. Basins of enhanced subsidence and deposition, as well as intervening platforms or arches ("uplifts" sensu lato), which subsided less than surrounding areas. These features were formed by spatially varying rates of lithospheric cooling related to the early synrift history of extension, and amplified by differential sediment loading.

2. Basin-margin fault systems in the northwestern and western segments of the Gulf of Mexico basin, due to flexing of the basin rim and uplift of adjacent provinces.

3. Structural basins and uplifts sensu stricto with resultant erosional unconformities and clastic wedges related to active tectonics: block faulting, epeirogenic doming, and the formation of fold-thrust belts.

4. Salt diapirs and related structures formed from flow of Jurassic salt that lies at the base of the sediment column. Different original salt thicknesses and different loading histories have created distinct salt-diapir provinces characterized by their style and age of diapirism. A "peripheral graben system" formed in the northern part of the basin because of downdip lateral movement of sediments overlying salt. Lateral movement of salt from beneath the rapidly prograding Cenozoic clastic packages has created large salt thrusts and related folds in the central part of the basin.

5. Syndepositional normal fault systems ("growth faults") formed by gravitational failure of the thick, mostly Cenozoic clastic sedimentary wedges. Rapid loading of these wedges and concomitant overpressuring of deeply buried sediments have resulted in both basinward sliding of sediments and deep flow of

shale. Other features such as shale ridges, shale diapirs, and shale-cored compressional folds were formed by the same processes.

Definitions

In the Gulf of Mexico Basin, the term "embayment" is frequently used. The term refers to a basin or subbasin, usually containing Middle Jurassic salt and Upper Jurassic sediments, that is more or less open toward the main Gulf of Mexico basin. It overlaps with the geomorphic use of the term, as Mesozoic or Cenozoic coastal plain sediments often form low-lying areas between cratonic uplifts. These "embayments," like many of the "basins" within the larger Gulf of Mexico basin, are passive features resulting from differential subsidence and sediment loading. Between these "embayments" or "basins" lie areas of lesser subsidence, which are variably known as "uplifts" (sensu lato), "arches," or, less commonly, "platforms." True structural "basins" and "uplifts" are also present, primarily in the western province.

"Diapir provinces" are areas with a concentration of salt diapirs and related features; they are usually rimmed by zones of salt pillows, and occur within a much larger region that is underlain by salt. "Syndepositional normal faults," where downthrown expansion of equivalent stratigraphic intervals occurs within thick clastic stratigraphic wedges, are commonly termed "growth faults" for convenience, even though not all of the intervals deformed by any particular fault will in fact expand across it. These faults are usually listric (flattening downward) in geometry and are inferred not to involve basement.

Provinces

The structural framework of the Gulf of Mexico basin can be subdivided into three major structural provinces (Fig. 1), which correspond to the three major lithofacies provinces that persist from the Late Jurassic to the Holocene.

1. The northwestern progradational margin (from northeastern Mexico to Alabama), which may be subdivided into an

Ewing, T. E., 1991, Structural framework, *in* Salvador, A., ed., The Gulf of Mexico Basin: Boulder, Colorado, Geological Society of America, The Geology of North America, v. J.

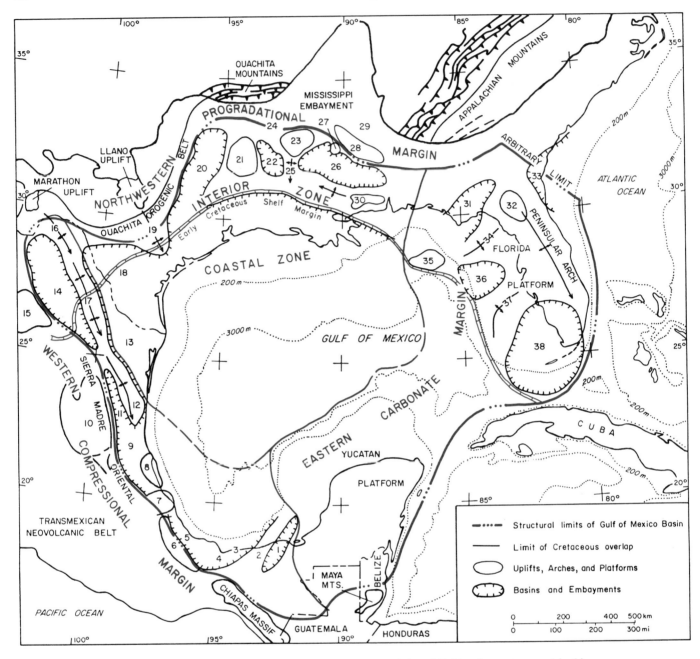

Figure 1. Index map showing major structural elements and the subdivision into sectors used in this chapter.

Interior Zone (mostly Mesozoic structures) and a Coastal Zone (mostly Cenozoic structures). On this margin, an early mixed clastic-carbonate shelf margin of Late Jurassic and Early Cretaceous age is affected by depositional and active-tectonic basins and uplifts, and salt diapirism in the "interior salt diapir provinces." In the Coastal Zone, the Mesozoic strata are buried beneath a thick wedge of Upper Cretaceous and Cenozoic coarse clastic sediments, which have prograded the shelf margin hundreds of kilometers seaward and generated "growth fault" systems and coastal and offshore salt diapir provinces.

2. The eastern carbonate margins (the Florida and Yucatan platforms). Here, broad and in general poorly known basins and uplifts are largely concealed by two major carbonate banks, the Florida and Yucatan platforms, which have persisted since Middle Jurassic time.

3. The western compressional margin (from the Isthmus of Tehuantepec to northeastern Mexico). In this area, the Mesozoic carbonate margin has been affected by Laramide (Late Cretaceous–Eocene) folding and thrusting, then covered by prograding fine-grained clastics and locally subjected to Neogene tectonics.

This chapter provides a brief description of the morphology, timing, and regional relations of each of these tectonic features as revealed both by surface geology and by oil and gas exploration in the subsurface. Chapter 4 discusses the crustal structure of the basin; Chapter 5 discusses in more detail the development of the various tectonic features in the Gulf of Mexico Basin.

FEATURES OF THE NORTHWESTERN PROGRADATIONAL MARGIN

This margin extends from the eastern edge of Cordilleran compressional deformation near the Texas-Mexico border eastward to southwestern Alabama and westernmost Florida. This is the best-known part of the entire Gulf of Mexico basin, as it has been the target of innumerable oil and gas exploration wells and the subject of many publications. It also contains the most complex assemblage of second-order structural features.

As mentioned above, the northwestern margin can be divided into two zones. They are separated by the trend of the Lower Cretaceous (Aptian-Albian) reefs from the Rio Grande north and east to the Mississippi delta (Fig. 1). Subsidence modeling by Buffler and Sawyer (1985) has indicated that this reef trend marks the approximate boundary between the interior, less extended, and the coastal, highly extended continental crust (see Chapter 4, this volume).

1. The *Interior zone* is mainly the product of Mesozoic tectonism and sedimentation. In this area, a broad rift complex of Late Triassic–Early Jurassic age was followed by later differential subsidence (Nunn and others, 1984). Salt was mobilized to form the interior diapir provinces. Late Mesozoic uplifts and related igneous activity interrupted the subsidence history. The area experienced only minor Cenozoic reactivation of structures, gentle uplift, and tilting.

2. In the *Coastal zone,* the Mesozoic history of extension and subsidence is concealed by a 10- to 15-km-thick Upper Cretaceous and Cenozoic clastic sequence. Progradation of this sequence into the deep Gulf of Mexico basin formed linear bands of syndepositional normal faults (growth faults) and related shale-cored, possibly mobile features, and destabilized thick salt deposits to form a number of diapir provinces.

Interior Zone

The inner margin of the Interior Zone—the margin of the Gulf of Mexico basin—is highly irregular, forming three embayments separated by positive areas or arches (Fig. 1, and Plate 2). The *Rio Grande embayment* in south Texas is a little-deformed basin showing faint signs of Laramide compression. The *San Marcos arch* to the northeast is a broad area of lesser subsidence that extends southeast from the Llano uplift of central Texas, where Precambrian rocks are exposed.

A very wide complex of embayments extends from the San Marcos arch eastward to Alabama. It is marked at its craton-

ward edges by a nearly continuous "peripheral graben system," which occurs just downdip of the updip limit of Jurassic salt. The broad, relatively positive *Sabine uplift* occurs in the midst of this embayment, subdividing it into the *East Texas basin* to the west, and the *North Louisiana Salt basin* and *Mississippi Salt basin* to the east. Each of these three basins contains many salt diapirs. To the north of the North Louisiana and Mississippi Salt basins is the *Mississippi embayment,* a broad spoon-shaped depression of Late Cretaceous and younger age, which extends north to Cairo, Illinois. It is separated from the southern basins by the *Monroe uplift,* mostly of Late Cretaceous age. South of the complex embayment is the *Wiggins uplift,* and related, relatively positive features, which lie immediately landward of the Early Cretaceous reef trend. South of the Sabine uplift, a marked increase in dip called the *Angelina-Caldwell flexure* (Anderson, 1979) occurs just landward of the reef trend. Elsewhere, the reef trend occurs on a gentle homocline dipping into the Gulf of Mexico basin.

Rio Grande embayment. This basin occupies the area northeast of the El Burro uplift and the Peyotes and Picachos arches (a zone of open Laramide folds in Cretaceous strata, which overlie less extended and high-standing basement blocks) and south of the basin-marginal Balcones fault zone (Fig. 2). It contains some Jurassic salt in its southern and southeastern portions, and a few salt domes (part of the Rio Grande diapir province) have formed landward of the reef trend, but salt tectonics is only a minor part of the basin's history. Jurassic and Cretaceous sedimentation was nearly continuous, and recorded a general subsidence and transgression in the Early Cretaceous, forming a paleobathymetric Maverick basin in Albian time behind the bounding Stuart City reef (Rose, 1972). The northern part of the basin is marked by Late Cretaceous volcanic and intrusive activity, which is associated with a minor structural high near Uvalde (Ewing, 1988a). Laramide compression developed a zone of northwest-trending low-amplitude folds in the western part of the basin.

Structural styles change along a linear zone, the Frio River line (Ewing, 1987), just northeast of the basin axis. The peripheral graben system characteristic of the northern Gulf of Mexico basin begins at this line as the "Charlotte-Jourdanton fault zone," and encircles several Mesozoic salt structures (Fig. 3). Also, the Balcones fault zone begins at this line and increases in displacement rapidly to the northeast. The line forms a hingeline (possibly with Paleozoic ancestry), where regional uplift, subsidence, and deformation histories have changed from Cretaceous time to the present.

San Marcos arch. The arch is a broad area of lesser subsidence between the Rio Grande embayment and the East Texas basin. It is marked by a southeastward bowing of outcrop patterns as young as Eocene, a southeastward bowing of the Jurassic subcrop and Paleozoic structural and stratigraphic trends, and by pronounced changes in the Cretaceous stratigraphy. Three major fault zones cross the San Marcos arch in a southwest-northeast direction (Fowler, 1956).

The *Balcones and Luling fault zones* form one broad area of

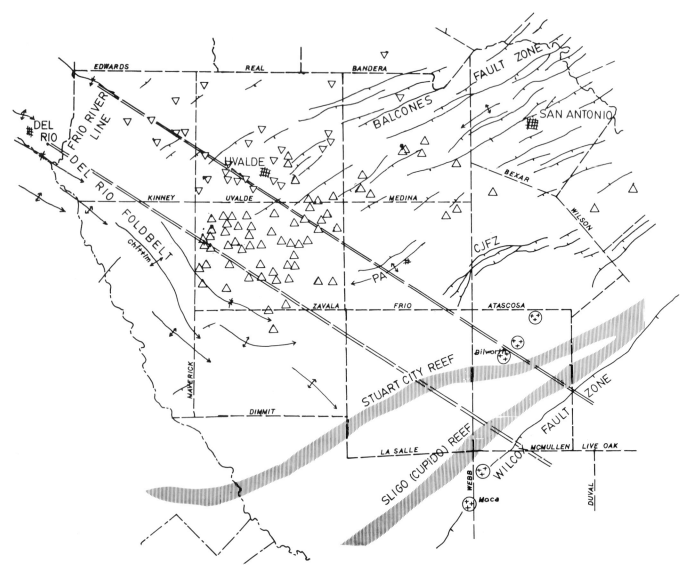

Figure 2. Structural elements of the Rio Grande embayment, from Ewing (1987). Triangles = Cretaceous igneous centers; PA = Pearsall arch; CJFZ = Charlotte-Jourdanton fault zone (part of the peripheral graben system).

basement-involved normal faults over a large region of central Texas between the Llano uplift and the peripheral graben system (Plate 2; Fig. 3). The Balcones fault zone is dominated by down-to-basin normal faults parallel to the trend of the buried Ouachita orogenic belt, with cumulative displacements up to 500 m. The faults have a pronounced physiographic expression where they juxtapose resistant Lower Cretaceous and less resistant Upper Cretaceous rocks, forming the Balcones Escarpment. The Luling fault zone to the southeast contains mostly up-to-basin normal faults with a cumulative throw of 300 to 600 m. Although faults of both senses occur throughout the area, the overall pattern is that of a graben about 50 km wide trending southwest-northeast across the San Marcos arch (Murray, 1961). Displacement of the faults decreases away from the axis of the arch in both directions. Both sets of faults displace the sub-Mesozoic unconformity and

the underlying Ouachita rocks. Their continuation at depth is not known, but the two fault zones overlie major crustal-scale faults, both within the Ouachita orogenic belt, and beneath its frontal portions (Fig. 3). The Balcones igneous province, consisting of mostly ultramafic, alkaline volcanoes and related intrusions of Late Cretaceous age, lies generally along the Balcones and Luling fault zones (see Chapter 6, this volume). In some areas, individual fault zones have aligned volcanic centers (Ewing and Caran, 1982).

The timing of movement on the Balcones and Luling fault zones is not well constrained, but Miocene unroofing of the Edwards Plateau northwest of the Balcones system is recorded in sediments of the Coastal Plain (Weeks, 1945). Such Miocene uplift is part of the general Neogene uplift of the western United States. An episode of Late Cretaceous movement has also been indicated (Sellards, 1932); earlier movement may also be possible. No Quaternary movement has yet been demonstrated on either fault zone.

The *Karnes fault zone* strikes northeast-southwest about 30 km northwest of the Early Cretaceous reef trend. It passes southwest through a set of en echelon right-stepping normal faults into the Charlotte-Jourdanton fault zone, and appears to pass northeast through en echelon left-stepping normal faults into the Milano fault zone (Ewing and others, 1990), making this fault zone a link in the peripheral graben system. Major normal faults define a graben about 6 to 8 km wide, with about 200 m of displacement at depth. Major movement on this fault zone began at the latest by the late Albian, as sediments of this age are thickened in the graben (Rose, 1972). Fault displacement has continued into the Tertiary; upper Eocene strata are displaced at the surface. Recent seismicity at the Fashing field due to depressurization of a gas reservoir indicates that aseismic creep may be occurring on many faults of this system (Pennington and others, 1988).

East Texas basin. This basin is bounded on the west and north by the peripheral graben of the Mexia-Talco fault system, on the northeast by a shallow syncline (Murray, 1961) that connects to the North Louisiana Salt basin, on the east by the Sabine Uplift, and on the south by a homocline that dips southward toward the Houston embayment. The northeast-trending basin axis is marked by a concentration of salt diapirs that forms the East Texas diapir province, which is ringed by belts of high-relief and low-relief salt pillows (Fig. 4).

The basin was the site of deposition of a thick late Middle Jurassic salt layer, followed (after an interval of carbonate deposition) by a large clastic influx during the Late Jurassic and Early

Cretaceous. Most of the salt features were initiated during the Late Jurassic, and grew into diapirs in the Early Cretaceous. In Early Cretaceous time the basin had substantial bathymetric expression (Rose, 1972). Minor subsidence continued into the Tertiary; strata of middle Eocene age are preserved in the center of the basin.

The *Mexia-Talco fault system* is a continuous zone of faulting that extends from central Texas to the Arkansas border. It consists of three segments of symmetric grabens about 8 to 12 km across (Fig. 5A): the Talco fault zone in northeast Texas, the Mexia fault zone in north-central Texas, and the Milano fault zone in central Texas. These segments are linked by zones of en echelon left-stepping down-to-basin normal faults (Jackson, 1982). The fault zone lies immediately downdip of the pinchout of the Jurassic salt; the graben segments form where the pinchout parallels regional strike. There is no conclusive evidence for basement involvement in the Mexia-Talco fault system, although this has been the subject of some controversy (Rodgers, 1984). Thick Upper Jurassic and Lower Cretaceous strata in the graben segments indicate that movement began in the Jurassic; surface sediments of Paleocene and Eocene age are also offset by the faults. Farther toward the center of the basin, a band of normal faults that were active in the Late Jurassic overlies a line of salt pillows.

The *Mount Enterprise fault zone* (Plate 2; Fig. 5B) is a roughly east-west–trending belt of normal faults on the southern and southeastern border of the East Texas basin proper, and extends east onto the Sabine uplift (Jackson, 1982). The fault zone varies in character, but includes down-to-the-north normal faults of both listric and straight geometry with associated antithetic faults. Major movement occurred in the Late Jurassic and

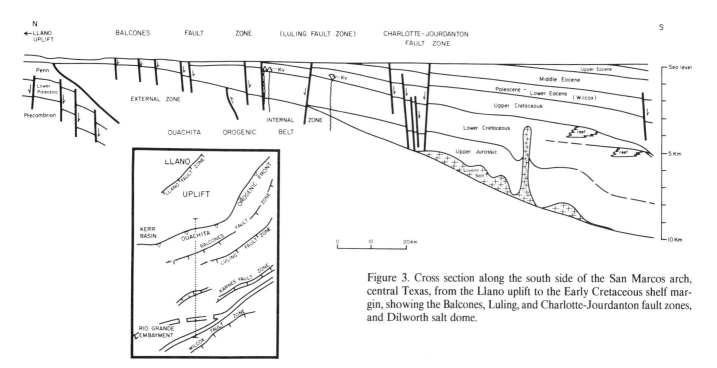

Figure 3. Cross section along the south side of the San Marcos arch, central Texas, from the Llano uplift to the Early Cretaceous shelf margin, showing the Balcones, Luling, and Charlotte-Jourdanton fault zones, and Dilworth salt dome.

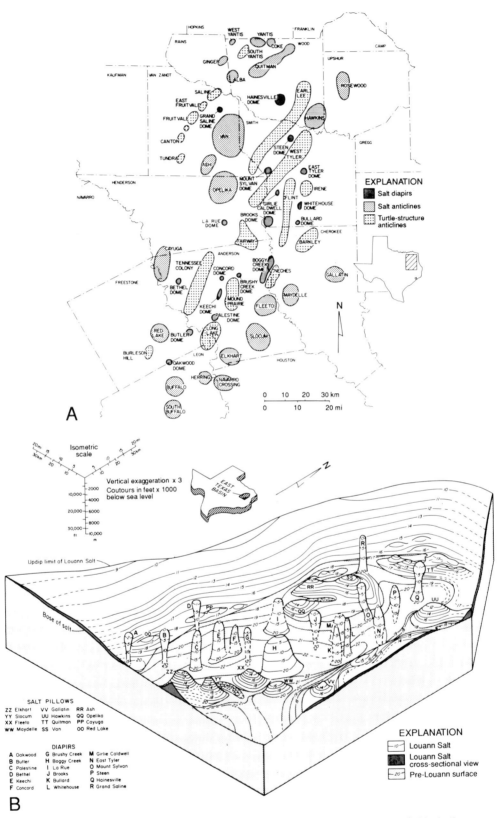

Figure 4. A, Distribution of salt structures, from Kreitler and others (1981). B. Isometric block diagram of the salt structures of the East Texas basin; from Jackson and Seni (1983).

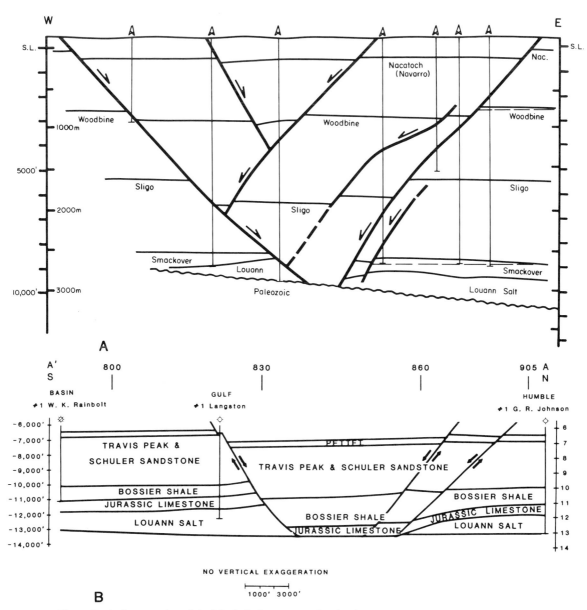

Figure 5. A, Cross section of the Mexia fault zone near Currie Field, based on well logs, from Locklin (1984). B, True-depth structural section of the Mt. Enterprise fault zone from seismic and borehole data; from Ferguson (1984).

Early Cretaceous, but surface strata are displaced and seismicity suggests continuing deformation.

Sabine uplift. The Sabine uplift lies between the East Texas and North Louisiana basins, and north of the Early Cretaceous shelf margin. Its present form is outlined at the surface by outcrops of Paleocene and lower Eocene strata (Wilcox Group) surrounded by younger rocks. The uplift has a complex history; its present form dates from the Late Cretaceous (Jackson and Laubach, 1988). Its boundaries are nearly everywhere gentle homoclines into the surrounding basins, with local fault and fracture zones on the north and south sides.

The uplift area in general has a thin layer of Jurassic salt that forms low-amplitude swells (Fig. 6), but there are "islands"

where salt and overlying Upper Jurassic carbonates are absent, notably the "Sabine Island" (Nicholas and Waddell, 1982). This area, at the south end of the Sabine uplift, may be part of an east-west series of salt-free basement highs, which includes the Wiggins uplift. Thicknesses of Upper Jurassic and Lower Cretaceous clastic rocks show little effect of the uplift.

In mid-Cretaceous (mid-Cenomanian) time the Sabine uplift area, as well as most of southern Arkansas and northern Louisiana, was uplifted to form the "Southern Arkansas uplift" (see Chapter 14, this volume). This large positive feature was subsequently eroded and partially covered with Woodbine or Tuscaloosa clastic sediments. The Sabine uplift was then raised again, and the Woodbine and various older strata were stripped

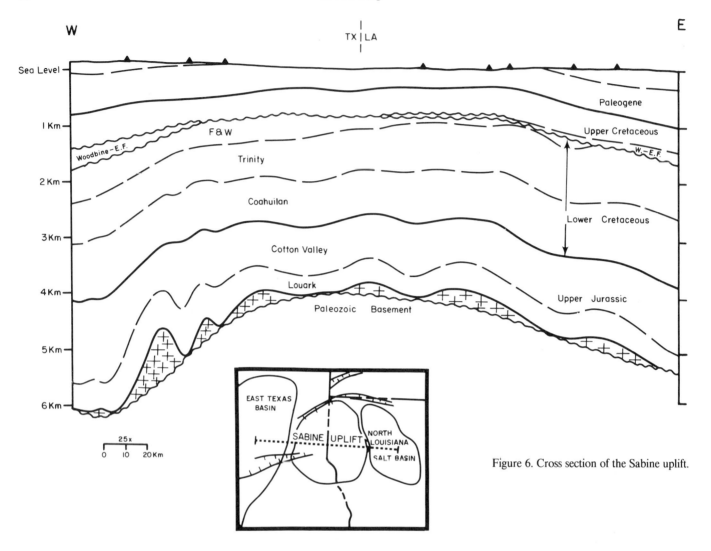

Figure 6. Cross section of the Sabine uplift.

and redeposited before deposition of the Austin Group chalk (Fig. 6; Granata, 1963; Halbouty and Halbouty, 1982). A third episode of uplift occurred in Eocene time, and is responsible for the present outcrop pattern (Jackson and Laubach, 1988). The southern margin of the uplift was sharply downflexed in Tertiary time, probably as a result of sediment loading in the Gulf of Mexico basin to the south. This resulted in the formation of the *Angelina-Caldwell flexure* (Anderson, 1979).

The *Rodessa fault zone* is a northeast-trending normal fault system on the northern flank of the Sabine uplift (Plate II). Little has been reported concerning its age and origin, but it displaces Lower Cretaceous over 100 m down to the north. Possibly related is the *Stateline fault zone,* which extends eastward along the Arkansas-Louisiana state line.

North Louisiana Salt basin. This is a northwest-southeast–trending basin of Jurassic through Tertiary age (Kupfer and others, 1976). Like the East Texas basin, it contains an axial salt-diapir province and flanking salt pillows. To the north is the *Southern Arkansas fault zone,* a peripheral graben system similar

to the Mexia-Talco fault system. To the east and southeast is the *LaSalle arch,* a very low arch marked chiefly by the absence of salt structures. To the northeast is the *Monroe uplift,* a Late Cretaceous feature described below.

The North Louisiana Salt basin has a Late Jurassic and Early Cretaceous history of subsidence and salt tectonics (Scardina, 1982; Lobao and Pilger, 1985), similar to the East Texas basin. In mid-Cenomanian time, the area formed part of the Southern Arkansas uplift. Much of the Lower Cretaceous sequence was eroded, and was preserved only in the deeper salt-withdrawal basins. Minor subsidence resumed in Late Cretaceous time and continued into the Tertiary. Fault movement on the Southern Arkansas fault zone continued at least as late as the middle Eocene.

Monroe uplift. This uplift is bounded by the North Louisiana Salt basin to the southwest, the Mississippi Salt basin to the south and southeast, and the Desha basin and Mississippi embayment to the north. The uplift is located astride the peripheral graben system, which it interrupts. It is marked by two or more

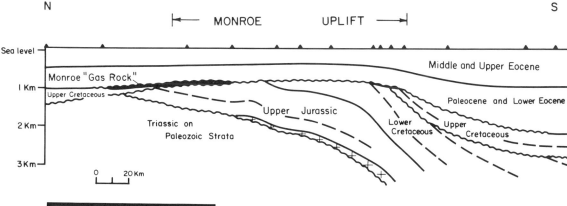

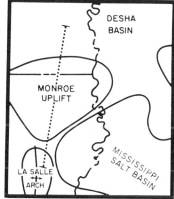

Figure 7. Cross section of the Monroe uplift, Louisiana and Arkansas. The pre–Upper Cretaceous unconformity represents the broad Southern Arkansas uplift; the post–Upper Cretaceous unconformity, the more local Monroe uplift.

coalescing unconformities representing major episodes of uplift and erosion (Johnson, 1958). Abundant igneous activity of Late Cretaceous age is intimately associated with the uplift.

There is no convincing evidence for the existence of the Monroe uplift as such before the Late Cretaceous. The area was near the center of the "Southern Arkansas uplift" of mid-Cretaceous age; units as old as Upper Jurassic subcrop beneath Upper Cretaceous strata over the northern part of the Monroe Uplift, with as much as 3,000 m of erosion indicated (Fig. 7; see Chapter 14, this volume). In Late Cretaceous time, a second, more local uplift formed the Monroe uplift proper, with a very pronounced angular unconformity between Campanian and Paleocene strata on its southern and southeastern flank (Fig. 7). The Monroe uplift formed a high ground on which the "Monroe gas rock" carbonate shoal of latest Cretaceous age was built. The area then subsided to receive lower Tertiary sediments. Continued slight uplift at the present time has been suggested by warping of Quaternary terraces and the results of repeated geodetic measurements (Burnett and Schumm, 1983; D. Trahan, personal communication, 1984).

Mississippi embayment; Desha basin. The Mississippi embayment is a southwest-plunging synclinal feature filled with sediments of Late Cretaceous, early Tertiary, and Quaternary ages, which overly unconformably a Paleozoic sedimentary section (Plate 3). The embayment occupies a pronounced physiographic embayment extending north-northeast from Louisiana to southernmost Illinois. It is located over an old graben system, the

Mississippi Valley graben (Hildenbrand and others, 1977), also known as the Reelfoot rift (Ervin and McGinnis, 1975). This graben system had a Cambrian origin, possibly formed along a Precambrian shear zone. It is, however, lined with large plutons of Late Cretaceous age (Hildenbrand and others, 1982; Byerly, this volume). Subsidence in most of the embayment began after this intrusive episode, reaching a maximum in Paleocene and Eocene time (Cushing and others, 1964). The superposition of the embayment over the old graben is asymmetric; the axis of post-Cretaceous subsidence lies on the eastern flank of the graben. The term "Desha basin" refers to the southern end of the Mississippi embayment, where it abuts the Monroe uplift to form a nearly closed basin (Murray, 1961).

Mississippi Salt basin. This basin extends from eastern Louisiana to westernmost Alabama. It contains numerous salt diapirs and related structures. It is bordered on the north and east by the peripheral graben system, which includes in this area the Pickens, Quitman, Gilbertown, and Pollard fault zones. On the south, it adjoins the Wiggins uplift and the "Adams County high"; to the west is the structurally low LaSalle arch. The history of the basin is similar to that of the East Texas and North Louisiana basins (Hughes, 1968), except that major clastic sedimentation persisted into the Late Cretaceous. The area of salt diapirs is larger than either of the other two interior diapir provinces, but little has been published about the regional distribution and timing of salt structures. Various local structures have been described, however; the most major of these is the north-trending Mobile graben in southwestern Alabama.

The *Pickens, Quitman, Gilbertown,* and *Pollard fault zones* bound the basin in central Mississippi and southwestern Alabama. Like the composite Mexia-Talco system, these zones include symmetric grabens about 8 to 12 km wide, often with great expansion of the Mesozoic clastic section within them (Murray, 1961). They are apparently associated with the cratonward pinchout of Jurassic salt. The faults reach the surface only near the Mississippi-Alabama state line (Quitman-Gilbertown area), where they displace strata as young as Miocene.

North of the peripheral grabens, a slight steepening of pre-Jurassic contours marks the "Central Mississippi deformed belt" (Plates 2 and 3). There is no published evidence that indicates this deformed belt was active during Gulf of Mexico basin evolution. Its chief definition has been as an area of shallow Paleozoic "basement" north of the Mississippi Salt basin and south of the (Paleozoic) Black Warrior basin.

The *Jackson dome* interrupts the homoclinal northern flank of the Mississippi Salt basin. It has a radius of about 19 km and uplifts Jurassic strata up to 2,000 m above the surrounding area. Its history is similar to that of the later stages of the Monroe uplift; abundant igneous activity in the Late Cretaceous and the deposition of latest Cretaceous reefal carbonates are associated with the uplift. Continued uplift has affected Tertiary strata, with up to 160 m of vertical uplift of exposed upper Eocene (Jackson Group) strata.

Wiggins uplift. This uplift lies in southernmost Mississippi and adjoining states. It is bounded on the north by the Mississippi Salt basin and on the south by the deep Cenozoic-filled basin of the Coastal Zone. Associated uplifts, possibly in en echelon position, occur farther west in southwestern Mississippi (the Adams County high), and may continue westward into the LaSalle arch. The Wiggins uplift is characterized by the absence of salt, local absence of the lower part of the Upper Jurassic sequence, and reduced sedimentation rates during some of the Cretaceous (Cagle and Khan, 1983). Slight present-day uplift is indicated by geodetic and geomorphic data (Burnett and Schumm, 1983). The uplift probably represents a block of thicker, possibly Paleozoic continental crust left behind during Late Triassic–Jurassic rifting, and may have fault-bounded margins. A Late Cretaceous volcanic center has been reported on its southern flank (Braunstein and McMichael, 1976).

Coastal Zone

South of the Early Cretaceous carbonate shelf margin, sub-basins that are due to Mesozoic rifting, differential salt deposition and deformation, subsidence, and uplift, if present, are buried by thick Cenozoic sediments and obscured by Cenozoic salt mobilization. The effects of some of these subbasins are detectable, however, in the distribution of salt diapirs. Salt features occur in distinct diapir provinces (Fig. 8), each characterized by spatial clustering and/or distinctive styles of salt occurrence. Between some of these provinces are areas with few or no salt features. These areas may represent old highs (horsts) where little or no salt was deposited in the Jurassic. The diapir provinces then may represent Jurassic subbasins that contained thick salt, which was later mobilized into diapiric structures. However, lateral (basin-ward) movement of salt as salt tongues or sheets due to shelf-margin loading and progradation, as documented for the Neogene (see below), may obscure the effects of Jurassic subbasins.

The major "growth-fault" (syndepositional normal fault) trends of Cenomanian through Quaternary age are, besides the salt structures, the distinguishing feature of the Coastal Zone. These fault trends were formed by gravitational instability at the rapidly prograding shelf margin, where large quantities of sand and mud were dumped on top of thick, undercompacted and overpressured marine mudstones (Winker and Edwards, 1983).

The embayments and arches of the Interior Zone have only a muted expression in the Coastal Zone, forming the Coastal Zone Rio Grande embayment (and its extension in northeastern Mexico, the Burgos basin), the Houston embayment, and the broad depocenter in southern Louisiana. However, these embayments are, in the main, load-induced depocenters—a response to the enhanced progradation of the shelf margin, created by the deltas of the large rivers that flow through the embayments of the Interior Zone.

Diapir provinces. Rio Grande diapir province. A few salt diapirs and inferred salt domes are known from south Texas coastal zone areas; they are contiguous with the Rio Grande embayment diapirs mentioned above. The extent of the province is indefinite; isolated salt structures are inferred to be present in several counties of south Texas (Posey, 1986). Salt structures in this area, unlike other coastal diapir provinces, do not disrupt the trend and style of the Paleogene growth fault zones.

Houston diapir province. About 60 salt diapirs, as well as a few known or inferred pillows, are reported from the Houston embayment of southeast Texas. The province is continuous with the South Louisiana province to the east, but is bounded on the south and west by areas free of significant salt structures. On the northwest, a small line of salt diapirs lying behind the Cretaceous reef trend connects the Houston diapir province with the East Texas basin, possibly representing a small connecting graben filled with thick salt.

The salt diapirs of the Houston embayment are generally shallow, with tops at less than 600 m depth (Halbouty, 1979), and form narrow cylinders with mushroom tops. In an area southeast of Houston, however, they may be as much as 4,000 m deep; this area roughly coincides with the late Oligocene depocenter. Salt diapirs are frequently aligned along major growth-fault trends, and had their maximum growth after the main phase of shelf margin progradation and attendant growth faulting (Ewing, 1983).

South Louisiana Shelf diapir province. This diapir province adjoins the Houston province on the west, and the Texas-Louisiana slope province on the south. It includes both onshore coastal south Louisiana and the wide continental shelf. Most diapirs in this province are small, cylindrical, and shallow, as in the Houston province; but more extensive salt ridges and large massifs are present, mostly in the areas of Pliocene and Pleistocene progradation (Murray, 1966). Areas of deeper-seated salt structures (probably diapiric) occur near Lafayette and New Orleans, associated with Oligocene and Miocene depocenters. An area with few salt structures (possibly an area of thinner salt originally) occurs south of New Orleans. The relation between

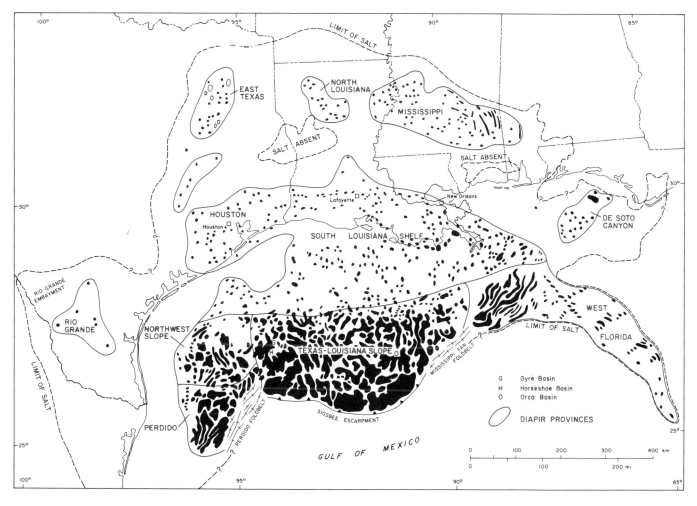

Figure 8. Salt diapir provinces of the northwestern progradational margin.

salt diapirism and shelf-margin progradation has not been fully worked out, but may be similar to the Houston area; however, more abundant intrusions and related crestal faulting and withdrawal basin formation have largely disrupted the growth-fault trends.

In the southern part of the province, evidence of horizontal movement of salt has been obtained, with penetration of salt along glide planes of fault systems and diapirs that appear to root in a relatively shallow horizon (Brooks, 1989). These occurrences appear to represent collapsed or deflated salt tongues. This area was probably much like the present continental slope during Pliocene and early Pleistocene time, but has since been covered by shelf-margin progradation.

Texas-Louisiana Slope diapir province. This province is contiguous with the South Louisiana Shelf diapir province; the De Soto Canyon diapir province to the east; and the Northwest Slope diapir province to the west. Salt structures in the slope areas have very prominent effects on bathymetry, allowing confident mapping of the major features (see Plate 1). Recent hydrocarbon exploration in the region is resulting in much improved

understanding of salt tectonism in this province, focusing particularly on the importance of lateral migration of salt "out from under" the prograding Neogene shelf margins (Worrall and Snelson, 1989).

At the base of the lower slope is the Sigsbee Escarpment, where still-coherent salt tongues or salt nappes are clearly imaged on seismic data. They override continental-rise sediments of Miocene and younger ages (Amery, 1969; Worzel and Burk, 1978; Humphris, 1978). The middle and lower slope in the major part of the province is characterized by very broad, gentle- to steep-sided diapiric massifs (Martin, 1980; Lee and others, 1989) with a random distribution and shape; these compose 50 to 90 percent of the slope (Plate 2). The massifs are separated by narrow, deep intraslope basins that commonly formed from truncated submarine canyons (Bouma and others, 1978); important ones include the *Gyre Basin* off southeast Texas, the nearby *Horseshoe Basin,* and the *Orca Basin* offshore south-central Louisiana (see Coleman and others, this volume). The large massifs are inferred to have developed from horizontal salt tongues, which have been disrupted by loading and consequent diapiric uplift (Humphris,

1978; Ray, 1988). Base of salt is occasionally visible on modern seismic reflection data (Lee and others, 1989). The upper slope contains relatively small, cylindrical salt stocks and salt ridges similar to the adjacent South Louisiana Shelf province. These may have formed by evolution from older midslope massifs under the higher sedimentation conditions near the Pleistocene shelf margin.

To the east, structures in the Mississippi Fan area have been described as linear salt ridges (Martin, 1980), but more recent work indicates that this linearity was an artifact of the data (R. Martin, personal communication, 1989). The salt features here appear to be salt tongues that have been covered and deformed by the rapidly deposited Pleistocene Mississippi Fan (see Chapter 12, this volume). A belt of salt-cored folds, the *Mississippi Fan foldbelt,* extends eastward into the West Florida diapir province. These are salt-cored, basinward-verging folds and thrust faults (Weimer and Buffler, 1989).

Northwest Slope diapir province. This province adjoins the Texas-Louisiana Slope diapir province to the east and the Perdido diapir province and foldbelt to the south. It is differentiated from them by a subdued bathymetry and reduced intensity of salt-tectonic activity. Along the upper slope, salt stocks similar to those to the east are known; but most of the slope is underlain by a shallow reflector inferred to be salt, which is gently deformed into anticlinal forms, uplifted, and faulted with only a few diapirs developed (Martin, 1980). This may represent a salt-intrusion complex of tongues coalescing into canopies (Hardin, 1989). The area's history is similar to the Texas-Louisiana Slope area, but it is less disrupted by sediment loading.

Perdido diapir province. This province adjoins the Northwest Slope diapir province. It is characterized by salt diapirs and narrow anticlines on the upper slope, large irregular massifs occupying most of the middle and lower slope, and by salt-cored anticlines of the *Perdido foldbelt* deforming abyssal-plain strata (Martin, 1984; Blickwede and Queffelec, 1988; Ray, 1988) that trend northeast along the base of the slope and upper rise. These salt-cored anticlines are coaxial with, and may pass southward into, the shale-cored anticlines of the Mexican Ridges (Buffler and others, 1979; Martin, 1980).

Growth-fault trends. Major strike-elongate zones of normal faulting, occurring entirely within the sedimentary column, form the most striking features of the Coastal Zone in those areas where salt tectonism is limited. These fault zones are intimately related in location and age to the prograding clastic shelf margins of Late Cretaceous. Tertiary, and Quaternary age. The faults can cause tremendous expansion of the upper-slope and shelf-margin marine clastic deposits associated with these trends. These faults are referred to for convenience as "growth faults," although they do not necessarily cause expansion ("growth") of section in all of the rock units that are displaced. Structures related to growth faulting form many of the major oil and gas traps in the region, and hence have been intensively studied (see Chapter 5, this volume).

Growth faults occur in a spectrum of structural styles (Fig.

9). In general, they may be classified as: (1) glide systems, in which a basal decollement is visible, into which overlying strata have been rotated, and in which lateral translation is greater than vertical motion; and (2) deep listric systems—the "differential compaction" faults of Bruce, 1973), in which faults flatten into a deep, diffuse detachment level (probably in ductile shale), and vertical subsidence is greater than horizontal displacement (Bruce, 1973; Ewing, 1988b).

Glide systems include intraformational slumps, "domino-style" detachments displacing a great thickness of sediments laid down before faulting began, and "escalator-style" detachments displacing highly expanded sediments, which form large pods in the upper plate (Ewing, 1988b). Glide-fault systems are inferred to represent failure of the progradational shelf margin on discrete low-strength decollement surfaces, with substantial basinward transport of thick sedimentary packages. This represents the "simple stratigraphic detachment" of Lopez (1989).

Deep listric systems reflect a dominance of vertical tectonics with moderate extension. Shelf-margin sediments sank down into highly overpressured and undercompacted slope shales, which probably flowed basinward in a ductile fashion. Features such as shale diapirs and ridges attest to the ductile (and buoyant) behavior of these thick shale sections. This represents the "detachment of ductile shale" of Lopez (1989).

Late motion on both glide and deep listric systems is typical; this phase of faulting with reverse drag is responsible for the large "rollover anticlines" that form major shallow oil and gas fields. Many of these late faults continue to the surface.

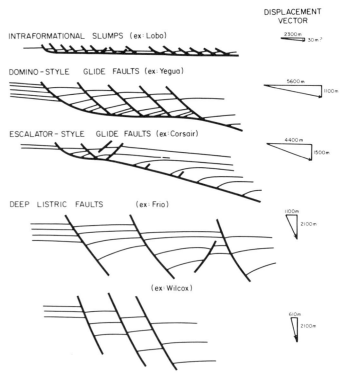

Figure 9. Four styles of intrasedimentary normal (growth) faulting, with estimates of relative horizontal and vertical displacements.

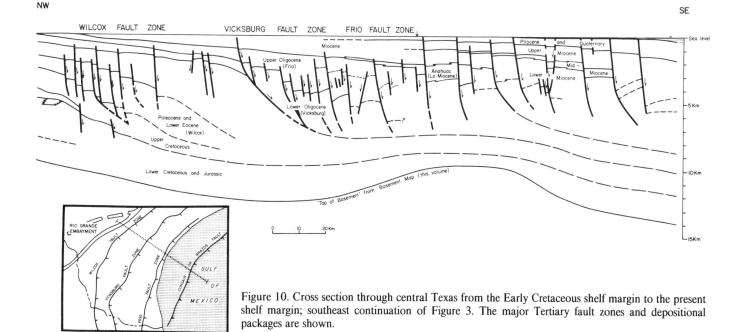

Figure 10. Cross section through central Texas from the Early Cretaceous shelf margin to the present shelf margin; southeast continuation of Figure 3. The major Tertiary fault zones and depositional packages are shown.

Lopez (1989) describes additional growth-fault styles involving salt withdrawal and salt-sill detachment. Salt withdrawal faults or "collapse faults" are localized around deep salt-withdrawal basins. Salt-sill detachments form in areas of loading and deflation of salt sills; this is observed primarily in the Neogene, as mentioned above.

The earliest major growth fault zone is the *Tuscaloosa fault zone* of early Late Cretaceous age. It occurs immediately south of the Early Cretaceous reef trend within the deltaic depocenters of the Tuscaloosa Group, primarily from central Louisiana southeastward into offshore Mississippi. Large rollover anticlines associated with several kilometers of expansion on this growth-fault system were reported by Smith (1985).

The next major shelf-margin progradation occurred in the Paleocene and early Eocene, when the large deltas of the Wilcox Group built out in south Texas and adjacent Tamaulipas, southeast Texas, and Louisiana. The resulting *Wilcox fault zone* consists of 5 to 10 closely spaced faults of deep-listric character, which expand the deltaic section as much as 3 to 5 times. In some areas, glide-fault systems with visible detachments have been noted in the more basinward part of the zone (Fig. 10). In the Houston embayment, these faults have been overprinted with diapir-related basins and uplifts (Ewing, 1983). In south Texas, a greater thickness of mobile, overpressured shale results in a complex structural style with shale ridges and large antithetic faults (Ewing, 1986). Behind the main fault trend in south Texas is the widespread Lobo intraformational slump that affects lowermost Wilcox and Midway strata.

Episodic progradation and faulting occurred in middle to late Eocene (Yegua) time, creating a *Yegua fault zone* (Ewing, 1989). This zone is best defined southwest of the Houston em-

bayment, where it consists of a "domino-style" glide-fault system downdip of a series of deep-listric or transitional faults, some of which may be inherited from Wilcox structures. Yegua-age faulting is also widespread in the Houston embayment and eastward into Louisiana, with both glide-fault and deep-listric geometries. Some earlier, middle Eocene (Sparta) expansion may also have taken place on Louisiana faults of this trend.

A major deltaic progradation in south Texas and adjacent Mexico in early Oligocene time created the *Vicksburg fault zone* (Stanley, 1970). The Vicksburg fault zone consists of a complex, escalator-style glide-fault system with secondary headwalls developed (Fig. 10); the Vicksburg section is expanded enormously across the fault. Abundant antithetic subsidiary faults and shale anticlines create complex structures (Ashford, 1972; Erxleben and Carnahan, 1983). The zone can be traced more or less continuously north into the central Texas Gulf Coast; to the northeast, it becomes more diffuse. Vicksburg-age faults can, however, be traced through southeast Texas, where they may in part occupy older Yegua fault traces.

Late Oligocene Frio progradation occurred throughout Texas and in Louisiana. The *Frio fault zone* is a broad deep-listric system, consisting of 5 to 10 rather sinuous major normal faults spaced 5 to 10 km apart, with intervening equant rollover anticlines and some up-to-basin faults (Fig. 10). Shale diapirs are abundant in the central Texas gulf coast (Plate 2; Bishop, 1977), while shale ridges frequently underlie major growth faults in south Texas (Ewing, 1986).

Extensive Miocene progradation along the entire Coastal zone created early, middle, and late Miocene fault systems. The single most conspicuous is the *Corsair fault zone,* an escalator-style glide-fault system of middle Miocene age (Fig. 11) that extends uninterrupted over 300 km parallel to the Texas coastline (Christiansen, 1983; Vogler and Robinson, 1987). The inboard early Miocene and outboard late Miocene fault zones are generally of a deeply listric character. Complex fault systems of these ages continue through the Miocene depocenters in southern Louisiana, but they are much disrupted by salt movement.

Plio-Pleistocene sedimentation induced growth-fault development near the present shelf edge. However, the major depocenter is located in the heart of the South Louisiana Shelf and the Texas-Louisiana Slope salt diapir provinces, where growth-fault trends are much disrupted by lateral and vertical salt mobility (Geitgey, 1988).

FEATURES OF THE EASTERN CARBONATE MARGINS

The present physiography of the eastern and southeastern margins of the Gulf of Mexico is dominated by the immense Florida and Yucatan carbonate platforms. The deep structure beneath these platforms is poorly known, giving an air of simplicity, which is almost certainly deceptive as major crustal boundaries run beneath both platforms (e.g., Klitgord and others, 1984). The southeastern Gulf of Mexico, which separates the platforms, has been studied by multichannel seismic and by DSDP drilling, and found to consist of a complex of rift blocks

(Phair and Buffler, 1983; Schlager and others, 1984); this complex may well continue to some extent beneath the surrounding platforms.

North Florida

Beneath the panhandle of Florida are several northeast-southwest–trending arches and basins (Barnett, 1975; Miller, 1982), notably the *Chattahoochee arch* and the *Apalachicola embayment*. These features are marked by seaward and landward swings in subcrop and strandline positions, especially in the Jurassic. The Apalachicola embayment overlies the southwestern extension of the South Georgia basin, a Triassic-Jurassic rift-related basin (Chowns and Williams, 1983; McBride and others, 1989). The Apalachicola embayment contains Jurassic and Cretaceous clastic sediments, with a Cretaceous thickness more than 2,000 m, overlain by 1,600 m of Tertiary strata.

Offshore, salt diapirs extend eastward from the Mississippi Fan into the *DeSoto Canyon diapir province* (Plate 2; Martin, 1980). More than 24 individual structures have been identified; they have grown in Cretaceous and Tertiary time. Included in the province is *Destin dome,* a salt-cored domal anticline up to 70 km long (Martin, 1980). South of the diapir province is the *Southern platform,* a structurally high area lacking salt. Possibly salt-related structures extend south of the Southern platform along the base of the Florida Escarpment.

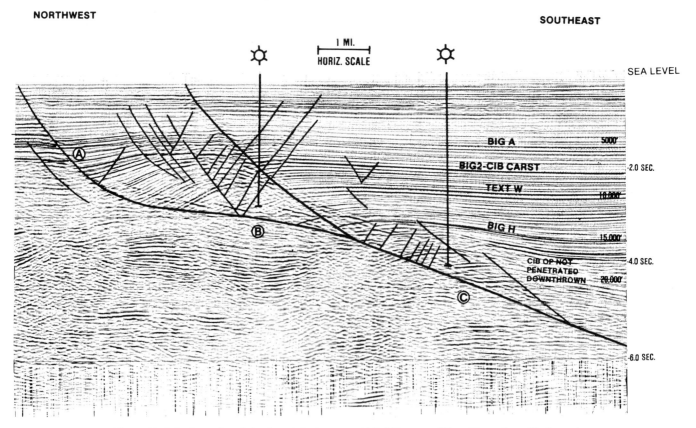

Figure 11. Seismic section of the Corsair fault zone, middle Miocene, offshore Texas; from Vogler and Robison (1987).

Onshore North Florida is dominated by the *Ocala uplift,* a broad uplift exposing Eocene limestone at the surface. The uplift passes southeastward into the *Peninsular arch,* which extends halfway down the peninsula; it is bounded to the east by the Southeast Georgia embayment; minor uplifts and basins detailed by Riggs (1984) affect the Miocene sediment patterns (see Chapter 16, this volume). In the Ocala uplift area, Lower Cretaceous sediments lie on Paleozoic sediments and on crystalline basement (see Chapter 7, this volume; Plate 3). Offshore west of the Peninsular arch are several lesser basins and uplifts, including the east- or northeast-trending *Middle Ground arch,* the *Tampa embayment,* and the *Sarasota arch.*

South Florida basin

This basin extends from the south end of Lake Okeechobee southward, southeastward, and westward to the shelf edge. Limited well control in the area indicates more than 8,000 m of uppermost Jurassic through Quaternary carbonate and evaporite strata in the basin. More than 1,600 m of the basin fill has accumulated in Tertiary times, indicating a long history of basin subsidence. The offshore area is essentially unknown, but the basin probably overlies in part the highly extended continental basement that was drilled by the Deep Sea Drilling Project (DSDP) in the southeastern Gulf of Mexico.

Southeast Gulf of Mexico

Today the area forms a deep-water entrance from the Caribbean and the Atlantic into the Gulf of Mexico (Plate I). Recent seismic work and DSDP drilling data indicate that it is underlain by rifted and attenuated continental crust (Plate 3; see Chapters 4 and 13, this volume). A northeast-trending normal fault system created a series of rotated blocks in the Late Triassic that have been modified by later northwest-trending normal faults in the Jurassic and episodically thereafter (Phair and Buffler, 1983; Schlager and others, 1984). The Catoche Tongue reentrant in the Yucatan platform overlies a block-faulted basin (Shaub, 1983b).

Yucatan platform

The platform includes the land area of the Yucatan Peninsula as well as the submerged Yucatan Shelf. At its northern and northwestern boundaries is the Campeche Escarpment and the Sigsbee Knolls diapir province; the Campeche Knolls diapir province is to the west. Basement lies beneath Lower Cretaceous carbonates and red beds of probable Late Jurassic–Early Cretaceous age, at depths ranging from 1,000 m in the east to more than 5,000 m offshore to the west (Viniegra O., 1981). Andesitic volcanic rocks have been reported within Upper Cretaceous rocks in the vicinity of Merida (Onate Espinosa, 1969; López Ramos, 1975; Peterson, 1983).

To the south, basement rises toward the La Libertad arch in Guatemala and to outcrop in the Maya Mountains in Belize. South of the La Libertad arch is the Peten basin, an eastern extension of the Isthmus Saline basin and the Sierra de Chiapas foldbelt (Plate 3; Bishop, 1980).

Sigsbee Knolls diapir province

Large salt diapirs, some coalesced diapirs, and salt ridges pierce several kilometers of abyssal plain strata in a band about 120 to 170 km northwest of the Campeche Escarpment, forming the Sigsbee Knolls (Plate 2; Martin, 1980). The band continues southwest to connect with the Campeche Knolls diapir province. Pillows of low and high amplitude are disclosed by seismic data to lie southeast of the diapirs up a sloping basement surface toward the Escarpment (see Chapter 13, this volume).

Campeche Knolls diapir province

This major diapir province is continuous northward with the Sigsbee Knolls province, and southward with the diapir province of the Isthmus Saline Basin and other basins of southern Mexico. It is divided into shelf and slope areas.

The shelf area contains very large salt stocks, pillows and salt-cored anticlines that uplift up to 5 km of Mesozoic and Cenozoic rocks (Martin, 1980). The domes do not generally rise to shallow depth, cresting from 1,000 to 3,000 m below the surface. The eastern area contains large domal swells aligned along northwest-trending fault zones that show reverse displacement (Santiago Acevedo and Mejia Dautt, 1980) and possible strike-slip motion (Viniegra O., 1981). There is also abundant dome-related normal faulting. The domal swells host the major marine Campeche oil fields. These structural complexities are similar in age and style to those of the onshore Villahermosa uplift to the southwest.

The slope area contains a variety of uplifts and basins, which are clearly imaged by bathymetry (Plate 1). Very large salt stocks and ridges are found on the north and west flanks. Large, closely spaced massifs are found on the upper slope, covered by 1 to 3 km of folded and faulted strata. In the center of the province is a broad area of uplifted strata with diapir-cored folds; the 1 to 4 km of section is cut by reverse faults that displace both salt and younger sediments (Watkins and others, 1978; Martin, 1980). On the east side of the province, there is a zone 30 to 60 km wide, of closely spaced anticlines and broad tilted blocks (Martin, 1980). On the west side of the province is a band of salt-cored anticlines trending north-south, separated by a swath of undeformed abyssal sediments from the subparallel Mexican Ridges.

The abundant diapiric activity in this area may be aided by folding and/or strike-slip faulting, as shown by broad areas of uplifted and reverse-faulted sediment (Martin, 1980). The folding may be related to the deformation of the Isthmian Embayment area, the "Maya tectonic event" of Miocene age (Watkins and others, 1978).

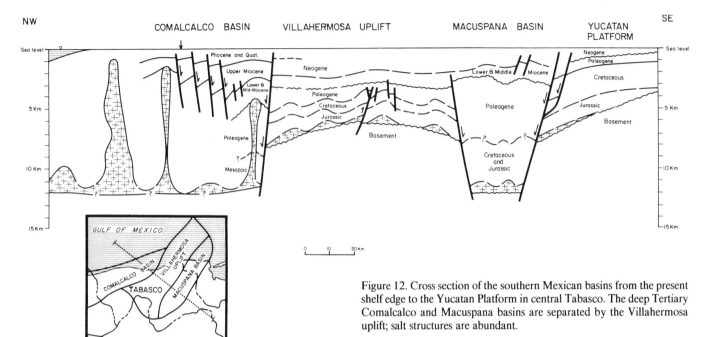

Figure 12. Cross section of the southern Mexican basins from the present shelf edge to the Yucatan Platform in central Tabasco. The deep Tertiary Comalcalco and Macuspana basins are separated by the Villahermosa uplift; salt structures are abundant.

FEATURES OF THE WESTERN COMPRESSIONAL MARGIN

The western (Mexican) margin of the Gulf of Mexico basin, though much narrower, shares some elements with the other margins, notably a variety of continental blocks or basement highs, outlined by Jurassic shoals or Cretaceous reefs. However, in this sector, compressional deformation of Mesozoic and Cenozoic age has profoundly affected the earlier structures. Also, Cenozoic clastic progradation is limited, mostly Neogene in age, and has not generated large synsedimentary structures.

The Mexican margin may be subdivided into several basins and embayments, separated by uplifts evident in the topography (Plates 1 and 2). From south to north, these are the basins of the Isthmus of Tehuantepec separated by the San Andrés Tuxtla volcanic highland from the *Veracruz basin,* the *Santa Ana massif,* and the *Tampico-Misantla basin.* To the north, a trend of positive elements and arches includes, from northwest to southeast, the *El Burro uplift* and the *Peyotes, Picachos,* and *Tamaulipas arches.* These separate the *Magiscatzin basin* and the *Sabinas basin* from the Rio Grande embayment and its Coastal zone extension into Mexico, the Burgos basin. Features of some of the embayments continue around the highland areas, mostly offshore. Subdivisions of the continental slope are based on structural style and the presence or absence of diapiric salt.

Isthmian embayment

This broad embayment, containing several basins and subbasins, has been the arena of much hydrocarbon exploration. From east to west, the major subdivisions are the *Macuspana basin,* the *Villahermosa uplift,* the *Comalcalco basin,* and the

Isthmus Saline basin. All of these subdivisions are underlain by salt and contain salt structures.

The *Macuspana basin* is a northeast-southwest–trending fault-bounded basin in which the sedimentary section is more than 12 km thick (Fig. 12, and Plate 3); exploration has been confined for the most part to the Tertiary (Santiago Acevedo and Mejia Dautt, 1980). There are numerous domal structures inferred to be caused by salt.

Northwest of the Macuspana basin is the *Villahermosa uplift* (Viniegra O., 1981), which is bounded on the northwest by the Comalcalco fault and on the east by the Frontera fault. Like the marine Bay of Campeche area into which the uplift passes, large salt-cored domal swells (at least one salt diapir is known) are complexly faulted with both normal and reverse faults of Miocene age. Crosscutting, northwest-trending faults have been postulated to be strike-slip faults (Viniegra O., 1981), but this is uncertain.

Northwest of the Comalcalco fault (which has 3 to 4 km of north-down displacement; Bishop, 1980) is the *Comalcalco basin.* The 4-km-thick Miocene section shows complex northeast-trending growth faulting with southeast-down displacement (Fig. 12; Gonzalez Alvarado, 1969).

To the west of the Comalcalco basin is the *Isthmus Saline basin,* with abundant salt diapirs (Contreras V. and Castillon B., 1968). The basin is naturally divided into two parts. The northern part contains salt diapirs and ridges aligned northeast-southwest, which pierce a thick Miocene succession. The southern section contains broad salt massifs where salt rises to within 60 m of the surface (Halbouty, 1979). The northwestern boundary of the Isthmian Saline Basin is a sharp linear, west of which salt structures are not present.

Some diapiric structures are found south of the elements

described above. They merge into the Sierra de Chiapas belt of folds and faults, composed of northwest-southeast–trending folds and strike-slip faults trending west-northwest–east-southeast and east-west, which are probably of Neogene age and developed on an evaporite basin.

A major change in structural style within Mexico takes place at the Isthmus of Tehuantepec, immediately south of the Isthmus Saline basin. To the west, an east-verging fold and thrust belt of Laramide (early Tertiary) age is prominently developed, continuing northward into the Sierra Madre Oriental. East of the isthmus, no such belt is present; instead, strike-slip faulting more or less parallel to the Chiapas massif is dominant. There is no evidence, however, for any Cenozoic structural break across the Isthmus; the Salina Cruz fault, shown crossing the Isthmus by many authors, has not been documented as a major fault by either surface mapping or by geophysical methods (Padilla y Sanchez, 1986; Salvador, 1988).

Veracruz basin

Northwest of the Isthmus Saline basin is the *Veracruz basin,* the eastern part of which is concealed by the Neogene to Quaternary volcanic rocks of the San Andrés Tuxtla "uplift" (López Ramos, 1979). The onshore basin is bounded on the west by thrusted Mesozoic carbonates of the Sierra Madre Oriental, which are onlapped by middle Miocene sediments. This thrusting has been linked to the early Tertiary Laramide compression that affects all of eastern Mexico north of the Isthmus of Tehuantepec. Near the shoreline, thick Eocene to lower Miocene sediments are mildly folded and overlain by a middle Miocene progradational sequence (Fig. 13; Mossman and Viniegra, 1976; Cruz Helú and others, 1977). Offshore, sediments as young as middle Miocene

are thrusted and overlain unconformably by the late Miocene to Quaternary prograding shelf margin (Fig. 13). This Miocene deformation is another part of the "Maya tectonic event" (Watkins and others, 1978).

The northern edge of the Veracruz basin is the *Santa Ana massif* where there are sparse exposures of Jurassic and Paleozoic rocks indicating a major uplift (Viniegra O., 1971). The older rocks are nearly covered by Pliocene to Holocene volcanic rocks. These volcanics are offset over a kilometer by down-to-basin faults, which offset rocks as young as 1.6 Ma (Cantagrel and Robin, 1979).

Tampico-Misantla basin

This basin extends north from the Santa Ana massif to the Tamaulipas arch north of Tampico. Its landward boundary is the Sierra Madre Oriental fold-and-thrust belt of early Tertiary age, which includes the Precambrian- and Paleozoic-cored *Huayacocotla anticlinorium* (Fig. 14 and Plate 2) and large eastward-transported thrust plates of Lower Cretaceous platform carbonates (Suter, 1987). Suter suggested that the Huayacocotla anticlinorium represents a strong fold above the frontal Sierra Madre Oriental thrust. Down-to-the-east normal faults with more than 2,000 m of displacement occur along the western side of the basin; they displace volcanic rocks as young as 3 Ma (Cantagrel and Robin, 1979). The major feature on the eastern side of the basin is the *Tuxpan platform,* a subtle basement feature upon which is prominent Early Cretaceous carbonate platform was constructed. At the northern margin of the basin is an arch and series of faults extending south from the Tamaulipas arch toward the Tuxpan platform (López Ramos, 1979).

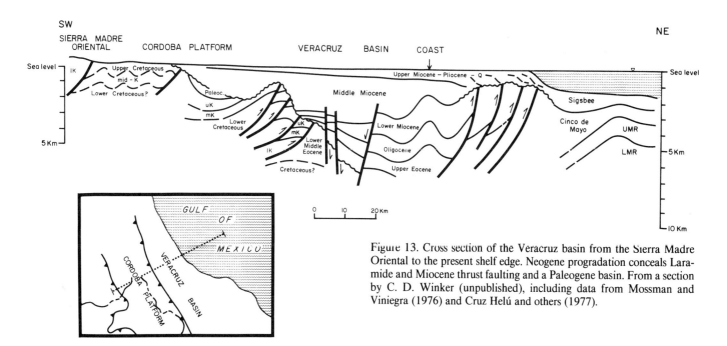

Figure 13. Cross section of the Veracruz basin from the Sierra Madre Oriental to the present shelf edge. Neogene progradation conceals Laramide and Miocene thrust faulting and a Paleogene basin. From a section by C. D. Winker (unpublished), including data from Mossman and Viniegra (1976) and Cruz Helú and others (1977).

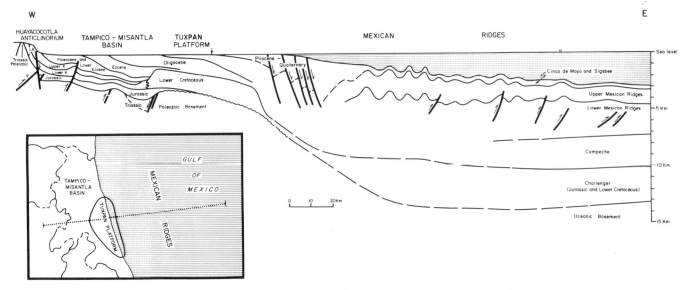

Figure 14. Cross section of Tampico-Misantla basin from the Huayacocotla anticlinorium through the Mexican Ridges. From a section by C. D. Winker (unpublished), with the Huayacocotla section added from Suter (1987).

The basin contains Jurassic and Cretaceous strata similar to other areas of the Gulf margin, overlying red beds of Late Triassic and Early Jurassic age. As much as 3,000 m of lower Tertiary rocks in the western part of the basin form a foredeep before the Sierra Madre Oriental; major early Tertiary paleocanyons extend east and southeast from the mountain front (Sansores Manzanilla and Girard Navarrete, 1969). This foredeep extends northward into the narrow *Magiscatzin basin* between the front of the Sierra Madre Oriental and the Tamaulipas arch.

Sabinas basin

The Sabinas basin (Humphrey, 1956) is a deformed sedimentary basin, best outlined by the presence of thick Upper Jurassic evaporites of the Olvido Formation. In Late Jurassic and Early Cretaceous time, coarse clastic sediments were shed into the basin from the Coahuila platform to the southwest, and from the El Burro uplift and the Peyotes and Picachos arches from the north and northeast (Alfonso Zwanziger, 1978). The boundaries of the Sabinas basin are probably fault controlled, and mark the transition from crust thinned by extension during the Jurassic, to thicker (less extended) continental blocks (McKee and others, 1984). The less extended blocks are in part those that contain Permo-Triassic granitic batholiths, which may have been more rigid and resistant to rifting. The basin subsided with all of northeastern Mexico and south Texas during the Early Cretaceous, leading to progressive landward retreat of the carbonate shelf margin and the establishment of open-marine conditions. Large amounts of Upper Cretaceous clastic sediment were shed off the Cordillera to the west; they are preserved today in the structural basins of the *Sabinas basin* and the *Parras basin*.

Laramide deformation (post-Maastrichtian, probably Eocene) created a northwest-southeast–trending system of tight,

evaporite-cored folds: the *Sabinas foldbelt* (Fig. 15). The northern boundary of this foldbelt is the *La Babia fault* (Smith, 1970), which may have some strike-slip component (Padilla y Sanchez, 1982). To the northeast, the Sabinas foldbelt grades into open, basement-involved folds that formed over the El Burro uplift and the Peyotes and Picachos arches (Roebeck and others, 1956). The southwest side of the foldbelt is in many places a southwest-verging fold-thrust pair against the Coahuila platform, which is itself only slightly folded (McKee and others, 1984). To the south and southeast in Nuevo Leon, the foldbelt narrows between the Picachos arch and the Sierra Madre Oriental; spectacular intrusive evaporite diapirs have formed in the cores of some of these folds (Wall and others, 1961). Compression across the foldbelt probably involved a left-lateral component, as evidenced by the sigmoidal terminations of major folds and en echelon geometries (Padilla y Sanchez, 1982; Charleston, 1981). Post-Laramide tectonism is not apparent; some mid-Tertiary intrusions were emplaced along east-west–trending lineaments.

El Burro Uplift, Peyotes and Picachos arches

A segmented northwest-southeast–trending basement high extends from the Serrania del Burro, in northernmost Coahuila, southeast to the Sierra de Picachos in northeastern Nuevo Leon. The northern segment, the El Burro uplift, is continuous to the northwest with the high country of west Texas. The middle and southern segments are the Peyotes and Picachos arches, marked by large open folds exposing Lower Cretaceous strata. The overall axis continues southeastward into the Tamaulipas arch and other basement highs of eastern Mexico, although there are gaps between some of the highs. The en echelon arrangement of positive features (the Peyotes, Picachos, and northern Tamaulipas arches) suggests early strike-slip faulting, probably pre-Oxfordian in age (Alfonso Zwanziger, 1978; Padilla y Sanchez, 1982). Upper Jurassic rocks are not present over the arch in its northern areas, and the Upper Jurassic strandline outlines the arch at its southeastern end (Chapter 8, this volume). The uplifts are probably fault bounded at depth, as described above. The arch forms the indefinite western boundary of the Rio Grande embayment of the Interior Zone, and of the Tertiary Burgos basin of the Coastal Zone.

The arch was deformed during the Laramide orogeny into broad, basement-cored domal folds that contrast with the Sabinas foldbelt to the southwest. To the northeast, low-amplitude folds with the same trend continue into the Rio Grande embayment and in the Burgos basin.

Tamaulipas arch

This high area extends south-southeast from the Sierra de San Carlos, southeast of Monterrey, to the Tampico area. On the west it dips into the Magiscatzin basin; on the east it is bounded by the *San Jose de las Rusias homocline* (Varela Hernandez, 1969) and a major Tertiary growth fault (Shaub, 1983a). Northeast of the Sierra de San Carlos is a sharp transition, probably faulted, into the Tertiary-filled and growth-faulted Burgos basin. Abundant Tertiary igneous rocks occur in the Sierras de San Carlos, Cruillas, and Tamaulipas, which are also the structurally highest areas of the Tamaulipas arch. The arch is more or less continuous with the Picachos and Peyotes arches and the Burro uplift to the northwest.

The Tamaulipas arch overlies and was formed by a basement high; several wells have penetrated Permo-Triassic granitic rocks beneath Lower Cretaceous and/or Upper Jurassic rocks (Varela Hernandez, 1969; Padilla y Sanchez, 1982; Chapter 7, this volume). Triassic and Jurassic red beds lie west of the uplift (López Ramos, 1979). The arch, exposed during the early Late Jurassic, was covered by the latest Late Jurassic and subsided strongly in the Early Cretaceous, when it was covered by deep-water sediments. The present expression of the arch was formed by broad Laramide folding of early Tertiary age and by development of the foredeep to the west, and was accentuated by Oligocene and Miocene igneous intrusion (Cantagrel and Robin, 1979).

Mexican Ridges

The steep continental slope of the western Gulf of Mexico is dominated by symmetrical to east-verging folds with some associated thrust faults developed in Miocene and Pliocene clastic strata (Fig. 14, and Plate 2). They are roughly collinear with the salt-cored anticlines of the Perdido foldbelt to the north, and were once considered to be salt-cored themselves (Massingill and others, 1973). These ridges are now understood to be shale-cored,

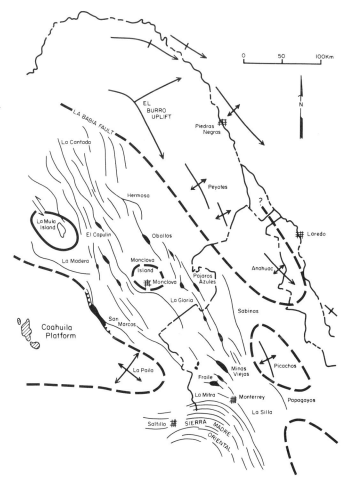

Figure 15. Map of anticlinal structures and diapirs in the Sabinas basin area. The open dashes mark the approximate boundaries of basement highs. Dark areas: exposures of Jurassic strata (mostly evaporites).

and formed by thin-skinned compression and decollement over mobile lower Tertiary shales (Buffler and others, 1979). The belt is 50 to 75 km wide, with folds 500 to 700 meters in relief, spaced every 10 km. The foldbelt is broken into two segments, separated by an area of low relief at 22°N; the northern segment trends north-northeast, and the southern segment north-northwest. Folding is Pliocene to Quaternary in age. Some or all of the documented compression may be due to basinward sliding (decollement) of the Cenozoic sedimentary wedge. The headwall of this slide is represented by the large growth faults detected at the shelf edge of eastern Mexico (Fig. 14, and Plate 2; Buffler and others, 1979; Winker, 1982; Fig. 3). These faults extend northward east of the Tamaulipas arch (Shaub, 1983a), and may be the southern continuation of some of the Tertiary growth faults of the Burgos basin.

REFERENCES CITED

Alfonso Zwanziger, J., 1978, Geologia regional del sistema sedimentario Cupido: Asociación Mexicana de Geólogos Petroleros Boletín, v. 30, no. 1–2, p. 1–55.

Amery, G. B., 1969, Structure of Sigsbee scarp, Gulf of Mexico: American Association of Petroleum Geologists Bulletin, v. 53, p. 2480–2482.

Anderson, E. G., 1979, Basic Mesozoic study in Louisiana, the northern coastal region, and the Gulf Basin province: Louisiana Geological Survey Folio Series 2, 46 p.

Ashford, T., 1972, Geoseismic history and development of Rincon field, Texas: Geophysics, v. 37, p. 797–812.

Barnett, R. S., 1975, Basement structure of Florida and its tectonic implications: Gulf Coast Association of Geological Societies Transactions, v. 25, p. 122–142.

Bishop, R. S., 1977, Shale diapir emplacement in south Texas; Laward and Sherriff examples: Gulf Coast Association of Geological Societies Transactions, v. 27, p. 20–31.

Bishop, W. F., 1980, Petroleum geology of northern Central America: Journal of Petroleum Geology, v. 3, no. 1, p. 3–39.

Blickwede, J. F., and Queffelec, T. A., 1988, Perdido Fold Belt; A new deepwater frontier in the western Gulf of Mexico [abs.]: American Association of Petroleum Geologists Bulletin, v. 72, p. 163.

Bouma, A. H., Moore, G. T., and Coleman, J. M., eds., 1978, Framework, facies, and oil-trapping characteristics of the upper continental margin: American Association of Petroleum Geologists Studies in Geology 7, 326 p.

Braunstein, J., and McMichael, C. E., 1976, Door Point; A buried volcano in southeastern Louisiana: Gulf Coast Association of Geological Societies Transactions, v. 26, p. 79–80.

Brooks, R. O., 1989, Horizontal component of Gulf of Mexico salt tectonics, in Gulf of Mexico salt tectonics; Associated processes and exploration potential; 10th Annual Research Conference Program and Extended Abstracts: Gulf Coast Section, Society of Economic Paleontologists and Mineralogists, p. 22–24.

Bruce, C. H., 1973, Pressured shale and related sediment deformation; A mechanism for development of regional contemporaneous faults: American Association of Petroleum Geologists Bulletin, v. 57, p. 878–886.

Buffler, R. T., and Sawyer, D. S., 1985, Distribution of crust and early history, Gulf of Mexico Basin: Gulf Coast Association of Geological Societies Transactions, v. 35, p. 333–344.

Buffler, R. T., Shaub, F. J., Watkins, J. S., and Worzel, J. L., 1979, Anatomy of the Mexican Ridges, southwestern Gulf of Mexico, in Watkins, J. S., and others, eds., Geological and geophysical investigations of continental margins: American Association of Petroleum Geologists Memoir 29, p. 319–327.

Burnett, A. W., and Schumm, S. A., 1983, Alluvial-river response to neotectonic deformation in Louisiana and Mississippi: Science (7 October 1983), v. 222, p. 49–50.

Cagle, J. W., and Khan, M. A., 1983, Smackover–Norphlet stratigraphy, south Wiggins Arch, Mississippi and Alabama: Gulf Coast Association of Geological Societies Transactions, v. 33, p. 23–29.

Cantagrel, J.-M., and Robin, C., 1979, K-Ar dating on eastern Mexican volcanic rocks; Relations between the andesitic and the alkaline provinces: Journal of Volcanology and Geothermal Research, v. 5, p. 99–114.

Charleston, S., 1981, A summary of the structural geology and tectonics of the State of Coahuila, Mexico, in Smith, C. I., and Katz, S. B., eds., Lower Cretaceous stratigraphy and structure, northern Mexico: West Texas Geological Society Publication 81-74, p. 28–36.

Chowns, T. M., and Williams, C. T., 1983, Pre-Cretaceous rocks beneath the Georgia Coastal Plain; Regional implications, in Studies related to the Charleston, South Carolina, earthquake of 1886; Tectonics and seismicity: U.S. Geological Survey Professional Paper 1313, p. L1–L42.

Christiansen, A. F., 1983, An example of a major syndepositional listric fault in Bally, A. W., ed., Seismic expression of structural styles; A picture and work atlas: American Association of Petroleum Geologists Studies in Geology 15, v. 2, p. 2.3.1-36–40.

Contreras V., H., and Castillón B., M., 1968, Morphology and origin of salt domes of Isthmus of Tehuantepec: American Association of Petroleum Geologists Memoir 8, p. 244–260.

Cruz Helú, P., Verdugo V., R., and Barcenas P., R., 1977, Origin and distribution of Tertiary conglomerates, Veracruz Basin, Mexico: American Association of Petroleum Geologists Bulletin, v. 61, p. 207–226.

Cushing, E. M., Boswell, E. H., and Hosman, R. L., 1964, General geology of the Mississippi Embayment: U.S. Geological Survey Professional Paper 448-B, 28 p.

Ervin, C. P., and McGinnis, L. D., 1975, Reelfoot Rift; Reactivated precursor to the Mississippi Embayment: Geological Society of America Bulletin, v. 86, p. 1287–1295.

Erxleben, A. W., and Carnahan, G., 1983, Slick Ranch area, Starr County, Texas, in Bally, A. W., ed., Seismic expression of structural styles; A picture and work atlas: American Association of Petroleum Geologists Studies in Geology 15, v. 2, p. 2.3.1-22–26.

Ewing, T. E., 1983, Growth faults and salt tectonics in the Houston diapir province; Relative timing and exploration significance: Gulf Coast Association of Geological Societies Transactions, v. 33, p. 83–90.

—— , 1986, Structural styles and structural evolution of the Wilcox and Frio growth-fault trends in Texas; Constraints on geopressured reservoirs: University of Texas at Austin Bureau of Economic Geology Report of Investigations 154, 86 p.

—— , 1987, The Frio River Line in south Texas; Transition from Cordilleran to northern Gulf tectonic regimes: Gulf Coast Association of Geological Societies Transactions, v. 37, p. 87–94.

—— , 1988a, The Uvalde Igneous field; A summary, in Ewing, T. E., and Rodgers, R. W., eds., Upper Cretaceous of southwest Texas: South Texas Geological Society Guidebook 89-3, p. 7–15.

—— , 1988b, Variation of "Growth Fault" structural styles in the Texas Gulf Coast Basin [abs.]: Gulf Coast Association of Geological Societies Transactions, v. 38, p. 579.

—— , 1989, The Downdip Yegua Trend; An overview: Gulf Coast Association of Geological Societies Transactions, v. 39, p. 75–83.

Ewing, T. E., and Caran, S. C., 1982, Late Cretaceous volcanism in south and central Texas; Stratigraphic, structural, and seismic models: Gulf Coast Association of Geological Societies Transactions, v. 32, p. 137–145.

Ewing, T. E., and others, 1990, Tectonic map of Texas: University of Texas at Austin Bureau of Economic Geology, scale 1:750,000.

Ferguson, J. D., 1984, Jurassic age salt tectonism within the Mt. Enterprise fault system, Rusk County, Texas, in Presley, M. W., ed., The Jurassic of east Texas: East Texas Geological Society, p. 157–161.

Fowler, P., 1956, Faults and folds of south-central Texas: Gulf Coast Association of Geological Societies Transactions, v. 6, p. 37–42.

Geitgey, J. E., 1988, Plio-Pleistocene evolution of central offshore Louisiana: Gulf Coast Association of Geological Societies Transactions, v. 38, p. 151–156.

Gonzalez Alvarado, J., 1969, Interpretación estructurl del area Encrucijada–Chontalpa, Tabasco, in Seminario sobre exploración petrolera, Mesa Redonda 5: Instituto Mexicano del Petroleo, 16 p.

Granata, W. H., 1963, Cretaceous stratigraphy and structural development of the Sabine Uplift area, Texas and Louisiana, in Report on selected North Louisiana and South Arkansas oil and gas fields and regional geology: Shreveport Geological Society, Reference Volume V, p. 50–95.

Halbouty, M. T., 1979, Salt Domes, Gulf Region, United States and Mexico, 2nd ed.: Houston, Gulf Publishing, 561 p.

Halbouty, M. T., and Halbouty, J. J., 1982, Relationships between East Texas Field region and the Sabine Uplift in Texas: American Association of Petroleum Geologists Bulletin, v. 66, no. 8, p. 1042–1054.

Hardin, N. S., 1989, Salt distribution and emplacement processes, northwest Gulf lower slope; A suture between two provinces, in Gulf of Mexico salt tectonics; Associated processes and exploration potential; 10th Annual Research Conference Program and Extended Abstracts: Gulf Coast Section, Society of Economic Paleontologists and Mineralogists, p. 54–59.

Hildebrand, T. G., Kane, M. F., and Stauder, W., 1977, Magnetic and gravity anomalies in the northern Mississippi Embayment and their spatial relationship to seismicity: U.S. Geological Survey Miscellaneous Field-Studies Map MF-914.

Hildebrand, T. G., Kane, M. F., and Hendricks, J. D., 1982, Magnetic basement in the upper Mississippi Embayment region; A preliminary report: U.S. Geological Survey Professional Paper 1236E, p. 39–53.

Hughes, D. J., 1968, Salt tectonics as related to several Smackover fields along the northeast rim of the Gulf of Mexico Basin: Gulf Coast Association of Geological Societies Transactions, v. 18, p. 320–329.

Humphrey, W. E., 1956, Tectonic framework of northeast Mexico: Gulf Coast Association of Geological Societies Transactions, v. 6, p. 25–35.

Humphris, C. C., Jr., 1978, Salt movement on continental slope, northern Gulf of Mexico, in Bouma, A. H., Moore, G. T., and Coleman, J. M., eds., Framework, facies, and oil-trapping characteristics of the upper continental margin: American Association of Petroleum Geologists Studies in Geology 7, p. 69–85.

Jackson, M.L.W., and Laubach, S. E., 1988, Cretaceous and Tertiary compressional tectonics as the cause of the Sabine Arch, east Texas and northwest Louisiana: Gulf Coast Association of Geological Societies Transactions, v. 38, p. 245–256.

Jackson, M.P.A., 1982, Fault tectonics of the East Texas Basin: University of Texas at Austin Bureau of Economic Geology Geological Circular 82-4, 31 p.

Jackson, M.P.A., and Seni, S. J., 1983, Geometry and evolution of salt structures in a marginal rift basin of the Gulf of Mexico, east Texas: Geology, v. 11, p. 131–135.

Johnson, O. H., Jr., 1958, The Monroe Uplift: Gulf Coast Association of Geological Societies Transactions, v. 8, p. 24–26.

Klitgord, K. S., Popenoe, P., and Schouten, H., 1984, Florida; A Jurassic transform plate boundary: Journal of Geophysical Research, v. 89, p. 7753–7772.

Kreitler, C. W., and others, 1981, Geology and geohydrology of the East Texas Basin: The University of Texas at Austin Bureau of Economic Geology Geological Circular 81-7, 207 p.

Kupfer, D. H., Crowe, C. T., and Hessenbruch, J. M., 1976, North Louisiana Basin and salt movements (Halokinetics): Gulf Coast Association of Geological Societies Transactions, v. 26, p. 94–110.

Lee, G. H., Bryant, W. R., and Watkins, J. S., 1989, Salt structures and sedimentary basins in the Keathley Canyon area, northwestern Gulf of Mexico; Their development and tectonic implications, in Gulf of Mexico salt tectonics associated processes, and exploration potential; 10th Annual Research Conference Program and Extended Abstracts: Gulf Coast Section, Society of Economic Paleontologists and Mineralogists, p. 90–93.

Lobao, J. J., and Pilger, R. H., Jr., 1985, Early evolution of salt structures in North Louisiana Salt Basin: Gulf Coast Association of Geological Societies Transactions, v. 35, p. 189–198.

Locklin, A. C., 1984, Currie field (Smackover), East Texas Basin, in Presley, M. W., ed., The Jurassic of east Texas: Tyler, Texas, East Texas Geological Society, p. 32–42.

Lopez, J. A., 1989, Distribution of structural styles in the northern Gulf of Mexico and Gulf Coast, in Gulf of Mexico salt tectonics, associated processes, and exploration potential: 10th Annual Research Conference Program and Extended Abstracts: Gulf Coast Section, Society of Economic Paleontologists and Mineralogists, p. 101–108.

López Ramos, E., 1975, Geological summary of the Yucatan Peninsula, in Nairn, A.E.M., and Stehli, F. G., eds., The ocean basins and margins; V. 3, The Gulf of Mexico and the Caribbean: New York, Plenum Press, p. 257–282.

——, 1979, Geología de México, 2nd edition, v. III, 446 p. (privately printed).

——, 1980, Geología de México, 2nd edition, v. II, 454 p. (privately printed).

Martin, R. G., 1980, Distribution of salt structures in the Gulf of Mexico: U.S.

Geological Survey Miscellaneous Field-Studies Map MF-1213, 2 sheets, scale 1:2,500,000.

Massingill, J. V., Bergantino, R. N., Fleming, H. S., and Feden, R. H., 1973, Geology and genesis of the Mexican Ridges: Journal of Geophysical Research, v. 78, p. 2498–2507.

McBride, J. H., Nelson, K. D., and Brown, L. D., 1989, Evidence and implications of an extensive early Mesozoic rift basin and basalt/diabase sequence beneath the southeast Coastal Plain; Geological Society of America Bulletin, v. 101, p. 512–520.

McKee, J. W., Jones, N. W., and Long, L. E., 1984, History of recurrent activity along a major fault in northeastern Mexico: Geology, v. 12, p. 103–107.

Miller, J. A., 1982, Structural control of Jurassic sedimentation in Alabama and Florida: American Association of Petroleum Geologists Bulletin, v. 66, p. 1289–1301.

Mossman, R. W., and Viniegra O., F., 1976, Complex fault structures in Veracruz province of Mexico: American Association of Petroleum Geologists Bulletin, v. 60, p. 379–388.

Murray, G. E., 1961, Geology of the Atlantic and Gulf Coastal Province of North America: New York, Harper and Brothers, 692 p. (especially p. 79–201).

——, 1966, Salt structures of the Gulf of Mexico Basin; A review: American Association of Petroleum Geologists Bulletin, v. 50, p. 439–478.

Nicholas, R. L., and Waddell, D. E., 1982, New Paleozoic subsurface data from the north-central Gulf Coast: Geological Society of America Abstracts with Programs, v. 14, p. 576.

Nunn, J. A., Scardina, A. D., and Pilger, R. H., Jr., 1984, The thermal evolution of the north-central Gulf Coast: Tectonics, v. 3, p. 723–740.

Onate Espinosa, R., 1969, Problemas de exploración en la zona sur; Problemas geofísicos de la Península de Yucatán: Instituto Mexicano del Petroleo, Mesa Redonda, v. 5, 18 p.

Padilla y Sanchez, R. J., 1982, Geologic evolution of the Sierra Madre Oriental between Linares, Concepción del Oro, Saltillo, and Monterrey, Mexico [Ph.D. thesis]: University of Texas at Austin, 217 p.

——, 1986, Post-Paleozoic tectonics of northeast Mexico and its role in the evolution of the Gulf of Mexico, in Urrutia Fucugauchi, J., ed., Dynamics and evolution of the lithosphere; Results and perspectives of geophysical research in Mexico, part A: Geofísica Internacional, v. 25, p. 157–206.

Pennington, W. D., Davis, S. D., Carlson, S. M., DuPree, J., and Ewing, T. E., 1988, The evolution of seismic barriers and asperities caused by the depressuring of fault planes in oil and gas fields of south Texas: Bulletin of the Seismological Society of America, v. 76, p. 939–948.

Peterson, J. A., 1983, Petroleum geology and resources of southeastern Mexico, northern Guatemala, and Belize: U.S. Geological Survey Circular 760, 44 p.

Phair, R. L., and Buffler, R. T., 1983, Pre-Middle Cretaceous history of the deep southeastern Gulf of Mexico, in Bally, A. W., ed., Seismic expression of structural styles; A picture and work atlas: American Association of Petroleum Geologists Studies in Geology 15, v. 2, p. 2.2.3-141–147.

Posey, J. S., 1986, The Louann Salt of the Gulf Coast Basin, with emphasis on south Texas, in Stapp, W. L., ed., Contributions to the geology of south Texas 1986: South Texas Geological Society, v. 440–446.

Ray, P. K., 1988, Lateral salt movement and associated traps on the continental slope of the Gulf of Mexico: Gulf Coast Association of Geological Societies Transactions, v. 38, p. 217–223.

Riggs, S. R., 1984, Paleoceanographic model of Neogene phosphorite deposition, U.S. Atlantic continental margin: Science, v. 223, no. 4632, p. 123–131.

Rodgers, D. A., 1984, Mexia and Talco fault zones, east Texas; Comparison of origins predicted by two tectonic models, in Presley, M. W., ed., The Jurassic of east Texas: Tyler, East Texas Geological Society, p. 23 31.

Roebeck, R. C., Pesquera V., R., and Ulloa A., S., 1956, Geología y depósitos de carbon de la region de Sabinas: Edo. de Coahuila, Mexico, 20th International Geological Congress, 103 p.

Rose, P. R., 1972, Edwards group, surface and subsurface, central Texas: University of Texas at Austin Bureau of Economic Geology Report of Investigations 74, 198 p.

Salvador, A., 1988, Reply *to* 'Late Triassic–Jurassic paleogeography and origin of Gulf of Mexico Basin': American Association of Petroleum Geologists Bulletin, v. 72, p. 1419–1422.

Sansores Manzanilla, E., and Girard Navarrete, R., 1969, Bosquejo geológico de la Zona Norte, *in* Seminario sobre exploración petrolera, Mesa Redonda No. 2, 36 p.

Santiago Acevedo, J., and Mejia Dautt, O., 1980, Giant fields in the southeast of Mexico: Gulf Coast Association of Geological Societies Transactions, v. 30, p. 1–31.

Scardina, A. D., 1982, Tectonic subsidence history of the North Louisiana Salt Basin: Louisiana State University Publications in Geology and Geophysics, Gulf Coast Studies 2, 33 p.

Schlager, W., Buffler, R. T., Angstadt, D., and Phair, R., 1984, Geologic history of the southeastern Gulf of Mexico, *in* Initial reports of the Deep Sea Drilling Project: Washington, D.C., U.S. Government Printing Office, v. 77, p. 715–738.

Sellards, E. H., 1932, Oil fields in igneous rocks in coastal plain of Texas: American Association of Petroleum Geologists Bulletin, v. 16, p. 741–768.

Shaub, F. J., 1983a, Growth faults on the southwestern margin of the Gulf of Mexico, *in* Bally, A. W., ed., Seismic expression of structural styles; A picture and work atlas: American Association of Petroleum Geologists Studies in Geology 15, v. 2, p. 2.3.3.-3 to 2.3.3.-15.

—— , 1983b, Origin of Catoche Tongue, *in* Bally, A. W., ed., Seismic expression of structural styles; A picture and work atlas: American Association of Petroleum Geologists Studies in Geology 15, v. 2, p. 2.2.3-129-140.

Smith, C. I., 1970, Lower Cretaceous stratigraphy, northern Coahuila, Mexico: University of Texas at Austin Bureau of Economic Geology Report of Investigations 65, 101 p.

Smith, G. W., 1985, Geology of the Deep Tuscaloosa (Upper Cretaceous) gas trend in Louisiana, *in* Perkins, B. F., and Martin, G. B., eds., Habitat of oil and gas in the Gulf Coast; Proceedings, 4th Research Conference: Gulf Coast Section, Society of Economic Paleontologists and Mineralogists, p. 153–190.

Stanley, T. B., Jr., 1970, Vicksburg fault zone, Texas, *in* Geology of giant petroleum fields: American Association of Petroleum Geologists Memoir 14, p. 301–308.

Suter, M., 1987, Structural traverse across the Sierra Madre Oriental fold-thrust belt in east-central Mexico: Geological Society of America Bulletin, v. 98, p. 249–264.

Varela Hernandez, A., 1969, Problemas de la exploración petrolera en el Homoclinal de San Jose de las Rusias, Tamaulipas, *in* Seminario sobre exploración petrolera, Mesa Redonda 2: Instituto Mexicano del Petroleo, 17 p.

Viniegra O., F., 1971, Age and evolution of salt basins of southeastern Mexico: American Association of Petroleum Geologists Bulletin, v. 55, p. 478–494.

—— , 1981, Great carbonate bank of Yucatan, southern Mexico: Journal of Petroleum Geology, v. 3, p. 247–278.

Vogler, H. A., and Robison, B. A., 1987, Exploration for deep geopressured gas; Corsair Trend, offshore Texas: American Association of Petroleum Geologists Bulletin, v. 71, p. 777–787.

Wall, J. R., Murray, G. E., and Diaz G., T., 1961, Geologic occurrence of intrusive gypsum and its effect on structural forms in Coahuila marginal folded province of northeastern Mexico: American Association of Petroleum Geologists Bulletin, v. 45, p. 1504–1522.

Watkins, J. S., Ladd, J. W., Buffler, R. T., Shaub, F. J., Houston, M. H., and Worzel, J. L., 1978, Occurrence of salt in the deep Gulf of Mexico, *in* Bouma, A. H., Moore, G. T., and Coleman, J. M., eds., Framework, facies, and oil-trapping characteristics of the upper continental margin: American Association of Petroleum Geologists Studies in Geology 7, p. 43–65.

Weeks, A. W., 1945, Oakville, Cuero, and Goliad Formations of Texas Coastal Plain between Brazos River and Rio Grande: American Association of Petroleum Geologists Bulletin, v. 29, p. 1721–1732.

Weimer, P., and Buffler, R. T., 1989, Structural geology of the Mississippi Fan Foldbelt, Deep Gulf of Mexico, *in* Gulf of Mexico salt tectonics, associated processes, and exploration potential; 10th Annual Research Conference Program and Extended Abstracts: Gulf Coast Section, Society of Economic Paleontologists and Mineralogists, p. 146–147.

Winker, C. D., 1982, Cenozoic shelf margins, northwestern Gulf of Mexico-Gulf Coast Association of Geological Societies Transactions, v. 32, p. 427–448.

Winker, C. D., and Edwards, M. B., 1983, Unstable progradational clastic shelf margins, *in* Stanley, D. J., and Moore, G. T., eds., The shelfbreak; Critical interface on continental margins: Society of Economic Paleontologists and Mineralogists Special Publication 13, p. 139–157.

Worrall, D. M., and Snelson, S., 1989, Evolution of the northern Gulf of Mexico, with emphasis on Cenozoic growth faulting and the role of salt, *in* Bally, A. W., and Palmer, A. R., eds., The Geology of North America; An overview: Boulder, Colorado, Geological Society of America, The Geology of North America, v. A, p. 97–138.

Worzel, J. L., and Burk, C. A., 1978, The margins of the Gulf of Mexico, *in* Geological and geophysical investigations of continental margins: American Association of Petroleum Geologists Memoir 29, p. 403–419.

MANUSCRIPT ACCEPTED BY THE SOCIETY JUNE 4, 1990

ACKNOWLEDGMENTS

This work was begun while the author was at the Texas Bureau of Economic Geology; the assistance of the staff there is gratefully acknowledged. A special debt is owed to Charles D. Winker for numerous discussions on Gulf of Mexico geology and for sections of the Mexican margins. The text has been reviewed by Ray G. Martin and Roberto Flores Lopez, whose numerous helpful comments are much appreciated.

The Geology of North America
Vol. J, The Gulf of Mexico Basin
The Geological Society of America, 1991

Chapter 4

The crust under the Gulf of Mexico basin

Dale S. Sawyer
Department of Geology and Geophysics, Rice University, P.O. Box 1892, Houston, Texas 77251
Richard T. Buffler
*Institute for Geophysics and Department of Geological Sciences, University of Texas, 8701 North Mopac Boulevard, Austin, Texas
78759*
Rex H. Pilger, Jr.*
Department of Geology and Geophysics, Louisiana State University, Baton Rouge, Louisiana 70803

INTRODUCTION

Although the sediments filling the shallow-water parts of the Gulf of Mexico basin have been extensively explored for hydrocarbons using seismic methods and are reasonably well known, the nature and distribution of the underlying crust and mantle are much less well known. Several characteristics of the basin's sediments have made deep-penetration observations difficult. Thick Cenozoic clastic sediments throughout the Gulf of Mexico basin attenuate seismic energy, while Jurassic to Upper Cretaceous carbonates and evaporites provide a large impedance contrast, through which seismic transmission is further limited. In parts of the basin, mobile salt forms pillows, domes, sills, and other structures, which focus and defocus seismic energy in ways that make imaging difficult. In the southeastern Gulf of Mexico, solution cavities and karst-like paleotopography in shallow-water carbonates severely scatter seismic signals. In spite of these difficulties, reflection seismic, refraction seismic, gravity, magnetic, and subsidence techniques have been used to resolve the gross characteristics of the crust under the Gulf of Mexico basin.

In all cases the terms "crust" and "basement" will be used as synonyms. In areas of normal or modified continental crust, the crust and basement are defined to include all rock lying beneath a widespread unconformity at the base of the marine Mesozoic section. This surface is overlain and onlapped by Middle Jurassic salt (or equivalent rocks) as well as younger sedimentary rocks. Included within "basement" are the Upper Triassic to Lower Jurassic rift sequences (see Chapter 8, this volume). Although these "red-bed" sequences should perhaps be considered a part of

the very early basin fill, they are often difficult to distinguish from rocks formed prior to the formation of the Gulf of Mexico basin. In addition, we suggest that the surface at the base of the marine Mesozoic section more accurately represents the overall configuration of the modern Gulf of Mexico basin. In areas identified as oceanic crust, the crust or basement is defined to include rocks of oceanic crust layer 2, usually composed of basalt pillows and dikes, and oceanic crust layer 3, usually composed of massive gabbro or serpentinite. In all areas the base of the crust or the Moho is defined using seismic refraction data as a layer of seismic velocity greater than 7.6 km/sec and usually between 8.0 and 8.5 km/sec.

Most of the issues addressed in this chapter will be examined in the context of the basement depth maps presented as Figure 1 and Plate 3, and a series of crustal-scale schematic cross sections presented as Figure 2. The crust under the Gulf of Mexico basin has been divided into four major types: oceanic, thin transitional, thick transitional, and continental (Figs. 1 and 2). These divisions reflect the manner in which crust was either created or modified by Mesozoic rifting. Continental crust, as used in this chapter, is crust that predated the formation of the Gulf of Mexico basin and was not significantly modified (i.e., extended, thinned, or intruded) by the more important Middle Jurassic and later rifting. Continental crust was, in some places, affected by a Late Triassic through Early Jurassic early phase of rifting, but the total amount by which the crust was extended during this phase was minimal. Continental crust is observed around the periphery of the basin. Transitional crust is crust that was originally continental, but was significantly extended and thinned, and probably intruded with

*Present address: ARCO Oil and Gas Co., 2300 West Plano Parkway, Plano, Texas 75075.

Sawyer, D. S., Buffler, R. T., and Pilger, R. H., Jr., 1991, The crust under the Gulf of Mexico basin, *in* Salvador, A., ed., The Gulf of Mexico Basin: Boulder, Colorado, Geological Society of America, The Geology of North America, v. J.

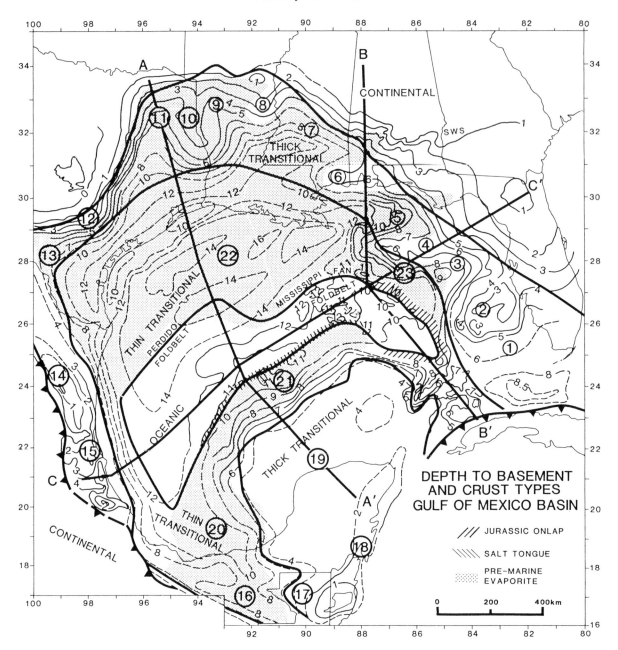

Figure 1. Map of the Gulf of Mexico basin showing (1) generalized depth to basement in kilometers; (2) distribution of four crust types: continental, thick transitional, thin transitional, and oceanic crust; and (3) known distribution of Middle Jurassic premarine evaporites (Louann Salt and equivalent rocks; stippled area). Basement includes oceanic crust plus all rocks lying below (older than) Middle Jurassic premarine evaporites. Solid contours indicate areas where top of basement is well constrained by wells or seismic-reflection data, whereas the dashed contours are areas where top of basement is more speculative, based on other geophysical data or extrapolation of shallower trends. Circled numbers refer to major named basement highs, lows, arches, basins, etc., as follows: (1) South Florida basin; (2) Sarasota arch; (3) Tampa embayment; (4) Middle Ground arch–Southern platform; (5) Apalachicola basin; (6) Wiggins uplift; (7) Mississippi salt basin; (8) Monroe uplift; (9) North Louisiana salt basin; (1) Sabine uplift; (11) East Texas (salt) basin; (12) San Marcos arch; (13) Rio Grande embayment– Burgos basin; (14) Tamaulipas arch; (15) Tuxpan platform; (16) Macuspana basin; (17) La Libertad arch; (18) Quintana Roo arch; (19) Yucatán block; (20) Campeche salt basin; (21) Sigsbee Salt basin; (22) North Gulf salt basin; (23) West Florida basin. This map is slightly different from Plate 3. The differences lie mostly in Mexico where the editor (Plate 3) had some additional data, and in the north-central Gulf of Mexico where the editor (Plate 3) chose to eliminate some speculative dashed contours retained in this figure.

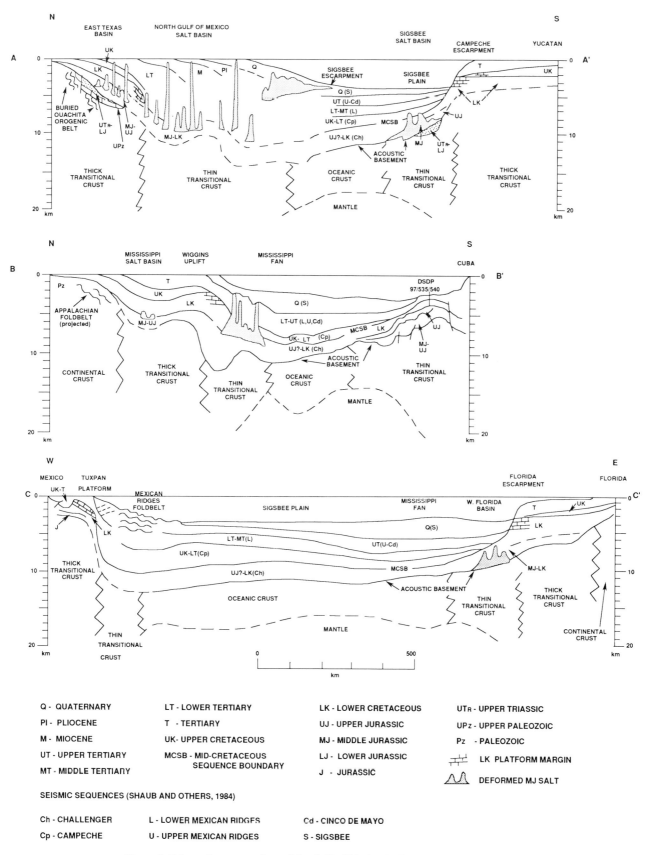

Figure 2. Schematic cross sections of the Gulf of Mexico basin (see Fig. 1 for locations).

magma, during Middle and Late Jurassic rifting. Transitional crust, in some places, also shows evidence of Late Triassic through Early Jurassic rifting. Transitional crust may be further divided into thick transitional crust, which was somewhat thinned during rifting, and thin transitional crust, which was dramatically thinned during rifting. Within the thick transitional crust there are many blocks that appear relatively unthinned but are surrounded by regions of greater thinning. Within the thin transitional crust, the crust was thinned fairly uniformly, without much lateral variation. Oceanic crust was formed under the deep Gulf of Mexico basin (Ibrahim and others, 1981; Ebeniro and others, 1988) during the Late Jurassic (Buffler and Sawyer, 1985). Sea-floor spreading probably continued for only about 5 to 10 m.y. It is not known whether this oceanic crust was like that forming today at sea-floor spreading centers. The transition from rifting to sea-floor spreading is not well understood at any passive margin. Since the cessation of sea-floor spreading, transitional crust and oceanic crust cooled and subsided, allowing deposition of the thick sedimentary sequences in the Gulf of Mexico basin.

The crust-type classification relates to the large-scale pattern of crustal modification that was the result of Late Triassic through Late Jurassic continental rifting related to the breakup of Pangea. The crust types do not in every case correspond exactly to the rock types or terranes that made up the crust of the region later to become the Gulf of Mexico basin. However, preexisting structures, rock types, thermal state, and crustal or lithospheric weaknesses probably affected, and even controlled, the rifting process.

The data used to constrain the depth to basement and distribution of crust type maps in the Gulf of Mexico basin are not evenly distributed (Fig. 3). The data around the edges of the basin, mostly the onshore data points, are wells that have penetrated basement, providing very precise depths to basement and in many cases log observations and samples of the crust, or they are wells that have penetrated rocks just above basement, providing minimum values for basement depth and guiding contouring (areas with solid contours in Fig. 1). There are also several Deep Sea Drilling Project wells in the deep southeastern Gulf of Mexico that penetrated basement. They tell us nothing about the lower crust or mantle. Most of the data points from the water-covered areas of the central and southeastern Gulf of Mexico basin are reflection seismic profiles that image basement. These are generally apparent in Figure 3 as linear sets of data points, and they provide an estimate of the depth to basement and indicate the large-scale texture of the basement surface (also areas of solid contours in Fig. 1). They also tell us nothing about the lower crust or mantle. A few of the data points, some on land and some in the water (e.g., the group in the northwestern Gulf of Mexico), are refraction seismic stations. These provide an estimate of the depth to basement (dashed contours) and the seismic velocity of the upper crustal rocks. Some refraction stations also provide an estimate of the depth to the Moho and the seismic velocity of the rocks in the lower crust and upper mantle. Over large areas of the basin, there is little or no control, and the depth-to-basement contours are estimates (dashed contours, Fig. 1).

The following sections describe first the total tectonic subsidence method, a new technique for identifying crust type and the location of crust-type boundaries, and then describe each of the crust types we identify in the Gulf of Mexico basin and the boundaries between them. Next we briefly summarize the sequence of events that produced the observed crust type distribution, and then conclude by mentioning several ongoing efforts to learn more about the crust and mantle under the Gulf of Mexico basin.

TOTAL TECTONIC SUBSIDENCE

In addition to more traditional methods of determining crust type, Buffler and Sawyer (1985) and Dunbar and Sawyer (1987) used the total tectonic subsidence analysis method (Sawyer, 1985) to identify the crust types and boundaries in the Gulf of Mexico basin. They used maps of bathymetry and basement depth in the Gulf of Mexico region to estimate and map total tectonic subsidence (TTS; Fig. 4). Dunbar and Sawyer (1987) used an earlier version of the basement depth data, shown on Figure 1 and Plate 3. The total subsidence is the water depth plus the sediment thickness at a point. The TTS is the total subsidence less the loading effect of the sediments. Dunbar and Sawyer (1987) assumed local isostatic equilibrium in calculating the loading effect of the sediment. Assumptions about sediment density were made using average compaction curves appropriate to the dominant sedimentary rock types in the various parts of the Gulf of Mexico basin. The use of the word "total" in these subsidence descriptions refers to the amount of subsidence from the time of rifting initiation to the present. Dunbar and Sawyer (1987) implicitly assumed that the basement surface was at sea level at the time of rifting initiation and, therefore, that the present depth of basement could be called total subsidence. TTS (Fig. 4) in the Gulf of Mexico basin ranges from zero on the periphery of the basin to more than 7 km in the western deep Gulf of Mexico.

Dunbar and Sawyer (1987) interpreted this total tectonic subsidence map in terms of crustal extension using a model by LePichon and Sibuet (1981). In that model, the amount of TTS resulting from crust extension is proportional to the amount of crustal thinning. The factor that relates the extension to the TTS is a function of the time since rifting, with TTS for a particular amount of extension increasing with time. Zero TTS is equivalent to zero extension (described by $\beta = 1$). For a basin of the age of the Gulf of Mexico basin (mid-Jurassic), 7.6 km of TTS maps into infinite extension (described by $\beta = \infty$). Values of TTS between zero and 7.6 km map to an extension of $\beta = (1 - TTS/7.6)^{-1}$. Dunbar and Sawyer's (1987) extension map (Fig. 5) shows that the extension contours parallel those of TTS. They interpreted extension of greater than $\beta = 4.5$ to correspond to oceanic crust, and thus have not contoured extension at higher values (Fig. 5). The region Dunbar and Sawyer (1987) identified as oceanic crust corresponds roughly with that identified on the basis of seismic observations (Fig. 5). Dunbar and Sawyer (1987) interpreted the oceanic crust to be somewhat wider in the eastern Gulf of Mexico basin than predicted using seismic data (Fig. 5).

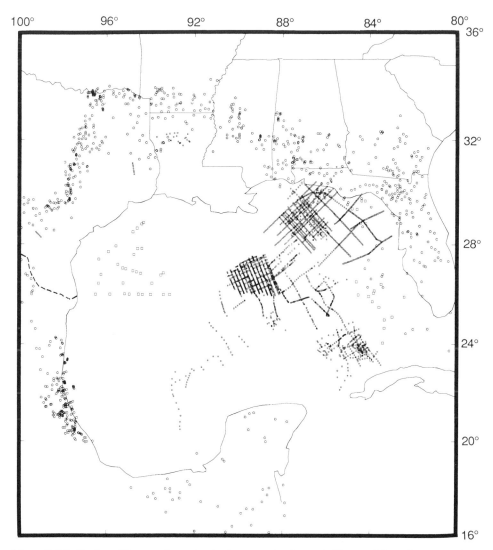

Figure 3. Distribution of data used to make the basement depth map in Figure 1. Most of the data points onshore are from wells drilled to or near basement (circles). Most of the lineated data points in the central and eastern Gulf of Mexico are reflection seismic lines, which are interpreted to image basement (smaller squares). Scattered points in the water (i.e., in the northeastern Gulf of Mexico) and on land come from seismic refraction experiments (larger squares). Note that the coverage is far from even. The source for these data is an unpublished industry-sponsored report on the Gulf of Mexico basin basement conducted by the University of Texas Institute for Geophysics.

The different approaches agree in the southwestern Gulf of Mexico basin, extending oceanic crust to the base of the steep basement slope. Dunbar and Sawyer (1987) extend oceanic crust farther north in the western Gulf of Mexico basin than shown in Figure 1.

The subsidence analysis of Dunbar and Sawyer (1987) may be used to estimate the amount of extension of the continental crust, and thus, its post-rifting thickness. Their extension map (Fig. 5) has been converted into maps of estimated crust thickness and Moho depth (Figs. 6 and 7). The crust thickness (Fig. 6) is determined by assuming an initial crust thickness of 40 km and dividing it by the amount of extension (Fig. 5). The Moho depth

(Fig. 7) is determined by adding the crust thickness just obtained to the basement depth. The crust thickness map (Fig. 6) shows clearly the asymmetry of the stretched crust in the Gulf of Mexico basin. The zone of extended crust is very wide in the north and east, while it is very narrow in the south and the west. The boundary of the oceanic crust identified using the subsidence analysis ($\beta = 4.5$; Fig. 5) lies just deeper than and parallel to the 10-km crust-thickness contour (Fig. 6). Crust thicknesses within thin transitional crust vary between about 10 and 20 km on the north side of the basin, and 10 to 30 km on the south side. Note that these thicknesses are estimates based on subsidence analyses rather than direct seismic observations.

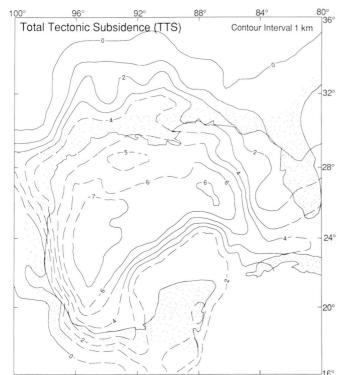

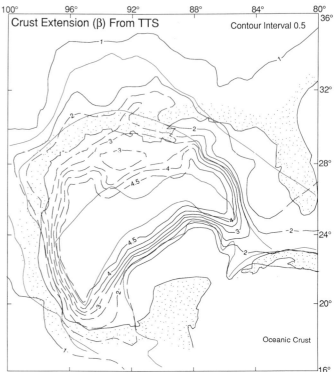

Figure 4. Map of total tectonic subsidence (TTS) by Dunbar and Sawyer (1987). TTS is the observed depth to basement at a point corrected for the loading effect of the sediment. It is calculated using the observed bathymetry, observed basement depth, and a model for the density of the sediments. It is interpreted as the total amount by which the basement would have subsided during the formation of the Gulf of Mexico basin if no sediment had been deposited.

Figure 5. Map of crust extension during the rifting that formed the Gulf of Mexico basin by Dunbar and Sawyer (1987). Extension is described using the extension parameter β. For example, $\beta = 2$ describes crust extension in which surface area is increased by a factor of 2. We also typically assume that increasing the surface area by a factor of 2, thins the crust to half its original thickness. $\beta = 1$ describes unextended crust. Estimates of crust extension are derived from the total tectonic subsidence (TTS) map in Figure 4. A model relating extension to subsidence by LePichon and Sibuet (1981) is used. The red lines are the crust type boundaries from Figure 1. We interpret $\beta = 4.5$ to correspond to the boundary between oceanic and transitional crust. Within oceanic crust, β has no obvious interpretation so we have not contoured it there.

The Moho depth determined from the subsidence is grossly similar to that determined using sparse refraction data (Fig. 7). Seismic observations in Texas agree very well with the subsidence estimates of the Moho depth. The seismic data from Louisiana show Moho to be about 9 km deeper. This is an area where, because of flexural loading due to the late Cenozoic Mississippi River sediments, the subsidence method is likely to overestimate the extension, and henceforth, underestimate the crustal thickness and Moho depth. Most of the rest of the refraction data points are out in the deep Gulf of Mexico. The subsidence analysis indicates that Moho depth there should be about 20 to 25 km. Most of the observations fall in the range 16 to 23 km. This level of agreement is as good or better than we would anticipate for the subsidence method. A map made in this way, though obviously not as good as one made by collecting many more refraction data, can nonetheless be useful in planning experiments and illustrating the general distribution of the crust.

The accuracy of the subsidence calculations is limited by knowledge of the basement depth. Current basement depth maps, including Figure 1 and Plate 3, tend to be poorly controlled, particularly in the northern Gulf of Mexico, where, due to thick sediments and mobilized salt, few seismic observations of base-

ment have been made (Fig. 3). As better data become available, some details of the maps will change, but the general framework probably will not.

CRUST TYPES AND BOUNDARIES

Oceanic crust

The area interpreted to be underlain by oceanic crust forms a convex-northwestward, arcuate band across the deep central Gulf of Mexico (Figs. 1 and 2). It is about 400 km wide in the western Gulf of Mexico basin but narrows considerably to the east. The oceanic crust in the western and central parts of the basin is interpreted to have been emplaced during the Late Jurassic as a result of rifting between the crust underlying the Yucatán Peninsula (the Yucatán block) and North America (Chapter 14, this volume). The oceanic crust in the eastern part of the basin may have been produced the same way, or it may have been produced by a different, possibly later, episode of plate separation (Chapter 14, this volume). The distribution of oceanic crust in the Gulf of Mexico basin does not appreciably discriminate between plate-reconstruction models for the following reasons: the details of its shape are not well known; it has relatively short conjugate margins to fit back together; one cannot be sure that the sea-floor spreading began synchronously along the whole margin (in fact it is unlikely that it did); and more than 50 percent of the total relative plate motion was accommodated by crustal extension rather than sea-floor spreading (Dunbar and Sawyer, 1987).

Oceanic crust in the Gulf of Mexico was first recognized in early refraction studies (M. Ewing and others, 1955; J. Ewing and others, 1960, 1962). Its distribution and character were better defined by later studies of seismic refraction and reflection data (Buffler and others, 1980, 1981; Ibrahim and others, 1981; Buffler and Sawyer, 1985; Ebeniro and others, 1988; see Trehu and others, 1989, for a thorough summary). The oceanic crust in the Gulf of Mexico basin is generally 5 to 6 km thick and is characterized by refraction velocities of 6.8 to 7.2 km/sec (Fig. 8). This velocity probably corresponds to oceanic layer 3 found in most normal ocean basins. In all but the most recent refraction data, oceanic layer 2 cannot be seen in the Gulf of Mexico basin because it is deeply buried and has velocities that are too similar to those of the overlying compacted clastic and carbonate sediments. Note that in Figure 8, which is an example of the best of the older seismic reflection data, the 4.8 km/s layer probably includes both deeply buried carbonates and oceanic layer 2. The top of this interval is the prominent Mid-Cretaceous Sequence Boundary (MCSB; Chapter 13, this volume). Over much of the eastern and central deep basin, acoustic basement is recognized on reflection data and is interpreted to be the top of the oceanic crust (Buffler and others, 1980, 1981; Ibrahim and others, 1981; Rosenthal, 1987). In most places it is an irregular surface on-lapped and filled with uniformly layered sedimentary sequences interpreted to be deep-water sediments (Fig. 8). This acoustic surface is interpreted to be the top of oceanic layer 2 and is believed to be composed of pillow basalts and basalt dikes (Ibrahim and others, 1981). The top of oceanic layer 3 is not typically observed in seismic reflection profile data (Fig. 8).

The best data to compare the velocity structure of the Gulf

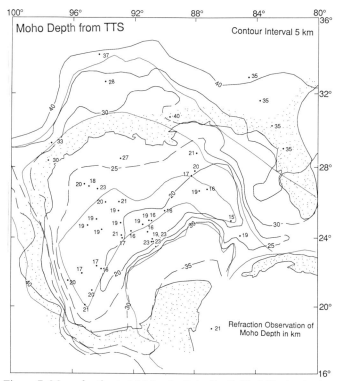

Figure 7. Map of estimated Moho depth in the Gulf of Mexico basin based on subsidence analyses (contours), and Moho depth observations made using seismic refraction, and in one area seismic reflection, methods (dots labelled with depth in kilometers). The subsidence estimates of Moho depth were derived by adding the estimated crust thickness (Fig. 6) to the basement depth (Fig. 1). Most of the refraction stations shown were taken from a compilation by Locker and Chatterjee (1984). Other refraction stations are from Sawyer and others (1986), Keller and others (1989), Nakamura and others (1988), and Nakamura (personal communication). The only reflection seismic data used in this map are located in southern Georgia and were obtained by Nelson and others (1985). The red lines are the crust type boundaries from Figure 1. TTS, total tectonic subsidence.

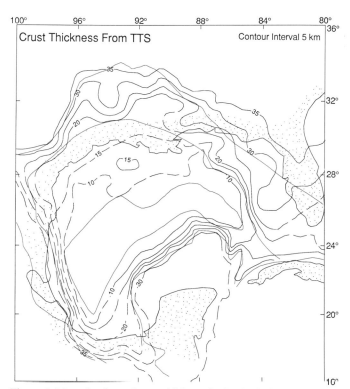

Figure 6. Map of estimated crust thickness in the Gulf of Mexico basin based on subsidence analyses. Estimates of crust thickness are derived from the map of crust extension (Fig. 5) and an assumption that the crust was originally 40 km thick. The oceanic crust is not contoured and is assumed to have normal oceanic crustal thickness of 6 to 8 km. The red lines are the crust type boundaries from Figure 1. TTS, total tectonic subsidence.

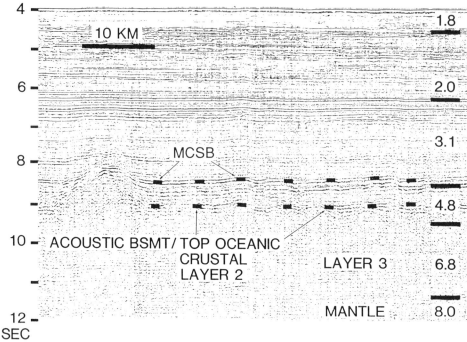

Figure 8. Portion of the University of Texas Institute for Geophysics (UTIG) seismic reflection line 16-2 with refraction observations of oceanic crust in the south-central Gulf of Mexico basin (modified from Ibrahim and others, 1981; see Fig. 11 for location). The refraction layer having velocity 4.8 km/s corresponds to the reflection layers identified as mainly carbonates below the Mid-Cretaceous Sequence Boundary (MCSB) plus oceanic layer 2. There is no reflection from the boundary between oceanic layers 2 and 3. There is, however, a change in refraction velocity at the layer 2 to layer 3 interface of 4.8 to 6.8 km/s. BSMT, basement.

of Mexico oceanic crust with that of a type oceanic section were reported by Ebeniro and others (1988). They used modern digital ocean-bottom seismographs to record large (4,000 in³), closely spaced (50 m) airgun shots at ranges up to 90 km in the northern Gulf of Mexico. Their Line 5 was located just south of the Sigsbee Escarpment (Figs. 9A and 10), the basinward edge of an extensive salt tongue(s) (Figs. 2 and 10) and within crust identified as oceanic (Fig. 1). These and data from two other ocean-bottom seismographs along Line 5 allowed Ebeniro and others (1988) to obtain a reasonably well-constrained velocity-depth structure for oceanic crust under Line 5 (Figs. 9B and 9C). The two curves labeled "extremal bounds" represent the minimum and maximum depths to rock of a particular velocity that are consistent with the observed seismic section (Fig. 9C). Note that the depth to rock of velocities in the 2- to 4-km/s range is constrained to within 1 km. As we go to higher velocities, and therefore deeper, the depth constraint becomes poorer, about ±3 km at 7 km/s and about ±4 km at 8 km/s. This method of presentation places more emphasis on the true information content and potential interpretation errors of the wide-angle seismic data than the usual method of showing layer boundaries. The Ebeniro and others (1988) interpretation shows an increase in velocity from about 4.5 km/s at the base of a carbonate section to about 5.2 km/s, typical of

oceanic layer 2 at a depth of 15 km below sea level (Fig. 9C). The boundary between carbonates and oceanic layer 2 seems to show up as a velocity increase in the upper-bound curve at about 15 km depth. The boundary between layer 2 and layer 3 probably shows up best as a velocity increase in the lower-bound curve at about 17 km depth. Below that, the velocity increases to greater than 8 km/s over about 6 km, from 17 to 23 km.

Ebeniro and others (1988) compared their solution with data from a well-studied area with unequivocal oceanic crust (Fig. 9C). The example they chose was a model by Spudich and Orcutt (1980) from young oceanic crust near the East Pacific Rise. Young oceanic crust is not an ideal comparison with Jurassic Gulf of Mexico oceanic crust, but it was used because the velocity structure there was known in great detail. The East Pacific Rise velocity-depth function was translated downward to simulate the greater burial of oceanic crust in the Gulf of Mexico. The character of the two curves is quite similar. The extremal-bounds solutions seem to show that the velocity steps upward at 15 km depth correspond to the top of oceanic layer 2, and those at about 17 km correspond to the top of oceanic layer 3. The maximum-depth extremal-bound curve agrees in slope with the velocity model within oceanic layer 3 down to the Moho. There is obviously not as much detail in the Ebeniro and others (1988) model, becuse in the Gulf of Mexico the oceanic crust is much more deeply buried. However, this comparison suggests that the crust in the central Gulf of Mexico basin is consistent in velocity character with oceanic crust elsewhere.

The seismic refraction results obtained by Ebeniro and others (1988), showed very thin (6 to 8 km) crust extending north of the Sigsbee Escarpment (Fig. 10). They even interpreted one ocean-bottom seismograph dataset from Line 3 (Figs. 9A and 10) to be underlain by oceanic crust. It is possible that this area is all oceanic, or that there were two loci of intense thinning of transitional crust, only one of which progressed to the stage of sea-floor spreading. In our maps we show this to be thin transitional crust rather than oceanic.

Rosenthal (1987), in a study of the oceanic crust in the eastern Gulf of Mexico using reflection seismic data, identified a series of ENE–trending linear basement highs and lows that are cut in one place by a prominent NNW–trending trough and are truncated to the east by an area of flat-lying basement (Fig. 11 and Plate 3). While along part of the boundary between rugged

and flat basement there is a change in reflection seismic data density that might at first glance explain the change, the northern part of the boundary seems adequately controlled and shows that the change is real. To the west (Fig. 11), the trend of the basement highs becomes more northeast. This change from east-northeast to northeast is matched by the change in the orientation of the boundary between oceanic and thin transitional crust (Fig. 1). The irregular acoustic basement marking the top of the volcanic oceanic crust is onlapped and filled with a uniformly layered sedimentary sequence interpreted to be mainly deep-water carbonates with some terrigenous clastic sediments (Challenger Unit of Shaub and others, 1984; Rosenthal, 1987). Rosenthal (1987) suggested that the two different areas of oceanic crust (highs-and-lows versus flat-lying) may have had different origins, and that the linear features may be spreading ridges related to a south-southeast opening of the central Gulf of Mexico. In that scenario, the NNW–trending boundaries could be transform faults or oceanic fracture zones. This sea-floor spreading orientation is most consistent with the plate reconstructions proposed by Klit-

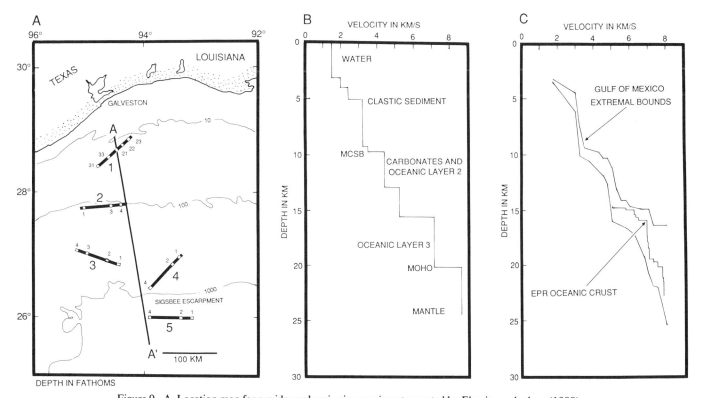

Figure 9. A, Location map for a wide-angle seismic experiment reported by Ebeniro and others (1988). Large numbers identify lines 1 through 5. Smaller numbers identify ocean-bottom seismograph locations along each line. The lines were 90 km long and large airgun shots were fired at an interval of 50 m along each line. Each instrument recorded all the shots on its line. A–A' is the location of a composite cross section shown as Figure 10. B, Horizontal, constant-velocity-layer solution and interpreted geology obtained by Ebeniro and others (1988) from Line 5 instrument 2. MCSB, Mid-Cretaceous Sequence Boundary. C, Comparison of extremal bounds solution obtained by Ebeniro and others (1988) from Line 5 instrument 2 with a detailed velocity-depth model obtained by Spudich and Orcutt (1980) for unequivocal oceanic crust near the East Pacific Rise (EPR). The two extremal bounds curves represent a seismic data inversion for the maximum and minimum depths at which rocks of each velocity will be found. The EPR data have been translated downward to account for the different water depth and sediment thickness.

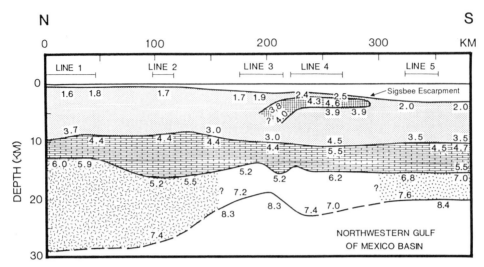

Figure 10. Generalized cross section of the northern Gulf of Mexico margin from Ebeniro and others (1988; see Fig. 9A for location of section line and data used). The interpreted horizons are dashed where not well constrained. Oceanic crust is indicated by the v-pattern shading on the right side of the figure. The velocities and greater thickness of the crust on the left are diagnostic of continental or modified continental crust. We suggest that it is all thin transitional crust. There seems to be a crust thickness minimum under Line 3 that may correspond to an alternate axis of crustal extension during the formation of the Gulf of Mexico basin that failed to proceed to sea-floor spreading.

gord and Schouten (1986), although their trends are more west-northwest.

Whether magnetic anomalies produced by sea-floor spreading are present in the Gulf of Mexico basin is unclear. The apparent absence of well-defined magnetic anomalies may be due to the great depth of burial and higher temperature of the oceanic crust in the Gulf of Mexico basin, to rapid sedimentation during ocean crust formation, or to incorrect identification of crust type. The magnetic anomalies are of small amplitude, roughly 100 nT, and existing data and maps do not show obvious lineations (Fig. 12; Hall and others, 1984). Shepherd (1983) suggested that some magnetic anomalies in the eastern Gulf of Mexico could have been produced by sea-floor spreading in a north-south direction, consistent with a pole to the east of the Gulf of Mexico and basin-opening models of Humphris (1978), Pindell and Dewey (1982), Pindell (1985), Salvador (1987), and Dunbar and Sawyer (1987). These anomaly orientations and identifications are quite tenuous; it seems likely that this is due largely to the poor data quality. When analysis of a recently flown aeromagnetic survey of the eastern Gulf of Mexico becomes available in 1991, it should become clearer whether sea-floor spreading magnetic anomalies are present there and what their orientations are. Then, if no anomalies are identified, it will not be for lack of dense, high-quality data.

Applying the oceanic crust subsidence studies of Parsons and Sclater (1977) to the Gulf of Mexico basin suggests that parts of the basin underlain by oceanic crust should have subsided by 6.2 km. The subsidence map of Dunbar and Sawyer (1987; Fig. 4) shows that most of the central Gulf of Mexico basin is consist-

ent with this prediction. There is a gentle increase in subsidence to the west within the region identified as oceanic crust. This subsidence maximum, about 7.1 km, corresponds with the location of a geoid low and geoid-derived gravity low (Haxby, 1987). The origin of these anomalies is unknown but suggests that some dynamic process is loading the crust in the western Gulf of Mexico basin.

The oceanic crust in the Gulf of Mexico basin was formed during a 5- to 10-m.y. span during the Middle to Late Jurassic (Buffler and Sawyer, 1985). This estimate is based on an assumed average spreading rate, the width of the oceanic crust, and the seismic stratigraphy. The water depth at the time is unknown, but we can guess it was about 2.5 km, approximately the present global average depth at mid-ocean ridges (Parsons and Sclater, 1977). In the absence of sedimentation, the oceanic crust had the potential to subside by about 3.9 km as it aged and cooled, for a total of about 6.4 km. Because the lithosphere is depressed isostatically during sediment loading, it takes about 3 km of sediment to fill 1 km of water depth. If sedimentation had kept pace with 6.4 km of subsidence, keeping the basin filled to sea level (which it probably never did), then the crust could have subsided to a depth of about 19 km (3 × 6.4 km sediment + 0 km water). The water depth in the central Gulf of Mexico basin is actually about 3.5 km (3.75 km maximum depth). The depth to the top of the crust appropriate for oceanic crust, therefore, is calculated to be about 12.2 km (3 × 2.9 km sediment + 3.5 km water). This is generally consistent with observations that show basement depths of from 10 to 14 km (Fig. 1 and Plate 3).

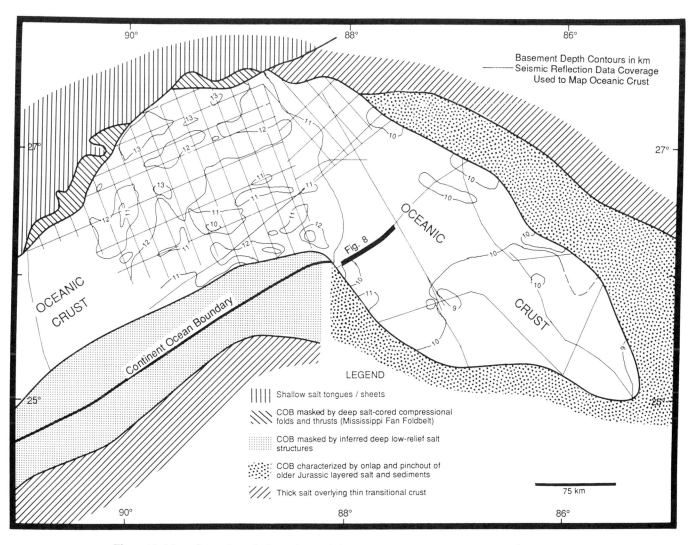

Figure 11. Map of crust boundaries and oceanic basement depth in the eastern Gulf of Mexico basin from Rosenthal (1987). The contours within oceanic crust are depth to basement obtained using a grid of seismic-reflection data. The finer straight lines are the grid of reflection data used to make the map. The shading patterns indicate the character of the boundary between oceanic and thin transitional crust observed. Note the apparent east-northeast–striking basement ridges in the left center of the figure. These lineations seem to be absent to the east. While some of this effect may be due to differences in data density, there does seem to be a clear change in the north where the data density is more uniform. The lineations seem to curve southward on the left side of the figure, roughly consistent with the shape of the edge of oceanic crust. COB is the continent-ocean boundary.

The boundary between oceanic and thin transitional crust

The boundary between the oceanic crust and thin transitional crust in the Gulf of Mexico basin is reasonably well defined by seismic refraction and reflection data. We assume that salt is only deposited over transitional crust and that, therefore, the edge of transitional crust is often marked by the edge of salt deposition. In the extreme northwest corner of the deep basin, the Perdido foldbelt apparently marks the seaward limit of the original salt deposition and the approximate boundary between oceanic and thin transitional crust (Figs. 1 and 13; Worral and Snelson,

1989). This deep, northeast-trending, late Tertiary foldbelt is inferred to be salt cored. The salt acted as a regional décollement zone that transmitted compressional stresses caused by updip sediment loading to the lower slope. The original distribution of salt overlying thin transitional crust but not oceanic crust, therefore, controlled the location of the foldbelt, which formed where the salt pinched out and the décollement surface ends. To the east in the vicinity of the Mississippi Fan, another deep, northeast-trending, late Tertiary fold and thrust belt (Mississippi Fan foldbelt; Figs. 1 and 11) again marks the seaward limit of salt and the boundary between oceanic and thin transitional crust (Rosenthal,

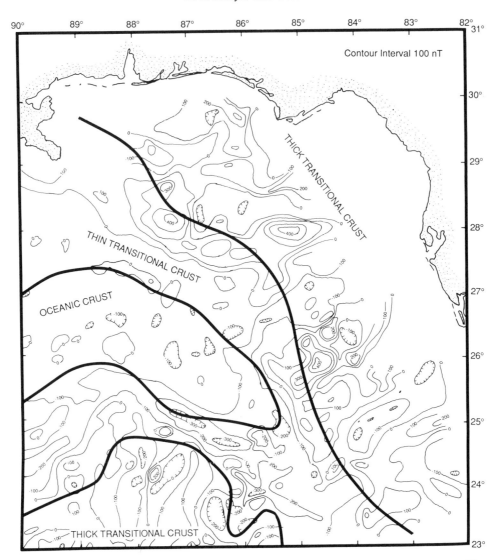

Figure 12. Map of magnetic anomalies in the eastern Gulf of Mexico from Hall and others (1984). The bold lines are the crust-type boundaries from Figure 1. The magnetic anomalies are generally more subdued over oceanic crust. Over thin transitional crust there is a general tendency for the anomalies to be oriented parallel to the crust boundaries. Over thick transitional crust there is a general tendency for the anomalies to be oriented perpendicularly to the crust boundaries.

1987). The two foldbelts are offset (Fig. 1). This zone of offset is characterized by a zone of deep-seated, northwest-trending salt structures along the lower slope (Lee and others, 1989). The northwest trend of these offsets and the salt trends are interpreted to represent a major transform related to the opening direction of the Gulf of Mexico basin.

In the eastern Gulf of Mexico basin the boundary is defined seismically by refraction data, a change in reflection character, and the termination of Jurassic salt and sediments (Buffler and others, 1981; Ibrahim and others, 1981; Phair and Buffler, 1983; Rosenthan, 1987; Figs. 11 and 14). These older Jurassic sediments represent rocks deposited on thin transitional crust prior to the emplacement of oceanic crust. The boundary here may be associated with a basement high.

There is also a significant change in the magnetic anomaly patterns in the eastern Gulf of Mexico basin that correlates reasonably well with the boundary between oceanic and thin transitional crust defined using seismic data (Fig. 12). Over oceanic crust, magnetic anomalies are quite small, while they are much larger over thin transitional crust. This is presumably due to the greater heterogeneity of transitional crust. If the magnetic anomaly data are used alone, the oceanic to thin transitional crust boundary probably would be drawn farther north on the south side of the eastern Gulf of Mexico basin and even more northward on the north side (Fig. 12). Within oceanic crust the anomalies show no apparent lineations, while within the thin transitional crust they appear to show along-strike lineation.

The boundary in the western Gulf of Mexico basin is in-

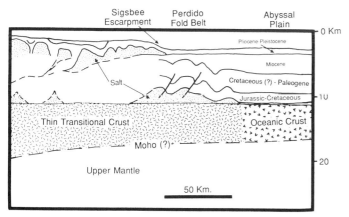

Figure 13. Cross section of the Perdido foldbelt in the northwestern Gulf of Mexico basin modified from Worral and Snelson (1989; Fig. 1). The faults and folds of the Perdido foldbelt are localized by the edge of Jurassic age salt deposition. As such they are an indirect indicator of the location of the edge of deposited salt and, by analogy to elsewhere in the Gulf of Mexico basin, the edge of oceanic crust. This manifestation of the edge of oceanic crust is also seen in the Mississippi Fan foldbelt to the east (Fig. 1).

ferred from only a few seismic refraction stations (Swolfs, 1967; Ibrahim and others, 1981) and the abrupt change in relief on the basement (Figs. 1 and 2C). A narrow NW-SE–trending zone of thin transitional crust is inferred based on the presence of relatively shallow basement in the subsurface of eastern Mexico and very deep basement inferred from the seismic data under the deep western Gulf of Mexico. Thin transitional crust is not directly observed here, but inferred. The exact nature of the boundary is not well constrained, but it is important in any prerifting plate reconstruction of the Gulf of Mexico region, as it may represent a major transform boundary during early Gulf of Mexico basin opening, particularly for models that move the Yucatán block south with respect to North America (Chapter 14, this volume). This is an area where future data acquisition could be valuable.

The entire boundary between thin transitional crust and oceanic crust along the northern and western margins of the Sigsbee Salt basin is obscured by a prominent high-amplitude reflector (Figs. 11 and 15), interpreted to be a salt tongue (Fig. 15) that migrated toward the oceanic crust sometime in the Late Jurassic to Early Cretaceous (Lin, 1984; Chapter 13, this volume). Seismic refraction data suggest that the actual boundary between thin transitional and oceanic crust occurs beneath the seaward edge of the salt tongue (Hall and others, 1982; Lin, 1984).

Thin transitional crust

Thin transitional crust flanks the area of oceanic crust on all sides and lies basinward of the Early Cretaceous carbonate margin (Figs. 1 and 2; Buffler and Sawyer, 1985). It occupies narrow belts to the east and west, a somewhat wider region to the south, and a broad zone to the north. This crust apparently extended

during Middle Jurassic time, as the African–South American Plate began pulling away from the North American Plate, but prior to the actual emplacement of oceanic crust (Chapter 14, this volume). Limited refraction data suggest that it ranges in thickness from 8 to 15 km and has velocities of 6.4 to 6.8 km/sec (Ibrahim and others, 1981; Ibrahim and Uchupi, 1983; Locker and Chatterjee, 1984; Ebeniro and others, 1988; Nakamura and others, 1988). No clear pattern of high-velocity lower crust has been shown for the margins of the Gulf of Mexico basin. Because it was highly extended, it subsided faster and to greater depths than the adjacent thicker crust landward of the Early Cretaceous carbonate margins (Figs. 1 and 2, and Plate 3). Thick salt was deposited in the more rapidly subsiding parts of the area as well as in the salt basins north of the Early Cretaceous margin. The thin transitional crust is overlain by Jurassic through Lower Cretaceous, nonmarine to deep-marine sedimentary rocks, deposited as the crust cooled and subsided prior to, during, and following, emplacment of oceanic crust (Fig. 2).

In the south-central deep Gulf of Mexico Sigsbee Salt basin area (Figs. 1 and 2A), a prominent, high-amplitude, basinward-dipping reflector/unconformity is interpreted to be the top of the crust (Fig. 15; Ladd and others, 1976; Long, 1978; Buffler and others, 1980; Buffler, 1983; Lin, 1984; Chapter 13, this volume). The reflector corresponds approximately with the top of a high-velocity crustal layer that has velocities of 6.4 to 6.8 km/sec (Fig. 15; Buffler and others, 1980; Ibrahim and others, 1981; Locker and Chatterjee, 1984). Over much of the area, the surface is relatively smooth (probably erosional), although in places it is offset by small faults (Lin, 1984). It is unconformably overlain and onlapped by the inferred Middle Jurassic salt as well as by progressively younger rocks, including Upper Jurassic and Cretaceous sediments (Fig. 15). The surface also truncates a thick, older sedimentary sequence interpreted to be a Late Triassic to Early Jurassic rift basin, similar to other rift basins that rim the northern and western Gulf of Mexico basin (Fig. 15). Late Triassic to Early Jurassic basins like this one are observed within thin transitional, thick transitional, and continental crust (as defined in this chapter). This sequence, however, could represent older Paleozoic sedimentary rocks. In either case, these older rocks are considered to be part of the basement as defined at the outset of this chapter.

In the deep southeastern Gulf of Mexico, the top of the thin transitional crust is recognized as acoustic basement in seismic profiles (Fig. 2B). Here it rises gradually to the south to relatively shallow depths over a prominent NE-SW–trending basement arch (Phair and Buffler, 1983; Schlager and others, 1984; Fig. 2B). This arch area is characterized by faulted blocky basement, in contrast to the unfaulted smooth crosional surface in the Sigsbee Salt basin area. Along the southern half of the arch, the basement blocks are very large. In one area, a large east-west–trending graben system is interpreted to be a Late Triassic to Early Jurassic rift basin. The northern flank of the arch is characterized by prominent tilted basement blocks forming a broad area of half-grabens filled with synrift sediments (Phair and Buffler,

1983; Schlager and others, 1984). The crust along the arch was interpreted to be transitional crust based on refraction data (Buffler and others, 1981; Ibrahim and others, 1981; Shaub and others, 1984). This was later confirmed by results of DSDP Leg 77 (Buffler and others, 1984). Two tilted basement blocks were drilled at sites 537 and 538, and metamorphic rocks were recovered that yielded early Paleozoic ages (about 500 Ma). At site 538 the metamorphic rocks were intruded by basic igneous rocks with ages of 190 and 160 Ma (Early to Middle Jurassic). These igneous rocks probably were emplaced during rifting and formation of the thin transitional crust.

The differences between the smooth, undeformed basement in the south-central Gulf of Mexico (Sigsbee Salt basin area) and the block-faulted basement in the southeastern Gulf of Mexico may be due to their different tectonic settings during the early evolution of the basin (Pilger, 1981; Klitgord and others, 1984; Buffler and Sawyer, 1985; Pindell, 1985; Salvador, 1987). The south-central Gulf of Mexico may have been more of a divergent

or pull-apart margin, parts of which were uplifted and eroded to give the smooth erosion surface along its southern flank. In contrast, the southeastern Gulf of Mexico probably was more of a transform margin, along which the Yucatán block moved out of the Gulf of Mexico, leaving foundered continental blocks along its edge.

In the eastern Gulf of Mexico beneath the West Florida basin, the top of the thin transitional crust is poorly imaged in the seismic data because of thick salt and sediments (Figs. 2 and 14). It is mainly observed along the western part of the West Florida basin.

Thin transitional crust is deeply buried in the northwestern Gulf of Mexico basin. Ebeniro and others (1988) used refraction data (Fig. 9) to construct a cross section (Fig. 10) of thin transitional and oceanic crust there. Within the thin transitional crust, which is only 6 to 12 km thick over a 200 km width, there is considerable thickness variation. They speculated that the crustal thin spot located at 200 km in Figure 10 may be a failed spread-

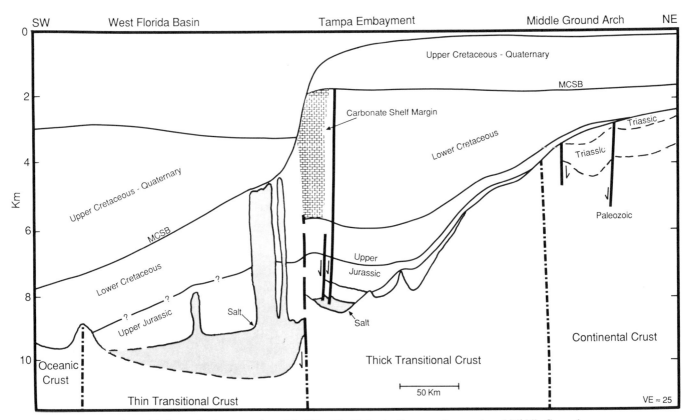

Figure 14. Generalized cross section of the West Florida basin, Tampa embayment, and Middle Ground arch (see Fig. 1, features 3, 4, 5, and 23, for general location) by Dobson (1990). This section spans all four crust types. The continental crust contains a Late Triassic basin, formed during the earliest phase of rifting which formed the Gulf of Mexico basin. These early basins also are seen within thick and thin transitional crust elsewhere. The boundary between continental crust and thick transitional crust corresponds to the onlap of Jurassic sedimentary rocks, a basement hinge zone, and the approximate truncation of Triassic basins. The boundary between thick and thin transitional crust is clearly marked by the common carbonate shelf margin buildup. In this area the boundary may also be the site of a basement fault. The boundary between thin transitional and oceanic crust is marked by the southwestward onlap of salt and Upper Jurassic sediments on basement, which may represent a basement high associated with the boundary. MCSB, Mid-Cretaceous Sequence Boundary.

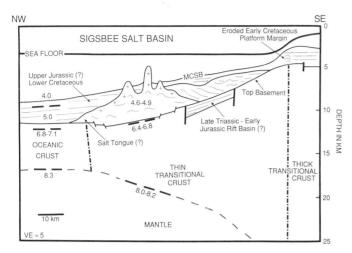

Figure 15. Generalized cross section of the Sigsbee Salt basin (see Fig. 1, feature 21, for general location) area of the southern Gulf of Mexico basin margin modified from Lin (1984). This section crosses oceanic, thin transitional, and thick transitional crust. The boundary between oceanic and thin transitional crust is the limit of salt deposition. There is typically some early salt movement across the boundary in the form of a salt tongue. A Late Triassic to Early Jurassic rift basin may be present within the thin transitional crust here. The boundary between thick and thin transitional crust here is a basement hinge overlain by an eroded Early Cretaceous carbonate margin. MCSB, Mid-Cretaceous Sequence Boundary.

ing axis. That possibility has not been included in our crust-type maps.

Magnetic data over thin transitional crust in the eastern Gulf of Mexico basin (Fig. 12) show that the structures are generally oriented along the strike of the basin margin. This pattern is particularly well developed in the thin transitional crust north of the Yucatan block in the eastern Sigsbee Salt basin and southwest of Florida in the West Florida Salt basin (Fig. 12). These are presumably structures produced during the extreme thinning that formed the thin transitional crust. Amplitudes of anomalies within thin transitional crust are often greater than 200 nT.

The zones of thin transitional crust on the north and south of the Gulf of Mexico basin are highly asymmetrical. If our crust-type identifications are correct, on the north it is roughly three times as wide as on the south (Figs. 1, 2, and 4). The thin transitional crust on the north is about 500 km wide, certainly among the widest zones of stretched crust on any passive margin. The mechanics of continental extension are not well understood, so it is not yet possible to be sure why the margins are so asymmetric. Wernicke (1985) has suggested that asymmetric margins may be produced when simple shear along subhorizontal detachments dominates the extension process. That process may have occurred in the Gulf of Mexico basin, but it seems unlikely that simple shear alone could produce the entire amount of asymmetry in this basin. We are also averse to automatically associating rift margin asymmetry to low-angle normal faulting. Dunbar and Sawyer (1989) reported variations in the amount of extension prior to the initiation of sea-floor spreading along the

margins of the Atlantic Ocean and Labrador Sea. They found some asymmetries as large as those in the Gulf of Mexico basin and attributed them to variations in the depth and orientation of preexisting weaknesses that served to localize the rifting process. In particular, broadly extended and asymmetric margins may be formed when no "conveniently oriented weaknesses"—those trending nearly normal to the direction of maximum tensional stress—are present in the continental crust. In this situation, rifting is diffuse, and the initiation of sea-floor spreading often occurs at one side or the other of the extended zone. This process is not inconsistent with the Wernicke (1985) model but can explain a wider range of margin configurations. It is not yet clear whether or how either of these models explains highly asymmetric crust extension at conjugate margins such as in the Gulf of Mexico basin.

The boundary between thin transitional and thick transitional crust

The boundary between the thin transitional crust and the thick transitional crust appears to correspond approximately to a major tectonic hinge zone in the basement (Plate 3; Figs. 1, 2, 14, 15, and 16). This hinge zone in many cases (Figs. 2, 15, and 16) is defined by an overall change in basement dip that separates more subsided basement from less subsided basement. In other cases it appears to be a faulted boundary (Fig. 14). The boundary was mapped on the basis of the distribution of the Early Cretaceous carbonate platform margin that rimmed the deep Gulf of Mexico basin during Early Cretaceous time (Figs. 1, 2, 14, 15, and 16). This change in differential basement subsidence apparently became the site of the Early Cretaceous carbonate margin as the entire basin subsided. The boundary is also identified by a change in the trend of basement features (Fig. 1). On the thin-transitional-crust side of the boundary, the basement blocks trend roughly parallel to the margin; on the thick-transitional-crust side of the boundary, the basement blocks trend roughly perpendicular to the margin.

The magnetic data in the eastern Gulf of Mexico basin show the trends of the basement blocks reasonably clearly (Fig. 12). At the boundary between thick and thin transitional crust, the orientation of the anomalies changes from parallel to the margin over thin transitional crust, to perpendicular to the margin over thick transitional crust (Fig. 12). This pattern in the trend of the magnetic anomalies and its association with the crust-type boundary is clear on the north and south sides of the eastern Gulf of Mexico basin. This pattern is not well developed elsewhere in the basin.

Thick transitional crust

The area of thick transitional crust generally lies landward of the Early Cretaceous carbonate margin (Figs. 1, 2, 14, 15, and 16). Thick transitional crust thickness varies from about 20 km up to normal continental crust thickness of about 35 to 40 km (Fig. 6), and is characterized by relatively shallow, well-defined basement highs with intervening lows (Fig. 1). The high areas overlie crust with thickness close to normal continental crust,

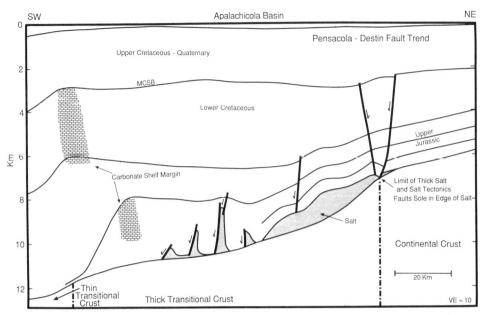

Figure 16. Cross section of the Apalachicola basin by Dobson (1990). See Figure 1, feature 5, for the location of the Apalachicola basin. Note that this section crosses continental, thick transitional, and thin transitional crust. The landward limit of thick salt deposition corresponds to the boundary between continental and thick transitional crust. A band of faults sole in the edge of that salt, providing surface manifestation of the boundary. The boundary between thick and thin transitional crust is marked by carbonate shelf margin buildups. Their location was probably controlled by differential subsidence due to the crust type difference. Thick salt is frequently present on both thick and thin transitional crust, although it is only seen over thick transitional crust in this section. MCSB, Mid-Cretaceous Sequence Boundary.

while the lows overlie thinner crust, probably extended continental crust. This peripheral area of the Gulf of Mexico basin apparently did not stretch as much as the central Gulf of Mexico, and therefore, did not subside as much or as rapidly. Thick Jurassic to Lower Cretaceous salt and sedimentary rocks were deposited in the lows or basins (Figs. 14 and 16), while these deposits are thinner or absent over the adjacent highs, indicating that the topography was present during rifting and the deposition of the salt and sediments.

The pattern of alternating basement highs and lows within the thick transitional crust is best illustrated in the northeastern Gulf of Mexico basin. There are few data to constrain its presence elsewhere. Proceeding from southern Florida toward the northwest, the highs are the Sarasota arch, the Middle Ground arch–Southern platform, the Wiggins uplift, the Monroe uplift, and the Sabine uplift (Fig. 1). The intervening basins are the South Florida basin, the Tampa embayment, the Apalachicola basin and embayment, the Mississippi Salt basin, the North Louisiana Salt basin, and the East Texas basin. Thomas (1988) speculated that this set of highs and lows formed in a broad transtensional zone. Within this zone, movement of the Yucatán block to the southeast with respect to North America produced a combination of tension and sinistral strike-slip movement. Many of the highs may represent continental blocks that have been rotated ball-bearing style in a counterclockwise direction within the transtensional zone. These blocks would not have experienced

much extension or internal deformation. The lows would then be basins formed due to greater crust extension between the continental blocks. The crust between the rotating blocks may have been highly deformed during the movement. Whatever their origin, the highs and lows now trend roughly perpendicularly to the margin of the Gulf of Mexico basin. This orientation contrasts with the structures within thin transitional crust, which tend to be parallel to the margin.

On a larger scale, Dunbar and Sawyer (1987) speculated that the whole Yucatán Block may have rotated counterclockwise in a ball-bearing style due to the motion of the South American Plate with respect to the North American Plate. Rotation of the Yucatan block ceased when the South American Plate moved south and was disengaged from the North American Plate. This mechanism might explain why the Gulf of Mexico basin only opened for a short period.

Magnetic data from thick transitional crust in the eastern Gulf of Mexico basin show anomalies that cross the strike of the margin (Fig. 12). This pattern is observed offshore of both Florida and the Yucatán Peninsula and is consistent with the observations of high and low basement blocks in thick transitional crust. The anomalies are relatively large, up to 450 nT, over thick transitional crust.

Tectonic-subsidence-history analyses in the North Louisiana Salt basin (Nunn and others, 1984) found that the subsidence history there is compatible with crustal extension of $\beta = 1.5$ to

2.0. If the crust was originally 40 km thick, this would predict present crustal thicknesses of 20 to 27 km, consistent with seismic refraction results. We expect that this is typical of the many salt basins overlying thick transitional crust. The intervening highs have probably seen little or no crustal extension.

The boundary between thick transitional crust and continental crust

The boundary between continental crust and thick transitional crust (Fig. 1) is drawn based on the distribution of mapped Mesozoic structures (fault trends and transforms) that are associated mainly with the Middle to Late Jurassic opening of the Gulf of Mexico basin (Thomas, 1988). In southern Florida, a major crustal boundary is inferred from gravity and magnetic data to correspond with the nrothwestward projection from the Atlantic of the Bahamas Fracture Zone (Klitgord and others, 1984, 1988; Klitgord and Schouten, 1986). The boundary also defines the northeastern edge of a zone of Mesozoic (Jurassic) mafic volcanic rocks described in wells (Chowns and Williams, 1983; Thomas, 1988; Thomas and others, 1989; Chapter 6, this volume). The boundary is projected northwestward across the West Florida Shelf on the basis of gravity and magnetic data (Klitgord and others, 1984, 1988) and a regional hinge zone or flexure of the basement surface as seen on the basement contour map (Fig. 1 and Plate 3). This hinge zone is also shown in the cross section of Figure 14, which shows regional thinning and onlap of the Jurassic sedimentary section of the Tampa embayment onto the hinge, and Figure 16, which shows regional onlap and thinning of the Louann Salt (premarine evaporites) at the hinge. The area of abrupt thinning of the salt became the locus for later faulting of the overlying Mesozoic section due to salt withdrawal. These faults are part of a more regional peripheral fault zone around the flank of the Apalachicola basin, also due to salt withdrawal.

The northwestward projection of the boundary between thick transitional crust and continental crust on land in the northern Gulf of Mexico basin corresponds to a zone of Mesozoic orthogonal faults mapped by seismic and well data (Thomas, 1988). This fault zone has been referred to as the Gulf Rim fault zone (Martin, 1978; Klitgord and Schouten, 1986) and more recently as the Alabama-Arkansas fault zone (Thomas, 1988). Movement along this fault zone was greatest from Late Triassic through Late Jurassic (Thomas, 1988). This zone controls the regional updip limit of the Louann Salt or premarine evaporites in the area (Fig. 1). Alignment of the Alabama-Arkansas fault zone with the projection of the Bahamas fracture zone suggests a large-scale left-slip system of transform and wrench faults that was active both during the Late Triassic–Early Jurassic rift phase as well as the later Middle Jurassic through Late Jurassic opening of the Gulf of Mexico basin (Thomas, 1988). The location of this boundary probably was inherited from the late Precambrian–early Paleozoic extensional margin with its associated transform faults, which framed a southeastward-projecting promontory

(Alabama Promontory) of the early North American craton (Thomas, 1988).

In southwest Arkansas the boundary between continental and thick transitional crust turns to the southwest across east Texas and then bends around the San Marcos arch along the trend of the Mesozoic fault system (Martin, 1978; Thomas, 1988; Chapter 3, this volume; Plate 2; Fig. 1). Again, movement along this fault system was during the Mesozoic, but some activity continued through the Cenozoic. This fault trend generally follows the trend of the underlying buried Ouachita orogenic belt and overlying Late Triassic–Early Jurassic basins described above, thus reflecting movements along inherited crustal weaknesses.

The origin of the hinge zone forming the boundary between thick transitional and continental crust, like that of the hinge zone on the U.S. Atlantic continental margin, is not well understood. Sawyer and Harry (1991) suggest that the Atlantic margin hinge zone may be the boundary between crust affected by two different but related phases of extension and subsidence. The first phase, lasting from about 225 to 195 Ma, formed the Triassic Newark-style basins, and was the result of crustal thinning localized by midcrustal-depth weaknesses. The second phase, lasting from about 195 to 175 Ma, formed the offshore basins, led to sea-floor spreading, and was localized by weaknesses in the upper mantle. The mantle weaknesses seem to have been displaced laterally or trend differently from the crustal weaknesses. A similar mechanism may be responsible for the formation of the hinge zone in the northern Gulf of Mexico basin, although a good picture is not yet available of what distribution of pre-rifting lithosphere weaknesses can explain the crustal configuration of the basin.

Continental crust

Continental crust, as used in this chapter, is the crust presumably unaffected by the Middle to Late Jurassic rifting, the main stage of rifting that led to the formation of the Gulf of Mexico basin. In places, the continental crust was affected by Late Triassic to Early Jurassic rifting that formed isolated basins. The basement depth in regions of continental crust varies from 0 km in Texas and the southern United States (basement outcrop) to 5 km below sea level on the West Florida Shelf (Fig. 1). It is typically 1 to 2 km deep. Five kilometers of sediment accumulation on normal continental crust is somewhat unusual. Here it has two causes. Part of the sediment accumulated due to small amounts of Late Triassic to Early Jurassic extension that thinned the continental crust and caused it to subside. The remainder probably accumulated as a result of the downward flexure of the continental crust resulting from thick sediment deposited in the deeper parts of the Gulf of Mexico basin. Some of this downward flexing along the West Florida Shelf may have been amplified by the late Cenozoic depocenter of the Mississippi River Delta. The large sediment load of the Cenozoic, and earlier, rocks is flexing the adjacent continental crust downward, allowing additional sediment deposition there.

The overall extension of the continental crust is probably very small. TTS analysis indicates that the crustal extension of continental crust is mostly less than $\beta = 1.25$ (25 percent extension), and everywhere less than $\beta = 1.5$ (50 percent extension; Fig. 5). However, even this is probably an overestimate, since some flexure-induced subsidence is incorrectly being attributed to extension in these areas. The continental crust is typically 35 to 40 km thick, with some areas in which, for the reasons just mentioned, subsidence analysis indicates it to be slightly thinner (Fig. 6).

That some extension of the continental crust took place during the Late Triassic and Early Jurassic is indicated by the presence of many fault-bounded basins of that age (Fig. 14). These are analogous to the Triassic basins along North America's Atlantic margin. Similar features are also observed in seismic data over transitional crust (Fig. 15). The presence of the older basins in areas affected by later intense extension (transitional crust) and areas unaffected by later intense extension (continental crust) suggests that during the early stage, rifting was more widespread than during the later stage. Current models of continental rifting predict that late-stage extension will be more laterally focused than early-stage extension (Dunbar and Sawyer, 1988, 1989; Sawyer and Harry, 1991). This is a consequence of the rheologies ascribed to the continental lithosphere. The Late Triassic and Early Jurassic basins generally follow the trend of the underlying late Paleozoic Appalachian-Ouachita orogenic belt, and therefore appear to represent the reactivation of older Paleozoic weaknesses in the crust.

All of the rocks in the Late Triassic to Early Jurassic rift basins are truncated by a prominent unconformity, which is the surface mapped in Figure 1 and Plate 3. This unconformity is probably equivalent to the post-rift unconformity of Falvey (1974) and Klitgord and others (1988). The formation and filling of the Late Triassic and Early Jurassic basins and the formation of the unconformity are believed to have preceded the main phase of extension that formed the transitional crust in the Middle and Late Jurassic.

EVOLUTION OF THE CRUST UNDER THE GULF OF MEXICO BASIN; A SUMMARY

The crust-type distribution suggests a model of the evolution of the crust under the Gulf of Mexico basin that includes: (1) a Late Triassic–Early Jurassic phase of early rifting; (2) a Middle Jurassic phase of rifting, crustal attenuation, and formation of a broad area of transitional crust; (3) a Late Jurassic period of oceanic crust formation in the deep central Gulf of Mexico; and (4) subsequent crustal subsidence modified by sediment deposition. Each of these phases is outlined briefly below.

Late Triassic–Early Jurassic, the early phase of rifting

This stage is characterized by the formation of linear zones of rifting of a brittle continental crust. Fault basins are characterized by large and small half-grabens bounded by listric normal faults. They are filled with nonmarine sediments and volcanics. Equivalent rift basins occur all along the east coast of North America, across northern Florida and southern Georgia, all along the northern Gulf of Mexico basin, and into Mexico. Similar rifts may occur beneath parts of the Sigsbee Salt basin in the deep southern Gulf of Mexico. All of these rifts represent the initial rift stage in the breakup of Pangea. Rifting apparently took place along older weaknesses and sutures inherited from earlier late Precambrian and Paleozoic phases of rifting and the later continental collision that formed Pangea. This early rift stage, though formed by relative movements between Africa, South America, and North America, did not significantly thin the continental crust throughout the Gulf of Mexico region.

Middle Jurassic, the main phase of crust attenuation

During the Middle Jurassic the entire Gulf of mexico basin region underwent rifting and significant attentuation of the crust. This resulted in formation of the areas of transitional crust and the associated basement highs and lows that form the Gulf of Mexico basin architecture we see today. During this period the Yucatán block began rotating out of the northern Gulf of Mexico basin region like a large ball bearing caught between broad, generally NW-SE–trending transtensional zones across Mexico and Florida. The continental crust in the outer periphery of the basin underwent only moderate thinning and remained relatively thick, forming the area of broad arches and basins that we refer to as thick transitional crust. The center, and perhaps originally hotter or weaker, part of the basin, however, underwent considerably more crust and lithosphere thinning and subsidence to form the large area of thin transitional crust. The pattern of extension was to gradually focus the crust thinning in a decreasing amount of the crust. Although the zone of thin transitional crust is now quite wide, it was formed from a piece of prerift continental crust that was quite narrow. Thick salt was deposited throughout the broad central area of thin crust as well as in many of the basins in the adjacent thick crust, as marine waters periodically spilled into the basin across sills probably located in central Mexico.

Late Jurassic oceanic crust is formed

The Late Jurassic is characterized by the emplacement of oceanic crust after sea-floor spreading began along a generally east-west–striking weakness in the thinning continental crust. The area of oceanic crust became wider to the west as Yucatán continued to rotate in a counterclockwise direction. This was accompanied by a general marine transgression into the Gulf of Mexico basin area as the basin began to cool and subside. The area of newly formed oceanic crust was covered by water several kilometers deep, while the area of transitional crust formed a continental shelf and slope. Gravity and sediment loading caused early basinward flow of salt tongues along the boundary between oceanic and thin transitional crust. The broad basement highs and lows continued to influence sedimentation around the periphery of the basin, and many of the highs remained emergent throughout most of the Late Jurassic (e.g., Yucatán). Some rifting and deposition of synrift sediments seems to have continued in the southeastern Gulf of Mexico. This is probably activity related to the eastward propagation of the initiation of sea-floor spreading. The driving forces seem to have slowed or stopped before sea-floor spreading could propagate further to join with the Atlantic oceanic crust.

Gulf of Mexico Basin subsidence phase

The Gulf of Mexico basin was certainly locked into its present configuration with respect to North America by Early Cretaceous time. Subsidence rates were probably highest for oceanic crust, somewhat less for thin transitional crust, and minimal for thick transitional crust and continental crust. Broad carbonate platforms with prominent rimmed margins became established along the tectonic hinge zones forming the boundary of differential subsidence between thin and thick crust. Sediment deposition amplifies the subsidence of the crust by about a factor of three. If sediment fills a part of the basin to sea level, the basement depth will be about three times what it would be in the absence of sedimentation. In the oceanic crust area, sedimentation was never rapid enough to fill the basin to sea level, however. Some of the thin transitional crust, particularly in the northwestern Gulf of Mexico basin, was filled to near sea level, first by carbonate sediment and then by clastic sediment. The areas of thick transitional crust and continental crust were probably continually filled to near sea level to the present.

ONGOING STUDIES

Considering the enormous economic value of the resources in the Gulf of Mexico basin and the intense exploration effort mounted to locate them, remarkably little is known about the crust under the basin. Several efforts now underway in the academic community promise to expand our knowledge somewhat. A detailed aeromagnetic survey in the eastern part of the deep Gulf of Mexico basin has been flown to identify sea-floor magnetic lineations and crustal boundaries, and the data should soon be available. New deep-penetration land seismic reflection data from the Llano uplift to the Gulf Coast in Texas were recently acquired by COCORP. The U.S. Geological Survey expects to acquire new deep penetration reflection profiles, with complementary ocean bottom, large offset seismic studies, in the eastern Gulf of Mexico in 1991. Continued cooperation between academic scientists and colleagues in the petroleum exploration industry will also produce new knowledge of the crust under the Gulf of Mexico basin as new well and seismic data become available, and with that knowledge should come improved understanding of the processes that formed the basin, its crust, and upper mantle.

REFERENCES CITED

Buffler, R. T., 1983, Structure and stratigraphy of the Sigsbee Salt Dome area, deep south-central Gulf of Mexico, *in* Bally, A. W., ed., Seismic expression of structural styles; A picture and work atlas: American Association of Petroleum Geologists Studies in Geology 15, v. 2, p. 2.3.2–2.3.56.

Buffler, R. T., and Sawyer, D. S., 1985, Distribution of crust and early history,

Gulf of Mexico basin: Gulf Coast Association of Geological Societies Transactions, v. 35, p. 333–344.

Buffler, R. T., Watkins, J. S., Worzel, J. L., and Shaub, F. J., 1980, Structure and early geologic history of the deep central Gulf of Mexico, *in* Pilger, R., ed., Proceedings of a Symposium on the Origin of the Gulf of Mexico and the Early Opening of the Central North Atlantic: Baton Rouge, Louisiana State University, p. 3–16.

Buffler, R. T., Shaub, F. J., Huerta, R., and Ibrahim, A. K., 1981, A model for the early evolution of the Gulf of Mexico basin, *in* Geology of Continental Margins Symposium, Proceedings of the 26th International Geological Congress, Paris, July 1980: Oceanologica Acta, p. 129–136.

Buffler, R. T., and others, 1984, Initial reports of the Deep-Sea Drilling Project: Washington, D.C., U.S. Government Printing Office, v. 77, 747 p.

Chowns, T. M., and Williams, C. T., 1983, Pre-Cretaceous rocks beneath the Georgia Coastal Plain; Regional implications: U.S. Geological Survey Professional Paper 1313-L, p. L1–L42.

Dobson, L., 1990, Seismic stratigraphy and geologic history of Jurassic rocks, northeastern Gulf of Mexico [M.A. thesis]: Austin, University of Texas, 105 p.

Dunbar, J. A., and Sawyer, D. S., 1987, Implications of continental crust extension for plate reconstruction; An example from the Gulf of Mexico: Tectonics, v. 6, p. 739–755.

—— , 1988, Continental rifting at pre-existing lithospheric weaknesses: Nature, v. 333, p. 450–452.

—— , 1989, Effects of continental heterogeneity on the distribution of extension and shape of rifted continental margins: Tectonics, v. 8, p. 1059–1078.

Ebeniro, J. O., Nakamura, Y., Sawyer, D. S., and O'Brien, W. P., Jr., 1988, Sedimentary and crustal structure of the northern Gulf of Mexico: Journal of Geophysical Research, v. 93, p. 9075–9092.

Ewing, J. I., Antoine, J., and Ewing, M., 1960, Geophysical measurements in the western Caribbean Sea and in the Gulf of Mexico: Journal of Geophysical Research, v. 65, p. 4087–4126.

Ewing, J. I., Worzel, J. L., and Ewing, M., 1962, Sediments and oceanic structural history of the Gulf of Mexico: Journal of Geophysical Research, v. 67, p. 2509–2527.

Ewing, M., Worzel, J. L., Ericson, D. B., and Heezen, B. C., 1955, Geophysical and geological investigations in the Gulf of Mexico, Part 1: Geophysics, v. 20, p. 1–18.

Falvey, D. A., 1974, The development of continental margins in plate tectonic theory: Australian Petroleum Exploration Association Journal, v. 14, p. 95–106.

Hall, D. J., Cavanaugh, T. D., Watkins, J. S., and McMillen, K. J., 1982, The rotational origin of the Gulf of Mexico based on regional gravity data, *in* Watkins, J. S., and Drake, C. E., eds., Studies in continental margin geology: American Association of Petroleum Geologists Memoir 34, p. 115–126.

Hall, S., Shepherd, A., Titus, M., and Snow, R., 1984, Magnetics, *in* Buffler, R. T., Locker, S. D., Bryant, W. R., Hall, S. A., and Pilger, R. H., Jr., eds., Gulf of Mexico: Woods Hole, Massachusetts, Marine Science International, Ocean margin Drilling Program Regional Atlas Series, Atlas 6, sheet 3.

Haxby, W. F., 1987, Gravity field of the world's oceans: Boulder, Colorado, National Oceanic and Atmospheric Administration National Geophysical Data Center, scale at Equator approximately 1:40,000,000.

Humphris, C. C., 1978, Salt movement on continental slopes, northern Gulf of Mexico, *in* Bouma, A. H., Moore, G. T., and Coleman, J. M., eds., Framework, facies, and oil-trapping characteristics of the upper continental margin: American Association of Petroleum Geologists Studies in Geology 7, p. 69–85.

Ibrahim, A. K., and Uchupi, E., 1983, Continental/oceanic crustal transition in the Gulf Coast geosyncline, *in* Watkins, J. S., and Drake, C. E., eds., Studies in continental margin geology: American Association of Petroleum Geologists Memoir 34, p. 155–165.

Ibrahim, A. K., Carye, J., Latham, G., and Buffler, R. T., 1981, Crustal structure in the Gulf of Mexico from OBS refraction and multichannel reflection data: American Association of Petroleum Geologists Bulletin, v. 65, p. 1207–1229.

Keller, G. R., and 6 others, 1989, Paleozoic continent-ocean transition in the Ouachita Mountains imaged from PASSCAL wide-angle seismic reflection-refraction data: Geology, v. 17, p. 119–122.

Klitgord, K. D., and Schouten, H., 1986, Plate kinematics of the central Atlantic, *in* Vogt, P. R., and Tucholke, B. E., eds., The western North Atlantic region: Boulder, Colorado, Geological Society of America, The Geology of North America, v. M, p. 351–378.

Klitgord, K. D., Popenoe, P., and Schouten, H., 1984, Florida; A Jurassic transform plate boundary: Journal of Geophysical Research, v. 89, p. 7753–7772.

Klitgord, K. D., Hutchinson, D. R., and Schouten, H., 1988, U.S. Atlantic margin; Structural and tectonic framework, *in* Sheridan, R. F., and Grow, J. A., eds., The Atlantic continental margin; U.S.: Boulder, Colorado, Geological Society of America, The Geology of North America, v. I-2, p. 19–55.

Ladd, J. W., Buffler, R. T., Watkins, J. S., Worzel, J. L., and Carranza, A., 1976, Deep seismic reflection results from the Gulf of Mexico: Geology, v. 4, p. 365–368.

Lee, G. H., Bryant, W. R., and Watkins, J. S., 1989, Salt structures and sedimentary basins in the Keathley Canyon area, northwestern Gulf of Mexico; Their development and tectonic implications, *in* Gulf of Mexico salt tectonics; Associated processes and exploration potential; 10th Annual Research Conference: Gulf Coast Section, Society of Exploration Paleontologists and Mineralogists Foundation Program and Extended and Illustrated Abstracts, p. 90–93.

LePichon, X., and Sibuet, J. C., 1981, Passive margins; A model of formation: Journal of Geophysical Research, v. 86, p. 3708–3720.

Lin, T., 1984, Seismic stratigraphy and structure of the Sigsbee salt basin, south-central Gulf of Mexico [M.A. thesis]: Austin, University of Texas, 102 p.

Locker, S. D., and Chatterjee, S. K., 1984, Seismic velocity structure, *in* Buffler, R. T., Locker, S. D., Bryant, W. R., Hall, S. A., and Pilger, R. H., Jr., eds., Gulf of Mexico: Woods Hole, Massachusetts, Marine Science International, Ocean margin Drilling Program Regional Atlas Series, Atlas 6, sheet 4.

Long, J. M., 1978, Seismic stratigraphy of part of the Campeche Escarpment southern Gulf of Mexico [M.A. thesis]: Austin, University of Texas, 105 p.

Martin, R. G., 1978, Northern and eastern Gulf of Mexico continental margin; Stratigraphic and structural framework, *in* Bouma, A. H., Moore, G. T., and Coleman, J. M., eds., Framework, facies, and oil-trapping characteristics of the upper continental margin: American Association of Petroleum Geologists Studies in Geology 7, p. 21–42.

Nakamura, Y., Sawyer, D. S., Shaub, F. J., MacKenzie, K., and Oberst, J., 1988, Deep crustal structure of the northwestern Gulf of Mexico: Gulf Coast Association of Geological Societies Transactions, v. 38, p. 207–215.

Nelson, K. D., and 8 others, 1985, New COCORP profiling in southeastern United States; Part 1, Late Paleozoic suture and Mesozoic rift basin: Geology, v. 13, p. 714–718.

Nunn, J. A., Scardina, A. D., and Pilger, R. H., Jr., 1984, Thermal evolution of the north-central Gulf Coast: Tectonics, v. 3, p. 723–740.

Parsons, B., and Sclater, J. G., 1977, An analysis of the variation of ocean floor bathymetry and heat flow with age: Journal of Geophysical Research, v. 82, p. 803–827.

Phair, R. L., and Buffler, R. T., 1983, Pre-Middle Cretaceous geologic history of the deep southeastern Gulf of Mexico, *in* Bally, A. W., ed., Seismic expression of structural styles; A picture and work atlas: American Association of Petroleum Geologists Studies in Geology 15, v. 2, p. 2.2.3–2.2.141.

Pilger, R. H., Jr., 1981, The opening of the Gulf of Mexico; Implications for the northern Gulf Coast: Gulf Coast Association of Geological Societies Transactions, v. 3, p. 377–381.

Pindell, J. L., 1985, Alleghenian reconstruction and subsequent evolution of the Gulf of Mexico, Bahamas, and proto-Caribbean: Tectonics, v. 4, p. 1–39.

Pindell, J. L., and Dewey, J. D., 1982, Permo-Triassic reconstructions of western Pangea and the evolution of the Gulf of Mexico/Caribbean region: Tectonics, v. 1, p. 179–211.

Rosenthal, D. B., 1987, Distribution of crust in the deep eastern Gulf of Mexico [M.A. thesis]: Austin, University of Texas, 149 p.

Salvador, A., 1987, Late Triassic–Jurassic paleogeography and origin of the Gulf of Mexico basin: American Association of Petroleum Geologists Bulletin, v. 71, p. 419–451.

Sawyer, D. S., 1985, Total tectonic subsidence; A parameter for distinguishing crust type at the U.S. Atlantic continental margin: Journal of Geophysical Research, v. 90, p. 7751–7769.

Sawyer, D. S., and Harry, D. L., 1991, Dynamic modeling of divergent margin formation; Application to the U.S. Atlantic Margin: Marine Geology (in press).

Sawyer, D. S., Ebeniro, J. O., O'Brien, W. P., Jr., Tsai, C. J., and Nakamura, Y., 1986, Gulf of Mexico seismic refraction study; Alaminos Canyon ocean bottom seismograph experiment: Austin, University of Texas Institute for Geophysics Technical Report 42, 26 p.

Schlager, W., Buffler, R. T., Angstadt, D., and Phair, R. L., 1984, Geologic history of the southeastern Gulf of Mexico, *in* Initial reports of the Deep Sea Drilling Project: Washington, D.C., U.S. Government Printing Office, v. 77, p. 715–738.

Shaub, F. J., Buffler, R. T., and Parsons, J. G., 1984, Seismic stratigraphic framework of deep central Gulf of Mexico basin: American Association of Petroleum Geologists Bulletin, v. 68, p. 1790–1802.

Shepherd, A., 1983, A study of magnetic anomalies in the eastern Gulf of Mexico [M.S. thesis]: Houston, Texas, University of Houston, 197 p.

Spudich, P., and Orcutt, J., 1980, Petrology and porosity of an oceanic crustal site; Results from wave-form modeling of seismic refraction data: Journal of Geophysical Research, v. 85, p. 1409–1433.

Swolfs, H. S., 1967, Seismic refraction studies in the southwestern Gulf of Mexico [M.A. thesis]: College Station, Texas A&M University, 42 p.

Thomas, W. A., 1988, Early Mesozoic faults of the northern Gulf Coastal Plain in the context of opening of the Atlantic Ocean, *in* Manspeizer, W., ed., Triassic-Jurassic rifting, continental breakup, and the origins of the Atlantic Ocean and passive margins, Part A: New York, Elsevier, p. 463–476.

Thomas, W. A., and 5 others, 1989, Pre-Mesozoic paleogeographic map of Appalachian-Ouachita orogen beneath Atlantic and Gulf Coastal Plains, *in* Hatcher, R. D., Jr., Thomas, W. A., and Viele, G. W., eds., The Appalachian-Ouachita orogen in the United States: Boulder, Colorado, Geological Society of America, The Geology of North America, v. F-2, Plate 6, scale 1:2,500,000.

Trehu, A. M., Klitgord, K. D., Sawyer, D. S., and Buffler, R. T., 1989, Atlantic and Gulf of Mexico continental margins, *in* Pakiser, L. C., and Mooney, W. D., eds., Geophysical framework of the continental United States: Boulder, Colorado, Geological Society of America Memoir 172, p. 349–382.

Wernicke, B., 1985, Uniform-sense normal simple shear of the continental lithosphere: Canadian Journal of Earth Sciences, v. 22, p. 108–125.

Worral, D. M., and Snelson, S., 1989, Evolution of the northern Gulf of Mexico, with emphasis on Cenozoic growth faulting and the role of salt, *in* Bally, A. W., and Palmer, A. R., eds., The geology of North America; An overview: Boulder, Colorado, Geological Society of America, The Geology of North America, v. A, p. 97–138.

MANUSCRIPT ACCEPTED BY THE SOCIETY DECEMBER 27, 1990

ACKNOWLEDGMENTS

We appreciate the efforts of William Thomas, Amos Salvador, Debbie Hutchinson, and Ian Norton, whose comments and criticisms have improved this chapter. Funding for DSS was provided by National Science Foundation Grants OCE-87-14004 and OCE-89-96263. Partial funding for work by RTB on this chapter came from the National Science Foundation Grant 84-17771 and the Texas Higher Education Coordinating Board Advanced Research Program Project 3513. We thank numerous industry and academic colleagues working in the Gulf of Mexico basin who, through the years, have shared their ideas with us. Special thanks to Laura Dobson who provided figures from her unpublished M.A. thesis on the Jurassic rocks in the northeastern Gulf of Mexico. This is University of Texas Institute for Geophysics Contribution No. 853.

The Geology of North America
Vol. J, The Gulf of Mexico Basin
The Geological Society of America, 1991

Chapter 5

Salt tectonics and listric-normal faulting

T. H. Nelson
Salt Tectonics International, Inc., 2203 Timberlock Place, Drawer 51, The Woodlands, Texas 77380

INTRODUCTION

Local structuring in the basins that rim the Gulf of Mexico is largely the result of gravity acting on sedimentary sections deposited on an unstable base of abnormally pressured shales and/or salt. The resulting deformation takes two primary forms, salt-flow structures and listric-normal faults that sole out at various levels above the basement.

Salt flow in this nonorogenic environment is the result of pressure gradients created by the sediments that overlie, or load, the salt. When differential loading occurs, pressures vary laterally within the salt layer, and the salt tends to move away from areas of higher pressure toward areas of lower pressure. Structures resulting from such flow have a wide variety of forms. These include low-relief anticlines and pillows, which simply deform the overlying beds; high-relief plugs and walls, which have a piercement relation with the sediments and often breach the surface; and extensive salt sheets, which have been emplaced laterally in clastic sediments deposited in a continental slope environment.

Listric-normal faults are the result of coherent, differential basinward movement of the sediments above some decollement layer. In the Gulf of Mexico basin, this layer may be either salt or abnormally pressured shale. The term "listric" comes from the Greek word for shovel and aptly describes the curved, three-dimensional geometry of the fault itself (Bally and others, 1981). The term was originally used to describe thrust faults; thus the adjective "normal" is needed to describe their extensional counterparts. The principal driving force for such faulting is gravity acting on a sloping surface, analogous to the force that drives glaciers. The form of structures created by listric-normal faulting alone is not as dramatically varied as those created by salt flow. They do, however, occur over a variety of geologic settings ranging from the onshore margins of the depositional basins within the region, to the thick sedimentary wedges that underlie the present continental shelf and slope.

The primary purpose of this chapter is to describe the manner in which salt structures and listric-normal faults form. To do this I have chosen to start with salt structures.

SALT FLOW

The term salt is often used to denote a limited group of minerals (see Table 1) precipitated during evaporation of natural waters. These include both sodium and potassium chlorides. In addition to their common origin, most of these minerals "flow" at relatively low levels of deviatoric stress. This ability to "flow" is inherent in the crystalline structure and high solubility of these minerals. Because they are less dense than most compacted sediments, the low-density elements of this group tend to flow upward relative to their source bed, often forming piercement structures. Of these low-density elements, halite (NaCl) is by far the most common and makes up between 90 and 98 percent of the salt in the Gulf of Mexico basin (Kupfer, 1989; Halbouty, 1979). Although pure halite has a specific gravity of 2.164 gm/cc, the effective density of rock salt is usually somewhat greater because of impurities disseminated through the salt. Anhydrite, which has a specific gravity of 2.960 gm/cc, is the most common of these impurities. Thus, for an average salt bed made up of 95 percent halite and 5 percent disseminated anhydrite, the bulk density would be about 2.2.

Laboratory studies of halite deformation over a wide range of temperatures, pressures, strain rates, and differential stresses show that in the absence of added water, flow is accomplished by the motion of lattice dislocations in the crystalline structure of the salt (e.g., Carter and Hansen, 1983). When water is added, however, flow may occur either by fluid-aided dislocation motion or by pressure solution and recrystallization (e.g., Spiers and others, 1990). This latter mechanism, termed solution transfer, is similar to the process of regelation by which glacial ice flows around an impediment at its base.

In addition to the effect that water has on the ability of salt to flow, the temperature of the salt is also an important factor. Again, laboratory tests (e.g., Carter and Heard, 1970; Carter and Hansen, 1983) show that the effective viscosity of salt decreases as temperature increases. Thus as salt becomes buried to ever greater depths, resistance to dislocation glide and climb within the salt decreases, and at some point, nominally dry salt may, under the right conditions of pressure and grain size, flow as easily as does wet salt.

Nelson, T. H., 1991, Salt tectonics and listric-normal faulting, *in* Salvador, A., ed., The Gulf of Mexico Basin: Boulder, Colorado, Geological Society of America, The Geology of North America, v. J.

TABLE 1. COMPOSITION AND SPECIFIC GRAVITY OF EVAPORITE MINERALS*

Mineral	Composition	Specific Gravity
NONRADIOACTIVE EVAPORITES		
Halite	NaCl	2.164
Anhydrite	$CaSO_4$	2.960
Gypsum	$CaSO_4 \cdot 2H_2O$	2.320
Trona	$Na_3(CO_3)(HCO_3) \cdot 2H_2O$	2.120
RADIOACTIVE EVAPORITES		
Sylvite	KCl	1.984
Carnallite	$KMgCl_3 \cdot 6H_2O$	1.610
Langbeinite	$K_2Mg_2(SO_4)_3$	2.830
Polyhalite	$K_2MgCa_2(SO_4)_4 \cdot 2H_2O$	2.780
Kainite	$MgSO_4 \cdot KCl \cdot 3H_2O$	2.130

*From Carmichael, 1984.

Intuitively, it is probable that the ease with which salt flows is also affected by the purity and thickness of the salt bed. Impurities such as sediments, anhydrite, and polyhalite included within the salt provide obstacles making flow more difficult. These impurities are concentrated close to and along the boundaries between salt crystals, and certainly adjacent to the sediments that encase the salt layer. Thus, given equal stress and temperature conditions, thick, pure salt layers should flow more easily than thin, dirty ones. In addition to simple flow considerations, a thick salt layer is also needed to supply the volume of salt required to grow a salt structure of any significant size.

Thick, pure salt needed to produce structures of the scale observed in the Gulf of Mexico basin is deposited in silled basins where a physical barrier (the sill) between the basin and the open ocean restricts water circulation, and arid climatic conditions create a high rate of surface evaporation (Dott and Rice, 1971). As is the case in the region of the Gulf of Mexico (see Chapters 8 and 14 of this volume), such basins are often a product of continental breakup. The best young example of this is the Red Sea region where, during the late stages of the rifting apart of Africa and Saudi Arabia, more than a thousand meters of salt was deposited, mostly during the middle Miocene (Lowell and Genik, 1972).

In the Gulf of Mexico basin, this process, which occurred in the late Triassic and Jurassic, resulted in salt deposition over a wide geographic area. In the north, there are two distinct sets of salt basins; the interior salt basins and the coastal and offshore salt basins (Martin, 1978, 1980). The interior basins include a small, isolated bit of salt in onshore south Texas, plus the extensive salt basins of northeast Texas, north Louisiana, southernmost Arkansas, Mississippi, Alabama, the Florida panhandle, and the northeast Gulf of Mexico (see Plate 2, this volume). The coastal and offshore basins lie to the south of the Early Cretaceous shelf margin and include the areas of onshore south Louisiana and

southeast Texas, and almost all of the Louisiana and Texas shelf and slope (Humphris, 1979; Martin, 1978, 1980). In the south, the salt basin extends from a position outboard of the Campeche Escarpment, through the Bay of Campeche, and then into onshore southern Mexico (see Plate, 2, and Chapters 3 and 8, this volume).

DRIVING FORCES

In the Gulf of Mexico basin, salt flow is driven primarily by gravitational forces related either to the slope of the surface, to differential sediment loading over the salt layer, or to buoyancy created by more dense sediments overlying the low-density salt (see Ramberg, 1968; Nettleton, 1934, 1943, 1955; Parker and McDowell, 1955). Kehle (1988) and Jackson and Galloway (1984) have discussed the driving forces for salt flow in some detail using a variety of models. For purposes of this chapter, based on this previous work, three simple and greatly idealized models have been chosen to illustrate the primary ways in which these driving forces are generated and the nature of the resulting salt flow.

Surface slope

The first of these models (Fig. 1a) depicts a driving force generated solely by gravity acting on a sloping upper sediment surface that is underlain by uniformly thick layers of salt and sediment. Because layer thicknesses are constant there is no pressure gradient induced within the salt by the overlying sediments. Thus, the only force acting on the salt is the component of gravity, which tends to drive the salt in the downslope direction. The magnitude of this driving force is proportional to the slope of the surface. Flow takes place when the magnitude of the driving force is great enough to overcome the total stresses resisting such flow.

Once flow is initiated, there are three types of motion that may occur in response to this force, and all tend to decrease the slope of the upper sediment surface. The three motion types are referred to here as pure salt flow, gravity gliding, and gravity spreading.

Pure salt flow occurs when the salt-sediment system is laterally constrained. In this circumstance, if the driving force is sufficient to overcome the resistance of the overlying sediments, salt flows out from under the upslope portion of the section, thinning the salt there, and into the downslope area, thickening the salt there (Fig. 1b). As salt thickness changes, the overlying sediments, in this simple case, are rotated and remain parallel to the top of the salt. The original salt volume is preserved within the area and is simply redistributed. Rotation of the overburden decreases the angle of slope of the surface and thus decreases the magnitude of the driving force acting on the salt. The system reaches equilibrium when the slope has flattened to the point where the driving force is too small to overcome the stresses resisting the flow.

Gravity gliding and gravity spreading occur when the system is not laterally constrained. Under these circumstances the salt and overlying sediments are both free to move in the downslope direction. As a result, the entire section is lengthened. The difference between gliding and spreading lies in how much lateral salt flow occurs. In gliding, little actual differential flow of salt is involved. The section above the glide surface simply slides downslope, and rotation of the section occurs primarily near the upslope end of the glide system (Fig. 2). When gravity spreading takes place, however, a significant volume of salt flows laterally over potentially long distances (Worral and Snelson, 1989). The surface slope is altered both by extensive rotation of the overlying sediments on normal faults that sole out in the salt, and by changes of thickness of the salt layer as salt is transferred in the downslope direction.

Since the driving force is primarily a function of surface slope, this type of salt flow can occur at a very early stage, under little or no sediment cover. Such early deformation is best preserved in areas where the salt layer is too thin to support later development of large piecement structures, which in the course of their growth, often deform the sediments to such a degree that evidence of early structuring is masked or obliterated.

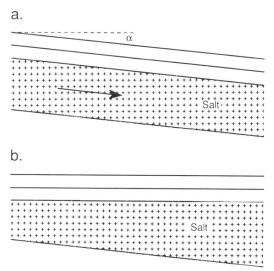

Figure 1. Model of salt flow driven by gravity acting on a sloping upper sediment surface. Initial conditions are shown in a. The driving force is proportional to sine α, and salt flows in the direction of the arrow. Equilibrium is illustrated in b.

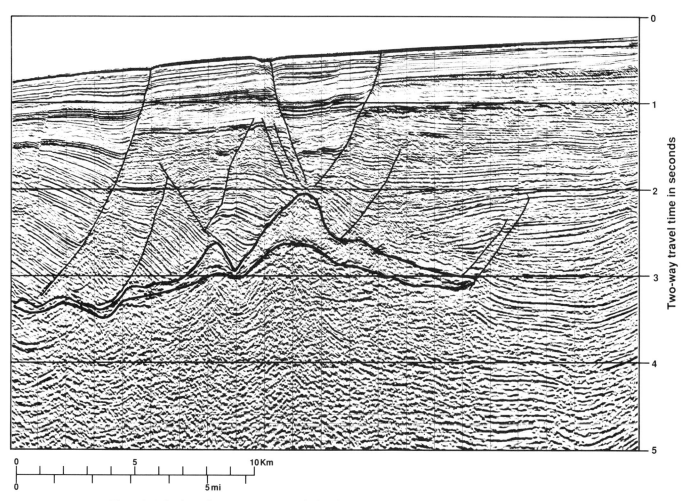

Figure 2. Seismic section across the updip limits of a thin salt sheet. The deformation is primarily due to gravity spreading of the salt accompanied by listric normal faulting of the overlying section.

a.

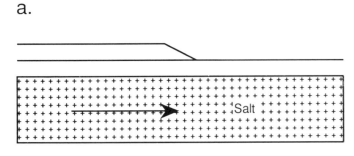

b.

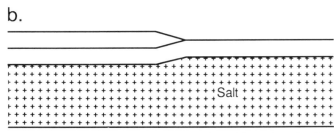

Figure 3. Model of salt flow driven by differential loading related to surface deposition. a, Initial conditions are shown with a wedge prograding over flat-lying layers of sediment and salt. Flow is in the direction of the arrow. b, As flow occurs the wedge subsides, and structure develops on the top of the salt.

Differential loading

Differential loading occurs whenever two adjacent vertical columns of sediment over a common level within a salt layer have unequal weight. Although this condition may be created by a variety of sediment densities and geometries, the driving mechanism is illustrated here by models of two simple end members.

The first of these models, illustrated in Figure 3, shows differential loading resulting from variations in surface deposition. Initially (Fig. 3a), a wedge of sediment overlies undeformed, horizontal layers of sediment and salt. Because of the added weight of the wedge, the pressure at any given level within the salt is greater beneath the wedge than it is beneath the area outside of the wedge. This pressure difference tends to drive salt out from under the wedge into the adjacent area of lower pressure. As a result, the salt layer beneath the wedge thins and the wedge subsides relative to the base of salt (Fig. 3b). If the section is laterally constrained, the salt layer beneath the nonwedge area thickens as salt is added, and the overlying sediments rise relative to the base of the salt. The net effect of this movement reduces the magnitude of the surface anomaly and hence the pressure difference within the salt. Equilibrium is reached when the pressure difference becomes too small to overcome the frictional stresses resisting salt flow.

The point at which equilibrium is reached, in terms of subsidence of the wedge, is dependent on the density of the sediments relative to the salt. If the sediments are less dense than the salt, the wedge, at equilibrium, will still have some positive topographic

or bathymetric expression. If sediment densities are equal to the salt, then, neglecting resisting stresses, the surface at equilibrium ideally will be flat, and no surface expression of the wedge, either positive or negative, will remain. If the sediments are more dense than the salt, wedge subsidence will form a surface depression.

In all of these cases, flow creates structure on the originally flat top of the salt. As sedimentation continues, further salt flow may be driven by differential loading that results solely from such structuring of the salt.

The model shown in Figure 4 illustrates the condition in which a simple structure is present on the top of the salt beneath a flat-lying surface. This geometry may produce differential loading of sufficient magnitude to drive salt flow. The direction of flow and the point at which equilibrium is reached are dependent on the density of the sediments relative to the salt.

Assuming that sediment densities in this model are horizontally stratified and are, therefore, strictly a function of depth, the direction of salt movement is determined by the density of the portion of the sedimentary section that is laterally equivalent to the local salt high (the interval labeled h in Fig. 4). There are three possible cases: the sediments in this interval are (1) less dense than the salt, (2) of the same denisty as the salt, or (3) more dense than the salt. Each case produces a different result.

When the sediments adjacent to the local salt high are less dense than the salt, the pressure at a common level within the salt is greater beneath the high than it is beneath the adjacent areas. This pressure difference tends to drive salt out from beneath the local high and into the surrounding salt bed. If such flow actually occurs, the high will decrease in amplitude and some form of collapse structure will form in the overlying sediments (Fig. 5). Inversion structures of this sort occur only when the salt layer is at very shallow burial depths and compaction has not progressed to the point where sediment densities are equal to or greater than the density of salt. It is most often observed above shallow salt sheets where highs on the tops of the sheets, created by the sheet emplacement process, cannot be supported in the low-density, near-surface environment once emplacement has ceased (Fairchild and Nelson, 1990).

Case 2, where salt and sediment densities are equal, produces no pressure difference within the salt. Thus, no salt flow occurs. For obvious reasons no separate figure is needed to illustrate this condition.

When the density of the sediments adjacent to the local salt high is greater than the density of the salt, the resulting pressure difference tends to drive salt out of the salt bed and into the local salt high. If such flow takes place, the top of the salt high rises relative to the base of salt (Fig. 6). It is this type of density-driven flow that produces the high-amplitude salt diapirs so prevalent in many areas of the Gulf of Mexico basin region.

The height to which a column of salt will rise is dictated by buoyant forces (Arrhenius, 1913, Nettleton, 1934) and is, therefore, also a function of the density of the sediments relative to the salt. Neglecting both internal and external resisting stresses, salt will rise to a level at which the weight of the salt column per unit

Surface

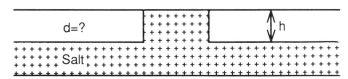

Figure 4. Model of differential loading related to structure on the top of the salt. Assuming sediment densities are horizontally stratified, flow direction of the salt is dependant on the density of the sediments in the interval labeled h.

Surface

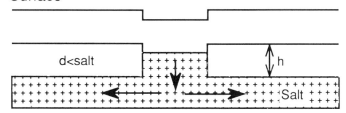

Figure 5. Model of damping of structural relief on top of the salt. This occurs when the sediments in interval h are less dense than the salt. Flow, shown by the arrows, is out of the salt high and into the source layer.

Surface

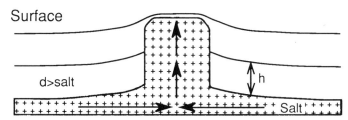

Figure 6. Model of diapirism. The sediments in interval h are more dense than the salt. This greater load drives salt out of the source layer into the initial salt high as indicated by the arrows.

area is equal to the per unit area weight of a column of laterally equivalent sediments. Since weight per unit area is a function of density, this point is reached when the average density of the sediments extending from the top of the source layer of salt upward to a depth equivalent to the top of the salt high is equal to the density of the salt.

Inherent in this model, in which density is the primary driving mechanism, is the implication that, given an upper sediment surface with little relief, continuous diapirism cannot occur until the sediments immediately above the salt are more dense than the

salt. The model also indicates that, once initiated, a diapir cannot reach the surface until the average density of the sediments between the surface and the source salt layer is equal to the density of the salt.

Exactly when these conditions are met varies widely from area to area. In parts of the interior salt basins of east Texas, north Louisiana, and Mississippi, Jurassic limestones overlie the salt. Such carbonate sediments often achieve densities equal to or greater than the salt at shallow burial depths. In such cases, conditions necessary for diapir initiation would have been established early (see Chapter 8 of this volume). The resulting piercements could, thus, have reached the surface early in their history and have grown through time by a process of passive downbuilding (Barton, 1933) in which the top of the diapiric salt remained always at or very near the surface while sediments were deposited around it. Such early growth of interior basin domes is well documented by seismic and well data (Labao and Pilger, 1985; Seni and Jackson, 1984). In addition, paleomagnetic data from the cap rock of the Winnfield dome in north Louisiana shows that this salt plug was near enough to the surface by late Jurassic time to have been subjected to solution (Gose and others, 1985, 1989).

In contrast, conditions in the areas of the Texas-Louisiana shelf and slope require significantly greater burial depths to achieve the necessary sediment densities. Here the section is composed of fine-grained clastic sediments deposited fairly rapidly in a deep-water environment. Well data from this area (Fig. 7) show that on average, sediments in this environment do not acquire a density greater than salt until they are buried to depths of about 1,500 m below the sea floor. In addition, integration of the density curve of Figure 7 indicates that an overburden thickness of about 3,600 m is required to support a diapir that rises from the source salt layer to the sea floor.

These relations between sediment densities, diapir initiation, and diapir height are strictly applicable only in an environment in which lateral forces of any significant magnitude are absent. The presence of large, coherent blocks of sediment which have moved basinward on decollement layers associated with major listric normal faults indicates that, in some of the Gulf of Mexico basin regions, significant lateral extensional forces have played a major role in the tectonics of the area. Space considerations also imply that some degree of shortening and lateral compression of the sedimentary section must also have occurred within and along the basinward ends of at least some of these blocks.

The effect of lateral extension when acting on salt layers is to spread the existing volume of salt and overlying sediment over a larger area, primarily in the downslope or basinward direction (Jackson and Cramez, 1989; Vendeville, 1989). This effect is in direct opposition to that associated with static loading, which tends to concentrate salt volume in vertical diapirs. The degree to which lateral extension affects a diapir depends on the rate of the extension relative to the rate of salt flow. Jackson and Cramez (1989) have suggested that when extension is rapid enough, existing diapirs caught up in the extension may become laterally un-

78 T. H. Nelson

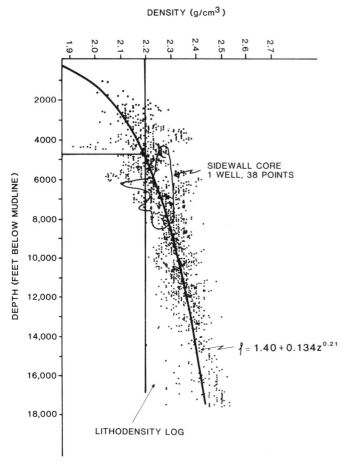

Figure 7. Plot of sediment densities versus burial depth. The data are from wells drilled on the Louisiana slope. The curve represents an average of the scatter of the log and sidewall core measurements.

Where lateral compression is small, however, its presence is difficult to detect. The only evidence may be cases in which diapirs appear to be initiated prior to establishment of a density inversion or where diapirs appear to reach the surface significantly in advance of burial depths required by a purely buoyant model.

I have seen little evidence of diapirs with such anomalous growth histories in the northern Gulf of Mexico. Beneath the outer shelf and slope of offshore Louisiana a significant number of diapirs have developed out of salt sheets that are now buried over a wide depth range. Examination of a number of these features shows that both the onset of vertical growth and the present height of the diapirs above their source sheets (Fig. 8) are in reasonable agreement with predictions based on salt flow driven solely by sediment densities.

DIAPIR EVOLUTION

Seismic sections from the Louisiana outer shelf, where salt sheets, emplaced primarily into sediments of Pliocene and early Pleistocene age, have been buried and remobilized, provide excellent examples of diapir evolution (Nelson, 1989). Although here the sediments consist of fine-grained, poorly consolidated sands and shales, the same basic growth sequence is probably applicable to diapirs in the interior salt basins where more competent section overlies the salt. In the interior salt basins lateral forces appear to be small or absent, and diapir growth appears to be driven primarily by differential loading and the contrast in density between the sediments and the salt.

The process of diapir evolution progresses through two distinctly different stages that are referred to here as active piercement and passive piercement.

Active piercement

An example of a diapir in the early active piercement stage of evolution is shown in Figure 9. The elements of this structure

constrained. As a result, a diapir may increase its diameter in the direction of extension to fill the gap between adjacent, divergent blocks. If the rate of lateral expansion of the diapir is greater than the rate of density-driven salt flow into the diapir from the source layer, then the amplitude of the diapir will decrease and the salt structure will deflate.

In contrast, the general effect of lateral compression acting on layered salt is to concentrate the initial salt volume and the overlying sediments into a smaller area. It does this by thickening the salt vertically and folding and sometimes thrusting it over the overlying sediments (e.g., Davis and Engelder, 1985). Thus lateral compression acts on layered salt in the same sense as does static loading. Where compression is large relative to the effects of differential loading the compressional overprint is evident in the form of the resulting structure (e.g., box folds, thrusting, etc.). In the basins of the northern part of the Gulf of Mexico region, such structuring is observed on a regional scale only in the Perdido and Mississippi Fan fold belts, which lie along the base of the continental slope off of Texas and Louisiana (see Plate 2, this volume; Blickwede and Queffelec, 1988; Weimer and Buffler, 1989).

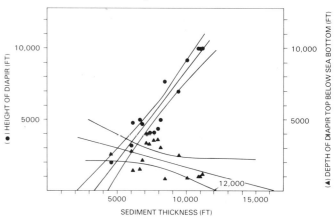

Figure 8. Relation between the thickness of sediment deposits above a source salt layer and the height of diapirs above that layer and the depth of diapir tops below the sea floor. Dots indicate diapir height, triangles indicate diapir depth below the sea floor. Straight lines are fit by linear regression, curved lines enclose the area of 95 percent confidence.

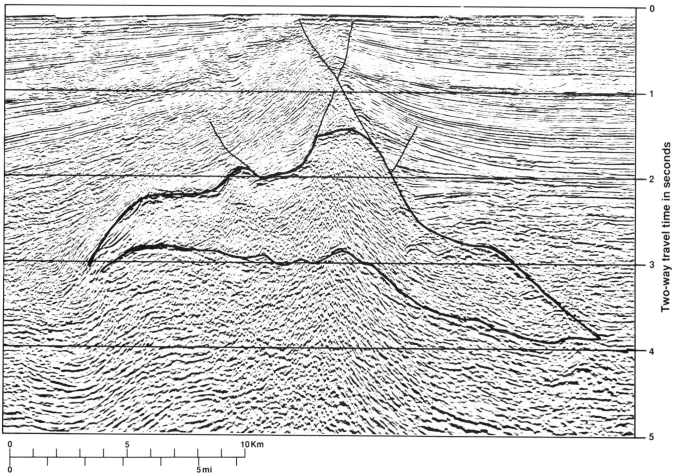

Figure 9. Seismic section across a salt structure in the early stage of active piercement. This structure is developing out of a salt sheet encased in Tertiary clastics.

characteristic of the early stage of diapirism are: (1) a relatively thick section of prediapiric sediments overlying the salt, (2) a major normal fault bounding one side of the rising salt body, and (3) the asymmetric nature of the overall structure.

The prediapiric section is most evident in the area to the left of the main fault that bounds the salt uplift. It extends from the top of the salt upward for an interval of slightly more than one second in time. Although the interval is complexly faulted in places, it appears to be rather evenly bedded. Some local thinning and thickening is present, indicating that early salt flow may have occurred. However, these thickness changes do not seem to be related to the growth of the present anticline. The overall thickness of the section, which was deposited prior to the initiation of this structure, reflects the amount of burial needed, in this depositional environment, to create a density inversion above the salt.

The major fault that bounds the salt structure on the right appears to have played a significant role in lowering the strength of the overburden and creating an early pressure gradient within the salt, thus controlling the initial location and early configuration of the feature. From the geometry of the beds on the downthrown side of this fault, it is evident that the fault predated

initiation of the diapir. These beds, which are the same age as the prediapiric section in the upthrown block, bend down and thicken into the fault in a pattern characteristic of an active listric normal fault. This fault may have either formed along or caused a structural depression on the top of the salt sheet. This resulted in differential loading of the salt, which once a density inversion was establisdhed, created a pressure gradient that drove salt out from under the low side of the fault and into the area immediately beneath the high side of the fault.

This pattern of salt flow has built a structure that is very asymmetric. The sediments overlying the salt to the left of the fault, where salt is being added, are uplifted and rotated back in the form of a flap. In contrast, sediments to the right of the fault move downward as salt is withdrawn to feed the growing structure. In addition, the relations at the salt/sediment interface are different for the two flanks. On the "flap" (upthrown) flank, the salt has uplifted but has not pierced the overlying section. On the "withdrawal" (downthrown) flank, however, the salt, with the help of the bounding fault, has risen stratigraphically through a portion of the section and, thus, has a piercement relation with that section. It is by this mechanism of moving up the high side of

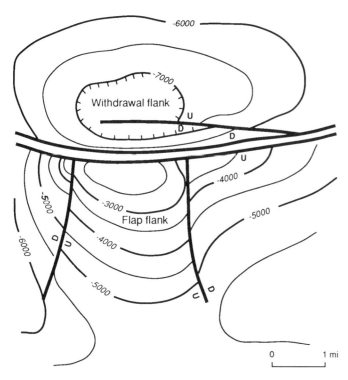

Figure 10. Schematic structure map of a sedimentary horizon over a salt feature in the early stage of active piercement.

an active normal fault that salt actively penetrates the overburden and gradually works its way to the surface.

The asymmetry, so well shown in cross section, is usually also expressed in map view (Fig. 10). The structure develops as a somewhat elongate anticline along the high side of the main bounding fault. The flap flank lies on the upthrown side of the fault, and the withdrawal flank lies along the downthrown side of the fault. Uplift of the flap flank is generally not of the same magnitude all along the lateral extent of the structure but instead is influenced by the presence of a secondary set of faults subperpendicular to the main bounding fault. These early structures, therefore, often develop local highs and lows along the main anticlinal crest. As the diapir becomes more mature, some of these secondary faults develop into radial faults so characteristic of classical shallow piercement plugs seen in the onshore and near-shore environment (see Fault systems around salt domes, this chapter). On the withdrawal flank the pattern of secondary faults is quite different. Here the faults are subparallel to the main bounding fault, and are either synthetic or antithetic to that fault.

Although most early-stage diapirs in this part of the outer continental shelf appear to have developed along preexisting normal faults, in some cases it is apparent that the bounding fault is actually the result of the initial upward movement of the salt. In these cases, initial uplift of the overburden as salt begins to rise places the section under tension, and normal faults begin to develop. Because of the influence of such factors as regional dip and variations in surface sedimentation, withdrawal of salt to feed the

growing structure generally comes from only one flank. This one-sided withdrawal in turn creates an asymmetry in the deformation of the sediments overlying the salt. As a result, faults that are down toward the withdrawal area usually develop more rapidly and become dominant. Once formed, these faults act in the same manner as does the main bounding fault described above, weakening the overburden and controlling the further development of the diapir. The difference between the two situations may simply be reflected in the rate of earliest growth of the salt structure. When no preexisting fault is present, there is nothing to lessen the strength of the overlying section; as a result, the stresses resisting salt flow are greater. This means that early development of the structure will occur at a slow rate. Once faults form and weaken the overburden, however, the growth rate accelerates and the structure grows rapidly.

Given an adequate supply of salt and continued progressive burial of the source-salt layer, the diapir will continue to actively pierce the overlying section and will eventually work its way to the surface. Figure 11 shows a seismic section across a diapir that is in the late stages of active piercement. The section deposited before this structure began to develop is clearly shown, although it is somewhat thicker than in the previous example. Uplift of the flap flank is of rather dramatic proportions, and as a result, the main bounding fault has a very large throw. Because of the deformation of the deep beds on the downthrown side of the fault, it is not clear whether the fault predated diapir initiation or formed as the result of initial upward movement of the salt.

The degree of uplift of the salt in this later stage of active piercement has also resulted in the creation of a second major fault, this one bounding the left side of the diapir. With the creation of this second fault, the diapir now has a piercement relation to the encasing sediments on both the withdrawal flank and the flap flank. This fault lowers the strength of the overlying section even further and facilitates the rapid, nearly vertical rise of the salt.

In plan view this diapir is a salt wall with its long axis parallel to the main bounding fault. The orientation of the salt wall may have been biased by the configuration of the salt sheet out of which it has grown. Since salt sheets appear to be driven by gravity acting on the sloping sea floor, they tend to flow down the local slope, forming elongate lobes perpendicular to the strike of that slope. In most cases this slope is to the south, toward the Sigsbee Abyssal Plain of the Gulf of Mexico.

It would also appear that this diapir is close to being in equilibrium with its surroundings, at least for the moment. In an earlier part of this chapter the point was made that a diapir will be in equilibrium when it has risen to the point where the average density of the sediments between the top of the source layer of salt and a depth equivalent to the top of the diapir is equal to the density of the salt. From the seismic data, the burial depth of the source salt layer to the west of the structure of Figure 11 is about 3,300 m. Using the depth-density relation of Figure 7 as representative of the density of the sediments overlying the salt in this area, and integrating the curve upward from the depth of the

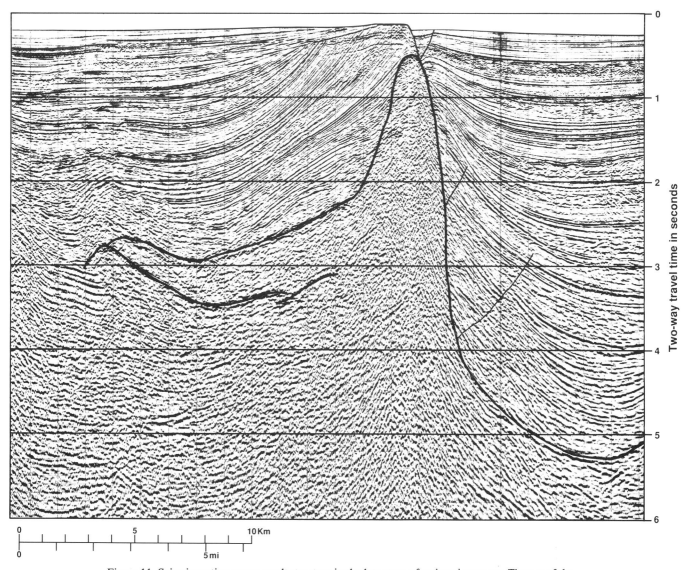

Figure 11. Seismic section across a salt structure in the late stage of active piercement. The top of the diapir, which is growing out of a salt sheet, has nearly reached the sea floor.

top of the source salt, gives an equilibrium diapir height of about 3,100 m. This would put the top of the diapiric salt only about 200 m beneath the sea floor, a number that is in very good agreement with the data on this structure.

In the foregoing examples of active stage piercements, I do not wish to imply that no salt flow occurred prior to the establishment of a density inversion. In fact, salt flow, driven either by gravity acting on the sloping sea floor or by differential surface loading, could well have taken place. Such flow would, however, have been predominantly lateral and not vertical. Once a density inversion was established, vertical flow of salt could be sustained, and the entire pattern of deformation changed dramatically.

Passive piercement

Once the salt reaches the surface of deposition, the active piercement process is complete, and the diapir enters what is

referred to here as the passive stage of piercement. During this stage of evolution, given an adequate supply of salt in the source layer, the top of the salt will remain at or above the level of the surrounding sea floor. Thus, as sediments continue to be brought into the area, they are deposited around the top of the diapir. Continued addition of sediments to the area around the diapir increases the pressure gradient within the source salt layer. This increased pressure gradient pushes the diapir even higher and allows it to keep up with sedimentation and stay at the surface. Only when either sedimentation rates are very high or when the supply of salt in the source layer begins to run out will the diapir become buried.

Because diapirs in this stage of evolution do not actively have to pierce any significant amounts of sediments, deformation of the beds adjacent to the salt is primarily the result of withdrawal of salt from the source layer and some drag due to

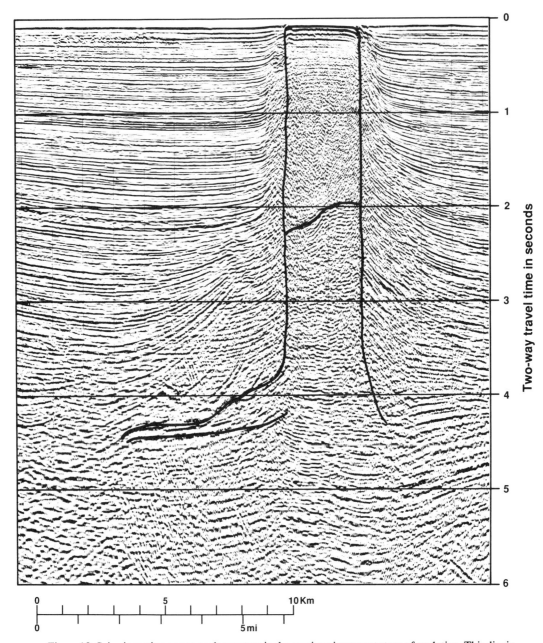

Figure 12. Seismic section across a salt structure in the passive piercement stage of evolution. This diapir has developed at the edge of a salt sheet. The steeply dipping beds to the left of the salt plug below 2.5 seconds represent deformation during the active piercement stage of evolution of this structure.

compaction of the sediments around the salt core. Thus, the dip of the beds deposited around a passive piercement diapir tend to be much less than the dips created during active piercement. The difference is well expressed on the diapir shown in Figure 12. The steeply dipping deep beds on the left flank of this structure reflect the active piercement stage and the more gently dipping shallow beds, which appear to abut the nearly vertical salt face with only a small area of upturn reflect the passive piercement stage.

With continued upward growth, a diapir tends to evolve from an elongate salt wall to a more circular salt plug. This change in form results from the fact that a salt wall often does not

grow at the same rate along its entire length. This creates secondary highs and lows along the crest of the wall. While the highs keep up with deposition the lows become progressively more buried. Eventually, one of these highs becomes dominant and continues to rise, forming a more cylindrical diapir seated on top of a deeper salt ridge.

Fault systems around salt domes

Faults around passive piercement salt domes generally extend outward from the salt/sediment interface at a high angle, forming a radial pattern (Fig. 13). These faults tend to be largest near the salt plug and die out down the flanks of the structure. They form in response to relative uplift of the sediments flanking the diapir. This relative uplift is due to a combination of salt withdrawal from the source layer, continued rise of the salt plug relative to the source layer, and compaction and subsidence of the sediments surrounding the salt plug. Around a more or less circular salt plug the extensional stress produced by this relative uplift tends to be oriented parallel to the salt/sediment interface. Thus, the faults that form in response to this stress tend to be perpendicular to that interface.

Since active salt withdrawal is usually more pronounced on one side of the dome, the sediments flanking the plug are subjected to different degrees of relative uplift. The radial faults reflect these differences. They are usually best developed on the nonwithdrawal flanks of the structure and, ideally, tend to be down toward the withdrawal flank.

In addition to the radial faults, a system of "keystone" grabens is often developed over the crest of passive piercements that have failed to keep pace with deposition and have become buried. These grabens form in response to bending of the sediments over the dome. This bending may be caused either by continued rise of the plug relative to its source layer or by compaction of the sediments flanking the salt stock.

In the interior salt basins of east Texas, north Louisiana, Mississippi, and Alabama, similar flexural grabens have developed over nonpiercement salt pillows (Seni and Jackson, 1984). Here the extensional stress that has produced the faults is due solely to the relative uplift produced by growth of the salt structure. Because of the limited volume of salt in these areas the rate of relative rise of the salt was slow, and piercement did not occur.

Cap rock

Some passive piercement salt plugs are mantled by layers of anhydrite, gypsum, and calcite referred to as "cap rock" (Halbouty, 1979). Cap rock forms when the top of a salt plug is subjected to solution primarily by meteoric waters (Murray, 1966; Posey and Kyle, 1988). Under these circumstances, the more soluble halite is taken into solution and removed while the less soluble impurities, mainly anhydrite, are left behind. As solution continues, progressively more anhydrite accumulates along the salt/cap rock interface in a process called underplating (Kyle and others, 1987). Where a supply of hydrocarbons is present, bacteria-associated chemical reactions convert the anhydrite to biogenic calcite and hydrogen sulfide (Feely and Kulp, 1957). Occasionally the hydrogen sulfide is converted to elemental sulfur, which may be deposited and preserved in the cap rock in commercial quantities (Posey and Kyle, 1988; Feely and Kulp, 1957; Halbouty, 1979; Chapter 16 of this volume).

Overhangs

Domes in the more mature stages of passive piercement sometimes develop significant overhangs. An overhang occurs

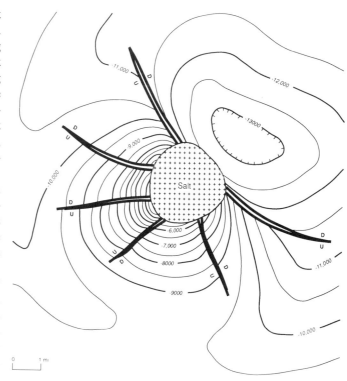

Figure 13. Schematic structure map of a sedimentary horizon around a typical passive piercement salt dome.

when the edge of the salt plug at some shallower level extends beyond the limits of the plug at some deeper level. Overhangs vary in size from minor protrusions of salt of limited thickness and extent to massive salt bulges more than a thousand meters thick and extending outward a mile or more from the main stock (Halbouty, 1979).

Large overhangs often occur on only one side of a salt plug. These one-sided overhangs have two possible origins. The first is that the plug as a whole is leaning in the direction of the overhang. Talbot (1977) showed, in model studies, that diapirs tend to grow with an orientation perpendicular to their source layer rather than rising vertically. If this is indeed the case in nature, then the axis of a salt plug growing out of a nonhorizontal salt layer would lean in the direction of dip of that source layer. If the angle of lean is significant, growth of the diapir would produce a pronounced overhang on the downdip side and a normally dipping flank on the updip side.

Bulging outward of the shallow portion of an otherwise vertical salt plug will also produce an overhang that, because of local or regional biases, may also be one sided. In the coastal and offshore areas of Louisiana and Texas the upper portions of shallow piercement salt plugs are encased in sediments that are less dense than the salt (Nelson, 1989). This density difference sets up a stress at the salt/sediment interface that tends to push the heavier salt outward into the lighter sediments. If this stress is large

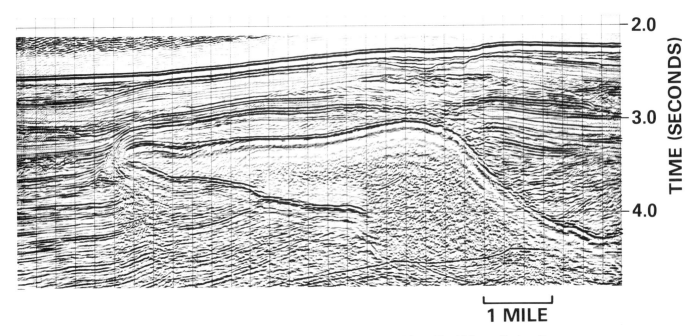

Figure 14. Seismic section across a salt sheet in offshore Louisiana. The high amplitude reflectors at approximately 3.2 seconds and 3.6 to 4.0 seconds record the top and base of the salt, respectively.

compared to the strength of the encasing sediments, the diameter of the shallow plug will increase and form a bulge. If no other forces are acting on the salt, this bulge should, ideally, form a symmetrical overhang completely around the dome. If there is any anisotropy in the system, however, the overhang may be one sided. For example, if the upper sediment surface is sloping, gravity, acting on that surface, will cause the bulge to develop preferentially on the downslope side of the salt plug.

Finally, a more symmetrical overhang may develop when a salt plug pinches off from its source layer. This phenomenon occurs only in the most mature stages of diapir evolution when the supply of salt in the source layer becomes depleted. Even with no deep salt source, buoyant forces may cause salt in the plug to continue to rise (Nettleton, 1934). Since the rising plug salt is no longer replenished from the original source layer, the diameter of the plug at depth decreases and an overhang is produced.

Salt sheets

Under some circumstances it appears that instead of continuing to move upward relative to its source layer, salt spreads laterally and forms salt sheets such as those from which the diapirs in the foregoing examples have grown. Although reflection seismic data often show both the top and base of the salt in these sheets, it rarely captures a good image of the geometry of the underlying sediments. Because of this lack of good subsalt seismic data and the paucity of wells that have drilled through such sheets, the origin of these features is still speculative. Centrifuge models (Talbot and Jackson, 1987; Jackson and Cornelius, 1987) using silicone putty of various densities and viscosities have

produced structural forms intriguingly similar to those observed beneath the slope of the northern Gulf of Mexico. The models are initially composed of layers of different density to simulate conditions found in nature. The deepest layer simulates the salt and has a moderate density. Immediately above this is a relatively high-density layer representing well-compacted sediments. The uppermost layer represents shallow, uncompacted sediments and has a low density. Under the strong gravitational force imposed on the model by the centrifuge, the "salt," driven by the load of the more dense overburden, rises diapirically toward the surface. Because the materials are viscous, that portion of the diapir which penetrates the shallow, low-density layer is laterally unconstrained. As a result, "salt" in this part of the diapir flows outward, intruding the encasing section. In some of the models the materials are Newtonian and have no strength; intrusion takes place where the outward stress acting on the "salt" is the greatest. This point occurs at the boundary between the shallow "uncompacted" layer and the deeper "well-compacted" layer. In other models using non-Newtonian materials, a stiff shallow layer produces similar outward flow of the "salt."

Data from the eastern part of the Louisiana slope (Fairchild and Nelson, 1989) suggest that here, sheets may well have formed in the manner indicated by the models of Jackson and Talbot. In this area the sheets, which Fairchild and Nelson refer to as sills, occur as relatively isolated bodies at rather shallow burial depths (Fig. 14). They have not been remobilized, and relations between the salt and the sediments both above and at the edges of the sills are clearly expressed on seismic data. From examination of these relations, Fairchild and Nelson have concluded that the sheets intruded the encasing slope sediments at depths of less than

300 m beneath the sea floor. This estimate of emplacement depth is based on seismic data that show onlap onto the bathymetric high created by intrusion of the sheet.

This intrusion depth is well above the level at which sediment densities equal the density of salt and indicates that below very shallow levels the sediments must have considerable strength. The magnitude of this strength is sufficient to prohibit salt intrusion at the density-equilibrium level where outward stress from the diapir toward the encasing sediments reaches a maximum value. The reason intrusion occurs at shallower levels is not, at present, clear. Limited well data (Fairchild and Nelson, 1989) indicate that sheets preferentially intrude a section that is nearly pure clay deposited in areas away from the location of an active depocenter. This suggests that material properties of the near-surface sediments encasing the top of the diapir may control sheet initiation.

Once initiated, the distance that the sheet extends outward from its feeding diapir is controlled by the volume of salt available and the presence or absence of impediments in its path. In the eastern portion of the northern Gulf of Mexico slope, rapid subsidence of the base of the sheets due to deeper salt withdrawal has limited the lateral extent of these bodies to an average of about 16 km. To the west, however, Worrall and Snelson (1989) show an example of a massive sheet that appears to extend continuously for a distance of about 160 km in the dip direction. These more westerly sheets described by Worrall and Snelson may have a different origin than those in the east. Gravity spreading in response to updip loading may have pushed the salt in these sheets into what is referred to as the Sygsbee nappe (Worrall and Snelson, 1989).

West (1989) has also proposed a model for the continued propagation of salt sheets down the slope. This model postulates that once a major sheet has initially formed, salt will, by progradational loading, be pushed out of the updip end of the sheet and, because of gravity acting on the sloping sea floor, will flow toward the downdip toe of the sheet. In this manner, the sheet will continue to extend downdip, gradually cutting up-section as it does so.

Gravity data and some seismic data indicate that, in many cases, the salt sheets no longer are connected to depth. In some instances it appears that the supply of deep salt has become depleted and the feeder stock, which spawned the sheet, has simply pinched off (Fairchild and Nelson, 1989). In other cases, the sheets have been separated from their feeders by mass movement of the section containing the sheet on listric normal faults (Wu and others, 1989).

The variety of models and ideas regarding the origin and growth of salt sheets are by no means mutually exclusive, nor does any one of them probably contain the whole truth about how such features evolve. There is some hope for the future, however, for these laterally extensive salt bodies hide thick sequences of sediments that may contain significant accumulations of oil and gas. Thus much effort will continue to be focused on salt sheets, including the drilling of some critical wells.

LISTRIC-NORMAL FAULTS

One of the major characteristics of the coastal and offshore basins of Louisiana and Texas is the presence of major growth faults (Shelton, 1984). Sediments deposited during the time these faults were most active increase several fold in thickness across the faults (Busch, 1975; Ocamb, 1961; Winker and Edwards, 1983; Galloway, 1986). The majority of these faults flatten with depth and are listric-normal faults that sole out within the sedimentary section (Bally and others, 1981). In onshore south Texas and the adjacent offshore basin, where little or no salt is present, these faults extend regionally along the strike of the basin and are subparallel to the present coastline. To the east, within the main salt basins, these listric growth faults have a different, more complex pattern. The extent of individual faults is more local, and their orientations often depart significantly from regional strike. The difference reflects the structural influence of salt diapirism in the area.

In their simplest form, listric-normal faults have the basic geometry shown in Figure 15. The dip of the fault decreases with depth and finally becomes a bedding-parallel decollement surface (Xiao and Suppe, 1989). Beds in the hanging wall of the fault typically bend down, or roll over, into the fault. Section deposited contemporaneous with fault movement thickens into the area of the downbend.

The downbend or roll of beds into the fault is the result of the gradual, vertical collapse of these beds to fill the gap created along the shallower, more steeply dipping part of the fault as the downthrown block moves away from the upthrown block along the deeper decollement surface (Cloos, 1968; Hamblin, 1965). The amount of downbend of a bed initially contiguous across the fault increases as the lateral displacement along the decollement surface increases. The downbend of the bed reaches a maximum when the bed in question impinges on the decollement surface. Thickening of beds into the downthrown side of the fault provides an indication of the timing of fault movement. The magnitude of relative lateral translation of the downthrown block is progressive through the life of the fault. Measurement of this translation requires careful reconstruction of each bed to its original position at the time of deposition.

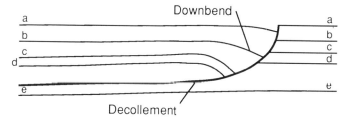

Figure 15. Diagrammatic cross section showing the basic characteristics of a simple listric-normal fault. The fault flattens with depth into a bedding parallel décollement surface (either salt or shale). Beds in the hanging wall downbend into the fault. Thickening the strata into the fault represents fault movement contemporaneous with deposition.

Initiation

Although often associated with salt basins, listric-normal faults are by no means restricted to areas underlain by salt. They are the result of differential basinward creep of the sediments above some decollement layer (Crans and others, 1980). They occur wherever a potential detachment layer is present within the section, and the section above that layer is laterally unconstrained, and there is a driving force of sufficient magnitude to overcome the frictional stresses within the detachment layer.

In the northern Gulf of Mexico basin, the decollement layer for a listric-normal fault may be either salt or a deep-water shale section (Bishop, 1973; Quarles, 1953; Bruce, 1973). The salt layer may be at the depositional level of the Jurassic Louann, or may be a salt sheet such as was described earlier in this chapter. The shale decollement may be related to instabilities created by abnormal pressure in general or by clay-mineral transitions, which release excess water into the system (Bruce, 1973).

Lateral movement of the section over the decollement layer is driven primarily by gravitational forces acting on the surface of depositon (or erosion). This driving force is the same as that described earlier in this chapter under salt flow, and is proportional to the sine of the slope of the surface (see Fig. 1).

Because listric-normal faults represent the boundary between blocks of sediment that are moving basinward at different rates, they occur either where the surface slope increases or resistance on a decollement layer decreases. Increases in surface slope may occur as a result of either processes of deposition or erosion, or underlying local structural movements.

Surface slope increases

The most obvious example of a change in surface slope created by deposition and related basin subsidence is the break between the continental shelf and slope offshore Texas and Louisiana. This feature, as it has migrated southward with the progradational infill of this margin of the Gulf, has been a persistent location for major listric-normal faults. These shelf-margin listric-normal faults are syndepositional, and the thickness of outer shelf and upper slope sediments increases significantly across them (Lopez, 1989). Because the shelf-slope break is a regional feature, such faults tend to extend for long distances along strike.

The best expressed and most widely known examples of listric-normal faults come from the set of faults created by the process of sediment fill and basin subsidence (Shelton, 1968, 1984). These are the major "flexure" systems of onshore south Texas and the adjacent offshore basin (Hardin and Hardin, 1961). In this area, differential movement of the section takes place either on a substrate of deep-water, abnormally pressured shale or, as suggested by Worrall and Snelson (1989), on the deep Louann Salt. The fault trends are believed to form initially at the shelf/slope break, and therefore, they parallel the strike of the basin. As sediments prograde over the area the locus of faulting shifts basinward, and a new, younger fault system develops.

Figure 16 shows a seismic section across one of these faults on the Texas shelf (Vogler and Robison, 1987). The listric nature of the fault and the characteristic downbending and thickening of the hanging-wall sediments are clearly evident on this section. The ultimate decollement layer lies well below the base of the seismic data, and the fault does not reach this layer (possibly Louann Salt) for some distance downdip from the end of this seismic section. The updip portion of the fault, however, flattens and, for some distance, becomes parallel to the underlying beds, forming a local decollement. Downdip, the fault again steepens and cuts across bedding. This latter, cross-cutting relation represents a fault ramp that effectively connects the shallow, local decollement with the deep, primary decollement.

Surface slope increases produced by underlying local structural movements are often created by withdrawal of thick salt. Described from many areas in the northern Gulf of Mexico basin (Quarles, 1953; Lopez, 1989), these faults tend to be more limited in their lateral extent. Although the faults generally sole out on salt and have a listric form, true lateral translation of section along the decollement surface is often less than that associated with faults created primarily by deposition and basin subsidence as discussed above. Most of the lateral translation occurs during the period when withdrawal of salt coincides with deposition of slope sediments. Two factors play a role in this relation. The first is that salt movement generally has a greater effect on bottom topography n a deep-water environment than it does in shallow water. Thus the magnitude of slope increase, and therefore driving force, coming into a withdrawal syncline tends to be greater. The second factor is that in the early stages of development of the withdrawal syncline, the salt, and therefore the decollement layer, is closer to the sea floor, and the sediments above the salt are laterally less well constrained. As the area is prograded and changes from deep-water slope to shallow-water shelf, the faults continue to be active but in response to flexural stresses related to vertical subsidence over the withdrawal area rather than to any significant lateral movement of the sediments.

Frictional resistance

Decreases in frictional resistance to lateral movement of sediments are caused by changes in the properties of the material at the level of detachment. One of the most common sites for listric-normal faults related to such changes is along the updip margins of subhorizontal salt bodies. In the interior salt basins of east Texas, north Louisiana, Mississippi, and Alabama, for example, a series of grabens composed of down-to-the-basin listric normal faults that sole out in the salt and up-to-the-basin antithetic faults have formed along the updip margins of the Louann salt (e.g., Mexia-Talco fault system, Arkansas Graben system, Pickens-Gilbertown-Polland fault system; Bishop, 1973). In this case, the change at the décollement level is obvious and is due to the presence of a ductile, low-friction salt layer downdip and the absence of such a layer updip. In the offshore, similar conditions exist along the updip margins of salt sheets (see Fig. 2).

The causes for decreases of frictional stresses along décollements that do not involve salt are difficult to document and are therefore more speculative. One of the most probable causes for such changes, however, is an increase in the amount of water present at the level of detachment (Bruce, 1973). Such décollements, whether they are primary or local, usually form in shale-prone, abnormally pressured slope sediments. At some point in their burial, water trapped in these shales bears enough of the overburden load to decrease frictional stresses within the shale layer and to permit detachment. This level may be associated with clay-mineral transitions that occur with burial and release significant quantities of free water into the system in a short time period (Bruce, 1973). Once detachment is initiated, the resulting listric normal fault probably acts as a conduit for at least some of this water to gradually escape from the system. In the updip portions of the fault, downbending brings progressively younger section into contact with the décollement layer (see Fig. 16). If this younger section is more sand prone, which is often the case, then the updip portion of the system will dewater more rapidly than will the downdip part of the décollement layer. With dewatering, frictional stresses along the updip portion of the system tend to increase, lateral motion of that portion of the hanging-wall block decreases, and a new fault forms downdip. This new fault may, if conditions are appropriate, sole out along the same décollement layer, and the process of downbending and dewater-ing begins at the new location as activity on the older fault gradually decreases.

Compensation

From a mechanical standpoint, the most perplexing aspect of listric-normal faulting involves the downdip compensation for all this extension of section. Clear evidence for large-scale compressional folding and thrusting has been observed only in the Perdido fold belt and the Mississippi Fan fold belt (Blickwede and Queffelec, 1988; Weimer and Buffler, 1989). These fold belts have been interpreted to be located over the downdip depositional limts of Jurassic salt and represent the compressional toes of décollement surfaces, which tie into some of the deeper rooted listric-normal faults updip. These fold belts, however, appear to account for only a part of the overall updip extension. Some compensating shortening of section may also be represented in shale uplifts located along the downdip ends of some faults. The internal structure of these possible toe structures is rarely visible on reflection seismic data. Thus, some degree of unrecognized folding and possibly thrusting could be present there (Mandl and Crans, 1981; Dailly, 1976).

Lateral compaction of sediments in the hanging-wall block may also compensate for some of the extension without producing recognizable compressional structures and certainly, as dis-

NORTHWEST   **SOUTHEAST**

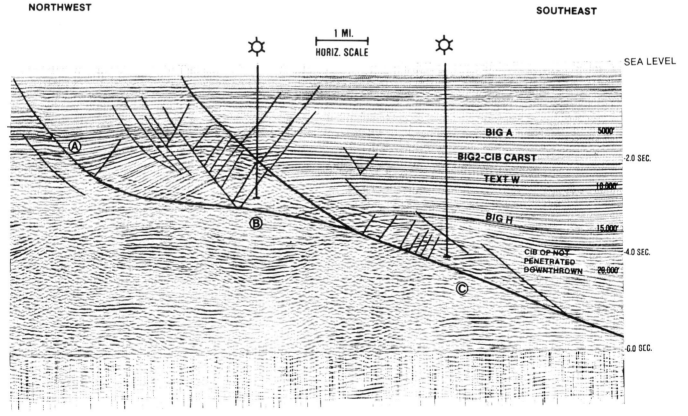

Figure 16. Seismic section across the Corsair fault trend, offshore Texas. The section illustrates the shallow, listric segment of the fault (A), a bedding parallel décollement surface on abnormally pressured shale (B), and a deep ramp as the décollement surface changes levels (C) (Vogler and Robison, 1987).

cussed earlier in this chapter, some small degree of lateral compressional deformation can be hidden within structurally complex salt and shale diapirs. Worrall and Snelson (1989) have also suggested that much of the updip extension along listric normal faults has been taken up by downdip emplacement of the thick, tabular salt sheets that underlie the Sygsbee Escarpment. In their model these salt sheets represent nappes of salt that have been moved up-section and downdip in response to updip sediment loading. Such movement of these large salt masses easily compensates for the lateral extension indicated by the existing listric normal fault systems present beneath the shelf and slope of the northern Gulf of Mexico basin.

CONCLUSION

The Gulf of Mexico basin region contains structural features that, for the most part, have been created by gravity acting on an unstable substrate in a nonorogenic environment. The basic principles for the formation of the two principal types of structures present, salt domes and listric normal faults, are relatively straightforward. Their interaction with each other and with depositional processes, however, creates an almost endless variety of features, for which this chapter provides only a taste.

This variety is reflected in the wide range of geologic conditions encountered in the large number of producing oil and gas fields within this region. In addition, the recent recognition of the presence of massive salt sheets encased in young Tertiary sediments beneath the outer continental shelf and slope of the northern Gulf of Mexico has provided a major new incentive to reexamine salt tectonics. These sheets not only provide a new and exciting field for geological investigation, but also represent a new frontier for petroleum exploration, a frontier that, conceptually at least, has significant economic potential.

REFERENCES CITED

Arrhenius, S., 1913, Zur Physik der Salzlagerstätten (On the physics of salt deposits): Meddelanden K. Vetenskapsak Ademius Nobelinstitut, v. 2, no. 20, p. 1–25.

Bally, A. W., Bernoulli, D., Davis, G. A., and Montadert, L., 1981, Listric normal faults, 26th International Geologic Congress (Paris), Colloque C3-Geology of Continental Margins: Oceanologica Acta, p. 87–101.

Barton, D. C., 1933, Mechanics of formation of salt domes with special reference to Gulf Coast salt domes of Texas and Louisiana: American Association of Petroleum Geologists Bulletin, v. 17, p. 1025–1083.

Bishop, W. F., 1973, Late Jurassic contemporaneous faults in north Louisiana and south Arkansas: American Association of Petroleum Geologists Bulletin, v. 57, p. 858–877.

Blickwede, J. F., and Queffelec, T. A., 1988, Perdido foldbelt; A new deep-water frontier in western Gulf of Mexico [abs.]: American Association of Petroleum Geologists Bulletin, v. 72, p. 163.

Bruce, C. H., 1973, Pressured shale and related sediment deformation; Mechanism for development of regional contemporaneous faults: American Association of Petroleum Geologists Bulletin, v. 57, p. 878–886 (also in GCAGS Transactions 1972, v. 22, p. 23–31.

Busch, D. A., 1975, Influence of growth faulting on sedimentation and prospect evaluation: American Association of Petroleum Geologists Bulletin, v. 59, p. 217–230.

Carmichael, R. S., ed., 1984, CRC handbook of physical properties of rocks, v. 3: Boca Raton, Florida, CRC Press, 340 p.

Carter, N. L., and Hansen, F. D., 1983, Creep of rocksalt; Tectonophysics, v. 92, p. 275–333.

Carter, N. L., and Heard, H. C., 1970, Temperature and rate dependent deformation of halite: American Journal of Science, v. 269, p. 193–249.

Cloos, E., 1968, Experimental analysis of Gulf Coast fracture patterns: American Association of Petroleum Geologists Bulletin, v. 52, p. 420–444.

Crans, W., Mandl, G., and Haremboure, J., 1980, On the theory of growth faulting; A geomechanical delta model based on gravity sliding: Journal of Petroleum Geology, v. 2, p. 265–307.

Dailly, G. C., 1976, A possible mechanism relating progradation, growth faulting, clay diapirism, and overthrusting in a regressive sequence of sediments: Canadian Petroleum Geology Bulletin, v. 24, p. 92–116.

Davis, D. M., and Engelder, T., 1985, The Role of salt in fold and thrust belts: Tectonophysics, v. 119, no. 4, p. 67–88.

Dott, R. H., and Rice, P., eds., 1971, Origin of evaporites: American Association of Petroleum Geologists Reprint Series 2, 208 p.

Fairchild, L. H., and Nelson, T. H., 1989, Emplacement and evolution of salt sills: Houston Geological Society Bulletin, v. 32, no. 1, p. 6–7 (extended abstract).

Feely, H. W., and Kulp, J. L., 1957, Origin of Gulf Coast salt dome sulfur deposits: American Association of Petroleum Geologists Bulletin, v. 41, p. 1802–1853.

Galloway, W. E., 1986, Growth faults and fault-related structures of prograding terrigenous clastic continental margins: Gulf Coast Association of Geological Societies Transactions, v. 36, p. 121–128.

Gose, W. A., Kyle, J. R., and Ulrich, M. R., 1985, Preliminary paleomagnetic investigations of the Winnfield salt dome cap rock, Louisiana: Gulf Coast Association of Geological Societies Transactions, v. 36, p. 97–106.

Gose, W. A., Kyle, J. R., and Ulrich, M. R., 1989, Direct dating of salt diapir growth by means of paleomagnetism: Gulf Coast Section Society of Economic Paleontologists and Mineralogists Foundation Tenth Annual Research Conference Program and Abstracts, p. 48–53.

Halbouty, M. F., 1979, Salt domes, Gulf Region, United States and Mexico, 2nd ed.: Houston, Texas, Gulf Publishing Company, 561 p.

Hamblin, W. K., 1965, Origin of reverse drag on the downthrown side of normal faults: Geological Society of America Bulletin, v. 76, p. 1145–1163.

Hardin, F. R., and Hardin, G. C., Jr., 1961, Contemporaneous normal faults of Gulf Cost and their relation to flexures: American Association of Petroleum Geologists Bulletin, v. 45, p. 238–248.

Humphris, C. C., 1979, Salt movement on continental slope, northern Gulf of Mexico: American Association of Petroleum Geologists Bulletin, v. 63, p. 782–798.

Jackson, M.P.A., and Cornelius, R. R., 1987, Stepwise centrifuge modeling of the effects of differential sedimentary loading on the formation of salt structures, in Lerche, I., and O'Brien, J. J., eds., Dynamical geology of salt and related structures: Orlando, Florida, Academic Press, p. 163–259.

Jackson, M.P.A., and Cramez, C., 1989, Seismic recognition of salt welds in salt tectonic regimes: Gulf Coast Section Society of Economic Paleontologists and Mineralogists Foundation Tenth Annual Research Conference Program and Abstracts, p. 66–71.

Jackson, M.P.A., and Galloway, W. E., 1984, Structural and depositional styles of Gulf Coast and Tertiary continental margins; Applications to hydrocarbon exploration: American Association of Petroleum Geologists Continuing Education Course Note Series 25, 226 p.

Kehle, R. O., 1988, The origin of salt structures, in Schreiber, B. C., ed., Evaporites and hydrocarbons: New York, Columbia University Press, p. 345–404.

Kupfer, D. H., 1989, Diapirism sequences as indicated by internal salt structures:

Gulf Coast Section Society of Economic Paleontologists and Mineralogists Foundation Tenth Annual Research Conference Program and Abstracts, p. 79–89.

Kyle, J. R., Ulrich, M. R., and Gose, W. A., 1987, Textural and paleomagnetic evidence for the mechanism and timing of anhydrite cap rock formation, Winnfield salt dome, Louisiana, *in* Lerche, I., and O'Brien, J., eds., Dynamical geology of salt and related structures: Orlando, Florida, Academic Press, p. 497–542.

Labao, J. J., and Pilger, R. H., 1985, Early evolution of salt structures in the North Louisiana salt basin: Gulf Coast Association of Geological Societies Transactions, v. 35, p. 189–198.

Lopez, J. A., 1989, Distribution of structural styles in the northern Gulf of Mexico and Gulf Coast: Gulf Coast Section Society of Economic Paleontologists and Mineralogists Foundation Tenth Annual Research Conference Program and Abstracts, p. 101–108.

Lowell, J. D., and Genik, G. J., 1972, Sea-floor spreading and structural evolution of southern Red Sea: American Association of Petroleum Geologists Bulletin, v. 56, p. 247–259.

Mandl, G., and Crans, W., 1981, Gravitational gliding in deltas, *in* McClay, K. R., and Price, N. J., eds., Thrust and nappe tectonics: Geological Society of London, p. 41–53.

Martin, R. G., 1978, Northern and eastern Gulf of Mexico continental margins; Stratigraphic and structural framework: American Association of Petroleum Geologists Studies in Geology 7, p. 21–42.

—— , 1980, Distribution of salt structures in the Gulf of Mexico; Map and descriptive text: U.S. Geological Survey Miscellaneous Field Studies Map MF-1213, scale 1:2,500,000.

Murray, G. E., 1966, Salt structures of the Gulf of Mexico Basin; A review: American Association of Petroleum Geologists Bulletin, v. 50, p. 439–478.

Nelson, T. H., 1989, Style of salt diapirs as a function of the stage of evolution and the nature of the encasing sediments [abs.]: Gulf Coast Section Society of Economic Paleontologists and Mineralogists Foundation Tenth Annual Research Conference Program and Abstracts, p. 109–110.

Nettleton, L. L., 1934, Fluid mechanics of salt domes: American Association of Petroleum Geologists Bulletin, v. 18, p. 1175–1204.

—— , 1943, Recent experimental and geophysical evidence of mechanics of salt-dome formation: American Association of Petroleum Geologists Bulletin, v. 27, p. 51–63.

—— , 1955, History of concepts of Gulf Coast salt dome formation: American Association of Petroleum Geologists Bulletin, v. 39, p. 2373–2383.

Ocamb, R. D., 1961, Growth faults of south Louisiana: Gulf Coast Association of Geological Societies Transactions, v. 11, p. 139–175.

Parker, T. J., and McDowell, A. N., 1955, Model studies of salt-dome tectonics: American Association of Petroleum Geologists Bulletin, v. 39, p. 2384–2470.

Posey, H. H., and Kyle, J. R., 1988, Fluid-rock interactions n the salt dome environment; An introduction and review: Chemical Geology, v. 74, p. 1–24.

Quarles, M., Jr., 1953, Salt-ridge hypothesis on origin of Texas Gulf Coast type of faulting: American Association of Petroleum Geologists Bulletin, v. 37, p. 489–508.

Ramberg, H., 1968, Instability of layered systems in the field of gravity, I and II: Physics of Earth and Planetary Interiors, v. 1, p. 427–474.

Sensi, S. J., and Jackson, M.P.A., 1984, Sedimentary record of Cretaceous and Tertiary salt movement, East Texas Basin; Times, rates, and volumes of salt flow and their implications for nuclear waste isolation and petroleum exploration: University of Texas at Austin Bureau of Economic Geology Report of Investigations 139, 89 p.

Shelton, J. W., 1968, Role of contemporaneous faulting during basin subsidence: American Association of Petroleum Geologists Bulletin, v. 52, p. 399–413.

—— , 1984, Listric normal faults; An illustrated summary: American Association of Petroleum Geologists Bulletin, v. 68, p. 901–915.

Spiers, C. J., and 5 others, 1990, Experimental determination of constituative parameters governing creep of rocksalt by pressure solution, *in* Knipe, R. J., and Rutter, E. H., eds., Deformation mechanisms; Geology and tectonics: Journal of Geological Society of London (in press).

Talbot, C. J., 1977, Inclined and asymmetric upward-moving gravity structures: Tectonophysics, v. 42, p. 159–181.

Talbot, C. J., and Jackson, M.P.A., 1987, Salt tectonics: Scientific American, v. 257, no. 2, p. 70–79.

Vendeville, B. C., 1989, Scaled experiments on the interaction between salt flow and overburden faulting during syndepositional extension: Gulf Coast Section Society of Economic Paleontologists and Mineralogists Foundation Tenth Annual Research Conference Program and Abstracts, p. 131–136.

Vogler, H. A., and Robison, B. A., 1987, Exploration for deep geopressured gas; Corsair trend, offshore Texas: American Association of Petroleum Geologists Bulletin, v. 71, p. 777–787.

Weimer, P., and Buffler, R. T., 1989, Structural geology of the Mississippi Fan Foldbelt, Gulf of Mexico: Gulf Coast Section Society of Economic Paleontologists and Mineralogists Foundation Tenth Annual Research Conference Program and Abstracts, p. 146–147.

West, D. B., 1989, Model for salt deformation of central Gulf of Mexico Basin: American Association of Petroleum Geologists Bulletin, v. 73, p. 1472–1482.

Winker, C. D., and Edwards, M. B., 1983, Unstable progradational clastic shelf margins, *in* Stanley, D. J., and Moore, G. T., eds., The shelfbreak; Critical interface on continental margins: Society of Economic Paleontologists and Mineralogists Special Publication 33, p. 139–157.

Worrall, D. M., and Snelson, S., 1989, Evolution of the northern Gulf of Mexico, with emphasis on Cenozoic growth faulting and the role of salt, *in* Bally, A. W., and Palmer, A. R., eds., The geology of North America; An overview: Boulder, Colorado, Geological Society of America, The Geology of North America, v. A, p. 97–138.

Wu, S., Cramez, C., Bally, A. W., and Vail, P. R., 1989, Evolution of allochthonous salt in the Mississippi Canyon area: Gulf Coast Section of the Society of Economic Paleontologists and Mineralogists Tenth Annual Research Conference Program and Abstracts, p. 161–165.

Xiao, H.-B., and Suppe, J., 1989, Role of compaction in listric shape of growth faults: American Association of Petroleum Geologists Bulletin, v. 72, p. 777–786.

MANUSCRIPT ACCEPTED BY THE SOCIETY SEPTEMBER 28, 1990

ACKNOWLEDGMENTS

I thank Neville Carter for his help in dealing with the micromechanics of salt flow and for his constructive suggestions on the initial manuscript. I also express my appreciation to Tom Hauge, Martin Jackson, Amos Salvador, Steve Seni, Sig Snelson, and Dan Worrall for their help and suggestions in preparing this chapter, and to Lee Fairchild for the many hours we have spent together discussing the evolution of salt structures. Finally, I thank TGS/GECO for supplying me with several of the excellent seismic sections that appear in this chapter, and Exxon for the opportunity to pursue salt structures for the 32 years I spent in their employ.

Printed in U.S.A.

Chapter 6

Igneous activity

Gary R. Byerly
Department of Geology and Geophysics, Louisiana State University, Baton Rouge, Louisiana 70803

INTRODUCTION

Some 220 million years ago, in Late Triassic time, the Gulf of Mexico basin began to form in the wake of the breakup of Paleozoic megacontinent Pangea, and the opening of the North Atlantic Ocean. Igneous processes played a major role in the formation of this basin, as the common occurrence of basaltic rocks in rift basins around the Gulf of Mexico margin indicates. Geophysical evidence indicates that the central basin is floored by oceanic crust, presumably similar to that of oceanic crust elsewhere. Igneous activity was, however, not confined to the early stages of evolution of the basin. During the late Mesozoic, major volcanic fields rimmed the northern margin of the basin, probably the result of intraplate stresses due to global plate reorganization or isostatic adjustment from increased sediment loads along this margin. The eastern margin of the basin may have been affected by a late Mesozoic–early Tertiary Caribbean magmatic arc complex. Throughout the Tertiary the western margin has had a complex history of igneous activity associated with subduction of the Pacific plate beneath the North American Plate. There are numerous volcanic fields on the coastal plain and presently three active submarine volcanoes within the Gulf of Mexico.

Throughout the evolution of the Gulf of Mexico basin, igneous rocks have also been a significant component of sediment being deposited within the basin. In the Late Cretaceous, local basins along the northern margin had adjacent volcanic sources, and during the early Tertiary uplift, these volcanic terrains were the source of epiclastic materials. Mid-Tertiary volcanism in northern Mexico, and perhaps renewed uplift along the northern margin, again resulted in abundant volcanic materials in local basins along the western and northwestern Gulf of Mexico basin margin.

EARLY MESOZOIC

Igneous activity in the Gulf of Mexico basin is well defined for the Late Triassic to Middle Jurassic. Some evidence suggests it may extend to the close of the Jurassic. The majority of this igneous activity may be resolved into periods centered at approximately 190 and 165 Ma. Evidence for younger Jurassic activity is not very sound, but will be discussed in this section. The wide range of isotopic age determinations throughout the Jurassic probably reflects alteration from original igneous compositions.

Most of the igneous rocks of early Mesozoic age are basalts and diabases associated with grabens and red beds similar to those of the Newark Supergroup of eastern North America. These grabens and volcanic rocks can be seen as continuous with those of eastern North America, and encircling much of the Gulf of Mexico. In south Florida, and possibly in Mexico, silicic and alkalic volcanic rocks occur with the basalts. In central Mississippi, granites are found in two locations.

Dike swarms in the southeastern United States

One or more major dike swarms are present in the southeastern United States. These dikes crosscut tectonic and metamorphic structures of the Piedmont and Blue Ridge Provinces, as well as post-orogenic granites as young as 264 Ma (Dooley and Wampler, 1983). They are not found in rocks younger than Early Jurassic. In Georgia there are more than 100 of these dikes. They are generally vertical, trend north-northwest, and range from less than 1 m to about 100 m in width (Dooley and Wampler, 1983).

Many workers have noted that early Mesozoic dike swarms may be associated with the initial breakup of Pangea (e.g., May, 1971). In the southernmost United States these dikes are subparallel and north-northwest in trend, while farther north, successively overlapping swarms trend north and north-northeast (Fig. 1a, location A). May (1971) suggests that in a pre-rift configuration, these swarms, plus those from adjacent parts of Africa and South America, form a radial pattern convergent on the Bahamas platform. This pattern, he concludes, is due to magma emplacement along lines of tension produced in the initial stage of continental breakup. It is not clear, however, that all the dikes formed at the same time. DeBoer and Snyder (1979) conclude from a comprehensive study of these dikes in eastern North America that significant regional variation exists in major- and trace-element compositions, magnetic susceptibilities, and probably ages. Dikes in the Carolinas have the highest FeO, La/Sm, and magnetic susceptibilities. DeBoer and Snyder (1979) suggest that these

Byerly, G. R., 1991, Igneous activity, *in* Salvador, A., The Gulf of Mexico Basin: Boulder, Colorado, Geological Society of America, The Geology of North America, v. J.

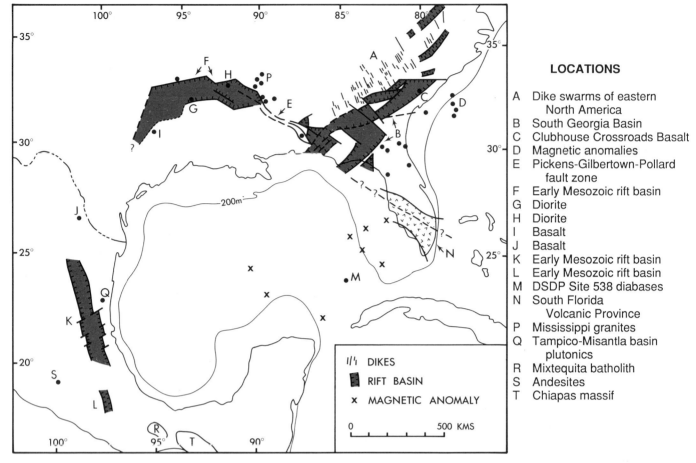

Figure 1a. Distribution of Triassic and Jurassic igneous rocks. Lettered locations are discussed individually in the text. Boundaries of the rift basins are generalized; detailed geology of these rifts is found in
Chapter 8, this volume. Most of these occurrences are from well cores and cuttings or defined by
geophysical anomalies.

patterns are reminiscent of the patterns over hot spots. They
conclude from thermoremanent magnetism of these dikes that
igneous activity began at the Carolina hot spot perhaps as early as
230 Ma (earliest Late Triassic) and lasted until 160 Ma (Middle
Jurassic), while farther away from the hot spot, igneous activity
began about 200 Ma (Early Jurassic) and lasted until 170 Ma
(Middle Jurassic).

Basalts and diabases of the South Georgia Basin

The South Georgia basin is defined in the subsurface from
South Carolina through Georgia and into the Gulf of Mexico
basin of south Alabama and north Florida (Gohn and others,
1978). The general boundaries of this basin, taken from Daniels
and others (1983), are shown in Figure 1a, location B. Basaltic
rocks are recognized from cores and inferred from magnetic
anomalies throughout the basin. The rocks apparently represent
dikes, flows, and sills that were contemporaneous with thick
deposits of red beds common in the basin. In addition, Daniels
and others (1983) report magnetic anomalies off the East Coast
that may represent the continuation of this basin. These and other

anomalies to the west are circular and may represent ring-dikes,
possibly more alkaline and younger than surrounding tholeiitic
rocks (Fig. 1a, location D). The basin fill attains a maximum
thickness in south Georgia, where it is at least 3.5 km thick. The
red beds are unconformably overlain by Upper Cretaceous sediments. Large numbers of faults subparallel to the northeast-
southwest axis of the basin have been postulated in the subsurface
of northern Florida (Barnett, 1975). A second major trend of
postulated faults is northwest-southeast and is subparallel to the
exposed Triassic and Jurassic dike swarms of north Georgia and
Alabama. Diabase sills appear to be associated with these faults.

The best-documented igneous rocks from this basin are
found in the subsurface near Charleston, South Carolina (Fig. 1a,
location C; Gottfried and others, 1983). A 256-m section of
basalt includes at least 7 lithologic units. Phillips (1983) recognizes additional subdivisions based on variations in magnetic in-
clinations and polarities. The presence of vesicular zones,
weathering zones, and several interbedded sedimentary units
clearly indicates that these units represent subaerial lava flows.
Two chemical types are recognized and are very similar to those
reported from other Triassic basins in eastern North America. A

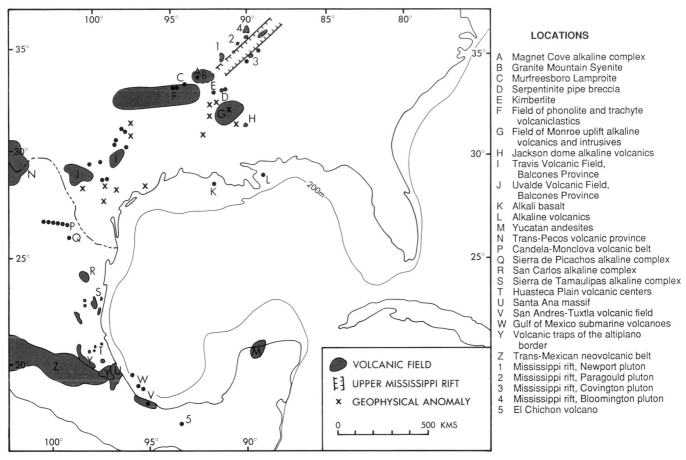

Figure 1b. Distribution of Upper Cretaceous igneous rocks. Lettered locations are individually discussed in the text. Boundaries of volcanic fields are generalized. No attempt is made to distinguish between subsurface and outcrop occurrences.

high TiO_2, quartz-normative tholeiite is fine grained, generally aphyric, but with rare clots of plagioclase and augite, and rare olivine. Plagioclase composition ranges from An_{60} to An_{70}. An olivine-normative tholeiite is 5 to 10 percent olivine-phyric with a coarse groundmass of plagioclase, An_{70}, augite, and chrome spinel. Unlike other basins where the quartz-normative flows clearly postdate the olivine-normative flows, those at Charleston are stratigraphically intercalated (Gottfried and others, 1983). The rocks are severely altered at the top of the sequence, and to a lesser degree at the tops of individual flow units, but the interiors of flows are remarkably well preserved. Clays, zeolites, carbonates, and silica have replaced much of the rock in the vesicular tops. The massive portions of the flows show only minor changes in H_2O, K_2O, Na_2O, and related trace elements. The other major elements, rare earths, P_2O_5, TiO_2, and related trace elements seem unaffected by alteration (Gottfried and others, 1983). The first age determinations reported for these rocks were 94 to 109 Ma by K-Ar techniques and probably reflect weathering related to the Late Cretaceous unconformity in this region. A second set of K-Ar age determinations ranges from 162 to 204 Ma. The most recent $^{40}Ar–^{39}Ar$ determination for the upper quartz-

normative flows is 184 ± 3 Ma (earliest Middle Jurassic) (Lanphere, 1983). This wide range of ages is most likely due to alteration effects. Similar difficulties have been encountered throughout the Gulf of Mexico basin in attempts to date subsurface basalts by K-Ar methods; these difficulties are discussed at length later in this section.

Elsewhere in the South Georgia basin, deep wells have encountered tholeiitic basalts and diabases, along with red beds. Because of their occurrence in the deep subsurface, dating either by paleontological or isotopic means has had only modest success. However, seemingly related dike swarms immediately north of the basin in Alabama and Georgia have provided somewhat better documentation of the ages for this igneous activity. Deininger and others (1975) report two discrete dike swarms in Alabama: (1) olivine-normative diabases with K-Ar ages of 184 to 193 Ma (mostly Early Jurassic), and (2) quartz-normative diabases with K-Ar ages of 161 to 168 Ma (mostly latest Middle Jurassic). Work on the Georgia dike swarms reveals that excess ^{40}Ar in many of these dikes results in variable and discordant K-Ar ages, but the most reliable ages for both olivine- and quartz-normative diabases are in the range of 190 to 195 Ma (Early

G. R. Byerly

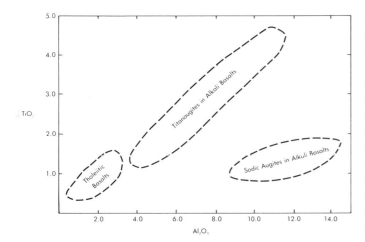

Figure 2. Pyroxenes in Triassic basalts, northern Florida. Many of the subsurface igneous samples have been altered to the extent that bulk rock analyses are unreliable indicators of original magmatic compositions. Compositions of preserved minerals in most rocks allow for a general assignment of the altered rock to a magmatic assemblage. In north Florida, alkalic and tholeiitic basalts are closely associated within the Triassic but easily distinguished based on pyroxene compositions.

Jurassic) (Dooley and Wampler, 1983). The intercalation of these two chemical types of flows in the Charleston wells clearly requires that at least some of these flow types were contemporaneous. Further work, especially ^{40}Ar-^{39}Ar, is needed to clarify the age relationship among these units.

Several workers have described rhyolite tuffs and flows below the red bed units of the South Georgia basin (for example, Neatherly and Thomas, 1975; Barnett, 1975). At present, only Paleozoic or older ages have been determined for these rocks (Mueller and Porch, 1983). Edick and Byerly (1981), however, report several occurrences of alkali basalts from north Florida. These rocks contain abundant olivine, chrome-spinel, and pleochroic pyroxenes rich in salite and titanangite components (Fig. 2). Though not isotopically dated, these occur in association with the local grabens and red beds of early Mesozoic age, and thus are a probable constituent of the Triassic igneous suite. Van Houten (1977) suggests that rare alkaline and andesitic rocks on either side of the North Atlantic may represent the earliest igneous activity associated with rifting, perhaps as early as about 215 Ma (Late Triassic).

Basalts and diabases of the south-central United States

From the termination of the South Georgia basin in south Alabama, the early Mesozoic basaltic province is offset 400 km northwest into Mississippi (see Fig. 1a, location F). This offset roughly coincides with the continuation of the Bahamas fracture zone and the Pickens-Gilbertown-Pollard fault zone in Mississippi (Fig. 1a, location E). Along this fault zone are found some Triassic sediments but no associated igneous rocks. Basalts and

diabases occur along a trend S80°W into northeast Texas in a series of poorly defined early Mesozoic basins or grabens. The northern and southern boundaries of this trend are modified by post-Jurassic uplift, erosion, and faulting, and thus may not accurately represent the original extent of early Mesozoic igneous activity in this region (Kidwell, 1951), although Scott and others (1961) suggest that normal faults of early Mesozoic age may define this basin. As much as 2 km of Eagle Mills Formation occurs in south Arkansas. A well near the southern margin of the basin penetrates 9 diabase horizons within the Eagle Mills; wells to the north penetrate progressively fewer diabases (Scott and others, 1961). In this basin, the Eagle Mills Formation is largely red shales with minor red sands and conglomerates. Plant fossils have allowed for fairly good correlation to the Upper Triassic–Lower Jurassic Newark Supergroup of eastern North America (see Chapter 8, this volume). Pebbles from the erosion of these diabases occur in the basal conglomerate of the Middle Jurassic Werner Formation (Scott and others, 1961). Kidwell (1951) suggests that regional differences in diabase composition are significant; those in Mississippi tend toward olivine-normative compositions, while those in Arkansas and Texas tend toward quartz-normative compositions. Little work has been done on the compositional variation in these rocks, although Edick and Byerly (1981) have shown that the compositional trends of the augites and pigeonites are clearly tholeiitic and similar to other diabases of the eastern North America province. Age determinations on these rocks, all by K-Ar, have been generally consistent with their occurrence in the Eagle Mills Formation (mostly Early Jurassic): 197 Ma (Baldwin and Adams, 1971) and 197, 184, 182, and 130 Ma (Denison and others, 1977). (The 130 Ma age will be discussed in a section on possible Late Jurassic igneous activity. It is tentatively considered not to reflect the igneous age of the rock.) Diorite is reported from south of the Sabine uplift in Texas (Flawn and others, 1961; see Fig. 1a, location G) and west of the Monroe uplift (Kidwell, 1951; see Fig. 1a, location H). Baldwin and Adams (1971) have determined an age of 196 Ma for the second diorite, but its geologic relation to the other diorite or to the more dominant diabases to the north is not well understood.

Igneous activity in the southwestern Gulf of Mexico Basin

Reports of early Mesozoic grabens and diabase/basalt are rare from the western Gulf of Mexico basin in the United States. Several explanations may exist: (1) the greater depth to lower Mesozoic strata in the southern Texas basins and the updip loss of these strata in central Texas with an erosional unconformity present a sampling problem; (2) recurrent tectonic activity along the western margin of the basin has obscured much of the detail of the early Mesozoic; and (3) crust of the western Gulf of Mexico basin did not respond to the early Mesozoic tectonic activity in the same way that the eastern part of the basin did.

Flawn and others (1961) report an isolated olivine basalt in red beds of early Mesozoic age in Freestone County, central

Texas (Fig. 1a, location I). No other west Texas occurrences are known. Dark amygdaloidal basalt or andesite flows are found with red beds in the subsurface near Nuevo Laredo, Nuevo Leon (Flawn and others, 1961; Fig. 1a, location J). Similar sediments are reported from across the Rio Grande in Texas.

A large north-trending graben system of early Mesozoic age occurs along the northeast coast of Mexico (see Chapter 8 of this volume). This graben system is at least 400 km long and averages 50 km wide, extending from north of Ciudad Victoria, Tamaulipas, to west of Tehuacan, Puebla (Fig. 1a, locations K and L). The Huizachal red beds fill the graben, along with minor diabase such as reported from near Ciudad Victoria. Although ages are not well constrained for these red beds, it seems likely that they are comparable to those of the Eagle Mills Formation and Newark Supergroup. Murray (1961) reports basalts in the red beds near Teziutlan, whereas López Ramos (1981) reports andesites and rhyolites in the slightly younger Jurassic red beds at Teziutlan and at Pueblo Viejo, Chiapas.

Igneous activity in the southeastern Gulf of Mexico Basin

Lyons (1957) notes 10 marine magnetic anomalies for this area, which he suggests are related to igneous intrusions (denoted by crosses in Fig. 1a). Work by the Deep Sea Drilling Project has helped to clarify the nature of these anomalies. Cores recovered during Leg 77, Hole 538A at Catoche Knoll (Buffler and others, 1984; and Fig. 1a, location M), contain two groups of diabase intrusives related to extensive early Mesozoic normal faulting. Dallmeyer (1982) reports ^{40}Ar-^{39}Ar ages for these to be 190 and 165 Ma, Early and Middle Jurassic respectively. Much of this region is apparently underlain by rifted transitional crust formed by Late Triassic to Jurassic extension and accompanied by periodic igneous activity (Buffler and others, 1982).

South Florida Volcanic Province

Barnett (1975) recognized major differences in the style and composition of early Mesozoic igneous activity between north and south Florida. The South Florida Volcanic Province (Fig. 1a, location N) is bounded on the north by a west-northwest-trending lineament that may have been a major strike-slip fault during the early Mesozoic. This fault may have been a continuation of the Bahamas fracture zone, perhaps the transform fault that connected spreading centers in the Gulf to those in the Atlantic (Klitgord and others, 1983). To the south the down-dip limit of these units is not known. Several wells up to 50 km offshore to the west have encountered volcanic rocks of early Mesozoic age. Abundant Upper Triassic to Middle Jurassic rhyolites and tuffs make south Florida unique (Barnett, 1975). Major- and trace-element compositions of some of the volcanic rocks of the province suggest that they are at least transitionally alkaline (Mueller and Porch, 1983).

A granophyre, probably a hypabyssal dacite porphyry, composed of plagioclase, quartz-orthoclase intergrowth, biotite,

and hornblende, has a K-Ar age of 170 ± 4 Ma, but an Rb-Sr age of 212 ± 30 Ma (Barnett, 1975). Although the Rb-Sr age is less precise, it is probably more accurate and may represent the earliest episode of igneous activity associated with the Late Triassic breakup of Pangea (Van Houten, 1977). Fine-grained rhyolite tuffs with phenocrysts of quartz, plagioclase, and sanidine have yielded a K-Ar age of 173 ± 4 Ma (Milton, 1972; Milton and Grasty, 1969). Milton also reports a sequence of interbedded tuffs and basalt. The basalt is olivine-normative and yields a K-Ar age of 183 ± 10 Ma. This would seem to confirm that some rhyolitic igneous activity overlapped with the basaltic activity, probably into Middle Jurassic time.

With few exceptions the alkaline and silicic igneous activity of south Florida is unique relative to other regions during the early Mesozoic. Minor andesites and trachytes in Virginia (Van Houten, 1977) and andesites and rhyolites near Veracruz (Lopez Ramos, 1981) are the only other examples of alkaline to silicic igneous activity. Many workers have used this fact to suggest that a major "hot spot" centered on south Florida resulted in this type of igneous activity. It is equally possible that this style of igneous activity characterized the earliest stage of rifting but, for the most part, is now represented by rocks on the continental shelves beyond our present sampling limits. This would conform to the style of continental rifting found in the East African rift and the probable complexity of the outer continental shelves known from seismic surveys (e.g., Sheridan, 1974; Buffler and Martin, 1982).

Granite intrusives of central Mississippi

Granitic rocks are not uncommon in the subsurface throughout the Gulf of Mexico basin, but they have generally been demonstrated on either stratigraphic or isotopic evidence to be Paleozoic or Precambrian. Denison and Muehlberger (1963) first documented the presence of early Mesozoic granites in Mississippi. They describe the granite as having volumetrically equal amounts of perthitic microcline, plagioclase, and quartz, with minor biotite, hornblende, zircon, apatite and magnetite. Stratigraphically, the granite lies beneath an apparent erosional unconformity with the Cretaceous Tuscaloosa Formation, with about 140 m of granite preserved above contact metamorphosed sandstone of unknown age. An Rb-Sr age of biotite of 183 ± 10 Ma (Middle Jurassic), with a similar Rb-Sr age of the microcline and K-Ar age on the biotite, are reported (Denison and Muehlberger, 1963). Two other occurrences of early Mesozoic granite from central Mississippi have K-Ar ages of 187 ± 7 Ma and 211 ± 20 Ma. These three occurrences (Fig. 1a, location P) are all within about 30 km of each other and may represent a single intrusive related to the Yocona dome, a small subsurface structure (Harrelson and Bicker, 1979). Petrographically these granites are clearly plutonic. They are not similar to the hypabyssal granophyres related to diabase sills. Denison and Muehlberger (1963) rule out the possibility of these ages reflecting either metamorphism or weathering.

The Jurassic Mississippi granites are at the eastern end of the

500-km-long series of east-west Triassic grabens of the south-central United States. Their genetic relationship to nearby basalts and diabases is not known (Kidwell, 1951). One possible analogy for these rocks comes from the White Mountains of New England (Zartman, 1977). The White Mountains contain a wide range of granitic intrusives that were apparently emplaced throughout the Mesozoic independently of the Mesozoic rifting of eastern North America.

Intrusive igneous rocks of Mexico

Intrusive granitic rocks were emplaced along the Cordilleran margin in Mexico throughout the Mesozoic. J. Jacobo-Albarran and M. Lopez-Infanzon (written communication, 1987) report that several wells in the Tampico-Misantla basin of east-central Mexico (Fig. 1a, location Q) have penetrated granites and granodiorites of Early to latest Triassic age (243 to 208 Ma) and Middle Jurassic (187 to 173 Ma) diorites and tonalites. They also report granites, granodiorites, and tonalites of Middle Triassic to Early Jurassic (235 to 192 Ma) from the Mixtequita batholith (Fig. 1a, location R) and granites, granodiorites, and tonalites of Early Triassic to earliest Cretaceous age (242 to 142 Ma) from the Chiapas massif in southern Mexico (Fig. 1a, location T).

Did basin-wide igneous activity continue through the Jurassic?

Igneous activity associated with the North American Cordillera may have continued throughout the Mesozoic. For example, bentonites and pillow lavas of Tithonian age are found about 90 km south-southwest of Mexico City interbedded with ammonite-bearing shales (Fig. 1a, location S). Elsewhere in the Gulf of Mexico basin, reports of Late Jurassic to Early Cretaceous igneous activity need critical evaluation. Reported isotopic ages for igneous rocks in the southeastern United States range throughout the Mesozoic (for example, Barnett, 1975). In many instances these ages do not conform to other geological interpretations, and new dating, especially by $^{40}Ar-^{39}Ar$ methods, is systematically reducing the range of probable ages for igneous activity in this region. Two other problems are encountered in the interpretation of some of these igneous rocks: (1) fine-grained intrusives may cut through many underlying units, and in wells they may be misidentified as flows; and (2) periodic sea-level changes during the Mesozoic resulted in several major erosional unconformities, providing an opportunity for older igneous rocks to be eroded, redeposited in younger sequences, and misinterpreted as younger igneous rocks. This section will examine some of these problems.

The first of these three problems is perhaps the easiest to deal with, since ultimately it should be possible to obtain new isotopic dates for samples whose ages now seem inconsistent with other geologic evidence. One example of this type of problem has already been discussed—those dates from the Charleston, South Carolina, subsurface. A similar problem is found with a basalt from Hardee County, Florida, which occurs in the subsurface at about 3,600 m depth. It is highly altered, and initial dating by K-Ar resulted in an age of 147 ± 7 Ma (Milton, 1972). Subsequent dating by $^{40}Ar-^{39}Ar$ yielded an age of 192 ± 7 Ma and 196 ± 6 Ma (Mueller and Porch, 1983). Basalt from St. Lucie County, Florida, occurs in the subsurface at about 3,900 m depth. The plagioclase is fresh, but groundmass and ferromagnesian minerals are chloritized. Milton (1972) reports an age of 89 ± 2 Ma based on K-Ar methods. This age is anomalous. Paleozoic granite and diorite are encountered immediately below this basalt. Triassic to Lower Jurassic volcanic rocks are found at similar depths in deep wells to the west (Milton, 1972). This basalt should be dated by $^{40}Ar-^{39}Ar$ to test its apparent young age. Denison and others (1977) report a 130-Ma whole-rock K-Ar age on diabase from 3,010 m deep in southwestern Arkansas. However, since this unit occurs within the Eagle Mills Formation (Kidwell, 1951) and all other local diabases are older, this age date is also a candidate for future $^{40}Ar-^{39}Ar$ work.

A quartz diabase in east Texas apparently intruded an anhydrite horizon that Moody (1949) considers Buckner in age, approximately 140 Ma. No isotopic age is available for this sample. However, anhydrite also occurs in the Eagle Mills (Kidwell, 1951) and in the Werner-Louann evaporite sequence (Chapter 8, this volume). Other possible examples of Upper Jurassic or Lower Cretaceous igneous rocks in the northern Gulf of Mexico basin may be interpreted equally well as products of erosion of older igneous rocks. Volcanic material is found in the Smackover Formation of central Mississippi, but detailed studies indicate that it is epiclastic material derived from Triassic basalts from the northern margin of the Mississippi Salt basin (Oxley and others, 1968).

Upper Cretaceous igneous rocks are also likely a source of confusion when they occur as fine-grained intrusives in underlying formations. This problem is especially acute for the Monroe uplift, where Upper Cretaceous intrusives and extrusives are abundant. Kidwell (1951) reports Cotton Valley Group (approximately 135 Ma) volcanic breccias from Ashley County, in southeast Arkansas. He later reinterprets these to be part of a younger diatreme containing much serpentine and other mixed lithologies (quoted in Stone and Sterling, 1964). Volcanic sediments from the Jackson dome are reported as Cotton Valley Group by Murray (1961) but as Late Cretaceous by Monroe (1954). Andesite found in the Late Jurassic Smackover Formation of Morehouse Parish in northeast Louisiana may best be interpreted as intrusive. Until new dating techniques are applied to these possible Upper Jurassic–Lower Cretaceous igneous rocks, their age assignments should be considered uncertain.

LATE MESOZOIC

After a probable hiatus in igneous activity of perhaps as much as 60 to 70 m.y., widespread, though relatively small-volume, volcanism took place throughout the western and northwestern Gulf of Mexico basin margin. The rocks are extremely alkaline in composition and are quite unlike the igneous

THE MAGNET COVE COMPLEX

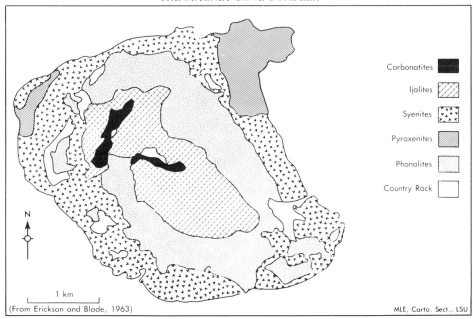

Figure 3. Geologic map of the Magnet Cove complex. This hypabyssal ring-dike complex of Late Cretaceous age is perhaps the best-known igneous feature of the northern Gulf of Mexico basin. The key has lithologies arranged by age from bottom to top. Based on geophysical anomalies many other similar complexes must be buried beneath coastal plain sediments.

rocks of the early Mesozoic Gulf of Mexico basin. Late Mesozoic igneous activity was also not associated with plate boundaries, although it may have been localized along ancient plate boundaries that acted as zones of weakness during subsequent global plate reorganizations or isostatic adjustments due to sediment loading. A large, though poorly understood, volcanic field lies in the subsurface of the northern Yucatan Peninsula.

Magnet Cove and related alkaline intrusives of central Arkansas

These rocks are well studied because most are at least partially exposed along the northern edge of the Gulf Coastal Plain. Including the subsurface extent beneath younger sediments, these intrusives are found in an area that is about 80 km east to west and 50 km north to south (Stone and Sterling, 1964; Morris, 1987). Six small intrusives and hundreds of related dikes occur along this trend from Hot Springs to Little Rock, Arkansas (Fig. 1b, circle about A and B). The larger bodies are primarily nepheline syenite, although often they occur with complex associations that include pyroxenites and carbonatites. The dikes are primarily either phonolite or lamprophyre.

Magnet Cove is the most often studied of these intrusives (Fig. 1b, location A). Although only about 4 km in diameter, it displays complex lithologies and intrusive relationships (Williams, 1891; Erickson and Blade, 1963). The overall structure is that of a series of ring dikes and plugs. It has a core of ijolite and carbonatite, an intermediate ring of trachyte and phonolite, and

an outer ring of nepheline syenite (Fig. 3). Two plugs of pyroxenite also cut the outer ring. Erickson and Blade (1963) suggest that, while all these intrusions are related, crosscutting relationships suggest the sequence (from oldest to youngest): (1) intermediate ring intrusives and breccia, (2) pyroxenites, (3) outer ring feldspathoidal syenites, (4) inner core ijolite, and (5) carbonatites and most dikes. The structure of this ring complex may be due to several episodes of cauldron subsidence into a relatively steep-sided magma chamber (Erickson and Blade, 1963).

Petrographic descriptions of the major lithologies are given below. They are from Erickson and Blade (1963) unless otherwise noted.

Phonolite and trachyte. These rocks are of variable composition, but all are finely crystalline and many are breccias. Many crustal xenoliths occur within these units. The breccias probably represent a constructional volcanic/hypabyssal stage of igneous activity. Altered phenocrysts of pyroxene and nepheline occur in a trachytic groundmass. Biotite and calcite seem to have deuterically replaced many of the original phases within the rock. Where pyroxenes are fresh, they are augite mantled by aegirine–augite. Where these rocks are freshest (younger?) they may contain anorthoclase, plagioclase, nepheline, aegirine-augite, and hornblende as major phases.

Pyroxenites. The dominant pyroxenite is jacupirangite, a magnetite-rich rock responsible for the magnetic highs of Magnet Cove. The rock is medium-grained with more than 50 percent pyroxene, up to 25 percent magnetite, and minor apatite and perovskite. Both sphene- and quartz-bearing varieties are found, but are apparently the result of assimilation of SiO_2-rich wallrock.

Feldspathoidal syenites. The most common lithology of Magnet Cove is the garnet-pseudoleucite syenite, which makes up about 21 percent of the igneous complex. It forms a nearly complete ring. It is typically medium grained and composed of pseudoleucite, sodic plagioclase and orthoclase, titanium-garnet, augite to aegirine, and nepheline. The common presence of miarolitic cavities suggests relatively low-pressure intrusion, as does the presence of leucite, whose stability is limited to the hypabyssal and volcanic environment. A sphene-nepheline syenite makes up about 7 percent of the igneous complex. It differs from the above syenite in lack of pseudoleucite and presence of hornblende and sphene.

Ijolites. The rocks of the inner core are quite varied lithologically and make up more than 12 percent of the complex. They are all principally nepheline, diopside, biotite, and titanium-garnet. Apatite, perovskite or sphene, and magmatic calcite are important minor phases.

Carbonatites. These represent about 2 percent of the complex. They occur as very coarse-grained calcite, with accessory minerals such as niobium-rich perovskite, zirconium garnet, apatite, and magnetite. The margins of these intrusives have apparently reacted with the silicate-rich rocks and are highly variable in their mineralogy. The carbonatite is clearly a late-stage intrusion into the complex. It contains inclusions of the syenite and ijolite and occurs as dikes crosscutting most of the other units.

Dikes within the complex display a wide range of composition from lamprophyres to syenites. These will be discussed with the other dikes that are abundant throughout this region.

Erickson and Blade (1963) note that a zone of metamorphosed sediments extends out 300 to 800 m from the intrusive contact. Near the contact the Stanley Shale is metamorphosed to a gneissic rock cut by aplitic dikes and composed of oligoclase, perthite, green hornblende, brown biotite, magnetite, and pyrite. The biotite zone of metamorphism extends out some 300 ft from the contact.

The average composition of the Magnet Cove igneous complex is melanocratic phonolite, but perhaps more significantly, the complex is strongly bimodal in composition with common leucocratic phonolites and pyroxenites (Erickson and Blade, 1963). Compared to most igneous rocks, these are richer in alkalis, calcium, and volatiles, especially CO_2, and poorer in silica. They are also richer in Ti, Zr, P, Nb, Ba, Sr, Y, and REE.

Economic deposits associated with Magnet Cove are primarily related to late-stage rutile-bearing veins. These have been mined for titanium, niobium, and vanadium. Iron ores have also been worked, mostly from magnetite-rich rocks.

The Potash Sulfur Springs complex occurs about 10 km west of Magnet Cove as an intrusive into Paleozoic sediments. It is less than half the size of Magnet Cove but has a similar ring structure and distribution of lithologies, although syenites are more abundant (Stone and others, 1982). Fenites, rocks produced by alkali metasomatism, are far more common than at Magnet Cove. Vanadium mineralization, apparently related to this fenitization, has produced several economic deposits. The "V" intrusive occurs several kilometers to the southwest. It is similar to the two previous examples, although perhaps somewhat more mafic in composition.

In Saline County (Fig. 1b, location A) several inliers of syenite are surrounded by early Tertiary sediments. Two lithologies are dominant: nepheline syenite and plagioclase-nepheline syenite (Williams, 1891). A great variety of dikes intrudes the syenites. These syenite occurrences are especially significant because during the early Tertiary they were exposed and subjected to lateritic weathering and erosion. Significant economic deposits of bauxite were produced (Stone and others, 1982). Several apparent explosion breccias have also been found in Saline County. Each is several hundred meters in diameter with a wide variety of clast lithologies (Stone and Sterling, 1964).

The syenites of Granite and Fourche Mountains occur just south of Little Rock (Fig. 1b, location A). There are two main lithologies here, according to Williams (1891). One, called the blue granite or pulaskite, is a medium- to coarse-grained nepheline syenite, somewhat porphyritic with orthoclase, nepheline, augite, arfvedsonite, biotite, sphene, and apatite. The second, called the grey granite, is a coarse-grained, trachytic nepheline syenite with microcline, nepheline, augite, aegirine, and biotite. The margins of these intrusives and associated dikes commonly contain miarolitic cavities, evidence for the relatively shallow emplacement of these plutons. Morris (1987) has found a minor olivine syenite phase that may be the oldest phase of the Granite Mountain intrusion.

Hundreds of related dikes and sills occur in this region. They are in greatest concentration near the larger intrusives. Kidwell (1951) notes that many of these strike east-west, but exceptions do occur. He has tabulated the number and composition as follows: nepheline syenite 24 percent, tinguaite 16 percent, ouachitite 32 percent, fourchite 22 percent, and monchiquite 6 percent. His definition of the latter three rocks, all lamprophyres, has been narrowed by recent workers to exclude plagioclase-bearing rocks that have been grouped as feldspathoidal gabbros (Erickson and Blade, 1963; Robinson, 1976). Thus, we can generalize that the three important groups of dike rocks are: (1) Feldspathoidal syenites. These range in composition from trachyte to phonolite, all with nepheline, some with pseudoleucite or sodalite. They have highly variable textures, from fine-grained trachytic to aplitic, porphyritic, or granitic. (2) Feldspathoidal gabbros. These are highly variable and are gradational into the lamprophyres based on the plagioclase content. Some contain analcime and nepheline. (3) Alkali lamprophyres. These are plagioclase-poor rocks with abundant mafic phenocrysts. Morris (1987) presents valuable new mineral and rock analyses from this suite.

The lamprophyres are especially characteristic of this province. They are all alkali-rich and silica-poor, but otherwise are highly variable in mineralogy and composition. Two of these were first named based on their occurrence in this province. Ouachitite contains phenocrysts of abundant titaniferous augite, biotite, and sometimes basaltic hornblende in a groundmass that ranges from glass to analcime (Kidwell, 1951). Fourchite contains phenocrysts of titaniferous augite, some with aegirine rims, minor barkevikite, and biotite in a groundmass dominated by analcime and small magnetite crystals (Kidwell, 1949). The other varieties of lamprophyres contain variable amounts of olivine, melilite, perovskite, nepheline, and leucite as phenocrysts. Robinson (1976) has shown that a compositional continuum exists between the feldspathoid gabbros and alkali lamprophyres. A plot of Cr-Ni-Co reveals a trend for all of these rocks very similar to the Skaergaard fractionation trend.

Magnetic anomalies and drilling logs indicate the existence of at least five other large bodies of alkalic to ultramafic composition within this region, but beneath the Gulf Coastal Plain. These seem to be similar to the exposed rocks of this province, although little work has been done on them.

Age relations for this suite of rocks are poorly constrained stratigraphically because they seem to intrude Paleozoic rocks, but the contact with the Mesozoic is missing, presumably eroded away. Thus, Tertiary rocks rest on an erosional unconformity with these intrusives. The following isotopic dates constrain the timing of the activity: (1) Magnet Cove trachyte 100, 101 ± 4 Ma whole-rock K-Ar (Baldwin and Adams, 1971); (2) Magnet Cove ijolite 95 ± 5 Ma biotite K-Ar (Zartman and others, 1967); (3) Magnet Cove carbonatite 103 ± 10 Ma apatite fission track (Scharon and Hsü, 1969); (4) Magnet Cove syenite 105 ± 10 Ma apatite fission track (Scharon and Hsü, 1969); (5) Potash Sulfur Springs syenite 100 ± 2 Ma U–Th–Pb on zircons (Zartman and Howard, 1985); and (6) Little Rock syenite 87 ± 4, 91 ± 5 Ma biotite K-Ar, and 86 ± 3 Ma biotite Rb-Sr (Zartman and others, 1967).

These early Late (and late Early) Cretaceous ages are consistent with other known igneous activity in the Gulf of Mexico basin. The consistency of the K-Ar, Rb-Sr, and fission-track ages is very good. The apatite age also indicates that cooling to below the blocking temperature for this mineral (100°C) was rapid. At Magnet Cove there seems to be no significant age difference between early (syenite) and late crosscutting (carbonatite and ijolite) intrusions, although these isotopic ages could have all been reset by the youngest intrusion. There may be a significant age difference between Magnet Cove and the syenites at Little Rock (Granite and Fouche Mountains).

Field, petrographic, compositional, and age relations among the rocks of this suite clearly indicate a genetic link. Most workers have called for a co-magmatic origin for these rocks from fractional crystallization of an alkali olivine basalt (for example, Erickson and Blade, 1963). Cumulus mushes, produced by settling of ferromagnesian minerals and floating of feldspars and feldspathoids, may best account for many of the rock types found in this region.

Diamond-bearing ultrapotassic rocks of Arkansas

The Murfreesboro lamproite occurs as an inlier within the Gulf Plain, just south of the Ouachita Mountains in Pike County,

Arkansas (Fig. 1b, location C). It is the best example of Late Cretaceous ultrapotassic igneous activity in the Gulf of Mexico Basin. This location is also known as the Prairie Creek or Crater of Diamonds intrusive. Cretaceous kimberlites occur to the north and west in Oklahoma and Kansas, and undated kimberlites are found in northern Arkansas. At least two subsurface occurrences of ultrapotassic rocks are known from southern Arkansas beneath the Gulf Plain sediments (Stone and Sterling, 1964): in Ashley County a breccia-filled volcanic pipe includes serpentinite with a variety of other lithologies (Fig. 1b, location D), and in Cleveland County a kimberlite contains olivine, augite, phlogopite, magnetite, and perovskite (Fig. 1b, location E). The Ashley County intrusive crosscuts Cotton Valley sediments and is therefore post–earliest Cretaceous. The Cleveland County intrusive crosscuts Paleozoic strata and is overlain by Upper Cretaceous sediments. Thus, both of these could be of similar age to the Murfreesboro lamproite. Miser and Ross (1922) describe the Murfreesboro locality as four small intrusions, each composed of massive, brecciated, and tuffaceous units. The main exposure is about 600 m in diameter. All are intrusive into the Lower Cretaceous and are overlain unconformably by the Upper Cretaceous Tokio Formation.

The massive units are probably the hypabyssal equivalents of the breccia and tuff (Miser and Ross, 1922). Where fresh, it is hard and dull black. Olivine, Fo_{90}, is the major phenocryst phase, although much of this has altered to serpentine. The groundmass is mostly phlogopite, often in poikilitic relationship to the remaining groundmass of amphibole, augite, spinel, and perovskite. Rare diamonds are also found, but the rock is notable in its lack of magnesian ilmenite, enstatite, and chrome diopside (Mitchell and Lewis, 1983). The BaO– and TiO_2–rich nature of the amphibole and phlogopite in these rocks allows for discrimination of lamproites from otherwise similar kimberlites (Scott Smith and Skinner, 1984). The breccias has a mineralogy similar to that of the massive unit (Miser and Ross, 1922), although diamonds may be somewhat more abundant (Stone and Sterling, 1964). It is usually weathered to clays. The rock fragments range from lithologies similar to the massive unit to aphanitic. Color variation in this unit is extreme: blue to green when first unearthed and yellow after long exposure. Bedding is developed in several locations with some size sorting. Dips in these beds are inclined toward the center of the intrusions, suggesting the infilling of an explosion pipe. This style is also more consistent with a lamproite diatreme (Scott Smith and Skinner, 1984). The tuff is blue-gray and of highly variable texture, although the original mineralogy was probably similar to that of the massive unit (Miser and Ross, 1922). Chlorite after phlogopite and serpentine after olivine are now the dominant minerals. No diamonds are found in the tuff (Stone and Sterling, 1964). Fragments of shale along with quartz and feldspar grains are abundant in some parts of the tuff, probably as the rsult of later reworking and mixing of the tuff with local sediments.

The age of the Murfreesboro lamproite is stratigraphically well constrained to be post-Aptian and pre-Coniacian. This is consistent with K-Ar isotopic ages on phlogopite from the massive unit, 97 ± 2 Ma, and the tuff, 106 ± 3 Ma (Zartman, 1977). This also makes these lamproites comparable in age to the Magnet Cove province 70 km to the northeast and to the phonolitic volcanic province that occurs immediately to the west of Murfreesboro.

Alkaline volcanics of southwestern Arkansas and adjacent areas

In Howard, Pike, and Sevier Counties of southwestern Arkansas, tuffaceous sediments constitute an important part of the Upper Cretaceous (Hazzard, 1939). These have sometimes been called the Centerpoint volcanics. Similar lithologies occur in adjacent parts of Texas, Oklahoma, and Louisiana (Fig. 1b, area circled around F; Ross and others, 1929). The outcrop area of these volcanic rocks runs east-west along the northern margin of the Gulf Coastal Plain for about 240 km into east Texas. Vent areas have been proposed in southwestern Arkansas (Fig. 1b, points above F; Ross and others, 1929). Several occurrences of similar lithologies have been observed in the subsurface in northwestern Louisiana, and in outcrop where the Prothro and Rayburn salt domes have pushed Upper Cretaceous sediments through the overlying Tertiary sediments (Ross and others, 1929). At Prothro dome, 33 m of tuff is exposed.

Most of these tuffs are apparently waterlaid (Ross and others, 1929). They are often crossbedded, and lithic clasts are well rounded. Individual units may be up to 40 m thick. Most often these beds are admixed with locally derived nonvolcanic sedimentary material, but in some beds are nearly pure volcanic material. The two major igneous lithologies observed are phonolite and trachyte. Both are undersaturated and feldspathoid-bearing. The phonolite most frequently occurs as a tuff or sand, but cobbles have been found locally. The cobbles are fine grained with a trachytic texture and phenocrysts of plagioclase and sanidine, with rare nepheline, augite or hornblende, and magnetite. The augite is strongly pleochroic and presumably Ti-rich. The trachyte occurs as a pumice tuff and as a lithic tuff. Sanidine is the only phenocryst phase. Groundmass biotite, sphene, magnetite, and a nonmagnetic black spinel are common.

Ross and others (1929) have also studied the diagenesis of these rocks in some detail. Fresh tuffs are olive to gray-green. Early alteration produced glauconite, followed by bentonite and finally kaolinite. The development of extensive early calcite cementation has arrested the alteration in various stages of completion.

A single K-Ar age of 98 ± 2 Ma has been obtained on a trachyte clast (A. Baksi, personal communication, 1988). These volcaniclastics are stratigraphically restricted to early Late Cretaceous (Ross and others, 1929). A Late Cretaceous unconformity may have resulted in some subsequent epiclastic deposits that may be difficult to distinguish from the original pyroclastic deposits.

Volcanism on the Monroe uplift

A structural platform created, or at least reactivated, during the Late Cretaceous occupies a broad area of northeast Louisiana, southeast Arkansas, and west-central Mississippi (Fig. 1b, circled area at G). Generally named the Monroe uplift, it is also referred to as the Sharkey platform, especially in Mississippi. The most comprehensive studies of igneous rocks of the Monroe uplift are those of Kidwell (1949, 1951) and Moody (1949). They describe over 50 subsurface occurrences of volcanic and hypabyssal rocks from the area. The rocks are generally similar in age and composition to those found in central and southwestern Arkansas.

Kidwell (1951) recognizes four major groups of igneous rocks within the Monroe uplift: (1) intermediate rocks such as trachytes, (2) alkaline rocks such as phonolites, (3) basalts, and (4) lamprophyres. The distribution of these rocks is remarkably nonuniform throughout the uplift. Lamprophyres, the most commonly encountered rock type, are mostly concentrated in a zone 20 km wide and 160 km long, trending N45°E on the southeast margin of the uplift. The phonolitic rocks are most abundant along the northeast margin of the uplift in Mississippi, and the trachytic rocks are most common along the northwest margin in Arkansas. The basalts are alkaline varieties, with salite the common pyroxene (Edick and Byerly, 1981).

Tuffs and breccias are very common and, although often interbedded with basalts and lamprophyres, they are usually phonolitic varieties. Kidwell (1949) suggests that these are probably analogous to the volcanic rocks that must have been associated with the alkaline intrusives of central Arkansas. Virtually all of the nepheline syenites reported have extremely weathered, flat-topped upper surfaces that lie unconformably beneath the Monroe gas rock, a local unit of Maastrichtian age.

The most apparent compositional difference between the Monroe uplift and central Arkansas intrusives is the dominance of amphibole over biotite in the lamprophyres, which makes amphibole monchiquite the common lamprophyre of this area (Kidwell, 1949). Basalt and associated feldspathoid-bearing basic volcanic rocks, including tephrite and nepheline, also occur exclusively in this area. However, this may be due to the preservation of upper level volcanics here that were eroded away from the central Arkansas region.

Spooner (1964) has studied volcaniclastic sediments in Louisiana south of the Monroe uplift. Oil-bearing stringer sands are rich in volcanic material. The Monroe uplift volcanoes probably supplied basins on all sides with a substantial amount of primary volcaniclastic material during Cenomanian time, and later, during uplift, erosion may have provided these basins with epiclastic volcanic material.

Only two isotopic ages are reported from this area. A K-Ar date on biotite from a biotite analcimite is 91 ± 3 Ma, and a K-Ar whole-rock age on a phonolite is 78 ± 3 Ma (Sundeen and Cook, 1977). The dated units occur in west-central Mississippi at 1,287 and 1,576 m depth, respectively. This would seem to indicate that

PILOT KNOB VOLCANO

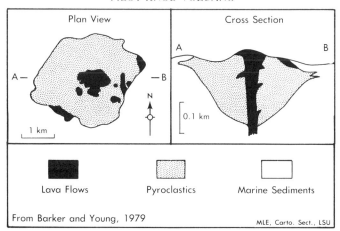

Figure 4. Interpretation of Pilot Knob Volcano based on surface mapping and geophysics. This large shallow-marine cindercone or tuff-ring was probably typical of the hundreds that rimmed the north margin of the Gulf of Mexico basin during Late Cretaceous time.

much of the volcanic activity here occurred during the Late Cretaceous prior to about 90 Ma. The younger age reflects either a cooling age, if this lower unit is intrusive, or an apparent age that is the result of alteration.

Numerous magnetic and gravity anomalies have been reported on the Monroe uplift (Lyons, 1957). These include Epps dome in Louisiana and the Midnight Volcano in Mississippi. These most likely reflect larger igneous complexes in the area.

Volcanism at Jackson dome

The Jackson dome in central Mississippi is a structural feature roughly 40 km in diameter that developed in Late Cretaceous time (Fig. 1b, location H). Monroe (1954) outlines the igneous activity that apparently accompanied this crustal warping. During Cenomanian time (91 to 97.5 Ma), volcanoes built a complex of islands that were periodically eroded until Santonian time (84 to 87.5 Ma). These volcanoes produced primarily phonolitic lavas, pyroclastics, and intrusives, along with minor lamprophyres. The Jackson gas rock, an Upper Cretaceous carbonate reef complex, lies unconformably on top of these volcanic units. Like many of the Upper Cretaceous igneous complexes in the northern Gulf of Mexico basin, the Jackson dome has pronounced magnetic and gravity maxima (Lyons, 1957).

The Balcones volcanic province

The Balcones volcanic province of south and central Texas is about 400 km long and 80 km wide. It follows closely the structural trend of the buried Paleozoic Ouachita orogenic belt and the Tertiary Balcones and Luling fault zones (Lonsdale, 1927; Spencer, 1969; Ewing and Caran, 1982). Most of the more

than 200 recognized sites of igneous activity are in one or two major fields on either side of the San Marcos arch (Fig. 1b, the Travis field location I and the Uvalde field location J). Volcanics crop out on the northwest side of the trend, but to the southeast are found only in the subsurface. Most of these subsurface occurrences have been located because of the gravity and magnetic anomalies associated with them. These include the "serpentine plugs," which are important small-scale hydrocarbon traps in the region.

A wide variety of extrusive and intrusive relationships can be identified in outcrops, especially in the Uvalde field and at Pilot Knob in the Travis field. Plugs, sills, and dikes are very fine grained and likely formed by shallow intrusion beneath local volcanic vents. Most occurrences are, however, tuffs or mixtures of volcanic debris with locally derived sediments, commonly carbonates (e.g., Barker and Young, 1979; Ewing and Caran, 1982). Lonsdale (1927) provides the most comprehensive discussion of the geology of the Balcones volcanic outcrops. At the Knippa aggregate quarry in Uvalde County he described a plug about 400 m in diameter intrusive into Austin chalk and with a carapace of "serpentine" produced by alteration of the intrusive. A more recent interpretation is that the massive igneous rock formed as a lava lake within a tuff ring (Ewing and Caran, 1982). The "serpentine" is thus reinterpreted as tuff, and Austin chalk as a mantling deposit on the volcanic platform. At Pilot Knob, detailed studies by numerous geologists and geophysicists have produced a fairly clear picture of the relatively short life of this Late Cretaceous volcano. Pilot Knob probably began in a relatively shallow sea, the result of numerous explosive eruptions producing ash, lapilli, and eventually a tuff mound. Subsequent explosions produced a crater, which filled with massive flow rock. Some flows may have breached the tuff mound and flowed down the flanks of the subaerially exposed volcano (Fig. 4). A total of approximately 1 km^3 of material was incorporated into this volcano, probably over a very short period of time (Barker and Young, 1979; Ewing and Caran, 1982).

Compositions found in the Balcones province are significantly different from those found elsewhere in the Upper Cretaceous of the Gulf of Mexico basin. Five major rock types and their approximate proportions are: melilite-olivine nepheline, 50 percent; olivine nephelinite, 30 percent; phonolite, 10 percent; alkali basalt, less than 10 percent; and nepheline basanite, less than 5 percent (Spencer, 1969). This is a much higher proportion of mafic-ultramafic compositions than the dominantly phonolitic compositions associated with central Arkansas or the Monroe uplift. A brief discussion of each of the lithologies is given below. Olivine nephelinite is the most mafic rock type found. It is generally fine grained, holocrystalline with brown titanaugite, olivine and nepheline, and minor apatite, biotite, and oxides. The melilite-olivine nephelinite has melilite and occasionally perovskite in addition to the above minerals. The nepheline basanite has plagioclase instead of melilite. Barker and Young (1979) provide many new mineral analyses and the observation that olivine clearly has a reaction relation to augite. In the alkali basalt the dominant minerals are titanaugite, plagioclase, and olivine, with minor biotite, calcite, apatite, and oxides. Analcime phonolite often shows porphyritic-trachytic texture and contains sanidine, aegirine-augite, and nepheline with minor olivine, titanaugite, sphene, apatite, and amphibole. A rare mafic variety of phonolite is also found. Spinel lherzolite and harzburgite xenoliths have been found in several locations.

Spencer (1969) has observed some geographic control on the distribution of the lithologies. He finds that all phonolites occur on the San Marcos arch and all basalts occur on or to the west of the arch. Little is known about the stratigraphic relations among these lithologies.

Ages for Balcones igneous activity are moderately well known from both stratigraphic and isotopic data. Volcaniclastic sediments occur from lower Cenomanian (Greenwood, 1956) to upper Maastrichtian (numerous examples reported in Spencer, 1969). This range from 100 to 69 Ma does, however, have a clear maximum in the Campanian, 74 to 84 Ma (Spencer, 1969). Baldwin and Adams (1971) report 20 whole-rock K-Ar ages from the Balcones Province. These ages range from 73 to 86 Ma for the mafic lithologies, but the two dated phonolites have ages of 63 Ma. These ages would seem to confirm that most of the igneous activity in the Balcones Province occurred after that in the Late Cretaceous of Arkansas, Louisiana, and Mississippi, although there may have been some overlap.

Spencer (1969) suggests that two primary magmas are required to generate the observed compositional variation found in the Balcones Province. The alkali basalts seem unrelated by fractionation to any of the other lithologies. However, the olivine nephelinite may be the primary magma to the melilite-olivine nephelinites and the basinite-phonolite group of rocks. This would require that the olivine nephelinite was near a liquidus thermal divide, which resulted in two opposing fractionation trends with slight variations in the composition of the primary magma or the P-T conditions of fractionation. Barker and Young (1979) further note that the primary olivine nephelinite magma was likely generated by a small degree of partial fusion under relatively thick, stable crust and then injected along a zone of long-term structural weakness. Barker and others (1987) consider that the bimodal Balcones suite may be the result of phonolite produced by high-pressure removal of kaersutitic amphibole, olivine, and clinopyroxene from primary olivine nephelinites, whereas the range of observed mafic lavas were produced by variable, but low, degrees of partial melting.

Eastern Mexico

Evidence of volcanic activity, common in the uppermost Jurassic of eastern Mexico, is also recorded in the Cretaceous section. Bentonitic horizons have been reported from Tamaulipas Inferior, Agua Nueva, and San Felipe Formations and andesitic pillow lavas, pyroclasts, and bentonite beds are present in the Xonamanca Formation, west of the Veracruz Basin. As in the latest Jurassic, the volcanic centers during the Cretaceous were probably located to the west, along the Pacific margin of Mexico.

Yucatan Peninsula and Campeche Bank

A large volcanic field is found in the subsurface of the Yucatan Peninsula near Merida (Fig. 1b, circle around M; López Ramos, 1981). Three wells were drilled on gravity and magnetic anomalies and yielded andesitic flows interbedded with tuffs, anhydrite, and limestone. A fourth well 150 km to the southeast of the main field also contains numerous pyroclastic beds interlayered with anhydrite and limestone. At least 500 m of flows and tuffs are in the main field. Nothing has been reported in the literature on the composition or mineralogy of the volcanic rocks. No isotopic ages are available, but stratigraphically these rocks are interbedded with Albian-Cenomanian and Turonian (88.5 to 97.5 Ma) fossiliferous limestones (López Ramos, 1981). If these rocks are alkaline, they might be related to the Late Cretaceous volcanism on the north rim of the Gulf of Mexico basin. If they are found to be calc-alkaline, this could have major implications for the tectonic evolution of the Yucatan since the nearest calc-alkaline rocks of similar age are from Cuba and form part of a Cretaceous Caribbean volcanic arc.

The Central Gulf of Mexico basin

Alkaline volcanic rocks have recently been discovered offshore along the Louisiana coast (Fig. 1b, location L; an offshore well, Braunstein and McMichael, 1976; and Fig. 1b, location K; Rezak and Tieh, 1984). Whole-rock K-Ar ages for these two samples are 82 ± 8 Ma and 76.8 ± 3.3 Ma, respectively. Little information is currently available on the composition, mineralogy, or structure of the well sample. The bottom sample was obtained from a marine outcrop that is part of the caprock of a large salt dome (Rezak and Tieh, 1984). This sample is basaltic with minor olivine, Ti-augite, and hornblende phenocrysts, and though altered, a rock composition that is alkaline in nature. Magnetic anomalies along the coast to the west (e.g., see Lyons, 1957) suggest that volcanism could have been as extensive along this belt as it was along the Balcones belt of similar age.

Igneous activity and the Upper Mississippi embayment

The northeasterly trending New Madrid fault zone, extending from Arkansas to southern Illinois, is marked by a high density of recent seismic activity. This zone has apparently been a locus of episodic rifting and igneous activity since Cambrian times. Kidwell (1951) and Moody (1949) both suggest that plutonic rocks recovered from deep wells to the northeast of the Magnet Cove and Little Rock plutons are likely part of a Late Cretaceous belt of alkaline igneous activity. Two occurrences of nepheline syenite and four of lamprophyre are reported from the rift. Ervin and McGinnis (1975) elaborate on this theme and call on rapid subsidence in the Gulf of Mexico basin to reactivate the rift during late Mesozoic time and bring on igneous intrusions along the border faults of the rift. Zoback and others (1980) present field mapping and seismic reflection profiles to support both Late Cretaceous and Cenozoic, probably late Eocene, igneous activity along the rift. This conclusion is based on stratigraphic thinning and arched reflectors over the subsurface Newport pluton (Fig. 1b, location 1). Isotopic ages for these intrusives are sparse. A mica peridotite from Lake County, Tennessee, yields a K-Ar date of 267 Ma on biotite, while three similar rocks at the northeast termination of the rift yield both K-Ar and Rb-Sr ages in the range 252 to 281 Ma (Zartman, 1977).

The recent USGS Professional Paper 1236 on the New Madrid earthquake region contains much geophysical information bearing on the igneous activity associated with the rift (e.g., Hildenbrand and others, 1982). Within the rift graben, four magnetic anomalies are observed. Three represent units of unknown age or composition, but appear to be intrusive into Paleozoic sediments. The fourth is a probable ring-dike structure, based on the anomaly pattern (Hildenbrand and others, 1982). Drill cores at this location recovered a Permian peridotite (Zartman, 1977). The margins of the rift graben are marked by a series of magnetic highs presumed to be mafic or ultramafic intrusions based on additional gravity modeling (Hildenbrand and others, 1982). The directions of the remanent magnetic field of 3 of these plutons (the Paragould, Covington, and Bloomington; Fig. 1b, locations 2, 3, and 4) suggest a post-early Mesozoic age (Hildenbrand and others, 1982).

CENOZOIC

Although some intraplate igneous activity in the Upper Mississippi embayment may have carried through into the Cenozoic, the majority of Cenozoic activity was confined to the western margin of the Gulf of Mexico basin. This activity can probably be related to Cenozoic subduction of various Pacific plates beneath the western margin of North America. A hiatus in subduction-related volcanism in the latest Cretaceous to early Tertiary (between 75 and 25 Ma) seen farther north is related to a change in dip of the plate subducted beneath North America (Engebretson and others, 1985). Minor subduction-related volcanism took place much farther inland from the earlier magmatic arc. Most of the igneous rocks of this suite are mildly alkaline, probably reflecting their petrogenesis at greater depths than calc-alkaline rocks to the west, although more complex tectonic models or combinations can be proposed. Young activity results in the large volcano San Martin on the shore of the Gulf and three submarine volcanoes south of Veracruz.

Northeastern Mexico and Trans-Pecos Texas

This suite of igneous rocks trends from west Texas to the southeast, roughly parallel with the Cenozoic southwestern plate margin of North America. It may be part of a chain of alkaline volcanics that runs along the eastern margin of the North American Cordillera from Montana to Central America (Barker, 1977; Robin and Tournon, 1978). Along this trend, individual volcanic centers tend to display younger ages to the southeast (Robin and Tournon, 1978).

The Trans-Pecos magmatic province extends from El Paso, Texas, for a distance of some 700 km to the southeast (Figure 1b, location N). Several hundred intrusives and extensive volcanic units occur within the province (Barker, 1977). The most abundant rocks are hawaiite and mugearite, with lesser amounts of nepheline-bearing trachyte and phonolite, all part of a single differentiation series that includes minor trachy-basalt and abundant peralkaline rhyolite (Barker, 1979). Relatively low initial ratios of strontium isotopes suggest a mantle origin for both series. Volcanic deposits associated with fissure eruptions have not been found, but the similarity in composition of lavas along strike for distances of several hundred kilometers suggests extensive magmatic reservoirs similar to modern fissure eruptions (Barker, 1977). Numerous cauldron complexes have been recently identified. They are associated with eruptions of major felsic lava flows and ash-flow tuffs (Henry and Price, 1984). Ages for igneous activity range from 46 to 16 Ma (middle Eocene to middle Miocene), but the most voluminous activity occurred in the much narrower interval of 38 to 30 Ma (K-Ar ages reviewed in Barker, 1979). The degree of fractionation of alkaline rocks emplaced during the compressional tectonic episode, 46 to 32 Ma, is clearly greater than that seen in lavas formed during the extensional tectonic episode, 24 to 17 Ma (Price and others, 1987).

Seven major volcanic centers lie on or near the Gulf Coastal Plain southeast of the Trans-Pecos volcanic province. The northernmost is the Cendela-Monclova volcanic belt, which trends east-west for 150 km across much of the Laramide structural trend of the region (Fig. 1b, location P). Diorite to syenite was intruded during late Eocene to early Oligocene time, possibly along a reactivated Triassic fault system (Bloomfield and Cepeda-Davila, 1973; Lopez Ramos, 1981). To the southeast lies the Sierra de Picachos alkaline complex (Fig. 1b, location Q). Approximately 100 km farther southeast the San Carlos complex is composed of numerous calc-alkaline intrusives (Fig. 1b, location R). The largest, about 139 km^2, also contains abundant lamprophyre dikes, and the crosscutting relations suggest that each progressively younger unit is more alkaline as well (Broomfield and Cepeda-Davila, 1973). Four K-Ar ages here have an average of 28.8 ± 0.6 Ma (late Oligocene). The Sierra de Tamaulipas (Fig. 1b, location S) lies some 150 km southeast and is composed of Miocene alkaline to peralkaline rhyolites, usually microphyric with sanidine, some quartz and acmite. Nepheline-bearing syenites contain sodic pyroxenes and amphiboles. Minor mafic volcanics include nephelinites, basanites, alkali basalts, and trachy-basalts, some as recent as Quaternary (Robin and Tournon, 1978). On the Huasteca plain (Fig. 1b, location T), 200 km southeast of Sierra de Tamaulipas, several alkaline centers occur with ages from Miocene to Pliocene (Robin and Tournon, 1978). At the Santa Ana massif (Fig. 1b, location U), also 200 km southeast of the Huasteca, Pliocene to Quaternary alkali basalts and trachybasalts mark the eastern termination of the otherwise predominantly calc-alkaline Trans-Mexican neovolcanic belt

(Robin and Tournon, 1978). The Tuxtla volcanic field (Fig. 1b, location V) lies about 200 km farther southeast, on the Gulf of Mexico coast. Volcanic activity began here in the late Oligocene and has continued until the present (López Ramos, 1981). The most recent eruption of Volcan de San Martin was in 1793 (Moore and del Castillo, 1974). There are 3 major volcanic centers and numerous small ones. The largest, including San Martin, are low shields, about 10 km in diameter, with summit calderas about 3.5 km in diameter. They are dominantly alkali basalt to hawaiite, typically olivine-clinopyroxene phyric, and rarely plagioclase phyric (Thorpe, 1977).

An apparently independent association of volcanic rocks also lies along the border between the altiplano and the coastal plain (Fig. 1b, location Y; Robin and Tournon, 1978). This zone is marked by a series of normal faults, which separate normal continental crust beneath the altiplano from the crust beneath the coastal plain, which is transitional to that of the oceanic crust of the deep Gulf basin. Here lava flows are dominantly basalts, alkali to tholeiite, hawaiites, and mugearites. Robin and Tournon (1978) report late Miocene ages of 7 to 8 Ma from the northern zone, 5 to 6 Ma for the central zone, and Pliocene ages of 2 to 3 Ma for the southern zone.

Several models have been proposed to account for the igneous activity of northeast Mexico and Trans-Pecos Texas. The most commonly advocated model is that of a continental rift. Comparisons and even connections are made with the Rio Grande rift to the north (Bloomfield and Cepeda-Davila, 1973; Barker, 1977; Robin and Tournon, 1978). The generally alkaline nature of these rocks and their association with normal faulting are regarded as consistent with a rifting model. Several workers have suggested that reactivation of older Gulf of Mexico basin structures was totally or partially responsible for this igneous activity (Bloomfield and Cepeda-Davila, 1973, for the east-west belt; Robin and Tournon, 1978, for the volcanics associated with the altiplano border faults; and Thorpe, 1977, for the recent Tuxtla and offshore volcanics). The faults oriented east-west are considered to be related to Triassic transform movement associated with the opening of the Gulf of Mexico basin. The altiplano border faults may be analogous to other faults concentric to the Gulf of Mexico basin and related to loading of the crust, e.g., the Balcones fault zone and igneous province. The probable relation to Cenozoic subduction along the southwestern margin of North America is discussed by most of the above authors. Barker (1979), however, presents a compelling case for a subduction origin, at least in the Trans-Pecos region. He notes that this association is part of a series of alkaline provinces that lie behind the North American Cordillera from Montana to Mexico, roughly parallel to the Cenozoic continental margin. The igneous activity began well before rifting began, and in some instances before Laramide compression had ceased (Price and Henry, 1984). Perhaps most compelling is the observation that the alkaline belt is but the easternmost of a series of parallel belts that are progressively more calc-alkaline toward the west. Thus, the increasing

alkalinity to the east would be related to increased depth of melting along the subduction zone.

Volcaniclastics throughout the northern and western margins of the Gulf of Mexico Basin

Hunter and Davis (1979) have reviewed the distribution of Cenozoic volcaniclastic sediments in the northern Gulf of Mexico basin. Input of at least some volcanic materials occurred throughout the Cenozoic, although a major contribution did not begin until late Eocene, about 40 Ma. The only Cenozoic vents found in this region are those in the Trans-Pecos and northeast Mexico volcanic fields. This implies a source over 1,000 km from some of the sites of deposition in Mississippi and Alabama, but these distal deposits are generally considered to be of airfall origin (Ross and others, 1929).

Probably the most important of these volcaniclastic units is the Catahoula Formation in Mexico, Texas, and Louisiana. Volcanic activity in late Oligocene to early Miocene time produced an extensive blanket of waterlaid and airfall material in south Texas and northeastern Mexico. The most common lithologies are tuffaceous clay, volcanic arenite or conglomerate, bentonite, and vitric tuff or ash (McBride and others, 1968). Within the conglomerates, the primary igneous lithologies are trachyte, trachyandesite, rhyolite, and welded tuff. In southeast Texas and Louisiana, substantial amounts of ash are found in the Catahoula Formation.

Most of the tuffaceous clays are unstratified to crudely bedded, with montmorillonite the dominant component. The tuff or ash units are also typically unstratified, with dominant glass shards and trace amounts of sanidine, anorthoclase, and quartz. Alteration products include montmorillonite, opal, and zeolite (McBride and others, 1968). The volcaniclasts in sandstones and conglomerates are primarily either trachyte-trachyandesite or rhyolite porphyry. The trachytic rocks generally have an altered groundmass with an overall amygdaloidal porphyritic texture. Abundant wormy to resorbed anorthoclase is the dominant phenocryst, while minor aegirine, sodic amphibole, and rare orthopyroxene and quartz also occur. The trachyandesites commonly contain augite (McBride and others, 1968). The rhyolite porphyry typically contains about 10 percent potassium feldspar phenocrysts.

The Catahoula Formation is composed primarily of nonmarine fluvial, mudflow, and airfall debris, but it clearly grades downdip into a marine facies (McBride and others, 1968). The San Marcos arch apparently restricted the fluvial and mudflow component to the Rio Grande embayment, whereas airfall deposits of volcanic ash covered a much broader area in east Texas and Louisiana (Galloway, 1977). Dokka (1982), however, has determined fission track ages for zircons from two south Louisiana samples that yield ages of 68 ± 10 Ma and 62 ± 9 Ma. He suggests these are likely derived from erosion of Late Cretaceous volcanic rocks. McBride and others (1968) consider the Trans-

Pecos volcanic field the source for both airfall and fluvial/mud-flow components, but more precise age determinations from the Trans-Pecos may not be consistent with this interpretation (Henry and Price, 1984). Nearly all the Trans-Pecos igneous activity took place prior to deposition of the Catahoula Formation. The Trans-Pecos could still be a major source of epiclastic volcanic materials, but northern Mexico may be a more likely source for the abundant airfall ash deposits.

The Tertiary section throughout the length of eastern Mexico also contains abundant evidence of volcanic activity: tuffaceous material and volcanic ash derived from numerous volcanic centers along the Pacific margin of Mexico.

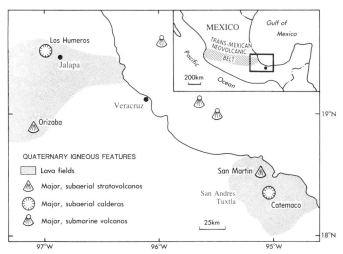

Figure 5. Quaternary igneous features near Veracruz, Mexico. The eastern termination of the Trans-Mexican neovolcanic belt may account for the presence of three submarine volcanoes offshore from Veracruz. These may also be aligned NNW along a trend with the Los Tuxtlas volcanic field and El Chichon Volcano.

Trans-Mexican neovolcanic belt of Mexico

In late Miocene time the subduction-related volcanic arc in Mexico was reoriented from its position parallel to the North American Cordillera to a roughly east-west configuration (Fig. 1b, location Z; López Ramos, 1981). This recent volcanic arc intersects the Gulf Coast about 50 km north of Veracruz (Figure 5), but apparently, igneous activity has never extended along this trend into the deep Gulf of Mexico basin (Moore and del Castillo, 1974). Three young sea-floor volcanoes lie 30 to 50 km off the Mexican coast (Fig. 1b, location W; Moore and del Castillo, 1974) but along a trend at a high angle to the Trans-Mexican neovolcanic belt and perhaps related to an offset of the arc that would include the Tuxtlas volcanic field and El Chichon Volcano (Fig. 1b, location 5) to the south. The submarine volcanoes are characterized by little sediment cover, large magnetic and gravity anomalies, and sharp topographic relief, all suggesting relatively recent activity. No information is available on the composition of

these seamounts. They may be similar to the predominantly andesitic rocks of the Trans-Mexican neovolcanic belt to the west, or they may be more closely related to the alkaline rocks of the Tuxtlas volcanic field. Both of these fields have volcanoes that have produced major historic eruptions (Moore and del Castillo, 1974).

In March 1982, a major explosive eruption took place at El Chichon volcano, Chiapas, Mexico (Luhr and Varekamp, 1984). El Chichon is offset from the Trans-Mexican neovolcanic belt even farther to the south than Tuxtlas, and like Tuxtlas it is compositionally distinct. The 1982 eruption produced primarily trachyandesite ash and pumice. In addition to being unusually alkaline, the El Chichon lavas are also unusually rich in sulfur. The 1982 pumices contain microphenocrysts of anhydrite.

AN OVERVIEW OF IGNEOUS ACTIVITY AND THE EVOLUTION OF THE GULF OF MEXICO BASIN

Throughout the evolution of the Gulf of Mexico basin, igneous processes played a significant role in forming the young basin, periodically modifying the margins, and contributing to the sediment load. Igneous processes may have also played a role in the thermal and uplift history. The earliest record of igneous activity comes from several large rift basins of Early Mesozoic age along the northern and western margin of the Gulf of Mexico basin (Fig. 6). Except for eastern Mexico, these basins are now covered by younger strata, but a fairly detailed picture of their geology exists because of petroleum exploration and production. Ages and styles of both igneous and sedimentary fill of these basins are very similar to that of analogous basins along the eastern margin of North America. Subaerial lavas, dikes, and sills of tholeiitic basalt are the dominant igneous materials found. This igneous activity is related to attenuation of continental crust about the margins of the Gulf of Mexico basin. In northern Florida, alkali basalts are interbedded with the tholeiitic basalts, and in southern Florida, silicic and alkalic volcanic rocks are common. This geographic distribution has been used to speculate that a hot spot centered on the Florida peninsula may have played an important role in the opening of the North Atlantic and Gulf of Mexico. Early Mesozoic intrusion of granites in central Mississippi cannot be easily explained as products of a continental margin rifting event.

During the Late Jurassic and Early Cretaceous the Gulf of Mexico basin probably experienced no igneous activity except for a minor component derived from western Mexico volcanic centers. The few reports of igneous rocks with ages in this interval represent either epiclastic igneous materials, younger intrusives misidentified as flows, or older igneous rocks with ages reset by low-temperature alteration.

In the Late Cretaceous, igneous activity resumed throughout the northern and northwestern margin of the Gulf of Mexico basin. Products of this activity are common in the subsurface from central Mississippi to west Texas and exposed in central Arkansas to west Texas. All of the igneous rocks from this inter-

G. R. Byerly

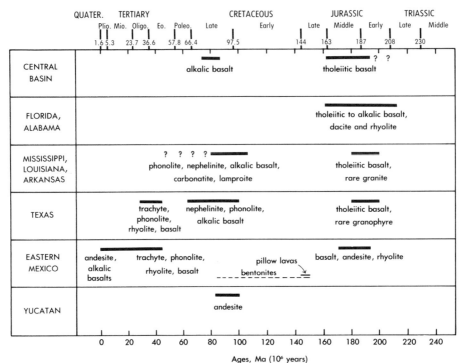

Figure 6. Summary of ages and geographical distribution of igneous activity for the Gulf of Mexico basin. Details are discussed in the text. Only ages that are consistent with other stratigraphic data have been included.

val are alkalic to nephelinitic, and although a wide spectrum of compositions is present, the suite is strongly bimodal with abundant phonolite and nephelinite. Rare carbonatites are associated with several intrusives of this suite, but the rare Upper Cretaceous lamproites also found in this region are not directly associated with the alkalic to nephelinitic suite. Some of this igneous activity may have been accompanied by local crustal warping, as seen in the Jackson dome. In general, however, the distribution of this type of igneous activity was controlled by the presence of older structural elements. Intraplate stress produced by global plate reorganization or regional isostatic adjustment from increased sediment load allowed the small-volume upper-mantle melts access into the upper crust by means of reactivated structures.

By about 40 Ma, subduction of a Pacific plate beneath North America was responsible for renewed igneous activity along the western and northwestern margin of the Gulf of Mexico basin. In Trans-Pecos Texas the earlier phase of activity 40 to 30 Ma was associated with Laramide tectonics and resulted in a wide spectrum of compositions from mafic to silicic, some also slightly alkaline; while a second phase from 30 to 20 Ma was associated with Basin and Range tectonics and is typically bimodal basalt and rhyolite. In Mexico this activity continues to the present and is represented by numerous Tertiary igneous complexes along the western margin of the Gulf. Several currently active volcanoes exist within the Gulf coastal plain or were submerged in the Gulf of Mexico along the Trans-Mexican neovolcanic belt.

REFERENCES CITED

Baldwin, O. D., and Adams, J.A.S., 1971, K/Ar ages of alkalic igneous rocks of the Balcones fault trend in Texas: Texas Journal of Science, v. 22, p. 223–231.

Barker, D. S., 1977, Northern Trans-Pecos magmatic province; Introduction and comparison with the Kenya rift: Geological Society of America Bulletin, v. 88, p. 1421–1427.

—— , 1979, Cenozoic magmatism in Trans-Pecos province; Relation to the Rio Grande rift, *in* Riecker, R. E., ed., Rio Grande rift; Tectonics and magmatism: American Geophysical Union, 438 p.

Barker, D. S., and Young, K. P., 1979, A marine Cretaceous nepheline basanite volcano at Austin, Texas: Texas Journal of Science, v. 31, p. 5–24.

Barker, D. S., Mitchell, R. H., and McKay, D., 1987, Late Cretaceous nephelinite

to phonolite magmatism in the Balcones province, Texas, *in* Morris, E. M., and Pasteris, J. D., eds., Mantle metasomatism and alkaline magmatism: Geological Society of America Special Paper 215, p. 293–304.

Barnett, R. S., 1975, Basement structure of Florida and its tectonic implications: Gulf Coast Association of Geological Societies Transactions, v. 25, p. 122–142.

Bloomfield, K., and Cepeda-Davila, L., 1973, Oligocene alkaline igneous activity in NE Mexico: Geology Magazine, v. 110, p. 551–555.

Braunstein, J., and McMichael, C. E., 1976, Door Point; A buried volcano in southeast Louisiana: Gulf Coast Association of Geological Societies Transactions, v. 26, p. 79–80.

Buffler, R. T., and Martin, R. G., 1982, Evolution of the Gulf of Mexico Basin:

Geological Society of America Abstracts with Programs, v. 14, p. 454.

Buffler, R. T., and others, 1984, Initial reports of the Deep Sea Drilling Project. Washington, D.C., U.S. Government Printing Office, v. 77, 747 p.

Dallmeyer, R. D., 1982, Pre-Mesozoic basement of the southeastern Gulf of Mexico: Geological Society of America Abstracts with Programs, v. 14, p. 471.

Daniels, D. L., Zietz, I., and Popenoe, P., 1983, Distribution of subsurface lower Mesozoic rocks in the southeastern United States, as interpreted from regional aeromagnetic and gravity maps: U.S. Geological Survey Professional Paper 1313-K, 24 p.

deBoer, J., and Snider, F. C., 1979, Magnetic and chemical variations in Mesozoic diabase dikes from eastern North America; Evidence for a hot-spot in the Carolinas?: Geological Society of America Bulletin, v. 90, p. 185–198.

Deininger, R. W., Dallmeyer, R. D., and Neathery, T. L., 1975, Chemical variations and K-Ar ages of diabase dikes in east-central Alabama: Geological Society of America Abstracts with Programs, v. 7, p. 482.

Denison, R. E., and Muehlberger, W. R., 1963, Buried granite in Mississippi: American Association of Petroleum Geologists Bulletin, v. 47, p. 865–867.

Denison, R. E., Burke, W. H., Otto, J. B., and Hetherington, E. A., 1977, Age of igneous and metamorphic activity affecting the Ouachita fold belt, *in* Symposium on the Geology of the Ouachita Mountains: Arkansas Geological Commission, p. 25–40.

Dokka, R. K., 1982, Implications of fission track ages from the Kaplan geothermal–geopressure zone, Vermilion Parish, Louisiana: Gulf Coast Association of Geological Societies, v. 32, p. 465–468.

Dooley, R. E., and Wampler, J. M., 1983, Potassium-argon relations in diabase dikes of Georgia; The influence of excess ^{40}Ar on the geochronology of early Mesozoic igneous and tectonic events: U.S. Geological Survey Professional Paper 1313-M, 24 p.

Edick, M. J., and Byerly, G. R., 1981, Post-Paleozoic igneous activity in the southeastern United States: Geological Society of America Abstracts with Programs, v. 13, p. 236.

Engebretson, D. C., Cox, A., and Gordon, R. G., 1985, Relative motions between oceanic and continental plates in the Pacific Basin: Geological Society of America Special Paper 206, 59 p.

Erickson, R. L., and Blade, L. V., 1963, Geochemistry and petrology of the alkalic igneous complex at Magnet Cove, Arkansas: U.S. Geological Survey Professional Paper 425, 95 p.

Ervin, C. P., and McGinnis, L. D., 1975, Reelfoot rift; Reactivated percursor to the Mississippi embayment: Geological Society of America Bulletin, v. 86, p. 1287–1295.

Ewing, T. E., and Caran, S. C., 1982, Late Cretaceous volcanism in south and central Texas; Stratigraphic, structural, and seismic models: Gulf Coast Association of Geological Societies Transactions, v. 32, p. 137–145.

Flawn, P. T., Goldstein, A., Jr., King, P. B., and Weaver, C. E., 1961, The Ouachita system: Austin, University of Texas Bureau of Economic Geology Publication 6120, 410 p.

Galloway, W. E., 1977, Catahoula Formation of the Texas Coastal Plain; Depositional systems, composition, structural development, groundwater flow history, and uranium distribution: Austin, University of Texas Bureau of Economic Geology Report of Investigation 87, 59 p.

Gohn, G. S., Gottfried, D., Lanphere, M. A., and Higgins, B. B., 1978, Regional implications of Triassic or Jurassic age for basalt and sedimentary red beds in the South Carolina Coastal Plain: Science, v. 202, p. 887–890.

Gottfried, D., Annel, C. S., and Byerly, G. R., 1983, Geochemistry and tectonic significance of subsurface basalts from Charleston, South Carolina; Clubhouse Crossroads test holes 2 and 3: U.S. Geological Survey Professional Paper 1313-A, 19 p.

Greenwood, R., 1956, Submarine volcanic mudflows and limestone dikes in the Grayson Formation (Cretaceous) of central Texas: Gulf Coast Association of Geological Societies Transactions, v. 6, p. 167–177.

Harrelson, D. W., and Bicker, A. R., Jr., 1979, Petrography of some subsurface igneous rocks of the Mississippi, Gulf Coast Association of Geological Societies Transactions, v. 29, p. 244–251.

Hazzard, R. T., 1939, The Centerpoint volcanics of southwest Arkansas; A facies of the Eagleford of northeast Texas: Shreveport Geological Society 14th Annual Field Trip Guidebook, p. 133–151.

Henry, C. D., and Price, J. G., 1984, Variations in caldera development in the Tertiary volcanic field of Trans-Pecos Texas: Journal of Geophysical Research, v. 89, p. 8765–8786.

Hildenbrand, T. G., Kane, M. F., and Hendricks, J. D., 1982, Magnetic basement in the upper Mississippi embayment region; A preliminary report: U.S. Geological Survey Professional Paper 1236-E, p. 39–53.

Hunter, B. E., and Davis, D. K., 1979, Distribution of volcanic sediments in the Gulf Coast province; Significance to petroleum geology: Gulf Coast Association of Geological Societies Transactions, v. 24, p. 147–155.

Kidwell, A. L., 1949, Mesozoic igneous activity in the northern Gulf Coastal Plain [Ph.D. thesis]: Chicago, Illinois, University of Chicago, 316 p.

——, 1951, Mesozoic igneous activity in the northern Gulf Coastal Plain: Gulf Coast Association of Geological Societies, v. 1, p. 182–199.

Klitgord, K. D., Dillon, W. P., and Popenoe, P., 1983, Mesozoic tectonics of the southeastern United States Coastal Plain and continental margin: U.S. Geological Survey Professional Paper 1313-P, 15 p.

Lanphere, M. A., 1983, ^{40}Ar/^{39}Ar ages of basalt from Clubhouse Crossroads test hole 2 near Charleston, South Carolina: U.S. Geological Survey Professional Paper 1313-B, 8 p.

Lonsdale, J. T., 1927, Igneous rocks of the Balcones fault region of Texas: Austin, University of Texas Bureau of Economic Geology Bulletin 2744, 178 p.

Lopez Ramos, E., 1981, Geologia de Mexico, 2nd ed., volume 3: Mexico City, 445 p.

Luhr, J. F., and Varekamp, J. C., 1984, Special issue on the 1982 eruption of El Chichon Volcano, Chiapas, Mexico: Journal of Volcanology and Geothermal Research, v. 23, 192 p.

Lyons, P. L., 1957, Geology and geophysics of the Gulf of Mexico: Gulf Coast Association of Geological Societies Transactions, v. 7, p. 1–10.

May, P. R., 1971, Pattern of Triassic–Jurassic diabase dikes around the North Atlantic in the context of predrift position of the continents: Geological Society of America Bulletin, v. 82, p. 1285–1292.

McBride, E. F., Lindemann, W. L., and Freeman, P. S., 1968, Lithology and petrology of the Gueydan (Catahoula) Formation in south Texas: Austin, University of Texas Bureau of Economic Geology Report of Investigation 63, 122 p.

Milton, C., 1972, Igneous and metamorphic basement rocks of Florida: Florida Bureau of Geology Bulletin, v. 55, 125 p.

Milton, C., and Grasty, R., 1969, "Basement" rocks in Florida and Georgia: American Association of Petroleum Geologists Bulletin, v. 53, p. 1483–1493.

Miser, H. D., and Ross, C. S., 1922, Diamond-bearing peridotite in Pike County, Arkansas: U.S. Geological Survey Bulletin 735, p. 279–322.

Mitchell, R. H., and Lewis, R. D., 1983, Peridotite-bearing xenoliths from the Prairie Creek mica peridotite, Arkansas: Canadian Mineralogist, v. 21, p. 59–64.

Monroe, W., 1954, Geology of the Jackson area, Mississippi: U.S. Geological Survey Bulletin 986, 113 p.

Moody, C. L., 1949, Mesozoic igneous rocks of the northern Gulf Coastal Plain: American Association of Petroleum Geologists Bulletin, v. 33, p. 1410–1428.

Moore, G. W., and del Castillo, L., 1974, Tectonic evolution of the southern Gulf of Mexico: Geological Society of America Bulletin, v. 85, p. 607–618.

Morris, E. M., 1987, The Cretaceous Arkansas alkalic province; A summary of petrology and geochemistry, *in* Moffis, E. M., and Pasteris, J. D., Mantle metasomatism and alkaline magmatism: Geological Society of America Special Paper 215, p. 217–234.

Mueller, P. A., and Porch, J. W., 1983, Tectonic implications of Paleozoic and Mesozoic igneous rocks in the subsurface of peninsular Florida: Gulf Coast Association of Geological Societies Transactions, p. 169–173.

Murray, G. E., 1961, Geology of the Atlantic and Gulf Coastal provinces of North America: New York, Harper and Brothers, 692 p.

Neatherey, T. L., and Thomas, W. A., 1975, Pre-Mesozoic basement rocks of the
 Alabama Coastal Plain: Gulf Coast Association of Geological Societies
 Transactions, v. 25, p. 86–99.
Oxley, M. L., Minihan, E., and Ridgway, J. M., 1968, A study of Jurassic
 sediments in portions of Mississippi and Alabama: Mississippi Geological
 Survey Bulletin, v. 109, p. 39–77.
Phillips, J. D., 1983, Paleomagnetic investigations of the Clubhouse Crossroads
 basalt: U.S. Geological Survey Professional Paper 1313-C, 18 p.
Price, J. G., and Henry, C. D., 1984, Stress orientations during Oligocene volcan-
 ism in Trans-Pecos Texas; Timing the transition from Laramide compression
 to Basin and Range tension: Geology, v. 12, p. 238–241.
Price, J. G., Henry, C. D., Barker, D. S., and Parker, D. F., 1987, Alkalic rocks of
 contrasting tectonic settings in Trans-Pecos Texas, in Morris, E. M., and
 Pasteris, J. D., Mantle metasomatism and alkaline magmatism: Geological
 Society of America Special Paper 215, p. 335–346.
Rezak, R., and Tieh, T. T., 1984, Basalt from Louisiana continental shelf: Geo-
 Marine Letters, v. 4, p. 69–76.
Robin, C., and Tournon, J., 1978, Spatial relations of andesitic and alkaline
 provinces in Mexico and Central America: Canadian Journal of Earth
 Sciences, v. 15, p. 1633–1641.
Robinson, E., 1976, Geochemistry of the lamprophyre rocks of the eastern Oua-
 chita Mountains, Arkansas [M.S. thesis]: Fayetteville, University of Arkan-
 sas, 147 p.
Ross, C. S., Miser, H. D., and Stephenson, L. W., 1929, Water-laid volcanic rocks
 of Upper Cretaceous age in southwestern Arkansas, southeastern Oklahoma,
 and northeastern Texas: U.S. Geological Survey Professional Paper 154-F,
 p. 175–202.
Scharon, L., and Hsü, I-C., 1969, Paleomagnetic investigation of some Arkansas
 alkali igneous rocks: Journal of Geophysical Research, v. 74, p. 2774–2779.
Scott, K. R., Hayes, W. E., and Fietz, R. P., 1961, Geology of the Eagle Mills
 Formation: Gulf Coast Association of Geological Societies Transactions,
 v. 11, p. 1–14.
Scott Smith, B. H., and Skinner, E.M.W., 1984, Diamondiferous lamproites:
 Journal of Geology, v. 92, p. 433–438.
Sheridan, R. E., 1974, Atlantic continental margin of North America, in Burk,
 C. A., and Drake, C. L., eds., The geology of continental margins: New
 York, Springer-Verlag, p. 391–407.
Spencer, A. B., 1969, Alkalic igneous rocks of the Balcones province, Texas:
 Journal of Petrology, v. 10, p. 272–306.
Spooner, H. V., Jr., 1964, Basal Tuscaloosa sediments, east-central Louisiana:
 American Association of Petroleum Geologists Bulletin, v. 48, p. 1–21.

Stone, C. G., and Sterling, P. J., 1964, Relationship of igneous activity to mineral
 deposits in Arkansas: Arkansas Geological Commission, 22 p.
Stone, C. G., Howard, J. M., and Holbrook, D. F., 1982, Field guide to the
 Magnet Cove area: Arkansas Geological Commission Guidebook 82-1,
 31 p.
Sundeen, D. A., and Cook, P. L., 1977, K-Ar dates from Upper Cretaceous
 volcanic rocks in the subsurface of west-central Mississippi: Geological So-
 ciety of America Bulletin, v. 88, p. 1144–1146.
Thorpe, R. S., 1977, Tectonic significance of alkaline volcanism in eastern Mexico: Tectonophysics, v. 40, p. T19–T26.
Van Houten, F. B., 1977, Triassic–Liassic deposits of Morocco and eastern North
 America: Comparison: American Association of Petroleum Geologists Bulletin, v. 61, p. 79–99.
Williams, J. F., 1891, The igneous rocks of Arkansas: Arkansas Geological Sur-
 vey Annual Report, v. 2, 391 p.
Zartman, R. E., 1977, Geochronology of some alkalic rock provinces in the
 eastern and central United States: Annual Review of Earth Science, v. 5,
 p. 257–286.
Zartman, R. E., and Howard, J. M., 1985, U-Th-Pb ages of large zircon crystals
 from the Potash Sulfur Springs igneous complex, Garland County, Arkansas:
 Geological Society of America Abstracts with Programs, v. 17, p. 198.
Zartman, R. E., Brock, M. R., Heyl, A. V., and Thomas, H. H., 1967, K/Ar and
 Rb/Sr ages of some alkalic intrusive rocks from central and eastern United
 States: American Journal of Science, v. 265, p. 848–870.
Zoback, M. D., and 15 others, 1980, Recurrent intraplate tectonism in the New
 Madrid seismic zone: Science, v. 209, p. 971–976.

Manuscript Accepted by the Society May 19, 1989

ACKNOWLEDGMENTS

Many colleagues at LSU read drafts of earlier versions of this manuscript and
provided valuable comments. Martha Edick and Rex Pilger were especially helpful. Dan Barker, Dave Dallmeyer, and Jonathan Price formally reviewed the
manuscript and suggested revisions that made the final product much more useful.
Jorge Jacobo-Albarran and Manuel Lopez-Infanzon of the Instituto Mexicano del
Petroleo supplied important unpublished data.

NOTES ADDED IN PROOF

New studies of Cenozoic (48-17 Ma) igneous activity of the Trans-Pecos
region of Texas and northern Mexico have been published in the last several years.
Henry and others (1991) review their field work on the stress evolution and
changes in magma compositions that reflect the transition from compressional
continental arc to extensional intraplate rift along the northwestern margin of the
Gulf of Mexico basin. The transition took place at about 31 Ma, although major
normal faulting did not begin until 24 Ma. Arc volcanism produced a continuous
suite of basalt to rhyolite, moderately alkaline, and low in Nb and Ta (James and
Henry, 1991). As rifting began, volcanism produced bimodal alkali basalt-
rhyolite, and after normal faulting began only alkali basalts formed. The rift-
related volcanic rocks are high in Nb and Ta. Several interesting new papers on
Trans-Pecos volcanic styles have also been published by this group (see Henry and
others, 1991, for references). A detailed field guide to Trans-Pecos volcanics is
also now available (Henry and others, 1989).

Heatherington and Mueller (1991) have published new isotopic and trace
element data on two basalts and a rhyolite of Mesozoic age from southwestern
Florida. Although Ar-Ar and Rb-Sr give Mesozoic ages, the model Nd-Sm ages
are all late Precambrian suggesting involvement of older lithosphere perhaps
similar to that found in northern Florida. High field-strength element compositions suggest a hot spot type of tectonic environment.

ADDITIONAL REFERENCES

Heatherington, A. L., and Mueller, P. A., 1991, Geochemical evidence for Triassic
 rifting in southwestern Florida: Tectonophysics, v. 188, p. 291–302.
Henry, C. D., Price, J. G., Parker, D. F., and Wolff, J. A., 1989, Mid-Tertiary
 silicic alkalic magmatism of Trans-Pecos Texas: Rheomorphic tuffs and
 extensive silicic lavas: New Mexico Bureau of Mines and Mineral Resources
 Memoir 46, p. 231–274.
Henry, C. D., Price, J. G., and James, E. W., 1991, Mid-Cenozoic Stress Evolu-
 tion and Magmatism in the Southern Cordillera, Texas and Mexico: Transi-
 tion from Continental Arc to Intraplate Extension: Journal of Geophysical
 Research, v. 96, p. 13545–13560.
James, E. W., and Henry, C. D., 1991, Compositional Changes in Trans-Pecos
 Texas Magmatism Coincident with Cenozoic Stress Realignment: Journal of
 Geophysical Research, v. 96, p. 13560–13575.

The Geology of North America
Vol. J, The Gulf of Mexico Basin
The Geological Society of America, 1991

Chapter 7

Pre-Triassic

R. D. Woods
4100 Jackson Avenue, Austin, Texas 78731
Amos Salvador
Department of Geological Sciences, The University of Texas at Austin, P.O. Box 7909, Austin, Texas 78713
A. E. Miles
Interpretation Consultants, Inc., 5402 Olympic Fields Lane, Houston, Texas 77069

INTRODUCTION

The thick Mesozoic-Cenozoic fill of the Gulf of Mexico basin was deposited on a floor of Paleozoic and older "basement" that is still poorly known. Fewer than 250 wells out in the basin, away from the structural rim, have penetrated pre-Mesozoic rocks; fewer than a dozen of these wells drilled as much as 1,000 m of this older section. Basinwide, the average penetration of these older rocks is only 220 m, and data from many of the wells are sketchy and incomplete. The pre-Triassic has long been considered "economic basement" by the oil industry, and stratigraphers have paid little attention to these rocks. Whereas a great part of the pre-Triassic rocks was subjected to strong deformation and metamorphism, it is now known that unaltered sediments of Cambrian(?) to Devonian age underlie the Mesozoic in the southeastern U.S. states of Alabama, Georgia, and Florida. Sediments of early Pennsylvanian to Permian age, some possibly older, are known to underlie the Mesozoic in the northwestern part of the basin, and seismic data suggest thicknesses of 5 km or more in some areas for these young Paleozoic sediments. Unmetamorphosed pre-Triassic rocks are also known from the western flank of the Gulf of Mexico basin, in eastern Mexico, and from its southern flank in southern Mexico, Guatemala, and Belize.

One of the earliest recorded penetrations of pre-Triassic "basement" out in the basin was reported in 1928. A well drilled in Marion County, Florida (location 1, Fig. 1), encountered rocks first called "basement schist and quartzite." Reexamination some two decades later revealed the rocks to be dark micaceous shales and quartzites, tentatively correlated on the basis of lithology to rocks dated elsewhere in Florida as early Ordovician in age. Because of the reported nature of the older rocks at the time it was drilled, this well confirmed the thinking of geologists about the pre-Triassic of the Gulf of Mexico basin as economic "basement." Another well, drilled a decade later, gave geologists reason to reconsider their thinking. The Union Producing Company's No. 1 Tensas Delta, Morehouse Parish, Louisiana (location 2, Fig. 1), located approximately 200 km south of the ex-

posed Ouachita structural front, penetrated 363 m of unaltered, relatively flat-lying marine clastic rocks with carbonate stringers instead of the expected metamorphic Ouachita facies. These sediments were first thought to be older Mesozoic but are now known to be Permian in age.

This chapter summarizes the limited and scattered information currently available on the pre-Triassic sedimentary, igneous, and metamorphic rocks that compose the pre-Triassic "basement" floor of the Gulf of Mexico basin. Distribution, lithology, thickness, and other characteristics of the various units currently recognized within the basin, and the basis for dating and correlating these units, are discussed. First, however, as reference and background, a brief summary of the pre-Triassic rock sequences that crop out in the uplifted areas around the structural rim of the Gulf of Mexico basin is included in this chapter. It is hoped this information will enhance the understanding and interpretation of the pre-Triassic "basement" underlying the basin. Some speculations regarding pre-Mesozoic history of the basin, limited by the very sparse geological and geophysical data available, are offered.

PRE-TRIASSIC OUTCROPS AROUND THE RIM OF THE GULF OF MEXICO BASIN

Pre-Triassic rocks are exposed along the northern, western, and part of the southern flanks of the Gulf of Mexico basin (Fig. 1). Precambrian rocks crop out only in the southern Appalachians, in the Llano uplift of central Texas, in two anticlinal structures in east-central Mexico—the Huizachal-Peregrina and Huayacocotla anticlinoriums—and in the southern Mexico state of Oaxaca. Most of the pre-Triassic "basement" rocks surrounding the basin and probably forming most of its "basement" are sedimentary sequences, part of the extensive Paleozoic cover of the North American craton, around the margins of which the cover graded into the sediments of peripheral troughs or basins.

Much of the pre-Triassic "basement" of the Gulf of Mexico

Woods, R. D., Salvador, A., and Miles, A. E., 1991, Pre-Triassic, *in* Salvador, A., The Gulf of Mexico Basin: Boulder, Colorado, Geological Society of America, The Geology of North America, v. J.

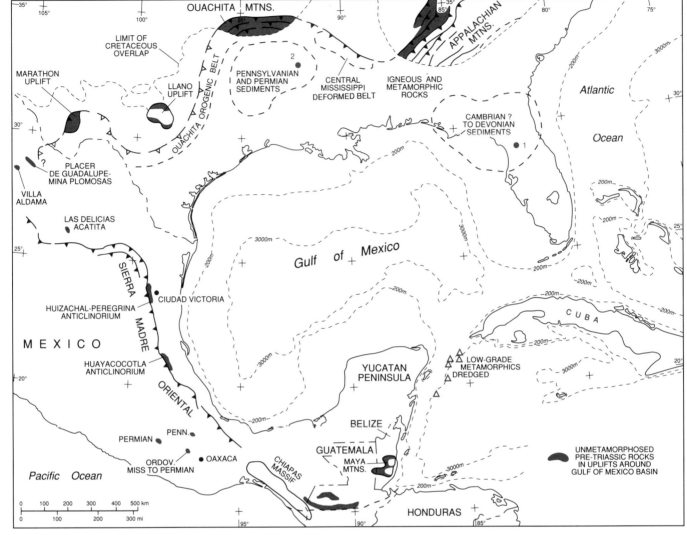

Figure 1. Unmetamorphosed pre-Triassic rocks cropping out in uplifts around the Gulf of Mexico basin.

basin is probably allochthonous, parts of thrust sheets emplaced during the collision of lithospheric plates that took place toward the end of the Paleozoic and that resulted in the formation of the supercontinent of Pangea.

Some lower to middle Paleozoic sediments may still exist beneath these late Paleozoic thrust sheets, but neither well control nor seismic data are adequate yet to determine this. With limited well control, clues to the distribution of lower Paleozoic rocks in the subsurface must be sought in exposures along the basin margin, in Mexico, Texas, Oklahoma, and Arkansas, in the Valley and Ridge outcrops of Alabama, and in a few exposures in Central America. Paleozoic outcrops are absent on the eastern flank of the basin, southeast of the Appalachians; there are no Paleozoic sediments currently recognized at the surface in the Yucatan platform or in the Greater Antilles (Khudoley and Meyerhoff, 1971). Principal outcrop areas of unmetamorphosed Paleozoic rocks in uplifts around the Gulf of Mexico basin are shown in Figure 1.

All systems of the Paleozoic are represented along the basin margin; equivalents should, therefore, be expected somewhere within the basin. Site of the present Gulf of Mexico basin during most of the Paleozoic has usually been interpreted as part of a deep basin or seaway flanking the North American craton to the south. Plate movements during the late Paleozoic increasingly restricted this sea and created uplifts that became sources of clastics for the depositional basins of the time.

Northern margin of the Gulf of Mexico basin

Along the northeast flank of the Gulf of Mexico basin, in Alabama and Georgia, the Piedmont province of the southern Appalachians exposes Precambrian high-grade metamorphics and granites, and the Valley and Ridge province exposes 8,000 to 10,000 m of Cambrian to Pennsylvanian rocks, folded and thrusted but not metamorphosed. A basal Cambrian clastic sequence of variable thickness, and 3,000 m of Cambrian-

Ordovician shelf carbonates and clastic rocks are in the southeastern part of the thrust belt; less than 300 m of Silurian-Devonian sandstones, limestones, and shales, and about 4,200 m of Carboniferous strata, principally clastic rocks containing coal and limestone beds. Limestones are predominant in the Mississippian to the northeast.

No surface exposures of Paleozoic rocks occur along the northern rim of the basin in central Mississippi. The more than 100 wells drilled into Paleozoic rocks along the Central Mississippi deformed belt have penetrated Cambrian to Devonian shelf sediments, principally carbonates, along the eastern half of the belt, and Carboniferous and possibly older, typically "Ouachita flysch facies" argillites and slates, along the western-northwestern part (Morgan, 1970; Thomas, 1972, 1985).

The Ouachita Mountains of southeastern Oklahoma and west-central Arkansas, and the Marathon uplift of west Texas are two large "windows" of the so-called Ouachita orogenic belt, an extensive late Paleozoic thrust belt, more than 1,800 km long, now largely buried below Mesozoic and Cenozoic rocks, that rims and forms the basement of the Gulf of Mexico basin in the subsurface of Texas, Oklahoma, Arkansas, and west-central Mississippi.

The Ouachita Mountains expose 15,000 to 17,000 m of Cambrian(?) to mid-Pennsylvanian rocks. The lower part of the section, of late Cambrian(?) to early Mississippian age, is 3,500 to 4,000 m thick and is composed predominantly of dark carbonates, black shales, cherts, and lesser amounts of sandstone. Siliceous rocks, chert and novaculite, 30 to 200 m thick, are particularly abundant and distinctive in the upper Silurian(?)–Devonian–lower Mississippian section (Arkansas Novaculite). The upper part of the section, of late Mississippian to mid-Pennsylvanian age, consists of as much as 12,000 to 13,000 m of generally thinly interbedded sandstones and shales, usually described as "flysch" ("Carboniferous flysch"). The contact between the Arkansas Novaculte and the "flysch" is sharp, but appears to be conformable, though it may represent a short stratigraphic hiatus.

In the Marathon uplift the exposed Paleozoic section is thinner, and includes units younger than those cropping out in the Ouachita Mountains—6,000 to 6,500 m of Cambrian to Permian rocks. From bottom to top the section is composed of 2,000 to 2,500 m of Cambrian-Ordovician carbonates, shales, and lesser amounts of sandstone; 100 to 330 m of cherts and novaculties of late Ordovician to early Mississippian age (the Maravillas Chert and Caballos Novaculite); and up to 4,000 m of upper Mississippian to mid-Pennsylvanian "flysch"—interbedded sandstones and shales, and including some limestones. The boundary between the Caballos Novaculte and the "flysch," as in the Ouachita Mountains, is sharp but essentially conformable. Approximately 150 m of Lower Permian (Wolfcampian) clastic rocks unconformably overlie the "flysch."

The environment in which the pre-"flysch" sediments of the Ouachita Mountains and the Marathon uplift were deposited—

deep water or shallow shelf—is still controversial (King, 1937, 1975; Flawn and others, 1961; Stone, 1977; McBride, 1978).

The buried part of the Ouachita orogenic belt between the Ouachita Mountains and the Marathon uplift is only known from information supplied by wells drilled along the belt; the information varies greatly in adequacy from one area to another (Flawn and others, 1961; King, 1975). The continuation of the Ouachita orogenic belt into Mexico has been the subject of considerable speculation, as is the nature of the Ouachita-Appalachian connection.

The Llano uplift of central Texas, on the craton edge of the Ouachita orogenic belt, half-way between the Marathon uplift and the Ouachita Mountains, exposes a foreland section more than 1,128 m thick. This section includes 365 m of upper Cambrian to Lower Ordovician sandstones and calcarenites; thin remnants of Upper Ordovician and Silurian limestones, and dark phosphatic Devonian shales including local chert breccias; 30 m of Mississippian dark phosphatic shales and crinoidal limestones; 305 m of lower Pennsylvanian dark, cherty limestones and dark shales; and 428 m of middle to upper Pennsylvanian shales, sandstones, and conglomerates, with thin limestone beds.

For a complete description of the Ouachita margin of the Gulf of Mexico basin, see Hatcher and others, 1989.

Northeastern Mexico

In northeastern Mexico, Paleozoic rocks crop out at the Placer de Guadalupe–Mina Plomosas and Villa Aldama areas, Chihuahua, and in the Las Delicias–Acatita area, Coahuila (Fig. 1).

In the Placer de Guadalupe–Mina Plomosas area, the Paleozoic section is predominantly composed of Ordovician to upper Pennsylvanian or Lower Permian marine limestones, shaly limestones, and shales more than 840 m thick overlain by up to 2,000 m of Permian siltstones, shales, and lesser amounts of shaly limestone ("flysch"). Chert is common, up to 50 percent, in the Devonian part of the section (Bridges, 1964, 1965; Bridges and DeFord, 1961; López Ramos, 1969). A thick section of the Permian "flysch" crops out in the Sierra del Cuervo, near Villa Aldama, about 60 km southwest of the Placer de Guadalupe–Mina Plomosas area. López Ramos (1969) reported a thickness of 2,600 m for this section, but Mellor and Breyer (1981) believed that this apparent large thickness is due to recumbent folding and that the section does not exceed 1,200 m.

In the Las Delicias–Acatita area, a thick, richly fossiliferous Pennsylvanian(?)-Permian section crops out in a window in the flat-lying Cretaceous sediments of the Coahuila Platform. It is composed of about 3,000 m of fine-grained to conglomeratic graywackes, and lesser amounts of shale and limestone. Several igneous bodies, probably sills, lava flows, and dikes are interlayered with or cut the section (King and others, 1944; Wardlaw and others, 1979; López Ramos, 1980). The upper Paleozoic sequence is intruded by Permian-Triassic granodiorites.

East-central Mexico

Farther south, in east-central Mexico, pre-Triassic rocks are exposed in two main areas: the cores of the Huizachal-Peregrina and Huayacocotla anticlinoriums (Fig. 1).

In the Huizachal-Peregrina anticlinorium, west of Ciudad Victoria, the oldest rocks exposed are the Novillo Gneiss of late Precambrian age (Fries and Rincón-Orta, 1965; Denison and others, 1971; Ortega-Gutiérrez, 1978a; Garrison and others, 1980). A Paleozoic sedimentary section as much as 2,000 m thick, first described by Carrillo Bravo (1961), is in fault contact with the Precambrian rocks. The lower 130 to 200 m of this section is unfossiliferous and has generally been assigned to the Cambrian and Ordovician on the basis of its stratigraphic position below fossiliferous Silurian sediments. From bottom to top, it is composed of the La Presa Quartzite (100 to 150 m) and the Naranjal Conglomerate (30 to 40 m). The Naranjal Conglomerate lies unconformably over the La Presa Quartzite and is composed of subrounded to rounded fragments of the underlying La Presa Quartzite, as well as of gneisses similar to those of the Novillo Gneiss. The unconformable relation between the La Presa Quartzite and the overlying Naranjal Conglomerate, the lack of evidence of metamorphism in the Naranjal Conglomerate, and its content of La Presa Quartzite pebbles, indicate that there is a considerable hiatus between these two units. In fact, some authors (e.g., Fries and others, 1962) have considered the La Presa Quartzite to be of Precambrian age.

Above the Naranjal Conglomerate is a marine section, for the most part fossiliferous, composed of sediments of latest Ordovician or Silurian to Permian age. The lowermost unit, the Victoria Limestone, is only 10 m thick, contains poorly preserved, partly recrystallized brachiopods undiagnostic for age determination, and could be of either latest Ordovician or early Silurian age.

Conformably overlying the Victoria Limestone is the Silurian Cañon de Caballeros Formation (80 to 90 m), composed of alternating limestones, shales, and thin sandstone beds; the Devonian La Yerba Formation (about 100 m), predominantly composed of siliceous sediments (chert and novaculite) interbedded with dark shales, sandstones, and some limestones; and the lower Mississippian Vicente Guerrero Formation, 160 to 200 m thick, composed of sandstones with conglomeratic bands and dark shales. The Ordovician(?)-Silurian to Mississippian section is apparently conformable. The siliceous Devonian section is lithologically similar to the Devonian section of the Placer de Guadalupe–Mina Plomosas area and to the section of the same age in the Marathon uplift area of west Texas.

The Vicente Guerrero Formation is overlain unconformably by 200 m of interbedded limestones, sandstones, and shales above a basal conglomerate 1 to 4 m thick—the Del Monte Formation of early Pennsylvanian age. The uppermost part of the Paleozoic section in the Huizachal-Peregrina anticlinorium is composed of more than 1,000 m of interbedded sandstones, siltstones, and shales ("flysch") of early Permian age, and has been called the Guacamaya Formation. Its contact with the underlying Del Monte Formation has nowhere been clearly observed. The Guacamaya Formation unconformably underlies the Upper Triassic–Lower Jurassic red beds of the La Boca Formation, described in Chapter 8 of this volume.

The core of the Huizachal-Peregrina anticlinorium also exposes a 500-m sequence of schists and associated grenstones and serpentinite—the Granjeno Schist—always in fault contact with other units and whose age has been the source of considerable controversy. Carrillo Bravo (1961), Fries and others (1962), and Fries and Rincón-Orta (1965) consider the Granjeno Schist to be of Precambrian age. De Cserna and others (1977) argue in favor of an Ordovician age on the basis of isotopic determinations, its metamorphic grade—lower than that of the Novillo Gneiss—and their observation that no components of the La Presa Quartzite seem to have been derived from the Granjeno Schist, whereas the Naranjal Conglomerate contains pebbles similar to the Granjeno. Denison and others (1971), Ramírez-Ramírez (1978), Garrison (1978), and Garrison and others (1980) assigned a Carboniferous-Permian age to the time of the metamorphism of the Granjeno Schist. De Cserna and others (1977), as well as Denison and others (1971), Ramírez-Ramírez (1978), Garrison (1978), and Garrison and others (1980), agreed, however, that the Granjeno Schist and the associated serpentinite are probably allochthonous to the Ciudad Victoria area and that they have been transported into the area as a result of regional orogenic events.

In the core of the Huayacocotla anticlinorium, 300 km south of the Huizachal-Peregrina anticlinorium, the pre-Triassic is represented by the Precambrian Huiznopala Gneiss (Fries and Rincón Orta, 1965) and the Permian Guacamaya Formation (Carrillo Bravo, 1965). The two units are always in fault contact. Older Paleozoic rocks do not crop out in this area but they probably underlie it.

The Guacamaya Formation in the Huayacocotla anticlinorium, as in the Ciudad Victoria area to the north, is predominantly composed of interbedded, generally thinly bedded conglomerates, sandstones, siltstones, and shales with occasional limestone beds ("flysch"). It attains a thickness of more than 3,345 m and has been dated as Early Permian, mainly on the basis of its contained fusuline fauna (Carrillo Bravo, 1965; Pérez Ramos, 1978).

Southern margin of the Gulf of Mexico basin

Metamorphic and intrusive igneous rocks of Precambrian and Paleozoic age border the Gulf of Mexico basin to the south, in the Mexican states of Puebla, Oaxaca, and Chiapas, and in Guatemala (Dengo and Bohnenberger, 1969; Kesler and others, 1970; Rodriguez-Torres, 1970; Ruiz-Castellanos, 1970; Kesler, 1971, 1973; Pantoja-Alor and others, 1974; Dengo, 1975; Ortega-Gutierrez, 1978b, 1981 [1984]; López Ramos, 1983) (Fig. 1). Low-grade metamorphic rocks—phyllite and marble—have been dredged from the lower part of the continental slope east of Yucatan (Fig. 1) (Vedder and others, 1973; Dillon and Vedder, 1973; Pyle and others, 1973). Unmetamorphosed Paleozoic sediments have been reported from southernmost Puebla, northeastern Guerrero, Oaxaca, Chiapas, Guatemala, and from the Maya Mountains of Belize (Fig. 1).

From southern Puebla, Silva Pineda (1970) described a Pennsylvanian fossil flora from a section about 600 m thick composed of sandstones and interbedded dark shales and coal beds that crops out 25 km south-southwest of Tehuacán.

In northeastern Guerrero, Corona-Esquivel (1981 [1983]) mapped an Upper Permian sequence, 450 to 775 m thick, composed of alternating sandstones, siltstones, and shales with occasional conglomerates and an 80–140-m limestone section in its middle part.

In Oaxaca, unmetamorphosed Paleozoic sedimentary rocks are known from two small areas about 70 km northwest of the city of Oaxaca (Pantoja-Alor and Robison, 1967; Robison and Pantoja-Alor, 1968; Pantoja-Alor, 1970). They overlie Precambrian or early Paleozoic metamorphics. An older Ordovician sequence, up to 200 m thick, is composed of a lower limestone unit and an upper unit of interbedded sandstones and shales. This unit is unconformably overlain by about 650 m of sandstones and shales with lesser amounts of limestone. This younger sequence, which becomes more shaly toward the top, has been dated as Mississippian, Pennsylvanian, and possible Permian in age.

More extensive exposures of unmetamorphosed or weakly metamorphosed Paleozoic rocks are known from the southeastern part of Chiapas (southern Mexico), from Guatemala, and from Belize (Fig. 1). The unmetamorphosed Paleozoic section in these three areas includes rocks of only late Paleozoic age (possible Mississippian, Pennsylvanian, and Permian) and is for the most part of fairly uniform lithologic composition from southern Mexico to Belize. It is assumed to rest on older Paleozoic metamorphic rocks, but the contact has not been observed anywhere.

The lower part of the section is composed of coarse-grained clastics—sandstones, siltstones with occasional conglomeratic horizons—and lesser amounts of shale. It may locally show low-grade metamorphism. Thicknesses between 1,000 and 3,000 m have been reported from Chiapas and Guatemala. This lower coarse clastic section has been assumed to be of Pennsylvanian age on the basis of its stratigraphic position below well-dated late Pennsylvanian to Permian sequences.

Overlying the lower coarse clastics is a section predominantly composed of shales and siltstones, with some sandstones and occasional, generally discontinuous, limestone beds. The limestone beds increase in number and thickness toward the top, while the sandstones become less abundant. This shaly section is about 500 to 1,300 m thick, and has been dated as of latest Pennsylvanian to Early Permian age on the basis of fusulines collected from the limestone beds.

The upper part of the Paleozoic section in Chiapas and Guatemala is composed of thick-bedded to massive, cliff-forming limestones. This unit appears to be absent or poorly represented in the Paleozoic section of the Maya Mountains of Belize. In Guatemala it ranges in thickness between 500 and 1,000 m, but it may reach close to 2,000 m in Chiapas. Principally on the basis of its contained fusulines, the upper limestone section has been assigned an early to middle Permian age. No upper Permian rocks have been reported from southern Mexico, Guatemala, or Belize.

In Guatemala and Belize, the entire upper Paleozoic section has been included in the Santa Rosa Group. No formation units have been proposed for the Santa Rosa Group in Belize; however, a sequence of acidic lavas and tuffs in the middle part of the group has been called the "Bladen volcanic member." For more detailed descriptions of the Paleozoic section of the Maya Mountains, see Dixon (1956), Bateson and Hall (1971, 1977), Hall and Bateson (1972), and Bateson (1972). In Guatemala, the Santa Rosa Group has generally been subdivided into four formations: the lower coarse-clastic part has been called the Chicol Formation, the middle shaly part has been included in the Tactic Formation below and the Esperanza Formation above, and the upper massive limestone unit has been called the Chochal Formation. For a more comprehensive discussion of these units, see Walper (1960), Clemons and Burkart (1971), Anderson and others (1973), and Clemons and others (1974).

In Chiapas, southern Mexico, the name "Santa Rosa" has only been applied to the lower coarse-clastic part of the section and to a thick low-grade metamorphic section of phyllites, slates, and sandstones, which underlies it unconformably and from which fossils of late Mississippian age have been reported (Hernández García, 1973; López Ramos, 1969, 1983). The lower metamorphic unit has been called the "Santa Rosa Inferior" ("Lower Santa Rosa"), and the upper clastic unit the "Santa Rosa Superior" ("Upper Santa Rosa"). The middle, predominantly shaly part of the upper Paleozoic section of Chiapas has been called the Grupera Formation, and the upper, massive limestone unit the Paso Hondo Formation (including at its base the La Vainilla Limestone of some authors). For a more detailed discussion of the upper Paleozoic section of southern Mexico, see Thompson and Miller (1944), Gutiérrez Gil (1956), Thompson (1956), Hernández García (1973), and López Ramos (1969, 1983).

The upper Paleozoic section of the Maya Mountains of Belize is intruded by granites of late Permian to mid-Triassic age. Similar granitic intrusive rocks in Guatemala are probably also of Permian-Triassic age.

PRE-TRIASSIC UNDERLYING THE GULF OF MEXICO BASIN

Most of the information about the pre-Triassic rocks underlying the Gulf of Mexico basin comes from the northeastern, northern, and western flanks of the basin, where most of the drilling reaching these rocks has taken place. Less information is available from the southern flank, in southern Mexico and Central America. For convenience, the discussion of the pre-Triassic rocks of the Gulf of Mexico basin will be divided into three parts: one dealing with the northeastern part of the basin (Alabama, Georgia, Florida, and adjacent parts of the Gulf of Mexico, roughly east of the projection into the basin of the Appalachians); another covering the subsurface of Mississippi, Arkansas, Louisiana, and Texas; and a third dealing principally with Mexico. Plate 3 of this volume (in pocket) shows the distribution of the pre-Triassic rocks underlying the Gulf of Mexico basin.

Subsurface of the northeastern Gulf of Mexico basin

Wells in the southeastern United States have penetrated a variety of pre-Triassic rocks, including relatively flat-lying Cambrian(?)-Ordovician to Devonian clastic strata, metamorphic rocks ranging from low-grade metasediments to gneiss, and igneous intrusions and extrusions dated from Precambrian(?) to late Paleozoic (Fig. 2). For additional information, see Milton and Grasty (1969), Milton (1972), and Puri and Vernon (1964).

Wells drilled into unmetamorphosed Paleozoic strata in the southeastern states are clustered principally in northern Florida, along the crest of the Peninsular arch, with a small group of wells in southeastern Georgia (Applin, 1951; Applin and Applin, 1965; Barnett, 1975). The unmetamorphosed Paleozoic sediments underlying the Cretaceous in northern Florida seem to form a gentle large syncline probably broken up by a complex fault system (Barnett, 1975). In addition, two wells on the northwest flank of the Apalachicola embayment have reached Paleozoic sediments, one in Houston County, southeastern Alabama, the other in Early County, southwestern Georgia, and Paleozoic sediments have been reported from a few wells in the eastern Florida panhandle. However, no well in the southeastern states has yet drilled a complete sequence of dated strata from Cambrian(?)–Lower Ordovician through Devonian. Maximum penetration has been close to 300 m, while average penetration of these older rocks in Florida is only 91 m, and 76 m in Alabama and Georgia. The total thickness of the section can, therefore, only be approximated. Wicker and Smith (1978), Smith (1982), and Chowns and Williams (1983), using both geophysical and well data, have estimated a total thickness of 2,000 to 2,500 m, of which approximately 1,900 m are of Ordovician age, and 600 m are Silurian and Devonian. A general summary of the stratigraphy, lithology, and environments of deposition of the Paleozoic sediments penetrated in the northeastern Gulf of Mexico Basin follows.

The Union-Kirkland well, Houston County, Alabama (location 1, Fig. 2), drilled in 1949, is one of the key wells in dating subsurface Paleozoic rocks of the southeastern states. Lower Ordovician graptolites and a Middle Ordovician trilobite (Pojeta and others, 1976) were recovered in this well from black shales and interbedded laminated sandstones. These sediments also contained *Skolithos* worm(?) borings, which have been reported to be widespread in Cambrian-Ordovician sediments in Europe, west Africa, and North America. This lithology—black shales and gray quartzitic, medium- to fine-grained sandstones with *Skolithos* borings—has been used to tentatively date 22 nonfossiliferous similar sections in wells in Florida and Georgia (Bridge and Berdan, 1951).

The Mont Warren–Chandler well, Early County, Georgia

(location 2), drilled in 1943, is another key well; the samples from this well have been studied by a number of specialists. On the basis of analysis of ostracodes, pelecypods, plant fragments, and acid-insoluble microfossils, ages ranging from Late Ordovician to early Pennsylvanian have been assigned to the section penetrated by this well (Applin, 1951; Swartz, 1945; McLaughlin, 1970; Pojeta and others, 1976). The most recent studies of this 219-m section of black shales, silty shales, and siltstones indicate that these beds are pre-Carboniferous in age, probably Devonian.

Paleozoic wells in southeastern Georgia are clustered in Clinch, Echols, Charlton, and Camden counties. Of the 11 wells in this area, 4 have paleontological evidence for dating the sections. The remainder have been tentatively assigned ages based on lithological similarities to the dated sections. Three of the Hunt–Superior Pine Products wells (location 3, Fig. 2) in Echols County contained Ordovician brachiopods and conodonts in black shales with interbeds of gray quartzitic sandstones. The thickness of the Paleozoic strata penetrated in these wells ranges from about 1 m to 105 m. The depositional environment for these sediments has been interpreted as offshore marine. The Humble-Bennett Langsdale well (location 4) in Echols County penetrated 23 m of dark shales with considerable interbedded sandstones beneath a well-developed weathered zone. The sandstones were cross-bedded with horizontal worm(?) tubes, unlike the vertical *Skolithos* borings characteristic of Lower Ordovician sediments elsewhere. Acid-insoluble microfossils have dated this section as of Late Silurian to Early Devonian age (Pojeta and others, 1976). The depositional environment has been interpreted as shallow shelf.

Paleontological data have been published for only 10 of about 70 wells in northern Florida where ages of Ordovician, Silurian, and Devonian have been assigned to the sediments. Approximately half of the "unfossiliferous" sections have been referred to the Lower Ordovician on the basis of lithological similarities to the paleontologically dated sections. Vertical *Skolithos* borings and fragments of linguloid brachiopods have been noted in several of these "undated" sections. Use of acid-insoluble microfossils has been limited in the analysis of the sections penetrated in these wells. It is possible that more sections could be dated if these techniques were more widely applied.

Among wells in northern Florida that have penetrated subsurface sections assigned to the Ordovician is the Hunt-Gibson 2, Madison County (location 5, Fig. 2), which was drilled through 231 m of predominantly black shales with interbedded gray or white, coarse-grained, quartzitic sandstones. The shales contained phosphatic brachiopods, *Conularia* sp., and a Middle Ordovician trilobite. Two other similar, but shorter, fossiliferous sections were penetrated in the Humble-Taylor well (location 6) and the Sun-Odom well (location 7), both in Suwannee County. The Humble-Taylor well yielded Middle to Late Ordovician conodonts; the Sun-Odom yielded similar conodonts and, in addition, phosphatic brachiopods and Middle Ordovician chitinozoans (Andress and others, 1969). The depositional environment for these sediments has been interpreted as shallow-water marine.

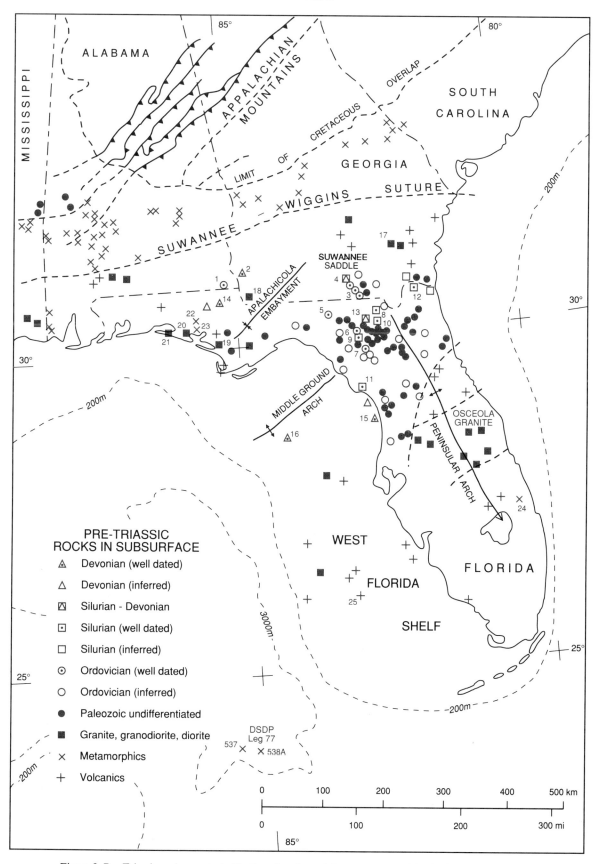

Figure 2. Pre-Triassic rocks penetrated in the subsurface of the eastern part of the Gulf of Mexico basin.

Wells penetrating fossiliferous Silurian strata in northern Florida include the Humble-Cone (location 8), the Gulf–Kie Vining (location 9), and the Sun-Sapp (location 10), all in Columbia County; the Coastal-Ragland in Levy County (location 11); the St. Mary's River Corp.–Hilliard Turpentine in Nassau County (location 12); and the Sun-Tillis, in Suwannee County (location 13). Maximum penetration of Silurian rocks to date, 293 m of dark gray to black shales and silty shales, was in the Humble-Cone well. Similar rocks were encountered in all the wells that are believed to have penetrated Silurian sections. The Silurian sediments have been interpreted to have been deposited in a shallow, nearshore to open-marine, low-energy environment adjacent to land areas of low relief. Chitinozoans found in all these wells indicate a middle to late Silurian age (Goldstein and others, 1969). Late Silurian pelecypods were found in the section penetrated by the Coastal-Ragland well. Other fossils reported include brachiopods, crinoid columnals, ostracodes, orthoconic cephalopods, acritarcs, algae, miospores, tasmanites, and arthropod fragments. A new species of eurypterid was recovered in the Gulf–Kie Vining well (Kjellesvig-Waering, 1950). Some aspects of the fauna from the Sun-Tillis well suggest Late Silurian or Early Devonian age. Fossils in the Humble-Cone well indicate a latest Silurian age, probably just below the Silurian-Devonian boundary (Pojeta and others, 1976).

The St. Mary's River Corp.–Hilliard Turpentine well, Nassau County, reportedly contained an ostracode which R. H. Bassler (*in* Schuchert, 1943, p. 454) identified as *Amphissites,* a genus found in Devonian to Mississippian rocks. This microfossil led Bassler to suggest a correlation of this section with the Chattanooga Shale. Esther Applin (letter to Schuchert, July 5, 1941), in a first look at the samples from this well, called the section "probably Pennsylvanian." Cole (1944) could find no authentic fossils in the samples and concluded that the section might be Triassic. J. W. Wells and R. D. Moore, both of whom also examined the samples, called the section unfossiliferous and concluded the "*Amphissites*" was probably a concretion. The age of these sediments remained uncertain until Cramer (1973) found Middle to Late Silurian chitinozoans in the samples. This brief history of the dating of one well is included to highlight the uncertainty of some of the age determinations and the need to apply newer techniques such as the examination of acid-insoluble microfossils in a restudy of all these older well samples.

A Devonian section was penetrated in the Humble-Tindel well, Jackson County, Florida (location 14, Fig. 2). This well drilled 245 m of reddish-brown to gray shale, siltstone, and cross-bedded sandstones, interpreted to be terrestrial deposits. Plant fragments of a type common to the Lower to Middle Devonian, and typical Devonian acid-insoluble microfossils date this section. The Mobil–State Lease 224A, Levy County (location 15), reportedly also had paleontological evidence for a Devonian age for the penetrated section. Devonian sandstones and shales were reported in Texaco's Block 252 well (location 16 in the gulf, offshore Florida) drilled on the Middle Ground arch to a total depth of 4,774 m.

No sediments younger than Devonian have yet been reported from the Paleozoic of the subsurface of Alabama, Georgia, and Florida, and none may be present. A widespread, peneplaned erosional surface over the complexly structured lower to middle Paleozoic rocks suggests that this part of the Gulf of Mexico basin area was undergoing erosion during the late Paleozoic. The Afro-European affinities of the fauna and flora of the lower and middle Paleozoic section, as well as the absence of upper Paleozoic deposits over the Florida Peninsula and other geologic features of the area have led a number of geologists to conclude that the Paleozoic of this region is a remnant of west Africa welded to North America in the late Paleozoic. The remnant was left behind when the North American and African continental plates separated and the Atlantic reopened in Late Triassic or Early Jurassic time (Wilson, 1966; Andress and others, 1969; Goldstein and others, 1969; Cramer, 1971, 1973; Rankin, 1975; King, 1975; Smith, 1982; Chowns and Williams, 1983; Nelson and others, 1985a, 1985b; Dallmeyer, 1987; Dallmeyer and others, 1987; Opdyke and others, 1987; Thomas and others, 1989; and many others). An Alleghenian suture of not-yet well-defined location has been proposed to separate the North American plate on the north from the welded terranes of African affinities to the south (Fig. 2) (Chowns and Williams, 1983; Nelson and others, 1985a; Thomas and others, 1989). It has been called the Suwanee-Wiggins suture by Thomas and others (1989).

Northwest of the area where unmetamorphosed Paleozoic sediments have been penetrated in the subsurface of southeastern Alabama, southern Georgia, and northern Florida, wells drilled below Mesozoic sediments have encountered metamorphic and igneous rocks similar to those known in the southern Appalachians.

In Alabama, about 60 wells are reported to have bottomed in metasedimentary rocks—slate, phyllite, schist, gneiss, mylonite, and amphibolite—as well as volcanics and granite (Fig. 2). In texture, structure, and lithology, these rocks resemble those exposed in the Appalachian Piedmont province and the flanking foreland folded and thrust-faulted belts of the southern Appalachians. Data on the subsurface pre-Triassic rocks in Alabama have been summarized by Thomas (1973), Neathery and Thomas (1975), and Thomas and others (1989), who grouped the penetrated rocks into four provinces that represent the southwestward extension of similar provinces of the Appalachian complex beneath the younger sediments of the Gulf coastal plain.

In Georgia, 29 wells are reported to have bottomed in metasedimentary rocks—schists and gneisses—granodiorites, rhyolites, and tuffs. Most of these are shallow and within a few kilometers of the exposed Piedmont province (Milton and Hurst, 1965; Chowns and Williams, 1983). Hurst (1960) grouped the pre-Cretaceous rocks in the subsurface of Georgia into four broad zones subparallel to the Piedmont. Zone 1, closest to the Piedmont, includes shallow penetrations of diorite, schist, gneiss, and similar igneous and metamorphic rocks. Wells in zone 2 typically bottom in red beds and diabase compositionally similar to Triassic rocks exposed along the eastern flank of the Appalachians (see

Chapter 8 of this volume). The deepest wells in zone 3 penetrate basalts, rhyolites, and tuffs that have been called "basement." Chowns and Williams (1983) assigned an age of "Proterozoic Z to lower Paleozoic" to this terrane of felsic volcanic rocks associated with granite plutons beneath the Suwannee saddle. Similar, possibly correlative rocks occur within the Apalachicola embayment and on the east side of the Peninsular arch in Florida. Wells in Hurst's zone 4 bottomed in Paleozoic sandstones and shales. Two wells in Pierce County, southeastern Georgia (location 17, Fig. 2), reported to have bottomed in "granite," may have penetrated altered intrusives. Milton and Hurst (1965) interpreted these rocks as either contact-metamorphosed arkose or hydrothermally altered intrusives. Chowns and Williams (1983), in a study of a similar "granite" in Seminole County, Georgia (location 18, Fig. 2), and a restudy of the Pierce County well samples, concluded, however, that all three of these wells penetrated true magmatic granites.

"Granites" penetrated in the western Florida panhandle (Gulf, Bay, and Walton counties, locations, 19, 20, and 21, Fig. 2) have been assigned an early Paleozoic (Cambrian) age by Barnett (1975), probably on the basis of their petrographic similarities to granitic rocks of this age in central Florida. The meta-arkoses and quartzites in Washington County (locations 22 and 23) have been considered to be of Cambrian or late Precambrian age. Arthur (1988) reported "an unpublished K-Ar (feldspar) age determination of 709 ± 25 Ma (Precambrian) from a granodiorite in Gulf County" (location 19, Fig. 2) that "suggests that the panhandle igneous complex may be older than that of central Florida."

South of the area underlain by unmetamorphosed Paleozoic sedimentary rocks, wells have penetrated a pre-Triassic "basement" below the Cretaceous section composed of igneous and metamorphoric rocks of variable composition (Fig. 2). To the north, the "basement" is composed of calc-alkalic, felsic volcanic, and metavolcanic rocks (andesites and rhyolites) that Barnett (1975) called the "volcanic and metamorphic complex," and to which he assigned a late Precambrian to early Cambrian age. Farther to the south, a felsic to intermediate plutonic body composed of granite, alaskite, and diorite is below the Cretaceous. It has been called the Osceola Granite. Isotopic dating has yielded a mid-Cambrian age (535 to 527 Ma) for this granitic body (Bass, 1969; Dallmeyer and others, 1987).

South of the Osceola Granite, at least one well (location 24, Fig. 2) cored metamorphic rocks (gneiss, schist, and amphibolite) below the Cretaceous section. They are overlain by felsic igneous rocks and Mesozoic basalt (Bass, 1969; Chowns and Williams, 1983; Thomas and others, 1989). Bass (1969) reported a Cambrian age (530 Ma) for the gneiss, suggesting that the metamorphic rocks (called the St. Lucie Metamorphic Complex) are roughly contemporaneous with the Osceola Granite. Other wells south of the Osceola Granite have penetrated felsic volcanic rocks similar to those farther north.

The relation between the Osceola Granite and adjacent volcanic and metamorphic rocks is uncertain (Bass, 1969). Barnett

(1975) considered the granite as intrusive into his "volcanic and metamorphic complex." Chowns and Williams (1983) believed it more likely that the felsic volcanic rocks are younger than the Osceola Granite and the St. Lucie metamorphics and that they are separated from them by either a northwest-dipping fault or an unconformity.

Wells drilled in the West Florida Shelf, west of the coast of southern Florida, are reported to have bottomed in volcanic rocks and granite. One of them, Tenneco's #672 (location 25, Fig. 2), drilled into a rhyolite porphyry isotopically dated at 316 ± 11 Ma (Pennsylvanian).

Conclusive evidence of the presence of Precambrian rocks under peninsular Florida and the West Florida Shelf has not yet been reported. The pre-Triassic igneous and metamorphic rocks forming the "basement" of these areas are overlain in their southern parts by an assemblage of diabases, basalts, and rhyolites of Early Jurassic age (see Chapter 6, this volume).

Farther south, in the southeastern Gulf of Mexico, between the Florida and Yucatan platforms, Deep Sea Drilling Project (DSDP) holes 537 and 538A reached "basement" rocks beneath the Lower Cretaceous section (Buffler and others, 1984) (Fig. 2). At DSDP site 538A, on the Catoche Knoll, the drill penetrated 64 m of gneiss and amphibolite intruded by partly serpentinized diabase dikes. Isotopic determinations by Dallmeyer (1984) yielded a Cambrian-Ordovician age (500 Ma) for the gneiss and an Early Jurassic age (190 Ma) for the diabase dikes. At site 537, the hole bottomed in 19 m of phyllite also dated as Cambrian-Ordovician (500 Ma).

Subsurface of the northwestern Gulf of Mexico basin (Mississippi, Arkansas, Louisiana, and Texas)

Information on pre-Triassic rocks under the northwestern Gulf of Mexico basin is limited. Fewer than 12 wells on the basinal flank of the Ouachita orogenic belt and the Ouachita Mountains, clustered in a few areas in Arkansas, Louisiana, and Texas, have penetrated these rocks. Only those wells located along the rim of the basin have drilled into pre-Pennsylvanian sediments. The nature of the lower to middle Paleozoic within the basin, if present, can only be surmised based on study of the sections encountered in the few deep wells drilled along the structural rim of the basin and of the limited and widely spaced outcrops of pre-Pennsylvanian rocks around its periphery.

Little is currently known about the pre-Triassic in southern Mississippi, south of the northwesterly trending Central Mississippi deformed belt (Fig. 3). Rainwater (1962) estimated the total Paleozoic section, Cambrian to Pennsylvanian, within the Black Warrior basin, north of the belt, to be more than 5,100 m thick. Wells drilled along the southeastern part of the Central Mississippi deformed belt have encountered Cambrian to Devonian carbonates and some Carboniferous clastics similar to those known from the Appalachians and the Black Warrior basin, and upper Paleozoic clastic rocks ("Ouachita facies") in its northwestern part (Morgan, 1970; Thomas, 1972, 1985). No transition between the

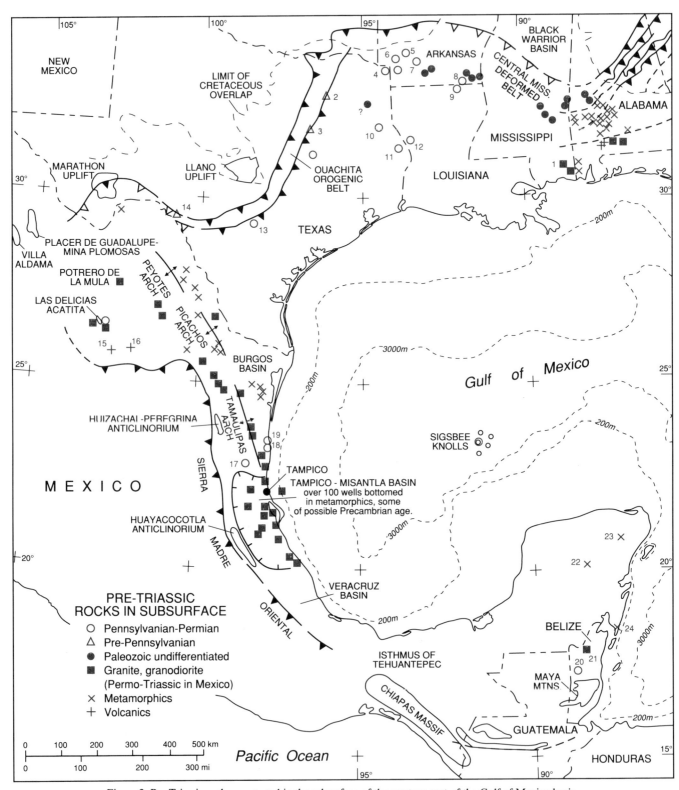

Figure 3. Pre-Triassic rocks penetrated in the subsurface of the western part of the Gulf of Mexico basin.

two has been reported. Equivalents of at least some parts of the Paleozoic section of the Central Mississippi deformed belt should be present south of the belt beneath the Mesozoic-Cenozoic cover. To the limit of control southwestward from the Central Mississippi deformed belt, the top of the Paleozoic dips basinward at a rate of approximately 100 m/km, and in southwestern Mississippi it may lie below 10 to 12 km.

Two wells in Jackson County, southeastern Mississippi, penetrated granite, possibly part of the buried Appalachian piedmont (location 1, Fig. 3). The granite in one of these wells, the Amoco-Saga No. 1 Cumbist, has been isotopically dated as Permian (272 ± 10 Ma) by Harrelson and Bicker (1979).

The Paleozoic subcrop west of Mississippi is known from only a few areas in north Louisiana, south Arkansas, and northeast Texas (Fig. 3).

Most of the wells that have reached pre-Triassic rocks in these areas are located along the basinal flank of the Ouachita orogenic belt and the Ouachita Mountains, near or within the Late Triassic–Early Jurassic graben system that parallels the basin rim. Several of them penetrated thick clastic sections of Upper Triassic–Lower Jurassic Eagle Mills Formation before drilling into Paleozoic rocks. The known pre-Triassic subsurface section comprises a sequence of lower Pennsylvanian to middle Permian clastics and carbonates, deposited in nearshore to middle-shelf, mostly marine, environments. Nicholas and Waddell (1982) described the basin in which these sediments were deposited as a shallow-marine successor, episutural basin developed on the Ouachita orogen. No well within the northwestern Gulf of Mexico basin has been drilled deep enough to encounter reliably dated pre-Pennsylvanian rocks. Only two wells drilled along the structural rim of the basin—the Hunt-Neely (location 2, Fig. 3) and the Shell-Barrett (location 3)—penetrated thick, slightly altered sections of Cambrian-Ordovician carbonates. Equivalents of these older rocks may underlie some of the upper Paleozoic section already drilled within the basin.

The Pennsylvanian-Permian section penetrated by wells in northeast Texas, southern Arkansas, and northern Louisiana is composed of about 1,100 m of middle to upper Pennsylvanian (Desmoinesian to Virgilian) dark marine clastics, biostromal carbonates, and sandstones; more than 930 m of lower Permian (Wolfcampian) dark gray shales, sandstones, and thin limestone stringers, with a few coal seams, and red beds near the top of the section; and about 360 m of middle Permian (Leonardian-Guadalupian) dark gray marine clastic rocks, including sandstones and thin dark gray carbonate layers. Unconformably below this section are more than 930 m of lower Pennsylvanian (Atokan and possibly older) clastics with a few thick beds of sandstone. This older section of dark clastics with a sparse fauna represents a deep-water deposit, probably a turbidite. This sequence of Pennsylvanian and Permian sediments has been reconstructed from stratigraphic information provided by the following wells.

The Wirick-Rochelle well (location 4, Fig. 3) in northeastern Bowie County, Texas, drilled a thin section of Upper Trias-

sic(?) clastics before penetrating 1,052 m of lower to middle Pennsylvanian strata. This section included 472 m of middle Pennsylvanian (Desmoinesian) dark gray to black shales, silty shales, siltstones, and gray, fine-grained sandstones, with some thin carbonate stringers, and 580 m of dark shales, siltstones, and sandstones, thought to be early Pennsylvanian (Morrowan-Atokan) in age.

The Humble-Royston well (location 5) in Hempstead County, Arkansas, approximately 65 km northeast of the Wirick-Rochelle well, penetrated 369 m of middle Pennsylvanian dark shales, sandy shales, sandstones, siltstones, and light-colored biostromal carbonates in the upper part of the section. Below this, the well drilled 102 m of black, hard "Ouachita-facies" shales and quartzitic sandstones with dips of 70°. Depositional environments of the middle Pennsylvanian section reflect a gradual basin subsidence with a concomitant marine transgression. Strata immediately above the unconformity on top of the "Ouachita-facies" rocks were mud-flat and marsh, black, carbonaceous shales with some coaly beds. These deposits gave way upward to nearshore beds, then to an open-marine facies. Beds of light gray and tan, mixed-grain, skeletal, pelletal, biostromal carbonates as much as 75 m thick, compose the upper part of the section. Calcite- and dolomite-filled fractures and vugs partly filled by calcite and dolomite cement occur within the carbonates. This upper marine section contains an abundant fauna, including fusulines that date the strata as middle Pennsylvanian (Desmoinesian). Also present are corals, algae, linguloid brachiopods, scolecodonts, gastropods, bryozoans, crinoid stems, sponge spicules, and echinoid spines. The abundant algae occur as masses, lumps, grains, and as encrusting-binding filaments. Acid-insoluble microfossils, leaf fragments, seeds, and plant stems of early Desmoinesian and possibly Atokan age are found in the lower part of the section.

The Ray-Cox well (location 6, Fig. 3), west of the Humble-Royston, drilled 382 m of nearshore middle to upper Pennsylvanian clastics, then penetrated 928 m of lower Pennsylvanian dark gray to black shales, siltstones, and gray fine-grained quartzitic sandstones exhibiting dips to 30° in some of the cores. The Chevron-Land well (location 7), in Nevada County, Arkansas, southeast of the Humble-Royston, penetrated 1,159 m of Upper Triassic–Lower Jurassic Eagle Mills red beds before drilling through 264 m of Lower Permian (Wolfcampian) marine shales and sandstones, underlain by 1,100 m of middle to upper Pennsylvanian (Desmoinesian to Virgilian) marine shelfal shales, limestones, and sandstones. Middle to upper Pennsylvanian carbonates have been observed in several other wells along the northwestern flank of the Gulf of Mexico Basin and as far south out into the basin as in the Shell-Templeton well (location 11), in Sabine County, Texas.

The Humble–Georgia Pacific well (location 8, Fig. 3) in Ashley County, Arkansas, some 175 km southeast of the Humble-Royston, penetrated 929 m of Lower Permian (Wolfcampian) strata lying unconformably over 1,283 m of middle Pennsylvanian (Desmoinesian) rocks. The Permian section comprised shallow-shelf, variegated, but mostly dark-colored clastics, and light to dark-colored shallow-water biostromal carbonates in beds up to 2.5 m thick, with an abundant fauna of crinoid stems, echinoid spines, algae, bryozoans, ostracodes, and spores. The Desmoinesian dark gray to black carbonaceous shales, siltstones, and tight, fine-grained sandstones contained only a sparse fauna and microflora. The few cores cut showed dips of 15° to 20°. Depositional environments from the lower part of the section reflect an upward change from nearshore, deltaic swamp, to shallow shelf. Several thin diabase dikes were noted in this lower section (Woods and Addington, 1973).

The Union–Tensas Delta well (location 9), in Morehouse Parish, Louisiana, located approximately 21 km south-southwest of the Humble–Georgia Pacific, penetrated 363 m of dark carbonaceous silty shales and siltstones, interbedded with gray sandy shales, some dark sandy and shaly limestones, and some red shales the upper part of the section. There were no apparent dips in the few cores cut, and the sediments were unaltered. This section was first though to be older Mesozoic on the basis of the presence of a sponge more similar to Mesozoic than to Paleozoic forms (Imlay, 1941). Examination of the macrofauna in more detail, mostly a study of external and internal molds of pelecypods and gastropods, led Imlay and Williams (1942) to conclude that the section was of late Paleozoic age, probably no older than Pennsylvanian. Initial examination of the microflora by Hoffmeister and Staplin (1954) prompted them to assign a middle to late Pennsylvanian age to this section, which has been designated the type section for the Morehouse Formation. In 1965, with more detailed information on palynomorphs and improved recovery procedures, Exxon palynologists concluded that the Desmoinesian spores noted by Hoffmeister and Staplin were reworked. A Permian (Leonardian-Guadalupian) microflora, not previously identified, permitted the designation of the upper two-thirds (264 m) of the section as Guadalupian and the lower third as Leonardian. Depositional environments within this section range upward from nearshore to shallow marine.

The wells just cited provide clues to the late Paleozoic depositional history along the northern margin of the Gulf of Mexico Basin. The Humble-Royston with middle Pennsylvanian (Desmoinesian) beds deposited on eroded "Ouachita facies," the Chevron-Land and Humble–Georgia Pacific wells, with Lower Permian (Wolfcampian) above middle to upper Pennsylvanian, and the Union–Tensas Delta well with its middle Permian section suggest a south-southwestward subsidence for this late Paleozoic basin with increasingly younger sediments deposited or preserved southward. Unconformities within these sections, such as between the Pennsylvanian and the Permian in the Humble–Georgia Pacific well, suggest continuing instability within this post-orogenic basin.

Some 225 to 280 km south-southwest of the wells just discussed are several others which may be located along the southeastern flank of this late Paleozoic basin: the Humble-Johnson well (location 10, Fig. 3), Panola County, and the Shell-Temple Industries well (location 11), Sabine County, both in Texas, and the Humble-Boise Southern well (location 12), Sabine Parish, Louisiana.

The Humble-Johnson well penetrated 108 m of Lower Permian, then 575 m of Pennsylvanian sediments below Upper Triassic red beds. Palynomorphs in the upper part of the section indicate a Lower Permian, probably Wolfcampian, age. Microflora in the lower section indicates a range in age from early Pennsylvanian (Morrowan) through middle Pennsylvanian (Atokan-Desmoinesian), to late Pennsylvanian (Virgilian). From oldest to youngest, these sediments reflect a regressive cycle from offshore marine deposits composed of dark gray claystones, silty claystones, and fine-grained sandstones with laminae generally dipping less than 10°, to prograding deltaic deposits of sparsely burrowed claystones, siltstones, and massive-bedded to laminated, fine-grained sandstones. The laminae were dominantly horizontal, containing small cross-beds and convoluted beds. These deposits in turn were succeeded upward by fluvial sediments, flood plain, point bar, and natural levee deposits of massive red and green mottled claystones and laminated fine to very fine-grained sandstones, characterized by large-scale cross-beds (Woods and Addington, 1973).

The Shell–Temple Industries well, located approximately 80 km southeast of the Humble-Johnson, penetrated 169 m of upper Pennsylvanian (Missourian) and 219 m of middle Pennsylvanian (Desmoinesian) shallow-marine packstones, grainstones, and wackestones, and fractured carbonates with some clastic rocks beneath Upper Jurassic Bossier shales. These strata contained an abundant fauna of brachipods, algae, pelecypods, ostracodes, and fusulines. Beneath this section the well penetrated 56 m of a rhyolite porphyry, dated at 255 ± 10 Ma (Late Permian) (Nicholas and Waddell, 1982). This date, however, may be too young because of low-temperature devitrification of the volcanics (Nicholas and Waddell, 1985, personal communication). The rhyolite may be part of the widespread late Paleozoic igneous activity recorded in northern Mexico, where lava flows within Permian sediments are common.

The Humble-Boise Southern well, approximately 95 km southeast of the Humble-Johnson and 26 km northeast of the Shell-Templeton well, penetrated 746 m of Lower Permian and a possibly older argillite (semi-slate) beneath Jurassic carbonates. Black micritic limestone with black shale, gray fine-grained sandstone, and siltstone laminae with near-vertical calcite-filled fractures were penetrated at the top of this section, along with a fossil rubble of ostracodes, bryozoans, crinoid stems, echinoid spines, brachiopod fragments, and late Pennsylvanian (Missourian) to Early Permian (Wolfcampian) fusulines. Most of the section below this shallow-marine calcarenite-micrite sequence was made up of dark shales and silty shales verging on argillite in alteration, with quartzitic, mostly poorly sorted "wacke-type" sandstones, occasionally graded and with slump structures. The section below the uppermost fossil rubble contained abundant plant debris, including leaf fragments, seeds(?), stems, and woody tissue, but very few other fossils. Near the bottom of the well, some dark gray to black micritic skeletal limestone and sandy fossil rubble similar to the uppermost section was observed, and a Wolfcampian fusuline, possibly out of place, was identified from the bottom sample (Woods and Addington, 1973). Nicholas and Waddell (1985, personal communication) reported finding Pennsylvanian (Missourian) fusulines in place in the bottom 40 m of section in the Humble–Boise Southern and concluded that this well penetrated upper Pennsylvanian through Lower Permian strata. The Shell-Boise Southern, 3 km northwest of the Humble–Boise Southern well, penetrated 240 m of shale, argillaceous sandstones, and some packstones and wackestones below Jurassic beds. An upper Pennsylvanian (Virgilian) fusuline was identified from a core taken at 3,962 m.

No south Texas well to date has drilled deep enough out in the basin to penetrate Paleozoic rocks. Two wells along the structural rim, however, provide clues to the possible section deeper in the basin in this area. These include the Pagenkopf-Bloom (location 13) in Bexar County, and the Shell-Stewart (location 14) in Val Verde County.

The Pagenkopf-Bloom well, located within the regional zone of microschist of the Ouachita orogenic belt, drilled 807 m of variegated clastics, unaltered to slightly metamorphosed to a semi-slate. This section included dark gray to black, locally maroon and green calcareous shales and claystones and gray to greenish-gray, medium- to fine-grained sandstones with conglomeratic sandstones in the lower 90 m of section. These clastics contained middle Pennsylvanian to Early Permian palynomorphs, suggesting the well drilled into a post-orogenic rift zone on the western margin of the basin. The Shell-Stewart well drilled, below the Cretaceous, 457 m of foreland-facies dolomites above 152 m of dolomitic quartzites with interbeds of sandy dolomite. Nicholas and Rozendal (1975) considered these metacarbonates to be Ellenburger and the dolomitic quartzites and sandy dolomites to probably be late Cambrian in age. Palmer and others (1984), on the basis of the discovery of Middle Cambrian boulders in the Marathon uplift, have suggested that the age of these carbonates may be Middle Cambrian rather than Early Ordovician. Below this section, the Shell-Stewart well drilled 1,737 m of metasediments with interbeds of metavolcanics and metaigneous rocks. Denison and others (1977), citing a K/Ar analysis of a metadacite and a metarhyolite from this well, interpreted the rhyolite to have been extruded at 700 Ma over even older igneous rocks. The Shell-Stewart well, with its thick section of Cambrian-Ordovician carbonates, suggests, as did the Shell-Barrett in Hill County (location 3, Fig. 3) and the Hunt-Neely in Lamar County (location 2), the possibility of equivalent carbonates extending out into the basin in south Texas.

Subsurface of the western and southern Gulf of Mexico basin (Mexico and Central America)

Pre-Triassic rocks underlying the western and southern flanks of the Gulf of Mexico basin have been penetrated by wells in northeastern and east-central Mexico and the Yucatan platform. No pre-Triassic rocks have been reported from the Veracruz basin or from the sedimentary basins of the Isthmus of Tehuantepec area and southeastern Mexico.

The pre-Triassic reached by wells in northeastern and east-central Mexico is predominantly represented by metamorphic rocks of probable Precambrian and/or early or mid-Paleozoic age and by Permian-Triassic granitic plutons. No pre-Pennsylvanian unmetamorphosed sediments have been reported from the subsurface of the western and southern flanks of the Gulf of Mexico basin, and only four wells, 100 to 130 km northwest and north of Tampico, and a few more in Belize are known to have penetrated unaltered upper Paleozoic (Pennsylvanian and Permian) sediments.

In the northeastern Mexico state of Coahuila and the northern part of the states of Nuevo León and Tamaulipas, numerous wells drilled in search of petroleum have reached pre-Triassic rocks, principally along the crests of the Peyotes, Picachos and Tamaulipas arches—low-grade metamorphics and intrusive granitic and granodioritic rocks (Fig. 3).

The low-grade metamorphics are most commonly composed of various types of schists and phyllites. Less common are metaquartzites and marbles (Flawn and Diaz G., 1959; Flawn and Maxwell, 1958; Flawn and others, 1961). The age of their latest metamorphism, as determined by a few isotopic analyses, seems to be Permian (Denison and others, 1971). Before undergoing low-grade metamorphism, these metamorphic rocks are thought to have been part of lower or mid-Paleozoic sedimentary rock sequences.

The granitic intrusives are mainly composed of granites and granodiorites and are generally considered to be Permian-Triassic in age. They extend in a northwest-southeast belt from central Coahuila (Potrero de la Mula) (Denison and others, 1971; Jones and others, 1984) to central Tamaulipas, along the trend of the Peyotes, Picachos, and Tamaulipas arches (Fig. 3). Granodiorite and granite have also been reported from the Las Delicias–Acatita area, where they intrude unmetamorphosed Permian sediments and associated volcanics. Wells Mayran-1 and Paila-1, in southern Coahuila (locations 15 and 16, respectively, Fig. 3), bottomed in what have been interpreted as late Paleozoic lava flows (Alfonso Zwanziger, 1978).

In east-central Mexico—southern Nuevo León, southern Tamaulipas, and northern Veracruz—more than 100 wells in the southern part of the Burgos basin and in the Tampico-Misantla basin have reached pre-Triassic rocks—high-grade as well as low-grade metamorphics, granites, and granodiorites. In addition, four wells have been reported to have reached unmetamorphosed sediments of late Paleozoic age.

Well Gonzalez-101, about 100 km northwest of Tampico (location 17, Fig. 3), penetrated 1,446.6 m of black, slightly calcareous shale interbedded with sandstones and conglomeratic sandstones below 908 m (Carrillo Bravo, 1961; López Ramos, 1980). Cores at 1,049 m and 2,213 m contained fusulines of early middle Pennsylvanian age. The lower part of the section exhibited some degree of metamorphism. A nearby well, Magiscatzin-1, may also have penetrated upper Paleozoic unmetamorphosed sediments (Sansores Manzanilla and Girard Navarrete, 1969). Two more wells, Zamorina-1 and Tepehuaje-1, 110 and 130 km, respectively, north of Tampico (locations 18 and 19, Fig. 3), also penetrated unmetamorphosed upper Paleozoic sediments. Zamorina-1 drilled 804 m of black, slightly sandy and carbonaceous shale similar to that of the Guacamaya Formation or early Permian age cropping out in the Huizachal-Peregrina anticlinorium (Toledo Toledo, 1969). Tepehuaje-1 penetrated only 56 m of an analogous section—black carbonaceous shales and interbedded dark siltstones and fine-grained sandstones. These wells may have drilled into the little-deformed eastern flank of a late Paleozoic orogenic belt that folded out of a Paleozoic geosyncline extending from western Coahuila southeastward through east-central Mexico and perhaps as far as the state of Chiapas in southern Mexico and Guatemala (López Ramos, 1969; Dengo, 1975). These wells, on the other hand, may have drilled into a shallow-marine successor basin east of the orogenic belt, possibly related to a similar basin in which the upper Paleozoic rocks penetrated in east Texas, northwest Louisiana, and southern Arkansas were deposited.

No conclusive evidence provided by isotopic determinations has been published to establish the presence of Precambrian rocks in the subsurface of east-central Mexico. However, a few wells in the Tampico-Misantla basin have bottomed in gneisses resembling the Huiznopala Gneiss and the Novillo Gneiss cropping out in the nearby Huayacocotla and Huizachal-Peregrina anticlinoriums, both of known late Precambrian age (Fig. 3). More than 100 other wells have reached lower-grade metamorphic rocks that, as in the case of the low-grade metamorphics of northeastern Mexico, are believed to be lower or mid-Paleozoic sedimentary sequences that underwent their last period of metamorphism during the late Permian of early Triassic.

In the Tampico-Misantla basin, about 100 wells have bottomed in medium- to fine-grained intrusives of granitic, granodioritic, and tonalitic composition (Fig. 3), indicating the presence in the subsurface of this basin of an intrusive body of considerable size extending along the eastern coast of Mexico (Quezadas Flores, 1961; López Ramos, 1972). Isotopic determinations indicate a Permian-Triassic age for the intrusive rocks. Parts of this intrusive body were apparently exposed as horst blocks that formed islands during the Late Jurassic and Early Cretaceous. Because no contact-metamorphic effects have been observed around these massive intrusives, the batholith may represent the roots of a late Paleozoic island-arc system that collided with the eastern continental margin of Mexico.

In the Yucatan Peninsula, north of the Maya Mountains of Belize, a number of wells have drilled into pre-Triassic rocks below the Cretaceous carbonate and evaporite section or below red beds of uncertain age. Those closer to the Maya Mountains, in northern Belize or northeastern Guatemala, have bottomed in upper Paleozoic Santa Rosa Group rocks. The largest penetration was probably in well Yalbac-1 in west-central Belize (locality 20, Fig. 3), which drilled through 1,370 m of Santa Rosa sediments. A well in northern Belize (Tower Hill-1, location 21) has been reported to have drilled into granite (Vedder and others, 1973; López Ramos, 1975). Farther north, in the Mexican part of the peninsula, two wells (Yucatan-1 and -4, locations 22 and 23, respectively) bottomed in low-grade metamorphic rocks; a third well in northeast Belize (Basil Jones-1, location 24) also encountered schists at total depth (López Ramos, 1975).

The pre-Triassic section in Yucatan-1 consisted of 19 m of rhyolite porphyry and schists. Bass and Zartman (1969) speculated on the similarity of the rhyolite with the Bladen volcanic member of the Santa Rosa Group in the Maya Mountains. López Ramos (1975, p. 267) reported an isotopic date of 410 Ma (Silurian) for the rhyolite and "the probable existence of a metamorphic event at about 300 m.y. (Mississippian)." Yucatan-4 encountered, at total depth, 8 m of slightly metamorphosed quartzite showing the effects of weathering.

Pre-Triassic rocks from the central Gulf of Mexico Basin

Upper Paleozoic sediments have been dredged from the flank of one of the Sigsbee Knolls within the central Gulf of Mexico (Fig. 3). These sediments were described as diagenetically altered glauconitic siltstones of Pennsylvanian age, dated at 318 Ma by K/Ar analysis (Pequegnat and others, 1971). They were interpreted to be marine deposits, transported by salt flow to their present position.

SUMMARY AND SPECULATIONS

Knowledge of the pre-Triassic rocks underlying the Gulf of Mexico basin is restricted by the limited number of wells that have reached these rocks. Most wells that have reached the pre-Triassic have seldom drilled more than a few tens of meters into it because the pre-Triassic rocks have been generally considered "economic basement" by the petroleum industry. Fewer than 12 wells in the basin have penetrated as much as 1,000 m of pre-Triassic rocks, and most of the wells that have reached them are located around the periphery of the basin. With few exceptions, therefore, it is nearly impossible to map the distribution of the pre-Triassic rocks under the basin and to establish their structural relations, or to determine the disposition of pre-Triassic paleogeographic provinces and structural belts. No evidence can be found in the pre-Triassic rocks of the Gulf of Mexico basin area, for example, that may indicate if the Mojave-Sonora megashear, recognized by Silver and Anderson (1974) in northwestern Mexico, extends into the area—the necessary documentation is not available. The lack of wells that have reached pre-Triassic rocks deeper in the basin also makes it difficult to interpret seismic data. Seismic records do indicate, however, that the upper Paleozoic section was subjected to compression near the northern, and perhaps the western margins of the basin, with resulting thrusting and overturned folds, whereas deeper in the basin the records show gentler folding and block faulting. The "basement" underlying the central part of the basin has been surmised to be composed of oceanic volcanic rocks on the basis of geophysical data (see chapters 4 and 13 of this volume).

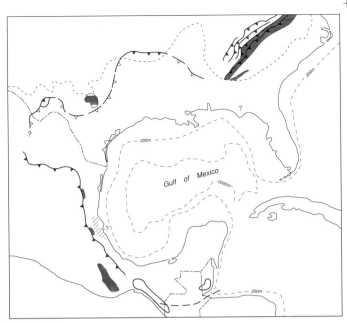

Figure 4. Distribution of Precambrian rocks in the Gulf of Mexico basin and surrounding uplifts.

Pre-Triassic rocks crop out in the uplifted regions surrounding the Gulf of Mexico basin—the Appalachian and Ouachita mountains, the Llano and Marathon uplifts in the United States, the Sierra Madre Oriental and the Sierra de Chiapas in Mexico, the mountains in central Guatemala, and the Maya Mountains of Belize (Fig. 1)—and some information about the pre-Triassic rocks under the Gulf of Mexico basin can be derived from their study. But even these outcrop areas are generally not extensive and are separated from each other by considerable distances.

In spite of this paucity of outcrop and subsurface data, it is possible to make some general statements and to attempt some broad speculations.

1. Within the Gulf of Mexico basin and the surrounding uplifted areas, rocks known to be of Precambrian age have only been reported from the Appalachian Mountains, from the Llano uplift of central Texas, from the cores of the Huizachal-Peregrina and Huayacocotla anticlinoriums in east-central Mexico, and from the Oaxaca area of southwestern Mexico. A few wells in the Florida panhandle and in the Tampico-Misantla Basin in east-central Mexico have penetrated igneous and metamorphic rocks that may also be Precambrian in age (Fig. 4).

2. Unmetamorphosed pre-Pennsylvanian rocks are known from the subsurface of northern Florida and adjacent parts of Georgia and Alabama, from the Appalachian and Ouachita mountains and parts of their possible connection under the Mesozoic sediments of the Gulf Coastal Plain, from the Llano and Marathon uplifts, from the Placer de Guadalupe area in Chihuahua, from the Huizachal-Peregrina anticlinorium in east-central Mexico, and from a small area in Oaxaca (Fig. 5). Metamorphic rocks in several areas around the periphery of the Gulf of Mexico basin, both in outcrop and in the subsurface, have been assigned

pre-Pennsylvanian ages, mostly on the basis of isotopic determinations or stratigraphic position.

The Paleozoic section in the subsurface of Florida, Georgia, and Alabama, 2,000 to 2,500 m thick, has been dated as Ordovician, Silurian, and Devonian on the basis of its contained fossils. From paleontological and other evidence it has been speculated that these Paleozoic sediments were once part of the African plate, sutured to the North American plate during the closing of the Iapetus Ocean during the late Paleozoic, and left there after the opening of the Atlantic Ocean during the Mesozoic. A weathered zone on top of these older Paleozoic rocks, estimated from seismic data to reach 300 m in places, suggests that the area remained essentially high during the late Paleozoic and that little or probably no deposition of upper Paleozoic sediments took place in the area.

3. Siliceous sediments (chert and novaculite) are common in the Devonian over a considerable area, from the Black Warrior basin of northern Mississippi west to the Ouachita Mountains, the Marathon uplift, and the Placer de Guadalupe–Mina Plomosas area of northeastern Mexico and from there south to the Huizachal-Peregrina anticlinorium of east-central Mexico (Fig. 6). This widespread occurrence of Devonian siliceous sediments suggests extensive volcanic activity in nearby areas at that time.

4. The Pennsylvanian and the Permian are much better represented and more widely distributed around the periphery of the Gulf of Mexico basin and surrounding uplifted areas (Fig. 7). It is not known, however, how far into the basin sediments of this age extend. Pennsylvanian and Permian rocks generally show no effect of regional metamorphism, and are considered to be post-orogenic deposits. They are occasionally associated with volcanic rocks, indicating another volcanic episode during the late Paleo-

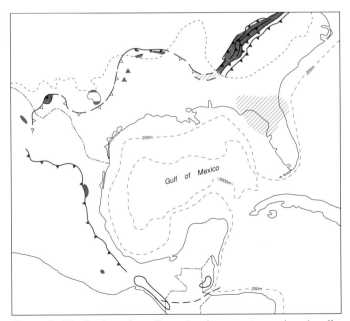

Figure 5. Distribution of pre-Pennsylvanian unmetamorphosed sediments in the Gulf of Mexico basin and surrounding uplifts.

zoic. No sediments of latest Permian age have been reported from the Gulf of Mexico basin and surrounding areas.

The Pennsylvanian-Permian section of the northern and western flanks of the Gulf of Mexico basin commonly includes a thick sequence of terrigenous clastics—interbedded, thinly bedded shales, siltstones, and sandstones—of the type that has often been referred to as "flysch" (Fig. 8). The "flysch" deposits of the U.S. part of the northern and northwestern flanks of the basin—those of the Ouachita Mountains and the Llano and Marathon uplifts—are of latest Mississippian and Pennsylvanian age, and those of northeast and east-central Mexico (Placer de Guadalupe-Mina Plomosas-Villa Aldama area, Las Delicias—Acatita area, and the Huizachal-Peregrina and Huayacocotla anticlinoriums) are of Permian age. The significance of this difference in age between the two seemingly similar stratigraphic sequences is not evident.

5. From the central part of the state of Coahuila, northeastern Mexico, to east-central Mexico, Guatemala, and Belize, the Paleozoic section is intruded by silicic plutonic rocks—granites, granodiorites, and tonalites (Fig. 9). No evidence of contact metamorphism of the Paleozoic rocks around the intrusions have been reported, but Paleozoic rocks are not known to overlie the intrusive bodies. Isotopic dates yield a majority of late Permian to early Triassic dates for these silicic batholiths. Two wells in southeastern Mississippi have penetrated a granitic intrusive body of Permian age.

6. Two major unconformities can be recognized within the pre-Triassic rock sequences in some parts of the Gulf of Mexico basin area. (1) A pre–Upper Mississippian–Pennsylvanian unconformity that separates thinner platform or metamorphosed sequences below from thicker generally unmetamorphosed se-

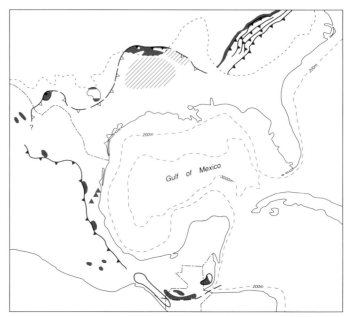

Figure 7. Distribution of Pennsylvanian and Permian unmetamorphosed sediments in the Gulf of Mexico basin and surrounding uplifts.

quences above, often including thick "flysch" deposits; and (2) A pre–Upper Triassic–Jurassic–Cretaceous unconformitty that records the termination of a long period of deformation, plutonic intrusion, and erosion that started during the late Permian and lasted in some areas until the Late Triassic, and in others until the Late Cretaceous.

The significance of the older unconformity is difficult to ascertain based solely on information from the Gulf of Mexico basin and immediately adjacent areas. The sudden influx of thick "flysch" sequences in late Mississippian, Pennsylvanian, or Permian time could be interpreted to indicate the onset of the collision of the African and South American Plates with the North American Plate, and the resulting folding, thrusting, and uplift, the appearance of an extensive provenance area, and the development of foredeep basins around the perimeter of the North American Plate. The younger unconformity is generally believed to record the latest events in the plate collision: the formation of the supercontinent of Pangea and the long period of erosion that followed.

With the meager information available today, it is futile to attempt to reconstruct the geologic history of the Gulf of Mexico basin area before the latest Mississippian or the Pennsylvanian. Even drawing information from regions far from the area, the interpretation of the early Paleozoic history of the Gulf of Mexico basin area would be illusory; we could only speculate.

There is, however, considerable evidence to suggest that during late Mississippian, Pennsylvanian, and early Permian time, the southern margin of the North American Plate changed from a passive to a convergent margin. During this span of time, major compressive tectonic activity took place along the northern

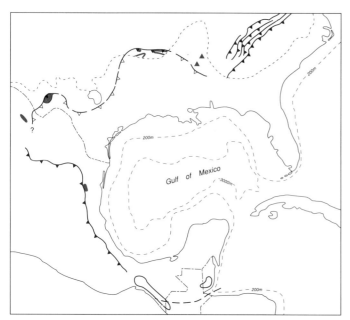

Figure 6. Distribution of siliceous sediments of Devonian age in the Gulf of Mexico basin and immediately surrounding areas.

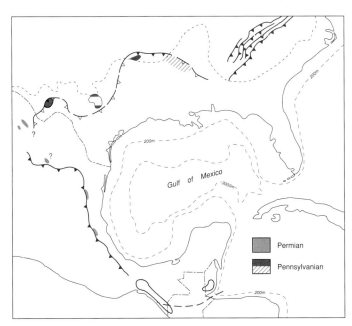

Figure 8. Distribution of Pennsylvanian and Permian "flysch" sequences in the Gulf of Mexico basin and surrounding uplifts.

and western flanks of the present basin. In the southern Appalachians areas of Alabama and in the Marathon Mountains area, it may have started sooner, during the late Mississippian. This compressive episode has been interpreted by many authors to be the result of the collision during the late Paleozoic of the North American and the African–South American continental plates, which resulted in the assemblage of the supercontinent of Pangea. Much stratigraphic and tectonic evidence supports this interpretation.

1. In the southeastern United States, the presence of a pre-Triassic terrane of plutonic, metamorphic, and sedimentary rocks with evident African affinities, unlike rocks of the same age in nearby areas, suggests that this part of the present North American Plate is a remnant of the African Plate, which was welded to the North American Plate when the two collided during the late Paleozoic. This fragment of the African Plate was left behind when the North American and African Plates separated at the time the North Atlantic Ocean opened, in the Late Triassic or Early Jurassic.

2. Plate convergence during the late Paleozoic is also indicated by extensive thrusting of this age toward the North American plate along the Appalachian Mountains, the Ouachita Mountains, the subsurface Ouachita orogenic belt, and the Marathon uplift. The continuation of this thrust belt into Mexico is still the subject of considerable controversy. The Granjeno Schist and associated serpentinite bodies of the Huizachal-Peregrina anticlinorium in east-central Mexico have been interpreted as allochthonous to the area, probably part of a westward-transported thrust slice. If this is the case, it could be speculated that a late Paleozoic thrust belt may underlie the

Mesozoic cover in east-central Mexico. The presence of unmetamorphosed Pennsylvanian and Permian sediments in wells Gonzalez 101, Zamorina-1, and Tepehuaje-1, southeast of the Huizachal-Peregrina anticlinorium, support this interpretation, because these upper Paleozoic sediments could be viewed as having been deposited in a successor basin behind the thrust belt. Whether such a thrust belt represents the long-sought continuation in Mexico of the Ouachita orogenic belt cannot be asserted on the basis of the limited information now available. If a late Paleozoic thrust belt is present in east-central Mexico, and could be recognized as the continuation of the Ouachita orogenic belt of Texas, it may be necessary to resort to large transcurrent displacement of the belt between the Marathon Mountains area and east-central Mexico, probably in pre–Late Triassic time.

3. The thick "flysch" sequences of Pennsylvanian and Permian age known along the northern, northwestern, and western flanks of the Gulf of Mexico basin also suggest uplift in these areas, probably the result of compressive tectonic episodes.

4. The belt of Permian-Triassic granitic plutons known from northern Central America to northeastern Mexico probably represents the record of plate collision and uplift along this belt during the late Paleozoic and early Mesozoic.

5. The absence of marine sediments of late Permian and Triassic age in the Gulf of Mexico basin and immediately surrounding areas indicates that these areas were indeed part of a large positive stable area subjected to erosion and peneplanation during the late Permian and most of the Triassic. As will be discussesd in chapters 8 and 14 of this volume, this long period of stability and erosion was not interrupted by new tectonic activity until late in the Triassic, when the development of complex systems of grabens and rifted basins marked the start of the separa-

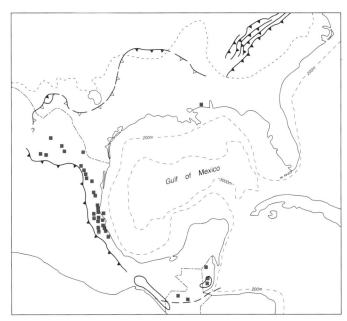

Figure 9. Distribution of Permian-Triassic granitic intrusive bodies in the Gulf of Mexico basin and immediately surrounding areas.

tion of the North American Plate from the African–South American Plate and the initial episodes of the formation of the Gulf of Mexico basin.

Many unanswered questions remain concerning the pre-Triassic history of the Gulf of Mexico basin area. Where is the continuation of the Ouachita orogenic belt in Mexico? What is the tectonic significance of the belt of Permian-Triassic granitic intrusions in northern Central America and Mexico? What is the nature of the connection between the Appalachian and the Ouachita orogenic belts? Are they continuous or discontinuous? Do they intersect each other or are they offset by a transcurrent fault system? What is the significance of the extensive Devonian deposits of chert and novaculite and the period of volcanic activity with which they are probably associated? Do the Mojave-Sonora megashear and other similar transform faults extend into the Gulf of Mexico basin area, and if so, when were they active? New, more detailed geological and geophysical studies and additional deep drilling in the Gulf of Mexico basin area are needed before attempts can be made to answer these questions with any hope of success.

ECONOMIC CONSIDERATIONS

Pre-Triassic unmetamorphosed sedimentary rocks can no longer be considered as "basement" except in the sense of having had an earlier, different, and more complex history than the overlying Mesozoic-Cenozoic fill. The oil industry in recent years has made an effort to explore this older section. No reliable figures are available on the areal extent or volume of the unmetamorphosed Paleozoic rocks, but estimates based on limited geophysical data have suggested an area of perhaps 200,000 to 300,000 km^2, with section thicknesses of 6,000 m or more in places. Fewer than 125 wells out in the basin have been drilled into this potentially large frontier; the average penetration has been about 220 m. Oil and gas shows have been reported in several areas, but to date no commercial accumulations have been discovered (Paine and Meyerhoff, 1970; Vernon, 1971; Woods and Addington, 1973; Valerius, 1978). However, the estimated large size of the area within the Gulf of Mexico basin underlain by unaltered pre-Triassic marine sediments and the considerable thickness these sediments attain in certain areas make them attractice as a potential target for oil and gas exploration.

No geothermal potential for this older section has been indicated, nor have any economic mineral deposits been reported, but no real effort has been made to explore for such resources. This deeper part of the Gulf of Mexico basin is truly one of the few remaining frontiers of North America.

REFERENCES CITED

Alfonso Zwanziger, J., 1978, Geología regional del sistema Cupido: Asociación Mexicana de Geólogos Petroleros Boletín, v. 30, p. 1–55.
Anderson, T. H., Burkart, B., Clemons, R. E., Bohnenberger, D. H., and Blount, D. N., 1973, Geology of the western Altos Cuchumatanes, northwestern Guatemala: Geological Society of America Bulletin, v. 84, p. 805–826.
Andress, N. E., Cramer, F. H., and Goldstein, R. F., 1969, Ordovician chitinozoans from Florida well samples: Gulf Coast Association of Geological Societies Transactions, v. 19, p. 369–375.
Applin, P. L., 1951, Preliminary report on buried pre-Mesozoic rocks in Florida and adjoining states: U.S. Geological Survey Circular 91, 28 p.
Applin, P. L., and Applin, E. R., 1965, The Comanche Series and associated rocks in the subsurface in central and south Florida: U.S. Geological Survey Professional Paper 447, 84 p.
Arden, D. D., Jr., 1974, Geology of the Suwannee Basin interpreted from geophysical profiles: Gulf Coast Association of Geological Societies Transactions, v. 24, p. 223–230.
Arthur, J. D., 1988, Petrogenesis of early Mesozoic tholeiite of the Florida basement and an overview of Florida basement geology: Florida Geological Survey Report of Investigation 97, 39 p.
Barnett, R. S., 1975, Basement structure of Florida and its tectonic implications: Gulf Coast Association of Geological Societies Transactions, v. 25, p. 122–142.
Bass, M. N., 1969, Petrography and ages of crystalline basement rocks of Florida; Some extrapolations: American Association of Petroleum Geologists Memoir 11, p. 283–310.
Bass, M. N., and Zartman, R. E., 1969, The basement of Yucatan Peninsula [abs.]: EOS American Geophysical Union Transactions, v. 50, p. 313.
Bateson, J. H., 1972, New interpretation of geology of Maya Mountains, British Honduras: American Association of Petroleum Geologists Bulletin, v. 56, p. 956–963.

Bateson, J. H., and Hall, I.H.S., 1971, Revised geologic nomenclature for pre-Cretaceous rocks of British Honduras: American Association of Petroleum Geologists Bulletin, v. 55, p. 529–530.
—— , 1977, The geology of the Maya Mountains, Belize: Institute of Geological Sciences Overseas Memoir 3, 47 p.
Bridge, J., and Berdan, J. M., 1951, Preliminary correlation of the Paleozoic rocks from test wells in Florida and adjacent parts of Georgia and Alabama: U.S. Geological Survey Preliminary Release, 89 p.
Bridges, L. W., 1964, Stratigraphy of Mina Plomosas–Placer de Guadalupe area, in Geology of Mina Plomosas–Placer de Guadalupe area, Chichuahua, Mexico: West Texas Geological Society 1964 field trip guidebook Publication 64–50, p. 50–59.
—— , 1965, Geología del area de Plomosas, Chihuahua; Part I, Estudios geológicos en el Estado de Chihuahua: Instituto de Geología, Universidad Nacional Autónoma de México Boletín 74, p. 1–134.
Bridges, L. W., and DeFord, R. K., 1961, Pre-Carboniferous Paleozoic rocks in central Chihuahua, Mexico: American Association of Petroleum Geologists Bulletin, v. 45, p. 98–104.
Buffler, R. T., and others, 1984, Initial reports of the Deep Sea Drilling Project, Volume 77: Washington, D.C., U.S. Government Printing Office, 747 p.
Carrillo Bravo, J., 1961, Geología del anticlinorio Huizachal-Peregrina al N-W de Ciudad Victoria, Tamps.: Asociación Mexicana de Geólogos Petroleros Boletín, v. 13, p. 1–98.
—— , 1965, Estudio geológico de una parte del anticlinorio de Huayacocotla: Asociación Mexicana de Geólogos Petroleros Boletín, v. 17, p. 73–96.
Chowns, T. M., and Williams, C. T., 1983, Pre-Cretaceous rocks beneath the Georgia coastal plain; Regional implications, in Gohn, G. S., ed., Studies related to the Charleston, South Carolina, earthquake of 1886; Tectonics and seismicity: U.S. Geological Survey Professional Paper 1313L, 42 p.

Clemons, R. E., and Burkart, B., 1971, Stratigraphy of northwestern Guatemala: Sociedad Geológica Mexicana Boletín, v. 32, p. 143–158.

Clemons, R. E., Anderson, T. H., Bohnenberger, O. H., and Burkart, B., 1974, Stratigraphic nomenclature of recognized Paleozoic and Mesozoic rocks of western Guatemala: American Association of Petroleum Geologists Bulletin, v. 58, p. 313–320.

Cole, W. S., 1944, Stratigraphic and paleontological studies of wells in Florida, No. 3: Florida Geological Survey Bulletin 26, p. 20–21.

Corona-Esquivel, R.J.J., 1981 [1983], Estratigrafía de la región de Olinalá-Tecocoyunca, noreste del Estado de Guerrero: Instituto de Geología, Universidad Nacional Autónoma de México, Revista, v. 5, p. 17–24.

Cramer, F. H., 1971, Position of the north Florida lower Paleozoic block in Silurian time; Phytoplankton evidence: Journal of Geophysical Research, v. 76, p. 4754–4757.

—— , 1973, Middle and Upper Silurian chitinozoan succession in Florida subsurface: Journal of Paleontology, v. 47, p. 279–288.

Cserna, Z. de, Graf, J. L., Jr., and Ortega-Gutierrez, F., 1977, Alóctono del Paleozoico inferior en la region de Ciudad Victoria, Estado de Tamaulipas: Instituto de Geología, Universidad Nacional Autónoma de México, Revista, v. 1, p. 33–43.

Dallmeyer, R. D., 1984, $^{40}Ar/^{39}Ar$ ages from a pre-Mesozoic crystalline basement penetrated at holes 537 and 538A of the Deep Sea Drilling Project Leg 77, southeastern Gulf of Mexico, *in* Buffler, R. T., and others, Initial reports of the Deep Sea Drilling Project, Volume 77: Washington, D.C., U.S. Government Printing Office, p. 497–504.

—— , 1987, $^{40}Ar/^{39}Ar$ age of detrital muscovite within Lower Ordovician sandstone in the coastal plain basement of Florida; Implications for West African terrane linkages: Geology, v. 15, p. 998–1001.

Dallmeyer, R. D., Caen-Vachette, M., and Villeneuve, M., 1987, Emplacement age of post-tectonic granites in southern Guinea (West Africa) and the peninsular Florida subsurface; Implications for origins of southern Appalachian exotic terranes: Geological Society of America Bulletin, v. 99, p. 87–93.

Dengo, G., 1975, Paleozoic and Mesozoic tectonic belts in Mexico and Central America, *in* Nairn, A.E.M., and Stehli, F. G., eds., The ocean basins and margins; Volume 3, The Gulf of Mexico and the Caribbean: New York, London, Plenum Press, p. 283–323.

Dengo, G., and Bohnenberger, O., 1969, Structural development of northern Central America, *in* Tectonic relations of northern Central America and the western Caribbean; The Bonacca expedition: American Association of Petroleum Geologists Memoir 11, p. 203–220.

Denison, R. E., Burke, W. H., Jr., Hetherington, E. A., and Otto, J. B., 1971, Basement rock framework of parts of Texas, southern New Mexico and northern Mexico, *in* Seewald, K., and Sundeen, D., eds., The geologic framework of the Chihuahua tectonic belt: West Texas Geological Society Publication 71-59, p. 3–14.

Denison, R. E., Burke, W. H., Jr., Otto, J. B., and Hetherington, E. A., 1977, Age of igneous and metamorphic activity affecting the Ouachita foldbelt, *in* Stone, C. G., ed., Symposium on the geology of the Ouachita Mountains, Volume 1: Arkansas Geological Commission, p. 25–40.

Dillon, W. P., and Vedder, J. G., 1973, Structure and development of the continental margin of British Honduras: Geological Society of America Bulletin, v. 84, p. 2713–2732.

Dixon, C. G., 1956, Geology of southern British Honduras, with notes on adjacent areas: Belize Government Printer, 85 p.

Flawn, P. T., and Diaz G., T., 1959, Problems of Paleozoic tectonics in north-central and northeastern Mexico: American Association of Petroleum Geologists Bulletin, v. 43, p. 224–230.

Flawn, P. T., and Maxwell, R. A., 1958, Metamorphic rocks in Sierra del Carmen, Mexico: American Association of Petroleum Geologists Bulletin, v. 42, p. 2245–2249.

Flawn, P. T., Goldstein, A., Jr., King, P. B., and Weaver, C. E., 1961, The Ouachita System: Austin, University of Texas Publication 6120, 401 p.

Fries, C., Jr., and Rincón-Orta, C., 1965, Nuevas aportaciones geocronológicas y tecnicas empleadas en el laboratorio de geocronometría: Instituto de Geología, Universidad Nacional Autónoma de México, Boletín, v. 73, p. 57–133.

Fries, C., Jr., Schmitter, E., Damon, P. E., Livingston, D. E., and Erickson, R., 1962, Edad de las rocas metamórficas en los cañones de La Peregrina y de Caballeros, parte centro-occidental de Tamaulipas: Instituto de Geología, Universidad Nacional Autónoma de México, Boletín, v. 64, p. 55–69.

Garrison, J. R., Jr., 1978, Reinterpretation of isotopic age data from the Granjeno Schist, Ciudad Victoria, Tamaulipas: Instituto de Geología, Universidad Nacional Autónoma de México, Revista, v. 2, p. 87–89.

Garrison, J. R., Jr., Ramírez-Ramírez, C., and Long, L. E., 1980, Rb-Sr isotopic study of the ages and provenance of Precambrian granulite and Paleozoic greenschist near Ciudad Victoria, Mexico, *in* Pilger, R. H., Jr., ed., The origin of the Gulf of Mexico and the early opening of the central North Atlantic Ocean: Baton Rouge, Louisiana State University, p. 37–49.

Goldstein, R. F., Cramer, F. H., and Andress, N. E., 1969, Silurian chitinozoans from Florida well samples: Gulf Coast Association of Geological Societies Transactions, v. 19, p. 377–384.

Gutierrez Gil, R., 1956, Bosquejo geológico del Estado de Chiapas, *in* Geología del Mesozoico y estratigrafía pérmica del Estado de Chiapas: Mexico, D. F., 20th International Geological Congress, Excursión C-15, p. 9–32.

Hall, I.H.S., and Bateson, J. H., 1972, Late Paleozoic lavas in Maya Mountains, British Honduras, and their possible regional significance: American Association of Petroleum Geologists Bulletin, v. 56, p. 950–956.

Harrelson, D. W., and Bicker, A. R., Jr., 1979, Petrography of some subsurface igneous rocks of Mississippi: Gulf Coast Association of Geological Societies Transactions, v. 29, p. 244–251.

Hatcher, R. D., Jr., Thomas, W. A., and Viele, G. W., eds., The Appalachian-Ouachita Orogen in the United States: Boulder, Colorado, Geological Society of America, The Geology of North America, v. F-2, 767 p.

Hernández García, R., 1973, Paleogeografía del Paleozoico de Chiapas, México: Asociación Mexicana de Geólogos Petroleros Boletín, v. 25, no. 1–3, p. 77–134.

Hoffmeister, W. S., and Staplin, F. L., 1954, Pennsylvanian age of the Morehouse formation in northeastern Louisiana: American Association of Petroleum Geologists Bulletin, v. 38, p. 158–159.

Hurst, V. J., 1960, Oil tests in Georgia: Georgia State Division of Conservation, Division of Mines, Mining, and Geology Information Circular 19, 14 p.

Imlay, R. W., 1941, Jurassic fossils from Arkansas, Louisiana, and east Texas: Journal of Paleontology, v. 15, p. 261–262.

Imlay, R. W., and Williams, J. S., 1942, Late Paleozoic age of the Morehouse formation of northeast Louisiana: American Association of Petroleum Geologists Bulletin, v. 26, p. 1672–1673.

Jones, N. W., McKee, J. W., Marquez D., B., Tovar, J., Long, L. E., and Laudon, T. S., 1984, The Mesozoic La Mula Island, Coahuila, Mexico: Geological Society of America Bulletin, v. 95, p. 1226–1241.

Kesler, S. E., 1971, Nature of ancestral orogenic zone in nuclear Central America: American Association of Petroleum Geologists Bulletin, v. 55, p. 2116–2129.

—— , 1973, Basement rock structural trends in southern Mexico: Geological Society of America Bulletin, v. 84, p. 1059–1064.

Kesler, S. E., Josey, W. L., and Collins, E. M., 1970, Basement rocks of western nuclear Central America; The Western Chuacús Group, Guatemala: Geological Society of America Bulletin, v. 81, p. 3307–3322.

Khudoley, K. M., and Meyerhoff, A. A., 1971, Paleogeography and geological history of the Greater Antilles: Geological Society of America Memoir 129, 199 p.

King, P. B., 1937, Geology of the Marathon region, Texas: U.S. Geological Survey Professional Paper 187, 148 p.

—— , 1975, The Ouachita and Appalachian orogenic belts, *in* Nairn, A.E.M., and Stehli, F. G., eds., The ocean basins and margins; Volume 3, The Gulf of Mexico and the Caribbean: New York, London, Plenum Press, p. 201–241.

King, R. E., and others, 1944, Geology and paleontology of the Permian area northwest of Las Delicias, southwestern Coahuila, Mexico: Geological Society of America Special Paper 52, 172 p.

Kjellesvig-Waering, E. N., 1950, A new Silurian eurypterid from Florida: Journal of Paleontology, v. 24, p. 229–231.

López Ramos, E., 1969, Marine Paleozoic rocks of Mexico: American Association of Petroleum Geologists Bulletin, v. 53, p. 2399–2417.

——, 1972, Estudio del basamento ígneo y metamórfico de las zonas Norte y Poza Rica: Asociación Mexicana de Geólogos Petroleros Boletín, v. 24, p. 266–323.

——, 1975, Geological summary of the Yucatan Peninsula, *in* Nairn, A.E.M., and Stehli, F. G., eds., The ocean basins and margins; Volume 3, The Gulf of Mexico and the Caribbean: New York, London, Plenum Press, p. 257–282.

——, 1980, Geología de México, Volume 2 (second edition), 454 p.

——, 1983, Geología de México, Volume 3 (third edition), 453 p.

McBride, E. F., 1978, The Ouachita trough sequence; Marathon region and Ouachita Mountains, *in* Tectonics and Paleozoic facies of the Marathon geosyncline, West Texas: West Texas Geological Society 1978 field conference guidebook Publication 78–17, p. 39–49.

McLaughlin, R. E., 1970, Palynology of core samples of Paleozoic sediments from beneath the coastal plain of Early County, Georgia: Georgia Geological Survey Information Circular 40, 27 p.

Mellor, E. I., and Breyer, J. A.,1981, Petrology of late Paleozoic basin-fill sandstones, north-central Mexico: Geological Society of America Bulletin, part I, v. 92, p. 367–373.

Milton, C., 1972, Igneous and metamorphic basement rocks of Florida: Florida Bureau of Geology Bulletin 55, 125 p.

Milton, C., and Grasty, R., 1969, "Basement" rocks of Florida and Georgia: American Association of Petroleum Geologists Bulletin, v. 53, p. 2483–2493.

Milton, C., and Hurst, V. J., 1965, Subsurface "basement" rocks of Georgia: Georgia Geological Survey Bulletin 76, 56 p.

Morgan, J. K., 1970, The Central Mississippi Uplift: Gulf Coast Association of Geological Societies Transactions, v. 20, p. 91–109.

Neathery, T. L., and Thomas, W. A., 1975, Pre-Mesozoic basement rocks of Alabama Coastal Plain: Gulf Coast Association of Geological Societies Transactions, v. 25, p. 86–98.

Nelson, K. D., and 8 others, 1985a, New COCORP profiling in the southeastern United States; Part I, Late Paleozoic suture and Mesozoic rift basin: Geology, v. 13, p. 714–718.

Nelson, K. D., and 5 others, 1985b, New COCORP profiling in the southeastern United States; Part 2, Brunswick and east coast magnetic anomalies, opening of the north-central Atlantic Ocean: Geology, v. 13, p. 718–721.

Nicholas, R. L., and Rozendal, R. A., 1975, Subsurface positive elements within the Ouachita foldbelt in Texas and their relation to Paleozoic craton margin: American Association of Petroleum Geologists Bulletin, v. 59, p. 193–216.

Nicholas, R. L., and Waddell, D. E., 1982, New Paleozoic subsurface data from the north-central Gulf Coast: Geological Society of America Abstracts with Programs, v. 14, p. 576.

Opdyke, N. D., and 5 others, 1987, Florida as an exotic terrane; Paleomagnetic and geochronologic investigation of lower Paleozoic rocks from the subsurface of Florida: Geology, v. 15, p. 900–903.

Ortega-Gutiérrez, F., 1978a, El Gneis Novillo y rocas metamórficas asociadas en los cañones del Novillo y de La Peregrina, area de Ciudad Victoria, Tamaulipas: Instituto de Geología, Universidad Nacional Autónoma de México, Revista, v. 2, p. 19–30.

——, 1978b, Estratigrafía del Complejo Acatlán en la Mixteca Baja, Estados de Puebla y Oaxaca: Instituto de Geología, Universidad Nacional Autónoma de México, Revista, v. 2, p. 112–131.

——, 1981 [1984], La evolución tectónica premisisípica del sur de México: Instituto de Geología, Universidad Nacional Autónoma de México, Revista, v. 5, p. 140–157.

Paine, W. F., and Meyerhoff, A. A., 1970, Gulf of Mexico Basin; Interactions among tectonics, sedimentation, and hydrocarbon accumulation: Gulf Coast Association of Geological Societies Transactions, v. 20, p. 5–44.

Palmer, A. R., DeMis, W. D., Muehlberger, W. R., and Robinson, R. A., 1984,

Geological implications of Middle Cambrian boulders from the Haymond Formation (Pennsylvanian) in the Marathon basin, west Texas: Geology, v. 12, p. 91–94.

Pantoja-Alor, J., 1970, Rocas sedimentarias paleozoicas de la region centro-septentrional de Oaxaca, *in* Segura, L. R., and Rodriguez-Torres, R., eds., Libro Guia de la Excursion México-Oaxaca: Sociedad Geológica Mexicana, p. 67–84.

Pantoja-Alor, J., and Robison, R. A., 1967, Paleozoic sedimentary rocks in Oaxaca, Mexico: Science, v. 157, p. 1033–1035.

Pantoja-Alor, J., Fries, C., Jr., Rincón-Orta, C., Silver, L. T., and Solorio-Munguia, J., 1974, Contribución a la geocronología del Estado de Chiapas: Asociación Mexicana de Geólogos Petroleros Boletín, v. 26, p. 205–223.

Pequegnat, W. E., Bryant, W. R., and Harris, J. E., 1971, Carboniferous sediments from Sigsbee Knolls, Gulf of Mexico: American Association of Petroleum Geologiss Bulletin, v. 55, p. 116–123.

Pérez Ramos, O, 1978, Estudio bioestrátigrafico del Paleozoico Superior del Anticlinorio de Huayacocotla en la Sierra Madre Oriental: Sociedad Geológica Mexicana Boletín, v. 39, no. 2, p. 126–135.

Pojeta, J., Jr., Kriz, J., and Berdan, J. M., 1976, Silurian-Devonian pelecypods and Paleozoic stratigraphy of subsurface rocks in Florida and Georgia and related Silurian pelecypods from Bolivia and Turkey: U.S. Geological Survey Professional Paper 879, 32 p.

Puri, H. S., and Vernon, R. O., 1964, Summary of the geology of Florida and a guidebook to the classic exposures: Florida Geological Survey Special Publication 5, 312 p.

Pyle, T. E., and 5 others, 1973, Metamorphic rocks from northwestern Caribbean Sea: Earth and Planetary Science Letters, v. 18, p. 339–344.

Quezadas Flores, A. G., 1961, Las rocas del basamento de la Cuenca de Tampico-Misantla: Asociación Mexicana de Geólogos Petroleros Boletín, v. 13, no. 9–10, p. 289–323.

Rainwater, E. H., 1962, Geologic history and oil and gas possibilities of Mississippi: Mississippi Geologic, Economic and Topographic Survey Bulletin 97, p. 77–86.

Ramírez-Ramírez, C., 1978, Reinterpretación tectónica del Esquisto Granjeno de Ciudad Victoria, Tamaulipas: Instituto de Geología, Universidad Nacional Autónoma de México, Revista, v. 2, p. 31–36.

Rankin, D. W., 1975, The continental margin of eastern North America in the southern Appalachians; The opening and closing of the proto-Atlantic Ocean: American Journal of Science, v. 275A, p. 298–336.

Robison, R. A., and Pantoja-Alor, J., 1968, Tremadocian trilobites from the Nochixtlan region, Oaxaca, Mexico: Journal of Paleontology, v. 42, p. 767–800.

Rodríguez-Torres, R., 1970, Geología metamórfica del area de Acatlán, Estado de Puebla, *in* Segura, L. R., and Rodriguez-Torres, R., eds., Libro Guia de la Excursión México-Oaxaca: Sociedad Geológica Mexicana, p. 51–54.

Ruiz-Castellanos, M., 1970, Reconocimiento geológico en el area de Mariscala-Amatitlán, Estado de Oaxaca, *in* Segura, L. R., and Rodriguez-Torres, R., eds., Libro Guia de la Excursión México-Oaxaca: Sociedad Geológica Mexicana, p. 55–66.

Sansores Manzanilla, E., and Girard Navarrete, R., 1969, Bosquejo geológico de la Zona Norte: Instituto Mexicano del Petróleo, Seminario sobre exploración petrolera, Mesa Redonda no. 2; Problemas de exploración de la Zona Norte, 36 p.

Schuchert, C., 1943, Stratigraphy of the eastern and central United States: New York, John Wiley and Sons, 1013 p.

Silva Pineda, A., 1970, Plantas del Pensilvánico de la región de Tehuacán, Puebla: Instituto de Geología, Universidad Nacional Autónoma de México, Paleontología Mexicana 29, 109 p.

Silver, L. T., and Anderson, T. H., 1974, Possible left-lateral early to middle Mesozoic disruption of the southwestern North American craton margin: Geological Society of America Abstracts with Programs, v. 6, p. 955.

Smith, D. L., 1982, Review of the tectonic history of the Florida basement: Tectonophysics, v. 88, p. 1–22.

Stone, C. G., editor, 1977, Symposium on the geology of the Ouachita Mountains; Volume 1, Stratigraphy, sedimentology, petrography, tectonics, and paleontology: Arkansas Geological Commission, 174 p.

Swartz, F. M., 1945, Mid-Paleozoic ostracode in exploratory well in Georgia; Muscle scars in *Leperditiiae* [abs.]: Geological Society of America Bulletin, v. 56, p. 1205.

Thomas, W. A., 1972, Paleozoic stratigraphy in Mississippi between Ouachita and Appalachian Mountains: American Association of Petroleum Geologists Bulletin, v. 56, p. 81–106.

——, 1973, Southwestern Appalachian structural system beneath the Gulf coastal plain: American Journal of Science, Cooper Volume, v. 273-A, p. 372–390.

——, 1985, The Appalachian-Ouachita connection; Paleozoic orogenic belt at the southern margin of North America: Annual Review of Earth and Planetary Sciences, v. 13, p. 175–199.

Thomas, W. A., and 5 others, 1989, The subsurface Appalachians beneath the Atlantic and Gulf Coast plains, *in* Hatcher, R. D., Jr., Thomas, W. A., and Viele, G. W., eds., The Appalachian-Ouachita Orogen in the United States: Boulder, Colorado, Geological Society of America, The Geology of North America, v. F-2 (in press).

Thompson, M. L., 1956, Rocas paleozoicas del sur de México, *in* Geología del Mesozoico y estratigrafía pérmica del Estado de Chiapas: Mexico, D. F., 20th International Geological Congress, Excursión C-15: p. 61–68.

Thompson, M. L., and Miller, A. K., 1944, The Permian of southernmost Mexico and its fusulinid faunas: Journal of Paleontology, v. 18, p. 481–504.

Toledo Toledo, M., 1969, Problemas de exploración de la plataforma continental de la Zona Norte: Instituto Mexicano del Petróleo, Seminario sobre exploración petrolera, Mesa Redonda No. 2; Problemas de exploración de la Zona Norte, 35 p.

Valerius, C. N., 1978, Extracts from published data pertaining to hydrocarbon possibilities in Paleozoic rocks in the Arkansas, Louisiana, Texas region with special emphasis on the Morehouse Formation: Shreveport Geological Society Special Publication, 77 p.

Vedder, J. G., McLeod, N. S., Lanphere, M. A., and Dillon, W. P., 1973, Age and tectonic implications of some low-grade metamorphic rocks from the Yucatan Channel: U.S. Geological Survey Journal of Research, v. 1, p. 157–164.

Vernon, R. C., 1971, Possible future petroleum potential of pre-Jurassic, western Gulf Basin, *in* Cram, I. H., ed., Future petroleum provinces of United States; Their geology and potential: American Association of Petroleum Geologists Memoir 15, v. 2, p. 954–979.

Walper, J. L., 1960, Geology of Cobán-Purulhá area, Alta Verapaz, Guatemala: American Association of Petroleum Geologists Bulletin, v. 44, p. 1273–1315.

Wardlaw, B. R., Furnish, W. M., and Nestell, M. K., 1979, Geology and paleontology of the Permian beds near La Delicias, Coahuila, Mexico: Geological Society of America Bulletin, part 1, v. 90, p. 111–116.

Wicker, R. A., and Smith, D. L., 1978, Re-evaluating the Florida basement: Gulf Coast Association of Geological Societies Transactions, v. 28, p. 681–687.

Wilson, J. T., 1966, Did the Atlantic close and then re-open?: Nature, v. 211, p. 676–681.

Woods, R. D., and Addington, J. W., 1973, Pre-Jurassic geological framework, northern Gulf Basin: Gulf Coast Association of Geological Societies Transactions, v. 23, p. 92–108.

ACKNOWLEDGMENTS

The authors are indebted to Exxon Company U.S.A. for the release of certain data used in the preparation of this chapter. They also appreciate and acknowledge the background information provided by C. Ramirez for the description of the pre-Triassic rocks in Mexico. R. L. Nicholas, W. A. Thomas, and G. W. Viele reviewed the manuscript and offered many excellent comments and suggestions. W. A. Thomas made available invaluable material on the pre-Triassic of the southeastern U.S. ahead of its publication in volume F-2, "The Appalachian-Ouachita Orogen in the United States," of this series (Hatcher and others, 1989).

MANUSCRIPT ACCEPTED BY THE SOCIETY MAY 22, 1989

Printed in U.S.A.

The Geology of North America
Vol. J, The Gulf of Mexico Basin
The Geological Society of America, 1991

Chapter 8

Triassic-Jurassic

Amos Salvador
Department of Geological Sciences, The University of Texas at Austin, P.O. Box 7909, Austin, Texas 78713-7909

INTRODUCTION

Geological and geophysical evidence—and confidence in the postulates of plate tectonics—indicates that in the dawn of the Mesozoic what would become the Gulf of Mexico basin area was part of a very large land mass, the supercontinent of Pangea. It is generally believed that toward the end of the Paleozoic and the beginning of the Mesozoic, Pangea grouped all the continental plates of the Earth. How all these plates were assembled is still a subject of considerable controversy, particularly how and where the North American and South American Plates fit together in the area that would eventually become the Gulf of Mexico basin and surrounding positive tectonic elements.

Many reconstructions of the Gulf of Mexico basin area during the late Paleozoic or earliest Mesozoic have been proposed (Bullard and others, 1965; Freeland and Dietz, 1971; Owen, 1976, 1983; Carey, 1958, 1976; Smith and Briden, 1977; Pilger, 1978; Salvador and Green, 1980; and many others). They all differ in the manner in which the authors interpreted how the plates were assembled, but agree, at least, in indicating that the North American and South American Plates were joined in some manner during the late Paleozoic and earliest Mesozoic as part of the emergent and stable supercontinent of Pangea.

The first recorded active events of the Mesozoic geologic history of the Gulf of Mexico basin area correspond with the beginning of the breakup of Pangea, probably during the Late Triassic, and the drifting of the North American Plate away from the African and South American Plates.

Two distinct tectono-stratigraphic periods can be recognized in the structural and stratigraphic evolution of the Gulf of Mexico Basin and surrounding areas during the Triassic and Jurassic.

1. A period of active rifting lasting from the Late Triassic to the end of the Middle Jurassic or the earliest part of the Late Jurassic. During this period, the stratigraphic events were closely controlled by the tectonic development of the area. The early part of this period of rifting is characterized by the accumulation of thick sequences of nonmarine clastics and associated volcanics in a complex system of rapidly subsiding grabens and rift basins. It started in the Late Triassic and continued until mid–Early Jurassic time. Contemporaneous marine sediments are only known from the western periphery of the Gulf of Mexico basin in east-central Mexico and adjacent areas to the west. The graben systems remained active, at least in east-central Mexico, during the late Early and early Middle Jurassic. They were filled with "red beds." No evidence of concurrent activity has been recognized in other parts of the Gulf of Mexico basin. The last part of the period of rifting, corresponding to the late Middle Jurassic (mainly the Callovian but possibly starting somewhat earlier and extending into the earliest Late Jurassic) resulted in the accumulation of extensive and thick salt deposits over large areas of the basin. The accumulation of thick sections of salt is best explained by assuming that the salt was formed in persistently subsiding grabens, some undoubtedly corresponding to those previously active. Thinner salt deposits were formed in the structurally higher blocks bounding the grabens, whereas no salt ever accumulated in the highest structurally positive areas.

2. The period of active rifting was followed, during most of the Late Jurassic, by a tectonically calm period characterized by the prolonged subsidence of the central part of the Gulf of Mexico basin. Around its subsiding central part, the basin was rimmed by broad, stable shelves and ramps on which the Upper Jurassic sequence, predominantly composed of shales and limestones, was deposited. During most of the Late Jurassic, the Florida and Yucatan platforms were emergent. Influx of coarser clastics came mainly from the north and northwest, limited at first, but copious during the latest stages of Late Jurassic deposition.

To decipher the origin and early development of the Gulf of Mexico basin it is essential to understand the stratigraphic and structural history of the Triassic and Jurassic, particularly the Late Triassic and the Early and Middle Jurassic. It was during this time that the basin was born and attained essentially the general geologic configuration that characterizes it today. The distribution and formation of the Upper Triassic–Lower Jurassic "red beds," the geometry of the rifts and basins in which they accumulated, and especially the environment, process, and time of formation of the extensive salt deposits are of particular interest because they can provide the best clues to interpret the earliest geologic events in the Gulf of Mexico basin. For this reason, the

Salvador, A., 1991, Triassic-Jurassic, *in* Salvador, A., The Gulf of Mexico Basin: Boulder, Colorado, Geological Society of America, The Geology of North America, v. J.

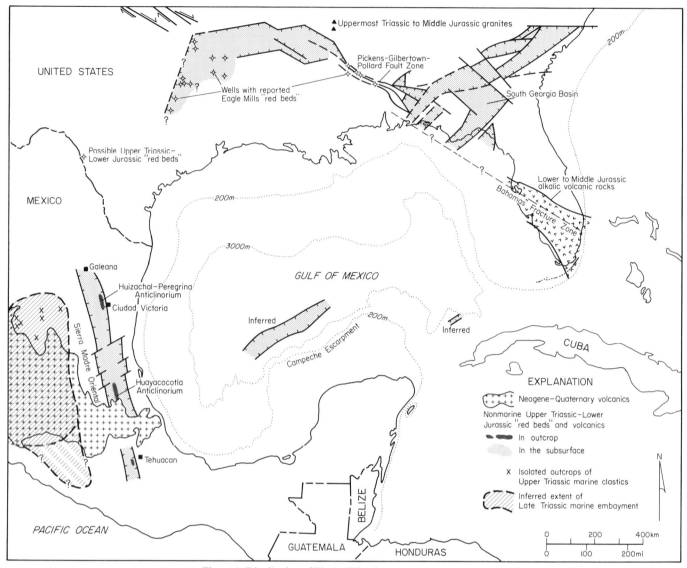

Figure 1. Distribution of Upper Triassic–Lower Jurassic rocks.

stratigraphy of the Triassic and Jurassic sediments will be described in perhaps more detail than warranted in a general synthesis of the geology of the Gulf of Mexico basin.

UPPER TRIASSIC–LOWER JURASSIC

The oldest post-Paleozoic rocks reported from the Gulf of Mexico basin and surrounding areas have been dated as Late Triassic and Early Jurassic in age. Marine rocks of this age are only known from central, east-central, and northwestern Mexico; other sequences assigned to the Upper Triassic–Lower Jurassic are composed of nonmarine clastics locally associated with volcanic rocks, and have been dated mainly on the basis of fossil plant remains, stratigraphic relations, and similarity to comparable but better dated sequences in nearby areas.

Nonmarine Upper Triassic–Lower Jurassic

Distribution. Nonmarine clastic sequences ("red beds") and associated volcanics of probable Late Triassic and Early Jurassic age are known from three main areas within the Gulf of Mexico basin (Fig. 1).

In the subsurface of South Carolina, Georgia, northwestern Florida, and southern Alabama, "red beds" have been penetrated by approximately 50 wells (Marine and Sipple, 1974; Gohn and others, 1978; Chowns and Williams, 1983). The "red beds" fill a trough that has been called the South Georgia basin. Northeast of this subsurface trough, Upper Triassic–Lower Jurassic "red beds" (the so-called Newark Supergroup) have long been known from surface exposures in elongate grabens the length of the northeastern part of the United States and eastern Canada.

In the subsurface of southern Arkansas and adjacent parts of east Texas and west-central Mississippi, "red beds" have also been penetrated by several dozen wells at depths from less than 300 m to more than 3,000 m (Scott and others, 1961; Vernon, 1971; Gawloski, 1983; Beju and others, 1986). Between the South Georgia basin and the south Arkansas area, the "red bed" sequence has been penetrated by a few wells in Mississippi and western Alabama along the Pickens-Gilbertown-Pollard fault zone. To the southwest of this fault zone, Eagle Mills "red beds" have not been reported from below the Louann Salt in the Mississippi salt basin. The Upper Triassic–Lower Jurassic "red bed" sequence crops out nowhere along the north flank of the Gulf of Mexico basin, in the southeastern United States.

In Mexico, Upper Triassic–Lower Jurassic "red beds" are exposed in several localities along a belt roughly corresponding to the front of the Sierra Madre Oriental from the vicinity of Galeana (southern Nuevo Leon) in the north to about 30 to 35 km southwest of Tehuacan (southern Puebla and northern Oaxaca) in the south. They have also been penetrated by numerous wells along the Gulf of Mexico coastal plain east of the Sierra Madre Oriental, in northern Veracruz and eastern San Luis Potosí. The geometry of the Late Triassic–Early Jurassic rift system in eastern Mexico is probably much more complex than shown in Figure 1, but additional subsurface control is needed to properly portray it.

On the basis of interpretation of seismic reflection lines, a graben filled with pre–Upper Jurassic "red beds" has been inferred to be present in the subsurface of the deep, central part of the Gulf of Mexico to the northwest of the Campeche Escarpment (the "Viejo seismic unit" of Ladd and others, 1976; the "rift basin sedimentary filling" of Buffler and others, 1980). A similar, but smaller, northeast-southwest–trending graben has been interpreted in the southeastern Gulf of Mexico, between the Florida and Yucatan platforms (Schlager and others, 1984).

Lithostratigraphy. In the United States, the Upper Triassic–Lower Jurassic nonmarine sequence has been called the Eagle Mills Formation (Shearer, 1938; Weeks, 1938). The term was first used to refer to all the nonmarine "red beds" and the salt deposits overlying upper Paleozoic beds and underlying the Smackover Formation. In today's terminology, the original Eagle Mills would include, from bottom to top, the Eagle Mills, Werner, Louann, and Norphlet Formations. The term "Eagle Mills" was later restricted to the "red bed" sequence underlying younger Jurassic evaporites or marine sediments. The Eagle Mills Formation is predominantly composed of red, purplish, greenish-gray, or mottled shales, mudstones, and siltstones; less abundant are sandstones, generally fine- to very fine-grained and gray, gray-green, or red. Coarse-grained sandstones and conglomerates are less common. Diabase dikes and sills are frequent.

In Mexico, the nomenclature of the Upper Triassic–Lower Jurassic nonmarine "red beds" has been confused by the presence of two "red bed" units in the Triassic-Jurassic section and needs clarification (see stratigraphic correlation chart, Fig. 2). The older of the two "red bed" units is presumably of Late Triassic–Early Jurassic age, and the younger unit, on the basis of its stratigraphic

position, has been assigned to the early Middle Jurassic. In the Huizachal-Peregrina anticlinorium, near Ciudad Victoria, the younger "red bed" unit unconformably overlies the older one, but the two were originally considered a single formation, the Huizachal Formation (Imlay and others, 1948). Farther south, both in the surface along the Huayacocotla anticlinorium and in the subsurface, the two "red bed" units are separated by a mostly marine unit of Early Jurassic age. In this southern area, Imlay and others (1948) did not recognize the older unit and they applied the name Huizachal to the younger "red beds." Erben (1956b) followed Imlay and others (1948). In 1959, Mixon, Murray, and Diaz set apart the two adjacent "red bed" units in the Huizachal-Peregrina anticlinorium; the older and thicker one was called the La Boca Formation and the younger and thinner unit was called the La Joya Formation. The two were included in the Huizachal Group. The terms "La Boca" and "La Joya" are now in common use in the Ciudad Victoria area, even though, as pointed out by Carrillo Bravo (1961), it would have been less confusing to retain the term Huizachal for only the Upper Triassic–Lower Jurassic unit. Farther south, where the two "red bed" units are separated by an intervening Lower Jurassic marine sequence, modern authors generally apply the term Huizachal to the older, Upper Triassic–Lower Jurassic "red beds," not to the younger unit, as originally done by Imlay and others (1948) and Erben (1956b). The younger, Middle Jurassic "red beds" are generally called the Cahuasas Formation.

The "red beds" of the La Boca Formation in the Huizachal-Peregrina anticlinorium include fluvial and alluvial fan sediments in which channel, inter-channel, bar, and debris-flow deposits have been recognized. They are composed of irregularly alternating conglomerates, sandstones, siltstones, and some volcanic rocks. The conglomerate clasts are generally angular and composed of metamorphic, sedimentary, and volcanic rocks. The predominant colors are gray, gray-green, and various shades of red. The Huizachal Formation farther south is also composed of thick-bedded conglomerates interbedded with reddish-brown, red, and greenish-gray sandstones, siltstones, and shales locally containing plant remains. Red sandy siltstone is probably the most common rock type. The conglomerate clasts are composed of quartz, igneous rocks, gneisses, schists, and various types of sandstones.

As described in detail in Chapter 6 of this volume, the Upper Triassic–Lower Jurassic "red beds" are associated throughout the Gulf of Mexico basin with dikes, flows, and sills of basalt and diabase. These intrusive and extrusive igneous bodies have been dated by isotopic methods, and range in age from 200 to 180 Ma—Early to Middle Jurassic (May, 1971; Van Houten, 1977; Chowns and Williams, 1983; Dooley and Wampler, 1983). Younger dates have been reported but are not considered reliable. Extensive dike swarms are common in the southeastern United States, both in the surface and in the subsurface. In Georgia, more than 100 of these dikes are known. They are generally vertical, 1 to 100 m in width, trend NNW–SSE and cut the Upper Triassic–Lower Jurassic "red beds," but not the overlying younger rocks. May (1971) interpreted these dike swarms to be related to the fracturing and initial rifting and breakup of Pangea.

In the subsurface of south Florida, more silicic and alkalic volcanic rocks—rhyolites and tuffs—occur with the basalts (Fig. 1). They have been dated from 212 to 170 Ma (latest Triassic to Middle Jurassic). This type of more acidic igneous activity is unique in the Triassic-Jurassic of the Gulf of Mexico basin area. It is not well understood but thought to be related to the intrusion of the dike swarms (see Chapter 6, this volume).

In central Mississippi, granite intrusions dated from 211 to 183 Ma (latest Triassic to Middle Jurassic) are known in two localities (Fig. 1). Their relation with the basalts and diabases of the same age is not known. Granites and granodiorites of Triassic age have also been reported from east-central Mexico (see Chapters 6 and 7, this volume).

Thickness. The Upper Triassic–Lower Jurassic "red bed" sequences, both in the United States and in Mexico, characteristically show abrupt lateral changes in thickness from more than 2,000 m to just a few meters or to nothing in very short distances.

Chowns and Williams (1983) reported that a well in central Georgia penetrated a complete section of "red beds" with a thickness of about 2,200 m. Other wells in the South Georgia basin have penetrated at least 1,220 m of Upper Triassic–Lower Jurassic rocks without reaching the base of the unit (Chowns and Williams, 1983). Daniels and others (1983) calculated thicknesses of about 3,500 m in east-central Georgia on the basis of

magnetic surveys. In northern Florida, Arden (1974a, 1974b) estimated a thickness of at least 1,830 m for the "red beds" from seismic data.

The thickest known section of Eagle Mills Formation in the southern Arkansas area is 2,124 m, penetrated between 558 and 2682 m (1,832 and 8,800 ft) by the Humble Oil & Refining Co. No. 1 G. B. Royston (Scott and others, 1961). Several other wells in this area have penetrated more than 1,200 m of Eagle Mills and were still in the formation at total depth.

In Mexico, the La Boca Formation attains a thickness of 2,380 m in the Cañon de la Boca in the Huizachal-Peregrina anticlinorium near Ciudad Victoria (Carrillo Bravo, 1961), but is known to be much thinner in several nearby localities. The Huizachal Formation is 1,300 to 1,400 m thick in the Huayacocotla anticlinorium, 350 km south of Ciudad Victoria, and 300 to 500 m in southern Puebla and northern Oaxaca. Wells in the east-central Mexico coastal plain have drilled through as much as 1,700 to 1,800 m of Upper Triassic–Lower Jurassic "red beds" without reaching the base.

Stratigraphic relations. In the South Georgia basin and in the southern Arkansas area, the Eagle Mills Formation unconformably overlies sediments of Paleozoic age. Pebbles and cobbles of Paleozoic limestone are common in the Eagle Mills basal section. The Eagle Mills is unconformably overlain in turn by various units ranging in age from Middle Jurassic to Late Cretaceous. In Mexico, the La Boca and Huizachal Formations show similar stratigraphic relations; they unconformably overlie Pa-

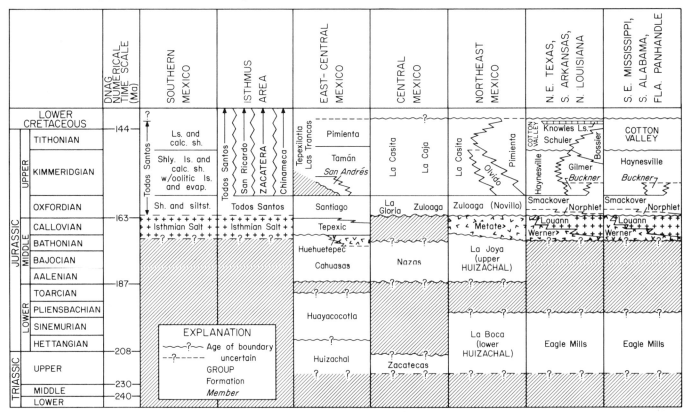

Figure 2. Simplified stratigraphic correlation chart (for more complete chart, see Plate 5 in pocket).

leozoic beds (Carboniferous, Permian) and are also unconformably overlain by Lower Jurassic to Lower Cretaceous sedimentary units.

Biostratigraphy and age. Megafossils are rare in the Upper Triassic–Lower Jurassic "red beds," and they consist almost exclusively of plant remains. Plant impressions found in cores cut in the Eagle Mills Formation in the No. 1 G. B. Royston well in southern Arkansas have been identified as *Macrotaeniopteris magnifolia* of Late Triassic to Early Jurassic age (Scott and others, 1961). Palynomorphs have been recovered from several wells in east Texas, southern Arkansas, and the South Georgia Basin (Gawloski, 1983; Beju and others, 1986; Moy and Traverse, 1986; Traverse, 1987). They denote a Late Triassic age for the Eagle Mills Formation.

Fossil plants collected in the upper part of the La Boca Formation in the Ciudad Victoria area have been identified as *Pterophyllum fragile, P. inaequale, Cephalotaxopsis carolinensis,* and *Podozamites* sp., which indicate a Late Triassic age for the upper part of the formation (Mixon and others, 1959; Carrillo Bravo, 1961). Also from the upper part of the La Boca Formation, plant remains have been identified as *Williamsonia netzahualcoyotlii* of probable Early Jurassic age (Carrillo Bravo, 1961). Clark and Hopson (1985) reported on discovering the skull of a mammal-like reptile from the middle part of the La Boca Formation, which they believed "suggests a Jurassic age for that part of the formation from which the fossil was collected."

In the Huayacocotla anticlinorium, plant remains collected in the lower part of the Huizachal Formation have been identified as *Todites carrilloi, Mertensides bullatus, Thaumatopteris* sp. cf. *T. kochibei, Pterophyllum longifolium, Stenopteris* sp. cf. *S. desmomera,* and *Asterotheca meriani* of probable Late Triassic age, and collections from the upper part of the formation included *Otozamites hespera, Otozamites reglei, Ptilophyllum acutifolium,* and *Williamsonia netzahualcoyotlii,* believed to be of Early Jurassic age (Silva-Pineda, 1963, 1979, 1981 (1983); Carrillo Bravo, 1965).

As mentioned earlier, isotopic dating of dikes and sills of diabase intruding the "red beds" and basalt flows interbedded in them indicate an Early to Middle Jurassic age (200 to 180 Ma), confirming the Late Triassic–Early Jurassic age for the "red beds."

Finally, a Late Triassic–Early Jurassic age for the Eagle Mills, La Boca, and Huizachal Formations is suggested by the similarity in lithologic composition and tectono-stratigraphic framework between these formations and the Newark Supergroup of eastern North America, which on the basis of spores, pollen, pelecypods, arthropods, reptiles, and fishes, has been dated as ranging in age from Late Triassic to Early Jurassic (Cornet and others, 1973; Cornet and Traverse, 1975; Manspeizer, 1982; Olsen and others, 1982).

Conditions of deposition—Paleogeography. The lithologic composition, the nature of its scarce fossils, the abrupt and extreme variation in thickness, and the stratigraphic relations of the Upper Triassic–Lower Jurassic "red beds," strongly suggest

that they represent the filling of grabens, half grabens, or rift basins that actively subsided during deposition and into which the products of the erosion of the bounding high-standing areas were deposited in the form of prograding alluvial fans, and fluvial, delta plain, or fresh-water-lake deposits. Coalescing alluvial fans containing coarse clastics were more common along the graben margins; finer clastics were deposited farther away from the source areas; swamps and shallow fresh-water lakes formed in the flat, central part of the troughs where subsidence was greatest. Organic-rich muds and local coal beds are characteristic of the deposits of these swamps and lakes. Deltas developed where rivers entered the lakes. Igneous activity in the form of dikes, sills, and lava flows was common along the rifts. Deposition on opposite sides of the active contemporaneous faults bounding the grabens accounts for the abrupt changes in thickness observed in the Upper Triassic–Lower Jurassic "red beds." The interpreted geographic distribution of the Late Triassic–Early Jurassic grabens and half grabens is shown in Figure 1.

Marine Upper Triassic

No Upper Triassic marine beds have been reported from the Gulf of Mexico Basin area. West of the basin, in the central Mexico states of Zacatecas and San Luis Potosí, marine, fossiliferous Upper Triassic clastics with occasional thin radiolarite beds and andesitic pillow lavas are known from a few small, isolated outcrops (Fig. 1). They were probably deposited in the eastern shallow shelf of an embayment of the Pacific Ocean, which did not reach the Gulf of Mexico basin area. These marine Upper Triassic beds of central Mexico and their fauna were first described by Burckhardt (1905) and by Burckhardt and Scalia (1906), and more recently by Cantú Chapa (1969b), Carrillo Bravo (1971), Martínez Pérez (1972), López Ramos (1980), and López Infanzón (1986).

Marine Lower Jurassic

Distribution. Lower Jurassic marine beds in the Gulf of Mexico basin area are restricted to east-central Mexico, in the eastern part of the state of Hidalgo and adjoining parts of the states of San Luis Potosí, Veracruz, and Puebla, where they crop out over a distance of about 130 km along the elongated and structurally complex Huayacocotla anticlinorium (Fig. 3). To the north and east of this foothill structure, Lower Jurassic marine beds have been penetrated by more than 40 wells drilled in search of petroleum. The known occurrences of the marine Lower Jurassic are in a belt about 100 km wide and 300 km long that corresponds roughly with the belt along the front of the Sierra Madre Oriental, where the Huizachal "red beds" also are found.

Lithostratigraphy. The Lower Jurassic marine section has been called the Huayacocotla Formation (Imlay and others, 1948) and is predominantly composed of dark gray to black shales and siltstones, commonly micaceous and carbonaceous and locally calcareous. The sandstone content of the formation

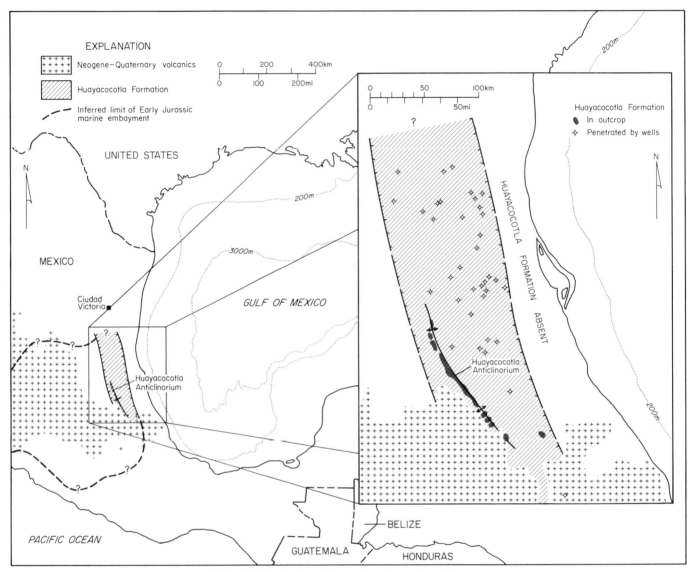

Figure 3. Distribution of Lower Jurassic marine sediments.

seems to increase from south to north, and upward in the section: in northern Puebla, the Huayacocotla Formation is almost entirely composed of dark shales and siltstones. Northward from this area, thin beds of fine- to medium-grained sandstone become more abundant in the section, particularly in its upper part. In the northernmost surface exposures, as well as in the subsurface, north and northeast of the outcrop area, the Huayacocotla Formation is composed predominantly of sandstones and siltstones, and its uppermost part is generally sandier, more carbonaceous, and contains abundant plant remains. The basal part of the formation, when exposed, is generally represented by a 20-m section of conglomerates and coarse-grained sandstones, overlain by an interval of sandy limestones and calcareous sandstones 40 to 50 m thick. The conglomerates contain clasts derived from the underlying Huizachal "red beds" (Carrillo Bravo, 1965).

Thickness. The Huayacocotla Formation seems to thin progressively from its outcrop area in the south toward the north

and east. It is absent over the Tuxpan platform. In the outcrop area, along the trend of the Huayacocotla anticlinorium, the formation has been reported to range between 900 and 1,500 m in thickness (Carrillo Bravo, 1965; Schmidt-Effing, 1980). Wells drilled along the coastal plain have penetrated complete sections of the formation ranging from 425 to 1,350 m in the south and west, and from 125 to 150 m toward the north and east. Other wells have drilled through Huayacocotla incomplete sections of up to 1,000 m without reaching the base of the formation.

Stratigraphic relations. The Huayacocotla Formation is generally unconformably underlain by the "red beds" of the Huizachal Formation, and unconformably overlain by the Middle Jurassic Cahuasas "red beds," or by various Upper Jurassic formations. The possibility cannot be discarded, however, that the Huayacocotla Formation grades north and east into nonmarine beds that are difficult to distinguish from the Huizachal sediments.

Biostratigraphy and age. On the basis of abundant fossils, the Huayacocotla Formation has been dated as Sinemurian in age, possibly extending upward into the early Pliensbachian. The base of the formation could be of Hettangian age, but no fossil evidence exists that permits the identification of this lowermost stage of the Lower Jurassic. The Hettangian, or part of it, may be represented in east-central Mexico by the lowermost part of the Huayacocotla Formation, by the uppermost beds of the Huizachal Formation, or by the hiatus indicated by the unconformity that separates these two units.

The shalier, more marine part of the Huayacocotla Formation contains a rich ammonite fauna characterized by numerous species of the genera *Coroniceras, Arnioceras, Euagassiceras, Oxynoticeras, Pleurechioceras, Vermiceras, Echioceras, Microderoceras,* and many others, all indicative of a Sinemurian age (Burckhardt, 1930; Erben, 1954, 1956a; Flores López, 1967, 1974). Sandier parts of the formation are not as rich in ammonites but contain a fauna of pelecypods and other shallow-water marine fossils. The Sinemurian ammonite fauna of the Huayacocotla Formation is closely similar or identical to contemporaneous faunas in Europe and elsewhere (Imlay, 1980).

Evidence for extending the age of the Huayacocotla to include the earliest Pliensbachian on the basis of ammonites is questionable (Burckhardt, 1930; Erben, 1956a; Imlay, 1980; Schmidt-Effing, 1980). Plant remains and small pelecypods collected from the uppermost, littoral part of the Huayacocotla Formation, however, are said to indicate an early Pliensbachian age (Burckhardt, 1930; Maldonado-Koerdell, 1950).

Conditions of deposition—Paleogeography. The lithologic composition and the fossils collected in the Huayacocotla Formation indicate that it was deposited in a marine environment in its southern area of occurrence (northern Puebla, eastern Hidalgo, and adjoining part of Veracruz), and that this marine environment passed gradually northward and eastward into a deltaic and fluvial environment of deposition. The fact that the belt of occurrence of the Huayacocotla marine beds corresponds closely with the belt along the front of the Sierra Madre Oriental, from where the underlying Huizachal "red beds" are known, suggests that the Huayacocotla Formation was deposited in the same graben system in which the Huizachal "red beds" had been deposited, and that this graben, active during the Late Triassic and earliest Jurassic, was still subsiding, though perhaps at a slower rate, during the middle part of the Early Jurassic (Sinemurian and at least early Pliensbachian). The occurrence of a marine section of Sinemurian age in east-central Mexico indicates that a connection with the ocean, most probably with the Pacific, as was the case during the Late Triassic, persisted during at least part of the Early Jurassic. Whereas during the Late Triassic the embayment of the Pacific Ocean extended only as far as the state of Zacatecas and the western part of San Luis Potosí, during the Early Jurassic, or at least during the Sinemurian, this ocean embayment spread eastward, reaching the NNW-SSE–trending graben in which only nonmarine sediments had been accumulating until then. The marine invasion from the Pacific extended

eastward only as far as the eastern bounding fault zone of the graben, and northward to eastern Hidalgo, southern San Luis Potosí, and adjoining parts of Veracruz. It did not seem to have reached the Ciudad Victoria area or to cover the Tuxpan platform. Southward, marine Lower Jurassic beds are only known as far south as northern Puebla; south of this area, older rocks are hidden below the thick Neogene-Quaternary volcanics of the Trans-Mexican neovolcanic belt. Lower Jurassic beds, however, have been reported from southwestern Mexico, in the states of Guerrero and Oaxaca, south of the Trans-Mexican neovolcanic belt (Imlay, 1953; Erben, 1956a). A connection between east-central Mexico and the Atlantic Ocean to the east, as proposed by Erben (1956a, 1957a) and Schmidt-Effing (1980), does not seem possible. Evidence indicates that the southern part of the North Atlantic did not open until late in the Middle Jurassic (Gradstein and Sheridan, 1983; Sheridan, 1983).

MIDDLE JURASSIC

In the Gulf of Mexico basin, a comprehensive section of Middle Jurassic beds is only known from east-central Mexico. It is composed of a lower sequence of nonmarine "red beds" (the Cahuasas and La Joya Formations), with local calcareous and evaporitic developments (the Huehuetepec Formation), and an upper marine interval of shallow-watr marine calcarenites and shales (the Tepexic Formation). The lower, nonmarine part of the section has not yielded identifiable fossils, and its age is only approximately known. It can be safely assumed, however, that it probably represents the lower part of the Middle Jurassic—parts or all of the Aalenian, Bajocian, and Bathonian stages. The calcarenitic unit has been conclusively dated as of Callovian age on the basis of its contained fauna. It is overlain by a sequence of marine dark shales, the base of which, at least in places, contains diagnostic latest Callovian ammonites. The major part of this shale unit—the Santiago Formation—is of Late Jurassic (Oxfordian) age (Fig. 2) and will be discussed in the following section of this chapter under "Upper Jurassic."

Extensive and locally very thick evaporite deposits, predominantly composed of halite, are known to occur over most of the Gulf of Mexico basin. Whereas the exact age of these salt deposits remains one of the most important unanswered questions concerning the stratigraphy of the basin, most authors concur today in assigning a late Middle Jurassic (Callovian) to perhaps earliest Late Jurassic age to the salt section. It will be discussed, therefore, in the following pages under "Middle Jurassic."

Correlation between the Middle Jurassic section in east-central Mexico and the widespread salt deposits of the Gulf of Mexico Basin is still speculative.

Lower Middle Jurassic nonmarine beds

Distribution. Nonmarine "red beds," probably representing the lower part of the Middle Jurassic, are known in east-central Mexico from a belt 400 km long and about 100 km wide along the front of the Sierra Madre Oriental, from the Ciudad Victoria area southward to the northern part of the state of Puebla. The distribution of the Middle Jurassic "red beds" corresponds closely with that of the Upper Triassic–Lower Jurassic beds in east-central Mexico. The Middle Jurassic "red beds" crop out along the Huizachal-Peregrina and Huayacocotla anticlinoriums, and have been penetrated by numerous wells along the coastal plain (Fig. 4). Nonmarine "red beds" are also known from several localities in north-central Mexico west of the front of the Sierra Madre Oriental (northern Zacatecas and eastern Durango). They have been mapped as the Nazas Formation, and even though their age and their stratigraphic relation with the "red beds" cropping out along the eastern foothills of the Sierra Madre Oriental have not been clearly established, they are believed to be of probable early Middle Jurassic age (López Infanzón, 1986). No sediments of equivalent age have been reported from any other sector of the Gulf of Mexico basin. Marine beds of

Middle Jurassic age (Bajocian, Bathonian, and Callovian) are known, however, from southwestern Mexico, in the states of Oaxaca and Guerrero (Erben, 1956b).

Lithostratigraphy. The Middle Jurassic "red beds" are now generally known as the La Joya Formation in the Huizachal-Peregrina anticlinorium, near Ciudad Victoria (Mixon and others, 1959), and as the Cahuasas Formation in the subsurface of the Tampico–Poza Rica area and along the Huayacocotla anticlinorium (Carrillo Bravo, 1965; Petrazzini and Basañez, 1978). Imlay and others (1948) and Erben (1956a) referred to the Cahuasas Formation of current use as Huizachal Formation. The correlation between the Cahuasas and the La Joya Formation is still tentative; it is based mostly on lithologic resemblance and similar stratigraphic relations. A 250-km gap exists, however, between the outcrops in the Huizachal-Peregrina and Huayacocotla anticlinoriums that has not been adequately bridged by subsurface information.

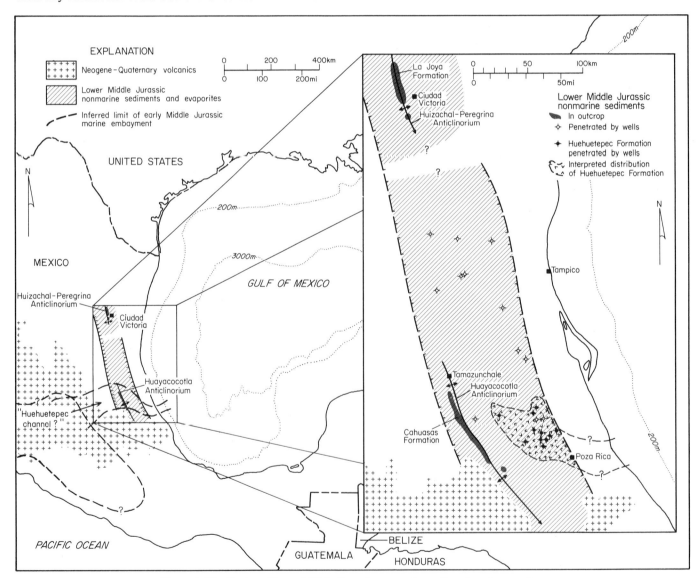

Figure 4. Distribution of lower Middle Jurassic nonmarine sediments and evaporites.

The Cahuasas Formation is composed of sandstones, conglomerates, siltstones, and shales of predominantly red color. Conglomerates are more common toward the base; they are poorly sorted, thick-bedded, and formed of subangular clasts of quartz and various kinds of sandstone. Cross-bedding is common in the sandstones and conglomerates.

The La Joya Formation is similarly composed of red to green or greenish-gray sandstones, conglomerates, and lesser amounts of shale and siltstone. The conglomerates are also more abundant at the base; they are poorly sorted, and contain clasts of igneous and metamorphic rocks as well as fragments of the underlying La Boca "red beds." A thin bed of dense, unfossiliferous fresh-water limestone, 1 to 4 m thick, often occurs above a basal conglomeratic section that varies in thickness between a few centimeters and 20 m. Laminated carbonates and calcareous mudstones are also common in the La Joya Formation (Bracken, 1984).

In the northernmost part of the state of Puebla and adjoining parts of Veracruz, northwest of Poza Rica (Fig. 4), more than 25 wells have penetrated as much as 260 m of interbedded marine limestones and calcarenites, evaporites, and "red beds" overlying the Cahuasas Formation. This section has been called the Huehuetepec Formation (González García, 1970). The formation is composed of two units. The lower unit, 20 to 30 m thick, is made up of micritic limestones, calcareous siltstones, fossiliferous limestones, and calcarenites formed predominantly by arenaceous foraminiferal tests, shell fragments, ostracodes, algae, and fecal pellets (*Favreina*). The fossils found in the limestones are principally shallow-water pelecypods, undiagnostic for age determination. The upper unit, 120 to 150 m thick, is composed of evaporites—salt and anhydrite—interbedded with red and grayish-green shales similar to the "red beds" of the underlying Cahuasas Formation. The anhydrite occurs in beds and nodules within the shales; the salt also occurs in beds and as crystals within the red shales.

The Huehuetepec Formation may be interpreted as a transitional unit between the underlying nonmarine "red beds" of the Cahuasas Formation and the overlying marine calcarenites of the Tepexic Formation.

Thickness. Thicknesses reported for the Cahuasas and La Joya Formations seldom exceed 100 m, but locally these formations may be much thicker. Thickness changes are generally abrupt and occur over very short distances. In the Huizachal-Peregrina anticlinorium, the La Joya Formation ranges between 50 and 150 m, most commonly between 60 and 100 m. Thicknesses ranging between 40 and 325 m have been measured for the Cahuasas Formation in the Huayacocotla anticlinorium. In the subsurface, wells have penetrated sections of Cahuasas Formation ranging from 20 to 75 or 80 m, occasionally reaching 120 m. An anomalously thick section of Cahuasas "red beds"—more than 1,200 m—has been reported by Carrillo Bravo (1965) in the canyon of the Amajac River, 20 km south-southwest of Tamazunchale, San Luis Potosí.

Stratigraphic relations. In the area of its southern occurrences, the Middle Jurassic "red beds" (the Cahuasas Formation) unconformably overlie the Huayacocotla Formation and underlie the Callovian Tepexic calcarenites or younger Jurassic formations. The contact between the Cahuasas and overlying formations has been interpreted to be unconformable in outcrops along the Huayacocotla anticlinorium, but may be conformable and even transitional farther east, where the Huehuetepec Formation occurs at the top of the Cahuasas Formation. Farther to the north, the La Joya Formation overlies, with marked angular unconformity, the La Boca Formation, as well as Paleozoic sediments and Precambrian igneous and metamorphic rocks (Mixon and others, 1959; Carrillo Bravo, 1961). It underlies the Upper Jurassic Zuloaga (Novillo) Formation—unconformably, according to some authors (Mixon and others, 1959; Carrillo Bravo, 1961), and conformably, according to others (Bracken, 1984).

Biostratigraphy and age. No fossils have been reported from the Middle Jurassic "red beds." Their age has to be deduced, therefore, from their position in the stratigraphic section; because the Cahuasas Formation is bracketed between the mid-Lower Jurassic and the Callovian, it has generally been considered to represent the early part of the Middle Jurassic—part or all of the Aalenian, Bajocian, and Bathonian stages, depending on how much missing time is postulated at the unconformities bounding the unit. The Huehuetepec Formation, if it can be considered a transition between the Cahuasas and the overlying Tepexic calcarenites, could represent the latest Bathonian and/or the earliest part of the Callovian.

The age of the La Joya Formation is much more loosely circumscribed: it overlies the Upper Triassic–Lower Jurassic(?) La Boca "red beds" and underlies the Upper Jurassic (Oxfordian) Zuloaga (Novillo) Formation. It may be restricted, however, as the Cahuasas is, to the lower part of the Middle Jurassic.

Conditions of deposition—Paleogeography. The lithologic composition, the absence of fossils, the areal distribution, the abrupt thickness changes, and the stratigraphic relations suggest that the Middle Jurassic "red beds" accumulated as prograding alluvial fans and as fluvial and lacustrine deposits in the same rift basin or graben system in which the Upper Triassic and Lower Jurassic sediments were deposited. The Middle Jurassic "red beds" were the product of the erosion of high-standing horst blocks bounding the graben or grabens. The presence of nonmarine "red beds" overlying the marine Huayacocotla Formation in the rift basin or basins of east-central Mexico is indicative of a retreat of the sea from the pre-existing marine embayment during the early part of the Middle Jurassic. Marine sequences of Bajocian, Bathonian, and Callovian age are known from southwestern Mexico, in the states of Oaxaca and Guerrero (Erben, 1956b), indicating that the embayment of the Pacific, which briefly advanced to east-central Mexico during the Early Jurassic, retreated westward during the early Middle Jurassic, leaving east-central Mexico exposed to erosion or as the site of nonmarine "red-bed" deposition. Locally, in northernmost Puebla and adjoining Vera-

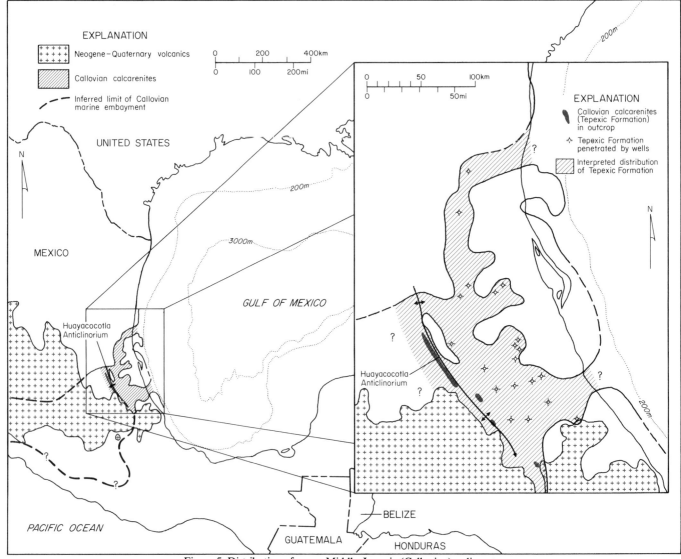

Figure 5. Distribution of upper Middle Jurassic (Callovian) sediments.

cruz, the late Bathonian may have witnessed the beginning of a new Pacific marine invasion over a flat coastal area (sahbka) in which the shallow-water carbonates, "red beds," and evaporites of the Huehuetepec Formation were deposited (Fig. 4).

Upper Middle Jurassic (Callovian) calcarenites

Distribution. Over parts of east-central Mexico, the lower Middle Jurassic "red beds" or the Huehuetepec evaporitic section are overlain by a thin sequence of calcarenites and marine shales. This Middle Jurassic (Callovian) calcarenite sequence crops out along the Huayacocotla anticlinorium in northern Puebla, eastern Hidalgo, and adjoining parts of Veracruz, and has been penetrated by numerous wells in northern Veracruz (Fig. 5). The Callovian calcarenites are absent in the Huizachal-Peregrina anticlinorium to the north and over the Tuxpan platform to the east.

Even though little has been published about the subsurface distribution of the Callovian calcarenites, some references in the literature suggest that the formation may not be continuous over the whole area of distribution, being locally absent due to erosion or nondeposition (González García, 1969; Toledo Toledo and Tavitas Galván, 1972).

Lithostratigraphy. The Callovian calcarenites have been called the Tepexic Formation (Erben, 1956a, 1956b). The unit is composed of fine- to coarse-grained, locally conglomeratic gray to dark gray or brown calcarenites interbedded with thinner beds of dark gray to black shales and calcareous shales. Calcareous sandstones and thin beds of oolitic limestone also have been reported from the formation. Occasionally, the Tepexic Formation contains thin quartz sandstone lenses, and quartz grains are not unusual in the calcarenites. Pelecypod shells are generally abundant; ammonites less so.

Thickness. The thickness of the Tepexic Formation rarely exceeds 100 m and generally ranges between 15 and 80 m. A maximum thickness of 225 m has been reported from a surface section. In the subsurface the formation ranges from 5 to 150 m (Erben, 1956a, 1956b; González García, 1969; Toledo Toledo and Tavitas Galván, 1972).

Stratigraphic relations. The Tepexic Formation generally unconformably overlies the Cahuasas "red beds." In northernmost Puebla and neighboring Veracruz, the Tepexic calcarenites may overlie the Cahuasas conformably, with a transition zone represented by the Huehuetepec carbonates, "red beds," and evaporites. The Tepexic underlies conformably, and in places transitionally, the Santiago Formation.

Biostratigraphy and age. The most common fossil in the Tepexic Formation is the pelecypod *Liogryphaea nebrascensis.* Oysters, echinoids, sponge spicules, and benthonic foraminifers are less common. Ammonites identified as *Neuqueniceras neogaeum, Erymnoceras,* cf. *E. mixtecorum,* and *Reineckeia* spp. have been collected from the dark shales interbedded with the calcarenites in the upper part of the formation (Erben, 1956a, 1956b; Cantú Chapa, 1971; Imlay, 1980).

This faunal assemblage indicates that the Tepexic Formation represents all or most of the Callovian Stage. *Liogryphaea nebrascensis* appears to range from the early Callovian to the early Oxfordian, whereas *Neuqueniceras neogaeum* and the unidentified species of *Reineckeia* indicate a mid-early to early late Callovian age for the upper part of the Tepexic Formation (Imlay, 1980); the lower part of the formation may represent most or all of the early part of the Callovian. In places, the age of the top of the formation may include the latest Callovian. The ammonite fauna of the Tepexic Formation is similar to that of Chile and other areas of the Pacific Realm (Imlay, 1980).

Conditions of deposition—Paleogeography. The Tepexic Formation was deposited in a shallow-water marine environment where recurrent periods of higher and lower energy conditions resulted in the formation of alternating beds of calcarenite and shale. The irregular distribution of the formation may indicate the existence of low islands separated by shallow tidal channels and arms of the sea in east-central Mexico during the Callovian. The Tepexic calcarenites and shales were probably deposited in these shallow channels and along the mainland coast as beach and shallow-shelf sediments (Fig. 5).

Upper Middle Jurassic(?) salt deposits

Large areas of the Gulf of Mexico basin are underlain by salt deposits. The salt has been deformed into diapiric pillows, walls, ridges, domal features, and spines, or it may form the nondiapiric core of salt-supported anticlines, swells, and extensive massifs. Whereas the base of the salt layer is often undeformed, the top nearly always shows flow to a lesser or greater extent. This distinctive type of deformation provides an important element to the structural character of those areas in the basin where significant salt deposits are present (see Chapter 5 of this volume).

Probably no other stratigraphic interval is as important in understanding the geology of the Gulf of Mexico basin as the Middle Jurassic salt. Information about the distribution of the salt in space and time is not only essential in understanding the distinctive structural character of the various sectors of the basin, but it also gives, as discussed in Chapter 14 of this volume, some of the best clues concerning the initial stages of the development of the basin.

Finally, salt-related structures—salt-cored anticlines, deep-seated salt domes, and piercement salt domes—provide the trapping mechanism for the accumulation of important volumes of oil and gas in the Gulf of Mexico basin.

For all these reasons, the salt deposits of the Gulf of Mexico basin—their distribution, thickness, composition, time and means of deformation, form of deposition, and age—have been the subject of numerous and varied studies. The literature on the subject is consequently extensive (Halbouty, 1979, for example, listed 77 pages of references).

Distribution. Subsurface control of the occurrence and distribution of the salt deposits in the Gulf of Mexico basin is limited for the most part to wells that have penetrated diapiric salt features which have risen to depths attainable by drill holes. The "mother salt layer" has only been penetrated along the northern rim of the basin, where it lies at shallower depths. Most of the information concerning the distribution and type of occurrence of the salt has been derived from seismic reflection data. Fortunately, due to the sharp change in acoustic impedance at the top and base of the salt, and when not masked by complex systems of diffractions, the boundaries of the salt can be recognized reliably in seismic sections. Diapiric salt features can generally be identified on the basis of their structural character, even though it is not always easy or possible to differentiate between salt and shale diapirs.

The extensive seismic surveys carried out in the Gulf of Mexico basin in the past two decades by the petroleum industry and various academic institutions have contributed a great amount of information concerning the salt deposits of the basin. More information is being added almost daily. Unfortunately, much of this geophysical information is proprietary, and not available at this time. Our knowledge of the distribution of the salt deposits in the Gulf of Mexico basin should, therefore, improve substantially during the next decades, as new nonproprietary seismic surveys are conducted, as some previously unavailable information becomes public, and as additional drilling takes place in new and deeper parts of the basin.

Available information indicates that salt deposits are known in two separate and extensive regions of the Gulf of Mexico basin (Fig. 6 and Plate 2): a northern region, including part of northeastern Mexico, the Gulf coastal plain of the states of Texas, Louisiana, Mississippi, and Alabama, the Florida panhandle, and adjacent parts of the continental shelf and slope of the Gulf of Mexico; and a southern region, including part of the deep Gulf of Mexico (Sigsbee Plain) immediately north of the Campeche Escarpment, the Bay of Campeche, parts of the Mexican states of

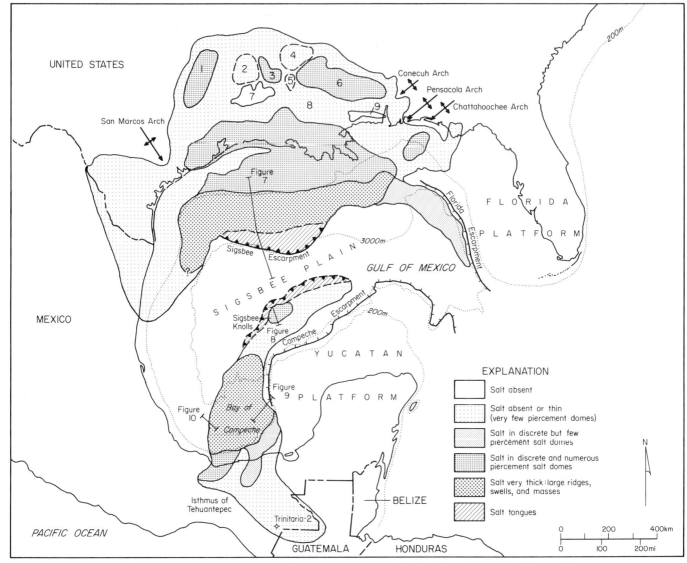

Figure 6. Distribution of upper Middle Jurassic salt deposits. 1: East Texas basin; 2: Sabine uplift;
3: North Louisiana salt basin; 4: Monroe uplift, 5: La Salle arch; 6: Mississippi salt basin; 7: Jasper arch;
8: Adams County high; 9: Wiggins uplift.

Veracruz, Chiapas and Tabasco, and a small part of northern
Guatemala. Between these two regions, salt seems to be absent
from the western flank of the basin in east-central Mexico, from
the central part of the western shelf and slope of the Gulf of
Mexico, from most of its deep central part, the Sigsbee Plain, and
from the southeastern part of the Gulf, between the Florida and
Yucatan platforms. The postulated absence of salt has been based
principally on the fact that the typical salt-flow structures, observed wherever salt is known to occur, have not been recognized
in the seismic reflection profiles across these areas.

The presence of a salt-free gap between the two separate
regions of salt occurrence raises a question: Is the salt in the
northern region the same age as that of the southern region? In
other words, was the salt in both regions formed essentially at the
same time during a single episode of extensive sea-water evaporation and salt precipitation? The assumption has been made here
that the salt deposits of the two areas described above, and shown
in Figure 6, are indeed the same age. This assumption, while
logical and supported by considerable evidence, as will be discussed later, is by no means irrefutable. It has been accepted as
valid, however, until new information proves it wrong.

Salt distribution is not uniform throughout the two regions
of salt occurrence; nor are the nature, size, and distribution of the
salt features.

In the northern region, south of the northern pinchout of the
salt, it is possible to recognize four distinct belts of salt occurrence
(Fig. 6).

1. The northernmost area of salt occurrence is characterized
by a series of basins containing numerous diapiric salt features
(the Interior salt basins), separated by structural highs where

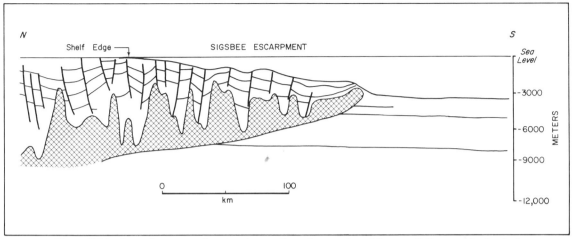

Figure 7. North-south cross section showing southward squeezing of Louann Salt caused by load of prograding Pliocene-Pleistocene sediments (modified after Humphris, 1978). For location see Figure 6.

diapiric salt features are not recognized and the salt is either thin or absent. From west to east, these alternating features are: the East Texas basin (1), the Sabine uplift (2), the North Louisiana salt basin (3), the Monroe uplift (4), the La Salle arch (5), and the Mississippi salt basin (6). It has been suggested (Andrews, 1960; Murray, 1961) that the northern limit of "thick" salt may be controlled by fault systems known from around the northern periphery of the Gulf of Mexico basin (see Plate 2 and Chapter 3 of this volume). On the other hand, it is possible that the fault systems may have been the result of the flow of the salt along its depositional pinchout.

2. South of the Interior salt basins is a structurally high belt where salt is thin or absent. It includes, from west to east, the Jasper arch (7), the Adams County high (8), and the Wiggins uplift (9).

3. This structurally high belt is, in turn, limited to the south by an area including coastal northeast Texas, southern Louisiana, and adjacent parts of the continental shelf of the Gulf of Mexico, where salt domes are numerous but are clearly separated and form discrete features.

4. Finally, in an area immediately south of the previous one that covers the northern slope of the Gulf of Mexico, the salt forms large stocks, ridges, walls, pillows, swells, and massifs rising above an almost continuous mass of salt. The salt features rise to relatively shallow depths and are clearly reflected in the bathymetry of the floor of the Gulf (see Plate 1), indicating an active and recent history of flow and uplift. The boundary between this area of almost continuous salt masses and the central, deep part of the Gulf of Mexico, where salt appears to be absent, is generally abrupt. However, along the Sigsbee Escarpment, at the base of the Texas-Louisiana continental slope, for a distance of almost 400 km (Figs. 6 and 7 and Plate 2), the salt seems to extend southward as if extruded or squeezed laterally over undisturbed sediments of Tertiary age (Amery, 1969; Humphris, 1978; Watkins and others, 1978; Foote and others, 1983; Ray, 1988). This belt of salt squeezing along the Sigsbee Escarpment has been explained as caused by the weight of the thick, southward-prograding Pliocene and Pleistocene sediments whose depocenters were located directly to the north of the Sigsbee Escarpment. As the Pliocene-Pleistocene sedimentary wedge prograded southward, it squeezed the highly ductile salt ahead of it, not unlike a rolling pin squeezing the dough as it is rolled over it.

West of these four distinctive salt provinces, the salt pinches out westward, is thin or absent over the San Marcos arch, and manifests itself in only half a dozen domes in south Texas. No salt domes have been reported from northeastern Mexico or the continental shelf bordering south Texas and northeast Mexico.

Within the southern area of salt occurrence, three distinct provinces can likewise be recognized.

1. A northern area located in the deep Sigsbee Plain, immediately north of the Campeche Escarpment where salt is relatively thin. In the central part of this area, however, numerous large salt domes penetrate the near-horizontal sedimentary layers. A few of these diapiric features rise to near the floor of the Gulf, forming conspicuous humps—the so-called Sigsbee Knolls—in the otherwise flat and featureless bathymetric surface (Martin, 1980; Foote and others, 1983) (Plate 1 and Fig. 8; for location see Fig. 6).

2. In the Bay of Campeche, the salt is present in large ridges, pillows, swells, and massifs rising above an almost continuous mass of salt. As in the Texas-Louisiana slope area, the salt features in the Bay of Campeche rise to very shallow depths, nearly to the floor of the Gulf, and are clearly reflected in the bathymetric surface (Plate 1 and Fig. 9; for location see Fig. 6), indicating that the salt has flowed and risen in very recent times or even that it may still be moving at present. The boundary of this area of nearly continuous salt features with the area to the west where no salt has been reported is abrupt—an area evidently containing a thick salt accumulation sharply juxtaposed with an area having no apparent salt (Fig. 10; for location see Fig. 6).

A. Salvador

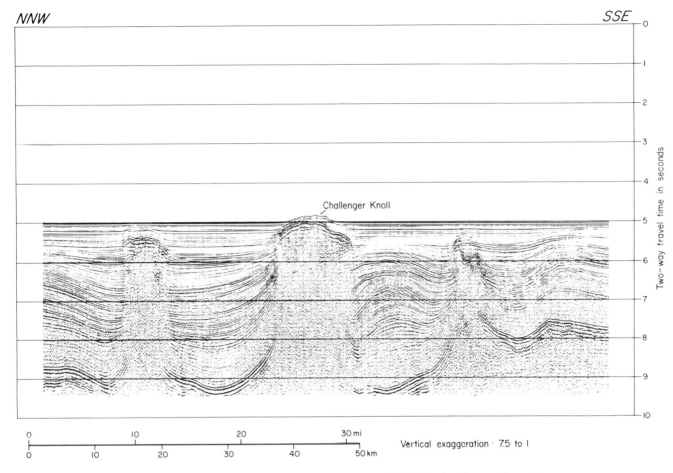

Figure 8. Seismic reflection profile across Challenger Knolls. For location see Figure 6.

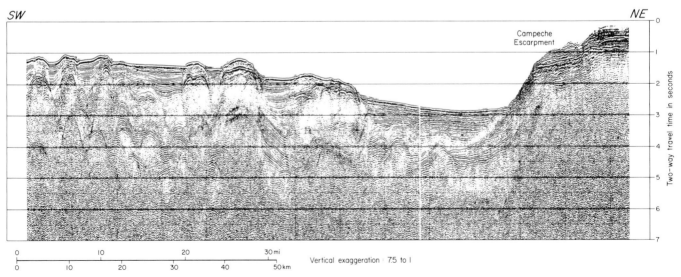

Figure 9. Seismic reflection profile across Bay of Campeche. For location see Figure 6.

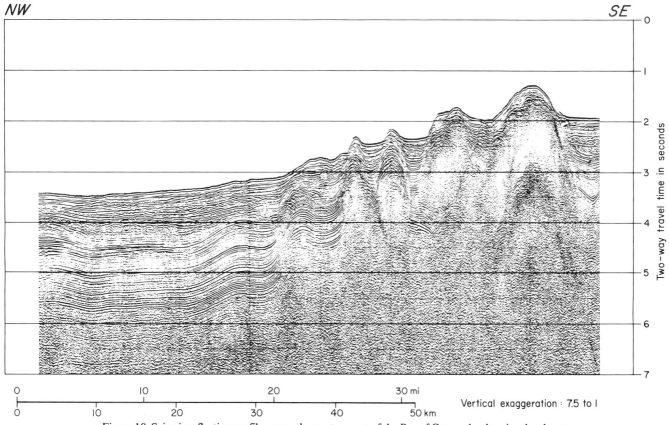

Figure 10. Seismic reflection profile across the western part of the Bay of Campeche showing the abrupt boundary between the area of thick salt occurrence and the area where no salt has been identified. For location see Figure 6.

3. South of the Bay of Campeche, in the states of Veracruz, Chiapas and Tabasco, the density of salt features decreases. Numerous but separate salt features are known; salt anticlines, deep-seated salt domes, and diapiric salt features are present in the area. Their frequency decreases southward toward northern Guatemala and adjoining parts of Mexico, where only a few isolated salt domes have been reported.

For a more thorough and complete description of the occurrence and distribution of the salt deposits in the Gulf of Mexico basin, the reader is referred to Martin (1978, 1980, 1981, 1984) and Martin and Foote (1981).

The present occurrence in the Gulf of Mexico basin of two distinct regions of salt distribution separated by an era, in the deeper parts of the Gulf and in its western shelf, where salt is apparently absent (Fig. 6), as well as the manifest similarity in the shape of the abrupt boundaries of the salt on either side of the salt-free central area, has led several authors (Humphris, 1978; Buffler and others, 1980; Salvador, 1987) to speculate that the two separate regions of salt occurrence are parts of a formerly single area of salt precipitation that was split and separated during the opening of the Gulf of Mexico basin. This hypothesis will be discussed in more detail in Chapter 14 of this volume.

Lithostratigraphy. The salt deposits that underlie the coastal plain of the southeastern United States and the adjacent continental shelf and slope (what has been described as the northern region of salt occurrence) have been generally called the Louann Salt or Louann Formation (Hazzard and others, 1947). In southern Arkansas, northern Louisiana, and adjacent parts of east Texas and Mississippi, a unit predominantly composed of anhydrite, commonly with conglomerates and red clastics at the base, has been recognized below the Louann Salt and named the Werner Anhydrite or Werner Formation (Hazzard and others, 1947). The Werner and Louann evaporites are believed to represent a single continuous cycle of salt precipitation.

The salt section in the salt-dome area of the Isthmus of Tehuantepec has been referred to as the Isthmian Salt, and that forming the Sigsbee Plain domes as the Challenger Salt (Ladd and others, 1976). The Minas Viejas Formation of northeastern Mexico, which commonly has been considered equivalent to the Louann Salt, is now known to be younger. It is discussed in the following section on the Upper Jurassic.

As known from well cores and from salt mines excavated into shallow salt domes in east Texas and Louisiana, the salt is almost entirely composed of medium to coarsely crystalline halite; anhydrite is the chief additional component but seldom forms more than 10 percent of the rock. Pyrite, quartz, dolomite, and potassium salts (sylvite) have also been identified, but only in minute amounts, except for rare local occurrences. The halite is white to light gray and is generally composed of interlocked equigranular, slightly elongated crystals, 0.5 to 1 cm in length. Banding is common; thicker bands of white or light gray halite alternate with darker gray bands richer in anhydrite. Bands of pure or nearly pure anhydrite are rare.

The Werner Formation is composed of a lower clastic section of red sandstones and sandy shales commonly conglomeratic at the base, overlain by white to grayish, pure, dense granular anhydrite, often interbedded with thin beds of dolomite in its lower part. The composition of the lower clastic section varies considerably depending on the nature of the terrain over which the Werner was deposited.

In northeastern Mexico, the Upper Jurassic sequence is underlain by an evaporitic unit principally composed of red to white, finely laminated anhydrite in updip areas, that grades basinward to a section in which the amount of halite seems to increase. This anhydritic unit has been called the Metate Formation.

Throughout the northern and northwestern rim of the Gulf of Mexico basin, the Louann Salt is not only underlain by an anhydrite unit, the Werner Formation, but seems also to grade laterally to anhydrite toward the margins of its area of distribution. Anderson (1979) mentioned occurrences of intertonguing of anhydrite and salt (halite) in wells drilled in central Mississippi and southwestern Alabama, and described the presence near the inland limits of the Louann Salt of massive anhydrite sections that reach a thickness several times that of the usual thickness of the Werner Formation. He interpreted these occurrences as evidence that the halite grades laterally to anhydrite toward the margin of the basin. Wells that have penetrated the Werner-Louann evaporitic section near its pinchout from northeastern Mexico to Alabama have often encountered the section and found it to be predominantly composed of anhydrite. Some authors have assigned this anhydritic interval to the Werner Formation and presumed that the overlying Louann Salt is absent due to beveling by post-Louann erosion along the margin of the basin. It is more likely, however, that the updip anhydritic interval represents not only the downdip Werner anhydrite, but also the lateral equivalent of at least part of the Louann halite section. This relation has also been recognized in northeast Mexico, where the Metate anhydrite grades basinward into a section progressively richer in halite.

Thickness. It is not easy to determine the thickness of the original "mother salt" except in areas where the salt is thin, undeformed, and in normal stratigraphic sequence: in these areas, the salt thickness can be estimated from well-bore and seismic information. This, unfortunately, is seldom the case. Where the salt has flowed, bulged, or diapirically penetrated the overlying sediments—and was to some extent dissolved in the process—the original thickness of the salt deposits cannot be accurately ascertained. A qualitative estimation can be made, however, on the basis of the density, shape, and size of the salt features: in areas where no salt features have been detected, as, for example, over and immediately south of the Sabine uplift, the salt can be

assumed to be thin or absent; the original salt deposit can be presumed to have been thicker in areas where a few scattered domes are present, as in south Texas, and thicker still where salt domes are numerous and closely spaced, as in the Interior salt basins (East Texas, North Louisiana, and Mississippi salt basins) or along the coastal part of southeast Texas and south Louisiana. The original "mother salt" deposit can be assumed to have been the thickest where the salt forms large ridges, pillows, swells, massifs, or nearly continuous masses. These areas of greatest salt thickness must have corresponded to areas of most rapid and persistent subsidence during the formation of the salt.

More quantitative attempts to determine the thickness of the original salt deposits by this author have resulted in estimates of about 1,000 m for the East Texas and North Louisiana salt basins, 1,200 to 1,500 m for the Mississippi salt basin and the belt along coastal southeast Texas and south Louisiana, more than 2,000 m in the Gulf of Campeche, and more than 3,000 m in the Texas-Louisiana continental slope. These estimates generally agree with those of other authors (Andrews, 1960; Oxley and others, 1967, 1968; Wilson, 1975; Anderson, 1979). Humphris (1984, personal communication) has estimated that the original salt deposits may have reached a thickness of more than 1,000 m in the East Texas, North Louisiana, and Mississippi salt basins, and as much as 4,000 m in the Texas-Louisiana Gulf of Mexico slope area.

In comparison with the Louann Salt, the Werner is generally thin; the lower clastic unit may range up to 30 m in thickness but may be absent over pre-Werner topographic highs. The updip anhydrite section ranges generally between 15 and 30 m, but thicknesses of as much as 200 m have been reported from northeastern Mexico in a well where the anhydrite section (Metate Formation) contains two intervals of halite 70 and 30 m thick.

Stratigraphic relations. When undeformed and in normal stratigraphic sequence along the northern periphery of the Gulf of Mexico basin, the Werner Anhydrite–Louann Salt sequence unconformably overlies the "red beds" of the Eagle Mills Formation or Paleozoic rocks, and underlies the Upper Jurassic Norphlet or Zuloaga Formations.

The nature of the contact between the Louann and overlying units is controversial; most authors have postulated a stratigraphic discontinuity at the top of the salt (Hazzard and others, 1947; Bishop, 1967; Shreveport Geological Society, 1968; Newkirk, 1971; Todd and Mitchum, 1977; Imlay, 1980; Tolson and others, 1983), but some believe that the contact may be conformable and even gradational (Nichols and others, 1968; McBride, 1981). Badon (1975) stated that the lower part of the Norphlet clastics grades laterally into the Louann in Clarke County, southeastern Mississippi. It is also possible that the salt/Norphlet contact may be conformable and transitional in the basin but disconformable around its margins.

Even those who favor a disconformable contact between the Werner-Louann evaporites and the Norphlet believe that the hiatus represented by the disconformity must be very small because there is no indication of extensive solution of the upper surface of the salt. A long period of exposure and dissolution between the end of the formation of the evaporites and the beginning of the progradation of the basal Upper Jurassic clastics would have resulted in the dissolution and removal of considerable volumes of salt.

In the southern part of the Gulf of Mexico basin, the salt is known in relative normal stratigraphic sequence along a belt directly north of the Campeche Escarpment, where, on the basis of seismic-reflection information, it has been interpreted to unconformably overlie Upper Triassic–Lower Jurassic(?) "rift-filling" beds (Buffler and others, 1980) or older "basement," and to underlie Upper Jurassic(?) sediments with apparent conformity. Well Trinitaria-2 drilled by PEMEX (Petroleos Mexicanos) near the Mexico-Guatemala border may have also penetrated the salt section in nondiapiric conformable relation with overlying sediments.

Over most of the Gulf of Mexico basin, the salt deposits are deformed into various types of diapiric features, where the salt has flowed and intruded the overlying sediments. The base of the salt, however, when recognizable in seismic-reflection sections, shows little deformation and unconformably overlies the underlying rocks: Upper Triassic–Lower Jurassic "red beds," Paleozoic sediments, or igneous-metamorphic "basement."

Biostratigraphy and age. The age of the salt deposits has long been one of the most important and controversial among the unresolved geologic problems in the Gulf of Mexico basin, particularly because the age of the salt is probably one of the fundamental pieces of information in establishing the time and dynamics of the early stages of development of the basin.

Early workers were divided between those who favored a Jurassic age for the Louann Salt (Hazzard, 1939; Imlay, 1940, 1943) and those who placed it in the Permian (Weeks, 1938; Hazzard and others, 1947). The arguments of those who favored a Jurassic age for the salt progressively gained the acceptance of Gulf Coast geologists, but as late as 1956, Halbouty and Hardin still favored a Permian age. In 1961, Scott and others, reported on the occurrence of plant impressions of probable Late Triassic to Early Jurassic age in the Eagle Mills Formation, confirming the Jurassic age of the overlying Louann Salt. The controversy concerning the age of the salt deposits of the Gulf of Mexico basin has not benefited by the occurrence in the basin of several evaporitic intervals, which at times, have not consistently been distinguished from each other.

No reliable paleontologic information is available to rigorously determine the age of the Jurassic salt—except for sparse palynomorphs, generally undiagnostic of a precise age, no fossils have been recovered from the salt. Kirkland and Gerhard (1971) have reported a palynomorph assemblage of Middle or Late Jurassic age from the calcite caprock of the Challenger Salt Dome in the Sigsbee Knolls area, which they believed was probably derived from the underlying salt beds. Jux (1961) had earlier described 24 palynomorph species from salt samples collected in five salt domes in east Texas and Louisiana, and had assigned a Late Triassic to Early Jurassic age to the salt on this basis. Jux's palynomorphs, however, are poorly preserved, greatly corroded,

and the samples show considerable contamination (Pflug, 1963). His age determination, therefore, cannot be considered reliable beyond indicating a Mesozoic age for the salt (Pflug, 1963).

Beds overlying the salt in PEMEX's well Trinitaria-2 in the southern Mexico State of Chiapas, near the Guatemalan border, contain a dwarfed mollusk fauna and poorly preserved palynomorphs that have given only inconclusive evidence for establishing the age of the salt.

In the absence of paleontological evidence, the age of the Jurassic salt deposits of the Gulf of Mexico basin must be established on the basis of evidence provided by its stratigraphic relations, or on indirect reasoning based on regional geological and paleogeographical considerations.

Along the north flank of the Gulf of Mexico basin, where the Werner-Louann evaporites are present in essentially normal stratigraphic sequence, they overlie the Eagle Mills "red beds" of Late Triassic to Early Jurassic age. The oldest well-dated interval above the salt is the lower part of the Smackover Formation, which has yielded late Oxfordian ammonites. The Norphlet Formation, between the Louann Salt and the Smackover, is unfossiliferous. On the basis of this direct evidence of stratigraphic bracketing, it may be said that the salt is post–Early Jurassic and pre–late Oxfordian in age. It is not possible at present to estimate the time represented by the Norphlet, but if it is taken into consideration that it attains significant thicknesses only in southern Alabama and adjoining parts of Mississippi and the Florida panhandle, where it probably represents rapidly deposited complexes of alluvial fans and fluvial and eolian sediments in places reworked by the shallow encroaching Smackover sea, it may be concluded that the top of the salt is not much older than the late Oxfordian, probably of late Callovian or early Oxfordian age.

Imlay had long favored an early Oxfordian age for the Louann Salt (1940, 1943, 1952, 1953). He reasoned that it should be older than late Oxfordian because it underlies beds of this age and must have been covered by the basal Upper Jurassic beds shortly after the end of salt formation in order to avoid solution and erosion. Further, Imlay believed, the Louann must be younger than Callovian because climatic conditions were unfavorable for salt deposition before latest Callovian time—the Lower and Middle Jurassic section in east-central and southeastern Mexico contains marine intervals as well as coal seams and plant-bearing beds indicative of a humid, tropical climate. Viniegra (1971) supported this line of evidence. Imlay (1980) has revised the age of the Louann Salt to range from late Callovian to early or early-middle Oxfordian, because part of the early Oxfordian (Divesian), as used by him in his early publications, is now placed in the late Callovian. He considered the Werner Formation as of early to middle Callovian age. Imlay (1980) added that a late Callovian to mid-Oxfordian age for the Louann Salt would imply more saline than normal marine water in the Gulf of Mexico at that time, and would explain the absence of fossils of this age in the lower part of the Santiago Formation of east-central Mexico and contemporaneous beds in other parts of the basin.

In the southern part of the Gulf of Mexico basin, the salt has not been penetrated by wells where it is present in normal stratigraphic sequence, with the exception of its occurrence in PEMEX's well Trinitaria-2. Everywhere else the salt has only been reached in diapiric structures. Numerous wells, however, have drilled into Kimmeridgian and upper Oxfordian beds without reaching the salt, indicating that the salt in this area is also pre-late Oxfordian in age (Santiago Acevedo and Mejía Dautt, 1980; Viniegra, 1981).

A more precise age for the salt deposits of the Gulf of Mexico basin may be established indirectly, based on regional stratigraphic and paleogeographic reasoning. As described in previous pages, the only Upper Triassic, Lower Jurassic, and Middle Jurassic marine beds in the Gulf of Mexico basin and neighboring areas are present in southwest, central, and east-central Mexico. The distribution, lithology, and stratigraphic relations of these beds support the interpretation that they were deposited in embayments of the Pacific Ocean of the time, and that these embayments did not reach the areas where salt is known in the basin today (Fig. 11, A, B, and C). The first indications that Pacific marine water may have extended across central Mexico and reached the areas of salt precipitation are found in the upper part of the late Bathonian or early Callovian Huehuetepec Formation of northern Puebla and bordering parts of Veracruz, and in the overlying Callovian-age Tepexic calcarenites (Fig. 11D). The upper part of the Huehuetepec Formation consists of 120 to 150 m of salt and anhydrite interbedded with red and greenish shales, a record of a period of evaporation of marine water with the resulting formation of salt deposits, probably in a coastal sabkha. It may be that the time of formation of the Huehuetepec evaporites corresponded roughly with the formation of at least part (the initial part?) of the larger and much thicker salt deposits in the rest of the Gulf of Mexico basin. The Tepexic calcarenites could also represent a time during which, in east-central Mexico, shal-

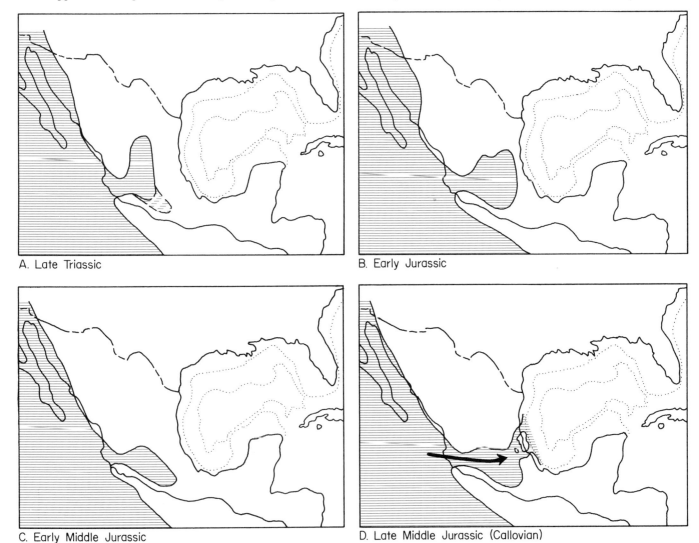

A. Late Triassic

B. Early Jurassic

C. Early Middle Jurassic

D. Late Middle Jurassic (Callovian)

Figure 11. Location of marine embayments of the Pacific Ocean during the Late Triassic and Early and Middle Jurassic, and possible intake of sea water into the ancestral Gulf of Mexico during the late Middle Jurassic (Callovian).

low channels between scattered islands connected the Pacific Ocean to the west with an ancestral Gulf of Mexico to the east (Figs. 4, 5, and 11D).

The Werner-Louann, Isthmian, and Challenger evaporite deposits would also have to be interpreted to be no older than late Middle Jurassic if the source for the sea water from which they precipitated was assumed to be the Atlantic Ocean instead of the Pacific, because, as mentioned earlier, the southern part of the North Atlantic is not believed to have opened until late in the Middle Jurassic (Gradstein and Sheridan, 1983; Sheridan, 1983).

In summary, direct evidence indicates that the Werner-Louann salt deposits of the northern part of the Gulf of Mexico basin and their probably contemporaneous salt deposits of the southern part of the basin—the Challenger Salt and the Isthmian Salt—are post–Early Jurassic and pre–late Oxfordian in age. On the basis of indirect evidence provided by regional stratigraphic and paleogeographic reasoning, the salt deposits may be considered as mainly of Callovian age, although their lower part may be late Bathonian and the uppermost part may extend into the early Oxfordian. It is also possible that the base of the Middle Jurassic evaporite deposits may be older (late Bathonian) only in east-central Mexico, where Pacific sea water first reached the Gulf of Mexico basin area (Huehuetepec Formation), and may be progressively younger (Callovian and earliest Oxfordian) away from this intake area. For the sake of conciseness, these evaporite deposits will be referred to as the "Middle Jurassic" or "upper Middle Jurassic" evaporites.

Conditions of deposition—Paleogeography. The extensive salt deposits of the Gulf of Mexico basin were most likely formed in very large shallow bodies of hypersaline water with limited communication with the ocean, where the evaporation exceeded the inflow of marine water. The climate was arid or semiarid, and the land around these bodies of water was probably low after a long period of emergence and erosion. No major rivers emptied into the hypersaline water bodies, and the supply of terrigenous clastics was, therefore, very limited.

The influx of marine water was most likely from the Pacific to the west, across central Mexico, and probably intermittent; at times of hurricanes or especially high tides, large volumes of Pacific sea water surged into the shallow water bodies in the Gulf of Mexico basin area. Between flooding periods, the connection with the Pacific would close or become restricted to a few of the deeper channels that communicated the Pacific with the ancestral Gulf of Mexico. The water in the shallow depressions evaporated, forming the extensive salt deposits that we know today. Halite precipitation predominated in the central parts of the large hypersaline water bodies, and anhydrite formed most of the evaporite deposits around their periphery. In order to account for the large thicknesses of salt, it is necessary to assume that the depressions in which the salt precipitated subsided gradually and more or less continuously during the time of salt formation. Some depressions subsided more rapidly and accumulated, therefore, thicker salt deposits; others subsided more slowly and include thinner sections of salt. Positive areas between the depressions were not covered by water and contain no salt, or may have been covered only at infrequent intervals and contain only a thin layer of salt. In the latter case, salt preservation would depend on protection from later erosion or dissolution.

Economic considerations. Salt domes provide traps for important oil and gas accumulations in the United States and Mexico, and the high thermal conductivity of the salt is believed to have favored the maturation of the organic matter in sediments near the salt diapiric features. Salt is mined or has been mined from shallow salt domes, and sulfur is produced from the caprock of some salt domes (see Chapter 16, this volume).

UPPER JURASSIC

Introduction

The period of extensive salt formation in the late Middle Jurassic was followed during the Late Jurassic by a widespread and prolonged marine transgression that covered most of the areas of the present Gulf of Mexico basin. The transgression was nearly continuous during the entire Late Jurassic, with only minor periods of regression or sea-level drop.

The Upper Jurassic section does not crop out in the U.S. part of the Gulf of Mexico basin. It is known, therefore, only from borehole and seismic information. The top of the Upper Jurassic section has been penetrated at depths ranging from approximately 300 m to more than 6,000 m. Excellent outcrops of Upper Jurassic beds are present, however, in northeastern and eastern Mexico, particularly along the Sierra Madre Oriental. The Upper Jurassic is also well known in the subsurface of the Gulf of Mexico Coastal Plain and shelf in eastern and southeastern Mexico. No Upper Jurassic sediments are known to be present in the Yucatan and Florida platforms except perhaps for southernmost Florida. In these two regions a Lower Cretaceous section predominantly composed of carbonates and evaporites directly overlies a variety of pre–Upper Jurassic rocks (see Plate 3).

Even though Upper Jurassic sediments probably underlie the entire Gulf of Mexico Basin west of the Florida platform and north of the Yucatan platform, direct knowledge of the Upper Jurassic section is restricted to the periphery of the basin, where it either crops out or is within reach of the drill. Over most of the basin, the Upper Jurassic is deeply buried below thick Cretaceous and Cenozoic sections, and its distribution and character can only be interpreted from seismic reflection information.

The northern limit of the deposition of Upper Jurassic sediments in the United States lies within the structural limits of the Gulf of Mexico basin. The present limit is for the most part crosional, but judging from the rates of thinning and the northward succession of lithofacies toward the basin margin, the original area of deposition is not believed to have extended much beyond the erosional limit as known today. In Mexico, on the other hand, Upper Jurassic sediments are known from many areas in the central and western parts of the country, well beyond the western structural limits of the Gulf of Mexico basin. The following discussion will only cover the description of the Upper Jurassic section within the limits of the basin; reference to occurrences of Upper Jurassic sequences outside these limits will only be made when necessary to the clear understanding of the Upper Jurassic stratigraphy and geologic history of the Gulf of Mexico basin.

The Upper Jurassic stratigraphic units generally represent a conformable and transgressive sequence, with each unit overstepping the preceding one and pinching out progressively farther landward. This stratigraphic relation has been interpreted to indicate coastal onlap due to a eustatic sea-level rise (Todd and Mitchum, 1977; Vail and others, 1984). Within this overall transgressive cycle, smaller regressive episodes have been recognized in the Upper Jurassic sequence.

The Upper Jurassic of the Gulf of Mexico basin is a predominantly marine section. Nonmarine fluvial and deltaic coarse clastics are present in the northern and northwestern margins of the basin. They reflect the presence in those areas of important rivers draining the continental interior and emptying into the Late Jurassic Gulf of Mexico. Of particular interest is the indication that a precursor of the Mississippi River was already contributing coarse clastics to the Gulf in the Late Jurassic and that its shifting deltas can be recognized in the sediments of this age in central Mississippi, northeastern Louisiana, and southeastern Arkansas. Nonmarine Upper Jurassic clastics are also known from southern Mexico and northwestern Guatemala.

Elsewhere in the Gulf of Mexico basin the Upper Jurassic is predominantly composed of shales, calcareous shales, and carbonates, for the most part indicative of shallow-water marine environments of deposition—shelves and ramps. Deeper water shales and fine-grained carbonates have been recognized in a more basinward position.

In contrast with the underlying Upper Triassic and Lower and Middle Jurassic sediments, which clearly reflect strong tectonic control of their environment of deposition—mostly related to actively subsiding grabens or troughs—the Upper Jurassic section does not show evidence of large-scale tectonic influence. Over most of the Gulf of Mexico basin, the Upper Jurassic is relatively uniform in lithologic composition and does not show abrupt changes in thickness, indicating a stable environment of deposition. However, on a smaller, more local scale, the deposition of the Upper Jurassic section was influenced by: (1) contemporaneous movement of regional tensional normal fault zones generally parallel to the periphery of the basin (Bishop, 1973; Miller, 1982); (2) the flow of the underlying Middle Jurassic salt (Andrews, 1960; Hughes, 1968; Wilson, 1975); (3) fluctuations of sea level (Todd and Mitchum, 1977; Vail and others, 1984); and (4) pre-existing topography, particularly in northeast Mexico, southern Alabama, and northwest Florida (Mancini and Benson, 1980; Miller, 1982; Tolson and others, 1983).

Presence of salt and the relative thickness of the salt exert a marked influence over the lithology and thickness of the Upper Jurassic section: where no salt underlies it, as in east-central

Mexico, the lithologic composition and the thickness of the Upper Jurassic section is fairly uniform; where appreciable thicknesses of salt were present under the Upper Jurassic, as in the East Texas, North Louisiana, and Mississippi salt basins, early deformation of the underlying salt materially influenced the sedimentary environment and, consequently, the thickness and lithofacies of the sediments being deposited during the time of salt deformation. This influence of the salt movement on the lithofacies and thickness of the Upper Jurassic section is particularly notable during the latest Oxfordian and the Kimmeridgian.

The influence of preexisting topography on the deposition of the Upper Jurassic section can be best illustrated in southern Alabama and the Florida panhandle (Wilson, 1975; Sigsby, 1976; Mancini and Benson, 1980; Miller, 1982; Tolson and others, 1983; Benson and Mancini, 1984). In this area, the lithologic composition and thickness of the Upper Jurassic section is clearly controlled by topographic ridges projecting into the Gulf of Mexico basin area from the southern Appalachians. In northeastern Mexico, the Peyotes-Picachos and Tamaulipas arches, the Coahuila platform, and other pre–Late Jurassic highs similarly controlled sedimentation of the Upper Jurassic section.

For the purpose of describing the Upper Jurassic of the Gulf of Mexico basin, the section has been broken down into three intervals that approximately correspond to standard global chronostratigraphic units (stages) and that also approximately correspond with generally recognized lithostratigraphic units or groups of units (formations or groups), and with distinct periods or cycles of sedimentation. They are listed, from younger to older, as follows (see Figs. 2 and 12).

Tithonian (including part of the Lower Cretaceous): Cotton Valley Group in the United States; upper part or entire La Casita, La Caja, Pimienta, and other time-equivalent formations in Mexico.

Kimmeridgian: Haynesville Formation (including Buckner and Gilmer) in the United States; in Mexico, Olvido and Taman Formations, lower part of the La Casita, and La Caja Formations, and time-equivalent named or unnamed units.

Oxfordian: Norphlet and Smackover Formations in the United States; in Mexico, Zuloaga, Novillo, La Gloria, and Santiago Formations, and time-equivalent named and unnamed units.

As would be expected, some of the lithostratigraphic units listed (formations and groups) do not exactly fit the chronostratigraphic framework; exceptions will be mentioned in the text: in certain areas, for example, the Santiago Formation may extend down into the uppermost Callovian, and the upper part of the Tamán Formation is of early Tithonian age; the Cotton Valley Group includes Lower Cretaceous beds in its upper part, and basinward it encompasses the Kimmeridgian and even part of the upper Oxfordian; the Buckner in the southern Arkansas–northern Louisiana area and the Haynesville Formation in southwestern Alabama and southeastern Mississippi include sediments of latest Oxfordian as well as Kimmeridgian age. The use of chronostratigraphic units (stages), however, facilitates regional treatment and synthesis of the stratigraphy and geologic history of the Upper Jurassic of the Gulf of Mexico basin. It also aids international understanding and communication.

The diagrams of Figure 12 illustrate in simplified form the stratigraphic relations of the Upper Jurassic lithostratigraphic units in various regions of the Gulf of Mexico basin and how they tie to the chronostratigraphic reference framework. As illustrated in this figure, around the periphery of the basin, in the U.S. Gulf coastal plain area and in northeastern Mexico, the lower part of the Upper Jurassic section is represented by nonmarine or littoral coarse-clastic sedimentary units, often separated by unconformities. Farther basinward, these units grade into nearshore and shallow-shelf sequences predominantly composed of carbonates, calcareous shales, and evaporites. The unconformities also tend to die in a basinward direction. The lithologic composition and the stratigraphic relations of the various lithostratigraphic units in the U.S. Gulf Coast area are very similar to those of northeast Mexico; only the names of the units are different: the Norphlet-Smackover-Haynesville sequence in the United States closely corresponds to the Zuloaga/Novillo/La Gloria–Olvido/lower La Casita in northeast Mexico, though the exact ages of these sequences may not precisely correspond (Fig. 12).

The upper part of the Upper Jurassic section is represented around the northern margin of the Gulf of Mexico basin by a more clastic section; the Schuler Formation of the Cotton Valley Group in the United States and the upper La Casita in northeast Mexico.

Both in the United States and in northeast Mexico, the Upper Jurassic grades farther basinward into a much shalier section, indicative of increasingly deeper water conditions of deposition. Away from the margins of the basin, for example, the Bossier Formation of the Cotton Valley Group in the United States, and the Pimienta Formation in northeast Mexico, expand at the expense of the overlying and underlying units; the Smackover Formation in the United States and its equivalents in Mexico, the Zuloaga and Novillo formations, become increasingly shaly and eventually can no longer be distinguished from the overlying Bossier or Pimienta shales on the basis of lithology. Beyond present-day control, the Upper Jurassic section can be assumed to thin progressively toward the deepest part of the basin, where it is probably composed of a relatively thin sequence of shales and calcareous shales deposited slowly and continuously in a deep-water marine environment.

In east-central and southern Mexico, the Upper Jurassic section within the Gulf of Mexico basin is characterized by fairly uniform lithology, marine dark argillaceous limestones and calcareous shales. Coarse clastics are rare. Around the periphery of the Late Jurassic land areas, shallow-water sections composed of grainstones, predominantly oolitic limestones, have been recognized.

Oxfordian

Distribution. Sediments of Oxfordian age are known from the northern, western, and part of the southern flanks of the Gulf of Mexico basin. Along the U.S. Gulf coastal plain, the Oxfordian does not crop out but it has been penetrated by hundreds of oil and gas wells along a belt extending from the Florida panhandle through southern Alabama, central Mississippi, northern Louisiana, southern Arkansas, and east Texas. Control over the San Marcos arch and in south Texas and northeastern Mexico is more sparse. Basinward from this belt, the Oxfordian section rapidly plunges below great thicknesses of younger sediments and is beyond the reach of the drill.

In northeastern and east-central Mexico, the Oxfordian crops out extensively in the core of large anticlinal structures in the Sabinas basin and along the Sierra Madre Oriental. It does not seem to extend northwestward to the state of Chihuahua. The Oxfordian has also been penetrated by many petroleum exploration wells in the Gulf coastal plain of east-central Mexico and in the shallow, nearshore part of the Gulf of Mexico shelf. The southernmost reported occurrence of reliably dated Oxfordian sediments in the Gulf of Mexico basin is in the Reforma petroleum-productive area, in northern Chiapas and Tabasco, and in the offshore Bay of Campeche area, where numerous wells drilled in search of oil and gas have reached the Oxfordian. South of these areas, the Oxfordian may be represented in the lower part of the Todos Santos "red bed" sequence.

No Oxfordian sediments are known from the Florida and Yucatan carbonate platforms where Lower Cretaceous or uppermost Jurassic sediments rest on "basement" rocks, indicating that these areas were emergent during the early part of the Late Jurassic.

In the central, deeper part of the Gulf of Mexico basin, the presence of Oxfordian and younger Upper Jurassic sediments has been postulated on the basis of interpretation of seismic reflection data (Ladd and others, 1976; Watkins and others, 1978; Buffler and others, 1980; Shaub and others, 1984; and many others). The Oxfordian, as interpreted from seismic data, appears to pinch out against the base of the Florida and Yucatan carbonate platforms. It is also believed to be absent from the deep part of the southeastern Gulf of Mexico, between these two platforms. Seismic information and the results of Leg 77 of the Deep Sea Drilling Project suggest that a pre–Lower Cretaceous section may be present in the area (Phair and Buffler, 1983; Schlager and others, 1984). How much of the Late Jurassic is represented in this section is not known. In the preparation of the maps for this chapter (Figs. 16 to 21), it has been assumed that the connection between the Gulf of Mexico and the Atlantic was not established until the late Kimmeridgian, at the time of an apparent increase of the extension of the Late Jurassic sea.

Lithostratigraphy. Two main distinct lithostratigraphic units have been recognized in the Oxfordian of the northern and northwestern parts of the Gulf of Mexico basin: a generally thin basal clastic unit, and an upper section predominantly composed of carbonates and shales, or almost entirely composed of shales, which makes up the major part of the Oxfordian section.

The basal Oxfordian coarse-clastic unit is known all along the northern and northwestern flanks of the Gulf of Mexico basin,

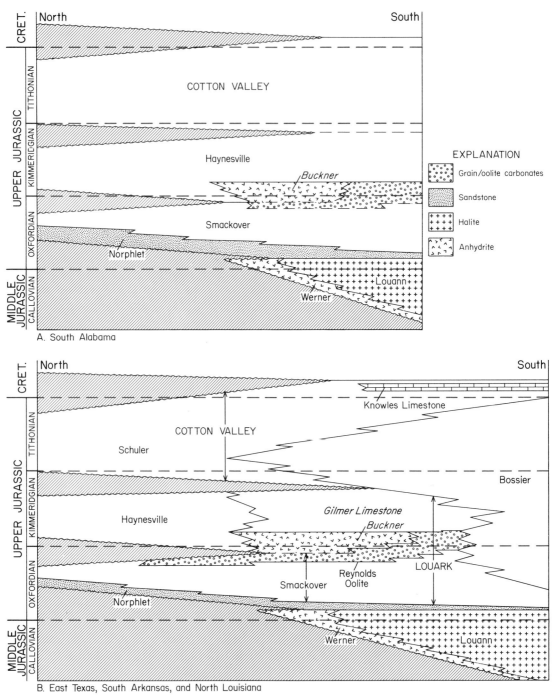

Figure 12. Nomenclature and stratigraphic relations of Upper Jurassic lithostratigraphic units, Gulf of Mexico basin.

from the Florida panhandle to northeastern Mexico and as far south as the Huizachal-Peregrina anticlinorium, near Ciudad Victoria. No similar clastics have been reported from most of the western and southern flanks of the basin in east-central and southern Mexico (Fig. 13). This basal coarse-clastic unit has been called the Norphlet Formation in the United States (Hazzard, 1939; Imlay, 1940; Hazzard and others, 1947), but no special name has been given to it in northeastern Mexico where, when

present, it has been considered to be the thin clastic basal part of the Oxfordian section.

The upper, carbonate-shale Oxfordian unit has been called the Smackover Formation in the U.S. part of the Gulf of Mexico Basin (Bingham, 1937; Shearer, 1938; Weeks, 1938; Imlay, 1940). In northeastern Mexico, beds of similar lithologic composition have been called Zuloaga Formation (Imlay, 1938) or Novillo Formation (Heim, 1940). Toward the land areas of the

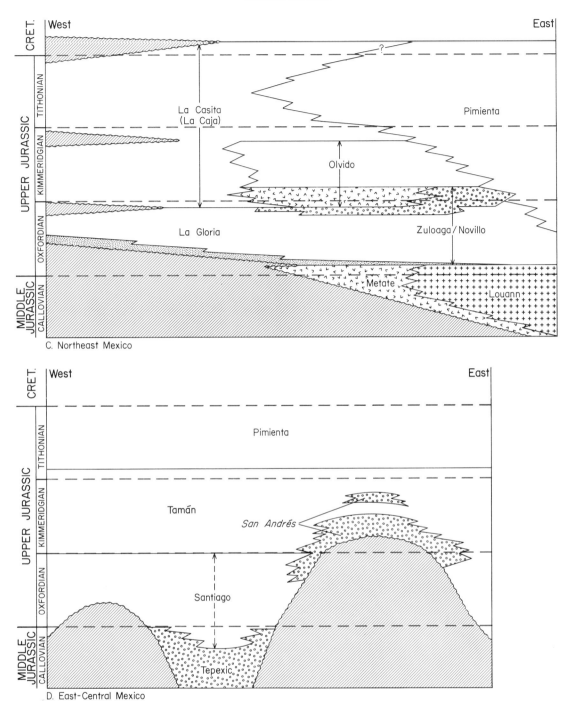

C. Northeast Mexico

D. East-Central Mexico

time, in northeastern Mexico, the Oxfordian section becomes increasingly sandy and has been included in the La Gloria Formation (Imlay, 1936). Southward, in east-central Mexico, the Oxfordian becomes more shaly and is represented by all or most of the Santiago Formation (Cantú Chapa, 1969a, 1971) or the equivalent shaly, lower part of the Tamán Formation of other authors. (The Santiago Formation locally may extend downward into the uppermost Callovian; see Fig. 12D.) In southern Mexico, the Oxfordian section, also predominantly composed of shale, has not been given a formational name.

The Oxfordian is fairly uniform in lithologic composition throughout the Gulf of Mexico basin. Lithologic similarity is particularly remarkable from northeastern Mexico to the Florida panhandle over a distance of more than 1,800 km.

The Norphlet Formation is predominantly composed of sandstones and conglomeratic sandstones of varied lithology, depending on the source of the clastics, the transporting agent, the relative position in the basin, the sedimentary environment in which it was deposited, and its time of deposition.

In southern Arkansas, northern Louisiana, Texas, and north-

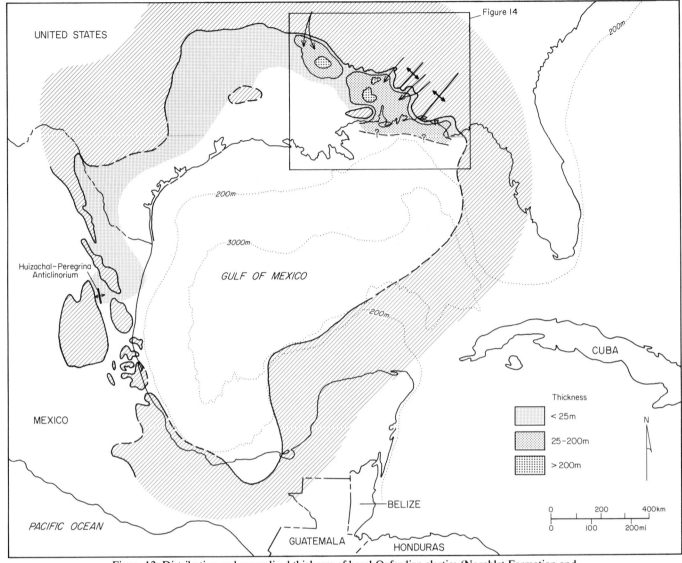

Figure 13. Distribution and generalized thickness of basal Oxfordian clastics (Norphlet Formation and equivalents).

east Mexico, the Norphlet Formation is composed of a thin section of immature, poorly sorted, feldspathic to lithic sandstones, which are red, brown, green, or greenish gray to gray and locally conglomeratic. The sandstones are interbedded with red, mottled or gray, often micaceous silty shales and siltstones. Low-angle cross-bedding is common. Conglomerates are more abundant updip, and the section becomes finer grained in a basinward direction. In this area, the Norphlet probably represents the initial deposits of a marine transgression.

In parts of central Mississippi, southwestern Alabama, and the westernmost part of the Florida panhandle, the Norphlet is considerably thicker and composed of fine- to medium-grained, well-sorted mature sandstones (Figs. 13 and 14). The sand grains are generally well-rounded and commonly coated with hematite. The sandstones have high-angle cross-bedding and do not contain detrital clay matrix. They have been interpreted as eolian sands

(Hartman, 1968; Badon, 1975; McBride, 1981; Pepper, 1982; Mancini and others, 1985; Marzano and others, 1988). In southwestern Alabama, the sandstones in the upper part of the Norphlet show evidence of reworking and are believed to have been deposited in a marine environment (Sigsby, 1976; Pepper, 1982; Mancini and others, 1985; Marzano and others, 1988).

In southwestern Alabama and the western Florida panhandle, the eolian sediments of the Norphlet Formation grade laterally updip (northeastward) to a section of siltstones, arkosic sandstones, and conglomeratic sandstones, predominantly red and generally poorly sorted. The Norphlet in this area has been interpreted to represent deposition in coalescing alluvial fans and fluvial systems (Sigsby, 1976; Mancini and others, 1985; Marzano and others, 1988). The eolian and alluvial-fluvial lithofacies are found in irregular alternations along a northwest-southeast belt (Fig. 14).

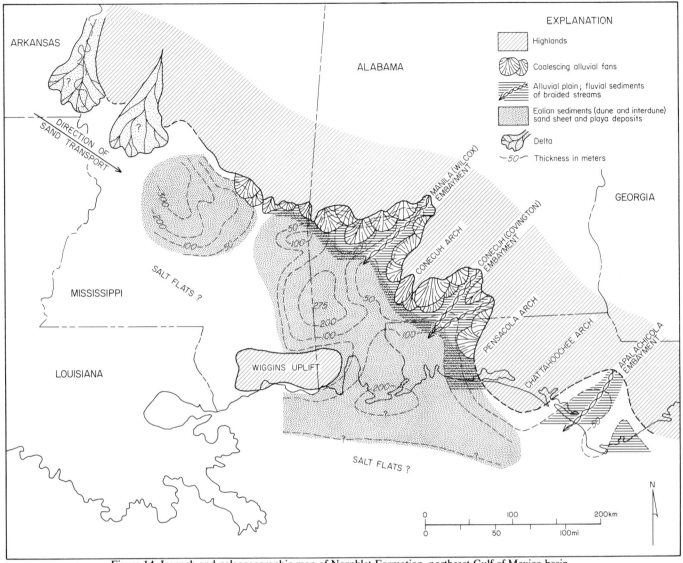

Figure 14. Isopach and paleogeographic map of Norphlet Formation, northeast Gulf of Mexico basin.

The Smackover Formation is predominantly composed of carbonates and calcareous shales and can easily be correlated by its lithology for considerable distances along its depositional strike. The more porous upper part of the Smackover Formation is the reservoir rock of many of the oil and gas fields around the northern rim of the Gulf of Mexico basin. The Smackover, consequently, is the unit of the Jurassic section in the Gulf of Mexico basin most thoroughly studied and described. Regional stratigraphic information on the Smackover may be found in Weeks (1938), Imlay (1943, 1952, 1980), Swain (1949), Vestal (1950), Thomas and Mann (1966), Oxley and others (1967, 1968), Bishop (1968, 1969), Dickinson (1968b, 1969), Dinkins (1968), Newkirk (1971), Ottmann and others (1973, 1976), Sigsby (1976), Wakelyn (1977), Mancini and Benson (1980), Budd and Loucks (1981), Loucks and Budd (1981), Benson and Mancini (1982), Miller (1982), Tolson and others (1983), Moore (1984), and Hancharik (1984).

Over most of the U.S. Gulf coastal plain area, from the Florida panhandle to south Texas, the Smackover Formation has been informally subdivided into two distinct units, often called "members": a lower unit composed of dark-colored carbonate mudstone and dense argillaceous limestone formed in a low-energy environment, and an upper unit typically composed of grain-supported carbonates—predominantly oolitic grainstones and packstones—formed in a high-energy shallow-water environment (Weeks, 1938; Bishop, 1968, 1969; Wakelyn, 1977, and many others). Some workers have subdivided the lower Smackover into two units, which they have called "lower" and "middle" Smackover (Dickinson, 1968a, 1968b; Budd and Loucks, 1981; Loucks and Budd, 1981). Although the subdivision of the Smackover into three "members" is feasible and useful in certain areas, for the purpose of a regional synthesis of the entire Gulf of Mexico Basin, the subdivision into two units is more practical and preferable.

Where the *lower Smackover* can be subdivided into two lithologic units, the lower unit is composed of finely laminated, dark gray or brown to black carbonate mudstone and argillaceous limestone rich in organic matter and pyrite; it contains occasional fine siltstone laminae, and is devoid of any evidence of burrowing or bioturbation (the so-called "laminated micrite," or the "lower member" of some authors). The upper section (the "middle" member of some authors) is made up of dark brown to black, dense argillaceous limestone (the so-called "Brown Dense Limestone" of Mississippi and Alabama), calcareous shale, and pelletoidal and oncolitic carbonate mudstone and wackestone. Burrowing is common in the upper part of the lower Smackover. Throughout the lower Smackover, nodular bands of secondary anhydrite have been occasionally reported.

Laminated mudstones are not always restricted to the basal part of the lower Smackover; they also occur higher in the section, interbedded with dense, argillaceous carbonates and calcareous shales. In such cases, the recognition of two distinct lithologic units in the lower Smackover (the "lower" and "middle" members of some authors) becomes difficult.

The *upper Smackover* is a regressive, upward-shoaling, higher energy section and is mainly composed of light to dark gray grainstones and packstones. Wackestones and mudstones are much less common. The carbonate grains are predominantly nonskeletal—ooids are most abundant, but pisolites, intraclasts, pellets, rhodolites, and oncolites are also present. Fecal pellets of the crustacean *Favreina* are frequent. Among the skeletal grains, echinoid fragments, whole or broken thick-shelled mollusks, and various kinds of algae are most common. The well-developed oolitic limestone section of the upper Smackover in south Arkansas has been called the "Reynolds Oolite" (Weeks, 1938; Akin and Graves, 1969).

The coarser, grain-supported carbonates of the upper Smackover often show cross-lamination. Dolomitization in various degree is common in the upper Smackover, particularly in its upper part. The occurrence of diagenetic dolomite appears to be related to the occurrence of an evaporitic section directly above the Smackover.

The boundary between the lithologic units into which the Smackover Formation has been subdivided is usually gradational.

Small sponge, algal, and coral patch reefs have been recently recognized in the upper part of the Smackover Formation from southwest Arkansas to the Florida panhandle (Baria and others, 1982; Crevello and Harris, 1984). They are commonly elongate, but not continuous linear shelf-margin reefs, and occur near the base of the upper Smackover in the eastern region but higher in the section in the Arkansas-Louisiana area. They are generally less than a kilometer by a kilometer or two in areal dimensions, but can attain a size of 8 by 16 km, and range in thickness between 4 and 30 m. The patch reefs are most commonly developed seaward of oolite shoals and over topographic highs within the shelf or ramp that reflect basement highs or salt-cored structures. In southern Alabama, patch reefs are commonly developed in the upper Smackover over the basement ridges that project into the basin from the southern Appalachians.

The previously described two-member lithologic sequence characteristic of the Smackover Formation in the U.S. part of the Gulf of Mexico basin is known to be somewhat modified in some areas.

In southern Alabama and the Florida panhandle, the upper Smackover higher energy section is predominantly composed of hardened pellets; oolites form only a minor part of the section (Ottmann and others, 1973, 1976; Sigsby, 1976; Bradford, 1984). This scarcity of oolites is probably due to the presence in the area during the Oxfordian of basement ridges and elongated salt highs, which gave rise to the formation between them of two shallow, protected embayments—the Manila or Wilcox embayment to the west and the Conecuh or Covington embayment to the east (Figs. 16, 17) (Sigsby, 1976; Mancini and Benson, 1980; Miller, 1982), in which restricted low-energy depositional environments did not favor the formation of oolitic grainstones. In these embayments the Smackover is predominantly composed of carbonate mudstones, wackestones, and packstones almost devoid of fossil constituents. High-energy grainstones are only known locally from the uppermost part of the Smackover; they are not uncommon in the thinner sections formed over the paleotopographic ridges and highs separating the embayments (Conecuh arch, Wiggins uplift, Fig. 16) (Sigsby, 1976; Benson and Mancini, 1984; Bradford, 1984).

In central Mississippi and adjoining parts of northeastern Louisiana and southeastern Arkansas, clean, fine- to medium-grained sandstones, often thick-bedded to massive, are found interbedded with carbonates and calcareous shales in both the lower and the upper parts of the Smackover Formation. They indicate the presence of a stream—the precursor of the Mississippi—draining the North American continental interior during the Oxfordian and supplying terrigenous clastics to the Late Jurassic Gulf of Mexico. These terrigenous clastics probably accumulated in a marine shelf or ramp in front of a fluvial and deltaic complex, and were spread by currents and waves. Those farther basinward have been interpreted as turbidites (Fig. 16) (Olsen, 1982) and as submarine fan complexes (Judice and Mazzulo, 1982).

A three-fold lithologic division of the Oxfordian section (the Zuloaga Formation) can also be recognized in northeastern Mexico: the lower unit is 1 to 20 m in thickness and composed of gray to brownish gray, fine- to medium-grained, occasionally conglomeratic, quartz sandstones. This unit is overlain by an interval of finely laminated, dark gray to brownish gray carbonate mudstones, argillaceous limestones, and pelletoidal wackestones rich in organic matter, pyrite, and anhydrite. The upper part of the Zuloaga Formation is predominantly made up of grain-supported carbonates—calcarenites, packstones, and oolitic limestones, which, as in the U.S. part of the basin, are often dolomitized. The carbonate grains are predominantly nonskeletal—ooids, grapestones, pisolites, and oncolites. The less-common skeletal grains are mainly fragments of corals, mollusks, brachiopods, and

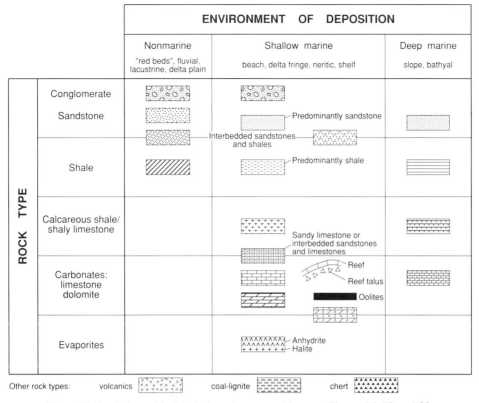

Figure 15. Symbols used in lithofacies-paleogeographic maps, Figures 16, 18, and 20.

sponge spicules (*Rhaxella*). Interbedded with the grainstones are dark gray to brown, dense, argillaceous mudstones, and pelletoidal micrites. They are generally burrowed and contain *Favreina* fecal pellets.

Northwest, toward the bordering land areas of the time in Mexico (the Coahuila platform, the El Burro uplift, and the Peyotes, Picachos, and Tamaulipas arches), the Oxfordian becomes increasingly sandy; the predominantly carbonate deeper water section typical of the Zuloaga grades into a section of interbedded nearshore limestones and sandstones, conglomeratic toward the base—the La Gloria Formation (Imlay, 1936, 1937).

Carbonates predominate in the Oxfordian section along the Sierra Madre Oriental as far south as the Ciudad Victoria area, where the Oxfordian section has been called Novillo Formation (Heim, 1940). It is predominantly composed of gray limestone, dolomitic limestone, and dolomite. Oolitic beds are common toward the top of the formation.

Farther south, the Oxfordian becomes increasingly shaly. The dark gray to black shales and calcareous shales with occasional thin layers of argillaceous limestone and calcareous nodules which represent the Oxfordian in east-central Mexico are included by some authors in the shalier lower part of the Tamán Formation (Sánchez López, 1961; Aguilera, 1972; De la Fuente and Moya Cuevas, 1972; Hermoso de la Torre and Martínez Pérez, 1972; Sánchez-Rosas, 1972; Toledo Toledo and Tavitas Galván, 1972). Cantú Chapa (1969a) excluded this lower shalier

section from the Tamán Formation and set it apart as a separate unit called the Santiago Formation.

In southern Mexico—the Isthmus area and the offshore Bay of Campeche—the Oxfordian section is composed of interbedded dark gray shales and siltstones with lesser amounts of limestone, light brown oolitic limestone, and very fine-grained sandstone (Santiago Acevedo, 1980; Santiago Acevedo and Mejía Dautt, 1980). No formational name has been given to this unit. The Todos Santos "red bed" sequence of southernmost Mexico may include sediments of Oxfordian age at its base.

Distinctive lateral lithologic gradations can generally be recognized in the Oxfordian section in a margin-to-basin direction. Along the margin of the basin, particularly in northeast Texas, Mississippi, and along parts of the northwestern rim, coarse clastics are common in the section, indicating proximity to adjacent land areas and/or discharge of streams into the basin. Basinward, clastics decrease and the section grades into sediments typical of a shallow shelf or gentle ramp—oolitic grainstones and packstones near the margin of the shelf, mudstones and local development of evaporites behind the oolitic bars. Farther basinward, calcareous shales and argillaceous limestones become progressively predominant in the section, making it more difficult to break the section down into distinct lithologic units. Each lithologic association forms concentric bands parallel to the rim of the Gulf of Mexico Basin, particularly along the northern and northwestern margins.

Thickness. The total Oxfordian section in the Gulf of Mexico basin ranges in thickness from a featheredge along parts of the periphery of the basin to as much as 400 to 500 m in more basinal areas (Figs. 13 and 17). Over most of the area of known occurrence, however, the Oxfordian seldom exceeds 300 m. To the west, in northeastern Durango, beyond the margin of the basin, the Oxfordian section has been reported to reach almost 1,000 m in thickness (Burckhardt, 1912, 1930; Oivanki, 1974). Structural complications in this area make these estimates of the thickness of the Oxfordian section questionable. In northern Zacatecas, the Oxfordian attains thicknesses of more than 700 m (Fig. 17).

Along the northwestern and northern margins of the basin, the Norphlet Formation and the equivalent basal Oxfordian sandstone unit in Mexico are thin, often just a few meters thick and generally less than 25 m (Fig. 13). Locally, the basal clastic section may reach a thickness of up to 100 m over limited and irregular areas. These local irregularities in the thickness of the basal Oxfordian clastic section reflect the topographic surface over which it was deposited. Thickness variations of the Norphlet

may also indicate either some solution and erosion or differential flow of the underlying salt deposits. The Norphlet is absent over the southern part of the Sabine uplift in east Texas and western Louisiana.

In the northeastern part of the Gulf of Mexico basin, in central Mississippi, southern Alabama, the western Florida panhandle, and adjoining parts of the Gulf of Mexico, the Norphlet section thickens greatly. Three distinct areas of thick Norphlet development may be recognized (Figs. 13 and 14; Marzano and others, 1988): one in central Mississippi, one centered in southwestern Alabama, and a third in the westernmost part of the Florida panhandle and adjacent parts of southernmost Alabama and the Gulf of Mexico shelf. The last two are separated by the Conecuh arch and the Wiggins uplift. As much as 360 m of the Norphlet Formation have been penetrated in central Mississippi and 250 m in southwestern Alabama. Farther to the east in the Apalachicola embayment, the central part of the Florida panhandle, the easternmost known occurrence of the Norphlet, the formation is again thin, attaining maximum thicknesses of 90 m.

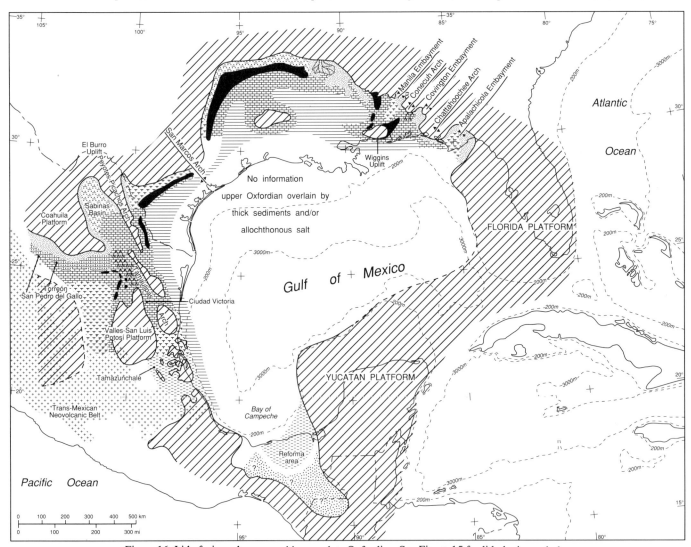

Figure 16. Lithofacies-paleogeographic map, late Oxfordian. See Figure 15 for lithologic symbols.

Thickness variations of the Norphlet Formation in the northeastern part of the basin are due in part to deposition over an irregular surface—ridges and embayments extending southwestward from the Appalachian highlands—and in part to the variable nature of the sedimentary section. The Norphlet is absent over the Wiggins uplift in southern Mississippi and southern Alabama.

The two thick accumulations in southern Alabama and the western part of the Florida panhandle probably represent either (1) fluvial valley fill and alluvial fans along the lower course of braided streams that flowed between the Appalachian ridges that projected southward into the broad coastal plain surrounding the newly formed proto–Gulf of Mexico, or (2) accumulations of eolian sands near these sources of clastic materials. The thick accumulation of eolian sands in central Mississippi may indicate proximity to the delta of the ancestral Mississippi River (Fig. 14).

The Oxfordian section overlying the basal clastic unit (the Smackover, Zuloaga, Novillo, Santiago, and coeval units) thickens basinward from its pinchout along the periphery of the Gulf of Mexico basin, but seldom attains thicknesses of more than

300 m (Fig. 17). Maximum thicknesses of 450 m have been reported from northern Louisiana and east Texas. Thicker sections may be present farther basinward, below the present reach of the drill, but beyond a certain point the Oxfordian should reach a maximum thickness and start thinning toward the central part of its depositional basin, due to increasing distance from the source of sediment.

The thickness of the Oxfordian section overlying the basal clastics is influenced locally by contemporaneous faults and by differential flow of the underlying salt.

Stratigraphic relations. As discussed before, the nature of the lower boundary of the Upper Jurassic section, particularly the contact between the basal Oxfordian clastics (Norphlet Formation and equivalent beds in northeastern Mexico) and the underlying evaporites, has been the subject of some controversy. It is not a trivial issue: whether this contact is conformable or unconformable and, if unconformable, what the duration of the hiatus between the units may be is important in estimating the age of the top of the salt.

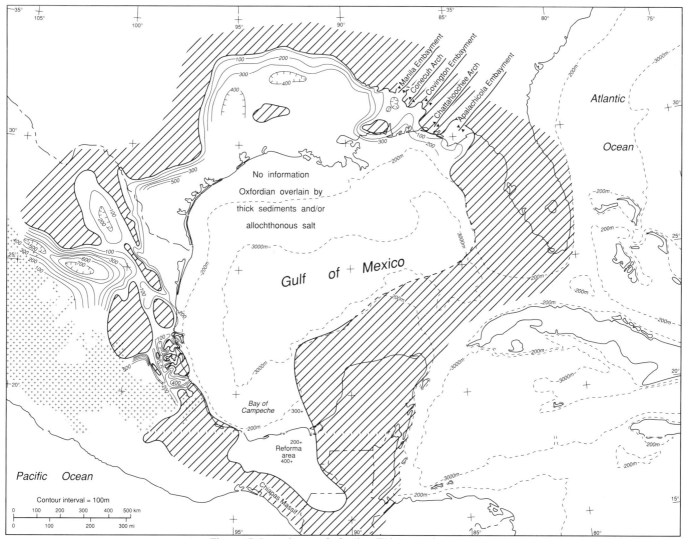

Figure 17. Isopach map, Oxfordian. Thicknesses in meters.

Evidence favors a conformable or, at most, a disconformable contact between the Norphlet Formation and equivalent basal Oxfordian clastics and the underlying Werner-Louann and Metate evaporites.

In the northeastern part of the basin, along the eastern part of the U.S. Gulf coastal plain, where the basal Oxfordian clastics are interpreted to be in part of alluvial/fluvial and in part of eolian origin, it is conceivable that these clastics could have begun to prograde from the north and northwest over the northern part of the extensive salt flats where the Werner-Louann and equivalent evaporites were being formed while salt was still accumulating to the south. The contact between the evaporites and the basal clastics should, in such a case, be essentially conformable and even transitional, both vertically and laterally. The change from evaporite formation to deposition of the basal Oxfordian clastics represents a change in paleogeography from extensive distribution of salt flats and little or no influx of clastics from the bordering lowlands, to the appearance of rivers draining these lands and contributing clastics to the ancestral Gulf of Mexico basin.

In the northwestern part of the basin, where the Norphlet appears to represent the initial clastic lag deposits of the Oxfordian transgression, the contact between this basal clastic section and the underlying Werner-Louann and Metate evaporites is probably also conformable. A long hiatus at the contact would be unlikely because, as mentioned before, such a long period of nondeposition would have resulted in extensive dissolution and erosion of the evaporites, for which there is no evidence.

Where the Norphlet and equivalent clastics overlie units older than the Werner-Louann evaporites—Eagle Mills, Huizachal, or La Joya "red beds," Paleozoic rocks, etc.—the contact is clearly unconformable with the Norphlet filling lows in the preexisting topographic surface. In such cases, the thickness of the basal clastics varies irregularly.

In east-central Mexico, where sedimentation had been nearly continuous during the late Middle Jurassic (Cahuasas-Tepexic sequence), the Oxfordian overlies older sediments conformably and often transitionally. The contact between the Middle Jurassic and the Oxfordian may, in fact, be present in some sections within the lower part of the marine shales of the Santiago Formation (See Fig. 12D). No clastic section is present at the base of the Oxfordian in this region.

In southern Mexico, the base of the Oxfordian has not been recognized in surface sections or reached in the subsurface. The Oxfordian may be present only in the more basinward parts of the Gulf of Mexico basin—the Reforma and Bay of Campeche areas.

The contact between the Norphlet, or equivalent clastics, and the overlying Oxfordian units—Smackover, Zuloaga, and Novillo formations—is usually sharp but generally recognized as conformable. Gradational contacts between the Norphlet and the overlying Smackover have also been reported in the literature (Oxley and others, 1967, 1968; Dinkins, 1968; Nichols and others, 1968; Tolson and others, 1983). Those authors who have postulated a disconformable contact between the Norphlet and

the Smackover Formation have given little direct evidence to support their interpretation (Newkirk, 1971; Todd and Mitchum, 1977; Budd and Loucks, 1981).

The basal Oxfordian clastics of the northwestern part of the basin appear to become progressively younger toward the margin and to grade laterally basinward into the lower part of the carbonate mudstones and argillaceous limestones of the overlying Smackover, Zuloaga, and Novillo Formations (Fig. 12).

In the deeper parts of the Gulf of Mexico basin, the Oxfordian is overlain conformably and in places gradationally by younger Upper Jurassic units. A disconformity may develop toward the periphery of the depositional basin. Over most of the basin, the top of the Oxfordian seems to closely correspond with the top of the Smackover, Zuloaga, Novillo, or Santiago Formations. However, down the slope of the ramps and near the edge of the shelf in which these units were deposited, oolitic grainstones and packstones, generally considered as the uppermost part of the Smackover and Zuloaga, extend upward into the Kimmeridgian. Farther basinward, the section becomes increasingly shaly, and it is often difficult to differentiate on the basis of lithologic composition between the shaly lateral equivalent of the Smackover, Zuloaga, and Novillo and the overlying shale units, the Bossier and Pimienta Formations (Fig. 12).

Biostratigraphy and age. No fossils have been recovered from the Norphlet and basal Oxfordian sands. The lowermost part of the overlying Smackover, Zuloaga, and Novillo Formations is generally unfossiliferous, although pelecypods of probable Oxfordian age have been reported from this interval (Imlay, 1943). The age of the basal Oxfordian clastic sequence, therefore, can only be postulated on the basis of its stratigraphic position between the Werner-Louann and equivalent evaporite deposits below, and the Smackover, Zuloaga, or Novillo Formations above. As stated earlier, the age of the evaporite section is not precisely known, but several lines of evidence point to a Callovian age, although its uppermost part may extend into the early Oxfordian. The lowermost age-diagnostic fossils in the section overlying the basal clastic units are ammonites of late Oxfordian age found 30 to 50 m above the base of the Smackover Formation. The Norphlet and equivalent units may be considered, therefore, to be early to middle Oxfordian in age.

Fossils diagnostic for age determination, particularly ammonites, are not common in the Oxfordian section of the Gulf of Mexico basin. From several surface sections and a few wells in east-central Mexico, south of the latitude of Tamazunchale, Cantú Chapa (1969a, 1971, 1979) has reported the occurrence in the upper part of the Santiago Formation of several species of *Ochetoceras*, *Dichotomosphinctes*, *Discosphinctes*, *Camplyllites*, and *Euaspidoceras*. They indicate a late Oxfordian age. No fossils have been reported from the lower part of the Oxfordian section of east-central Mexico, with two possible exceptions: Cantú Chapa (1971) has described a specimen of *Fehlmannites* sp. from a deformed calcareous nodule in a highly fractured shale of the Santiago Formation, and Erben (1957b) reported the occurrence of *Creniceras renggeri* and *Peltoceras (Parapeltoceras) annulare* in the basal beds of the Tamán Formation, now included in the Santiago Formation. There fossils have been interpreted to indicate an early Oxfordian age (Cantú Chapa, 1971; Erben, 1957b), although Imlay (1980) recommended caution in accepting this interpretation. As mentioned earlier, the lowest part of the Santiago Formation may extend in some areas into the Callovian, as indicated by the occurrence of species of *Reineckeia* (Cantú Chapa, 1971) and *Erymnoceras* (Erben, 1956b, 1957b).

In northeastern Mexico, no ammonites have been reported from the Zuloaga Formation, except in the San Pedro del Gallo area, 80 km west of Torreón (Burckhardt, 1912). The age of the Zuloaga has been established on the basis of its stratigraphic position below beds containing early Kimmeridgian fossil assemblages. The formation, however, often contains a distinctive fauna of pelecypods (*Nerinea, Pholadomya, Pleuromya, Trigonia, Astarte*, etc.), which led Burckhardt (1906, 1912, 1930) to give it the name "Calcaires à Nérinées."

In the U.S. part of the Gulf of Mexico basin, the lower part of the Smackover Formation is sparsely fossiliferous, and fragmentary ammonites of late Oxfordian age, similar to those described from east-central Mexico, have been recovered from cores cut in the lower, middle, and upper parts of the formation in wells drilled in northeast Texas and northern Louisiana, where the Smackover represents open marine conditions of deposition. They include several species of *Perisphinctes (Dichotomosphinctes)*, *P. (Discosphinctes)*, *P. (Orthosphinctes)*, *Euaspidoceras*, *Ochetoceras*, *Taramelliceras*, *T. (Proscaphites)*, and *Idoceras* (Imlay, 1943, 1945, 1980; Imlay and Herman, 1984). Mollusks (Imlay, 1941) and corals (Wells, 1942) have been described from the Smackover Formation, but they only indicate a Middle or Late Jurassic age.

In other sectors of the U.S. part of the Gulf of Mexico basin, such as southern Alabama, the Smackover generally is only sparsely fossiliferous, containing foraminifers and some pelecypods.

The ammonite faunas collected in the Oxfordian sequences of the Gulf of Mexico basin contain the same genera found in the beds of the same age in the Mediterranean area. They also include specimens similar to species in California and southwest Oregon (Imlay, 1980).

The scarcity of fossils in the Oxfordian section of the Gulf of Mexico basin may be attributed to several causes. The Oxfordian sediments represent the initial phase of the Late Jurassic transgression, and were deposited in littoral or marginal and shallow-water marine, often restricted, environments, with poor communication with the open ocean. Only where the Oxfordian represents unrestricted marine conditions of deposition are fossils more plentiful, and only where they represent open marine conditions have ammonites been found.

The absence of a rich megafauna in the lower part of the Oxfordian (and uppermost Callovian) section has been attributed by Imlay (1980) to the high salinity of the sea water filling the early Gulf of Mexico; he postulated that the Werner-Louann and other contemporaneous evaporite deposits were being formed at that time. Such a high-salinity environment would not be conducive to the well being of normal marine faunas.

Conditions of deposition—Paleogeography. Toward the beginning of the Late Jurassic, a marked change in the paleogeography took place in the ancestral Gulf of Mexico basin area. During the late Middle Jurassic (Callovian), most of the area was covered by large shallow-marine embayments only intermittently connected with the Pacific Ocean through central Mexico. A widespread and locally very thick evaporite section was deposited in these shallow-water bodies. Evaporite formation probably persisted into the early part of the Late Jurassic (Oxfordian), at which time terrigenous clastics began to enter the basin from the north along its northeastern periphery. They accumulated first as nonmarine alluvial, fluvial, and eolian sediments, and later, along the entire north flank of the basin, as the basal deposits of the widespread Late Jurassic marine invasion.

The lower Upper Jurassic (Oxfordian) terrigenous clastics of the northeastern part of the basin (the Norphlet Formation of central Mississippi, southwestern Alabama, and the western part of the Florida panhandle) were contributed by streams flowing south from the continental interior and the southern Appalachians.

In southwestern Alabama and the westernmost part of the Florida panhandle, clastics entered the basin through valleys between prominent ridges projecting southwestward from the Appalachian highlands (Fig. 14). The climate was still arid, and along the valleys the basal Oxfordian clastics represent the deposits of coalescing alluvial fan complexes and of shifting braided streams. Southwest of the area of valleys and ridges, where the streams reached the plains surrounding the incipient Gulf of Mexico, the sediments they carried accumulated and were subsequently redistributed, in part by wind, to form extensive dune fields. Dune, interdune, sand sheet, and playa deposits are, therefore, prevalent in this area (Mancini and others, 1985; Marzano and others, 1988). Formation of evaporite deposits was probably still taking place farther to the south.

The Norphlet section of south-central Mississippi is also of eolian origin (Hartman, 1968; Badon, 1975; McBride, 1981; Marzano and others, 1988). No alluvial-fan and fluvial deposits equivalent to those of southern Alabama and the western Florida panhandle, which could have been the source of these eolian sediments, have been reported from Mississippi, although highlands are known to have existed to the north. It is possible, however, that the clastic material was supplied by a stream that drained the interior of the continent and began to accumulate its sediment load in west-central Mississippi, perhaps in the form of a delta, shortly before and during the deposition of the eolian Norphlet sands (Fig. 14). The clastics supplied by such a stream would then have been winnowed and transported toward south-central Mississippi by winds, which at that time blew predominantly toward the east, southeast, and south (Gary Kocurek, 1987, personal communication). Evidence of a stream emptying into the early Gulf of Mexico in west-central Mississippi, though not available for the time of deposition of the Norphlet, is manifest in the upper Oxfordian sediments of the Smackover Formation. If the presence in pre-Smackover time of the mouth or delta of a stream in west-central Mississippi could be confirmed, it would be the first indication of the existence of the precursor of the Mississippi River.

While terrigenous clastics were entering the northeastern Gulf of Mexico basin area and were accumulating as alluvial, fluvial, and eolian sediments, the northwestern part of the basin, west of the present Mississippi River, lacking a source of terrigenous clastics, was still the site of evaporite formation in vast shallow-water bodies.

A widespread marine transgression took place in the Gulf of Mexico basin after the deposition of the alluvial, fluvial, and eolian basal Oxfordian clastic section of the northeastern part of the basin. It was during this marine transgression that the thin Norphlet lag deposits of the northwestern part of the basin accumulated, from southern Arkansas and northern Louisiana to northeastern Mexico. These deposits represent the reworking and spreading out by a transgressive sea of the products of a long period of erosion of the lowlands bordering the shallow evaporite basins. They are generally composed of the residue of weathering of the rocks cropping out in the lowlands and were deposited filling irregularities in the topographic surface over which the sea transgressed. In the northeastern part of the basin, the marine invasion is first reflected in the uppermost beds of the Norphlet Formation of southern Alabama and the western Florida panhandle, which have been interpreted as shallow-water marine deposits, probably the product of the reworking of the pre-existing sediments by the transgressing sea (Sigsby, 1976; Pepper, 1982; Mancini and others, 1985; Marzano and others, 1988).

It appears, therefore, that the initial Oxfordian clastic deposition in the Gulf of Mexico basin took place in the northeastern part of the basin, mostly in the form of extensive eolian deposits along the coastal plains, and as the deposits of coalescing alluvial fans and braided streams along the valleys and foothills of the highlands to the northeast. A stream draining the continental interior may have entered the ancestral Gulf of Mexico in west-central Mississippi, furnishing additional terrigenous clastics to the basin. No source of clastics is known to have existed at this time west of this hypothetical stream. Deposition of the basal Oxfordian terrigenous clastics in this western area did not take place, therefore, until the transgressive Oxfordian sea reached the area.

Though mapped under the common name of "Norphlet Formation" both east and west of the present Mississippi River, the basal Oxfordian clastic section is of a somewhat different age and represents different depositional processes in the northeastern and northwestern parts of the Gulf of Mexico basin.

As the Oxfordian transgression progressed over the shallow ramps or shelves surrounding the early Gulf of Mexico, the supply of coarse terrigenous clastics seems to have diminished and the basal sands were covered by the finer grained carbonate mudstones and micritic limestones of the lower part of the Smackover, Zuloaga, and Novillo Formations. The climate grad-

ually became more humid, perhaps the outcome of the Late Jurassic marine invasion which resulted in the formation of a widespread expanse of normal-marine water over the Gulf of Mexico basin region, where an arid lowland and shallow restricted hypersaline water bodies had previously existed.

Along the western margin of the basin, in east-central Mexico, marine deposition continued without interruption from the late Middle Jurassic to the Late Jurassic. No significant influx of coarser clastics has been recognized in this part of the basin. In southern Mexico, the sedimentary environments and paleogeography of the early Oxfordian are not known because rocks of this age have not yet been recognized.

During the remaining part of the Oxfordian, coarse clastics were restricted to two main areas (Fig. 16). One is in central Mississippi and adjoining parts of northeastern Louisiana and southeastern Arkansas, where sandstones are found interbedded with carbonates and calcareous shales throughout the Smackover Formation. The areal distribution of these sandstones has been interpreted as evidence of the presence of the mouth of a major stream entering the ancestral Gulf of Mexico from the north, probably the paleo-Mississippi River. Basinward from this source of terrigenous clastics, sands may have been deposited in deeper water as turbidites (Olsen, 1982) and/or as submarine fan complexes (Judice and Mazzullo, 1982).

The second area of coarser clastic influx into the Gulf of Mexico basin during the Oxfordian is in northeastern Mexico, where the Zuloaga Formation, a predominantly carbonate and shale unit, becomes increasingly sandy in a westward and northwestward direction. The sandier lateral-equivalent section has been called the La Gloria Formation.

The lower carbonate mudstone and limestone part of the Smackover and Zuloaga Formations, immediately overlying the basal clastics, was deposited under low-energy, severely restricted conditions, in a shallow sea slowly transgressing over a broad ramp gently dipping basinward. Paucity of fossils or other indications of life have been attributed to high salinity or toxic sea-bottom conditions. The laminated carbonate mudstones are thought to represent sedimentation in intertidal mudflats or low supratidal coastal areas in which development of algal mats alternated with deposition of carbonate muds.

The Oxfordian transgression probably reached its maximum extent toward the end of the deposition of the lower carbonate mudstone and argillaceous limestone section of the Smackover and Zuloaga Formations. Conditions at this time were no longer severely restricted, and the section contains the fossil remains of a varied marine fauna, as well as indications of burrowing. This upper part of the "lower" Smackover and Zuloaga represents low-energy subtidal conditions of deposition.

The upper part of the Oxfordian section in the northern and northwestern parts of the Gulf of Mexico basin was deposited under high-energy conditions during a stillstand or moderate lowering of the sea level. For the most part, the margins of the basin were still broad, regional ramps sloping gently toward the basin. These ramps were progressively modified by the formation of a widespread shoal system in which grainstones and packstones, particularly oolitic grainstones, formed offshore bars and beaches. These linear shoals separated a landward area of intertidal coastal plains and littoral lagoons from a deeper and quieter shelf or ramp and a basinal area where micritic limestones, carbonate muds, and shales were deposited (Fig. 16).

Along the trend of the shoals the water was shallow, clear, generally agitated by waves, and favorable for the formation and accumulation above wave base of grain-supported carbonates and similar high-energy sediments in offshore bars, tidal bars, spits, spillover lobes, and beaches. These grain-supported carbonates, generally composed of numerous shoaling-upward sedimentary cycles, were periodically exposed subaerially and often eroded. Exposure to erosion may have been caused by salt deformation and uplift, by sea-level lowering, or by a combination of both. Patch reefs grew locally over sea-floor highs, and occasional islands and embayments developed along the trend of the shoals.

Grainstone shoals seem to have been particularly well developed during the latest Oxfordian and earliest Kimmeridgian. Such a grainstone shoal system resulted in the formation of protected hypersaline lagoons and sabkhas landward from the shoals. In these lagoons and sabkhas, evaporites, low-energy carbonates, and some fine-grained terrigenous clastics accumulated. This lagoonal sequence, called the Buckner Formation or Member, was more distinctly developed during the early part of the Kimmeridgian and will be described in more detail in the following section of this chapter.

Through the late Oxfordian, the land-to-basin sequence of depositional environments—coastal plain/lagoon/sabkha to shoal to shelf to basin—prograded basinward, and toward the end of the Oxfordian the gentle ramps may have evolved, at least in some areas, into shelves with well-developed shelf margins, where high-energy grain carbonates accumulated. In southern Alabama and the Florida panhandle, the Oxfordian sediments accumulated in a much more restricted environment, protected from the open sea and segmented into smaller barred basins by "basement" ridges projecting into the basin from the southern Appalachians and by salt-swells and other "basement" highs (Sigsby, 1976; Mancini and Benson, 1980; Tolson and others, 1983).

What caused the progressive change in depositional environment during the Oxfordian from gentle basinward-sloping ramps to a more complex system of shoals and eventually to a shelf, is still the subject of speculation. It may be explained by rapid production of shallow-water carbonates which, unless counteracted by faster subsidence, could cause the natural evolution of a ramp profile into a shelf profile. Contemporaneous growth faults known to have developed around the periphery of the basin have also been suggested as a cause, as have changes in sea level, and the early flow of the underlying Louann Salt. The growth faults could be related to the movement of the salt.

Evidence of local salt movements concurrent with deposi-

tion of the upper Oxfordian sediments is convincing (Hughes, 1968). These movements are reflected not only in the physiography of the basin margins, but also in the distribution of the lithofacies and thickness of the upper Oxfordian sediments (upper Smackover in particular). As the salt flowed into bulges and ridges it formed anticlines and synclines; grainstones formed in the highs, generally as oolitic tidal bars, whereas finer grained carbonates and calcareous shales accumulated in the lows.

In east-central Mexico, the conditions of deposition and paleogeography were considerably different from those in the northern and northwestern margins on the basin. The Oxfordian section was not underlain by a salt interval and does not, therefore, show the effects of the early flow and deformation of such an incompetent and ductile layer. No coarse clastics have been reported from the Oxfordian section, indicating very limited or no influx of such clastics and the absence, therefore, of important streams draining into the western margin of the basin. The Oxfordian section of this western margin, on the other hand, seems to still reflect the horst and graben structural framework that controlled deposition during Late Triassic to Middle Jurassic time (Fig. 17). A thicker and more marine Oxfordian, rich in organic matter and representative of a low-energy (euxinic?) environments, has been reported from the areas formerly occupied by the grabens in which the Upper Triassic, Lower and Middle Jurassic sediments were deposited; thinner and more littoral sediments are found in the structurally high areas.

Not much is known now about the distribution, thickness, environment of deposition, and paleogeography of the Oxfordian section in southern Mexico and neighboring parts of the Gulf of Mexico—the Reforma area and the Bay of Campeche—or how far inland the Oxfordian marine transgression extended. It is possible that the Oxfordian may be represented in the basal part of the Todos Santos Formation "red beds" of southern Mexico and northwestern Guatemala (Fig. 16). The Oxfordian of southern Mexico is known to be underlain by Middle Jurassic salt and should, therefore, reflect the contemporaneous deformation of the salt.

As discussed earlier, the Oxfordian Gulf of Mexico was probably not yet connected with the recently opened North Atlantic Ocean; its only marine communication was with the Pacific through central Mexico, as during the Early and Middle Jurassic. A thick Oxfordian section with ammonite remains in the San Pedro del Gallo area of eastern Durango could be interpreted to indicate that another connection with the Pacific existed during the Oxfordian west of Torreón. In northeastern Mexico, a belt of islands, shoals, and shallow-water shelves extending from southwest Texas to east-central Mexico separates the early Gulf of Mexico to the east from a smaller basin in central Mexico west of the Valles–San Luis Potosí platform (eastern Durango, southern Coahuila, northern Zacatecas, and western San Luis Potosí). The two basins probably join under the thick Neogene-Quaternary volcanics of the Trans-Mexican neovolcanic belt and connect with the Pacific Ocean (Fig. 16). It is not possible to ascertain how deep the central part of the proto-Gulf of Mexico was during the Oxfordian, because rocks of this age are only known from the peripheral margins of the basin.

Economic considerations. Both the Norphlet and Smackover Formations are important producers of oil and gas in the U.S. Gulf Coast area. Production from the Smackover is, for the most part, from oolitic grainstones, dolomitized grain-supported carbonates (grainstones, packstones), and wackestones in the upper part of the formation. Accumulation is in both structural and stratigraphic traps, principally in east Texas, southern Arkansas–northern Louisiana, southern Alabama, and the Florida panhandle.

The organic-rich carbonate mudstones of the lower part of the Smackover are considered to be the source of the oil and gas in the Smackover and Norphlet reservoirs (Mancini and Benson, 1980, Chapter 15, this volume).

In northeastern Mexico, the Zuloaga formation is a minor oil- and gas-producing unit. No production from Oxfordian sediments has been reported from southern Mexico, where the section is predominantly composed of shale and calcareous shale. However, the Oxfordian, as well as the rest of the overlying Upper Jurassic section, is also considered an excellent source rock of petroleum in east-central and southern Mexico.

Kimmeridgian

Distribution. The geographic distribution of the Kimmeridgian sediments in the Gulf of Mexico basin is similar to that of the Oxfordian beds that they overlie. The Kimmeridgian generally oversteps the Oxfordian and extends farther landward. The distance of this overstepping is nowhere sizeable, with the exception of northeastern and southern Mexico. In northeastern Mexico, the Oxfordian seems to be restricted to the Sabinas basin and surrounding areas, whereas Kimmeridgian sediments have been recognized the length of the "Chihuahua Trough" as far north as the El Paso area, considerably beyond the known extent of Oxfordian sediments (Salvador, 1987). In southern Mexico, marine sediments of Kimmeridgian age are present bordering the northern part of the Chiapas Massif, whereas marine sedimentation during the Oxfordian was apparently restricted to areas closer to the present-day Gulf of Mexico.

As in the case of the Oxfordian, the Kimmeridgian nowhere crops out along the northern rim of the basin, where it is only known from wells drilled in search of oil and gas. It is known at the surface, however, along the Sierra Madre Oriental in eastern Mexico, and in the foothills bordering the Chiapas Massif in southern Mexico. Kimmeridgian sediments have been identified in wells along the entire Gulf coastal plain of Mexico, from the U.S. border to as far south as the Reforma area, in northern Chiapas and Tabasco, and the Bay of Campeche shelf. No Kimmeridgian sediments are known from the Florida and Yucatan carbonate platforms.

Lithostratigraphy. The lithostratigraphic terminology of the Kimmeridgian section in the Gulf of Mexico basin is not as clearly defined or as generally accepted as that of the Oxfordian. This may be due to the fact that, unlike the Oxfordian, which maintains remarkable lithologic uniformity over most of the basin, the Kimmeridgian shows considerable vertical and lateral lithologic variability.

In the U.S. Gulf coastal plain and northeast Mexico, the Kimmeridgian section is composed of a variety of clastics, carbonates, and evaporites. Over wide areas a basal evaporite unit, predominantly composed of anhydrite but including locally thick halite sections, can be recognized (Fig. 18). It is restricted to those areas on the shelves where proper barriers made the formation of evaporites possible. The term "Buckner" has generally been given to this lower evaporitic section in the United States (Shearer, 1938; Weeks, 1938), and the overlying section has been called "Haynesville" when predominantly composed of terrigenous clastics (Philpott and Hazzard, 1949), or "Gilmer Limestone" (Forgotson and Forgotson, 1976) when made up of carbonates. Some authors have used these terms as units of formational rank (Dickinson, 1968a, 1968b, 1969; Imlay, 1980), whereas others prefer to apply the term "Haynesville" to the entire section between the top of the Smackover Formation and the base of the overlying Cotton Valley Group, and consider the Buckner evaporite section to be a member of the Haynesville Formation (Philpott and Hazzard, 1949; Oxley and others, 1968). For the purpose of a regional geologic synthesis it seems more appropriate and advantageous to use the term "Haynesville" in the broader sense, and to consider the Buckner and Gilmer as members of the Haynesville Formation (see Fig. 12).

In northeastern Mexico, a Kimmeridgian section lithologically similar to that of the U.S. Gulf coastal plain area has been called the Olvido Formation (Heim, 1940). A lower evaporitic unit and an upper sandstone and/or carbonate unit can also be recognized within the Olvido Formation. In the neighborhood of Monterrey, the evaporitic lower unit of the Olvido Formation has been called the Minas Viejas Formation and mistakenly correlated with the older Werner-Louann anhydrite and salt section (Humphrey, 1956; Wall and others, 1961; Weidie and Wolleben, 1969; and many others). Landward, the Olvido grades into the lower part of a sequence predominantly composed of terrigenous clastics that extends from the top of the Zuloaga Formation to the lowermost Cretaceous carbonate units. It has been called the La Casita Formation (Imlay, 1936, 1937; Vokes, 1963). In these landward areas, therefore, the Kimmeridgian is represented by the lower part of the La Casita Formation (see Figs. 2 and 12).

Both in the U.S. Gulf coastal plain and in northeastern Mexico, the Kimmeridgian, as well as the underlying and overlying parts of the Upper Jurassic section, becomes increasingly shaly in a basinward direction and grades into a thick, predominantly shale unit that represents most of the Upper Jurassic. This thick shale interval has been called Bossier Formation in the United States and Pimienta Formation in northeastern and eastern Mexico (see Fig. 12). In most of the U.S. Gulf coastal plain area and northeast Mexico, the Buckner and lower Olvido evaporites are separated from the more marine, basinal, shaly

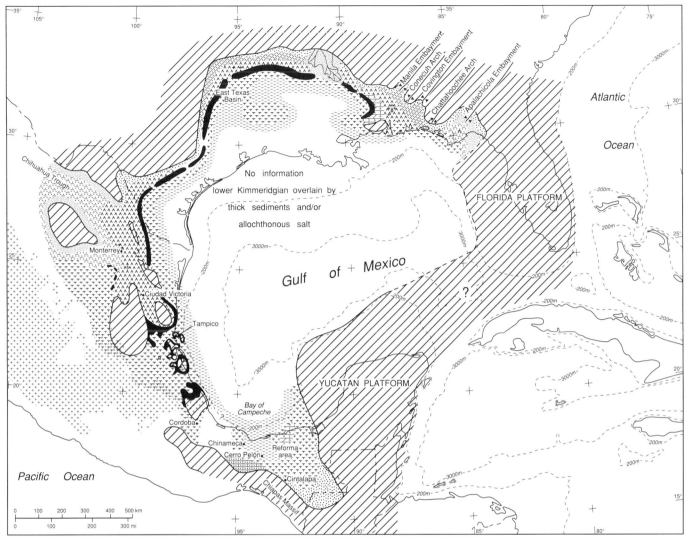

Figure 18. Lithofacies-paleogeographic map, early Kimmeridgian. See Figure 15 for lithologic symbols.

Kimmeridgian section by a belt of oolitic grainstones, which generally have been considered as part of the upper Smackover-Zuloaga. In such cases, the Smackover-Zuloaga should be considered to extend into the lower part of the Kimmeridgian (see Fig. 12).

The Kimmeridgian part of the Upper Jurassic section in east-central Mexico is represented by part or all of the predominantly carbonate section of the "restricted" Tamán Formation (Heim, 1926) or by all or part of the upper part of the Tamán of those authors who do not consider the shaly lower part of the formation as the Santiago Formation (Aguilera, 1972; De la Fuente and Moya Cuevas, 1972; Hermoso de la Torre and Martínez Pérez, 1972; Sánchez-Rosas, 1972; Toledo Toledo and Tavitas Galván, 1972). The lower part of the Las Trancas Formation of the states of Querétaro and Hidalgo (Segerstrom, 1961, 1962) and the lower part of the Tepexilotla Formation of the Córdoba (Veracruz) area (Mena Rojas, 1960, 1962; Viniegra, 1965; Carrasco and others, 1975) are also Kimmeridgian in age.

In southern Mexico, the lower part of the Chinameca Limestone, which crops out in several localities of the Isthmus region (Chinameca dome, Cerro Pelón) has been assigned to the Kimmeridgian; the upper part of the unit is said to be of Early Cretaceous age (Burckhardt, 1930; Benavides, 1950; Santiago Acevedo, 1962; Contreras and Castillón, 1968). No formational names have been proposed for the Upper Jurassic section in the subsurface of southern Mexico. Quezada Muñetón (1984) has described an Upper Jurassic section (the Zacatera Group) near the northeastern termination of the Chiapas massif, along the trans-isthmian highway; the middle part of the section is composed of oolitic and oncolitic grainstones interbedded with carbonate mudstones and wackestones; he assigned this middle part of the section to the Kimmeridgian on the basis of its contained microfauna. Similar carbonates have been mapped bordering to the north, the Chiapas massif at least as far southeast as the Cintalapa area. They are probably also Kimmeridgian in age, but no paleontological evidence to support such an age has been published. Southeastward from Cintalapa, as far as northwestern Guatemala, the Kimmeridgian may be represented in the "red beds" of the Todos Santos Formation or San Ricardo Formation of Richards (1963).

The lithology of the Kimmeridgian section in the Gulf of Mexico basin varies considerably. Along the northern and northwestern flanks it generally can be divided into a lower evaporitic unit and an upper unit composed of different proportions of terrigenous clastics and carbonates. In the western and southern flanks, the Kimmeridgian is predominantly composed of carbonates and calcareous shales, with little or no coarse clastics.

The lower evaporitic section of the northern and northwestern flanks of the basin—the Buckner Member of the Haynesville Formation in the United States and the lower part of the Olvido Formation in northeastern Mexico—is characterized by anhydrite and "red beds." The most common lithologic component is white, pink, or gray, massive or nodular anhydrite in thick massive beds interbedded with thinner beds of dolomite, dark to light gray argillaceous limestone, anhydritic limestone, and anhydritic or dolomitic brown or red mudstone. Anhydrite is more abundant toward the base. Locally, halite is present in the section, generally in thin stringers but occasionally in units several tens of meters thick (northeast Mexico, northeast Texas, Manila embayment of southern Alabama). Within the lower Kimmeridgian, it is not uncommon to be able to recognize a gradation from more abundant halite in the central, thicker parts of evaporite bodies, to predominant anhydrite and eventually carbonates toward the rims. Grainstone beds are rare in the lower part of the Kimmeridgian. Red, green, or greenish gray terrigenous clastics—shales, siltstones, and sandstones—are commonly found interbedded with the evaporitic intervals. They generally increase in abundance landward and toward the top of the section. The Buckner Member is also noticeably sandier in central Mississippi and northeastern Louisiana.

The lower evaporitic unit of the Haynesville Formation (Buckner Member) and its equivalent lower part of the Olvido can be traced around the periphery of the Gulf of Mexico basin, with only occasional interruptions, from the western part of the Florida panhandle to the Ciudad Victoria area in northeastern Mexico (Fig. 18).

Downdip, the Buckner evaporitic section has been observed to grade to oolitic grainstones similar to those of the upper part of the Smackover Formation. This grainstone section has generally been included in the latter unit. The oolitic grainstones have been interpreted to represent barrier islands or shoals that restricted water circulation and contributed to the development of depositional environments favorable for the formation of evaporitic and sabkha sequences between them and the land areas. Some authors have related the formation of the barrier islands and shoals to the growth of salt-cored anticlines.

The upper part of the Haynesville Formation is a lithologically variable sequence of terrigenous clastics and carbonates with occasional occurrences of anhydrite. Eastward from northeast Texas, the sandstones and shales predominate, particularly in northeastern Louisiana, southeastern Arkansas, and central Mississippi, where the section is almost entirely composed of calcareous and anhydritic fine-grained sandstones and dark red calcareous and anhydritic shale. The upper Haynesville clastics are generally red, thinly bedded, and range in grain size from shales to coarse-grained sandstones and conglomerates.

In the Apalachicola embayment of northwestern Florida, the Haynesville Formation is composed of gray and red, often calcareous shales, and lesser amounts of fine-grained sandstones and thin beds of micritic limestone. The Buckner basal evaporitic section has limited distribution in this area (Mitchell-Tapping, 1982).

In the East Texas basin, the upper part of the Haynesville Formation, between the Buckner Member and the overlying Cotton Valley Group, is composed of gray to brown oolitic and pelletal grainstones and packstones interbedded with argillaceous limestones and carbonate mudstones—the Gilmer Limestone (Forgotson and Forgotson, 1976; Ahr, 1981), also known in the literature as the "Cotton Valley Limestone" (Collins, 1980). The grainstones are composed predominantly of nonskeletal grains, are often cross-bedded, and occasionally graded. They indicate deposition as tidal bars in very shallow, open-marine water under conditions of considerable current and wave agitation. The Gilmer Limestone is best developed in both flanks of the East Texas basin. It grades landward into an increasingly more clastic section. On the other hand, the Gilmer, as well as the underlying Buckner evaporitic interval, grade basinward laterally into a carbonate section difficult to differentiate from the Smackover. The term "Louark Group" has been used to name this undifferentiated, predominantly carbonate Upper Jurassic section above the Louann Salt, the Norphlet Formation, or older rocks, and below the Cotton Valley Group.

In south Texas and northeastern Mexico, the Kimmeridgian section is similar to that of east Texas: it is composed of a lower evaporite–red bed unit (Buckner or lower Olvido) and an upper carbonate unit (upper part of the Haynesville or upper Olvido). The upper carbonate unit is composed of calcareous mudstones, bioclastic grainstones, oolitic grainstones, calcarenites, wackestones, and occasional evaporitic intervals. Whole or broken gastropods, bryozoans, calcareous algae, sponge spicules, foraminifers, and ostracodes are common. Landward, to the west, the Olvido grades into a more clastic section—sandstones, siltstones, and shales with lesser amounts of limestone—which has been included in the lower part of the La Casita Formation.

In east-central Mexico, south of the Ciudad Victoria area, the Kimmeridgian is represented by a section composed of dark gray to black, medium- to thick-bedded, well-bedded, argillaceous limestones interbedded with dark gray to black shales and calcareous shales—the "restricted" Tamán Formation. The limestone generally predominates, but the proportion of limestone to shale in the section varies considerably from area to area. Smaller amounts of coarser terrigenous clastics, black chert in beds and lenses, and bands of calcareous concretions have been reported from some localities (Carrillo Bravo, 1965).

Locally, the argillaceous limestone and calcareous shale sequence characteristic of the Tamán Formation grades laterally to skeletal and oolitic grainstones and calcarenites, indicative of high-energy, shallow-water deposition in shoals parallel to the coastline or surrounding islands (Fig. 18). This grain carbonate section has been called the San Andrés Formation (Carrillo Martínez, 1960) or, preferably, the San Andrés Member of the Tamán Formation (Cantú Chapa, 1971). The San Andrés Member shows considerable vertical and lateral variation in lithologic composition and thickness. The most common components are oolitic and bioclastic grainstones and pelletoid grainstones. Siltstones and sandstones are much less frequent. Most prevalent skeletal grains are whole shells or fragments of echinoids, crinoids, mollusks, corals, calcareous algae, and benthonic foraminifers, whereas oolites, pseudoolites, and pelletoids form the bulk of the nonskeletal grains (Carrillo Martínez, 1960; Cantú Chapa, 1971; Aguilera, 1972; Hermoso de la Torre and Martínez Pérez, 1972; Sánchez-Rosas, 1972; Toledo Toledo and Tavitas Galván, 1972).

The San Andrés Member generally ranges in thickness from less than a meter to around 250 m, but thicknesses of up to 470 m have been reported. The unit, as a whole, varies considerably in thickness over short distances, as do its often lenticular individual beds.

In the northern part of the Isthmus of Tehuantepec, the Kimmeridgian is represented as the lower part of the Chinameca Limestone, a sequence of alternating thin-bedded gray to dark gray argillaceous limestones and calcareous shales that apparently ranges in age from the Kimmeridgian to the Early Cretaceous (Burckhardt, 1930; Benavides, 1950; Santiago Acevedo, 1962; Contreras and Castillón, 1968).

In the subsurface of the Reforma area, the Kimmeridgian is composed of dark gray, partly dolomitized pellet, oolite, and bioclastic packstones with lesser amounts of anhydrite and mudstone. In the Bay of Campeche fields, the upper part of the Kimmeridgian is predominantly argillaceous with minor amounts of oolitic limestone; the lower part is composed of interbedded shales, dolomites, bentonites, and a few anhydrite beds (Santiago Acevedo, 1980; Santiago Acevedo and Mejía-Dautt, 1980). In the southern part of the Isthmus of Tehuantepec, Quezada Muñetón (1984) has assigned to the Kimmeridgian, on the basis of its contained microfauna, a section composed of oolitic and oncolitic grainstones with interbedded carbonate mudstones and packstones, indicative of a shallow-water marine environment.

Thickness. The Kimmeridgian of the Gulf of Mexico basin thickens from its pinchout along the periphery of the basin to more than 500 m in the more basinward areas of southern Alabama and Mississippi, east Texas, south Texas, and northeastern and east-central Mexico. In southern Mexico, the Kimmeridgian section has been reported to range from 150 to 300 m (Fig. 19). The Kimmeridgian section shows sharp variations in thickness, particularly along the northern margin of the basin, that may be attributed to local structural growth due to contemporaneous

fault movement and to the deformation of the underlying Louann Salt. Abrupt lateral changes in thickness are particularly common in the Buckner and lower Olvido evaporites, which are known to attain thickness of several hundred meters in some areas.

As the Oxfordian, the Kimmeridgian should reach a maximum thickness some distance beyond present control and then thin toward the central part of the basin.

Stratigraphic relations. The lower boundary of the Kimmeridgian section in the Gulf of Mexico basin is conformable in more basinal sections but often unconformable toward the margins of the basin. Where the Kimmeridgian oversteps the Oxfordian and lies over older rocks, the contact is unconformable.

In the United States, the Haynesville Formation and, when present, its lower evaporitic member, conformably and often transitionally overlie the Smackover Formation almost everywhere. In downdip basinal sequences, the Oxfordian-Kimmeridgian boundary occurs within a homogeneous shale section generally included in the Bossier Formation. Similar conformable and often transitional relations have been recognized in northeastern Mexico in downdip sections where the marine Oxfordian underlies marine Kimmeridgian sequences.

Some authors have postulated locally unconformable contacts between the Buckner evaporites and the overlying upper part of the Haynesville Formation (Goebel, 1950; Forgotson and Forgotson, 1976). Local unconformities within the Haynesville section should be expected as a result of equally local deformation of the underlying salt section, uplift, and erosion.

An unconformity has also been recognized in certain areas between the Kimmeridgian and the overlying Tithonian sequences. Forgotson and Forgotson (1976) reported "a major unconformity" between their Gilmer Limestone and the overlying Cotton Valley Group in east Texas. Todd and Mitchum (1977) confirmed this unconformable relation based on their seismic-stratigraphy study. They reported that in east Texas the Bossier Formation "strongly onlaps the Gilmer Limestone surface and displays obvious cycle terminations against it. . . ." In south Texas, they added, "the onlap is not as strong. . . ." Dinkins (1968) has also recognized an unconformity at the top of the Haynesville in Mississippi. Todd and Mitchum (1977) have interpreted this unconformity as the result of the eustatic lowering and subsequent rise of the sea level. Local tectonic influence, however, cannot be discarded.

Except for the Tamaulipas-Constituciones oil field, a few kilometers north of Tampico, where Stabler (1972) reported that the Tithonian Pimienta Formation onlaps the underlying beds from west to east, no other author has recognized an unconformity between the Kimmeridgian and overlying beds in Mexico. Most report a conformable and often transitional boundary.

Biostratigraphy and age. In the more marine, fossiliferous, and well-exposed sections of the Tamán Formation in east-central Mexico, Cantú Chapa (1969a, 1971, 1979) has recognized four biozones. They are characterized, from youngest to oldest, by species of the following ammonite genera:

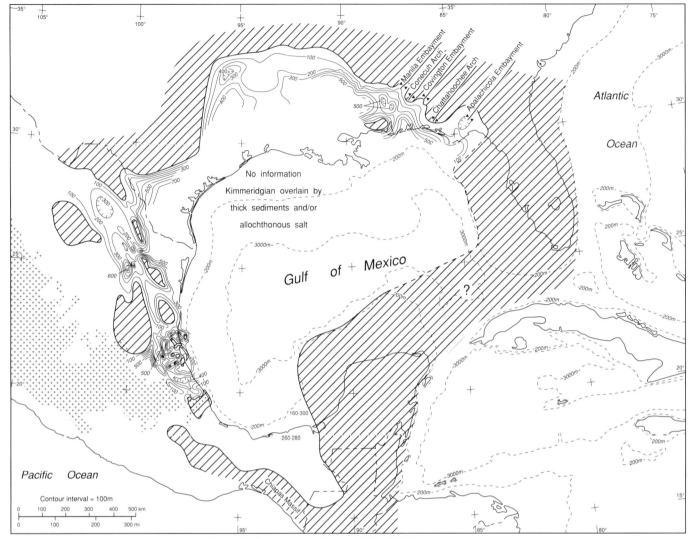

Figure 19. Isopach map, Kimmeridgian. Thicknesses in meters.

● *Virgatosphinctes mexicanus* (and the locally very abundant pelecypod *Aulacomyella neogeae*) in addition to *Taramelliceras (Metahaploceras)* and *Subdichotomoceras*.

● *Glochiceras* of the *G. fialar* group. The pelecypods *Buchia concentrica* and *B. mosquensis* have also been reported from this interval (Imlay, 1980), as well as species of *Ochetoceras*, *Glochiceras (Lingulaticeras)*, *Taramelliceras (Metahaploceras)*, *Aspidoceras*, and *Subdichotomoceras*.

● *Idoceras* spp., in addition to rare occurrences of *Aspidoceras* and *Nebrodites*.

● *Ataxioceras (A.) aff. subinvolutum* associated with *Rosenia (Involuticeras)* and *Taramelliceras (Metahaploceras)*.

The lower three biozones indicate a Kimmeridgian age, and the uppermost can be assigned to the earliest Tithonian.

Thin-shell pelecypods, echinoids, sponge spicules *(Rhaxella),* and radiolaria are also common in the Tamán Formation.

Many of these forms have also been reported in northeast Mexico from the lower part of the La Casita Formation *(Idoceras* spp., *Glochiceras*, etc.) and its more marine equivalent unit, the

La Caja Formation (Imlay, 1938), denoting a Kimmeridgian to earliest Tithonian age for this part of the section (Burckhardt, 1930; Imlay, 1939, 1943, 1952, 1953, 1980).

In southern Mexico, the occurrence of sediments of Kimmeridgian age has been indicated by the recognition of species of *Aspidoceras* and *Idoceras* in the lower part of the Chinameca Limestone (Burckhardt, 1930) and *Nebrodites* in well Chac-1 in the Gulf of Campeche (Cantú Chapa, 1977).

In the United States, a few ammonites of Kimmeridgian age have been recovered in northwest Louisiana and east Texas from cores cut in the lower, basinal part of the Bossier shales, in the Haynesville Formation and in its Gilmer Member. Most common are species of *Idoceras*, *Ataxioceras*, and *Nebrodites* (early Kimmeridgian), *Glochiceras*, *Taramelliceras*, and *T. (Metahaploceras)* (early to late Kimmeridgian), identical to species identified from Kimmeridgian sections in east-central and northeast Mexico (Imlay, 1941, 1943, 1945, 1952, 1980; Imlay and Herman, 1984).

Like the Oxfordian ammonite faunas, those of the Kimmeridgian sequences of the Gulf of Mexico basin contain genera typical of the Mediterranean region and similar to specimens found in California and Oregon (Imlay, 1980).

In summary, the Buckner and lower Olvido evaporites have not yielded any age-diagnostic fossils; they have been assigned an early Kimmeridgian or latest Oxfordian age on the basis of stratigraphic position. No age-diagnostic fossils have been reported, either, from the predominantly clastic updip upper part of the Haynesville Formation; it has been assumed to represent the remaining part of the Kimmeridgian, also on the basis of stratigraphic position and correlation with more marine downdip equivalent sequences, from which age-diagnostic fossils have been recovered. Kimmeridgian ammonites have been recovered from the more marine sections of the upper Haynesville. In Mexico, the presence of sedimentary sequences of Kimmeridgian age is better controlled, particularly in the east-central part of the country, where rich ammonite faunas have been collected and studied. Less marine sections of the La Casita Formation in northeastern Mexico have been dated on the basis of scattered occurrences of ammonites, stratigraphic position, and regional correlations.

Conditions of deposition—Paleogeography. The trend toward shallower water conditions initiated during the late Oxfordian continued during the early Kimmeridgian. Over all the Gulf of Mexico basin, the lower Kimmeridgian section appears to reflect less marine or shallower marine environments of deposition than the underlying Oxfordian. However, after a short period of shoaling and retreat of the sea, the Late Jurassic transgression continued during the Kimmeridgian. As discussed earlier, no evidence exists at present to support the interpretation that the ancestral Gulf of Mexico was in communication with the Atlantic during the Kimmeridgian, or to say that such a connection did not exist. In the preparation of the maps for this chapter (Figs. 18 and 19) it has been assumed that a connection between the Gulf of Mexico and the Atlantic was not established until after the period of restricted marine condition that characterized the early Kimmeridgian.

Throughout the U.S. Gulf coastal plain, the occurrence of the Buckner evaporites overlying the high-energy shallow-water marine upper Smackover indicates the presence of hypersaline coastal lagoons and/or sabkhas where marine conditions had previously existed. Similar changes in the conditions of deposition occurred in northeastern Mexico, where the evaporitic lower part of the Olvido Formation overlies the marine carbonates of the Zuloaga or Novillo Formations. Farther south, in east-central Mexico, the Tamán Formation is also considered to have been formed under shallower water marine conditions than the underlying Santiago Formation. In southeastern Mexico, the limited information available only allows one to postulate an open marine environment in the area during the entire Late Jurassic, except in the areas bordering the Yucatan platform and the Chiapas massif.

The supratidal mudflats, hypersaline lagoons, and sabkhas in which the Buckner and lower Olvido evaporitic section was

formed graded landward into nonmarine fluvial and deltaic environments in which a predominantly coarse clastic section was deposited. Basinward, the evaporites grade into oolitic grainstones (uppermost Smackover and Zuloaga) that were deposited as offshore bars and shoals and provided the restrictive barriers behind which evaporites could be formed in an environment protected from waves and currents (Fig. 18). The evaporitic basins occupied a long belt parallel to the margin of the basin. Anhydrite is the most common of the evaporitic deposits, but halite was formed in the largest and/or deepest among these basins.

Basinward from the oolitic barriers, the Kimmeridgian section is represented by predominantly shaly sections deposited in an increasingly deeper water marine environment (lower part of the Bossier and Pimienta formations).

The upper part of the Kimmeridgian section in the eastern part of the U.S. Gulf coastal plain was deposited in very shallow-water marine to intertidal and supratidal littoral environments (intertidal flats, tidal deltas, beaches). Terrigenous clastics predominate in the section, particularly in an area of clastic input in northeastern Louisiana, southeastern Arkansas, and central Mississippi, where the presence of a delta complex—the delta of the ancestral Mississippi River—can again be inferred. On both sides of the delta, terrigenous clastics derived from the north filled the lagoons and covered the sabkhas. Only in more protected environments, like in southwestern Alabama, did evaporite formation persist intermittently during the later part of the Kimmeridgian. In Texas and northeast Mexico, the influx of terrigenous clastics was not as dominant, and the upper Kimmeridgian section is characterized by oolitic and pelletal grainstones and packstones (the Gilmer Limestone and upper Olvido) deposited in shallow, open-marine shelves under high-energy conditions.

In east-central Mexico the Kimmeridgian section (Tamán and time-equivalent formations) was deposited in an open-marine shelf into which the influx of terrigenous clastics was minimal. The coast was probably irregular in shape, and numerous islands dotted the offshore. Interbedded limestones and shales were deposited in the deeper parts of the shelf, and oolitic and skeletal grainstones and calcarenites were formed in shallower and higher energy environments as bars, banks, and shoals along the coast and around the islands (San Andrés Member). Between the shoals and the coast, littoral lagoons were common.

The Kimmeridgian section known in southern Mexico from the Gulf coastal plain and the Gulf of Mexico shelf was also deposited in an open-marine shelf free of terrigenous clastics. Some of the sections—the lower part of the Chinameca Limestone and the subsurface section penetrated under the Bay of Campeche—seem to represent deposition in the deeper part of the shelf or a short distance beyond the edge of the shelf; others, like that penetrated in the Reforma area, were probably deposited in shallower open-marine environments—the shallow part of the shelf or offshore shoals. Sections assigned to the Kimmeridgian along the margin of the Chiapas Massif indicate deposition in a very shallow marine environment, probably in close proximity to the coast. In southernmost Mexico and northwestern Guatemala, the Kimmeridgian is represented by nonmarine "red beds."

Economic considerations. The more porous and permeable oolitic and skeletal grainstones of the Gilmer and San Andrés members provide the reservoirs for commercial oil and gas accumulations in the United States and Mexico.

Tithonian

Distribution. The geographic distribution of the Tithonian sediments in the Gulf of Mexico basin is similar to that of the underlying Kimmeridgian section. The Tithonian generally oversteps the Kimmeridgian, but the distance of this overstepping is nowhere sizable.

As in the case of the underlying part of the Upper Jurassic section, the Tithonian nowhere crops out along the northern flank of the basin. It is well exposed, however, along the Sierra Madre Oriental and in the Sierra de Chiapas in eastern and southern Mexico. Tithonian sediments have been penetrated by numerous wells along the length of the Gulf coastal plain of the United States and Mexico, from the Florida panhandle to the Reforma petroleum district and the Campeche shelf in southern Mexico. No Tithonian sediments are known from the Yucatan platform, but in the extreme southern part of the Florida peninsula, the lowermost part of the sedimentary section has been dated provisionally as of probable latest Tithonian age.

Lithostratigraphy. The Tithonian section is present in the Gulf of Mexico basin in two contrasting lithofacies: along the northern and parts of the northwestern flanks of the basin, the Tithonian is represented by a thick wedge of coarse terrigenous clastics, whereas in the western and southwestern flanks, the Tithonian section is much thinner, predominantly composed of shales, calcareous shales, and argillaceous limestones, and essentially devoid of coarse clastics.

The northern predominantly coarse-clastic section has been referred to in the subsurface of the U.S. Gulf coastal plain as the Cotton Valley Group (Shearer, 1938; Weeks, 1938; Hazzard, 1939; Swain, 1944, 1949). In east Texas, northern Louisiana, and southern Arkansas, where the Cotton Valley was first described, it includes in its upper and landward part a nonmarine or littoral coarser unit, the Schuler Formation (Shearer, 1938; Swain, 1944), and in its lower and basinward part, a marine predominantly shale unit, the Bossier Formation (Swain, 1944) (see Figs. 2 and 12). These two formations have been further subdivided into a number of formal (members) and informal local units on the basis of lithologic composition, color, or other physical properties (Swain, 1944; Mann and Thomas, 1964; Coleman and Coleman, 1981). In the East Texas basin and adjacent parts of northern Louisiana, a very distinctive limestone unit, the Knowles Limestone (Mann and Thomas, 1964), occurs at or near the top of the Cotton Valley Group. Even though the Knowles Limestone has only a limited distribution, it is an important unit because it establishes that the age of the Cotton Valley Group, at least in the area where the Knowles is present, ranges up into the Early Cretaceous.

In south Texas and northeast Mexico, the Tithonian is a predominantly shaly section in the more basinal areas, but becomes progressively more sandy in a landward direction—the upper part of the La Casita Formation (Imlay, 1936, 1937). A unit intermediate in lithologic composition has been called La Caja Formation (Imlay, 1938). Along the Gulf coastal plain of eastern and southern Mexico and the foothills of the Sierra Madre Oriental, the Tithonian is represented by the upper part of the Tamán Formation and by a generally thin unit composed of argillaceous limestones and calcareous shales called the Pimienta Formation (Heim, 1926; Cantú Chapa, 1969a, 1971; Aguilera, 1972; De la Fuente and Moya Cuevas, 1972; Hermoso de la Torre and Martínez Pérez, 1972; Sánchez-Rosas, 1972; Stabler, 1972; Toledo Toledo and Tavitas Galván, 1972). The term "Pimienta" has also been applied by some authors to the shaly basinal Tithonian section of northeastern Mexico (Varela Hernández, 1969; Alfonso Zwanziger, 1978; Cantú Chapa, 1982, among others). In the states of Querétaro and Hidalgo, the Las Trancas Formation (Segerstrom, 1961, 1962) probably includes beds of Tithonian age. In southern Mexico, the Tithonian is represented in the upper part of the Tepexilotla Formation of the Córdoba area (Mena Rojas, 1960, 1962; Carrasco and others, 1975) and the middle part of the Chinameca Limestone farther south (Benavides, 1950; Santiago Acevedo, 1962).

The Cotton Valley Group in east Texas, northern Louisiana, and southern Arkansas is a thick sequence of terrigenous clastics, in general coarser in an updip, landward direction and upward in any given section (Swain, 1944; Forgotson, 1954; Mann and Thomas, 1964; Thomas and Mann, 1966; Dickinson, 1968b; Collins, 1980; McGowen and Harris, 1984) (Figs. 20 and 21). The lithology of the Cotton Valley varies considerably, however, in a vertical and lateral direction. This variability accounts for the numerous formal and informal units into which the Cotton Valley has locally been subdivided. Growth of salt-cored structures has been recognized during the Tithonian, but it has considerably less influence in the lithologic character and thickness distribution of the Cotton Valley section than it had in the deposition of the underlying Upper Jurassic units.

The lower and down-dip, finer grained, basinal part of the Cotton Valley Group (the Bossier Formation) is composed of dark gray to black, marine, fossiliferous shales and calcareous shales with occasional thin beds of fine-grained sandstone. The lower part of the formation is generally more calcareous than the upper part and may contain argillaceous limestone beds. Both the lower and upper boundaries of the Bossier are time transgressive (see Fig. 12B). In its most basinward occurrence, the formation comprises most of the Upper Jurassic section and is the lateral equivalent of at least the upper part of the Smackover, the Haynesville, and most of the Cotton Valley Group. Landward, the Bossier becomes increasingly sandy and thins as it gradually changes to other lithostratigraphic units; it eventually pinches out as a formation, and grades into the coarser grained, updip, Cotton Valley—the Schuler Formation (see Fig. 12B).

The upper and updip, coarser grained part of the Cotton

Valley Group (the Schuler Formation) is composed of an irregular interbedding of a great variety of terrigenous clastics—mudstones, shales, siltstones, sandstones, and conglomerates. Grain size increases updip, where conglomerates and coarse sandstones predominate. Basinward, the Schuler becomes increasingly shaly and progressively grades into the upper part of the Bossier Formation. The color of the finer grained clastics—mudstones, shales, and siltstones—varies from red, maroon, and purple, varicolored and mottled in the updip sections to more greenish and grayish tones downdip. The sandstones and conglomerates are generally light greenish-gray, gray, or white, but may show shades of red in the more landward sections. Bedding is irregular, and bed lenticularity and cross-bedding are common in the sandstones. Quartz is the most abundant component. Carbonaceous material and thin coal beds have often been reported, particularly from the updip, nonmarine or nearshore part of the formation.

The distinctive Knowles Limestone of east Texas and northern Louisiana is a regionally persistent unit of dark gray limestone or interbedded argillaceous limestone and shale. It occurs at or near the top of the Cotton Valley Group, grades landward into the coarse clastics of the upper Schuler, and reaches a maximum thickness of 360 m (Mann and Thomas, 1964; Coleman and Coleman, 1981; Cregg and Ahr, 1984; Finneran and others, 1984). The Knowles Limestone has been described as composed predominantly of fossiliferous wackestones, skeletal wackestones, pelletal and oolitic packstones, and grainstones. Algal boundstones with stromatoporoids and corals also occur in the Knowles Limestone, indicating that part of this unit represents the development of patch reefs (Cregg and Ahr, 1984; Finneran and others, 1984).

In central Mississippi, the shaly Bossier Formation has not been recognized, and the Cotton Valley Group is composed predominantly of coarse terrigenous clastics—white, red, and pink, fine- to coarse-grained sandstones, and conglomeratic sandstones interbedded with lesser amounts of dark red, brown, or mottled silty shales, and occasional thin lignite beds. The section becomes finer grained toward the top, where lighter colored shales, siltstones, fine-grained sandstones, and occasional thin limestones

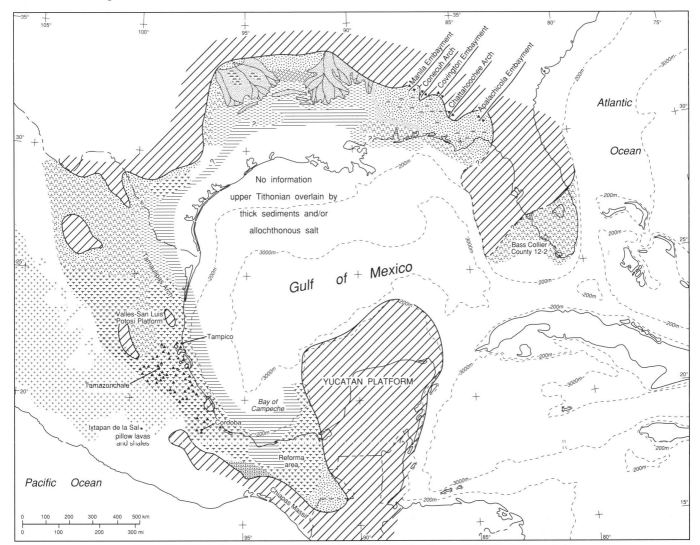

Figure 20. Lithofacies-paleogeographic map, late Tithonian. See Figure 15 for lithologic symbols.

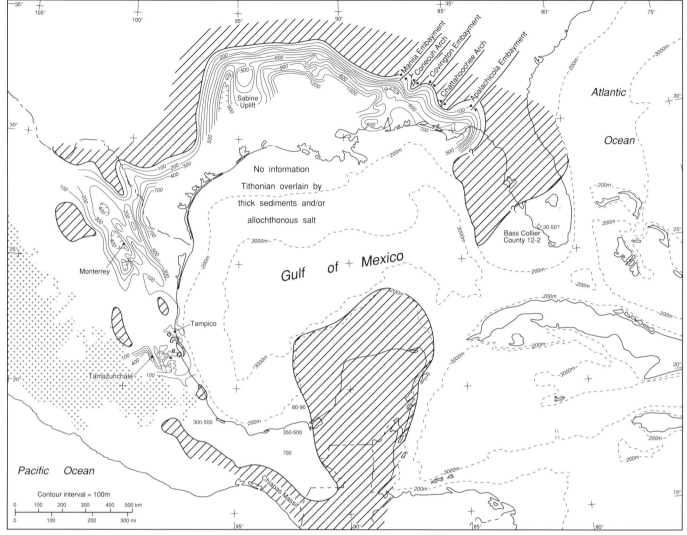

Figure 21. Isopach map, Tithonian. Thicknesses in meters.

are characteristic of the Cotton Valley. In west-central Missis-sippi, the coarse-clastic Tithonian section continues to reflect the presence in the area of the delta of the ancestral Mississippi River (Dinkins, 1968; Oxley and others, 1967, 1968) (Fig. 20).

In southern Alabama, the Cotton Valley is also predomi-nantly composed of coarse-grained terrigenous clastics (Tolson and others, 1983). Red, reddish gray, or light gray, coarse-grained sandstones, conglomeratic sandstones, and conglomerates prevail in updip sections. Thin coal beds are common toward the top. The clastics become finer grained downdip, shale beds increase in number and in thickness, and colors grade from reds and reddish grays to darker gray tones. Occasional thin limestone beds occur in the more basinward sections.

In southernmost Florida, a few wells have penetrated di-rectly above the basement a 100- to 200-m-thick section com-posed of a 75-m-thick basal sequence of red arkosic sandstones and calcareous sandstones interbedded with red and varicolored shales, and an upper part composed of skeletal, oolitic, and mic-ritic limestones, dolomitic limestones, and dolomites. Lesser

amounts of shale and anhydrite are found interbedded with the carbonates. Applin and Applin (1965) called this unit the Fort Pierce Formation and tentatively assigned a latest Jurassic age to its lowermost part, based on the presence of microfossils in the lower part of the upper carbonate section; similar microfossils were described by Bronnimann (1955) from Upper Jurassic (Tithonian) rocks in Cuba. How much of the Fort Pierce Forma-tion is of latest Jurassic age is not known.

Applegate and others (1981) included the section directly above the "basement" in south Florida in their Wood River Formation. They described it as predominantly composed of anhydrite and dolomite with a 120-m basal section of shale and fine- to coarse-grained varicolored sandstone. The Wood River Formation is probably equivalent to the Fort Pierce Formation of Applin and Applin (1965). In the type well (Bass Collier County 12-2), the Wood River Formation overlies a rhyolite porphyry for which, according to Applegate and others (1981), an isotopic age of 189 Ma (Early Jurassic) has been determined. They as-signed a "late Jurassic(?)" age to the entire formation, but emphasized that this age was based partly on circumstantial evidence, primarily on long-distance regional correlation of the Wood River with the Cotton Valley Group of the northern flank of the Gulf of Mexico basin and the fact that if all or most of the Wood River Formation were of Early Cretaceous age, too large a gap would be present between this formation and the underlying rhyolite porphyry. A palynological study of the Bass Collier County 12-2 well indicates, however, that of the 640 m that the Wood River Formation attains in this well, only the lowermost 30 to 50 m may be of possible Tithonian age (Salvador, 1987).

In south Texas and northeastern Mexico, the Tithonian section is represented by a basinward section predominantly composed of dark gray marine shale (Bossier Formation in south Texas and Pimienta Formation in northeastern Mexico) that grades landward, toward the northwest and west, into an increasingly sandy section—the upper part of the La Casita Formation of northeastern Mexico and an equivalent lithologic unit in south Texas.

The La Casita Formation (Imlay, 1936, 1937) includes a lithologically very variable section composed of black to gray shales, sandstones, conglomeratic sandstones, and occasional limestone beds. Its upper part (Tithonian) is generally sandier that the lower part (Kimmeridgian).

Farther south, in east-central and southern Mexico, the Tithonian section becomes more calcareous and is represented by the uppermost part of the Tamán Formation and a generally thin, widely distributed unit of extremely uniform and homogeneous lithology—the Pimienta Formation (Heim, 1926). The Pimienta is composed of thinly interbedded, well-bedded, black to dark gray dense limestone, argillaceous limestone, calcareous shale, and dark shale rich in organic matter. Characteristic of this unit are black chert in thin beds, lenses, and nodules, and gray or green bentonites and bentonitic shales. Bands of calcareous concretions are present occasionally, more commonly toward the base. The uniform and persistent lithology of the Pimienta Formation and its abundant and age-diagnostic fossil content have made it possible to subdivide the formation into smaller units recognizable over long distances, based either on lithologic or on biostratigraphic characteristics.

In the Isthmus area of southern Mexico, the Tithonian is also made up of interbedded, thin-bedded argillaceous limestones and calcareous shales—the middle part of the Kimmeridgian to Early Cretaceous Chinameca Limestone and upper part of the Tepexilotla Formation. The Tithonian maintains a similar lithologic composition in the Reforma area and under the Campeche shelf. The upper part of the section is more shaly and contains some thin sandstone beds (Santiago Acevedo, 1980; Santiago Acevedo and Mejía Dautt, 1980).

Along the northern foothills of the Chiapas massif, in the Isthmus of Tehuantepec area, the Upper Jurassic is predominantly composed of shallow-water carbonates (the Zacatera Group). Its upper part has been assigned to the Tithonian on the basis of the occurrence of species of the ammonites *Salinites* and *Micracanthoceras* (Quezada Muñetón, 1984). Grainstones, oo-

litic limestones, and calcarenites predominate in the lower part of the section assigned to the Tithonian; argillaceous limestones and calcareous shales predominate in the upper part. The Upper Jurassic carbonate section is underlain in the Isthmus area by "red beds" of uncertain age, and is overlain by a marine Lower Cretaceous section. Southeastward, toward Guatemala, along the northeastern foothills of the Chiapas massif, the marine Upper Jurassic section grades laterally into a "red bed" sequence—the Todos Santos Formation of Richards (1963)—whose age is still controversial.

Near the town of Ixtapan de la Sal, about 90 km south-southwest of Mexico City, Campa and others (1974) have reported the occurrence of Tithonian ammonites and tintinnids from a sequence of alternating, slightly metamorphosed volcanic and sedimentary rocks (Fig. 20). The metavolcanic rocks are andesitic-dacitic lavas, often showing pillow structures, tuffs, and volcanic agglomerates. The metasedimentary rocks are predominantly shales, sandstones, and limestones. Among the Tithonian ammonites reported from this locality are species of *Wichmanniceras* and *Micracanthoceras*.

Thickness. The Tithonian section has its thickest development along the northern flank of the Gulf of Mexico basin where the coarse-clastic section of the Cotton Valley Group reaches thicknesses of 1,000 to 1,200 m in northern Louisiana and west-central Mississippi (Fig. 21). Even though the present northern pinchout of the Tithonian is in part due to erosion, thickness and lithologic distribution indicate that this erosional edge does not differ appreciably from the original depositional edge. The Cotton Valley Group thins over the Sabine uplift, locally over salt-cored structures that grew during the Tithonian, and on the upthrown side of contemporaneous faults (Bishop, 1973).

The Tithonian is thinner along the northwestern and western flanks of the basin. Maximum recorded thicknesses in south Texas and northeastern Mexico range between 500 and 700 m. It reaches a thickness of 400 m in the Monterrey area, but thins to less than 100 m to the west. In east-central Mexico, the Tithonian section (uppermost Tamán and Pimienta) is generally less than 200 m thick and thicknesses of as much as 300 to 500 m have been reported only in the area west of Tamazunchale. The formation is very thin or absent over a few emergent highs due to either erosion or nondeposition. In southern Mexico, thicknesses of as much as 500 to 700 m of Tithonian have been penetrated in the Reforma area.

Unlike the northern flank of the Gulf of Mexico basin, where the Tithonian section wedges out within the present boundary of the structural basin, the limits of the Tithonian depositional basin extent westward beyond the western structural boundary in eastern Mexico. The Tithonian should be represented by a very thin shaly section in the central part of the basin.

Stratigraphic relations. Along the northern flank of the Gulf of Mexico basin, the coarse-clastic littoral or nearshore Cotton Valley section (the Schuler Formation) rests unconformably over the Haynesville Formation or older rocks. Basinward, however, the unconformity dies out, and the boundary becomes conformable and transitional. Todd and Mitchum (1977) illustrated this relation as it can be observed in seismic-reflection profiles from northeast and south Texas. In both areas, first the Bossier and then the Schuler lap over the underlying Upper Jurassic section and, along the periphery of the basin, lie over older rocks. The same stratigraphic relation has been reported from other parts of the northern flank of the Gulf of Mexico Basin all the way east to the Apalachicola embayment (Mitchell-Tapping, 1982).

The contact between the Cotton Valley and the overlying Lower Cretaceous coarse clastics is generally sharp and unconformable, except in more basinal areas where the boundary between the two sequences may be conformable (Weeks, 1938; Swain, 1944; Forgotson, 1954; Todd and Mitchum, 1977).

In northeastern Mexico, no evident unconformities have been reported at the base or top of the Tithonian section except at the Tamaulipas-Constituciones oil field, a few kilometers north of Tampico, where Stabler (1972) reported a progressive onlap of the Pimienta Formation over older Upper Jurassic beds. Disconformities may develop, however, toward the rim of the depositional basins, particularly at the top of the La Casita Formation. Southward, in east-central and south Mexico, where the uppermost Jurassic and lowermost Cretaceous sections represent openmarine conditions of continuous deposition, the Tithonian has been reported to overlie the older Upper Jurassic section and to underlie the Lower Cretaceous, both conformably and transitionally.

Biostratigraphy and age. The more marine Tithonian sections of east-central and northeast Mexico have yielded a rich fauna of ammonites and tintinnids that have allowed precise age determinations.

Imlay (1952, 1953, 1980), and Cantú Chapa (1967, 1971, 1979, 1982) have recognized ammonite assemblages that range in age from the earliest to the latest Tithonian. Imlay subdivided the Tithonian into "lower" and "upper," whereas Cantú Chapa recognized a threefold division: "lower," "middle," and "upper." The general sequence of the occurrence of the most common ammonites in the Tithonian of Mexico seems to be as follows:

● latest Tithonian: *Parodontoceras*, *Substeueroceras*, *Proniceras*, and *Protacanthodiscus*;

● somewhat older Tithonian: *Suarites*, *Acevedites*, *Kossmatia*, *Durangites*, *Pseudolissoceras*, *Haploceras*, *Whichmanniceras*, and *Corongoceras*;

● early Tithonian: *Hybonoticeras*, *Virgatosphinctes*, *Mazapilites*, and *Aulacosphinctoides*.

Many more species of ammonites have been reported from the abundantly fossiliferous Tithonian sections of Mexico. These ammonite assemblages contain many species common with the faunas collected in Cuba and the Mediterranean area, as well as with those in California and Oregon. Some of the Tithonian ammonite genera of the Gulf of Mexico basin area are found only, or mainly, in areas bordering the Pacific Ocean south of Oregon (Imlay, 1980).

The upper part of the Pimienta Formation is also rich in tintinnids (*Calpionella eliptica*, *C. alpina*, *Calpionellites darderi*, *Tintinnopsella oblonga*, *T. longa*, and others: Cantú Chapa, 1967; Velasco Torres and Sepúlveda de Leon, 1973) that indicate a latest Tithonian to possible Early Cretaceous (Berriasian-Valanginian) age.

Several species of *Inoceramus*, other mollusks, and radiolarians have also been described from the Tithonian sections of Mexico.

Fossil evidence to establish the age of the section assigned to the Tithonian along the northern margin of the Gulf of Mexico basin is not adequate. The coarse-grained nonmarine or nearshore Cotton Valley (the Schuler Formation) is generally unfossiliferous. From its more marine, basinward time equivalent (the upper part of the Bossier Formation), Imlay and Herman (1984) have reported an ammonite fauna containing species of *Virgatosphinctes*, *Salinites*, *Durangites?*, *Proniceras*, and *Substeueroceras*, which indicates a Tithonian age for that part of the section. Cooper and Shaffer (1976) have confirmed the late Tithonian age of part of the Bossier in northern Louisiana and southern Arkansas on the basis of a distinctive calcareous nannofossil assemblage.

The upper part of the Cotton Valley Group has been found to extent into the Early Cretaceous. Finneran and others (1984) and Scott (1984) reported upper Berriasian-Valanginian tintinnids (*Tintinnopsella carpathica*, *Lorenziella hungarica*, *Calpionella alpina*, *Calpionellopsis oblonga*, *C. simplex*, *Calpionellites darderi*), foraminifers, and corals from the Knowles Limestone. Imlay and Herman (1984) mentioned "middle Neocomian ammonites, such as *Neocomites*, *Maderia* and *Leopoldia*" from the uppermost Schuler in a well in south Texas. Fournier (1977) contributed palynological evidence for an Early Cretaceous age for the upper part of the Bossier Shale, as did Cooper and Shaffer (1976) on the basis of calcareous nannofossils.

Conditions of deposition—Paleogeography. The Late Jurassic transgression reached its peak during the Tithonian or earliest Cretaceous. During this time the sea advanced over the surrounding lands and covered most of the islands still remaining around the periphery of the ancestral Gulf of Mexico. The Tithonian sea may also have covered the southern part of the Florida platform, and evidence exists that the Gulf of Mexico was at this time in communication with the Atlantic Ocean.

During the Tithonian, and continuing into the early part of the Early Cretaceous, the northern flank of the Gulf of Mexico basin witnessed a great influx of terrigenous clastics contributed by several major streams draining areas to the north and northwest—the Appalachians, the Ouachitas, and the continental interior of the United States. The delta of the ancestral Mississippi River can also be recognized in the Tithonian section of northeastern Louisiana and west-central Mississippi (Forgotson, 1954; Oxley and others, 1967, 1968; Mann and Thomas, 1964). Other important and complex deltas, often coalescing with each other, were present in northeast Texas (McGowen and Harris, 1984) (Fig. 20).

The Tithonian section of the northern flank of the basin

reflects the continuing transgression and an extensive progradation of a thick wedge of terrigenous clastics, not necessarily connected with a regression. The terrigenous clastics of this wedge were worked over by waves along the fronts of the deltas and redistributed along the coast by east-to-west longshore currents. Pre–Late Jurassic paleotopography and salt-induced deformation still influenced sedimentation.

The updip part of the Tithonian section, near its northern pinchout, is characterized by nonmarine beds, mainly fluvial "red beds." These grade downdip (basinward) into deltaic complexes that prograde over an extensive shallow shelf or ramp. Farther downdip, the coarse-clastic deltaic sequences (Schuler) are replaced by shalier prodelta and basinal sequences (Bossier). Barrier beaches, bars, barrier islands, and lagoons were locally developed along the coast, principally in northern Louisiana (Mann and Thomas, 1964).

The Sabine uplift area seems to have been a shallow platform on which clastics were winnowed and deposited as bars. Deep marine depocenters were present on both sides of the Sabine uplift area in east Texas and in northern Louisiana. These depocenters contain particularly thick sections of Bossier shales and were the site, during the early part of the Cretaceous, of the deposition of the Knowles Limestone in a shallow-water marine shelf.

In east-central Mississippi, Alabama, and the Florida panhandle, the Tithonian section is predominantly composed of nonmarine, littoral, or deltaic sequences, which grade basinward to progressively more marine sections. The Wiggins uplift was covered by the sea.

The stable Florida platform, emergent during the Triassic and most of the Jurassic, was invaded by the sea, at least in part, during the latest Tithonian or Early Cretaceous. The sea covered first the southern part of the platform; as it advanced it reworked the products of a long period of erosion and deposited a diachronous basal clastic unit—progressively younger in a northward and eastward direction—as nearshore accumulations.

The Tithonian transgression is also reflected in the sedimentary section of the western flank of the basin. The sea advanced over the land areas, and most of the previously emergent high areas were submerged (Fig. 20). In northeastern Mexico, the Tamaulipas arch was almost completely covered by the sea and shales characterize the Tithonian section in the downdip, open-marine areas. To the west, evidence of an increase in the supply of terrestrial clastics is found in the coarser nature of the predominantly clastic upper part of the La Casita Formation.

Farther south, in east-central and southern Mexico, most of the Valles–San Luis Potosí platform seems to have been covered by the sea during the Tithonian, as were the majority of the other positive elements which were emergent earlier in the Late Jurassic. The sediments deposited in east-central and southern Mexico during the Tithonian–the Pimienta Formation and time-equivalent units—reflect a basinal, low-energy environment, free from the influx of coarse terrigenous clastics. Only west of Tamazunchale (westernmost outcrops of the Las Trancas Formation), south of Córdoba, and in the southern part of the Isthmus of Tehuantepec, along the northern limit of the Chiapas massif, have coarser clastics or carbonate grainstones of very shallow marine origin been reported from alleged uppermost Jurassic sections. They may indicate proximity to a western landmass or landmasses.

The presence of numerous bentonite beds and black chert in beds, lenses, and nodules in the Pimienta Formation, and of volcaniclastic sediments in the Las Trancas Formation, indicates the existence of volcanic activity at the time, probably in western Mexico. The occurrence near Ixtapan de la Sal of Tithonian ammonites in slightly metamorphosed shales interbedded with lava flows confirms this interpretation.

Economic considerations. The shales and calcareous shales of the Pimienta Formation, rich in organic matter, have been considered as excellent source rocks for the oil in the many fields of east-central and southern Mexico. The sandstones of the Cotton Valley Group, particularly the barrier-island sandstone complexes in its upper part, are the reservoirs for important accumulations of oil and gas along the northern flank of the basin.

REFERENCES CITED

Aguilera, H. E., 1972, Ambientes de depósito de las formaciones del Jurásico Superior en la región Tampico-Tuxpan: Asociación Mexicana de Geólogos Petroleros Boletín, v. 24, nos. 1–3, p. 129–163.

Ahr, W. M., 1981, The Gilmer Limestone; Oolite tidal bars on the Sabine Uplift: Gulf Coast Association of Geological Societies Transactions, v. 31, p. 1–6.

Akin, R. H., Jr., and Graves, R. W., Jr., 1969, Reynolds Oolite of southern Arkansas: American Association of Petroleum Geologists Bulletin, v. 53, p. 1909–1922.

Alfonso Zwanziger, J., 1978, Geología regional del sistema Cupido: Asociación Mexicana de Geólogos Petroleros Boletín, v. 30, nos. 1–2, p. 1–55.

Amery, G. B., 1969, Structure of Sigsbee Scarp, Gulf of Mexico: American Association of Petroleum Geologists Bulletin, v. 53, p. 2480–2482.

Anderson, E. G., 1979, Basic Mesozoic study in Louisiana, the northern coastal region, and the Gulf Province: Louisiana Geological Survey Folio Series no. 3, 58 p.

Andrews, D. I., 1960, The Louann Salt and its relationship to Gulf Coast salt domes: Gulf Coast Association of Geological Societies Transactions, v. 10, p. 215–240.

Applegate, A. V., Winston, G. O., and Palacas, J. G., 1981, Subdivision and regional stratigraphy of the pre-Punta Gorda rocks (lowermost Cretaceous-Jurassic?) in south Florida: Gulf Coast Association of Geological Societies Supplement to Transactions, v. 31, p. 447–453.

Applin, P. L., and Applin, E. R., 1965, The Comanche Series and associated rocks in the subsurface in central and south Florida: U.S. Geological Survey Professional Paper 447, 84 p.

Arden, D. D., Jr., 1974a, A geophysical profile in the Suwannee Basin, northwestern Florida, in Stafford, L. P., ed., Symposium on petroleum geology of the Georgia coastal plain: Georgia Geologic Survey Bulletin 87, p. 111–122.

—— , 1974b, Geology of the Suwannee Basin interpreted from geophysical profiles: Gulf Coast Association of Geological Societies Transactions, v. 24, p. 223–230.

Badon, C. L., 1975, Stratigraphy and petrology of Jurassic Norphlet Formation,

Clarke County, Mississippi: American Association of Petroleum Geologists Bulletin, v. 59, p. 377–392.

Baria, L. R., Stoudt, D. L., Harris, P. M., and Crevello, P. D., 1982, Upper Jurassic reefs of Smackover Formation, United States Gulf Coast: American Association of Petroleum Geologists Bulletin, v. 66, p. 1449–1482.

Beju, D., Guillory, R., Wood, G. D., and Finneran, J. M., 1986, Palynomorphs from Upper Triassic Eagle Mills Formation of northeastern Texas, U.S.A. [abs.]: Palynology, v. 10, p. 244.

Benavides G., L., 1950, El anticlinal del Cerro Pelón, Municipio de Minatitlan, Veracruz: Asociación Mexicana de Geólogos Petroleros Boletín, v. 2, no. 10, p. 599–616.

Benson, D. J., and Mancini, E. A., 1982, Petrology and reservoir characteristics of the Smackover Formation, Hatter's Pond Field; Implications for Smackover exploration in southwestern Alabama: Gulf Coast Association of Geological Societies Transactions, v. 32, p. 67–75.

———, 1984, Porosity development and reservoir characteristics of the Smackover Formation in southwest Alabama, in Ventress, P. S., and others, eds., The Jurassic of the Gulf Rim, Proceedings, 3rd Annual Research Conference: Society of Economic Paleontologists and Mineralogists, Gulf Coast Section, p. 1–17.

Bingham, D. H., 1937, Developments in Arkansas-Louisiana-Texas area, 1936–1937: American Association of Petroleum Geologists Bulletin, v. 21, p. 1068–1073.

Bishop, W. F., 1967, Age of pre-Smackover Formations, north Louisiana and south Arkansas: American Association of Petroleum Geologists Bulletin, v. 51, p. 244–250.

———, 1968, Petrology of Upper Smackover Limestone in North Haynesville Field, Claiborne Parish, Louisiana: American Association of Petroleum Geologists Bulletin, v. 52, p. 92–128.

———, 1969, Environmental control of porosity in the upper Smackover Limestone, North Haynesville Field, Claiborne Parish, Louisiana: Gulf Coast Association of Geological Societies Transactions, v. 19, p. 155–169.

———, 1973, Late Jurassic contemporaneous faults in north Louisiana and south Arkansas: American Association of Petroleum Geologists Bulletin, v. 57, p. 858–877.

Bracken, B., 1984, Environments of deposition and early diagenesis, La Joya Formation, Huizachal Group Red Beds, northeast Mexico, in Ventress, P. S., and others, eds., The Jurassic of the Gulf Rim, Proceedings, 3rd Annual Research Conference: Society of Economic Paleontologists and Mineralogists, Gulf Coast Section, p. 19–26.

Bradford, C. A., 1984, Transgressive-regressive carbonate facies of the Smackover Formation, Escambia County, Alabama, in Ventress, P. S., and others, The Jurassic of the Gulf Rim, Proceedings, 3rd Annual Research Conference: Society of Economic Paleontologists and Mineralogists, Gulf Coast Section, p. 27–39.

Bronnimann, P., 1955, Microfossils incertae sedis from the Upper Jurassic and Lower Cretaceous of Cuba: Micropaleontology, v. 1, p. 28–51.

Budd, D. A., and Loucks, R. G., 1981, Smackover and Lower Buckner formations, south Texas; Depositional systems on a Jurassic carbonate ramp: University of Texas at Austin Bureau of Economic Geology Report of Investigations 112, 38 p.

Buffler, R. T., Watkins, J. S., Shaub, F. J., and Worzel, J. L., 1980, Structure and early geologic history of the deep central Gulf of Mexico Basin, in Pilger, R. H., Jr., ed., The origin of the Gulf of Mexico and the early opening of the central North Atlantic Ocean: Baton Rouge, Louisiana State University, p. 3–16.

Bullard, E. C., Everett, J. E., and Smith, A. G., 1965, The fit of the continents around the Atlantic: Royal Society of London Philosophical Transactions, ser. A, v. 258, p. 41–51.

Burckhardt, C., 1905, La faune marine du Trias Superieur de Zacatecas: Instituto Geológico de México Boletín no. 21, 41 p.

———, 1906, La faune Jurassique de Mazapil, avec un appendice sur les fossiles du Crétacique Inférieur: Instituto Geológico de México Boletín no. 23, 216 p.

———, 1912, Faunes jurassiques et crétaciques de San Pedro del Gallo: Instituto Geológico de México Boletín no. 29, 264 p.

———, 1930, Etude synthétique sur le Mésozoique Mexicain, Premiére partie: Société Paléontologique Suisse Mémoires, v. 49, p. 1–123.

Burckhardt, C., and Scalia, S., 1906, Géologie des environs de Zacatecas: Mexico, D. F., 10th International Geological Congress, Guide to Excursion 16, 26 p.

Campa, M. F., Campos, M., Flores, R., and Oviedo, R., 1974, La secuencia mesozoica volcánico-sedimentaria metamorfizada de Ixtapan de la Sal, Méx.-Teloloapan, Gro.: Sociedad Geológica Mexicana Boletín, v. 35, p. 7–28.

Cantú Chapa, A., 1967, El límite Jurásico-Cretácico en Mazatepec, Puebla (México), in Estratigrafía del Jurásico de Mazatepec, Puebla (México): Instituto Mexicano del Petróleo Monografia no. 1, p. 3–24.

———, 1969a, Estratigrafía del Jurásico Medio-Superior del subsuelo de Poza Rica, Veracruz (Area de Soledad-Miquetla): Instituto Mexicano del Petróleo Revista, v. 1, no. 1, p. 3–9.

———, 1969b, Una nueva localidad del Triásico Superior marino en México: Instituto Mexicano del Petróleo Revista, v. 1, no. 2, p. 71–72.

———, 1971, La Serie Huasteca (Jurásico Medio-Superior) del centro este de México: Instituto Mexicano del Petróleo Revista, v. 3, no. 2, p. 17–40.

———, 1977, Las amonitas del Jurásico Superior del Pozo Chac 1, norte de Campeche (Golfo de México): Instituto Mexicano del Petróleo Revista, v. 9, no. 2, p. 38–39.

———, 1979, Biostratigrafía de la Serie Huasteca (Jurásico Medio y Superior) en el subsuelo de Poza Rica, Veracruz: Instituto Mexicano del Petróleo Revista, v. 11, no. 2, p. 14–24.

———, 1982, The Jurassic-Cretaceous boundary in the subsurface of eastern Mexico: Journal of Petroleum Geology, v. 4, p. 311–318.

Carey, S. W., 1958, The tectonic approach to continental drift, in Carey, S. W., convener, Continental drift; A symposium: Hobart, University of Tasmania Geological Department, p. 177–355.

———, 1976, The expanding Earth; Developments in Geotectonics 10: New York, Elsevier Scientific Publishing Company, 488 p.

Carrasco, B., Flores, V., and Godoy, D., 1975, Tobas del Cretácico Inferior del área de Fortin-Zongolica, Estado de Veracruz: Instituto Mexicano del Petróleo Revista, v. 7, no. 4, p. 7–27.

Carrillo Bravo, J., 1961, Geología del Anticlinorio Huizachal-Peregrina al N-W de Ciudad Victoria, Tamps: Asociación Mexicana de Geólogos Petroleros Boletín, v. 13, nos. 1–2, p. 1–98.

———, 1965, Estudio geológico de una parte del Anticlinorio de Huayacocotla: Asociación Mexicana de Geólogos Petroleros Boletín, v. 17, nos. 5–6, p. 73–96.

———, 1971, La Plataforma Valles-San Luis Potosí, S.L.P.: Asociación Mexicana de Geólogos Petroleros Boletín, v. 23, nos. 1–6, p. 1–113.

Carrillo Martínez, P., 1960, Estudio geológico de los campos petroleros de San Andrés, Hallazgo y Gran Morelos, Estado de Veracruz: Asociación Mexicana de Geológos Petroleros Boletín, v. 12, nos. 1–2, p. 1–73.

Chowns, T. M., and Williams, C. T., 1983, Pre-Cretaceous rocks beneath the Georgia coastal plain; Regional implications: U.S. Geological Survey Professional Paper 1313-L, 42 p.

Clark, J. M., and Hopson, J. A., 1985, Distinctive mammal-like reptile from Mexico and its bearing on the phylogeny of the Tritylondontidae: Nature, v. 315, p. 398–400.

Coleman, J. L., Jr., and Coleman, C. J., 1981, Stratigraphic, sedimentologic and diagenetic framework for the Jurassic Cotton Valley Terryville massive sandstone complex, northern Louisiana: Gulf Coast Association of Geological Societies Transactions, v. 31, p. 71–79.

Collins, S. E., 1980, Jurassic Cotton Valley and Smackover reservoir trends, east Texas, north Louisiana, and south Arkansas: American Association of Petroleum Geologists Bulletin, v. 64, p. 1004–1013.

Contreras V., H., and Castillón B., M., 1968, Morphology and origin of salt domes of Isthmus of Tehuantepec, in Braunstein, J., and O'Brien, G. D., eds., Diapirism and diapirs: American Association of Petroleum Geologists Memoir 8, p. 244–260.

Cooper, W. W., and Shaffer, B. L., 1976, Nannofossil biostratigraphy of the Bossier Shale and the Jurassic-Cretaceous boundary: Gulf Coast Association of Geological Societies Transactions, v. 26, p. 178–184.

Cornet, B., and Traverse, A., 1975, Palynological contribution to the chronology and stratigraphy of the Hartford Basin in Connecticut and Massachusetts: Geoscience and Man, v. 11, p. 1–33.

Cornet, B., Traverse, A., and McDonald, N. G., 1973, Fossil spores, pollen, and fishes from Connecticut indicate Early Jurassic age for part of the Newark Group: Science, v. 21, p. 1243–1247.

Cregg, A. K., and Ahr, W. M., 1984, Paleoenvironment of an upper Cotton Valley (Knowless Limestone) patch reef, Milan County, Texas, *in* Ventress, P. S., and others, eds., The Jurassic of the Gulf Rim, Proceedings, 3rd Annual Research Conference: Society of Economic Paleontologists and Mineralogists, Gulf Coast Section, p. 41–56.

Crevello, P. D., and Harris, P. M., 1984, Depositional models for Jurassic reefal buildups, *in* Ventress, P. S., and others, eds., The Jurassic of the Gulf Rim, Proceedings, 3rd Annual Research Conference: Society of Economic Paleontologists and Mineralogists, Gulf Coast Section, p. 57–102.

Daniels, D. L., Zietz, I., and Popenoe, P., 1983, Distribution of subsurface lower Mesozoic rocks in the southeastern United States, as interpreted from regional aeromagnetic and gravity maps: U.S. Geological Survey Professional Paper 1313–K, 24 p.

De la Fuente, I., and Moya Cuevas, F., 1972, Programa de investigación estratigráfica y discusion generalizada de los sedimentos del Jurásico Superior en la porción central de la Cuenca de Chicontepec: Asociación Mexicana de Geólogos Petroleros Boletín, v. 24, nos. 1–3, p. 15–27.

Dickinson, K. A., 1968a, Petrology of the Buckner Formation in adjacent parts of Texas, Louisiana, and Arkansas: Journal of Sedimentary Petrology, v. 38, p. 555–567.

——, 1968b, Upper Jurassic stratigraphy of some adjacent parts of Texas, Louisiana, and Arkansas: U.S. Geological Survey Professional Paper 594–E, 25 p.

——, 1969, Upper Jurassic carbonate rocks in northeastern Texas and adjoining parts of Arkansas and Louisiana: Gulf Coast Association of Geological Societies Transactions, v. 19, p. 175–187.

Dinkins, T. H., Jr., 1968, Jurassic stratigraphy of central and southern Mississippi, *in* Jurassic stratigraphy of Mississippi: Mississippi Geologic Economic and Topographic Survey Bulletin 109, p. 9–37.

Dooley, R. E., and Wampler, J. M., 1983, Potassium–argon relations in diabase dikes in Georgia; The influence of excess [40]Ar on the geochronology of early Mesozoic igneous and tectonic events: U.S. Geological Survey Professional Paper 1313–M, 24 p.

Erben, H. K., 1954, Nuevos datos sobre el Liásico de Huayacocotla, Veracruz: Sociedad Geológica de México Boletín, v. 17, no. 2, p. 31–40.

——, 1956a, El Jurásico Inferior de México y sus amonitas: Mexico D.F., 20th International Geological Congress, 393 p.

——, 1956b, El Jurásico Medio y el Calloviano de México: Mexico, D.F., 20th International Geological Congress, 140 p.

——, 1957a, Paleogeographic reconstructions for the Lower and Middle Jurassic and for the Callovian of Mexico, *in* El Mesozoico del Hemisferio Occidental y sus correlaciones mundiales: Mexico, D. F., 20th International Geological Congress Transactions, Section II, p. 35–41.

——, 1957b, New biostratigraphic correlations in the Jurassic of eastern and south-central Mexico, *in* El Mesozoico del Hemisferio Occidental y sus correlaciones mundiales: Mexico, D. F., 20th International Geological Congress Transactions, Section II, p. 43–52.

Finneran, J. M., Scott, R. W., Taylor, G. A., and Anderson, G. H., 1984, Lowermost Cretaceous ramp reefs; Knowles Limestone, southwest flank of the East Texas Basin, *in* Ventress, P. S., and others, eds., The Jurassic of the Gulf Rim, Proceedings, 3rd Annual Research Conference: Society of Economic Paleontologists and Mineralogists, Gulf Coast Section, p. 125–133.

Flores López, R., 1967, La fauna Liásica de Mazatepec, Puebla (Mexico), *in* Estratigrafía del Jurásico de Mazatepec, Puebla (México): Instituto Mexicano del Petróleo Monografia no. 1, p. 25–30.

——, 1974, Datos sobre la bioestratigrafía del Jurásico Inferior y Medio del subsuelo de la región de Tampico, Tamps.: Instituto Mexicano del Petróleo Revista, v. 6, no. 3, p. 6–15.

Foote, R. Q., Martin, R. G., and Powers, R. B., 1983, Oil and gas potential of the Maritime Boundary region in the central Gulf of Mexico: American Association of Petroleum Geologists Bulletin, v. 67, p. 1047–1065.

Forgotson, J. M., and Forgotson, J. M., Jr., 1976, Definition of Gilmer Limestone, Upper Jurassic formation, northeastern Texas: American Association of Petroleum Geologists Bulletin, v. 60, p. 1119–1123.

Forgotson, J. M., Jr., 1954, Regional stratigraphic analysis of Cotton Valley Group of Upper Gulf Coastal Plain: American Association of Petroleum Geologists Bulletin, v. 38, p. 2476–2499 (also in Gulf Coast Association of Geological Societies Transactions, v. 4, p. 143–154).

Fournier, G. R., 1977, Palynological evidence of Early Cretaceous age for part of the Bossier Shale and the role of Norris' "C" assemblage in determining the Jurassic-Cretaceous boundary [abs.]: Palynology, v. 1, p. 173.

Freeland, G. L., and Dietz, R. S., 1971, Plate tectonic evolution of Caribbean-Gulf of Mexico region: Nature, v. 232, no. 5305, p. 20–23.

Gawloski, T., 1983, Stratigraphy and environmental significance of the continental Triassic rocks of Texas: Waco, Texas, Baylor Geological Studies Bulletin 41, 47 p.

Goebel, L. A., 1950, Cairo Field, Union County, Arkansas: American Association of Petroleum Geologists Bulletin, v. 34, p. 1954–1980.

Gohn, G. S., Gottfried, D., Lanphere, M. A., and Higgins, B. B., 1978, Regional implications of Triassic or Jurassic age for basalt and sedimentary red beds in the South Carolina Coastal Plain: Science, v. 202, p. 887–890.

González García, R., 1969, Areas con posibilidades de producción en sedimentos del Jurásico Superior (Caloviano-Titoniano): Instituto Mexicano del Petróleo, Seminario sobre Exploración Petrolera, Mesa Redonda No. 3, Problemas de Exploración del Distrito Poza Rica, 11 p.

——, 1970, La Formación Huehuetepec, nueva unidad litoestratigráfica del Jurásico de Poza Rica: Ingeniería Petrolera, v. 10, no. 7, p. 5–22.

Gradstein, F. M., and Sheridan, R. E., 1983, On the Jurassic Atlantic Ocean and a synthesis of results of Deep Sea Drilling Project Leg 76, *in* Sheridan, R. E., Gradstein, F. M., and others, eds., Initial reports of the Deep Sea Drilling Project, Volume 76: Washington, D.C., U.S. Government Printing Office, p. 913–943.

Halbouty, M. T., 1979, Salt domes; Gulf region, United States and Mexico (second edition): Houston, Texas, Gulf Publishing Co., 561 p.

Halbouty, M. T., and Hardin, G. C., Jr., 1956, Genesis of salt domes of Gulf Coastal Plain: American Association of Petroleum Geologists Bulletin, v. 40, p. 737–746.

Hancharik, J. M., 1984, Facies analysis and petroleum potential of the Jurassic Smackover Formation, western and northern areas, East Texas Basin, *in* Presley, M. W., ed., The Jurassic of East Texas: Tyler, Texas, East Texas Geological Society, p. 67–78.

Hartman, J. A., 1968, The Norphlet sandstone, Pelahatchie Field, Rankin County, Mississippi: Gulf Coast Association of Geological Societies Transactions, v. 18, p. 2–11.

Hazzard, R. T., 1939, Notes on the Comanche and Pre-Comanche? Mesozoic formations of the Ark-La-Tx area, and a suggested correlation with northern Mexico: Shreveport Geological Society Guide Book to 14th Annual Field Trip, p. 155–178.

Hazzard, R. T., Spooner, W. C., and Blanpied, B. W., 1947, Notes on the stratigraphy of the formations which underlie the Smackover Limestone in south Arkansas, northeast Texas, and north Louisiana, *in* 1945 reference report on certain oil and gas fields of north Louisiana, south Arkansas, Mississippi, and Alabama, volume 2: Shreveport Geological Society, p. 483–503.

Heim, A., 1926, Notes on the Jurassic of Tamazunchale (Sierra Madre Oriental, Mexico): Eclogae Geologicae Helvetiae, v. 20, p. 84–87.

——, 1940, The front ranges of Sierra Madre Oriental, Mexico, from Ciudad Victoria to Tamazunchale: Eclogae Geologicae Helvetiae, v. 33, p. 313–352.

Hermoso de la Torre, C., and Martínez Pérez, J., 1972, Medición detallada de formaciones del Jurásico Superior en el frente de la Sierra Madre Oriental: Asociación Mexicana de Geólogos Petroleros Boletín, v. 24, nos. 1–3, p. 45–63.

Hughes, D. J., 1968, Salt tectonics as related to several Smackover fields along the northeast rim of the Gulf of Mexico Basin: Gulf Coast Association of Geological Societies Transactions, v. 18, p. 320–330.

Humphrey, W. E., 1956, Tectonic framework of northeast Mexico: Gulf Coast Association of Geological Societies Transactions, v. 6, p. 25–35.

Humphris, C. C., Jr., 1978, Salt movement on continental slope, northern Gulf of Mexico, *in* Bouma, A. H., Moore, G. T., and Coleman, J. M., eds., Framework, facies, and oil-trapping characteristics of the upper continental margin: American Association of Petroleum Geologists Studies in Geology 7, p. 69–85 (also in American Association of Petroleum Geologists Bulletin, v. 63, p. 782–798).

Imlay, R. W., 1936, Evolution of the Coahuila Peninsula, Mexico; Part 4, Geology of the western part of the Sierra de Parras: Geological Society of America Bulletin, v. 47, p. 1091–1152.

—— , 1937, Evolution of the Coahuila Peninsula, Mexico; Part 5, Geology of the middle part of the Sierra de Parras: Geological Society of America Bulletin, v. 48, p. 587–630.

—— , 1938, Studies of the Mexican geosyncline: Geological Society of America Bulletin, v. 49, p. 1651–1694.

—— , 1939, Jurassic ammonites from Mexico: Geological Society of America Bulletin, v. 50, p. 1–78.

—— , 1940, Lower Cretaceous and Jurassic formations of southern Arkansas and their oil and gas possibilities: Arkansas Geological Survey Information Circular 12, 64 p.

—— , 1941, Jurassic fossils from Arkansas, Louisiana, and eastern Texas: Journal of Paleontology, v. 15, p. 256–277.

—— , 1943, Jurassic formations of Gulf region: American Association of Petroleum Geologists Bulletin, v. 27, p. 1407–1533.

—— , 1945, Jurassic fossils from the southern states, no. 2: Journal of Paleontology, v. 19, p. 253–276.

—— , 1952, Correlation of the Jurassic formations of North America, exclusive of Canada: Geological Society of America Bulletin, v. 63, p. 953–992.

—— , 1953, Las formaciones Jurásicas de México: Sociedad Geológica Mexicana Boletín, v. 16, no. 1, p. 1–65.

—— , 1980, Jurassic paleobiogeography of the conterminous United States in its continental setting: U.S. Geological Survey Professional Paper 1062, 134 p.

Imlay, R. W., and Herman, G., 1984, Upper Jurassic ammonites from the subsurface of Texas, Louisiana and Mississippi, *in* Ventress, P. S., and others, eds., The Jurassic of the Gulf Rim; Proceedings, 3rd Annual Research Conference: Society of Economic Paleontologists and Mineralogists, Gulf Coast Section, p. 149–170.

Imlay, R. W., Cepeda, E., Alvarez, M., and Diaz, T., 1948, Stratigraphic relations of certain Jurassic formations in eastern Mexico: American Association of Petroleum Geologists Bulletin, v. 32, p. 1750–1761.

Judice, P. C., and Mazzullo, S. J., 1982, The Gray Sandstones (Jurassic) in Terryville Field, Louisiana; Basinal deposition and exploration model: Gulf Coast Association of Geological Societies Transactions, v. 32, p. 23–43.

Jux, U., 1961, The palynologic age of diapiric and bedded salt in the Gulf Coastal Province: Department of Conservation, Louisiana Geological Survey Geological Bulletin No. 38, 46 p.

Kirkland, D. W., and Gerhard, J. E., 1971, Jurassic salt, central Gulf of Mexico, and its temporal relation to circum-Gulf evaporites: American Association of Petroleum Geologists Bulletin, v. 55, p. 680–686.

Ladd, J. W., Buffler, R. T., Watkins, J. S., Worzel, J. L., and Carranza, A., 1976, Deep seismic reflection results from the Gulf of Mexico: Geology, v. 4, p. 365–368.

López Infanzón, M., 1986, Estudio petrogenético de las rocas ígneas de las formaciones Huizachal y Nazas: Sociedad Geológica Mexicana Boletín, v. 47, no. 2, p. 1–42.

López Ramos, E., 1980, Geología de Mexico, Volume 2 (second edition), 454 p.

Loucks, R. G., and Budd, D. A., 1981, Diagenesis and reservoir potential of the

Upper Jurassic Smackover Formation of south Texas: Gulf Coast Association of Geological Societies Transactions, v. 31, p. 339–346.

Maldonado-Koerdell, M., 1950, Los estudios paleobotánicos en México, con un catálogo sistemático de sus plantas fosiles (excepto Tallophyta y Bryophyta): Instituto Geológico de México Boletín, v. 55, no. 8, 72 p.

Mancini, E. A., and Benson, D. J., 1980, Regional stratigraphy of Upper Jurassic Smackover carbonates of southwest Alabama: Gulf Coast Association of Geological Societies Transactions, v. 30, p. 151–165.

Mancini, E. A., Mink, R. M., Bearden, B. L., and Wilkerson, R. P., 1985, Norphlet Formation (Upper Jurassic) of southwestern and offshore Alabama; Environments of deposition and petroleum geology: American Association of Petroleum Geologists Bulletin, v. 69, p. 881–898.

Mann, C. J., and Thomas, W. A., 1964, Cotton Valley Group (Jurassic) nomenclature, Louisiana and Arkansas: Gulf Coast Association of Geological Societies Transactions, v. 14, p. 143–152.

Manspeizer, W., 1982, Triassic-Liassic basins and climate of the Atlantic passive margins: Geologische Rundschau, v. 71, p. 895–917.

Marine, Z. W., and Siple, G. E., 1974, Buried Triassic basin in the central Savannah River area, South Carolina and Georgia: Geological Society of America Bulletin, v. 85, p. 311–320.

Martin, R. G., 1978, Northern and eastern Gulf of Mexico continental margin; Stratigraphic and structural framework, *in* Bouma, A. H., Moore, G. T., and Coleman, J. M., eds., Framework, facies, and oil-trapping characteristics of the upper continental margin: American Association of Petroleum Geologists Studies in Geology 7, p. 21–42.

—— , 1980, Distribution of salt structures in the Gulf of Mexico; Map and descriptive text: U.S. Geological Survey Map MF–1213, scale 1:2,500,000.

—— , 1981, Regional geology of the Gulf of Mexico, *in* Powers, R. B., ed., Geologic framework, petroleum potential, petroleum resource estimates, mineral and geothermal resources, geologic hazards, and deep-water drilling technology of the Maritime Boundary region in the Gulf of Mexico: U.S. Geological Survey Open-File Report 81–265, p. 19–29.

—— , 1984, Diapiric trends in the deep-water Gulf Basin, Research Conference, December 1984: Society of Economic Paleontologists and Mineralogists, Gulf Coast Section, p. 60–62.

Martin, R. G., and Foote, R. Q., 1981, Geology and geophysics of the Maritime Boundary assessment areas, *in* Powers, R. B., ed., Geologic framework, petroleum potential, petroleum resource estimates, mineral and geothermal resources, geologic hazards, and deep-water drilling technology of the Maritime Boundary Region in the Gulf of Mexico: U.S. Geological Survey Open-File Report 81–265, p. 30–67.

Martínez Pérez, J., 1972, Exploración geológica del área El Estribo–San Francisco, S.L.P. (Hojas K-8 y K-9): Asociación Mexicana de Geólogos Petroleros Boletín, v. 24, nos. 7–9, p. 325–402.

Marzano, M. S., Pense, G. M., and Andronaco, P., 1988, A comparison of the Jurassic Norphlet Formation in Mary Ann Field, Mobile Bay, Alabama to onshore regional Norphlet trends: Gulf Coast Association of Geological Societies Transactions, v. 38, p. 85–100.

May, P. R., 1971, Pattern of Triassic-Jurassic dikes around the North Atlantic in the context of pre-drift positions of the continents: Geological Society of America Bulletin, v. 82, p. 1285–1292.

McBride, E. F., 1981, Diagenetic history of Norphlet Formation (Upper Jurassic), Rankin County, Mississippi: Gulf Coast Association of Geological Societies Transactions, v. 31, p. 347–351.

McGowen, M. K., and Harris, D. W., 1984, Cotton Valley (Upper Jurassic) and Hosston (Lower Cretaceous) depositional systems and their influence on salt tectonics in the East Texas Basin, *in* Ventress, P. S., and others, eds., The Jurassic of the Gulf Rim; Proceedings, 8th Annual Research Conference: Society of Economic Paleontologists and Mineralogists, Gulf Coast Section, p. 213–253.

Mena Rojas, E., 1960, El Jurásico marino de la región de Córdoba: Asociación Mexicana de Geólogos Petroleros Boletín, v. 12, no. 7–8, p. 243–252.

—— , 1962, Geología y posibilidades petrolíferas del Jurásico marino en la región de Córdoba, Ver.: Asociación Mexicana de Geólogos Petroleros Boletín, v. 14, nos. 3–4, p. 77–84.

Miller, J. A., 1982, Structural control of Jurassic sedimentation in Alabama and Florida: American Association of Petroleum Geologists Bulletin, v. 66, p. 1289–1301.

Mitchell-Tapping, H. J., 1982, Exploration analysis of the Jurassic Apalachicola Embayment of Florida: Gulf Coast Association of Geological Societies Transactions, v. 32, p. 413–425.

Mixon, R. B., Murray, G. E., and Diaz G., T., 1959, Age and correlation of Huizachal Group (Mesozoic), State of Tamaulipas, Mexico: American Association of Petroleum Geologists Bulletin, v. 43, p. 757–771.

Moore, C. H., 1984, The upper Smackover of the Gulf Rim; Depositional systems, diagenesis, porosity evolution, and hydrocarbon production, *in* Ventress, P. S., and others, eds., The Jurassic of the Gulf Rim; Proceedings, 3rd Annual Research Conference: Society of Economic Paleontologists and Mineralogists, Gulf Coast Section, p. 283–307.

Moy, C., and Traverse, A., 1986, Palynostratigraphy of the subsurface Eagle Mills Formation (Triassic) from a well in east-central Texas, U.S.A.: Palynology, v. 10, p. 225–234.

Murray, G. E., 1961, Geology of the Atlantic and Gulf Coastal Province of North America: New York, Harper and Brothers, 692 p.

Newkirk, T. F., 1971, Possible future petroleum potential of Jurassic, western Gulf Basin, *in* Cram, I. H., ed., Future petroleum provinces of the United States; Their geology and potential: American Association of Petroleum Geologists Memoir 15, v. 2, p. 927–953.

Nichols, P. H., Peterson, G. E., and Wuestner, C. E., 1968, Summary of subsurface geology of northeast Texas, *in* Beebe, B. W., and Curtis, B. F., eds., Natural gases of North America: American Association of Petroleum Geologists Memoir 9, v. 1, p. 982–1004.

Oivanki, S. M., 1974, Paleodepositional environments in the Upper Jurassic Zuloaga Formation (Smackover), northeast Mexico: Gulf Coast Association of Geological Societies Transactions, v. 24, p. 258–278.

Olsen, P. E., McCune, A. R., and Thompson, K. S., 1982, Correlation of the Early Mesozoic Newark Supergroup by vertebrates, principally fishes: American Journal of Science, v. 282, p. 1–44.

Olsen, R. S., 1982, Depositional environments of Jurassic Smackover sandstones, Thomasville Field, Rankin County, Mississippi: Gulf Coast Association of Geological Societies Transactions, v. 32, p. 59–66.

Ottmann, R. D., Keyes, P. L., and Ziegler, M. A., 1973, Jay Field; A Jurassic stratigraphic trap: Gulf Coast Association of Geological Societies Transactions, v. 23, p. 146–157.

—— , 1976, Jay Field, Florida; A Jurassic stratigraphic trap, *in* Braunstein, J., ed., North American oil and gas fields: American Association of Petroleum Geologists Memoir 24, p. 276–286.

Owen, H. C., 1976, Continental displacement and expansion of the Earth during the Mesozoic and Cenozoic: Philosophical Transactions of the Royal Society of London, ser. A, v. 281, p. 223–291.

—— , 1983, Atlas of continental displacement; 200 million years to the present: Cambridge University Press, 159 p.

Oxley, M. L., Minihan, E., and Ridgway, J. M., 1967, A study of Jurassic sediments in portions of Mississippi and Alabama: Gulf Coast Association of Geological Societies Transactions, v. 17, p. 24–48.

—— , 1968, A study of Jurassic sediments in portions of Mississippi and Alabama, *in* Jurassic stratigraphy of Mississippi: Mississippi Geologic Economic and Topographic Survey Bulletin 109, p. 39–77.

Pepper, F., 1982, Depositional environments of the Norphlet Formation (Jurassic) for southwestern Alabama: Gulf Coast Association of Geological Societies Transactions, v. 32, p. 17–22.

Petrazzini, G., and Basañez, M. A., 1978, Sedimentación del Jurásico Medio-Superior en el Anticlinorio de Huayacocotla-Cuenca de Chicontepec, Estados de Hidalgo y Veracruz, México: Instituto Mexicano del Petróleo Revista, v. 10, no. 3, p. 6–25.

Pflug, H. D., 1963, The palynological age of diapiric and bedded salt in the Gulf Coastal Province, by Ulrich Jux: American Association of Petroleum Geologists Bulletin, v. 47, p. 180–181.

Phair, R. L., and Buffler, R. T., 1983, Pre-middle Cretaceous geologic history of the deep southeastern Gulf of Mexico, *in* Bally, A. W., ed., Seismic expression of structural styles; A picture and work atlas: American Association of Petroleum Geologists Studies in Geology 15, v. 2, p. 2.2.3-141–2.2.3-147.

Philpott, T. H., and Hazzard, R. T., 1949, Preliminary correlation chart of upper Gulf coast, *in* Guidebook 17th Annual Field Trip: Louisiana, Shreveport Geological Society, Fig. 5.

Pilger, R. H., Jr., 1978, A closed Gulf of Mexico, pre-Atlantic Ocean plate reconstruction and the early rift history of the Gulf and North Atlantic: Gulf Coast Association of Geological Societies Transactions, v. 28, p. 385–393.

Quezada Muñetón, J. M., 1984, El Grupo Zacatera del Jurásico Medio-Cretácico Inferior de la Depresion Istmica, 20 km al Norte de Matias Romero, Oax.: Sociedad Geológica Mexicana, VII Convención Nacional, Memorias, p. 40–59.

Ray, P. K., 1988, Lateral salt movement and associated traps on the continental slope of the Gulf of Mexico: Gulf Coast Association of Geological Societies Transactions, v. 38, p. 217–223.

Richards, H. G., 1963, Stratigraphy of earliest Mesozoic sediments in southeastern Mexico and western Guatemala: American Association of Petroleum Geologists Bulletin, v. 47, p. 1861–1870.

Salvador, A., 1987, Late Triassic-Jurassic paleogeography and origin of Gulf of Mexico Basin: American Association of Petroleum Geologists Bulletin, v. 71, p. 419–451.

Salvador, A., and Green, A. R., 1980, Opening of the Caribbean Tethys (origin and development of the Caribbean and the Gulf of Mexico), *in* Aubouin, J., and others, coordinators, Geology of the Alpine chains born of the Tethys; 26th International Geological Congress, (Paris), Symposium C5: Bureau de Recherches Géologiques et Minières Mémoire 115, p. 224–229.

Sánchez López, R., 1961, Aplicación de la paleogeografía a la búsqueda de yacimientos petrolíferos en el área Misantla-Ebano-Pánuco: Asociación Mexicana de Geólogos Petroleros Boletín, v. 13, nos. 11–12, p. 361–376.

Sánchez-Rosas, G. J., 1972, Estudio estratigráfico del área Los Cues-Salinas, SE de Tamaulipas y N de Veracruz: Asociación Mexicana de Geólogos Petroleros Boletín, v. 24, nos. 1–3, p. 65–91.

Santiago Acevedo, J., 1962, Estructuras de la porción occidental del frente de la Sierra de Chiapas: Asociación Mexicana de Geólogos Petroleros Boletín, v. 14, nos. 5–6, p. 111–134.

—— , 1980, Giant fields of the southern zone; Mexico, *in* Halbouty, M. T., ed., Giant oil and gas fields of the decade 1968–1978: American Association of Petroleum Geologists Memoir 30, p. 339–385.

Santiago Acevedo, J., and Mejía Dautt, O., 1980, Giant fields in the southeast of Mexico: Gulf Coast Association of Geological Societies Transactions, v. 30, p. 1–31.

Schlager, W., Buffler, R. T., Angstadt, D., and Phair, R., 1984, Geologic history of the southeastern Gulf of Mexico, *in* Buffler, R. T., Schlager, W., and others, eds., Initial reports of the Deep Sea Drilling Project, Volume 77: Washington, D.C., U.S. Government Printing Office, p. 715–738.

Schmidt-Effing, R., 1980, The Huayacocotla aulacogen in Mexico (Lower Jurassic) and the origin of the Gulf of Mexico, *in* Pilger, R. H., Jr., ed., The origin of the Gulf of Mexico and the early opening of the central North Atlantic Ocean: Baton Rouge, Louisiana State University, p. 79–86.

Scott, K. R., Hayes, W. E., and Fietz, R. P., 1961, Geology of the Eagle Mills Formation: Gulf Coast Association of Geological Societies Transactions, v. 11, p. 1–14.

Scott, R. W., 1984, Significant fossils of the Knowles Limestone, Lower Cretaceous, Texas, *in* Ventress, P. S., and others, eds., The Jurassic of the Gulf Rim; Proceedings, 3rd Annual Research Conference: Society of Economic Paleontologists and Mineralogists, Gulf Coast Section, p. 333–346.

Segerstrom, K., 1961, Geology of the Bernal-Jalpan area, Estado de Querétaro, Mexico: U.S. Geological Survey Bulletin 1104–B, p. 19–86.

—— , 1962, Geology of south-central Hidalgo and northeastern México, Mexico: U.S. Geological Survey Bulletin 1104–C, p. 87–162.

Shaub, F. J., Buffler, R. T., and Parsons, J. G., 1984, Seismic stratigraphic framework of deep central Gulf of Mexico Basin: American Association of Petroleum Geologists Bulletin, v. 68, p. 1790–1802.

Shearer, H. K., 1938, Developments in south Arkansas and north Louisiana in 1937: American Association of Petroleum Geologists Bulletin, v. 22, p. 719–727.

Sheridan, R. E., 1983, Phenomena of pulsation tectonics related to the breakup of the eastern North American continental margin, *in* Sheridan, R. E., Gradstein, F. M., and others, eds., Initial reports of the Deep Sea Drilling Project Volume 76: Washington, D.C., U.S. Government Printing Office, p. 897–909.

Shreveport Geological Society, 1968, Stratigraphy and selected gas-field studies of north Louisiana, *in* Beebe, B. W., and Curtis, B. F., eds., Natural gases of North America: American Association of Petroleum Geologists Memoir 9, v. 1, p. 1099–1175.

Sigsby, R. J., 1976, Paleoenvironmental analysis of the Big Escambia Creek–Jay–Blackjack Creek Field area: Gulf Coast Association of Geological Societies Transactions, v. 26, p. 258–278.

Silva-Pineda, A., 1963, Plantas del Triásico Superior del Estado de Hidalgo: Instituto de Geología, Universidad Nacional Autónoma de México, Paleontología Mexicana No. 18, 12 p.

——, 1979, La Flora Triásica de México: Instituto de Geología, Universidad Nacional Autónoma de México, Revista, v. 3, p. 138–145.

—— 1981 (1983), *Asterotheca* y plantas asociadas de la Formación Huizachal (Triásico Superior) del Estado de Hidalgo: Instituto de Geología, Universidad Nacional Autónoma de México, Revista, v. 5, p. 47–54.

Smith, A. G., and Briden, J. C., 1977, Mesozoic and Cenozoic paleocontinental Maps: Cambridge, Cambridge University Press, 63 p.

Stabler, C. L., 1972, Secuencia de una trampa estratigráfica en el Jurásico Superior: Campo Tamaulipas-Constituciones: Asociación Mexicana de Geólogos Petroleros Boletín, v. 24, nos. 1–3, p. 165–197.

Swain, F. M., 1944, Stratigraphy of Cotton Valley beds of northern Gulf Coast Plain: American Association of Petroleum Geologists Bulletin, v. 28, p. 577–614.

——, 1949, Upper Jurassic of northeastern Texas: American Association of Petroleum Geologists Bulletin, v. 33, p. 1206–1250.

Thomas, W. A., and Mann, C. J., 1966, Late Jurassic depositional environments, Louisiana and Arkansas: American Association of Petroleum Geologists Bulletin, v. 50, p. 178–182.

Todd, R. G., and Mitchum, R. M., Jr., 1977, Seismic stratigraphy and global changes of sea level; Part 8, Identification of Upper Triassic, Jurassic, and Lower Cretaceous seismic sequences in Gulf of Mexico and offshore west Africa, *in* Payton, C. E., ed., Seismic stratigraphy; Applications to hydrocarbon exploration: American Association of Petroleum Geologists Memoir 26, p. 145–163.

Toledo Toledo, M., and Tavitas Galván, J. E., 1972, Analisis estratigráfico del Jurásico Superior en pozos perforados al norte del Río Pánuco: Asociación Mexicana de Geólogos Petroleros Boletín, v. 24, nos. 1–3, p. 93–127.

Tolson, J. S., Copeland, C. W., and Bearden, B. L., 1983, Stratigraphic profiles of Jurassic strata in the western part of the Alabama coastal plain: Geological Survey of Alabama Bulletin 122, 425 p.

Traverse, A., 1987, Pollen and spores date origin of rift basins from Texas to Nova Scotia as early Late Triassic: Science, v. 236, p. 1469–1472.

Vail, P. R., Hardenbol, J., and Todd, R. G., 1984, Jurassic unconformities, chronostratigraphy, and sea-level changes from seismic stratigraphy and biostratigraphy, *in* Schlee, J. S., ed., Interregional unconformities and hydrocarbon accumulation: American Association of Petroleum Geologists Memoir 36, p. 129–144.

Van Houten, F. B., 1977, Triassic-Liassic deposits of Morocco and eastern North America; Comparison: American Association of Petroleum Geologists Bulletin, v. 61, p. 79–99.

Varela Hernández, A., 1969, Problemas de la exploración petrolera en el homocli-
nal de San José de las Rusias, Tamaulipas: Instituto Mexicano del Petróleo, Seminario sobre Exploración Petrolera, Mesa Redonda No. 2: Problemas de Exploración de la Zona Norte, 17 p.

Velasco Torres, J. J., and Sepúlveda de León, G., 1973, Biostratigrafía de la Formación Pimienta en el Distrito Poza Rica: Ingeniería Petrolera, v. 13, no. 3, p. 99–108.

Vernon, R. C., 1971, Possible future petroleum potential of pre-Jurassic, western Gulf Basin, *in* Cram, I. H., ed., Future petroleum provinces of the United States; Their geology and potential: American Association of Petroleum Geologists Memoir 15, v. 2, p. 954–979.

Vestal, J. H., 1950, Petroleum geology of the Smackover Formation of southern Arkansas: Arkansas Resources and Development Commission, Division of Geology Information Circular 14, 19 p.

Viniegra, F., 1965, Geología del Macizo de Tezuitlán y la cuenca cenozoica de Veracruz: Asociación Mexicana de Geólogos Petroleros Boletín, v. 17, nos. 7-12, p. 101–163.

Viniegra O., F., 1971, Age and evolution of salt basins of southeastern Mexico: American Association of Petroleum Geologists Bulletin, v. 55, p. 478–494.

——, 1981, Great carbonate bank of Yucatan, southern Mexico: Journal of Petroleum Geology, v. 3, p. 247–278.

Vokes, H. E., 1963, Geology of the Cañon de la Huasteca area in the Sierra Madre Oriental, Nuevo Leon, Mexico: New Orleans, Louisiana, Tulane Studies in Geology, v. 1, p. 126–148.

Wakelyn, B. D., 1977, Petrology of the Smackover Formation (Jurassic), Perry and Stone counties, Mississippi: Gulf Coast Association of Geological Societies Transactions, v. 27, p. 386–408.

Wall, J. R., Murray, G. E., and Diaz G., T., 1961, Geologic occurrence of intrusive gypsum and its effect on structural forms in Coahuila marginal folded province of northeastern Mexico: American Association of Petroleum Geologists Bulletin, v. 45, p. 1504–1522.

Watkins, J. S., Ladd, J. W., Buffler, R. T., Shaub, F. J., Houston, M. H., and Worzel, J. L., 1978, Occurrence and evolution of salt in deep Gulf of Mexico, *in* Bouma, A. H., Moore, G. T., and Coleman, J. M., eds., Framework, facies, and oil-trapping characteristics of the upper continental margin: American Association of Petroleum Geologists Studies in Geology 7, p. 43–65.

Weeks, W. B., 1938, South Arkansas stratigraphy with emphasis on the older coastal plain beds: American Association of Petroleum Geologists Bulletin, v. 22, p. 953–983.

Weidie, A. E., and Wolleben, J. A., 1969, Upper Jurassic stratigraphic relations near Monterrey, Nuevo León, Mexico: American Association of Petroleum Geologists Bulletin, v. 53, p. 2418–2420.

Wells, J. W., 1942, Jurassic corals from the Smackover Limestone, Arkansas: Journal of Paleontology, v. 16, p. 126–129.

Wilson, G. V., 1975, Early differential subsidence and configuration of the northern Gulf Coast Basin in southwest Alabama and northwest Florida: Gulf Coast Association of Geological Societies Transactions, v. 25, p. 196–206.

MANUSCRIPT ACCEPTED BY THE SOCIETY MAY 24, 1989

ACKNOWLEDGMENTS

This chapter is respectfully dedicated to Ralph W. Imlay, whose pioneering studies of the Mesozoic section of the southern United States and Mexico have been the foundation upon which all subsequent work in the region has been based.

Raúl González, George Herman, Curtis C. Humphris, and Ray G. Martin carefully reviewed the manuscript and offered many thoughtful comments and suggestions. Discussions with Gary Kocurek about the Norphlet Formation were extremely useful. Their invaluable contributions for significant improvements of the manuscript are gratefully acknowledged.

The Geology of North America
Vol. J, The Gulf of Mexico Basin
The Geological Society of America, 1991

Chapter 9

Lower Cretaceous

Edward McFarlan, Jr.
3131 West Alabama Street, Suite 531, Houston, Texas 77098
L. Silvio Menes
Petroleos Mexicanos, Mexico, D.F., Mexico

INTRODUCTION

Throughout most of Early Cretaceous time, the Gulf of Mexico basin was a major site of continental and marine deposition surrounded by the Appalachian and Ouachita uplands on the north, the Llano and Marathon uplifts on the northwest and the Chiapas massif and Maya Mountains to the south. During this time there were marine connections to the Pacific Ocean to the west and to the Atlantic Ocean to the southeast. The Gulf of Mexico basin was tectonically stable except for continuing slow subsidence of its central part, growth faulting on the margins of some depocenters, and local deformation related to underlying Jurassic salt. Shallow-marine water covered its rims and peripheral shelves, and progressively deeper waters its slope and abyssal plain.

Lower Cretaceous rocks form a continuous disc of sediments, which thin and pinch out updip along the periphery of the Gulf of Mexico basin. Lower Cretaceous sequences crop out along the northwestern, western, southwestern rims of the basin. No Lower Cretaceous outcrops are known east of the Mississippi River nor in the Florida and Yucatán Peninsulas (Fig. 1).

The Lower Cretaceous sediments are primarily carbonates and evaporites on the circum-Gulf shelves, and carbonates in the bathyal areas. Continental and shallow-marine terrigenous clastic sediments occur primarily around the northern and northwestern rims of the basin, from northeastern Mexico to the Florida panhandle. They are most prevalent in the lower part of the Lower Cretaceous section (Berriasian to Barremian), and represent the sediment load of rivers draining the continental interior and the uplands limiting the basin to the north.

The Lower Cretaceous section reaches thicknesses of 2,700 to 3,700 m (8,000 to 11,000 ft) along the outer margin of the northern Early Cretaceous shelf (McFarlan, 1977). Within the shelf, Lower Cretaceous deposits thicken locally to more than 2,700 m (8,000 ft) in shallow-water shelfal basins (Fig. 1). Many

of these shelfal basins are associated with crustal depressions that contained previously significant thicknesses of underlying Jurassic salt. Much of the salt has shifted into domes and pillows that penetrate Lower Cretaceous deposits.

Beneath the continental slope and continental rise in the northern Gulf of Mexico basin, the Lower Cretaceous thicknesses are unknown. In this part of the basin, the Lower Cretaceous strata have been downwarped to depths beyond the drill. Salt-dome growth and down-to-the-basin faults have disturbed the layering; hence, correlation on seismic data is tenuous and uncertain.

The Lower Cretaceous section thins from the shelfal areas toward the central part of the Gulf of Mexico basin, where the section is probably less than 2,000 m (6,000 ft) thick.

On the southern flank of the Gulf of Mexico basin (Fig. 1), Lower Cretaceous sediments thicken gulfward from pinchouts along the trend of upland massifs in southern Mexico and Yucatán (López Ramos, 1975; Vinegra O., 1981; Carrillo Bravo, 1965; Schlager and Buffler, 1984). In northern and central Mexico, relatively thin Lower Cretaceous deposits were laid down continuously across most of the craton, forming a broad shelf 160 to 640 km (100 to 400 mi) wide (López Ramos, 1982; Carrillo Bravo, 1965; Smith, 1970; Wilson and Pialli, 1977; Cook and Bally, 1975). Several distinct platforms—the Coahuila, Valles–San Luis Potosí, Tuxpan, and Córdoba platforms—maintained a more positive position within the shelf and were surrounded by areas of deeper water (see Plate 2). Reef complexes are present along some of the margins of these platforms. They reached thicknesses of 2,500 to 3,000 m (8,300 to 9,900 ft).

Even though the Early Cretaceous was a time of considerable tectonic stability in the Gulf of Mexico basin region, a number of positive and negative topographic features were present in the basin that had been inherited from Jurassic and earlier periods of structural deformation. The general subsidence of the basin, which had started in the Late Jurassic, and differential sedimen-

McFarlan, E., Jr., and Menes, L. S., 1991, Lower Cretaceous, *in* Salvador, A., ed., The Gulf of Mexico Basin: Boulder, Colorado, Geological Society of America, The Geology of North America, v. J.

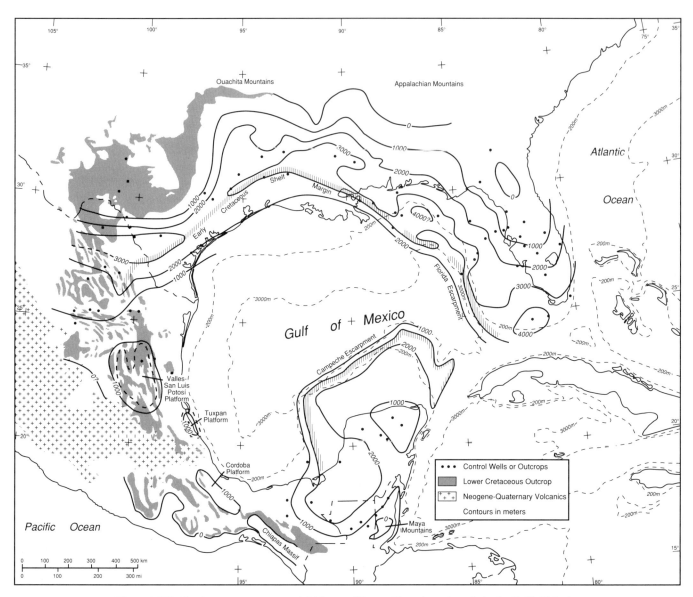

Figure 1. Distribution, outcrop areas, and thickness of Lower Cretaceous deposits in the Gulf of Mexico basin.

tary compaction caused these features to influence Early Creta-ceous deposition.

Among the more prominent pre–Early Cretaceous positive features are the Peninsular arch of Florida, the Coahuila, Valles–San Luis Potosí (Carrillo Bravo, 1971), Tuxpan (Viniegra O. and Castillo-Tejero, 1970; Viniegra O., 1981) and Córdoba platforms in Mexico and the Maya Mountains in Belize (Bateson, 1972). Negative features included areas of thicker Jurassic salt deposi-tion, which showed persistent subsidence of their central parts and growth faulting along their peripheries during the Early Cre-taceous. Thicker-than-normal Lower Cretaceous sedimentary sections are found in these structural sags.

STRATIGRAPHIC SUBDIVISION AND TERMINOLOGY

Three major sedimentary provinces can be recognized in the Lower Cretaceous section of the Gulf of Mexico basin: to the north, from east Texas to southern Alabama, carbonates and coarse-grained terrigenous clastic sediments predominate along a region into which rivers draining the continental interior and the Appalachian and Ouachita uplands deposited their load. To the west, from central Texas through northeastern Mexico to south-ern Mexico, the section is primarily composed of carbonates with only thin intervals of fine-grained terrigenous clastic sediments—

shales and calcareous shales. The southeastern province includes the Florida and Yucatán platforms where interbedded carbonates and evaporites characterize the Lower Cretaceous section.

Some of the earliest descriptions of the Lower Cretaceous rocks are those of Hill (1887a, b, 1891, 1893a, b, and 1894) from outcrops in north-central Texas. Hill subdivided the upper part of the Lower Cretaceous section of the area, from bottom to top, into the Trinity, Fredericksburg, and Washita groups. Similar sections were also described in northeastern Mexico by Dumble (1895). An older section was later recognized in northeastern Mexico below the Trinity Group by Bosc (1923), Bose and Cavins (1927), and Imlay (1936, 1937, 1938). This lower part was named the Coahuila Group by Imlay (1940b). During the following decades, descriptions of outcropping Lower Cretaceous strata in Texas and their fauna were provided by Adkins (1933), Lozo (1959), Lozo and Stricklin (1956), Lozo and Smith (1964), Rose (1972), Bay (1977), and many others. In Arkansas, Louisiana, Mississippi, Alabama, and Florida, the updip pinchout of Lower Cretaceous strata in the subsurface required studies of well samples by many geologists, including Weeks (1938), Hazzard (1939), Hazzard and others (1947), Murray (1948), and Forgotson (1957, 1963).

Early investigations of surface and subsurface Lower Cretaceous strata were undertaken simultaneously in Mexico. Some of these early studies included those by Bose and Cavins (1927), Burckhardt (1930–1931), Muir (1936), Humphrey (1949), Diaz (1952), Smith (1970), de Cserna (1970), and many others.

Several symposia have taken place over the years to discuss the stratigraphy of the Lower Cretaceous rocks of Texas and Mexico (Lozo and others, 1959; Hendricks, 1967; Perkins, 1974; Bebout and Loucks, 1977). The stratigraphy of the Lower Cretaceous over a large part of the northern flank of the Gulf of Mexico basin has been summarized by Anderson (1979).

Continuing investigations of an increasing number of fossil groups have provided a firm basis for dating the Lower Cretaceous section. Imlay (1944, 1945) demonstrated how the outcropping strata could be traced into the subsurface by means of ammonites, and be correlated with microfossil zones recognized in well cuttings. Many additional studies of ammonites by Young (1967, 1972, 1974, 1977, 1978) provided a basis for improved correlations in Texas and Mexico. Young's ammonite zonation of the Lower Cretaceous sediments of Texas and Mexico is shown in Table 1.

Exploration for oil and gas throughout the Gulf of Mexico basin resulted in the need to identify microscopic fossils in samples from cores and well cuttings for more accurate dating and correlation. To this end, many biostratigraphers, including Bonet (1952, 1956), Trejo (1960, 1973, and 1975), Cabrera and Menes (1973), and others, used samples from type sections and outcrops dated by ammonites to zone the Lower Cretaceous section with planktonic microfossils.

Useful zonations have been developed for the Taraises Formation (Berriasian-Hauterivian) based on calpionellids and nannoconids (Blauser and McNulty, 1980), and for the Albian,

Cenomanian, and Turonian based on foraminifera and calcispheres (Ice and McNulty, 1980). Michael (1972) used planktonic foraminifera to zone the Lower Cretaceous sections of Texas. The results of these studies, together with those of Bonet and Trejo mentioned above, provided useful application of planktonic foraminifera, calpionellids, coccoliths, and nannoconids to the dating of the Lower Cretaceous strata of the United States and Mexico.

The base and top of the Lower Cretaceous section in the Gulf of Mexico basin do not everywhere coincide to clear lithostratigraphic boundaries or to regional unconformities or hiatuses (Fig. 2 and Plate 5).

The Jurassic–Lower Cretaceous boundary in the southern Florida carbonate platform is probably within the lower part of the Wood River Formation; over most of the Florida Peninsula, the Lower Cretaceous section unconformably overlies the "basement." Along the northern flank of the basin the boundary has been placed in the upper part of the Cotton Valley Group, which is now known to range into the Lower Cretaceous (see Chapter 8, this volume). In northern and east-central Mexico, the Jurassic–Lower Cretaceous boundary has generally been placed at the top of the La Casita and Pimienta Formations, although these two units may range into the Lower Cretaceous; in the southern part of the country, the base of the Lower Cretaceous falls within the Chinameca and Todos Santos Formations and the Zacatera Group.

The boundary between the Lower Cretaceous and the Upper Cretaceous, as recognized along the northern flank of the Gulf of Mexico basin, from Texas to the Florida panhandle, does not correspond with the internationally accepted boundary between these two chronostratigraphic series. In this region, an important regional unconformity of mid-Cenomanian age marks the end of a cycle of widespread carbonate deposition and the beginning of another cycle, during which terrigenous clastic sediment become more abundant. The boundary between the Lower and Upper Cretaceous has traditionally been placed at this unconformity even though the uppermost beds below it are of Late Cretaceous (early Cenomanian) age. To be consistent with international usage, the boundary should be placed at or near the base of the Grayson and Del Rio Formations and age-equivalent units (Fig. 2). In Mexico, the Lower Cretaceous–Upper Cretaceous boundary would fall within the Cuesta del Cura, El Abra, and Tamaulipas Superior Formations, the Sierra Madre Group, and contemporaneous units. Common in Mexico, however, is the recognition of a "Middle Cretaceous," which includes beds of Albian and Cenomanian age.

The Lower Cretaceous rocks that crop out extensively along the northwestern and western flanks of the Gulf of Mexico basin in the United States and Mexico have been carefully studied for over 100 years. The section, particularly its upper part (of Aptian and Albian age), has been subdivided into numerous groups, formations, and members, many of them of only local significance. From the surface, these units have been carried into the subsurface to the large number of wells drilled in search of petro-

TABLE 1. AMMONITE ZONATION OF THE LOWER CRETACEOUS SEDIMENTS OF TEXAS & MEXICO (AFTER YOUNG, 1972, 1974)

Stage	Zone	C = Cosmopolitan E = Endemic M = Mixed Division	Division
LOWER CENOMANIAN	Budaiceras hyatti	E	WASHITA
LOWER CENOMANIAN	Graysonites	C	WASHITA
LOWER CENOMANIAN	Plesioturrilites brazoensis		WASHITA
UPPER ALBIAN	Drakeoceras drakei	M	WASHITA
UPPER ALBIAN	Mortoniceras wintoni	M	WASHITA
UPPER ALBIAN	Drakeoceras lasswitzi	C	WASHITA
UPPER ALBIAN	Pervinquieria equidistans	C	WASHITA
UPPER ALBIAN	Eopachydiscus brazoensis	C	WASHITA
UPPER ALBIAN	Adkinsites bravoensis	C	WASHITA
UPPER ALBIAN	Manuaniceras powclli	E	FREDERICKSBURG
MIDDLE ALBIAN	Manuaniceras carbonarium	M	FREDERICKSBURG
MIDDLE ALBIAN	Oxytropidoceras salasi	C	FREDERICKSBURG
MIDDLE ALBIAN	Metengonoceras hilli	E	FREDERICKSBURG
MIDDLE ALBIAN	Endemic engonocerids	E	UPPER TRINITY
MIDDLE ALBIAN	Ceratostreon weatherfordense	M	UPPER TRINITY
LOWER ALBIAN	Hypacanthoplites comalensis	E	UPPER TRINITY
LOWER ALBIAN	Douvilleiceras "mammillatum"	C	UPPER TRINITY
LOWER ALBIAN	Hypacanthoplites craginl		UPPER TRINITY
LOWER ALBIAN	Hypacanthoplites mayfieldensis	E	UPPER TRINITY
APTIAN	Kazanskyella spathi	C	MIDDLE TRINITY
APTIAN	Dufrenoyia justinae	E	MIDDLE TRINITY
APTIAN	Cheloniceras spp.	C	MIDDLE TRINITY
NEOCOMIAN		?	LOWER TRINITY
NEOCOMIAN	Leopoldia victoriensis	C	LOWER TRINITY

leum, at times on the basis of their distinctive lithology, but all too often by means of wire-line-log correlation, regardless of lithologic characteristics. The section identified in the wells has been traced toward the center of the basin, and into areas of poor or no well control, by reflection-seismic methods, many times with difficulty due to the disturbances caused by the deformation and intrusion of the Jurassic salt.

For the purpose of the following discussion, and to avoid the possible confusion that the proliferation of lithostratigraphic terms may cause, the Lower Cretaceous section of the Gulf of Mexico basin has been subdivided into three chronostratigraphic units, in part bounded by unconformities. Reference to chronostratigraphic units of the standard global chronostratigraphic scale makes easier the regional treatment of the Lower Cretaceous stratigraphy of the basin and the interpretation of its geologic history. The three units are, from oldest to youngest: (1) Berriasian and Valanginian; (2) Hauterivian, Barremian, and Aptian; and (3) Albian and lower Cenomanian.

The lower unit (Berriasian and Valanginian) records a period of clastic sedimentation in coastal and nearshore areas of the northern and northeastern Gulf of Mexico basin, and of carbonate deposition in the shelfal areas. In east-central Mexico, car-

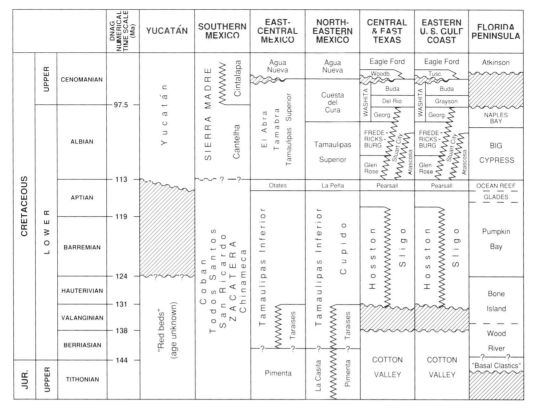

Figure 2. Simplified stratigraphic correlation chart, Lower Cretaceous of the Gulf of Mexico basin (for more complete chart, see Plate 5 in pocket). Group names are shown in all capitals, formations have only first letter capitalized.

bonate deposition predominated on the many platforms and shelfal areas. Throughout the basin the Berriasian and Valanginian deposits grade gulfward into basinal micrites and pelagic shales.

The base of the Berriasian is generally gradational and conformable with underlying Upper Jurassic rocks. In contrast, the top of the Valanginian is marked by an erosional unconformity across the northern and northeastern parts of the Gulf of Mexico basin. Elsewhere, the Valanginian-Hauterivian boundary has been reported to be conformable.

The Berriasian and Valanginian section also records the early stages in the development of the distinct paleogeographic features that characterize the Gulf of Mexico basin area during the Early Cretaceous: coastal plain, shelf, shelf margin, distinct stable platforms, and basin. During the Berriasian and Valanginian the shelves and platforms were narrow compared to those that formed later and contrast with the ramps common during the Late Jurassic.

The middle unit (Hauterivian-Barremian-Aptian) records a transgressive episode in its lower part and progradation in the middle and upper sections. The basal transgressive Hauterivian sediments rest unconformably on the eroded surface of the Berriasian shelfal rocks in the northern part of the basin. This boundary is conformable farther gulfward where Hauterivian sediments overlie the Valanginian section. In Mexico, as mentioned

above, Hauterivian carbonates apparently rest conformably on Valanginian limestones in most shelfal and platform areas.

In the northern and southern rims, basal Hauterivian transgressive clastic rocks and limestones grade upward into medium- and fine-grained terrigenous clastic sediments and then into a regressive upper section of carbonates. These carbonates are capped along the northern part of the basin by a widespread marine shaly unit of latest Aptian age (the Pearsall, La Peña, and Otates Formations) that records a landward encroachment of the shoreline across a broad flat shelf and an influx of fine-grained terrigenous clastic sediments from the north and northeast. To the southeast, lesser amount of clastic materials derived from the Maya Mountains and the Chiapas massif are included in the section. Along the eastern flank of the basin, rocks of Hauterivian to Aptian age are predominantly aggrading carbonates deposited far from sources of terrigenous clastic sediments.

The regressive and aggrading development of shelfal and platform carbonates of Hauterivian, Barremian, and Aptian age resulted directly in the broadening of the circum-Gulf shelves and platforms and in the formation of a more distinct shelf/basin margin. The outer shelf margin in most places was the site of basin-rimming carbonate banks.

The upper unit (Albian–lower Cenomanian) is composed of a prograding and aggrading sequence of predominantly shelfal carbonates with shelf-margin bank sediments that indicate a marked increase in slope where the banks grade into basin-margin micrites and pelagic shales. Subsidence and a eustatic rise in sea level were factors influencing deposition of this upper unit.

Albian time began with the deposition of regressive carbonates deposited conformably on the underlying upper Aptian shales around most of the northern, western, and southern margins of the Gulf of Mexico basin area. On the Yucatán and western Florida shelves, Albian carbonates and evaporites overlie, also conformably, Aptian carbonates.

The Albian period of prograding and aggrading carbonate shelf and shelf-margin bank development ended with an influx of fine-grained terrigenous clastic sediments around the northern and northwestern rims in early Cenomanian time (the Grayson and Del Rio Formations). Along the western, eastern, and southern rims, carbonate deposition continued, apparently without interruption, from late Albian to early Cenomanian time.

Figures 3 to 7 illustrate the stratigraphic relations, subdivisions, and terminology of the Lower Cretaceous section of the Gulf of Mexico basin.

BERRIASIAN-VALANGINIAN

Distribution

Berriasian-Valanginian sediments are present around the periphery of the Gulf of Mexico basin (Figs. 8, 9). Along the northern margin of the basin, thin coastal and nearshore terrigenous clastics grade gulfward into much thicker shallow-water shelfal carbonates and beyond into basinal micrites and calcareous shales (Hazzard, 1939; Hazzard and others, 1947; Lozo and Stricklin, 1956; Murray, 1961; and many others). In northeastern and central Mexico, rocks of these ages have been identified and described from excellent outcrops all the way from the state of Coahuila to the Trans-Mexican neovolcanic belt, the whole length of the Sierra Madre Oriental (Fig. 1; Imlay, 1940b, 1944; Humphrey, 1949, 1956; Smith, 1970; Conklin and Moore, 1977; Enos, 1983; Lopez Ramos, 1982). Numerous wells also have penetrated beds of Berriasian-Valanginian age along the coastal plains of northeastern and east-central Mexico. In the southern part of the Gulf of Mexico basin, in southern Mexico, the

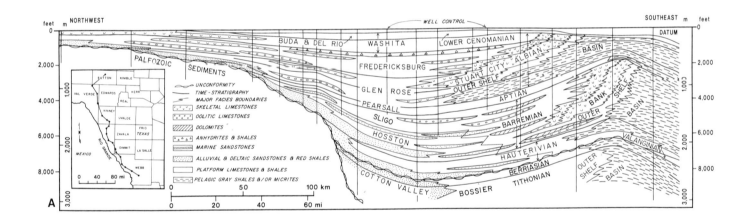

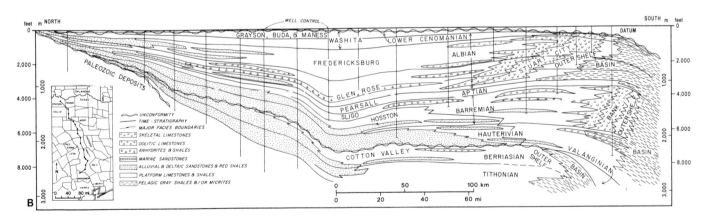

Figure 3. A, Generalized regional dip section of Lower Cretaceous deposits across south Texas. B, Generalized regional dip section of Lower Cretaceous deposits across east Texas.

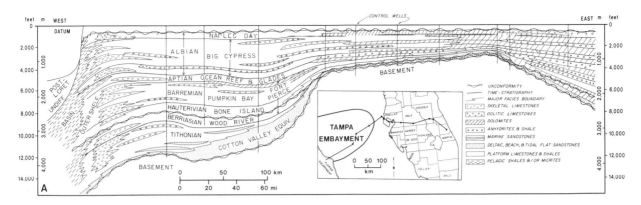

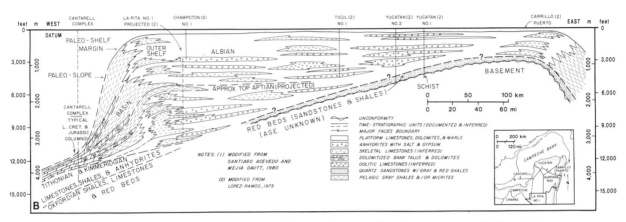

Figure 4. A, Generalized regional section of Lower Cretaceous deposits across south Florida and the West Florida shelf. B, Generalized regional section of Lower Cretaceous deposits across the Yucatan platform (modified from Lopez Ramos, 1975; Santiago Acevedo and Mejia Dautt, 1980; Viniegra O., 1981).

Berriasian-Valanginian section is represented by "red beds" along the front of the Chiapas massif; they grade basinward into a section of marine limestones and shales that has been penetrated by many wells in the petroleum-producing districts of southern Mexico, both onshore and offshore (Viniegra O., 1971; Lopez Ramos, 1981; Viniegra O., 1981). No well-dated sediments of this age have been reported from the subsurface of the northern part of the Yucatan Peninsula.

Lithostratigraphy

In the subsurface of the northern U.S. Gulf Coastal Plain, Berriasian sediments form a basinward-thickening wedge of coarse terrigenous clastics that interfinger downdip with marine shales and, in places, with limestones (Fig. 9). To the southwest, the coarse clastics change to a thinner section of shales and argillaceous limestones. These Berriasian clastics constitute the uppermost part of the Cotton Valley Group described by Weeks (1938), Hazzard (1939), Swain (1944), Forgotson (1954), and others (see Chapter 8, this volume).

Updip in east Texas, northern Louisiana, and southern Ar-

kansas, the Cotton Valley Group is represented by terrigenous clastic sediments of the Schuler Formation. They consist of nonmarine aluvial, fluvial, and nearshore red and white sandstones, siltstones, and conglomerates interbedded with red, maroon, and purple mudstones and shales. These beds grade gulfward into the marine gray and black fossiliferous shales of the Bossier Formation.

In east Texas and west-central Louisiana, the distinctive Knowles Limestone, composed of oolitic and skeletal grainstones, occurs near the top of the Cotton Valley Group (Mann and Thomas, 1964).

In Mississippi and Alabama, the Cretaceous part of the Cotton Valley Group is composed of white and red, fine-to-coarse sandstones and conglomerates with red shales and beds of argillaceous limestones near the top (Tolson and others, 1983).

From south Texas to Alabama, the Valanginian deposits are limited primarily to basinal micrites and calcareous shales. These Valanginian basinal deposits have only been recognized in deep, downdip wells. Valanginian time is represented in the updip areas by an erosional unconformity on the top of the Cotton Valley (Berriasian) deposits.

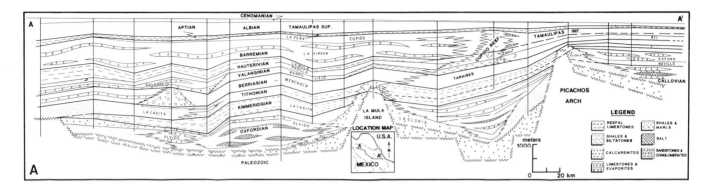

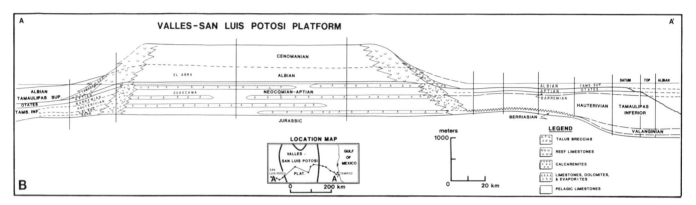

Figure 5. A, Generalized section of Lower Cretaceous deposits from the northern Sabinas basin to the Picachos arch, northeastern Mexico. B, Generalized section of Lower Cretaceous deposits across the Valles–San Luis Potosí platform and western part of the Tampico-Misantla basin, east-central Mexico.

Eastward around the rim of the basin, the Berriasian and Valanginian sediments of Florida and the West Florida shelf form the basal part of the Fort Pierce Group of Applin and Applin (1965) and the Wood River and lower part of the Bone Island Formations of Applegate and others (1981). The Wood River Formation consists of a 520-m-thick section of interbedded dense brown limestones and dense brown dolomites with anhydrite layers. These beds grade downward into a 120-m basal clastic unit of reddish brown shale with layers of fine-to-coarse vari-colored sandstone. The basal "red beds" are probably Jurassic in age (Applegate and others, 1981). These "red beds" rest on rhyo-lite basalt dated at 189 Ma.

In northeastern Mexico, the Coahuila platform and the El Burro uplift supplied the terrigenous clastics of the lower part of the Taraises Formation. Along the northeastern and southern borders, respectively, of these positive features the Berriasian-Valanginian section is composed of the conglomerates, red sand-stones, siltstones, and shales of alluvial fans, which farther away from the positive elements, grade to alluvial coastal plain deposits (de Cserna, 1970; Fig. 9).

In the Sabinas basin depression between the Coahuila plat-form and the El Burro uplift, dark-colored argillaceous lime-stones, shales, and nodular marls of the Menchaca Formation

were deposited (Fig. 5A). These sediments contain an abundant fauna of ammonites and a microfauna of tintinnids and calpionel-lids, indicating deposition on the outer shelf. Conformably overly-ing this sequence are the shales and siltstones of the Barril Viejo Formation of Valanginian age, which represent shallow-water conditions of deposition. They grade toward the southeast to the dark-colored, slightly dolomitized argillaceous limestones of the Taraises Formation. This unit is rich in calpionellids and nanno-conids, indicating deposition on a platform margin that extended in a broad arc surrounding the Coahuila Platform. From this arc, water depth increased progressively toward the east and southeast where the section is composed of the light gray argillaceous lime-stones with black chert nodules of the lower part of the Tamaulipas Inferior Formation, which occurs over a large area of east-central Mexico including the Tampico-Misantla, and Vera-cruz basins.

Locally, the basin-margin Taraises deposits grade into shallow-water facies in the areas surrounding the Valles–San Luis Potosí and Tuxpan platforms (Fig. 9). Along the margins of the former, the initial deposits of the Lower Cretaceous are composed of biogenic calcarenites and patch reefs (Moya Cuevas, 1974). Along the margins of the Tuxpan platform, skeletal calcarenites and bentonite-bearing limestones and dolomites grade laterally into the argillaceous limestones in the lower part of the Tamaulipas Inferior Formation (Viniegra O. and Castillo-Tejero, 1970). These reefs around the margin of the platforms restricted the flow of seawater, and as a result, evaporites, limestones, and dolomites were deposited in their central parts. Toward the basinal areas from these platforms, the platform-margin sediments grade upward and gulfward into Tamaulipas Inferior limestones deposited in relatively deep water (Fig. 5B). The lower part of the Tamaulipas Inferior section probably also includes sediments of Valanginian age.

No Berriasian-Valanginian rocks have been reported from the Veracruz basin, in part because extensive dolomitization and strong structural deformation has made difficult their identification. West of this basin, on the Córdoba platform, about 100 km west-southwest of Veracruz, a Berriasian-Valanginian section, the Xonamanca Formation, has been described by Carrasco and others (1975). It overlies conformably and transitionally the Upper Jurassic Tepexilotla Formation (see Chapter 8, this volume) and is composed of 400 to 600 m of graywackes, siltstones, shales, and micritic limestones interbedded with dacitic tuffs and bentonites, which can form as much as 50 to 60 percent of the section. Farther to the southeast on the Córdoba platform, in northern Oaxaca, an exploratory well penetrated Berriasian sediments composed of light-colored limestone containing a fauna of

tintinnids and calpiopnellids indicating deposition on an outer platform. Along the western margin of the platform, the Berriasian is represented by dark-colored argillaceous limestones containing a fauna of tintinnids and calpionellids, indicating deposition in deep water (Figs. 6B and 9).

Along the northeastern front of the Chiapas massif, the continental reddish sandstones, siltstones, and conglomerates of the Todos Santos Formation were deposited in Berriasian and Valanginian times (Fig. 7). This terrigenous sequence grades northward in the Sierra de Chiapas first to coastal sediments and platform dolomitized limestones, then to bentonite-bearing argillaceous limestones that mark the transition to the open-marine microcrystalline limestones and dark shales that crop out in the northern part of the Sierra de Chiapas and have been penetrated by numerous wells in the coastal plains of the Isthmus of Tehuantepec area and in the Bay of Campeche (Viniegra O., 1971, 1981; Vargas, 1978; Lopez Ramos, 1981). Both the argillaceous and the microcrystalline limestones of this Lower Cretaceous section contain abundant ammonites, tintinnids, calpionellids, and nannoconids of Berriasian-Valanginian age (Trejo, 1960, 1973).

Andesitic lavas have been reported along the eastern foothills of the Chiapas massif. They have been dated by isotopic methods as of latest Jurassic and earliest Cretaceous age.

In the southeastern part of the Gulf of Mexico basin, a

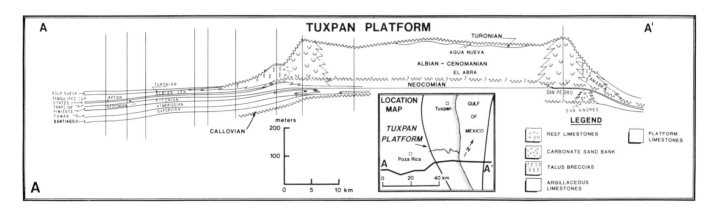

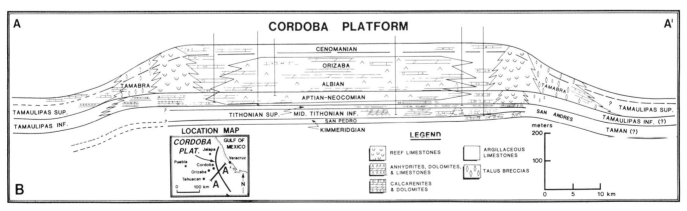

Figure 6. A, Generalized section of Lower Cretaceous deposits across the Tuxpan platform, east-central Mexico. B, Generalized section of Lower Cretaceous deposits across the Córdoba platform, southern Mexico.

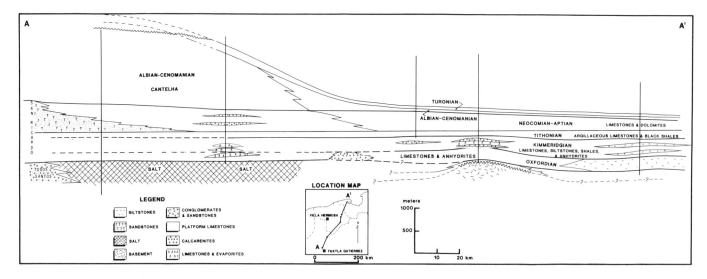

Figure 7. Generalized strike section of Lower Cretaceous deposits of southern Mexico.

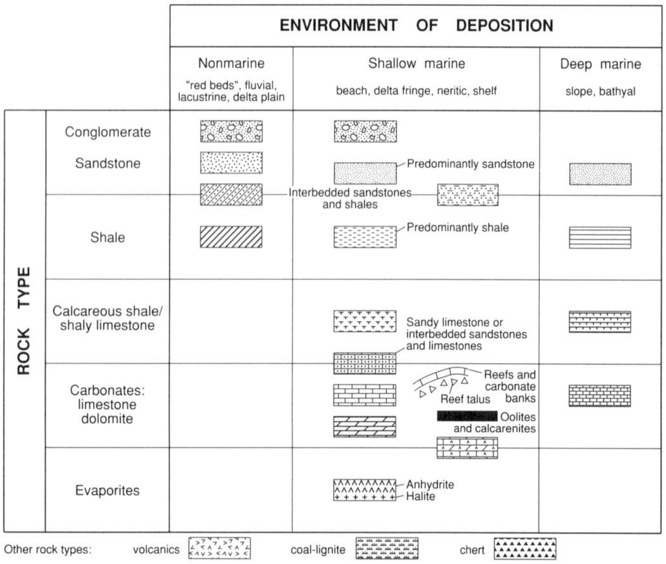

ENVIRONMENT OF DEPOSITION		
Nonmarine "red beds", fluvial, lacustrine, delta plain	Shallow marine beach, delta fringe, neritic, shelf	Deep marine slope, bathyal

Figure 8. Symbols used in lithofacies-paleogeographic maps, Figures 9, 10, and 11.

carbonate platform with evaporitic influence developed in Berriasian and Valanginian times (Fig. 9). It was limited to the southwest by the Chiapas massif and to the northeast by the emergent part of the Yucatan block, forming an embayment that extended southeastward to Guatemala. In the central and eastern parts of this platform, subsurface information has revealed the existence of a restricted area in which interbedded evaporites and dolomites (Cobán Formation) were deposited (Viniegra O., 1971; Lopez Ramos, 1975). Even though the limited information now available does not indicate the presence of reefal barriers or carbonate banks along the margin of the platforms, the presence of such deposits cannot be entirely discounted. Such barriers or

banks would explain the occurrence of the carbonate-evaporite sequence in the southeastern part of the platform. To the north and west of the emergent Yucatan landmass, shallow-water carbonates grade to the deep-water sediments known from the basins of southeastern Mexico, where they contain turbidite intercalations and talus material derived from the platform (Fig. 4B).

Along the northeastern margin of the Yucatan landmass, red arkosic sandstones and conglomerates overlain by white, shallow-water skeletal and oolitic limestones of Berriasian-Valanginian age were recovered at sites 535, 537, and 538 of Leg 77 of the Deep Sea Drilling Project (Schlager and others, 1984a, 1984b). The arkosic sandstones and conglomerates were interpreted to indicate a nearby source of clastic sediments. In Berriasian time, the source was probably uplands along the northeastern coast of the Yucatan landmass. The overlying shallow-water carbonates suggest a 16- to 32 km northward progradation of the Yucatan platform. Schlager and others (1984) note, however, that the sediments could be interpreted as deep-water talus deposits. If so, the Straits of Florida was a major opening to the Atlantic Ocean at this time.

Thickness

Berriasian sediments in the northern flank of the Gulf of Mexico basin thicken from a featheredge updip to 300 to 700 m along the shelf margin, 100 to 300 km downdip (Fig. 3A and B). In the eastern part of the basin, these deposits are about 300 m thick on the shelf margin of the West Florida shelf; they thin to about 30 m across the axis of the Florida Peninsular arch (Fig. 4A). The Valanginian sediments pinch out at the Berriasian shelf margin along the northern Gulf of Mexico basin, and thicken gradually gulfward (Fig. 3A and B).

In the western flank of the basin, the Berriasian deposits are

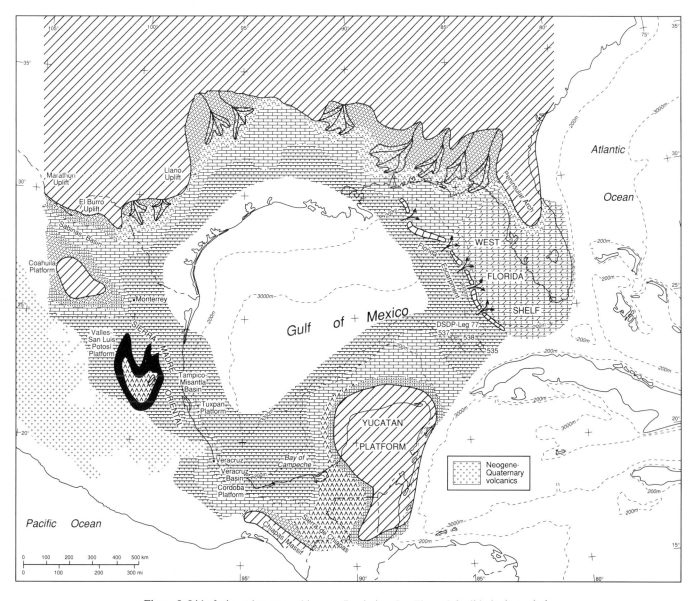

Figure 9. Lithofacies-paleogeographic map, Berriasian. See Figure 8 for lithologic symbols.

less than 200 m thick on the shelf margin, thinning inland to about 100 m on the Valles–San Luis Potosí platform (Fig. 5B). Their pinchout is obscured by the tectonic deformation in the Sierra Madre Oriental. Subsurface data indicate thin Berriasian sediments on the Tuxpan Córdoba platforms. On the western shelf margin off Yucatan, the Berriasian section thickens from its pinchout to about 100 m at the shelf margin. Valanginian beds are recognized throughout the western flank of the Gulf of Mexico basin in Mexico, where the sequence seldom reaches a thickness of more than 400 m.

Stratigraphic relations

The Berriasian clastic and carbonate sediments rest conformably on uppermost Upper Jurassic sediments (McFarlan, 1977;

Santiago Acevedo, 1980). The Jurassic-Cretaceous boundary is established on faunal analyses and not on some distinctive depositional event. In general, inland clastic sediments and shelfal carbonates were being deposited continually during a highstand of sea level in a progradational pattern throughout latest Tithonian and Berriasian times.

An erosional unconformity is recognized at the top of the Berriasian deposits throughout the shelfal areas of the northern Gulf of Mexico basin (McFarlan, 1977; Todd and Mitchum, 1977). This unconformity is marked by: (1) the truncation of Upper Cotton Valley beds along this surface, (2) the juxtaposition of overlying strandplain sandstones with underlying shelfal deposits, and (3) evidence of oxidation and leaching in areas with good core control such as east Texas. Information from wells and seismic surveys indicates that along the northern Gulf of Mexico basin, basinward from the shelf margin, in the deeper parts of the basin, clastic and carbonate deposition was continuous through the Valanginian and into the Hauterivian. This suggests uplift of the basin margin and/or a lowering of sea level during the Valanginian.

In the more basinal areas of Mexico, the sequence of Berriasian, Valanginian, and Hauterivian strata seems to be conformable, indicating continuous sedimentation during this time also in these areas. It is possible that these basinal areas were subsiding and that the possible effect of sea-level lowering, therefore, cannot be recognized in their Lower Cretaceous sections. In the platform areas, the Berriasian-Valanginian sediments are thin (Figs. 5 and 6), suggesting that possible hiatuses or disconformities of Valanginian age may as yet remain unrecognized.

Biostratigraphy and age

Throughout the Gulf of Mexico basin, the Berriasian and Valanginian open-marine deposits are dated primarily with planktonic microfossils. The more common forms used in the United States and Mexico include several species of *Nannoconus* and of the calpionellids *Calpionella, Calpionellopsis, Calpionellites Tintinnopsella, Remaniella, Lorenziella,* and *Crassicollaria.* Distinctive forms are *Calpionella alpina, C. elliptica, Calpionellopsis oblonga, C. simplex, Calpionellites darderi, Tintinnopsella carpathica, T. longa,* and *Lorenziella hungarica.*

In the platform carbonate sections, fossils are less abundant and generally nondiagnostic of age. The Berriasian-Valanginian faunas of northern Mexico were described by Imlay (1940b).

Conditions of deposition and paleogeography

During earliest Cretaceous time (Berriasian and Valanginian) the sea covered most of the area of the present Gulf of Mexico basin (Fig. 9). This extensive seaway had broad connections with the Pacific Ocean to the west and with the Atlantic Ocean to the southeast. To the north, the water-covered area was bordered by the emergent continental North America, with prominent uplands and intervening lowlands, along the latter, rivers draining the continental interior flowed into the ancestral Gulf of Mexico. To the south, the area of the present Yucatan Peninsula and much of the surrounding shelf, as well as the Chiapas massif, remained emergent. The continental shelves varied in width, and at their margins the sea floor sloped more steeply toward the deeper parts of the basin.

The continental shelf on the northern flank ranged from 80 to more than 480 km in width. In Mexico, the shelfal areas ranged in width from 80 to 200 km. For central and northeastern Mexico, the western extent of the shelf is uncertain due to the structural complexities of these areas (Lopez Ramos, 1982).

On the broad northern rim, the shelf was bordered to the north by a low-relief coastal plain 80 to 320 km wide. Streams

draining the continental interior and the Appalachian and Ouachita uplands carried clastic sediments to alluvial valley, deltaic, and interdeltaic environments of the coastal plain (Fig. 9). Where these streams reached the shoreline, the inner part of the shelf, to a water depth of 30 m, was primarily an area of clastic deposition. Beyond the influence of streams, carbonate sedimentation predominated on the outer shelf. The middle part of the shelf was marked by water depths that increased to more than 60 m, with the deeper water in the center of some local basins within the shelf. On the outer part of the shelf, bank-forming rudists, corals, and molluscs formed shoals that rose to water depths of 10 to 30 m. Greater water depths probably existed between discontinuous banks where tidal channels and passes were present (Fig. 9).

Along the northern and northwestern flanks of the present Gulf of Mexico basin, the boundary between the outer shelf margin and the inner slope was recognized not only by presence of the bank shoals, but also by a marked increase in dip into the basin. The slope changed abruptly from about 12 m per kilometer to about 45 m per kilometer. Along the gulfward margins of the West Florida and Yucatan shelves, the slope into the basin formed steep escarpments.

Berriasian clastic deposition in Mexico was related closely to emergent areas to the northwest and southeast of the basin (Fig. 9). The northwestern land areas included the Coahuila platform and the El Burro uplift. These land areas were surrounded by continental sediments and these, in turn, by shallow-water littoral sediments. The Sabinas basin filled with moderately deep-water shelfal sediments which graded basinward into deep basinal deposits to the east. In east-central Mexico, a similar pre-Cretaceous exposed land area was the foundation over which the Valles–San Luis Potosí platform was built. East of this platform, near the present coast of the Gulf of Mexico, the Tuxpan platform evolved over an older Permian-Triassic granitic pluton. Each of these platforms became an isolated area of shelfal carbonate aggradation. In the southeastern and southern part of the Gulf of Mexico basin, the Córdoba platform, west of the Veracruz basin, was the site of deposition of platform carbonates. It may have extended to the south as a narrow platform to the north of the Chiapas massif. The occurrence of extensive dacitic tuffs and bentonites in the Berriasian-Valanginian section of the Córdoba platform area and bentonites in other areas of southern Mexico indicate extensive volcanic activity probably along the present Pacific coast of Mexico, an area where there is considerable evidence of pre-Albian volcanism.

To the southeast, a northwest-southeast-trending shelfal basin that extended as far as Guatemala was bounded on the southwest by the Chiapas massif, and to the northeast by the Yucatan emerged landmass. In this basin, shelfal limestones and evaporites accumulated.

The Berriasian and Valanginian stratigraphy reflects a time of considerable tectonic stability. Subsidence, particularly of the central part of the basin, which started in Late Jurassic, continued during earliest Cretaceous time.

Economic considerations

The Cotton Valley Group contains predominantly gas accumulations along the northern flank of the Gulf of Mexico basin (see Chapter 15 of this volume). The amount of these resources that can be assigned to the upper, Lower Cretaceous, part of the group is probably of only modest importance. No other significant oil and gas accumulations in Berriasian and Valanginian sediments have been identified in the Gulf of Mexico basin.

HAUTERIVIAN-BARREMIAN-APTIAN

Distribution

Hauterivian to Aptian rocks are present all around the periphery of the Gulf of Mexico basin even though they only crop out in the northwestern and western flanks of the basin (Fig. 1). No reliably dated rocks of this age have been reported from the subsurface of the northern part of the Yucatan Peninsula, and the age of the lower part of the Cretaceous section in the area remains controversial.

Lithostratigraphy

In the northern flank of the Gulf of Mexico basin, the Hauterivian, Barremian, and lower part of the Aptian are represented by the subsurface Hosston Formation. This formation consists of terrigenous clastic sediments named from wells near the town of Hosston in Caddo Parish, Louisiana (Imlay, 1940a). The section includes a lower part of white, pink, red, and green silts and fine-to-coarse sandstones that grade into an upper part of red, gray, and green micaceous shale and silt. To the east, in Mississippi and Alabama, where large streams drained the continental interior and the Appalachian uplands, the Hosston Formation becomes more coarse-grained and graveliferous (McFarlan, 1977). To the west in Texas, the Hosston clastic section becomes finer grained and includes a larger amount of chert. In the East Texas basin, the Hauterivian-Barremian section is predominantly formed by fine- to coarse-grained clastic sediments contributed by streams flowing into the area from the northwest (Bushaw, 1968; McGowen and Harris, 1984). The Hosston has been locally called the Travis Peak Formation in the East Texas basin.

Throughout the northern Gulf of Mexico basin, the transgressive Hosston clastic section interfingers gulfward with, and is overlain by, the argillaceous and fossiliferous limestones of the Sligo Formation (Fig. 3A and B). These limestones aggrade and prograde the shelfal areas. The Sligo Formation was first recognized in the Sligo Field, Bossier Parish, Louisiana, as a gray-to-brown argillaceous limestone overlying the Hosston red and brown clastics (Hazzard, 1939; Imlay, 1940a; Martin and others, 1954). To the east and gulfward, in Louisiana, Mississippi, and Alabama, the Sligo Formation becomes a more massive, shallow-water, fossiliferous shelfal limestone. On the shelf margin, discontinuous rudist banks developed. In shoal areas on the tip of

structurally positive areas and around the rims of shelfal basins, oolitic limestones and rudist banks formed. Across the entire crest of the Sabine uplift, oolitic limestones were deposited in many layers interbedded with fossiliferous limestones (McFarlan, 1977; Fig. 10). This oolitic limestone section in east Texas and western Louisiana is a significant reservoir rock; in the East Texas basin it is known by the term "Pettet Limestone."

The deposition of Hosston clastic sediments and Sligo carbonates was continuous through Hauterivian, Barremian, and early Aptian times across the sequence updip, in an aggradational series of beds along the outer shelfal area, and as a progradational sequence along the outer shelf margin where rudist banks predominated (Fig. 3A and B; Bebout, 1977; Wooten and Dunaway, 1977; Bebout and others, 1981).

In late Aptian time, carbonate deposition along the northern Gulf of Mexico basin was succeeded by deposition of the predominantly terrigenous clastic section of the Pearsall Formation (Bushaw, 1968; Loucks, 1977). This unit, composed of shale and thin limestone layers, covered the Sligo carbonates from south Texas to northern Florida. The lower part of the Pearsall, the Pine Island Shale, was recognized and described from well data in the Pine Island Field, Caddo Parish, Louisiana (Weeks, 1938; Imlay, 1940a). It consists of about 120 m of dark, splintery shale interbedded with thin, dense gray limestones. The unit becomes sandy updip and more arenaceous and argillaceous in Mississippi and Alabama. In Texas, shale predominates in the Pine Island, which is also known locally as the Hammett Shale (Lozo and Stricklin, 1956).

Overlying the Pine Island Shale is the James Limestone. It is present from south Texas, where it is often called Cow Creek (Lozo and Stricklin, 1956; Stricklin, 1973), to north Florida. The James Limestone was first described from the subsurface of Union Parish, Louisiana, and consists of oolitic, coquinoidal, sandy gray limestone (Weeks, 1938; Imlay, 1944). To the east, in Mississippi and Alabama, this formation includes thin, fine-grained sand layers interbedded with the grain limestone. In Texas, the James Limestone is a dense, nonporous gray limestone interbedded with shale. Overlying the James and forming the upper part of the Pearsall Formation is the Bexar Shale.

In Florida and in the West Florida shelf, the Hauterivian through Aptian sediments, as known from numerous wells, consist of a thick sequence of carbonates and some interbedded evaporites that onlap the gently sloping flanks of the Peninsular arch. Generally, these beds thicken gulfward and pinch out against the arch (Applin and Applin, 1944, 1965). Near their pinchout the carbonate-evaporite section grades into a laterally equivalent clastic sequence (Fig. 10).

The Hauterivian part of the section, composed of tan oolitic and gray micritic limestones interbedded with anhydrite and dolomite, constitutes the Bone Island Formation (Applegate and others, 1981). The lower part of this thick formation is judged to be Valanginian in age. The Pumpkin Bay Formation overlies the Bone Island Formation; it has been considered to be of Barremian to early Aptian age. The Bone Island and lower part of the

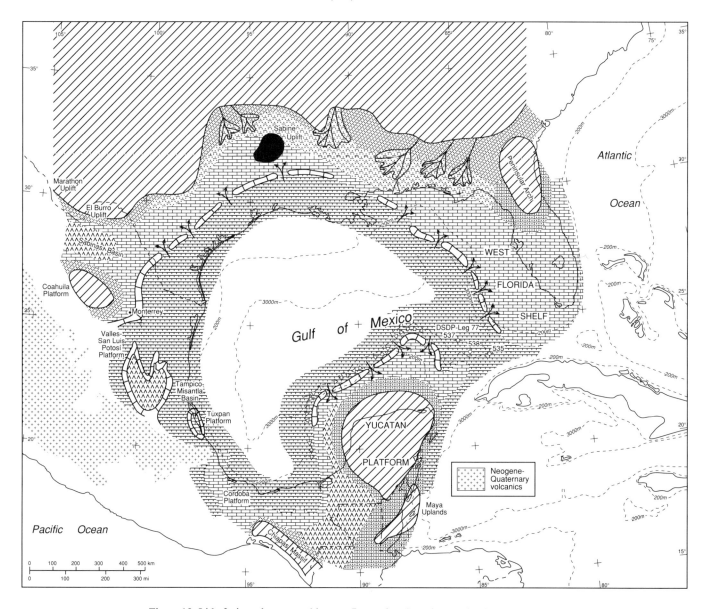

Figure 10. Lithofacies-paleogeographic map, Barremian. See Figure 8 for lithologic symbols.

Pumpkin Bay are equivalent to most of the Fort Pierce Formation of Applin and Applin (1965; Fig. 4A). The Pumpkin Bay Formation consists of tan micritic and oolitic limestones interbedded with vugular dolomite and anhydrite (Applegate and others, 1981). The Glades Group (Winston, 1971, 1976), of mid-Aptian age, overlies the Pumpkin Bay Formation and is divided in ascending order into the Lehigh Acres Formation and the Punta Gorda Anhydrite. According to Applegate and others (1981), the basal West Felda Shale Member of the Lehigh Acres Formation is regionally persistent and equivalent in age to the Pine Island Shale of the northern flank of the Gulf of Mexico basin. This shale is overlain by interbedded gray micritic limestones, brown, generally porous, dolomite and white anhydrite. The Punta Gorda Anhydrite is a 180-m-thick white anhydrite

with interbedded thin beds of fossiliferous limestones, dolomites, and shale (Applin and Applin, 1965). The Punta Gorda has often been correlated on the basis of lithologic similarity and nearly equivalent age with the Ferry Lake Anhydrite of the northern flank of the basin.

However, Braunstein and others (1988), in the COSUNA (Correlation of Stratigraphic Units of North America) stratigraphic correlation chart for the Gulf Coast Region, have shown the West Felda Shale to be older than the Pine Island Shale, and the Punta Gorda Anhydrite to be also older than the Ferry Lake Anhydrite. This interpretation has been accepted in this chapter and has been shown in Plate 5 of this volume. More recently, Scott (1990) favored the West Felda/Pine Island and Punta Gorda/Ferry Lake correspondence. He bases this correlation mainly on lithologic similarity and on his assignation of an early Albian age to the Sunniland Formation, which overlies the Punta Gorda Anhydrite. This age is supported by the presence in the Sunniland of *Paracoskinolina sunnilandensis* and *Orbitolina texana*.

The upper part of the Aptian in south Florida is the Ocean Reef Group (Winston, 1971, 1976), which is subdivided in ascending order into the Sunniland, Lake Trafford, and Rattlesnake Hammock Formations. The Sunniland Formation, overlying the Punta Gorda Anhydrite, includes dense, chalky, and fossiliferous limestones with porous bioclastic limestones and thin beds of dolomite and shale (Applin and Applin, 1965; Winston, 1976; Scott, 1990). The Lake Trafford Formation is primarily a gray, sparsely fossiliferous limestone with thin beds of anhydrite. These beds are overlain by a regionally persistent anhydrite with thin layers of limestone and dolomite; the Rattlesnake Hammock Formation.

In northeastern Mexico, the Hauterivian-Barremian-Aptian sequence is well known in the Sabinas basin (Fig. 10) where a large part of the section crops out (Fig. 1). Gulfward, many wells drilled in search of petroleum have also penetrated this section (Fig. 5A). The sequence is comparable to the time-equivalent section in the United States except for the absence of thick clastic sediments associated with large rivers.

Toward the rim of the Hauterivian depositional basin the section begins with the Patula arkosic red clastics, which onlap the Paleozoic metamorphic rocks (Smith, 1970). These beds grade basinward to the thin-bedded limestones of the Padilla Formation, which in turn, grade laterally into the thick-bedded limestones of the Cupido Formation (Bishop, 1970; Alfonso Zwanziger, 1978; Selvius and Wilson, 1985; Fig. 5A). Along the shelf margin, rudist carbonate banks are present in the Cupido section. These carbonate banks continued to build a prograding margin during Hauterivian, Barremian and early Aptian time with aggrading Cupido limestones on the landward side, and basinal micrites and pelagic shales of the Tamaulipas Inferior Formation on the gulfward side (Winker and Buffler, 1988). The shelf-margin and basin carbonates of the Cupido and Tamaulipas Inferior Formations are well known from outcrop studies near Monterrey where the progradational relations are clearly exposed (Conklin and Moore, 1977; Wilson and Pialli, 1977).

As the Cupido shelf carbonates accumulated along the margin of the shelf, increased erosion of the Coahuila platform and the El Burro uplift resulted in the deposition of the La Mula red silty shales and calcareous mudstones over the Patula and Padilla formations along the inner part of the shelf. The La Mula interfingers gulfward with the fossiliferous outer-shelf Cupido limestones (Smith, 1970; Bishop, 1970).

In the subsurface of the central part of the Sabinas basin, interbedded shallow-water limestones and evaporites of the La Virgen Formation were deposited on top of the La Mula shales and limestones (Fig. 5A) in late Barremian time. These beds interfinger gulfward with the shelfal Cupido limestones.

The uppermost Aptian section is composed of the shales and brown calcareous mudstones of the La Peña Formation, laterally equivalent to the Pearsall Formation of the northern flank of the basin. The La Peña indicates a significant influx of fine clastic sediments into the Sabinas basin and a measureable stratigraphic onlap over the inland parts of the basin (Smith, 1970; Bishop, 1970).

Subsurface data cross the Valles–San Luis Potosí platform reveal the deposition over the platform during Hauterivian, Barremian, and most of Aptian times of the thin, shallow-water, shelfal carbonates of the Guaxcamá Formation. These shelfal carbonates grade gulfward into the light brown, thick-bedded, open-marine limestones of the Tamaulipas Inferior of the Tampico-Misantla basin (Figs. 5B and 10). On the platform margins, fossiliferous grainstone banks developed, which formed a broad pear-shaped lagoon more than 200 km wide and 300 km in a north-south direction. The restriction of sea-water circulation in this lagoon resulted in the deposition of dolomites and evaporites interbedded with thin-bedded shallow-water limestones (Moya Cuevas, 1974).

The uppermost part of the Aptian section is characterized in the Tampico-Misantla basin and the adjacent Valles–San Luis Potosí platform by a thin interval of shales and calcareous shales, the Otates Formation (Muir, 1936), equivalent to the La Peña and Pearsall Formations of northeastern Mexico and the United States.

During Hauterivian through Aptian times, the Tuxpan and Córdoba platforms were the sites for the accumulation of thin shallow-water limestones and evaporites surrounded by platform-margin reefs (Fig. 6A and B). These reefal beds grade laterally into the basinal limestones of the Tamaulipas Inferior Formation of the Tampico-Misantla and Veracruz basins.

In the Isthmus of Tehuantepec area, basinal shales and calcareous shales of the upper part of the Chinameca Formation grade southward to platform limestones (upper part of The Zacatera Group), and then, along the foothills of the Chiapas Massif, to the continental and coastal "red beds" of the Todos Santos Formation (Fig. 10). In northern Guatemala and adjacent regions of southern Mexico, the Hauterivian-Aptian section is characterized by a sequence of interbedded evaporites, dolomites, and limestones called the Cobán Formation.

How much of the present Yucatan Peninsula was the site of deposition of carbonates and evaporites during Hauterivian to Aptian times is still a controversial subject. In the absence of published fossil evidence, the age of the basal part of the Cretaceous section in the area is not known. Lopez Ramos (1975) believes that the northern part of the peninsula was not covered by the sea until the Albian or Cenomanian. Viniegra O. (1971, 1981), on the other hand, believes that the whole area of the Yucatan Peninsula was covered by shallow water during most of the Early Cretaceos, and postulates that deposition of carbonates and evaporites in the area started as early as the Neocomian (Berriasian-Barremian).

The northern part of the present Yucatan Peninsula and adjacent parts of the shelf are interpreted here as having an exposed flat upland during Hauterivian to Aptian time (Fig. 10). A narrow belt of coastal and tidal-flat sandstones and siltstones constitutes the shoreline deposits of the emergent area. These coastal sediments grade seaward into shelf-lagoon shallow-water limestones with layers of anhydrite. On the 50- to 75-km wide shelf, fossiliferous carbonate banks developed along the shelf margin. These shallow-water limestones grade gulfward into deep-water limestones with dolomitic talus debris (Viniegra O., 1981).

Between the Yucatan and Florida platforms, alternating gray laminated to bioturbated marly limestones, interpreted to be Hauterivian-Aptian in age and to have been deposited in a deep-water basinal environment, were recovered at sites 535, 537, and 538 of Leg 77 of the Deep Sea Drilling Project (Schlager and Buffler, 1984).

Thickness

The Hauterivian, Barremian, and Aptian sediments generally pinch out updip along the rims of the Gulf of Mexico basin. Downdip these sediments thicken to a maximum along the shelf margin that varies from 1,200 to 1,500 m in the United States (Figs. 3A, B, 4A, B). In northeastern Mexico, these sediments attain a thickness of about 1,700 m along the marine shelf (Fig. 5A). In the southeastern evaporite basin, thicknesses are more than 1,500 m (Lopez Ramos, 1975).

In the platform areas of Mexico, the structurally high settings during deposition caused the Hauterivian through Aptian section to be thin when compared to the shelf margins. Thicknesses on platforms range from about 150 m on the Tuxpan platform to about 600 m on the Córdoba platform (Figs. 6A, B).

Stratigraphic relations

Along the updip part of the northern flank of the Gulf of Mexico basin, the Hauterivian deposits are separated from the underlying Berriasian coastal and shelfal deposits by an erosional unconformity. Strandplain Hauterivian sandstones rest on Berriasian shallow shelfal clastics and carbonates along an irregular surface. Basinward, beyond the shelf margin, a wedge of Valanginian deposits is present between the Berriasian and the Hauterivian, and the unconformity at the base of the Hauterivian grades to a conformable and gradational contact.

In Mexico, the Hauterivian sediments have been reported to lie conformably on Valanginian deposits in both shelfal and basinal areas. However, the reduced thickness of the Lower Cretaceous deposits on many platforms in Mexico may obscure possible unconformable relations at the base of the Hauterivian section in these platforms.

The Hauterivian to Aptian section appears to lack major internal unconformities and hiatuses over most of the Gulf of Mexico basin. The contact with the overlying Albian sediments seems everywhere conformable. A marked change from the fine-grained clastic sediments of the Aptian to the overlying Albian carbonates occurs in the shelfal areas of the northern and northwestern flanks of the Gulf of Mexico basin, but the contact appears to be gradational and conformable.

Biostratigraphy and age

Marine Hauterivian to Aptian sediments of the Gulf of Mexico basin have been dated with the help of an abundant and diagnostic fauna (Alencaster, 1984). Particularly abundant are ammonites, nannoconids, and foraminifera.

In Mexico, the occurrence of numerous species of *Nannoconus* and planktonic foraminifera have been particularly useful in dating the open-marine Hauterivian to Aptian section in the subsurface. Diagnostic of the Hauterivian to Aptian section are *Nannoconus steinmanni, N. kamptneri, N. bermudezi, N. colomi, N. boneti, N. wassalli, N. bucheri,* and *N. globulus* (Trejo, 1975). Planktonic foraminifera are more abundant and useful in dating the Aptian sediments, among them several species of *Globigerinelloides, Hedbergella,* and *Leupoldina.* Also diagnostic of the Aptian is the calpionellid *Colomiella mexicana.*

In the platform carbonate sections, species of *Choffatella* and *Orbitolina* have been used mainly to date the Hauterivian to Aptian sediments.

In outcrop, the Hauterivian to Aptian beds commonly have been dated by means of their contained ammonite fauna (Imlay, 1944, 1945; Young, 1974). Table 1 shows the ammonite zonation of the Aptian section proposed by Young (1972, 1974).

Conditions of deposition and paleogeography

After Valanginian time there was a transgression of the shoreline along the northern flank of the Gulf of Mexico basin, as the sea covered again the former coastal plain. Basal Hauterivian clastic sediments are associated with continued regional subsidence and a slow eustatic rise in sea level. The clastic sediments were derived from the adjacent upland areas to the north and deposited in deltaic, interdeltaic, beach, and shelfal marine environments. In later Hauterivian and early Barremian times, regional subsidence and eustatic sea-level rise continued. Coastal plains became drowned, and the quantity of clastic sediments was reduced in nearshore areas. Gulfward in the shelfal areas there was a change from clastic to carbonate deposition with development of limestone banks along an increasingly distinct shelf margin (Fig. 10). The continental-slope margin continued to be dominated by the deposition of pelagic limestones, calcareous mudstones, and gray shales.

This depositional cycle ended in Aptian time with the cessation of carbonate deposition probably caused by the influx of the fine-grained terrigenous clastic sediments of the Pearsall, La Peña, and Otates Formations around the northern and northwestern margins of the basin. The marked change in deposition is attributed, in part, to regional inland uplift and shelfal downwarping, and in part to a fluctuation in eustatic sea level (Smith, 1970; McFarlan, 1977).

Except for the northern area, which received a continuing

influx of terrigenous clastic sediments, the rims of the Gulf of Mexico basin were primarily areas of carbonate deposition during Hauterivian, Barremian, and Aptian times (Fig. 10). Most of the carbonates were deposited in broad, shallow-water shelfal environments and deeper-water basinal settings. Regional subsidence and a slow eustatic rise in sea level continued, causing an onlapping of sedimentary units inland. On the shelves, carbonate deposition was rapid enough to maintain shallow-water deposition. In fact, the accumulation of shelfal carbonates and the continuous development of reef-bank organisms at the shelf margin resulted in 20 to 60 km of gulfward progradation on the northern and western rims. On the West Florida shelf and the Yucatán shelf, aggradation of carbonates caused a nearly vertical escarpment to develop along the outer shelf margin.

In Mexico, the initial stages of shelf, reef, and lagoon deposition occurred on the shelfal platforms. The Hauterivian through Barremian sediments record such carbonate deposition on the Córdoba, Tuxpan, and Valles–San Luis Potosí platforms. Evaporites were deposited in the large lagoonal area behind the reefs along the margins of the Valles–San Luis Potosí platform. Water depths in the shelfal basins of southern Mexico were so shallow and water circulation so restricted by reefs and banks that thick evaporites accumulated over wide areas (Fig. 10).

In northeastern Mexico, shelf carbonates grade inland into clastic deposits derived from the adjacent uplands. In Hauterivian and Barremian time, inland uplift was accelerated and there was a large influx of the terrigenous clastic sediments of the Hosston, Patula, and La Mula Formations (Fig. 10). Along the northern rim, the deposition of clastics in alluvial valley, deltaic, prodelta, and interdeltaic environments formed a broad coastal plain 160 to 400 km wide crossed by many stream systems. Along the southern, western, and eastern rims of the basin, smaller stream systems on uplands brought a lesser influx of terrigenous clastics to the sea. As a result the coastal beach, tidal, deltaic, and interdeltaic depositional environments were narrow, about 50 to 100 km wide.

The reduction in clastic influx in Barremian time suggests that inland streams around the Gulf of Mexico basin were partly drowned by continued regional subsidence and sea-level rise. In addition, a period of inland structural stability probably led to reduced runoff of the streams entering the Gulf of Mexico.

In late Aptian time, the widespread deposition of dark fossiliferous Pearsall, La Peña, and Otates marine shales indicates a rejuvenation of clastic influx and a slight deepening of the paleo-water depths on the shelves. The accumulation of carbonates and the development of reefs and banks stopped abruptly, probably due to the influx of the terrigenous clastic sediments, except in west Florida, Yucatán, and other shelves far removed from clastic sedimentation.

Economic considerations

Porous Hosston nearshore sandstones and Sligo and Cupido shelfal and reefal carbonate grainstones are known to be reservoir rocks for small amounts of petroleum. Most of these accumula-

tions are in lower Hosston sandstones on the Sabine uplift and on the crests and flanks of positive Jurassic salt features. The Sligo and Cupido shelf-margin reef trend of Aptian age contains small oil and gas fields in Texas and northeastern Mexico. The trend of small fields in south Florida produces oil from the skeletal limestones of the Sunniland Formation of latest Aptian or early Albian age.

ALBIAN–LOWER CENOMANIAN

Distribution

As in case of the older Lower Cretaceous, the Albian–lower Cenomanian sediments crop out only along the northwestern and western flanks of the Gulf of Mexico basin (Fig. 1). Rocks of this age, however, have been reported to occur in wells from all around the periphery of the basin. Albian and lower Cenomanian beds cover the entire Yucatan and Coahuila platforms, extend northwestward to the Texas and Oklahoma panhandles, eastern New Mexico, southeastern Colorado and Kansas, and are believed to be present over the central, deeper part of the basin.

Lithostratigraphy

Across the shelfal areas of northeastern Mexico, Texas, Louisiana, Arkansas, Mississippi, and Alabama, the basal Albian beds are represented by the limestones of the Glen Rose Formation, originally named by Hill (1891) for the town of Glen Rose, Somervell County, Texas. In the Sabinas basin, these rocks are often referred to as the Tamaulipas Superior Formation, and consist of outer-shelf marls, thin-bedded limestones, and gray skeletal limestones with an abundant marine fauna, which includes ammonites, echinoids, rudists, molluscs, and foraminifera (Smith, 1970). Northward in south Texas, the Glen Rose changes to a gray argillaceous dolomite and micrite with anhydrite layers (Rose, 1972). In the East Texas basin and the shelfal basins of Louisiana, Mississippi, and Alabama, the Glen Rose is subdivided in ascending order into the Rodessa Limestone, the Ferry Lake Anhydrite, and the Rusk (Mooringsport) Limestone (Murray, 1961). The Rodessa Limestone is a gray, arenaceous, argillaceous, oolitic, skeletal limestone interbedded with thin sandstone and shale layers (Weeks, 1938; Hazzard, 1939; Imlay, 1940a). The massive, white Ferry Lake Anhydrite (Imlay, 1940a), interbedded with thin limestones and dolomites, overlies the Rodessa Formation. In turn, the Ferry Lake is overlain by the Rusk (Mooringsport) Formation, which is composed of oolitic limestones interbedded with shale, sandstone, and marl (Imlay, 1940a).

All of the Glen Rose shelfal carbonates described above interfinger updip with terrigenous clastics that onlap pre-Cretaceous rocks (Stricklin and others, 1971). Generally, these sandstones and shales are red, brown, and gray, and nonfossiliferous.

Downdip and gulfward, the shelfal Glen Rose beds interfinger with prograding shelf-margin reefal limestones containing rudists, corals, molluscs, and other shallow-water bank fauna. This lithology is generally continuous, distinctive, and mappable. The name Stuart City Formation has been applied throughout the northern rim of the Gulf of Mexico basin to this reefal unit (Bebout and Loucks, 1974; Scott, 1990).

Across the northern rim of the Gulf of Mexico basin, the Glen Rose limestones are overlain unconformably by the Fredericksburg Group predominantly composed of shelfal limestones with a transgressive onlapping pattern. In parts of the Sabinas basin of northeastern Mexico, beds equivalent to the Fredericksburg Group include the Telephone Canyon clayey marls and nodular limestones and the overlying West Nueces miliolid and turritellid limestones (Smith, 1981). In south Texas and on the San Marcos arch of central Texas, the Fredericksburg Group includes the Edwards Limestone, which is subdivided into the transgressive West Nueces limestones, and the younger McKnight evaporites to the south and the Kainer massive dolomitic micrites and the Person biomicrites and rudist grainstones to the north (Rose, 1972). The McKnight evaporites and limestones occupy a subsiding shelfal basin that developed in northeastern Mexico and southern Texas (Fig. 11). These lagoonal evaporites interfinger around the western and northern rims of this shelfal basin with the Devils River caprinid mounds and miliolid limestones and on the eastern and southern rims with the Stuart City shelf-margin rudist-reef trend (Rose, 1972; Bebout, 1974; Wooten and Dunaway, 1977; Cook, 1979; Smith, 1981).

Along the northern shelfal area from east Texas to west-central Mississippi, the Fredericksburg Group is divided in ascending order into the Paluxy, Goodland, and Kiamichi Formations. The Paluxy Formation is composed of nonfossiliferous, red, brown, and gray sandstones and shales (Hill, 1894). This clastic section is overlain by the white chalky limestones and pelletal marls of the Goodland Formation (Hill, 1891; Adkins, 1933). The Kiamichi Formation is generally a dark calcareous clay interbedded with yellow, thin-bedded limestones (Nunnally and Fowler, 1954). The Fredericksburg Group grades updip into red and brown mudstones, shales, and fine-to-medium-grained sandstones. Downdip the entire group interfingers with the Stuart City rudist-reef trend along the shelf margin, which in turn, grades farther gulfward with the basinal micrites and shales of the Atascosa Formation (Winter, 1961a, b).

Fredericksburg time was terminated across the northern rim of the Gulf of Mexico basin by gradual inland uplift and shoreline regression at the beginning of Washita time, as indicated by some updip evidence of exposure and erosion of Fredericksburg deposits (Rose, 1972). Downdip deposition persisted as sea level continued to rise slowly and the Washita Group of sediments accumulated. The Washita Group includes, from older to younger, the Georgetown, Del Rio, and Buda Formations. The Georgetown is of latest Albian age, and the Del Rio and Buda are early Cenomanian. In the Sabinas basin, the Georgetown equivalents are the Devils River, the Salmon Peak, and the Las Pilas Formations. The caprinid mounds and miliolid limestones of the Devils River are replaced gulfward by the Salmon Peak calcareous mudstones, which grade farther gulfward into the Las Pilas caprinid mounds and skeletal grainstones along the shelf margin (Smith, 1981).

After the period of updip exposure during late Fredericksburg time, Washita flooding of the northern rim was recorded by the deposition of the Georgetown limestones. The outcrop studies and original descriptions are reviewed by Adkins (1933). In general, brown, gray, argillaceous micrites and skeletal limestones with *Gryphea washitensis* in south and central Texas (Rose, 1972) change to gray and yellow micritic limestones with gray shales and marls and, in places, with cream-colored chalky limestones in the shelfal basins of the northern rim of the basin in Louisiana and Mississippi (Murray, 1961).

Across the entire northern rim, the Georgetown limestones grade basinward into the skeletal limestones of the Stuart City reef trend along the outer shelf margin.

A period of inland uplift resulted in the deposition of primarily marine shales across the entire northern rim in middle Washita time (earliest Cenomanian). In northeastern Mexico as well as in south and central Texas, this marine-shale section, the Del Rio Formation, is a thin gray calcareous shale with molluscs and foraminifera (Rose, 1972). In east Texas and eastward into Louisiana and Mississippi, an equivalent thin unit of flaggy, dark gray shale has been called the Grayson Formation (Adkins, 1933).

The Del Rio–Grayson shales are overlain across the northern rim of the Gulf of Mexico Basin by the onlapping Buda limestones marking a brief period of flooding near the end of Washita time. In northern Mexico, the Buda is a white calcareous mudstone and nodular wackestone (Smith, 1970), which changes to a dark brown argillaceous, fossiliferous micrite in south and central Texas (Rose, 1972).

Farther to the east, the Buda Formation is a light gray, organic-rich nodular limestone that is fossiliferous in places (Adkins, 1933).

The youngest Washita deposits on the northern rim of the Gulf of Mexico basin are the bronze- and copper-colored shales of the Maness Formation of the East Texas basin (Hazzard and others, 1947) and the Dantzler red shales and sands that overlie the Washita limestones in west-central Mississippi and southern Alabama (Hazzard and others, 1947). Both of these formations represent deposition in basinal areas not affected by mid-Cenomanian uplift and/or sea-level drop, and subsequent erosion. (The Del Rio–Grayson, Buda, Maness, and Dantzler Formations are discussed in Chapter 10, Upper Cretaceous, of this volume).

In south Florida and the West Florida shelf, the Albian was a period of uninterrupted carbonate and evaporite deposition. Deposits of this age are known only in the subsurface. The section is divided, into the lower Big Cypress Group and the overlying Naples Bay Group (Winston, 1971, 1976). They have been dated based on fauna identified by Applin and Applin (1965). The Big Cypress Group includes, from oldest to youngest, the Marco Junction dense, chalky limestones with layers of dolomite and anhydrite, the Gordon Pass anhydrite with dense micritic limestone, and the Dollar Bay alternating limestone, anhydrite, and dolomite.

The overlying Naples Bay Group consists of the following formations, in ascending order: The Panther Camp pelletal micrite with thin anhydrite layers; the Rookery Bay anhydrite with pelletal micrite and chalky limestone beds; and the Corkscrew Swamp alternating limestones, anhydrites, and dolomites. All of the formations are mappable throughout the Florida platform (Winston, 1971 1976).

A possible erosional unconformity between the Naples Bay Group and the overlying Atkinson Formation, of mid-Cenomanian to Turonian age, has been generally recognized (Gohn, 1988). This unconformity is interpreted to represent an erosional episode of early to mid-Cenomanian age.

Along the margin of the West Florida shelf, a reefal trend rims the shelf and forms the Florida Escarpment (Bryant and others, 1969).

In the southern part of the Sabinas basin and the Burgos basin, the Tamaulipas Superior and Cuesta del Cura Formations (Albian and lower Cenomanian) are represented by a shallowing-upward sequence of pelagic limestones, dense limestones, dolomites with anhydrites, and miliolid limestones wioth biostromal elements (Bishop, 1970). The younger units grade laterally along the margins of the Coahuila platform to reefal limestones composed of rudistid carbonates (Garza Gonzalez, 1973).

In east-central Mexico, the Tuxpan platform was the site of a large Early Cretaceous atoll of Albian and early Cenomanian age resting on a Late Jurassic positive element (Viniegra O. and Castillo-Tejero, 1970). The El Abra Formation, more than 1,000 m thick, consists of massive, light gray rudist limestone with abundant fauna. This limestone is fractured, has irregular bedding and exhibits a karst topography. The reef lithofacies interfingers toward the center of the platform with a lagoonal sequence composed primarily of miliolid limestones with layers of mudstone and anhydrite (Cabrera and Menes, 1973). A fore-reef section is absent. The platform slopes to the southeast, and the reefal section has an eroded crest on which beds as young as Oligocene rest. Upper Cretaceous and Eocene beds are missing (Viniegra O. and Castillo-Tejero, 1970).

At the foot of the steep reef flank lies the Tamabra Formation (Barnetche and Illing, 1956; Becerra, 1970; Enos, 1977), which includes caprinid and rudist limestones with abundant pelecypods and corals. This may be reef debris or an indigenous early reef facies that did not persist in Albian time. A normal fault has been postulated between the El Abra reef and the Tamabra Formation (Coogan and others, 1972).

The El Abra and Tamabra Formations interfinger away from the Tuxpan platform with the basinal, dark, thin-bedded, calcareous mudstones and micrites of the Tamaulipas Superior Formation.

A reefal complex composed of biogenic limestones with rudistids (caprinids), corals, and coraline algae, interbedded with biogenic calcarenites developed along the margins of the Valles–San Luis Potosí platform (Carillo Bravo, 1969, 1971; Carrasco,

1970, 1971, 1977; Enos, 1974). In the center of the platform, gray-to-cream-colored limestones containing abundant miliolids, pelecypods, ostracods, and gastropods were deposited together with beds of evaporites. In the forereef zone, the products of reef destruction and erosion accumulated as talus deposits composed of biogenic fragments. The fragments occur as gravity-slide and turbidite-type deposits in the Tamabra Formation. Such debris sequences interfinger with the deep-water argillaceous limestones in the basal part of the Tamaulipas Superior Formation (Fig. 5B).

In the southwestern part of the Gulf of Mexico basin, the Córdoba platform also contains a carbonate sequence similar to that of the Valles–San Luis Potosí and Tuxpan platforms (Gonzalez Alvarado, 1976). In the central part of the platform, the Albian section contains lagoonal sediments including thick carbonates with miliolids and ostracods, and a sequence of anhydrites, dolomites, and dolomitized limestones (Fig. 6B). The eastern margin of the Córdoba platform is poorly known due to structural complications.

Albian–lower Cenomanian sediments predominantly composed of limestones and dolomites (lower part of the Sierra Madre Group) are present throughout the Sierra Madre del Sur and the Sierra de Chiapas (Mora and others, 1975). Toward the north these platform carbonates grade to open marine, dark brown and black, laminated argillaceous limestones.

In the Yucatán platform, the Albian and lower Cenomanian are represented by the dolomites, evaporites, and limestones with abundant rudists, miliolids, valvulinids, and ostracods of the lower part of the Yucatán Formation (Viniegra O., 1981). This platform section interfingers gulfward with deep-water micrites and pelagic shales with abundant planktonic foraminifera. The Albian carbonate sequence along the Campeche Escarpment is probably composed of reefal barriers similar to those developed along the margins of the Valles–San Luis Potosí, Tuxpan, and Florida platforms (Bryant and others, 1969).

Thickness

Albian and lower Cenomanian deposits thicken from their pinchout along the rims of the Gulf of Mexico basin, across the shelves to the reef complexes and shelf-margin banks of the time. Along the shelf margins, the beds range in thickness from 1,500 m off Yucatán on the Campeche shelf to 900 m on the West Florida shelf and to 600 m on the Texas shelf. In the shelfal basins, 500 m of Albian sediments were deposited in the basins of southern Mexico (Fig. 7), 600 m in the Sabinas basin (Fig. 5A), and 800 to 1,100 m in the shelfal basins of the northern flank of the basin. With regional subsidence and rise in sea level, platforms in Mexico received thicker-than-normal deposits in Albian time. More than 800 m of carbonates and evaporites accumulated on the Tuxpan and Córdoba platforms (Figs. 6A and 6B). On the Yucatan platform, Albian carbonates reach thicknesses of about 1,900 m. Seismic data indicate the presence of thin Albian sediments in the central part of the Gulf of Mexico basin.

Stratigraphic relations

Except in the Yucatan Peninsula, where a carbonate-evaporite section believed to be Albian in age rests unconformably over older "red beds" and "basement" rocks, the Albian section overlies conformably the Aptian throughout the Gulf of Mexico basin. The boundary between the Albian and the overlying Cenomanian section is also conformable and transitional.

The Albian-Cenomanian boundary is also the boundary between the Lower Cretaceous and the Upper Cretaceous, as recognized internationally. However, as mentioned earlier, along the northern and northwestern flanks of the Gulf of Mexico basin, the Lower Cretaceous–Upper Cretaceous boundary has been arbitrarily placed at an important regional unconformity within the Cenomanian section—at the base of the Woodbine, Tuscaloosa, Eagle Ford, and Atkinson Formations.

Biostratigraphy and age

Marine Albian rocks have been well dated on the basis of their abundant contained fossils (Alencaster, 1984). Ammonites have been used predominantly in surface sections (Imlay, 1944, 1945; Young, 1974). Table 1 shows the ammonite zonation of the Albian section of Texas and Mexico proposed by Young (1974) on the basis of ammonites. Common forms are *Drakeoceras drakei, Mortoniceras wintoni,* and *Adkinsites bravoensis.*

In the subsurface the Albian section has been dated most frequently by means of planktonic foraminifera (Longoria and Gamper, 1977), and calpionellids. Common among the planktonic foraminifera are several species of *Hedbergella, Globigerinoides, Favusella,* and *Ticinella;* several species of the calpionellid *Colomiella* are also common in the Albian section of the Gulf of Mexico basin.

Rich and diverse rudist faunas have been most useful in dating the Lower Cretaceous reefs and banks in the Gulf of Mexico basin, particularly the extensive and well-developed Albian reefs (Coogan, 1977).

The biostratigraphy of the Albian-Cenomanian boundary and of the lower Cenomanian section is discussed in Chapter 10 of this volume.

Conditions of deposition and paleogeography

Albian sediments in the Gulf of Mexico basin record an apparent rise in sea level. The Yucatan and Coahuila platforms were covered by shallow water, and the ancestral Gulf of Mexico was part of a broad seaway with connections not only with the Atlantic and Pacific Oceans, but also with the Western Interior Seaway through north Texas, eastern New Mexico, and southeastern Colorado (Fig. 11). The Albian section all around the periphery of the present Gulf of Mexico basin is made up predominantly of carbonates. Evaporites are found interbedded with the carbonates in the stable Florida, Yucatan, Córdoba, Tuxpan, Valles–San Luis Potosí, and Coahuila platforms, as well as in the

Maverick basin. All these areas were covered by very shallow water where banks, reefs, and shoals restricted marine circulation. Evaporitic sections were particularly well developed in the Yucatan Platform where its central area became a large evaporite lagoon surrounded by shelfal areas of shallow-water limestone deposition. Terrigenous clastic sediments were still being contributed to the basin along its northern flank by a few major streams draining the continental interior and the Appalachian uplands, even though the rise in sea level had drowned smaller rivers and the lower courses of the larger streams. A large alluvial and deltaic plain was present from central Mississippi to southern Georgia and the Florida panhandle. The delta of the ancestral Mississippi was a prominent feature of this deltaic plain. Some of the deltas advanced across the shelf to a position close to the shelf margin. Westward-flowing long-shore currents carried the clastic load along the shelf away from the mouths of the rivers for several tens of kilometers.

The maximum extent of development of carbonate platforms in the Gulf of Mexico basin took place during Albian time. Important rudist and other mollusk reefs and carbonate banks developed along the margins of the carbonate shelves and platforms. Depending on their environment of deposition and tectonic associations, these carbonate buildups developed a variety of styles: aggradation, progradation, or a combination of these processes (Winker and Buffler, 1988; Scott, 1990). Reefs and carbonate banks were less well developed or absent in major embayments of the shelf, as in southern Mexico, where high-energy conditions were lacking. On sea-floor mounds caused by underlying structural uplifts, oolitic and skeletal limestones formed shallow shoals.

Beyond the shelf and platform margins, the deeper parts of the basin received fine-grained limestone and pelagic shale during the Albian. The slope into the deeper part of the basin was greatest along the Florida and Campeche Escarpments. Rudist-bank limestones of Albian age have been dredged from the surface of the Florida Escarpment (Bryant and others, 1969). Along the Campeche Escarpment, shelf-margin carbonate debris flowed down the slope (Viniegra O., 1981).

The conditions of deposition and paleogeographic elements during the early Cenomanian were similar to those of the Albian. The sea covering the present area of the Gulf of Mexico basin still extended northwestward toward the Western Interior Seaway, and terrigenous clastic sediments continued to be contributed to the basin from the north. A strong pulse of fine-grained clastic material deposited the Del Rio–Grayson sequence during earliest Cenomanian over most of the northern flank of the basin, possibly contributing to the demise of the Stuart City reef. (For further discussion of the geologic developments during the early Cenomanian, see Chapter 10 of this volume.)

Economic considerations

Albian–lower Cenomanian shelf-margin and platform-margin porous reef limestones constitute important reservoirs in Mexico. The oil fields of the Golden Lane trend rimming the

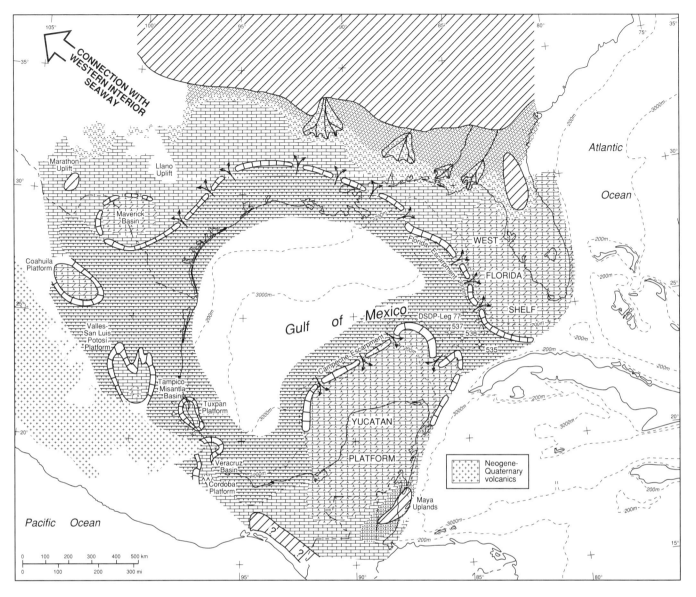

Figure 11. Lithofacies-paleogeographic map, Albian. See Figure 8 for lithologic symbols.

Tuxpan platform and the Poza Rica field are considered among the world's giant oil fields. They produce respectively from the porous rudistid-reef limestones in the El Abra Formation and from the Tamabra Formation (see Chapter 15, this volume). In southern Mexico, Albian–lower Cenomanian carbonates are also important reservoirs. In the Reforma area the A. J. Bermudez, Cactus/Nispero, Sitio Grande, and numerous other important fields produce oil and gas from fractured dolomites on large faulted domes probably underlain by salt (Santiago Acevedo, 1980; Santiago Acevedo and Mejia Dautt, 1980; Chapter 15, this volume).

In the United States, Albian rudist-bank skeletal limestones produce gas in the Stuart City field, La Salle County, Texas, along the outer margin of the Albian shelf.

REFERENCES CITED

Adkins, W. S., 1933, The Mesozoic systems in Texas, *in* The geology of Texas; V. 1, Stratigraphy: University of Texas Bulletin 3232, p. 239–517.

Alencaster, G., 1984, Late Jurassic–Cretaceous molluscan paleogeography of the southern half of Mexico, *in* Westermann, G.E.G., ed., Jurassic-Cretaceous biochronology and paleogeography of North America: Geological Association of Canada Special Paper 27, p. 77–88.

Alfonso, Zwanziger, J., 1978, Geología regional del sistema sedimentario Cupido: Asociación Mexicana de Geólogos Petroleros Boletín, v. 30, nos. 1–2, p. 1–55.

Anderson, E. G., 1979, Basic Mesozoic study in Louisiana, the northern coastal region, and the Gulf Basin province: Louisiana Geological Survey Folio Series 3, 58 p.

Applegate, A. V., Winston, G. O., and Palacas, J. G., 1981, Subdivision and regional stratigraphy of the pre-Punta Gorda Rocks (lowermost Cretaceous-

Jurassic?) in south Florida: Gulf Coast Association of Geological Societies Transactions, v. 31, p. 447–453.

Applin, P. L., and Applin, E. R., 1944, Regional subsurface stratigraphy and structure of Florida and southern Georgia: American Association of Petroleum Geologists Bulletin, v. 28, p. 1673–1753.

——, 1965, The Comanche Series and associated rocks in the subsurface in central and south Florida: U.S. Geological Survey Professional Paper 447, 84 p.

Barnetche, A., and Illing, L. V., 1956, The Tamabra Limestone of the Poza Rica Oilfield: 20th International Geological Congress, Mexico, 38 p.

Bateson, J. H., 1972, New interpretation of the geology of Maya Mountains, British Honduras: American Association of Petroleum Geologists Bulletin, v. 56, p. 956–963.

Bay, T. A., Jr., 1977, Lower Cretaceous stratigraphic models from Texas and Mexico, in Bebout, D. G., and Loucks, R. D., eds., Cretaceous carbonates of Texas and Mexico; Applications to subsurface exploration: University of Texas at Austin Bureau of Economic Geology Report of Investigations 89, p. 12–30.

Bebout, D. G., 1974, Lower Cretaceous Stuart City shelf margin of south Texas; Its depositional and diagenetic environments and their relationship to porosity: Gulf Coast Association of Geological Societies Transactions, v. 24, p. 138–159.

——, 1977, Sligo and Hosston depositional patterns, subsurface of south Texas, in Bebout, D. G., and Loucks, R. D., eds., Cretaceous carbonates of Texas and Mexico; Applications to subsurface exploration: University of Texas at Austin Bureau of Economic Geology Report of Investigations 89, p. 79–96.

Bebout, D. G., and Loucks, R. G., 1974, Stuart City trend, Lower Cretaceous, south Texas; A carbonate shelf-margin model for hydrocarbon exploration: University of Texas at Austin Bureau of Economic Geology Report of Investigations 78, 80 p.

——, eds., 1977, Cretaceous carbonates of Texas and Mexico; Applications to subsurface exploration: University of Texas at Austin Bureau of Economic Geology Report of Investigations 89, 332 p.

Bebout, D. G., Budd, D. A., and Schatzinger, R. A., 1981, Depositional and diagenetic history of the Sligo and Hosston Formations (Lower Cretaceous) in south Texas: University of Texas at Austin Bureau of Economic Geology Report of Investigations 109, 70 p.

Becerra, H. A., 1970, Estudio biostratigráfico de la Formación Tamabra del Cretácico en el Distrito de Poza Rica: Instituto Mexicano del Petroleo Revista, v. 2, no. 3, p. 21–39.

Bishop, B. A., 1970, Stratigraphy of Sierra de Picachos and vicinity, Nuevo Leon, Mexico: American Association of Petroleum Geologists Bulletin, v. 54, p. 1245–1270.

Blauser, W. H., and McNulty, C. L., 1980, Calpionellids and nannoconids of the Taraises Formation (Early Cretaceous) in Santa Rosa Canyon, Sierra de Santa Rosa, Nuevo Leon, Mexico: Gulf Coast Association of Geological Societies Transactions, v. 30, p. 263–272.

Bonet, F., 1952, La Facies Urgoniana del Cretácico Medio de la Region de Tampico: Asociación Mexicana de Geólogos Petroleros Boletín, v. 4, p. 153–262.

——, 1956, Zonificación microfaunística de las calizas cretácicas del Este de Mexico: Asociación Mexicana de Geólogos Petroleros Boletín, v. 8, nos. 7–8, p. 389–487.

Bose, E., 1923, Vestiges of an ancient continent in northeast Mexico: American Journal of Science, 5th ser., v. 206, p. 127–136, 196–214, 310–337.

Bose, E., and Cavins, O. A., 1927, The Cretaceous and Tertiary of southern Texas and northern Mexico: University of Texas Bulletin 2748, p. 7–142.

Braunstein, J., Huddlestun, P., and Biel, R., coordinators, 1988, Correlation chart of Gulf Coast Region: American Association of Petroleum Geologists, Correlation of Stratigraphic Units of North America (COSUNA) Project.

Bryant, W. R., and 5 others, 1969, Escarpments, reef trends and diapiric structures, eastern Gulf of Mexico: American Association of Petroleum Geologists Bulletin, v. 53, p. 2506–2542.

Burckhardt, C., 1930–1931, Etude synthétique sur le Mésozoique mexicain, Sec-

onde partie: Société Paleontologique Suisse Memoires, v. 50, p. 125–280.

Bushaw, D. J., 1968, Environmental synthesis of the east Texas Lower Cretaceous: Gulf Coast Association of Geological Societies Transactions, v. 18, p. 416–435.

Cabrera, C. R., and Menes, L. S., 1973, Applicación de importancia económica de las microfacies de la Formación El Abra: Asociación Mexicana de Geólogos Petroleros Boletín, v. 25, nos. 7–9, p. 237–307.

Carrasco, B., 1970, La Formación El Abra (Formación El Doctor) en la Plataforma Valles–San Luis Potosí: Instituto Mexicano del Petroleo Revista, v. 2, no. 3, p. 97–99.

Carrasco V., B., 1971, Litofacies de la Formación El Abra en Plataforma de Actopan, Hgo.: Instituto Mexicano del Petroleo Revista, v. 3, no. 1, p. 5–26.

——, 1977, Albian sedimentation of submarine autochthonous and allochthonous carbonates, east edge of the Valles–San Luis Potosí Platform, Mexico, in Cook, H. E. and Enos, P., eds., Deep-water carbonate environments: Society of Economic Paleontologists and Mineralogists Special Publication 25, p. 263–272.

Carrasco, B., Flores, V., and Godoy, D., 1975, Tobas del Cretácico Inferior del area Fortin-Zongolica, Estado de Veracruz: Instituto Mexicano del Petroleo Revista, v. 7, no. 4, p. 7–27.

Carrillo Bravo, J., 1965, Estudio geológico de una parte del anticlinorio de Huayacocotla: Asociación Mexicana de Geólogos Petroleros Boletín, v. 17, p. 73–96.

——, 1969, Exploración geológica y posibilidades petroleras de la plataforma Valles–San Luis Potosí: Instituto Mexicano del Petroleo, Seminario sobre exploración petrolera, Mesa Redonda no. 6, 20 p.

——, 1971, La plataforma Valles–San Luis Potosí: Asociación Mexicana de Geólogos Petroleros Boletín, v. 23, nos. 1–6, p. 1–113.

Conklin, J., and Moore, C., 1977, Paleoenvironmental analysis of the Lower Cretaceous Cupido Formation, northeast Mexico, in Bebout, D. G., and Loucks, R. G., eds., Cretaceous carbonates of Texas and Mexico; Applications to subsurface exploration: University of Texas at Austin Bureau of Economic Geology Report of Investigations 89, p. 302–323.

Coogan, A. H., 1977, Early and Middle Cretaceous Hippuritacea (Rudists) of the Gulf Coast, in Bebout, D. G., and Loucks, R. G., eds., Cretaceous carbonates of Texas and Mexico; Applications to subsurface exploration: University of Texas at Austin Bureau of Economic Geology Report of Investigations 89, p. 32–70.

Coogan, A. H., Bebout, D. G., and Maggio, C., 1972, Depositional environments and geologic history of Golden Lane and Poza Rica Trend, Mexico; An alternate view: American Association of Petroleum Geologists Bulletin, v. 56, p. 1419–1447.

Cook, T. D., 1979, Exploration history of South Texas Lower Cretaceous carbonate platform: American Association of Petroleum Geologists Bulletin, v. 63, p. 32–49.

Cook, T. D., and Bally, A. W., 1975, Stratigraphic atlas of North and Central America: Princeton, New Jersey, Princeton University Press, 272 p.

de Cserna, Z., 1970, Mesozoic sedimentation, magmatic activity, and deformation in northern Mexico, in Seewald, K., and Sundeen, D., eds., The geologic framework of the Chihuahua Tectonic Belt: Midland, West Texas Geological Society Symposium in honor of Professor Ronald K. DeFord, p. 99–117.

Diaz, T., 1952, Geología estructural del Anticlinal Peyotes: Asociación Mexicana de Geólogos Petroleros Boletín, v. 4, p. 117–147.

Dumble, E. T., 1895, Cretaceous of western Texas and Coahuila, Mexico: Geological Society of America Bulletin, v. 6, p. 375–388.

Enos, P., 1974, Reefs, platforms, and basins of Middle Cretaceous in northeast Mexico: American Association of Petroleum Geologists Bulletin, v. 58, p. 800–809.

——, 1977, Tamabra Limestone of the Poza Rica trend, Cretaceous, Mexico, in Cook, H. E., and Enos, P., eds., Deep-water carbonate environments: Society of Economic Paleontologists and Mineralogists Special Publication 25, p. 273–314.

——, 1983, Late Mesozoic paleogeography of Mexico, in Reynolds, M. W., and Dolly, E. D., eds., Mesozoic paleogeography of the west-central United

States: Rocky Mountain Section, Society of Economic Paleontologists and Mineralogists Paleogeography Symposium 2, p. 133–157.

Forgotson, J. M., Jr., 1954, Regional stratigraphic analysis of Cotton Valley Group of upper Gulf Coast plain: American Association of Petroleum Geologists Bulletin, v. 38, p. 2476–2499.

——, 1957, Stratigraphy of Comanchean Cretaceous Trinity Group: American Association of Petroleum Geologists Bulletin, v. 41, p. 2328–2363.

——, 1963, Depositional history and paleotectonic framework of Comanchean Cretaceous Trinity Stage, Gulf Coast area: American Association of Petroleum Geologists Bulletin, v. 47, p. 69–103.

Garza Gonzalez, R., 1973, Model sedimentario del Albiano-Cenomaniano en la porción sureste de la Plataforma de Coahuila (Prospecto Parras, Edo. de Coahuila): Asociación Mexicana de Geólogis Petroleros Boletín, v. 25, nos. 7–9, p. 309–339.

Gohn, G. S., 1988, Late Mesozoic and early Cenozoic geology of the Atlantic Coastal Plain; North Carolina to Florida, *in* Sheridan, R. E., and Grow, J. A., eds., The Atlantic continental margin; U.S.: Boulder, Colorado, Geological Society of America, The Geology of North America, v. I-2, p. 107–130.

Gonzalez Alvarado, J., 1976, Resultados obtenidos en la exploración de la Plataforma Córdoba y principales campos productores: Sociedad Geológica Mexicana Boletín, v. 37, p. 53–59.

Hazzard, R. T., 1939, Notes on the Comanche and pre-Comanche(?) Mesozoic formations of the Arkansas, Louisiana, and Texas area, and a suggested correlation with northern Mexico: Shreveport Geological Society Guidebook, 14th Annual Field Trip, p. 155–164.

Hazzard, R. T., Blanpied, B. W., and Spooner, W. C., 1947, Notes on correlations of the Cretaceous of east Texas, south Arkansas, north Louisiana, Mississippi, and Alabama: Shreveport Geological Society 1945 Reference Report, v. 2, p. 471–480.

Hendricks, L., ed., 1967, Comanchean (Lower Cretaceous) stratigraphy and paleontology of Texas: Permian Basin Section, Society of Economic Paleontologists and Mineralogists Publication 67-8, 410 p.

Hill, R. T., 1887a, The Texas section of the American Cretaceous: American Journal of Science, series 3, v. 34, p. 287–309.

——, 1887b, The topography and geology of the Cross Timbers and surrounding regions in northern Texas: American Journal of Science, series 3, v. 33, p. 291–303.

——, 1891, The Comanche Series of the Texas-Arkansas region: Geological Society of America Bulletin, v. 2, p. 503–528.

——, 1893a, The Cretaceous formations of Mexico and their relations to North American geographic development: American Journal of Science, series 3, v. 45, p. 307–324.

——, 1893b, Paleontology of the Cretaceous formations of Texas; The invertebrate paleontology of the Trinity Division: Biologic Society of Washington Proceedings, v. 8, p. 9–40.

——, 1894, Geology of parts of Texas, Indian Territory, and Arkansas adjacent to Red River: Geological Society of America Bulletin, v. 5, p. 297–338.

Humphrey, W. E., 1949, Geology of the Sierra de los Muertos area Mexico: Geological Society of America Bulletin, v. 60, p. 89–176.

——, 1956, Tectonic framework of northeast Mexico: Gulf Coast Association of Geological Societies Transactions, v. 6, p. 25–35.

Ice, R. G., and McNulty, C. L., 1980, Foraminifers and calcispheres from the Cuesta del Cura and lower Agua Nueva Formations (Cretaceous) in east-central Mexico: Gulf Coast Association of Geological Societies Transactions, v. 30, p. 403–414.

Imlay, R. W., 1936, Evolution of the Coahuila Peninsula, Mexico; Part 4, Geology of the western part of the Sierra de Parras: Geological Society of America Bulletin, v. 47, p. 1091–1152.

——, 1937, Geology of the middle part of the Sierra de Parras, Coahuila, Mexico: Geological Society of America Bulletin, v. 48, p. 587–630.

——, 1938, Studies of the Mexican geosyncline: Geological Society of America Bulletin, v. 49, p. 1651–1694.

——, 1940a, Lower Cretaceous and Jurassic formations of southern Arkansas and their oil and gas possibilities: Arkansas Geological Survey Information Circular 12, 64 p.

——, 1940b, Neocomian faunas of northern Mexico: Geological Society of America Bulletin, v. 51, p. 117–190.

——, 1944, Cretaceous formations of central America and Mexico: American Association of Petroleum Geologists Bulletin, v. 28, p. 1077–1195.

——, 1945, Subsurface Lower Cretaceous formations of south Texas: American Association of Petroleum Geologists Bulletin, v. 29, p. 1416–1469.

Longoria, J. F., and Gamper, M. A., 1977, Albian planktonic foraminifera from the Sabinas Basin of northern Mexico: Journal of Foraminifera Research, v. 7, p. 196–215.

Lopez Ramos, E., 1975, Geological summary of the Yucatan Peninsula, *in* Nairn, A.E.M., and Stehli, F. G., eds., The ocean basins and margins; v. 3, The Gulf of Mexico and the Caribbean: New York, Plenum Press, p. 257–282.

——, 1981, Geología de México, Tomo III, 2nd Edition: México, D.F., 446 p.

——, 1982, Geología de México, Tomo II, 3rd Edition: Mexico, D.F., 454 p.

Loucks, R. G., 1977, Porosity development and distribution in shoal-water carbonate complexes; Subsurface Pearsall Formation (Lower Cretaceous), south Texas, *in* Bebout, D. G., and Loucks, R. G., eds., Cretaceous carbonates of Texas and Mexico; Applications to subsurface exploration: University of Texas at Austin Bureau of Economic Geology Report of Investigations 89, p. 97–126.

Lozo, F. E., 1959, Stratigraphic relations of the Edwards Limestone and associated formations in north-central Texas, *in* Lozo, F. E., Nelson, H. F., Young, K., Sherburne, O. B., and Sandridge, J. R., eds., Symposium on Edwards Limestone in central Texas: University of Texas at Austin Bureau of Economic Geology Publications 5905, p. 1–19.

Lozo, F. E., and Stricklin, F. L., Jr., 1956, Stratigraphic notes on the outcrop basal Cretaceous, central Texas: Gulf Coast Association of Geological Societies Transactions, v. 6, p. 67–78.

Lozo, F. E., and Smith, C. I., 1964, Revision of Comanche Cretaceous stratigraphic nomenclature, southern Edwards Plateau, southwest Texas: Gulf Coast Association of Geological Societies Transactions, v. 14, p. 285–306.

Lozo, F. E., Nelson, H. F., Young, K., Sherburne, O. B., and Sandridge, J. R., eds., 1959, Symposium on Edwards Limestone in central Texas: University of Texas at Austin Bureau of Economic Geology Publication 5905, 235 p.

Mann, C. J., and Thomas, W. A., 1964, Cotton Valley Group (Jurassic) nomenclature, Louisiana and Arkansas: Gulf Coast Association of Geological Societies Transactions, v. 14, p. 143–152.

Martin, J. L., Hough, L. W., Reggio, D. L., and Sandberg, A. E., 1954, Geology of Webster Parish: Louisiana Geological Survey Bulletin 29, 252 p.

McFarlan, E., Jr., 1977, Lower Cretaceous sedimentary facies and sea-level changes, United States Gulf Coast, *in* Bebout, D. G., and Loucks, R. G., eds., Cretaceous carbonates of Texas and Mexico; Applications to subsurface exploration: University of Texas at Austin Bureau of Economic Geology Report of Investigations 89, p. 5–11.

McGowen, M. K., and Harris, D. W., 1984, Cotton Valley (Upper Jurassic) and Hosston (Lower Cretaceous) depositional systems and their influence on salt tectonics in the East Texas Basin, *in* Ventress, P. S., Bebout, D. G., Perkins, B. F., and Moore, C. H., eds., The Jurassic of the Gulf Rim: Gulf Coast Section, Society of Economic Paleontologists and Mineralogists Third Annual Research Conference Proceedings, p. 213–253.

Michael, F. Y., 1972, Planktonic foraminifera from the Comanchean Series (Cretaceous) of Texas: Journal of Foraminiferal Research, v. 2, p. 200–220.

Mora, J. C., Schlaepfer, C. J., and Rodriguez, E. M., 1975, Estratifrafía y microfacies de la Sierra Madre del Sur, Chiapas: Asociación Mexicana de Geólogos Petroleros Boletín, v. 27, nos. 1–3, p. 1–103.

Moya Cuevas, F., 1974, Estudio sedimentario del Cretácico Medio en el area margen oriental de la Plataforma Valles–San Luis Potosí: Asociación Mexicana de Geólogos Petroleros Boletín, v. 26, nos. 10–12, p. 200–220.

Muir, J. M., 1936, Geology of the Tampico region, Mexico: American Association of Petroleum Geologists, 280 p.

Murray, G. E., 1948, Geology of De Soto and Red River Parishes, Louisiana: Louisiana Geological Survey Bulletin 25, 312 p.

——, 1961, Geology of the Atlantic and Gulf Coast Province of North America: New York, Harper and Brothers Geoscience series, 680 p.

Nunnally, J. D., and Fowler, H. F., 1954, Lower Cretaceous stratigraphy of Mississippi: Mississippi Geological Survey Bulletin 79, 45 p.

Perkins, B. F., ed., 1974, Aspects of Trinity Division geology: Geoscience and Man, v. 8, 228 p.

Rose, P. R., 1972, Edwards Group, surface and subsurface, central Texas: University of Texas at Austin Bureau of Economic Geology Report of Investigations 74, 198 p.

Santiago Acevedo, J., 1980, Giant fields of the southern zone; Mexico, in Halbouty, M. T., ed., Giant oil and gas fields of the decade 1968-1978: American Association of Petroleum Geologists Memoir 30, p. 339–385.

Santiago Acevedo, J., and Mejia Dautt, O., 1980, Giant oil fields in the southeast of Mexico: Gulf Coast Association of Geological Societies Transactions, v. 30, p. 1–31.

Schlager, W., and 15 others, 1984a, Deep Sea Drilling Project, Leg 77, southeastern Gulf of Mexico: Geological Society of America Bulletin, v. 95, p. 226–236.

Schlager, W., Buffler, R. T., Angstadt, D., and Phair, R., 1984b, Geologic history of the southeastern Gulf of Mexico, in Buffler, R. T., and others, eds., Initial reports of the Deep Sea Drilling Project: Washington, D.C., U.S. Government Printing Office, v. 77, p. 715–738.

Scott, R. W., 1990, Models and stratigraphy of mid-Cretaceous reef communities, Gulf of Mexico, in Concepts of sedimentology and paleontology, v. 2: Society of Economic Paleontologists and Mineralogists (Society of Sedimentary Geology), 102 p.

Selvius, D. B., and Wilson, J. L., 1985, Lithostratigraphy and algal-foraminiferal biostratigraphy of the Cupido Formation, Lower Cretaceous, northeast Mexico, in Perkins, B. F., and Martin, G. B., eds., Habitat of oil and gas in the Gulf Coast: Gulf Coast Section, Society of Economic Paleontologists and Mineralogists 4th Annual Research Conference, p. 285–311.

Smith, C. I., 1970, Lower Cretaceous stratigraphy, northern Coahuila, Mexico: University of Texas at Austin Bureau of Economic Geology Report of Investigations 65, 101 p.

——, 1981, Review of geologic setting, stratigraphy, and facies distribution of the Lower Cretaceous in northern Mexico, in Lower Cretaceous stratigraphy and structure, northern Mexico: West Texas Geological Society Field Trip Guidebook, Publication 81-74, p. 1–27.

Stricklin, F. L., Jr., 1973, Environmental reconstruction of a carbonate beach complex: Cow Creek (Lower Cretaceous) Formation of central Texas: Geological Society of America Bulletin, v. 84, p. 1349–1368.

Stricklin, F. L., Jr., Smith, C. I., and Lozo, F. E., 1971, Stratigraphy of Lower Cretaceous Trinity deposits of central Texas: University of Texas at Austin Bureau of Economic Geology Report of Investigations 71, 63 p.

Swain, F. M., 1944, Stratigraphy of Cotton Valley beds of northern Gulf Coastal Plain: American Association of Petroleum Geologists Bulletin, v. 28, p. 577–614.

Todd, R. G., and Mitchum, R. M., Jr., 1977, Seismic stratigraphy and global changes of sea level; Part 8, Identification of Upper Triassic, Jurassic, and Lower Cretaceous seismic sequences in Gulf of Mexico and offshore west Africa, in Payton, C. E., ed., Seismic stratigraphy; Applications to hydrocarbon exploration: American Association of Petroleum Geologists Memoir 26, p. 145–164.

Tolson, J. S., Copeland, C. W., and Bearden, B. L., 1983, Stratigraphic profiles of Jurassic strata in the western part of the Alabama Coastal Plain: Geological Survey of Alabama Bulletin 122, 425 p.

Trejo, M., 1960, La Familia Nannoconidae y su alcance estratigráfico en America (protozoa, incertae saedis): Asociación Mexicana de Geólogos Petroleros Boletín, v. 12, nos. 9–10, p. 259–314.

——, 1973, Tintínidos Mesozoicos de México [Doctoral thesis]: Escuela Nacional de Ciencas Biologicas, México, 67 p.

——, 1975, Zonificación del limite Aptiano-Albiano de México: Instituto Méxicano del Petroleo Revista, v. 7, no. 3, p. 6–29.

Vargas, A. F., 1978, A paleosedimentary study of Reforma area: Petroleo Internacional v. 26, p. 44–48.

Viniegra O., F., 1971, Age and evolution of salt basins of southeastern Mexico: American Association of Petroleum Geologists Bulletin, v. 55, p. 478–494.

——, 1981, Great carbonate bank of Yucatan, southern Mexico: Journal of Petroleum Geology, v. 3, p. 247–278.

Viniegra O., F., and Castillo-Tejero, C., 1970, Golden Lane fields, Veracruz, Mexico, in Halbouty, M. T., ed., Geology of giant petroleum fields: American Association of Petroleum Geologists Memoir 14, p. 309–325.

Weeks, W. B., 1938, South Arkansas stratigraphy with emphasis on the older coastal plain beds: American Association of Petroleum Geologists Bulletin, v. 22, p. 953–983.

Wilson, J. L., and Pialli, G., 1977, A Lower Cretaceous shelf margin in northern Mexico, in Bebout, D. G., and Loucks, R. G., eds., Cretaceous carbonates of Texas and Mexico; Applications to subsurface exploration: University of Texas at Austin Bureau of Economic Geology Report of Investigations 89, p. 286–294.

Winker, C. D., and Buffler, R. T., 1988, Paleogeographic evolution of early deep-water Gulf of Mexico and margins, Jurassic to Middle Cretaceous (Comanchean): American Association of Petroleum Geologists Bulletin, v. 72, p. 318–346.

Winston, G. O., 1971, Regional structure, stratigraphy, and oil possibilities of the South Florida Basin: Gulf Coast Association of Geological Societies Transactions, v. 21, p. 15–29.

——, 1976, Six proposed formations in the undefined portion of the Lower Cretaceous section in south Florida: Gulf Coast Association of Geological Societies Transactions, v. 26, p. 69–72.

Winter, J. A., 1961a, The Atascosa Group of the Lower Cretaceous of South Texas: South Texas Geological Society Bulletin, v. 1, no. 11, p. 15–21.

——, 1961b, Stratigraphy of the Lower Cretaceous (subsurface) of south Texas: Gulf Coast Association of Geological Societies Transactions, v. 11, p. 15–24.

Wooten, J. W., and Dunaway, W. E., 1977, Lower Cretaceous carbonates of central south Texas; A shelf-margin study, in Bebout, D. E., and Loucks, R. G., eds., Cretaceous carbonates of Texas and Mexico; Applications to subsurface exploration: University of Texas at Austin Bureau of Economic Geology Report of Investigations 89, p. 71–78.

Young, K., 1967, Comanche series (Cretaceous), south-central Texas, in Hendricks, L., ed., Comanchean (Lower Cretaceous) stratigraphy and paleontology of Texas: Permian Basin Section, Society of Economic Paleontologists and Mineralogists Publication 67-8, p. 9–29.

——, 1972, Cretaceous paleogeography; Implications of endemic ammonite faunas: University of Texas at Austin Bureau of Economic Geology Geological Circular 72-2, 13 p.

——, 1974, Lower Albian and Aptian (Cretaceous) ammonites of Texas, in Perkins, B. F., ed., Aspects of Trinity Division geology: Geoscience and Man, v. 8, p. 175–228.

——, 1977, Middle Cretaceous rocks of Mexico and Texas, in Bebout, D. G., and Loucks, R. D., eds., Cretaceous carbonates of Texas and Mexico; Applications to subsurface exploration: University of Texas at Austin Bureau of Economic Geology Report of Investigations 89, p. 325–332.

——, 1978, Lower Cenomanian and late Albian (Cretaceous) ammonites, especially Lyelliceridae, of Texas and Mexico: Texas Memorial Museum Bulletin 26, 99 p.

MANUSCRIPT ACCEPTED BY THE SOCIETY DECEMBER 26, 1990

ACKNOWLEDGMENTS

Special appreciation is expressed to Exxon Company, USA, and Petroleos Mexicanos for granting permission to prepare and publish the information contained in this chapter. The manuscript was reviewed by C. I. Smith, James L. Wilson, and Amos Salvador. Their many constructive comments and suggestions are gratefully acknowledged.

Printed in U.S.A.

The Geology of North America
Vol. J, The Gulf of Mexico Basin
The Geological Society of America, 1991

Chapter 10

Upper Cretaceous

Norman F. Sohl
U.S. Geological Survey, 970 National Center, Reston, Virginia 22092
Eduardo Martínez R., Pedro Salmerón-Ureña, and Fidel Soto-Jaramillo
Instituto Mexicano del Petroleo, Eje Centrál Lázaro Cardenas 152, Mexico 14, D.F., Mexico

INTRODUCTION

In the Gulf of Mexico basin and contiguous areas the Late Cretaceous history is marked as a general time of oceanic high stand. The Late Cretaceous here begins with a short period, characterized over the northern margin, by basin fill. This interval was followed by major transgression, during which marine waters inundated the basin margins and eventually linked the Gulf of Mexico with the great epicontinental Western Interior Seaway. The following is a summary of these and the subsequent events that transpired during the Late Cretaceous in the Gulf of Mexico basin region.

Extent, thickness, and source

Upper Cretaceous rocks form a virtually continuous blanket over the Gulf of Mexico basin. Overall, the inner edge of the northern outcrop parallels the deeply buried Paleozoic Ouachita orogenic belt. Conspicuous projecting features, such as the Rio Grande and Mississippi embayments, are situated on pronounced salients of the older Ouachita orogenic belt. At the updip pinchout, the present-day outcrop pattern follows a trebly arcuate path (Fig. 1). The easternmost Gulfward-trending arc extends from Georgia westward, wrapping around the southwest end of the Appalachian trend across Alabama, and thence northward through Mississippi, Tennessee, and Kentucky, to terminate at the head of the Mississippi embayment in southern Illinois. Because of local overlap by Tertiary deposits, the second and complementary outcrop arc is intermittent, but wraps around the Arkansas platform through Missouri and Arkansas, then trends westward and terminates north of Dallas, Texas. From this point a third outcrop arc trends first south, then southwestward across Texas, generally paralleling the Balcones Escarpment, and then westward to the head of the Rio Grande embayment. Southward because of complex structure and post-Cretaceous erosion, outcrops become intermittent through eastern Mexico to the state of Chiapas.

The Upper Cretaceous sedimentary package thickens downdip in most areas to a shelf edge that is postulated to correspond to the trend of the Early Cretaceous "reef" track. The trend generally follows the edge of the Florida Escarpment, striking into the subsurface of southern Louisiana and then arcuately following the path of the "Stuart City Reef" line south across Texas to the Rio Grande embayment. This concentrically arranged Late Cretaceous shelfal area is broadest on the northern margins and narrows southward through east-central Mexico (Fig. 1).

Late Cretaceous sedimentation history has a strong overprint of eustasy. Cyclic sea-level fluctuation affected a broad, generally low relief, shelfal area. Sources of sediment varied throughout the Late Cretaceous. Nearby extra-shelfal sources, such as the Appalachian and Ouachita trends, provided much of the terrigenous sediment during the earliest depositional episodes. During the middle part of the Late Cretaceous, biogenic sedimentation becomes especially prominent in many areas, and in others, terrigenous input is finer-grained than for earlier times. In the northeastern Gulf of Mexico basin areas, terrigenous deposits appear to be dominantly derived from adjacent Appalachian areas throughout the Late Cretaceous. However, in the northwestern parts of the basin, from the Coniacian through the Maastrichtian, increasing amounts of terrigenous material appear to be derived from western sources. In Mexico, Late Cretaceous terrigenous influx from the west is even more marked. This shift in source may be linked to far removed Laramide events (Weidie and others, 1972). Superimposed on this general pattern of external source of sediment are within-shelf, local, tectonically generated sediment sources. Periodic uplift and subsequent erosional stripping of such features as the Sabine uplift provided terrigenous clastics to margining basins within the shelf. In the central and deeper parts of the Gulf of Mexico basin, the post–mid-Cenomanian marks a period of deep-marine sedimentation consisting of very fine-grained clastics in the west and central areas and pelagic deposits in the east (Schaub and others, 1984).

Some significant contrasts exist between the types of lithologies common in the Lower Cretaceous and those that are common to the Upper Cretaceous of the Gulf of Mexico basin. Prime among these is the post-Turonian deposition of chalk and chalky marl as shelfal carbonates across the northern Gulf Coastal Plain

Sohl, N. F., Martínez R., E., Salmerón-Ureña, P., and Soto-Jaramillo, F., 1991, Upper Cretaceous, *in* Salvador, A., ed., The Gulf of Mexico Basin: Boulder, Colorado, Geological Society of America, The Geology of North America, v. J.

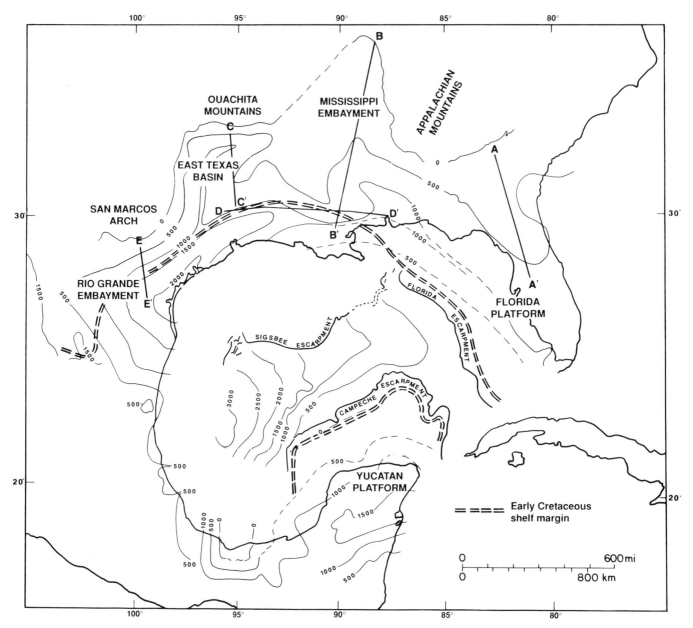

Figure 1. Isopachous map of the Upper Cretaceous rocks of the Gulf of Mexico basin. Lines A-A′ through E-E′ show positions of stratigraphic cross sections illustrated in Figures 3 to 7. Thicknesses in meters.

areas. A second contrast is that the rudist "reef" and associated carbonate platform facies, which are so prominent a component of Lower Cretaceous deposits, are displaced far to the south in deposits of the Late Cretaceous in Mexico and Guatemala where they developed mainly on platformal margins (Alencaster, 1984; Young, 1983).

The full extent of the landward encroachment of marine Upper Cretaceous rocks is not fully expressed by the present outcrop. The most obvious discontinuity is the lack of outcrop in the area that once connected the Gulf with the Western Interior Seaway (Figs. 9 through 12). In the eastern Gulf Coastal Plain, isolated faulted inliers well north of the present outcrop line in

Alabama (Monroe, 1941, Plate 1) and areal lithofacies, biofacies, and thickness patterns, which are truncated by the outcrop line (Sohl and Koch, 1986), bear witness to the once-greater extent of Upper Cretaceous rocks in this area. Southward in Mexico, post-Cretaceous uplift, folding, faulting, and subsequent erosion have significantly altered the original limits of Cretaceous deposits.

Although the Upper Cretaceous deposits of the Gulf of Mexico basin are predominantly marine, there were episodes of widespread deposition of nonmarine, fluvial and deltaic, coarsely terrigenous clastics. For example, mid-Cenomanian uplift in the northern areas of the basin provided the source of the terrigenous fluvial and delta-plain deposits found in the Woodbine and Tuscaloosa Formations (mid-Cenomanian to Turonian) of the northeast Texas to Alabama area. Subsequent to Cenomanian uplift, significant deltaic deposits are more localized, as for example, the McNairy Sand in the northern part of the Mississippi embayment (Pryor, 1960). Throughout the Late Cretaceous, terrigenous deposits were most abundant in the northern part of the Gulf of Mexico basin, while carbonates and calcareous shales were predominant in much of Mexico. Upper Cretaceous, platformal carbonates, at times interbedded with evaporites, dominate the sequences in the most southerly areas in peninsular Florida and the Yucatan platform.

The thickest Upper Cretaceous sections are found in such major salients as the Mississippi and Rio Grande embayments, and in structural basins like the East Texas basin (Fig. 1). In other areas, deposits generally thin over structural features such as the San Marcos arch in central Texas. In the northern Gulf of Mexico basin the general trend is for thickening of the sediment wedge downdip toward the Late Cretaceous shelf break with as much as 1,800 m of accumulation in areas such as the Rio Grande embayment (Fig. 1). Beyond the shelf break, deposits thin toward the deeper parts of the Gulf of Mexico. Upper Cretaceous sediments in the deep part of the basin are poorly known. Toward the north, they lie beneath exceptionally thick Cenozoic deposits and are beyond the reach of the drill. Southward, along the margins of the Yucatan Shelf, thin sequences of pelagic Upper Cretaceous sediments have been encountered in cores recovered by drilling during DSDP Leg 10 (McNeeley, 1973). Seismic data suggest that the central part of the basin has a blanket of Upper Cretaceous to lower Tertiary deposits that are thickest in the west and thinner toward the east (Schaub and others, 1984).

Contact relations

The base of the Upper Cretaceous occurs with no pronounced physical break within rocks of the Washita Group and equivalents. However, when discussing the "Upper Cretaceous" deposits of the U.S. Gulf Coastal Plain, most North American geologists think of the mid-Cenomanian through Maastrichtian section that has been called the provincial "Gulfian Series." The main reason for such usage is that the contact between the "Gulfian Series" and the underlying rocks is the most profound physical break in the Cretaceous sedimentary record of the northern Gulf of Mexico basin. Throughout this area, post-lower Cenomanian, "Gulfian" strata rest with unconformity upon lowermost Cenomanian or older strata. For example, in the area from Georgia to the Mississippi embayment, "Gulfian" rocks overlap older Cretaceous rocks and rest upon a range of Jurassic, Triassic, and Paleozoic rocks (see Chapter 14, this volume). Imlay (1940), Anderson (1979), and others have shown that similar unconformable relations exist in Arkansas and well into the subsurface of Louisiana. To the west, in Texas, a disconformity of lesser time range is present in outcrop and commonly encompasses only a part of the Cenomanian stage. Relations in the subsurface of Texas are variable. Bailey and others (1945) sug-

gested that there was transition between the "Gulfian" section and older rocks in the center of the East Texas basin. In contrast, Hazzard and others (1947) interpreted the deposits of this interval as a terminal, regressive, basin-center fill phase of the "Comanchean." Lozo (1951) reviewed the evidence relating to the problem and concluded that insufficient data were available to answer the question. Subsequent evidence from wells suggests that deposition was transitional in deeper parts of the basin.

The pronounced unconformity at the base of the "Gulfian" strata cannot be universally traced southward into the Gulf of Mexico basin in Mexico, suggesting that during this part of the mid-Cretaceous the structural history of the northern part of the basin differed from that of the southern part. Only locally, on such elevated platform areas as the Tuxpan platform and the Valles–San Luis Potosí platform, is there a break suggesting this event. Similarly, along the margins of the Yucatan platform the boundary between the Challenger and Campeche seismic units has been recognized as a distinctive unconformity (see Chapter 13 of this volume).

Historically, the great majority of workers have assumed that the Cretaceous-Tertiary boundary in the Gulf of Mexico basin region is represented by a major hiatus, a period during which there was "emergence of the land, or a lowering of sea level . . . such that the strand line retreated to the relatively steep outer slope of the Continental Shelf, throughout the length of the Atlantic and Gulf Coastal Plain" (Stephenson, 1941, p. 33). The universality of the Cretaceous-Tertiary disconformity in the Gulf of Mexico basin has increasingly been brought into question in recent years. In the northeastern Gulf Coastal Plain the basal Tertiary units overlap the Cretaceous in many areas, and a demonstrable erosional unconformity exists. The resultant hiatus varies in magnitude from place to place, but in general, as current biostratigraphic resolution and geochronometric techniques have improved, the "lost interval" has commonly been reduced in span. In some sequences, such as the Brazos River (Jiang and Gartner, 1986) and the Rio Grande embayment (Cooper, 1973) sections in Texas, or the Parras basin of Mexico (Weidie and Murray, 1967), evidence has been presented that suggests that the Cretaceous and Tertiary are transitional.

Notes on the biostratigraphic framework

In spite of the mass of paleontologic data generated by work on the Cretaceous rocks of the Gulf of Mexico basin, many problems of correlation still exist. For example, within the Gulf of Mexico basin, correlation of rocks containing a Tethyan (tropical) biota with those of the warm temperate region is hampered by problems of provinciality. Few larger invertebrates are common to the biota of the two faunal provinces (Sohl, 1971). Ammonites and planktonic foraminiferans common to the terrigenous lithofacies of the northern areas of the Gulf of Mexico basin region are especially rare in the Tethyan shallow-water carbonate bank deposits common to the Mexican platform areas. Such carbonate bank deposits are dominated by faunas associated with framework structures composed of rudist pelecypods that along with large foraminifers have often served for biostratigraphic zonation. In addition, especially in Mexico, Bonet (1956) and other workers have proposed zones utilizing calcispheres and other microfossil groups as markers. Fortunately, there are areas that contain faunas that are transitional between the two provinces. That is, sections may contain faunal mixtures that allow recognition of stratigraphic tie points that facilitate, at least broadly, the integration between the zonations of the two faunal provinces.

Precision of correlation of the Upper Cretaceous section of the Gulf of Mexico basin with the standard section in Europe is variable. Ammonite and inoceramid bivalve occurrences early provided a general framework. However, their rarity in many lithofacies, compounded by the problems of provinciality discussed previously, resulted in the erection of a system of provincial stages (Murray, 1961). Subsequently, investigations by Pessagno (1969) and a host of other workers on planktonic foraminifers, nannofossils, and other microfossil groups of wide distribution, have made giant steps toward correlation with the standard sections and total abandonment of the provincial stage nomenclature.

STRATIGRAPHY

In the following discussion of the Upper Cretaceous stratigraphy of the Gulf of Mexico basin, the section has been divided into five chronostratigraphic units. Usually these units do not precisely correspond to either lithostratigraphic unit boundaries or the proposed provincial stage boundaries. Within any single geographic subdivision of the basin, major lithostratigraphic units or groups of units might form a more convenient frame for discussion because they commonly relate to major cyclic depositional events (Young, 1982). Such unconformity-bounded units lend themselves nicely to discussion for the Texas area. Unfortunately, they may not coordinate with events in another part of the Gulf of Mexico basin. Thus, the chronostratigraphic approach forms the most consistent basis for discussion of such a broad area. The subdivisions used here are as follows, from older to younger:

Lower Cenomanian. This interval marks a time of widespread transgression in the Gulf of Mexico basin that caused the drowning of the previously existing reef trends. Late in the interval, the northern parts of the basin underwent tilting accompanied by regression.

Middle Cenomanian–Turonian. Rocks of this time interval are discussed together because they represent the initial great episode of post-early Cenomanian shelf inundation (Fig. 9).

Coniacian–Santonian. This interval includes most of the period of great chalk and carbonate deposition in the northwestern Gulf of Mexico basin and the Western Interior Seaway (Fig. 10).

Campanian. During this interval the site of carbonate deposition shifted to the northeastern part of the basin. In Texas and Mexico the source of much of the terrigenous sediment influx

shifted to the west, reflecting the early stages of the Laramide orogeny (Fig. 11).

Maastrichtian. This stage begins as a continuance of the Campanian depositional pattern followed by an increase in coarseness of the terrigenous component. Along the northern margin of the basin, late in the interval, there is a return to finer-grained sediments. Locally the deposition of marly chalk represents the last extensive period of Cretaceous shelf inundation. This episode is terminated by eustatic sea-level drop over much of the basin (Fig. 12).

Lower Cenomanian

Distribution. From the Florida platform westward through Alabama and Mississippi to Arkansas, lower Cenomanian deposits are known only from the subsurface. Outcrops occur from northeast Texas southward through Mexico until, in the Yucatan area, such beds are known only in the subsurface. In the deeper parts of the basin, lower Cenomanian pelagic deposits have been recovered by drilling in the area between the Florida and the Campeche Escarpments at DSDP site 97 of Leg 40 (Worzel and others, 1973, see chapter 13 of this volume). In addition, the upper part of the Challenger seismic unit of the central basin is presumed to include lower Cenomanian sediments.

The distribution of lower Cenomanian deposits is sporadic along the northern rim of the Gulf of Mexico basin. This distribution pattern is the result of erosion of these and earlier deposits during and subsequent to the mid-Cenomanian uplift of the northern margin of the basin.

Lithostratigraphy. Lower Cenomanian lithostratigraphic units (Fig. 2) are assigned to the upper part of the Washita Group. In areas where formational units are not distinguished, lower Cenomanian strata form the upper part of what regional reports refer to as "Washita undifferentiated" (Plate 5, in pocket).

In subsurface sections in Florida, lower Cenomanian beds may be absent or may occur in units designated by Applin and Applin (1965, p. 63), as "Washita age" or simply "undifferentiated limestone and shale" (Fig. 3). Such units are mainly calcitic dolomite to chalky limestones that commonly bear abundant miliolid foraminifers. In central Florida, such carbonates are overlain by a thin unit of green shale. Sparsity of diagnostic fossils prevents an estimation of how much of these Washita-equivalent sections are lower Cenomanian. Updip, in the Florida panhandle, the upper part of the undifferentiated Fredricksburg-Washita sequence is composed of red and variegated clay shale and fine to coarse-grained quartzose sandstone that is considered to be a marginal marine lithofacies (Applin and Applin, 1965). Fossils contained in rare intercalations of fossiliferous sandy limestones indicate that some of the beds are lower Cenomanian (Buda Formation equivalents). Except for such rare occurrences of fossils, there is no lithic distinction of beds of this age from older units in this lithofacies. Westward in southern Mississippi, however, the lower Cenomanian deposits take on a tripartite character (Fig. 4), which in the deeper parts of the basin, can be recognized into Texas. These divisions and their outcrop equivalents are, from base to top, as follows: a lower shale, marl or clay (= Grayson or Del Rio); limestone (= Buda); shale (= Dantzler or Maness). The latter, predominantly shale units, are restricted to the center of basins or near shelf areas where there is probable continuous deposition into the overlying units (Figs. 5 and 6).

Throughout the surface and subsurface of east Texas, the lowermost Cenomanian unit is the Grayson Formation, named for the type locality at Grayson Bluff in northeast Texas (Winton, 1925). The Grayson is primarily fossiliferous, gray to brownish gray, argillaceous marl and calcareous mudstone with subsidiary nodular limestones. Toward the south the Grayson grades into the Del Rio Formation, a unit that is recognized from central Texas to northeastern Mexico (Fig. 7). It is a gray, fossiliferous, calcareous clay much like the Grayson to the north, but generally is less calcareous in its middle part. In northeastern Mexico, the Del Rio clay becomes more calcareous and contains thin-bedded, gray, argillaceous limestone.

In most areas of both Texas and northeastern Mexico, the Buda Limestone lies conformably upon the Grayson–Del Rio.

Two subdivisions of the Buda have been recognized in central Texas; a lower glauconitic biomicrite and an upper part consisting of fossiliferous intramicrite (Martin, 1967).

In north Texas, the full thickness of the Buda is present only in the deeper part of the East Texas basin, where it consists mainly of a white, finely crystalline limestone that becomes more marly toward the top, and is overlain by the Maness Shale (Figs. 5 and 6). The Maness is a laminated to massive, calcareous, clayey shale that is commonly bronze to copper colored (Bailey and others, 1945). In the subsurface of south Texas, the Buda is a persistent unit, but may pinch out locally over the Stuart City reef trend (Winter, 1962).

Southward in Mexico, carbonate deposits dominate the sequences of most of the basins. On the Tamaulipas arch, in the Burgos basin and the southeastern part of the Sabinas basin, the upper part of the Cuesta del Cura Formation is Cenomanian. The unit is composed of dark gray, microcrystalline, thin- to medium-bedded limestones that contain nodules and bands of gray to black chert and some intercalations of dark gray to black, finely laminated, calcareous shale. In central Mexico, from the front

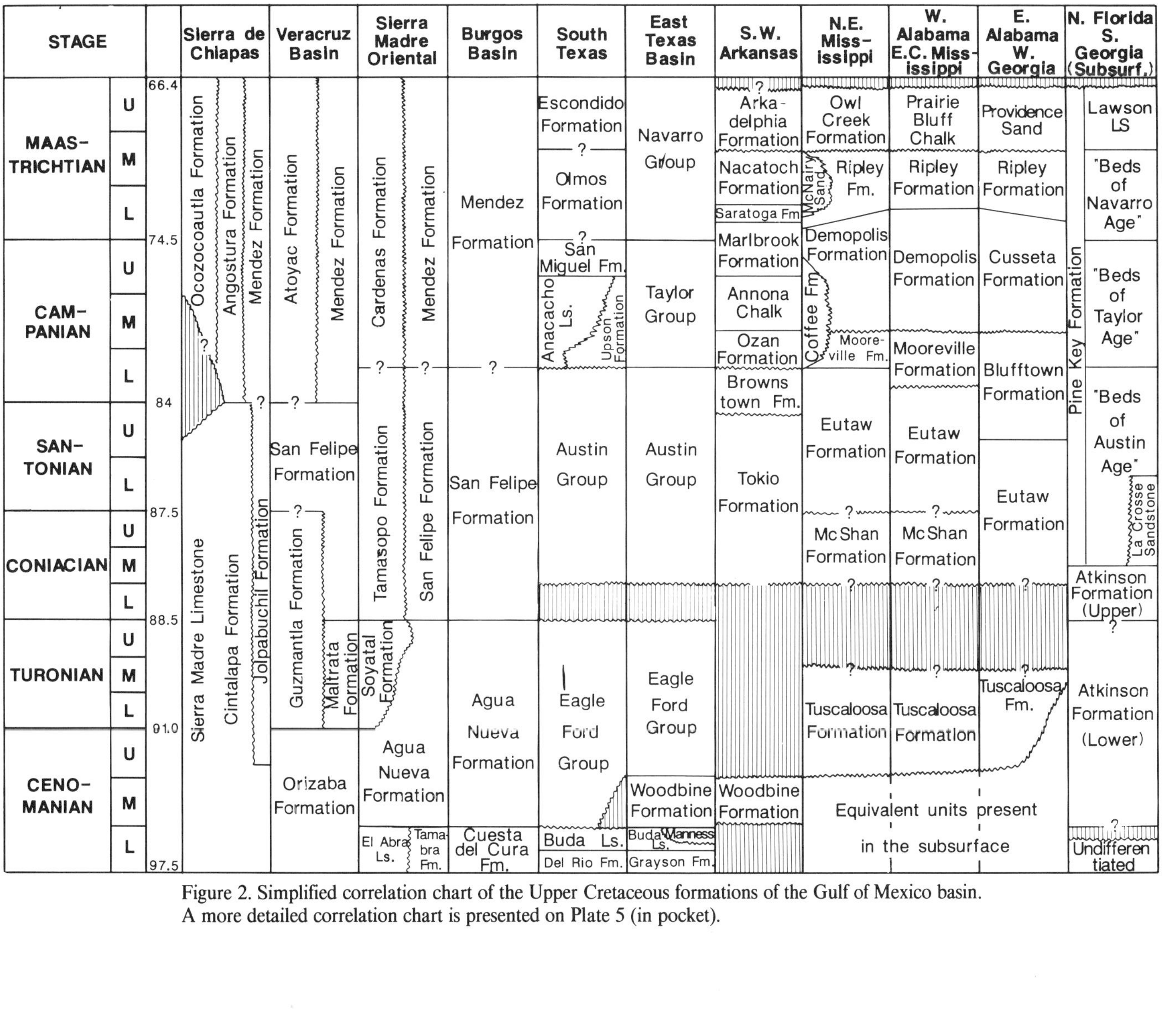

STAGE			Sierra de Chiapas	Veracruz Basin	Sierra Madre Oriental	Burgos Basin	South Texas	East Texas Basin	S.W. Arkansas	N.E. Mississippi	W. Alabama E.C. Mississippi	E. Alabama W. Georgia	N. Florida S. Georgia (Subsurf.)
MAAS-TRICHTIAN	U	66.4	Ocozocoautla Formation	Mendez Formation	Cardenas Formation	Mendez Formation	Escondido Formation	Navarro Group	Arka-delphia Formation	Owl Creek Formation	Prairie Bluff Chalk	Providence Sand	Lawson LS
	M		Angostura Formation	Atoyac Formation			? Olmos Formation		Nacatoch Formation	Ripley Fm.	Ripley Formation	Ripley Formation	"Beds of Navarro Age"
	L	74.5		Mendez Formation		Mendez Formation	San Miguel Fm.		Saratoga Fm. Marlbrook Formation	Demopolis Formation	Demopolis Formation	Cusseta Formation	
CAM-PANIAN	U			San Felipe Formation			Anacacho Ls. Upson Formation	Taylor Group	Annona Chalk	Mooreville Fm.	Mooreville Formation	Blufftown Formation	"Beds of Taylor Age"
	M								Ozan Formation				
	L	84	? ?	? ?	? ?	?			Browns town Fm.	Eutaw Formation	Eutaw Formation		"Beds of Austin Age"
SAN-TONIAN	U		Sierra Madre Limestone	San Felipe Formation	Tamasopo Formation / San Felipe Formation	San Felipe Formation	Austin Group	Austin Group	Tokio Formation			Eutaw Formation	
	L	87.5								McShan Formation	McShan Formation		La Crosse Sandstone
CONIACIAN	U		Cintalapa Formation	Guzmantla Formation									Atkinson Formation (Upper)
	M									?	?	?	
	L	88.5	Jolpabuchil Formation	Maltrata Formation / Soyatal Formation									
TURONIAN	U						Eagle Ford Group	Eagle Ford Group				Tuscaloosa Fm.	Atkinson Formation (Lower)
	M					Agua Nueva Formation				Tuscaloosa Formation	Tuscaloosa Formation		
	L	91.0											
CENO-MANIAN	U			Agua Nueva Formation		Agua Nueva Formation	Eagle Ford Group						
	M			Orizaba Formation				Woodbine Formation	Woodbine Formation	Equivalent units present in the subsurface			
	L	97.5		El Abra Ls. / Tambra Fm.	Cuesta del Cura Fm.	Buda Ls.	Buda Ls. Del Rio Fm.	Buda Maness Grayson Fm.					Undifferentiated

Figure 2. Simplified correlation chart of the Upper Cretaceous formations of the Gulf of Mexico basin. A more detailed correlation chart is presented on Plate 5 (in pocket).

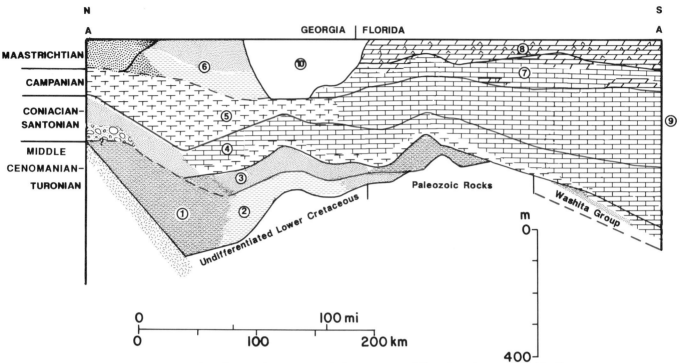

Figure 3. Cross section of Upper Cretaceous rocks between central Georgia and central Florida. See Figure 1 for location of section and Figure 8 for lithologic symbols. Numbers on section represent the following: 1, "Tuscaloosa Undifferentiated" of Applin and Applin (1944); 2, Lower Member of the Atkinson Formation; 3, Upper Member of the Atkinson Formation; 4, "Beds of Austin Age" (part); 5, "Beds of Taylor Age"; 6, units UK5 and UK6 (Prowell and others, 1985); 7, Lower member of the Lawson Limestone; 8, Upper Member of the Lawson Limestone; 9, Pine Key Formation; 10, absence of Maastrichtian deposits in Suwanee saddle. The red lines indicate the boundaries of the chronostratigraphic intervals indicated to the left of the diagram.

range of the Sierra Madre Oriental eastward into the Tampico-Misantla basin, several formations have been delineated that, in their upper parts, are of Cenomanian age (Fig. 2 and Chapter 9 of this volume). In the center of the basin, the Tamaulipas Superior consists of dark gray to white, finely crystalline limestone with some interbeds of black, silty, calcareous shale and marl. Black to white chert beds may be common locally (Muir, 1936). Both eastward and westward, the dark limestones give way to the dolomitic to pure limestones of the upper part of the Tamabra Formation. A few thin shale beds occur in this unit, but for the most part, it consists of a fine, skeletal, limy silt, derived from the destruction of rudist bivalves (Barnetche and Illing, 1956). The Tamabra, in turn, grades into the El Abra Limestone, which is usually a cavernous to dense, hard, gray, brown, or white, fossiliferous limestone. The El Abra consists of a variety of lithic types ranging from bioclastic limestone to beds composed predominantly of whole rudist valves. Locally the unit may grade to oolitic bank or anhydrite lithofacies (Coogan and others, 1972).

In the Veracruz basin, El Abra–like lithofacies, consisting of rudist or miliolid-rich, dark to light gray limestones with interbeds of microolites, is termed the Orizaba Formation (Benavides, 1956). The middle part of this limestone unit is assumed to be early Cenomanian equivalent. In southern Mexico, in the State

of Chiapas, a thick, monotonous sequence of limestones occurs that is called the Sierra Madre Limestone. Currently, most authors treat this unit as a group subdivided into formations. The lower unit, the Cantelha Formation consists of between 400 and 900 m of dolomite and dolomitic limestone with occasional beds of fossiliferous limestone that are considered low-energy platform deposits (Sanchez Montes de Oca, 1969). The orbitolinid fauna indicates that the unit is mainly Albian. The Cantelha grades upward into the Cenomanian through Turonian beds of the Cintalapa Formation. The Cintalapa consists of as much as 750 m of light-colored, medium-bedded limestones with occasional intervals of dolomitic limestone. Locally, rudist bivalves are sufficiently abundant to indicate development of patch reefs. The Cintalapa appears to be a shallow-water platform deposit in the Tuxtla Gutierrez region of Chiapas. To the north, in the Macuspana and Isthmus Saline basins, sequences of mudstones and dolomites occur, the middle parts of which are presumed to be early Cenomanian. On the Yucatan Peninsula, carbonate and anhydrite deposits of the Yucatan Formation are, in part, believed to include Albian-Cenomanian equivalents, but thickness and character are poorly known (Lopez Ramos, 1975).

Thickness. The lower Cenomanian section in the Gulf of Mexico basin ranges as much as 400 m thick in basinal se-

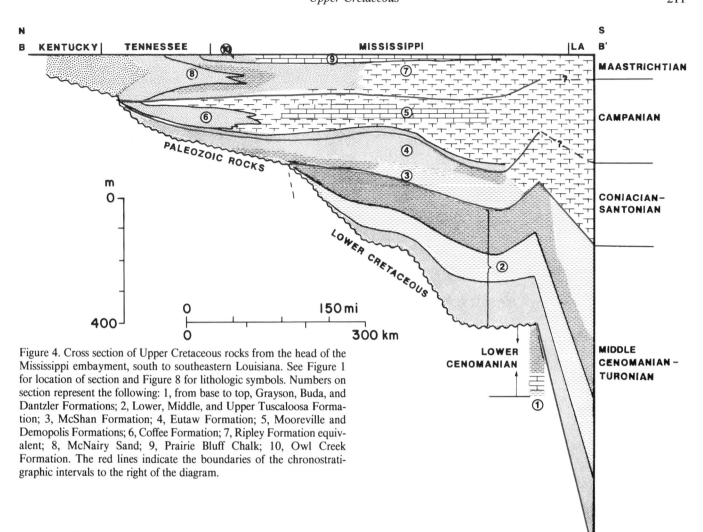

Figure 4. Cross section of Upper Cretaceous rocks from the head of the Mississippi embayment, south to southeastern Louisiana. See Figure 1 for location of section and Figure 8 for lithologic symbols. Numbers on section represent the following: 1, from base to top, Grayson, Buda, and Dantzler Formations; 2, Lower, Middle, and Upper Tuscaloosa Formation; 3, McShan Formation; 4, Eutaw Formation; 5, Mooreville and Demopolis Formations; 6, Coffee Formation; 7, Ripley Formation equivalent; 8, McNairy Sand; 9, Prairie Bluff Chalk; 10, Owl Creek Formation. The red lines indicate the boundaries of the chronostratigraphic intervals to the right of the diagram.

quences. Thicker sections may be present, for example, in the Yucatan Peninsula, but there, as in many other areas, sufficient biostratigraphic information is lacking to indicate how much of a given interval belongs to the lower Cenomanian.

Along the northern rim of the basin, post–early Cenomanian erosion and truncation of beds results in significant variation in thickness. On the outcrop the Grayson, Buda, and equivalent units total 30 m or less and are commonly represented only by erosional remnants. The thickest lower Cenomanian deposits in the northern Gulf of Mexico basin are found in southwest Mississippi, where the Dantzler Formation ranges between 300 and 400 m, and in the East Texas basin where as much as 200 m of deposit are present in areas of Maness Shale occurrence (Anderson, 1979). Farther south, in the Rio Grande embayment, Del Rio–Buda beds attain a thickness of about 160 m. Winter (1962) has noted irregularities in thickness distribution of the Del Rio in the subsurface of southwestern Texas. The unit thins over structural highs, thickens on the downthrown sides of faults and thins over the Stuart City reef trend.

In Mexico, units equivalent to those of Texas are present, but thickness is poorly known. In the states of Tamaulipas and Veracruz, the upper 300 m of the Tamaulipas Superior appears to

be Cenomanian in age, but how much of the maximum recorded 2,537 m of the El Abra Limestone is Cenomanian is unknown. Areas farther south in Mexico have thick units of carbonates that similarly are dated only as ranging from Albian through Cenomanian.

In sum, thicknesses of the units in the basin margin sections show the effects of mid-Cenomanian erosion in most areas of the northern Gulf of Mexico basin margin. Locally, pinchout may occur over arches or highs, and very locally, as in central Texas, thickness may vary slightly because of infilling of preexisting topography. In most of these areas, downdip thickening is gradual, and few pronounced thick basinal fills are evident.

Stratigraphic relations. In all areas of the Gulf of Mexico basin, lower Cenomanian beds appear transitional with those of the Albian except for local basin margin sections. For example, in Texas both the Grayson and Del Rio are conformable upon underlying units except where the Del Rio feathers out on the flank of the San Marcos arch (Young, 1986b). In this case, beds of the middle Buda lie disconformably upon older units. In most areas, deposition was continuous through the lower Cenomanian.

EAST TEXAS BASIN

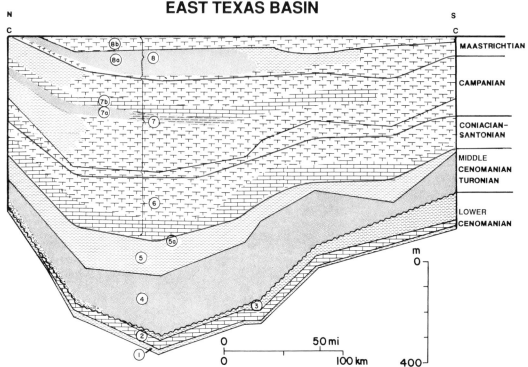

Figure 5. Cross section of Upper Cretaceous rocks in the East Texas basin. See Figure 1 for location of section and Figure 8 for lithologic symbols. Numbers on section represent the following; 1, Grayson Formation; 2, Buda Limestone; 3, Maness Shale; 4, Woodbine Group; 5, Eagle Ford Group; 5a, Sub-Clarksville beds of the Eagle Fort Group; 6, Austin Group; 7, Taylor Group; 7a, Wolfe City Sand; 7b, Pecan Gap Chalk; 8, Navarro Group, 8a, Nacatoch Sand; 8b, Corsicana Marl and Kemp Clay. The red lines indicate the boundaries of the chronostratigraphic intervals shown to the right of the diagram.

However, from central to north Texas a disconformity separates the Del Rio–Grayson from the overlying Buda.

Lower Cenomanian deposits are unconformably overlain by younger deposits throughout the northern part of the Gulf of Mexico basin except in the center of the East Texas basin and perhaps in the deep subsurface areas of deposition of the Dantzler Formation in southern Mississippi. Hazzard and others (1947) have maintained that a disconformity exists at the top of the lower Cenomanian even in the East Texas basin, but additional well data suggest transitional deposition.

Southward in Mexico, deposition in the basin sequences was continuous, but on the Tuxpan and Valles–San Luis Potosi platforms and through much of the Sierra Madre Oriental, there is karst development at the top of the lower Cenomanian El Abra deposits (Coogan and others, 1972). In these areas, overlying beds range from the Turonian to the Oligocene.

Biostratigraphy and age. Interpretation of the Albian-Cenomanian boundary in the Gulf of Mexico basin remains controversial. The two most used groups, foraminifers and ammonites, both provide usable boundary markers, but they indicate somewhat different stratigraphic levels. Young (1986a) suggests the Albian-Cenomanian boundary be placed at the base of the ammonite zone of *Plesioturrilites brazoensis* (Roemer), with the remainder of the lower Cenomanian encompassed within the

successive ascending zones of *Graysonites adkinsi, G. lozoi,* and *Budaiceras hyatti.* In this scheme the base of the Cenomanian would lie within the Georgetown Formation. Mancini (1979) has reviewed the planktonic foraminifer evidence of Pessagno (1969) and Michael (1972) and concluded that the boundary lies within the Grayson Formation at the first appearance of *Rotalipora evoluta* and *Praeglobotruncana delrioensis.* According to Mancini, this boundary would coincide with the boundary between the ammonite zones of *P. brazoensis* and *G. adkinsi.* Whichever boundary is used, the disparity is not great.

In southern Florida, the miliolid *Nummoloculina heimi* Bonet, has been found in several deep wells (Applin and Applin, 1965). Thus, Washita-equivalent beds are presumed present, but only the report of the pelecypod *Pecten (Neithea) texanus* can be considered to be indicative of the Cenomanian part of the Washita. *N. heimi* has also been reported from deep wells in southern Alabama and Mississippi, but published biostratigraphic data are sparse. In the outcrop and shallow subsurface of Texas there is abundant evidence from both microfossils and megafossils to substantiate the early Cenomanian age of the Grayson, Del Rio, and Buda Formations. The first occurrence of such species of planktonic foraminifers as *Rotalipora evoluta* and *Praeglobotruncana delrioensis* are used as markers for this interval (Mancini, 1979), along with other species cited by Pessagno (1969). Among ammonites, species of *Plesioturrilites, Graysonites,* and *Budaiceras* have proved valuable for zoning the lower Cenomanian in Texas and northern Mexico (Young, 1977, 1986a).

In Mexico, south of the states of Coahuila and northern Zacatecas, ammonites are extremely rare; most age assignments are based on microfossils and, to a minor extent, rudist pelecypods where their stratigraphic ranges have been documented. Böse (1923) described an early Cenomanian ammonite fauna from the Cuesta del Cura Formation that included species of *Tetragonites, Turrilites,* and *Anisoceras.* Locally, this formation also contains rudist bivalves and planktonic foraminifers, such as *Rotalipora evoluta,* indicative of the early Cenomanian (Pessagno, 1969), and other species indicative of the *Rotalipora gandolfi* Zone as defined by Ice and McNulty (1980). In the carbonate units of central and south Mexico—for example, in the Tampico-Misasntla and Veracruz basins—correlation in reefal units of the lower Cenomanian, like the El Abra, is mainly based on larger foraminifer species of the *Nummoloculina heimi* Subzone. In addition, the pectenoid pelecypod *Neithea roemeri,* a Buda species, occurs in the upper part of the formation (Muir, 1936). Coogan (1977) described species of rudists belonging to *Caprinuloidea* and *Mexicaprina* from similar stratigraphic levels. In peri-reefal facies of the Orizaba Formation, calpionellids, such as *Pithonella trejoi,* are biostratigraphically useful. On the Yucatan platform, shallow-water carbonates contain the elements of the *Nummoloculina heimi* Subzone. Deep-water pelagic sediments encountered in the Deep Sea Drilling Project cores from site 97, Leg 10, located on the Campeche Escarpment, yielded an early Cenomanian planktonic foraminifer assemblage that includes *Rotalipora appenninica, R. evoluta,* and *Praeglobotruncana delrioensis* (McNeeley, 1973).

Conditions of deposition and paleogeography. The patterns of sediment source, depositional basins, "reef" trends, and structural highs that controlled the conditions of deposition of the lower Cenomanian units are a continuation of those seen for the Lower Cretaceous outlined in the previous chapter. However, the foundering of the "Stuart City Reef" trend along the northern and northwestern rim of the Gulf of Mexico basin during the close of the early Cenomanian (Winter 1961; Young, 1972) set the stage for new controls on patterns of deposition. During later parts of the Late Cretaceous, major changes occurred, new embayments and positive structures appeared, and the onset of the Laramide orogeny changed the depositional patterns so long in place. Thus, the lower Cenomanian deposits represent the last episode of the long history of the "Comanche Series" major cycle of deposition on the northern margins of the

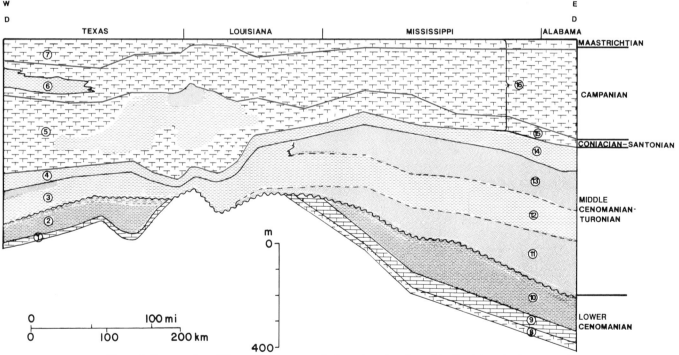

Figure 6. Cross section of Upper Cretaceous rocks from the subsurface of the East Texas basin, eastward along the southern flank of the Sabine uplift to Mobile Bay, Alabama. Mainly after Anderson (1979). See Figure 1 for location of section and Figure 8 for lithologic symbols. Numbers on section represent the following: 1, Grayson and Buda Formations; 2, Maness Shale; 3, Woodbine Group; 4, Eagle Ford Group; 5, Austin Group (part); 6, Upper part of Austin Group and Taylor Group; 7, Navarro Group; 8, Grayson equivalent; 9, Buda equivalent; 10, Dantzler Formation; 11, Lower Tuscaloosa; 12, Middle Tuscaloosa; 13, Upper Tuscaloosa; 14, Eagle Ford equivalent; 15, Eutaw-McShan equivalent; 16, Selma Chalk undifferentiated. The red lines indicate the boundaries of the chronostratigraphic intervals shown to the right of the diagram.

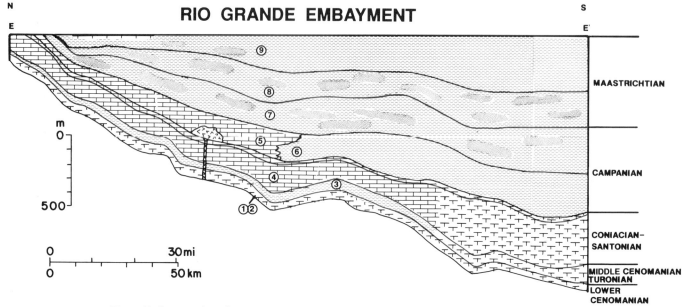

Figure 7. Cross section of Upper Cretaceous rocks in the Rio Grande embayment of southern Texas. See Figure 1 for location of section and Figure 8 for lithologic symbols. Numbers on section represent the following: 1 and 2, Del Rio Formation and Buda Limestone; 3, Eagle Ford Group, 4, Austin Group; 5, Anacacho Formation; 6, Upson Clay; 7, San Miguel Formation; 8, Olmos Formation; 9, Escondido Formation. The red lines indicate the boundaries of the chronostratigraphic intervals shown to the right of the diagram.

Gulf of Mexico basin. Southward into Mexico, however, depositional patterns continued in much the same fashion for a considerable time.

Along the northeastern margin of the Gulf of Mexico basin, much of the record of early Cenomanian deposition is clouded by poor biostratigraphic control and probable mid-Cenomanian erosion. It is possible that carbonates continued to be deposited in southern Florida, but deposition may have ceased during this time in central Florida. In the Florida Panhandle and southeastern Alabama, a few fossils suggest that mainly marginal marine conditions prevailed during the early Cenomanian. Westward, in southwestern Mississippi fluvial to marginal marine conditions also prevailed in which red terrigenous clastics grade downdip into a unit that is a marine limestone (Buda?) at its base, that upsection grades to red to gray, commonly carbonaceous sandstone, and shale (Dantzler). In south-central Louisiana, reefal or carbonate bank equivalents of the Grayson-Buda interfinger landward into interbedded shale and brown limestone (Anderson, 1979, Plate 4). Presumably, these downdip carbonates are the eastern extension of the Stuart City Reef trend that extends westward through Texas and southward into Mexico.

Along the northwestern margins of the Gulf of Mexico basin, deposition during the early Cenomanian occurred in marine, inner neritic sites in north Texas. Here the Grayson marls and shales represent a period of increased influx of terrigenous deposits compared to the underlying Georgetown (Mancini, 1977). Southward, in central Texas, these "blanket clays" become less calcareous (Del Rio Formation); some of them were

deposited in hypersaline waters. Hayward and Brown (1967) suggest that the lower part of the Del Rio represents a regressive phase that is followed by a transgression in the upper part. This transgression led to a return to carbonate deposition represented by the Buda. Young (1986b), in contrast, pointed to the feathering out of the total Del Rio section on the flanks of the San Marcos arch and suggested that the Del Rio represented only a regression. The Buda Limestone represents the final transgressive phase of the lower Cenomanian. It is only irregularly present over the Del Rio through much of the outcrop because of mid-Cenomanian erosion, but it is recognized widely in the subsurface as a period of carbonate deposition. Martin (1967) has suggested that the division of the Buda in south-central Texas into two units represents separate transgressive-regressive episodes. He suggests the lower and more glauconitic biomicritic unit was deposited in a great shallow, semi-protected lagoonal environment and that the upper, more fossiliferous intramicrite represents open-marine, higher energy conditions of deposition. In almost all areas, the Buda shows a very low but regular amount of thickening into the subsurface, which indicates stable conditions of carbonate deposition over a broad shelf (Fig. 6). In Texas, the youngest deposits of the lower Cenomanian are known only in the deeper parts of the East Texas basin where terrigenous, reddish to lignitic clays and interbedded fine sands of the Maness Formation overlie the Buda limestones and represent a final basin-fill event of the early Cenomanian.

Young (1983) indicates that the early Cenomanian seas covered all of Mexico, except for parts of the present-day western margin. He presents evidence for a rapid early Cenomanian transgression over the Coahuila platform that drowned out reef building at the platform margin. The effects of this transgression are not so marked to the south where reef building continued.

In northeastern Mexico, the Cuesta del Cura Formation may reflect poor circulation or, in part, the Cenomanian anoxic event noted in many other areas. In any case, the carbonates of this unit probably represent sedimentation in deeper water than the platform deposits to the west and are probably representative of an outer platformal or basinal environment. In east-central Mexico (Tampico-Misantla basin), rudist banks or reefs occur at the margins of platforms (Valles–San Luis Potosi and Tuxpan platforms). The back-reef, lagoonal, platformal areas were covered by shallow water where oolitic banks developed locally and in some areas supratidal evaporites formed. Miliolid foraminiferal communities were common on the lagoonal floor, and rudist patch reefs, margined by reefal debris, were also frequently developed (El Abra). A coral-rudist barrier reef formed at the platformal margins with both back and fore-reefal talus grading away from the reefal margins (Tamabra). Basinward, the deposits of reefal debris interfinger with, and grade into, micritic limestone and some dark shales that contain abundant planktonic foramin-

ifers that were deposited in deeper open-marine waters (Upper Tamaulipas or Cuesta del Cura; see Chapter 9, this volume). In Chiapas, thick carbonate deposits contain beds composed of radiolitid rudist pavements, suggesting a shallow carbonate-shelf setting that is replaced gulfward in Isthmian Mexico by mudstone to dolomitic units. Shallow-water carbonates and evaporites accumulated on the Yucatan platform, a condition that was to continue throughout the remainder of the Late Cretaceous. The poor record of lower Cenomanian deposits in the center of the Gulf of Mexico basin is sufficient only to show that, where known, pelagic deposition prevailed.

Economic considerations. In Mexico, Albian to Cenomanian rocks have been especially productive of oil. The reefal facies of the El Abra on the Tuxpan platform is a world-famous reservoir. The Tamabra Limestone is also a major producer, especially in the Poza Rica District.

Middle Cenomanian–Turonian

Distribution. Deposits of middle Cenomanian through Turonian age occur widely over the Gulf of Mexico basin (Fig. 9). Along the northern rim of the basin, rocks of this age crop out

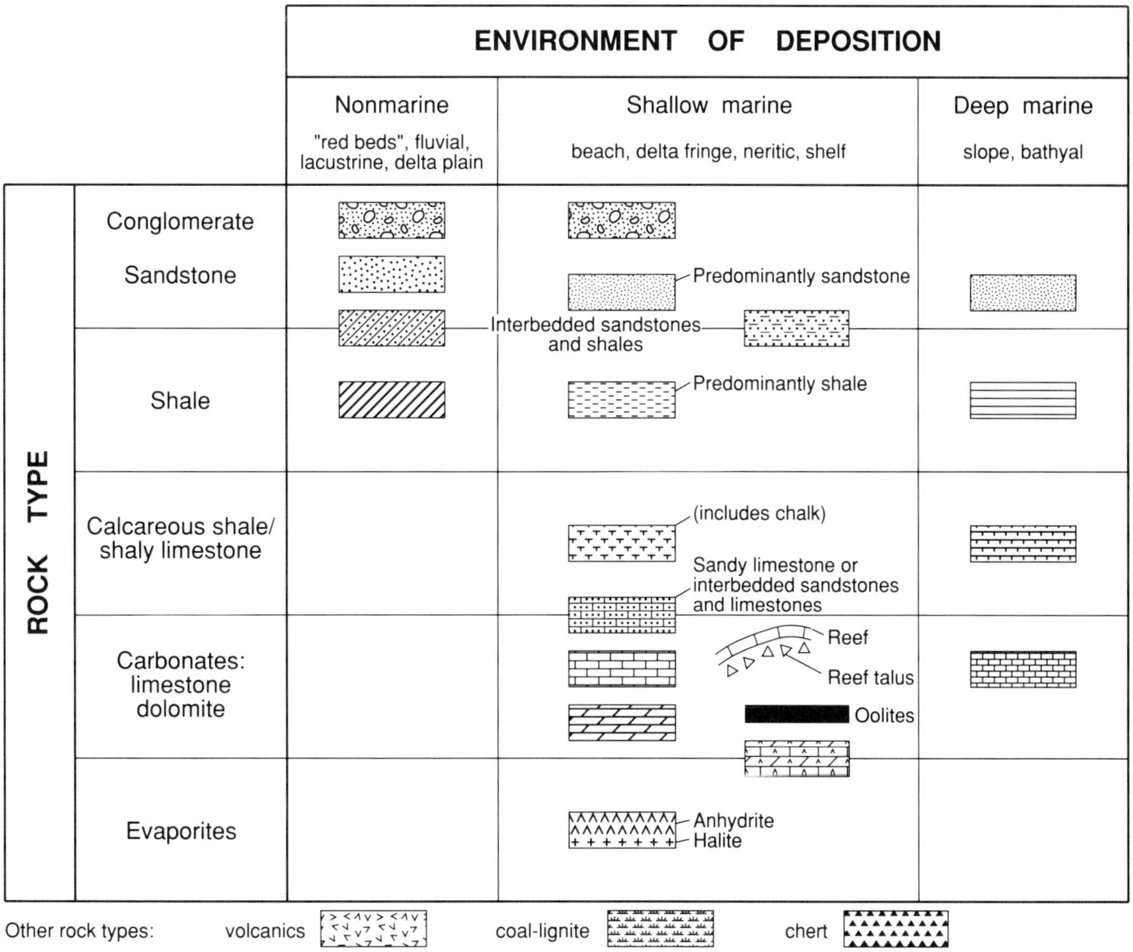

Figure 8. Symbols used in cross sections, Figures 3 to 7, and in lithofacies maps, Figures 9 to 12.

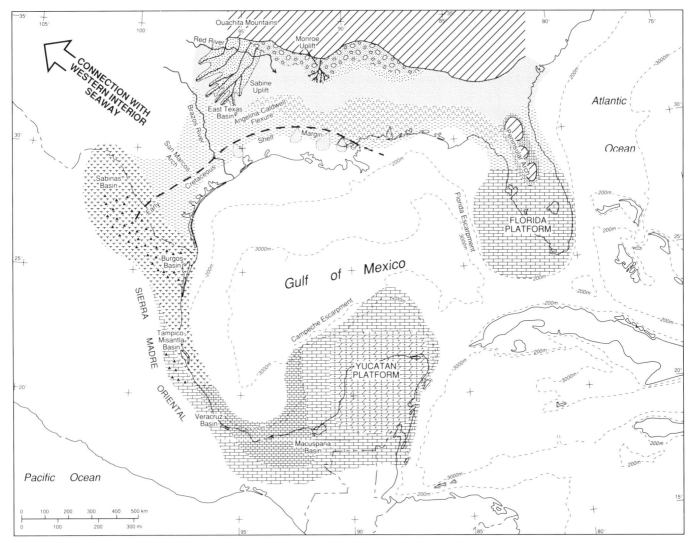

Figure 9. Lithofacies-paleogeographic map, middle Cenomanian through Turonian. See Figure 8 for lithologic symbols.

continuously from eastern Alabama westward into east-central Mississippi where the section is overlapped by younger deposits at about the latitude of the Arkansas-Louisiana border. Outcrops appear again in southwestern Arkansas and continue through northeast Texas. Southward through central Texas, outcrops are virtually continuous. In south-central Texas, these rocks are intermittently exposed, but from south Texas to northern Mexico, the outcrop belt is continuous. In Mexico, exposures occur along the front of the Sierra Madre Oriental southward into Chiapas. Basinward, middle Cenomanian to Turonian rocks are known from abundant well data; they extend counterclockwise, from the Florida carbonate platform, across the Late Cretaceous shelf to the shelf break around the margins of the basin, to the Yucatan carbonate platform. Basinward of the shelf break, little is known about rocks of this age. Some deltaic spillover deposits are locally recorded from deep drilling along the northern Gulf of Mexico basin margin. No mid-Cenomanian to Turonian deposits have

been encountered in the DSDP cores recovered on legs 10 or 77 in the southeastern part of the basin. Middle or late Cenomanian fossils recovered from these cores are all considered to have been reworked and derived from materials transported from the nearby carbonate platforms. In the central Gulf of Mexico basin, mid-Cenomanian to Turonian equivalents are presumed present within the Campeche seismic unit (see Chapter 13, this volume).

Lithostratigraphy. In general, middle Cenomanian through Turonian units are predominantly terrigenous across the northern rim of the Gulf of Mexico basin, but carbonate deposits dominate the sequences on the Florida and Yucatan carbonate platforms and along the central and southern parts of the Gulf Coastal Plain in eastern Mexico (Fig. 9).

In the subsurface of northern Florida and southern Georgia, the varied sequence of terrigenous clastics, which grade downdip into carbonates (Fig. 3), is termed the Atkinson Formation (Applin and Applin, 1965, 1967). Only the lower member of the formation is Cenomanian to Turonian (Gohn, 1988). Sand units of the Atkinson occupy the updip area of Georgia, but become finer grained southward and grade into dark, carbonaceous shale in north Florida. In central Florida, the lower member becomes more calcareous, with oolitic units grading to dense limestones toward the south.

From the panhandle of Florida, westward to the deep subsurface of southern Louisiana, coeval rocks are included in the Tuscaloosa Group or Tuscaloosa Formation of varied usage. In the type area of western Alabama the Tuscaloosa has been divided into two formations (Drennen, 1953), but in the subsurface, informal designations have prevailed. The lower unit, the Coker Formation, is a micaceous, commonly, crossbedded sand interbedded with varicolored clays and gravel. The overlying Gordo Formation is composed of lenticular beds of gravel, reddish sand, and purplish red clay. Volcanic materials occur as sedimentary adjuncts in many subsurface sections in Alabama and southwestern Mississippi. In general, the terrigenous deposits of the Tuscaloosa become finer grained, darker colored, and more marine in character in downdip sections (Fig. 4). The lower Tuscaloosa of the subsurface is composed of massive, medium- to coarse-grained sands with minor thin beds of reddish to gray shale (= massive sand of eastern and central Mississippi). The basal Tuscaloosa becomes younger updip to the northwest. In southern Louisiana, the middle Tuscaloosa is mainly a gray, fossiliferous, micaceous shale with interbedded calcareous, glauconitic sands that thicken downdip. The upper Tuscaloosa, in central Mississippi to northeastern Louisiana, consists mainly of varicolored mudstones, and gray, medium- to coarse-grained sandstone with some gravels. In southernmost Louisiana, immediately south of the shelf break, thick sequences of terrigenous clastics are termed the "expanded Tuscaloosa." The section consists of upward-coarsening successions of gray shale, siltstone, and fine-grained argillaceous sandstones, which are capped by massive, clean, well-sorted, medium-grained sandstones of shallow-water origin.

The Woodbine Formation or Woodbine Group of the western U.S. Gulf Coastal Plain, as should be expected when dealing with any fluvio-deltaic to marginal-marine sequence, is lithologically highly variable and complicated by recurrence of sediment type in any succession. In the type area in east Texas, the lower part of the Woodbine consists of irregularly bedded, generally gray to tan, noncalcareous sands and sandstones that are interbedded with lesser amounts of shale. Locally, the basal unit may be gravelly (Fig. 5). Upsection, shale becomes more common, may be carbonaceous, and contains interbedded sandstone lenses that occasionally bear a low-diversity marginal-marine fauna. The upper Woodbine consists of massive to lenticular, medium-grained, sometimes glauconitic, fossiliferous sand and sandstone that is interbedded with carbonaceous shales of various shades of gray. The sands are commonly tufaceous with lenticular tufaceous sandstone most common to the north and east in the Red River Valley of Texas. In adjacent areas of Oklahoma, the

tufaceous character is also very apparent, and in southwestern Arkansas, the Woodbine includes a thick basal gravel composed of novaculite, quartzite, and chert that is overlain by volcanic sandstone and conglomerates ("Centerpoint volcanics"). South of the type area, the Woodbine thins greatly, accompanied by a decrease in sand content until, south of the Brazos River, the outcrop consists totally of black, generally noncalcareous shales (Pepper Formation) that are representative of the upper part of the Woodbine. Locally, a few fossiliferous, limy seams occur in the shale. The shale thins southward, until in south-central Texas, all Woodbine-equivalent beds pinch out on the flank of the San Marcos arch. In the south Texas outcrop, any Woodbine equivalents have been included in the Eagle Ford Group (Fig. 2).

In the subsurface of the East Texas basin, the Woodbine is as lithically variable as in the outcrop, but terrigenous clastics dominate all sections. Overall, sand is dominant in the northern part of the East Texas basin, with an increase in shale toward the south (Stehli and others, 1972). Most workers recognize a twofold subdivision of the subsurface Woodbine. The lower part is mainly clean sand with subsidiary amounts of silty clay, and beds of gravel. The upper part consists of red to gray clays, tuffaceous sands, and fossiliferous sandy clays. Basinward of the shelf break, the Woodbine consists of turbiditic slope deposits associated with submarine-fan development (Foss, 1979; Fig. 9).

The Eagle Ford Group represents the upper part of the mid-Cenomanian through Turonian period of deposition in the U.S. western Gulf Coastal Plain. These deposits correspond to the fine-grained phase of terrigenous deposition that, in north Texas, began with the coarse-grained deposits of the Woodbine sequence. In the type area near Dallas, Texas, the Eagle Ford is divisible into three parts (Surles, 1979): (1) a lower 5 to 7 m of gray to brown sandstone, siltstone, and shale that contains a basal reworked zone of phosphatic nodules, sandstone or alunite pebbles, and other exotic components; (2) 90 m of brownish, calcareous mudstone interbedded with limestone and siltstone beds and common seams of bentonite; (3) 30 m of gray, calcareous mudstone with a few thin siltstone or limy beds, abundant calcareous concretions, and in the upper part, two distinctive "fish beds" (McNulty, 1965). In the subsurface, the uppermost Eagle Ford deposit is a sand-dominated unit called the "Sub-Clarksville," which may not be represented in the outcrop of the type area. However, to the northeast toward the Texas-Oklahoma border, it is probably represented by outcrops consisting of a lower sandstone unit overlain by laminated shale. The Eagle Ford thins toward southwestern Arkansas, where it consists of blue calcareous shale with common interbeds of red clay. Downdip, in northwestern Louisiana, the Eagle Ford is a thin unit of dark, fossiliferous shale interbedded with tuffaceous sands and greenish bentonitic shale (Hazzard, 1939). Southward from the type area, the Eagle Ford becomes less silty, thins dramatically, and in central Texas, consists of a lower calcareous montmorillonitic clay with limestone interbeds that is overlain by black, blocky, noncalcareous clay in which thin bentonite beds occur at various levels (Pessagno, 1969). South and west of the

San Marcos arch, the Eagle Ford consists primarily of black shale that, farther westward, becomes a sequence of limy flagstones, siltstones, mudstones, and chalky limestones (Boquillas Formation). In Mexico, the Eagle Ford is present in the Sabinas basin and in the northeastern part of the Burgos basin. In these areas it is composed of laminated, fissile, calcareous, and carbonaceous shales that alternate with beds of laminated dark, argillaceous limestones.

Mid-Cenomanian through Turonian rocks in eastern Mexico are called the Agua Nueva Formation. The formation is present mainly along the foothills of the Sierra Madre Oriental and in the subsurface at the Gulf Coastal Plain, from the eastern part of the Burgos basin southward through the Tampico-Misantla basin to the Macuspana basin of the Bay of Campeche. In the northern areas, the formation is composed of microcrystalline limestone with chert nodules that is interbedded with cryptocrystalline argillaceous limestone. Intercalations of carbonaceous laminated black shale, bentonite, bentonitic marls, and calcareous shales also occur in minor amounts. The limestones present a laminated or banded appearance in outcrop. Where the formation overlies the El Abra Limestone, as on the Tuxpan platform, a basal breccia composed of fragments derived from the underlying unit is found at the base of the Agua Nueva. To the south, in the Macuspana Basin, the formation is composed of dark, cherty, argillaceous, microcrystalline limestone alternating with calcarenites rich in planktonic foraminifers and radiolaria.

The Soyatal Formation occurs along the Sierra Madre Oriental west of the Tampico-Misantla basin. In the type area, the lower part of the 200-m-thick formation is a generally unfossiliferous, gray, laminated, argillaceous limestone intercalated with mudstones and beds of limestone that contain primary chert nodules. The upper part of the formation is dominantly a gray to pale yellow, calcareous mudstone with lesser beds of black limestone.

In the Veracruz basin, the upper part of the Orizaba Formation (previously discussed) is considered middle to upper Cenomanian. This unit is conformably overlain by the Guzmantla Formation in the subsurface of the basin and in outcrops to the west. The Guzmantla ranges to as much as 1,600 m thick and is lithologically variable. The formation is mainly medium- to thick-bedded limestone that is occasionally dolomitized. Locally, brecciated limestone lenses may occur. Biogenic components vary significantly through the sequence, giving rise to the term "Pelagic Guzmantla" for the lower part, which contains abundant planktonic microfossils. Another, but more areally restricted, Turonian unit occurs in the subsurface of the Veracruz basin and in outcrops of the adjacent southeastern part of the Sierra Madre Oriental, the Maltrata Formation. This unit conformably overlies the Orizaba and underlies post-Turonian parts of the Guzmantla (Velarde, 1969). It ranges to as much as 1,100 m thick westward in outcrop, but thins to a maximum of 225 m in the subsurface. It is mainly a dark gray, cryptocrystalline to argillaceous limestone with lenses and nodules of black chert. Intercalations of laminated, dark gray, argillaceous limestones occur in the middle part

of the formation, and intraclasts as much as 5 cm in size are present in the upper part of the unit.

In the state of Chiapas, middle Cenomanian through Turonian deposits are assigned to the Cintalapa Formation of the Sierra Madre Limestone (previously discussed) that intergrades with the Jolpabuchil Formation. This latter unit consists of alternating gray to tan, fossiliferous, fine-grained, marly limestone and thin beds of lithographic limestone. The Cintalapa represents the internal platform facies and the Jolpabuchil represents the deeper water, quieter environment of the external platform (Fig. 2). On the Yucatan platform, deposition of evaporites and carbonates continued during the Cenomanian, but during the Turonian, deposition of microcrystalline limestone in the west-central part of the platform suggests that more normal-marine conditions prevailed.

Thickness. Mid-Cenomanian uplift along the northern margin of the Gulf of Mexico basin, coupled with the foundering of the Early Cretaceous reefal shelf margin, provide a setting in which one would expect considerable regional variation in sediment accumulation during the earliest stages of the Late Cretaceous. Uplifted areas provided a source for abundant terrigenous clastics, and thus, the lower part of the middle Cenomanian to Turonian interval records thick accumulation of fluvio-deltaic sediments.

In the northeastern parts of the Gulf of Mexico basin, in Georgia and northern Florida, only some 60 m of sand and clay are assignable to this interval, with local absence of deposits along the Peninsular arch in central Florida. Westward, in the deeper subsurface of Alabama, the Tuscaloosa ranges to near 550 m thick, but thins basinward over the position of the Early Cretaceous carbonate-bank edge. The thickest known Tuscaloosa deposits occur in the basinward sections of the "Gas Trend" in southeastern Louisiana, where up to 1,030 m of section has been penetrated (Christina and Martin, 1979). Thinning of the Tuscaloosa and equivalents occurs over structural highs such as the Monroe uplift in southern Arkansas.

In northeast Texas, the Woodbine–Eagle Ford may total as much as 210 m in outcrop. Thickness is greatest toward the northern source of clastics and generally corresponds to areas of deltaic accumulation. Thus, in the subsurface of the East Texas basin, as much as 600 m of Woodbine and Eagle Ford deposits have been noted (Stehli and others, 1972; Surles, 1979). Both units thin eastward and southeastward and are absent over parts of the Sabine uplift to the east in the Texas-Louisiana border area. Thinning southward is also noted over the Angelina-Caldwell flexure, but gulfward of this point, slope deposits associated with submarine-fan development thicken to 420 m. Southward, along the outcrop in Texas, both units thin or are locally absent over the San Marcos arch (Young, 1986b). In south Texas, mid-Cenomanian through Turonian deposits thicken only slightly into the subsurface in many areas and show a rather uniform maximum of about 75 m (Bebout and Schatzinger, 1977). This trend is followed by thinning to 45 m over the old reef trend and then gulfward thickening to 150 to 180 m.

To the south, in Mexico, in the Sabinas and part of the Burgos basins, as much as 320 m of mid-Cenomanian to Turonian deposits are recorded. Along the front of the Sierra Madre Oriental and in the subsurface of the Tampico-Misantla and more southern basins, coeval Turonian units generally range in thickness from 30 to 300 m. Greater thicknesses may be present in the Guzmantla Formation, which may range to 1,600 m in thickness, but what portion of that total is represented by mid-Cenomanian through Turonian rocks is not precisely known. Along the western part of the Yucatan platform, about 700 m of limestone and evaporites of the Yucatan Formation are assigned to the Turonian, but there is a general trend toward thinning both to the east and to the south across the peninsula (Lopez Ramos, 1975).

Stratigraphic relations. Throughout the northern margin of the Gulf of Mexico basin, middle Cenomanian to Turonian rocks rest disconformably upon rocks ranging in age from Paleozoic to early Cenomanian. In a regional sense, the surface of contact is a low-magnitude angular unconformity. Because the mid-Cenomanian uplift in the northern areas was accompanied by a eustatic sea-level drop that exposed the shelf, virtually to the shelf break in many areas (Anderson, 1979, Plate 4), the subsequent transgressive deposits may be of different ages at their base in different areas. Such factors make the time magnitude of the unconformity variable from place to place. For example, in the deep subsurface of western Alabama, Tuscaloosa beds rest disconformably upon Lower Cretaceous units, but in updip areas the Tuscaloosa overlaps onto rocks of the Paleozoic. Positive structures show individual histories: in the Arkansas and Louisiana area, the Monroe uplift was not covered until the Turonian, but in contrast, the Sabine uplift was rejuvenated during the Turonian with attendant stripping of Woodbine deposits.

Deposition was continuous in the center of the East Texas basin, but toward the basin margins, sands of the middle Cenomanian part of the Woodbine Group generally lie upon rocks of the upper Albian. South of the San Marcos arch, in both surface and subsurface, the Woodbine lithologies are missing, and the Eagle Ford lies on the Buda Formation (Fig. 7).

Through most areas along the northern margin of the Gulf of Mexico basin, younger units lie disconformably upon rocks of the Tuscaloosa and Eagle Ford Groups. Such contacts are usually marked by reworked beds at the base of the overlying unit and piping of younger sediments down in burrows into the subjacent strata. In Texas, this hiatus is also marked by the absence, in outcrop, of the Subclarksville beds. The presence or absence of a disconformity in subsurface sections is not well documented. The presence of increased section, such as the Subclarksville beds, in deeper parts of the East Texas basin, indicates at least a diminution in magnitude of the hiatus, if not a conformable transition in downdip sections.

The marked disconformity present between older rocks and those of the middle Cenomanian to Turonian, so widely present along the northern margin of the Gulf of Mexico basin, has not been recognized in the basinal sequences in Mexico. In many instances this interval falls within formational units or, if coinci-dent with formational boundaries, the units appear to be conformable or transitional. Only rarely, as in the case of the Agua Nueva Formation where it overlies the El Abra Formation in the platforms flanking the Tampico-Misasntla basin, is a demonstrable disconformity present. In other platform areas a disconformity may be present, but it is masked by occurring between lithically similar reefal limestones.

Addy and Buffler (1984) and Shaub and others (1984) have interpreted seismic data to indicate that a disconformity is present at the top of the Washita beds in the deeper parts of the Gulf of Mexico, but more recently Faust (1990) has shown that such a disconformity is not present in the area.

Biostratigraphy and age. Age assignment of the lower part of the Atkinson Formation of south Georgia is based on scattered and sparse data. Data from several wells drilled through the Atkinson Formation have yielded pollen of the *Complexiopollis-Atlantopollis* Assemblage Zone (Christopher, 1982a, b), the ostracode *Rehacythereis eaglefordensis* (Hazel, 1969), planktonic foraminifers of the *Rotalipora greenhornensis* assemblage, and such pelecypods as *Mytiloides "labiatus"* and *Exogyra wollmani.* In sum, the taxa suggest an age range of mid-Cenomanian to Turonian. In Alabama, subsurface strata assigned to the Tuscaloosa Group have yielded planktonic foraminifers of the middle to late Cenomanian (Sohl and Smith, 1980; Mancini and others, 1987). Christopher (1982a, b) presented evidence that the outcropping basal Tuscaloosa of the type section in Alabama, and the entire Tuscaloosa Group of eastern Alabama and western Georgia, is assignable to the *Complexiopollis-Atlantopollis* Assemblage Zone (= late Cenomanian and early Turonian). No information is currently available as to the age of the upper part of the Tuscaloosa type section. At least some of the beds mapped as Tuscaloosa in northern Mississippi and Georgia are much younger than the type Tuscaloosa (Russell, and others, 1983; Christopher, 1982a). In sum, the lower Tuscaloosa of the deep subsurface is the correlative of the Woodbine Group of Texas, but the outcropping Tuscaloosa appears to be an equivalent of all or part of the Eagle Ford Group.

In the outcrop areas of the western U.S. Gulf Coastal Plain, there are abundant biostratigraphic data dealing with the Woodbine and Eagle Ford Groups. Based on the work of many authors, Young and Powell (1976) divided the middle Cenomanian through Turonian rocks of Texas and northeastern Mexico into seven zones, as follows, in ascending order: (Cenomanian) *Conlinoceras tarrantense; Eucalycoceras bentonianum; Acanthoceras alvaradoensis; Kanabiceras septemseriatim;* (Turonian) *Metoicoceras whitei; Coilopoceras eaglefordensis; Prionocyclus hyatti.* Planktonic foraminifers have been studied and zoned for these same rocks by Pessagno (1967), who recognized the *Rotalipora cushmani-greenhornensis* Zone, *Marginotruncana sigali* Zone, and *Whitienella archaeocretacea* Zone. Smith (1981) and Valentine (1984) have provided information on the Turonian nannofossils, and Christopher (1982a, b) and Brown and Pierce (1962), a zonation of the palynomorphs. Different workers have placed tee Cenomanian-Turonian boundary at different positions within the Eagle Ford Group. Current consensus is that the boundary lies near the base of the upper third of the group in a fully developed sequence. Both Adkins and Lozo (1951) and Pessagno (1969) considered that the lower Turonian was missing in north Texas, but Young and Powell (1976) believe indicative foraminiferal faunas are present in a thin interval that represents a condensed zone and that, if present at all, the hiatus is small. Latest Turonian faunas are missing in northeast and north-central Texas, a fact probably related to the regression at that time, but such faunas are present in the remainder of Texas and south into northeast Mexico.

Documentation of the faunas of the Gulf Coastal Plain in Mexico is sporadic at best. In the Sabinas basin and in parts of the Burgos basin, the Eagle Ford contains a fauna similar to that in Texas (Böse and Cavins, 1927). Ammonite faunas are most common in the western part of this area and rare toward the east, but common inoceramid pelecypods of the *Mytiloides "labiatus"* lineage indicate equivalency. To the south, the widespread Agua Nueva Formation contains the Cenomanian ammonites *Mantelliceras* (Muir, 1936) and *Eucalycoceras* (Young, 1977) and Turonian inoceramid pelecypods of the *Mytiloides labiatus* lineage. Ice and McNulty (1980) have reported early middle Cenomanian through middle Turonian planktonic foraminifers from the unit. Along the trend of the Sierra Madre Oriental, the Soyatal Formation is generally sparsely fossiliferous, but Segerstrom (1961) has reported *Mytiloides labiatus* from the Soyatal in the state of Hidalgo, and de Cserna and Bello-Barradas (1963) found specimens of the rudist pelecypod *Hippurites resectus mexicanus* in San Luis Potosí. Both species indicate that the containing beds are Turonian.

In the Veracruz basin and the Sierra Madre Oriental front, the Maltrata and Guzmantla Formations are Turonian, but also contain some microfossils (*Valvulammina picardi*) of the Coniacian and possibly the Santonian (Bonet, 1969). The Guzmantla may possibly range as low as the early Cenomanian zone of *Dicyclina schlumbergeri*. In addition, the Guzmantla contains a varied macrofauna of rudist bivalves of Senonian aspect, but their stratigraphic ranges are poorly understood (Mullereid, 1947).

Conditions of deposition and paleogeography. The middle Cenomanian was a time of major change in paleogeography, especially in the northern Gulf of Mexico basin (Fig. 9). Uplift along this northern margin, coupled with eustatic sea-level drop, exposed the shelf virtually to its edge. The uplifted margin provided an abundant source of terrigenous materials to the basins on the shelf. Large fluvio-deltaic systems developed in Texas and areas eastward to Mississippi and Alabama. Thus, throughout this northern area, terrigenous clastics dominate in the sections. These deposits were spread across the coastal plain and shelf to the shelf break that lay at the position of the Early Cretaceous rudist reef trend (Stuart City). Late Cenomanian to early Turonian transgression extended well inland to the point where marine deposits overlapped those of the earlier Cretaceous and a connection was made with the waters of the Western Interior Seaway (Kauffman and others, 1977). This connection

was maintained through the remainder of the Cretaceous until some time in the Maastrichtian. The late Turonian represents a regressive phase in many of the north Gulf of Mexico basin areas.

In southern Georgia and north Florida, middle Cenomanian through Turonian deposits, derived from a piedmont source, show a progression from marginal-marine to marine, quartzose, terrigenous clastics and bioclastic nearshore units in the north, to fine-grained, terrigenous shelfal deposits in the south. Locally, along the trend of the Peninsular arch in Florida, erosion of a series of islands provided a source for rims of calcareous sands (Fig. 9). Southernmost Florida remained a carbonate platform. Through the panhandle area of Florida, the sediments are nearshore deposits of interdeltaic facies.

In the near-coast subsurface of Alabama to Louisiana, the mid-to-upper Cenomanian (lower Tuscaloosa) is composed of a basal, fluvial, deltaic sand overlain by a series of prodeltaic shales with interbedded, deeper water, turbiditic sand bodies that grade landward into a complex of barrier-island to delta-margin sands (Funkhouser and others, 1980; Berg and Cook, 1968). These deposits thin over the Early Cretaceous carbonate-bank edge and subsequent distributary systems breech the drowned reef, spilling sands over the edge. The net result is the development of a series of fore-bank deltaic splays extending into the Gulf of Mexico basin (Fig. 9). Updip, the mid-Cenomanian sequence displays massive sands of a fluvial meander system that is overlain by deposits of a deltaic distributary system. This sequence suggests upslope or shoreward migration of the deltaic front. All these deposits are overlain by inner neritic shale representing the late Cenomanian–early Turonian transgression (T6 of Kauffman, 1977). It is the deposits of this transgressive phase that are the basal units of the outcropping Tuscaloosa at the type section in western Alabama, but laterally they grade to fluvio-deltaic deposits.

In the western U.S. Gulf Coastal Plain, mid-Cenomanian deposits were also deltaic. The main source of these terrigenous clastics was from the north, in the Paleozoic and metamorphic terranes of the Ouachita Mountains in Arkansas and southern Oklahoma. Within the shelf, the Monroe uplift of southeast Arkansas and northeast Louisiana seems to have been a positive structure until it was covered by Turonian deposits. In the southern part of the East Texas basin, Foss (1979, 1980) has interpreted the mid-Cenomanian (Woodbine) deposits to represent those of a prograding shelfal margin with submarine fan sediments developed basinward of the Early Cretaceous shelf margin. Behind this margin, the sequence begins in fluvial channel and overbank deposits that grade upward into thin-bedded, nonbioturbated sands and are followed by thick-bedded, bioturbated units that are capped by Eagle Ford shale. Updip, in the south and southwest parts of the East Texas basin, gravel locally forms the basal mid-Cenomanian unit, but otherwise the lower part is a transgressive marine shale (Oliver, 1971). To the northeast, there was concurrent development of fluvio-deltaic progradation (Dexter Fluvial System and Freestone Delta System). With time, these coarser terrigenous clastics progressively overrode the transgressive marine shales to the south. Through time, the feeder streams of the deltaic complex migrated eastward until in the early late Cenomanian, much of the area previously occupied by the fluvial system because the site of strandline deposition.

During the latter part of the middle Cenomanian and early part of the late Cenomanian, rejuvenation occurred in the sediment source area of southern Oklahoma and southeastern Arkansas (Oliver, 1971). Toward the north and northeast of the Sabine Uplift, the contribution of detrital volcanic materials increases greatly (Centerpoint volcanics). Some 17 mid-Cretaceous volcanoes have been identified in this area (see Chapter 6, this volume). This interval is also marked by erosion in many updip areas and was followed by crustal downwarping and transgression that drowned out the fluvio-deltaic systems. In the East Texas basin, initial Eagle Ford deposits represent a complex of mud-dominated deltas, with the main sediment source areas remaining to the north. Maximum transgression, and concomitant greatest extent of the Eagle Ford sea, occurred during the latter part of the late Cenomanian into the earliest Turonian. Mud and sand were brought into this sea by prograding deltas, but limestones were deposited in marginal embayments and interdistributary areas. This period was followed by an increased supply of terrigenous material, especially sand and mud, and a decrease in limestone formation. During the early Turonian there was activation of the Sabine uplift. This resulted in erosion of both Woodbine and lower Eagle Ford deposits from the uplifted area. The products of the stripping were redeposited as a sequence of fringing sands and westerly prograding fan deltas (Oliver, 1971; Halbouty and Halbouty, 1982). Terrigenous clastics were also spread, mainly by turbidity currents, southward from the Sabine uplift across the Angelina-Caldwell flexure and the site of the Early Cretaceous shelf margin, into the deeper part of the Gulf of Mexico basin, forming a series of slope-fan lobes that grade basinward into distal turbidites (Siemers, 1978). Late Turonian deposition in east Texas ended with a flood of northern-derived, coarser terrigenous sediments prograding into the regressing shallow Eagle Ford sea (Subclarksville sands).

From central Texas and southward from the flanks of the San Marcos arch, updip outcrop and shallow subsurface areas show intermixtures of flaggy limestones and silty shales with occasional bentonite seams that are interpreted to have been deposited in lagoonal to inner neritic environments. Locally, marginal sections composed predominantly of volcanic detritus are present. Downdip, the Turonian deposits become a calcareous, silty shale that grades to less calcareous and less silty shales gulfward. These shales were deposited under low-energy conditions on a generally shallow shelf.

Throughout northeastern Mexico, middle Cenomanian through Turonian deposits are mainly terrigenous (Eagle Ford), consisting primarily of carbonaceous to calcareous muds in the Sabinas basin and northeastern part of the Burgos basin. These muds were deposited in shallow-marine waters on the outer part of a platform under conditions of restricted circulation. To the south, equivalent deposits become more calcareous. In the

Tampico-Misantla basin, argillaceous carbonates alternate with carbonaceous carbonates and shales (Agua Nueva). To the south and southeast, deposition of mudstones and dark argillaceous carbonates compose most basin fills, but the Turonian beds appear to have a lesser terrigenous component than the Cenomanian. Bentonites and siliceous material, generally in the form of chert, are common to most areas of these southern basins. Most of these dark carbonaceous muds and limestones suggest deposition in a reducing environment and the abundance of planktonic microfauna suggest open-marine waters. Upward-shallowing is suggested in some sections, such as that of the Maltrata Formation of the Veracruz basin, wherein shallow-water benthonic microfaunas progressively increase toward the upper part of the unit. This pattern may correlate with the general regressive phase of the upper Turonian seen in the northern part of the Gulf of Mexico basin and in the Western Interior Seaway.

Throughout the Cenomanian and Turonian, the Yucatan platform was covered by shallow waters and remained a site of deposition of limestones and evaporites. Drilling in the Gulf along the margins of the Yucatan Shelf during the course of DSDP Leg 10 failed to encounter any in situ middle Cenomanian through Turonian deposits. However, some middle Cenomanian pelagic deposits were encountered in coreholes drilled during the course of DSDP Leg 77 in the area between the Campeche and Florida Escarpments (Watkins and McNulty, 1984).

Economic considerations. The abundance of terrigenous clastics and organic-rich shales deposited during the middle Cenomanian through Turonian in the northern part of the Gulf of Mexico basin provided excellent hydrocarbon source and reservoir rocks. The East Texas Field is the second largest oil field in North America. It is estimated that the field has yielded more than 6 billion barrels of oil. The main producing level is the Woodbine sands where they are truncated along the western flank of the Sabine uplift, forming a simple stratigraphic trap. The dark organic shales of the Eagle Ford are considered to be a major source rock and a seal, both for this and other lesser oil accumulations. In southern Louisiana, numerous gas fields have been developed in the "expanded" section of Tuscaloosa sands and shales that lie gulfward of the Early Cretaceous carbonate-bank edge. Most of the traps are associated with down-to-the-south faulting. Eastward, the Tuscaloosa sands are also producers of oil and gas. In south Alabama, significant production in the South Carlton and Pollard fields is from Tuscaloosa sands (Pilot Sand) that have been interpreted as marine bars. In other areas, such as the Mallalieu field in southwestern Mississippi, the producing sands have been interpreted to be point-bar sandstones and narrow channel-fill sandstones, all part of the fluvial system of the lower Tuscaloosa.

Coniacian-Santonian

Distribution. In the Gulf of Mexico basin, Coniacian and Santonian rocks crop out over a wider area than those of the underlying Cenomanian and Turonian (Fig. 10). The outcrop

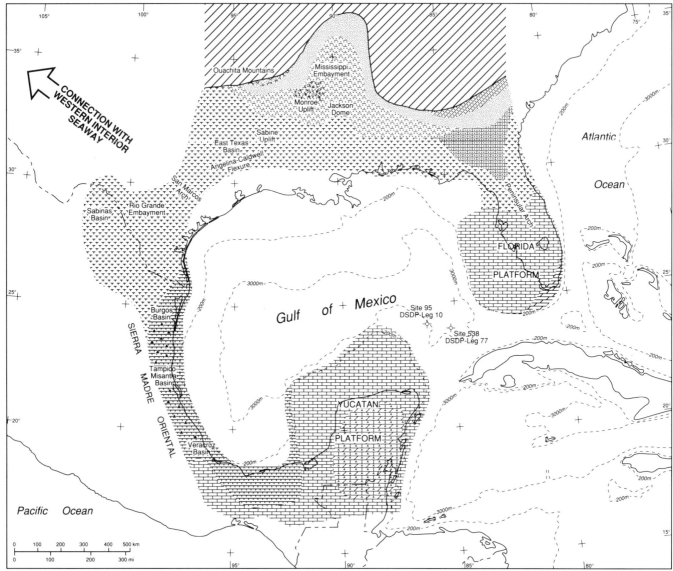

Figure 10. Lithofacies-paleogeographic map, Coniacian and Santonian. See Figure 8 for lithologic symbols.

line is interrupted only in such areas as the Mississippi embayment, where they are overlapped by younger deposits, and in central Texas, where the units not only thin over the San Marcos arch, but are disrupted by faulting. In Mexico, structural complications also disrupt the outcrop line, and in Yucatan, coeval beds are known only from the subsurface. During the Coniacian and Santonian, areas such as those along the Peninsular arch in Florida and the Sabine uplift in Texas and Louisiana were finally inundated; these areas were either never covered by Cenomanian and Turonian deposits or had them removed by postdepositional erosion. Beyond the shelf break, in the deeper parts of the Gulf of Mexico basin, Santonian deep-water deposits have been recovered in cores from site 95 of DSDP Leg 10, at the northeast extremity of the Yucatan Shelf and at site 538, Leg 77, on the Catoche Knoll in the southeastern Gulf. Knowledge of the distri-

bution of Coniacian and Santonian rocks in the deeper part of the basin is scanty, but they are probably represented within the Campeche seismic unit and were, most likely, pelagic (see Chapter 13, this volume).

Lithostratigraphy. Coniacian to Santonian rocks contrast most strikingly with those of the middle Cenomanian through Turonian, primarily by the change to carbonate deposition in many areas where deposition of terrigenous clastics had previously prevailed. Deposition of carbonates was dominant during the Coniacian and Santonian in peninsular Florida. The lower part of the 800-m-thick Pine Key Formation is Coniacian and Santonian and consists of alternating hard and soft, light-colored limestone and chalky limestone. In the southeastern tip of the peninsula and into the Florida Keys, the Pine Key Formation grades into the Card Sound Dolomite. The latter is about 450 m thick and consists of tan to gray, fine- to medium-crystalline dolomite that is a shallow-water reef-bank deposit. To the north, the Pine Key grades into a thinner sequence of shallow-water carbonate and terrigenous deposits (Meyerhoff and Hatten, 1974). In northern Florida and southern Georgia, these deposits consist of the uppermost part of the Atkinson Formation, "Beds of Austin Age" (Applin and Applin (1967), and the La Crosse Sandstone. The lower part of this interval consists of delta front to prodelta, fine sands and clays that coarsen and become thicker updip. The upper part consists of light gray, finely textured, chalky limestone that, in its lower part, contains interbeds of marly shale. The La Crosse Sandstone occurs in north-central Florida and consists of a maximum of about 80 m of generally light tan, fine to medium-grained quartzose sand (Babcock, 1969). The sand is underlain by shales of the Atkinson and overlain by chalk of the "undifferentiated Austin."

Westward, through the subsurface of southwestern Alabama and Mississippi, equivalent strata are assigned to the Eutaw Formation. It is generally divided into a lower and an upper unit. The lower part consists of almost 100 m of laminated, carbonaceous, dark gray shales and sands that commonly are glauconitic. The upper unit is mainly massive calcareous to noncalcareous, glauconitic sand and micaceous shale. In coastal subsurface sections of south and southwestern Mississippi the upper 60 m or more of the Eutaw becomes a marly to dense hard chalk (Fig. 4). Bentonites are common in many sections and locally, especially in the Monroe uplift area of western Mississippi, there is an abundance of pyroclastic sediment (Merrill, 1983).

In outcrop, in Alabama and Mississippi, the Coniacian-Santonian section is divided into two formations. The lower unit is called the McShan Formation and consists of as much as 60 m of thin- to thick-bedded, massive to cross-bedded, glauconitic sands; laminated sand and clay; and thick lensing clays. The Eutaw Formation unconformably overlies the McShan. The Eutaw consists of a lower glauconitic, fine-grained, micaceous sand and thin-bedded clay, which upsection becomes mainly a massive or thick-bedded, glauconitic micaceous, fossiliferous, fine quartz sand.

In southern Arkansas and Louisiana, updip Coniacian to Santonian deposits are mainly terrigenous clastics that become finer-grained downdip and grade to calcareous shales or carbonates in the deep subsurface. In the outcrop in Arkansas, the lower unit is called the Tokio Formation, which rests disconformably on older units. The Tokio consists of a maximum of 90 m that includes a basal gravel overlain by cross-bedded to thick-bedded, sometimes glauconitic, fine- to coarse-grained, quartz sand. The gravels are absent in the shallow downdip sections, and gray, calcareous, often tuffaceous clays become increasingly common. In the deeper subsurface of central Louisiana, the lower 30 m of the Tokio grades into white to gray, alternately hard and soft, chalky shale with some interbedded limestones that are correlated with the Ector Chalk of the surface sections of east Texas. The Brownstown Marl is disconformable upon the underlying Tokio in outcrop (Dane, 1929), but is considered by most workers to be

conformable in the subsurface. The upper part of the Brownstown appears to range upward into the lower Campanian. The Brownstown consists of gray to brown, micaceous, calcareous clays, sandy clays, and subordinate sand beds that are about 40 m thick in the outcrop of the type area in Arkansas. Downdip the unit becomes mainly a dark, calcareous clay with some thin limestone beds, but in northern Louisiana, coarser terrigenous clastics again dominate the section. These sands are derived from erosion of the Monroe uplift.

Westward, into northeast Texas, the Tokio-Brownstown interval becomes increasingly calcareous. As new chalky units appear in the section, formational nomenclature changes, and the units are included in the Austin Group. The Austin Group is unconformity-bounded in many areas and ranges from Coniacian into the lower part of the Campanian. The Coniacian-Santonian interval in northeast Texas, on the outcrop, appears cyclic, with a lower chalk unit (Ector) that grades upward to gray clay and finally a sandy unit.

In the subsurface, the Coniacian to Santonian parts of the Austin show a general thickening, from the Sabine uplift to the west and southwest into the East Texas basin. This trend is paralleled by an increase in carbonates westward and southward (Stehli and others, 1972; Fig. 6). The southward increase in carbonates correlates with the "chalk" development in the classic Austin Group type sequence of central Texas. In this area, the Austin Group is almost 100 m thick and has been subdivided into five formations (Durham, 1957; Young, 1985). The Coniacian through Santonian section is disconformable on the Eagle Ford Shale and begins in a unit of grayish massive limestone. These limestones grade upward into a thick interval of whitish, hard to soft chalk that contains an abundance of inoceramid pelecypods. The chalks are overlain, usually conformably, by marly, fragmental limestone and capped by chalks. Northward these units grade to more terrigenous sections in the East Texas basin, and southward they thin on the flanks of the San Marcos arch (Young, 1986b). In south Texas, the undifferentiated Austin thickens to as much as 300 m toward the Rio Grande embayment, and consists of a lower chalk and fragmental limestone that becomes calcareous shale and limestone above. Undifferentiated Austin is also recognized in northeastern Mexico, in the Sabinas basin and in the north-northeastern part of the Burgos basin. In these areas the Austin is mainly a limestone with minor amounts of interbedded calcareous shales in its lower part, but becomes increasingly shaly and marly in the upper part.

From the southern part of the Burgos basin to the basins of southeastern Mexico, the sequence of Coniacian to lower Campanian limestone and shale that conformably overlies the Agua Nueva Formation, is called the San Felipe Formation. The unit ranges from about 40 to 150 m in thickness and consists of greenish gray, generally thin-bedded, crystalline to argillaceous limestones that are interbedded with thin layers of green shale and a few sandstone beds. Chert or pyrite nodules and an abundant planktonic fauna occur in the limestones. Tuffaceous materials are common at some levels, and there is a trend toward increased terrigenous content upward in the sequence, toward the gradation into the overlying Mendez Shale. Along the front of the Sierra Madre Oriental the Tamasopo Formation is a platformal, reefal facies equivalent of the basinal San Felipe (Muir, 1936). Basinal San Felipe limestone and shale are encountered in the Veracruz and Isthmian basins. Southward in Chiapas, parts of the Sierra Madre Limestone have been considered Coniacian to Santonian. Rocks of that age are certainly present, but their assignment to that unit has been questioned by Martínez (1972). In Yucatan, the Coniacian to Santonian part of the Yucatan Formation differs from other parts of this limestone and evaporite-dominated unit in "the relative abundance of rudists, the presence of oolitic and pseudo-oolitic limestone, and the presence of cryptocrystalline limestone" (Lopez Ramos, 1975, p. 274).

Thickness. In most areas, Coniacian and Campanian shelfal deposits are 250 m or less thick, and basin sections are about 400 m. Occurrence of rocks of this age in the deep central part of the basin is limited to the 0.3 m of sediment encountered in cores of hole A, site 538, DSDP Leg 77 from the southeastern part of the basin between the Yucatan and Florida platforms.

In south Florida, accumulations of Coniacian-Santonian carbonates probably exceeds 400 m, but coeval updip terrigenous units of south Georgia do not exceed 100 m in thickness. In northern Florida, along the crest of the Peninsular arch, beds of the upper Atkinson thin or are locally absent. Westward through Alabama and Mississippi to Arkansas, outcrop thickness may range from 30 to near 100 m; the section thickens to 250 m in subsurface sections, but thins again toward the shelf edge. Thicker sections are seen in the Mississippi embayment between the Jackson dome and the Monroe uplift. Here wells have penetrated as much as 325 m of Coniacian to Santonian rocks, but these units are truncated on the flanks of the structures themselves. Westward into Texas, there is overall thickening toward the East Texas basin. Average thicknesses in this basin range from 200 to 250 m, but in certain areas, they may reach 700 m thick (Stehli and others, 1972). Southward in Texas, the Austin reaches 100 m thick in basin areas, but thins upon the intervening highs. Over the San Marcos arch, Young (1985) cites localities above the Balcones Escarpment where the Coniacian and Santonian parts of the Austin are absent. Downdip, in southeast Texas, the Coniacian to Santonian part of the Austin thickens to 200 m on the shelf, but increases to more than 300 m into the Rio Grande Embayment.

In northeastern Mexico, the Austin ranges between 100 and 250 m. Southward, along the front of the Sierra Madre Oriental, the San Felipe may range up to 225 m (Peregrina Canyon). Eastward, into the Tampico-Misantla basin, the San Felipe is relatively thin, attaining a maximum thickness of about 150 m in the northern part of the basin. However, in the southern part of the basin there is a marked thinning to as little as 20 m. Such reduced thicknesses are especially apparent in areas where the San Felipe overlies the carbonate bank facies of the El Abra Limestone. This may reflect thinning over the older Early Cretaceous rudist-reef trend.

On the Yucatan platform, the thickest accumulation of Coniacian and Santonian rocks is in the north-central part of the peninsula (890 m), with thinner sections to the east and southwest.

Stratigraphic relations. Throughout the northern margin of the Gulf of Mexico basin, updip units of the Coniacian rest disconformably upon Turonian or, where these are overlapped, older rocks. From south Texas into the basins of Mexico, with local exceptions, conformity between the Turonian and Coniacian rocks seems to be the rule. The Santonian-Campanian boundary lies within stratigraphic units in many areas, but locally the boundary may be contained within a hiatus.

In Florida to southern Georgia, Gohn (1988) has considered beds of upper Turonian to lowermost Coniacian to be absent. Thus, the Card Sound Dolomite, the Pine Key, and the upper Atkinson all lie disconformably on lower Turonian beds of the lower member of the Atkinson Formation. In the outcrop in Alabama and Mississippi the transgressive, marginal-marine, carbonaceous and gravelly clays of the basal part of the McShan Formation rest disconformably upon the fluvio-deltaic deposits of the Tuscaloosa Group. The regressive deposits of the upper part of the McShan are eroded and disconformably overlain by marine beds of the Eutaw Formation (Russell and others, 1983). Northward in the subsurface of the Mississippi embayment the Eutaw overlaps the McShan and Tuscaloosa deposits and lies upon Paleozoic strata (Fig. 4).

In southwest Arkansas, the Tokio Formation rests disconformably on older units. This discontormity carries well into the subsurface of northern Louisiana, being especially noticeable where the Tokio covers the truncated Eagle Ford and Woodbine deposits, on the flank of the Sabine uplift. The unconformity at the base of the Austin Group in Texas is recognized to the deepest part of the East Texas basin (Stehli and others, 1972), as well as being apparent at the outcrop from north to southwest Texas (Young, 1985). The magnitude of time represented by this disconformity varies greatly along the outcrop line. Disconformities also occur between units within the Austin Group. These are usually developed where one of the units feathers out on the flank of a structural high. In such instances the hiatus may be lost when followed into the deeper parts of the basin (Young, 1985).

The regional disconformity at the base of the Coniacian deposits has not been recognized southward in Mexico, where in the Sabinas basin and the northeastern part of the Burgos basin, both the base and the top of the Austin are said to be conformable. In the basins southward in Mexico, both the lower and upper contacts of the San Felipe are described as conformable in most areas. However, where the San Felipe overlies the El Abra Formation, as in the southern part of the Tampico-Misantla basin, either a pronounced disconformity must be present or the San Felipe has a significantly greater age range here than elsewhere. In the Veracrus basin, deposition of the Guzmantla Formation was continuous across the Turonian-Coniacian boundary and through the Santonian. The nature of its upper contact is in dispute. Some workers support conformable contacts, but others maintain that a disconformity is present with a hiatus that includes most of the Campanian. This same dispute extends to the sections in Chiapas, where the presence of Campanian beds versus Coniacian and Santonian is queried. Continuous deposition prevailed on the Yucatan platform.

Biostratigraphy and age. Published work on the biostratigraphy of Coniacian through Santonian rocks of the Gulf of Mexico basin is regionally uneven. Both for macrofossil and microfossil groups, more work has been done on the sequences in Texas than for any other area.

Little published information is available on the fossils of the carbonate sequence of peninsular Florida except for rather long-ranging foraminifers. The more terrigenous facies into which they grade northward in southern Georgia have yielded pollen assemblages definitive of the Coniacian to Santonian *Complexiopollis exigua–Santalacites minor* through *?Pseudoplicapollis cuneata–Semioculopollis verrucosa* Zones (Christopher, 1982a, b). Westward, the McShan and Eutaw Formations of Alabama and western Georgia have yielded, in addition to the same palynomorph assemblages, the bivalve *Sphenoceramus pachti* and the nannofossils *Lithastrinus floralis, Marthasterites furcatus,* and *Tetralithus obscurus,* which are indicative of late Coniacian through late Santonian equivalence. In Mississippi, lower Campanian ammonites, nannofossils, and planktonic foraminifers have been recovered (Smith and Mancini, 1983) from the upper Eutaw, but the lower part is of the same age as to the east. The Tokio Formation in Arkansas has yielded Coniacian to upper Santonian ostracodes of the *Cythereis dallasensis* and *Veenia quadrialira* Zones from its upper beds (Hazel and Brouwers, 1982). The overlying Brownstone is, in part at least, late Santonian, but mainly Campanian based on both ammonite and ostracode occurrences. Many authors consider the Arkansas Brownstown to be totally Campanian (Pessagno, 1967; Young, 1963; Hazel and Brouwers, 1982). Part of the evidence usually cited in suport of such a view is the report of the occurrence of the Campanian ammonite *Scaphites hippocrepis* from the formation (Dane, 1929). Cobban (1969) assigned the Brownstown specimen to *Scaphites leei* I, a form restricted to Santonian rocks.

Young (1963, 1985) has provided a zonation of the Austin Group of Texas that is based on ammonites as well as stratigraphic range data of other biostratigraphically useful ostreid and inoceramid pelecypods. His ammonite zones *Peroniceras haasi, Peroniceras westphalicum,* and *Prionocycloceras gabrielensis,* in ascending order, serve to divide the Coniacian. The lowest Coniacian is presumed absent because indications of the presence of the *Barroisiceras petrocoriense* Zone are lacking. His Santonian zones, in ascending order, are: *Texanites stangeri, Texanites texanus texanus, T. texanus gallica,* and *Behavites behavensis.* The first three are lower Santonian, and the last zone is assigned to the upper Santonian. The Santonian-Campanian boundary, as defined by Young, is well below that of Pessagno (1967) and Longoria (1984), who would assign the upper part of the Austin Group to the Santonian on the basis of planktonic foraminifers. In contrast, Hazel and Brouwers (1982) proposed the concurrent

last appearance of the foraminifer *Sigalia deflaensis* and first appearance of *Globotruncana elevata* as marking the boundary. Nannofossil workers recognize the first appearance of *Aspidolithus parcus* as the boundary marker. These last three markers occur in the lower part of the upper Austin, and their recognition as boundary indicators has the added advantage of being closely coordinated with the boundary based on ammonites and ostracodes.

In Mexico, the San Felipe Formation varies widely in age in different areas. However, in the areas considered here, e.g., in the Tampico-Misantla basin and along the front of thc Sierra Madre Oriental, the San Felipe contains planktonic foraminifer assemblages of the Coniacian through lower part of the Campanian (Pessagno, 1967; Gamper, 1977a; Soto, 1980).

Bonet (1969) cites Coniacian microfossils as present in the Guzmantla Formation of the Veracruz basin. Farther south, in the state of Chiapas, Martínez (1972) has recognized Coniacian and Santonian microfossil assemblages in rocks here considered part of the Sierra Madre Limestone, but which Martínez considers a separate but unnamed unit that lies between the Sierra Madre Limestone and the Ocozocuautla Formation.

Beyond the shelf, in the deeper part of the basin, Santonian planktonic foraminifers diagnostic of the *Dicarinella concavata* and *D. asymetrica* Zones occur in a thin and condensed section at DSDP site 538, Leg 77, on Catoche Knoll. Otherwise, in-place and well-dated Coniacian and Santonian deposits are unknown in the deep basin.

Conditions of deposition and paleogeography. In many areas of the Gulf of Mexico basin, the late Turonian was a time of regression. Coniacian rocks rest with marked disconformity upon Turonian or Cenomanian rocks in many areas of the northern part of the basin, a relation that extends well into the deeper basin of deposition within the shelf. Southward in Mexico, deposition was continuous in most platform sections and in basinal sequences.

In the east part of the basin, in southern Florida and the Keys, shallow-water, reef-bank carbonate deposits grade westward into deeper-water carbonates (Pine Key Formation). These deeper-water carbonates are bounded on the north by inner-shelf to prodelta deposits in southern Georgia. Locally, along the Peninsular arch, erosion of highs provided a source for areally restricted quartzose clastics (La Crosse Sandstone). Westward in Georgia and into eastern Alabama, bar and lagoonal deposits prevailed. In western Alabama and Mississippi, Coniacian and Santonian sediments represent two cycles of deposition. The earliest cycle (McShan) transgressed and overlapped the fluvio-deltaic deposits of the underlying Tuscaloosa and is represented by terrigenous sediments that were derived from Piedmont and Appalachian sources and deposited in near-shore, low-energy environments. The second transgressive pulse (Eutaw) is represented by deposits of higher energy, commonly with coarse, gravelly, reworked sediments at the base that overlap the underlying unit. Toward the shelfal margin, Coniacian and Santonian sediments become more calcareous and grade to limestone. In

addition to land-derived terrigenous sediments, abundant volcanic detritus is found in the Coniacian to Santonian rocks of the
area. Tuffaceous sands and water-laid volcanics are especially
prevalent in west-central and northwestern Mississippi in proximity to the suggested vents, such as the Midnight Volcano (Mellen,
1958), associated with the Monroe uplift.

To the west, across Arkansas, Louisiana, and into northeast
Texas, terrigenous supply during the Coniacian and Santonian
continued to come from the north in the Ouachita trend. Volcanic detritus is especially prominent early in this time interval
and had its main source in several volcanos in the Murfreesboro,
Arkansas, area (Ross and others, 1929). Such water-laid perido-
tite tuffs, kaolinites, and bentonites are common to the Tokio
Formation and to equivalent units, both westward along the out-
crop and well into the subsurface. Lithofacies similar to those of
the eastern Gulf of Mexico basin are found in Arkansas and
Louisiana. The Tokio, like the McShan, is a transgressive deposit.
The gravels at its base were mainly reworked from the underlying
Woodbine by the advancing sea. The Tokio overlaps the older
Woodbine and northward rests upon Lower Cretaceous rocks.
The Brownstown in updip areas rests upon the eroded upper
Tokio much in the same fashion as the Eutaw of the east lies on
the McShan. Westward in northeast Texas, Coniacian-Santonian
deposition of carbonates increases, and terrigenous contributions
from the northeasterly sources in the Ouachita uplift decrease.
On the platform areas to the west, the carbonates (Austin Group)
exhibit massive, resistant, chalk beds that alternate with intervening softer marly or argillaceous chalk units that suggest some
climatic control on the influx of terrigenous material into the
areas of carbonate deposition. Positive areas between depositional basins received the thinnest sections of carbonates. In
general, the Coniacian to Santonian chalks contain faunas and
such features as rip-up clasts, which suggest deposition on a shallow, open-marine shelf well removed from the shoreline. Scattered thin bentonite beds indicate periodic ash fall, but are not as
prevalent as in the areas to the northeast. In most areas, the chalks
grade downdip to dark, calcareous, shelfal shale in the basins and
across the shelf, but sections thin and change to chalk or limestone at the shelf edge (Angelina-Caldwell flexure), which was a
more positive area of higher energy. Farther west and north from
the present area of outcrop in Texas, chalk deposition was probably continuous with the epicontinental sea of the Western Interior, wherein the coeval Niobrara Chalk was being deposited
over a vast area.

In Mexico, the Coniacian and Santonian seas had contracted eastward 50 to more than 100 km from their greatest
extent during the Turonian, but inundated all of the area of the
present Gulf of Mexico basin. A pattern of heavily terrigenous
deltaic deposition prevailed to the west of the present coastal
plains, but as in the Cenomanian and Turonian, shallow-water
carbonate deposition prevailed on the platforms and in the intervening basins of eastern Mexico. Increased terrigenous input into
the depositional system with time is noticeable even in the basins
of eastern Mexico and is expressed in an upward increase in

terrigenous material toward the top of such units as the San
Felipe Formation. On the platform areas, carbonate-bank facies
are well developed, and rudist patch reefs are distributed in such
units as the Guzmantla. In total, the picture for most of the
Coniacian and Santonian in Mexico was not significantly different from the later Cenomanian to Turonian.

Economic considerations. Coniacian and Santonian rocks
are productive of oil in several areas of the northern part of the
Gulf of Mexico basin. Sands of the Eutaw Formation are reservoir rocks in Mississippi and southwestern Alabama. Much of the
accumulation there is in faulted traps associated with salt dome
structures. Although a minor amount of production has come
from Austin Group rocks in the East Texas basin, these rocks are
most productive in south Texas where, in such places as the
Pearsall field, the oil resides in fracture systems within the chalks
(Stapp, 1977).

Campanian

The Campanian (Fig. 11) marks an interval during which
deposition of shelfal carbonates became more widespread along
the northeastern part of the Gulf of Mexico basin. To the west, in
central and south Texas, there is a pronounced increase in volcanic activity, and during the late Campanian, there is an increased influx of fine terrigenous sediments from western sources,
which probably reflects early stages of the Laramide orogeny. In
Mexico, there is continued progradation of the shoreline eastward
and increased supply of terrigenous clastics to the basins.

Distribution. The outcrop of Campanian rocks follows the
same trend as those of the Coniacian and Santonian interval and
are interrupted only where they are overlapped by younger deposits in the upper Mississippi embayment. In Mexico, Campanian deposits do not extend as far west as do those of the Santonian
and Coniacian. In the deeper parts of the Gulf of Mexico basin,
Campanian rocks are presumed to be represented by an undetermined part of the Campeche seismic unit of Schaub and others
(1984).

Lithostratigraphy. In the subsurface of parts of the U.S.
Gulf Coastal Plain, Campanian rocks are often alluded to as
"Taylor undifferentiated" or "rocks of Taylor Age." In outcrop,
numerous formations have been proposed that are totally or
partly Campanian (Fig. 2). In the keys of southern Florida, the
lower part of the Rebecca Shoal Dolomite is considered to be
partly Campanian. It is a gray to white, crystalline dolomite that
interfingers with the Pine Key Formation to the north in the
South Florida basin (Winston, 1978). The Pine Key Formation is
partly Campanian and represents continuation of deposition of
chalky limestones and interbedded dolomite. In north Florida,
these limestones grade into a sequence of marine clays and marls
that are overlain by micaceous, carbonaceous, glauconitic, clayey
silts and fine sands, which are designated "Beds of Taylor Age."
In the central Georgia subsurface, the whole "Taylor" sequence
becomes silt and sand.

In western Georgia and eastern Alabama, outcropping
Campanian units are composed predominantly of terrigenous

rocks. In Georgia these deposits are divided into the Blufftown and Cusseta Formations, but in Alabama the Cusseta is generally considered a member of the Ripley Formation. The uppermost part of the Cusseta ranges into beds of the lower part of the Maastrichtian. The Blufftown Formation is about 180 m thick and contains three cyclic sedimentary packages. Each consists of a coarse-grained, reworked basal bed that is followed above by a lower, glauconitic, commonly calcareous, fine sand or micaceous clay that grades upward into interbedded silts and clays—which may be carbonaceous—and is capped by shoreface, commonly cross-bedded sands. All the cyclic units represent deposition on the inner shelf. The Cusseta Formation disconformably overlies the Blufftown and, in the updip outcrop, consists of about 50 m of fine-grained, clayey sands of the inner shelf, overlain by increasingly coarse, commonly cross-bedded sand that may contain

abundant *Ophiomorpha*. Downdip, both the Blufftown and the Cusseta become finer grained and grade to clay and chalky deposits. Surface nomenclature has not been applied; these beds are usually referred to informally as "Beds of Taylor Age" or, if dominated by carbonates, they become the lower part of the Selma Group. Laterally, along the outcrop line, toward central Alabama, both formations become finer grained and more calcareous. Thus, to the west the Blufftown grades into the coeval parts of the Mooreville Chalk, and the Cusseta grades into the Demopolis Formation. The Mooreville Chalk of Alabama consists primarily of marl some 60 to 70 m thick. The lower part of the Mooreville is commonly sandy and variably glauconitic with local thin beds of limestone or bentonite. In many sections the basal bed is coarsely glauconitic and quartzose, and commonly contains concentrations of phosphatic molds of molluscs and ver-

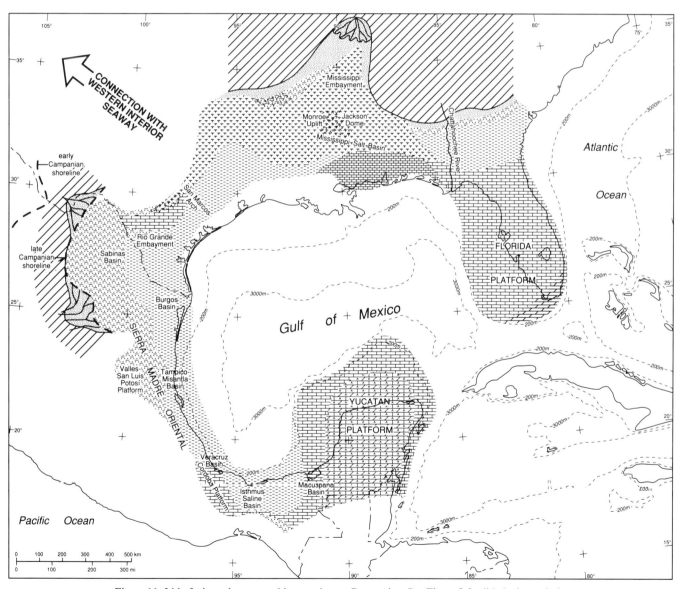

Figure 11. Lithofacies-paleogeographic map, lower Campanian. See Figure 8 for lithologic symbols.

tebrate remains. In the upper part of the Mooreville, terrigenous components increase in abundance, and bedding is less massive. The top of the Mooreville is marked by one to several beds of calcisphere-rich limestone. The overlying Demopolis Chalk attains a thickness of about 135 m. The basal Demopolis is generally a marl that may contain quartzose clastics and concentrations of phosphatic internal molds. This marl is commonly piped down in burrows into the underlying limestone. The middle part of the formation consists of alternating hard and softer beds of relatively pure chalk, but there is an upward gradation to marls as terrigenous components make up a greater part of the rock. The upper contact is gradational with the overlying Ripley Formation. In the deeper subsurface, the Mooreville and Demopolis are usually not separately distinguished and become a part of the Selma Chalk, a unit that includes beds of both the Campanian and the Maastrichtian. Lithically, the subsurface sections of Alabama and Mississippi are dominated by marls and chalks, but locally, from western Alabama through west-central Mississippi, pyroclastic debris is to be found in the Demopolis.

Northward along the outcrop in Mississippi, the carbonate-dominated Mooreville and Demopolis grade into the terrigenous Coffee Formation. The Coffee attains a thickness of almost 60 m. The lower part of the formation consists of thick to massive-bedded, fossiliferous, silty and clayey, glauconitic sand. The upper part of the Coffee is mainly composed of loose, white, quartz sands that are interbedded with laminated to thin-bedded, brownish to gray, commonly carbonaceous clay. These shelfal marine sands and barrier-bar deposits grade northward in Tennessee into a deltaic complex (Russell and others, 1983). From north-central Mississippi westward into the subsurface, the Coffee sands grade into chalk of the undifferentiated Selma Group (Fig. 4).

From southwestern Arkansas to northeastern Texas, Campanian rocks have been divided into a number of formational and member units. In southwestern Arkansas, the upper part of the Brownstown Formation is Campanian and is overlain by the Ozan Formation, which consists mainly of about 75 m of tan to gray, calcareous, micaceous shale that is interbedded with variably glauconitic sand. The lower 1 to 5 m of the formation is commonly a very glauconitic, sandy marl that may contain concentrations of phosphatic internal molds, pebbles of black chert, sharks teeth, and abundant shells. In the subsurface, the Ozan grades eastward in the deeper parts of the Mississippi embayment, and southward in Louisiana into the lower part of the undifferentiated Selma Chalk. The next succeeding unit is the Annona Chalk. It consists mainly of as much as about 90 m of massive to thick-bedded, hard, white chalk that becomes marly in its upper part. The Anonna is widely recognized in the subsurface of eastern Arkansas and Louisiana. Laterally, into northeast Texas, terrigenous components increase, and the unit becomes an impure chalk facies recognizable as far as central Texas (Young, 1985). The uppermost Campanian unit in Arkansas is the Marlbrook Formation, which consists of as much as 60 m of blue to gray calcareous shales and marly chalk.

In northeast Texas, new units and names are introduced for the Campanian section, mainly in response to lithofacies changes. The lowermost unit is the Gober Chalk, which ranges in thickness to 120 m, and is composed of argillaceous chalk, chalky clay, or marl in its lower part, and of a well-indurated limestone in its uppermost part. Eastward, the Gober thins and overlies various parts of the Brownstown, into which it grades. The Gober is overlain by clays of the Ozan Formation, which in turn, are overlain by as much as 60 m of calcareous, fossiliferous, gray to brown sand or sandstone, interbedded with sandy marls. This sand unit is coarser-grained in the northern outcrop belt, thins in the subsurface, and may be lost on the outcrop to the south (Beall, 1964). A 15-m unit of blue-gray, argillaceous, sandy chalk that commonly contains interbeds of marl in its middle part disconformably overlies the sands and is a correlative of the Annona Chalk. The unit thickens both southward to central Texas and southeastward in the subsurface of the East Texas basin, where it becomes a less argillaceous chalk. Upward the chalk grades into the Marlbrook Formation through a transition of chalky marl.

In central Texas, Campanian rocks have been subdivided into numerous formational units (Young, 1965, 1982, 1985). The lowest part consists of about 15 m of chalk and clayey limestone assigned to the upper part of the Austin Group. The overlying Taylor Group consists of a lower 21 m of gray marl and chalk, deposited far from shore in a shallow shelf environment. These beds are overlain by 90 m of greenish to brownish, somewhat calcareous claystone that is disconformably overlain by 15 m of argillaceous chalk, which correlates with the Annona Chalk of Arkansas. The uppermost Campanian of central Texas ranges to 120 m in thickness and consists of greenish to brownish gray clay or claystone, which is transitional to the underlying chalk. Most of the abovementioned outcrop divisions have not been recognized in the deeper subsurface where equivalent Campanian beds are generally referred to as the Taylor Formation or Taylor Group. Other, more localized, formal lithostratigraphic units have been proposed for sequences associated with such features as the Pilot Knob Volcano near Austin, Texas (Young, 1975, 1985).

In south Texas, south of the San Marcos arch, Campanian rocks consist of a succession that begins with carbonate deposits and ends with terrigenous sediments, some of which were deposited under deltaic influence. The section thickens southwestward into the Rio Grande embayment. The lower unit is the equivalent of the Campanian part of the Austin Chalk. The Austin is disconformably overlain by the Anacacho Limestone, a mainly argillaceous, bioclastic limestone that is interbedded with subsidiary amounts of clay and marl. The Anacacho ranges as much as 250 m thick (Spencer, 1965). In outcrop, the Anacacho appears to represent a shallow-water carbonate bank built upon sea-floor elevations that were created by intrusive igneous activity. The limestone is mainly composed of skeletal debris (Wilson, 1983) and downdip it grades into the terrigenous shelf mudstones of the Upson Formation and the lower part of the San Miguel (Fig. 5).

The middle Campanian Upson Formation consists of as much as 150 m of fossiliferous, dark to greenish gray clay. Outcrop exposures arc arcally limited, in part by pre-Navarroan erosion of the updip area, but the unit is widely recognized in the subsurface where it is conformably overlain by the San Miguel Formation. The San Miguel Formation in outcrop consists of as much as 120 m of fossiliferous sands and sandy limestones that are interbedded with gray clays, which generally conformably overlie the Anacacho carbonates. Similar lithologic content is present in the subsurface San Miguel; thicknesses as much as 350 m have been recorded (Lewis, 1977). Weise (1979) and Tyler and others (1986) interpreted the San Miguel as a product of deposition in a wave-dominated deltaic system.

In northeastern Mexico, the Campanian section is as thick as in southern Texas. The Campanian part of the Austin Formation contains alternating thin beds of shale and chalky limestone in about equal proportions. The overlying Upson Formation outcrops more widely than in Texas, occurring in the Sabinas basin and in the northern part of the Burgos basin. It is composed of about 400 m of friable shales and siltstones with sporadic and thin intercalations of fine-grained, calcareous, sandstone beds that contain coquinoid layers. The Upson is considered to be a shallow-water shelfal deposit transitional to the overlying San Miguel Formation. The San Miguel Formation in northern Mexico consists of as much as 300 m of thin- to medium-bedded, sometimes cross-bedded, calcite-cemented sandstones that are interbedded with gray shales. To the south, the unit shows evidence of deposition under deltaic influence, and some subaerial plain deposits reputedly contain fresh-water fossils (Weidie and others, 1972). The upper part of the formation becomes increasingly shaly and is transitional to the overlying Maastrichtian Olmos Formation.

Southward in Mexico, the upper part of the previously described San Felipe Formation is Campanian and is conformably overlain by the Mendez Formation, whose age ranges through the Maastrichtian. It is present in outcrop and in the subsurface, in the eastern and southeastern parts of the Burgos basin, and in the Tampico-Misantla and southeastern basins. The Mendez is principally gray and greenish marls that are interbedded with shale and thin greenish beds of bentonite, but westward, toward the Sierra Madre Oriental, sandy beds become more common. Locally, as in the Tampico-Misantla and Veracruz basins, breccias derived from the El Abra or Tamabra Formations may occur at the base. Overall, the Mendez represents deposition in a low-energy shelfal or basin setting. Where the Mendez rests upon beds of the San Felipe, the contact is conformable, but elsewhere it lies unconformably on such older units as the Lower Tamaulipas, Tamabra, or El Abra. Westward, on the Valles San Luis Potosí platform, the foothills of the Sierra Madre Oriental, and in the west-central part of the Tampico-Misantla basin, the Cardenas Formation appears to be the stratigraphic equivalent of the Mendez Formation. The Cardenas Formation ranges in thickness from 200 to 1,100 m, and like the Mendez, is probably Campanian only in its lower part. Terrigenous sediments dominate in the

Cardenas sequence, and shale is predominant in the lower part of the formation, followed by alternating proportions of interbedded shale, siltstone, sandstone, and bioclastic limestone (Myers, 1968). Locally, the presence of mud cracks and oyster beds, and an abundance of bioclasts attest to the shallow-water nature of much of the deposit. In the type area, the Cardenas lies in fault contact with the underlying Tamasopo Formation. Sansores and Navarrete (1969) and Young (1983) have suggested that the Tamasopo may be a lateral equivalent of the Cardenas.

To the south, in the subsurface of the Veracruz basin, the Mendez Formation grades laterally into a "reefal" unit named the Atoyac Formation, which consists of massive to thick-bedded, platformal, calcarenitic to calciruditic limestone that locally may be dolomitized. In the subsurface the Atoyac is generally about 200 m thick, but in the areas of outcrop of the Cordoba platform, thicknesses are close to 350 m (Viniegra, 1965). Lenses of limestone rich in rudist bivalves occur sporadically, but true reefal frameworks are not present, and the preponderance of fauna favors accumulation as a carbonate bank in very shallow marine waters. To the south, the Mendez Formation is widespread in the Isthmus Saline and Macuspana basins, but southward, in the state of Chiapas, the section changes significantly as the clays of the Mendez are replaced by the deposits of the Angostura and Ocozocoautla Formations. How much of these units should be assigned to the Campanian is currently in dispute (Martínez, 1972; Sanchez Montes de Oca, 1969). Thus, the two above-named units will be discussed along with the Maastrichtian formations.

On the Yucatan platform, the Campanian to Maastrichtian part of the Yucatan Formation consists mainly of anhydrites, but with limestones becoming dominant to the west and north. Wells drilled in the northwestern part of the Yucatan Peninsula have penetrated a body of andesites and andesitic tuffs within the Campanian-Maastrichtian section.

Thickness. The Campanian deposits in the Gulf of Mexico basin may be absent over structural highs, but elsewhere range from a few meters near the basin margin to as much as 700 m in deeper parts of embayments. Throughout the northern margin, Campanian deposits are relatively thin, commonly measuring less than 100 m. This is especially true of the areas dominated by deposition of terrigenous sediments in Georgia, Alabama, and the upper parts of the Mississippi embayment. Greater thicknesses, in excess of 300 m, are found in the carbonates of the southern part of peninsular Florida and along the axis of the Mississippi salt basin, in southern Mississippi and southwestern Alabama. Within this later basin, local variation in thickness is related to thinning of deposits over the crest of salt domes and thickening in the "rim synclines" that margin the domes. Thinning or pinchout of lithic units on the flanks of the Jackson dome and Monroe uplift of the Mississippi embayment reflects positive movement of these structures during the Campanian. Westward, in the East Texas basin, there is less pronounced thickening of Campanian units in the subsurface than for earlier deposits; a maximum thickness of about 270 m is recorded. The greatest thickness of deposits of this age in Texas occurs in the Rio Grande embayment where, in the subsurface, more than 600 m of Campanian rocks have been penetrated. Minor thickening of units, such as the Anacacho, is related to deposition around volcanic vents or structures related to intrusive bodies, but the greatest thicknesses are related to deltaic deposition. Such units as the San Miguel Formation contain as much as 360 m of terrigenous clastics accumulated in a deltaic complex. Similar thicknesses and lithologies of these units occur in northeastern Mexico. Farther to the south in Mexico, determination of thicknesses of Campanian rocks is less precise because of controversy as to the age limits of the various units. Thicknesses of more than 1,000 m have been recorded rarely for such basinal units as the Mendez, but only perhaps as much as one-half of that thickness can be attributed to Campanian rocks. Units deposited on the platform areas are generally thinner: for the Campanian component, commonly on the order of 90 to 180 m. In Yucatan, Campanian rocks are estimated to range from about 450 to 900 m thick.

Stratigraphic relations. In most areas of the Gulf of Mexico basin, the Santonian-Campanian boundary falls within lithostratigraphic units. The Campanian-Maastrichtian boundary has been cited as disconformable in some areas, but in most it lies within conformable sequences. In the eastern U.S. Gulf Coastal Plain, especially in the updip and outcrop area, Campanian deposits represent cyclic transgressive-regressive, disconformity-bounded, sedimentary packages. Such cycles can be discerned both in the terrigenous clastic and the carbonate-dominated sequences. Thus, in the Chattahoochee River Valley of Georgia and Alabama, the contact between the Blufftown and Cusseta Formations is marked by a distinct bed of coarse sand containing worm and disoriented shell debris, concentrations of bone fragments, bored cobbles, and phosphatic pebbles (Monroe, 1941). This bed lies at the base of the Cusseta and is overlain by shelfal marls or sands. This same coeval break in section is also evidenced at the base of the Demopolis Chalk of central Alabama and east-central Mississippi. In these areas, phosphatic internal molds are commonly concentrated in the lower part of the unit, with the marls of the Demopolis penetrating, in burrows, the underlying Mooreville Formation. This break in sedimentation within the Campanian of the eastern U.S. Gulf Coastal Plain section may be coordinate with the disconformity or omission surface between the Ozan Formation and Annona Chalk in Arkansas (Dane, 1929; Bottjer, 1981). In Texas, this break may correlate with the disconformity at the base of the chalk unit of the middle part of the Taylor (Ellisor and Teagle, 1934; Young, 1985, 1986b). If this middle Campanian break is coeval in all these areas, it records a major event in the Gulf of Mexico basin eustatic history. Other breaks of lesser or only more local significance may be discerned in the Campanian sequence. Within the Campanian part of the Blufftown Formation of Alabama and Mississippi, two cycles of terrigenous clastic sedimentation are present and separated by a disconformity. In the coeval Mooreville Formation of central Alabama, two cycles can also be distinguished (King, 1983), but in these downdip chalkier sections, the physical break

of the basin margin section has not been recognized and is reflected only by cyclic repetition of sediment types.

In the subsurface of the Mississippi embayment, both the Jackson dome and the Monroe uplift show positive movement during the Campanian that is reflected by thinning or absence of such units as the Anonna Chalk over the structural summit.

In Arkansas, the disconformity at the base of the Annona has already been mentioned, but a second disconformity (lower Campanian) exists between the Ozan Formation and the subjacent Brownstown Formation (Dane, 1929). This break is present in northeast Texas, between the Gober Chalk and the Ozan Formation and within the Taylor Group of central Texas (Hazel and Brouwers, 1982). Also in central Texas, various Taylor units are truncated on the flanks of the San Marcos arch (Young, 1986b).

During the late Campanian or early Maastrichtian, there was uplift and erosion of the northern margin of the Rio Grande embayment, which resulted in the development of a disconformity at the top of the Anacacho. In some areas of south Texas, upper Maastrichtian beds overlie the Anacacho, and the intervening San Miguel, Upson, and Olmos Formations are missing (Brown, 1965). Within the Rio Grande embayment and in the Sabinas basin of northeastern Mexico, the Campanian section is reputedly a conformable sequence. Southward, the San Felipe Formation and Mendez Shale are conformable where they are in contact, but in the Tampico-Misantla basin, the Mendez contains limestone breccias in its base, derived from older formations, and lies disconformably upon the lower Tamaulipas, Tamabra, or El Abra Formations. This break represents the mid-Cretaceous unconformity.

To the south in Chiapas, the relations of Campanian units are in dispute. There is agreement that a disconformity exists within the Coniacian to Campanian platformal sequence. Sanchez Montes de Oca (1969) maintains that Coniacian and Santonian deposits are absent from much of the platform. In contrast, Martínez (1972) presents faunal evidence that Coniacian and Santonian rocks are present, but that the Campanian is absent. More work needs to be done before the true relations and magnitude of this hiatus can be resolved. To the east and northeast of Chiapas, the Campanian sequence has not been studied.

Biostratigraphy and age. As used here, the Santonian-Campanian boundary is placed at the last occurrence datum of the planktonic foraminifer *Sigalia deflaensis* (Sigal) and the first appearance of *Globotruncana elevata* (Brotzen). This has the convenience of being near the base of the *Alatacythere cheethami* Zone (ostracode) of Hazel and Brouwers (1982) and thus has the utility of transcending most marine-facies boundaries. In addition, this placement conforms quite closely to the ammonite boundary (Young, 1963) by being near the base of the *Submortoniceras tesquequitensis* Zone. All of the above criteria lie stratigraphically lower than Pessagno's (1969) boundary placement at the top of his *Globotruncana fornicata* Subzone.

The boundaries of Campanian rocks are poorly known in the subsurface of Florida, especially in the carbonate sequences of the Pine Key Formation. Lists of foraminifers from wells indicate the presence of Campanian species in both Florida and southern Georgia, but data on occurrence are sparse. In contrast, there is considerable faunal information for the outcropping units of the northern part of the Gulf of Mexico basin. In the terrigenous clastic facies of eastern Alabama and western Georgia, the upper two-thirds of the Blufftown Formation contains the ammonites *Scaphites hippocrepis* and *Delawarella delawarensis* (Sohl and Smith, 1980) and ostracodes of the *Alatacythere cheethami* and *Ascetoleberis plummeri* Zones (Hazel and Brouwers, 1982). These fosils indicate that the upper two cycles of the Blufftown range from the lower Campanian through the lower part of the middle Campanian. The Cusseta Formation (= lower Ripley of Alabama) contains such molluscs as *Anomia tellenoides* in its upper part and late Campanian ostracodes below. Thus, the formation ranges from the upper part of the middle Campanian into the lower Maastrichtian.

In eastern Mississippi, both ammonite data (Young, 1963) and planktonic foraminifers and nannofossils (Smith and Mancini, 1983) suggest that the Santonian-Campanian boundary lies in the upper part of the Eutaw Formation. The limestone at the top of the Mooreville contains the middle Campanian ammonite *Trachyscaphites spiniger*. The overlying Demopolis Chalk contains *Trachyscaphites spiniger porchi* in its basal part, and *Atreta cretacea* about 20 m above the base. Both species are also to be found elsewhere in the Arkansas to Texas region in such units as the Annona Chalk and Anacacho Formations and may be assigned to the upper part of the middle Campanian. The Campanian-Maastrichtian boundary lies in the upper part of the Demopolis above beds containing *Globotruncana calcarata* and below the occurrence of *Exogyra cancellata*.

In Arkansas, the Santonian-Campanian boundary probably lies within the Brownstown Formation. The overlying Ozan Formation is virtually coordinate with the lower to middle Campanian ostracode zone of *Ascetoleberis plummeri* (Hazel and Brouwers, 1982). The overlying Annona contains the distinctive foraminifer *Globotruncana elevata*, the bivalve *Atreta cretacea*, and the echinoid *Echinocorys texanus*, all species of wide occurrence in the upper middle or upper Campanian. The Marlbrook contains late Campanian fossils, but the report by Dane (1929) of the occurrence of *Exogyra cancellata* in the formation suggests that some part is lower Maastrichtian. In northeastern Texas, the Gober Chalk contains such ammonites as *Delawarella danei* and *Submortoniceras vanuxemi*, and ostracodes of the *Alatacythere cheethami* Zone of the lower Campanian. In central Texas, Young (1963, 1986a) has proposed a Campanian ammonite zonation consisting of six zones. The lower three zones of *Submortoniceras tequesquetensis, Delawarella delawarensis*, and *Delawarella sabinalensis*, encompass the lower Campanian. The middle Campanian consists of the *Hoplitoplacenticeras marrotti* Zone, while the *Placenticeras meeki* and *Manambolites ricensis* zones represent the upper Campanian. Pessagno (1969) reports late Campanian foraminifers such as *Globotruncana calcarata* from the "upper Taylor Clays" from several areas in Texas. In the

Rio Grande embayment sections, the age relations are poorly known for the San Miguel and Upson Clay. On the basis of macrofossil content, Stephenson (1931) assigned the formations to the upper Campanian. Planktonic foraminifers are sparse in the San Miguel, more diverse in the Upson, but in neither formation do they appear especially diagnostic.

In Mexico, south of the Rio Grande embayment, most sections are conformable, and stage boundaries fall within formations. The upper part of the San Felipe Formation in the Tampico-Misantla basin and along the front of the Sierra Madre Oriental is early Campanian, based on both planktonic foraminifers (Pessagno, 1969; Gamper, 1977a) and ammonites (Young, 1983). The overlying and widely occurring Mendez Shale is dated primarily on the basis of planktonic foraminifers. Pessagno (1969) suggested that older Mendez beds are present in the northern area of outcrop, but most areas contain beds ranging from the upper part of the lower Campanian to uppermost Maastrichtian. To the south in Chiapas, most authors have considered parts of the Angostura and Ocozocouatla Formations to be Campanian, but Martínez (1972) presented microfossil evidence that he interpreted as indicating that the Campanian was absent. An abundant macrofauna exists in these beds, but it is dominated by rudist bivalves (Alencaster, 1971) whose species ranges, though imprecisely known, are assumed to be Maastrichtian. More work is needed to resolve the age relations of the Chiapas Campanian section.

Conditions of deposition and paleogeography. In all Gulf of Mexico basin areas the preserved record of Campanian deposition is marine and the shoreline lay outside the areas of existing outcrop. However, in some areas such as southern Mexico, there is evidence that shoreline displacement was several hundred kilometers to the east of its Santonian position (Young, 1983).

Early Campanian transgression is expressed in many areas of the northern and western margins of the Gulf of Mexico basin. To the northeast, inner-shelf terrigenous clastics grade cyclically upward into barrier-strandplain deposits (Blufftown Formation) in the outcrop. Deltaic deposition of terrigenous clastics (Coffee Sand) began during the early Campanian in the northern part of the Mississippi embayment and, with time, prograded southward into northern Mississippi (Dockery and Jennings, 1988). The streams feeding this delta system had headwaters to the northeast in the Piedmont region of the southern Appalachians (Potter and Pryor, 1961). Downdip, these sections of terrigenous clastics grade to hemipelagic calcareous clays and marls (Mooreville), but even these sequences show upward increase in terrigenous components through the lower Campanian. Similar conditions prevailed in northeastern Texas, where short-lived influx of deltaic terrigenous clastics, having a westerly or northwesterly source, developed late in the early Campanian. Early Campanian transgression is expressed by the inundation of the San Marcos arch by the upper part of the Austin Group. Elsewhere through central Texas, shallow-water shelfal clays and carbonates dominate the lower Campanian deposits, with major sources of terrigenous clastics far removed to the north. Much of the montmorillonitic clays may be volcanic dust derived from westerly sources.

Of special note along the northwestern margin of the Gulf of Mexico basin is the abundance of volcanic-derived deposits. During the deposition of the Austin, Taylor, and Navarro Group equivalent units, volcanic activity occurred at more than 200 sites along an arcuate belt some 400 km long extending from central through south Texas (see Chapter 6 of this volume). An extensive literature dealing with these "serpentine plugs" exists and is summarized by Ewing and Caran (1982) and Matthews (1986). It is envisioned that the contact of rising magma with seawater caused an initial phase of violent explosion, causing cratering and expulsion of glass shards and rock fragments. Inward-slumping and additions by later tuff falls commonly filled the crater, and the accreted tuff mound may have risen to 90 m above the paleo–sea floor. Such tuff mounds formed a platform for construction of carbonate mounds. Adjacent to such volcanic-island masses are progressive sequences of sedimentary environments. First are margining lagoonal deposits, then a shoal complex that includes beach deposits and patch reef biostromes. Debris from this shallow-water complex was shed downslope and into the shelf-margin mudstones (Young, 1975; Luttrell, 1977).

In northeastern Mexico, the lower Campanian shoreline lay well to the west of the Sabinas and Burgos basins (Weidie and others, 1972). In general, these basins show increased terrigenous clastic content upward through the Campanian. Lower Campanian, shallow-shelf, chalky limestones and shales become increasingly silty upward in the section. The basins to the south in Mexico were sites of mainly shale deposition, with net terrigenous clastic increase to the west. On platformal areas, such as the Valles–San Luis Potosí platform, terrigenous deposits predominate, with shallow-water siltstones, sandstones, and bioclastic limestones as common lithologies (lower Cardenas or Tamasopo). In the Veracruz basin and on the Córdoba platform, massive platform limestones were deposited that are calcarenitic, calciruditic, miliolid or rudist rich (Atoyac). Farther south in the state of Chiapas, the age status of rocks of the platform areas are in doubt. These rocks grade northeastward into the evaporitic and carbonate sections of the Yucatan platform.

The middle and late Campanian is initiated by a second major transgressive phase that affected the northern and western margins of the Gulf of Mexico basin in a pattern similar to that of the early Campanian transgression. Initially, deposition of rather pure, open-shelf, pelagic chalks became widespread (Demopolis–Annona–mid-Taylor Group) from Alabama to central Texas. In the latter area it is marked again by inundation of the San Marcos arch. Upward these chalks grade into chalky marl or calcareous clay, but the regressive phase is not completed until the Maastrichtian. Terrigenous clastic deposition dominated in a few areas along the northern margin, such as in eastern Alabama and western Georgia, where shelfal clayey sands grade into nearshore cross-bedded units (Cusseta). The upper part of the Mississippi embayment saw continued deposition of the deltaic sands of the Coffee Formation.

A connection with the Western Interior Seaway persisted during most of the Campanian. In south Texas, middle Campanian deposits consist of shallow-water carbonate-bank deposits (Anacacho) that are related to sea-floor highs developed over intrusive rocks. Mudstones (Upson) were deposited basinward of these limestone-shelf areas. Beginning in the late Campanian, this area received a great influx of terrigenous sediment that is recorded by the sands, sandy limestones, and clay of the San Miguel Formation deltaic complex. The source of these terrigenous materials is believed to have come from uplifted areas to the northwest and west, and marginal to the Rio Grande embayment. Such deposits are also present in the Sabinas and Burgos basins of northeastern Mexico (Weidie and others, 1972).

Farther southward in Mexico, the lower Campanian shales grade upward into thick sequences of interbedded marl shale and sandstone that form the basinal deposits, while platform areas received silts and sand or developed carbonate platform deposits.

Economic considerations. Significant amounts of oil have been produced from Campanian rocks, especially in central and south Texas. Principal production in many areas is associated with the occurrence of "serpentine plugs." Reservoirs are generally stratigraphic traps in porous calcarenitic units of the upper Austin in central Texas, but in some fields, traps are related to fractures along fault zones that were created by the intrusions. In south Texas, similar stratigraphic traps occur both in the Anacacho carbonates marginal to the tuff mounds and in the overlying sands of the San Miguel, where they thin or drape the mounds. Pinchout of the deltaic sands creates traps within the San Miguel that are major producers in such South Texas fields as Big Wells.

Maastrichtian

Distribution. In general, Maastrichtian rocks crop out in a band parallel to and gulfward of those of the Campanian (Fig. 12). Locally, in areas such as western Alabama and the western margins of the Mississippi embayment, Tertiary units may overlap part or all of the Maastrichtian section. Terrigenous clastic deposits dominate the outcrop sequences and, especially in the western Gulf of Mexico basin, reflect rejuvenated source areas that are related to Laramide events. In Mexico, Maastrichtian marine deposits reflect continued migration of the shoreline eastward to a point, in some areas, as much as 200 km from that of the Campanian margin. Maastrichtian rocks blanket the shelf and platform areas from Florida to Yucatan. In the deeper parts of the basin, Maastrichtian rocks are presumed to be represented by an undetermined part of the Campeche seismic unit of Schaub and others (1984).

Lithostratigraphy. In the Florida Keys and southern peninsular Florida, the Maastrichtian saw continued deposition of the upper parts of the carbonate Rebecca Shoals Dolomite and the Pine Key Formation (chalky limestones and interbedded dolomite). In north Florida, the subtidal carbonates of the Pine Key grade into the peritidal to subtidal Lawson Limestone (Applin and Applin, 1944, 1967). The Lawson is divisible into a lower interbedded chalk and dolomitic chalk, overlain by an algal-rich nodular limestone that contains abundant orbitoid foraminifera and clasts of rudist bivalves. The unit is 120 m or more thick in north Florida, but at the Georgia-Florida border, Maastrichtian deposits are thin to locally absent (Fig. 3). This border area, commonly termed the Suwannee saddle or Suwannee strait, separates the carbonate deposits of south Florida from the terrigenous open-marine to delta-plain deposits of central Georgia (Prowell and others, 1985).

Westward in the Chattahoochee River Valley of Georgia and Alabama, the lowermost Maastrichtian unit is the upper part of the Cusseta Formation (or member of some usage). The Cusseta is composed of about 60 m of interbedded micaceous silt and clay that updip becomes coarser-grained, sometimes cross-bedded, upper-shoreface sands. The Cusseta grades laterally, to-

ward central Alabama, into the Demopolis Chalk, the upper part of which is early Maastrichtian. This upper 10 to 15 m of the Demopolis is siltier and sandier than the underlying marls and chalks and commonly contains a more diverse molluscan fauna; it is traceable across central Alabama, through Mississippi, and into southern Tennessee, where it grades into the sands of the lower part of the Ripley Formation (Sohl, 1960; Russell and others, 1983). The Ripley Formation is a lower to middle Maastrichtian terrigenous unit that is recognizable from western Georgia, across Alabama, into northeastern Mississippi, and northward into Tennessee. It consists of about 25 to 75 m of massive to medium-bedded, bioturbated, fossiliferous, micaceous, glauconitic, fine-grained sand. Clayey silt or clay beds and indurated sandstones are common in some sections. Cyclic fluctuations can be seen in the alternation between the massive sands

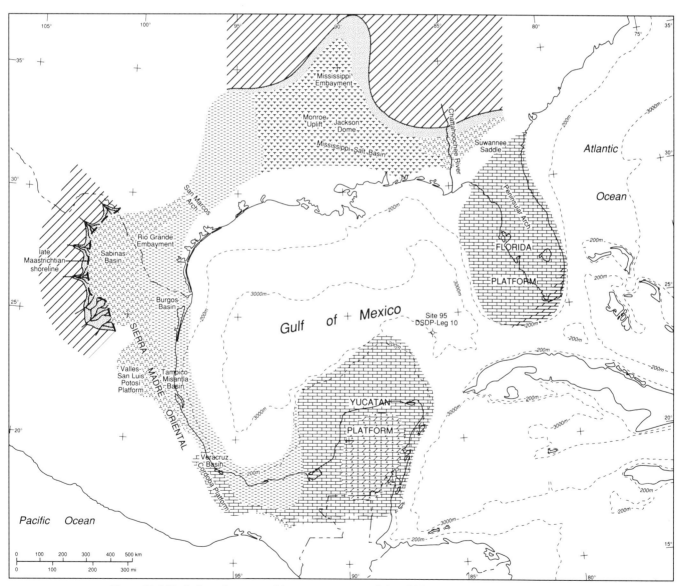

Figure 12. Lithofacies-paleogeographic map, upper part of the middle Maastrichtian. See Figure 8 for lithologic symbols.

and the units of thin-bedded sand and clay. In outcrop, it represents deposition on a shallow shelf and is the regressive phase of the cycle begun with the deposition of the Demopolis Formation. Downdip, the Ripley becomes increasingly clayey and marly, and is commonly included in the undifferentiated Selma Group of the subsurface (Fig. 4). Northward in Tennessee, the marine Ripley sands progressively lose their identity and are replaced by the mainly Maastrichtian, deltaic deposits of the McNairy Sand (Pryor, 1960). This sand-clay complex ranges from 30 m in Kentucky to almost 180 m in Tennessee and eastern Arkansas near the head of the Mississippi embayment.

The uppermost Maastrichtian unit in western Georgia is the Providence Sand. It rests disconformably upon the Ripley Formation. The disconformity is marked by a reworked lag unit of grit, quartz granules, and phosphatic pebbles. The overlying 50 m of the Providence Sand, in the downdip or river valley sections of the outcrop and shallow subsurface, is primarily a dark gray, fossiliferous, micaceous clay to fine sand. These shallow shelfal deposits grade updip into marginal-marine to nonmarine, crossbedded sands, which are interstratified sand and clay and white sandy clay (Donovan, 1986). Westward, from the Chattahoochee River Valley to central Alabama, the sands of the Providence gradually are replaced by the Prairie Bluff Chalk (late middle to late Maastrichtian). The Prairie Bluff rests disconformably on the Ripley below, with the basal beds generally being a chalky clay that contains various amounts of coarse quartz sand and granules, glauconite, bone clasts, phosphatic molds, and phosphatized clasts to cobble size, derived from underlying lithologies. Upward, the unit becomes a hard, brittle, white to bluish white, chalky marl. The upper beds are commonly a fine sandy clay. In the outcrop, the Prairie Bluff varies from 4 to 12 m in thickness. The upper contact with the Tertiary is commonly a scoured surface. Northward, in northeastern Mississippi, the Prairie Bluff becomes more argillaceous and grades into the Owl Creek Formation (Fig. 2). The Owl Creek consists of almost 15 m of dark gray, glauconitic, fossiliferous, silty, clayey, fine sand. Equivalents of the Owl Creek extend to near the head of the Mississippi embayment, but are not present in southern Illinois, possibly because of post-Cretaceous erosion.

In the subsurface, in both the gulfward direction and into the Mississippi embayment, the sandy units of the outcropping Maastrichtian sections rapidly become chalky and indistinguishable from underlying and overlying chalky deposits (Fig. 4). Thus, they are lumped into the Selma Group, a unit that includes all post-McShan deposits. Of special note, in the subsurface of the Mississippi embayment sections, is the presence of the informally named Jackson Gas Rock and Monroe Gas Rock units. The Gas Rock is a thin unit over the crest of the Jackson Dome, but thickens to as much as 500 m on the margin of the structure (Fig. 12). The Gas Rock of the Monroe Uplift rarely exceeds 30 m. The units consist mainly of hard to soft, porous, crystalline limestone, chalky limestone, and chalk. The Jackson Gas Rock rests unconformably upon the truncated beds of Jurassic deposits (Cotton Valley) at the crest of the structure, then rests toward the margin of the uplift, overlies truncated deposits of successively younger Lower and Upper Cretaceous deposits, and finally cuts Campanian rocks. The Gas Rock is disconformably overlain by Paleocene (Midway) deposits (Braunstein, 1950; Monroe, 1954).

Maastrichtian deposits of the Arkansas to Louisiana and northeastern Texas area are generally divided into three or four units. However, throughout the area they conform to a threefold sedimentological division: a lower fine-grained unit, a middle coarser-grained unit, and a return to fine-grained deposition. These three divisions represent parts of two transgressive-regressive cycles of deposition. In Arkansas, the lowest unit (early Maastrichtian) is the Saratoga Chalk. It consists of about 9 to 18 m of sparingly glauconitic, sandy chalk; marly chalk; and calcareous sandstone. The lower bed of the Saratoga represents a condensed zone containing abundant phosphatic molds and coarse grains of quartz (Bottjer, 1978, 1981). The Saratoga maintains its shallow, inner-shelf, chalky character well into the subsurface of Louisiana, where it thickens only slightly; but westward along the outcrop into northeastern Texas, it becomes less calcareous and grades to a fossiliferous, gray, sandy clay. The middle Maastrichtian Nacatoch Formation conformably overlies both the Saratoga Chalk of Arkansas and its equivalent clay in northeast Texas. It consists of 30 to 90 m of medium- to fine-grained, dark, clayey sand that grades upward to a massive glauconitic sand and an overlying, commonly cross-bedded quartz sand with *Ophiomorpha* burrows. Fossiliferous calcareous sandstone concretions and clay lenses occur sporadically. In the subsurface, it both thins and grades to dark calcareous clay. The Nacatoch is continuously exposed in outcrop from southwestern Arkansas to northern Texas, where it is overlapped by late Maastrichtian deposits (Stephenson, 1941). Young (1982) notes that isolated outcrops of sandstone as far south as central Texas lie in a stratigraphic position, suggesting that they may be Nacatoch equivalents.

The uppermost Maastrichtian unit in Arkansas is the Arkadelphia Formation. Updip it consists of as much as 45 m of a lower, gray, calcareous clay that is overlain by gray marls with subsidiary thin beds of glauconitic chalk and soft marls. In the subsurface of Louisiana, the formation is mainly gray, fossiliferous chalk; calcareous clays; and interbedded, gray, calcareous, micaceous sandstone. The contact with the underlying Nacatoch sand is disconformable (Dane, 1929) with a basal bed of marly, pebbly sand that contains clasts derived from underlying units. In the subsurface, most workers have considered the contact to be conformable. Contact with the overlying Midway is disconformable. Westward, into northeastern Texas, the Arkadelphia is replaced by the units of the upper Navarro Group, which consists of claystone and chalky marl, and subsidiary sandy beds. These latter units carry south through central Texas.

Character of the Maastrichtian deposits changes rapidly south of the San Marcos arch, toward the Rio Grande embayment. In this area, the increased influx of quartzose terrigenous sediments that began in the upper Campanian continued to dominate the Maastrichtian sections (Figs. 6 and 12). The lowest unit is the Olmos Formation, which in outcrop, consists of mainly nomarine, irregularly bedded clays, shales, and sandstones along with lesser adjuncts of seams of coal, lignite, fire clay, and carbonaceous shales. The unit ranges from 120 to 150 m in thickness and is transitional with the overlying Escondido (Cooper, 1971), but to the northeast, an erosional disconformity exists between the two units (Tyler and Ambrose, 1986). The terrigenous wedge of Olmos deltaic sediments thickens to more than 400 m in the deeper subsurface of the Rio Grande embayment. Southward, the Olmos is recognized in the Sabinas basin and northwestern part of the Burgos basin in Mexico. The uppermost Cretaceous unit of the Rio Grande embayment is the Escondido Formation. The formation has been divided into a number of members (Cooper, 1970, 1971). The lower three-quarters of the formation consists mainly of bioturbated mudstones interbedded with medium- to fine-grained, subarkosic sandstones and *Crassostrea* coquina beds. Some sandstones are cross-bedded, others bear ripple marks or contain burrows of *Ophiomorpha*, indicating their shallow-water origin. The upper one-quarter of the formation consists of sublittoral, inner-shelf deposits of calcareous, glauconitic, sandy mudstone, siltstone and impure limestone. Southward, in Mexico, the formation is present in the Sabinas basin and the northwestern part of the Burgos basin. The outcropping Escondido is about 250 m thick in Texas, but thickens gulfward in the embayment to almost 400 m. In Mexico, the formation ranges between 150 and 500 m. The basal contact is transitional in southern Texas and northern Mexico (Cooper, 1971). The upper contact of the Escondido has been considered disconformable by Stephenson (1915), but transitional by Cooper (1973).

In northeastern Mexico, the Maastrichtian marks a time of great influx of terrigenous clastics from west to southwesterly sources (Weidie and others, 1972). To the east, however, deposition of claystones of the Mendez Formation continued. Southward, on the Valles–San Luis Potosí platform and along the front of the Sierra Madre Oriental, deposition of the terrigenous sediments and bioclastic limestones of the Cardenas Formation, which began in the Campanian, continued throughout the Maastrichtian. Similarly, the calcarenitic platform limestones of the Atoyac Formation continued to accumulate in the Veracruz basin and on the Córdoba platform.

South of the sites of Mendez Shale deposition in southeastern Mexico, on the platform region of central Chiapas, terrigenous and carbonate-bank deposits accumulated throughout the Maastrichtian. In west-central Chiapas the coarser clastic unit is called the Ocozocoautla Formation. Gutierrez Gil (1956) and Chubb (1959) proposed subdivisions of the formation, but Sanchez Montes de Oca (1969) considered the subdivisions to reflect only local lithofacies changes of no regional significance. According to Sanchez Montes de Oca (1969), the formation is almost 800 m thick in the type area near the village of Ocozocoautla. The lower part of the formation is predominantly a silica-cemented, reddish quartz gravel that rests disconformably upon the older Sierra Madre Limestone and contains worn, re-

worked specimens of rudist bivalves. These gravels are overlain by fossiliferous, calcareous sandstones and brownish shale, with the upper part of the formation dominated by alternating fine sands, siltstones, and mudstones. The abundance of orbitoid foraminifers, algae, actaeonellid and nerineid gastropods, and rudist pelecypods, coupled with the absence of true framework structures, suggests that much of the formation represents a shallow-water lagoonal deposit. Eastward, the unit loses much of its quartzose sand, becomes more calcareous, and interfingers with the Angostura Formation. In its type area southeast of Tuxtla Gutierrez, the Angostura ranges to 1,300 m (Sanchez Montes de Oca, 1969). It consists mainly of fossiliferous, muddy limestones, coarsely calcarenitic limestone, calcareous mudstone, and beds of recrystallized, brecciated limestone. Some units contain abundant rudist pelecypods, whereas others are rich in gastropods, algae, or various foraminifers. The unit is considered a predominantly perireefal deposit. To the east-northeast, the formation is presumed to grade to basinal deposits of the Mendez Formation.

In the subsurface of the Yucatan Peninsula, Maastrichtian equivalents in the central part of the peninsula are mainly evaporites with occasional thick beds of brecciated limestone and cryptocrystalline limestone. To the northwest, the amount of anhydrite decreases and limestone increases, until in the far north of the area, the section consists totally of limestones and marly limestones. North of the peninsula, at the northeastern margin of the Yucatan Shelf, middle to late Maastrichtian, pelagic, chalky ooze was encountered in core 19 of site 95, DSDP Leg 10.

In the deeper parts of the Gulf of Mexico basin, Maastrichtian deposits are assumed to make up an unknown part of the Campeche seismic unit. The sediments are probably pelagic oozes in the esatern part of the basin, much like those encountered at the margins of the Yucatan Shelf mentioned above. However, one might expect terrigenous components to increase closer to the Mexican coast.

Thickness. In the Gulf of Mexico basin, Maastrichtian vary widely in thickness, reaching a maximum of more than 600 m in delatic deposits of the Rio Grande embayment. In Florida, along the trend of the Peninsular arch, there is pronounced thinning of the carbonate units from south to north. This thinning culminates in the vicinity of the Florida-Georgia boundary where there is total or near loss of Maastrichtian section (Applin and Applin, 1944, 1967); the thinning separates the areas of carbonate deposition to the south from those of terrigenous deposition to the north (Figs. 3 and 12). The origin of the area of thinning or loss has been ascribed to scour by ocean currents or to slow deposition or nondeposition in a structural low (Suwanee saddle) that served virtually as a starved basin, located far from a sediment source area. Elsewhere, along the northern margin of the Gulf of Mexico basin, from Georgia to Arkansas, most updip outcrop areas of Maastrichtian deposits have thicknesses that range from a few to 180 m. Generally there is thinning of Maastrichtian deposits from the outcrop into the subsurface. Thus, in areas such as the Mississippi salt basin that were sites of thick accumulations during earlier periods of the Upper Cretaceous, Maastrichtian units are only 30 to 60 m thick. Thick accumulations of sediment are to be found only associated with such structures as the Jackson dome, where the Gas Rock ranges from 19 m thick on the crest of the dome to 550 m on the northwestern flank. Similar trends, but greater thicknesses, are to be seen in the surface to subsurface section of northeast Texas. In the outcrop of central Texas, there is local variation in thickness caused by erosion, nondeposition, or overlap by later deposits, but most of these breaks in sequence are lost in the subsurface. South of the San Marcos arch, in southern Texas, terrigenous clastic deposition prevailed throughout the Maastrichtian, with nearly 650 m of sands and clays accumulating in the Rio Grande embayment. Within units such as the Olmos, there is local variation in thickness because of thinning by compaction over volcanic plugs. In the Sabinas and northwestern part of the Burgos basins of northeastern Mexico, Maastrichtian deposits range from about 300 to 750 m thick. Southward in Mexico, platform deposits of terrigenous mudstones, shales, sands, and interbedded limestones such as the Cardenas Formation, are as much as about 900 m thick, but there is much local variation. In general, such Maastrichtian platform deposits grade basinward into sequences of terrigenous muds, such as the Mendez Shale, that range from a few meters to, in some places like the southern part of the Burgos basin, as much as 1,375 m thick. However, such thick Mendez sequences include some unknown part that is composed of older beds. As much as 1,400 m of post-Turonian, Upper Cretaceous anhydrites and limestone occur on the Yucatan platform, but how much is Maastrichtian is unknown.

Stratigraphic relations. In most areas of the Gulf of Mexico basin, deposition was continuous from the Campanian into the Maastrichtian. Disconformities and overlap between units within the Maastrichtian Stage are widely documented. For example, in the subsurface of north Florida, Applin and Applin (1967) indicate that the lower member of the Lawson Limestone is overlapped by the upper member, which then lies on Campanian deposits. Along the eastern part of the Gulf of Mexico basin, the major break in the sedimentary sequence occurs in the upper Maastrichtian. From western Georgia to Tennessee this break is represented by the disconformity between, successively, the coeval Providence Sand, Prairie Bluff Chalk, and Owl Creek Formations and the underlying Ripley Formation (Sohl and Koch, 1986). In Alabama and Georgia, this disconformity has been traced well into the subsurface by Donovan (1986). The same break may be present in Arkansas at the base of the Arkadelphia (Dane, 1929) and in northeastern Texas at the base of the upper Navarro Group beds above the Nacatoch Formation. In this latter case, the middle Maastrichtian Nacatoch Formation thins southward until it is represented by only a few isolated outcrops; in central Texas it is lost. There, upper Maastrichtian beds rest upon those of the upper Campanian.

In the Rio Grande embayment the Olmos Formation conformably overlies the San Miguel Formation and, in the subsurface, is conformable with the overlying Escondido. In updip areas, along the northern margin of the embayment, the Olmos is eroded and a disconformity exists between it and the overlying Escondido. In this area, the period of post-Olmos erosion resulted in the Escondido unconformably overlying progressively older beds: first the Olmos and then the San Miguel in an updip direction. Southward and into Mexico, the Escondido resumes its conformable relation with the Olmos (Cooper, 1973).

To the south in Mexico, the stratigraphic relations of the Mendez Formation are variable. Where the formation overlies the San Felipe, the contact is gradational. In such areas as the platforms around the Tampico-Misantla basin, the Mendez unconformably overlies the lower Tamaulipas, Tamabra, or El Abra Formations, commonly with a basal breccia derived from these underlying units. In the Veracruz basin, the carbonate-bank deposits of the Altoyac Formation may disconformably overlie locally the Guzmantla Formation, and the hiatus represents part of the Campanian. The Atoyac grades laterally into the Mendez and in places conformably overlies San Felipe or Mendez beds, or underlies beds of the upper Mendez (Viniegra, 1965). In Chiapas, the Ocozocoautla and Angostura Formations laterally intergrade, and in central Chiapas, both rest disconformably upon the units of the Sierra Madre Limestone. The magnitude of this hiatus is a subject of dispute and ranges from the total Campanian to absence of the Turonian through Santonian. To the southwest, near the Guatemala border, Chubb (1959) reports that the rocks of the "Ocozocuautla Series" overlap the Sierra Madre limestone and lie upon Lower Cretaceous (Todos Santos) or Paleozoic rocks. To the northeast, on the Yucatan platform, deposition was continuous from the Campanian through the Maastrichtian and into the Paleocene.

Along the northern margin of the Gulf of Mexico basin, Tertiary deposits rest disconformably upon Maastrichtian beds. In Georgia and eastern Alabama, Donovan (1986) has shown that the upper contact of the marginal-marine facies of the Providence Sand shows evidence of subaerial weathering and regional truncation. Westward, through Alabama and Mississippi to Missouri, the basal Paleocene beds commonly contain an abundance of reworked Cretaceous fossils that are intermixed with those of the basal Tertiary (Stephenson, 1955; Sohl, 1960). This Cretaceous-Tertiary disconformity is stated to be present everywhere in the subsurface of the Mississippi embayment (Mississippi Geological Society, 1941). In Arkansas the contact between the Arkadelphia and the overlying Paleocene deposits is disconformable, and the magnitude of this break increases in the subsurface of Louisiana on the south flank of the Sabine uplift (Lloyd and Hazzard, 1939; Granata, 1963).

Throughout Texas, Tertiary deposits rest disconformably upon those of the Cretaceous, with the exception of the Brazos River sequence of central Texas (Jiang and Gartner, 1986; Hansen and others, 1987) and the Rio Grande embayment (Cooper, 1973; Young, 1982) where continuous deposition has been postulated.

In Mexico, the Cretaceous-Tertiary boundary has been viewed as conformable in most basinal and some platform sequences. For example, the Mendez Formation is usually conformable with overlying Tertiary units such as the Velasco Formation; both Hay (1960) and Pessagno (1969), however, report local disconformities, and the platformal Cardenas Formation of San Luis Potosí has an eroded upper contact overlain by deposits of unknown age (Myers, 1968). Elsewhere in the platform and basin sections of Chiapas and Yucatan, deposition was continuous.

Biostratigraphy and age. For the purposes of this chapter, the Campanian-Maastrichtian boundary is placed at the last occurrence of the planktonic foraminifer *Globotruncana calcarta.* Such a use has the advantage of being stratigraphically close to recognizable datums in other widely used groups of organisms. For example, it is very close to the last appearance datum of the nannofossil *Eiffelithus eximius* and the first appearance of the ostreid bivalve *Exogyra cancellata.* If a three-part division of the Maastrichtian is used, the lowest occurrence level of the foraminifer *Gansserina gansseri* is commonly used as the base of the middle division, and the lowest appearance of the foraminifer *Abathomphalus mayaroensis* as the marker for the upper division. The absence of *A. mayoroensis* from the sections of the northern margin of the Gulf has resulted in a number of foraminifer workers assuming that late Maastrichtian deposits are missing from the outcrop of the northern Gulf Coastal Plain (Smith and Pessagno, 1973). Others, basing their conclusions on nannofossil occurrence, maintain that uppermost Maastrichtian deposits are present at a number of localities.

In northern Florida, cores from the Lawson Limestone have yielded a diverse fauna that includes planktonic foraminifers. The foraminifers suggest a Maastrichtian date for the formation, but are inconclusive as to how much of the stage is represented (Applin and Applin, 1967). Among the terrigenous clastic units in western Georgia and eastern Alabama, the upper part of the Cusseta Formation contains such lower Maastrichtian species as the bivalve *Anomia tellenoides,* a common mollusk in the *Exogyra cancellata* Zone. These sands grade westward into the upper part of the Demopolis Formation that, from central Alabama to Tennessee, yields both macro- and microfossils of the lower Maastrichtian. The overlying Ripley ranges from the upper part of the lower Maastrichtian through most of the middle Maastrichtian. *Exogyra cancellata* and early Maastrichtian foraminifers of the *Rugotruncana subcircumnodifer* Zone are found in the lower part of the formation, and the normal form of *Flemingostrea subspatulata* and planktonic foraminifers of the *Gansserina gansseri* Zone occur at higher, middle Maastrichtian levels (Sohl and Smith, 1980; Smith and Mancini, 1983). The overlying Providence Sand ranges from upper middle to late Maastrichtian and is entirely within the *Haustator bilira* Assemblage Zone (Sohl and Koch, 1986), as are its coeval westward equivalents, the Prairie Bluff Chalk and the Owl Creek Formation. Some authors maintain that only middle Maastrichtian beds of the *G. gansseri* Zone are present in the Prairie Bluff Chalk (Smith and Mancini, 1983), but in some sections, nannofossils of the *Nephrolithus frequens* Zone have been recorded

from the upper part of the formation (Worlsey, 1974; Thierstein, 1981). Such occurrences are considered by most workers to equate with a position well into the late Maastrichtian foraminifer *Abathomphalus mayaroensis* Zone. The Prairie Bluff macrofossils are indicative of the *Haustator bilira* Assemblage Zone and include ammonites such as *Sphenodicus spp., Coahuilites sp., Discoscaphites conradi,* and *Baculites columna.*

West of the Mississippi embayment, in southwestern Arkansas, the upper part of the Marlbrook Formation and the overlying Saratoga Chalk contain *Exogyra cancellata* and foraminifers of the *Rugotruncana subcircumnodifer* Subzone of the early Maastrichtian (Pessagno, 1969). The biostratigraphic limits of the Nacatoch Formation are not precisely known in Arkansas, but it does have the ostreid *Flemingostrea subspatulata* (normal form of Sohl and Smith, 1980) that ranges through the lower part of the foraminifer zone of *G. gansseri.* The Arkadelphia Formation contains microfossils of the late middle Maastrichtian (Pessagno, 1969).

In Texas, the macrofossils of the Navarro Group have been extensively documented by Stephenson (1941), but this work was placed within the broad zonation of the exogyrene ostreids and resulted in no finer biostratigraphic subdivisions than did the recording of local occurrence by formation. Cooper (1971) and Young (1982, 1986b) have utilized mainly sphenodiscid ammonites to erect a four-fold zonation of the Maastrichtian rocks of Texas that is restricted primarily to southern Texas because of limited occurrence of ammonites in rocks of the Navarro Group in northeastern Texas. Beds of the lower part of the Navarro Group of northeastern Texas yield *E. cancellata, Anomia tellinoides,* and foraminifers of the early Maastrichtian *R. subcircumnodifer* Zone. Microfossils are poorly represented in the terrigenous clastics of the Nacatoch, but macrofossils suggest that it ranges from the upper part of the early Maastrichtian into the middle Maastrichtian. In north and central Texas, the age range of the highest units of the Navarro Group is disputed. All workers agree that beds of the middle Maastrichtian (*G. gansseri* Zone) are present, but the absence of the name-bearing species of the planktonic foraminifer zone of *Abathomphalus mayoroensis* suggested to Smith and Pessagno (1973) that no late Maastrichtian beds were present. Jiang and Gartner (1986) have reported nannofossils from a studied section on the Brazos River that yielded nannoplankton of the *Predisocsphaera quadripunctata* Acme Zone of the late Maastrichtian *Micula mura* Zone. Among the macrofossils, these upper Navarro Group beds contain an assemblage assignable to the *Haustator bilira* Assemblage Zone containing such ammonites as *Discoscaphites conradi, Sphenodiscus pleurisepta,* and *Baculites columna,* which elsewhere are considered to occur in either late middle or late Maastrichtian deposits.

In the Rio Grande embayment sequence of southern Texas and northeastern Mexico, little biostratigraphic information exists on the outcropping Olmos Formation. The overlying Escondido contains a very restricted foraminifer fauna that Pessagno (1969) assigned to the middle Maastrichtian (*G. gansseri* Zone). The unit has also been zoned on sphenodiscid ammonites and by Böese and Cavins (1927) and Cooper (1971). In addition, Sohl and Koch (1984) record occurrences of such other molluscs as *Haustator bilira, Camptonectes bubonis, Striarca cuneata,* and *Liopeplum cretaceum,* which are stratigraphically restricted to late middle to late Maastrichtian units on other parts of the coastal plain.

South of the Rio Grande embayment in Mexico, the basinal facies of the Mendez contains planktonic foraminifer assemblages assignable to the *R. subcircumnodifer, G. gansseri,* and *A. mayoroensis* Zones, indicating a full range of the Maastrichtian. In addition, Gamper (1977b) cites the presence of *Globigerina eugubina* as evidence that the deposition of the Mendez is continuous with the lower Paleocene. In the platform sections, biostratigraphic reliance has been placed on the occurrence of rudist bivalves, other associated molluscs, and larger foraminifers. For example, Myers (1968), in his study of the Cardenas Formation, identified a diverse molluscan assemblage that contains such species as *Exogyra costata,* a form that in the northern Gulf Coastal Plain ranges from lower into, if not through, upper Maastrichtian beds. Farther southeast, in the Córdoba Platform area, Bonet (1969) has recorded numerous orbitoid foraminifers and planktonic foraminifers from the carbonates of the Atoyac Formation. Based on the sum of stratigraphic ranges of these various taxa, he suggested that the Atoyac ranges from the upper Campanian through lower Maastrichtian. Less certainty exists as to the age of the terrigenous and carbonate clastic sections of Chiapas. An abundance of macrofossils, mainly rudist bivalves, has been known from these sections for many years (Alencaster, 1971). Much of the controversy about the age relations between the Ocozocoautla Formation and the underlying Sierra Madre Limestone, has been discussed by Chubb (1959). Using a combination of foraminifer and rudist bivalve species that occur in other better-known sections, he concluded that the upper part of the Sierra Madre Limestone was Turonian and was overlain by the Ocozocoautla that contained Campanian foraminifers at its base. The upper part of the Ocozocoautla Formation contained the rudists *Titanosarcolites* and *Chiapasella* and large foraminifers that Chubb believed were Maastrichtian because of their presumed stratigraphic range in Jamaica. Sanchez Montes de Oca (1969) came to the same conclusion as to the relation of the Ocozocoautla Formation. In addition, he cited the presence of such orbitiod foraminifers and calcispheres as *Pseudorbitoides rutteni, Calcisphaerula innominata, Vaughanina cubensis,* and others in the Angostura Formation, which in sum, suggested assignment of the formation to the Campanian and Maastrichtian. Martínez (1972) accepted the presence of a disconformity at the base of the Ocozocoautla-Angostura, but interpreted its significance differently. He described the presence of a basal conglomerate that incorporates a microfauna, including planktonic foraminifers, that range from at least Campanian through lower Maastrichtian. He interpreted this information to indicate that the disconformity represented a hiatus extending from perhaps the late Santonian to the early Maastrichtian, and he limits

the range of the Ocozocoautla and Angostura Formations to the Maastrichtian and, perhaps, to only the middle and late parts thereof. Such a conclusion seems at odds with current concepts of the stratigraphic ranges of the various rudist species that occur in common between these formations and the units in the islands of the Greater Antilles and Trinidad. The cited differences need to be resolved by further work.

Conditions of deposition and paleogeography. The dominance of terrigenous deposition that began in Campanian times continued, on most of the margins of the Gulf of Mexico basin, into the early Maastrichtian. Deposition of limestones and chalky limestones was mainly limited to such areas as southern Florida and the Yucatan platform (Fig. 12).

Along the northeastern margin of the basin, lower Maastrichtian deposits generally consist of terrigenous sediments deposited in a shallow-shelf environment. With minor bathymetric oscillation, such conditions continue into the middle Maastrichtian and represent the overall regressive phase of the cycle begun with the deposition of the late Campanian part of the Demopolis Formation. Sea-level drop is recorded by an erosional disconformity at the top of the Ripley Formation that can be traced throughout the eastern Gulf Coastal Plain well into the subsurface (Fig. 2). A major lower Maastrichtian episode of influx of terrigenous material is recorded by deposits in the northern part of the Mississippi embayment. Beginning in the late part of the early Maastrichtian, progradation of fluvio-deltaic environments (McNairy) spread sand, silt, and clay southward and westward into the Mississippi embayment, where they grade into deposits of shelfal environments (Pryor, 1960). Progradation reached its maximum during the earliest part of the middle Maastrichtian and was followed by a transgressive episode that is marked by a northward overriding of the deltaic sands by barrier and shoreface sands (upper Ripley) that reached as far north as Tennessee. Termination of this transgressive event is marked by a disconformity between the Ripley and the overlying Owl Creek Formation (Stephenson and Monroe, 1940).

From Arkansas to northeastern Texas, early Maastrichtian shelfal marl and chalk deposition gave way to a major influx of coarser terrigenous sediments (Nacatoch), which lasted into early middle Maastrichtian. The source of these sediments is streams entering the basin from the north in Arkansas and from both the north and northwest in northeastern Texas (McGowen and others, 1980). These deltaic and interdeltaic deposits grade basinward, into shelfal sandstones and mudstones.

The influx of terrigenous clastics into the Rio Grande embayment of southern Texas and northeastern Mexico, which began in the Campanian, continued into the Maastrichtian (Weidie and others, 1972). The source of these clastics is to the west and northwest of the embayment, and deposition took place on a broad, shallow shelf. During the early Maastrichtian, probable uplift in the Laramide orogenic belt provided an increased sediment supply. This increased supply provided the material for development of lobate to elongate, prograding deltaic systems (Olmos). Locally, along the northern margin of the Rio Grande embayment, this period of deltaic development was brought to an end by uplift and erosion that, in updip areas, truncated the Olmos and some older deposits. The exact timing of this event is unclear, but seems most likely middle Maastrichtian. The connection of the Gulf of Mexico basin with the Western Interior Seaway was probably cut during the Maastrichtian.

In Mexico, the lower Maastrichtian shoreline had moved considerably to the east of its Campanian position (Fig. 12). In the east-central states of Tamaulipas and Veracruz, fine clastic deposits (Mendez) accumulated in basins far from the western Laramide source areas. In many areas, deposition of these basinal shales was continuous from the late Campanian through the Maastrichtian. On platform areas to the west of these basinal sequences, such as those in the states of San Luis Potosí and Chiapas, terrigenous deposits dominate the Maastrichtian section. Conglomeratic units farther to the west indicate close proximity to the shoreline. During intervals of lowered terrigenous clastic input, carbonate platform and rudist bank facies developed. These platformal sequences grade basinward to deeper-water mudstone deposits. On the Yucatan Platform, evaporites were being deposited in the center of the peninsula, while mainly limestone accumulated to the north. This was an area receiving little or no terrigenous input during the Maastrichtian, and this pattern continued uninterrupted into the Paleocene.

In the basin center, the few data points for Maastrichtian deposits indicate the basin was receiving only thin pelagic sediments; only nearer the present Mexican coastline was there significant input of terrigenous clastics.

Upper middle Maastrichtian and lower upper Maastrichtian transgressive deposits rest disconformably on and overlap the regressive deposits of the earlier Maastrichtian throughout most areas of the northern and western margins of the Gulf of Mexico basin. This terminal Cretaceous cycle is represented updip in Georgia by prograding, nonmarine, deltaic sequences that grade downdip into delta-front and prodelta deposits. Downdip and westward into Alabama, these deltaic deposits grade to inner-shelf, fossiliferous, muddy sand and silty clay. From central Alabama to Mississippi, this cycle consists of a lower transgressive unit of chalk or chalky marl grading upward to sandy clays or shelfal mud and fine sand. Similar conditions and sediments prevailed through Arkansas, northeastern Texas, and southern to central Texas. In the latter area, the section of middle Maastrichtian calcareous marls marks the last Cretaceous inundation of the San Marcos arch (Young, 1986b). In the Rio Grande embayment, middle to late Maastrichtian deposits transgress the truncated surface of the Olmos Formation. The regressive depositional phase of this cycle continued in this area into the Tertiary, according to Cooper (1970). Similarly, to the south in Mexico, Gamper (1977b) suggests that in some areas Mendez deposition continued into the Paleocene. Farther south in Mexico, terrigenous deposits alternating with rudist bank sediment covered platform areas, while calcareous muds continued to accumulate in basinal sites during the late Maastrichtian.

The hypothesis that throughout the Gulf of Mexico basin there was a major sea-level drop and retreat of the waters to the shelf margin at the end of the Cretaceous is no longer tenable. Deposition appears continuous in parts of Mexico and at least nearly continuous in a few other areas of the western and northern margins of the basin, where the hiatus between the Cretaceous and Tertiary is demonstrably small.

Economic considerations. Maastrichtian rocks have produced oil in a number of areas of the Gulf Coastal Plains. In Louisiana and Mississippi, significant gas and lesser amounts of oil have been recovered from the "Gas Rock" facies of the Monroe uplift in Louisiana and from the Jackson dome and Tinsley field in Mississippi. The Nacatoch Sand has been a good source of shallow production in a number of small fields, especially in the East Texas basin. Production there is mainly from the shelfal sand facies, especially in areas of structural closure related to salt dome rim structures and fields associated with the Mexia fault system. In the Rio Grande embayment, modest production of oil and gas have been recovered from the deltaic sandstones of the Olmos Formation. Updip traps are related to sandstones truncated by the post-Olmos unconformity and sealed by Escondido mudstones. Lithofacies pinchouts, volcanic mound structures, and fault-bounded structures have also been found to be productive. In addition to oil and gas, the Olmos Formation in the Sabinas basin in Mexico contains workable beds of coal (see Chapter 16, this volume).

SUMMARY OF GULF UPPER CRETACEOUS MARINE TRANSGRESSION HISTORY

To date, there has been no all-encompassing study of the pattern of transgressions and regressions present through the Late Cretaceous in the Gulf of Mexico basin. However, Young (1986b) has presented a study dealing with the timing of transgressions on the San Marcos Platform in central Texas. Figure 13 presents an attempt to integrate Young's data with that from other parts of the basin. In addition, comparisons are made with the transgressive history of the Atlantic Coastal Plain and the Western Interior.

In general, patterns of transgression and regression are most easily discernible along the margins of the Gulf of Mexico basin, where breaks in section, caused by erosion or nondeposition, are accentuated. Any of these hiatuses may diminish in magnitude or even disappear when traced into the basin. Within an area such as the Gulf of Mexico basin, comparisons of the transgressive history between different regions may be complicated by activation of structural elements. These may superimpose a local transgressive or regressive phase that differs from that of contiguous areas. The effects of the late Cenomanian to early Turonian growth of the Sabine uplift, or the periodic raising of the Monroe uplift, serve as examples of tectonic control on the sedimentation in surrounding areas. Comparisons to areas outside the Gulf of Mexico basin are even more suspect, especially when, as Young

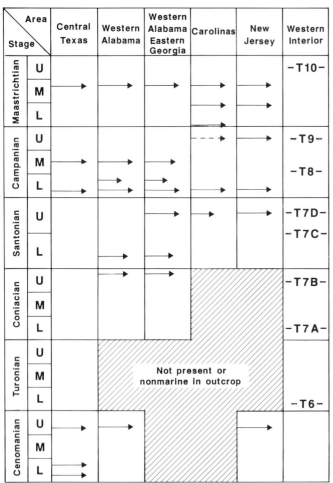

Figure 13. Comparison of times of major Late Cretaceous transgressions among selected areas of the northern Gulf of Mexico basin with such outside areas as the Atlantic Coastal Plain and Western Interior region. Arrows indicate times of peak transgression. Number designations for the Western Interior follow Kauffman (1977); data for central Texas from Young (1986b); data for the Atlantic Coastal Plain from Owens and Sohl (1969), Gohn (1988) and Sohl and Owens (1990).

(1986b, p. 134) has pointed out, correlation of the events is made by using different biostratigraphic scales.

In Figure 13, it is obvious that, at a few levels, there appears to be reasonable correspondence between certain transgressive events in different regions. The late Cenomanian transgression, evident in Texas (lower Eagle Ford), appears to correspond to that in the Tuscaloosa Group of western Alabama, and to the section in northern New Jersey (Woodbridge Member, Raritan Formation). This same transgressive pulse is recognized in the

subsurface of the Atlantic Seaboard as far south as Charleston, South Carolina. In the Western Interior, transgression T7 (Kauffman, 1977) is slightly younger. The transgressive overlap of the McShan Formation in Mississippi and Alabama is imprecisely dated, but overlying Eutaw beds dated as early Santonian suggest that it is late Coniacian. Although this placement is close to T7b of the Western Interior, it would be foolhardy to correlate the two on such evidence. The early Santonian pulse of the eastern part of the Gulf of Mexico basin finds no corollary elsewhere, but the late Santonian, lower cycle of the Blufftown Formation of the Chattahoochee River Valley is close to the Middendorf event in the Carolinas and the Cliffwood Beds (Magothy Formation) of the northern New Jersey Coastal Plain, and it approximates T7d of the Western Interior. The early Campanian transgression in central Texas seems to equate to the lower cycle of the Mooreville Chalk in Alabama, the middle cycle of the Blufftown Formation in Georgia, the lower Tar Heel of the Carolinas, and the Merchantville Formation of Delaware and New Jersey. A second, but younger, early Campanian cycle occurs in the upper part of the Mooreville and Blufftown Formations of the eastern Gulf Coast and is restricted thereto unless they equate with T8 of the Western Interior. A middle Campanian transgression is widespread in the northern Gulf Coastal Plain and is represented in such units as the middle Taylor, Annona, Demopolis, and Cusseta, but is not found elsewhere. The late Campanian through early Maastrichtian transgressions, evident on the Atlantic margin, have no recognizable counterpart in the Gulf of Mexico basin. However, the late middle Maastrichtian transgression, reflected in the late Navarro Group, Owl Creek, Prairie Bluff, and Providence of the northern part of the basin, seems correlative to that in the upper Peedee of the Carolinas and the Tinton Sand of New Jersey. How this transgression relates to T10 of the Western Interior is unknown at this time. Presumably, all connection with the interior seaway had vanished by the late Maastrichtian. Correlation of the transgressive events that occurred in the northern part of the Gulf of Mexico basin with sections in Mexico is highly speculative at this time. Some workers have considered the Upper Cretaceous of Mexico to be a single regressive event. This is obviously a simplistic view. However, the area of Mexico covered in this volume is mainly the eastern basinal sections where stratigraphic interruptions are not as evident as on the seaway margins. In addition, correlation of the northern Gulf of Mexico basin sections with Mexican shallow-water platform sections is imprecise because of the contrast in sedimentary depositional environments and the accompanying distinctive nature of the faunas that occur in those environments. For these reasons, such comparisons are not made here.

REFERENCES CITED

Addy, S. K., and Buffler, R. T., 1984, Seismic stratigraphy of shelf and slope, north-eastern Gulf of Mexico: American Association of Petroleum Geologists Bulletin, v. 68, no. 11, p. 1782–1789.

Adkins, W. S., and Lozo, F. E., 1951, Stratigraphy of the Woodbine and Eagle Ford, Waco area, Texas: Dallas, Texas, Southern Methodist University Press Fondren Scientific Series, no. 4, p. 101–161.

Alencaster, G., 1971, Rudistas del Cretacico Superior de Chiapas: Universidad Nacional Autónoma de Mexico, Paleontología Mexicana, no. 34, 91 p.

——, 1984, Late Jurassic–Cretaceous molluscan paleogeography of the southern half of Mexico, *in* Westermann, G.E.G., ed., Jurassic-Cretaceous biochronology and paleogeography of North America: Geological Association of Canada Special Paper 27, p. 76–88.

Anderson, E. G., 1979, Basic Mesozoic study in Louisiana; The northern Coastal Plain region and the Gulf Basin Province: Louisiana Geological Survey Folio Series 3, 58 p.

Applin, P. L., and Applin, E. R., 1944, Regional subsurface stratigraphy and structure of Florida and southern Georgia: American Association of Petroleum Geologists Bulletin, v. 28, p. 1673–1753.

——, 1965, The Comanche Series and associated rocks in the subsurface in central and south Florida: U.S. Geological Survey Professional Paper 477, 84 p.

——, 1967, The Gulf Series in the subsurface in northern Florida and southern Georgia: U.S. Geological Survey Professional Paper 524-G, 34 p.

Babcock, C., 1969, Geology of the Upper Cretaceous clastic section northern peninsular Florida: Florida Geological Survey Information Circular 60, 44 p.

Bailey, T. L., Evans, F. G., and Adkins, W. S., 1945, Revision of stratigraphy of part of Cretaceous of Tyler Basin, northeast Texas: American Association of Petroleum Geologists Bulletin, v. 29, p. 170–186.

Barnetche, A., and Illing, L. V., 1956, The Tamabara Limestone of the Poza Rica oil field, Veracruz, Mexico: 20th International Geological Congress, Mexico, 38 p.

Beall, A. O., Jr., 1964, Stratigraphy of the Taylor Formation, Upper Cretaceous, east-central Texas: Waco, Texas, Baylor Geological Studies Bulletin 6, 34 p.

Bebout, D. G., and Schatzinger, R. A., 1977, Regional Cretaceous cross sections, south Texas, *in* Bebout, D. G., and Loucks, R. G., eds., Cretaceous carbonates of Texas and Mexico: Austin, University of Texas Bureau of Economic Geology Report of Investigations 89, p. 4.

Benavides, G. L., 1956, Notas sobre la geología petrolera de México, *in* Guzman, E. J., ed., Symposium sobre Yacimentos de petroleo y gas: 20th Congreso Geológico Internacional México, v. 3, p. 351–362.

Berg, R. R., and Cook, B. C., 1968, Petrography and origin of lower Tuscaloosa sandstone, Mallalieu field, Lincoln County, Mississippi: Gulf Coast Association of Geological Societies Transactions, v. 18, p. 242–255.

Böse, E., 1923, Algunas faunas Cretácicas de Zacatectas, Durango y Guerrero: Instituto de Geología (Mexico) Boletin 42, 219 p.

Böse, E., and Cavins, O. A., 1927, The Cretaceous and Tertiary of southern Texas and northern Mexico: Austin, University of Texas Bureau of Economic Geology Bulletin 2748, p. 7–142.

Bonet, F., 1956, Zonificación microfaunística de las calizas Cretácicas del este de Mexico: Asociación Mexicana de Geólogos Petroleros Boletín, v. 8, no. 7-8, p. 389–488.

——, 1969, Microfacies de las calizas Cretácicas de la región Córdoba-Orizaba, *in* Seminario sobre exploracion petrolear, Mesa Redonda 4; Problemas de exploracion de la Cuenca de Papaloapan: Mexico City, Instituto Mexicano del Petroleo, 24 p.

Bottjer, D. J., 1978, Paleoecology, ichnology, and depositional environments of Upper Cretaceous chalks (Annona Formation); Chalk member of Saratoga Formation, southwestern Arkansas [Thesis]: Terre Haute, Indiana University, 424 p.

——, Structure of Upper Cretaceous Chalk benthic communities, southwestern Arkansas: Palaeogeography, Palaeoclimatology, Palaeoecology, v. 34, p. 225–256.

Braunstein, J., 1950, Subsurface stratigraphy of the Upper Cretaceous in Mississippi: Mississippi Geological Society 8th Field Trip Guidebook, p. 13–21.

Brown, C. W., and Pierce, R. L., 1962, Palynologic correlations in Cretaceous Eagle Ford Group, northeast Texas: American Association of Petroleum Geologists Bulletin, v. 46, p. 2133–2147.

Brown, N. K., Jr., 1965, Stratigraphy of the Upper Cretaceous beds in the vicinity of d'Hanis, Medina County, Texas: Corpus Christi Geological Society Annual Field Trip, p. 23–30.

Christina, C. C., and Martin, K. G., 1979, The lower Tuscaloosa trend of south-central Louisiana; You ain't seen nothing till you've seen the Tuscaloosa: Gulf Coast Association of Geological Societies Transactions, v. 29, p. 37–41.

Christopher, R. A., 1982, Palynostratigraphy of the basal Cretaceous units of the eastern Gulf and southern Atlantic Coastal Plains, *in* Arden, D. D., Beck, B. F., and Morrow, E., eds., Proceedings, Second Symposium on the Geology of the southeastern Coastal Plain: Georgia Geological Survey Information Circular 53, p. 10–23.

——, 1982b, The occurrence of the Complexiopollis-Atlantopolis zone (palynomorphs) in the Eagle Ford Group (Upper Cretaceous) of Texas: Jornal of Paleontology, v. 56, p. 525–541.

Chubb, L. J., 1959, Upper Cretaceous of central Chiapas: American Association of Petroleum Geologists Bulletin, v. 43, p. 725–756.

Cobban, W. A., 1969, The Late Cretaceous ammonites *Scaphites leei* Reeside and *Scaphites hippocrepis* (DeKay) in the western interior of the United States: U.S. Geological Survey Professional Paper 619, 29 p.

Coogan, A. H., 1977, Early and Middle Cretaceous Hippuritacea (rudists) of the Gulf Coast, *in* Bebout, D. G., and Loucks, R. G., eds., Cretaceous carbonates of Texas and Mexico; Applications to subsurface explorations: Austin, University of Texas Bureau of Economic Geology Report of Investigations 89, p. 32–70.

Coogan, A. H., Bebout, D. G., and Maggio, C., 1972, Depositional environments and geologic history of Golden Lane and Poza Rica Trend, Mexico; An alternative view: American Association of Petroleum Geologists Bulletin, v. 56, p. 1419–1447.

Cooper, J. D., 1970, Stratigraphy and paleontology of Escondido Formation (Upper Cretaceous), Maverick, County, Texas, and northern Mexico [Thesis]: Dallas, University of Texas, 287 p.

——, 1971, Maestrichtian (Upper Cretaceous) biostratigraphy, Maverick County, Texas, and northern Coahuila, Mexico: Gulf Coast Association of Geological Societies Transactions, 21, p. 57–65.

——, 1973, Cretaceous-Tertiary transition; Rio Grande outcrop section: American Journal of Science, v. 273, p. 431–443.

Dane, C. H., 1929, Upper Cretaceous formations of southwestern Arkansas: Arkansas Geological Survey Bulletin 1, 215 p.

de Cserna, E. G., and Bello-Barradas, A., 1963, Geología de la parte central de la Sierra de Alvarez, Municipoio de Zaragoza, Estado de San Luis Potosí: Universidad Nacional Autónoma de México Instituto Geología Boletin, v. 71, p. 23–63.

Dockery, D. T., III, and Jennings, S. P., 1988, Stratigraphy of the Tupelo Tongue of the Coffee Sand (Upper Campanian), northern Lee County, Mississippi: Mississippi Geology, v. 9, no. 1, p. 1–7.

Donovan, A. A., 1986, Sedimentology of the Providence Formation, *in* Reinhardt, J., ed., Stratigraphy and sedimentology of the continental nearshore and marine Cretaceous sediments of the eastern Gulf Coastal Plain: Society of Economic Paleontologists and Mineralogists Annual Meeting Field Trip 3, p. 29–56.

Drennen, C. W., 1953, Reclassification of outcropping Tuscaloosa group in Alabama: American Association of Petroleum Geologists Bulletin, v. 37, p. 522–538.

Durham, C. O., Jr., 1957, The Austin Group in central Texas [thesis]: New York,

Columbia University, 130 p.

Ellisor, A. E., and Teagle, J., 1934, Correlation of Pecan Gap Chalk in Texas: American Association of Petroleum Geologists Bulletin, v. 18, p. 1506–1536.

Ewing, T. E., and Caran, S. C., 1982, Late Cretaceous volcanism in south and central Texas; Stratigraphic, structural, and seismic models Gulf Coast Association of Geological Societies Transactions, v. 32, p. 137–145.

Faust, M. J., 1990, Seismic stratigraphy of the mid-Cretaceous unconformity (MCU) in the central Gulf of Mexico Basin: Geophysics, v. 55, p. 868–884.

Foss, D. E., 1979, Depositional environment of Woodbine Sandstones, Polk County, Texas: Gulf Coast Association of Geological Societies Transactions, v. 29, p. 83–94.

—— , 1980, Depositional environment of Woodbine Sandstones, Polk County, Texas, in 1st Annual Research Conference Program and Abstracts: Gulf Coast Section: Society of Economic Paleontologists and Mineralogists, p. 12–14.

Funkhouser, L. W., Bland, F. X., and Humphris, C. C., 1980, The deep Tuscaloosa gas trend of south Louisiana: Oil and Gas Journal, v. 78, no. 36, p. 96–98, 100, 101.

Gamper, M. A., 1977a, Estratigrafía y microfacies Cretácicas del anticlinorio Huizachal-Peregrina (Sierra Madre Oriental): Sociedad Geologica Mexicana Boletín, v. 38, no. 2, p. 1–17.

—— , 1977b, Acerca del limite Cretácico–Terciario en Mexico: Universidad Nacional Autónoma de México Instituto de Geologia Revista, v. 1, p. 23–27.

Gohn, G. S., 1988, Late Mesozoic and early Cenozoic geology of the Atlantic Coastal Plain; North Carolina to Florida, in Sheridan, R. E., and Grow, J. A., eds., The Atlantic Continental Margin: Boulder, Colorado, The Geology of North America, v. I-2, p. 107–130.

Granata, W. H., Jr., 1963, Cretaceous stratigraphy and structural development of the Sabine uplift area, Texas and Louisiana, in Hermann, L. A., ed., Report on selected north Louisiana and south Arkansas oil and gas fields and regional geology: Shreveport Geological Society Reference Report, v. 5, p. 50–96.

Gutierrez Gil, G. R., 1956, Geología del Mesozoico y Estratigrafía Permica del Estado de Chiapas: Congreso Geologico Internacional, 20th Session, Mexico Guide Book, Excursion C-15, p. 1–82.

Halbouty, M. T., and Halbouty, J. J., 1982, Relationships between East Texas field region and Sabine Uplift in Texas: American Association of Petroleum Geologists Bulletin, v. 66, p. 1042–1054.

Hansen, T., Farrand, R. B., Montgomery, H. A., Billman, H. G., and Blechschmidt, G., 1987, Sedimentology and extinction patterns across the Cretaceous-Tertiary boundary interval in East Texas: Cretaceous Research, v. 8, p. 229–252.

Hay, W. W., 1960, The Cretaceous-Tertiary boundary in the Tampico Embayment Mexico: 21st International Geological Congress, Copenhagen, Proceedings, part 6, p. 70–77.

Hayward, O. T., and Brown, L. F., 1967, Comanchean (Cretaceous) rocks of central Texas, in Hendricks, L., ed., Comanchean (Lower Cretaceous) stratigraphy and paleontology of Texas: Permian Basin Section, Society of Economic Paleontologists and Mineralogists Publication 67-8, p. 31–48.

Hazel, J. E., 1969, *Cythereis eaglefordensis* Alexander, 1929; A guide fossil for deposits of latest Cenomanian age in the Western Interior and Gulf Coast regions of the United States: U.S. Geological Survey Professional Paper 650-D, p. D155–D158.

Hazel, J. E., and Brouwers, E. M., 1982, Biostratigraphic and chronostratigraphic distribution of ostracodes in the Coniacian-Maastrichtian (Austinian-Navarroan) in the Atlantic and Gulf Coastal Province, in Maddocks, R. E., ed., Texas ostracoda; A guidebook of excursions and related papers for the 8th International Symposium on Ostracods: Houston, Texas, University of Houston, p. 166–198.

Hazzard, R. T., 1939, Notes on the Comanche and pre-Comanche? Mesozoic formations of the Arkansas-Louisiana-Texas area and a suggested correlation with northern Mexico, in Upper and Lower Cretaceous of southwest Arkansas: Shreveport Geologic Society Guidebook 14th Annual Field Trip, p. 155–165.

Hazzard, R. T., Blanpied, B. W., and Spooner, W. D., 1947, Notes on correlation of the Cretaceous of east Texas, south Arkansas, north Louisiana, Mississippi and Alabama: Shreveport Geological Society 1945 reference report on certain oil and gas fields of north Louisiana, south Arkansas, Mississippi, and Alabama, v. 2, p. 472–481.

Ice, R. G., and McNulty, C. L., 1980, Foraminifers and calcispheres from the Cuesta del Cura and lower Agua Nueva Formations (Cretaceous) in east-central Mexico: Gulf Coast Association of Geological Societies Transactions, v. 30, p. 403–425.

Imlay, R. W., 1940, Lower Cretaceous and Jurassic formations of southern Arkansas and their oil and gas possibilities: Arkansas Geological Survey Information Circular 12, 64 p.

Jiang, M. J., and Gartner, S., 1986, Calcareous nannofossil succession across the Cretaceous-Tertiary boundary in east-central Texas: Micropaleontlogy, v. 32, p. 232–255.

Kauffman, E. G., 1977, Geological and biological overview; Western Interior Cretaceous Basin, in Kauffmann, E. G., ed., Cretaceous facies, faunas, and paleoenvironments across the Western Interior Cretaceous Basin: The Mountain Geologist, v. 14, p. 75–99.

Kauffmann, E. G., Hattin, D. E., and Powell, J. D., 1977, Stratigraphic, paleontologic, and paleoenvironmental analysis of the Upper Cretaceous rocks of Cimarron County, northwestern Oklahoma: Geological Society of America Memoir 149, 150 p.

King, D. T., Jr., 1983, Shelf sedimentary facies, their cycles, and correlation; Lower Selma Chalk of Montgomery County, Alabama, in Carrington, T. J., ed., Current studies of Cretaceous formations in eastern Alabama and Columbus, Georgia: Alabama Geological Society 20th Annual Field Trip Guidebook, p. 21–25.

Lewis, J. O., 1977, Stratigraphy and entrapment of hydrocarbons in the San Miguel sands of southwest Texas: Gulf Coast Association of Geological Societies Transactions, v. 27, p. 90–98.

Lloyd, A. M., and Hazzard, R. T., 1939, North-south cross-section from the Paleozoic outcrops in Howard County, Arkansas, to Beauregard Parish, Louisiana: Shreveport Geological Society 14th Annual Field Trip Guidebook, p. 89–90.

Longoria, J. F., 1984, Cretaceous biochronology from the Gulf of Mexico region based on planktonic microfossils: Micropaleontology, v. 30, p. 225–242.

Lopez Ramos, E., 1975, Geological summary of the Yucatan Peninsula, in Nairn, A.E.M., and Stehli, F. G., The ocean basins and margins: Volume 3, The Gulf of Mexico and the Caribbean: New York, Plenum Press, p. 257–282.

Lozo, F. E., 1951, Stratigraphic notes on the Maness (Comanche Cretaceous) shale, in Woodbine and adjacent strata: Dallas, Texas, Southern Methodist University Press Fondren Scientific Series 4, p. 65–91.

Luttrell, P. E., 1977, Carbonate facies distribution and diagenesis associated with volcanic cones; Anacacho Limestone (Upper Cretaceous), Elaine field, Dimmit County, Texas, in Bebout, D. G., and Loucks, R. G., eds., Cretaceous carbonates of Texas and Mexico: Austin, University of Texas Bureau of Economic Geology Report of Investigations 896, p. 160–285.

Mancini, E. A., 1977, Depositional environment of the Grayson Formation (Upper Cretaceous) of Texas: Gulf Coast Association of Geological Societies Transactions, v. 27, p. 334–351.

—— , 1979, Late Albian and early Cenomanian Grayson ammonite biostratigraphy in north-central Texas: Journal of Paleontology, v. 53, p. 1013–1022.

Mancini, E. A., Mink, R. M., Payton, J. W., and Bearden, B. L., 1987, Environments of deposition and petroleum geology of Tuscaloosa Group (Upper Cretaceous), South Carlton and Pollard fields, southwestern Alabama: American Association of Petroleum Geologists Bulletin, v. 71, p. 1128–1142.

Martin, K. J., 1967, Stratigraphy of the Buda Limestone, south-central Texas, in Hendricks, L., ed., Comanchean (Lower Cretaceous) stratigraphy and paleontology of Texas: Permian Basin Section, Society of Economic Paleontologists and Mineralogists Publication 67-8, p. 287–299.

Martínez, R. E., 1972, Presencia del Turoniano, Coniaciano, Santoniano y ausencia del Campaniano en el Mesozoico de Chiapas: Intituto Mexicano del Petroleo Revista, v. 4, no. 4, p. 5–15.

Matthews, T. F., 1986, The petroleum potential of "Serpentine Plugs" and associated rocks, central and south Texas: Dallas, Texas Baylor Geological Studies Bulletin 44, 43 p.

McBride, E. F., Weidie, A. E., and Wolleben, J. A., 1974, Stratigraphy and structure of the Parras and La Popa Basins, northeastern Mexico: Geological Society of America Bulletin, v. 85, p. 1603–1622.

McGowen, M. K., Agagu, O. K., and Lopez, C. M., 1980, Depositional systems in the Nacatoch Sand (Upper Cretaceous) east Texas Basin and southwest Arkansas: Gulf Coast Association of Geological Societies Transactions, v. 30, p. 173.

McNeeley, B. W., 1973, Biostratigraphy of the Mesozoic and Paleogene pelagic sediments of the Campeche Embankment, *in* Worzel, J. L., ed., Initial reports of the Deep Sea Drilling Project: Washington, D.C., U.S. Government Printing Office, v. 10, p. 679–695.

McNulty, C. L., 1965, Lithology of the Eagle Ford–Austin contact in northeastern Texas: American Association of Petroleum Geologists Bulletin, v. 50, p. 375–379.

Mellen, F. F., 1958, Cretaceous shelf sediments in Mississippi: Mississippi Geological Survey Bulletin 85, 112 p.

Merrill, R. K., 1983, Source of the volcanic precursor to upper bentonite in Monroe County, Mississippi: Mississippi Geology, v. 3, no. 4, p. 1–7.

Meyerhoff, A. A., and Hatten, C. W., 1974, Bahamas Salient of North America; Tectonic framework, stratigraphy, and petroleum potential: American Association of Petroleum Geologists Bulletin, v. 58, p. 1201–1239.

Michael, F. Y., 1972, Planktonic foraminifera from the Comanchean Series (Cretaceous) of Texas: Journal of Foraminiferal Research, v. 2, p. 200–220.

Mississippi Geological Society, 1941, Subsurface sections of central Mississippi, chiefly Cretaceous: Mississippi Geological Society, p. 1–6.

Monroe, W. H., 1941, Notes on deposits of Selma and Ripley age in Alabama: Geological Survey of Alabama Bulletin 48, 150 p.

——— , 1954, Geology of the Jackson area, Mississippi: U.S. Geological Survey Bulletin 986, 113 p.

Muir, J. M., 1936, Geology of the Tampico region, Mexico: American Association of Petroleum Geologists, 280 p.

Mullereid, F.K.G., 1947, Paleogeología de la caliza de Coreloba y Orizaba: Instituto de Biología, Mexico, Anales, v. 18, no. 2, p. 361–462.

Murray, G., 1961, Geology of the Atlantic and Gulf Coastal Province of North America: New York, Harper and Brothers, 692 p.

Myers, R. L., 1968, Biostratigraphy of the Cardenas Formation (Upper Cretaceous), San Luis Potosi, Mexico: Universidad Nacional Autónoma de Mexico, Paleontologia Mexicana 24, 89 p.

Oliver, W. B., 1971, Depositional systems in the Woodbine Formation (Upper Cretaceous), northeast Texas: Austin, University of Texas Bureau of Economic Geology Report of Investigations 73, 28 p.

Owens, J. P., and Sohl, N. F., 1969, Shelf and deltaic paleoenvironments in the Cretaceous-Tertiary formations of the New Jersey Coastal Plain, *in* Subitzky, S., ed., Geology of selected areas in New Jersey and eastern Pennsylvania: New Brunswick, New Jersey, Rutgers University Press, p. 235–278.

Pessagno, E. A., Jr., 1967, Upper Cretaceous planktonic foraminifera from the western Gulf Coastal Plain: Palaontographica Americana, v. 5, no. 37, p. 245–445.

——— , 1969, Upper Cretaceous stratigraphy of the western Gulf Coast area of Mexico, Texas, and Arkansas: Geological Society of America Memoir 111, 139 p.

Potter, P. E., and Pryor, W. A., 1961, Dispersal centers of Paleozoic and later clastics of the upper Mississippi Valley and adjacent areas: Geological Society of America Bulletin, v. 72, p. 1195–1250.

Prowell, D. C., Christopher, R. A., Edwards, L. E., Bybell, L. M., and Gill, H. E., 1985, Geologic section of the undip Coastal Plain from central Georgia to western South Carolina: U.S. Geological Survey Miscellaneous Field Studies Map MF-1737.

Pryor, W. A., 1960, Cretaceous sedimentation in upper Mississippi Embayment: American Association of Petroleum Geologists Bulletin, v. 44, p. 1473–1504.

Ross, C. S., Miser, H. D., and Stephensoin, L. W., 1929, Water-laid volcanic rocks of early Upper Cretaceous age in southwestern Arkansas, southeastern Oklahoma, and northwestern Texas; U.S. Geological Survey Professional Paper 154-F, p. 175–202.

Russell, E. E., Keady, D. M., Mancini, E. A., and Smith, C. C., 1983, Upper Cretaceous lithostratigraphy and biostratigraphy in northeast Mississippi, southwest Tennessee, and northwest Alabama; Shelf chalks and coastal clastics: Gulf Coast Section, Society of Economic Paleontologists and Mineralogists Spring Field Trip, 72 p.

Sanchez, Montes de Oca, R., 1969, Estratigrafía y paleogeografía del Mesozoico de Chiapas, *in* Seminario sobre exploracion petrolera, Mesa Redonda 5; Problemas de exploracion de la Zona Sur: Mexico City, Instituto Mexicano del Petroleo, 31 p.

Sansores Manzanilla, E., and Navarrete, R., 1969, Bosquejo geológico de la zona norte, *in* Seminario sobre exploracion petrolera, Mesa Redonda 2; Problemas de exploracion de la zona norte: Mexico City, Instituto Mexicano del Petroleo, 37 p.

Schaub, F. J., Buffler, R. T., and Parsons, J. G., 1984, Seismic stratigraphic framework of deep central Gulf of Mexico Basin: American Association of Petroleum Geologists Bulletin, v. 68, p. 1790–1802.

Segerstrom, K., 1961, Geología del suroeste del Estado de Hildago y del noreste del Estado de México: Associación Mexicana de Geólogos Petroleros Boletin, v. 13, p. 147–168.

Siemers, C. T., 1978, Submarine-fan deposition of the Woodbine–Eagle Ford interval (Upper Cretaceous), Tyler County, Texas: Gulf Coast Association of Geological Societies Transactions, v. 28, p. 483–533.

Smith, C. C., 1981, Calcareous nannoplankton and stratigraphy of late Turonian, Coniacian, and early Santonian Age of the Eagle Ford and Austin Groups of Texas: U.S. Geological Survey Professional Paper 1075, 98 p.

Smith, C. C., and Mancini, E. A., 1983, Calcareous nannofossil and planktonic foraminiferal biostratigraphy, *in* Russell, E. E., Keady, D. M., Mancini, E. A., and Smith, C. C., Upper Cretaceous lithostratigraphy and biostratigraphy in northeast Mississippi, southwest Tennessee, and northwest Alabama; Shelf chalks and coastal clastics: Gulf Coast Section, Society of Economic Paleontologists and Mineralogists Spring Field Trip, p. 16–28.

Smith, C. C., and Pessagno, E. A., Jr., 1973, Planktonic foraminifera and stratigraphy of the Corsicana Formation (Maastrichtian) north-central Texas: Cushman Foundation for Foraminiferal Research Special Publication 12, 68 p.

Sohl, N. F., 1960, Archaeogastropods, mesogastropods, and stratigraphy of the Ripley, Owl Creek, and Prairie Bluff Formations: U.S. Geological Survey Professional Paper 331A, 151 p.

——— , 1971, North American biotic provinces delineated by gastropods: Proceedings, North American Paleontological Convention, Chicago 1969, v. 2 part 1, p. 1610–1638.

Sohl, N. F., and Koch, C. F., 1984, Upper Cretaceous (Maestrichtian) larger invertebrate fossils from the *Haustator bilira* Assemblage Zone in the west Gulf Coastal Plain: U.S. Geological Survey Open-File Report 84-687, 282 p.

——— , 1986, Molluscan biostratigraphy and biofacies of the *Haustator bilira* Assemblage zone (Maastrichtian) of the east Gulf Coastal Plain, *in* Rienhardt, J., ed., Stratigraphy and sedimentology of continental nearshore and marine Cretaceous sediments of the Eastern Gulf Coastal Plain: Society of Economic Paleontologists and Mineralogists Annual Meeting Field Trip 3, p. 45–56.

Sohl, N. F., and Owens, J. P., 1990, Cretaceous stratigraphy of the Carolina Coastal Plain, *in* Horton, J. W., Jr., and Zullo, V. A., eds., The geology of the Carolinas: Carolina Geological Society 50th Anniversary Volume, p. 191–220.

Sohl, N. F., and Smith, C. C., 1980, Notes on Cretaceous biostratigraphy, *in* Frey, R. W., ed., Excursions in southeastern geology; Field trip guidebook for the Annual Meeting of the Geological Society of America, Atlanta, Georgia: American Geological Institute, v. 2, field trip 20, p. 392–402.

Soto Jaramillo, F., 1980, Zonificación microfaunística de parte de los estratos cretácicos del cañon de la Borrega, Tamaulipas: Sociedad Geológica

Mexicana, 5th Convención Geológica Nacional Resumenes, p. 42–43.

Spencer, A. B., 1965, Alkalic igneous rocks of Uvalde County, Texas: Corpus Christi Geological Society Annual Field Trip, p. 13–21.

Stapp, W. L., 1977, The geology of the fractured Austin and Buda Formations in the subsurface of South Texas: Gulf Coast Association of Geological Societies Transactions, v. 27, p. 208–229.

Stehli, F. G., Creath, W. B., Upshaw, C. E., and Forgotson, J. M., Jr., 1972, Depositional history of Gulfian Cretaceous of East Texas Embayment: American Association of Petroleum Geologists Bulletin, v. 56, p. 38–56.

Stephenson, L. W., 1915, The Cretaceous–Eocene contact in the Atlantic and Gulf Coastal Plain: U.S. Geological Survey Professional Paper 90-J, p. 155–182.

——— , 1931, Taylor age of the San Miguel Formation of Maverick County, Texas: American Association of Petroleum Geologists Bulletin, v. 15, p. 793–800.

——— , 1941, The larger invertebrate fossils of the Navarro Group of Texas: Austin, University of Texas Publication 4101, 641 p.

——— , 1955, Owl Creek (Upper Cretaceous) fossils from Crowleys Ridge, southeastern Missouri: U.S. Geological Survey Professional Paper 374-E, p. 97–140.

Stephenson, L. W., and Monroe, W. H., 1940, The Upper Cretaceous deposits: Mississippi State Geological Survey Bulletin 40, 296 p.

Surles, M. A., Jr., 1979, Stratigraphy of the Eagle Ford Group (Upper Cretaceous) and its source-rock potential in the East Texas Basin: Dallas, Texas, Baylor Geological Studies Bulletin 45, 57 p.

Thierstein, H. R., 1981, Late Cretaceous nannoplankton and the change at the Cretaceous-Tertiary boundary, *in* Warme, J. E., Douglas, R. G., and Winterer, E. L., eds., The Deep Sea Drilling Project; A decade of progress: Society of Economic Paleontologists and Mineralogists Special Publication 32, p. 355–394.

Tyler, N., and Ambrose, W. A., 1986, Depositional systems and oil and gas plays in the Cretaceous Olmos Formation, South Texas: Austin, University of Texas Bureau of Economic Geology Report of Investigations 152, 42 p.

Tyler, N. Gholston, J. C., and Ambrose, W. A., 1986, Genetic stratigraphy and oil recovery in an Upper Cretaceous wave-dominated deltaic reservoir, Big Wells (San Miguel) field, south Texas: Austin, University of Texas Bureau of Economic Geology Report of Investigations 153, 38 p.

Valentine, P. C.,1984, Turonian (Eagle Fordian) stratigraphy of the Atlantic Coastal Plain and Texas: U.S. Geological Survey Professional Paper 1315, 21 p.

Velarde, P., 1969, Posibilidades petroliferas de la formaciones Mesozoicas en la cuenca de Veracruz, *in* Seminario sobre exploracion petrolera, Mesa Redonda 4; Problemas de exploracion de la cuenca del Papaloapan: Instituto Mexicano del Petroleo, 12 p.

Viniegra, F., 1965, Geologia del Macizo de Teziutlan y de la cuenca cenozoica de Veracruz: Asociacion Mexicana de Geologos Petroleros Boletin, v. 17, no. 7-12, p. 101–163.

Watkins, D. K., and McNulty, C. L., 1984, Paleontological synthesis, Leg 77, *in* Buffler, R. T., and others, eds., Initial reports of the Deep Sea Drilling Project: Washington, D.C., U.S. Government Printing Office, v. 77, p. 703–714.

Weidie, A. E., and Murray, G. E., 1967, Geology of Parras Basin and adjacent areas of northeastern Mexico: American Association of Petroleum Geologists Bulletin, v. 51, p. 678–695.

Weidie, A. E., Wollenben, J.A., and McBride, E. F., 1972, Late Cretaceous depositional systems in northeastern Mexico: Gulf Coast Association of Geological Societies Transactions, v. 22, p. 323–329.

Weise, B. R., 1979, Wave-dominated deltaic systems of the Upper Cretaceous San Miguel Formation, Maverick Basin, South Texas: Gulf Coast Association of Geological Societies Transactions, v. 29, p. 202–214.

Wilson, D., 1983, Paleoenvironments of the Upper Cretaceous Anacacho Formation in southwest Texas: South Texas Geological Society Bulletin, v. 24, no. 2, p. 22–28.

Winston, G. O., 1978, Rebecca Shoal reef complex (Upper Cretaceous and Paleocene) in south Florida: American Association of Petroleum Geologists Bulletin, v. 62, p. 121–127.

Winter, J. A., 1961, Stratigraphy of the Lower Cretaceous (subsurface) of south Texas: Gulf Coast Association of Geological Societies Transactions, v. 11, p. 15–24.

——— , 1962, Fredericksburg and Washita strata (subsurface Lower Cretaceous), southwest Texas, *in* Stapp, W. L., ed., Contributions to the geology of south Texas: San Antonio, South Texas Geological Society, p. 81–115.

Winton, W. M., 1925, The geology of Denton County: Austin, University of Texas Bulletin 2544, p. 1–86.

Worsley, T., 1974, The Cretaceous–Tertiary boundary event in the ocean *in* Hay, W. W., ed., Studies in paleo-oceanography: Society of Economic Paleontologists and Mineralogists Special Publication 20, p. 94–125.

Worzel, J. L., and others, 1973, Initial reports of the Deep Sea Drilling Project: Washington, D.C., U.S. Government Printing Office, v. 10, 747 p.

Young, K., 1963, Upper Cretaceous ammonites from the Gulf Coast of the United States: Austin, University of Texas Publication 6304, 373 p.

——— , 1965, A revision of Taylor nomenclature, Upper Cretaceous, Central Texas: Austin, University of Texas Bureau of Economic Geology Geological Circular 65-3, p. 1–11.

——— , 1972, Cretaceous paleogeography; Implications for endemic ammonite faunas: Austin, University of Texas Bureau of Economic Geology Geological Circular 72-2, 12 p.

——— , 1975, Pilot Knob; A marine Cretaceous volcano, *in* Young, K., Barker, D. S., and Jonas, E. C., eds., Stratigraphy of the Austin Chalk in the vicinity of the Pilot Knob; Guidebook for the South-Central Section of the Geological Society of America 9th Annual Meeting: Austin, University of Texas Bureau of Economic Geology, p. 8–20.

——— , 1977, Middle Cretaceous rocks of Mexico and Texas, *in* Bebout, D. G., and Loucks, R. G., eds., Cretaceous carbonates of Texas and Mexico: Austin, University of Texas Bureau of Economic Geology Report of Investigations 89, p. 325–331.

——— , 1982, Cretaceous rocks of central Texas; Biostratigraphy and lithostratigraphy, *in* Maddocks, R. F., ed., Texas ostracoda; Guidebook of excursions and related papers for the 8th International Symposium on Ostracoda: Houston, Texas, University of Houston, p. 111–125.

——— , 1983, Mexico, *in* Moullade, M., and Nairn, A. E., eds., The Phanerozoic geology of the world; Volume 2, The Mesozoic: Amsterdam, Elselvier, p. 61–88.

——— , 1985, The Austin division of central Texas, *in* Young, K., and Woodruff, C. M., Jr., eds., Austin Chalk in its type area; Stratigraphy and structure: Austin Geological Society Guidebook 7, p. 3–52.

——— , 1986a, The Albian-Cenomanian (Lower Cretaceous–Upper Cretaceous) boundary in Texas and northern Mexico: Journal of Paleontology, v. 60, no. 6, p. 1212–1219.

——— , 1986b, Cretaceous marine inundations of the San Marcos Platform, Texas: Cretaceous Research, v. 7, p. 117–140.

Young, K., and Powell, J. D., 1976, Late Albian correlations in Texas and Mexico: Annales du Museum d'historie Naturelle de Nice, v. 4, p. xxvl–xxv36.

MANUSCRIPT ACCEPTED BY THE SOCIETY JULY 27, 1990

ACKNOWLEDGMENTS

I owe a debt of gratitude to numerous individuals for discussion of problems pertaining to the Upper Cretaceous geology of the Gulf of Mexico Basin. Special thanks are due to those who contributed much to this effort through their careful review of the manuscript. These reviewers are as follows: C. W. Copeland and C. C. Smith of the Alabama Geological Survey; Amos Salvador and Keith Young of the Department of Geological Sciences, University of Texas at Austin; T. M. Scott of the Florida Geological Survey; and L. G. Wingard of the U.S. Geological Survey. However, the responsibility for any errors of fact or interpretation is solely mine.

The Geology of North America
Vol. J, The Gulf of Mexico Basin
The Geological Society of America, 1991

Chapter 11

Cenozoic

W. E. Galloway
The University of Texas at Austin, Department of Geological Sciences, P.O. Box 7909, Austin, Texas 78713
D. G. Bebout and W. L. Fisher
The University of Texas at Austin, Bureau of Economic Geology, Austin, Texas 78713
J. B. Dunlap, Jr.
Paleo-Data Inc., 6619 Fleur de Lis Drive, New Orleans, Louisiana 70124
R. Cabrera-Castro
Petroleos Mexicanos, Poza Rica, Veracruz, Mexico
J. E. Lugo-Rivera
Petroleos Mexicanos, Coatzacoalcos, Veracruz, Mexico
Thomas M. Scott
Florida Geological Survey, 903 West Tennessee, Tallahassee, Florida 32304-7700

INTRODUCTION

The Cenozoic Gulf of Mexico basin

While carbonate and evaporite deposition continued over the stable Florida and Yucatan Platforms, terrigenous clastic deposition dominated the rest of the Gulf of Mexico basin during the Cenozoic.

The position of the Cretaceous shelves and platforms determined to a great extent the shape and size of the basin at the beginning of the Cenozoic. This stratigraphic and structural framework was modified during the Cenozoic by the vast influx of terrigenous clastic sediments from the north and west, and by the structural impact of the Laramide orogeny during the Paleocene and Eocene. Subsidence of the central part of the basin continued during the Cenozoic, but it was the result more of sedimentary loading than of the thermal cooling of the oceanic crust.

The immense volumes of terrigenous clastic sediments that entered the Gulf of Mexico basin, particularly along its northern and northwestern margins, caused rapid basinward migration of shoreline deposition across the shelves, ultimately to positions considerably beyond the trend of the Cretaceous shelf margins. When large volumes of terrigenous clastics began to accumulate basinward of the Cretaceous shelf margins, an offlapping depositional style was developed along the northern and northwestern Gulf of Mexico basin that characterizes the Cenozoic. Very thick sedimentary sections began to accumulate over the continental slopes, and to fill the deeper parts of the basin, loading and depressing the attenuated continental crust and the oceanic crust.

Stratigraphic-structural provinces

The same three distinct stratigraphic-structural provinces in existence during the Mesozoic can still be recognized during the Cenozoic around the center of the Gulf of Mexico basin: a northern and northwestern province characterized by dominantly terrigenous clastic sedimentation and basinward progradation; a western "Laramide-modified" province; and the carbonate-evaporite province of the Florida and Yucatan platforms to the east and southeast. The central part of the Gulf of Mexico remained under deep water, a fourth stratigraphic-structural province.

Northern and Northwestern Progradational Province. The depositional patterns of the Northern and Northwestern Progradational Province are dominated by an abundant supply of terrigenous clastics and an overall prominent progradational depositional style. The gulfward-dipping and thickening wedges of sandstone and shale reached a maximum thickness of approximately 12,000 m just offshore of Texas and Louisiana. The offlapping nature of the section is well illustrated by regional-dip cross sections constructed from logs of deep wells, which show the maximum thickening of each major wedge being gulfward of that of the previous, or older, wedge (Fig. 1). The amount of continental margin progradation varied from stratigraphic unit to unit and, geographically, within a single stratigraphic unit as the focus of continental drainage shifted among the major deltaic depocenters.

Despite the fact that salt tectonics, which is a prominent feature of the Houston embayment, plays only a minor role in the

Galloway, W. E., Bebout, D. G., Fisher, W. L., Dunlap, J. B., Jr., Cabrera-Castro, R., Lugo-Rivera, J. E., and Scott, T. M., 1991, Cenozoic, *in* Salvador, A., ed., The Gulf of Mexico Basin: Boulder, Colorado, Geological Society of America, The Geology of North America, v. J.

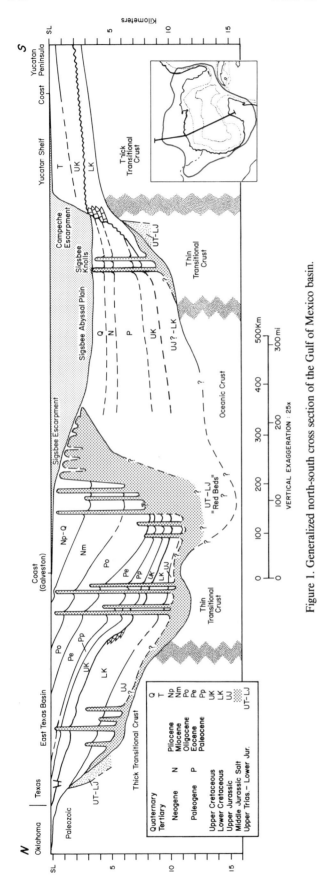

Figure 1. Generalized north-south cross section of the Gulf of Mexico basin.

structural development of the Burgos basin, the Rio Grande embayment, and the South Texas shelf, the four areas show many similarities. Paleogene units (Wilcox through Jackson groups) built across the subsurface Cretaceous shelf to a continental margin defined by the underlying Early Cretaceous reef trends. The Wilcox and Yegua cycles show moderate-scale growth faulting where they prograded beyond the Early Cretaceous shelf margin and onto the continental slope. Displacement across faults is typically 1 km (3,300 ft) or less, and stratigraphic thickening on the downthrown side is moderate. Beginning with the Oligocene units (Vicksburg and Frio Formations), deposition advanced into the deep Gulf of Mexico basin and onto thick Paleogene slope and basinal muds underlain by oceanic basement. Thick sequences of interbedded paralic sands and shales accumulated on the downthrown sides of major listric normal growth faults and in interdiapiric salt-withdrawal sags. Up to 2 km (6,600 ft) of sandy sediment may be largely, or even completely isolated structurally from its much thinner updip equivalent. Depositional loading of the underlying thick sequences of slope mud resulted in the development of large shale ridges and associated gently dipping listric-normal faults, which may displace sand bodies as much as tens of kilometers laterally and several kilometers vertically from their genetic equivalents on the upthrown side of the fault.

Western Laramide-modified Province. The western and southwestern Cenozoic margin of the Gulf of Mexico basin differs dramatically from its northern counterpart. In this province, the older Mesozoic margin was structurally modified by Laramide compression in early Tertiary time. Paleocene and younger marine strata onlap folded and thrusted Mesozoic carbonate rocks. Principal sedimentary/tectonic depocenters include the Tampico-Misantla and Veracruz basins, and the basins of southeastern Mexico. The prograded coastal plain and shelf were comparatively narrow, rarely exceeding 150 km (93 mi) from the Cenozoic onlap limits to the present continental margin; bypassing characterized the deposition in these shelf sequences. During the early Cenozoic, canyons were cut and later filled, and progressive offlap of Miocene and younger clastics produced the present shelf. Despite the presence of an adjacent tectonic upland, Cenozoic sediments are commonly impoverished in siliciclastic sand. Carbonate conglomerates and thick carbonate mudstone units reflect the dominance of Mesozoic carbonates and shales in the source terranes of the Sierra Madre Oriental.

Florida and Yucatan Platform Province. The aggradational depositional style characteristic of the Florida and Yucatan platforms is the result of the combination of a relatively low subsidence rate and accumulation of carbonates and evaporites sufficient only to maintain vertical growth. Progradation of large quantities of sediment off the stable platforms into the more rapidly subsiding central part of the Gulf of Mexico basin was prevented by the lack of a consistent source of terrigenous clastic sediments. The Cenozoic comprises a maximum of about 2,000 m of carbonates, evaporites, and some terrigenous clastics on the Florida platform and a carbonate-evaporite section of similar thickness on the Yucatan Platform.

Central Gulf of Mexico basin. More than 4,000 m of pelagic sediments were accumulated on the central part of the Gulf of Mexico during the Cenozoic. These sediments are composed of tests of planktonic organisms indigenous to the Gulf waters and of fine-grained to locally coarse, sandy terrigenous material transported across the northwestern and western shelves and into the center of the basin where it was deposited as bottomsets of the offlap clastic wedges. The units of the deep basin thicken dramatically against the toes of these offlap wedges.

STRATIGRAPHIC SUBDIVISION AND TERMINOLOGY

Overview

The stratigraphy of the Gulf of Mexico basin, while documented in great detail for local areas and depocenters, becomes quite confused when a synthesis of the entire basin is attempted.

The Cenozoic stratigraphic section over most of the northern and northwestern flanks of the basin is dominantly composed of very thick, laterally variable, and monotonous sequences of interbedded sandstones, siltstones, and shales difficult to break down into distinctive lithostratigraphic units (groups, formations, and members) that can be recognized and mapped over appreciable areas. The boundaries, both vertical and lateral, between recognized units are generally transitional, and in most cases vague, subjective, and the topic of durable controversies, the result of the recurrence of closely similar and related environments that persisted in the Gulf of Mexico basin during the entire Cenozoic. In east-central and southern Mexico, the Neogene section is composed of a similar alternation of terrigenous clastics, while the Paleogene is represented by a uniform shale section; neither one lends itself to a clear subdivision on the basis of diagnostic lithologic criteria.

For these reasons, it should not be surprising that early work on surface exposures resulted in the recognition of a myriad of formations, members, lentils, tongues, and beds, which commonly changed at geographical rather than geological boundaries. With the discovery of oil in the Gulf of Mexico basin, the stratigraphy of the region rapidly became economically as well as academically important, and the addition of subsurface information brought a growing number of problems not apparent on the surface. Drilling revealed that the sediments thicken rapidly in the subsurface. Formations that are nonmarine at the outcrop are marine in the subsurface. Thick marine wedges with no surface outcrop were recognized, and projecting surface nomenclature into the subsurface became increasingly difficult. Present subsurface Cenozoic nomenclature in the Gulf of Mexico basin is almost completely separate from the surface exposures, particularly in Neogene strata. Economic stratigraphy has largely replaced the rules of stratigraphic nomenclature, and many "formations," biozones, and lithofacies zones lack adequate definitions.

The establishment of a much-needed practical and operational stratigraphic classification and terminology, lacking a sound lithostratigraphic basis, has had to rely heavily on biostratigraphic zonation. Subdivision of the Cenozoic terrigenous clastic sequences of the northern and western Gulf of Mexico basin, and correlation between the major depocenters of the basin are now largely dependent on paleontology; specifically, benthonic and planktonic foraminifers, calcareous nannoplankton, and to some degree, ostracods. In some areas, parts of the Cenozoic section have been referred to as "Miocene," "Pliocene," or "Pleistocene."

Each depocenter is unique in its position relative to the ancient shelf-slope break, source of sediments, and time of most active deposition. Sandstone-shale sections more than a thousand meters thick in one depocenter may be represented by only 100 to 200 meters in a contemporaneous, but less active depocenter. Changing rates of deposition, rapid thickening along paleogeographic and structural flexures, shifting deltaic sources, and changes in paleobathymetry combine to make lithostratigraphic correlations unreliable.

The carbonate content of Cenozoic sediments in the Gulf of Mexico basin increases steadily from the Mississippi River eastward, and from the Isthmus of Tehuantepec northeastward. The sediments of the Florida and Yucatan platforms are predominantly carbonates and evaporites. In contrast with the active clastic depocenters of the northwestern and western Gulf of Mexico basin, the carbonate sections are thin and relatively undisturbed by the downwarping and other structural complications typical of the active clastic depocenters. Correlation of carbonates is based both on lithology and paleontology and, although consistent within a carbonate province, the carbonate units are nearly impossible to project directly into the clastic provinces. As the carbonate content of the sediments increases, faunal changes also occur, which prevent the interprovince correlation of biozones. Larger scale paleontologic correlations are recognized between the northern clastic and carbonate provinces. Most of these paleontological correlations have been made between surface outcrops that can be traced east and west of the transition between the two provinces in Mississippi, Alabama, and the panhandle of Florida.

The conceptual understanding of the complex structural and depositional history of the Cenozoic fill of the Gulf of Mexico basin has been built, therefore, upon four cornerstones: (1) regional biostratigraphic zonation and chronostratigraphic correlation between depositional provinces, (2) paleontologic and sedimentologic calibration of paleobathymetry, (3) recognition of regional cycles or depositional episodes, and (4) application of genetic depositional models based on Quaternary depositional systems.

A simplified stratigraphic correlation chart (Fig. 2) shows the nomenclature of the main Cenozoic lithostratigraphic units most commonly used in the Gulf of Mexico basin. The biostratigraphic zonation of the Cenozoic section based on benthonic and planktonic foraminifera and calcareous nannoplankton is shown in Table 1.

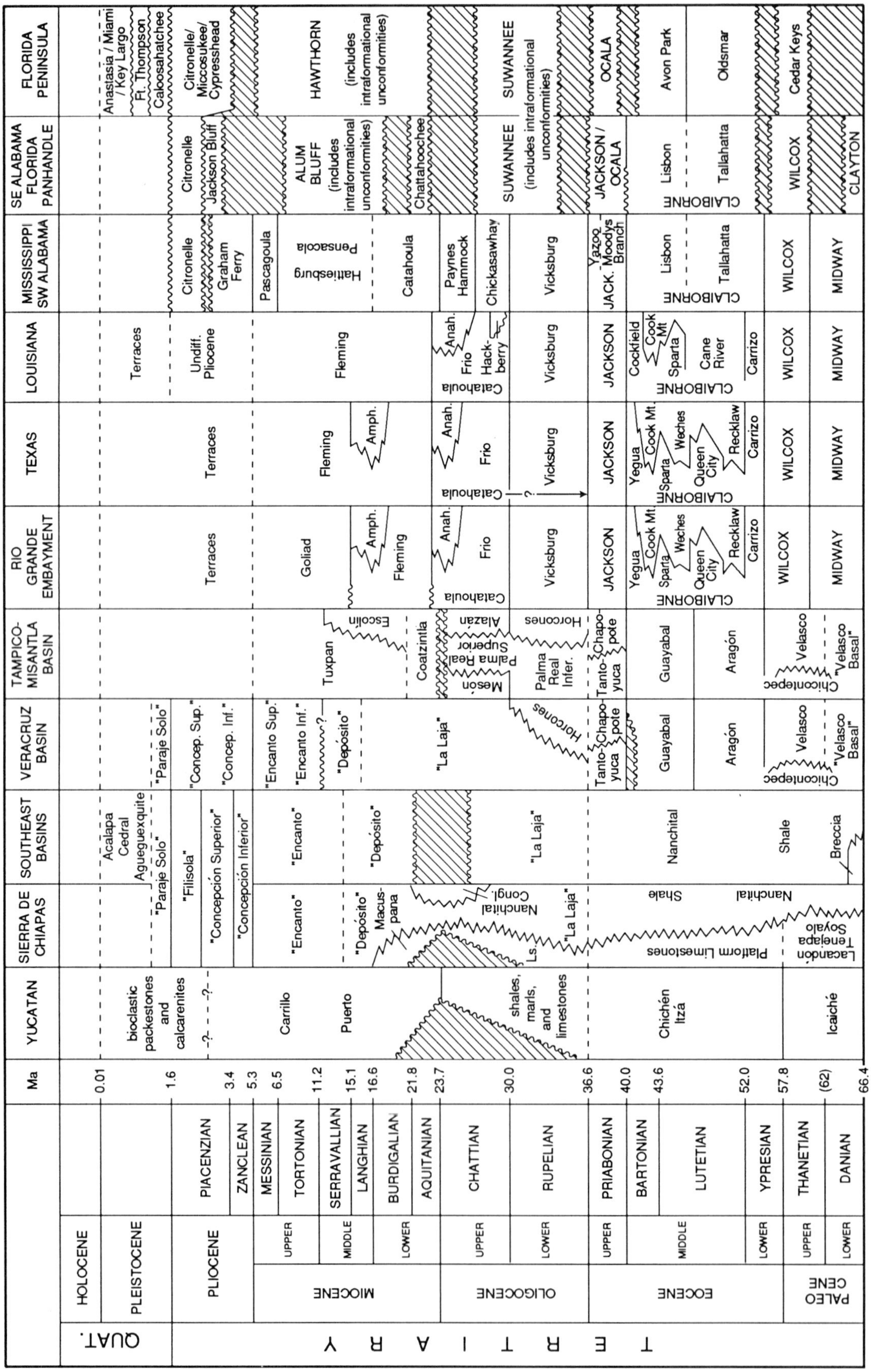

Figure 2. Simplified stratigraphic correlation chart of the Cenozoic of the Gulf of Mexico basin (for more complete chart, see Plate 5). Groups are shown in capitals, formations in lowercase.

TABLE 1.

CHRONOSTRATIGRAPHIC UNITS			DNAG NUM. TIME SCALE (Ma.)	CENOZOIC ZONATION–NORTHERN AND NORTHWESTERN GULF OF MEXICO BASIN		
				BENTHONIC FORAMINIFERA (Last Occurrence)[1]	PLANKTONIC FORAMINIFERA (Last Occurrence)[1]	CALCAREOUS NANNOPLANKTON (Last Occurrence)[1,2]
QUATERNARY	HOLOCENE		0.01		Globorotalia menardii flexuosa acme	
	PLEISTOCENE			Sangamon Fauna	Globorotalia truncatulinoides [D/S] Globorotalia tosaensis	Pseudoemiliania lacunosa A Pseudoemiliania lacunosa B
				Trimosina denticulata (Trimosina A)		Pseudoemiliania lacunosa C
				Hyalinea balthica Trifarina holcki (Angulogerina B)	Sphaeroidinella dehiscens acme A	Helicosphaera sellii Calcidiscus macintyrei
				*	Sphaeroidinella dehiscens acme B	Discoaster brouweri
			1.6	Cristellaria S		
PLIOCENE	PIACENZIAN			Lenticulina 1	Globorotalia menardii [S/D] Globorotalia miocenica	Discoaster pentaradiatus
						Discoaster surculus
				**	Globorotalia multicamerata	Discoaster tamalis
			3.4		Globoquadrina altispira	Sphenolithus abies A
	ZANCLEAN			Buliminella basispinata (Buliminella 1)	Globorotalia margaritae Globigerina nepenthes	Sphenolithus abies B Amaurolithus spp.
					Globigerinoides mitra	Ceratolithus acutus
				Textularia X	Globorotalia menardii [D/S]	Discoaster A
						Discoaster B
			5.3			
MIOCENE	UPPER	MESSINIAN		Robulus E	Globoquadrina dehiscens	Discoaster berggrenii
			6.5	Bigenerina A Cristellaria K	Globorotalia acostaensis [D/S] Sphaeroidinellopsis seminulina	Discoaster neohamatus
		TORTONIAN		Cyclammina 3 Discorbis 12		Discoaster prepentaradiatus
				Spiroplectammina barrowi (Textularia L)		Discoaster bollii
			11.2			Catinaster spp. Discoaster hamatus
				Bigenerina 2–Cibicides carstensi	Globorotalia mayeri	Coccolithus miopelagicus
	MIDDLE	SERRAVALLIAN		Uvigerina 3	Globorotalia fohsi robusta	
				Textularia stapperi (Textularia W)	Globorotalia fohsi fohsi	
				Bigenerina humblei	Globorotalia fohsi barisanensis	Sphenolithus heteromorphus
			15.1	Lenticulina cristi (Cristellaria I)	ORBULINA DATUM [3]	Discoaster deflandrei acme
		LANGHIAN		Cibicides opima		
				Amphistegina chippolensis (Amphistegina B) Robulus macomberi (Robulus L, 43)	Praeorbulina glomerosa	Helicosphaera ampliaperta
			16.6		Globigerinatella insueta	
	LOWER	BURDIGALIAN		Camerina 1–Robulus mayeri Gyroidina 9 Cristellaria A–Discorbis B	Catapsydrax stainforthi	Sphenolithus belemnos
			21.5	Marginulina ascensionensis	Catapsydrax dissimilis	
		AQUITANIAN		Siphonina davisi Planulina palmerae		Cyclicargolithus abisectus
				Lenticulina hanseni Cristellaria R	Globorotalia kugleri	Dictyococcites bisectus
			23.7	Robulus A–Discorbis "restricted"	Globigerina ciperoensis Globigerina sellii	
TERTIARY	OLIGOCENE	UPPER	CHATTIAN	Discorbis gravelli Heterostegina sp.		Sphenolithus distentus
				Cibicides jeffersonensis		
				Bolivina perca	Globorotalia opima opima	
				Marginulina idiomorpha Marginulina vaginata Marginulina howei		
				Textularia 14 Camerina A Miogypsinoides A Cibicides hazzardi		
				Marginulina texana		Sphenolithus predistentus
			30.0	Bolivina mexicana		
		LOWER	RUPELIAN	Nonion struma Nodosaria blanpiedi Textularia seligi Anomalina bilateralis		
				Textularia warreni	Globigerina ampliaperatura	Reticulofenestra umbilica Ericsonia subdistichus
				Textularia mississippiensis Loxostoma delicata	Pseudohastigerina micra	
				Cibicides pippeni		Discoaster saipanensis
			36.0		Globorotalia cerroazulensis (S.L.) Hantkenina alabamensis	
	EOCENE	UPPER	PRIABONIAN	Marginulina cocoaensis Textularia hockleyensis	Globorotalia cerroazulensis pomerali	
				Textularia dibollensis Camerina moodysbranchensis	Globigerinatheka semiinvoluta Globigerinatheka barri	
			40.0	Discorbis yeguaensis	Truncorotaloides rohri	Chiasmolithus grandis
		MIDDLE	BARTONIAN	Eponides yeguaensis	Truncorotaloides topilensis	
				Ceratobulimina eximia	Orbulinoides beckmanni	Chiasmolithus solitus
					Globorotalia pentacamerata	
			43.6	Operculinoides sabinensis		
				Anomalina B	Globorotalia aragonensis	Chiasmolithus gigas
			LUTETIAN	Textularia smithvillensis Cyclammina caneriverensis Discocyclina advena Bifarina B		
			52.0		Globigerina soldadoensis soldadoensis	Rhabdosphaera inflata
		LOWER	YPRESIAN		Globorotalia quetra Globigerina soldadoensis angulosa	
					Globorotalia rex	
					Globorotalia formosa formosa	
					Globorotalia wilcoxensis Globorotalia marginodentata	
					Globorotalia acuta	
			57.8		Globorotalia velascoensis	
	PALEOCENE	UPPER	THANETIAN	Cytheridea sabinensis (ostracod)		Discoaster multiradiatus Tribrachiatus contortus
				Discorbis washburni	Globorotalia pseudomenardii	Discoaster diastypus
						Heliolithus kleinpelli
				Trifarina herberti Clavulinoides midwayensis	Globorotalia angulata	
				Vaginulina robusta Vaginulina longiforma	Globorotalia pseudobulloides	
			(62)	Vaginulina midwayana	Globorotalia uncinata	
					Globorotalia incostans	
		LOWER	DANIAN	Polymorphina cushmani Vaginulina gracilis	Globorotalia trinidadensis	
					Globigerina daubjergensis	
				Robulus pseudocostatus	Globigerina fringa Globigerina eugubina	
			66.4			

Chart prepared by D. O. LeRoy, Exxon Company U.S.A., Houston, Texas and J. B. Dunlap Jr., New Orleans, Louisiana

NOTES: (1) "Last occurrence" defines extinction levels in geologic time.
(2) Calcareous nannoplankton not discussed in text as text predates chart.
(3) "ORBULINA DATUM" defines first occurrence in geologic time.
4. Relationship between zones based on benthonic foraminifers and planktonic foraminifers not firmly established.
5. Relationship between zones based on calcareous nannoplankton and planktonic foraminifers not firmly established.
** Base of Pleistocene–Global Time Scale.
** Base of Pleistocene–U. S. Gulf Coast.

[D/S] Coiling direction change in planktonic foraminifera
D—Dextral (right) coiled
S—Sinistral (left) coiled

Biostratigraphic zonation of the Cenozoic of the northern Gulf of Mexico basin

More than 60 years of investigations of Cenozoic sediments of the northern Gulf of Mexico basin have resulted in the establishment of a biostratigraphic zonation over the entire marine section (Table 1). The first recorded use of paleontology for subsurface correlation in the United States Gulf Coast area was in 1919. By 1928 there were several hundred micropaleontologists being utilized by the petroleum industry in the Gulf of Mexico basin. In areas of nonmarine and shallow or restricted marine conditions, vertebrate paleontologists, paleobotanists, palynologists and malacologists have contributed age and stratigraphic determinations to allow improved correlation between the outcrop and the subsurface.

Fossil protozoans (foraminifera) were the first to be recognized as useful in subsurface correlation. They were small enough to be recovered from well cuttings undamaged by the drill bit, were common in the marine sections, and evolved relatively quickly, allowing correlative extinction points to be established. Foraminifera continue to be the primary tool for paleontologic correlations in the Gulf of Mexico basin, but calcareous nannoplankton, palynomorphs, and other ultramicroscopic fossil groups are also used to verify, supplement, and expand the older zonations based solely on foraminifera. Zonations continue to be refined as deeper wells are drilled and use of planktonic foraminifera and calcareous nannoplankton is increased.

In addition to petroleum industry paleontologists studying calcareous nannoplankton in drilled sections throughout the northern and northwestern Gulf of Mexico basin, investigators such as Bybell (1982) and Siesser (1983) have contributed greatly to understanding the stratigraphic position of Paleogene formations of the eastern part of the region, and to tie them to the internationally recognized calcareous nannoplankton biozones of Martini (1971).

The stratigraphic framework, as well as most of the subsurface nomenclature of the United States Gulf Coast region, is based upon the extinction horizons of various benthonic foraminifera. Many of these biostratigraphic zones have never been formally described and delineation of zonal boundaries has been reached only through long usage.

Biostratigraphic zonation based primarily on benthonic foraminifera presents several problems. It has long been recognized that most benthonic foraminifera are affected by various ecologic factors that may cause them to be time transgressive in deeper marine sections. This is a very real problem as sedimentary environments can change from middle shelf to middle slope within a few kilometers in the downdip direction. A related problem that can exist in any marine section is the correlation along a lithofacies instead of a time horizon. Paleoecologic mapping and careful differentiation of ecologically and temporally sensitive species can help resolve these common problems.

With deep drilling, "deep water" lithofacies, which were once considered unique to the major onlapping shale wedges,

have been identified downdip for each stratigraphic unit. Depositional environments in these sequences often reached bathyal and even abyssal depths. Shallow-water benthonic marker faunas are normally not found in these deep-water sections and it is then necessary to use planktonic foraminifera to establish correlations. Planktonic zones can usually be directly related to the traditional benthonic zonation in paleoshelf margin lithofacies. For example, in a mid-dip position the planktonic *Globorotalia fohsi fohsi* occurs with the benthonic marker *Bigenerina humblei.* In lower-slope and abyssal environments *Bigenerina humblei* is either very rare or completely absent and *Globorotalia fohsi fohsi* is used as a top for the zone.

The use of planktonic foraminifera in correlating and subdividing strata has increased tremendously over the past 40 years. Planktonic foraminifera and calcareous nannoplankton are affected by certain ecologic factors, primarily temperature, and their use is restricted to open marine conditions, preferably middle to outer shelf and deeper environments. They are generally useful in the mid and low latitudes, and are well developed in all the marine units of the United States Gulf Coast region. The Deep Sea Drilling Program and its ongoing successor, the Ocean Drilling Program, have greatly enhanced our knowledge of the stratigraphic ranges of planktonic foraminifera and calcareous nannoplankton; these fossil groups now provide reliable correlations in areas of abrupt lithofacies changes and deep-water deposition, and define time lines for regional mapping over a wider geographical area than possible with most of the traditional benthonic markers. Currently micropaleontologists can correlate the Cenozoic section of the Gulf of Mexico basin with any zonation based on temperate to subtropical planktonic foraminifera and calcareous nannoplankton. Chronostratigraphic correlations with standard type sections must wait until the revision of these sections is complete.

The biostratigraphic zonation for the northern Gulf of Mexico basin is summarized in Table 1. Key references for the identification and ranges of planktonic foraminifera in the Gulf of Mexico basin are Postuma (1971), and Stainforth and others (1975). The calcareous nannoplankton shown on Table 1 are not discussed in this text as their extinction datums and the relationship of these datums to both planktonic and benthonic foraminiferal biostratigraphic zonation is still being investigated.

Biostratigraphic zonation of the Cenozoic of Mexico

The bases of the Cenozoic stratigraphy of the western and southwestern flanks of the Gulf of Mexico basin were established during the initial years of petroleum exploration in the Tampico and Isthmus of Tehuantepec areas, beginning in the latter years of the last century but occurring principally during the first four decades of this century. During this early exploratory period, many lithostratigraphic units—groups, formations, members— were proposed as a result of surface mapping by the oil industry, but the original descriptions of these units often did not fulfill the necessary requirements for their formalization. These original descriptions included, in addition to the diagnostic lithologic composition, the fossiliferous content of particular interest for the determination of the age and stratigraphic position of the units.

Subsequent drilling of numerous wells in search of petroleum along the Gulf Coastal Plain in east-central and southern Mexico, showed that the units recognized in the subsurface stratigraphic section did not correspond with those described from surface sections.

As in the northern Gulf of Mexico basin, subdivision of the Cenozoic stratigraphic section into distinctive lithostratigraphic units was not always easy, and, as in the northern part of the basin, biostratigraphic zonation soon became the fundamental approach to the stratigraphic breakdown of the Cenozoic section in east-central and southern Mexico. Only through the use of fossils is it possible to subdivide the monotonous terrigenous clastic sequences that characterize the Cenozoic in these regions. Biostratigraphic zonation not only supplied the basis for a sound and reliable stratigraphy, but allowed to establish local as well as regional correlations.

Some of the pioneering work in applying the study of foraminifera to petroleum exploration was done in Mexico, with benthonic foraminifera used first. More recently, the study of planktonic foraminifera has enabled paleontologists to overcome the shortcomings of the benthonic forms that are often more influenced by ecologic (environmental) factors than by chronology. Both benthonic and planktonic foraminifera are now most commonly used in biostratigraphic studies of the Cenozoic in the Mexican part of the Gulf of Mexico basin. Like in the northern Gulf of Mexico basin, subdivisions of the Cenozoic section have been designated locally as "Eocene," or "Oligocene," or "Miocene," as an alternative to biostratigraphic zonation. In southern Mexico some biostratigraphic units have, unfortunately, been given geographic names, conveying the wrong impression that they are lithostratigraphic units. The biostratigraphic zonation for the Cenozoic of Mexico as it may compare to that of the northern Gulf of Mexico basin is shown on Table 2.

ORGANIZATION OF THIS CHAPTER

The nature of the Cenozoic stratigraphic sequences in the Gulf of Mexico basin makes it impractical, and indeed unworkable, to describe it following the approach used in previous chapters. Description of the lithologic composition of unit after unit—either lithostratigraphic or chronostratigraphic—would be extremely repetitious because of the uniform character of the section. It has been thought preferable, therefore, to place more emphasis on the sedimentary processes governing the deposition of the Cenozoic sequence as a whole, and devote less space to the formal stratigraphic subdivision and nomenclature.

Each of the three stratigraphic-structural provinces around the central part of the Gulf of Mexico basin is divided into three primary areas of discussion: (1) an overview of the stratigraphic and structural style, and various related geologic factors; (2) a detailed review of the biostratigraphic zonation in common use

within the province and comments on paleoecology and paleobathymetry; and (3) detailed discussions of representative depositional sequences. Additional subjects will be discussed, as necessary. The fourth province, the central, deep part of the Gulf of Mexico, is discussed in Chapter 13 of this volume.

THE NORTHERN AND NORTHWESTERN PROGRADATIONAL PROVINCE

Overview of depositional and structural styles

Depositional style. In general, there has been a progressive gulfward shift of the sandstone/shale depocenters from older to younger Cenozoic units (Fig. 1). However, progressive spatial and temporal variations to this trend have resulted, depending on the interaction of regional subsidence, sea-level changes, and sediment supply. For example, during periods of high sediment supply, the main sand depocenter shifted to the continental shelf edge, where large quantities of sediment were delivered onto the slope and basin floor. Conversely, during periods of maximum shelf flooding and reduced sediment supply, the main sand depocenters shifted landward, with little coarse sediment reaching the basin. Various combinations of these controls resulted in an episodic pattern of sedimentation and resultant lithostratigraphy (Fig. 3).

Each depositional episode was marked by progradation and subsequent retrogradation of the sandy shore zone and associated coastal-plain lithofacies sequences by several tens of kilometers. Successive transgressions usually did not extend as far landward as their precursors. In contrast, the shelf edge was built basinward during offlap, but commonly remained as a permanent record of successive steps in basin filling (Fig. 4); it was merely submerged more deeply and prone to modest regrading during transgression and subsequent periods of maximum flooding. Thus, the sedimentary record in the Gulf of Mexico basin documents two aspects of episodicity: (1) the overall facies tract (most readily defined by the position of the shore zone) has oscillated widely, reflecting progradation followed by retrogradation and marine flooding of much of the depositional platform (Fig. 3); and (2) the continental margin (defined by the shelf edge) has alternated between periods of active outbuilding and periods of relative stability or minor retrogradation.

Although subsidence of this divergent oceanic basin margin was primarily induced by sedimentary loading, large-scale variation of sediment supply is required to explain the distinct pulses of shelf-edge progradation. By the late Neogene, the formation of increasing ice volumes to the north also significantly lowered eustatic sea level, and changes in ice volume provided a mechanism for rapid eustatic fluctuations. Climatic conditions and extrabasinal tectonic events largely control the supply of sediment to a basin. In the northwest shelf of the Gulf of Mexico basin, Winker (1981, 1982) recognized three distinct periods of major sediment influx and rapid progradation and correlated them with tectonic events in the western United States and Mexico (Fig. 5): (1) late Paleocene to early Eocene of east Texas (lower Wilcox)

TABLE 2. COMPARISON BETWEEN CENOZOIC PLANKTONIC FORAMINIFERAL ZONATION OF STAINFORTH AND OTHERS (1975) AND CENOZOIC ZONATION OF NORTHERN AND NORTHWESTERN GULF OF MEXICO BASIN

| CHRONOSTRATIGRAPHIC UNITS | | | DNAG NUM. TIME SCALE (Ma.) | CENOZOIC PLANKTONIC FORAMINIFERAL ZONATION (Stainforth and others, 1975) | DNAG NUM. TIME SCALE (Ma.) | CENOZOIC ZONATION – NORTHERN AND NORTHWESTERN GULF OF MEXICO BASIN | |
						PLANKTONIC FORAMINIFERA (Last Occurrence)[1]	CALCAREOUS NANNOPLANKTON (Last Occurrence)[1,2]
QUATERNARY	HOLOCENE		0.01		0.01	Globorotalia menardii flexuosa acme	
	PLEISTOCENE			Globorotalia truncatulinoides		Globorotalia truncatulinoides / Globorotalia tosaensis	Pseudoemiliania lacunosa A / Pseudoemiliania lacunosa B
						Sphaeroidinella dehiscens acme A	Pseudoemiliania lacunosa C
						Sphaeroidinella dehiscens acme B	Helicosphaera sellii / Calcidiscus macintyrei
			1.6		1.6		Discoaster brouweri
PLIOCENE	PIACENZIAN					Globorotalia menardii / Globorotalia miocenica	Discoaster pentaradiatus
							Discoaster surculus
				Pulleniatina obliquiloculata		Globorotalia multicamerata / Globoquadrina altispira	Discoaster tamalis
			3.4		3.4	Globorotalia margaritae / Globigerina nepenthes	Sphenolithus abies A
	ZANCLEAN			Globorotalia margaritae		Globigerinoides mitra	Sphenolithus abies B / Amaurolithus spp.
						Globorotalia menardii	Ceratolithus acutus / Discoaster A / Discoaster B
			5.3		5.3	Globoquadrina dehiscens	Discoaster berggrenii
MIOCENE	UPPER	MESSINIAN	6.5	Globorotalia acostaensis	6.5	Globorotalia acostaensis / Sphaeroidinellopsis seminulina	Discoaster neohamatus / Discoaster prepentaradiatus
		TORTONIAN					Discoaster bollii / Catinaster spp. / Discoaster hamatus
			11.2	Globorotalia menardii	11.2	Globorotalia mayeri	Coccolithus miopelagicus
	MIDDLE	SERRAVALLIAN		Globorotalia siakensis		Globorotalia fohsi robusta	
				Globorotalia fohsi robusta		Globorotalia fohsi fohsi	
				Globorotalia fohsi lobata		Globorotalia fohsi barisanensis	Sphenolithus heteromorphus
			15.1	Globorotalia fohsi fohsi / Globorotalia fohsi peripheroronda	15.1	ORBULINA DATUM[3]	Discoaster deflandrei acme
		LANGHIAN		Praeorbulina glomerosa		Praeorbulina glomerosa	Helicosphaera ampliaperta
			16.6		16.6	Globigerinatella insueta	
	LOWER	BURDIGALIAN		Globigerinatella insueta		Catapsydrax stainforthi	Sphenolithus belemnos
			21.5	Catapsydrax stainforthi	21.5	Catapsydrax dissimilis	
		AQUITANIAN		Catapsydrax dissimilis		Globorotalia kugleri	Cyclicargolithus abisectus
			23.7	Globorotalia kugleri	23.7	Globigerina ciperoensis / Globigerina sellii	Dictyococcites bisectus
TERTIARY	OLIGOCENE	UPPER		Globorotalia ciperoensis		Globorotalia opima opima	Sphenolithus distentus
		CHATTIAN					
			30.0	Globorotalia opima opima	30.0		Sphenolithus predistentus
		LOWER					
		RUPELIAN		Globigerina ampliaperatura		Globigerina ampliaperatura	Reticulofenestra umbilica / Ericsonia subdistichus
				Cassigerinella chipolensis		Pseudohastigerina micra	
			36.0	Pseudohastigerina micra	36.0	Globorotalia cerroazulensis (S.L.) / Hantkenina alabamensis	Discoaster saipanensis
EOCENE	UPPER	PRIABONIAN		Globorotalia cerroazulensis (S.L.)		Globorotalia cerroazulensis pomerali	
			40.0	Globigerinatheka semiinvoluta	40.0	Globigerinatheka semiinvoluta / Globigerinatheka barri / Truncorotaloides rohri	Chiasmolithus grandis
	MIDDLE	BARTONIAN		Truncorotaloides rohri		Truncorotaloides topilensis	
				Orbulinoides beckmanni		Orbulinoides beckmanni / Globorotalia pentacamerata	Chiasmolithus solitus
			43.6	Globorotalia lehneri	43.6	Globorotalia aragonensis	
		LUTETIAN		Globigerinatheka subconglobata			Chiasmolithus gigas
			52.0	Hantkenina aragonensis	52.0	Globigerina soldadoensis soldadoensis / Globorotalia quetra / Globigerina soldadoensis angulosa	Rhabdosphaera inflata
	LOWER	YPRESIAN		Globorotalia pentacamerata		Globorotalia rex / Globorotalia formosa formosa	
				Globorotalia aragonensis		Globorotalia wilcoxensis / Globorotalia marginodentata	
			57.8	Globorotalia formosa formosa / Globorotalia subbotinae / Globorotalia edgari	57.8	Globorotalia acuta	
PALEOCENE	UPPER	THANETIAN		Globorotalia velascoensis		Globorotalia velascoensis	Discoaster multiradiatus / Tribrachiatus contortus
						Globorotalia pseudomenardii	Discoaster diastypus / Heliolithus kleinpelli
				Globorotalia pseudomenardii		Globorotalia angulata / Globorotalia pseudobulloides	
			(62)	Globorotalia angulata	(62)	Globorotalia uncinata / Globorotalia incostans	
	LOWER	DANIAN		Globorotalia uncinata			
				Globorotalia trinidadensis / Globorotalia pseudobulloides		Globorotalia trinidadensis / Globigerina daubjergensis / Globigerina fringa	
			66.4	Globigerina eugubina	66.4	Globigerina eugubina	

NOTES: (1) "Last occurrence" defines extinction levels in geologic time.
(2) Calcareous nannoplankton not discussed in text as text predates chart.
(3) "ORBULINA DATUM" defines first occurrence in geologic time.
4 Relationship between zones based on calcareous nannoplankton and planktonic foraminifers not firmly established.
✱ Base of Pleistocene–Global Time Scale.
✱✱ Base of Pleistocene–U.S. Gulf Coast.

DL / TS Coiling direction change in planktonic foraminifera
D—Dextral (right) coiled
S—Sinistral (left) coiled

Chart prepared by D. O. LeRoy, Exxon Company
U.S.A., Houston, Texas and J. B. Dunlap Jr.,
New Orleans, Louisiana

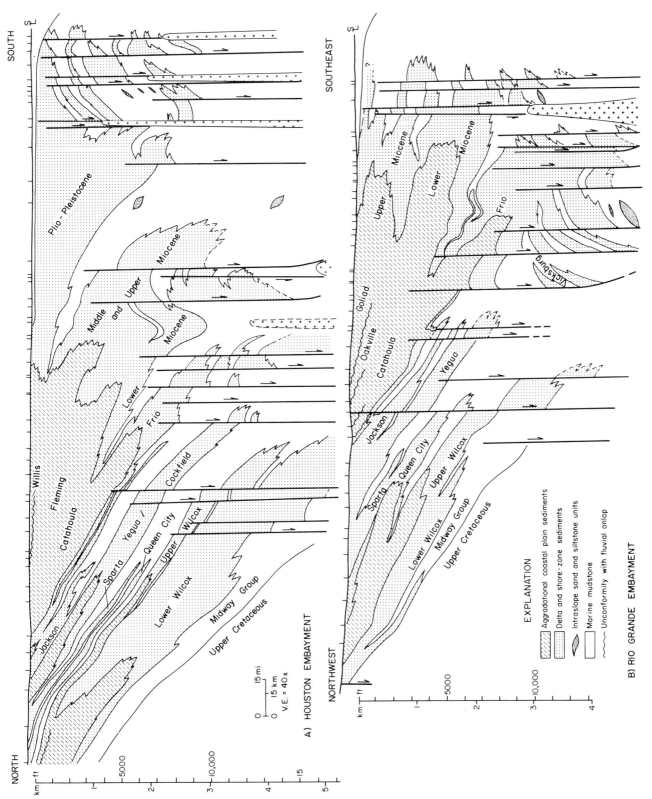

Figure 3. Regional stratigraphic cross sections of Cenozoic strata, northern Gulf of Mexico basin. A. Houston embayment; B. Rio Grande embayment.

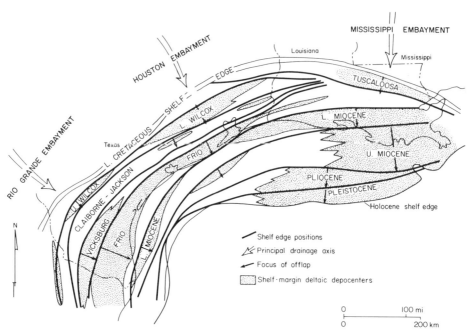

Figure 4. Cenozoic paleoshelf margins of the northwestern Gulf of Mexico basin. After Winker (1981).

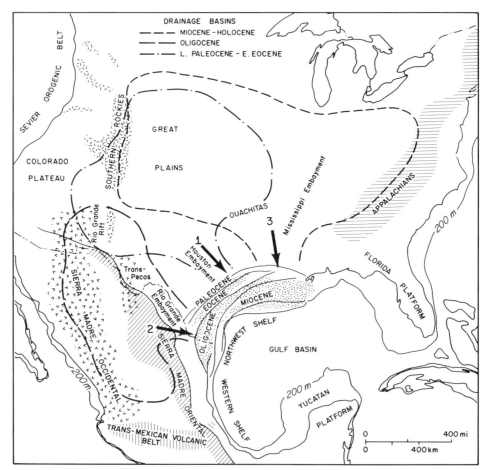

Figure 5. Probably Tertiary drainage basins and source areas for the major shelf-margin depocenters. After Winker (1981).

from early orogeny in the southern Rocky Mountains; (2) Oligocene (Vicksburg/Frio) of south Texas from calc-alkalic volcanism in the Sierra Madre Occidental; and (3) Miocene to Quaternary of coastal and offshore Louisiana related to reactivation of the southern Rocky Mountains, uplift of the Colorado Plateau, eastward tilting of the Great Plains, and renewed uplift of the southern Appalachians. Each of the major influxes focussed sedimentation into one of the three broad structural sags or "embayments" that flank the northwest Gulf of Mexico basin. In these shifting depocenters, large shelf-margin delta systems actively prograded the continental margin (Fig. 5).

Structural style. Variations in thickness and depositional patterns of the Cenozoic sediments are affected by a number of basement-controlled structures (Chapter 3 and Plate 3 of this volume) including the Rio Grande embayment, East Texas basin, Mississippi embayment, and South Florida basin. These embayments and basins are generally areas of more rapid Cenozoic subsidence and sediment accumulation. Complementary arches and uplifts between these embayments and basins have subsided less and have accumulated a thinner sediment section. The major positive structures are the San Marcos arch and the Sabine uplift. At any particular time these structural features were very subtle and probably not recognizable on the surface; but the cumulative effect throughout the Cenozoic was considerable.

Other major gravity tectonic structures have developed throughout the Cenozoic contemporaneously with sedimentation of sequences deposited beyond the shelf edge in huge progradational wedges. Growth faults, salt domes, and shale diapirs and ridges formed in areas where large masses of sediment were deposited rapidly on the unstable, muddy, prograding continental margin. Slumping and mass movement of sediment at the toe of large shelf-edge deltaic and continental slope systems initiated growth faults that ultimately extended upward through the entire sedimentary section as progradation progressed gulfward. Loading also mobilized underlying geopressured, plastic shale and salt beds into massifs and domes that rose contemporaneously with deposition around them. Patterns of faulting and diapirism reflect the interplay of depositional and gravity tectonic processes (Stude, 1978; Galloway, 1986b; Worrall and Snelson, 1989).

Pressure, temperature, and salinity environments. The northern shelf of the Gulf of Mexico basin is well known for the high-pressure zone, generally called the geopressured zone, which occurs in the subsurface at depths as shallow as 2,100 m and extends into the underlying Mesozoic slope and basinal deposits. The geopressured zone occurs where huge volumes of sediment were deposited gulfward of the stable Cretaceous shelf; subsidence was rapid and thick sections of sand and mud accumulated. Growth faults formed very early and aided in the entrapment of pore fluids within the structurally and stratigraphically isolated sands and mud. With further burial, pore pressures increased as trapped water bore an increasing proportion of the overburden load (Bredehoeft and Hanshaw, 1968; Bethke and others, 1988) and pressure gradients increased above that of hydrostatic (0.456 psi/ft). These sediments with high pore pressures are referred to as geopressured. Dewatering of shale and maturation of organic matter to hydrocarbons are believed by some to have contributed to the high pore pressure.

The depth to the top of the geopressured zone (Fig. 6) varies considerably depending largely on the amount of sandstone in the deep section; in areas of high sandstone content the top of geopressure is deep, and where shale is dominant the top of geopressure is shallow. In South Texas, the top ranges from 2,100 to 3,600 m in depth, and in Louisiana it ranges from 2,700 to 5,500 m in depth. Zones of shallow and deep geopressures trend along strike parallel to the coast. In Texas, the zones of deep geopressure represent the high-sandstone trends of the Eocene (Wilcox; inner band) and Oligocene (Frio); in Louisiana, the zones of deep geopressure reflect the high-sandstone trend of the Cretaceous (Tuscaloosa) and Miocene. The greatest depth to the top of geopressure occurs in southeast Louisiana in the sandstone-rich Upper Miocene section. The geopressured zone extends beneath the continental shelf and occurs in sediments as young as Pleistocene. In each of these major Gulf Coast Cenozoic formations there is a change from massive sandstones updip, interbedded sandstones and shale mid-dip, and massive shale downdip, containing, in many areas, fan sand deposits. The top of the geopressured zone generally occurs within the interbedded sandstones and shales deposited at the transition from delta and shore-zone to marine shelf and slope lithofacies (Fig. 3). The high pressures occur in this mid-dip position because of the presence of adequate shale to retard the escape of pore fluids during loading and subsidence.

Other properties of the rocks and fluids, such as temperature, salinity, porosity/permeability, and clay mineralogy change from the hydropressured zone into the geopressured zone. For example, in the thoroughly studied USDOE/General Crude geothermal test well, the observed temperature gradient increases from 1.0 to 1.5°F/100 ft in the hydropressured zone to 1.8 to 3.5°F/100 ft in the geopressured zone (Fig. 7). The fluid temperature at the top of the geopressured zone generally ranges from 200 to 250°F. The fluid salinity generally ranges from 100,000 to 250,000 ppm in the hydropressured zone and from 50 to 150,000 ppm in the geopressured zone. Regionally, porosity decreases with depth (Fig. 8); however, an increase as a result of secondary leaching near the top of the geopressured zone is commonly observed (Fig. 7). The dominant clay mineral within the hydropressured zone is montmorillonite (Fig. 7), with illite increasing with depth and becoming the dominant clay mineral in the geopressured zone (Freed, 1982). Thermal maturation of organic matter in sediments occurs mainly within the geopressured zone. While oil occurs almost exclusively within the hydropressured zone, major gas reserves are present in both the hydropressured and the geopressured zones, though the greatest gas volumes lie also in hydropressured reservoirs. The primary controls of these rock and fluid properties appear to be temperature, pressure, and lithology. Growth faults are essential in that they inhibit lateral fluid movement and dewatering at a very shallow depth, leading to the development of high pressure during later subsidence. Ad-

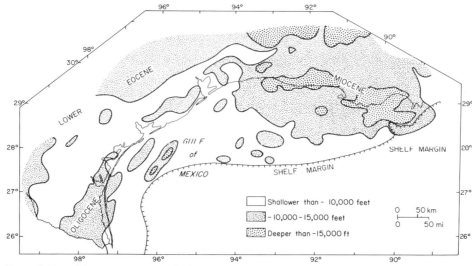

Figure 6. Top of the geopressure zone in the northwest Gulf of Mexico basin. From unpublished maps in the files of the Texas Bureau of Economic Geology and the Louisiana Geological Survey.

jacent fault blocks commonly retain markedly different fluid pressure and salinity regimes.

Depositional episodes and depositional systems. With the systematic analysis of subsurface data generated by petroleum exploration, the United States Gulf Coast Cenozoic sequence was recognized as an offlapping succession of progressively younger depositional units. Further, both subsurface and outcrop stratigraphy indicated a cyclical alternation of progradational, sandstone-rich units separated by comparatively thin, fossiliferous, widespread marine units. This fill configuration and cyclical character was early recognized and outlined by Fisk (1940), Bornhauser (1947), Lowman (1949), and Stenzel (1952) and later discussed by Fisher (1964), Rainwater (1964b), Coleman and Gagliano (1964), Dixon (1965), Curtis and Picou (1980), Dodge and Posey (1981), and Galloway (1987). A conceptual framework for interpretation of these Cenozoic fill cycles was provided by Frazier (1974) with introduction of the concept of *depositional episodes.*

Frazier's model of a depositional episode and the resulting depositional complex can, with modification, be applied to many terrigenous clastic basins. With its origins in the Quaternary of the northern Gulf Coast, it is particularly applicable to the older clastic succession of this province (Galloway, 1989b). Several sedimentologic principles underlie the depositional-episode model:

1. Terrigenous clastic sediments are allochthonous.

2. Basins are filled with repetitive alternation of depositional and nondepositional intervals. At any one time, active deposition is concentrated in a limited portion of the basin, and negligible amounts of sediment accumulate elsewhere.

3. The time interval represented by these nondepositional, or "hiatal" surfaces, which bound the active depositional units, varies areally; however, each surface contains at least one isochron throughout its areal extent.

4. Each episode of regional deposition incorporates numerous depositional events, which are localized cycles of deposition separated from underlying and overlying depositional events by short-duration, transgression-related hiatal intervals. The product of a depositional event is now commonly called a parasequence.

Further, each depositional episode deposits three primary lithofacies assemblages (Fig. 9A). Slope progradational and base-of-slope aggradational phase deposits progressively fill the basin from the margin, producing a wedge of sediment that commonly thickens basinward. Shallow water and coastal deposits prograde onto the offlapping slope depositional platform. Terrestrial aggradational phase deposits cap the progradational platform, and are thickest in the landward direction. Termination of the depositional pulse, along with continued subsidence, results in a transgressive phase of reworking and deposition of a thin, commonly discontinuous veneer of marine sediment across the submerged portion of the cycle. Upper slope and shelf margin mass wasting and erosion may deposit an onlapping wedge of marine sediment at the base of the slope. Thus, each depositional episode produces a composite facies progression recording initial marine offlap, terrestrial aggradation, and terminal transgression and shelf inundation. As shown on the time-distance diagram in Figure 9A, each depositional episode is bounded by intervals of slow or nondeposition as the locus of active deposition shifts about the basin margin.

Multiple depositional events combine to produce a depositional episode and its resultant genetic stratigraphic sequence and component parasequences. The genetic stratigraphic sequence is the assemblage of facies sequences deposited during a period of relative climatic, tectonic, and geomorphic stability in the basin and its surrounding source terranes. The genetic stratigraphic sequences are the major transgressively bounded offlapping wedges commonly recognized in the Gulf Coast Cenozoic. Each se-

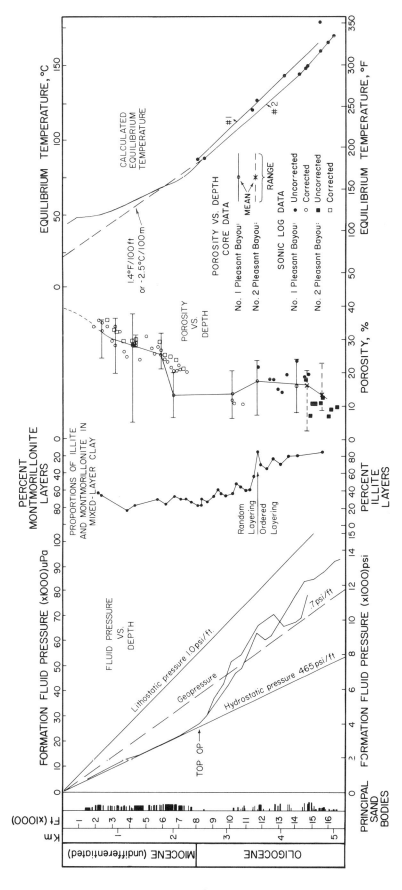

Figure 7. Typical interrelationships among lithology, pressure gradient, clay mineralogy, sandstone porosity, and temperature gradient, northwest Gulf of Mexico basin. Data from U.S. Department of Energy—General Crude Pleasant Bayou No. 1 and 2, Brazoria County, Texas.

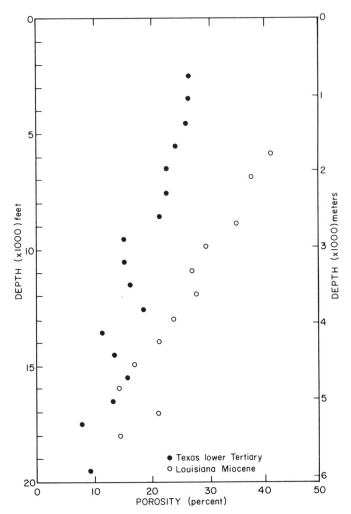

Figure 8. Mean sandstone porosity versus depth from whole-core analyses of lower Tertiary (Paleocene-Oligocene) formations along the Texas Gulf Coast (Loucks and others, 1986) and Miocene sandstones from Louisiana (unpublished data, Core Laboratories, Inc.) Only values from sandstones without matrix were used. Data were averaged over 300-m intervals.

quence is overlain bsainward by a regional marine shale wedge and bounded by associated hiatal surfaces stratigraphically manifested as condensed intervals (Fig. 9B). Each genetic sequence potentially contains a major subaerial hiatal surface or condensed interval that represents increasing spans of time landward. Hiatal surfaces, condensed intervals, and transgressive units all provide physical stratigraphic markers widely used in subsurface and surface correlation (including thin, widespread marls or impure limestone beds, richly glauconitic sandstone or mudstone sheets, fossiliferous or burrow-churned mudstone, veneers of relict, marine-reworked sand or mud, and subaqueous erosion surfaces). The genetic coherence of facies and related surfaces produced by the Frazier episode led Galloway (1989a, b) to propose the concept of flooding-bounded genetic stratigraphic sequence as a distinctly different and useful alternative to subaerial

unconformity-bounded sequences as defined by Vail and others (1984).

Recognition of genetic sequences and their component facies associations has two important applications. First, they provide regionally mappable genetic stratigraphic units. Second, the model predictively relates progradational, aggradational, and transgressive deposits in the temporal and spatial framework of the basin margin.

A second important concept in genetic stratigraphic analysis of the offlapping cycles in the Cenozoic of the Gulf of Mexico basin is that of *depositional systems*. First applied by Boyd and Dyer (1964) in their analysis of the south Texas Frio, the concept was further defined and deployed by Fisher and McGowen (1967) in their regional analysis of the Texas lower Wilcox. In their usage, depositional systems were fabricated principally from integration of genetic facies, and involved recognition of large-scale genetic units comparable to modern depositional systems, such as fluvial systems or delta systems. Thus, depositional systems—three dimensional, genetically defined, physical stratigraphic units consisting of process-related sedimentary facies (Galloway and others, 1979)—are in turn the physical geographic elements (Chorley and Kennedy, 1971) of a depositional complex.

Depositional systems are recognized and interpreted by: (1) their position within the basin, (2) their bulk lithologic composition, (3) the three-dimensional geometry of framework facies (usually sand bodies), and (4) the internal sedimentary features, bedding relationships, vertical textural sequences, and minor but diagnostic components.

Delineation and description of depositional systems is particularly applicable to basins, such as the Gulf of Mexico basin, in which subsurface data are abundant. Work subsequent to the middle 1960s by many authors has applied the concept of depositional systems to interpretation of most of the major Cenozoic stratigraphic units of the northern Gulf of Mexico basin.

Genetic stratigraphic framework and evolution during the Cenozoic. Throughout the Cenozoic, the northern and northwestern margin of the Gulf of Mexico basin has been an ongoing depocenter for thick successions of terrigenous clastic sediment. Sequential accumulation of deltaic, shore-zone, shelf, and slope systems has prograded the continental margin of northern Mexico, Texas, and Louisiana approximately 300 km basinward of the reef-delimited Cretaceous carbonate shelf edge (Fig. 4). Along any segment of the basin margin this history of coastal-plain offlap produced several well-defined, sand-rich genetic stratigraphic sequences, bounded on their basinward margins by widespread marine shale wedges or tongues.

Depositional architecture of the genetic sequence can be further characterized as slope offlap or platform aggradation. Principal depositional episodes resulted in progradation of major delta systems beyond the previously extant shelf margin, loading the thick, muddy continental slope wedge and underlying crust. Such offlapping episodes are characterized by greatly thickened delta-front and shore-zone lithofacies sequences, large-scale

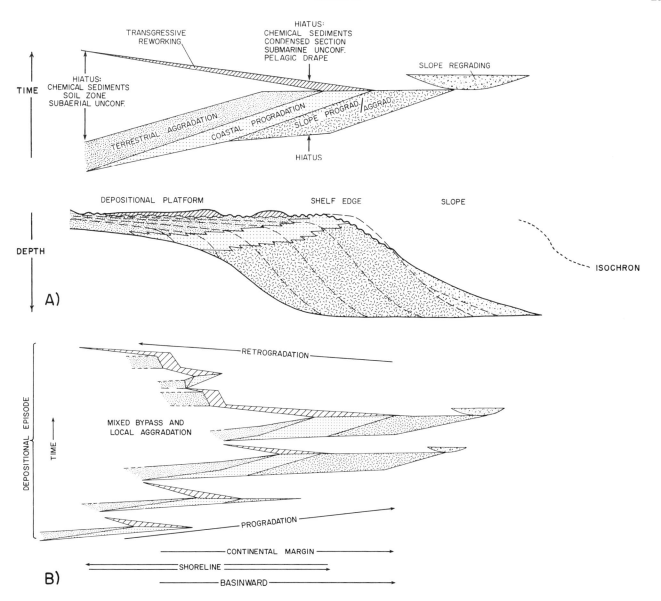

Figure 9. Idealized facies architecture of a depositional episode and resultant genetic stratigraphic sequence. A, Time-space diagram (episode) and depth-space diagram (genetic stratigraphic sequence) showing temporal and stratigraphic relationships of facies associations of a simple depositional episode. B, Schematic representation of the complex history of outbuilding typical of a northwest Gulf of Mexico basin depositional episode. Smaller cycles of progradation and transgression punctuated the long-term episode of regional offlap and retrogradation. From Galloway (1989a).

growth faulting, diapirism, and extensive development of geopressure (for a review, see Winker and Edwards, 1983). In contrast, platform aggradation occurred when less extensive or mud-dominated sediment input resulted in deltaic and shore-zone accumulation upon the comparatively shallow, drowned continental platform. Typically, such minor depositional episodes followed periods of extensive transgression and shoreline retreat and built out onto fluvio-deltaic coastal-plain deposits of earlier offlapping episodes.

Principal depositional episodes of the northern Gulf of Mexico basin are listed and classified in Table 3. Each is apparent on one or both of the regional subsurface cross sections (Fig. 3). Six offlapping episodes, the lower Wilcox (Paleocene–early Eocene), upper Wilcox (early Eocene), Vicksburg/Frio (Oligocene), early-to-middle Miocene, late Miocene, and Pliocene-Pleistocene episodes, are responsible for most of the Cenozoic accretion of the continental platform.

Three-dimensional analysis of the successive genetic strati-

TABLE 3. CONSTRUCTIONAL SAND-RICH DEPOSITIONAL
EPISODES OF THE NORTHERN AND NORTHWESTERN
GULF COAST BASIN

Depositional Episode	Depositional Architecture
Plio-Pleistocene*	Slope offlap
Upper Miocene	Slope offlap
Middle Miocene	Local slope offlap
Lower Miocene	Slope offlap
Frio	Slope offlap
Vicksburg	Local slope offlap
Jackson-Yazoo	Platform aggradation, muddy slope offlap
Yegua-Cockfield	Local slope offlap
Sparta	Platform aggradation
Queen City-Cane River	Platform aggradation
Upper Wilcox-Carrizo	Slope offlap
Lower Wilcox	Slope offlap

Subdivision of eustatic episodes within Pliocene and Quaternary
deposits is not prominent in their regional physical stratigraphy.

graphic sequences, which has been made possible by the extensive knowledge of the section obtained by drilling and reflection seismology, provides a unique insight into the dynamics of deposition along a divergent margin (McGookey, 1975; Curtis and Picou, 1980; Galloway, 1987). Several generalizations emerge:

1. Each of the sand-rich genetic sequences consists of one or more major delta systems separated or flanked by interdeltaic wave-dominated shore-zone systems and associated shelves. The deltaic headlands define the points of maximum continental platform expansion in the slope offlap episodes; here, delta systems extended onto the upper continental slope, supplying sediment to the lower continental slope and deep basin.

2. The location of deltaic depocenters has consistently occupied one of the major embayments. However, through time successive deltaic depocenters have shifted across the northern and northwestern shelf (Martin, 1978; Winker, 1982). Throughout the Paleocene and Eocene, largest delta systems lay along the axis of the Houston embayment (Fig. 4). In Oligocene time the major continental drainage element shifted to the Rio Grande embayment, and the largest of the Vicksburg and Frio deltaic systems prograded far beyond the earlier Eocene platform margin, which is delineated by the Vicksburg fault zone. In Miocene time, a final dramatic drainage shift to the Mississippi embayment focused continental platform offlap along the south Louisiana Coastal Plain and adjoining continental shelf. These shifts are believed to be related to continental tectonism and consequent reorganization of drainage patterns (Winker, 1982).

3. The large delta systems were fed by fluvial systems of continental proportions. The nature of the individual fluvial systems varied, reflecting the caliber of the sediment load and paleoclimate. Early Tertiary fluvial systems, which are represented at outcrop by units such as the Simsboro and Carrizo Sandstones (lower Wilcox and lower Claiborne equivalents), are interpreted

to be mixed-load and bed-load systems that traversed a tropical to subtropical coastal plain (Fisher and McGowen, 1967; Hamlin, 1983). By Oligocene time, the relatively uniform subtropical coastal climate was replaced by a pronounced climatic zonation from subarid to arid in northern Mexico and south Texas to humid subtropical in Louisiana and eastward (Galloway, 1977). Primarily bed-load fluvial systems traversed south Texas and the Rio Grande embayment (Galloway, 1977, 1981); as major continental drainage shifted to the Mississippi embayment in the late Miocene, drainage elements in the western part of the basin became extremely flashy to ephemeral (Hoel, 1982). Fluvial systems of the Houston and Mississippi embayments were dominantly of mixed- to suspended-load types (Galloway, 1981).

4. During major offlap episodes, distal delta, shore-zone, and shelf deposits prograded directly onto laterally unconfined, thick, muddy, underconsolidated, and geopressured sediments of the continental slope (Bruce, 1973). Large-scale plastic deformation, which was induced at the base of the slope by depositional loading, was manifested in the upper slope, shelf, and delta section as basinward sloping, listric growth faults (Fig. 10). Because deformation was greatest at the locus of maximum deposition, the shelf margin, greatest fault offset and expansion of the section lie within delta fringe, prodelta, or shelf, and upper slope facies. Thus, successive growth fault zones closely correspond to the basinward pinch-out of the progradational delta and shore zone sands into deep-water marine mud. This vertically persistent facies change, combined with the lateral offset of sands across faults, exerts a dominant control on the stratigraphic position of the top of the geopressured zone (Winker and Edwards, 1983; Fig. 10).

5. Large-scale isostatic subsidence of the crust in response to sediment loading produced ongoing coastward tilting of successively older depositional sequences. Each depositional complex has a hinge line, or more correctly, a hinge zone, which parallels the basin margin (the focus of crustal loading) and is a line of zero vertical displacement (Winker, 1979). Landward from the hinge zone, strata are uplifted slightly. Increased tilting with time is reflected in the increasing slope of Quaternary depositional surfaces of the lower coastal plain and in the increasing, though gentle dip of successively older Tertiary units. The hinge zone lies far inland of the continental margin, which is the focus of crustal loading; consequently only the most updip portions of each episode are subjected to uplift and subsequent erosional truncation. Several extremely low-angle discordant surfaces and formational onlaps occur along the Gulf Coastal Plain. Examples include the truncation and overlap of the Catahoula Formation by the Goliad Formation, which is so prominent on geologic maps of the Rio Grande embayment, and by the Miocene Oakville Formation across the San Marcos arch. This process of load-induced crustal subsidence and peripheral uplift has been modeled by Quinlan and Beaumont (1984) and recognized around the margin of the Quaternary depocenter (Fisk, 1944; Jurkowski and others, 1984).

6. Thin, widespread, highly fossiliferous marine shale

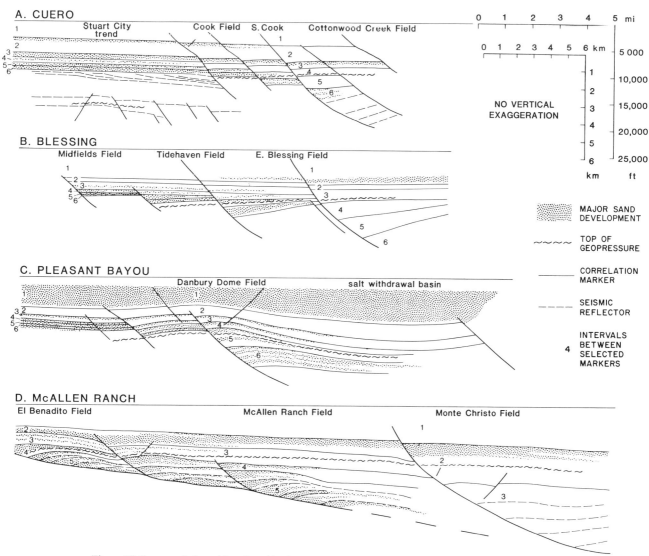

Figure 10. Structural, depositional, and hydrogeologic features of growth-faulted Cenozoic continental-margin deposits, northwestern Gulf of Mexico basin. For location of sections, see Figure 12. From Winker (1982).

wedges or tongues separate downdip portions of the genetic sequences. Updip equivalents of these regionally transgressive units include fossiliferous, glauconitic sandstones exhibiting features indicative of shallow-marine or shore-zone reworking and deposition. Studies of evolutionary rates (Stenzel, 1949; Fisher and others, 1964) show that such thin units are condensed sections that record extended periods of geologic time. More recent work (Loutit and others, 1983, 1988) notes the association of less obvious geochemical anomalies associated with such condensed sections. In the subsurface, equivalent fossiliferous shales contain a diverse inner to outer neritic fauna that includes many of the prominent index forms (Fig. 11).

7. Laterally restricted shale wedges or prominent submarine canyon fills containing an outer neritic to bathyal microfauna replace the Cenozoic offlap sequences at several stratigraphic levels. Best-known examples include the Oligocene Hackberry and Miocene Abbeville embayments, the Paleogene Yoakum and Lavaca paleocanyons, and the Timbalier and associated Pleistocene canyons (Fig. 12; Hoyt, 1959; Paine, 1971; Chuber and Begeman, 1982). Deep drilling and improved seismic resolution have revealed many additional canyons and erosionally based embayments that also enclose deep marine shale sections. The embayments were recognized early as noteworthy because they contained no obvious, updip, shallow-water equivalent and exhibit erosional relationships to underlying shallow-water facies. Although some authors have proposed a subaerial origin for the embayments and canyons, the preponderance of evidence favors combinations of rapid local subsidence and submarine slumping and incision (Coleman and others, 1983; Ewing and Reed, 1984; Dingus and Galloway, 1990; Galloway and others, 1988, 1991).

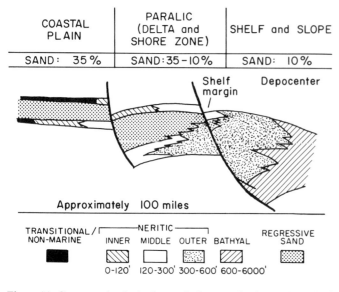

Figure 11. Conceptual paleobathymetric framework of a northern Gulf of Mexico basin Cenozoic genetic sequence. Bounding neritic (shelf) mudrocks extend updip from the continental margin depocenter defined by greatest expansion across contemporaneous growth faults and interfingering of outer neritic and bathyal faunas. Modified from Curtis and Picou (1978).

Paleocanyons exhibit a range of morphologies. The broad, open embayments display comparatively minimal erosional modification of the basal contact with older paralic sediments and contain onlapping fill. More obviously erosional features, such as the Lavaca paleocanyon, formed along rapidly prograding, unstable continental margins. Multiple distinct canyons may be stacked to form a slope-canyon depositional system. Such "immature" canyons are characterized by low length/width ratios and broad, flat floors; they extend at most a few tens of kilometers across the contemporaneous shelf platform, but excavated large volumes of prodelta and slope sediment, largely by slumping. Canyon fills correlate with local marine shale units that separate contemporaneous sandy deltaic and shore-zone facies sequences. Well-developed, isolated paleosubmarine canyons occur as widely separated features commonly associated with retrogradational or transgressive facies sequences (Dingus and Galloway, 1990). They exhibit high length/width ratios, extend many tens of kilometers updip of the contemporaneous shelf edge, and were excavated deeply, as much as 1 km, into underlying paralic and coastal plain deposits. Significantly, the canyon fills correlate with widespread mudstone units deposited during periods of maximum shelf flooding, and, like modern delta-associated canyons (Burke, 1972), commonly occur at flanks of major delta systems. Reasons for the specific temporal and paleogeographic localization of the features remain enigmatic, and are likely diverse.

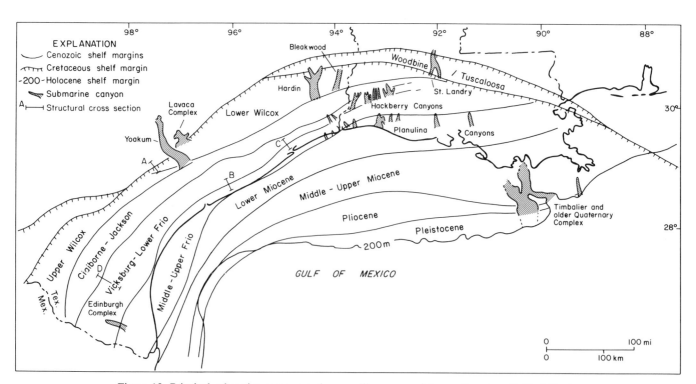

Figure 12. Principal submarine canyons and canyonlike gorges of the northwestern Gulf of Mexico basin Cenozoic wedge. Best known canyons are associated with the Wilcox, Frio, and Quaternary shelf edges. Modified from Winker (1982) with additions from numerous published and unpublished sources. Location of cross sections A to D (Fig. 10) also shown.

Biostratigraphy and paleoecology

Biostratigraphic zonation. (Table 1, Fig. 2, and Stratigraphic Correlation Chart, Plate 5, in pocket). (The Paleocene through Oligocene part of the following discussion is based to a great extent on the work of H. L. Tipsword, 1962).

Paleocene: Midway Group.

1. *Kincaid and Clayton Formations.* The lower Midway contains an abundant foraminiferal fauna, which is recognized throughout the Gulf of Mexico basin and is readily distinguished from that of the underlying Cretaceous beds. This fauna was first described in a classic paper by Plummer (1926). Kellough (1959) has shown the distribution of the foraminifera of the lower Midway of east Texas. More than 200 species have been described from these beds.

2. *Wills Point Formation.* The upper Midway is generally less fossiliferous than the lower portion. The top of the Wills Point Formation is marked by the extinction of *Robulus midwayensis* and *Vaginulina longiforma.*

As is typical of all Paleocene formations in the Gulf of Mexico basin, those of the Midway Group were deposited in a slope environment in the deeper portions of the basin. These environments are not favorable for the traditional updip marker benthonic foraminifera and it is necessary to use planktonic forms. In these deep-water environments the top of the Midway can be reliably identified on the extinction of *Globorotalia velascoensis* and *Globorotalia pseudomenardii.*

3. *Naheola, Nanafalia, and Tuscahoma Formations.* The Naheola, Nanafalia, and Tuscahoma Formations of southern Alabama, previously considered to be Eocene in age, are now known to be of Paleocene age based on planktonic faunas (Mancini and Oliver, 1981).

Paleocene-Eocene: Wilcox Group. The Wilcox Group represents one of the major Cenozoic progradational episodes in the Gulf of Mexico basin. Very thick beds of fluvial, deltaic, and shore-zone sequences make up the section in the central and northwestern United States Gulf Coast. With the exception of the Sabine uplift and the south Texas areas, the outcrops of the Wilcox Group contain, therefore, few marine fossils. In Mississippi and Alabama, on the other hand, the section is largely marine. In the subsurface of the northern Gulf of Mexico basin, the Wilcox Group becomes marine and fossiliferous; it is, however, difficult to separate it from the overlying Claiborne Group. There is no lithologic break, and the contact is not marked by a distinctive change in the benthonic foraminifera. The contact is apparent, however, in the changes in the planktonic fauna. The extinction of *Globigerina soldadoensis* is considered to be the top of the Wilcox Group in the downdip marine section (Stainforth and others, 1975).

1. *Hatchetigbee Formation.* The Bashi marl member of the upper Wilcox Hatchetigbee Formation of Mississippi and Alabama contains an abundant microfauna that has been described by Cushman and Ponton (1932), Cushman and Garrett (1939), and Cushman (1944). The partially equivalent Sabinetown Formation of Louisiana and Texas contains a nondiagnostic shallow-marine foraminiferal fauna; however, it does contain a distinctive ostracod, *Cytheridea sabinensis.* The top of the subsurface Wilcox Group in this area can be identified most reliably on the extinction of the planktonic foraminifer *Globorotalia wilcoxensis.*

Eocene: Claiborne Group.

1. *Reklaw and Tallahatta Formations.* The lower Claiborne Tallahatta Formation of Alabama and Mississippi and the Reklaw Formation of Texas do not contain diagnostic foraminifera.

2. *Weches, Cane River, Winona, and lower Lisbon Formations.* The Weches Formation is a thin, transgressive, glauconitic unit containing an abundant foraminiferal fauna described by Cushman and Thomas (1929), Cushman and Ellisor (1933), and Garrett (1941). Its top is marked by the extinction of *Textularia smithvillensis.* The Cane River Formation of Louisiana has a well-developed fauna described by Hussey (1943, 1949). The Winona Formation of Mississippi and the lower part of the Lisbon Formation of Alabama contain faunas similar to those of the Weches and Cane River Formations.

3. *Cook Mountain Formation and upper Lisbon Formation.* The guide species to these formations is *Ceratobulimina exima,* which is widespread over the entire clastic province. *Operculinoides sabinensis* occurs in the lower portion. The fauna of these formations has been described by Howe (1939), Garrett (1941), and Cushman and Todd (1945a).

4. *Yegua and Cockfield Formations.* The Yegua and Cockfield Formations of Louisiana and Texas are nonmarine or lagoonal at the outcrop. In the subsurface these formations are recognized by the presence of *Discorbis yeguaensis* and *Eponides yeguaensis,* described from the Yegua Formation of east Texas by Weinzierl and Applin (1929). These two species occur throughout most of the clastic province. *Nonionella cockfieldensis* has been listed as a guide species to the Yegua and Cockfield; however, it was described by Cushman and Ellisor (1933) from basal Jackson beds in Sabine County, Texas. In deep-water sections the top of the Claiborne Group is marked by the extinction of the planktonic foraminifer *Truncorotaloides rohri.*

Eocene: Jackson Group.

1. *Moodys Branch Formation.* The basal Jackson Moodys Branch Formation is present from east Texas to the Florida panhandle. In Texas and Louisiana the Moodys Branch contains *Camerina moodysbranchensis.* In the Mississippi-Alabama-Florida section *Camerina jacksonensis* becomes the more common index fossil. In downdip, deeper-water sections the Moodys Branch cannot be picked reliably on fossils. Cushman and Todd (1945b) described the foraminifera from the type Moodys Branch of Jackson, Mississippi. Bergquist (1942) published descriptions of Moodys Branch species from Scott County, Mississippi, and Gravell and Hanna (1935) described the larger foraminifera from Texas, Louisiana, and Mississippi.

2. *Caddell, Wellborn, Manning (McElroy), and Whitsett Formations.* Ellisor (1933) divided the Jackson Group of Texas from older to younger into the Caddell, McElroy, and Whitsett Formations and described their foraminiferal faunas. (At present

the Jackson Group is commonly subdivided, also from older to younger, into the Caddell, Wellborn, Manning or McElroy, and Whitsett Formations.) The diagnostic species of the Caddell Formation is *Textularia dibollensis*. The subsurface McElroy Formation is normally picked on *Textularia hockleyensis* even though this species ranges into the overlying Caddell Formation at the outcrop. The overlying Whitsett Formation (and the top of the Jackson Group) is recognized by the occurrence of *Massilina pratti*. The mid-dip, slightly deeper water portion of the Whitsett Formation contains *Marginulina cocoaensis* and *Uvigerina jacksonensis*. In downdip deep-water sections the Jackson Group is not differentiated and the top is picked on the extinction points of either *Globorotalia cerroazulensis* (*Globorotalia cocoaensis*) or *Hantkenina alabamensis*, both planktonic foraminifera.

Papers dealing with Jackson Group foraminifera of Texas include Cushman and Applin (1926), Cushman and Ellisor (1933), and Stuckey (1960).

3. *Yazoo Formation*. The Yazoo Formation overlies the Moodys Branch Formation in western Alabama, Mississippi, and Louisiana. Composed mainly of marine clays with a rich foraminiferal fauna, it has been extensively studied and subdivided. Cushman described the Jackson foraminifera of the southeastern United States Gulf Coast (1924, 1926, 1931, 1933, 1935a) and also devoted three additional papers (1925, 1928, 1946) to species attributed to the Cocoa Sand (a sandstone unit within the Yazoo), but which are actually from the Shubuta Clay above the Cocoa Sand. Other publications concerned with Jackson species of the central and eastern Gulf Coast are by Monsour (1937) and Bandy (1949).

Textularia dibollensis occurs in the lower Yazoo throughout much of the United States Gulf Coast Province. *Textularia hockleyensis* occurs above the *Textularia dibollensis* Zone and it also has wide distribution. The western Alabama–Mississippi outcrop section is a somewhat deeper water facies than that of the Jackson Group outcrop of the Louisiana-Texas section but contains similar benthonic forms.

As in Texas, the downdip, deep-water facies of the Jackson Group is not subdivided and the top of the Jackson is recognized on the extinction points of either *Globorotalia cerroazulensis* or *Hantkenina alabamensis*.

Oligocene: Vicksburg Group.

1. *Red Bluff Formation*. The Red Bluff Formation occurs at the base of the Vicksburg from Texas to the panhandle of Florida. In Louisiana the Red Bluff is recognized only in the subsurface. Cushman (1922a, 1923) described species of foraminifera from the type locality of the Red Bluff Formation at Hiwannee, Mississippi, and Cushman and Garrett (1938) described the fauna from the Red Bluff of Alabama. As most foraminifera that occur in the Red Bluff are not restricted to that unit it is recognized only on the basis of assemblages.

2. *Vicksburg Formation*. The Vicksburg Formation of Texas is known only from the subsurface. The top of the shallow-marine Vicksburg in Texas and Louisiana may be recognized on the extinction of *Textularia warreni* but this species does not occur in the eastern Gulf Coast. The top of the downdip Vicksburg in Texas is identified by the presence of *Loxostoma delicata* (B). Cushman and Ellisor (1931) and Ellisor (1933) described species occurring in the Vicksburg of Texas.

The Vicksburg is exposed in a narrow belt in central Louisiana and contains a typical Vicksburg fauna. In the subsurface the top of the Vicksburg can be recognized by the extinction of *Textularia mississippiensis* where *Textularia warreni* is not developed. The top of the Vicksburg Formation in deeper-water sections over the entire northern Gulf of Mexico basin can be picked at the extinction point of the planktonic foraminifer *Globigerina ampliapertura*.

3. *Mariana Formation*. The Mariana Formation is a limestone formation, which overlies the Red Bluff Formation in western Florida, Alabama, and Mississippi. Westward across Mississippi, it grades into a glauconitic, fossiliferous sand and marl, the Mint Springs Formation or Member. Both the Mariana and the more clastic Mint Springs contain abundant foraminifera. Cushman (1922b, 1923) described foraminifera from the type localities of the Mint Springs at Vicksburg, Mississippi, and from the Mariana Formation at Mariana, Florida. Cole and Ponton (1930) also described the foraminifera from the Mariana of Florida. *Lepidocyclina mantelli* is the guide species for the Mariana Formation.

4. *Byram Formation*. In Alabama and Mississippi this formation is a glauconitic limestone, overlain by an argillaceous limestone or marl, and glauconitic sand. The lower limestone has been called both the Glendon Formation and a member of the Byram Formation. Publications by Cushman (1922a, 1923) have descriptions of foraminifera from the Byram Formation of Mississippi. Additional species from the Byram of Mississippi and Alabama are listed by Cushman (1929, 1935b), Cushman and McGlamery (1938, 1939), and Cushman and Todd (1946).

Oligocene: Catahoula Group or Formation.

1. *Frio Formation of Texas and Louisiana*. The Frio Formation has been defined and redefined many times since the surface beds were originally described by Dumble along the Frio River in Live Oak County, Texas, in 1894. The original outcrop described by Dumble is now believed to correlate in part to both the Vicksburg and the lowest Catahoula-Frio of the deep subsurface (Holcomb, 1964; Galloway and others, 1982). As currently used in Texas, the Frio Formation is a series of deltaic and marginal marine sands and shales, which are the downdip equivalents of the nonmarine Catahoula.

In 1957, A. D. Warren formally proposed that the subsurface Frio beds of Louisiana be designated as the Frio Formation with its type locality in the China field in Jefferson Davis Parish, Louisiana. Although this would necessitate invalidating the older usage for the Texas outcrop section, it conforms to the current definition of the subsurface Frio in Texas and Louisiana.

From top to bottom the subsurface Frio of the Texas and Louisiana area is zoned on extinctions of the several benthonic foraminifera. Where it ranges above the Vicksburg, the extinction of *Anomalina bilateralis* can be used as a correlation point. It normally occurs below *Nonion struma* and *Textularia seligi* (Table 1). The foraminifera of the Frio Formation have been described from various localities (Garrett and Ellis, 1937; Ellis, 1939; Stuckey, 1946).

In southeast Texas and southwest Louisiana an onlap deep-water shale unit termed the "Hackberry facies" occurs in the middle Frio. The Hackberry-Frio relationship has been discussed by Stuckey (1964), and Berg and Powers (1980). The Hackberry fauna occurs between *Marginulina texana* and *Nonion struma* (Table 1); however, in downdip areas of south Louisiana a deep-water "pseudo-Hackberry fauna" is found above *Marginulina texana*. Garrett (1938) described the Hackberry assemblage from wells in Cameron Parish, Louisiana, and Jefferson County, Texas. This fauna is an upper slope to abyssal assemblage.

2. *Chickasawhay and Paynes Hammock Formations of southern Mississippi and southwestern Alabama.* The Chickasawhay Formation consists of arenaceous and argillaceous limestones with marly and silty clays. The overlying Paynes Hammock Formation contains similar sediments but is richer in sandstones. The Chickasawhay and the lower part of the Paynes Hammock Formation are the equivalents of the subsurface Frio of Texas and Louisiana. The outcrop of the Chickasawhay contains the subsurface Frio markers *Nonion struma* and *Nodosaria blanpiedi,* whereas *Cibicides hazzardi* occurs both in the upper Chickasawhay and in the lower Paynes Hammock.

Poag (1966) found no distinct change between the Paynes Hammock and Chickasawhay beds at the outcrop, which suggests that these two formations are a single biostratigraphic unit. A distinctive calcareous unit in the upper part of the Paynes Hammock Formation contains a benthonic fauna that Poag (1966) believed to be an updip equivalent of the *Heterostegina* Zone of the Anahuac Formation in the subsurface of Texas and Louisiana. Further investigations (Poag, 1972) of the planktonic fauna of the Chickasawhay and Paynes Hammock Formations supported his earlier conclusion that they are equivalent to the *Heterostegina* through *Nodosaria blanpiedi* Zones of the subsurface Anahuac and Frio Formations. Based on planktonic foraminifera Poag (1972) assigned a late Oligocene age to these beds.

3. *Anahuac Formation.* The Anahuac Formation is found in the subsurface of Texas, Louisiana, and southwestern Mississippi. It is primarily a marine shale wedge deposited following a relative rise in sea level that generally correlates with an earliest Miocene eustatic rise (Haq and others, 1987). The Anahuac overlaps the regressive Frio Formation in downdip areas and is overlain by the progradational sands of the lower Miocene. In local areas the Anahuac Formation contains carbonate bank and reefal deposits, commonly found on salt domes, which were bathymetric highs.

Although the foraminiferal zones of the Anahuac were first described in 1925 by Applin, Ellisor, and Kniker, it was not until 1944 that Ellisor formally defined it as a formation from the subsurface of Chambers County, Texas.

The Anahuac is divided into three major paleontologic subdivisions, from top to bottom the *Discorbis, Heterostegina,* and *Marginulina* Zones. The *Discorbis* Zone is characterized, from top to bottom, by *Discorbis* "restricted," *Discorbis nomada,* and *Discorbis gravelli.* In downdip areas the *Discorbis* fauna may or may not be well developed and the extinction of *Robulus* A is considered as a reliable marker for the top of the *Discorbis* Zone. This is a species that has not been formally named but is widely recognized by petroleum-industry paleontologists. (The use of unpublished species becomes more common in the zonation of Miocene and younger beds due primarily to restrictions placed by the petroleum industry on publication of information and zonations that are considered to be of possible use to competitors. These unpublished species will be referred to by the designation commonly used in the petroleum industry.)

The *Heterostegina* Zone underlies the *Discorbis* Zone and is characterized by species of *Heterostegina.* These species of foraminifera are confined to relatively shallow water; downdip, this zone may be recognized but not defined by the presence of *Cibicides jeffersonensis. Bolivina perca* occurs in the lower portion of the zone. It is in the *Heterostegina* Zone that carbonate bank and reef deposits are found.

The *Marginulina* Zone is the oldest zone of the Anahuac and contains the extinction points of three easily recognizable species of *Marginulina.* From youngest to oldest they are *Marginulina idiomorpha, Marginulina mexicana* var. *vaginata,* and *Marginulina howei.*

The foraminifera of the Anahuac Formation have been described by Gravell and Hanna (1937), Garrett and Ellis (1937), Garrett (1939), Ellisor (1944), and Cushman and Ellisor (1945).

"Miocene" and Fleming Group or Formation of Texas or Louisiana. In Texas the Fleming Group consists of the Oakville and Lagarto Formations. Although nonmarine at the outcrop, these formations become marine and thicken rapidly basinward, becoming thickest under the present continental shelf. In Mississippi the equivalent section includes the surface Catahoula and Fleming Formations of Miocene age. However, the Catahoula Formation of Louisiana is not equivalent to the upper part of the Oakville Formation and the Lagarto Formation of Texas. The confusion of surface formation and group names between Texas and Louisiana is due to isolated surface mapping and dates back many years. It is unfortunate but persistent.

The subsurface Miocene of Louisiana, like that of Texas, becomes marine downdip and very thick in southern Louisiana and under the continental shelf of the Gulf of Mexico. The great thickness, geographic extent, intense structural and diapiric activity, shifting depocenters, both major and minor transgressions and regressions, together with depositional environments varying from nonmarine to abyssal, combine to make subdivision and paleontologic zonation of the subsurface Miocene difficult. The Miocene has been zoned using foraminifera from drill cuttings and cores recovered during petroleum exploration. Many zonations have been developed independently by paleontologists working for the various companies, and many names, numbers, letters, and codes have been attached to the same marker species. Paleontologic zonations were, and in many cases are, considered proprietary information. As a result many well-known marker species have never been formally named and their informal letter or number designation has become entrenched in the literature. In some cases, even after a species has been formally designated, the informal name is preferred and used by many workers.

In general, the subdivisions of the Miocene of the Texas and Louisiana subsurface utilize the same index foraminifera. In the thicker sections of south Louisiana and the adjoining Gulf of Mexico more subzones have been utilized and the deeper marine shales of the various local depocenters have their own subdivisions within the regional zonation.

Traditionally the Miocene of Texas and Louisiana has been divided into "Lower," "Middle," and "Upper" units. McLean (1957) proposed stage names for these subdivisions, but they have never received wide acceptance. As used in the following discussion, the subdivision of the Miocene into "Lower," "Middle," and "Upper" conforms closely, but not exactly, with the recommendations of the IUGS Regional Committee on Mediterranean Neogene Stratigraphy in the Proceedings of their 6th Congress held in Bratislava in 1975 (Senes, 1975–1976). The reason for possible discrepancies is that the boundaries recommended by the Committee are based on *first appearance upward* in the section of certain species, a procedure difficult to follow in subsurface work where the section is penetrated from top to bottom and paleontological identification is made from drill cuttings, often contaminated by recirculation of previously drilled sediments and material sloughed from the walls of the drill hole.

1. *Lower Miocene.* The Lower Miocene is zoned (from youngest to oldest) based on a succession of extinction points beginning with *Amphistegina chipolensis* B and ending with *Cristellaria R* (Table 1).

The Lower Miocene below the *Siphonina davisi* Zone grades downdip into a deep-water facies generally termed the "Abbeville facies." This facies contains an assemblage of foraminifera typical of slope deposition and is common in downdip areas throughout the northwestern Gulf of Mexico basin. Like all other lithofacies, the Abbeville becomes progressively younger downdip.

The lower boundary of the Miocene appears to be conformable and its placement in the Gulf of Mexico basin is based on the local benthonic subdivision (Table 1). This contact is close to the one recommended by the IUGS Regional Committee on Mediterranean Neogene Stratigraphy (Senes, 1975–1976); however, the boundary as recommended is the *initial* appearance of *Praeorbulina glomerosa,* which cannot be placed accurately when using drill cuttings.

2. *Middle Miocene.* The Middle Miocene is zoned (from youngest to oldest) based on the extinction points of *Cibicides carstensi* or *Bigenerina nodosaria* var. *directa* ("2") through *Cibicides carstensi opima* (Table 1). The rapid thickening of the Middle Miocene, and the presence of thick sections of sediments deposited in a continental slope environment often necessitate using planktonic foraminifera for correlations in downdip areas. These sediments contain a series of planktonic foraminifera

(Table 1), which can be reliably correlated not only in the Gulf of Mexico basin but also on a worldwide basis.

The downdip lithofacies of the Middle Miocene are known as the "Breton Sound facies" for the deep-water equivalent of the *Cibicides carstensi* section and the "Harang facies" (Echols and Curtis, 1973) as the equivalent of the *Lenticulina cristi* section. Both of these lithofacies contain fauna characteristic of a slope environment and are confined to southeast Louisiana.

3. *Upper Miocene.* The contact between the Middle and the Upper Miocene is problematical for the same reason that the Middle and Lower Miocene contact is. The IUGS Regional Committee on Mediterranean Neogene Stratigraphy (Senes, 1975–1976) recommends this boundary be placed at the initial appearance of *Globorotalia acostaensis,* which cannot be recognized using drill cuttings. The contact between the *Cibicides carstensi* Zone and the *Spiroplectammina barrowi* Zone (Table 1) approximates this occurrence. The contact of the Miocene with the overlying Pliocene is likewise based on the initial appearance of the planktonic *Globorotalia margaritae.*

The Upper Miocene is zoned (from youngest to oldest) on the extinction points of *Robulus E* through *Spiroplectammina barrowi* (*Textularia* L; Table 1). Downdip portions of the Upper Miocene are difficult to zone. Many benthonic markers do not survive in the deeper-water environments and there are no useful planktonic extinctions. Local markers and assemblage zones are used, but the accurate downdip zonation of the Upper Miocene remains a problem.

Papers dealing with the foraminifera of the Texas-Louisiana Miocene include those by Howe and McDonald (1938), Cushman and Ellisor (1939), Ellisor (1940), Garrett (1942), Pope and Smith (1948), Akers (1955), Pope (1955), Akers and Drooger (1957), Butler (1960), and Skinner (1972).

Pliocene-Pleistocene. Examination of microfaunas recovered by extensive drilling on the present continental shelf and upper slope has resulted in a firm biostratigraphic zonation of the thick Pliocene-Pleistocene section in the Gulf of Mexico basin. Working with benthonic and planktonic foraminifera along with calcareous nannoplankton, the faunal zonation is, in its broader aspects, standard for most workers in the region. Delineation of warm- and cold-water cycles present in the Pliocene-Pleistocene section is also a routine procedure (Beard, 1969). What is not firmly established is the placement of the Pliocene-Pleistocene (Neogene-Quaternary) boundary, or the relation of the warm and cool cycles recognized in the marine section to the supposedly Pleistocene terraces mapped in central Louisiana, or to the classical four-stage glacial terminology of the midcontinent region. This Pliocene-Pleistocene boundary problem does not affect the biostratigraphic zonation of the marine Pliocene-Pleistocene section of the Gulf of Mexico basin, and resolution of the problem probably will not be found within the province.

The problem of defining the Pliocene-Pleistocene boundary in the Gulf of Mexico basin involves two basic issues. The first and most fundamental is the correlation of the United States Gulf Coast faunal events with events present in the boundary-stratotype section in Italy. The second issue is whether the boundary should be based entirely upon paleontological evidence or should be placed at what is believed to be the initial onset of glacial influence in the Gulf of Mexico basin. Fundamental to both these issues is the lack of isotopic dating and paleomagnetic data for the marine section in the Gulf of Mexico basin. Only one isotopic date (Beard and others, 1976) and no paleomagnetic data have been published to date.

Three different paleontological horizons have been proposed for the Pliocene-Pleistocene boundary in the Gulf of Mexico basin. The youngest is the extinction point of *Discoaster brouweri,* which approximates the extinction point of the benthonic foraminifer *Trifarina (Uvigerina) holcki (Angulogerina B)* (see Poag, 1971). The extinction of *D. brouweri,* which is favored by many paleontologists, occurs at the approximate top of the Olduvai magnetic polarity zone. At one time this boundary was recommended by the IUGS Regional Committee on Mediterranean Neogene Stratigraphy. However, the level of extinction of *Discoaster brouweri* occurs above (or younger than) the accepted Pliocene-Pleistocene boundary-stratotype at Vrica, Calabria, Italy (Beard and others, 1982).

The second proposed Pliocene-Pleistocene boundary for the Gulf of Mexico basin is at the extinction of the planktonic foraminifer *Globorotalia miocenica.* This extinction point is a widespread event wherever warm-water conditions existed and is the boundary most generally accepted by United States Gulf Coast industry workers. *Globorotalia miocenica,* however, has not been found in the Vrica section.

The Pliocene-Pleistocene boundary as proposed by Beard and others (1982) is at the extinction of the planktonic foraminifer *Globoquadrina altispira,* which is dated at approximately 2.8 Ma. The first cool-water cycle in the late Pliocene or early Pleistocene occurs just after the extinction of *Globoquadrina altispira.* This event is presumed to reflect the onset of continental glaciation and to be of Pleistocene age on that basis. The occurrence of *Globoquadrina altispira* in the Mediterranean region has been documented (DSDP Sites 125 and 132) and because it has not been found (to date) in the Vrica section, this proposed boundary is considered to be below the Pliocene-Pleistocene boundary at Vrica.

Eight *major* cycles of alternating sequences of warm-water and cold-water foraminiferal faunas have been recognized in the Gulf of Mexico basin above the extinction of *Globoquadrina altispira* (Beard, 1969). Many attempts have been made to equate these events with terrace deposits cropping out onshore along the shores of the Gulf of Mexico and with the classical four-stage glacial terminology of the midcontinent region. Investigations of cores from the Gulf of Mexico basin indicate that a continental glacial stage can encompass several marine cycles of warm and cool water (Kennett and Huddlestun, 1972). It is apparent that

glaciation was initiated about 3 m.y. ago in the Northern Hemisphere (Berggren, 1972), and this date correlates with the first of the eight cool/warm cycles documented by Beard and others (1982) in the Gulf of Mexico basin. Given the complexity of the marine record, it appears unlikely that direct correlation of the marine and nonmarine section will be obtained without the addition of more isotopic dates.

Correlation of the marine Pliocene-Pleistocene of the Gulf of Mexico basin with the exposed nonmarine units that crop out around the entire Gulf is problematic. In central Louisiana, Fisk (1952) mapped four terraces and their coastwise equivalents, which he considered to be expressions of higher sea levels during the interglacial stages of the Pleistocene. Fisk designated the Pleistocene terraces from oldest to youngest as the Williana, Bentley, Montgomery, and Prairie. He discounted the concept of the widespread Citronelle Formation as a Pliocene formation. Instead, he regarded these deposits as the "basal graveliferous beds" of the several terraces. Doering (1956, 1958) disagreed with Fisk's mapping and presented a revised version of the stratigraphy of the terrace deposits but did not disagree with Fisk's conclusion that they were produced by eustatic changes in sea level. Doering (1958) concluded that the Citronelle Formation was preglacial, although he considered it as early Pleistocene rather than Pliocene. The Citronelle has now been dated as Pliocene on the basis of vertebrate fauna (Isphording and Lamb, 1971), although this does not contribute to either prove or disprove its preglacial origin.

Both Fish and Doering based their interpretations on evidence from the west side of the Mississippi Valley. Their nomenclatural designations east of the river are the result of interpretative projections across the wide valley. More recent work (Durham and others, 1967) shows only post-Citronelle terraces on the east side of the Mississippi River in the vicinity of Baton Rouge. Previously mapped "terraces" are now known to be fault scarps. As a result, Saucier (1974) and others remapped the topographic features of south Louisiana and published their findings on the 1984 Louisiana Geological Map. This study resulted in the delineation of three major Pleistocene coast-wise terraces, from oldest to youngest: the High, Intermediate, and Prairie terraces.

Detailed stratigraphic and geomorphic studies by Otvos (1972, 1975, 1976) have shown that the oldest mappable Pleistocene formation along the northeastern rim of the Gulf of Mexico basin is the Prairie Formation and its equivalents (the Biloxi and Gulfport Formations). The Prairie Formation in southern Mississippi, Alabama, and in the Florida Panhandle coastal zone is part of a single discontinuous chain of Pleistocene barrier ridges also identified along the Texas and southwest Louisiana coast (Ingleside–Live Oak system). These late Pleistocene deposits are believed to be Sangamonian in age (100,000 to 170,000 yr B.P.; Otvos, 1972).

Otvos (1981) also presented evidence that the ridges and

scarps along the northeastern rim of the Gulf of Mexico, which have been interpreted as evolving during pre-Sangamon Pleistocene high stands of sea level, are structural lineaments and that tectonism was probably responsible for the coastal escarpments.

The subsurface subdivision of the Pliocene-Pleistocene section beneath the Continental Shelf has concerned many investigators. Akers and Holck (1957) studied the biostratigraphy in detail, using sidewall cores from a well in South Pass Block 41 at the mouth of the Mississippi River. They demonstrated that the Pliocene-Pleistocene section could be divided into four units, each composed of a deep-water topstratum and a shallow-water substratum. Their interpretation was supported on a cross-section network within the Mississippi deltaic plain.

In 1988, McFarlan and LeRoy presented a detailed surface to subsurface stratigraphic study utilizing lithologic, paleontologic, electric-log, and seismic correlations from over 380 core holes, water wells, and petroleum industry wells across Louisiana into the offshore area. They concluded that the Quaternary deposits could be divided into six major time-stratigraphic units, three of which correlate to the outcropping terraces of Saucier (1974); two occur as updip subcrop pinchouts; and one correlates to the modern Mississippi deltaic and chenier plain. Using historical Gulf Coast nomenclature, the units are, from oldest to youngest: Nebraskan-Aftonian = High Terrace; Kansan-Yarmouthian = Intermediate Terrace; early Illinoian = (subcrop pinchout); late Illinoian–Sangamonian = (subcrop pinchout); early Wisconsinian = Prairie Terrace; and late Wisconsinian = Mississippian deltaic and chenier plain.

The entire problem of identification, origin, differentiation, and long-distance correlation of terraces has yet to be resolved (see Isphording, 1981; and Chapter 2, this volume).

The Pleistocene-Holocene boundary is defined in the northern Gulf of Mexico basin as the last occurrence of *Globorotalia inflata* (cool) and an increase in abundance of *Globorotalia menardii* (warm). On the basis of C^{14} and oxygen isotopes this boundary is dated at approximately 11,000 yr B.P. (Flower and Kennett, 1990).

From younger to older the following extinctions and events are used to zone the Pliocene-Pleistocene of the Gulf of Mexico basin:

- Disappearance of *Globorotalia inflata* (planktonic).
- Disappearance of *Globorotalia tumida flexuosa* (planktonic).
- Extinction of *Trimosina denticulata* (benthonic).
- An increase in the abundance of *Sphaeroidinella dehiscens* (planktonic).
- Disappearance of *Hyalinea balthica* (benthonic).
- Extinction of *Trifarina holcki* (benthonic) and extinction of discoasters (Pliocene-Pleistocene boundary of Poag [1971], estimated to be 1.65 Ma. Berggren and others [1985], using new data, suggest this boundary must be as old as 1.90 Ma).
- Change in preferred coiling direction of *Globorotalia cultrata* from predominantly left to predominantly right.

- Extinction of *Lenticulina 1* (benthonic) and *Globorotalia miocenica* (planktonic) (Pliocene-Pleistocene boundary that some Gulf Coast workers estimate to be 2.2 Ma).
- Extinction of *Globorotalia multicamerata* (planktonic).
- Extinction of *Globoquadrina altispira* (planktonic) (Pliocene-Pleistocene boundary of Beard and others, 1982, estimated to be 2.8 Ma).
- Extinction of *Globorotalia margaritae* (planktonic).
- Extinction of *Globigerina nepenthes* (planktonic).
- Extinction of *Buliminella basispinata* (benthonic).
- Extinction of *Textularia X* (benthonic).

Biostratigraphic zonation of the Pliocene-Pleistocene section is difficult. The thick Pliocene-Pleistocene section creates long intervals between biostratigraphic markers even though the actual time span is geologically short. Most of the Pliocene-Pleistocene fauna are living today and there are very few reliable benthonic extinctions for zonations of shelfal environments. Repeated rises and falls of sea level during the late Pliocene and the entire Pleistocene have repeatedly flooded and exposed much of the continental shelf producing subtle regional unconformities that further complicate biostratigraphic zonation.

Additional references to the biostratigraphy of the Pliocene-Pleistocene of the northern Gulf of Mexico basin include Phleger and Parker (1951), Andersen (1961), Butler (1962), Akers and Dorman (1964), Akers (1972), Lamb and Beard (1972), Skinner (1972), LeRoy and Levinson (1974), and Kohl (1980).

Paleoecology and paleobathymetry. An increasingly important aspect of the stratigraphic work on the thick Cenozoic sequences of the Gulf of Mexico basin is the establishment of the paleoecology and paleobathymetry of the sediments making up the successive intervals of these sequences.

Paleoecology is the study of fossil organisms in relation to the environments in which they lived. The basis for most paleoecologic studies is a strict application of the principle of uniformitarianism; that is, the ecology of present organisms is the key to that of past organisms.

Although early studies on the ecology of modern foraminifera were done by R. D. Norton in 1930 and M. L. Natland in 1933 off California, it was not until after World War II that the use of modern foraminifera to construct models for paleoenvironments attracted the attention of the petroleum industry. Landmark papers by Stainforth (1948) on the Tertiary section of Trinidad and by Lowman (1949) on the Tertiary of the Gulf of Mexico basin showed the potential of paleoecology in petroleum exploration. In 1955, Crouch delineated six paleoecologic zones in the Miocene of south Louisiana, and paleoecology was recognized as a viable exploration tool. A paper by Bandy and Arnal in 1960, "Concepts of Foraminiferal Paleoecology," immediately became a classic reference and strongly influenced most future work in the field. In 1966, the Gulf Coast Section of the Society of Economic Paleontologists and Mineralogists published two papers that summarized the paleoecologic zones used by the petroleum industry (see Albers, 1966; Tipsword and others, 1966).

The most widespread application of paleoecology in the Gulf of Mexico basin region is determination of paleobathymetry. The water-depth distribution of modern foraminifera is well known and this information provides a basis for accurate paleobathymetric interpretations. The paleobathymetry of a sedimentary unit is determined by comparing the foraminiferal fauna of that unit with the bathymetric distribution of similar fauna in the present-day Gulf of Mexico. This direct method relies upon the assumption that the ecologic response of certain benthonic foraminifera to numerous physical and chemical factors which vary with depth has not changed significantly since the Oligocene. This assumption has been supported by Boltovskoy (1980) in a study of the benthonic foraminifera in 837 samples from upper Cenozoic deposits of nine Deep Sea Drilling Project (DSDP) cores located in the middle bathyal zone of the South Atlantic, South Pacific, and Indian Oceans. He concluded that the Quaternary and Oligocene assemblages consisted of nearly the same species, and that the Holocene benthonic fauna developed no later than the Oligocene. The paleoenvironments of sediments older than Oligocene are based upon faunal associations and the assumption that all associated species lived under similar conditions. The faunal elements having modern counterparts provide the paleoenvironment for the entire assemblage.

The distribution of living foraminifera in the Gulf of Mexico has been extensively studied and their bathymetric zonation is well documented (Phleger, 1951; Phleger and Parker, 1951; Parker, 1954; Pflum and Frerichs, 1976; Poag, 1981; Culver and Buzas, 1981). Benthonic species that are considered to be isobathyal form the framework of this zonation.

The bathymetric zonation is also calibrated by the succession of species and the morphologic gradations of single species. Utilizing this data base, comparisons are made between patterns of Holocene and older foraminiferal distributions that permit interpretation of the older environments in relation to a known frame of reference. Paleoenvironments ranging from inner neritic to abyssal can be recognized and are conventionally defined as shown in Figure 13.

The northwest Gulf of Mexico basin, containing a complete Cenozoic section which can be correlated from continental or shallow marine to deep marine sequences, provides an almost

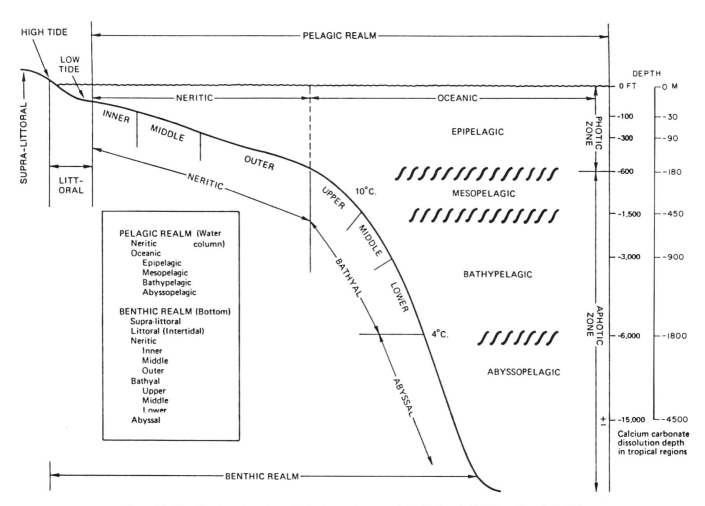

Figure 13. Classification of marine paleobathymetric zones. After Hedgpeth (1957) and Lamb (1981).

ideal setting for paleobathymetric studies. Major sand-rich depositional cycles display characteristic paleobathymetric relationships as defined by the interbedded and bounding shales (Curtis and Picou, 1980). Updip, coastal-, and delta-plain deposits are bounded by thin shale tongues containing inner to middle neritic forms (Fig. 11). Downdip, the highly thickened deltaic and shore-zone sand and shale complex, which was deposited in growth-fault- or diapir-bounded subbasins, grades basinward into outer neritic and bathyal shales. These deep-water argillaceous rocks are commonly geopressured.

Paleobathymetry is essential in determination of depositional environments and the sedimentary history of the resulting rock bodies. Reconstruction of ancient sea floors gives clues to sediment source, transportation, and depositional processes, and thus to distribution and geometry of sedimentary bodies. Hydrocarbon occurrences plotted on paleobathymetric maps outline trends of favorable environments. Detailed paleobathymetric mapping shows time of growth on structures, unconformities, faulting, and large bodies of transported sediments (Stude, 1978). Geologic interpretations based on seismic stratigraphy are greatly enhanced by a knowledge of the paleobathymetry of the section under study.

Oxygen-isotope analysis of the planktonic foraminifera tests provides another tool for both correlation and paleoecology. This method can give precise correlations and is especially useful in the study of deep-sea cores (Emiliani, 1969). Oxygen-isotope records provide evidence of glacial and interglacial cycles and glacio-eustatic sea-level changes. Work by Emiliani and others (1975), Kennett and Shackleton (1975), Malmgren and Kennett (1976), and Brunner and Keigwin (1981) in the Gulf of Mexico basin has shown that oxygen-isotope data from the Pleistocene can be correlated with worldwide events and used to construct a "type section" of oxygen-isotope data for the entire basin. Oxygen-isotope stratigraphy has become an important tool for worldwide correlations, especially for the Pleistocene, and will become increasingly important in the Gulf of Mexico basin as studies continue.

Representative genetic depositional sequences

Lower Wilcox Formation: Paleocene offlap. The lower Wilcox of the Gulf of Mexico basin, of late Paleocene age, marks the first Cenozoic episode of major deltaic offlap (Figs. 14 and 15). Principal delta systems were developed in the northern part of the basin (chiefly in Louisiana and Mississippi) and in its northwestern part (in Texas, largely east of the Guadalupe River). Generally contemporaneous with these episodes of deltation are: an extensive strandline-shelf complex in south Texas and northeastern Mexico; a tectonic, deep-basin system in the southwestern part of the basin along the east Mexican coast; and a broad carbonate-evaporite platform system in the Florida and Yucatan platforms. Eastern Mexico and the Florida and Yucatan platforms will be discussed in following sections of this chapter.

The major delta systems were responsible for substantial basin loading leading to the first Cenozoic episode of growth faulting and salt mobilization. These kinds of depositional-

tectonic events were to be repeated several times during the Cenozoic history of the Gulf of Mexico basin.

Component delta systems. Two major delta systems, with associated updip fluvial systems and downdip deeper-water slope systems, have been recognized and delineated in the upper Paleocene section of the Gulf of Mexico basin (Fig. 15). These are the Rockdale delta system in the northwestern part of the basin, in Texas (Fisher and McGowen, 1967), and the Holly Springs delta system of the Louisiana and Mississippi northern Gulf of Mexico basin (Galloway, 1968). Updip in the outcrop and shallow subsurface, at least parts of the feeding fluvial systems of the delta systems are preserved. These fluvial systems were predominantly meandering streams, both fine-grained and coarser-grained varieties. Associated downdip, deep-water slope systems have been penetrated only locally but can be interpreted from deep-penetration seismic data.

Both delta systems are extensive, each covering about 100,000 km^2. Each shows three main regional phases of progradation and abandonment. The Rockdale delta system developed six major lobal complexes, and the Holly Springs system, four. The analogous Holocene Mississippi delta system developed five lobal complexes as mapped by Frazier (1967), the currently active lobe being part of one of the lobal complexes.

Both the Rockdale and Holly Springs delta systems developed extensive growth faulting, notably where they prograded basinward of the underlying structural shelf edge of Cretaceous carbonates (Fig. 16). Both systems significantly mobilized underlying salt and associated shale deposits, with major salt-withdrawal basins constituting preferred and persistent deltaic depocenters. The Rockdale delta system is the thicker of the two (Fig. 15), reaching about 760 m on the prograded platform landward of zones of growth faulting, and as much as 3,000 m basinward. The Holly Springs system thickness ranges from a maximum of 450 to 1,500 m in comparable structural positions. Depositional architecture of the delta systems on the platform and on the basin contrast. Where developed above the underlying Cretaceous carbonate platforms, the progradational depositional facies of the system (delta front and prodelta) are relatively thin and the bulk of the systems consists of aggradational delta plain facies (Fig. 16). Downdip or basinward of the underlying Cretaceous shelf edge, progradation facies (delta front and prodelta) are stacked and generally associated with growth faults and are both thick and predominant.

The strong analogy of these delta systems to the fluvially dominated Holocene Mississippi delta system has been shown by Fisher and McGowen (1967) and by Galloway (1968), among others. Similarities include scale, size (not thickness), distribution, depositional facies composition, magnitude and caliber of feeding fluvial systems, and lobe geometry, style, and superposition.

Principal component facies. The lower Wilcox delta systems consist of four major depositional facies associations: three constructional facies—delta plain, delta front, and prodelta—and a fourth, thin, though regionally extensive destructional facies developed upon local abandonment (Fig. 17).

1. *Delta-plain facies.* Immediately downdip and basinward of feeding fluvial systems is the most landward component of the delta, the delta-plain facies. The sandstone skeletal framework of the facies is a complex of elongated, relatively straight, but commonly thick, distributary-channel deposits. These channel sandstones are generally uniform in texture, have sharp lower boundaries (erosional) and sharp upper boundaries (rapid abandonment through stream evulsion). Hence they give a characteristic box-shaped electric-log pattern. Associated laterally with these distributary-channel sands are levee silt deposits; local and thin, interdistributary sand splays; extensive interdistributary or overbank muds; and, most characteristically, numerous lignite seams. Composition of the lower Wilcox delta-plain facies is remarkably similar to the Holocene Mississippi delta-plain facies, though in the Holocene analogue, peats take the place of the lower Wilcox lignites.

In the platform portions of the lower Wilcox deltas, the stacked, aggrading delta-plain facies, the predominant facies, is associated with several basinward delta progradations. This relationship is broken only locally by thin transgressive units corresponding to periods of lobal or delta abandonment basinward.

2. *Delta-front facies.* The delta-front facies (channel-mouth bars, frontally splayed sands, and varying amounts of wave-reworked sands) is the progradational sand facies of the lower Wilcox delta system. On the platform, this facies is relatively thin and underlies the thick, aggradational facies of the delta plain; off the platform, the facies is stacked and thick and becomes the downdip equivalent of the landward delta-plain facies. Basinward, it grades into the predominantly shale facies of the prodelta. Both the delta-front and prodelta facies are intimately involved in growth faulting. Individual sequences of the delta-front facies show upward coarsening in texture and increasing frequency of sands, giving a characteristic progradational electric-log pattern. Thickness of individual sequences varies but commonly is 60 to 150 m; progradational sequences are separated by thin, commonly glauconitic, delta-destructional units. Aggregate thickness of the stacked, growth-faulted, progradational delta-front facies is as much as 1,500 m.

Distribution of the sandy delta-front facies defines the external geometry of prograded delta lobes. In the lower Wilcox delta systems, as in the comparable Holocene Mississippi delta system, two basic geometries are developed: a relatively narrow, elongated type, and a more arcuate or lobate type. These compare to the barfinger and shoal-water delta lobes of Fisk (1961). The elongated lobes generally are developed over relatively thick prodelta muds, show little or no wave reworking of sands and apparently involved rapid differential subsidence of prograding sands into underlying water-saturated prodelta muds. A lobate geometry commonly develops where progradation was over underlying abandoned delta lobes, which provided a relatively stable substrate. This delta geometry is commonly stratigraphically higher in the delta sequences. Further, it shows a significant amount of marine reworking and coalescing of sand bodies, giving it the lobate or more rounded geometry. The different geometries do not necessarily imply a change in wave activity or marine energy but rather the period of time that the prograded front was subject to wave modification as a function of rate of differential subsidence and degree of consequent sand storage.

3. *Prodelta facies.* The prodelta is the most basinward and thickest depositional facies of the lower Wilcox delta system. It, in turn, grades basinward into deeper-water sediments of the continental slope off the platform. The lower Wilcox prodelta facies is chiefly shale, generally dark and organic, with scant marine fossils.

It was primarily the excessive loading of the delta front and particularly the unstable prodelta muds that triggered extensive growth faulting associated with these facies of lower Wilcox deltas. Further, the loading served to extensively mobilize underlying salt and associated shale. Mobilization, commonly diapiric, was toward the thinner interdeltaic areas and especially basinward of the depocenter to an area largely coextensive with the area of slope deposition. A comparable diapiric field is associated with continental slope off Texas and Louisiana, basinward of the loading of Pliocene-Pleistocene delta systems (Lehner, 1969; Stuart and Caughey, 1977; Martin, 1978). Because the lower Wilcox was a major salt mobilizer, salt structures penetrating the lower Wilcox, as chiefly drilled, are in the reservoir-poor interdeltaic areas. As a result, little or no oil or gas production in the lower Wilcox is related to salt structures; the principal zones of production are delta-front sands associated with growth-fault traps. It was not until these mobilized salt structures of the Wilcox interdeltaic and slope areas were embanked by younger systems with favorable reservoir facies, chiefly the Frio sequences, that their contribution to oil and gas traps came into play. Such are the complementary roles of mobilizing and embanking systems.

Associated late Paleocene systems. Although the major deltas of the lower Wilcox of the northwestern and northern parts of the Gulf of Mexico basin are volumetrically significant, have been major hydrocarbon producers, and have attracted major exploration, other depositional systems developed in the basin contemporaneously with the lower Wilcox deltation. One of them—the south Texas–northeast Mexico strandline-shelf system—is a significant hydrocarbon producing area.

1. *Lavaca submarine canyon system.* The Lavaca submarine canyon system consists of at least three large erosional features that are nested to slightly offset and that formed along the rapidly prograding, unstable lower Wilcox (Paleocene) continental margin. The canyon system (Fig. 18) has maximum dimensions of 49 km in width, 41 km in length, and contains as much as 850 m of compacted muddy fill. Decompaction of canyon fill shows an original thickness (canyon depth) of up to 1,300 m.

Regionally, the Lavaca submarine canyon system formed within the Garwood subembayment of the Lower Wilcox depositional complex, and was located on the southwest flank of the Colorado lobe complex of the Rockdale delta system (Fig. 15). Facies analysis revealss that the canyons formed on a muddy, progradational delta flank and were subsequently filled and buried by prodelta muds and delta-front sands. Submarine mass wasting

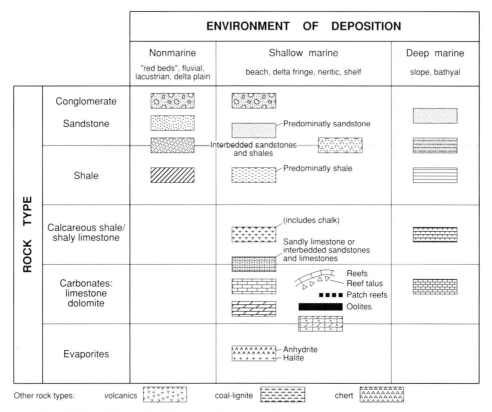

Figure 14. Symbols used in lithofacies-paleogeographic maps, Figures 15, 19, 21, 25, and 28.

caused headward erosion of the lower Wilcox shelf edge and was primarily responsible for canyon excavation. The canyon exhibits onlap fill, capped by offlapping, progradational seismic reflectors. Marine mudstones with few, isolated sandstone bodies comprise the dominant fill within the canyon. Mudstones are predominately massive with zones of contorted, slumped, and irregular bedding; sandstone units consist of slumped masses of shallow-water sands that grade into distal turbidites. These facies reflect submarine-current-modified slump deposits.

Interpretation of initial canyon cutting suggests that erosion was due to shelf-edge failure following active depositional loading of the continental margin. A likely control was a sudden shift in local depocenters, possibly lobe switching in the Rockdale delta system. This resulted in temporary sediment starvation at the shelf edge, during which headward slumping, and to a lesser extent current scour and turbidity flows, enlarged the canyon boundaries updip. Comparison of the Lavaca canyon system with the middle Wilcox Yoakum canyon, which is located in the same subembayment, indicates that the magnitude of headward erosion for this type of canyon was constrained by the limited duration of sediment interruption.

2. *The south Texas–northeast Mexico strandline-shelf system.* During the late Paleocene, contemporaneous with and to

the south of lower Wilcox deltation, an extensive structural shelf, largely coextensive with underlying Cretaceous carbonate platforms, was the site of a major barrier-bar/strandline system with an associated updip lagoonal system and basinward mud-rich continental shelf and slope systems (Fig. 15). The strike-trending sand system was sourced by marine reworking of prograding delta lobes to the northeast with transport by prevailing southwest longshore currents. The barrier bar system (Cotulla system of Fisher and McGowen, 1967) has an aggregate thickness of up to 350 m, grades updip to predominantly lagoonal mud facies (Indio lagoon system) and downdip to predominantly shelf muds. Although there is a principal axial facies trend of the barrier bar system, several, less common, strike-fed sand deposits are preserved basinward. These mark more basinward shoreline positions apparently related to a period of maximum progradation of the sourcing delta system to the northwest. Neither growth faulting nor salt-shale mobilization was a common adjunct of this strandline-shelf system. Certain sands near the outer structural edge of the south Texas shelf were considered by Fisher and McGowen (1967) as lower Wilcox "shelf-edge" sands. These have subsequently been placed in the upper Wilcox by Edwards (1981) and interpreted to be prograded deltas.

The south Texas–northeast Mexico strike-oriented systems

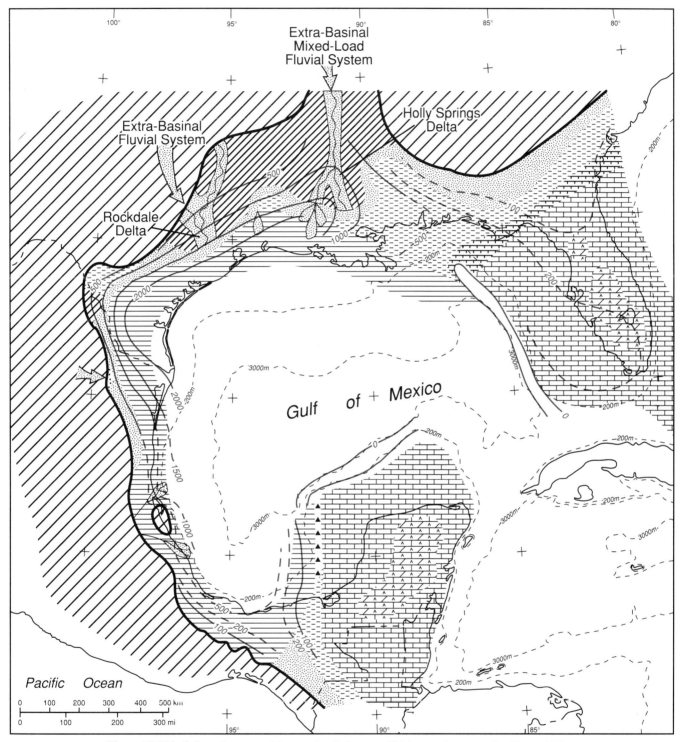

Figure 15. Lithofacies-paleogeographic map, late Paleocene. Thicknesses in meters. See Figure 14 for lithologic symbols.

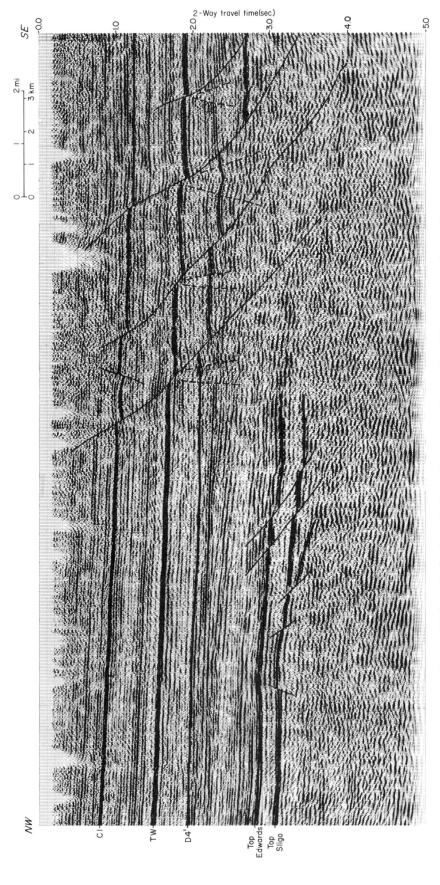

Figure 16. Migrated reflection seismic line (for interpretation see Fig. 10A) showing the buried Cretaceous shelf margin and development of coastward-dipping growth faults in the lower Wilcox, the first Tertiary genetic sequence to overstep the Mesozoic margin and to prograde into the deep Gulf of Mexico basin. Updip of the growth fault zone, shingled clinoform reflections at the base of the Wilcox record progradation of the delta system across the deep Paleocene shelf platform. Cl, Claiborn; Tw, Wilcox; D4, local lower Wilcox marker. For location see Figure 12.

ROCKDALE DELTA SYSTEM, LOWER WILCOX GROUP (EOCENE) TEXAS

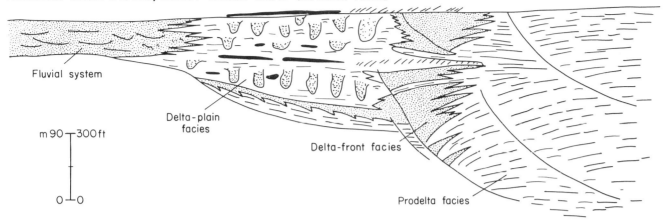

MISSISSIPPI DELTA SYSTEM (MODERN and PRE-MODERN) LOUISIANA

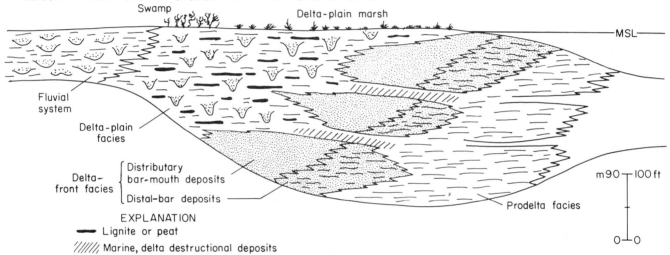

EXPLANATION
━━ Lignite or peat
///// Marine, delta destructional deposits

Figure 17. Comparison of constituent facies, Paleocene lower Wilcox of Texas (modified from Fisher and McGowen, 1967), and Holocene Mississippi delta system (modified from Coleman and Gagliano, 1964).

are the paleogeographic analogues of Holocene strike systems (south Louisiana chenier plain and Texas barrier-bar system) related to Holocene Mississippi River deltation.

Jackson Group: Eocene platform aggradation. The Jackson Group, which is of late Eocene age, is one of the volumetrically sandstone-poor depositional complexes of the northern Gulf of Mexico basin (Fig. 3). Terrigenous clastics were deposited on the submerged shelf produced primarily by the earlier Wilcox offlap and modified by deposition of several subsequent cycles. The Yegua Formation, which directly underlies the Jackson Group, locally extended the continental shelf by deltaic progradation in the Houston embayment before onset of the Jackson platform aggradational episode. Because Jackson Group sandstones were deposited upon the depositional foundation of older sequences and not on an actively prograding continental slope, syndepositional structural features are largely inherited from the

older salt and shale mobilizing episodes. Thus Jackson sediments are only mildly deformed or offset along the Wilcox growth-fault belt and are locally intruded by shallow piercement salt domes. Outer neritic and upper slope Jackson mudstones lie below most well penetration, but are locally uplifted by large-scale diapirism triggered by subsequent basin loading during deposition of the overlying Vicksburg and Frio sediments.

The Jackson depocenter is located in the Texas portion of the northwestern Gulf of Mexico basin (Fig. 19), where several formations, including, in ascending stratigraphic order, the Caddell, McElroy, and Whitsett Formations, are defined and regionally mapped. To the east, the Jackson Group is divided, also from bottom to top, into the Moodys Branch and Yazoo Formations. Terrigenous clastics grade across Alabama into the carbonates of the Ocala Limestone.

By late Eocene time active volcanism in areas to the west

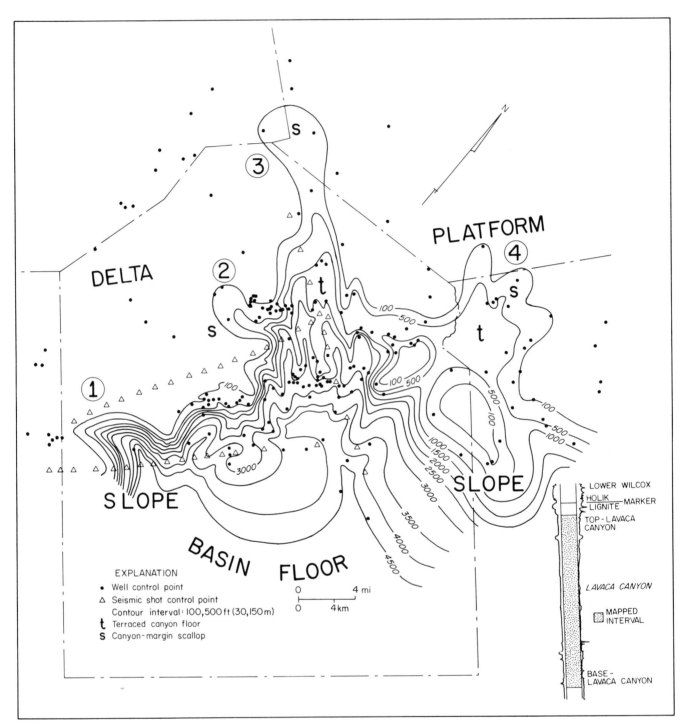

Figure 18. Decompacted isopach map of the fill of the Lavaca submarine canyon that was excavated into the prograding flank of the lower Wilcox Rockdale delta system.

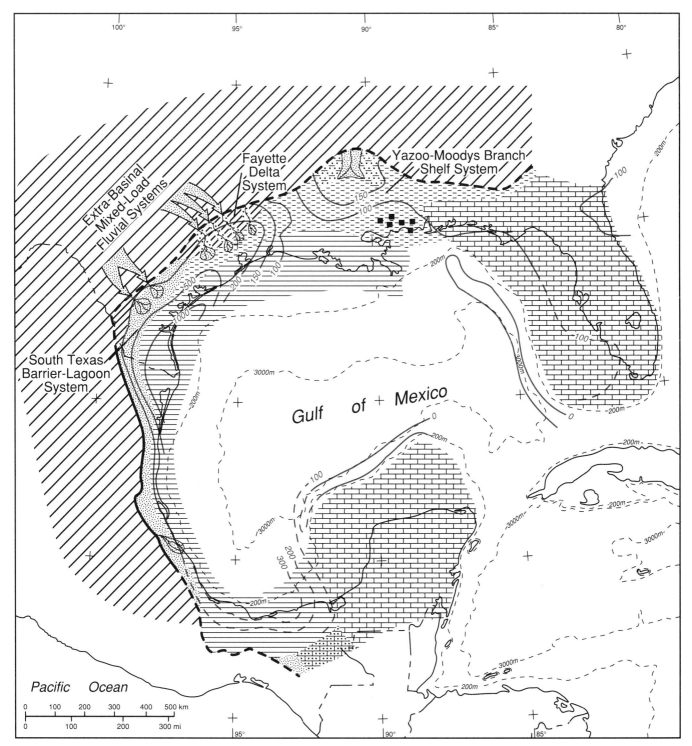

Figure 19. Lithofacies-paleogeographic map, late Eocene. Thicknesses in meters. See Figure 14 for lithologic symbols.

and southwest introduced abundant air-fall volcanic debris both directly into shallow-water and coastal-plain environments and into the catchment basins of streams draining into the Gulf of Mexico. Vitric ash and bentonite, bentonitic claystone, and mudstone beds are abundant in the Jackson Formation.

Principal depositional systems of the northwestern Gulf of Mexico basin were outlined by Fisher and others (1970). They include (Fig. 19) the Fayette delta system and associated distal fluvial facies, the South Texas barrier-lagoon system, and the Moodys Branch-Yazoo shelf system.

Fayette delta system. The Fayette delta system prograded across the older Eocene platform in a broad, ill-defined area of the Houston embayment. Original mapping of sandstone distribution in the total Jackson Group by Fisher and others (1970) and later mapping of sandstone distribution within the deltaic facies only by Kaiser and others (1978) reveal a fluvial-dominated delta system exhibiting several major lobes. The generalized progradational sequence consists of basal prodelta mudstones overlain by sandy delta-front and channel-mouth bar facies, in turn overlain by interbedded, lithologically diverse delta-plain deposits. The entire deltaic complex is generally less than 300 m thick. A relatively thin sequence of sandy fluvial deposits, the Whitsett Formation, caps the delta deposits at the outcrop and extends into the shallow subsurface (Fisher and others, 1970).

The Fayette delta system is mud-rich. Sandstones are fine to very fine; some medium sand occurs in the fluvial sandstones of the Whitsett Formation. Lignite beds, up to a few feet in thickness, occur in delta-plain facies, particularly of the Manning Formation (Kaiser and others, 1978).

South Texas Barrier-Lagoon system. The Jackson Group of the central and southern Texas Coastal Plain contains a framework of strongly strike-oriented sandstone bodies, which align in a narrow belt that extends from the southwestern margin of the Fayette delta system into Mexico (Fig. 16). The sandstones are flanked updip by lagoonal and coastal mudstones containing brackish or terrestrial faunas. Basinward, sand units grade into gray, fossiliferous marine shales (Fisher and others, 1970).

Thickness of the barrier-lagoon sequence, which is comparable to that of the delta system, and presence of local dip-oriented sandstone axes connecting across the lagoonal mudstone belt to the strike-parallel sandstone trend (Fig. 20), document the importance of local sediment input along the wave-dominated shore zone. Large suspended-load channel fills containing thick point-bar sequences have been exposed by uranium mining at the north end of the barrier-lagoon axis (Galloway and others, 1978). In all, the barrier-lagoon system extends more than 400 km from the flank of the delta system into northeastern Mexico.

Along its northern segment, the South Texas barrier-lagoon system has been well exposed by open-pit uranium and, more recently, lignite mining, revealing a complex mosaic of component facies (Galloway and others, 1979). Upward-coarsening shoreface deposits consist of fine to very fine sand, silt, and mud. Many barrier-bar sand bodies have relatively sharp bases, suggesting a transgressive origin, or progradation onto a very shallow shelf. Barrier-core sands are interrupted along strike by inlet-fill tuffaceous sands and associated spit-accretion deposits. Back-barrier facies include washover-fan deposits, small bay-head and cuspate deltas, lagoon and bay mudstones and ash beds, and thin, impure barrier-flat lignites. Coastal mudflat and marsh deposits form the landward component of the complete facies tract.

Commercial resources of the Jackson barrier-lagoon system include lignite (Kaiser and others, 1978), uranium (Galloway and others, 1978), petroleum (Fisher and others, 1970; Galloway and others, 1983), and bentonitic clay and zeolite. The South Texas barrier-lagoon system hosts one of the few major

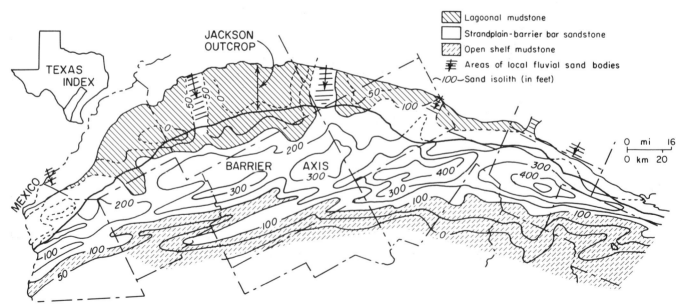

Figure 20. South Texas barrier/lagoon system of the Jackson Group (Eocene) sequence. Modified from Fisher and others (1970).

stratigraphic petroleum plays in the United States Gulf Coast Cenozoic basin fill.

Moodys Branch-Yazoo shelf system. Extending eastward from the eastern limit of the Fayette delta system is a broad expanse of glauconitic, fossiliferous mudstone and marl. In southern Alabama, terrigenous clastics grade into limestone and marl. The inferred depositional environments range from inner shelf and strandline to open shelf. The basal formation of the Jackson Group, the Moodys Branch, contains abundant glauconite and is interpreted to be in part a shelf unit capping the underlying Yegua-Cockfield depositional episode. The glauconitic zone extends from beneath the center of the Fayette delta system across Louisiana into Alabama; however, its thickness rarely exceeds 3 m (Fisher and others, 1970).

Frio Formation: Construction of a continental platform. A major Oligocene depositional episode began with early Oligocene Vicksburg deposition and culminated with deposition of the *Heterostegina-Marginulina* sands and associated transgressive deposits of the marine Anahuac shale tongue. Thus, the Vicksburg/Frio genetic stratigraphic sequence lies between the top of the *Marginulina cocoaensis* Zone and the top of the *Discorbis* Zone, and encompasses all Oligocene strata. Relationships between the Frio, recognized largely in the subsurface of the northwestern Gulf of Mexico basin, and its outcrop equivalents, have been obscured by a long history of definition and redefinition of its nomenclature (Murray, 1961). More recent detailed correlations of the outcrop section with the subsurface (Galloway, 1977; Baker, 1978; Coleman and Galloway, 1990) confirm the equivalence of the subsurface Vicksburg and Frio with the Catahoula Formation of the Texas Coastal Plain (but not apparently with the redefined Catahoula type section in Louisiana).

The Frio Formation together with the underlying Vicksburg Formation is one of the great progradational wedges of the northwestern Gulf of Mexico basin (Figs. 3 and 4). Frio deposition, which extended from northeastern Mexico to Mississippi, deposited as much as 4,500 m of sediment, and prograded the continental platform basinward as much as 80 km. The resultant clastic wedge extends from the Burgos basin, across the Rio Grande, to the Houston and Mississippi embayments and grades eastward into a thin (about 100 m or in maximum thickness), aggradational sequence consisting of shallow shelf shale, limestone, and glauconitic marl. In the Houston and Mississippi Embayments, the Frio was deposited on extensive salt-diapir fields initially mobilized by Wilcox slope offlap, inheriting the older dome distribution patterns, causing extensive renewed diapirism, and creating complex syndepositional structure within sand-rich deltaic sequences.

In addition to scale, the Vicksburg-Frio episode is distinguished by the dramatic shift of continental drainage into the Rio Grande embayment (Fig. 4), the climax of volcanic sediment influx into the northern Gulf of Mexico basin (Galloway, 1977), the onset of coastal-plain climatic deterioration, and an abrupt basinward shoreline shift due to an earliest Oligocene eustatic sea-level fall (Coleman and Galloway, 1990).

Frio depositional elements of the north-central Gulf of Mexico basin include four major clastic systems, which were delineated and described by Galloway and others (1982b, c), and Galloway (1986a). Two large delta systems, the Norias and Houston delta systems, occupied the Rio Grande and Houston Embayments (Fig. 21). Between them, extending across the San Marcos arch, was the extensive Frio barrier/strandplain system, the Greta/Carancahua barrier/strandplain system, described initially by Boyd and Dyer (1964). To the east, in Louisiana, the extensive, onlapping Hackberry slope canyon system lies within an erosionally based embayment within the Frio section (Paine, 1971). Subsidiary depositional elements include (Fig. 21) (1) delta systems in northeastern Mexico, fed by the Norma Conglomerate fluvial axis; and, in Louisiana, (2) the Gueydan bed-load fluvial system, Chita-Corrigan mixed-load fluvial system, and intervening coastal streamplain (Galloway, 1977) and (3) shallow clastic shelf, carbonate platform and local reef/bank complexes, and deep-water carbonate ramp in fore-barrier and marginal platform areas. Deeply buried, largely undrilled clastic slope and basinal systems complete the Frio facies tract.

Norias delta system. The Norias delta system constitutes the principal Frio depocenter (Fig. 21). It was fed by a single, large fluvial system of continental proportions. The drainage basin of this fluvial system extended across Trans-Pecos Texas, as evidenced by the abundance of volcanic rock fragments and plagioclase feldspar in Gueydan and Frio conglomerates and sandstones (McBride and others, 1968; Galloway, 1977; Loucks and others, 1986). Expansion and vertical accumulation of thick deltaic sequences was initiated where the fluvial system crossed the underlying Vicksburg wedge onto the continental slope. Here, along the so-called Frio flexure, extensive syndepositional growth faults were activated, producing the first in a series of fault-defined depositional subbasins.

Maximum width of the delta system exceeds 200 km, and sand-rich deltaic facies are inferred from seismic data to extend offshore from Kenedy County, the apex of the system. Downdip, and at depth, deltaic facies grade into deep-water continental slope deposits of the offlapping platform. Regional sand distribution patterns as well as local sand-body configurations indicate deposition of the Norias delta system as a mosaic of largely wave-dominated delta lobes. The expansive offlap of the Norias delta system was arrested by the regional Anahuac transgression, but the rate of sediment supply was sufficient to severely limit the extent of marine-shale onlap. The upper Frio *Heterostegina-Marginulina* deltaic complexes continued to prograde across the deltaic platform in the face of the regional transgression.

Component facies of the Norias delta system include delta-plain, delta-front, prodelta and upper slope, and delta-flank deposits. Delta-plain deposits include distributary channel fill, crevasse splay, and destructional bar and beach-ridge facies. Distributary channel fills form coastward-branching framework

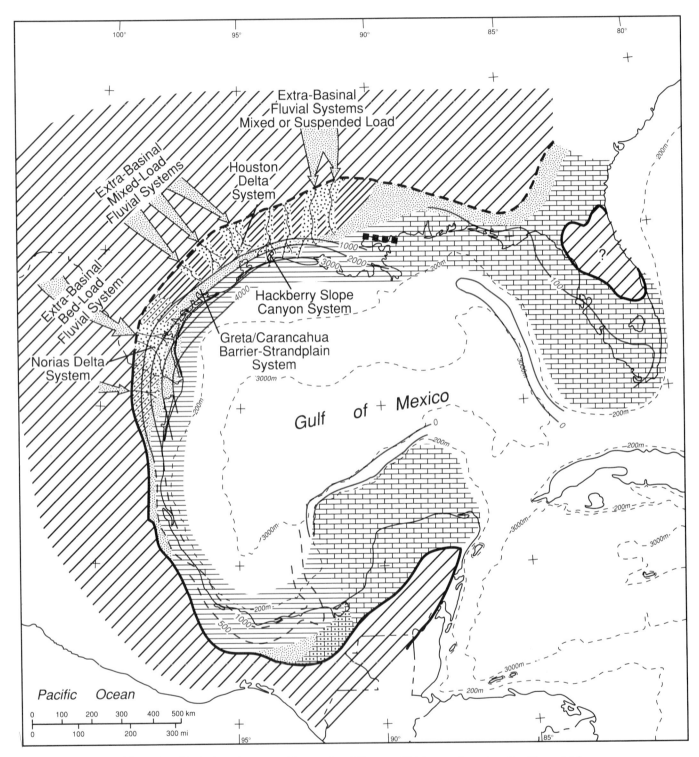

Figure 21. Lithofacies-paleogeographic map, late Oligocene. Thicknesses in meters. See Figure 14 for lithologic symbols.

sandstone bodies (Nanz, 1954). Bedded organic debris is notably absent from Norias delta-plain facies; this, along with the common red and brown colors of deltaic mudstones, is likely a result of the increasing aridity of the coastal-plain climate. Delta-front facies include the dominantly progradational, upward-coarsening sequences of channel mouth bar and laterally equivalent wave-reworked delta-margin sands. Sandstones cap prodelta mudstone and interbedded frontal splay deposits. In deeper water, more distal settings where the first delta lobes prograded directly onto continental slope and into rapidly subsiding growth-fault subbasins, the overall delta-front sequence increases to as much as a few hundreds of meters, and cyclic progradation of successive lobes accumulated composite sequences of delta-front deposits as much as a few thousands of meters thick.

Delta-flank facies include thick, massive sandstone bodies that trend parallel to depositional strike and grade laterally into the massive barrier sandstones of the interdeltaic coastline.

Throughout, the structural style and, in turn, the depositional fabric of the Norias delta system is dominated by progradation of a mud-rich continental platform and associated extensional shale tectonics.

In the Burgos basin, a second, though similar delta system prograded into the Gulf of Mexico. The major fluvial system feeding this delta complex is represented by the Norma Conglomerate. Quartzitic-clast composition indicates that this fluvial/delta complex was a separate depositional element that was probably sourced in northeastern Mexico. Contemporaneous growth-fault development shows that Oligocene continental margin offlap characterized northeastern Mexico (Busch, 1973).

Houston delta system. The Houston delta system is typified by fluvial-dominated digitate to lobate delta lobes (Gernant and Kesling, 1966; Bebout and others, 1978; Tyler and Han, 1982). Although its lateral extent is comparable to that of the Norias delta system, both the thickness of the Frio section and the extent of offlap are less (Fig. 21). The Houston delta system is interpreted to be the product of two or possibly three moderate-sized streams, comparable to the modern Colorado or Brazos Rivers in the same geographic area. The overall facies tract of the Houston delta system is similar to that of the Norias. However, delta-destructional and flank deposits appear to be more extensively preserved.

The Houston delta system prograded onto the salt-intruded Wilcox delta and its associated continental slope. Consequently, a very complex structural framework, including deeply buried salt anticlines, shallow salt diapirs and associated faults, local salt-withdrawal basins, and growth faults, segments the deltaic deposits into numerous subbasins. Further, because of inherent density contrasts, salt mobilization continued to affect the sand-rich delta-plain facies, as well as the delta-front and prodelta facies.

Detailed study of lower Frio delta lobes of the Houston delta system by Gernant and Kesling (1966) provides an excellent example of the integration of physical stratigraphy and paleoecology in environmental description. The distribution of paleoenvironments as defined by foraminiferal assemblages con-

forms closely with the digitate, fluvial-dominated delta lobe geometry outlined by an isolith map of individual sand units (Fig. 22A). A dip cross section through the map area (Fig. 22B) illustrates the basinward increasing paleo-water-depth in the lower to middle Frio interval. A major growth fault expands the sand-poor middle to inner shelf (prodelta) muds and thin sands, and outer-shelf fauna appears in the lower part of the section only on the downthrown side.

Greta/Carancahua barrier-strandplain system. The two major Frio deltaic headlands were separated by a wave-dominated shore zone along which was deposited a linear sandstone belt that separates fossiliferous, marine mudstones downdip from restricted lagoonal or coastal mudstones updip (Fig. 21). The locus of sand accumulation shifted 15 to 30 km basinward during Frio offlap. A well-developed barrier-bar and lagoon system that consists largely of sediment reworked laterally from the Norias delta system lies to the south. Northward, the sandstone axis becomes less well-defined, and the presence inland of deposits of numerous small coastal-plain streams suggests a sandy strandplain, which grades into delta-flank deposits of the Houston delta system.

Barrier and strandplain framework sandstone bodies consist of massive, thick, stacked shoreface and aggradational barrier-core sands that may persist through intervals as much as 2,500 m thick (Fig. 23). The sandstone units pinch out abruptly both landward and gulfward into lagoonal, bay, coastal mudflat, or shelf mudstones. Aprons of sand and silt deposited in lower shoreface and shelf settings by storm-generated waves and currents are incorporated in barrier-front sequences that may exceed 300 m in thickness in rapidly subsiding growth-fault-bounded subbasins.

Structural grain is dominated by growth-fault zones, which primarily affect shoreface, shelf, and upper slope facies (Fig. 23). The greatest vertical stacking and amalgamation of sandstone occurs between the Vicksburg and Frio flexures; however, the axis of the system crosses to the down-dropped side of the Frio flexure as it approaches the Norias deltaic headland.

Hackberry slope canyon system. The Hackberry embayment is an erosionally floored embayment carved by submarine processes into the lower and middle Frio continental platform margin and filled with onlapping deep-water slope sands. Subsequent progradational mudstones and sandstones of later Frio and Miocene offlap cover the upper Hackberry mudstones. Deep-water sediments of the wedge or "embayment" are as much as 1,000 m thick and pinch out updip along a highly serrate erosional edge, suggesting submarine canyon and gorge cutting (Paine, 1971). The sandy lower Hackberry fill consists of onlapping gorge-confined submarine channel, levee, and turbidite lobe sandstones and interbedded mudstones (Fig. 24) (Paine, 1968; Benson, 1971; Berg and Powers, 1980; Ewing and Reed, 1984) overlain by upper Hackberry massive mudstones bearing the characteristic deep-water microfauna.

Topography of the Hackberry slope was likely highly complicated by ongoing salt diapirism. Maximum thicknesses of fan-

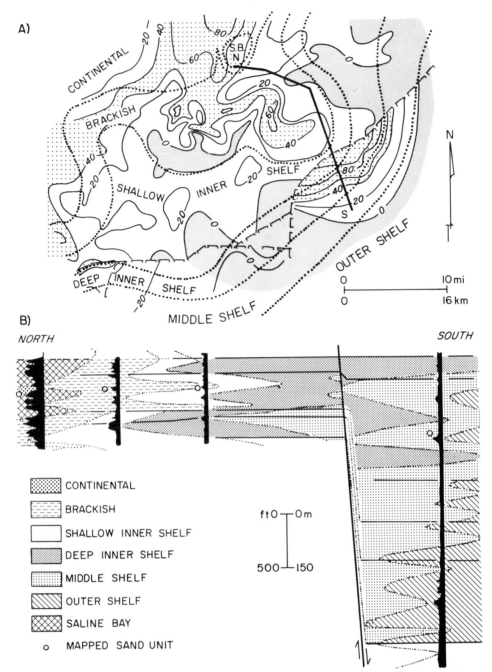

Figure 22. Net sand isopach map (A) and SP-log cross section (B) showing the paleoecologic setting of an individual lobe complex of the Frio Houston delta system. Paleoecologic zones are based on benthonic foraminifera. A major growth fault expands the distal delta-fringe and prodelta facies. Modified from Gernant and Kesling (1966).

channel deposits lie in interdomal (and presumably bathymetric) lows (Paine, 1971; Curtis and Echols, 1985), as is typical of submarine-slope deposits.

The Hackberry slope system lies in the interdeltaic area between the Houston and Louisiana deltaic axes, and postdates the postulated middle Oligocene eustatic sea-level drop of Haq and others (1987). Incision extends far downdip, cutting paleo-

slope deposits well below reasonable limits for eustatic sea-level fall (Pitmann and Golovchenko, 1983). Along the landward fringe of the Hackberry embayment, older paralic Frio sediments are overlain by deep-water gorge fill. Excavation appears to have been dominantly or exclusively a submarine event readily explained by sediment starvation and rapid local subsidence foundering of the lower Frio contienntal margin (Ewing and Reed,

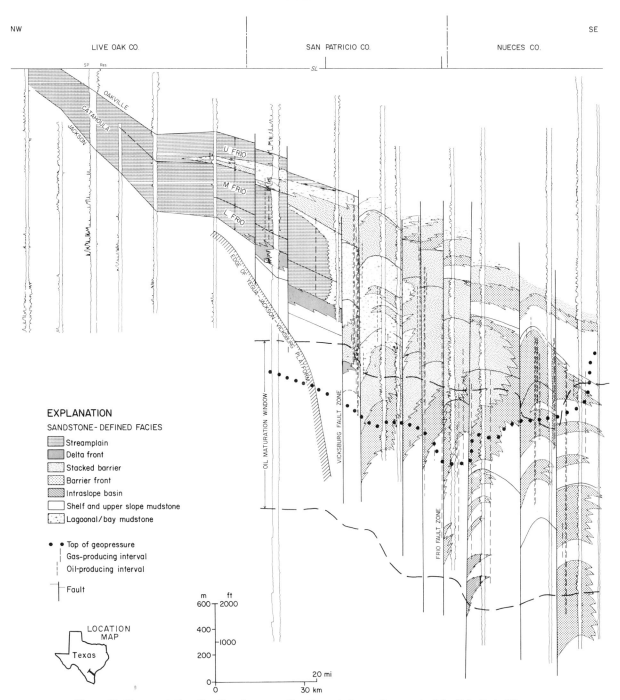

Figure 23. Representative dip-aligned cross section through the southern part of the Frio Greta/Caran-cahua barrier-strandplain system. Features shown include the position of the underlying paleocontinental margin, major growth faults, generalized facies assemblages, depths of the oil-generation window, top of the geopressured zone, and stratigraphic distribution of oil and gas production in major fields lying along the line of section. From Galloway and others (1982b).

1984). Headward erosion of Hackberry submarine canyons tapped inner shelf and shoreface long-shore drift (Fig. 21), providing a ready supply of sand that backfilled the canyons and created uniquely prolific deep-water hydrocarbon reservoirs.

Synthesis: The Frio offlap episode. With the onset of Frio

deposition, the northwestern shelf of the Gulf of Mexico assumed an amazingly modern physical geography that duplicated many attributes of Quaternary drainage patterns, coastal land forms, and climatic gradient. Further, as shown on the dip cross sections such as Figure 23, the Frio illustrates the close interrelationships

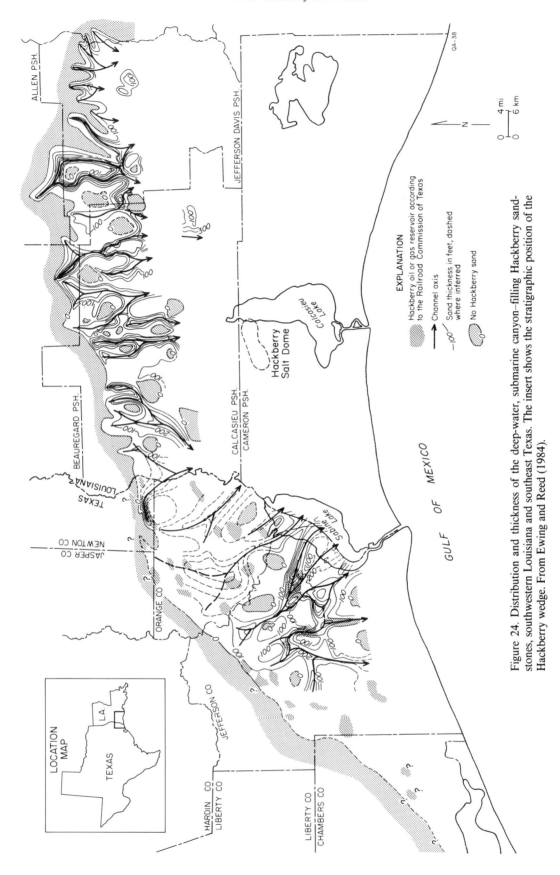

Figure 24. Distribution and thickness of the deep-water, submarine canyon–filling Hackberry sandstones, southwestern Louisiana and southeast Texas. The insert shows the stratigraphic position of the Hackberry wedge. From Ewing and Reed (1984).

among depositional facies, structural style, hydrogeology, thermal maturity, and distribution of "world-class" hydrocarbon resources. Facies affected by syndepositional structural growth are primarily upper-slope, shelf, and distal-delta or barrier sequences. The top of the geopressured zone, marking the depth of nearly complete hydrologic isolation of the deep-basin fill, typically lies within the transition zone between the main body of the deltaic and shore zone systems and these muddier distal facies suites. Further, isotherms tend to parallel the top of the geopressure zone, delineating zones of liquid and gaseous hydrocarbon generation that lie significantly below the actual depth of occurrence of most oil and gas pools. Inescapably, one must conclude that considerable vertical migration of fluids, including petroleum, from sources in shelf, prodelta, and slope facies into comparatively shallow reservoirs has occurred despite the evidence of partial hydrologic isolation of the source section.

Natural resources. The Frio Formation and its shallow stratigraphic equivalent, the Catahoula Formation, are a major repository of natural resources. For Texas alone, the Frio is calculated to have contained known recoverable reserves totaling approximately 8 billion barrels of oil and 70 trillion cubic feet (Tcf) of natural gas (Galloway and others, 1982a, b). Uranium reserves lie primarily in the Gueydan fluvial system and are conservatively estimated to exceed 45,000 metric tons of uranium oxide. Substantial uranium and petroleum reserves are also found in the Norma and Frio deposits of the Burgos basin. In addition, fluvial facies are major sources of fresh ground water in the northern Gulf Coastal Plain. Deep, geopressured waters in the Frio have been a target for investigation of potential geothermal and dissolved-gas resources (see Chapter 16 of this volume).

Miocene offlap: Emergence of the Paleo-Mississippi drainage basin. Along the northwestern Gulf of Mexico basin, from northeastern Mexico to the Florida panhandle, thick offlapping lower- and middle-upper Miocene sequences of terrigenous clastics were deposited gulfward from the older Frio shelf margin (Figs. 4, 11, and 25). Deposition of each sequence was punctuated by subsidiary coastal transgressive pulses (*Marginula* A through *Textularia* W shale tongues). Each genetic sequence grades from fluvial to deltaic to marine from updip to downdip. The maximum thickness of the Miocene is approximately 1,200 m in Texas and 2,400 m in Louisiana updip and onshore; the Miocene is reported to exceed 6,000 m (Rainwater, 1964a) to 7,600 m (Meyerhoff, 1968) in thickness in offshore Louisiana.

The more than five-fold thickening of the Miocene from updip to downdip is attributed mainly to deposition on and basinward of the unstable substrate of the underlying Frio Formation and thick salt; huge growth faults developed during deposition and provided space for the accumulation of the thick, structurally segmented sections of sediment (Fig. 26; Leutze, 1972). The growth faults were initiated and remained active mainly during deltaic deposition and seldom can be traced into the overlying fluvial section (Curtis, 1970; Curtis and Picou, 1980). Miocene sandstone and shale units also thickened into salt-withdrawal basins, which formed contemporaneously with salt movement into nearby salt domes. The salt migration into the domes is believed to be the result of loading by sediment during the gulfward progradation. Deposition coincident with salt withdrawal and contemporaneous with growth-fault movement resulted in the accumulation of thick, highly expanded sections of sandstone and shale.

The considerably greater thickness of the Miocene section in Louisiana is the result of a major shift of feeder systems from Texas in the Oligocene, to southeast Texas (Kiatta, 1971) and western Louisiana in the early Miocene (Sloane, 1971), and to eastern Louisiana in the middle and late Miocene (Fig. 4). The west-to-east shift of depocenters was accompanied by more than 80 km of seaward migration of major deltaic complexes between the early and late Miocene; the late Miocene deltas occupied the position of the present-day Mississippi Delta. The shift of terrigenous-clastic depocenters in Louisiana is also reflected in the character of the Miocene sediments of Florida; in the lower Miocene episode, when the major terrigenous depocenters were located in western Louisiana, carbonates were deposited throughout the Florida peninsula and panhandle. However, during the middle-upper Miocene episode, after the depocenters shifted to eastern Louisiana, sandstone, shale, and marl were deposited throughout the peninsula, with the exception of south Florida where limestone was continuously deposited.

The updip portion of the Miocene depositional wedges represents fluvial and delta-plain environments. The fluvial systems deposited sand and mud on the stable shelf constructed by the underlying Tertiary units. Up to 80 percent sand accumulated in areas of major feeder systems, and 20 to 40 percent in other areas. The dominant dip orientation of the feeder systems is well displayed in the middle Texas Gulf Coast where they have been mapped in detail in the Lower Miocene (Solis, 1981; Galloway and others, 1986). Sediments transported along numerous dip-oriented feeders were deposited in strike-oriented wave-dominated deltas, barrier bars, and strandplains in the area of the present-day shoreline (Doyle, 1979; Galloway and others, 1986; Morton and others, 1988). Detailed study of the updip Miocene (Catahoula and Fleming Formations) in Rapides Parish, Louisiana, indicated that the sands there were deposited mainly as channels and point bars, and that the fine-grained terrigenous clastics accumulated on levees and in back swamps and lakes on a deltaic plain (Glowacz and Horne, 1971). A major Late Miocene depocenter offshore of southeast Louisiana indicates a shift of drainage to the Mississippi River system.

In Louisiana, a few major fluvial feeder systems supplied large quantities of sand and mud to huge delta systems at the shelf hingeline. On sandstone-percent maps (Fig. 27), the demarcation between the fluvial/delta-plain systems (20 to 40 percent sandstone) is clearly evident. Individual deltas range up to 65 by 120 km in dimensions (Curtis, 1970). Major exceptions to the low sandstone percentage in the delta proper do occur but are not common; in these areas, continuous sandstone units 150 to 335 m thick occur on the downdip side of huge down-to-the-coast growth faults.

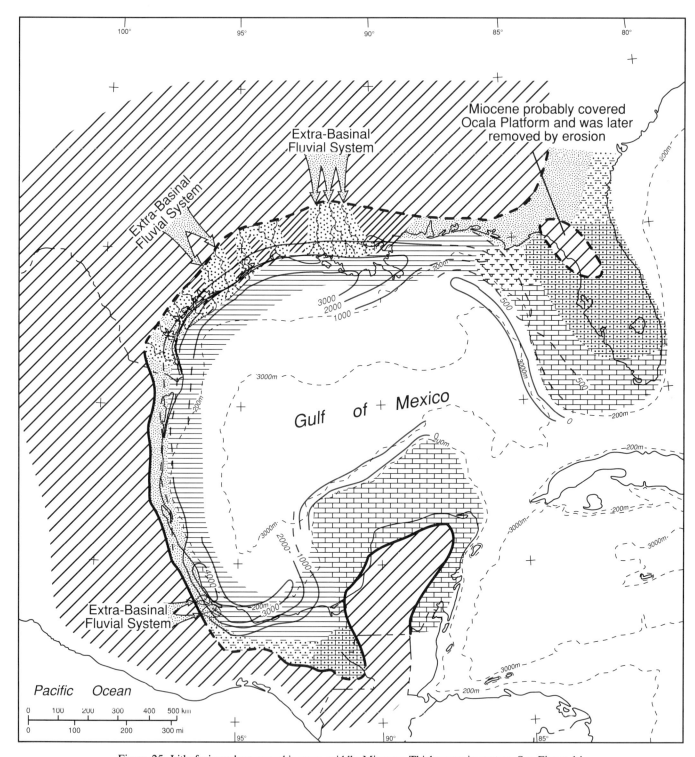

Figure 25. Lithofacies-paleogeographic map, middle Miocene. Thicknesses in meters. See Figure 14 for lithologic symbols.

NW   SE

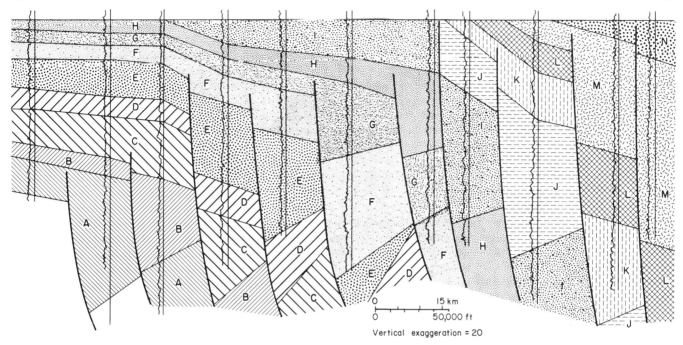

Figure 26. Miocene growth faulting in southeast Louisiana. The cross section illustrates the dramatic progressive expansion of deltaic depositional increments (A to N) across a succession of contemporaneous faults. Note the nearly complete structural segmentation of increments such as I. From Curtis and Picou (1978).

The middle and upper Miocene of offshore Texas is greater than 3,000 m thick. The offshore Texas Miocene is shale-dominated, but sandstone-prone areas are coincident with the present-day courses of the Sabine, Brazos, and Rio Grande Rivers. Of particular note was the progradation of a subsidiary middle to upper Miocene delta system across the central Texas shelf and local deposition directly onto the older Miocene slope. The resultant sedimentary loading of the unstable margin produced a prominent, highly listric normal fault that trapped and preserved a depositional complex of distal delta and slope sediments in the down-dropped block (Rainwater, 1964a; Meyerhoff, 1968; Morton and others, 1985, 1988). This combined structural and depositional feature has evolved into the well-known Corsair gas-producing trend of offshore Texas (Vogler and Robison, 1987).

Economically significant hydrocarbon accumulations occur mainly in the deltaic and inner and middle neritic marine Miocene sandstones associated with delta systems (Meyerhoff, 1968); oil is found mainly in areas of salt-dome penetration of the Miocene deltaic systems. This relationship is best illustrated by comparing the maps showing the belts of production for the Lower, Middle, and Upper Miocene across south Louisiana with the gulfward edge of the sandstone-rich section of fluvial deposits, as indicated on the sandstone-percentage maps (Fig. 27). The areas

of production are also coincident with the various positions of the major delta lobes. Major uranium deposits are hosted by Lower Miocene Oakville bed-load fluvial channel facies in south Texas (Galloway and others, 1982a).

Pliocene-Pleistocene offlap: Building the modern shelf and slope. The major Pliocene-Pleistocene depocenters of the northern Gulf of Mexico basin lie beneath the outer continental shelf and upper slope of Texas and Louisiana (Figs. 4 and 28). Covering a time span of only 5.3 m.y. the Pliocene-Pleistocene is the shortest and last major depositional episode in the Gulf of Mexico basin. During this time the locus of maximum deposition shifted over 300 km southwestward from just west of the present mouth of the Mississippi River to the edge of the continental shelf south of the present shoreline at the Texas-Louisiana border (Fig. 4). This shift was accompanied by 80 km of progradation of the continental shelf edge to its present position near the 200-m isobath (Woodbury and others, 1973).

The sedimentary processes that filled the Pliocene-Pleistocene depocenters were a continuation of the processes of deltaic sedimentation and shelf-edge progradation which had been continuously active in the Gulf of Mexico basin since the end of the Cretaceous. The distinguishing characteristics of this period of deposition are (1) a change in tempo and depositional responses of the sedimentary regimes to numerous large, eustatic changes in

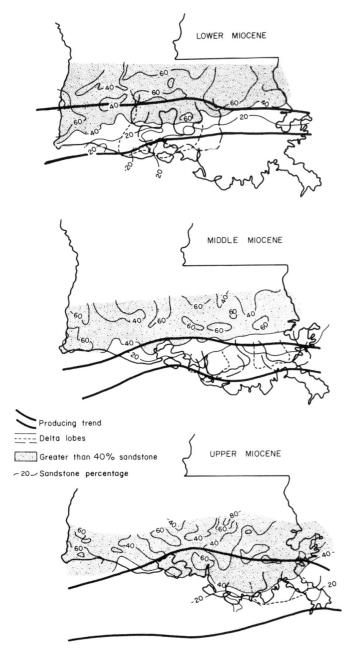

Figure 27. Miocene interval sandstone percentages (Louisiana Geological Survey), hydrocarbon producing trends (Limes and Stipe, 1959), and delta lobes (Curtis, 1970).

sea level that began about 2.5 to 3 Ma and continued through the late Pliocene and entire Pleistocene (McFarlan and LeRoy, 1988; Armentrout and Clement, 1990); and (2) a resurgence of sediment supply in response to epeirogenic uplift and glaciation of central North America. It appears likely that the southwestward shift in the major locus of sedimentation was in response to a westwardly shift in the major sources of sediment at the time of the initial drop in sea level in the late Pliocene.

The onset of North American continental glaciation at ap-

proximately 3 Ma and the eustatic drop in sea level in the Gulf of Mexico of an estimated 150 m was the first of eight major cycles of sea level changes in the Gulf of Mexico basin (Beard and others, 1982; Fig. 29). During periods of falling and low sea level, a series of delta lobes and interdeltaic depositional facies expanded rapidly across the exposed continental shelf until the sedimentary load was being deposited on the outermost shelf and finally on the upper continental slope (Figs. 28, 30, and 31). Large valleys were incised into the exposed continental shelf (Suter and Berryhill, 1985). The development of shelf-margin deltas and deposition of sediment within the shelf-slope transition zone were accompanied by active growth faulting. Sediment overload and consequent subsidence of the shelf edge triggered uplifts of diapiric structures on the upper continental slope. The active diapirs in turn strongly influenced sediment dispersal patterns on the slope (Lehner, 1969; Woodbury and others, 1976; Bouma, 1982).

The gradual rise of sea level during interglacial periods shifted the zone of active deposition landward. Shelf-edge delta systems were partially reworked and the marine transgression caused a regional hiatus over the drowned fluvial-deltaic systems, which had now become the continental shelf (Suter and Berryhill, 1985). Repeated exposure of the continental shelf during glacial stages and gradual drowning during interglacial stages created hiatuses at both the top and bottom of each glacial-interglacial cycle in that portion of the cycle deposited in a shelf environment.

The present-day continental slope of the northern Gulf of Mexico is essentially the relict Pleistocene depositional surface of the latest Pleistocene glacial stage (18,000 to 15,000 yr B.P.; McIntyre and Kipp, 1976) modified somewhat by the Holocene transgression. The complex topography of the present slope (Fig. 32) is largely the result of syndepositional diapirism, active growth faulting, and the gravity transport of sediments produced by the loading of sediments at the shelf/slope break during the periods of lowered sea level (Lehner, 1969; Martin, 1976). Detailed multidisciplinary studies of the present-day slope have provided the background necessary to understand the depositional processes that have filled the Pleistocene depocenter (see Bouma and others, 1976; Trippet, 1981; Chapters 2 and 12 of this volume).

The process of shelf-slope progradation by deltaic overload includes extensive "mass-wasting" on the upper continental slope (Woodbury and others, 1976; Booth, 1979; Coleman and others, 1983). Such gravity-driven resedimentation involves sediment masses as much as several tens to hundreds of meters thick (Fig. 32). Booth (1979) has shown that the slumps and slides involved in mass-wasting on the upper continental slope are the result of high rates of deposition. These high rates of deposition (3 mm/yr), because of underconsolidation and excess pore pressure, can trigger mass-wasting on slopes of 1 to 2° without depending on such external causes as storms or earthquakes. Rapid mass-wasting has created the most recent family of submarine canyons that head the Mississippi Fan (Coleman and others, 1983; Chapters 2 and 12 of this volume).

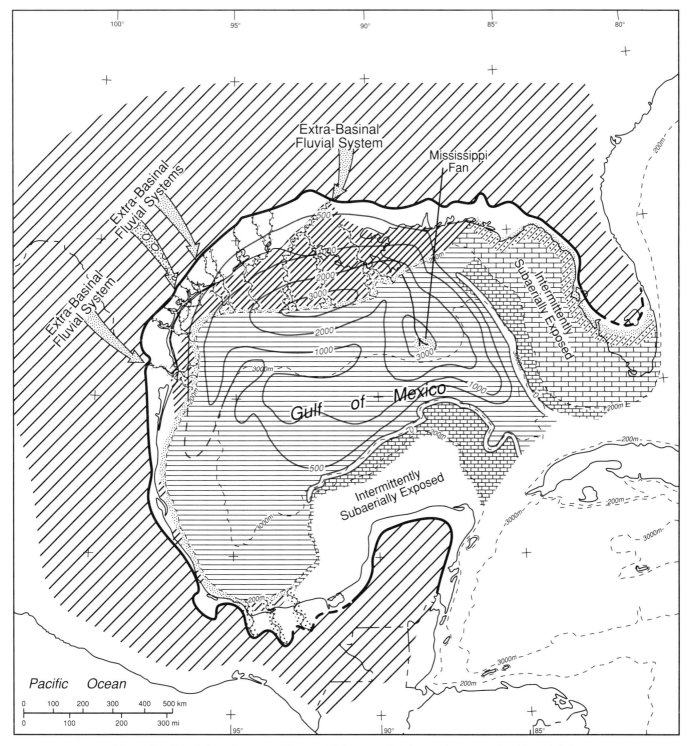

Figure 28. Lithofacies-paleogeographic map, Pleistocene. Thicknesses in meters. See Figure 14 for lithologic symbols.

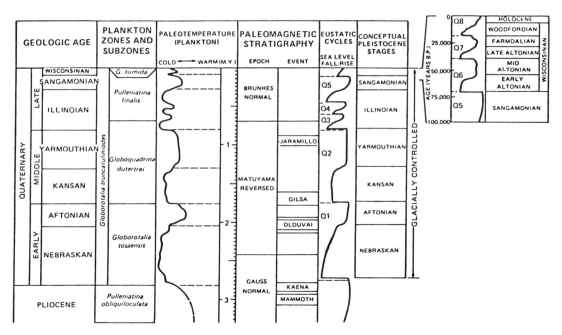

Figure 29. Pleistocene cycles and chronostratigraphy. From Beard and others (1982).

Economic significance. Due to unconformities and structural deformation, it is difficult to estimate the total thickness of Pleistocene sediments under the northern Gulf of Mexico. It is probably in excess of 5,000 m. Before the initiation of active drilling in the youngest portion of the Pleistocene depocenter underlying the outer continental shelf of Texas and Louisiana, there was considerable speculation that hydrocarbons would not be present in commercial amounts in sediments so young. Soon after the initial discoveries, primarily gas, improved seismic data were able to lead the way to what has become a major gas, and, more recently, a deep-water oil trend.

Although arguments still exist over the source of hydrocarbons present in the Pleistocene (see Chapter 15 of this volume), there remains no doubt that rapid deposition of suitable reservoir facies in zones of structural activity capable of producing traps, and interfingered with deltaic and continental slope shales, has provided a Pleistocene hydrocarbon province. The relative lack of time for the process of indigenous maturation would explain the presence of mostly biogenic gas rather than oil in the young sediments (Rice, 1980). The source of the deep-water (flexure trend) oils is speculative.

Hydrocarbon production in the Pleistocene depocenter is primarily from sediments deposited in outer-neritic and slope environments—the zone of maximum deposition and structural activity. Principal traps are associated with diapiric structures, domal structures (probably deep diapiric structures), roll-over features associated with growth faults, and faulted anticlines. The proprietary nature of most industry data concerning the Pleistocene depocenter has permitted publication of very few field studies. Nevertheless, the structural traps are familiar since they are

similar to the better-documented structures in the Pliocene and Miocene depocenters to the north and east (Roberts, 1982).

An excellent example of a Pleistocene gas accumulation on the upper continental slope is in High Island South Addition, Blocks A560 and A561 (Lund and others, 1978). Utilization of a multidisciplinary approach involving seismolog, subsurface geology, paleontology, petrology, and electric-log and dipmeter interpretation allowed delineation of sand-body geometry, transport directions, and depositional environments. Blocks A560 and A561 are located 130 km south-southeast of Galveston, Texas, in 80 m of water. The 200-m bathymetric contour is 8 km south of the lease. The depositional model of Lund and others (1978) is based on data from the initial eight exploratory wells drilled by 1975. Most of the wells penetrated the entire Pleistocene section and bottomed in the upper Pliocene (*Globoquadrina altispira* Biozone).

The controlling structure is a large growth fault with 460 m of throw, which was active during much of the Pleistocene. Hydrocarbon presence was anticipated by interpretation of seismic "bright spots." Paleontological information indicates that all of the productive interval was deposited on the upper continental slope. Petrological analyses of sidewall cores, including grain size, sorting, and grain and matrix composition, resulted in the determination that the sand bodies were composed of detrital material transported by turbidity currents and deposited as part of one or more intraslope submarine fan systems.

Characteristics that Lund and others (1978) used to distinguish submarine-fan deposits included: (1) presence of glauconite; (2) poor to good sorting depending on location in the fan; (3) common presence of rock fragments, feldspar, and mica grains; and (4) high content of original depositional matrix. However, these criteria are not uniquely diagnostic of fan deposition. Changes in the vertical distribution of grain sizes were used to distinguish channel deposits feeding the submarine fans from the fans themselves. A downward increase in grain size was taken to indicate deposition in a submarine channel. An upward-coarsening sequence was taken to be typical of sand deposited in a submarine fan lobe. Submarine channels supplied sediments from a generally northeast direction, and the submarine-fan system created a complex system of overlapping sand bodies. Fan geometry is exhibited by both individual fan members and by the aggregate of all sands in the fan system (Lund and others, 1978).

The present-day continental slope, essentially an exposed Pleistocene depositional surface, provides a model for the processes that filled not only the Pleistocene but also the Pliocene and Miocene depocenters of the northern Gulf of Mexico basin. By use of this model, combined with our increasing knowledge of shelf-slope structural and depositional patterns, many previously poorly understood aspects of the Cenozoic depositional history of the Gulf Coast can be interpreted.

Gulf Coast depositional episodes: An evolving interplay among tectonism, sediment supply, and eustasy

The Cenozoic history of the northern and northwestern Gulf of Mexico basin was characterized by rapid, episodic sediment input and resultant deposition of thick, prograding depositional sequences exhibiting considerable gravity deformation. Each depositional episode was defined by progradation and subsequent retrogradation of the shoreline by several tens of kilometers, although successive transgressions usually did not ex-

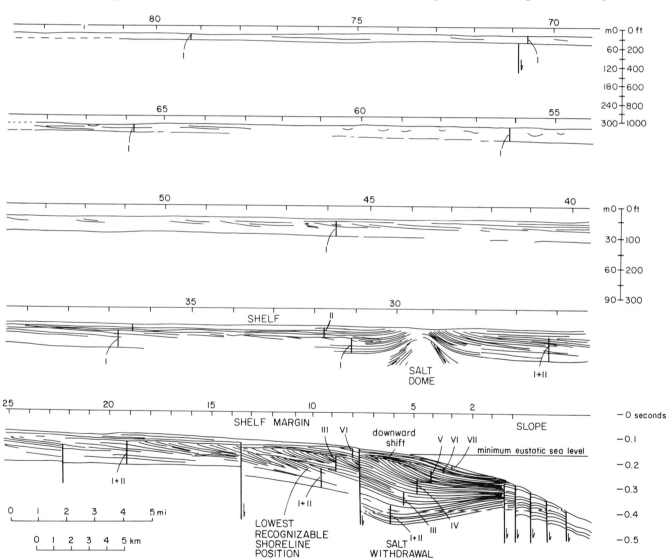

Figure 30. Line drawing based on a dip-oriented sparker cross section of the Texas shelf and upper slope in the vicinity of Galveston. The depositional and structural regimes of late Pleistocene deposition, as well as interpreted depositional cycles (Roman numerals) and seismic sequence boundaries at the shelf break are evidenced by bedding architecture. Modified from Winker (1979).

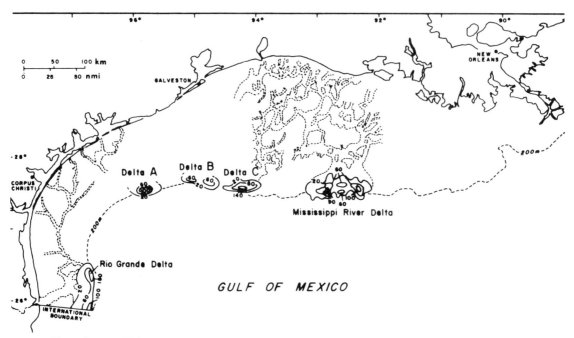

Figure 31. Late Pleistocene valley axes and shelf-margin deltas deposited during the last lowstand of sea level. Thickness of deltaic sediments in meters. From Suter and Berryhill (1985).

tend as far landward as their precursors. In contrast, the shelf edge was built basinward during offlap, but commonly remained as a permanent record of successive steps in basin filling. It was merely submerged more deeply and prone to modest regrading during transgression and subsequent periods of maximum flooding. Thus, the sedimentary record in the basin documents two aspects of episodicity. The overall facies tract (most readily defined by the position of the shore zone) has oscillated widely, reflecting progradation followed by retrogradation. The continental margin (defined by the bathymetric shelf edge) has alternated between periods of active outbuilding and periods of relative stability or minor retrogradation.

The principal Cenozoic depositional episodes that resulted in continental margin outbuilding of the northwest Gulf of Mexico basin are summarized in Figure 33. Although subsidence of this divergent oceanic basin margin was primarily induced by sedimentary loading, large-scale variation of sediment supply is required to explain the distinct pulses of shelf-edge progradation. By the late Neogene, increasing ice volumes also significantly lowered eustatic sea level, and changes in ice volume provided a mechanism for rapid eustatic fluctuations.

Simple comparison of the episode history with the proposed eustatic curve of Haq and others (1987) shows that correlation is indeed good during the Pliocene and Pleistocene. For Miocene time, the cycles still show correlation, but the relative magnitudes and exact timing of progradational and bounding transgressive events as well as the multitude of proposed sea-level fluctuation are increasingly disparate. Correlation of the Oligocene Vicksburg-Frio episode with the Oligocene sea-level curve is compli-

cated by uncertainties in the paleontologic dating of these two units, but poses problems regardless of the specific chronology accepted. The early Vicksburg-Frio depositional episode does, however, coincide closely with a well-documented reorientation in the crustal stress regime (Price and Henry, 1984). Style of basin-margin subsidence changed markedly at this time as well. As shown in Figure 3, Oligocene and younger sequences are characterized by prominent basin-margin thinning. Within the Paleogene, correspondence between the eustatic and episode curves is poor to fair at best. Again, age determinations using standard planktonic zonation pose some uncertainties; discrepancies in published dates may be interpreted to mean that offlap events are not perfectly synchronous around the basin margin or simply reflect remaining difficulties in tying the conventional benthonic zones to a standard planktonic framework (see Table 1). Analysis of depositional rates (Galloway, 1990) shows good correlation of major episodes with times of high sediment supply. In contrast, periods of regional Paleogene flooding are times of minimal sediment input.

Comparison of episode history with the onset or duration of major tectonic events of the central and western North American Plate reveals some compelling associations. The terminal Laramide deformation phase extended from the southwestern United States (late Paleocene–early Eocene) into northern Mexico (middle Eocene; Chapin and Cather, 1981; Dickinson, 1981; Eaton, 1986). Pulses of deformation and uplift centered in the middle and then the southern Rocky Mountains correspond directly to outbuilding of the lower and upper Wilcox continental margins (Fig. 33). As the locus of uplift moved south into Mexico, supply

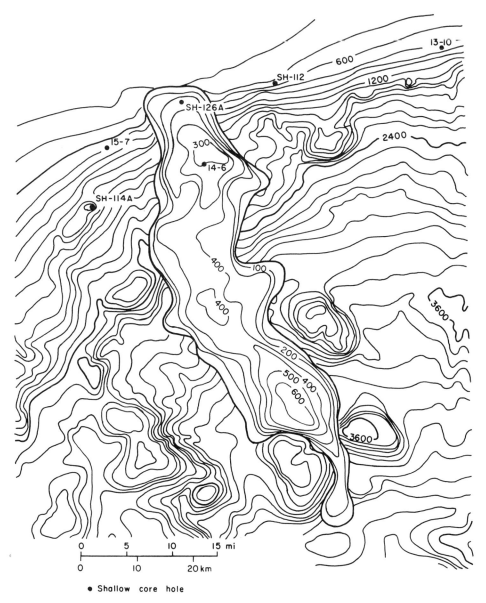

Figure 32. Isopach map of slump-displaced late Pleistocene sediments, Texas continental margin. From Woodbury and others (1976).

of sediment to the northwest Gulf of Mexico basin decreased, and the continental margin was generally flooded. By late Eocene time, tectonic quiescence dominated and the southern Rocky Mountains were beveled, forming a regional erosional surface (Epis and Chapin, 1975). Meanwhile, volcanism began in the southwestern United States and northern Mexico. Concomitantly, the modest Yegua and Jackson sequences were deposited. An important episode of explosive rhyolitic volcanism and crustal heating spread from west Texas into northern Mexico during late Eocene and Oligocene time (McDowell and Clabaugh, 1979). A contemporaneous surge of sediment, notable for its content of volcanic rock fragments, entered the Gulf of Mexico basin. Be-

ginning in late Oligocene, the initial subsidence along the Rio Grande rift occurred (Chapin, 1979), culminating in large-scale graben formation 27 to 20 m.y. ago. Rapid subsidence of this north-trending feature (Fig. 5) created an immense sediment trap, beheading the southwestern United States drainage system that had played a prominent role in Oligocene continental margin outbuilding. By the end of early Miocene time, a major reorganization of the intraplate stress regime initiated regional extension across western North America. This Basin and Range episode of normal faulting extended as far east as the inner margin of the northwest United States Gulf of Mexico Coastal Plain, where the Balcones fault was reactivated. Regional epeirogenic uplift of the

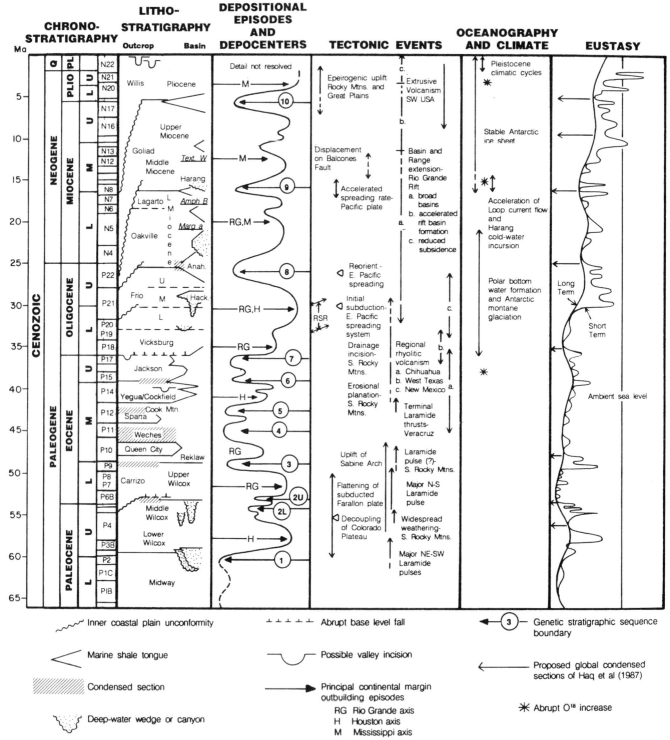

Figure 33. Comparative temporal history of United States Gulf Coast Cenozoic depositional episodes, proposed eustatic sea-level changes, oceanographic evolution in response to Cenozoic climatic cooling, and tectonic events of western North America. Major continental-margin outbuilding episodes and their depocenters are shown by excursions to the right on the episode curve. The principal lithostratigraphic elements (including basin-margin unconformities and submarine erosion features) are tabulated according to most recently published and unpublished dates based on planktonic foraminifera. The chart indicates progressive evolution from input-dominated sequences in the Paleogene to increasingly eustatic-dominated sequences in the Neogene. The eustatic curve is that of Haq and others (1987). Modified from Galloway (1989b).

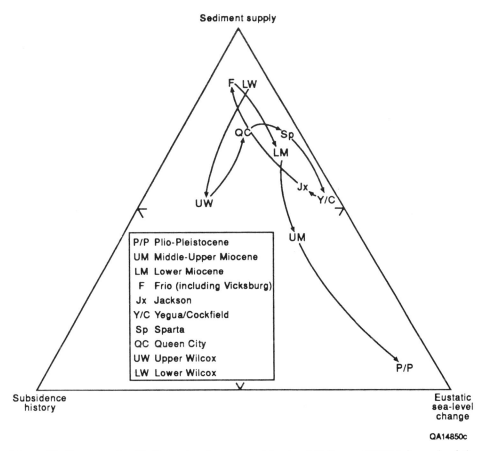

Figure 34. Sequence classification using the process triangle of Galloway (1989b) for each of the Cenozoic depositional episodes of the northwest Gulf of Mexico basin.

Rocky Mountains and adjacent western Midcontinent by more than 1,000 m occurred in the Pliocene (Chapin, 1979; Dickinson, 1981). At the same time major pulses of continental margin outbuilding occurred (Fig. 33), primarily in the northern Gulf of Mexico basin (Fig. 5).

This cursory review of comparative tectonic and depositional histories suggests several generalizations and allows a provisional qualitative classification of the principal northern Gulf of Mexico basin genetic stratigraphic sequences on a process triangle (Fig. 34).

1. Major depositional episodes (Wilcox, Frio, early Miocene) responsible for continental margin progradation correspond to principal tectonic events within the North American Plate. These events are in turn related to the evolution of the active plate margin along the western rim of the continent.

2. Details of sequence development may also reflect a eustatic sea-level overprint, particularly during the Neogene when polar ice caps began to affect ocean circulation and volume. The middle to upper Miocene and Pliocene sequences are likely bracketed by eustatic transgressions.

3. Finally, as pointed out by Winker (1982), tectonic his-

tory readily explains the shifting position of the major deltaic depocenters along the northern margin of the Gulf of Mexico basin (Fig. 5). In the early Paleocene, sediment was derived primarily from the southern Rocky Mountain terrane and directed into the closest depocenter—the Houston embayment. Volcanism and regional uplift of Trans-Pecos Texas and the Sierra Madre Occidental in northwestern Mexico sent a considerable surge of sediment into the adjacent Rio Grande embayment. This drainage axis was partly beheaded in early Miocene time by faulting and subsidence of the Rio Grande rift. Late Neogene epeirogenic uplift of the western interior of the continent diverted drainage far to the east into the midcontinent. Here it was collected by the paleo–Mississippi River and transported southward into the Mississippi embayment on the north-central rim of the Gulf of Mexico basin. This pattern continues today. Although it can be argued that eustatic events defined the age and duration of the Cenozoic sequences, only tectonics of the intracontinental source and transport areas can explain the shifting of depocenters.

It is of interest to compare the relative time intervals of progradation and retrogradation within the northern Gulf of Mexico basin Cenozoic episodes. For the well-dated Oligocene

Frio and lower and middle-upper Miocene sequences retrogradation occupies from as little as 15 to as much as 50 percent of the
total duration of the episode (Galloway, 1987). Duration of the
episodes is variable, but each encompasses several million years.
Thus, the concept of geologically instantaneous transgression or
offlap is certainly not supported by the sedimentary record of the
northern margin of the Gulf of Mexico basin.

Perspectives and research opportunities

The Cenozoic fill of the northern Gulf of Mexico basin offers
a unique opportunity for four-dimensional study of depositional
processes and their products. Widespread distribution of mineral
fuels has resulted in extensive penetration of the shallower section, and remote methods, particularly reflection seismology,
have extended the three-dimensional perspective to greater
depths. Active depositional, structural, diagenetic, and hydrologic
processes may be studied in the younger Cenozoic or modern
sediments and extended back into the stratigraphic record. Modern sedimentary processes and facies can be documented, and results applied directly to the interpretation of Cenozoic paleogeography. Consequently, the Gulf of Mexico basin has become a
natural laboratory in which much has been accomplished but in
which many questions and research opportunities remain.

The causes of depositional cyclicity and the processes of the
shelf to slope transition in an actively prograding continental
margin remain only qualitatively understood. The complex and
variable stratigraphic and structural evolution and styles of unstable clastic shelf margins still generate much debate and hypotheses. Syntheses of modern geophysical and conventional
subsurface data with new stratigraphic and structural concepts
offer opportunities for improved understanding of this dynamic
setting.

The youth of the basin fill permits examination of active
hydrologic and thermal processes as well as their products, including diagenesis of both the sands and muds and of organic
matter, that are important in the early consolidation history of
large basins. Ironically, despite the large volume of hydrocarbons
that has been produced from the Cenozoic strata of the Gulf of
Mexico basin, processes of origin and migration of oil and gas
remain more speculative than documented. The vast bulk of the
petroleum productive section contains, according to conventional
criteria, thermally immature, marginal to poor source rocks. Despite this, the Cenozoic of the Gulf of Mexico basin is one of the
world's great hydrocarbon provinces (see Chapter 15 of this volume). Much observational and circumstantial data indicate the
importance of mass and energy transfer from the ultradeep
(greater than 6 km) Cenozoic and Mesozoic foundation of the
Cenozoic fill. Only systematic deep drilling and sampling can
confirm the speculative role of these basinal sediments and their
expelled fluids in diagenesis, hydrochemistry, and mineral resource genesis within the overlying section.

THE WESTERN LARAMIDE-MODIFIED PROVINCE

Introduction

The broad seaway, which during the early Late Cretaceous
extended from the Pacific Ocean, across Mexico to the Gulf of
Mexico region, began to be restricted during the late Late Cretaceous by the uplifting of uplands along the Pacific margin of
present Mexico. Uplift progressed eastward and culminated during the Paleocene and early Eocene with the formation of the
Sierra Madre Oriental of eastern Mexico. This orogenic episode,
the Laramide orogeny, folded, faulted, and uplifted the Cretaceous carbonate section and developed the complex thrust belt
and large folds of the Sierra Madre Oriental.

The Laramide orogeny drastically modified the paleogeography of the western flank of the present Gulf of Mexico basin,
blocking the communication between the Pacific Ocean and the
Gulf of Mexico, except perhaps across the present Isthmus of
Tehuantepec, and causing the progressive displacement of the
western coast of the Gulf during the late Late Cretaceous and
early Tertiary—toward the east in northeastern and east-central
Mexico, and toward the north in the southeastern part of the
country. The Laramide uplift with the resulting formation of the
Sierra Madre Oriental was accompanied by the development of a
series of depressions roughly parallel to the orogenic belt that are
known today as the Tertiary basins of eastern Mexico: the Burgos, Tampico-Misantla, and Veracruz basins, and the so-called
"Southeastern Tertiary basins"—the Isthmus Saline, Comalcalco,
and Macuspana basins (see Plate 2 in pocket).

The *Burgos basin* is located in northeastern Mexico; it is the
southern extension into Mexican territory of the Rio Grande
embayment. It is limited to the north by the Rio Grande (Rio
Bravo), to the south-southeast by the Tamaulipas arch, to the
west by the Cretaceous/Tertiary boundary, and to the east by the
present coast of the Gulf of Mexico. The Burgos basin has an area
of approximately 40,000 km^2. The stratigraphic section is composed of terrigenous sediments distributed in bands roughly parallel to the present coast of the Gulf; the oldest sediments
(Paleocene) crop out to the west and the youngest to the east. The
Cenozoic section, with a maximum thickness of 10,000 m, is
composed of a sequence of alternating sandstones and shales rich
in foraminifera. The variations in lithology are closely related to
the environment of deposition and to the location of growth faults
active during the Cenozoic. Sedimentation was eminently regressive, only interrupted by short transgressive episodes, in an environment that ranged from brackish to bathyal.

Five depositional episodes can be distinguished in the Burgos basin, each with its own corresponding systems—deltaic,
fluvial, and littoral (see Figs. 15, 19, 21, 25, and 28).

The *Tampico-Misantla basin* comprises the northern part of
the State of Veracruz, the southern part of the State of Tamaulipas, the easternmost parts of San Luis Potosí and Hidalgo, and
northern Puebla. Geologically it is limited to the north by the

Tamaulipas arch, to the south by the Santa Ana (Teziutlan) massif, to the west by the foothills of the Sierra Madre Oriental, and to the east by the Gulf of Mexico coast line.

The stratigraphic column is composed of sandstones, shales, conglomerates, and rarely by reefal limestones. Planktonic and benthonic faunas are abundant in the sediments. Upper Tertiary volcanic rocks occur locally intruding and associated with the sedimentary section. The Cenozoic stratigraphic section was deposited in environments that range from neritic to bathyal. The average thickness of this terrigenous section in the Tampico-Misantla basin is 6,000 m.

The formation of the basin is the result of the subsidence of the Tamaulipas arch during the Laramide orogeny, toward the end of the Cretaceous and the beginning of the Tertiary. The area, in addition, was subject to two tectonic events, one in late Eocene–early Oligocene time, and another during the Miocene. Five depositional episodes are also distinguished, during which distinctive depositional features were formed: the La Flor Ayotoxco submarine fan (late Paleocene), the Chicontepec and Nautla canyons (early Eocene), and the Cazones and Tecolutla paleodeltas (late Eocene). They will be discussed later in this chapter.

The *Veracruz basin,* as developed during the Tertiary, is located to the east of the Córdoba platform. It covers part of the Gulf Coastal Plain and extends to the subsurface of the continental shelf of the Gulf of Mexico. It originated during the early Paleocene when a thick sequence of shales, sandstones, and conglomerates was deposited. The Cenozoic section reaches a thickness of 8,000 m in the center of the basin and thins toward its flanks. The stratigraphic column ranges from the Paleocene to the Pliocene. Several unconformities can be recognized in the section. The main ones are the middle Eocene–upper Eocene and the middle Miocene–upper Miocene unconformities in the central and eastern parts of the basin. In the western flank of the basin, a third unconformity is present at the base of the upper Miocene, which rests unconformably over beds ranging in age from early Cretaceous to late Eocene.

Miocene volcanic rocks cover part of the basin. They are andesitic and basaltic lavas and pyroclastic beds derived from nearby volcanic centers that form the eastern part of the Trans-Mexican Neovolcanic Belt (Pico de Orizaba, Cofre de Perote) and the San Martin Volcano in the San Andrés Tuxtla (Los Tuxtlas) uplift. In the Veracruz basin, four depositional episodes can be interpreted. They are more evident during the Eocene when fan-shaped paleocanyons were developed.

The *southeastern Tertiary basins* cover the southeastern part of the State of Veracruz and nearly all of the State of Tabasco. Three basins are recognized in this area: the Isthmus Saline basin, the Comalcalco basin, and the Macuspana basin. They are limited to the north by the continental slope of the Gulf of Mexico, to the south by the northern front of the Sierra de Chiapas, to the west by the Veracruz basin, and to the east by the Yucatan platform.

In the Comalcalco and Macuspana basins the Cenozoic section is composed of a sequence of up to 16,000 m of shales and sandstones, which commonly contains in its upper part bentonites and volcanic ash horizons. The Cenozoic environments of deposition range from neritic to bathyal. Locally, the Eocene and Miocene sections contain some calcarenitic limestone beds. Of the total Cenozoic section in these basins, as much as about 4,000 m may locally correspond to the Miocene-Pliocene. This anomalously large thickness of the upper Tertiary section is due to the development in southeastern Mexico during the late Tertiary of an extensional regime that resulted in the evolution of large syndepositional growth faults (Palizada, Macuspana, and Comalcalco faults, among others) that created deep troughs or grabens into which the terrigenous sediments were deposited from the late Tertiary to the Quaternary. Reverse faults known in the southeastern basins have resulted from early and late Miocene diastrophic movements and from diapiric intrusion of the Middle Jurassic salt.

In the southeastern basins, two main unconformities have been recognized as well as four depositional episodes related to the orogenic processes of the Laramide, Chiapaneca, and Cascadian orogenies.

Deposition in the Tertiary basins of eastern and southeastern Mexico was dominated by terrigenous clastics—conglomerates, sandstones, and shales—derived from the positive areas to the west and south.

Carbonate deposition in the Mexican part of the Gulf of Mexico basin during the Cenozoic was restricted to the Yucatan platform and to an area in the Sierra de Chiapas, between the Chiapas massif and the southeastern basins, where the Yucatan Tertiary carbonate section grades to terrigenous clastic sequences.

Numerous submarine canyons were formed along the western shelf and slope of the Tertiary Gulf of Mexico, perhaps triggered by Laramide orogenic movements. These Tertiary canyons were subsequently filled by turbiditic sediments transported by density currents.

Biostratigraphy and paleoecology

Biostratigraphic zonation. Unlike the northern and northwestern Gulf of Mexico basin province where planktonic foraminifera are not abundant in the predominantly fluvial and deltaic sequences of the Cenozoic, much of the equivalent section in the Tampico-Misantla, Veracruz, and southeastern basins in Mexico, particularly the Paleocene, Eocene, and Oligocene, is composed of deep-water marine sequences containing a rich planktonic foraminiferal fauna that has allowed a detailed subdivision and precise dating of the section (Barker and Blow, 1976).

The stratigraphic units mentioned in the following discussion of the biostratigraphic zonation of the Cenozoic of the western Gulf of Mexico basin are illustrated in Figure 2, a simplified correlation chart of the Cenozoic of the basin; in Plate 5, in the pocket, a more detailed correlation chart; and in Table 2, which includes the planktonic foraminifera biostratigraphic zonation of Stainforth and others (1975) generally followed in the study of the Cenozoic section in Mexico, and which may in general correlate to the planktonic foraminiferal and calcareous nannoplankton biostratigraphy of the northern and northwestern Gulf of Mexico basin.

Paleocene. Sedimentary sequences of Paleocene age are widely distributed along the entire coastal plain of the western Gulf of Mexico basin. The Paleocene is characterized by a section of shales containing lesser amounts of interbedded sandstones and conglomerates. Planktonic foraminifera vary in abundance from rare to abundant.

The Cretaceous/Paleocene boundary is easily placed at the change from marls with globotruncanids to shales with globigerinids. In the Tertiary basins it has been possible to recognize the majority of the planktonic foraminifera of the eight biostratigraphic units into which the Paleocene has been subdivided by Stainforth and others (1975; Table 2).

In the Burgos basin the Paleocene contains *Globigerina eugubina* Luterbacher and Premoli Silva, and *Globorotalia compressa* (Plummer) associated with *Rzehakina epigona* (Rzehak) and *Clavulinoides (Tritaxia) midwayensis* (Cushman). These species are diagnostic of the shaly sediments of the Midway Formation. The depositional environment of this unit is deep water, and its average thickness about 500 m. The Paleocene part of the overlying Wilcox Formation includes an interval ranging from the *Globorotalia angulata* Biozone to the *Globorotalia velascoensis* Biozone, and is composed of shales and sandstones characteristic of barrier bars.

In the Tampico-Misantla and Veracruz basins, the lower Paleocene includes the *Globigerina eugubina, Globorotalia pseudobulloides,* and *Globorotalia trinidadensis* Biozones. The planktonic foraminifera of these biozones are diagnostic of the shaly, deep-water sediments of the "Basal Velasco" Formation. The thickness of this unit is generally less than 600 m.

The middle Paleocene in these same two basins occurs from the southern Magiscatzin basin in the north to the northern and central parts of the Veracruz basin. It is about 2,000 m thick. Two laterally equivalent lithofacies can be recognized: one composed of deep-water shales (Velasco Formation) with abundant planktonic foraminifera; and another characterized by a turbiditic sequence of interbedded shales, sandstones, and conglomerates (Chicontepec Formation). The lower part of these two different lithofacies has been dated by the presence of fossils of the *Globorotalia uncinata* and *Globorotalia pusilla pusilla* Biozones, and corresponds to the "Velasco Inferior" and "Chicontepec Inferior" Formations. The upper Paleocene is represented by the "Velasco Medio" and "Chicontepec Medio" Formations, dated by *Globorotalia pseudomenardii* Bolli and *Globorotalia velascoensis* (Cushman).

In southeastern Mexico, the Paleocene sequence is also composed of deep-water shales with abundant arenaceous and calcareous benthonic foraminifera such as *Gavelinella rubiginosa* (Cushman), *G. velascoensis* (Cushman), *G. becariiformis*

(White), *Nuttallinella florealis* (White), *Osangularia velascoensis* Cushman, and many more. Also present are the majority of the planktonic species that characterize the Paleocene biostratigraphic units from the *Globorotalia trinidadensis* to the *Globorotalia velascoensis* Biozones. In the southeastern basins, the Paleocene corresponds to the lower part of the Nanchital Shale and has an average thickness of 330 m. In the eastern part of the Gulf of Campeche, a distinctive dolomitized calcareous talus breccia represents part of the Upper Cretaceous (126 m) and the basal Paleocene (28 m). It is overlain by a sequence of argillaceous limestones with intercalated shales which in turn is overlain by a shale section. The total thickness of the Paleocene section overlying the breccia is 100 to 200 m.

In the eastern and northeastern parts of the Sierra de Chiapas, the Paleocene section includes the platform carbonates of the Lacandon Formation, which rest conformably over the platform carbonates of the Upper Cretaceous (Maastrichtian) Angostura Formation (Quezada Muñetón, 1987 [1990]). Some of the benthonic foraminifera identified in the Paleocene section include *Storrsella haastersi* (van den Bold), *Elphidium nassauensis* (Applin and Jordan), *Pseudophragmina* sp., and milliolids. The reported average thickness of the Lacandon Formation is 307 m. In the central part of the Sierra de Chiapas, the Paleocene section in an outer platform lithofacies has been described under the name "Tenejapa Formation." Described from this unit are *Elphidium nassauensis* (Applin and Jordan) and planktonic foraminifera distinctive of the Paleocene *Globorotalia trinidadensis* to *Globorotalia pusilla pusilla* Biozones. The reported average thickness of the Tenejapa Formation is 300 m. Finally, in the same Sierra de Chiapas region, to the west and north of the area where the Paleocene is represented by platform and outer platform sediments, the Paleocene is composed of a flysch section called Soyalo Formation, which contains abundant planktonic foraminifera of the interval from *Globorotalia trinidadensis* Biozone to *Globorotalia velascoensis* Biozone. The reported average thickness of the Soyalo Formation is 208 m.

Eocene. In the northeastern part of the Burgos basin, the Lower Eocene is characterized biostratigraphically by the presence of *Globorotalia aequa* Cushman and Renz, *Globorotalia rex* Martin, *Globigerina gravelli* Bronnimann, *Globorotalia pseudotopilensis* (Subbotina), and *Pseudohastigerina wilcoxensis* (Cushman and Ponton). The Lower Eocene is a progradational unit of shales and sandstones indicative of coastal deposition. This lithologic unit is part of the Wilcox Formation and reaches an average thickness of 1,000 m.

The Middle Eocene of the Burgos basin is represented by shelf and basin sediments characterized by sandstone-shale sequences. The lower part is shaly, formed in a middle neritic to outer neritic-bathyal environment. Its basal part is known as the Recklaw Formation and is dated by the occurrence of *Globorotalia broedermanni* Cushman and Bermudez, *Globorotalia crassata* (Cushman), *Asterigerina tatumi* Hussey, and *Haplophragmoides tallahattensis* (Bandy). It has a thickness that ranges from 700 to 800 m. The middle part, the Queen City Formation, is a sandstone-shale sequence; the sandstone bodies represent coastal-bar systems. No diagnostic faunal assemblages have been reported from the Queen City Formation; it has a thickness of 400 to 500 m.

Resting over the Queen City Formation is the Weches Formation characterized by a transgressive shaly section formed in a middle neritic to outer neritic environment. Planktonic foraminifera are rare. It is identified by the extinction of *Textularia smithvillensis* (Cushman and Ellisor). The Weches, in turn, is overlain by the shale-sandstone sequence of the Cook Mountain Formation. It represents a moderate regression, and a shallow neritic environment. Its fauna is characterized by *Truncorotaloides rohri* Bronnimann and Bermudez, associated with *Ceratobulimina eximia* Rzehak.

The uppermost part of the Middle Eocene is composed of sandy shales with interbedded sandstone beds, representing a continental or littoral environment. The microfauna is scarce and includes *Discorbis yeguaensis* Weinzierl and Applin and *Eponides yeguaensis* Weinzierl and Applin. These species characterize the Yegua Formation.

The Upper Eocene (Jackson Formation) consists of a transgressive lower part composed of sandstones and shales deposited in shallow water, and an upper part composed predominantly of shales with interbedded isolated sandstone beds deposited in an inner neritic to bathyal environment. The diagnostic species of this prograding, regressive unit are *Hantkenina alabamensis* Cushman and *Globorotalia cerroazulensis* (Cole), associated with the benthonic forms *Textularia hockleyensis* Cushman and Applin and *Textularia dibollensis* Cushman and Applin.

In the Tampico-Misantla and Veracruz basins, the Lower Eocene is found in two laterally equivalent lithofacies dated by *Globorotalia rex* Martin, *Globorotalia formosa formosa* Bolli, and *Globorotalia palmerae* (Cushman and Bermudez). One lithofacies, represented by the "Velasco Superior" Formation, is composed of deep-water shales, and the second, of turbiditic origin, corresponds to the shale-sandstone-conglomerate sequences of the upper part of the "Chicontepec Superior."

Overlying the above-described units in the Tampico-Misantla and Veracruz basins is the Aragón Formation composed also of deep-water shales rich in planktonic foraminifera, among them *Globorotalia aragonensis* and *Hantkenina aragonensis* associated with *Globorotalia collatea* (Finlay) and *Globigerina soldadoensis* Bronnimann. These species date the Aragón Formation as of Early Eocene to earliest Middle Eocene age.

The Middle Eocene Guayabal Formation is generally shaly and rich in planktonic foraminifera. In the northern part of the Tampico-Misantla basin it becomes sandier and the planktonic microfossils become much less abundant. In the Tuxpan platform, the Middle Eocene is not represented as it pinches out from west to east over the Cretaceous rocks. The paleontologic and lithologic characters of the Guayabal Formation indicate that it was deposited in a relatively deep and quiet sea. Four planktonic-foraminifera zones are recognized in the Guayabal Formation—from the *Globigerinatheka subconglobata* to the *Truncoro-*

taloides rohri Biozones. They are associated with the benthonic form *Eponides guayabalensis* Cole.

In the Tampico-Misantla and Veracruz basins, the upper Eocene is also represented by two lithofacies, one representative of deep-water deposition and composed of shales containing an abundant planktonic microfauna (Chapopote Formation), and the second of shallow-water origin (Tantoyuca Formation) rich in larger foraminifera and poor in planktonic forms. The lithology of the Tantoyuca Formation is characteristic of molasse sequences composed of alternating conglomerates, sandstones, and shales. The deposition of this unit took place during the culmination of the Laramide tectonic event.

The Chapopote Formation is dated by containing the *Globorotalia cerroazulensis* Biozone in addition to *Hantkenina alabamensis* Cushman, *Globorotalia centralis* Cushman and Bermudez, and *Globigerapsis semiinvolute* (Keijzer). Its average thickness is 200 m. The Tantoyuca Formation contains a higher proportion of benthonic fossils. Most abundant among them are *Operculina* sp. and *Lepidocyclina* sp. and rare specimens of the planktonic species *Globorotalia cerroazulensis* (Cole).

In the southeastern basins, the Eocene rocks form the upper part of the Nanchital Shale. This unit is very fossiliferous and contains many of the planktonic foraminifera that characterize all Eocene biozones from *Globorotalia subbotinae* to *Globorotalia cerroazulensis.*

In the Isthmus Saline basin and in the western part of the Sierra de Chiapas, a unit composed of conglomerates with intercalations of sandstones and shales—the Uzpanapa Conglomerate—has been described at the base of the Eocene section. The components of the conglomerates and sandstones are igneous and metamorphic rocks, recycled sandstones, recrystallized limestones, and Cretaceous fossiliferous limestones. The cement is calcareous and siliceous. This unit has been interpreted as a wild-flysch. The intercalated shales contain planktonic foraminifera representative of early and middle Eocene biozones. Conglomerate lenses of similar lithologic composition have been described from the upper Eocene sequence.

Toward the southern part of the Isthmus Saline basin and in the western part of the Sierra de Chiapas, the Eocene ranges in thickness between 300 and 1,200 m. In the Chiapas-Tabasco area, to the north, the thickness ranges between 300 and 900 m, and in the Gulf of Campeche it averages only 368 m.

The Eocene of the Sierra de Chiapas represents a transition between the carbonate section of the Yucatan platform to the north and the terrigenous-clastic sequences of the southeastern basins. It consists predominantly of sandstone-and-shale sequences, deposited in marine platform environments, that grade toward the Chiapas massif to littoral and continental reddish conglomerates and sandstones. Argillaceous and bioclastic limestones become more common toward the Yucatan platform. The Eocene sequences in the Sierra de Chiapas contain abundant fossils of the genera *Pseudophragmina, Rhapydionina, Raadshoovenia, Nummulites, Lepidocyclina,* and *Helicostegina,* as well as coraline algae, mollusks, and abundant other foraminifera.

Oligocene. In the Burgos basin, the lower part of the Oligocene is represented by the Vicksburg Formation, which contains a foraminiferal fauna distinctive of the *Globigerina ampliapertura* Biozone, associated with *Heterolepa mexicana* (Nuttall) and *Anomalina bilateralis* Cushman. The Vicksburg is composed of a sandstone-shale sequence deposited in an outer neritic to bathyal environment. It has an average thickness of 1,200 to 1,500 m.

Overlying the Vicksburg is the "Marine Frio" Formation, which contains foraminifera representative of the *Globorotalia opima opima* Biozone with *Globigerina ciperoensis angulisuturalis* Bolli, associated with *Nodosaria blanpiedi* Ellis. This unit is composed of shales and sandy shales deposited in brackish and mixed environments. It has a thickness of 1,500 m. Over the "Marine Frio" are 800 to 1,500 m of interbedded sandstones and shales that make up the "Nonmarine Frio" Formation. No age-diagnostic fossils have been identified in this unit. Over the "Nonmarine Frio" is a sequence, 400 m thick, also composed of nonmarine conglomerates with limestone fragments, the Norma Conglomerate. These nonmarine sequences are characteristic of the upper part of the Oligocene in the northwestern part of the Burgos basin. They grade basinward to a section composed of shales and sandstones deposited in a neritic environment—the Anahuac Formation—that contains abundant benthonic foraminifera such as *Discorbis gravelli* Garrett, *Heterostegina antillea* Cushman, *H. texana* Gravell and Hanna, and *Marginulina mexicana* Cushman, as well as planktonic foraminifera representative of the *Globigerina ciperoensis* Biozone.

In the Tampico-Misantla and Veracruz basins, the extinction of *Globorotalia cerroazulensis* (Cole) and *Hantkenina alabamensis* Cushman marks the base of the Oligocene section. Its lower part is represented by the Horcones and "Palma Real Inferior" Formations in which the most common forms are *Globigerina parva* Bolli, *Cassigerinella chipolensis* (Cushman and Ponton), and *Pseudohastigerina micra* (Cole), all diagnostic of the *Globigerina ampliapertura* Biozone. The Horcones and "Palma Real Inferior" Formations are composed of globigerinid oozes, deep-water shales, and sandstones. The Horcones Formation, in contrast to the "Palma Real Inferior," is recognized by the exclusive occurrence of *Rotalialina mexicana* Cushman and the planktonic fossils previously mentioned. The Horcones, less than 500 m thick, is local in distribution and difficult to map.

Overlying the Horcones and "Palma Real Inferior" are the laterally equivalent Mesón, "Palma Real Superior," and Alazán Formations. They contain faunas representative of the *Globorotalia opima opima* and *Globigerina ciperoensis* biozones, diagnostic of the upper part of the Oligocene. The Mesón Formation is composed of thick-bedded sandstones rich in *Lepidocyclina* sp., coelenterates, echinoderms, and mollusks, and deposited in a shallow-marine environment. The "Palma Real Superior" is composed of slightly sandy shales, sandstones, and occasional reef limestones. It was deposited in a somewhat deeper marine environment, basinward from the Mesón Formation. Doubtful forms of *Globorotalia kugleri* Bolli have been reported from this unit.

The Alazán Formation is predominantly composed of deep-water shales containing a rich planktonic foraminiferal fauna.

In the Veracruz basin, the Oligocene is represented by the bathyal sandstone-shale sequences of the lower part of the La Laja Formation. Its contained faunas are diagnostic of the *Globorotalia opima opima* and *Globigerina ciperoensis* biozones.

In the southeastern basins, the Oligocene is composed of shales and sandy shales with intercalations of tuffs and bentonites that make up the La Laja Formation. It contains abundant benthonic foraminifera such as *Trochamminoides* sp., *Haplophragmoides* sp., *Bathysiphon* sp., *Pleurostomella alternans* Schwager, *Chilostomella oolina* Schwager, *Vulvulina pennatula* Bastsch, and many more. It was deposited in a lower bathyal environment. Planktonic foraminifera are also abundant, including many species characteristic of the *Cassigerinella chipolensis–Pseudohastigerina micra* to *Globigerina ciperoensis* Biozones.

Toward the southern part of the Isthmus Saline basin and the western part of the Sierra de Chiapas, a sequence of conglomerates, sandstones, and shales has been described. The conglomerates have a similar composition to those of the Eocene Uzpanapa Conglomerate. The shales interbedded with the conglomerates have yielded planktonic foraminifera characteristic of the Oligocene biozones.

Frost and Langenheim (1974) have described several carbonate units from the Sierra de Chiapas. They contain shaly and sandy intercalations rich in larger foraminifera such as *Lepidocyclina (Eulepidina) undosa* Cushman, *Nummulites panamensis* Cushman, and others, as well as coraline algae, corals, gastropods, and pelecypods that indicate shallow-water environments of deposition. These sequences have been differentiated locally with the name "Mompuyil Formation."

Miocene. In the northwestern part of the Gulf of Mexico basin (the Burgos basin), the Miocene starts with the development of a complete regressive-transgressive cycle. The regressive stage corresponds to the Catahoula Formation, formed in a continental or transitional environment. It contains a fauna of mollusks and charophytes.

During the middle and late Miocene and the earliest Pliocene, the shale-sandstone sequence of the Oakville Formation covered the Catahoula. It contains a fauna representative of the *Globorotalia foshi* to *Globorotalia margaritae* Biozones. Associated to these planktonic species are the benthonic forms *Textularia panamensis* Cushman and *Bigenerina humblei* Cushman and Ellisor.

In the Tampico-Misantla basin, the lower Miocene is represented by the lower part of the sandstone-shale sequence of the Tuxpan Formation of the Tampico area, and the Coatzintla Formation of the Poza Rica area. The thickness of these units is less than 500 m. They contain species of *Catapsydrax* in addition to *Globigerina rohri* Bolli, *Globigerinoides triloba altiaperturus* Bolli, *Globigerina ciperoensis angustiumbilicata* Bolli, and associated abundant specimens of *Siphogenerina* sp. and *Uvigerina* sp.

The middle and upper parts of the Tuxpan Formation, of Middle and Late Miocene age, are characterized by the *Globorotalia fohsi, Globorotalia mayeri, Globorotalia menardii,* and *Globorotalia acostaensis* biozones. Lithologically the section is composed of interbedded sandy shales and sandstones, which in some cases are barren of fossils.

The Escolín Formation, of only local distribution, is identified in the Poza Rica area by the presence of the genera *Sorites* and *Miogypsina,* which have a range comprised between the *Globigerinatella insueta* and the base of the *Globorotalia acostaensis* biozones. These forms are associated with *Globoquadrina dehiscens* (Chapman, Parr and Collins), *Globigerinoides conglobatus* (Brady), and *Globorotalia tumida* (Brady). Lithologically, the Escolín Formation is composed of sandstones and sandy shales with occasional intercalations of conglomerates.

In the Veracruz basin, the shale-sandstone sequences of the upper part of the La Laja Formation are dated by the extinction of the *Catapsydrax* group and *Globorotalia fohsi.* This unit extends to the Middle Miocene, represents deposition in a bathyal environment, and reaches a maximum thickness of 2,500 m. (In the Veracruz basin, as well as in the southeastern basins, the upper Cenozoic section has been subdivided into a number of stratigraphic units characterized and identified principally by their contained faunas. Even though these units are in reality biostratigraphic units, they have been assigned geographic names, appearing to be, therefore, lithostratigraphic units. The names of these units will be distinguished by placing them between quotation marks.)

The La Laja Formation is overlain by the deep-water argillaceous sediments of the "Deposito Formation," distinguished by the appearance of *Globorotalia mayeri* Cushman and Ellisor and *Globorotalia menardii* (d'Orbigny). It represents the upper part of the Middle Miocene and has a thickness of up to 4,000 m.

Overlying the "Deposito Formation" is the "Encanto Formation" composed of bentonitic shaly sandstones corresponding to the *Globorotalia acostaensis* Biozone. It also contains numerous species of benthonic foraminifera, and represents deposition in inner neritic to bathyal environments. Its maximum thickness is less than 600 m.

The Miocene section of the southeastern basins has been extensively studied in the subsurface. It consists of a sequence of interbedded fossiliferous shales and sandstones, deposited in a bathyal environment. The Early Miocene in the western part of the basins is represented in the upper part of the La Laja Formation, rich in deep-water, calcareous and arenaceous benthonic foraminifera. This upper part of the La Laja can be dated by the recognition of the *Catapsydrax dissimilis* and *Catapsydrax stainforthi* biozones. Toward the eastern part of the region the top of the La Laja Formation corresponds with the top of the Oligocene, because there is an unconformity between the Upper Oligocene and the Lower Miocene.

The biostratigraphic unit called "Deposito" that characterizes a large part of the central and eastern regions of the Cenozoic basins of southeastern Mexico, is identified in the subsurface by the extinction of *Anomalinoides cicatricosa* (Schwager) at its top, and the abundance of *Gyroidinoides broeckhiana* (Karrer). Planktonic foraminifera are abundant and include species characteristic of the interval from the *Globigerinatella insueta* to the *Globorotalia fohsi fohsi* biozones. The "Deposito" is composed of alternating shales and sandstones, deposited in a bathyal environment from perhaps the Early Miocene to the early part of the Middle Miocene.

During the late Middle Miocene and the Late Miocene, deposition of interbedded sandstones and shales continued. The sandstone content increases upward in the section. This unit has been called "Encanto" and is well recognized in the subsurface by the extinction at its top of *Siphouvigerina auberiana* d'Orbigny and the abundance of *Planulina filisolaensis* Nuttall, *Uvigerina peregrina* Cushman, *Sigmoilopsis schlumbergeri* Silvestri, and many other diagnostic species. Planktonic foraminifera are also very abundant and include the species characteristic of the biozones from *Globorotalia foshi lobata-robusta* to *Globorotalia acostaensis.* This unit represents a middle to upper bathyal environment of deposition.

In the Sierra de Chiapas, Frost and Langenheim (1974) mapped several Miocene formations on the basis of their lithology and fauna. It will be necessary, however, to carry out additional stratigraphic and sedimentological studies to better determine the depositional episodes and systems they represent. These authors report the presence in the section of several genera of larger foraminifera such as *Heterostegina, Lepidocyclina, Sorites, Miogypsina,* and *Archaias,* as well as corals and algae.

The Miocene section of the northern part of the Sierra de Chiapas and the southeastern part of the region of the southeastern basins, includes the Macuspana Formation composed of platform limestones. In the northwestern part of the Sierra de Chiapas are exposures of a thick conglomerate section, containing fragments of igneous rocks and sandstones, that is known as the Nanchital Conglomerate.

Pliocene-Pleistocene. The Early and Middle Pliocene are represented in the Veracruz and southeastern basins by the biostratigraphic units called "Concepción Inferior," and "Concepción Superior" (Akers, 1979, 1981, 1984; Machain Castillo, 1986). The top of the "Concepción Inferior" is recognized in the subsurface by the extinction of *Marginulinopsis marginulinoides* (Göes), *M. mesinae* (Souaya), and *Amphicoryna hirsuta* (d'Orbigny). The "Concepción Inferior" was deposited in a middle- to outer-neritic environment.

The top of the "Concepción Superior" is recognized in the subsurface of the southeastern basins by the extinction of *Astacolus vaughani* (Cushman), and has been interpreted to have been deposited in a middle-neritic environment. The planktonic foraminifera present in this unit include *Globigerinoides obliquus extremus* Bolli and Bermúdez, *Globigerinoides conglobatus* (Brady), *Globorotalia crassaformis* (Galloway and Wissler), and many more.

Overlying the shale-sandstone sequence of the "Concepción Superior," is a sedimentary sequence composed of interbedded shales and sandstones that has been subdivided into the "Filisola," "Paraje Solo," "Agueguexquite," and "Cedral" stratigraphic units. Foraminifera are scarce in these units in comparison with those that underlie them. Several authors have described the different lithofacies of the deltaic systems represented in these units (Pérez Matus and Valle González, 1974; Pérez Matus and Barbosa, 1975; Ricoy, 1989).

Pérez Matus (1978) has postulated the presence of a probable submarine canyon in Miocene-Pliocene sediments in the eastern part of the Isthmus Saline basin.

Paleoecology and paleobathymetry. The drilling of a large number of exploratory and development wells in the northwestern and western Gulf of Mexico basin has made it possible to undertake, also in Mexico, paleoecologic and paleobathymetric studies of the Cenozoic sedimentary section. The paleoecologic interpretation is based on the comparison between modern and fossil biofacies. The studies used as references are those of Bandy and Arnal (1957, 1960), Bradshaw (1959), Phleger (1960), and many others.

The paleoecologic and paleobathymetric interpretations mentioned in the previous sections are based on the analysis of countless surface and well samples. Paleoenvironments ranging from inner neritic to bathyal have been recognized; their definition follows the conventions shown in Figure 13. The faunal assemblages or cenozones discussed are based on the following considerations:

1. The planktonic and benthonic fossils have been studied separately as they reflect different environments.

2. The total populations have been determined and the relationship between the various components studied.

3. An absolute and relative quantitative analysis of the fossils contained in the samples was carried out.

The paleobathymetry thus obtained was essential to the interpretation of the sedimentologic models (deltas, channels, etc.) that can be recognized in the Cenozoic stratigraphic sequences which surround and roughly parallel the present Gulf of Mexico.

Genesis and stratigraphic evolution of the Cenozoic of the western Gulf of Mexico basin

The stratigraphic information supplied by surface and subsurface geologic studies of the Cenozoic rocks of the western Gulf of Mexico basin have allowed the identification of distinct depositional episodes. The Laramide, Chiapaneca, and Cascadian orogenies undoubtedly played an important role in the genesis of these episodes, in the processes that formed the Cenozoic basins, and in regulating many of the stratigraphic aspects and features recognized in the Cenozoic sequence of the basin (deltas, channels, etc.). These features have remained as witnesses of the sedimentary episodes. Following is a description of the main genetic features for each basin or province.

Burgos basin. Toward the end of the Cretaceous, the Laramide orogeny folded and uplifted the Mesozoic rocks of the Sierra Madre Oriental and the Sabinas basin, to the west of the Burgos basin. It resulted in the formation of a depression (foreland basin), dipping toward the east, over which a thick clastic sedimentary sequence of multiple origin was deposited. Sedimentation was predominantly in shallow-marine environments and of cyclical transgressive-regressive character, with westward transgressions, and eastward regressions and progradation.

The last stages of the Laramide orogeny can be recognized in the Paleocene and Eocene stratigraphic section. During the Oligocene and Miocene, tectonic activity was less intense. This activity caused epeirogenic movements and tilting that favored the formation of north-south–trending growth faults contemporaneous with the deposition. These faults are associated to each sedimentary episode and are represented in the Paleocene, Eocene, Oligocene, and Miocene belts (Fig. 35).

After the lowering of the sea level that marked the end of the Cretaceous, a major westward transgression took place in the Burgos basin during the Paleocene. It resulted in the deposition of a predominantly shale sequence (the Midway Formation), which in updip areas covered disconformably the eroded Cretaceous surface. The Oasis-Pandura fluvial-deltaic system was formed at this time in the northwestern part of the Burgos basin southsoutheast of the city of Nuevo Laredo (Figs. 15 and 36). It covers an area of approximately 400 km^2, and it may represent the ancestral Rio Salado or Rio Grande (Rio Bravo), which flowed from northwest to southeast and deposited a sequence of sands ranging in thickness from 4 to 13 m.

In the Burgos basin, the Eocene was witness to four progradational depositional episodes—Wilcox, Queen City, Cook Mountain, and Yegua—and to three transgressive episodes—Recklaw, Weches, and Jackson. A further transgressive-regressive cycle occurred during the latest Eocene, the Jackson cycle (Fig. 2).

A series of intensive uplifting movements were active in the uplands west of the Burgos basin during the early Oligocene; they resulted in a large influx of clastic sediments into the basin. An initial regressive event favored the development of a system of growth faults of widespread regional distribution and considerable displacement. The development of the growth faults, in turn, strongly influenced sedimentation; sandstones are more abundant in the down-thrown side of the growth faults (Fig. 37). The structural activity, in combination with frequent oscillations of the coast line, resulted in the formation of a sedimentary sequence deposited in environments ranging from outer neritic to bathyal (Vicksburg Formation). Renewed uplift of the positive areas to the west during the late Oligocene resulted in another strong influx of clastics into the basinal areas. These clastics were transported and distributed by fluvial currents, creating a complex sedimentary system deposited in littoral to inner and middle neritic environments, the "Marine Frio" and the "Nonmarine Frio" Formations. The Frio was covered to the west by the Norma Conglomerate, and to the east by the transgressive marine shaly section of the Anahuac Formation.

A further regressive-transgressive cycle took place during the

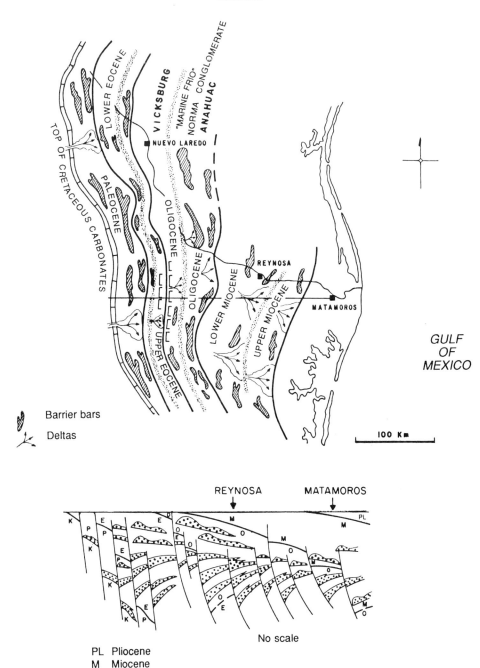

Figure 35. Tertiary depositional systems, Burgos basin.

Miocene—the Catahoula and Oakville Formations. It was followed in the Pliocene by an extended eastward regressive and prograding episode, which has continued to the present.

Tampico-Misantla basin. The Tampico-Misantla basin was formed during the latest Cretaceous or early Tertiary when the Laramide orogeny uplifted and folded the Mesozoic sedimentary rocks of the so-called "Mexican Geosyncline" to form the Sierra Madre Oriental. The present expression of the basin is the result of the interaction between the western Laramide thrusting and folding, and the rigid positive basement blocks to the east that

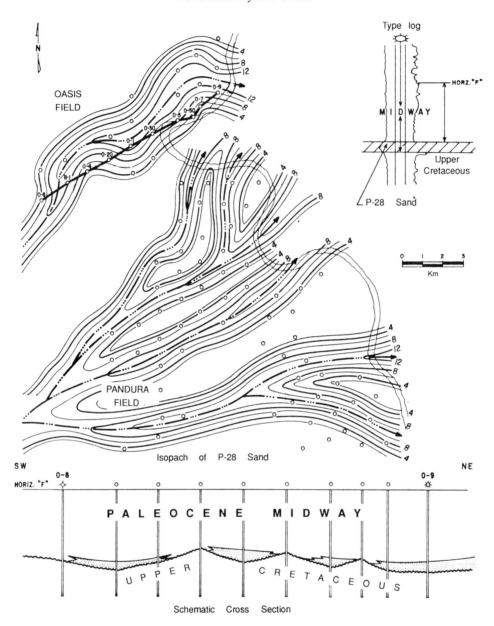

Figure 36. The Oasis-Pandura fluvial-deltaic system (early Paleocene), Burgos basin.

extended southward from the Tamaulipas arch to the Tuxpan platform (Plate 2). The new basinal area thus formed became the site of continuous Cenozoic sedimentation.

The sedimentary evolution of the Tampico-Misantla basin was governed by intermittent uplift to the west, and progressive subsidence or tilting to the east. The relationship between land and sea was invariably unstable, determined by the successive and alternating transgressive and regressive cycles (Fig. 38). These sedimentary cycles are recognized in the stratigraphic sequence by a series of contrasting sedimentary features.

The uplift of the Sierra Madre Oriental toward the end of the Cretaceous created a fluvial drainage system along which

large volumes of fine-grained sediments (the "Basal Velasco" Formation) were transported during the early Paleocene and deposited in a deep-water trough located in the southern part of the present Tampico-Misantla basin, the so-called "Chicontepec basin."

During the rest of the Paleocene, deposition took place in two contrasting contemporaneous lithofacies: the first, to the south and east, represented by the shaly sediments of the Velasco Formation, containing a deep-water fauna, and the second, to the north and west, characterized by a turbiditic sequence of alternating sandstones and shales, containing a faunal association indicative of both shallow- and deep-water environments. This

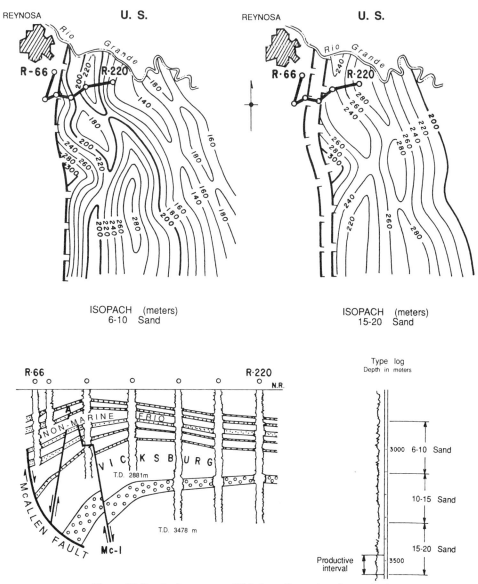

Figure 37. Barrier-bar systems, Vicksburg Formation, Burgos basin.

lithofacies relationship suggests the development of a foreland basin in the northwestern part of the present Tampico-Misantla basin.

Toward the end of the Paleocene, the southern part of the basin was uplifted to form the Santa Ana (Teziutlan) massif. The sediments derived from the erosion of most of the Mesozoic section in the massif cut submarine channels or canyons (the Bejuco-La Laja, Chicontepec, and Nautla paleocanyons; Fig. 39) through the Cretaceous and part of the Jurassic rocks of the region. The erosional limits of the channels were roughly parallel to the Laramide structures formed toward the end of the Cretaceous. During the early Eocene, these paleocanyons were filled with clastic sediments derived, for the most part, from the western uplands and from the Tuxpan platform to the east, which was an emergent positive area at the time. The material filling the paleocanyons is composed of as much as 700 m of alternating sandstones and shales, typical of a flysch sequence. Reworked faunas of Cretaceous to Eocene age have been recovered from it.

Outside of the erosional canyons, the deep-water bathyal shales of the Aragón and Guayabal Formations compose the early and middle Eocene section of the Tampico-Misantla basin.

The Chicontepec paleocanyon covers an area of 3,100 km^2 (Figs. 39 and 40) and was cut by submarine currents that flowed in a direction roughly parallel to the Sierra Madre Oriental and the Tuxpan platform. In its eastern part the canyon is cut down to the Upper Jurassic; in its northwestern part the section at the bottom of the canyon is Paleocene (Busch and Govela, 1978).

The Nautla paleocanyon is located a short distance to the

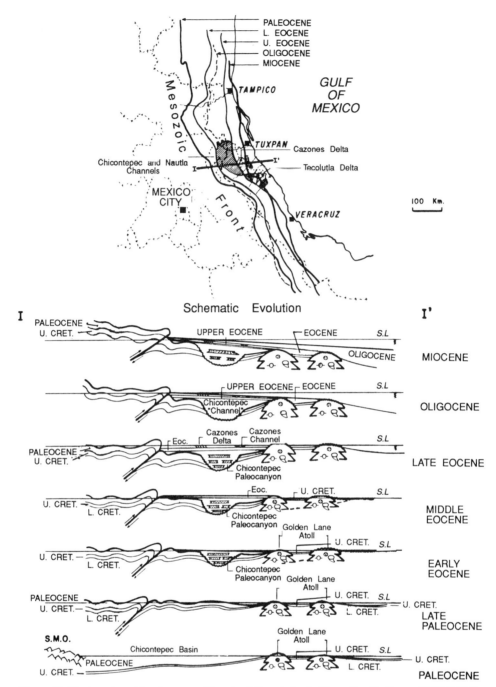

Figure 38. Tertiary depositional systems, Tampico-Misantla basin. From Carrillo Bravo, 1980. S.M.O., Sierra Madre Oriental.

south of the Chicontepec paleocanyon (Figs. 39 and 41), to which it has great stratigraphic, sedimentologic, and genetic similarity. It has a length of 75 km, an average width of 25 km, covers an area of about 1,900 km², and it was cut down to Cretaceous and Jurassic rocks. The canyon is filled with lower Eocene sandstones and shales. Net-sand thickness reaches 180 m in the deeper part of the canyon.

The last thrusting and folding episodes of the Laramide orogeny, during the middle Eocene, gave final shape to the Sierra Madre Oriental west of the Tampico-Misantla basin and resulted in the deposition of the molassic sequence of the Tantoyuca Formation, which includes the Cazones and Tecolutla deltaic complexes. Eastward, the Tantoyuca grades to the deep-water shaly section of the Chapopote Formation.

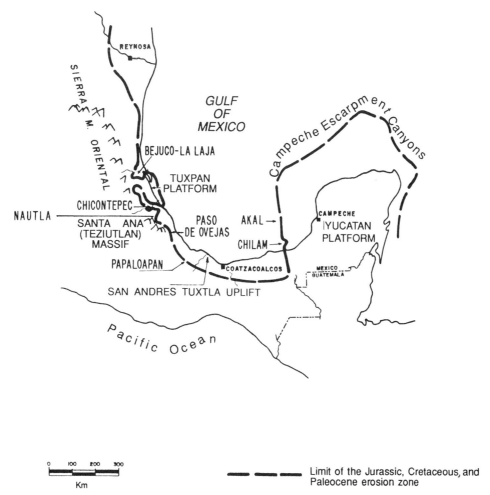

Figure 39. Tertiary submarine paleocanyons of the coastal plain and continental shelf of the western Gulf of Mexico. From Carrillo Bravo, 1980.

The Cazones delta is located in the southern part of the Tampico-Misantla basin (Figs. 38 and 42) and has a length of 25 km. The sand content reaches about 140 m in its western, proximal part, and decreases progressively eastward to nothing in the distal areas. Three well-defined lithofacies have been recognized in the Cazones delta: the fluvial lithofacies, to the west, averages 14 m in thickness and is characterized by claystones, coarse-grained sandstones, and conglomerates that show upward gradation to finer grained sediments. The section is cut by distributary channels that make difficult the correlation between wells. The delta-front lithofacies is characterized by stream-mouth sand bars that average 15 m in thickness. Finally, the prodelta lithofacies is composed predominantly of shale with only occasional thin sandstone beds.

Deposition of deep-water shales prevailed during the early Oligocene, but during the late Oligocene, vertical uplift of the Sierra Madre Oriental resulted in a renewed supply of large amounts of clastic sediments to the western part of the Tampico-Misantla basin. Regressive coarse-clastic wedges were deposited in shallow water (the "Palma Real Superior" and Mesón Formations); they grade eastward to the deeper-water shales of the Alazán Formation.

The late Oligocene regression was followed by a transgression during the early Miocene, and a new regression during the early and middle Miocene (the Coatzintla and Escolin Formations of the Poza Rica area, and the lower part of the Tuxpan Formation of the Tampico area). The Miocene closed with a transgression recorded in the interbedded sandstone-shale section of the upper part of the Tuxpan Formation.

After the late Miocene transgression, the region of the Tampico-Misantla basin was affected by a gentle but continuous emergence and the waters of the Gulf of Mexico retreated gradually toward the east. Deposition during the Pliocene and Quaternary was restricted to the present shelf and slope of the Gulf.

Veracruz basin. The tectonic and sedimentary evolution of the Cenozoic Veracruz basin is closely tied to the genesis and

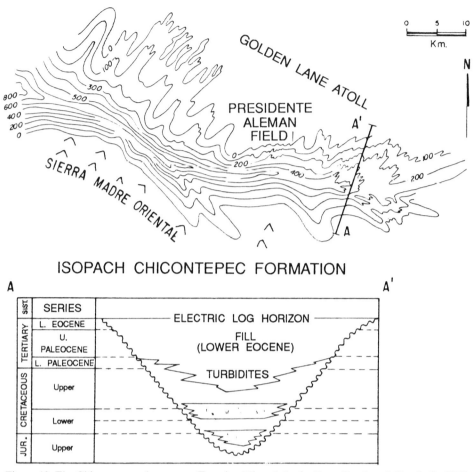

ISOPACH CHICONTEPEC FORMATION

Figure 40. The Chicontepec paleocanyon, Tampico-Misantla basin. From Busch and Govela S., 1978. For location see Figure 39.

evolution of the Sierra Madre Oriental, the main source of sediments to the basin. The Veracruz basin was formed during the early Tertiary starting at the time the Cretaceous sediments along the western margin of the present basin were uplifted and deformed. The resulting foreland basin received the sediments derived from the erosion of the area to the west. The influx of the terrigenous clastic sediments ended a period of predominantly carbonate sedimentation in the region.

At the end of the Middle Eocene, the Córdoba platform (to the north and west of the Veracruz basin) was tilted due to gravitational effects. Acting as a rigid block, the platform was pushed toward the basin resulting in a series of compressional structures (folds associated with reverse faults) that define the deformational style of the western margin of the basin (Fig. 43). This deformation resulted in a sequence of fluctuations of the coast line, transgressions and regressions associated with the persistent uplift of the source of the sediments, and the continuous subsidence of the depositional areas. In general, from the Paleocene to the Middle Eocene regression was dominant and was followed by transgressive events during the Late Eocene and the

Oligocene. The sedimentary sequence deposited during the late Eocene transgression rests unconformably along the western margin of the basin over sediments ranging in age from Cretaceous to Middle Eocene.

As in the Tampico-Misantla basin, deposition of deep-water shaly sediments predominated during the Early Paleocene (the "Basal Velasco" Formation). During the Late Paleocene and Early Early Eocene, two isochronous lithofacies were developed: one composed of deep-water shales (the Velasco Formation), and another of interbedded sandstones and shales, occasionally conglomeratic, typical of turbiditic deposits (the Chicontepec Formation).

During the late Early Eocene and Middle Eocene, a shaly section, rich in planktonic foraminifera (the Aragón and Guayabal Formations), was deposited in a deep and quiet environment, indicating that diastrophism had diminished along the Sierra Madre Oriental to the west. The renewal of tectonic activity during the early Late Eocene resulted in the uplift of the western part of the Veracruz basin area. Cretaceous to Middle Eocene rocks were exposed and eroded, and a dendritic drainage system

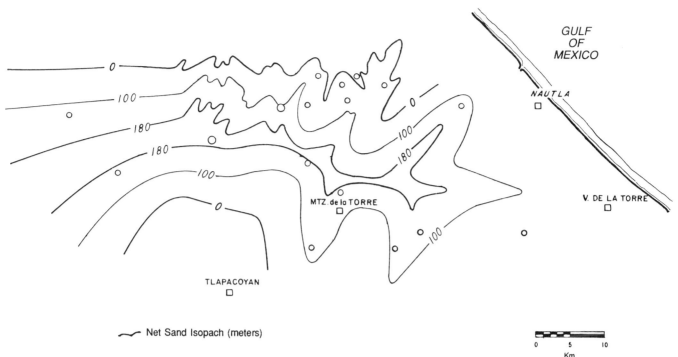

Figure 41. The Nautla paleocanyon, Tampico-Misantla basin. For location see Figure 39.

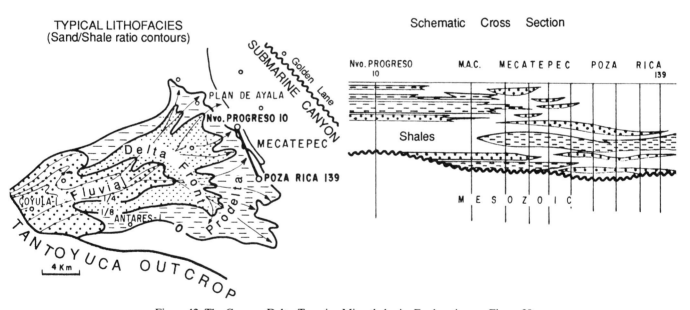

Figure 42. The Cazones Delta, Tampico-Misantla basin. For location see Figure 38.

was developed over the eroded surface (Cruz Helú and others, 1977). The eroded surface was covered by the molasse sequences of the Tantoyuca Formation of Late Eocene age (Fig. 44). It was during this time that fluvial-deltaic systems were accumulated as part of a regressive sedimentary episode. Toward the east, the

deltaic sequences graded to the deep-water shales of the Chapopote Formation, rich in planktonic foraminifera.

One drainage channel, corresponding to the present course of the Papaloapan River, had an abrupt gradient and steep walls, forming what has been called the Papaloapan paleocanyon (Figs.

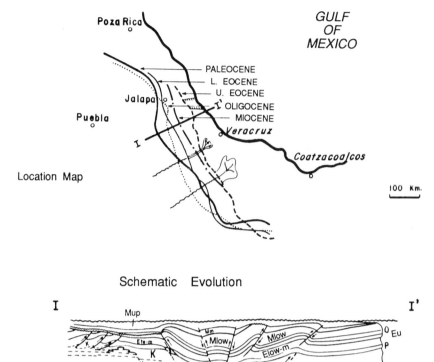

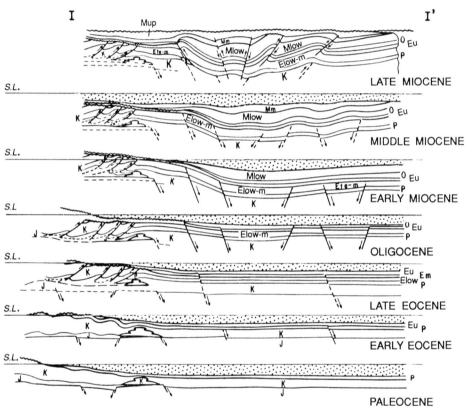

Figure 43. Tertiary depositional systems, Veracruz basin.

39 and 44). It has a length of 40 km, a width of 20 km, and is filled in its eastern part by a great thickness of conglomerates, sandstones, and shales deposited in a bathyal environment by turbidity currents. The canyon was initiated in the early late Eocene and remained active until the end of the Miocene.

A transgressive regime dominated deposition in the Veracruz Basin during the Oligocene when the deep-water sandstones

and shales of the lower part of the La Laja Formation accumulated in the western and central parts of the basin while the predominantly shale section of the Horcones Formation was deposited more basinward, toward the east.

A strong influx of clastics took place again during the Early Miocene. The isostatic disequilibrium caused by the weight of the thick sedimentary wedge, or by extension phenomena related to

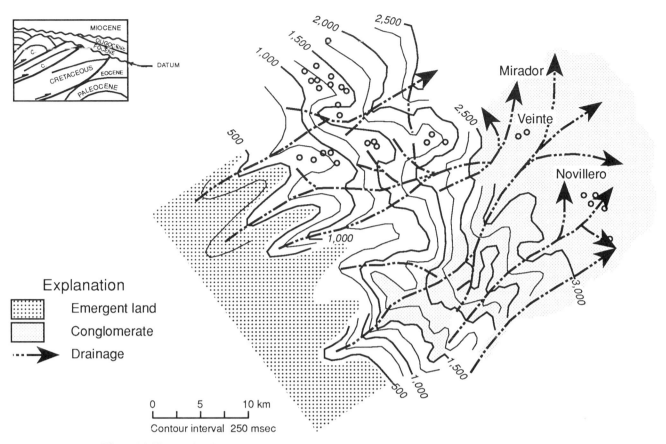

Figure 44. The pre–late Eocene eroded surface and the Papaloapan paleocanyon, Veracruz basin. From Cruz Helú and others, 1977.

horizontal-displacement faults, resulted in subsidence of the basin, rapid sedimentation, and the formation of growth faults in the Veracruz basin. At the same time, the eastern part of the basin was uplifted by the initial episodes of volcanic activity along the eastern part of the Trans-Mexican Neovolcanic Belt, and a deep depocenter was formed.

The Lower and Middle Miocene section of the Veracruz basin is composed of the interbedded sandstone-and-shale sequences of the upper part of the La Laja and the "Deposito" Formations. Uplift of the eastern part of the Veracruz basin and subsidence of its central part continued during the Late Miocene. The "Encanto" sandstone-shale section, rich in benthonic foraminifera, was deposited at the time unconformably over older beds.

The Early Pliocene is represented by the Middle neritic to upper bathyal bentonitic shales and sandstones of the "Concepción Inferior" Formation, and the Middle Pliocene by the shallower, middle neritic, sandstone-shale section of the "Concepción Superior." They were overlain, during the Pleistocene, by a lithologically similar sequence, the "Paraje Solo" Formation.

Volcanic activity along the Trans-Mexican Neovolcanic Belt accelerated during the Pliocene resulting in the uplift of the land area to the west, the termination of the Cenozoic sedimentary

cycle in the Veracruz basin, and the development of the present appearance of the coast line.

Southeastern basins. In southeastern Mexico, three Cenozoic basins have been recognized, from west to east: the Isthmus Saline, the Comalcalco, and the Macuspana basins (see Plate 2). Cenozoic rocks crop out in the Sierra de Chiapas, to the southeast (Quezada Muñetón, 1987 [1990]). Subsurface geological information as well as information provided by seismic surveys, permit the recognition in the southeastern basins of four genetically related sedimentary sequences (Ricoy, 1989).

The first sequence (I, Fig. 45) includes the Paleocene, Eocene, and Oligocene sediments, derived from the erosion of the rocks of the Chiapas massif and the Sierra de Chiapas during the tectonic events of the early Cenozoic. In the southeastern basins the Paleocene to Oligocene section is predominantly composed of shales (the Nanchital Shale and the La Laja Formation). Along the western front of the Sierra de Chiapas the Eocene-Oligocene sequence contains lenticular conglomerates probably deposited as submarine fans by turbidity currents. In the central part of the Sierra de Chiapas the Paleocene is composed of sandstones and shales (flysch), that grade northeastward to platform limestones. The Eocene is composed of platform limestones with shale inter-

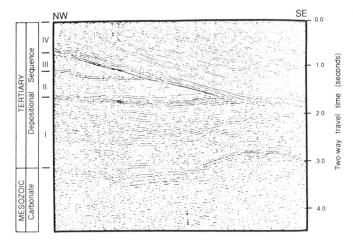

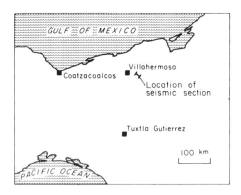

Figure 45. Seismic section showing Tertiary sedimentary sequences, Comalcalco Basin. From Ricoy, 1989.

calations; toward the southeast, the section grades to red beds. The Oligocene is poorly represented in the Sierra de Chiapas and is only locally identified as platform limestones with some shale intercalations in the central and northern part of the area.

In the Gulf of Campeche a dolomitized limestone breccia is present straddling the Cretaceous/Paleocene boundary. It extends southward to part of the states of Chiapas and Tabasco. The breccia is transitionally covered by argillaceous limestones and marls, which, in turn, grade upward, to fine-grained terrigenous sediments—shales and bentonitic shales of middle and late Paleocene age. The overlying Eocene is composed of bentonitic greenish-gray shales with pyrite grains. In the northeastern part of the Gulf of Campeche it contains calcarenitic intercalations deposited by turbidity currents flowing from the western margin of the Yucatan platform. The Oligocene is predominantly composed of shales, partly bentonitic, containing variable amounts of sand-size clastics and occasional lenticular beds of conglomerate.

The second sequence (II, Fig. 45) is related to an early Miocene tectonic event (the "Chiapaneca orogeny"), and comprises the "Deposito" and "Encanto" units of Miocene age. These units are generally composed of a monotonous alternation of sandstones and bluish gray, fossiliferous shales, probably depos-

ited as submarine fans in the bathyal zone. Interbedded in these units it is common to find tuffaceous horizons that provide evidence of contemporaneous volcanic activity. In the eastern part of the southeastern basins, along the northern part of the Sierra de Chiapas, the sequence is composed of marly shales, sandstones, bioclastic limestones, and sandy limestones with an abundant benthonic fauna (the Macuspana Limestone). In the Macuspana basin and in the Gulf of Campeche, the second sequence is predominantly composed of gray, fossiliferous shales with some intercalations of fine- to medium-grained sandstones.

The third sequence (III, Fig. 45) is associated to the Cascadian tectonic event and includes the so-called "Concepción Inferior," "Concepción Superior," and "Filisola" units, deposited during the Pliocene. This sedimentary sequence is composed of gray, fossiliferous shales with abundant intercalations of micaceous quartz sandstones deposited in a platform environment. Also characteristic of these units is their content of bentonitic material and volcanic-ash beds.

The fourth sequence (IV, Fig. 45) includes the "Paraje Solo," "Agueguexquite," and "Cedral" units, deposited during the Pleistocene. In general, they are composed of a gradational sequence of shales and sandstones with occasional conglomerates; they prograde from southeast to northwest in the Isthmus of Tehuantepec area. This sequence is related to fluvial and deltaic deposits of the present Coatzacoalcos, Uzpanapa, Grijalva, Usumacinta, and other rivers of the Isthmus region.

THE FLORIDA AND YUCATAN PLATFORMS

During the Cenozoic, as before during the Mesozoic, carbonate deposition prevailed in the Florida and Yucatan stable platforms. Evaporites are an important component of the Paleocene and lower Eocene sections of both platforms, but infrequent above the middle Eocene. Also common to the Florida and Yucatan platforms are two distinct stratigraphic hiatuses and unconformities, one between the Cretaceous and the Tertiary, and another between the Oligocene and Miocene sequences. As a result of the second unconformity, the Oligocene has a restricted distribution in both platforms due to either nondeposition or erosion.

The main difference between the Florida and Yucatan platforms is the lithologic composition of their upper Cenozoic stratigraphic section: while the Yucatan platform, located a long distance from possible sources of terrigenous clastic sediments, contains an upper Cenozoic section entirely composed of carbonates, the equivalent section in the Florida platform is dominated by siliciclastic sequences derived from the southern Appalachian Mountains.

Summary descriptions of the stratigraphy of the Cenozoic of the Florida platform have been published by Toulmin (1955), Puri and Vernon (1964) and Rainwater (1971). In spite of their somewhat outdated stratigraphic nomenclature, their summaries provide a valuable overview of the Cenozoic section of the Florida platform. Murray and Weidie (1962), Bonet and Butterlin

(1962), and Lopez Ramos (1973, 1975) have provided the most complete summaries of the Cenozoic stratigraphy of the Yucatan Peninsula.

The Florida Platform

Introduction. The Cenozoic stratigraphy of the Florida platform is characterized by marine deposition on a shallow-water platform. Much of the Florida platform remained isolated from the influx of significant amounts of terrigenous sediments during the Paleogene. Currents traversing the Suwanne straits/ Gulf trough (Figs. 46 and 47) effectively isolated the platform by transporting siliciclastic sediments away from the carbonate-producing environments of peninsular Florida, resulting in a vast carbonate platform (Figs. 15, 19, and 21). West of the straits, in the central and western Florida panhandle, terrigenous siliciclastic sediments mixed with the carbonates were deposited; they grade westward into less carbonate-bearing siliciclastics.

By the late Paleogene, carbonate sedimentation occurred over the entire Florida platform in response to a decreased influx of siliciclastic sediments (Fig. 21). These carbonate sediments function as the state's major aquifer system, the Floridan aquifer system (see Chapter 17 of this volume).

Near the end of the Paleogene, renewed uplift of the Appalachians dramatically altered the siliciclastic sediment supply reaching the Florida platform (Stuckey, 1965; Scott, 1988). The Gulf trough was filled by siliciclastics and these sediments began to infringe on the carbonate-depositing environments of the platform. During the early Neogene, siliciclastic-dominated sedimentation slowly migrated southward, replacing the vast areas of carbonate sedimentation (Fig. 25). The influx of the siliciclastic sediments proceeded more rapidly along the east coast of Florida under the influence of the Atlantic coastal processes than along the Gulf of Mexico coast. By the end of the Miocene, most of the Florida platform was covered by siliciclastic sediments leaving only limited areas of primarily carbonate deposition.

The siliciclastic sediment supply diminished in the latter Neogene and into the Quaternary. As a result, carbonate sedimentation was reestablished in southernmost Florida and persists to the present.

Paleogene. Paleogene sediments lie unconformably on Upper Cretaceous sediments throughout the Florida platform as a result of the lowering of sea level at the end of the Cretaceous that exposed the platform (see Chapter 10 of this volume). The topography of the pre-Cenozoic unconformity reflects the structures that controlled the deposition of the lower Cenozoic sediments. These structures included a major positive feature, the Peninsular arch, surrounded by three negative features: the South Florida basin to the south, the Southeast Georgia embayment to the northeast, and the Apalachicola embayment to the northwest (Fig. 46).

According to Chen (1965) the Paleocene to Middle Eocene sediments of the Clayton Formation and the Wilcox and Claiborne Groups (Fig. 2) in much of the Florida panhandle consist of variably fossiliferous, calcareous shales and sandstones, and argillaceous and arenaceous carbonates (Fig. 15). Glauconite is often present. Based on limited paleontologic data, Huddlestun and others (1988) indicated that unconformable contacts are present within the Paleogene siliciclastic section. The thickness of the siliciclastic Paleocene to Middle Eocene section ranges from approximately 300 to nearly 1,000 m; Chen, 1965; Huddlestun and others, 1988). Late Paleogene carbonates overlie and in part are laterally equivalent to the mixed siliciclastic-carbonate sediments in much of the panhandle.

In the eastern panhandle and the Florida Peninsula, carbonates and associated evaporites comprise the entire Paleogene section. The Cedar Keys Formation and the lower portion of the Oldsmar Formation (the Paleocene and lower portion of the Lower Eocene; Fig. 2), consist primarily of dolostone with minor amounts of limestone (Fig. 15). Anhydrite as pore-fillings and beds is present within the section indicating periods of restricted circulation on the Florida platform (Chen, 1965). Thick anhydrite beds may be present locally (Miller, 1986). The upper portion of the Oldsmar Formation and the Avon Park Formation (the upper part of the Lower and the Middle Eocene section) are composed of interbedded limestone and dolostone with varying occurrences of anhydrite and gypsum (Miller, 1986). Regionally, the ratios of dolostone to limestone vary significantly. This section is often abundantly fossiliferous, containing both microfossils and macrofossils (Applin and Applin, 1944; Cole, 1944; Applin and Jordan, 1945; Vernon, 1951; Miller, 1986). The upper part of the Avon Park Formation often contains carbonized marine grass specimens and unidentifiable carbonized fragments with occasional, thin, lignitic beds.

The distribution of anhydrite and gypsum in the Paleogene sediments is highly variable. The evaporites occur most widely in the Cedar Keys and Oldsmar Formations. Occurrences in the Avon Park Formation are common but not widespread. In many areas, the anhydrite and gypsum have been dissolved as the freshwater lens of the Floridan aquifer system expanded. Anhydrite and gypsum are rarely reported from sequences younger than Middle Eocene.

Depositional environments during the Paleocene through Middle Eocene ranged from subtidal to supratidal (Randazzo, 1976), producing grainstones to mudstones. Minor sea-level fluctuations drastically altered the depositional environments and developed numerous diastems. The major formational units are separated by paraconformities. Post-depositional diagenesis extensively affected the sediments (Randazzo and Zachos, 1974; Randazzo, 1976; Thayer and Miller, 1984). Dolomitization has been extensive resulting in very fine grained, relatively impermeable dolostones to very coarse grained, often vugular to cavernous, extremely permeable dolostones. The induration of the sediments varies dramatically from very poorly indurated to very hard and dense. Chert is sporadically present in these sediments. Thickness of the Paleocene through Middle Eocene section ranges from 500 to 1,400 m (Chen, 1965).

By the end of the Early Eocene time, the Peninsular arch no

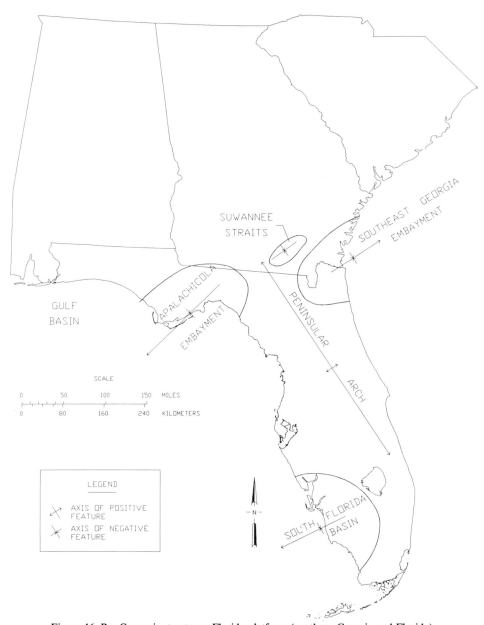

Figure 46. Pre-Cenozoic structures, Florida platform (southern Georgia and Florida).

longer affected sediment deposition in peninsular Florida. The negative areas surrounding it continued to be basins although somewhat modified. Younger positive features began to affect deposition. These include the Chattahoochee anticline in the central panhandle and, in the peninsula, the Ocala platform (formerly called Ocala uplift), and the Sanford high with its associated extensions to the north and south (Fig. 47). Upper Middle Eocene carbonates, the oldest sediments exposed in Florida, crop out on the crest of the Ocala platform.

By the end of the Middle Eocene, the carbonate depositional environments had spread north and west in response to a diminished siliciclastic supply (Fig. 19). Upper Eocene and Oligocene

carbonates are predominantly fossiliferous limestones ranging from grainstones to wackestones. Dolostones are locally dominant. Siliciclastics are only a minor part of the section, they are most abundant in the Oligocene section, which becomes increasingly rich in siliciclastics westward in the Florida panhandle (Miller, 1986). Chert is a very common component of the Upper Eocene and Oligocene carbonates in and near their outcrop areas. Induration is highly variable and is often related to the strata's position relative to the present surface.

The Upper Eocene and Oligocene carbonates are often highly fossiliferous; most common are foraminifera, echinoids, and mollusks (Cushman, 1921; Vaughan, 1928; Gravell and

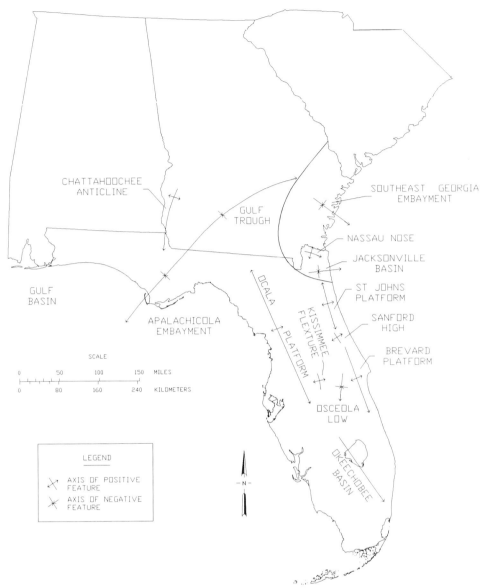

Figure 47. Mid- to late Eocene structures, Florida platform (southern Georgia and Florida).

Hanna, 1938; Cole, 1934, 1942; Vernon, 1942; Applin and Jordan, 1945; Puri, 1957; Akers and Drooger, 1957; Yon and Hendry, 1972). Deposition of the Upper Eocene and Oligocene carbonates occurred in open to marginal marine, subtidal environments (Cheetham, 1963; Randazzo, 1972; Miller, 1986). Their thickness ranges from zero to more than 250 m (Miller, 1986).

The Upper Eocene carbonates are known as the Ocala Group, those of the Oligocene as the Suwannee Limestone. Both the Upper Eocene and the Oligocene sedimentary sections were most likely deposited over virtually the entire Florida platform (Figs. 19 and 21). They were later eroded, however, during a period of exposure and erosion of the platform during the late

Oligocene or early Miocene. The Upper Eocene Ocala Group was removed from the crest of the Ocala Platform and the Sanford high; the Suwannee Limestone is absent, also due to erosion, from much of the northern two-thirds of the Florida Peninsula. The Miocene beds deposited above this important unconformity, therefore, covered sediments ranging in age from late Middle Eocene on the crest of the Ocala platform to Oligocene on its flanks and in the southern part of the peninsula.

Neogene. Dramatic changes in the depositional environments of the Florida platform occurred at the end of the Paleogene or beginning of the Neogene. Siliciclastic sediments rapidly encroached upon the carbonate-producing environments creating mixed carbonate-siliciclastic sediments (Fig. 25). Carbonates dominated in the Lower Miocene Chattahoochee and St. Marks Formations and the basal part of the Hawthorn Group (Fig. 2). During the time of maximum siliciclastic supply, virtually all sediments deposited over the Florida platform contained a considerable siliciclastic fraction or were principally siliciclastic deposits. From late Early Miocene through Pliocene, siliciclastic sedimentation dominated in the Alum Bluff and Hawthorn Groups and in the Jackson Bluff, Citronelle, Miccosukee, and Cypresshead Formations (Fig. 2).

Another significant change in the Florida platform during the Miocene included the onset of phosphogenesis. Riggs (1984; Chapter 16 of this volume) defined the phosphogenic episode of the southeastern United States. Virtually all the Neogene sediments in much of the eastern Florida panhandle and the peninsular area contain phosphate grains, occasionally in economically important concentrations (see Plate 4). The diagenetically formed phosphate grains were reworked extensively and deposited in the Neogene and Quaternary sediments with the greatest accumulations occurring in sediments of the Hawthorn Group (Scott, 1988; Compton and others, 1990).

During the Neogene, sediments deposited in environments ranging from subtidal to supratidal and deltaic resulted in an extremely variable lithologic framework (Goodell and Yon, 1960). Plate 5 shows this variable lithologic composition. Diastems are very common and paraconformities separate the major lithostratigraphic units. Lithologies range from pure clays to pure carbonates although the end-member lithologies are not common. Neogene sediments are often very fossiliferous, containing abundant mollusks, echinoids, corals, barnacles, foraminifera, and others.

The distribution of Neogene sediments reflects the depositional and erosional controls exerted by Florida's Cenozoic structures as well as by numerous fluctuations of sea level. Erosion removed the Neogene sediments deposited on the Chattahoochee anticline and the Ocala platform and at least partially from the Sanford high (Fig. 47). Neogene sediments extend across much of southern Florida and erosionally offlap the positive features in the northern peninsula and central panhandle. The Neogene sediment thickness ranges from 0 to more than 300 m in the peninsula and from 0 to more than 400 m in the panhandle.

Neogene siliciclastic sediments often exhibit a calcareous to dolomitic matrix. They contain clay to gravel-sized grains, range from very poorly sorted to well sorted and vary from unindurated to well indurated. The highlands in the panhandle and the peninsula are composed of late Neogene siliciclastic sediments of the Citronelle, Miccosukee, and Cypresshead Formations.

Carbonates deposited during the Neogene generally contain a significant siliciclastic fraction. In the Hawthorn Group, dolostone predominates in the northern two-thirds of the peninsula grading southward and westward into limestones. Pervasive diagenesis of the Hawthorn Group carbonates and carbonate-bearing sediments resulted in the destruction of most fossils in the Miocene section of much of the Florida Peninsula and eastern-

most panhandle. The Pliocene carbonates of the Tamiami Formation are characteristically composed of limestone exhibiting varying degrees of recrystallization and may be very fossiliferous. The dolostones and limestones range from very fine to very coarse grained and are poorly to well indurated. The limestones vary from mudstones to grainstones.

Quaternary. Quaternary sediments cover much of Florida. Marine siliciclastic sediments continued to be deposited in the coastal areas of the panhandle and the northern and central peninsula with mixed siliciclastic and carbonate sediments in the southern peninsula and carbonates along the southern end of the peninsula and in the keys (Fig. 28). Alluvial and eolian agents reworked and transported siliciclastic sediments in areas above sea level. The marine and eolian agents are responsible for the development of numerous dune fields and beach-ridge plains (see Chapter 2, this volume).

Although it is known that sea levels fluctuated through the Quaternary, the amount of fluctuation above present sea level is not easily determined. Early investigators speculated that Quaternary sea levels reached as high as 65 m above the present level (Vernon, 1942; Cooke, 1945). More recent investigations have suggested that Quaternary sea levels did not exceed 30 m above the present level (Altshuler and Young, 1960; Alt and Brooks, 1965). Haq and others (1987) indicate short-term eustatic sea-level fluctuations in the Quaternary of less than 25 m. Fossiliferous Quaternary sediments in peninsular Florida have not been identified from elevations exceeding 20 m.

The Quaternary sediments are often not differentiated into formational units in much of Florida, except along the east coast and in southern Florida. Where undifferentiated, the sediments are unfossiliferous siliciclastics ranging from clays to gravels in parts of the panhandle, and clays to medium-grained sands in much of the peninsula. Mineralogically, the siliciclastic sediments are primarily composed of clay minerals and quartz.

In southern Florida and along the east coast, the lithologies of the Quaternary are quite varied and often have a significant fossil content. The lithostratigraphic units recognized in this area include the Caloosahatchee, Fort Thompson, and Anastasia Formations and the Miami and Key Largo Limestones (Fig. 2). Sediments comprising these Quaternary units include clean sands, fossiliferous sands, and sandy, fossiliferous to unfossiliferous limestones. Ooids and fossil coral patch reefs are important components of the southern Florida Quaternary carbonate units. Fossils present include principally mollusks, echinoids, bryozoans, corals, and foraminifera.

Depositional environments of the marine Quaternary sediments ranged from very shallow water to the shoreface. The shallow-water environments of southern Florida allowed the development of reefs, carbonate muds, and very fossiliferous carbonates (Hoffmeister, 1974). Northward along the east coast the carbonates grade into fossiliferous siliciclastic sediments. In the panhandle, Quaternary sediments are generally unfossiliferous siliciclastics. Numerous diastems occur as a result of the sea-level fluctuations, currents, and storms. Huddlestun and others (1988) noted a number of unconformities separating the major identified Quaternary formations.

The Quaternary sediments in Florida range in thickness from a thin veneer to more than 50 m. However, the average thickness is generally less than 20 m.

Offshore Florida platform: Gulf of Mexico

The exposed Florida Peninsula constitutes less than one-half of the Florida platform. A narrow, submerged shelf and slope exist on the Atlantic Ocean side of the platform, while a broad shelf and slope occur on the Gulf of Mexico side. The Gulf side formed as the result of Late Cretaceous subsidence from a shallow-water shelf to a deep deep-water basin (Mitchum, 1978).

The investigation of the West Florida Slope by Mitchum (1978) reveals some correlations with the sediments underlying the exposed portion of the platform. The Cenozoic sediments of the West Florida Slope are deep-water deposits while those on-shore are shallow-water deposits. The sediments under the slope reflect the geologic history of the sediments beneath the exposed portion of the platform, with predominantly carbonate sediments occurring throughout the Paleogene, and with increasing amounts of siliciclastic sediments in the Neogene and Quaternary. Within the post-Paleocene sediments there is a trend of decreasing siliciclastic content from north to south. It is interesting to note that, according to Mitchum (1978), the influx of siliciclastic sediments that occurred during the Early Miocene in the shallow-water sediments first affected the deep-water sediments in the Middle Miocene.

The rate of sediment accumulation across the Florida platform shows a great variability. The Paleogene carbonate sequence under the West Florida Slope varies from approximately 300 to 600 m while the carbonate buildup in the shallow water depositional environments deposited 500 to more than 1,600 m of carbonates. Offshore, the Neogene and Quaternary sediment thickness ranges from 0 to 1,500 m while onshore the maximum thickness is approximately 450 m.

The Yucatan platform

The stratigraphy of the Cenozoic section of the Yucatan platform is not as well known as that of the Florida platform. Only a few wells have been drilled through the section in the Yucatan Peninsula and information on them is scarce.

The Cenozoic of the Yucatan platform ranges in thickness from less than 100 m in the southern part of the Yucatan Peninsula, in northern Guatemala and Belize, to 1,000 m in the northern part of the peninsula, and to about 2,000 m along the western margin of the Yucatan shelf, under the waters of the Gulf of Mexico. It is predominantly composed of carbonates with lesser amounts of interbedded evaporites, particularly in the Paleocene and Eocene. Terrigenous clastics are entirely absent from the Cenozoic section.

The Paleocene is represented by the lagoonal carbonates and evaporites of the Icaiché Formation. There is little information pertaining to the foraminiferal content and thickness of this unit. Toward the northern part of the Yucatan Peninsula, where the Paleocene becomes shalier, it is possible to recognize locally the *Globorotalia velascoensis* Biozone in a sequence of argillaceous limestone with interbedded marls and shales. The Paleocene was deposited over an irregular surface of eroded Upper Cretaceous rocks, indicating the presence, also in the Yucatan platform, of a stratigraphic hiatus between the Cretaceous and the Tertiary that has been interpreted to have been caused by a lowering of the sea level.

The Eocene has been referred to as the Chichén Itzá Formation, a section of shallow-water, light-colored, massive and fossiliferous limestones which has been subdivided into three members, from bottom to top:

- Xbacal Member, composed of limestones and marls with abundant *Nummulites catenula* (Cushman and Jarvis), *Discocyclina (Discocyclina) cristensis* (Vaughan), *Coskinolina elongata* Cole, and numerous miliolids of Early Eocene age.
- Pisté Member, composed of thick-bedded limestones with calcareous algae, foraminifera of the genera *Coskinolina, Dictyoconus,* and *Amphistegina,* and abundant miliolids of Middle Eocene age.
- Chumbec Member, composed of massive limestones with *Nummulites wilcoxi* (Heilprim), *Lepidocyclina (Pliolepidina) pustulosa* Douville, *Amphistegina* cf. *lopeztrigoi* Palmer, and other miliolids representative of the upper part of the Middle Eocene and of the Upper Eocene.

The Oligocene has an irregular distribution in the Yucatan platform and has not been given a formational name. It is composed of a sequence of light-colored, thick-bedded, poorly stratified calcarenites with intercalations of marls and shales that contain *Lepidocyclina (Eulepidina) undosa* Cushman.

The Miocene-Pliocene section, the Carrillo Puerto Formation, overlies unconformably the Eocene or, when present, the Oligocene. The occurrence in the Yucatan platform of an unconformity at the base of the Miocene, also recognized in the Florida platform, may indicate a lowering of the sea level during the late Oligocene or Early Miocene.

The Carrillo Puerto Formation is composed of light-colored, fossiliferous limestones, argillaceous limestones, and marls containing a shallow-water fauna composed of *Peneroplis proteus* d'Orbigny, *Archaias angulatus* (Fichtell and Moll), *Gypsina pilans* d'Orbigny, *Sorites* sp., and *Amphistegina* sp.

The Upper Pliocene-Pleistocene section is characterized by cream-colored bioclastic packestones and calcarenites.

REFERENCES CITED

Akers, W. H., 1955, Some planktonic foraminifera of the American Gulf Coast and suggested correlations with the Caribbean Tertiary: Journal of Paleontology, v. 29, p. 647–664.

——, 1972, Planktonic foraminifera and biostratigraphy of some Neogene formations, northern Florida, and Atlantic Coastal Plain: Tulane Studies in Geology and Paleontology, v. 9, 139 p.

——, 1979, Planktic foraminifera and calcareous nannoplankton biostratigraphy of the Neogene of Mexico; Part I, Middle Pliocene: Tulane Studies in Geology and Paleontology, v. 15, no. 1, p. 1–32.

——, 1981, Planktic foraminifera and calcareous nannoplankton biostratigraphy of the Neogene of Mexico; Addendum to Part I, Some additional Mid-Pliocene localities and further discussion on the Agueguexquite and Concepcion Superior beds: Tulane Studies in Geology and Paleontology, v. 16, nos. 3–4, p. 145–148.

——, 1984, Planktic foraminifera and calcareous nannoplankton biostratigraphy of the Neogene of Mexico; Part II, Lower Pliocene: Tulane Studies in Geology and Paleontology, v. 18, nos. 1–2, p. 21–36.

Akers, W. H., and Dorman, J. H., 1964, Pleistocene foraminifera of the Gulf Coast: Tulane Studies in Geology, v. 3, no. 1, p. 1–93.

Akers, W. H., and Drooger, C. W., 1957, Miogypsinids, planktonic foraminifera, and Gulf Coast Oligocene-Miocene correlations: American Association of Petroleum Geologists Bulletin, v. 41, p. 656–678.

Akers, W. H., and Holck, A.J.J., 1957, Pleistocene beds near the edge of the continental shelf, southeastern Louisiana: Geological Society of America Bulletin, v. 68, p. 983–992.

Albers, C. C., 1966, Foraminiferal ecologic zones of the Gulf Coast: Gulf Coast Association of Geological Societies Transactions, v. 16, p. 345–354.

Alt, D., and Brooks, H. K., 1965, Age of the Florida marine terraces: Journal of Geology, v. 73, p. 406–411.

Altschuler, Z. S., and Young, E. J., 1960, Residual origin of the "Pleistocene" sand mantle in central Florida uplands and its bearing on marine terraces and Cenozoic uplift: U.S. Geological Survey Professional Paper 400-B, p. 202–207.

Andersen, H. V., 1961, Genesis and paleontology of the Mississippi River mudlumps; Part II, Foraminifera of the mudlumps, lower Mississippi River delta: Louisiana Department of Conservation Geological Bulletin 35, 208 p.

Applin, P. L., and Applin, E. R., 1944, Regional subsurface stratigraphy and structure of Florida and southern Georgia: American Association of Petroleum Geologists Bulletin, v. 28, p. 1673–1753.

Applin, E. R., and Jordan, L., 1945, Diagnostic foraminifera from subsurface formations in Florida: Journal of Paleontology, v. 19, p. 129–148.

Applin, E. R., Ellisor, A. E., and Kniker, H. T., 1925, Subsurface stratigraphy of the Coastal Plain of Texas and Louisiana: American Association of Petroleum Geologists Bulletin, v. 9, p. 79–122.

Armentrout, J. M., and Clement, J. F., 1990, Biostratigraphic calibration of depositional cycles; A case study in High Island–Galveston–East Breaks area, offshore Texas, *in* Sequence stratigraphy as an exploration tool, concepts and practice in the Gulf Coast: Society of Economic Paleontologists and Mineralogists, Gulf Coast Section, Eleventh Annual Research Conference Program and Abstracts, p. 21–51.

Baker, E. T., Jr., 1978, Stratigraphic and hydrogeologic framework of part of the Coastal Plain of Texas: U.S. Geological Survey Open-File Report 77-712, 32 p.

Bandy, O. L., 1949, Eocene and Oligocene foraminifera from Little Stave Creek, Clarke County, Alabama: Bulletins of American Paleontology, v. 32, no. 131, 210 p.

Bandy, O. L., and Arnal, R. E., 1957, Distribution of recent foraminifera off west coast of Central America: American Association of Petroleum Geologists Bulletin, v. 41, p. 2037–2053.

——, 1960, Concepts of foraminiferal paleoecology: American Association of Petroleum Geologists Bulletin, v. 44, p. 1921–1932.

Barker, R. W., and Blow, W. H., 1976, Biostratigraphy of some Tertiary forma-

tions in the Tampico-Misantla Embayment, Mexico: Journal of Foraminiferal Research, v. 6, p. 39–58.

Beard, J. H., 1969, Pleistocene paleotemperature record based on planktonic foraminifers, Gulf of Mexico: Gulf Coast Association of Geological Societies Transactions, v. 19, p. 535–553.

Beard, J. H., Boellstorff, J., Menconi, L. C., and Stude, J. R., 1976, Fission-track age of Pliocene volcanic glass from the Gulf of Mexico: Gulf Coast Association of Geological Societies Transactions, v. 26, p. 156–163.

Beard, J. H., Sangree, J. B., and Smith, L. A., 1982, Quaternary chronology, paleoclimate, depositional sequences, and eustatic cycles: American Association of Petroleum Geologists Bulletin, v. 66, p. 158–169.

Bebout, D. G., Loucks, R. G., and Gregory, A. R., 1978, Frio sandstone reservoirs in the deep subsurface along the Texas Gulf Coast, their potential for the production of geopressured geothermal energy: The University of Texas at Austin, Bureau of Economic Geology Report of Investigations 91, 100 p.

Benson, P. H., 1971, Geology of the Oligocene Hackberry trend, Gillis English Bayou–Manchester area, Calcasieu Parish, Louisiana: Gulf Coast Association of Geological Societies Transactions, v. 21, p. 1–14.

Berg, R. R., and Powers, B. K., 1980, Morphology of turbidite-channel reservoirs, lower Hackberry (Oligocene), southeast Texas: Gulf Coast Association of Geological Societies Transactions, v. 30, p. 41–48.

Berggren, W. A., 1972, Late Pliocene-Pleistocene glaciation, *in* Initial reports of the Deep Sea Drilling Project, v. XII: Washington, D.C., U.S. Government Printing Office, p. 953–963.

Berggren, W. A., Kent, D. V., and Van Couvering, J. A., 1985, Neogene geochronology and chronostratigraphy, *in* Snelling, N.J., ed., Geochronology and the geologic time scale: Geological Society of London Memoir 10, pt. 2, p. 211–260.

Bergquist, H. R., 1942, Scott County fossils, Jackson foraminifera and ostracoda: Mississippi Geological Survey Bulletin 49, p. 1–146.

Bethke, C. M., Harrison, W. J., Upton, C., and Altaner, S. P., 1988, Supercomputer analysis of sedimentary basins: Science, v. 239, p. 261–267.

Boltovskoy, E., 1980, On the benthonic bathyal-zone foraminifera as stratigraphic guide fossils: Journal of Foraminiferal Research, v. 10, p. 163–172.

Bonet, F., and Butterlin, J., 1962, Stratigraphy of the northern part of the Yucatan Peninsula, *in* Guidebook to field trip to Yucatan Peninsula: New Orleans Geological Society, p. 52–57.

Booth, J. S., 1979, Recent history of mass-wasting on the upper continental slopes, northern Gulf of Mexico, as interpreted from the consolidation states of the sediment, *in* Doyle, L. J., and Pilkey, O. H., Jr., eds., Geology of continental slopes: Society of Economic Paleontologists and Mineralogists Special Publication 27, p. 153–164.

Bornhauser, M., 1947, Marine sedimentary cycles of Tertiary in Mississippi Embayment and central Gulf Coast area: American Association of Petroleum Geologists Bulletin, v. 31, p. 698–712.

Bouma, A. H., 1982, Intraslope basins in northwest Gulf of Mexico; A key to ancient submarine canyons and fans, *in* Watkins, J. S., and Drake, C. L., eds., Studies in continental margin geology: American Association of Petroleum Geologists Memoir 34, p. 567–581.

Bouma, A. H., Moore, G. T., and Coleman, J. M., eds., 1976, Beyond the shelf break: American Association of Petroleum Geologists Marine Geology Committee Short Course, New Orleans, Louisiana, v. 2.

Boyd, D. B., and Dyer, B. F., 1964, Frio barrier bar system of South Texas: Gulf Coast Association of Geological Societies Transactions, v. 14, p. 309–322.

Bradshaw, J. S., 1959, Ecology of living planktonic foraminifera in the north and equatorial Pacific Ocean: Contributions from the Cushman Laboratory for Foraminiferal Research, v. 10, part 2, p. 25–64.

Bredehoeft, J. D., and Hanshaw, B. B., 1968, On the maintenance of anomalous fluid pressures; I, Thick sedimentary sequences: Geological Society of America Bulletin, v. 79, p. 1097–1106.

Bruce, C. H., 1973, Pressured shale and related sediment deformation; Mechanism for development of regional contemporaneous faults: American Association of Petroleum Geologists Bulletin, v. 57, p. 878–886.

Brunner, C. A., and Keigwin, L. D., 1981, Late Neogene biostratigraphy and stable isotope stratigraphy of a drilled core from the Gulf of Mexico: Marine Micropaleontology, v. 6, p. 397–418.

Burke, K., 1972, Longshore drift submarine canyons, and submarine fans in development of Niger Delta: American Association of Petroleum Geologists Bulletin, v. 56, p. 1975–1983.

Busch, D. A., 1973, Oligocene studies, northeast Mexico: Gulf Coast Association of Geological Societies Transactions, v. 23, p. 136–145.

Busch, D. A., and Govela S., A., 1978, Stratigraphy and structure of Chicontepec turbidites, southwestern Tampico-Misantla Basin, Mexico: American Association of Petroleum Geologists Bulletin, v. 62, p. 235–246.

Butler, E. A., 1960, Paleontology of the L. L. & E. et al. Well Unit 1-L, No. 1: Louisiana Department of Conservation Folio Series 1.

—— , 1962, *Bigenerina humblei* and the Humble H. J. Ellender No. 1, Lirette Field, Terrebonne Parish, Louisiana: Gulf Coast Association of Geological Societies Transactions, v. 12, p. 271–282.

Bybell, L. M., 1982, Late Eocene to early Oligocene calcareous nannofossils in Alabama and Mississippi: Gulf Coast Association of Geological Societies Transactions, v. 32, p. 295–302.

Carrillo Bravo, J., 1980, Paleocañones terciarios de la planicie costera del Golfo de México: Asociación Mexicana de Geólogos Petroleros Boletín, v. 32, p. 27–55.

Chapin, C. E., 1979, Evolution of the Rio Grande Rift; A summary, *in* Riecker, R. E., ed., Rio Grande Rift; Tectonics and magmatism: Washington, D.C., American Geophysical Union, p. 1–5.

Chapin, C. E., and Cather, S. M., 1981, Eocene tectonics and sedimentation in the Colorado Plateau–Rocky Mountains area, *in* Dickinson, W. R., and Payne, W. D., eds., Relations of tectonics to ore deposits in the Southern Cordillera: Arizona Geological Society Digest, no. 14, p. 199–213.

Cheetham, A. H., 1963, Late Eocene zoogeography of the eastern Gulf Coast region: Geological Society of America Memoir 91, 113 p.

Chen, C. S., 1965, The regional lithostratigraphic analysis of Paleocene and Eocene rocks of Florida: Florida Geological Survey Bulletin 45, 105 p.

Chorley, R. J., and Kennedy, B. A., 1971, Physical geography, a systems approach: London, Prentice-Hall International, Inc., 370 p.

Chuber, S., and Begeman, R. L., 1982, Productive lower Wilcox stratigraphic traps from an entrenched valley in Kinkler field, Lavaca County, Texas: Gulf Coast Association of Geological Societies Transactions, v. 32, p. 255–262.

Cole, W. S., 1934, Oligocene orbitoids from near Duncan Church, Washington County, Florida: Journal of Paleontology, v. 8, p. 21–28.

—— , 1942, Stratigraphic and paleontologic studies of wells in Florida, No. 2: Florida Geological Survey Bulletin 20, p. 1–89.

—— , 1944, Stratigraphic and paleontologic studies of wells in Florida; No. 3, City of Quincy Water Well St. Mary's River Oil Corporation, Hilliard Turpentine Company No. 1 Well: Florida Geological Survey Bulletin 26, p. 11–168.

Cole, W. S., and Ponton, G. M., 1930, The foraminifera of the Marianna Limestone of Florida: Florida Geological Survey Bulletin 5, p. 19–69.

Coleman, James M., and Gagliano, S. M., 1964, Cyclic sedimentation in the Mississippi River deltaic plain: Gulf Coast Association of Geological Societies Transactions, v. 14, p. 67–80.

Coleman, James M., Prior, D. B., and Lindsay, J. F., 1983, Deltaic influences on shelfedge instability processes, *in* Stanley, D. J., and Moore, G. T., eds., The shelfbreak; Critical interface on continental margins: Society of Economic Paleontologists and Mineralogists Special Publication 33, p. 121–138.

Coleman, Janet M., and Galloway, W. E., 1990, Sequence stratigraphic analysis of the lower Oligocene Vicksburg Formation of Texas, *in* Society of Economic Paleontologists and Mineralogists Foundation, Gulf Coast Section, Eleventh Annual Research Conference Programs and Abstracts, p. 99–112.

Compton, J. S., Snyder, S. W., and Hodell, D. A., 1990, Phosphogenesis and weathering of shelf sediments from the southeastern United States; Implications of Miocene $\delta^{13}C$ excursions and global cooling: Geology, v. 18, p. 1227–1230.

Cooke, C. W., 1945, Geology of Florida: Florida Geological Survey Bulletin 29, 339 p.

Crouch, R. W., 1955, A practical application of paleontology in exploration: Gulf Coast Association of Geological Societies Transactions, v. 5, p. 89–96.

Cruz Helú, P., Verdugo V., R., and Barcenas P., R., 1977, Origin and distribution of Tertiary conglomerates, Veracruz Basin, Mexico: American Association of Petroleum Geologists Bulletin, v. 61, p. 207–226.

Culver, S. J., and Buzas, M. A., 1981, Distribution of recent benthic foraminifera in the Gulf of Mexico: Smithsonian Contributions to Marine Science, v. 1, p. 1–412, v. 2, p. 413–898.

Curtis, D. M., 1970, Miocene deltaic sedimentation, Louisiana Gulf Coast, *in* Morgan, J. P., ed., Deltaic sedimentation, modern and ancient: Society of Economic Paleontologists and Mineralogists Special Publication 15, p. 293–308.

Curtis, D. M., and Echols, D. J., 1985, Habitat of oil and gas in the middle Frio (Oligocene) Hackberry, *in* Perkins, B. F., and Martin, G. B., eds., Habitat of oil and gas in the Gulf Coast: Society of Economic Paleontologists and Mineralogists, Gulf Coast Section, Fourth Annual Research Conference Proceedings, p. 263–274.

Curtis, D. M., and Picou, E. B., Jr., 1978, Gulf Coast Cenozoic; A model of the application of stratigraphic concepts to exploration on passive margins: Gulf Coast Association of Geological Societies Transactions, v. 28, p. 103–120.

—— , 1980, Gulf Coast Cenozoic; A model for the application of stratigraphic concepts to exploration of passive margins: Canadian Society of Petroleum Geologists Memoir 6, p. 243–268.

Cushman, J. A., 1921, American species of Operculina and Heterostegina: U.S. Geological Survey Professional Paper 128-E, p. 125–137.

—— , 1922a, The Byram Calcareous Marl of Mississippi and its foraminifera: U.S. Geological Survey Professional Paper 129-E, 43 p.

—— , 1922b, The foraminifera of the Mint Springs Calcareous Marl Member of the Marianna Limestone: U.S. Geological Survey Professional Paper 129F, 30 p.

—— , 1923, The foraminifera of the Vicksburg Group: U.S. Geological Survey Professional Paper 133, p. 11–71.

—— , 1924, A new genus of Eocene foraminifera: U.S. National Museum Proceedings, v. 66, p. 1–4.

—— , 1925, Eocene foraminifera from the Cocoa Sand of Alabama: Contributions from the Cushman Laboratory for Foraminiferal Research, v. 1, pt. 1, p. 65–68.

—— , 1926, Some new foraminifera from the upper Eocene of the southeastern coastal plain of the United States: Contributions from the Cushman Laboratory for Foraminiferal Research, v. 2, pt. 2, p. 29–36.

—— , 1928, Additional foraminifera from the upper Eocene of Alabama: Contributions from the Cushman Laboratory for Foraminiferal Research, v. 4, pt. 3, p. 73–79.

—— , 1929, Notes on the foraminifera of the Byram Marl: Contributions from the Cushman Laboratory for Foraminiferal Research, v. 5, pt. 2, p. 40–48.

—— , 1931, Three new upper Eocene foraminifera: Contributions from the Cushman Laboratory for Foraminiferal Research, v. 7, pt. 3, p. 58–60.

—— , 1933, New foraminifera from the upper Jackson Eocene of the southeastern coastal plain region of the United States: Contributions from the Cushman Laboratory for Foraminiferal Research, v. 9, pt. 1, p. 1–21.

—— , 1935a, Upper Eocene foraminifera of the southeastern United States: U.S. Geological Survey Professional Paper 181, p. 1–81.

—— , 1935b, New species of foraminifera from the lower Oligocene of Mississippi: Contributions from the Cushman Laboratory for Foraminiferal Research, v. 11, pt. 2, p. 25–39.

—— , 1944, A foraminiferal fauna of the Wilcox Eocene, Bashi Formation from near Yellow Bluff, Alabama: American Journal of Science, v. 242, p. 7–18.

—— , 1946, A rich foraminiferal fauna from the Cocoa Sand of Alabama: Contributions from the Cushman Laboratory for Foraminiferal Research Special Paper 16, 40 p.

Cushman, J. A., and Applin, E. R., 1926, Texas Jackson Foraminifera: American Association of Petroleum Geologists Bulletin, v. 10, p. 154–189.

Cushman, J. A., and Ellisor, A. C., 1931, Some new Tertiary foraminifera from Texas: Contributions from the Cushman Laboratory for Foraminiferal Research, v. 7, pt. 3, p. 51–58.

——, 1933, Two new Texas foraminifera: Contributions from the Cushman Laboratory for Foraminiferal Research, v. 9, pt. 4, p. 95–96.

——, 1939, New species of foraminifera from the Oligocene and Miocene: Contributions from the Cushman Laboratory for Foraminiferal Research, v. 15, pt. 1, p. 1–14.

——, 1945, The foraminiferal fauna of the Anahuac Formation: Journal of Paleontology, v. 19, p. 545–572.

Cushman, J. A., and Garrett, J. B., 1938, Three new rotaliform foraminifera from the lower Oligocene and upper Eocene of Alabama: Contributions from the Cushman Laboratory for Foraminiferal Research, v. 14, pt. 3, p. 62–66.

——, 1939, Eocene foraminifera of Wilcox age from Woods Bluff, Alabama: Contributions from the Cushman Laboratory for Foraminiferal Research, v. 15, pt. 4, p. 77–89.

Cushman, J. A., and McGlamery, W., 1938, Oligocene foraminifera from Choctaw Bluff, Alabama: U.S. Geological Survey Professional Paper 189-D, p. 101–119.

——, 1939, New species of foraminifera from the lower Oligocene of Alabama: Contributions from the Cushman Laboratory for Foraminiferal Research, v. 15, pt. 3, p. 45–49.

Cushman, J. A., and Ponton, G. M., 1932, An Eocene foraminiferal fauna of Wilcox age from Alabama: Contributions from the Cushman Laboratory for Foraminiferal Research, v. 8, pt. 3, p. 51–72.

Cushman, J. A., and Thomas, N. L., 1929, Abundant foraminifera of the east Texas Greensands: Journal of Paleontology, v. 3, p. 176–184.

Cushman, J. A., and Todd, R., 1945a, A foraminiferal fauna from the Lisbon Formation of Alabama: Contributions from the Cushman Laboratory for Foraminiferal Research, v. 21, pt. 1, p. 11–21.

——, 1945b, Foraminifera of the type locality of the Moodys Marl Member of the Jackson Formation of Mississippi: Contributions from the Cushman Laboratory for Foraminiferal Research, v. 21, pt. 4, p. 79–105.

——, 1946, A foraminiferal fauna from the Byram Marl at its type locality: Contributions from the Cushman Laboratory for Foraminiferal Research, v. 22, pt. 3, p. 76–102.

Dickinson, W. R., 1981, Plate tectonic evolution of the southern Cordillera: Arizona Geological Society Digest, v. 14, p. 113–135.

Dingus, W. F., and Galloway, W. E., 1990, Morphology, paleogeographic setting, and origin of the middle Wilcox Yoakum Canyon, Texas Coastal Plain: American Association of Petroleum Geologists Bulletin, v. 74, p. 1055–1076.

Dixon, L. H., 1965, Cenozoic cyclic deposition in the subsurface of central Louisiana: Louisiana Geological Survey Bulletin 42, 124 p.

Dodge, M. M., and Posey, J. S., 1981, Structural cross sections, Tertiary formations, Texas Gulf Coast: The University of Texas at Austin, Bureau of Economic Geology, Cross sections 1-22 and A-D.

Doering, J., 1956, Review of Quaternary surface formations of Gulf Coast region: American Association of Petroleum Geologists Bulletin, v. 40, p. 1816–1862.

——, 1958, Citronelle age problem: American Association of Petroleum Geologists Bulletin, v. 42, p. 764–786.

Doyle, J. D., 1979, Depositional patterns of Miocene facies, middle Texas Coastal Plain: The University of Texas at Austin, Bureau of Economic Geology Report of Investigations 99, 28 p.

Dumble, E. T., 1894, The Cenozoic deposits of Texas: Journal of Geology, v. 2, p. 549–567.

Durham, C. O., Jr., Moore, C. H., Jr., and Parsons, B., 1967, An agnostic view of the terraces, Natchez to New Orleans, *in* Mississippi Alluvial Valley and Terraces, Field Trip Guidebook, 1967 Geological Society of America Annual meeting: Baton Rouge, Louisiana State University, Part E, 22 p.

Eaton, G. P., 1986, A tectonic redefinition of the Southern Rocky Mountains: Tectonophysics, v. 132, p. 163–193.

Echols, D. J., and Curtis, D. M., 1973, Paleontologic evidence for mid-Miocene refrigeration, from subsurface marine shales, Louisiana Gulf Coast: Gulf Coast Association of Geological Societies Transactions, v. 23, p. 422–426.

Edwards, M. B., 1981, The Upper Wilcox Rosita Delta System of south Texas; Record of growth-faulted shelf-edge deltas: American Association of Petro-

leum Geologists Bulletin, v. 65, p. 54–73.

Ellis, A. D., Jr., 1939, Significant foraminifera from the Chickasawhay Beds of Wayne County, Mississippi; Journal of Paleontology, v. 13, p. 423–424.

Ellisor, A. C., 1933, Jackson Group of formations in Texas with notes on the Frio and Vicksburg: American Association of Petroleum Geologists Bulletin, v. 17, p. 1293–1350.

——, 1940, Subsurface Miocene of southern Louisiana: American Association of Petroleum Geologists Bulletin, v. 24, p. 435–475.

——, 1944, Anahuac Formation: American Association of Petroleum Geologists Bulletin, v. 28, p. 1355–1375.

Emiliani, C., 1969, A new paleontology: Micropaleontology, v. 15, p. 265–300.

Emiliani, C., and 7 others, 1975, Paleoclimatological analysis of late Quaternary cores from the northeastern Gulf of Mexico: Science, v. 189, p. 1083–1088.

Epis, R. C., and Chapin, C. E., 1975, Geomorphic and tectonic implications of the post-Laramide, late Eocene erosion surface in the southern Rocky Mountains, *in* Curtis, B. F., ed., Cenozoic history of the southern Rocky Mountains: Geological Society of America Memoir 144, p. 45–74.

Ewing, T. E., and Reed, R. S., 1984, Depositional systems and structural controls of Hackberry sandstone reservoirs, *in* Southeast Texas: The University of Texas at Austin, Bureau of Economic Geology Geological Circular 84-7, 48 p.

Fisher, W. L., 1964, Sedimentary patterns in Eocene cyclic deposits, northern Gulf Coast region: Kansas Geological Survey Bulletin 169, p. 151–170.

Fisher, W. L., and McGowen, J. H., 1967, Depositional systems in the Wilcox Group of Texas and their relationship to occurrence of oil and gas: Gulf Coast Association of Geological Societies Transactions, v. 17, p. 105–125.

Fisher, W. L., Rodda, P. U., and Dietrich, J. W., 1964, Evolution of *Athleta petrosa* stock (Eocene gastropoda) of Texas: The University of Texas Publication 6413, 117 p.

Fisher, W. L., Proctor, C. V., Jr., Galloway, W. E., and Nagle, J. S., 1970, Depositional systems in the Jackson Group of Texas; Their relationship to oil, gas, and uranium: The University of Texas at Austin, Bureau of Economic Geology Geological Circular 70-4, 28 p.

Fisk, H. N., 1940, Geology of Avoyelles and Rapides Parishes: Louisiana Geological Survey Bulletin 18, 240 p.

——, 1944, Geological investigation of the alluvial valley of the Lower Mississippi River: U.S. Army Corps of Engineers, Mississippi River Commission, 78 p.

——, 1952, Geological investigation of the Atchafalaya Basin and the problem of Mississippi River diversion: U.S. Army Corps of Engineers, Waterways Experiment Station, v. 1, 145 p.

——, 1961, Bar-finger sands of the Mississippi Delta, *in* Geometry of Sandstone Bodies; A Symposium: American Association of Petroleum Geologists, Tulsa, Oklahoma, p. 29–52.

Flower, B. P., and Kennett, J. P., 1990, The Younger Dryas cool episode in the Gulf of Mexico: Paleoceanography, v. 5, p. 949–961.

Frazier, D. E., 1967, Recent deltaic deposits of the Mississippi River; Their development and chronology: Gulf Coast Association of Geological Societies Transactions, v. 17, p. 287–315.

——, 1974, Depositional episodes; Their relationship to the Quaternary stratigraphic framework in the northwestern portion of the Gulf Basin: The University of Texas at Austin, Bureau of Economic Geology Geological Circular 74-1, 28 p.

Freed, R. L., 1982, Clay mineralogy and depositional history of the Frio Formation in two geopressured wells, Brazoria County, Texas: Gulf Coast Association of Geological Societies Transactions, v. 32, p. 459–464.

Frost, S. H., and Langenheim, R. L., Jr., 1974, Cenozoic reef biofacies; Tertiary larger foraminifera and scleractinian corals from Chiapas, Mexico: De Kalb, Northern Illinois University Press, 388 p.

Galloway, W. E., 1968, Depositional systems of the lower Wilcox Group, north-central Gulf Coast Basin: Gulf Coast Association of Geological Societies Transactions, v. 18, p. 275–289.

——, 1977, Catahoula Formation of the Texas Coastal Plain; Depositional systems, composition, structural development, ground-water flow history, and uranium distribution: The University of Texas at Austin, Bureau of Economic Geology Report of Investigations 87, 59 p.

——, 1981, Depositional architecture of Cenozoic Gulf Coastal Plain fluvial systems, *in* Ethridge, F. G., and Flores, R. M., eds., Nonmarine depositional environments; Models for exploration: Society of Economic Paleontologists and Mineralogists Special Publication 31, p. 127–155.

——, 1986a, Depositional and structural framework of the distal Frio Formation, Texas coastal zone and shelf: The University of Texas at Austin, Bureau of Economic Geology Geological Circular 86-8, 16 p.

——, 1986b, Growth faults and fault-related structures of prograding terrigenous clastic continental margins: Gulf Coast Association of Geological Societies Transactions, v. 36, p. 121–128.

——, 1987, Depositional and structural architecture of prograding clastic continental margins; Tectonic influence on patterns of basin filling: Norsk Geologisk Tidsskrift, v. 67, p. 237–251.

——, 1989a, Genetic stratigraphic sequences in basin analysis I; Architecture and genesis of flooding-surface bounded depositional units: American Association of Petroleum Geologists Bulletin, v. 73, p. 125–142.

——, 1989b, Genetic stratigraphic sequences in basin analysis II; Application to northwest Gulf of Mexico Cenozoic basin: American Association of Petroleum Geologists Bulletin, v. 73, p. 143–154.

——, 1990, Paleogene depositional episodes, genetic stratigraphic sequences, and sediment accumulation rates, NW Gulf of Mexico Basin, *in* Sequence stratigraphy as an exploration tool, concepts and practice in the Gulf Coast: Society of Economic Paleontologists and Mineralogists, Gulf Coast Section, Eleventh Annual Research Conference Program and Abstracts, p. 165–176.

Galloway, W. E., Kreitler, C. W., and McGowen, J. H., 1978, Depositional and ground-water flow systems in the exploration for uranium: The University of Texas at Austin, Bureau of Economic Geology, 267 p.

Galloway, W. E., Finley, R. J., and Henry, C. D., 1979, South Texas uranium province, geologic perspective: The University of Texas at Austin, Bureau of Economic Geology Guidebook 18, 81 p.

Galloway, W. E., Henry, C. D., and Smith, G. E., 1982a, Depositional framework, hydrostratigraphy, and uranium mineralization of the Oakville Sandstone (Miocene), Texas Coastal Plain: The University of Texas at Austin, Bureau of Economic Geology Report of Investigations 113, 51 p.

Galloway, W. E., Hobday, D. K., and Magara, K., 1982b, Frio Formation of Texas Gulf Coastal Plain: Depositional systems, structural framework, and hydrocarbon distribution: American Association of Petroleum Geologists Bulletin, v. 66, p. 649–688.

——, 1982c, Frio Formation of the Texas Gulf Coast Basin; Depositional systems, structural framework, and hydrocarbon origin, migration, distribution, and exploration potential: The University of Texas at Austin, Bureau of Economic Geology Report of Investigations 122, 78 p.

Galloway, W. E., Ewing, T. E., Garrett, C. M., Jr., Tyler, N., and Bebout, D. G., 1983, Atlas of major Texas oil reservoirs: The University of Texas at Austin, Bureau of Economic Geology, 139 p.

Galloway, W. E., Jirik, L. A., Morton, R. A., and DuBar, J. R., 1986, Lower Miocene (Fleming) depositional episode of the Texas coastal plain and continental shelf; Structural framework, facies, and hydrocarbon resources: The University of Texas at Austin, Bureau of Economic Geology Report of Investigations 150, 50 p.

Galloway, W. E., Dingus, W. F., and Paige, R. E., 1988, Depositional framework and genesis of Wilcox submarine canyon systems, northwest Gulf Coast [abs.]: American Association of Petroleum Geologists Bulletin, v. 72, p. 187–188.

——, 1991, Seismic and depositional facies of Paleocene-Eocene Wilcox Group submarine canyon fills, northwest Gulf Coast, U.S.A., *in* Weimer, P., and Link, M., eds., Seismic facies and sedimentary processes of submarine fans and turbidite systems: New York, Springer-Verlag (in press).

Garrett, J. B., 1938, The Hackberry assemblage; An interesting foraminiferal fauna of post-Vicksburg age from deep wells in the Gulf Coast: Journal of Paleontology, v. 12, p. 309–317.

——, 1939, Some middle Tertiary smaller foraminifera from subsurface beds of

Jefferson Co., Texas: Journal of Paleontology, v. 13, p. 575–579.

——, 1941, New Middle Eocene foraminifera from southern Alabama and Mississippi: Journal of Paleontology, v. 15, p. 153–156.

——, 1942, Some Miocene foraminifera from subsurface strata of coastal Texas: Journal of Paleontology, v. 16, p. 461–63 (Correction, Journal of Paleontology, v. 24, p. 506).

Garrett, J. B., and Ellis, A. D., Jr., 1937, Distinctive foraminifera of the genus *Marginulina* from middle Tertiary beds of the Gulf Coast: Journal of Paleontology, v. 11, p. 629–633.

Gernant, R. E., and Kesling, R. V., 1966, Foraminiferal paleoecology and paleoenvironmental reconstruction of the Oligocene middle Frio in Chambers County, Texas: Gulf Coast Association of Geological Societies Transactions, v. 16, p. 131–158.

Glowacz, M. F., and Horne, J. C., 1971, Depositional environments of the early Miocene as exposed in the Cane River diversion canal, Rapides Parish, Louisiana: Gulf Coast Association of Geological Societies Transactions, v. 21, p. 379–386.

Goodell, H. G., and Yon, J. W., Jr., 1960, The regional lithostratigraphy of the post-Eocene rocks of Florida, *in* Puri, H. S., ed., Late Cenozoic stratigraphy and sedimentation of central Florida: Southeastern Geological Society, 9th Field Trip, p. 75–113.

Gravell, D. W., and Hanna, M. A., 1935, Larger foraminifera from the Moodys Branch Marl, Jackson Eocene of Texas, Louisiana, and Mississippi: Journal of Paleontology, v. 9, p. 327–340.

——, 1937, The *Lepidocyclina texana* horizon in the Heterostegina Zone, Upper Oligocene of Texas and Louisiana: Journal of Paleontology, v. 11, p. 517–529.

——, 1938, Subsurface Tertiary zones of correlation through Mississippi, Alabama, and Florida: American Association of Petroleum Geologists Bulletin, v. 22, p. 984–1013.

Hamlin, C. S., 1983, Fluvial depositional systems of the Carrizo–upper Wilcox in south Texas; Gulf Coast Association of Geological Societies Transactions, v. 33, p. 281–287.

Haq, B. J., Hardenbol, J., and Vail, P. R., 1987, Chronology of fluctuating sea levels since the Triassic: Science, v. 235, p. 1156–1166.

Hedgpeth, J. W., 1957, Classification of marine environments, *in* Hedgpeth, J. W., ed., Treatise on marine ecology and paleoecology: Geological Society of America Memoir 67, v. 1, p. 17–28.

Hoel, H. D., 1982, Goliad Formation of the south Texas Gulf Coastal Plain; Regional genetic stratigraphy and uranium mineralization [M.A. thesis]: The University of Texas at Austin, 173 p.

Hoffmeister, J. E., 1974, Land from the sea: Coral Gables, Florida, University of Miami Press, 143 p.

Holcomb, C. W., 1964, Frio Formation of southern Texas: Gulf Coast Association of Geological Societies, v. 14, p. 23–33.

Hoyt, W. V., 1959, Erosional channel in the middle Wilcox near Yoakum, Lavaca County, Texas: Gulf Coast Association of Geological Societies Transactions, v. 9, p. 41–50.

Howe, H. V., 1939, Louisiana Cook Mountain Eocene foraminifera: Louisiana Geological Survey Bulletin 14, 122 p.

Howe, H. V., and McDonald, S. M., 1938, Two new species of the foraminiferal genus *Marginulina*: Louisiana Geological Survey Bulletin 13, p. 209–211.

Huddlestun, P., Braunstein, J., and Biel, R., coordinators, 1988, Gulf Coast region correlation chart: American Association of Petroleum Geologists, Correlation of Stratigraphic Units of North America.

Hussey, K. M., 1943, Distinctive new species of foraminifera from the Cane River Eocene of Louisiana: Journal of Paleontology, v. 17, p. 160–167.

——, 1949, Louisiana Cane River Eocene foraminifera: Journal of Paleontology, v. 23, p. 109–144.

Isphording, W. C., 1981, Neogene geology of southeastern Mississippi and southwestern Alabama, *in* Field trip guidebook for southern Mississippi: Southeastern Geological Society Publication 2, p. 1–25.

Isphording, W. C., and Lamb, G. M., 1971, Age and origin of the Citronelle Formation in Alabama: Geological Society of America Bulletin, v. 82,

p. 775–780.

Jurkowski, G., Ni, J., and Brown, L., 1984, Modern uparching of the Gulf Coastal Plain: Journal of Geophysical Research, v. 89, p. 6247–6255.

Kaiser, W. R., Johnston, J. E., and Bach, W. N., 1978, Sand-body geometry and the occurrence of lignite in the Eocene of Texas: The University of Texas at Austin, Bureau of Economic Geology Geological Circular 78-4, 19 p.

Kellough, G. R., 1959, Biostratigraphic and paleoecologic study of Midway foraminifera along Tehuacana Creek, Limestone County, Texas: Gulf Coast Association of Geological Societies Transactions, v. 9, p. 147–160.

Kennett, J. P., and Huddlestun, P., 1972, Late Pleistocene paleoclimatology, foraminiferal biostratigraphy and tephrochronology, western Gulf of Mexico: Quaternary Research, v. 2, p. 38–69.

Kennett, J. P., and Shackleton, N. J., 1975, Laurentide ice sheet meltwater recorded in Gulf of Mexico deep-sea cores: Science, v. 188, p. 147–150.

Kiatta, H. W., 1971, The stratigraphy and petroleum potential of the lower Miocene, offshore Galveston and Jefferson Counties, Texas: Gulf Coast Association of Geological Societies Transactions, v. 21, p. 257–270.

Kohl, B., 1980, The lower Pliocene benthic foraminifers from the Isthmus of Tehuantepec, Mexico [Ph.D. thesis]: New Orleans, Louisiana, Tulane University, Graduate School, Department of Paleontology, 483 p.

Lamb, J. L., 1981, Marine environmental terminology and depth-related environments: Gulf Coast Association of Geological Societies Transactions, v. 31, p. 329–337.

Lamb, J. L., and Beard, J. H., 1972, Late Neogene planktonic foraminifers in the Caribbean, Gulf of Mexico, and Italian stratotypes: The University of Kansas Paleontological Contributions, Article 57 (Protozoa-8), 67 p.

Lehner, P., 1969, Salt tectonics and Pleistocene stratigraphy on continental slope of northern Gulf of Mexico: American Association of Petroleum Geologists Bulletin, v. 53, p. 2431–2479.

Leutze, W. P., 1972, Stratigraphy of Cibicides carstensi Zone, Miocene of Louisiana: American Association of Petroleum Geologists Bulletin, v. 56, p. 775–789.

LeRoy, D. O., and Levinson, S. A., 1974, A deep-water Pleistocene microfossil assemblage from a well in the northern Gulf of Mexico: Micropaleontology, v. 20, p. 1–37.

Limes, L. L., and Stipe, J. C., 1959, Occurrence of Miocene oil in south Louisiana: Gulf Coast Association of Geological Societies Transactions, v. 9, p. 77–90.

Lopez Ramos, E., 1973, Estudio geológico de la Península de Yucatán: Asociación Mexicana de Geólogos Petroleros Boletín, v. 25, nos. 1–3, p. 23–76.

—— , 1975, Geological summary of the Yucatan Peninsula, in Nairn, A.E.M., and Stehli, F. G., eds., The ocean basins and margins; v. 3, The Gulf of Mexico and the Caribbean: New York and London, Plenum Press, p. 257–282.

Loucks, R. G., Dodge, M. M., and Galloway, W. E., 1986, Sandstone consolidation analysis to delineate areas of high-quality reservoirs suitable for production of geopressured geothermal energy along the Texas Gulf Coast: The University of Texas at Austin, Bureau of Economic Geology Report of Investigations 149, 78 p.

Loutit, T. S., Baum, G. R., and Wright, R. C., 1983, Eocene-Oligocene sea-level changes as reflected in Alabama outcrop sections [abs.]: American Association of Petroleum Geologists Bulletin, v. 67, p. 506.

Loutit, T. S., Hardenbol, J., Vail, P. R., and Baum, G. R., 1988, Condensed sections; The key to age determination and correlation of continental margin sequences, in Wilgus, C. K., Hastings, B. S., Kendall, C. G., Posamentier, H. G., Ross, C. A., and van Wagoner, J. C., eds., Sea-level changes; An integrated approach: Society of Economic Paleontologists and Mineralogists Special Publication 42, p. 183–216.

Lowman, S. W., 1949, Sedimentary facies in Gulf Coast: American Association of Petroleum Geologists Bulletin, v. 33, p. 1939–1997.

Lund, J. W., King, J. S., Berlitz, R., and Gilreath, J. A., 1978, Pre-platform exploration of High Island Blocks A-560 and A-561: Gulf Coast Association of Geological Societies Transactions, v. 28, pt. 1, p. 273–294.

Machain Castillo, M. L., 1986, Ostracode biostratigraphy and paleoecology of the

Pliocene of the Isthmian Salt Basin, Veracruz, Mexico: Tulane Studies in Geology and Paleontology, v. 19, nos. 3–4, p. 123–139.

Malmgren, B., and Kennett, J. P., 1976, Principal component analysis of Quaternary planktonic foraminifera in the Gulf of Mexico; Paleoclimatic applications: Marine Micropaleontology, v. 1, p. 299–306.

Mancini, E. A., and Oliver, G. E., 1981, Planktonic foraminifera from the Tuscahoma Sand (Upper Paleocene) of southwest Alabama: Micropaleontology, v. 27, p. 204–225.

Martin, R. G., 1976, Geologic framework of northern and eastern continental margin, Gulf of Mexico, in Bouma, A. H., Moore, G. T., and Coleman, J. M., eds., Beyond the shelf break: American Association of Petroleum Geologists Marine Geology Committee Short Course, v. 2, p. A1–A28.

—— , 1978, Northern and eastern Gulf of Mexico continental margin; Stratigraphic and structural framework, in Bouma, A. H., Moore, G. T., and Coleman, J. M., eds., Framework, facies, and oil-trapping characteristics of the upper continental margin: American Association of Petroleum Geologists Studies in Geology 7, p. 21–42.

Martini, E., 1971, Standard Tertiary and Quaternary calcareous nannoplankton zonation, in Farinacci, A., ed., Proceedings, Second Planktonic Conference, Rome, 1970: Rome, Edizioni Technoscienza, p. 739–783.

McBride, E. F., Lindemann, W. L., and Freeman, P. S., 1968, Lithology and petrology of the Gueydan (Catahoula) Formation in south Texas: The University of Texas at Austin, Bureau of Economic Geology Report of Investigations 63, 122 p.

McDowell, F. W., and Clabaugh, S. E., 1979, Ignimbrites of the Sierra Madre Occidental and their relation to the tectonic history of western Mexico, in Chapin, C. E., and Elston, W. E., eds., Ash-flow tuffs: Geological Society of America Special Paper 180, p. 113–124.

McFarlan, E., Jr., and LeRoy, D. O., 1988, Subsurface geology of the late Tertiary and Quaternary deposits, coastal Louisiana and the adjacent continental shelf: Gulf Coast Association of Geological Societies Transactions, v. 38, p. 421–433.

McGookey, D. P., 1975, Gulf Coast Cenozoic sediments and structure; An excellent example of extracontinental sedimentation: Gulf Coast Association of Geological Societies Transactions, v. 25, p. 104–120.

McIntyre, A., and Kipp, N. G., 1976, Glacial North Atlantic 18,000 years ago, in Cline, R. M., and Hays, J. D., eds., Investigation of late Quaternary paleoceanography and paleoclimatology: Geological Society of America Memoir 145, p. 43–76.

McLean, C. M., 1957, Miocene geology of southeastern Louisiana: Gulf Coast Association of Geological Societies Transactions, v. 7, p. 241–245.

Meyerhoff, A. A., ed., 1968, Geology of natural gas in south Louisiana, in Beebe, B. W., ed., Natural gases of North America: American Association of Petroleum Geologists Memoir 9, v. 1, p. 376–581.

Miller, J. A., 1986, Hydrogeologic framework of the Floridan aquifer system in Florida and parts of Georgia, Alabama and South Carolina: U.S. Geological Survey Professional Paper 1403-B, 91 p.

Mitchum, R. M., Jr., 1978, Seismic stratigraphic investigation of West Florida Slope, Gulf of Mexico, in Bouma, A. H., Moore, G. T., and Coleman, J. M., eds., Framework, facies, and oil-trapping characteristics of the upper continental margin: American Association of Petroleum Geologists Studies in Geology 7, p. 193–223.

Monsour, E., 1937, Micropaleontological analysis of Jackson Eocene of eastern Mississippi: American Association of Petroleum Geologists Bulletin, v. 21, p. 80–86.

Morton, R. A., Jirik, L. A., and Foote, R. Q., 1985, Depositional history, facies analysis, and production characteristics of hydrocarbon-bearing sediments, offshore Texas: The University of Texas at Austin, Bureau of Economic Geology Geological Circular 85-2, 31 p.

Morton, R. A., Jirik, L. A., and Galloway, W. E., 1988, Middle-Upper Miocene depositional sequences of the Texas coastal plain and continental shelf: University of Texas at Austin, Bureau of Economic Geology Report of Investigations 174, 40 p.

Murray, G. E., 1961, Geology of the Atlantic and Gulf Coastal Province of North

America: New York, Harper and Brothes, 692 p.

Murray, G. E., and Weidie, A. E., Jr., 1962, Regional geologic summary of Yucatan Peninsula, *in* Guidebook to field trip to Yucatan Peninsula: New Orleans Geological Society, p. 5–51.

Nanz, R. H., Jr., 1954, Genesis of Oligocene sandstone reservoir, Seeligson field, Jim Wells and Kleberg Counties, Texas: American Association of Petroleum Geologists Bulletin, v. 38, p. 96–117.

Natland, M. L., 1933, The temperature and depth distribution of some Recent and fossil foraminifera in the southern California region: Scripps Institute of Oceanography Bulletin, Technical Services, v. 3, p. 225–230.

Norton, R. D., 1930, Ecologic relations of some foraminifera: Scripps Institute of Oceanography Bulletin, Technical Services, v. 2, p. 311–388.

Otvos, E. G., Jr., 1972, Mississippi Gulf Coast Pleistocene beach barriers, and the age problem of the Atlantic–Gulf Coast "Pamlico"-"Ingleside" beach ridge system: Southeastern Geology, v. 14, p. 241–250.

——, 1975, Late Pleistocene transgressive unit (Biloxi Formation), northern Gulf Coast: American Association of Petroleum Geologists Bulletin, v. 59, p. 148–154.

——, 1976, Post-Miocene geological development of the Mississippi-Alabama coastal zone: Journal of the Mississippi Academy of Sciences, v. 21, p. 101–114.

——, 1981, Tectonic lineaments of Pliocene and Quaternary shorelines, northeast Gulf Coast: Geology, v. 9, p. 398–404.

Paine, W. R., 1968, Stratigraphy and sedimentation of subsurface Hackberry wedge and associated beds of southwestern Louisiana: American Association of Petroleum Geologists Bulletin, v. 52, p. 322–342.

——, 1971, Petrology and sedimentation of the Hackberry sequence of southwest Louisiana: Gulf Coast Association of Geological Societies Transactions, v. 21, p. 37–55.

Parker, F. L., 1954, Distribution of the foraminifera in the northeastern Gulf of Mexico: Havard College, Museum of Comparative Zoology Bulletin, v. 111, p. 453–588.

Pérez Matus, J. D., 1978, Probable sistema de cañón submarino en la subprovincia de Agua Dulce, Ver.: Ingenieria Petrolera, v. 18, no. 3, p. 5–18.

Pérez Matus, J. D., and Barbosa, H. R., 1975, Sistemas de barras de barrera en el área de Tupilco, Distrito de Comalcalco, Tab.: Ingenieria Petrolera, v. 15, no. 11, p. 419–450.

Pérez Matus, J. D., and Valle González, V., 1974, Estudio de los medios ambientes de depositación de una arena del Campo Magallanes, Zona Sur: Ingenieria Petrolera, v. 14, no. 11, p. 437–449.

Pflum, C. E., and Frerichs, W. E., 1976, Gulf of Mexico deep-water foraminifers: Cushman Foundation for Foraminiferal Research Special Publication 14, 125 p.

Phleger, F. B., 1951, Ecology of foraminifera, northwest Gulf of Mexico; Part 1, Foraminiferal distribution: Geological Society of America Memoir 46, p. 1–88.

——, 1960, Ecology and distribution of recent foraminifera: Baltimore, Johns Hopkins Press, 297 p.

Phleger, F. B., and Parker, F. L., 1951 Ecology of foraminifera, northwest Gulf of Mexico; Part 2, Foraminifera species: Geological Society of America Memoir 46, p. 1–64.

Pitmann, W. C., and Golovchenko, X., 1983, The effect of sealevel change on the shelf edge and slope of passive margins: Society of Economic Paleontologists and Mineralogists Special Publication 33, p. 41–58.

Plummer, H. J., 1926, Foraminifera of the Midway Formation in Texas: University of Texas at Austin Bulletin 2644, 206 p.

Poag, C. W., 1966, Paynes Hammock (Lower Miocene) foraminifera of Alabama and Mississippi: Micropaleontology, v. 12, p. 393–440.

——, 1971, A reevaluation of the Gulf Coast Pliocene-Pleistocene boundary: Gulf Coast Association of Geological Societies Transactions, v. 21, p. 291–305.

——, 1972, Planktonic foraminifera of the Chickasawhay Formation, United States Gulf Coast: Micropaleontology, v. 18, p. 257–277.

——, 1981, Ecologic atlas of benthic foraminifera of the Gulf of Mexico: Woods Hole, Massachusetts, Marine Science International, p. 1–174.

Pope, D. E., 1955, Comparison of the Harang and Hackberry facies in south Louisiana: Gulf Coast Association of Geological Societies Transactions, v. 5, p. 153–163.

Pope, D. E., and Smith, D. J., 1948, The Harang fauna of Louisiana: Louisiana Geological Survey Bulletin 26, 80 p.

Postuma, J. A., 1971, Manual of planktonic foraminifera: New York, Elseiver, 422 p.

Price, J. G., and Henry, C. D., 1984, Stress orientations during Oligocene volcanism in Trans-Pecos Texas; Timing the transition from Laramide compression to Basin and Range tension: Geology, v. 12, p. 238–241.

Puri, H. S., 1957, Stratigraphy and zonation of the Ocala Group: Florida Geological Survey Bulletin 38, 248 p.

Puri, H. S., and Vernon, R. O., 1964, Summary of the geology of Florida and a guidebook to the classic exposures: Florida Geological Survey Special Publication 5, 255 p.

Quezada Muñetón, J. M., 1987(1990), El Cretácico Medio-Superior, y el límite Cretácico Superior-Terciario inferior en la Sierra de Chiapas: Asociación Mexicana de Geólogos Petroleros Boletín, v. 39, no. 1, p. 3–98.

Quinlan, G. M., and Beaumont, C., 1984, Appalachian thrusting, lithospheric flexure, and the Paleozoic stratigraphy of the Eastern Interior of North America: Canadian Journal of Earth Sciences, v. 21, p. 973–996.

Rainwater, E. H., 1964a, Regional stratigraphy of the Gulf Coast Miocene: Gulf Coast Association of Geological Societies Transactions, v. 14, p. 81–124.

——, 1964b, Transgressions and regressions in the Gulf Coast Tertiary: Gulf Coast Association of Geological Societies Transactions, v. 14, p. 217–230.

——, 1971, Possible future petroleum potential of peninsular Florida and adjacent continental shelves, *in* Cram, I. H., Future petroleum provinces of the United States—Their geology and potential: American Association of Petroleum Geologists Memoir 15, v. 2, p. 1311–1341.

Randazzo, A. F., 1972, Petrography of the Suwannee Limestone: Florida Bureau of Geology Bulletin 54, Part II, 13 p.

——, 1976, Petrographic and geohydrologic model of aquifer limestones in Florida: Water Resources Research Center Publication 35, University of Florida, 51 p.

Randazzo, A. F., and Zachos, L. G., 1974, Classification and description of dolomitic fabrics of rocks from the Floridan aquifer, USA: Sedimentary Geology, v. 37, p. 151–162.

Rice, D. D., 1980, Chemical and isotopic evidence of the origins of natural gases in offshore Gulf of Mexico: Gulf Coast Association of Geological Societies Transactions, v. 30, p. 203–213.

Ricoy, J. M., 1989, Tertiary terrigenous depositional systems of the Mexican Isthmus basins [Ph.D. thesis]: The University of Texas at Austin, 145 p.

Riggs, S. R., 1984, Paleoceanographic model of Neogene phosphorite deposition, U.S. Atlantic continental margin: Science, v. 223, p. 123–131.

Roberts, W. H., III, 1982, Gulf Coast magic: Gulf Coast Association of Geological Societies Transactions, v. 32, p. 205–216.

Saucier, R. T., 1974, Quaternary geology of the Lower Mississippi Valley: Arkansas Archeological Survey Research Series, no. 6, 26 p.

Scott, T. M., 1988, The lithostratigraphy of the Hawthorn Group (Miocene) of Florida: Florida Geological Survey Bulletin 59, 148 p.

Senes, J., ed., 1975–1976, Proceedings of the VI Congress of the Regional Committee on Mediterranean Neogene Stratigraphy (IUGS Commission on Stratigraphy), Bratislava, 1975, v. 1: Bratislava, VEDA Publishing House, 453 p., v. 2, 69 p.

Siesser, W. G., 1983, Paleogene calcareous nannoplankton biostratigraphy; Mississippi, Alabama, and Tennessee: Mississippi Department of Natural Resources Bureau of Geology Bulletin 125, 61 p.

Skinner, H. C., ed., 1972, Gulf Coast stratigraphic correlation methods with an atlas and catalogue of principal index foraminiferida: New Orleans, Louisiana Heritage Press, 213 p.

Sloane, B. J., 1971, Recent developments in the Miocene Planulina gas trend of south Louisiana: Gulf Coast Association of Geological Societies Transactions, v. 21, p. 199–210.

Solis I., R. F., 1981, Upper Tertiary and Quaternary depositional systems, central Coastal Plain, Texas; Regional geology of the coastal aquifer and potential liquid-waste repositories: The University of Texas at Austin, Bureau of Economic Geology Report of Investigations 108, 89 p.

Stainforth, R. M., 1948, Description, correlation and paleoecology of Tertiary Cipero Marl Formation, Trinidad, B.W.I.: American Association of Petroleum Geologists Bulletin, v. 33, p. 1293–1330.

Stainforth, R. M., Lamb, J. L., Luterbacher, H., Beard, J. L., and Jeffords, R. M., 1975, Cenozoic planktonic foraminiferal zonation and characteristics of index forms: The University of Kansas Paleontological Institute Contributions, Article 62, 425 p.

Stenzel, H. B., 1949, Successional speciation in paleontology; The case of the oysters of the *sellaeformis* stock: Evolution, v. 3, p. 34–50. (Reprinted as The University of Texas at Austin, Bureau of Economic Geology Report of Investigations 3, 16 p.).

—— , 1952, Boundary problems: Mississippi Geological Society Guidebook, 9th Field Trip, p. 11–13.

Stuart, C. J., and Caughey, C. A., 1977, Seismic facies and sedimentology of terrigenous Pleistocene deposits in northwest and central Gulf of Mexico, *in* Payton, C. E., ed., Seismic stratigraphy—applications to hydrocarbon exploration: American Association of Petroleum Geologists Memoir 26, p. 249–275.

Stuckey, C. W., Jr., 1946, Some Textulariidae from the Gulf Coast Tertiary: Journal of Paleontology, v. 20, p. 163–165.

—— , 1960, A correlation of the Gulf Coast Jackson: Gulf Coast Association of Geological Societies Transactions, v. 10, p. 285–298.

—— , 1964, The stratigraphic relationship of the Hackberry, Abbeville, and Harang faunal assemblages: Gulf Coast Association of Geological Societies Transactions, v. 14, p. 209–216.

Stuckey, J. L., 1965, North Carolina; Its geology and mineral resources: North Carolina Department of Conservation and Development, 550 p.

Stude, G. R., 1978, Depositional environments of the Gulf of Mexico South Timbalier Block-54 salt dome and salt dome models: Gulf Coast Association of Geological Societies Transactions, v. 28, p. 627–646.

Suter, J. R., and Berryhill, H. L., Jr., 1985, Late Quaternary shelf-margin deltas, northwest Gulf of Mexico: American Association of Petroleum Geologists Bulletin, v. 69, p. 77–91.

Thayer, P. A., and Miller, J. A., 1984, Petrology of lower and middle Eocene carbonate rocks, Floridan aquifer, central Florida: Gulf Coast Association of Geological Societies Transactions, v. 34, p. 421–434.

Tipsword, H. L., 1962, Tertiary foraminifera in Gulf Coast petroleum exploration and development, *in* Geology of the Gulf Coast and central Texas and guidebook of excursions: Houston Geological Society, p. 16–57.

Tipsword, H. L., Setzer, F. M., and Smith, F. L., 1966, Interpretation of depositional environment in Gulf Coast petroleum exploration and related stratigraphy: Gulf Coast Association of Geological Societies Transactions, v. 16, p. 119–130.

Toulmin, L. D., 1955, Cenozoic geology of southeastern Alabama, Florida, and Georgia: American Association of Petroleum Geologists Bulletin, v. 39, p. 207–235.

Trippet, A. R., 1981, Characteristics of diapirs on the outer continental shelf–upper continental slope boundary, northwest Gulf of Mexico: Gulf Coast Association of Geological Societies Transactions, v. 31, p. 391–397.

Tyler, N., and Han, J. H., 1982, Elements of high constructive deltaic sedimentation, lower Frio Formation, Texas: Gulf Coast Association of Geological Societies Transactions, v. 32, p. 527–540.

Vail, P. R., Hardenbol, J., and Todd, R. G., 1984, Jurassic unconformities, chronostratigraphy, and sea-level changes from seismic stratigraphy and biostratigraphy, *in* Schlee, J. S., ed., Interregional unconformities and hydro-

carbon accumulation: American Association of Petroleum Geologists Memoir 36, p. 129–144.

Vaughan, T. W., 1928, New Species of *Operculina* and *Discocyclina* from the Ocala Limestone: Florida Geological Survey, 19th Annual Report, p. 155–165.

Vernon, R. O., 1942, Geology of Holmes and Washington Counties, Florida: Florida Geological Survey Bulletin 21, 161 p.

—— , 1951, Geology of Citrus and Levy Counties, Florida: Florida Geological Survey Bulletin 33, 256 p.

Vogler, H. A., and Robison, B. A., 1987, Exploration for deep geopressured gas; Corsair trend, offshore Texas: American Association of Petroleum Geologists Bulletin, v. 71, p. 777–787.

Warren, A. D., 1957, The Anahuac and Frio sediments in Louisiana: Gulf Coast Association of Geological Societies Transactions, v. 7, p. 221–237.

Weinzierl, L. L., and Applin, E. R., 1929, The Claiborne Formation on the Coastal Domes: Journal of Paleontology, v. 3, p. 384–410.

Winker, C. D., 1979, Late Pleistocene fluvial-deltaic deposition: Texas Coastal Plain and shelf [M.A. thesis]: The University of Texas at Austin, 187 p.

—— , 1981, Cenozoic shelf margins, northwestern Gulf of Mexico, *in* Perkins, B. F., Ventress, W.P.S., and Edwards, M. B., eds., Recognition of shallow-water versus deep-water sedimentary facies in growth-structure-affected formations of the Gulf Coast Basin: Society of Economic Paleontologists and Mineralogists, Gulf Coast Section, 2nd Annual Research Conference, Program and Abstracts, p. 74–82.

—— , 1982, Cenozoic shelf margins, northwestern Gulf of Mexico: Gulf Coast Association of Geological Societies Transactions, v. 32, p. 427–448.

Winker, C. D., and Edwards, M. B., 1983, Unstable progradational clastic shelf margins, *in* Stanley, D. J., and Moore, G. T., eds., The shelfbreak; Critical interface on continental margins: Society of Economic Paleontologists and Mineralogists Special Publication 33, p. 139–157.

Woodbury, H. O., Murray, I. B., Jr., Pickford, P. J., and Akers, W. H., 1973, Pliocene and Pleistocene depocenters, Outer Continental Shelf, Louisiana and Texas: American Association of Petroleum Geologists Bulletin, v. 57, p. 2428–2439.

Woodbury, H. O., Spotts, J. H., and Akers, W. H., 1976, Gulf of Mexico continental slope sediments and sedimentation, *in* Bouma, A. H., Moore, G. T., and Coleman, J. M., eds., Beyond the shelf break: American Association of Petroleum Geologists Marine Geology Committee Short Course, v. 2, p. C-1–C-28.

Worrall, D. M., and Snelson, S., 1989, Evolution of the northern Gulf of Mexico, with emphasis on Cenozoic growth faulting and the role of salt, *in* Bally, A. W., and Palmer, A. R., eds., The Geology of North America; An overview: Boulder, Colorado, Geological Society of America, The Geology of North America, v. A, p. 93–138.

Yon, J. W., Jr., and Hendry, C. W., Jr., 1972, Suwannee Limestone in Hernando and Pasco Counties, Florida: Florida Bureau of Geology Bulletin 54, Part I, 42 p.

ACKNOWLEDGMENTS

Duane O. LeRoy and Edward B. Picou, Jr., carefully reviewed the manuscript. Their many constructive comments and suggestions for its improvement are gratefully acknowledged. Duane LeRoy's efforts in preparing Tables 1 and 2 are particularly appreciated.

MANUSCRIPT ACCEPTED BY THE SOCIETY JULY 29, 1991

The Geology of North America
Vol. J, The Gulf of Mexico Basin
The Geological Society of America, 1991

Chapter 12

Late Quaternary sedimentation

James M. Coleman and Harry H. Roberts
Coastal Studies Institute, School of Geoscience, Louisiana State University, Baton Rouge, Louisiana 70803
William R. Bryant
Department of Oceanography, Texas A & M University, College Station, Texas 77843

INTRODUCTION

The Gulf of Mexico basin (Fig. 1) is the largest semi-enclosed depositional basin in North America and has been the site of extensive hydrocarbon exploration and exploitation since the turn of the century. Since Late Jurassic times, the drainage basin of the Mississippi River system has been delivering sediments to the Gulf of Mexico (Worzel and Burke, 1978; Chapter 8, this volume). Mesozoic and Cenozoic deposits are estimated to have attained a total thickness in excess of 15 km (Martin and Bouma, 1978; Bouma and others, 1978a). Thus, the river system has been operative over relatively long periods of time, constantly feeding sediments to the receiving basin and building a thick Jurassic, Cretaceous, Tertiary, and Quaternary sequence of interfingering deltaic, nearshore coastal brackish water, and marine sediments, which have prograded the coastal plain shoreline seaward. Relatively little sediment yield has occurred during the Quaternary from the southern rim of the Gulf of Mexico basin. Through time, depocenters have shifted within the northern flank of the basin, forming a relatively thick sequence of Tertiary and Quaternary clastic sediments. The zone of maximum thickness trends roughly east-west near the present-day coastal plain of Louisiana and west toward Texas. Rapid subsidence associated primarily with sediment loading and salt and shale diapirism has been responsible for unusually thick, localized sedimentary accumulations and for the complex bathymetry on the continental slope (Fig. 1; Plate I, this volume). Throughout the Tertiary and Quaternary, minor and major transgressions and regressions have occurred, although the major depositional component has been a net basinward progradation (Fig. 2). This diagram shows a generalized cross section of the coastal plain, continental shelf, and continental slope off Louisiana and the major characteristics of the Gulf of Mexico basin.

The modern depositional pattern in the Gulf of Mexico basin, where terrigenous deposits dominate on the northern and western shelf areas and carbonates occur on the broad platforms of the eastern and southern Gulf, has persisted since Late Jurassic–Early Cretaceous time. Before the Late Jurassic, there was only deposition of the Louann Salt in late Middle Jurassic, and mostly nonmarine red beds before that. It was during the latest Jurassic that the terrigenous clastic influx from tectonically elevated northern and western continental interiors began to overwhelm the mainly carbonate environments that encircled the Gulf during Late Jurassic times (Garrison and Martin, 1973). Thus the modern-day environments represent good analogs for interpreting the Cretaceous and Tertiary section in the Gulf of Mexico basin. The bulk of the sediment was delivered to the northern margin of the Gulf during the Cenozoic and prograded the shelf as much as 300 km from the margin of the Cretaceous platform deposits to the present shelf edge. This progradation rate is exceptionally high, averaging 5 to 6 km of shelf-edge progradation per million years. Such rapid progradation contrasts sharply with that of the eastern margin of the United States where the average shelf-edge progradation is measured in fractions of a kilometer per million years.

The central portion of the northern Gulf of Mexico basin is dominated by the Mississippi embayment, which extends over 800 km north from south Louisiana to Illinois. This exceptionally large downwarp, which influences the coastal plain from Alabama to east Texas, consists of Mesozoic and Cenozoic rocks. It has been proposed that variations in alignment correlate with basement fractures, which also control the form and shape of the Mississippi Valley and River (Fisk, 1944). However, Saucier (1974) suggested that the configuration of the Lower Mississippi Valley is more related to preglacial drainage than to fault control. The Mississippi embayment, like most of the western part of the Gulf of Mexico basin, has been subject to vertical movement through the formation of salt domes and development of near-surface faults. The major fault systems are deep seated and originated in Mesozoic and even earlier times; faulting is a process that continues to the present and influences the development of specific surface forms (see Chapter 5, this volume).

THE QUATERNARY

The Quaternary deposits of the northern Gulf of Mexico basin are unusually thick. As much as 3,600 m have accumulated beneath the present shelf in offshore Louisiana and Texas, and up

Coleman, J. M., Roberts, H. H., and Bryant, W. R., 1991, Late Quaternary sedimentation, *in* Salvador, A., The Gulf of Mexico Basin: Boulder, Colorado, Geological Society of America, The Geology of North America, v. J.

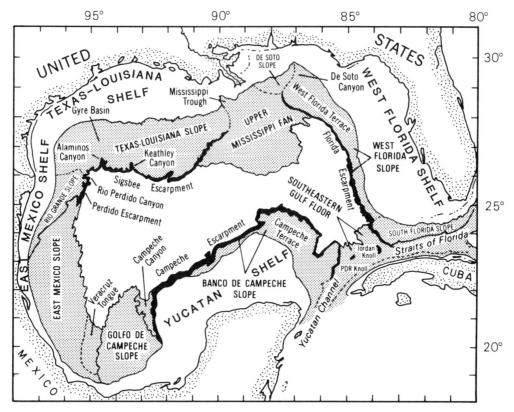

Figure 1. Physiographic provinces in the Gulf of Mexico basin. After Martin and Bouma (1978).

to 3,000 m have accumulated in the deep Gulf of Mexico basin in the vicinity of the present Mississippi Fan. The thickness of similar-age deposits along the western and southern rim of the basin is virtually unknown as little research or drilling has been conducted in this area. On the northern continental slope, much of the Quaternary has been deposited in salt-withdrawal basins (intraslope basins), and sediments often accumulate to considerable thicknesses. Morphology and sedimentation patterns were strongly influenced by drastic and relatively frequent changes in climate (temperature, precipitation, wind), vegetation, drainage patterns, discharge characteristics, and sea level throughout the Quaternary.

Although the precise history of the Quaternary is poorly known, the general effects of advancing and retreating ice on climate and vegetation are becoming clearer through research in palynology, isotope stratigraphy, and other related sciences. The major effect of these climatic changes has been the fluctuation of sea level throughout the Quaternary. Lowering of sea level with waxing glaciers exposed vast areas of former continental shelf; it also lowered the base level of streams draining into the Gulf, which in turn, led to entrenchment of river channels across the continental shelf. Lowering sea levels also caused rapid seaward progradation of river deltas, which rapidly built out the continental margin. With a rise in sea level as glaciers waned, vast areas of former coastal plain were submerged, the base level of streams was raised, and river valleys rapidly aggraded. Estimates of the

present shelf (and slope) area actually exposed during various times in the Quaternary, even during late Wisconsinan time, vary depending on the extent of sea-level lowering accepted.

Mechanisms responsible for the high-frequency eustatic sea-level changes have long been debated, but research by Hays and others (1976), Imbrie and Imbrie (1980), Pisias and Shackleton (1984), and many others has emphasized the importance of orbital control of the Quaternary ice ages. Research on Quaternary sediments, as well as on ancient sedimentary sequences, has seen a growing body of evidence linking sedimentary cycles with perturbations in the Earth's orbit, climatic variations, and sea-level fluctuations (see van Tassell, 1987, for summary). These studies have emphasized the importance of eccentricity (100,000- and 400,000-yr periods), tilt (41,000-yr period), and precession (19,000- and 22,000-yr periods) on controlling global climatic changes and, hence, changing sea levels.

Thus, frequencies, rates, and magnitude of sea-level changes are poorly known for the Quaternary. Although absolute values of sea-level change are still the subject of debate, isotopic studies in the Gulf of Mexico basin have shown the presence of high-frequency fluctuations within the sedimentary column. Over 50 isotope stages have been defined for the past 2 m.y., and these studies, in conjunction with foraminiferal and nannofossil datums, offer a stratigraphy capable of resolving stratigraphic correlations within 10,000 to 20,000 yr in the late Pleistocene and 40,000 to 50,000 yr in the early Pleistocene (Williams and

Trainor, 1986). These high-frequency changes in sea level, melt-water discharge, and changing sites of deposition have strongly influenced the sedimentation styles in the northern Gulf of Mexico basin.

Subaerial and subaqueous near-surface sediments, primarily late Wisconsinan (latest Pleistocene) and Holocene in age, have been strongly influenced by the last major fall and subsequent rise in sea level. The last major low sea-level stand occurred approximately 18,000 yr ago. The level of the sea during this period of time is also subject to much debate, but Bloom (1983) suggested that the current range of estimates of 60 to 120 m for the late Wisconsinan sea-level change is probably an accurate expression of the extreme complexity that exists when trying to specifically define such an elusive variable. It is clear that the relatively straightforward changes in sea level that result primarily from changes in global climate must be considered with the changes due to regional and local subsidence, compaction, tectonic movement, isostatic adjustment, and changes in the geoid (Morner, 1981). Thus, it is now accepted that sea-level change along every coast has probably been unique (Bloom, 1983).

GULF OF MEXICO BASIN

Late Wisconsinan and Holocene sediments of the Gulf of Mexico basin display high variability and record varying process controls. The subaerial deposits of the eastern and southern Gulf are dominated by carbonate deposition, gradually merging with the quartz-rich sands of the barrier island–lagoon complex of Alabama and Mississippi. The central part of the northern Gulf is dominated by the modern Mississippi River delta plain and its marginal chenier plain. Westward, along the Texas and north-eastern Mexican coast, barrier islands and lagoons are typical of the coastal plain.

The continental shelf off west Florida is composed primarily of carbonate, which gradually merges into the relict sand sheet that fronts most of peninsular Florida, Alabama, and Mississippi. Studies by Gould and Stewart (1956), Doyle and Sparks (1980), and Doyle (1981) along the west Florida shelf show that facies patterns tend to parallel the adjacent coastline. Coarse carbonates (grainstones) mixed with quartz sands are typical of the nearshore portions of the shelf, while these facies grade into carbonate mudstones near the shelf edge. Off the Mississippi Delta and western Louisiana, the shelf is variable in width and relatively complex in its pattern of sedimentation. The broad shelf off Texas displays little topography and consists mostly of relict sediments.

The northern continental slope of the Gulf of Mexico is a region of relatively gentle gradients interrupted by a complex series of basins and highs formed by salt and shale diapirs. It is blanketed with fine-grained pelagic and hemipelagic sediments, which are interrupted by carbonate-rich sediments on topographic highs such as diapirs. The continental slope off Louisiana is characterized by the presence of the only major submarine fan in the Gulf, the Mississippi Fan.

Subaerial environments

Beaches. Beaches of the northern Gulf of Mexico can be divided into three general types: (1) the quartz-rich barrier islands–lagoons of peninsular Florida, Alabama, and Mississippi; (2) the chenier plain of western Louisiana; and (3) the barrier island–lagoon complex of coastal Texas and northeastern Mexico.

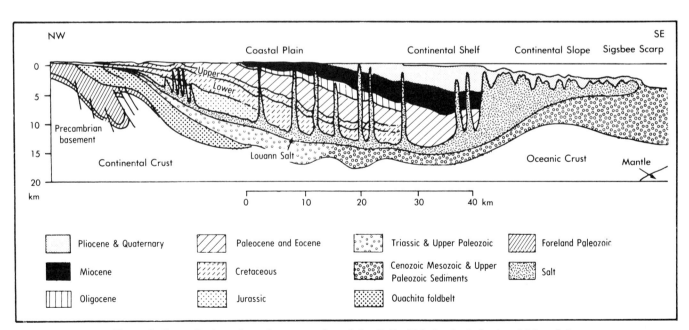

Figure 2. Generalized north-south cross section of the Gulf of Mexico basin in the vicinity of the Louisiana coast. After Buffler (1981).

The Mississippi-Alabama-Florida barrier islands are nearly continuous along the shoreline from peninsular Florida to the Mississippi Delta. In some instances, the barrier islands are Holocene in age and are separated from late Wisconsinan sediments of the mainland by either narrow elongate lagoons or broad, shallow estuaries with small rivers feeding into the heads of the water bodies. In other cases, the Holocene beach deposits are welded to older Wisconsinan-age beach and eolian deposits (Otvos, 1984). The one major characteristic of these beaches is that they are composed of clean, fine- to medium-grained quartz sand. In some instances, quartz content approaches 99 percent. Heavy-mineral analysis indicates that the source is the Appalachians and the sediments were delivered to the coast by the numerous small streams that drain the southern end of the Appalachian Mountains. During falling and lower sea stands, these streams extended and entrenched themselves across the continental shelf, delivering primarily small amounts of predominantly bedload sediments to the shelf edge. When sea level began its transgression, wave processes reworked the fluvial sediments, winnowing out the fines and concentrating the more resistant heavy minerals and quartz. The region experiences relatively low subsidence rates; thus, this reworking process has occurred multiple times during the late Pleistocene, resulting in clean, well-sorted, multicycle, quartz-rich sediments. The northern shelf reaches its narrowest width (less than 30 km) just offshore from the Alabama-Florida boundary, and resultant wave energy is higher along this stretch of the Gulf than in any other region. With a lowered sea level, the full impact of wave energy would be concentrated on reworking processes and would not be dissipated by a low, sloping continental shelf. Today, little sediment is supplied to these beaches except for erosion along some headland areas. Erosion and lateral migration of the barriers are the most common processes. Most barriers began to form 5,000 to 6,000 yr ago when the rate of sea-level rise diminished. During this period of time the barrier sands attained a thickness ranging from 10 to 18 m. Often the beaches display relatively thick transgressive eolian sands overlying the barrier islands.

Chenier plain. The chenier plain (Fig. 3) is located in southwestern Louisiana, immediately west of the Mississippi River delta plain. This marshy region is some 25 to 30 km wide and nearly 200 km in length and represents a Holocene progradational wedge of sediments composed primarily of stranded beach ridges separated by broad, fresh- and brackish-water marshes. This seaward-thickening wedge of coastal sediments was deposited during the latter part of the early Holocene sea-level rise as a result of the shifting sites of deltaic deposition (Russell and Howe, 1935; Byrne and others, 1959; Gould and McFarlan, 1959). When the Mississippi discharged its sediment load along the western margin of the delta plain, westward littoral currents blanketed the nearshore with an abundance of muds and fine silts. Rapid coastal progradation took place, and marshes capped the prograding mudflats. However, when the delta site changed to a more easterly course, sediment supply diminished along the western Louisiana coast and erosional processes became dominant.

During this time, reworked fine-grained sands and shell debris accumulated along the strand. With continued erosion and transgression, narrow linear beach ridges were formed. Since the process of delta switching was a multicycle process, this mudflat progradation and transgressive reworking occurred repeatedly, and a net seaward coastal progradation resulted. Thus, the chenier plain formed in a period of slightly less than 4,500 yr. The oldest stranded beach ridge or chenier is approximately 2,800 yr old (Byrne and others, 1959; Gould and McFarlan, 1959). Figure 3 illustrates the alternating stranded transgressive beaches separated by marsh deposits that overlie mudflat deposits. Transgressive, brackish to marine deposits containing shell and peat lag deposits are found at the base of this chenier plain sediment wedge. These sediments, overlying late Wisconsinan weathered zones, were laid down 6,000 to 4,500 yr ago (Coleman, 1966). They represent shallow shelf muds that are overlain by a series of mudflat, bay bottom, and marsh deposits that account for the overall net progradation of the coastline during the past 4,000 yr. Stranded within this fine-grained progradational wedge are the reworked transgressive beach deposits. These "chenier deposits" consist primarily of sand and shell with minor amounts of silt and clay. Average composition is 71 percent sand, 23 percent shell, and 6 percent silt, clay, and organic debris. Shell lags occur as distinct layers as well as disseminated within the fine sands. Burrowing can be intensive, often sufficient to mask most primary stratification. The beach ridges attain thicknesses of up to 6 m, but most are less than 3 m in thickness; widths are variable, but rarely exceed 1 km.

Texas barrier island coast. A barrier island–lagoon complex stretches along the entire Texas coast and continues southward into Mexico. The narrow, predominantly Holocene barriers are separated from the mainland by narrow lagoons and wide estuaries along the central and eastern Texas coast and by broad hypersaline flats and lagoons along the western Texas coast. On the northern side of many of these bays and lagoons, a series of late Pleistocene barriers marking previous high sea-level stands can be recognized. Many of the bays have small deltas building into and partially filling the estuaries (Guadalupe, Nueces, and Colorado Rivers). Most of the bays are relatively shallow (less than 5 m deep) and are floored with highly bioturbated silty and

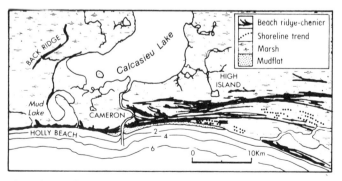

Figure 3. Map illustrating the stranded Holocene beach ridges in western Louisiana coastal chenier plain. After Byrne and others (1959).

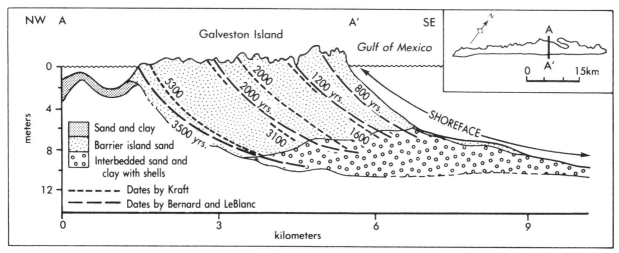

Figure 4. Cross section of a regressive barrier island, Galveston Island, Texas. The form lines with age dates indicate the seaward buildout of the barrier island during the latter part of the Holocene. The ages are based on radiocarbon dates. From Bernard and others (1962).

sandy clays. In the western and southwestern segments of the Texas shoreline, the climate becomes more arid, and the water bodies landward of the barrier islands are hypersaline in nature. Laguna Madre and Baffin Bay in southern Texas are examples of this setting. Salinity ranges from 40 to >80 ppt, and the bay bottoms are floored with highly organic silty muds that have a high methane content. Evaporite flats, algal flats, pelleted muds, and carbonates are common to the tidal flats surrounding the shallow water bodies. West of these hypersaline lagoons is the extensive South Texas sand sheet, an eolian deposit that covers in excess of 7,000 km². Much of the sand sheet is presently nearly stabilized by vegetation, although active dunes up to 15 m high are presently migrating. This sand sheet has undoubtedly been accumulating for a considerable length of time, at least since the late Wisconsinan. The eolian deposits range in thickness from 3 to 70 m and overlie late Wisconsinan lagoonal clays and evaporites.

Sediments of the Texas barrier islands consist predominantly of relatively well-sorted, fine- to medium-grained sands containing a relatively high percentage of quartz. Most of the barriers display small eolian dunes and evidence of numerous washover fans constructed during hurricane washovers. In the more arid part of the barrier system, cemented eolian deposits (eolianites) are present. The barrier island sands are variable in thickness, ranging from only a few meters to 15 m. These barrier islands are regressive in nature as shown in Figure 4, a cross section of the Galveston Island barrier. Shoreface sediments commenced seaward progradation when the rate of sea-level rise decreased, approximately 5,500 B.P. Thicker sand deposits are associated with the infilling of the migratory inlets, and in these instances, sands can attain a thickness on the order of 50 m.

Mississippi River alluvial valley. All of the rivers draining into the Gulf of Mexico Basin were drastically influenced by the changing sea levels associated with the retreat and advance of

the Quaternary glaciers. Most, if not all, of the rivers entrenched themselves as sea level lowered and their courses extended across the newly exposed continental shelf. These erosional valleys were subsequently filled during the rise in sea level. The most spectacular example of this process is illustrated by the Mississippi River alluvial valley. During the last low sea level, the Mississippi River reexcavated an alluvial valley that had been eroded, filled, and rescoured many times during the Quaternary; each valley cutting and filling was a result of changes in sea level. The modern lower Mississippi River Valley displays widths ranging from 60 to 170 km and has a length of 780 km from Cairo, Illinois, to the Gulf of Mexico. Although depths to the scoured base of the valley are highly variable, they range from 125 m in the southern end to 60 m in the vicinity of the Louisiana-Arkansas border (Fisk, 1944, 1947, 1952). The pioneering and classical work of Fisk (1944) outlined the major attributes of the valley cutting and filling episodes and the basic geometry of the various infilling depositional units. The processes of valley cutting and the initial fill as sea level began to rise are shown in the block diagrams of Figure 5. Since this initial work, many additions and concepts concerning the sequence of events, chronology, and processes have been proposed (Saucier, 1974).

Depth to the scoured base of the valley floor, especially near the present coastline, is much deeper than some estimates of the last lowering of sea level. This apparent anomaly indicates that subsidence since late Wisconsinan time has drastically affected the southern part of the valley. The infilling sequence originally described by Fisk (1944) consists of two major facies: a lower coarse, graveliferous sequence referred to as substratum (Fig. 5B) and a finer-grained upper sequence referred to as topstratum. The substratum deposits represent a wide range of environments such as braided channels (the bulk of the substratum deposits), lacustrine, and overbank deposits (crevasse splays, natural levee, swamp, and marsh). However, the bulk of this coarse-grained sediment in the lower part of the valley (south of Vicksburg, Mississippi) was deposited during the latter part of the late Wisconsinan low sea level and the early part of the Holocene rise in sea level. These sediments, based on correlations from offshore borings where isotope chronology is available, are dated from approximately 20,000 to 12,000 B.P. During this period, no evidence of major deltaic sequences has been found on the continental shelf off Louisiana. This is based on approximately 700 offshore borings (Coleman and Roberts, 1988a, b). If major deltaic sequences existed on the shelf during this period of time, they should have been evident in the boring data. Thus, considerable evidence tends to indicate that as sea level commenced its latest rise, the bulk of the sediment load carried by the Mississippi River was being deposited within the valley, resulting in rapid aggradation. This aggradation is especially rapid in the valley fill in off-shore Louisiana and beneath the coastal plain now occupied by the late Holocene Mississippi River deltas. The fill is extremely thick, up to 70 m near the present shoreline. The upper finer-grained sediments, referred to as topstratum by Fisk (1944), represents the various meander belts and overbank sediments deposited within the past 10,000 to 12,000 yr. It is during this time that relatively thick and well-developed delta sequences can be recognized from borings offshore and onshore in the deltaic plain.

The meandering pattern, as shown in the satellite image of Figure 6, that characterizes the modern Mississippi River Valley results from the complex interaction of valley slope, sediment

Figure 6. ERTS satellite image of the lower Mississippi River alluvial valley showing the modern river meander belt. Note the numerous abandoned channel scars and oxbow lakes within the meander belt.

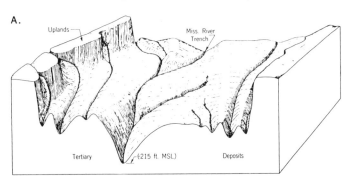

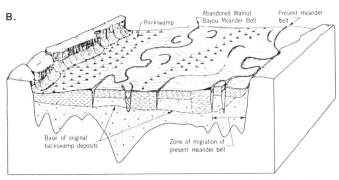

Figure 5. Block diagrams illustrating the evacuated Mississippi River alluvial valley during lowered sea level (A) and the initial coarse "substratum" fill during the initial phase of sea-level rise (B). After Fisk (1944).

load, discharge characteristics, channel hydrology, and the nature of bed and bank materials. River migration within well-defined limits results in the formation of meander belts that consist of the coarse reworked sediments referred to as point bar deposits. The complexity of the meander belts is illustrated in Figure 7, a line drawing illustrating the various lateral migrations of the modern Mississippi River meander belt just north of Baton Rouge, Louisiana. The morphology and internal geometry of the meander belt deposits are extremely complex as can be seen in the figure. The deposits consist of coarse sandy channel lags at the base, overlain by relatively clean, cross-bedded and cross-laminated sands that formed the point bar during channel migration. Capping the coarser sands are finer-grained sands and silts that represent overbank deposition during periods of high water. These deposits are usually well laminated and cyclic in nature. The uppermost capping unit consists generally of highly rooted and bioturbated clays and organic-rich clays that often display early diagenetic cementation by carbonates. The meander belt deposits vary considerably in thickness, ranging from 10 m (generally upstream) to 70 m (downstream). Widths also vary, ranging from about 30 km upstream to less than 10 km downstream. The general tendency is for the meander belt to thicken and become narrower downstream, and the highest internal complexity is generally found upstream. Figure 8 is a cross section of the modern Mississippi River meander belt just south of Baton Rouge, Louisiana (Conrad Point). At this point, the meander belt is relatively narrow, less than 8 km in width. The point bar deposits average 50 m in thickness and display a fining-upward vertical

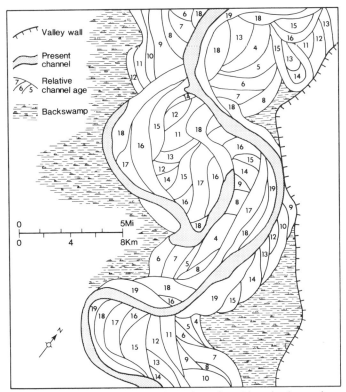

Figure 7. Map depicting the various channel positions in the modern meander belt of the Mississippi River just north of Baton Rouge, Louisiana. Numbers are relative ages ranging from oldest (1) to youngest. Modified from Fisk (1944).

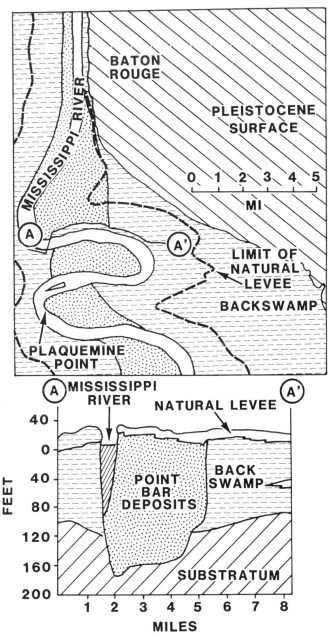

Figure 8. Cross section of Conrad Point, Mississippi River. This point bar is relatively far basinward in the alluvial valley and is relatively narrow in comparison with those upstream.

sequence. The coarsest sediments (medium-grained sands) are located at the base, and rip-up clasts and contorted bedding are common. Above the coarser, largely cross-bedded sands are well-laminated sands and soils containing an abundance of transported organic debris. Capping the point bar sequence are finer-grained deposits that display an abundance of ripple-drift and convoluted laminations. Within the meander belt deposits, linear masses of poorly sorted clays and silts, up to 40 m thick, are often present that represent infilling of abandoned channels and swales. Between the meander belts are generally finer-grained deposits representing overbank sedimentation. These deposits are highly variable, ranging from natural levee silts, crevasse splay sands, silts, clays, organic rich lacustrine clays, lacustrine delta sands and silts, and peaty and organic-rich backswamp clays.

Mississippi River delta. Deltaic sedimentation has dominated the northern Gulf of Mexico basin since Late Jurassic time (Chapters 8 to 11, this volume). This is the major mode of sediment transport and deposition that has been responsible for the huge volume of sediment found in this basin. Within the basin, deltas occur in many different settings and vary greatly in size. Many are very small, especially those forming on the landward sides of lagoons. Some lagoonal deltas, e.g., that of the Colorado River (Texas), have grown to such an extent that they

virtually filled their lagoons during the Holocene and have continued to grow into the open ocean. Another common type of delta in the province is the one that forms at the head of an estuary. The creation of estuaries along the Texas and Mississippi-Alabama coast during the last rise of sea level provided an ideal situation for estuarine delta formation. In estuaries, partly because of limited areal extent, relatively shallow water depths, and reduced marine destructive processes, delta formation proceeds rapidly. These deltas may display typical deltaic morphology, but

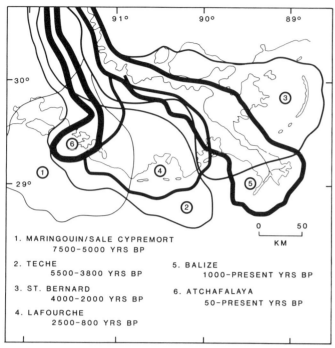

Figure 9. Outlines of the Holocene delta lobes of the Mississippi River delta plain. After Coleman (1988).

generally consist of thin depositional sequences. They do not account for a large volume of deltaic sediment.

The Holocene Mississippi River delta plain (Fig. 9) is not only one of the most studied but is the most important supplier of sediment to the Gulf of Mexico basin. It is one of the world's largest deltas, with an areal coverage of 28,600 km^2, of which 4,700 km^2 (16 percent) is subaqueous. The bulk of the subaerial portion is composed of marshes, interdistributary brackish-water bays, overbank splays, and bay fills. Marine processes, such as shoreline wave power, tides, and littoral currents, are extremely low, and river discharge is high; thus, riverine processes dominate and rapid deltaic progradation is common. Fine-grained sediments are spread far into the Gulf of Mexico; the suspended sediment load is sufficiently high that deposition occurs, and a large, relatively thick subaqueous delta platform is constructed. These marine sediments are rapidly deposited, have a high organic content, and are accompanied by abundant biochemical methane gas production. The sediments are underconsolidated, and excess pore water pressures exist, all of which lead to the creation of various types of subaqueous sediment instability.

The present deltaic plain is mainly the result of deltaic progradation and shifting delta lobes occurring since relative stillstand was reached in Holocene time. During the last low stand of sea level 18,000 yr ago, the Mississippi River was delivering sediment and forming deltas near the present-day shelf edge (Suter and Berryhill, 1985). As sea level began to rise, sediment yield to the coastline diminished, as most of the sediment was being deposited within the rapidly aggrading alluvial valley that had been deeply

eroded during low sea levels. Alluvial valley infilling continued as sea level rose. Commencing approximately 12,000 yr ago, valley infilling was nearly complete, and deltas began to actively prograde. During the period 18,000 to 12,000 B.P., the Louisiana shelf deposits record a relatively thin pelagic clay and a thin sandy shell horizon that mark this hiatus in deltaic development (Coleman and Roberts, 1988a).

By approximately 7,000 to 9,000 yr ago a series of progradational lobes began to develop (Kolb and Van Lopik, 1966; Coleman, 1988). This deltaic activity forms the modern subaerial delta plain of coastal Louisiana (Fig. 9). It is generally accepted that there are seven major deltaic lobes; each displays an active progradational period of 1,000 to 1,500 yr. Internally, these deltaic lobes are quite complex; Frazier (1967) has identified 17 individual delta lobes. The most recent of these lobes, known as the Balize Delta (Fig. 9), is almost a textbook example of the birdfoot-type delta. It has an area of about 600 km^2, which is less than one-fourth the average of 2,700 km^2 for the older lobes. In contrast, it is five to six times as thick (100 to 120 m versus 20 m).

Progradation rates vary among specific distributaries from less than 50 m/yr to more than 100 m/yr. The seaward progradation of Southwest Pass, as determined from historic map comparisons, is illustrated in Figure 10. During the period 1764 to 1979 (195 yr), the mouth of Southwest Pass prograded seaward a distance of 14 km. The width of the distributary mouth bar is on the order of 8 to 10 km, and the thickness of the sand body varies from 20 to 80 m. Thus, in less than 200 yr, a sand body 8 to 10 km wide, 14 km long, and 20 to 80 m thick has formed; this is indeed rapid sedimentation. Slopes are also variable; near the

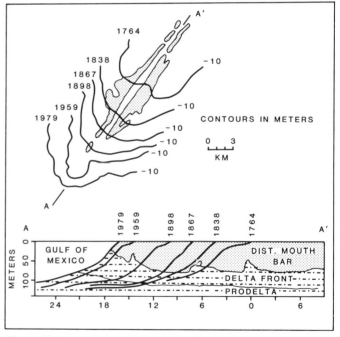

Figure 10. Progradation of Southwest Pass, Mississippi River delta, based on historic maps. Modified and updated from Fisk (1947).

shelf break they average 1.7 to 2.2 degrees, whereas in the inter-distributary bays they are seldom more than 0.2 degrees.

Compaction in the Mississippi River delta generally ranges between 30 and 100 cm/100 yr, although in the vicinity of distributary mouths, where sands are being deposited on weak clays, local subsidence may reach 200 cm/yr (Coleman and Prior, 1980). Also of significance is the consolidation that results from dewatering and degassing of the subaqueous delta deposits. Environments of the subaerial delta include distributary channels, small natural levees, interdistributary bays, overbank splays, bay fills, and marshes. The subaqueous delta includes distributary-mouth bars, delta front or distal bar, and prodelta depositional environments.

Subaerial delta. The distributary channels in the Mississippi River delta range in width from a few meters to 1 km and in depth from 1 or 2 m to more than 30 m. The deeper distributaries often scour through the sandy distributary mouth bar deposits into the underlying finer-grained marine clays. This reduces the tendency for channel migration, and the channels become more stable and display a relatively straight course (Russell, 1967).

The subaerial deltaic plain displays numerous abandoned distributaries. Distributary abandonment may be caused by a number of processes, including storm surges, channel alterations upstream, reduction in channel gradient, and logjams. The channel-fill deposits consist mostly of fine-grained sediments. Often, a coarser channel lag is present near the scoured base of a channel. Well-laminated silts and clays typically form the middle part of the fill, while homogeneous or bioturbated organic-rich clays cap the channel fill.

The interdistributary bays of the Mississippi River delta are normally brackish most of the year, but during periods of high flood, salinity will diminish significantly. These bays range in area to more than 200 km^2 and have irregular shapes. Deposition, which usually occurs only at times of high river flooding, consists mainly of fine-grained sediments, which are usually highly bioturbated. During large floods, crevassing or overbank splays often develop. This process is responsible for the infilling of interdistributary bays and the creation of some of the major land areas of the lower delta. The larger interdistributary bays are filled by a similar process, but generally require a few hundred years for complete infilling. Each bay fill forms initially as a break in the major distributary channel during flood stage, gradually increases in flow through successive floods, reaches a peak of deposition, wanes, and becomes inactive. As a result of subsidence, the crevasse system is inundated by marine waters, reverting to a bay environment, and thus completing a sedimentary cycle (Coleman, 1976). This process is repeated multiple times during the life of a delta lobe, and these bay fills often result in the typical cyclic sedimentary units that characterize the subaerial delta. The historic development of a modern bay fill, Cubits Gap, is illustrated in Figure 11. The initial break occurred in 1860; prior to that time, the bay (Bay Rondo) displayed water depths as much as 12 m. The break occurred in 1860, and by 1884 (Fig. 11), a small complex channel pattern was evident on the historic maps. Pro-

gradation continued until approximately 1946, at which point the channel pattern was extremely complex and a period of deterioration commenced. The map of 1971 (Fig. 11) shows that the marsh cover was deteriorating, and by 1988 most of the marsh cover had been lost, and compaction allowed a bay to encroach across the older delta surface.

The modern Balize lobe of the Mississippi River delta complex is approaching 800 yr in age, and the process of delta switching is already beginning. The Atchafalaya River began to capture discharge from the Mississippi River approximately 100 yr ago. While capturing less than 30 percent of the flow, the Atchafalaya River filled its basin with fine-grained sediment, and some 15 yr ago began a process of building a major subaerial delta lobe (Fig. 9; Roberts and others, 1980). Building of this presently small delta is perhaps the most important event in the Mississippi delta complex within historical times because it represents the initial stage of river diversion to a new site of delta building. At present, the delta (Figs. 9 and 12) is filling Atchafalaya Bay, and along with another small delta from an artificial cut (western delta lobe), approximately 40 km^2 of new deltaic marshland has been developed since subaerial delta emergence in 1973. These small deltas are developing in much the same way as bay fills in the modern Mississippi delta lobe. Although delta growth has been rapid, Mississippi River flow down the Atchafalaya distributary has been artificially controlled at 30 percent since 1963. Left completely to natural processes and discharge, the mechanism of delta switching would be well on its way by now, and the modern Balize lobe would be rapidly undergoing abandonment. Artificial controls will slow down the natural process, but eventually, a major delta lobe will exist along the central Louisiana coast.

Subaqueous delta. The subaqueous portion of the Mississippi River deltaic plain consists of that area of the Gulf of Mexico that is offshore of the Mississippi River delta, below low-tide level, and is actively receiving riverborne sediments. Major subenvironments generally radiate from the mouths of the distributary channels. Three major depositional environments are present: the distributary mouth bar, the delta front, and the prodelta (Fig. 13). The sediment-laden fresh riverwater debouches through the distributary mouth, spreading as a buoyant plume. Water velocity decelerates with distance from the river mouth, and the coarser clastics are rapidly dropped out of suspension. The deposition of these coarser sediments produces the distributary mouth bar, which, being in relatively shallow water, is subject to reworking by marine processes. These deposits also contain large quantities of transported organic matter, commonly referred to as "coffee grounds." The distributary mouth bar deposits display well-developed cross-bedding and cross-laminations, particularly the presence of climbing ripples. Farther offshore, deposition of fine sands, silts, and minor amounts of clay forms the delta front or distal bar environment. These deposits display well-developed reverse-graded and normal-graded laminations. Farthest offshore, deposition is characterized by only fine-grained clays, referred to as prodelta deposits.

Subaqueous slumping and downslope mass movement of

334 J. M. Coleman and Others

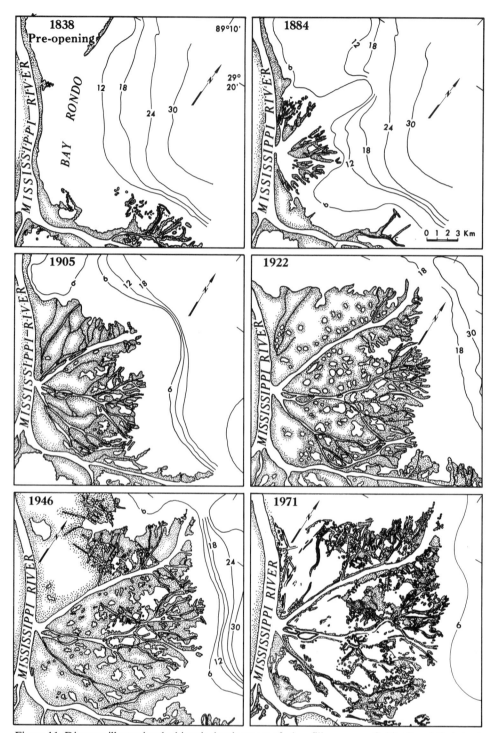

Figure 11. Diagram illustrating the historic development of a bay fill sequence, Cubits Gap, Mississippi River delta. After Coleman (1988).

sediments, although only recently recognized as normal processes, are very important to subaqueous morphology off the Mississippi River delta. Recent research has shown that low-angle slopes at the delta front are unstable and that large amounts of sediment are transported from shallow to deep water in a variety of ways (Coleman, 1976). Factors influencing the stability of bottom sed-

iments include rapid deposition and sedimentary loading, biochemical degradation of organic material and methane gas production, underconsolidation of fine-grained deposits, and cyclic loading induced by passage of winter storms and hurricanes (Coleman and Prior, 1980).

Rapid deposition of the coarser distributary mouth bar sands

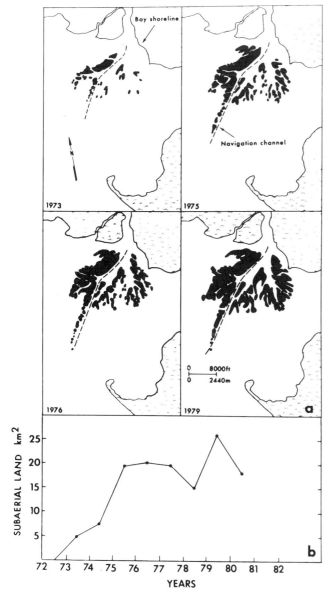

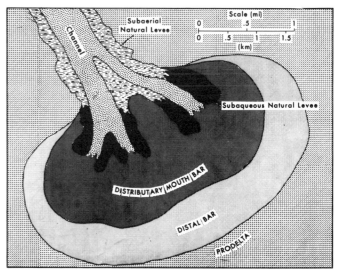

Figure 13. Schematic diagram illustrating the distribution of environments in the subaqueous Mississippi River delta. After Coleman (1988).

Figure 12. Map illustrating the recent development of the Atchafalaya River delta into Atchafalaya Bay. See Figure 9 (delta lobe #6) for location.

over the weaker prodelta clays leads to diapiric intrusions of clay into and through the sands (Fig. 14). Such forms are usually called mud lumps or mud diapirs. Other instability forms include collapse depressions, peripheral rotational slides, mudflow gullies, depositional lobes, and shelf-edge slumps. These types of instabilitics arc schematically illustrated in Figure 14. Detailed description of processes of failure, geometry, and distribution of these features can be found in Coleman (1976) and Coleman and others (1983).

Marshes. The coastal wetlands of the northern Gulf of Mexico basin represent the most extensive marshland development in the continental United States. The largest expanses of

marshes are associated with the broad plains of the Mississippi River delta and its adjacent downdrift (westward) region of western Louisiana and eastern Texas. Salt marsh grasses, such as *Spartina alterniflora,* form a relatively thin seaward zone that is regularly flooded by low tides and storm washovers. The brackish marsh is the most extensive of marsh types and contains dense stands of marsh rushes (*Spartina patens, Distichlis spicata,* and *Juncus roemerianus*). Tidal channel patterns are not well developed and generally consist of a high density of relatively small channels that do not tend to show the pronounced dendritic pattern so characteristic of the tidal channels found in higher tide regions. Tidal ranges are low, rarely exceeding 0.5 m. Within the Mississippi River delta region, the marshes are subjected to high rates of subsidence and compaction, which results in a constantly changing marsh landscape. In many instances, biomass production (marsh accretion) is barely able to maintain pace with rising sea level and subsidence (Hatton, 1981). In the most landward part of the delta plain, freshwater marshes are found. In many instances, these fresh to slightly brackish marshes consist of a relatively thin floating mat of vegetation overlying a highly turbid water and organic ooze. These marshes are referred to locally as flotant (Russell, 1942). He proposed that flotant results when the rooted living marsh mat separates from the substrate (often former delta deposits) because of rapid subsidence. These marshes are prone to rapid deterioration because of the extreme water-level variations that occur during the passage of hurricanes and tropical storms and the variety of destruction caused by wildlife such as nutria, muskrat, and waterfowl (O'Neil, 1949; Chabreck, 1982).

Marsh deposits are characterized by extremely high organic content, often forming relatively thick peat deposits, especially in fresh marsh settings that have been bypassed by clastic deposition.

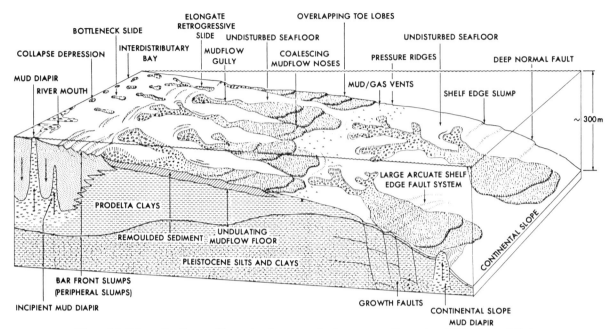

Figure 14. Schematic diagram illustrating the types of subaqueous sediment instabilities in the Mississippi River delta. After Coleman (1988).

The organic clays are highly rooted, and bioturbation is generally so intense that all primary stratification is destroyed. Rapid biochemical processes are highly active, and a wide variety of early diagenetic products are common, including calcium carbonate and iron carbonate nodules and laminations, and replacement of rootlets by vivianite and pyrite.

Subaqueous environments

The submerged portion of the Gulf of Mexico basin can be divided into several major regions or provinces: the west Florida Shelf and Slope, the Mississippi Fan, the Texas-Louisiana Shelf and Slope, the east Mexican Shelf and Slope, the Golfo de Campeche Slope, the Yucatan Shelf, and the Campeche Terrace (Fig. 1).

The sedimentation patterns and nearsurface structure are poorly known for many of these regions; data are available in the north-central Gulf because of the large number of drill sites acquired for offshore petroleum platforms. The modern continental shelf and slope, especially in the north-central Gulf, have been dominated by the huge sediment yields of the Mississippi River and its shifting sites of deposition during the late Wisconsinan and Holocene. Areas in the immediate vicinity of the deltas receive large volumes of sediment, resulting in seaward building of the shelf edge, rapid subsidence associated with sediment loading, and vertical aggradation. Areas removed from the site of active sedimentation accumulate only thin veneers of hemipelagic and pelagic sediment. In some instances, particularly along the eastern and southern margins of the Gulf, carbonate deposition dominates. Another major factor controlling sedimentation patterns

and modern shelf morphology has been changing sea levels associated with the advance and retreat of continental glaciers during the latter part of the Pleistocene and Holocene. When sea level is lowered, sites of active nearshore sedimentation migrate seaward, causing a rapid build-out of the shelf edge in front of the prograding delta lobes or exposure to subaerial erosional processes in areas removed from the sites of active deltaic sedimentation.

Continental shelves. Morphology of modern continental shelves of the Gulf of Mexico basin has been controlled primarily by sea-level fluctuations during the Pleistocene, location of the site of active sedimentation of the Mississippi River, and tectonics associated with active diapirism. Depositional environments on the continental shelf can generally be categorized as terrigenous on the northern and western shelf areas and carbonate on the broad platforms of the eastern and southeastern Gulf. This pattern has persisted since Late Jurassic–Early Cretaceous time (Garrison and Martin, 1973). Most sediment was delivered to the northern margin of the Gulf during the Cenozoic and prograded the shelf as much as 300 km from the margin of the Cretaceous platform deposits to the present shelf edge.

Western Florida Shelf. The continental shelf in this region is extremely broad, often exhibiting widths of nearly 200 km. The surface is relatively smooth and displays little morphologic variability. Most of the sediment zones are oriented parallel to the shelf contours and are thus parallel to the present shelf edge (Fig. 15). The shelf break is rather abrupt in places, and the bottom gradient of the continental slope is extremely steep and dissected by canyons. However, this shelf-slope area is a modern example of a nonrimmed carbonate margin that is much less steep than a typical reef-dominated rimmed margin. The only major bathymetric relief on the western Florida Shelf is formed by bedforms in the mollusc-rich sand and coralline algae ridges (Doyle, 1981; Doyle and Holmes, 1985).

Northeastern Gulf Shelf. The shelf break off westernmost Florida, Alabama, and Mississippi ranges in depth from 60 to 100 m and is characterized by a relict topography covered by a thin sand and mud sheet deposited during the last eustatic sea-level rise (Fig. 16). The shelf is rather narrow because of the large reentrant referred to as the DeSoto Canyon. This large reentrant is not a true submarine canyon, but is formed at the point where the Tertiary clastic sediments lap against the carbonate facies of the northeastern Gulf. The sediments underlying the thin Holocene sediment cover consist of fluvial sands and gravels deposited during the last low sea-level stand. Most of the inner shelf consists of a clean quartz-rich sand. The sand is multicycle, having been reworked numerous times during the changing sea levels associated with the late Pleistocene. Linear shoals, probably representing relict nearshore topography (beach and barrier systems), are present within this sand sheet. Small bedforms are actively migrating on the shelf, indicating modern-day reworking of these relict topographic features. The outer rim of the shelf consists of a lime-mud facies, a mixture of calcium carbonate, quartz, and terrigenous clays. Offshore of Mississippi and Alabama (Fig. 16), the same lime-mud facies exists, but contains several zones of carbonate buildups and inter-carbonate facies. The buildups occur in two depth zones, one at 65 to 80 m and another at 97 to 110 m deep. The carbonate buildup zone is characterized by algal limestone pinnacles, some attaining relief in excess of 15 m. In the inner-reef areas, molluscan shell debris and other carbonate debris in a muddy matrix is the most common lithology. Separating the sand facies from the lime-mud facies is a transition zone consisting of calcareous muddy sands and silts.

Offshore Louisiana Shelf. The shelf off Louisiana is highly variable in width, generally less than 20 km off the active mouths of the Mississippi River delta and in excess of 180 km off the western Louisiana coast. The narrow shelf in front of the Mississippi River is probably the most dynamic area in the submerged Gulf of Mexico margins. Sedimentation rates are extremely rapid, generally in excess of 1 m/yr off the immediate mouths of the Mississippi River. High rate of sediment accumulation results in underconsolidated sediments and inherent sediment instability even though the shelf gradient is generally less than 0.5°. The most common type of instability is the mudflow gullies (Fig. 14). Submarine failures result in a radiating pattern of channels, with seafloor relief in excess of 5 m, that crease the narrow shelf to the edge of the continental slope. It is estimated that as much as 50 percent of the sediment annually delivered to the river mouth is displaced seaward near the shelf edge by these instability processes. At the shelf edge, large-scale shelf-edge failures and growth faults control the shelf-edge morphology (Coleman and others, 1983).

West of the active Mississippi River delta, the shelf is extremely broad and mud covered, displaying little topographic relief. Holocene muds, which blanket the shelf, vary in thickness

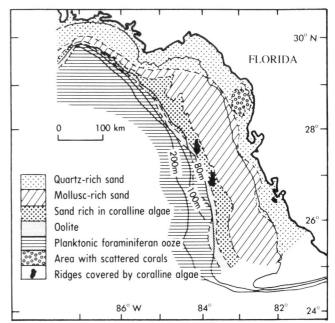

Figure 15. Distribution of sediments on the West Florida Shelf. After Hine (1983).

from a few meters to 10 m and were derived from the Mississippi River by the slow westward drift that characterizes this portion of the northern Gulf. Off central Louisiana, there are a few shoals, the remnants of former deltas that formed during the Holocene transgression and are now stranded on the mid- and inner shelf. These shoals consist of sands reworked from the transgressed delta facies. They form relatively low-relief structures and are generally thin, less than a few meters thick (Boyd and Penland, 1981).

The abundance of offshore foundation borings off the Louisiana coast has resulted in documenting the complex nature of the cyclic deposits associated with changing sea levels during the Wisconsinan (Coleman and Roberts, 1988a, b). The borings reveal that major differences exist in sedimentation patterns during rising and high sea levels as contrasted to falling and low sea levels. Sedimentation during periods of high sea level is characterized by: thin, slowly accumulated depositional sequences, referred to as condensed sections; calcareous deposits, including hemipelagics and shell hashes; wide lateral continuity; and high-amplitude acoustic response. Sedimentation during periods of low sea level is characterized by: variable-thickness, rapidly accumulated sequences, referred to as expanded sections; coarse-grained clastic deposits, including abundant sands and gravels; well-defined depositional trends; and a wide variety of seismic responses (Coleman and Roberts, 1988a, b).

Beneath the Holocene blanket of hemipelagic muds rich in calcareous microfaunal tests are eroded channel systems (Fig. 17) that formed during the last low sea level as the channels of both the Mississippi River and coastal streams entrenched themselves

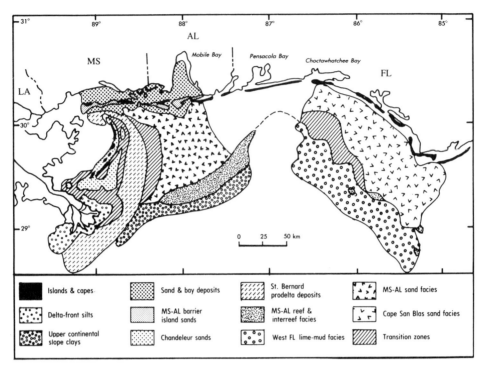

Figure 16. Distribution of surficial sediments on the Florida-Alabama-Mississippi Shelf. After Ludwick (1964).

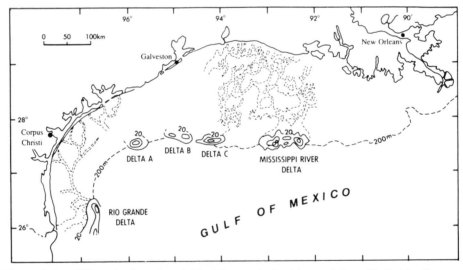

Figure 17. Late Wisconsin channels and delta lobes on the Louisiana and Texas Shelf. After Suter and Berryhill (1985).

in response to the lowered sea level (Suter and Berryhill, 1985). Offshore borings near the outer shelf indicate a considerable volume of fluvial and deltaic sands associated with these low-sea-level riverine systems. Sand thicknesses on the order of 70 to 100 m are present. Figure 18 illustrates three borings taken on the Louisiana continental shelf. The boring in Vermilion Block 124 (Fig. 18A) illustrates a coarse-grained graveliferous channel de-

posit that was scoured across the shelf during low sea level in early Wisconsinan times. The total sand thickness is on the order of 80 m. The boring in West Cameron Block 216 (Fig. 18B) shows two stacked channel sequences separated by a thin condensed section (#5, Fig. 18B). Most of these channel deposits display a blocky nature, but some show the typical fining-upward vertical sequence (unit 4, Fig. 18B). The boring in Main Pass

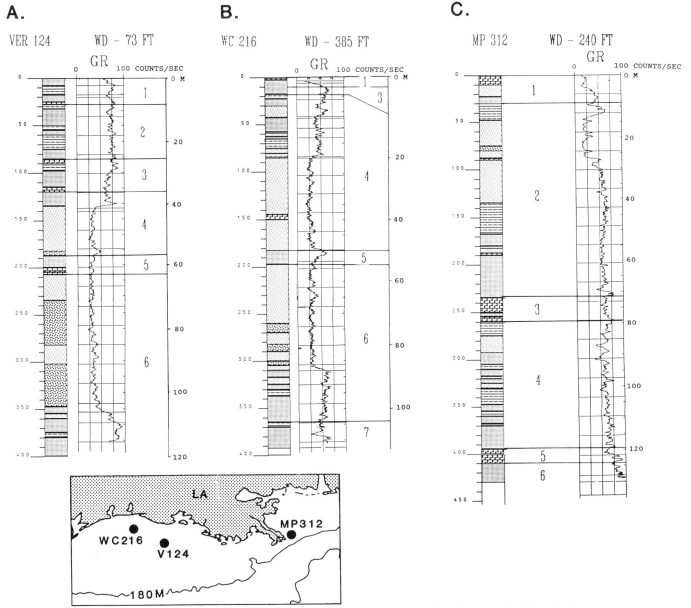

Figure 18. Lithology, gamma-ray trace, and condensed (even numbers) and expanded (odd numbers) sections of borings on the Louisiana continental shelf. A. Vermilion Block 124. B. West Cameron Block 216. C. Main Pass Block 312. Depth in feet. The numbers in the "Stage" column represent correlations to ^{18}O isotope stages. The odd numbers are periods of high sea-level stand and the even numbers are periods of low sea-level stand. After Coleman and Roberts (1988a).

Block 312 (Fig. 18C) illustrates the vertical sequence associated with a falling- and low-sea-level shelf-edge delta. Unit 3 (Fig. 18C) is the condensed section laid down during the previous high sea level, whereas unit 2 (Fig. 18C) displays the classical coarsening-upward vertical sequence typical of deltaic facies. Figure 19 is a high-resolution seismic record run across a shelf-edge delta (seismic line is in vicinity of the boring illustrated in Fig. 18C). Note the wedge shape of the delta unit and the presence of the fore-set seismic reflectors. Note also that the uppermost delta unit (between 0.3 and 0.4 sec) overlies an older delta

unit that is poorly defined on the seismic line. The shelf-edge deltas in this region are generally 50 to 60 m in thickness.

Offshore Texas Shelf. The shelf offshore of Texas is relatively broad, displaying only minor topographic relief. Shelf sediments contain only a thin cover of Holocene deposits overlying an eroded late Pleistocene subaerial fluvial plain. In many instances, upper Pleistocene sediments crop out on the sea floor. Entrenched stream channels, similar to those off the Louisiana Shelf, are common across the shelf, but have been completely infilled during the Holocene rising-sea-level stage (Suter and Ber-

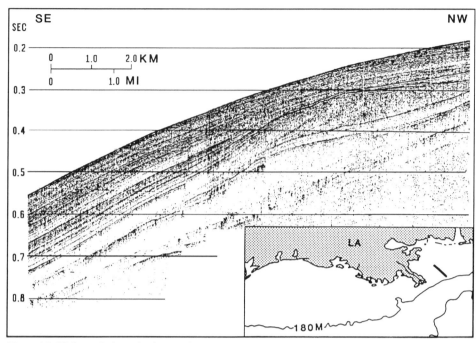

Figure 19. High-resolution seismic record run across a shelf-edge delta in the Main Pass area, Mississippi River delta. After Coleman and Roberts (1988a).

ryhill, 1985). The shelf edge consists of a series of prograded late Pleistocene shelf-edge deltas that formed in response to the lowering of sea level (Suter and Berryhill, 1985). Little information is available on types of sediment composing these delta lobes and the thicknesses of the sequences. Much of the shelf edge is characterized by a crenulated topography, the result of sediment instability processes during the late Pleistocene and by diapiric intrusion of salt and shale masses. Both deep-seated and shallow compaction faults are common across the shelf, but show little topographic expression because of truncation during the last eustatic rise of sea level (Berryhill, 1980, 1981a, b). Along the outer shelf edge, carbonate banks are common, similar to those off the Louisiana Shelf. Those off Texas, however, do not tend to have much Holocene sediment cover. The most prominent of these carbonate banks are the West and East Flower Garden Banks off Galveston, Texas, which have developed into true coral-algal reefs. These reefs display active carbonate-producing faunal and floral growth. The reefs have developed on diapiric structures near the shelf edge. Many of the carbonate banks cap diapiric highs and were actively growing during the lowered late Pleistocene sea level. Today, only a few display active growth because of Holocene sediment cover or lack of sunlight penetration because of water depth.

Continental Slope. The continental slope of the Gulf of Mexico is a region of gently sloping sea floor that extends from the shelf edge, or roughly at the 200-m isobath, to the upper limit of the continental rise, at a depth of 2,800 m. The slope occupies more than 500,000 km^2 of prominent escarpments, smooth and gently sloping surfaces, knolls, small-scale pinnacles, intraslope basins, ridge and valley topography, and submarine channels (Martin and Bouma, 1978; Bouma and others, 1978b; Chapter 2 and Plate I, this volume). Martin and Bouma (1978) described 9 distinctive subprovinces and many individual features (Fig. 1). The factors that have controlled the present-day morphology include: reef building on the Florida and Yucatan carbonate platforms; erosion, nondeposition, and faulting in the Straits of Florida and Yucatan Channel; diapirism and differential sedimentation on the slopes off Texas and Louisiana; tectonic uplift and diapirism in the Golfo de Campeche; rapid accumulation of terrigenous sediment of Pliocene and Pleistocene age in offshore Louisiana; and the folding of a thick blanket of sediment due to décollement on the slope off eastern Mexico.

Texas-Louisiana Slope. The most complex province is the Texas-Louisiana Slope, containing a 120,000 km^2 area of knoll and basin sea floor (Plate I, this volume). The average gradient of the slope is slightly less than 1°, but slopes greater than 20° are not uncommon around the many knolls and basins. Steep-sided knolls, enclosed intraslope basins, and canyon-like topography characterize the eastern two-thirds of the slope, whereas the occasional knolls and low-relief noses mark the otherwise featureless slope of the western sector (Martin and Bouma, 1978; Bouma and others, 1978a). The extreme topographic relief of the slope is a product of salt diapirism (Figs. 1 and 2) and salt withdrawal beneath the basins. Intraslope basins, such as the Gyre Basin (Bouma and others, 1978b) and the Orca Basin (Trabant and Presley, 1978), are flanked and commonly surrounded by salt domes and contain exceptionally thick sections of Tertiary sediments (Bouma and Coleman, 1986). The basins are directly re-

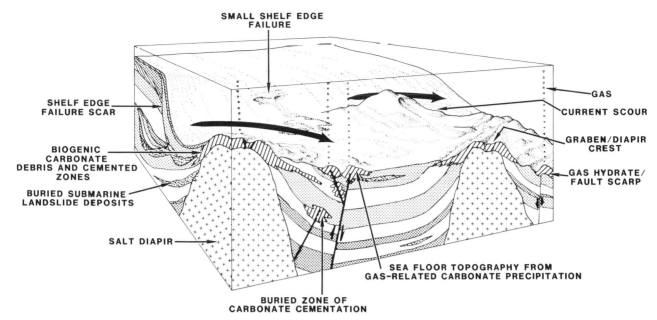

Figure 20. Schematic diagram illustrating the major types of features on the Texas-Louisiana continental slope.

lated to the growth of the adjacent salt spines, which blocked active submarine channels or coalesced to create sea-floor expression in noncanyon areas.

The Sigsbee Escarpment (Fig. 1) is the most prominent feature at the base of the slope. This escarpment is nearly continuous along the entire base of the slope from the western Gulf to the DeSoto Canyon. The scarp is the expression of the lobate frontal edge of the northern Gulf diapiric province and is underlain throughout its length by a complex system of salt ridges, overthrust tongues, and steep-sided massifs (Humphris, 1978; Martin, 1978, 1984). The continuity of the escarpment is broken locally by diapiric outliers and large, pronounced reentrants of several large interlobal canyons such as the Alaminos and Keathley Canyons (Fig. 1).

Mapping of high-resolution seismic and side-scan sonar data on the upper continental slope of Texas-Louisiana reveals a wide range of near-surface geologic features. The principal feature associations are shown schematically in Figure 20, which depicts a typical, outer continental shelf–upper continental slope setting. Although Figure 20 is schematic in nature, a map depicting two salt diapirs and related near-surface geological features (Green Canyon offshore area) is shown in Figure 21. A variety of complex faults, carbonate banks, seafloor erosion, gassy sediments, and displaced sediments are common features on the continental slope.

The shelf edge is commonly marked by a distinctive break in slope, but the geometry and gradients are dependent on the presence or absence of several factors: faulting, diapirism, local low-sea-level sediment accumulations (shelf-edge deltas), fluvial channels, the development of bioherms, or shelf-edge erosion. Figure 20 depicts a series of prograding inclined strata represent-

ing delivery of sediment to the shelf edge during lowered sea level. Because of the rapid accumulation, these sediments show post-depositional and syn-depositional features. Typically, growth faults tend to accentuate the shelf-edge break. Some of these faults extend through the near-surface sediments to the sea floor and have associated gas seeps, indicating their deep origin within the sedimentary sequence. Associated with some of the

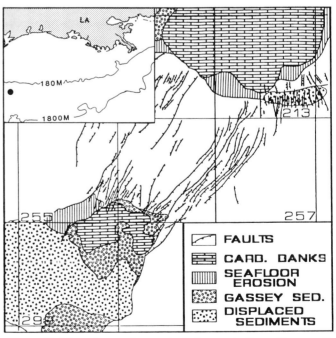

Figure 21. Map of nearsurface geologic features in Green Canyon offshore blocks.

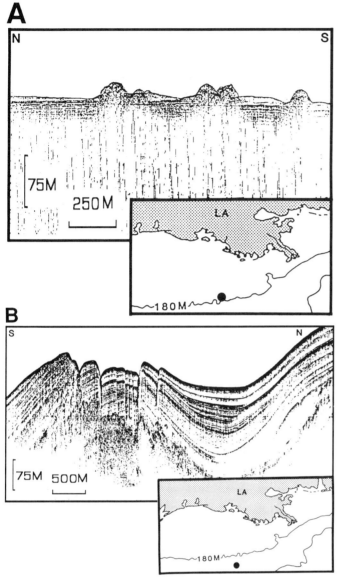

Figure 22. High-resolution seismic records illustrating geologic features on the Louisiana upper continental slope. A. Pinger record illustrating carbonate buildups. B. Air gun record showing complex faulting on salt diapiric crest and the adjacent salt-withdrawal basin.

growth fault systems are both large-scale and localized landslides. Sediments overlying such features typically display differential compaction or may inherit geometries from the underlying landslide morphology.

Because of changes in sea level, banks and bioherms are common occurrences at the shelf-edge break and on top of upper-slope diapiric spines. Figure 22A illustrates a pinger record run across several small bioherms on the upper continental slope. The carbonate buildups are generally less than 0.5 km in diameter and have relief on the order of 8 to 10 m. They generally occur in relatively well defined bands as shown in Figure 21. Their distri-

bution at the sea floor depends on sedimentation and erosion patterns during subsequent rise in sea level. Sea-floor erosion on the shelf and shelf break often represents relict features, but in some cases may be continuing to the present in response to oceanographic factors.

Diapiric ridges, spines, and intraslope basins are common features on the upper and middle continental slope. Figure 22B is a high-resolution air gun record run across a small salt diapir and the adjacent salt-withdrawal basin. Surface expression and relief is complex, being dominated by faulting associated with differential growth of the diapir. Complex fault patterns exist across the diapir's crest (Fig. 22B), often extending into adjacent basins and forming intricate and widespread fault systems (Fig. 21). The intraslope basins are often the site of extremely thick accumulations of terrigenous sediments. Drilling of two intraslope basins (Orca and Pigmy basins) during DSDP Leg 96 showed that the intraslope basins received large volumes of sediment during both high and low sea levels (Bouma and others, 1986; Bouma and Coleman, 1985; Bouma and Stelting, 1986). Cyclicity is observed on high-resolution seismic records, such as shown in Figure 23, a seismic record acquired across Pigmy basin. Drilling in Pigmy basin during Leg 96 indicated that these cyclic seismic responses were the result of differential sediment accumulation during changes in sea level. Accumulation rates during the Holocene high sea level was 166.7 cm/1,000 yr; the rate was 84.6 cm/1,000 yr for the last low-sea-level period. Thin-bedded turbidites and hemipelagic sediments dominated the sedimentary section. Displaced sediments, likely from the adjacent steep walls of the diapiric flanks, were common within the borings.

In addition to the major faults associated with diapiric structures, complex down-to-the-basin faults are common on the upper continental slopes. Figure 24A illustrates these types of complex fault systems. Such faulting provides pathways for gas migration, both thermogenic and biogenic, into the near-surface sediments. Chemosynthetic organisms, rapid diagenesis and cementation, and clathrate development are associated with these gas seeps, the latter exclusively with deep-source thermogenic gas (Figs. 20 and 21). All these features can result in acoustic wipe-outs on high-resolution geophysical data, and their interpretation (together with chaotic landslide deposits) is difficult. Gas seeps are especially common; Figure 24B illustrates a multi-channel seismic record run across a rather large gas seep on the middle continental slope off Louisiana. Erosion across the diapir crests often exposes older and greater-consolidated sediments, leading to highly diverse acoustic images. Growth of diapirs and associated oversteepening of their flanks can promote localized and large-scale near-surface, as well as deeply buried, landslide features.

The Mississippi Trough or Canyon (Fig. 1) is a broad submarine channel that formed by retrogressive slumping during late Pleistocene times and has been partially infilled during the Holocene rise in sea level (Coleman and others, 1983). At the base of the slope, in the upper part of the Mississippi Fan, sediments 2,400 m thick have been deposited in the past 2.1 m.y.

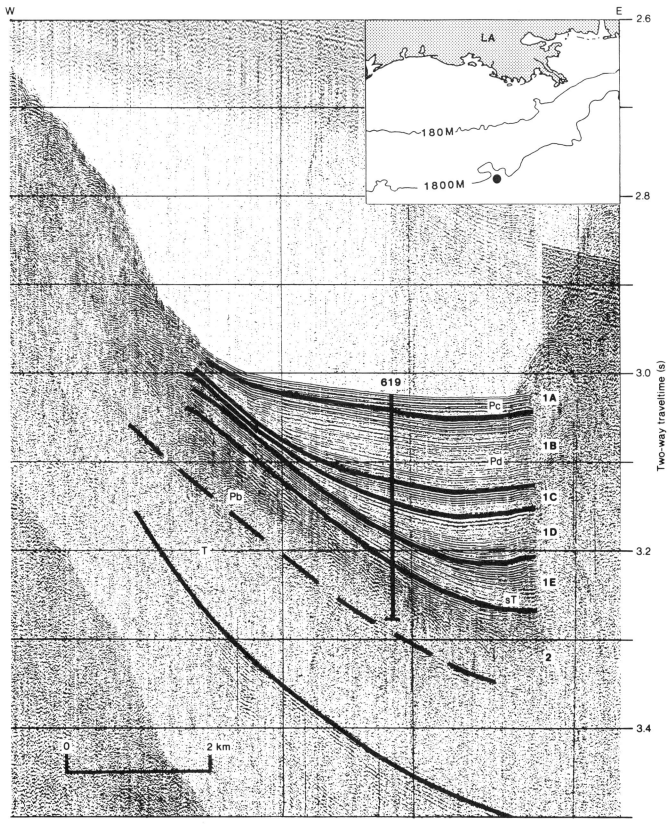

Figure 23. U.S. Geological Survey line 117 across Pigmy Basin, Louisiana continental slope. This minisparker seismic profile across an intraslope salt-withdrawal basin illustrates the seismic cyclicity on the continental slope. After Bouma and Coleman (1986).

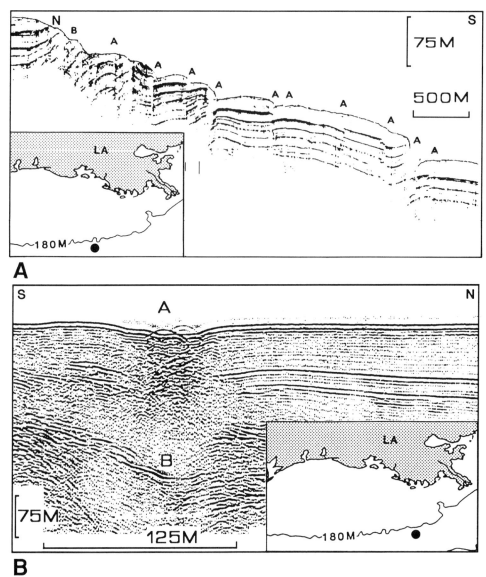

Figure 24. High-resolution seismic records illustrating geologic features on the Louisiana upper continental slope. A. Pinger record illustrating complex down-to-the-basin faulting. B. Multi-channel seismic trace illustrating a large gas seep.

Mississippi Fan. The Mississippi Fan covers an area in excess of 300,000 km^2 and has a volume of about 290,000 km^3 (Stuart and Caughey, 1976; Moore and others, 1979). The canyon-fan system commences on the continental shelf southwest of the modern Mississippi River where the Mississippi Canyon cuts across the continental slope (Fig. 25A). At approximately 1,500 m water depth, the canyon widens significantly, and overbanking from the incised channel can be observed on seismic data. Leg 96 of the Deep Sea Drilling Project cored several sites on the Mississippi Fan (Fig. 25B).

The Quaternary Mississippi Fan consists of several fan lobes, each having an elongate shape. These lobes are connected to an incised submarine canyon cut into the continental shelf and slope. The fan-lobe complex is basically composed of channel-overbank deposits that can be subdivided into four regions (Fig. 25B): (1) a canyon that was probably formed by massive slope failure; (2) the upper fan, which terminates near the base of the slope (the main channel acts as a conduit for sediments delivered to the more distal parts of the system); (3) the middle fan, which is an aggradational unit with a convex upward surface and a sinuous aggradational channel (graveliferous and sand-rich deposits form the basal portion of the channel and are capped by fine-grained sediments of the passive fill); and (4) the lower fan, which is also aggradational in nature and shows evidence of having numerous small channels that have switched through time. Near the ends of each channel, broad sand sheets have been deposited. Late Pleistocene sedimentation rates were extremely high during glacial low sea-level stands and minimal during inter-

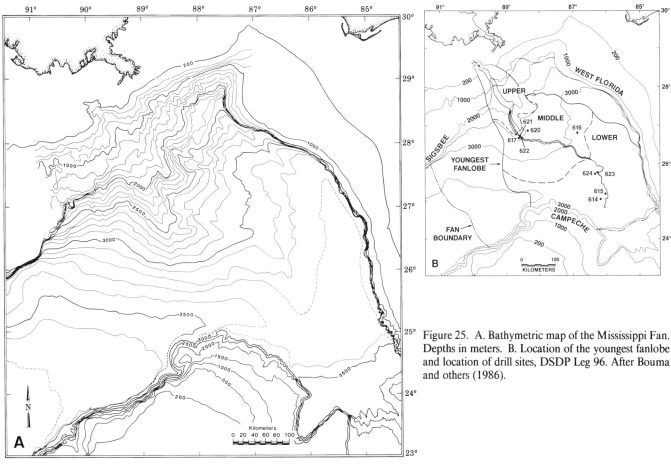

Figure 25. A. Bathymetric map of the Mississippi Fan. Depths in meters. B. Location of the youngest fanlobe and location of drill sites, DSDP Leg 96. After Bouma and others (1986).

glacial high sea levels. In the youngest fanlobe, during the late Wisconsinan glacial stage, accumulation rates ranged from 12 m/1,000 yr for the middle fan to 6 m/1,000 yr for the lower fan. During the interglacials, sedimentation rates rarely attained a few tens of centimeters per thousand years. The Mississippi Fan is a complex of overlapping fan lobes that were deposited during low sea-level stands in the Pleistocene (Moore and others, 1978; Bouma and Coleman, 1985). Drilling on the Mississippi Fan (Fig. 25B) during DSDP Leg 96 (Bouma and others, 1986; Feeley and others, 1985; Stelting and others, 1985; Coleman and others, 1985; Roberts and Thayer, 1985) documented the sedimentologic and faunal characteristics of these deposits.

The middle fan (Fig. 25B) begins in water depths that range from 2,200 to 2,300 m. It is marked by a change in gradient that corresponds to the base of the slope. The middle fan is lenticular in cross section, and the youngest fan lobe (late Wisconsinan in age) is more than 400 m thick. Figure 26 illustrates a map constructed from the SeaMARC side-scan sonar images. The prominent, sinuous channel varies in width from slightly greater than 3 km to less than 1 km, and displays relief from the channel thalweg to the top of the overbank of 40 to 50 m. Adjacent to the channels are a series of ridges and swales and abandoned scars. This channel complex (meander belt) is approximately 10 to 24 km in width and covers a much broader area than the modern bathymetric channel. Seismic data indicate that the base of the

eroded channel is approximately 420 m below the sea floor. The sinuosity of the channel and other morphologic characteristics, such as ridges and swales, suggest a migratory nature of the mid-fan channel. The drilling results indicated that the presence of an acoustic high-amplitude zone at the base of the channel fill represents a "coarse-grained" channel lag deposit (Stelting and others, 1985). Figure 27 illustrates a high-resolution seismic line (upper) and an interpreted section (lower) acquired across this channel-overbank sequence. The interpreted illustration attempts to show that the channel migrated laterally and aggraded through time, leaving behind a sequence of coarse-grained channel lag deposits that are represented by high-amplitude reflectors in seismic data.

Figure 28 illustrates the lithologic log, gamma-ray trace, and a core from DSDP drill site 621 (see Fig. 25B for location) on the mid-fan channel. The channel fill is composed almost entirely of redeposited sediments that contain shallow-water microfauna. The vertical sequence displays a fining-upward trend, and the basal channel lag deposits are graveliferous in nature (Fig. 28).

The lower fan commences in water depths of approximately 3,100 to 3,200 m (Fig. 25A). Four drill sites (Fig. 25B) were acquired in this part of the Mississippi Fan. Figure 29 was constructed from SeaMARC side-scan sonar and depicts the major morphologic characteristics of the lower fan region (Bouma and others, 1986). The main channel on the central part of the young-

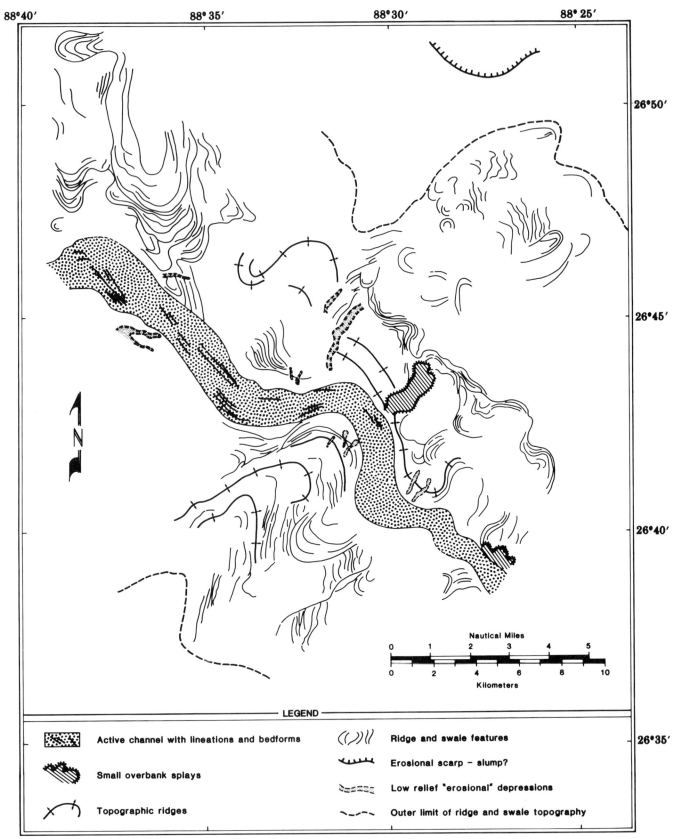

Figure 26. Morphology of the middle fan as mapped from Gloria and SeaMARC side-scan sonar data. After Bouma and others (1986).

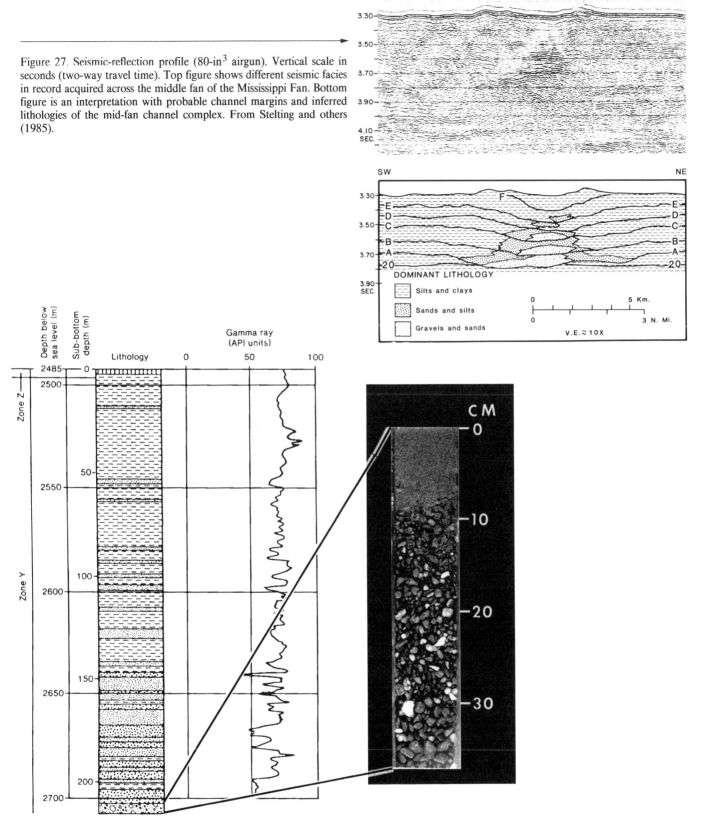

Figure 27. Seismic-reflection profile (80-in^3 airgun). Vertical scale in seconds (two-way travel time). Top figure shows different seismic facies in record acquired across the middle fan of the Mississippi Fan. Bottom figure is an interpretation with probable channel margins and inferred lithologies of the mid-fan channel complex. From Stelting and others (1985).

Figure 28. Lithologic log, gamma-ray trace, and core photograph of DSDP site 621, mid-fan channel drill site of the Mississippi Fan. See Figure 25B for location of boring. Modified from Bouma and others (1986).

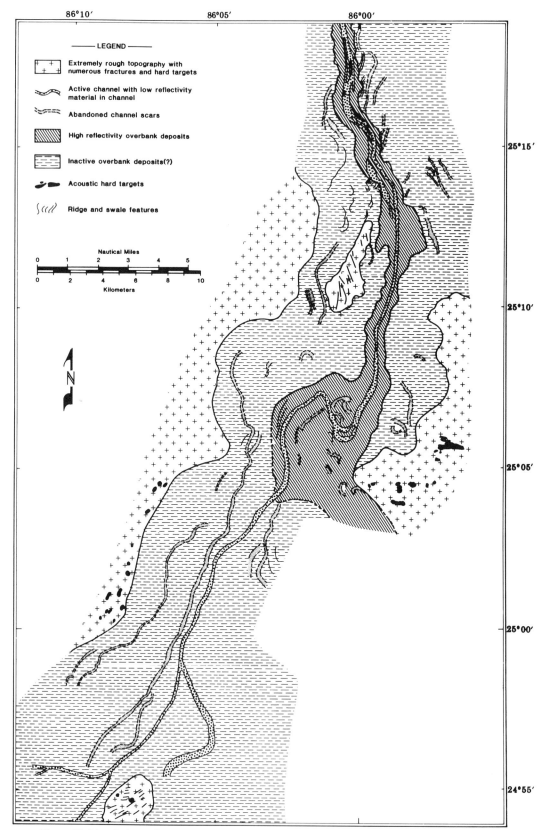

Figure 29. Morphology of a portion of the lower Mississippi Fan as mapped from SeaMARC side-scan sonar data. From Bouma and others (1986).

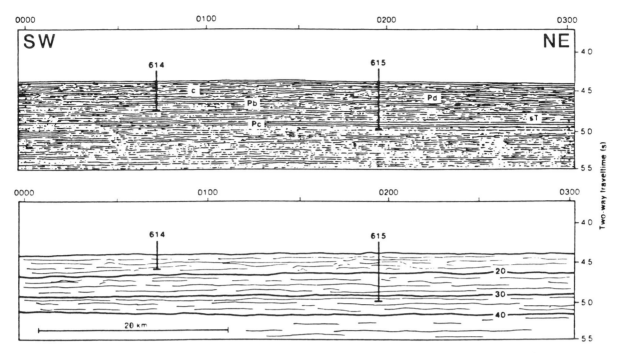

Figure 30. Seismic trace (upper) and line drawing (bottom) through sites 614 and 615, DSDP Leg 96, Mississippi Fan. From Bouma and others (1986).

est fan lobe has narrowed significantly, and the channel is not as sinuous as on the middle fan. In addition, the zone of abandoned meander scars is not apparent; instead there appear to be numerous, relatively straight, abandoned channels adjacent to the latest channel. The acoustic high-amplitude zone at the base of the channel that was so prominent in the middle fan becomes less prominent and gradually disappears downfan. A seismic section and line drawing near the downfan terminus of the channel is illustrated in Figure 30. The acoustic reflectors are more continuous and generally exhibit less divergence than in those in the middle fan. The channel, so prominent in the mid-fan seismic lines, has broken up into numerous smaller channel-overbank sequences that are relatively hard to discern. Correlation of the seismic records with the borings indicates that the higher-amplitude, slightly discontinuous seismic reflectors could be correlated with the sandier units that have been inferred to represent sheet-sands deposited at the terminus of the channels. Individual reflectors show low lateral continuity, but packages of reflectors can be correlated over considerable distances (O'Connell and others, 1985).

Drilling in the lower fan indicated the presence of numerous sheet sands of considerable thickness deposited primarily during periods of falling and low sea level. Figure 31 illustrates the lithologic log, wire-line traces, and a core from DSDP drill site 615 acquired on the lower fan (Fig. 25B). Two fan lobes were penetrated in the 500-m cored boring; the lower fan lobe contained 65 percent net sand, and the upper fan lobe contained 47 percent net sand. Between the sheet sands, thin-graded sand beds dominated the deposits.

The DeSoto Slope and Canyon (Fig. 1) lies between the eastern limit of the upper Mississippi Fan and the West Florida Terrace. This section of the slope is relatively smooth and is underlain by a thick sequence of conformably bedded sediment that is deformed by minor monoclinal folds and isolated small salt domes. The most conspicuous physiographic element in this area is the broad valley formed by the depositional convergence of the terrigenous slope with the northernmost exposure of the Florida Escarpment.

Mississippi-Alabama-Florida Slope. The northern portion of the continental slope off Mississippi, Alabama, and western Florida is relatively smooth and unbroken. Thick sequences of salt and shale are not present in this region, and the clastic wedge overlying the carbonate platform is relatively thin and undeformed. Small and large erosional gullies, possibly resulting from the late Wisconsinan lowering of sea level or relatively minor mass wasting, crease this continental slope.

The continental slope along the eastern margin of the Gulf of Mexico is extremely complex and is defined by a double reef trend. The shallower reef lies between 130 and 150 m, whereas the deeper reef trend ranges between 210 and 300 m water depth (Doyle and Holmes, 1985). Late Pleistocene and Holocene sediments composed of foraminifera-coccolith ooze accumulated at an exceedingly rapid rate, averaging 30 cm/1000 yr (Doyle and Holmes, 1985). At a water depth between 1,000 and 2,000 m, the slope increases significantly, in places exceeding 20 degrees, and forms the West Florida Escarpment. The escarpment is erosional in nature and is composed of sediments deposited in a shallow-water, back-reef, and lagoonal facies (Freeman-Lynde,

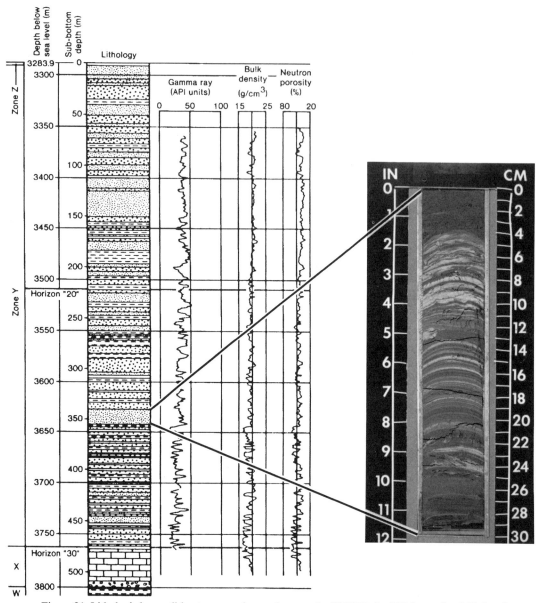

Figure 31. Lithologic log, well-log traces, and core photograph of DSDP site 615, lower fan drill site of the Mississippi Fan. See Figure 25B for location of boring. Modified from Bouma and others (1986).

1983). On the upper slope, at a water depth of approximately 500 m, a terrace, believed to be composed of outcropping Miocene sediments, is a prominent morphologic element. A series of gullies and small canyons crease the middle and upper slope. Some of the most striking features of the West Florida upper continental slope are the irregular morphology associated with mass movement processes. The mass wasting features range from creep to massive slides to gravity-induced folds tens of kilometers long (Doyle and Holmes, 1985).

CONCLUSIONS

The entire Gulf of Mexico basin is bounded to the east and south by carbonate platforms, while the remaining regions of the basin are characterized by terrigenous clastics. The late Quaternary deposits of the northern Gulf of Mexico basin show a complex distribution pattern that has been strongly influenced by changing eustatic sea levels during the Quaternary and shifting sites of deposition of the main sediment supplier, the Mississippi River. Most of the sedimentary sequences are characterized by cyclic sedimentation resulting from the constant alongshore and on/offshore shifting of the depositional sites, and by major transgressions and regressions associated with eustatic sea-level changes. The outer continental shelf and continental slope have been drastically modified by a wide variety of sediment instability processes, faulting, and diapiric intrusion. The complexities of the sedimentation patterns have been deciphered only because the sediments are young enough to yield a relatively precise chronostratigraphic framework. It is highly probable that similar depositional patterns have existed within the Gulf of Mexico basin throughout the late Mesozoic and Tertiary, but lack of a high-resolution chronostratigraphic framework does not allow the precise differentiation of this thick sedimentary wedge.

REFERENCES CITED

Bernard, H. A., Leblanc, R. J., and Major, C. F., 1962, Recent and Pleistocene geology of southeast Texas, field excursion no. 3, *in* Rainwater, E. H., and Zingula, R. P., eds., Geology of the Gulf Coast and central Texas; Geological Society of America Guidebook of Excursions: Houston Geological Society, p. 174–224.

Berryhill, H. L., Jr., 1980, Map showing paleogeography of the continental shelf during the low stand of sea level, Wisconsin glacial epoch, Port Isabel Quadrangle, Texas: U.S. Geological Survey Map I-1254-E, scale 1:250,000.

—— , 1981a, Map showing paleogeography of the continental shelf during the low stand of sea level, Wisconsin glacial epoch, Beeville Quadrangle, Texas: U.S. Geological Survey Map I-1288-E, scale 1:250,000.

—— , 1981b, Map showing paleogeography of the continental shelf during the low stand of sea level, Wisconsin glacial epoch, Corpus Christi Quadrangle, Texas: U.S. Geological Survey Map I-1287-E, scale 1:250,000.

Bloom, A. L., 1983, Sea level and coastal morphology of the United States through the late Wisconsin glacial maximum, *in* Wright, H. E., Jr., ed., Late-Quaternary environments of the United States: Minneapolis, University of Minnesota, v. 1, p. 215–229.

Bouma, A. H., and Coleman, J. M., 1985, Mississippi fan; Leg 96 program and principal results, *in* Bouma, A. H., Normark, W. R., and Barnes, N. E., eds., Submarine fans and related turbidite systems: New York, Springer-Verlag, p. 247–252.

—— , 1986, Interslope basin deposits and potential relation to continental shelf, northern Gulf of Mexico: Gulf Coast Association of Geological Societies Transactions, v. 36, p. 419–428.

Bouma, A. H., and Stelting, C. E., 1986, Seismic stratigraphy and sedimentary processes in Orca and Pigmy basins, *in* Bouma, A. H., Coleman, J. M., Meyer, A. W., and others, eds., Initial reports of the Deep Sea Drilling Project: Washington, D.C., U.S. Government Printing Office, v. 96, p. 563–576.

Bouma, A. H., Moore, G. T., and Coleman, J. M., 1978a, Framework, facies, and oil-trapping characteristics of the upper continental margin: American Association of Petroleum Geologists Studies in Geology no. 7, 326 p.

Bouma, A. H., Smith, L. B., Sidner, B. R., and McKee, T. R., 1978b, Interslope basin of northwest Gulf of Mexico, *in* Bouma, A. H., Moore, G. T., and Coleman, J. M., eds., Framework, facies, and oil-trapping characteristics of the upper continental margin: American Association of Petroleum Geologists Studies in Geology no. 7, p. 289–302.

Bouma, A. H., Coleman, J. M., Meyer, A. W., and others, 1986, Initial reports of the Deep Sea Drilling Project: Washington, D.C., U.S. Government Printing Office, v. 96, 824 p.

Boyd, R., and Penland, S., 1981, Washover of deltaic barriers on the Louisiana coast: Gulf Coast Association of Geological Societies Transactions, v. 31, p. 243–248.

Buffler, R. T., 1981, Seismic stratigraphy and geologic history of the Gulf of Mexico basin: Gulf Coast Association of Geological Societies 1981 meeting, Corpus Christi, Texas: American Association of Petroleum Geologists Short Course, 172 p.

Byrne, J. V., LeRoy, D. O., and Riley, C. M., 1959, The chenier plain and its stratigraphy, southwestern Louisiana: Gulf Coast Association of Geological Societies Transactions, v. 9, p. 237–270.

Chabreck, R. H., 1982, The effect of coastal alteration on marsh plants, *in* Boesch, D. F., ed., Proceedings of the Conference on Coastal Erosion and Wetland

Modification in Louisiana; Causes, consequences, and options: U.S. Department of the Interior FWS/OBS–82/59, p. 92–98.

Coleman, J. M., 1966, Recent coastal sedimentation; Central Louisiana coast: Baton Rouge, Louisiana State University Coastal Studies Institute Technical Report 29, 73 p.

—— , 1976, Deltas: processes of deposition and models for exploration: Champaign, Illinois, Continuing Education Publication, 102 p.

—— , 1988, Dynamic changes and processes in the Mississippi River delta: Geological Society of America Bulletin, v. 100, p. 999–1015.

Coleman, J. M., and Roberts, H. H., 1988a, Sedimentary development of the Louisiana continental shelf related to sea level cycles; Part 1, Sedimentary sequences: Geo-Marine Letters, v. 8, p. 63–108.

—— , 1988b, Sedimentary development of the Louisiana continental shelf related to sea level cycles; Part 2, Seismic response: Geo-Marine Letters, v. 8, p. 109–119.

Coleman, J. M., and Prior, D. B., 1980, Deltaic sand bodies: American Association of Petroleum Geologists Continuing Education Course Note Series 15, 171 p.

Coleman, J. M., Prior, D. B., and Lindsay, J., 1983, Deltaic influences on shelf-edge instability processes, *in* Stanley, D. J., and Moore, G. T., eds., The shelfbreak; Critical interface on continental margins: Society of Economic Paleontologists and Mineralogists Special Publication 33, p. 121–137.

Coleman, J. M., Bouma, A. H., Roberts, H. H., Thayer, P. A., and DSDP Leg 96 Scientific Party, 1985, X-ray radiography of Mississippi Fan cores, *in* Bouma, A. H., Normark, W. R., and Barnes, N. E., eds., Submarine fans and related turbidite systems: New York, Springer-Verlag, p. 311–318.

Doyle, L. J., 1981, Depositional systems of the continental margin of the eastern Gulf of Mexico west of peninsular Florida; A possible modern analog to some depositional models for the Permian Delaware Basin: Gulf Coast Association of Geological Societies Transactions, v. 31, p. 279–282.

Doyle, L. J., and Holmes, C. W., 1985, Shallow structure, stratigraphy, and carbonate sedimentary processes of West Florida upper continental slope: American Association of Petroleum Geologists Bulletin, v. 69, p. 1133–1144.

Doyle, L. J., and Sparks, T. N., 1980, Sediments of the Mississippi, Alabama, and Florida (MAFLA) continental shelf: Society of Economic Paleontologists and Mineralogists Journal, v. 50, p. 905–916.

Feeley, M. H., Buffler, R. T., and Bryant, W. R., 1985, Depositional units and growth patterns of the Mississippi Fan, *in* Bouma, A. H., Normark, W. R., and Barnes, N. E., eds., Submarine fans and related turbidite systems: New York, Springer-Verlag, p. 253–257.

Fisk, H. N., 1944, Geological investigations of the alluvial valley of the lower Mississippi River: Vicksburg, Mississippi, U.S. Army Engineers Mississippi River Commission, 78 p.

—— , 1947, Fine-grained alluvial deposits and their effects on Mississippi River activity: Vicksburg, Mississippi, U.S. Army Engineers Waterways Experiment Station, v. 1 and 2, 82 p.

—— , 1952, Geological investigation of the Atchafalaya Basin and the problem of Mississippi River diversion: Vicksburg, Mississippi, U.S. Army Corps of Engineers Waterways Experiment Station, v. 1, 145 p.

Frazier, D. E., 1967, Recent deltaic deposits of the Mississippi River; Their development and chronology: Gulf Coast Association of Geological Societies Transactions, v. 17, p. 287–315.

Freeman-Lynde, R. P., 1983, Cretaceous and Tertiary samples dredged from the Florida escarpment, eastern Gulf of Mexico: Gulf Coast Association of Geological Societies Transactions, v. 33, p. 91–99.

Garrison, L. E., and Martin, R. G., 1973, Geologic structures in the Gulf of Mexico basin: U.S. Geological Survey Professional Paper 773, 85 p.

Gould, H. R., and McFarlan, E., Jr., 1959, Geologic history of the chenier plain, southwestern Louisiana: Gulf Coast Association of Geological Societies Transactions, v. 9, p. 237–270.

Gould, H. R., and Stewart, R. H., 1956, Continental terrace sediments in the northeastern Gulf of Mexico: Society of Economic Paleontologists and Mineralogists Special Publication 3, p. 2–20.

Hatton, R. S., 1981, Aspects of marsh accretion and geochemistry; Barataria Basin, Louisiana [M.S. thesis]: Baton Rouge, Louisiana State University, 116 p.

Hays, J. D., Imbrie, J., and Shackleton, N. J., 1976, Variations in the Earth's orbit; Pacemaker of the ice ages: Science, v. 194, p. 1121–1132.

Hine, A. C., 1983, Modern shallow-water carbonate platform margins, *in* Cook, H. E., Hine, A. C., and Mullins, H. T., eds., Platform margins and deepwater carbonates: Society of Economic Paleontologists and Mineralogists Short Course no. 12, part 3, p. 1–100.

Humphris, C. C., Jr., 1978, Salt movement on continental slope, northern Gulf of Mexico, *in* Bouma, A. H., Moore, G. T., and Coleman, J. M., eds., Framework, facies, and oil-trapping characteristics of the upper continental margin: American Association of Petroleum Geologists Studies in Geology no. 7, p. 69–85.

Imbrie, J., and Imbrie, Z. L., 1980, Modeling the climatic response to orbital variations: Science, v. 207, p. 934–952.

Kolb, C. R., and Van Lopik, J. R., 1966, Depositional environments of the Mississippi River deltaic plain, southeastern Louisiana, *in* Shirley, M. L., and Ragsdale, J. A., eds., Deltas in their geologic framework: Houston, Texas, Houston Geological Society, p. 17–61.

Ludwick, J. C., 1964, Sediments in northeastern Gulf of Mexico, *in* Miller, R. L., ed., Papers in marine Geology: New York, Macmillan, p. 204–238.

Martin, R. G., 1978, Northern and eastern Gulf of Mexico continental margin; Stratigraphic and structural framework, in Bouma, A. H., Moore, G. T., and Coleman, J. M., eds., Framework, facies, and oil-trapping characteristics of the upper continental margin: American Association of Petroleum Geologists Studies in Geology no. 7, p. 21–42.

—— , 1984, Diapiric trends in the deep-water gulf Basin: Austin, Texas, Gulf Coast Section of Society of Economic Paleontologists and Mineralogists Foundation Research Conference, p. 60–62.

Martin, R. G., and Bouma, A. H., 1978, Physiography of Gulf of Mexico, in Bouma, A. H., Moore, G. T., and Coleman, J. M., eds., Framework, facies, and oil-trapping characteristics of the upper continental margin: American Association of Petroleum Geologists Studies in Geology no. 7, p. 3–19.

Moore, G. T., Starke, G. W., Bonham, L. C., and Woodbury, H. O., 1978, Mississippi fan, Gulf of Mexico; Physiography, stratigraphy, and sedimentation patterns, *in* Bouma, A. H., Moore, G. T., and Coleman, J. M., eds., Framework, facies, and oil-trapping characteristics of the upper continental margin: American Association of Petroleum Geologists Studies in Geology no. 7, p. 155–191.

Moore, G. T., Woodbury, H. O., Worzel, J. L., Watkins, J. S., and Starke, G. W., 1979, Investigations of the Mississippi fan, Gulf of Mexico: American Association of Petroleum Geologists Memoir 29, p. 393–402.

Morner, N. A., 1983, Sea levels, *in* Gardner, R., and Scoging, H., eds., Megageomorphology: Oxford, Clarendon Press, p. 73–91.

O'Connell, S., and 9 others, and DSDP Leg 96 Shipboard Scientists, 1985, Drilling results on the lower Mississippi fan, *in* Bouma, A. H., Normark, W. R., and Barnes, N. E., eds., Submarine fans and related turbidites, *in* Frontiers in sedimentary geology 1: New York, Springer-Verlag, p. 291–298.

O'Neil, T., 1949, The muskrat in the Louisiana coastal marshes: New Orleans, Louisiana Department Wildlife and Fisheries, 152 p.

Otvos, E. G., 1984, Barrier platforms; Northern Gulf of Mexico: Marine Geology, v. 63, p. 285–305.

Penland, S., and Boyd, R., 1981, Shoreline changes on the Louisiana barrier coast:

Institute of Electrical and Electronic Engineers, Oceans, v. 81, p. 209–219.

Pisias, N. J., and Shackleton, N. T., 1984, Modeling the global climate response to orbital forcing and atmospheric carbon dioxide change: Nature, v. 310, p. 757–759.

Roberts, H. H., and Thayer, P. A., 1985, Petrology of Mississippi fan depositional environments, in Bouma, A. H., Normark, W. R., and Barnes, N. E., eds., Submarine fans and related turbidite systems: New York, Springer-Verlag, p. 331–339.

Roberts, H. H., Adams, R. D., and Cunningham, R. W., 1980, Evolution of sand-dominant subaerial phase; Atchafalaya Delta, Louisiana: American Association of Petroleum Geologists Bulletin, v. 64, p. 264–279.

Russell, R. J., 1942, Flotant: The Geographical Review, v. 32, p. 74–98.

—— , 1967, Origins of estuaries, *in* Lauff, G. H., ed., Estuaries: American Association for the Advancement of Science Publication 83, p. 93–99.

Russell, R. J., and Howe, H. V., 1935, Cheniers of southwestern Louisiana: Geographical Review, v. 25, p. 449–461.

Saucier, R. T., 1974, Quaternary geology of the Lower Mississippi Valley: Arkansas Archaeological Survey Research Series no. 6, 25 p.

Stelting, C. E., and 9 others, and DSDP Leg 96 Shipboard Scientists, 1985, Drilling results on the middle Mississippi Fan, *in* Bouma, A. H., Normark, W. R., and Barnes, N. E., eds., Submarine fans and related turbidite systems: New York, Springer-Verlag, p. 275–282.

Stuart, C. J., and Caughey, C. A., 1976, Form and composition of the Mississippi Fan: Gulf Coast Association of Geological Societies Transactions, v. 26, p. 333–343.

Suter, J. R., and Berryhill, H. L., Jr., 1985, Late Quaternary shelf-margin deltas, Northwest Gulf of Mexico: American Association of Petroleum Geologists Bulletin, v. 69, p. 77–91.

Trabant, P. K., and Presley, B. J., 1978, Orca Basin, anoxic depression on the continental slope, northwest Gulf of Mexico, *in* Bouma, A. H., Moore, G. T., and Coleman, J. M., eds., Framework, facies, and oil-trapping characteristics of the upper continental margin: American Association of Petroleum Geologists Studies in Geology no. 7, p. 303–311.

van Tassell, J., 1987, Upper Devonian Catskill Delta margin cyclic sedimentation: Brallier, Scherr, and Foreknobs Formations of Virginia and West Virginia: Geological Society of America Bulletin, v. 99, p. 414–426.

Williams, D. F. and Trainor, D., 1986, Application of isotope chronostratigraphy in the northern Gulf of Mexico: Gulf Coast Association of Geological Societies Transactions, v. 36, p. 589–600.

Worzel, J. L., and Burke, C. A., 1978, Margins of the Gulf of Mexico: American Association of Petroleum Geologists Bulletin, v. 62, p. 2290–2307.

MANUSCRIPT ACCEPTED BY THE SOCIETY MAY 25, 1989

ACKNOWLEDGMENTS

The authors wish to thank colleagues in the Coastal Studies Institute and in the Departments of Geology at Louisiana State University and Texas A & M who have been part of much of the research in the Gulf of Mexico and who have freely discussed their findings over the years. The database on which much of the interpretations have been based has been acquired through the years from numerous petroleum companies, and the authors thank them for sharing their data. The manuscript was reviewed by Arnold H. Bouma and Whitney Autin; their reviews resulted in significant improvement to the manuscript.

The Geology of North America
Vol. J, The Gulf of Mexico Basin
The Geological Society of America, 1991

Chapter 13

Seismic stratigraphy of the deep Gulf of Mexico basin and adjacent margins

Richard T. Buffler

Institute for Geophysics, University of Texas at Austin, 8701 Mopac Boulevard, Austin, Texas 78759

INTRODUCTION

The geology of the deep Gulf of Mexico basin can be interpreted only by means of geophysical data, since few direct geologic data are available. The purpose of this chapter, therefore, is to review the seismic stratigraphy and geologic setting of the deep Gulf of Mexico basin and adjacent margins as inferred mainly from the interpretation of seismic reflection data. The deep Gulf of Mexico basin as used in this chapter is bathymetrically the deepest part of the basin (Fig. 1). The area is also part of the structurally deepest part of the basin and is underlain by a thick section (as much as 9 to 10 km) of generally undeformed sedimentary rocks overlying basement (Fig. 2). The term deep, therefore, refers both to the structural configuration of the basin as well as the depth of water. The margins of the deep basin are defined either by areas of deformed sedimentary rocks or steep escarpments that disrupt the seismic record and limit the correlation of strata in the deep basin with the better defined geology of the adjacent shallower parts of the basin (Figs. 1 and 2). The areas of deformation include the Campeche-Sigsbee Knolls to the southwest, the Mexican Ridges to the west, and the Sigsbee Escarpment to the north, which marks the southern limit of the extensively deformed Texas-Louisiana Slope (Fig. 1). The steep escarpments include the Florida Escarpment to the east and the Campeche Escarpment to the south (Figs. 1 and 2). In the southeastern Gulf the deep basin extends between the two escarpments to the Cuban margin.

Little was known about the geology of the deep Gulf of Mexico basin prior to the early 1960s, when seismic reflection and refraction data first revealed a thick sedimentary section overlying an acoustic basement consisting of either oceanic crust or thinned continental crust (transitional crust). During the mid-1970s the University of Texas Institute for Geophysics (UTIG, formerly the Marine Science Institute) began a long-term project to collect, process, and interpret multifold seismic reflection and refraction data in the deep Gulf and adjacent margins. This project continues today with over 40,000 km of data collected (Fig. 3). Interpretation of these data, as well as older singlefold

data, Deep Sea Drilling Project (DSDP) holes, and available industry data, provide the basis for the review of the seismic stratigraphy that follows.

EARLY SEISMIC STUDIES

During the late 1950s and early 1960s, singlefold seismic profiling and seismic refraction experiments in the deep ocean basins, including the Gulf of Mexico, were carried out mainly by marine academic institutions. The first major description of the sedimentary section, velocities, and structure of the deep Gulf of Mexico basin was by investigators from Lamont-Doherty Geological Observatory (Ewing and others, 1960, 1962). This was followed by several studies that described the location of major salt structures (Ewing and Antoine, 1966; Worzel and others, 1968) and discussed the velocity of the sedimentary layers (Houtz and others, 1968). At about the same time, Texas A & M University began extensive profiling in the deep Gulf, which resulted in discussions of the salt structures (Antoine and Bryant, 1969), the Mexican Ridges foldbelt (Jones and others, 1967; Bryant and others, 1968), and the escarpments around the Gulf (Bryant and others, 1969). Woods Hole Oceanographic Institution also conducted seismic surveys in the Gulf (Uchupi and Emery, 1968). Antoine (1972) and Wilhelm and Ewing (1972) summarized much of this early seismic data and presented a broad view of the deep basin structure and geologic history as it was known at that time. These seismic data also were used to plan two early DSDP drilling legs in the Gulf of Mexico: Leg 1 (Sites 1 to 3) (Ewing and others, 1969) and Leg 10 (Sites 85 to 97) (Worzel and others, 1973).

Another significant early geophysical cruise in the Gulf of Mexico was that of the USNS *Kane* in 1969, jointly sponsored by the U.S. Geological Survey and the U.S. Naval Oceanographic Office. Using a 160,000-joule sparker system, this cruise collected more than 25,000 km of high-quality singlefold reflection profiles. It was the first comprehensive seismic reflection survey of

Buffler, R. T., 1991, Seismic stratigraphy of the deep Gulf of Mexico basin and adjacent margins, *in* Salvador, A., ed., The Gulf of Mexico Basin: Boulder, Colorado, Geological Society of America, The Geology of North America, v. J.

354   *R. T. Buffler*

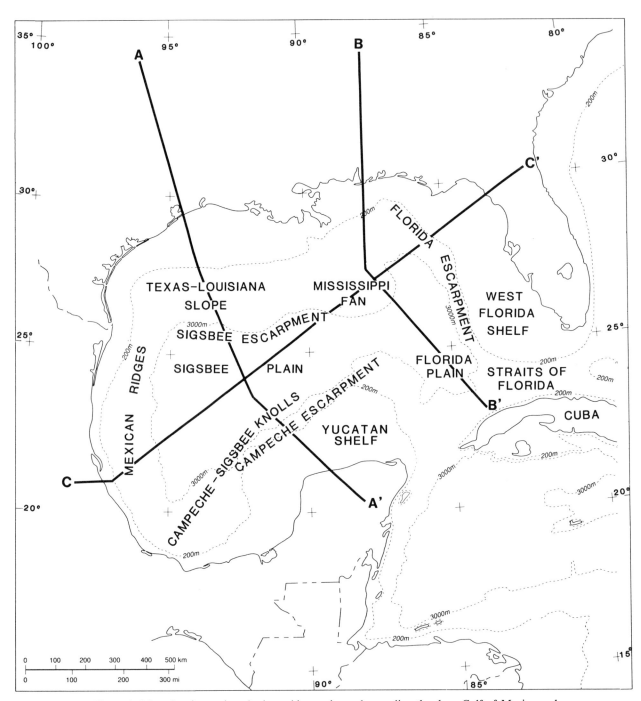

Figure 1. Map showing major physiographic provinces that outline the deep Gulf of Mexico and adjacent margins, and location of cross sections in Figure 2.

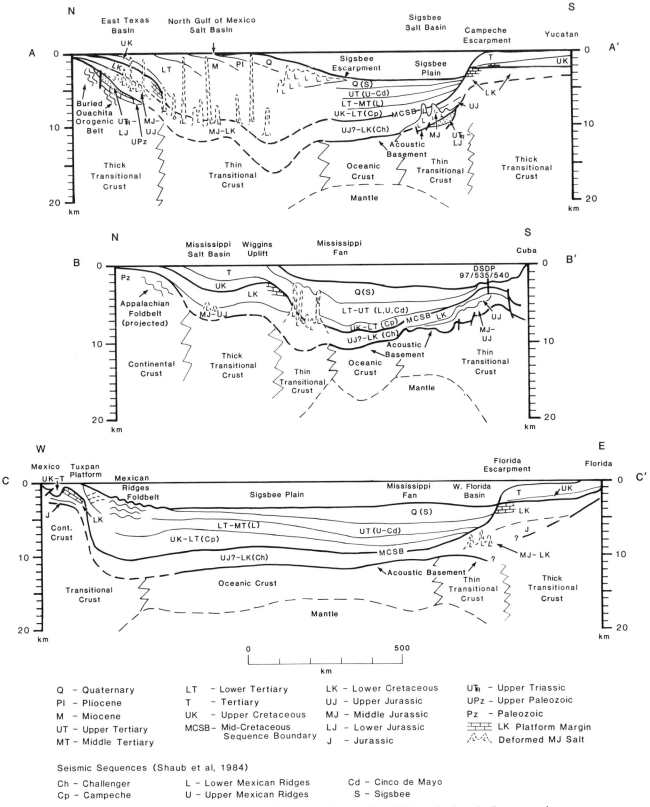

Figure 2. Generalized regional cross sections across the deep Gulf of Mexico basin and adjacent margins showing major crustal types, major structural elements, and overlying stratigraphic units by age (see legend). Abbreviations in parentheses are seismic sequences in the deep central Gulf discussed in the text. Figure 1 shows location of sections.

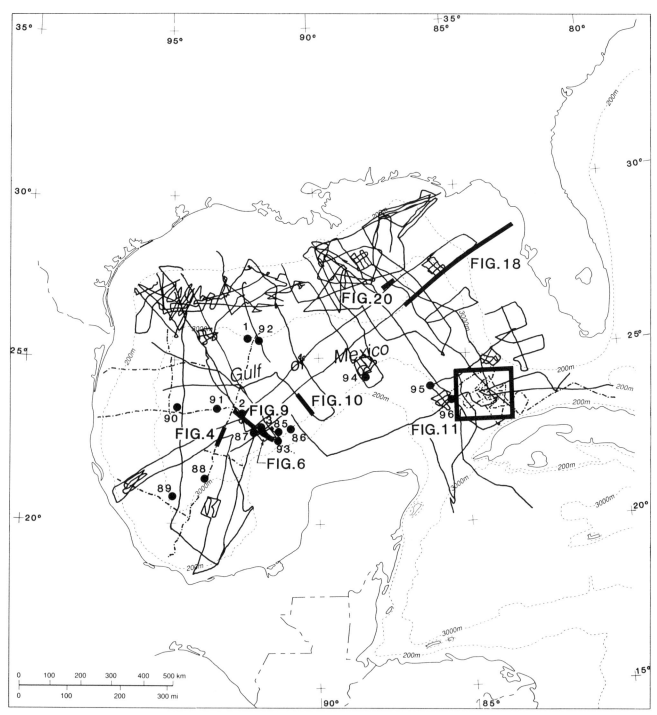

Figure 3. Map showing location of multifold seismic reflection data collected in the deep Gulf of Mexico basin and adjacent margins since 1975 by the University of Texas (UT). Dotted lines in western Gulf represent early National Science Foundation (NSF) sponsored cruises, while dotted lines in southeastern Gulf represent an NSF sponsored Deep Sea Drilling Project (DSDP) site survey cruise for DSDP Leg 77. Solid lines are cruises sponsored by industry. Heavy bars are locations of seismic lines and cross sections used as figures in text. Box outlines detailed area shown on Figure 11. Dots indicate location of DSDP sites.

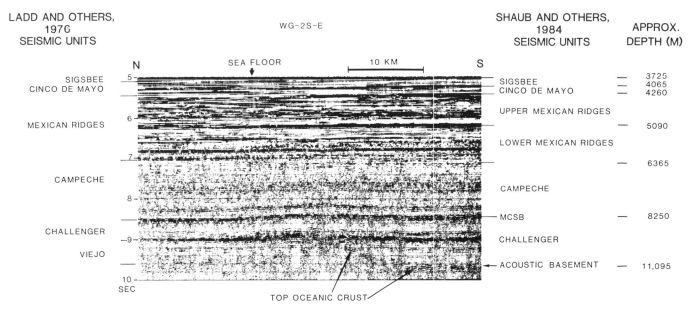

Figure 4. Portion of UT seismic line WG-2 in western Gulf of Mexico showing seismic units originally defined by Ladd and others (1976) and redefined seismic sequences by Shaub and others (1984). See Figure 3 for location. Modified from Shaub and others (1984).

the Gulf and was the best seismic data collected to that time. A summary of the seismic lines and a discussion of geologic structures was presented by Garrison and Martin (1973). The data also were used extensively in a discussion of the Pleistocene sediments in the deep basin (Stuart and Caughey, 1977), and they are still widely used today as a valuable supplemental source for interpreting the geologic history of the deep basin.

Other extensive geophysical cruises recording singlefold seismic data were conducted in the early 1970s, including: (1) the cruises of the R/V *Uribe* (1970) and the R/V *Unitedgeo* I (1971) in the southwestern Gulf sponsored by the U.S. Geological Survey and the Universidad Nacional Autónoma de México (Garrison, 1972; Moore and del Castillo, 1974), and (2) the cruises of the Texas A & M University R/V *Alaminos* (1973) sponsored by the National Science Foundation and the USNS *Keathley* (1970) in the northern Gulf (Shih and others, 1977). A track chart including the location of all the early seismic lines in the Gulf is shown in Martin (1980).

These early profiles presented a fairly comprehensive picture of the stratigraphy, structure, and geologic history of only the shallower part of the basin, as the energy sources available at the time did not allow for deep penetration of the thick sedimentary section and resolution of the deep structure and stratigraphy. Only refraction data provided any clues to the details of the deep stratigraphy and structure. These studies, therefore, set the scene for the use of multifold seismic reflection profiling.

UTIG MULTIFOLD SEISMIC STUDIES

In 1974 and 1975 UTIG collected the first nonproprietary multifold seismic data in the Gulf of Mexico using the R/V *Ida*

Green. These cruises were made possible mainly through the efforts of Joel Watkins and Joe Worzel. Approximately 3,600 km were collected in the western Gulf, a project sponsored by the National Science Foundation (Fig. 3, dotted lines). This was the beginning of a long-term project to better understand the geologic history and tectonic framework of the deep basin and surrounding margins. Some of the first results of these early surveys were published by Watkins and others (1975a, b). Ladd and others (1976) discussed the structure, stratigraphy, and geologic history of the deep western Gulf and showed for the first time the entire thick sedimentary section lying on an acoustic basement inferred to be the top of the crust (Fig. 4). Ladd and others (1976) defined six seismic units on the basis of reflection characteristics and basinwide continuity (Fig. 4). These seismic units and the early seismic lines in the western Gulf were further discussed and refined in later papers by Watkins and others (1976) and Watkins and others (1978).

Between 1975 and 1978, UTIG conducted additional multifold surveys in the deep basin sponsored by the petroleum industry, collecting long regional lines as well as grids at selected margins (Fig. 3). All of these new data, plus the old, were used for further descriptions of the structure and stratigraphy of the entire basin (Worzel and Burk, 1979), as well as specific areas and topics such as the Sigsbee Escarpment (Buffler and others, 1978), the Mexican Ridges foldbelt (Buffler and others, 1979), the Mississippi Fan (Moore and others, 1979), regional salt structures (Martin, 1980), and the maritime boundary region in the Gulf (Foote and others, 1983). They also were used to help develop models for the deep structure and early history of the basin (Buffler and others, 1980, 1981; Ibrahim and others, 1981; Buffler and Sawyer, 1985).

In addition, several University of Texas (UT) Master's theses were produced using the UTIG seismic lines, including seismic stratigraphic studies of the Sigsbee Escarpment area (Seekatz, 1977), the northwest Campeche Escarpment area (Long, 1978), the northeast Campeche Escarpment area (Huerta, 1980), the western Gulf of Mexico basin (Bertagne, 1980), the Sigsbee Salt basin area (Lin, 1984), and the Mississippi Fan–Mississippi Canyon area (Walters, 1985). Partial results of Bertagne (1980) were published in the American Association of Petroleum Geologists (AAPG) Bulletin (Bertagne, 1984). Pew (1982) produced a Master's thesis on the southern Mexican Ridges foldbelt based on the multifold seismic lines plus the singlefold seismic lines available in the region. The UTIG data also were used to map, describe, and interpret a prominent seismic reflector generally referred to as the "mid-Cretaceous unconformity" (MCU) in the deep basin (Faust, 1984). Students from other universities also used the data to study the deep western Gulf (Usmani, 1980, University of Houston) as well as the West Florida Basin in the deep eastern Gulf (Lord, 1986, Rice University).

Regional seismic lines collected in the southeastern Gulf of Mexico plus additional site surveys conducted in 1980 (dotted lines, Fig. 3) were used to design a DSDP program, which was drilled during DSDP Leg 77 in December–January, 1980–81. Results of this drilling are discussed in Schlager and others (1984a) and Buffler and others (1984b). Two UT Master's theses incorporated the drilling results of this leg into seismic stratigraphic studies of the Lower Cretaceous section (Phair, 1984) as well as the Upper Cretaceous–Cenozoic section (Angstadt, 1983; Angstadt and others, 1985). As part of the latter work, an Eocene unconformity was investigated in detail (Angstadt and others, 1983). Examples of seismic lines at each drill site are included in the site chapters in the DSDP volume (Buffler and others, 1984b), while a summary of the seismic stratigraphy and geologic history of the southeastern Gulf and the regional implications is included as a separate chapter (Schlager and others, 1984).

The UTIG seismic data have been used in other recent studies of the deep Gulf of Mexico basin. Examples are included with a series of short papers in two *Studies in Geology* volumes edited by A. W. Bally and published by the American Association of Petroleum Geologists (AAPG). The papers summarize several different areas, including the Mexican Ridges foldbelt (Buffler, 1983b), the East Mexico shelf and slope (Shaub, 1983b), the Sigsbee Knolls area (Buffler, 1983c), the Sigsbee Escarpment (Buffler, 1983a), the Campeche and Florida Escarpments (Locker and Buffler, 1983), the Catoche Tongue area (Shaub, 1983a), and the deep southeastern Gulf (Phair and Buffler, 1983). Regional lines across the entire basin are included in the *Ocean Margin Drilling Program Atlas of the Gulf of Mexico* (Buffler and others, 1984a). The UTIG seismic lines were used to construct the various isopach, structure, and lithofacies maps for the deep basin found in this atlas.

Examples of the seismic data were used to illustrate various shelf margins around the Gulf of Mexico (Winker, 1982, 1984; Winker and Edwards, 1983; Winker and Buffler, 1988), as well

as the seismic stratigraphy of the northeastern Gulf, which provides one of the few correlations of seismic horizons between the deep and shallow Gulf (Addy and Buffler, 1984). A more regional paper by Shaub and others (1984) redefines the seismic stratigraphic framework of the deep Gulf (Figs. 4 and 5), presents isopach maps of the various seismic units, and discusses the regional geologic history. In addition, lines collected from the southern West Florida shelf were used to develop a preliminary stratigraphic and structural framework (Shaub, 1984), while data from the eastern Gulf were used extensively in a Texas A & M University Ph.D. dissertation involving a comprehensive seismic stratigraphic study of the Mississippi Fan (Feeley and others, 1984). Most recently, data collected in 1983 in the Mississippi Fan area and the northeastern Gulf, supplemented by an extensive grid of industry seismic data, were used in an analysis of oceanic crust in the deep eastern Gulf (Rosenthal, 1987), an analysis of the northwest Florida Early Cretaceous carbonate platform and its margin (Corso, 1987), and a detailed study of the Mississippi Fan itself (Weimer, 1989).

SEISMIC STRATIGRAPHY

Generalized geologic setting

An acoustic basement is recognized on the seismic data over much of the central and eastern deep Gulf of Mexico (Figs. 2, 4, and 6). It has different characteristics in different places—in the central deep Gulf it is interpreted to be the top of oceanic crust (Figs. 2, 4, and 7), while in surrounding areas it is interpreted to be the top of stretched continental crust (transitional crust; Figs. 2, 6, and 7) and could consist of crystalline continental basement, Paleozoic sedimentary or metasedimentary rocks, or early Mesozoic rift sequences. The distribution, seismic character nature, and origin of the acoustic basement and crust in the deep basin are discussed in detail in Chapter 4 (this volume).

Overlying the crust in the deep Gulf is a thick sedimentary section (Figs. 2, 4, 5, and 6) representing a long history beginning in the Middle Jurassic. This rock record is separated into two main sequences by a prominent high-amplitude reflector that can be mapped easily on the basis of the seismic data throughout the deep basin. This reflector corresponds with a major unconformity along the southern and eastern margins of the deep basin, but in the central Gulf, beds above appear conformable with beds below (Faust, 1990) (Figs. 4 and 6). This reflector/unconformity is inferred to be mid-Cretaceous (mid-Cenomanian?) in age and generally has been referred to informally as the "mid-Cretaceous unconformity" or "MCU" (Buffler and others, 1980; Buffler and Sawyer, 1985; Faust, 1990; Schlager and others, 1984b; Shaub and others, 1984). Since this reflector does not everywhere represent an unconformity and has the characteristics of a major sequence boundary as defined by Vail and others (1977), it is more appropriate to refer to this boundary as the "mid-Cretaceous sequence boundary" or MCSB. This new terminology is adopted throughout this chapter. The MCSB apparently marks a major turning point in deep basin sedimentation and history and is discussed in more detail in a later section.

The pre-MCSB sequence represents a long and complex history involving the early evolution of the deep Gulf of Mexico basin (Buffler and Sawyer, 1985). There now exists support among many geologists working in the area for a general interpretation of the early history of the deep basin, which includes; (1) a Late Triassic to Middle Jurassic stretching of continental crust and formation of transitional crust, culminating with the widespread deposition of evaporites (of probable Callovian age); (2) a Late Jurassic period of oceanic crust formation in the deep central Gulf accompanied by a widespread marine transgression (Oxfordian); (3) a Late Jurassic through Early Cretaceous period of cooling and subsidence of the crust and buildup of extensive

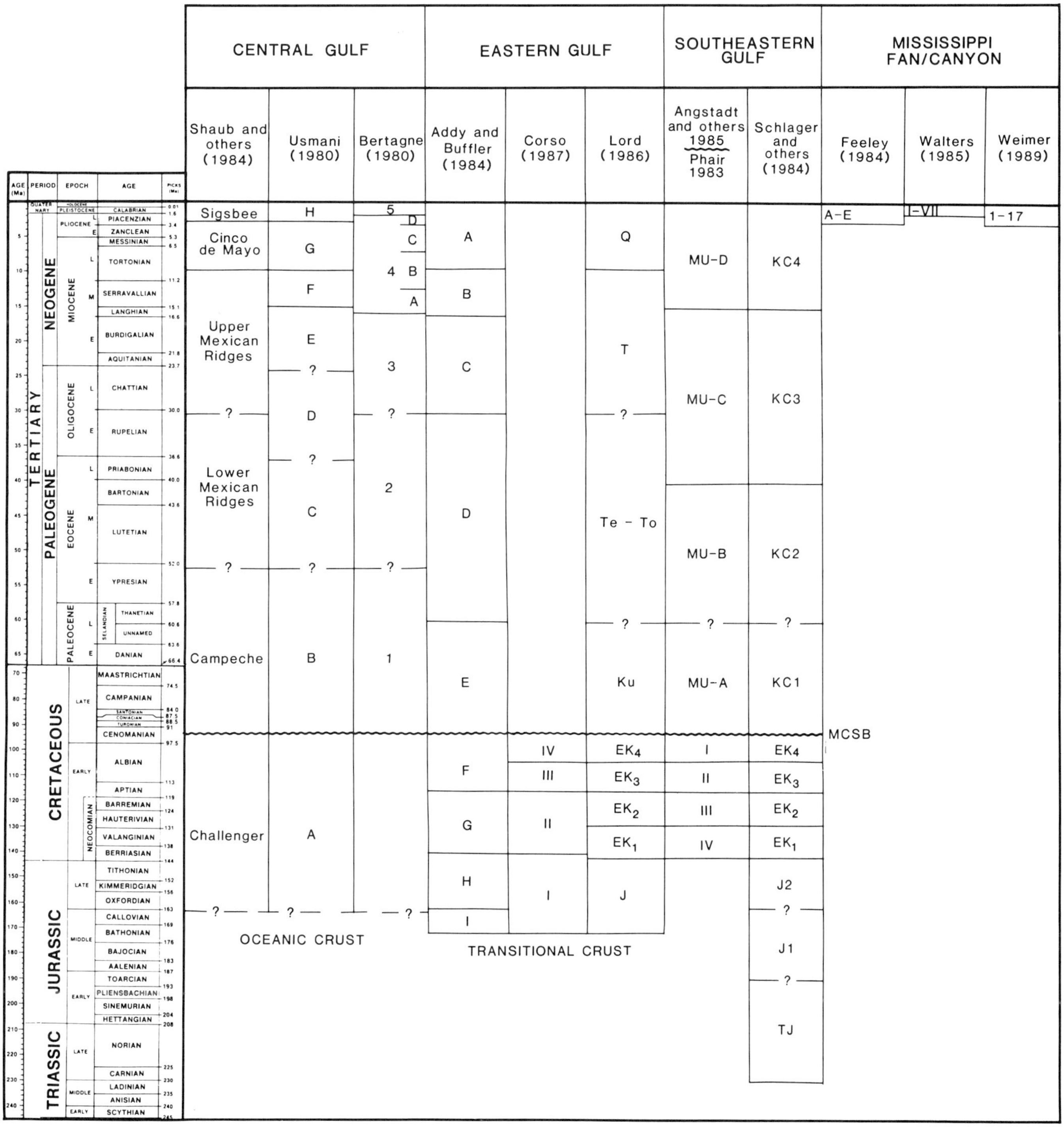

Time scale (Period / Epoch / Age)	CENTRAL GULF — Shaub and others (1984)	Usmani (1980)	Bertagne (1980)	EASTERN GULF — Addy and Buffler (1984)	Corso (1987)	Lord (1986)	SOUTHEASTERN GULF — Angstadt and others 1985 / Phair 1983	Schlager and others (1984)	MISSISSIPPI FAN/CANYON — Feeley (1984)	Walters (1985)	Weimer (1989)
Quaternary / Pliocene (Calabrian–Piacenzian–Zanclean)	Sigsbee	H	5 / D	A		Q			A–E	I–VII	1–17
Miocene (Messinian–Tortonian)	Cinco de Mayo	G	C				MU-D	KC4			
Miocene (Serravallian–Langhian)		F	4 B / A	B							
Miocene (Burdigalian–Aquitanian)	Upper Mexican Ridges	E	3	C		T					
		?									
Oligocene (Chattian–Rupelian)	?	D	?			?	MU-C	KC3			
Eocene (Priabonian–Bartonian)	Lower Mexican Ridges	?	2	D		Te – To					
Eocene (Lutetian)		C					MU-B	KC2			
Eocene (Ypresian)	?	?	?								
Paleocene (Thanetian–Danian)	Campeche	B	1			?	?	?			
Cretaceous Late (Maastrichtian–Cenomanian)				E		Ku	MU-A	KC1			
MCSB											
Cretaceous Early (Albian)	Challenger	A		F	IV	EK$_4$	I	EK$_4$			
Cretaceous Early (Aptian)					III	EK$_3$	II	EK$_3$			
Cretaceous Early (Barremian–Hauterivian)				G	II	EK$_2$	III	EK$_2$			
Cretaceous Early (Valanginian–Berriasian)						EK$_1$	IV	EK$_1$			
Jurassic Late (Tithonian–Kimmeridgian)				H	I	J		J2			
Jurassic (Oxfordian–Callovian)	?	?	?	I				?			
Jurassic Middle (Bathonian–Bajocian)	OCEANIC CRUST			TRANSITIONAL CRUST				J1			
Jurassic Early (Aalenian–Hettangian)								?			
Triassic Late (Norian–Carnian)								TJ			

Figure 5. Correlation chart of various seismic stratigraphic units mapped in the deep Gulf of Mexico basin and adjacent margins.

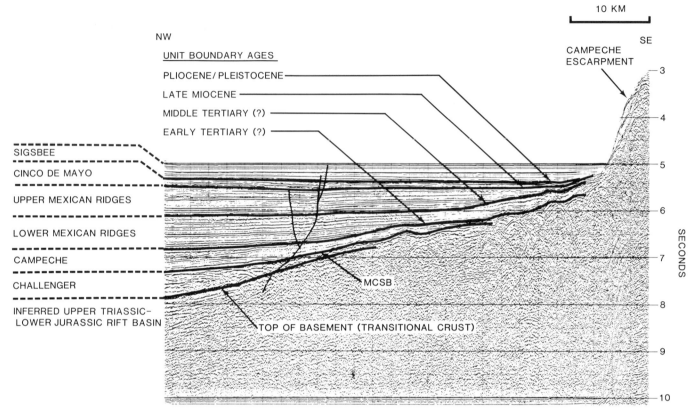

Figure 6. Portion of UT seismic line CE-2 showing names, ages, and stratigraphic relations of deep Gulf of Mexico seismic sequences along base of Campeche Escarpment as defined by Shaub and others (1984). See Figure 3 for location. Modified from Shaub and others (1984).

carbonate platforms surrounding a deep basin; and (4) establishment of the widespread mid-Cretaceous (mid-Cenomanian?) reflector/unconformity or sequence boundary (MCSB). The model is based on many lines of evidence, including gravity, magnetic, paleomagnetic, seismic reflection, and seismic refraction data, as well as depth-to-basement maps, total tectonic subsidence analyses, plate reconstructions, the distribution of Jurassic salt and sediments, and stratigraphic studies from the periphery of the basin (Buffler and Sawyer, 1985; Chapter 4, this volume).

The post-MCSB history consists primarily of filling the deep basin with thick wedges of siliciclastic sediments, first from the northwest in the Late Cretaceous–early Cenozoic (Fig. 2C), and then from the north (ancestral Mississippi River drainage) in the late Cenozoic (Fig. 2A). Most of the deep southeastern Gulf remained an area relatively starved of sediments, except in the extreme southeast just north of Cuba, where a large wedge of sediments derived from Cuba accumulated in a foredeep during Late Cretaceous through Eocene time (Fig. 2B). Deep shelf (pelagic) carbonates accumulated on the adjacent outer margins of the Yucatan and Florida platforms following a post-MCSB stepback or retreat of the platform margins.

This brief summary sets the scene for a more detailed discussion of the seismic stratigraphy of each of the two major sequences in specific parts of the deep basin and adjacent margins.

These discussions are derived mainly from the literature and the unpublished theses discussed previously. Most of the interpretations are based on the UTIG multifold seismic data supplemented by DSDP results, the earlier singlefold seismic data, available industry seismic data, and other regional information.

Pre-MCSB (Challenger Unit)

The early UTIG multifold seismic surveys in the deep Gulf of Mexico identified a 2- to 3-km-thick seismic stratigraphic sequence that overlies acoustic basement and is capped by the prominent mid-Cretaceous reflector/unconformity (now the MCSB) (Ladd and others, 1976; Watkins and others, 1976, 1978; Buffler and others, 1980) (Figs. 2, 4, 5, and 6). This sequence can be mapped throughout the entire deep Gulf of Mexico basin and represents a long Middle Jurassic through Early Cretaceous history, beginning with the deposition of thick evaporites and culminating with the establishment of the MCSB. In the deep Gulf this sequence is known as the Challenger Unit, as redefined by Shaub and others (1984) (Figs. 2, 4, 5, and 6; Table 1). The Challenger Unit has different characteristics and represents a somewhat different sedimentary history in various parts of the deep basin, as discussed below.

Area of oceanic crust. Over most of the area underlain by oceanic crust in the deep central Gulf (Fig. 7) the Challenger Unit

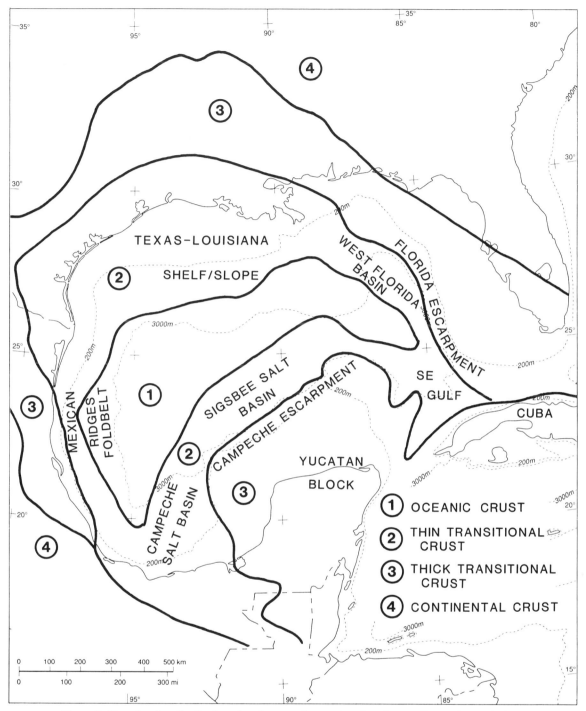

Figure 7. Map showing distribution of major crustal types (numbered) as well as major geologic or tectonic provinces discussed in text.

averages about 2 to 3 km thick (Shaub and others, 1984) (Figs. 2 and 4). It is characterized by uniform, low- to high-amplitude, continuous reflections that often have a high degree of internal ringing or reverberation (Fig. 4). The lower part of the section onlaps and infills around highs in acoustic basement, which is inferred to be the top of the oceanic crust (Fig. 4). The top of the

unit is the prominent MCSB. In the central area of oceanic crust the beds above and below the MCSB surface are conformable (Fig. 4) (Buffler and others, 1980; Faust, 1984; Shaub and others, 1984; Rosenthal, 1987).

The age of the Challenger Unit in the deep central Gulf overlying oceanic crust is inferred to be Late Jurassic through

TABLE 1. SUMMARY OF DEEP GULF OF MEXICO SEISMIC SEQUENCES*

Sequence	Age	Typical Reflection Characteristics	Suggested Depositional Environment Depocenter/Source
Sigsbee	Pleistocene	Mid-fan; variable, but generally high amplitude and frequency; complex, even chaotic reflection configurations. Lower fan: high amplitude, high frequency; continuous, parallel and horizontal; in places wavy or distorted with channels. Western and southwestern continental rise; generally acoustically transparent.	Abyssal submarine fan and other northern-source mass-transport deposits in eastern two-thirds of Gulf, contributed by Pleistocene Mississippi River, reflection configurations in mid-fan area interpreted as channels, levees, channel-fill, interchannel and overbank deposits; mostly suspension deposits in west; some fine-grained turbidites also derived from Mexican rivers in western basin. Local carbonate debris from escarpments. Local depocenter in Bay of Campeche (southwestern Gulf).
Cinco de Mayo	Late Miocene-Pliocene	Generally acoustically transparent; otherwise, variable amplitude and frequency; parallel and subhorizontal.	Abyssal terrigenous sediments and biogenic ooze. No major depocenter in deep Gulf; sediments thicken slightly in northern and southwestern Gulf. Most of clastic supply to deep basin may be trapped by sedimentary deformation along northern and western margins.
Upper Mexican Ridges	Middle Tertiary(?)-Late Miocene	High amplitude, high frequency; continuous, parallel, horizontal/subhorizontal; minor channels and clinoforms near depocenters.	Predominantly deep marine distal sandstones, siltstones, and mudstones. Progradation continues from western margin. A northeastern depocenter is also established in the study area for the first time and is attributed to the ancestral Mississippi River. Smaller depocenter occurs in southwestern Gulf.
Lower Mexican Ridges	Early Tertiary(?)-Middle Tertiary(?)	Moderate amplitudes and frequencies; generally continuous; commonly parallel and subhorizontal.	Predominantly deep marine distal siliciclastic sediments. Broad western depocenter attributable to ancestral Texas and Mexican rivers mainly in Rio Grande Embayment region. Considered a continuation and progradation of the sedimentation pattern of the underlying Campeche unit. Sediments are distributed throughout the entire deep basin.
Campeche	Mid-Cretaceous–Early Tertiary(?)	Low amplitude, low frequency; continuous parallel, horizontal or gently dipping.	Predominantly deep marine distal siliclastics and carbonates in western two-thirds of study area; pelagic in east. Western depocenter source was probably Rio Grande Embayment.
Challenger	Middle Jurassic(?)-mid-Cretaceous	Moderate amplitudes, low frequency; generally continuous, parallel and sub-horizontal in central Gulf; discontinuous and gently dipping, deformed and even chaotic around the margins of the deep basin.	Unit immediately overlies acoustic basement (oceanic and transitional crust). Predominantly deep marine sediments in central Gulf. Evaporites, shallow then deep marine sediment around margins of deep basin. Eastern depocenter (West Florida basin) source probably was Tampa Embayment; central Gulf depocenter (Sigsbee Salt basin) source was Yucatan block to south.

*Modified from Shaub and others, 1984.

Early Cretaceous. This age is inferred, in part, based on the model for the early evolution of the basin discussed above. This model assumes that the emplacement of oceanic crust took place following deposition of Callovian salt in subsiding basins, which are now separated by the oceanic crust (Buffler and Sawyer, 1985; Salvador, 1987). This assumes that no salt was deposited over oceanic crust, as suggested by the seismic data (Rosenthal, 1987). If the salt around the margins of the deep Gulf is late Middle Jurassic (Callovian?) in age as suggested by Salvador (1987, and Chapter 8, this volume), the sediments overlying oceanic crust must be younger. Seismic data in the eastern and southeastern Gulf indicate that much of the inferred Jurassic sedimentary section pinches out or thins drastically basinward in the area of transitional crust and does not overlie the area of oceanic crust (Phair and Buffler, 1983; Lord, 1986; Rosenthal, 1987). This suggests, therefore, that much of the Challenger Unit over oceanic crust in the eastern Gulf is Early Cretaceous in age. The upper age limit assumes a mid-Cretaceous (mid-Cenomanian?) age for the MCSB as discussed in the following section.

Based on its continuous seismic reflection character, its basinal setting overlying oceanic crust, and extrapolations from DSDP sites in the southeastern Gulf, the Challenger Unit is interpreted to consist mainly of deep-water, fine-grained sediments. Rocks of similar age around the periphery of the basin are primarily shallow-water carbonates and terrigenous clastics (Chapters 8 and 9, this volume), while fine-grained, deep-water Lower Cretaceous carbonates were drilled on Leg 77 in the southeastern Gulf (Buffler and others, 1984b). This evidence plus the relatively high refraction velocities for the unit (4.7 to 5.2 km/s; Ibrahim and others, 1981), suggests that much of the unit consists of fine-grained carbonates. In places, however, the unit may contain significant siliciclastic rocks (shales), such as in the northeastern Gulf where siliciclastics reached far into the basin during Late Jurassic–Early Cretaceous time (Chapters 8 and 9, this volume). The onlap and fill of the lower part of the unit onto the irregular oceanic crust (Fig. 4) indicates deposition by uniform, slow-moving current systems and distal turbidity currents, rather than strictly by drape of pelagic sediments.

Sigsbee Salt basin. The Sigsbee Salt basin, located just north of the Campeche Escarpment (Fig. 7), is a large Mesozoic structural downwarp or basin filled with up to 3 km of Middle Jurassic through Lower Cretaceous sediments (Figs. 2A, 6, 8, 9, and 10). The basin was first identified on multifold seismic data by Ladd and others (1976) and Watkins and others (1976) and was later described in more detail by Long (1978), Buffler and others (1980), and Buffler (1983c). More recently the basin was studied and described as part of a Master's thesis and was given the name Sigsbee Salt basin (Lin, 1984).

The Challenger Unit in the Sigsbee Salt basin unconformably overlies thin transitional crust (Chapter 4, this volume) (Figs. 2A, 6, 7, 8, 9, and 10). The base of the unit is a prominent, relatively smooth, high-amplitude reflector/unconformity that dips basinward and truncates an inferred Late Triassic–Early Jurassic rift basin considered as part of basement (Chapters 4 and 8,

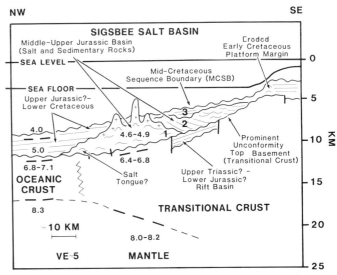

Figure 8. Schematic cross section across central Sigsbee Salt basin showing important pre-MCSB stratigraphic and tectonic features discussed in text. Numbers represent average refraction values (km/s) for the area. Large numbers (1, 2, and 3) are informal seismic sequences.

this volume) (Figs. 6, 8, and 9). The top of the unit is the MCSB, which here is a prominent unconformity. In this region the Challenger Unit consists of three major seismic sequences: Challenger 1—a basal, generally transparent sequence that is deformed in places; Challenger 2—an overlying layered sequence that also is deformed; and Challenger 3—a younger undeformed sequence that onlaps the other two (Figs. 8 and 10).

The basal transparent sequence (Challenger 1) is interpreted to be thick evaporites, mainly salt, deposited directly on the older transitional crust. This interpretation is based on its deformational character as large pillows and domes (Figs. 8 and 9), typical of salt structures in other parts of the Gulf of Mexico basin. A high-amplitude reflector at the top of the deformed sequence also is typical of salt. In addition, cap-rock was drilled at DSDP hole 2 on the top of Challenger Knoll (Fig. 9). The salt thickness is distributed asymmetrically; it is thicker along the northern and northwestern margins of the basin. Here the thick salt is deformed extensively into pillows and domes, some of which penetrate into the overlying Cenozoic section and reach the floor of the Gulf (Fig. 9). This distribution suggests differential subsidence of the basin just prior to or during deposition of the salt (mid-Jurassic). This period of subsidence in the Sigsbee Salt basin followed the period of inferred Late Triassic–Early Jurassic rifting, as suggested by the rift basins lying below the southern part of the basin and truncated by the prominent unconformity (Lin, 1984) (Figs. 6 and 8).

The entire northern and northwestern margin of the Sigsbee Salt basin is characterized by a persistent seismic geometry consisting of a northward-thinning transparent zone capped by a

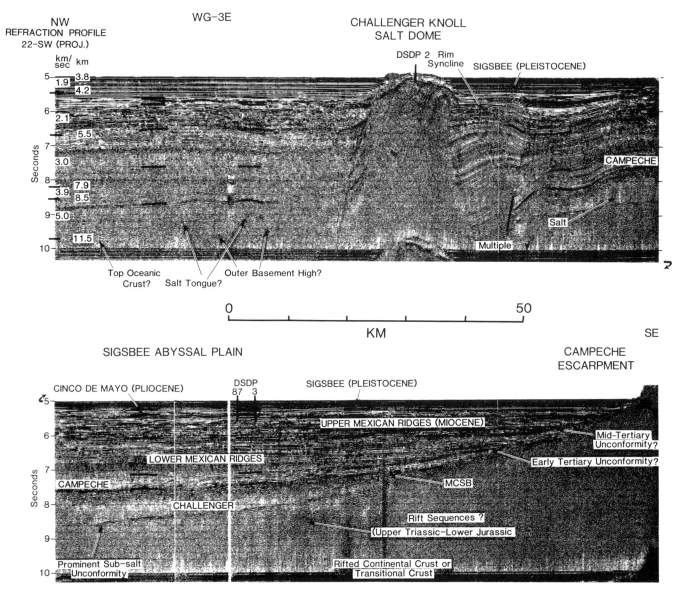

Figure 9. Portion of UT seismic line WG-3 across Sigsbee Salt basin showing major structural features and seismic stratigraphic sequences discussed in text. See Figure 3 for location. Modified from Buffler (1983c).

high-amplitude reflection, which in turn is onlapped by the pre-MCSB oceanic sediments described above (Figs. 8 and 9). This early topographic feature is continuous with the area of thick salt and is interpreted to be a salt tongue that flowed basinward during the Late Jurassic (Figs. 8 and 9). The salt probably flowed in response to gravity as the adjacent, newly formed oceanic crust to the north began to cool and subside. It also may have been responding, in part, to differential loading by a thick Upper Jurassic–Lower Cretaceous sedimentary section that accumulated in a depocenter just to the south (Fig. 8). Based on the seismic data, as well as gravity and magnetic data, the boundary between oceanic and transitional crust is interpreted to be at or just below the northern edge of this salt tongue (Lin, 1984) (Figs. 8 and 9).

Along the southern margin of the Sigsbee Salt basin, the salt is less deformed, thins by onlap against basement, and finally pinches out depositionally at the base of the Campeche Escarpment (Figs. 6 and 9). Similar stratigraphic relations occur along the updip parts of other peripheral Gulf salt basins such as in the northern Gulf of Mexico basin (e.g., Jackson and Seni, 1983; Vail and others, 1977). An exception to this relation is in the central part of the Sigsbee Salt basin, where a thick salt and sediment section extends right up to the edge of the Campeche Escarpment (Fig. 10), as discussed below.

Based on (1) preliminary ages from DSDP hole 2 (Kirkland and Gerhard, 1971); (2) the basin's setting similar to other salt basins in the northern Gulf of Mexico basin (such as the East

Texas basin; Jackson and Seni, 1983); and (3) models for the early evolution of the basin (e.g., Humphris, 1978; Buffler and others, 1981; Buffler and Sawyer, 1985; Salvador, 1987, and Chapter 8, this volume), the salt is inferred to be equivalent in age to the Louann Salt of the northern Gulf of Mexico basin. This would make the salt approximately late Middle Jurassic in age, possibly Callovian (Salvador, 1987; Chapter 8, this volume).

Overlying the salt in the central Sigsbee Salt basin and located just landward of the major salt structures is a thick sequence of generally deformed sedimentary rocks (Challenger 2; Figs. 8 and 10) (Buffler and others, 1980). Here the southeast-dipping Challenger 2 sedimentary section thickens to the southeast and downlaps directly onto the basal unconformity, indicating sedimentation contemporaneous with deformation due to the withdrawal and northward flow of salt into the large salt structures and salt tongue. This thick sedimentary section possibly represents a large local depocenter located within the basin. It extends right up to the base of the Campeche Escarpment with no apparent depositional thinning (Fig. 10), suggesting the presence of a local embayment that extends under the Campeche Escarpment (Lin, 1984). Elsewhere in the area, the equivalent section is thin and pinches out just basinward of the escarpment, similar to the salt (Figs. 6 and 9). Based on its position above the inferred Middle Jurassic salt and its overall setting, Challenger 2 is probably Late Jurassic to Early Cretaceous in age, possibly the equivalent of the Upper Jurassic section (Smackover to Cotton Valley rocks) in the northern Gulf of Mexico basin. The sequence probably consists of shallow-water carbonates as well as nonmarine to shallow-marine clastics, whose source may have been the embayment to the south.

The third and youngest sequence making up the Challenger Unit in the Sigsbee Salt basin area consists of a relatively thin, undeformed unit that onlaps and infills above the two older deformed sequences (Challenger 3; Figs. 8 and 10). It is equivalent to the deep-water sediments that overlie the adjacent oceanic crust and onlap the salt tongue (Fig. 8). This sequence probably mainly represents deep-water carbonates and shales equivalent in age to the sediments of the adjacent Early Cretaceous carbonate platforms. The platform margins were established along a tectonic hinge zone (flexure in basement) at the Campeche Escarpment (Fig. 8). Over much of the area of the Sigsbee Salt basin this sequence is too thin to be distinguished and mapped separately from the underlying post-salt sequence (Figs. 6 and 9), suggesting relatively starved conditions during the Early Cretaceous. Overall, the two post-salt sedimentary sequences (Challenger 2 and 3) probably represent an upward transition from shallow- to deep-water conditions as the basin underwent post-rift subsidence.

Southeastern Gulf of Mexico. A thick Jurassic–Lower Cretaceous (pre-MCSB) sequence occurs throughout the deep southeastern Gulf, lying between the Florida and Campeche Escarpments (Figs. 2B, 5, and 7). This area has been studied extensively and was the site of DSDP Leg 77 (Phair and Buffler, 1983; Phair, 1984; Shaub, 1983a: Schlager and others, 1984a; Buffler and others, 1984b; Schlager and others, 1984) (Figs. 11 and 12). In this area the basement rises to the south and the thick Mesozoic section (2 to 4 km) is either exposed along the sea floor or is at relatively shallow depths (Figs. 2B and 12). This makes it an ideal place for both seismic studies and drilling, and it represents the only place in the deep Gulf where Mesozoic sediments have been penetrated by drilling.

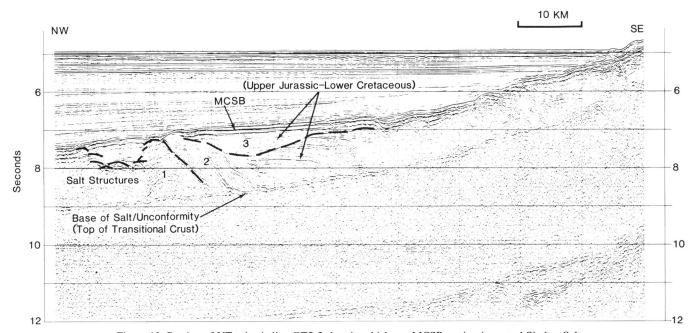

Figure 10. Portion of UT seismic line GT2-3 showing thick pre-MCSB section in central Sigsbee Salt basin informally subdivided into three sequences (Challenger 1, 2, and 3). See text for discussion of sequences. See Figure 3 for location. Modified from Lin (1984).

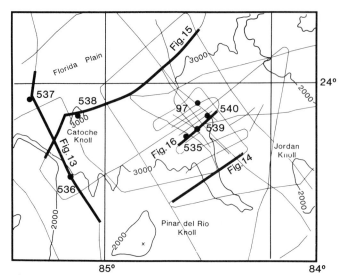

Figure 11. Map showing location of UT seismic lines in southeastern Gulf of Mexico, DSDP Leg 10 and 77 sites, and location of seismic examples shown in Figures 13 to 16. See Figure 3 for location of area. Depth in meters.

The thick pre-MCSB sedimentary section overlies thin transitional crust as discussed in Chapter 4 (this volume) (Figs. 2B, 7, and 12) and has been interpreted to represent a complex history involving the upward transition from nonmarine rocks to shallow-marine and then to deep-marine rocks as the area subsided. This section has been subdivided into four major sequences (TJ, J1, J2, and EK), that represent four major periods in the geologic history of the region (Figs. 5 and 12; Table 2) (Schlager and others, 1984b). Each of these sequences is described in more detail in Phair and Buffler (1983), Phair (1984), and Schlager and others (1984b) and is summarized below and in Table 2. The sequence designations (J1, J2, etc.) adopted by Schlager and others (1984b) are also used in this chapter for continuity.

The oldest sequence (TJ) fills a northeast-trending rift basin in the western part of the area (Figs. 12 and 13). Reflectors within this sequence dip to the southeast and are truncated by a prominent unconformity (Fig. 13). Because of its structural setting and its stratigraphic relation, this sequence is interpreted to be equivalent to the rocks filling Late Triassic–Early Jurassic rift basins along the east coast of the United States (Newark Supergroup), under northern Florida, southern Georgia, and the northern Gulf coastal plain (Eagle Mills Formation), in east-central Mexico (La

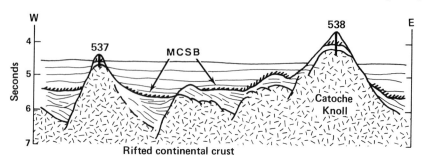

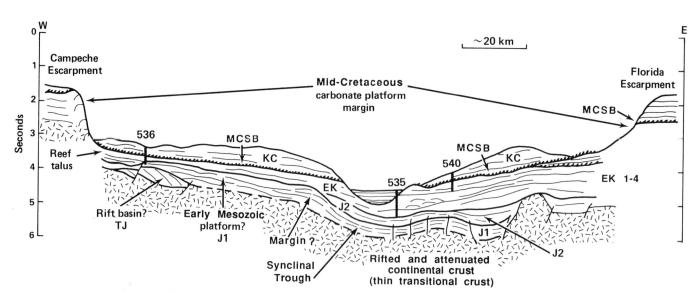

Figure 12. Schematic cross sections (two-way traveltime in seconds) across the southeastern Gulf of Mexico showing major geologic features, seismic sequences, and location of DSDP Leg 77 sites. See text and Table 2 for discussion of sequences. Modified from Schlager and others (1984b). See Figure 11 for location of DSDP sites.

Boca and Huizachal Formations), and possibly under the Sigsbee Salt basin as discussed earlier. Because of its inferred age it is designated TJ, and because of its inferred origin, it probably consists of nonmarine clastic and volcanic rocks. Alternatively, it could consist of unmetamorphosed upper Paleozoic sedimentary rocks, such as occur to the south in Belize (Chapter 7, this volume). This sequence is considered as part of the basement beneath the Gulf as defined in Chapter 4 (this volume), and as shown on Plate 3.

The next younger sequence (J1, Table 2) is quite widespread throughout the entire southeastern Gulf of Mexico. In the southwest it forms a broad undeformed platform overlying basement, while to the southeast it fills in broad basins lying between large uplifted basement blocks (Figs. 12, 13, and 14). These two areas are separated by a prominent NW-SE–trending structural depression or synclinal area (Figs. 12 and 14). A tentative tie between sequence J1 and shallow-marine dolomite drilled at the bottom of DSDP hole 536 (Fig. 12) suggests that part of this sequence may be composed of shallow-marine sediments in this area. The northern edge of the broad, undeformed platform probably is faulted (Fig. 13), while the eastern edge may represent the margin of a carbonate platform (Figs. 12 and 14). Alternatively, this eastern margin could simply represent structural flexuring and faulting (Fig. 14).

To the north, sequence J1 fills a series of half-grabens and probably consists of synrift rocks (probably nonmarine; Figs. 13 and 15). Thus, overall, sequence J1 probably represents a general upward transition from nonmarine to shallow-marine deposition,

although details are yet obscure. Because of its stratigraphic relation overlying the inferred Upper Triassic–Lower Jurassic rift sequence (TJ) and underlying inferred deeper marine Jurassic (J2) (described below), sequence J1 is inferred to be Middle Jurassic in age. It may be partly equivalent to the Werner-Louann evaporites in the northern Gulf coastal plain (Chapter 8, this volume), the lower part of the San Cayetano Group in Cuba to the south (Pszczolkowski, 1978), and a thick Middle Jurassic section buried beneath the Atlantic margin to the east (Klitgord and others, 1988).

The third major sequence (J2, Table 2) is restricted to the eastern half of the area, as it onlaps and pinches out to the west against the older, relatively high-standing platform in the synclinal area (Figs. 12 and 14), as well as high-standing fault blocks (Fig. 15). Its position just below the Lower Cretaceous rocks drilled during DSDP Leg 77 suggest a Late Jurassic age. The more uniform seismic facies of this sequence, similar to the overlying marine Lower Cretaceous (Figs. 14 and 15), suggests that it also represents marine sediments. The occurrence in the southeastern Gulf of this inferred marine sequence implies that a Late Jurassic seaway had developed between the Gulf of Mexico and a proto-Caribbean Sea to the south (Phair and Buffler, 1983; Phair, 1984; Schlager and others, 1984b). Deeper marine conditions apparently occurred in structural troughs, while shallow, possibly platform, conditions occurred on adjacent structural highs (Phair, 1984). Sequence J2 probably is approximately equivalent to Upper Jurassic marine rocks in the circum-Gulf region (Smackover to Cotton Valley and equivalent sequences in Mexico) (Sal-

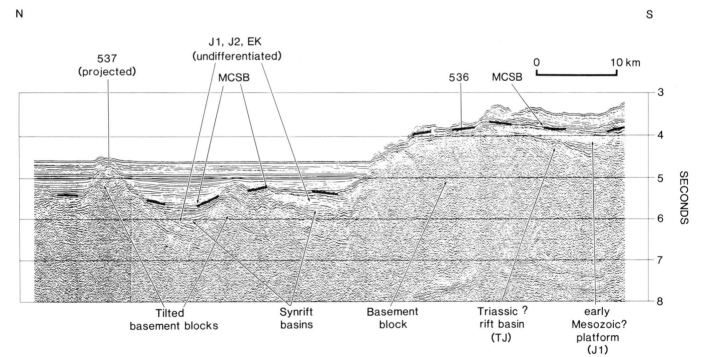

Figure 13. Portion of UT seismic line GT3-75 near the base of the Campeche Escarpment showing regional setting and stratigraphic relations at DSDP sites 536 and 537. See Figure 11 for location. Note change from broad platform in south to tilted basement blocks in north.

TABLE 2. MAJOR SEISMIC SEQUENCES IN THE SOUTHEASTERN GULF OF MEXICO*

Sequence	Age	Seismic facies and regional variation	Equivalent in DSDP holes	Interpretation
KC	Late Cretaceous-Cenozoic on basis of DSDP Holes 97, 536, 537, 538A, and 540	In the south, the lower part (Cretaceous-Eocene) form wedges that thicken considerably toward Cuba, characterized by continuous to discontinuous, divergent facies. Upper part (Eocene-Recent) is a thin blanket with many unconformities. Unit thins in central area around DSDP sites and then thickens again into Gulf of Mexico basin farther to the north. Subdivided into four mapping units (Table 4).	Drilled thin distal part of unit at Holes 540 and 97, approximately 270 m of carbonate ooze, chalk, and marl. Lower part disturbed by creep and slumping. Lower 56 m are gravity-flow deposits.	Eocene-Recent consists of pelagic drape in southern and central part of area. Onlapped by distal turbidites of Gulf basin from the north. Cenomanian-Eocene forms a thick wedge of turbidites and gravity-flow deposits in foredeep of Cuban orogen. Local carbonate cebris shed from Florida Escarpment (Angstadt, 1983; Angstadt and others, 1985).
EK	Early Cretaceous on basis of DSDP Holes 535, 536, 537, 538A, and 540	Widespread throughout northern and eastern part of area. Up to 2 to 3 km thick. Thickens to east along base of Florida Escarpment. Thin or absent to south and west owing to both non-deposition on high-standing areas and post-mid-Cretaceous erosion. Subdivided into four mapping units (Table 3). Relatively uniform parallel continuous facies over most of the area. More discontinuous and hummocky along Florida Escarpment. Prominent N-S distributed hummocky facies in middle unit.	Drilled almost entire unit in Holes 535 and 540; approximately 750 m of pelagic limestone–marly limestone cycles with episodic interbeddings of shallow-water carbonate sand and mud. Thin equivalents drilled at Holes 537 and 538A (clastics and shallow-water limestones overlain by pelagic limestones) and at Hole 536 (talus deposits).	Deep-water carbonate deposition in northern and eastern part of study area. Main source was Florida platform to east and planktonic carbonate production (Phair, 1984).
J2	Probably Late Jurassic on basis on position below lowermost Cretaceous drilled in Hole 535	Uniform, variable-amplitude, continuous reflections. Widespread over most of area. Deformed some in depressions between horsts. Absent or thin on high-standing blocks.	Not reached.	Seismic character similar to Cretaceous above, suggesting deep-marine deposition in central part of area. Possible low-relief shelf margins in places suggest transition to shallow-marine around periphery of area and on high-standing blocks. Probably represents major marine transgression and establishment of seaway in area (Phair and Buffler, 1983).

TABLE 2. MAJOR SEISMIC SEQUENCES IN THE SOUTHEASTERN GULF OF MEXICO* (continued)

Sequence	Age	Seismic facies and regional variation	Equivalent in DSDP holes	Interpretation
J1	Inferred to be Jurassic, possibly Middle Jurassic, on basis of superposition	Widespread unit in south with relatively uniform thickness (several km) and seismic character (high-amplitude, discontinuous). Onlaps broad basement highs. Undeformed in south except down-dropped along prominent NW-SE graben system and along broad trough north of Cuba. To north unit fills half-grabens between tilted fault blocks.	May have just reached top of unit in bottom of Hole 536. Shallow-water dolomite, Jurassic or Permian on basis of 87/86Sr ratio.	Seismic character and setting to north suggests nonmarine synrift sediments (alluvial fans, lacustrine deposits, volcanics, evaporites). Upper part of unit in south may be shallow-marine platform. Lower part to south may be nonmarine (Phair and Buffler, 1983). Possible equivalent to San Cayetano Formation on Cuba.
TJ	Inferred to be Late Triassic–Early Jurassic on basis of setting and stratigraphic relations	Southeast-dipping parallel reflections filling NE-SW–trending graben system. Onlaps basement. Truncated by prominent unconformity.	Not reached.	Inferred to represent equivalent of Upper Triassic–Lower Jurassic rift sequences observed elsewhere around margins of North Atlantic and Gulf of Mexico. Probably nonmarine sediments and volcanics (Phair and Buffler, 1983).
Basement	Lower Paleozoic metamorphic rocks (500 ± Ma) intruded by early Mesozoic dikes and sills (160 to 190 Ma)	Acoustic basement over much of the area. In places contains some low-amplitude reflections. Occurs as broad uplifted blocks and lows to south and high-relief tilted fault-blocks in northern part of area.	Drilled phyllite in Hole 537 and gneiss-amphibolite intruded by basic dikes and sills in Hole 538A.	Top of rifted and attenuated continental crust (Buffler and others, 1981; Ibrahim and others, 1981). Probably mostly lower Paleozoic (Pan-Africa) metamorphic rocks, possibly with Paleozoic sedimentary rock in places. Dissected by Jurassic intrusives. Good example of "transitional" crust.

*Modified from Schlager and others, 1984b.

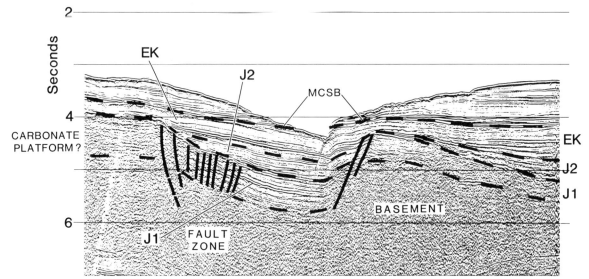

Figure 14. Portion of UT seismic line SF-11 across prominent northwest-trending structural depression or synclinal area showing stratigraphic relations of pre-MCSB Mesozoic sequences. Western edge of depression may represent faulted carbonate platform margin. See Figure 11 for location.

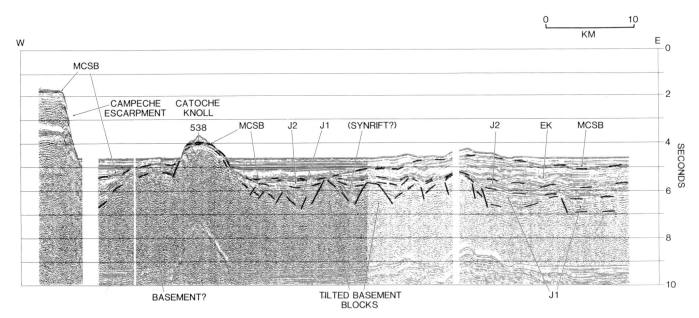

Figure 15. Portion of UT seismic line SF-2 showing stratigraphic relations of pre-MCSB rocks in area of tilted fault blocks to the north. Sequence J1 probably represents nonmarine synrift sediments, which are overlain by marine strata (J2 and EK). DSDP hole 538 on Catoche Knoll drilled Neocomian shallow-water carbonates overlying early Paleozoic metamorphic basement, that had been intruded by Early Jurassic diabase dikes and sills. Note truncation of upper EK section by MCSB. See Figure 11 for location.

vador, 1987; Chapter 8, this volume). It also probably is equivalent to Upper Jurassic shallow-marine rocks that are now uplifted in western Cuba to the south (Pszczolkowski, 1978).

The final major pre-MCSB sequence (EK, Table 2) is widespread across most of the southeastern Gulf region. Rocks equivalent to this sequence were drilled during Leg 77 at all of the holes, 535 through 540 (Fig. 12) (Schlager and others, 1984a, b; Buffler and others, 1984b). At holes 535 and 540, an almost complete Lower Cretaceous section was sampled (Figs. 12 and 16) consisting of deep-water cyclic carbonates. Seismic line SF-15 through both of these holes (Fig. 16) shows their local geologic setting. At hole 536 along the base of the Campeche Escarpment, a thin Aptian-Albian talus wedge was drilled (Figs. 12 and 13), while at holes 537 and 538A, drilled on the top of two basement fault-blocks (Figs. 12, 13, and 15), a thin, lowermost Cretaceous (Berriasian-Valanginian) shallow-water limestone section was penetrated overlain by a pelagic Aptian-Albian section. Information from these sites and the seismic data, therefore, document a complex Early Cretaceous history beginning with widespread shallow-water platforms located on basement highs and deeper water embayments in structural lows (Phair, 1984; Schlager and others, 1984b). By Aptian-Albian time the area had continued to collapse and the entire deep southeastern Gulf represented a deep-water environment. Steep carbonate margins had developed along the adjacent Campeche and Florida Escarpments (Fig. 12).

Only in the eastern half of the deep southeastern Gulf is the Lower Cretaceous section (EK) thick enough to be further subdivided into seismic sequences and studied in more detail (Phair, 1984). This detailed seismic stratigraphic analysis was combined with the DSDP Leg 77 results to document the Early Cretaceous geologic history (Schlager and others, 1984b). The section was subdivided into four seismic sequences (EK1 through EK4, Table 3) that were mapped over the entire eastern half of the area. Line SF-15 (Fig. 16) shows the four units, their bounding unconformities, and their ties to drill holes 535 and 540. Ages of each unit are given in Table 3.

Isotime maps document the filling of the eastern basin (Fig. 17). Unit EK1 (Berriasian-Hauterivian) reflects a major deep-water depocenter in the northwestern part of the area. To the south a narrow, deeper seaway extended between shallow-water banks located on basement highs. The three overlying units (EK2, EK3, and EK4; Hauterivian through Cenomanian) indicate a major influx of sediment into a deep-water basin from the adjacent shallow-water Florida carbonate platform to the east and only a minor contribution from the Yucatan platform to the west (Figs. 12, 15, 16, and 17; Map for EK4 not included in Fig. 17).

West Florida basin. The thick Jurassic–Lower Cretaceous sequence in the southeastern Gulf can be traced northward into the deep eastern Gulf, where it fills another large Mesozoic structural depression or basin lying just west of the Florida Escarpment (Figs. 2C, 7, and 18). This basin has a setting similar to the

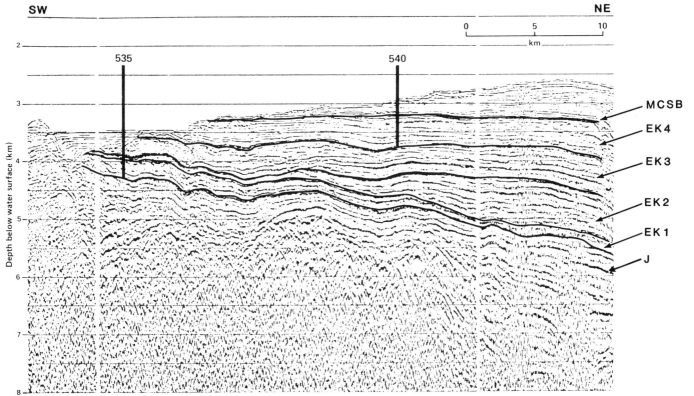

Figure 16. Depth-converted UT seismic line SF-15 showing Lower Cretaceous seismic sequences EK1 through EK4 near DSDP sites 535 and 540 (modified from Schlager and others, 1984b). See Figure 11 for location.

TABLE 3. EARLY CRETACEOUS AGE (EK) SEISMIC SEQUENCES, SOUTHEASTERN GULF OF MEXICO*

Sequence	Age	Seismic facies and regional variation	Equivalent in DSDP Holes	Interpretation
EK4	Late Albian–Early Cenomanian	Parallel, continuous with moderately high amplitude; toward Florida Escarpment become discontinuous and unit thickens rapidly; restricted to eastern part of study area.	Hole 540: 417 m of hemiplagic limestone–marly limestone cycles with some shallow-water detritus.	EK2-4: Mainly pelagic carbonates in SW, grading rapidly into hemipelagic deposits with much redeposited platform material toward Florida Escarpment; rates and loci of platform input changed with time, creating sequence boundaries.
EK3	Early Albian-late Albian	Parallel-subparallel, continuous with moderate amplitude; grade westward into hummocky clinoforms, suggesting westward shift of depocenter with time; thin or absent in the SW and largely eroded near Cuba.	Hole 535: 233 m of subdued cycles of limestone and marly limestone, frequent interbeds of shallow-water calcarenites (and mud?). Equivalent at Hole 536 is platform talus; at 537 and 538A, pelagic chalk and limestone.	
EK2	Late Hauterivian-late Aptian (several hiatuses)	Parallel, continuous; high-amplitude in SW changing to low-amplitude discontinuous toward Florida Escarpment; rapid thickening in same direction; thin or absent in SW.	Hole 535: 80 m of cyclic alternation of pelagic white limestone and marly limestone rich in organic matter. Equivalent at Hole 537 and 538A is condensed section of pelagic chalk and limestone.	
EK1	Late Berriasian-Hauterivian	Parallel, continuous with moderate amplitude; thickens at foot of northern Florida Escarpment; thin or absent in SW; thickness highly variable in block-faulted area to N.	Hole 535: 248 m of pelagic limestone in cyclic alternation with marly limestone rich in organic matter; hardgrounds (increasing in frequency downhole). Equivalent to shallow water carbonate platform sediments at Hole 537 and 538A.	Mainly pelagic carbonates; minor sediment contribution from low-relief platform in NE (Florida); in W shallow water platform affected by syn-sedimentary block faulting.

*Modified from Schlager and others, 1984b, after Phair, 1984.

Sigsbee Salt basin (i.e., lying between oceanic crust and the Early Cretaceous platform margin). In the northeast part of the basin the sequence contains a thick salt layer overlying basement and is capped everywhere by the MCSB (Figs. 18 and 19A). Thus, its age also probably ranges from Middle Jurassic through Early Cretaceous. The basin is bounded on the west by oceanic crust and on the east and northeast by the Florida Escarpment (Fig. 18). Because of its location, this basin is designated the West Florida basin (Lord, 1986, 1987). Its previous designation as the West Florida Salt basin (Buffler and Sawyer, 1985; Martin, 1984) is not entirely appropriate because salt is obvious only in the north and northwest part of the basin (Fig. 19A). The basin has been studied recently by Lord (1986, 1987) and Rosenthal (1987).

Seismic data in the area show a section of layered sedimentary rocks as much as 4 to 6 km thick along the base of the Florida Escarpment (Lord, 1986, 1987) (Fig. 18). True basement is difficult to observe over most of the area, and acoustic basement is often simply the top of inferred salt pillows or just the lowermost coherent reflection. Thus, the thicknesses may represent just minimum values. This area of maximum thickness abuts abruptly the Florida Escarpment. Even though reflections cannot

be carried below the steep escarpment, the thick sequence probably is correlative with the thick Mesozoic section that lies beneath the West Florida shelf to the east and northeast (Corso, 1987) (Fig. 18). Salt may extend east of the escarpment into the Tampa embayment, a prominent basement low lying between the Sarasota arch to the south and the Middle Ground arch to the north.

The overall sequence in the West Florida basin thins to the west by onlap onto basement inferred to be the top of thin transitional crust (Fig. 18) (Chapter 4, this volume). This relation was used by Rosenthal (1987) to define the oceanic/transitional crust boundary throughout the eastern Gulf of Mexico (Figs. 18 and 20). The place of maximum onlap marks the boundary (COB). The older sequence (Jurassic) shown in Figure 20 (2) is inferred to have been deposited on an eastward-dipping surface of transitional crust (1) (outer high?) prior to the emplacement of the adjacent oceanic crust (4) and the deposition of the overlying oceanic sediments (3 and 5). As the onlapping sequence appears to contain salt to the east, it is inferred to be, in part, Middle to Late Jurassic in age. This agrees with the earlier correlation of inferred Jurassic rocks from the southeastern Gulf by Phair and Buffler (1983) and Schlager and others (1984b), who showed the Jurassic section thinning and pinching out by onlap to the north

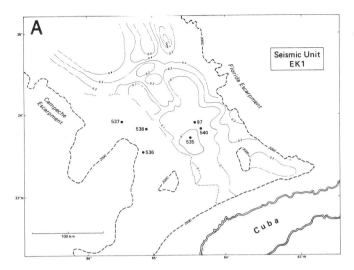

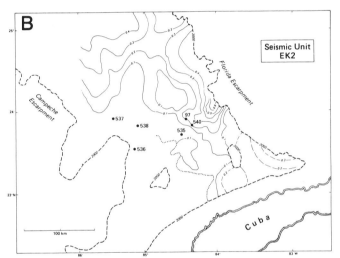

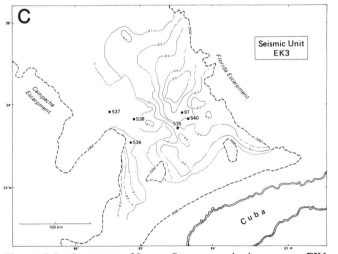

Figure 17. Isochron maps of Lower Cretaceous seismic sequences EK1, EK2, and EK3 (A, B, C) in southeastern Gulf of Mexico (two-way traveltime in seconds; from Schlager and others, 1984b). Water depth in meters (heavy lines). EK1 shows thickening to the north, while EK2 and EK3 show thickening to the east, suggesting a source from the Florida Escarpment area.

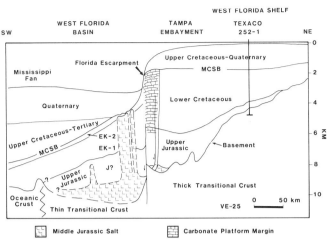

Figure 18. Regional northeast-southwest cross section across West Florida basin, Florida Escarpment and Tampa Embayment showing stratigraphic relations of Mesozoic rocks. See Figure 3 for location. Section based on interpretation of UT, USGS, and industry seismic data (from L. Dobson, personal communication, 1990).

and not overlying oceanic crust. Again, this further suggests that most of the sediments overlying oceanic crust are Early Cretaceous in age in this area.

Martin (1980, 1984) first described the distribution of salt in the basin, while Lord (1986, 1987) described for the first time the entire West Florida basin using the available UTIG seismic data. These analyses showed that the salt is characterized generally by low-relief pillows, domes, and diapirs. The salt structures affect mainly the pre-MCSB sequence; only a few actually penetrate the post-MCSB sections (Fig. 18). This suggests that most of the salt movement and deformation took place during the Late Jurassic–Early Cretaceous. The lack of later remobilization here may be due to thin original salt and/or lack of significant later overburden, as most of the Cenozoic section is quite thin in this part of the Gulf (Fig. 18) (Shaub and others, 1984).

Lord (1986, 1987) developed a preliminary seismic stratigraphic framework for the basin by correlating seismic units from the southeastern Gulf north along two regional seismic lines (Fig. 5). Correlations of the older units were difficult due to basement relief, and units J1 and J2 were combined into one inferred Jurassic unit J(?). Units EK1 and EK2 were correlated into the West Florida basin with confidence, while EK3 and EK4 are truncated in the southeastern Gulf by the MCSB and are absent or too thin to map in the West Florida basin.

An isopach of unit J(?) shows a northeast thickening to more than 2 km along the base of the Florida Escarpment (Fig. 19B). The map shows minimum thicknesses, as the base of this unit is not well defined. This northeast-thickening trend continues with unit EK1, which also shows a depocenter with over 2 km of sediments near the central part of the Florida Escarpment (Fig. 19C). EK2 is much thinner and more uniform in thickness, with

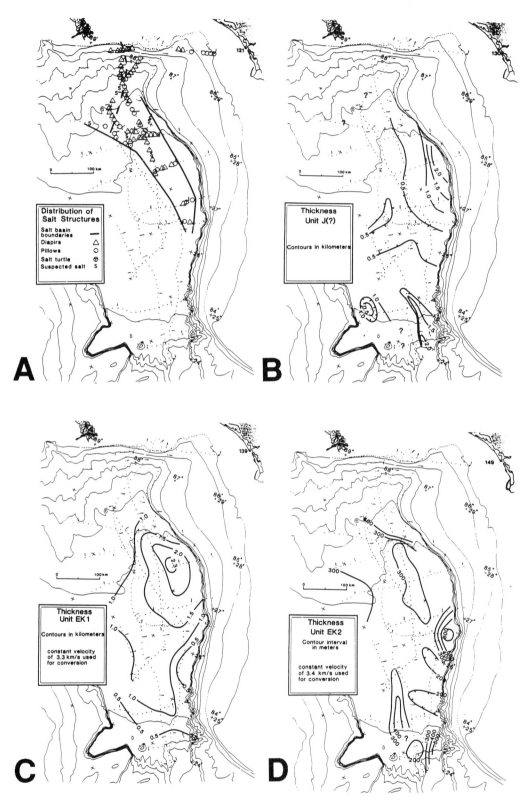

Figure 19. Maps showing distribution of salt in West Florida basin (A) as well as isopach maps (km) of three pre-MCSB seismic units mapped in the West Florida basin (B, C, and D; from Lord, 1986, 1987). Thickening of units suggests a source to the east. Salt structures mapped include diapir (Δ), pillows (O), and suspected structures (S). Heavy dashed line separates area to west where diapirs penetrate MCSB from area to east where diapirs generally do not penetrate MCSB.

only a small thickening to more than 400 m along the central escarpment (Fig. 19D).

The depositional setting for these units is unknown. The J(?) unit probably represents a general transition upward from shallow marine to deep marine following deposition of the salt. By EK1 time the basin definitely was a deep-water setting as a carbonate platform margin had become established along the Florida Escarpment (e.g., Corso, 1987) (Fig. 18). Also, EK1 represents deep-water carbonates at hole 540 in the southeastern Gulf as discussed above. In the area of the thick depocenter, the EK1 unit is characterized by large overlapping lobes, suggesting deposition of deep-sea fans. As mentioned earlier the thick J(?) and EK1 sections probably do not end abruptly at the escarpment but continue beneath the escarpment into the Tampa embayment, although they may be offset by faulting (Fig. 18). The overlying thinner EK2 unit apparently reflects the establishment in Aptian–early Albian time of a well-defined rimmed platform margin at the escarpment, which reduced the amount of sediment getting to the basin.

Mid-Cretaceous sequence boundary (MCSB)

The Lower Cretaceous and older sequences are separated from the Upper Cretaceous and Cenozoic sequences by a prominent reflector in the deep basin and by a major unconformity along the adjacent margins (Buffler and others, 1980; Faust, 1990; Schlager and others, 1984b; Shaub and others, 1984). The unconformity is easily recognized along the southern, eastern, and southeastern flanks of the deep basin as a surface that truncates beds below and is onlapped by the later basin fill (Figs. 6, 8, 9, 10, 12, 13, 14, 15, 18, and 20) (Faust, 1990). Along the base of the Campeche Escarpment, basinward-trending channels that erode into Lower Cretaceous rocks suggest erosion by turbidites, while along the Florida Escarpment, channels trending parallel to the escarpment suggest contour currents. In the southeastern Gulf of Mexico, erosional channels appear to be controlled by deeper structure.

In the central Gulf the unconformity becomes a conformable surface, which is recognized as a prominent high-amplitude reflector (Fig. 4). At this reflector there is a major downward increase in refraction velocity, which probably reflects a major lithology change from deep-water, fine-grained, terrigenous clastics above to predominantly deep-water carbonates below. Thus, it represents a turning point in the sedimentation patterns and depositional history of the basin (Shaub and others, 1984; Schlager, 1989).

The MCSB has been tentatively dated as mid-Cretaceous (Fig. 5) and tentatively correlated with a mid-Cenomanian drop in sea level (Vail and others, 1977; Buffler and others, 1980; Shaub and others, 1984); thus, its previous designation as "mid-Cretaceous unconformity" or MCU. As mentioned earlier, this surface has the characteristics of a major sequence boundary, and thus, its new designation as the "mid-Cretaceous sequence boundary" or MCSB. This age is based on correlations with well

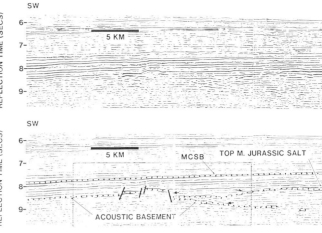
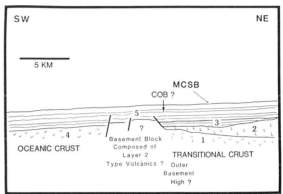

Figure 20. Uninterpreted (upper) and interpreted (middle) seismic profile plus interpreted section (lower) across inferred boundary between oceanic crust (4) and transitional crust (1). Older (Jurassic?) salt and sedimentary rocks (2) onlap transitional crust (outer basement high?). Basement block may represent volcanics extruded at continent-ocean boundary (COB). Older rocks (1, 2, 4) overlain by Upper Jurassic?–Lower Cretaceous deep-water sedimentary rocks (3 and 5; from Rosenthal, 1987).

control in the northeastern Gulf (Addy and Buffler, 1984), with the adjacent shallow-water carbonate banks (Locker and Buffler, 1983), and with DSDP Leg 77 drill holes in the southeastern Gulf (Buffler and others, 1984b; Schlager and others, 1984a; Phair, 1984; Schlager and others, 1984b) (Fig. 4).

Along the margins of the deep southern, eastern, and southeastern Gulf of Mexico basin, both above and below the Campeche and Florida Escarpments, the seismic data and DSDP results suggest that the hiatus represented by the unconformity may span much of the Late Cretaceous. Lower Cretaceous rocks are overlain by Paleocene rocks with no or only thin Upper Cretaceous present in these areas (Worzel and others, 1973; Mitchum, 1978; Locker and Buffler, 1983; Buffler and others, 1984b; Schlager and others, 1984b). The surface, therefore, represents a major period of erosion as well as nondeposition or starved conditions along the margins, although slumping on tectonically steepened slopes also has been suggested as a mechanism to explain the extreme reduction in the Upper Cretaceous section (Schlager and others, 1984b). This period of starvation apparently followed a widespread mid-Cretaceous drowning of the Early Cretaceous margins.

A likely primary cause of this major break in Gulf of Mexico sedimentation was a major mid-Cenomanian drop in sea level (Vail and others, 1977; Haq and others, 1987). This drop could have interrupted carbonate sedimentation along the Florida and Yucatan carbonate platform margins and concentrated turbidity currents and contour currents along the base of the slope, thus causing the widespread erosion. The subsequent rise in sea level could have terminally drowned the outer carbonate platforms, causing the margins to "step back" and become reestablished in a more landward position during the Late Cretaceous. Long-term climatic changes, opening and closing of seaways, and "nutrient" poisoning also could have contributed to this drowning event (Winker and Buffler, 1988; Schlager, 1989). Thus, even though the hiatus developed due to the drowning and general starvation of the deep margins, the major event that initiated the erosion and starvation was mid-Cretaceous in age. It is believed, therefore, that the term "mid-Cretaceous" is still appropriate for this major sequence boundary.

By mid-Cenomanian time, the Gulf of Mexico was a deep ocean basin surrounded on most sides by broad carbonate platforms with steep margins (Winker and Buffler, 1988). The major deep-water outlet was through the southeastern Gulf, although shallower seaways extended across Mexico into the Pacific. Simple reconstructions of the basin at this time, compensating for both sediment loading and tectonic cooling, suggest that the central Gulf was approximately 4 to 5 km deep over the area of oceanic crust (Winker and Buffler, 1988). Maximum relief along the margins, particularly along the Florida and Campeche Escarpments, was 1 to 2 km, and there was a marked asymmetry of the carbonate margins. Steep, rimmed bypass margins occurred all along the Florida Escarpment, the Campeche Escarpment, and the Tuxpan Platform of Mexico, while more gentle margins had developed along the northern Gulf as well as in the southwestern Gulf (Fig. 2). This asymmetry could be due to a combination of (a) an influx of siliciclastics from the north, (b) the distribution of paleocurrent systems, (c) variable paleoclimates, and (d) proximity to more rapidly subsiding areas of oceanic crust in the central Gulf. This picture of the early Gulf of Mexico basin now sets the scene and provides a framework for the post-MCSB filling of the deep basin.

Post-MCSB

The post-mid-Cretaceous (post-MCSB) section reaches maximum thicknesses in two areas of the deep Gulf of Mexico: the central part of the basin, and the far southeastern Gulf just north of Cuba. The second area of deposition represents the filling of a narrow foredeep in front of Cuba during the early Tertiary. Separating the two areas is a structural high capped by relatively thin, pelagic carbonate sediments of Tertiary age. Each of the areas is described briefly below.

Central Gulf of Mexico. The post-MCSB section in the deep Gulf is up to 5 or 6 km thick under much of the basin (Fig. 2). The section thins to the south, east, and southeast, and pinches out by onlap along the base of the Campeche and Florida Escarpments as well as over the structural high in the southeastern Gulf (Figs. 2, 6, 8, 9, 10, 13, 15, and 18). The section in the western Gulf of Mexico originally was divided into four seismic units (Fig. 4) (Ladd and others, 1976) based mainly on seismic characteristics. More recently, the post-MCSB seismic stratigraphy was redefined and the section subdivided into five major seismic units—from older to younger, Campeche, Lower Mexican Ridges, Upper Mexican Ridges, Cinco de Mayo, and Sigsbee (Shaub and others, 1984) (Table 1; Figs. 4, 5, and 6). The new units were defined as seismic sequences; their boundaries are the major unconformities along the margin of the deep basin where the units thin and pinch out (e.g., base of the Campeche Escarpment, Figs. 6 and 9). Ages for the top of the Campeche and Lower Mexican Ridges units are not well constrained, and are based only on extrapolation (Table 1; Fig. 6). The tops of the overlying two units are dated by correlations with DSDP holes 87, 90, and 91 in the deep basin (Fig. 3).

The five seismic units have been mapped throughout the deep central basin (Shaub and others, 1984). Isopach maps of each unit document the progressive filling of the basin (Figs. 21, 22, and 23). The Campeche and Lower Mexican Ridges units show major Late Cretaceous–Paleogene depocenters filling the entire deep western and northwestern Gulf (Fig. 21). They reflect erosion of the "Laramide" orogenic belt of northern Mexico as well as the supply of sediments by an ancestral Rio Grande system. By Upper Mexican Ridges time (mainly Miocene; Fig. 22A), shelf margins and major depocenters had shifted to the north and sediments were being supplied mainly from the ancestral Mississippi River, although there was still some western input from local rivers. During the Pliocene (Cinco de Mayo time; Fig. 22B), major sedimentation in the deep Gulf was restricted by deformed zones that had developed to the west (Mexican Ridges foldbelt) and north (Texas-Louisiana salt province). By Pleistocene time, however (Fig. 23), sediments from the Mississippi River and other smaller rivers broke through the northern deformed area, and sediments were carried to the slope and deep basin through canyons and deposited as deep-sea fans in the northern and eastern Gulf during low stands of sea level (Sigsbee Unit). More detailed local studies in the deep western Gulf (Bertagne, 1980, 1984; Usmani, 1980) have resulted in the subdivision of the post-MCSB section into up to eight seismic sequences (Fig. 5). These studies mapped the distribution of seismic facies and documented the history of the region, which was characterized by the deposition of hemipelagic sediments and turbidites.

The seismic stratigraphic studies and the limited DSDP data (down through middle Miocene) suggest that the thick post-MCSB sequence in the deep Gulf represents the more distal parts of large wedges of siliciclastic sediments deposited seaward of well-defined shelf margins. Through time, the major depocenters and equivalent shelf margins shifted from the west to the north

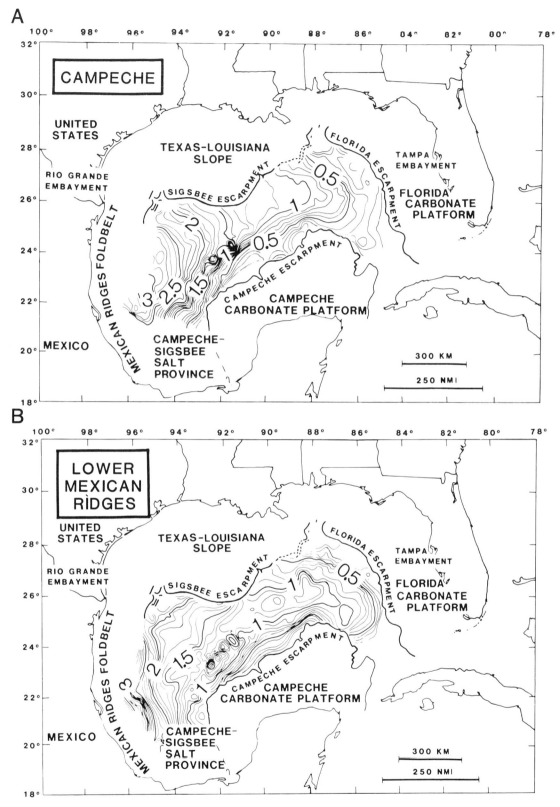

Figure 21. Isopach maps of mid-Cretaceous–early Tertiary(?) Campeche (A) and early(?) to mid-Tertiary(?) Lower Mexican Ridges (B) seismic sequences in the deep Gulf of Mexico (from Shaub and others, 1984). Contour interval is 0.1 km. Small anomalies in central Gulf are due to salt tectonics.

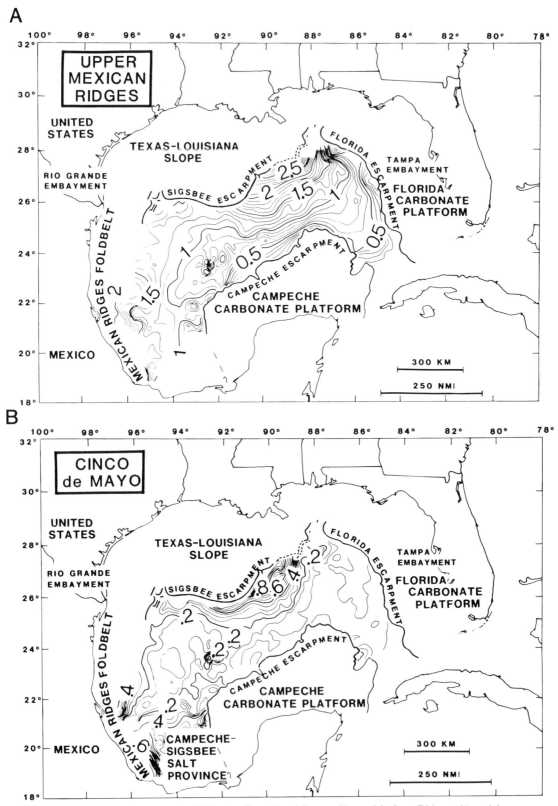

Figure 22. Isopach maps of mid-Tertiary(?) to late Miocene Upper Mexican Ridges (A) and late Miocene through Pliocene Cinco de Mayo (B) seismic sequences in the deep Gulf of Mexico (from Shaub and others, 1984). Contour interval is 0.1 km. Small anomalies in central Gulf are due to salt tectonics.

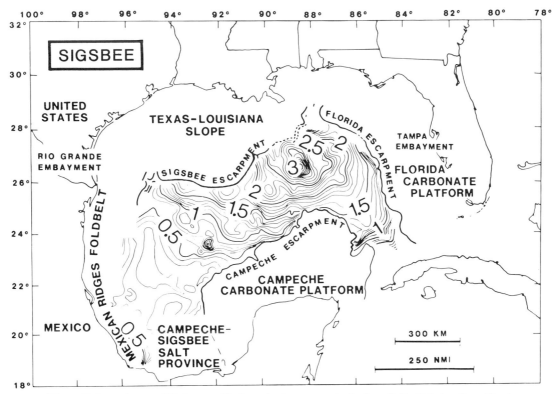

Figure 23. Isopach map of Pleistocene Sigsbee seismic sequence in the deep Gulf of Mexico (from Shaub and others, 1984). Contour interval is 0.1 km. Small anomalies in central Gulf are due to salt tectonics.

(Figs. 21, 22, and 23). The sediments are interpreted to consist mainly of fine-grained pelagic and hemipelagic sediments interbedded with fine- and coarse-grained turbidites. The thicker parts of the depocenters are often characterized by seismically defined channels and probably represent deep-sea fan systems.

Studies need to be made comparing Cenozoic sedimentation in the deep Gulf with that of the surrounding margins. A first attempt at this was made through the compilation of geological and geophysical data for a limited part of the deep basin and adjacent margins as part of an Ocean Margin Drilling Program Atlas (Buffler and others, 1984a). This study was limited by poor correlations between the post-MCSB units in the deep Gulf basin and equivalent units in surrounding regions. The lack of good age control, particularly for the lower units, limits a direct age comparison, while the deformed margins and steep escarpments surrounding the deep Gulf preclude direct ties using the seismic data. The only direct seismic tie is in the northeastern Gulf of Mexico and the Mississippi Fan/Canyon area, where a few seismic boundaries can be carried onto the shelf and tied to wells (Addy and Buffler, 1984; Walters, 1985) (Fig. 5). More deep well control in the deep Gulf is needed, as well as long regional, deep penetration seismic lines from the shelf to the deep basin, before a complete picture of the Cenozoic evolution of the entire central Gulf of Mexico basin can be developed. A few deep industry test wells have now been drilled on the upper Mississippi Fan, but information from these wells is still proprietary.

Mississippi Fan. The Sigsbee Unit in the eastern Gulf has been the subject of considerable interest during the past few years (Fig. 23). Here the unit contains the Mississippi Fan, a large deep-sea fan system up to 3 km thick that fills much of the eastern Gulf. The fan is just part of a larger accumulation of sediments deposited by the Mississippi River system during the late Pliocene and Pleistocene along the outer shelf, slope, and deep basin, during glacial low-stands of sea level (Chapter 12, this volume).

Early regional seismic studies described only in general terms the geometry, seismic stratigraphy, facies, and processes of the fan itself (Moore and others, 1978, 1979; Stuart and Caughey, 1977; Shaub and others, 1984). More detailed seismic stratigraphic analyses of the fan area, based on UTIG seismic lines and other seismic data, defined up to eight seismic sequences as well as seismic facies for the Sigsbee Unit on the fan (Feeley and others, 1984) and five seismic units across the adjacent slope (Walters, 1985; Fig. 5). These studies document the complex shifting of depocenters for each unit, the lateral distribution of seismic facies, and the cyclic nature of vertical sequences within each unit, which are related to the rise and fall of sea level during the Pleistocene. The complex internal nature of seismic facies within the units reflect deposition by a wide variety of fan processes, including turbidites, debris flows, slumps, as well as channel-overbank. Primary fan deposition apparently is related to submarine canyons that cut across the slope and funneled sediments to the deep basin. The Mississippi Canyon is the youngest of these canyon systems.

A regional survey of the fan using the long-range side-scan sonar system known as *Gloria* discovered a large, sinuous channel system that traverses the entire fan from the Mississippi Canyon to the Florida Plain (Garrison and others, 1983). The channel was studied in more detail using seismic data and detailed side-scan sonar techniques (Kastens and Shor, 1985, 1986). This channel and the associated modern fan lobe became the focus for DSDP Leg 96 (Bouma and others, 1984a, b). Drill holes at the upper fan channel sites found sand and gravel in the core of the channel flanked by fine-grained overbank deposits and slump deposits. The channel sands and gravels correspond to distinct high-amplitude reflection packages on the seismic data. Holes drilled at the distal end of the channel found sheet sands that had been deposited 400 to 500 km from their source, the Mississippi Canyon area. Results of DSDP Leg 96, including more detailed seismic stratigraphic studies of the drill sites and

channel systems, are published in the official DSDP Initial Reports volume for Leg 96 (Bouma and others, 1986).

The availability of a relatively dense grid of industry multifold seismic data has allowed for a much more detailed analysis of the middle part of the fan lying just south of the deformed slope (Weimer, 1989, 1990a, b; Weimer and Buffler, 1988, 1989b) (Figs. 5, 24, and 25). In this area, 17 sequences have been defined and mapped, each characterized by channel/levee/overbank systems. Sequence boundaries are regionally continuous reflections interpreted to be condensed intervals representing periods of starved deposition on the fan. An exaggerated east-west cross section across the fan (Fig. 24) shows the overlapping nature of the individual sequences and associated channels and the overall shift in deposition from west to east through time. A map of the channel valleys (Fig. 25) containing the high-amplitude channel facies also shows a very complex pattern. The channels show extreme variability in sinuosity, aggradation, and lateral migration, reflecting a very complex depositional history. Besides

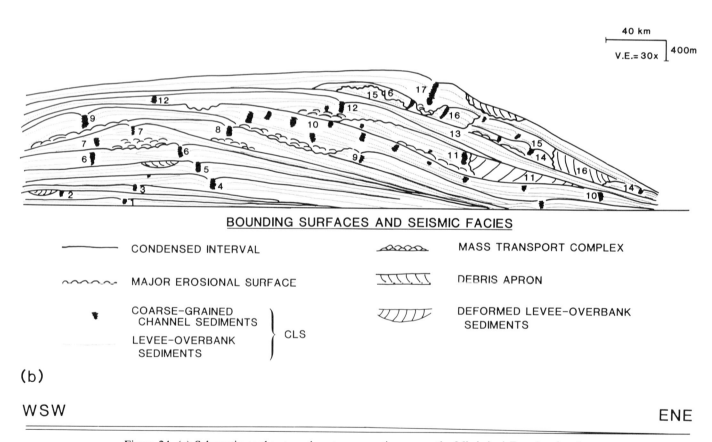

Figure 24. (a) Schematic northeast-southwest cross section across the Mississippi Fan showing the distribution of 17 sequences, with their bounding surfaces and depositional environments (from Weimer, 1989). Note the extreme vertical exaggeration. See Figure 25 for lateral distribution of channels within individual channel/levee systems (CLS). (b) Section of fan with no vertical exaggeration.

the well-defined channel systems, broader bands of high-amplitude seismic facies as well as units of hummocky and chaotic seismic facies at the base of many sequences suggest deposition by a variety of disorganized mass-transport processes, including broad channel systems, slumps, slides, and debris flows (Fig. 24).

The cyclic pattern within each sequence probably reflects deposition in response to Pleistocene sea-level cycles. The lower, less organized seismic facies represent sediment eroded from submarine canyons during the initial fall in sea level, while the more organized channel/levee/overbank facies represents material funneled directly down the canyons to the deep basin during low stands and the following rise.

Southeastern Gulf of Mexico. The units described above in the deep Gulf of Mexico thin and lap onto a high-standing block in the deep southeastern Gulf lying between the Florida and Campeche Escarpments (Fig. 2B). Here the post-MCSB section is only several hundred meters thick (Fig. 26). It was drilled during DSDP Leg 10 (holes 96 and 97) and Leg 77 (holes 536 through 540) (Figs. 11 and 12) and consists mainly of pelagic ooze and chalk with numerous unconformities (Worzel and oth-

ers, 1973; Buffler and others, 1984b; Schlager and others, 1984a). It represents an area that experienced periodic erosion and non-deposition, probably due to deep-sea currents that flowed through the narrow straits between the Florida and Campeche Escarpments (Angstadt, 1983; Angstadt and others, 1983, 1985).

Further south toward Cuba the post-MCSB section thickens rapidly to almost 4 km along the north coast of Cuba (Fig. 26). Here the post-MCSB seismic section has been studied in detail and divided into nine seismic sequences and four seismic mapping units (KC1 to KC4; Table 4; Figs. 5 and 26) (Angstadt, 1983; Schlager and others, 1984b; Angstadt and others, 1985). These units have been mapped throughout the area and tied as best as possible to DSDP holes 97 and 540 (Fig. 26), where the section is quite thin and abbreviated, making correlations difficult.

Isotime and seismic facies maps for each unit document the post-MCSB history of the area (Angstadt, 1983; Schlager and others, 1984b; Angstadt and others, 1985) (Fig. 27). The thick, wedge-shaped units KC1 and KC2 represent the filling of a tectonic foredeep in front of the Cuban orogen, probably formed by

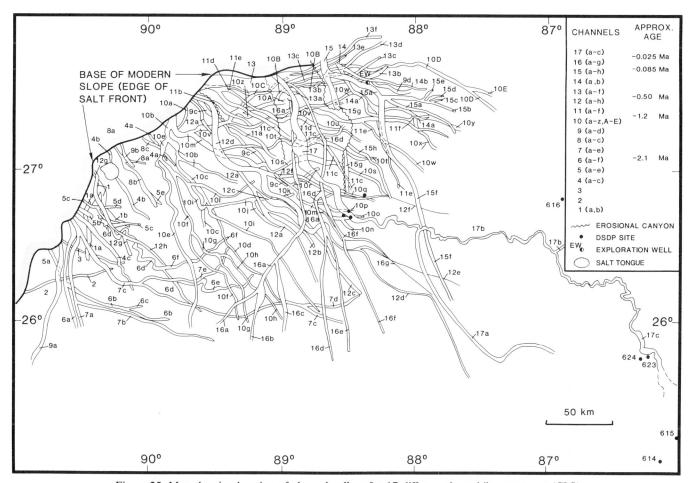

Figure 25. Map showing location of channel valleys for 17 different channel/levee systems (CLS) identified in the Mississippi Fan (from Weimer, 1989). Channels are numbered from 1 to 17 (oldest to youngest). Within any one CLS, different channel ages are lettered from oldest to youngest (a, b, c, etc.). Approximate ages are from Walters (1985).

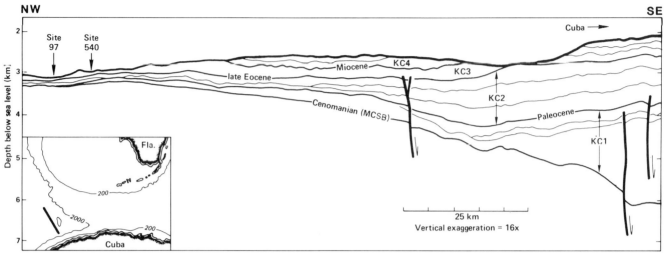

Figure 26. Regional northwest-southeast cross section across southeastern Gulf of Mexico based on depth-converted seismic profiles (from Schlager and others, 1984b). Shown are seismic sequences and major mapping units KC1 through KC4. Note southward thickening of units KC1 and KC2 into Cuban foredeep. Units KC3 and KC4 have been truncated by deep-sea erosion. See inset for location.

TABLE 4. LATE CRETACEOUS-CENOZOIC AGE (KC) SEISMIC SEQUENCES IN SOUTHEASTERN GULF OF MEXICO*

Sequence	Age	Seismic facies and regional variation	Equivalent in DSDP Hole 540	Interpretation
KC4	Middle Miocene-Recent	Parallel, continuous with numerous unconformities; forms discontinuous drape.	52 m of calcareous ooze and chalk.	Pelagic drape, partly eroded by turbidity currents and contour currents.
KC3	Late Eocene-early Middle Miocene	Parallel, continuous grading into discontinuous facies near Cuba; reflection continuity degrades to SW due to slumping and valley-cutting.	191 m of chalk and marly chalk deformed by sliding (inclined bedding, folds, microfaults, deformed burrows).	Mostly pelagic drape, complicated by slumping on (tectonically?) over-steepened slopes.
KC2	Late Paleocene-early Late Eocene	Divergent, continuous, thickening toward Cuba; hummocky along base of Florida Escarpment; top boundary is major unconformity with considerable relief and consistent truncation.	29 m of alternating chalk and dark marly limestone, deformed by creep or slumping. Top boundary corresponds to late Eocene hiatus.	Mainly turbidites in the south (toward Cuba) with increasing pelagic component to north.
KC1	Mid-Cenomanian-Late Paleocene	Strongly divergent toward Cuba, with concomitant decrease in continuity; minor thickening and decrease in continuity toward Florida Escarpment.	56 m of sediment gravity flows (pebbly chalks overlain by graded conglomerates and sandstones), possibly thin interbeds of hemipelagic limestone. Sedimentation discontinuous; major hiatuses. Section represents only small part of correlative seismic unit.	Mainly slumps and sediment gravity-flows shed from Cuba, with minor source along Florida Escarpment.

*Modified from Schlager and others, 1984b, after Angstadt, 1983. Also see Angstadt and others, 1985.

the loading of thrust sheets during the Late Cretaceous through Eocene "Laramide" deformation of Cuba (Figs. 26 and 27; Table 4). This deformation apparently was caused by the collision of an island-arc system with the North American margin (e.g., Pindell and Dewey, 1982). Sediments that onlap and fill the foredeep probably represent mainly turbidites and other mass-transport deposits eroded from the deforming tectonic front. Equivalent-age rocks are exposed in thrust sheets in northwestern Cuba just to the south (Angstadt and others, 1985).

A prominent late Eocene unconformity separating units KC2 from KC3 represents a major turning point in the sedimentation of the area, probably corresponding to the cessation of tectonic activity in Cuba (Fig. 26). The two overlying post–late Eocene units (KC3 and KC4) are thinner, more tabular and more widespread than the underlying units (KC1 and KC2), and they contain numerous unconformities (Figs. 26 and 27; Table 4). They represent mainly pelagic-hemipelagic sedimentation that was interrupted by periods of erosion and/or nondeposition due to deep-sea currents funnelling through the relatively narrow Straits of Florida north of Cuba (Fig. 26) (Angstadt, 1983; Schlager and others, 1984b; Angstadt and others, 1985).

Adjacent margins

A discussion of the deep central Gulf of Mexico basin, as defined in this chapter, would not be complete without a description of the adjacent margins. As mentioned earlier, the deep basin is bounded on the north, west, and southwest by deformed slopes, while on the east and south, it is bounded by the steep Florida and Campeche Escarpments, respectively (Figs. 1 and 7). Each of these margins is discussed briefly below.

To the west and southwest the deep Gulf seismic units show extensive gravity-induced late Cenozoic deformation. Most of this area is known as the Mexican Ridges foldbelt (Figs. 1, 2C, and 7), which extends along the entire length of the Mexican margin (Bryant and others, 1968; Watkins and others, 1976; Buffler and others, 1979; Pew, 1982; Buffler, 1983b). The deep Gulf seismic units can be carried through the more gentle folds of the outer foldbelt, but they cannot be carried up onto the shelf through the more intensely folded part of the foldbelt or through a large growth fault system at the shelf edge. The only exception is the MCSB, which can be followed almost to the end of the seismic data at the shelf break. The MCSB extends relatively undeformed below a deformed zone or décollement in lower Tertiary (probably geopressured) shales, which forms the base of the fold system (Buffler and others, 1979; Pew, 1982; Buffler, 1983b) (Fig. 2C). This gravity-induced deformation began in Oligocene-Miocene time and continues today. Farther to the south the Mexican Ridges foldbelt terminates abruptly against the southern end of the Campeche–Sigsbee Knolls province, another large area characterized by late Cenozoic salt deformation that totally disrupts the deep Gulf section (Watkins and others, 1978; Pew, 1982; Shaub, 1983b). It is the southern extension of the Sigsbee Salt basin described earlier.

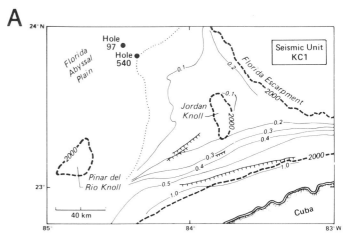
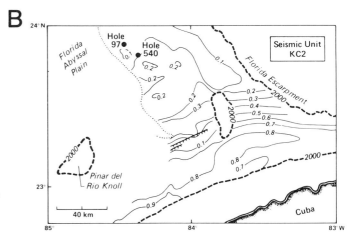
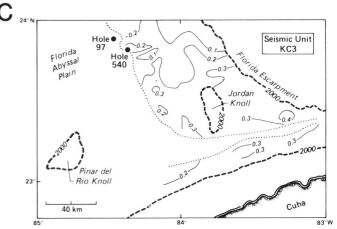

Figure 27. Isochron maps (two-way traveltime in seconds) of Late Cretaceous–Cenozoic seismic mapping units KC1 through KC3 (A through C). Unit KC4 not shown because of limited areal extent due to erosion at the sea floor. Water depth in meters (heavy dashed lines).

In the northwestern part of the Gulf of Mexico the deep basin is bordered by still another foldbelt, the Perdido foldbelt, characterized by deep, subparallel box folds that affect the entire sedimentary section, including the pre-MCSB Mesozoic section (Foote and others, 1983; Blickwede and Queffelec, 1988). The major folding here occurred about Oligocene-Miocene time, most likely the result of sediment loading along the outer shelf/upper slope. Stresses probably were transmitted to the lower slope along a deep décollement surface in the Jurassic salt and were buttressed against the thick sedimentary section overlying oceanic crust where no salt is present.

The northern margin of the deep basin is bounded by the Texas-Louisiana Slope, a broad, irregular area of basins and domes caused by extensive salt and shale deformation (Buffler and others, 1978; Watkins and others, 1978; Martin, 1980, 1984) (Figs. 1, 2A, and 7). In the central area the lower part of the deformed slope bulges out into the Gulf as the Sigsbee Escarpment, a prominent but low-relief topographic feature. This feature is formed by the leading edge of a series of salt tongues or sills that are thrust out over the deep Gulf sediments (Amery, 1969; Buffler and others, 1978; Humphris, 1978; Watkins and others, 1978; Martin, 1984) (Fig. 2A). Deep-basin sediments extend undeformed under the tongues for at least 10 to 20 km and possibly up to 90 km in places (Fig. 2A). The tongues have moved seaward mainly due to the differential loading of sediment updip and gravity.

These shallow salt structures continue to the east where they intrude sediments of the upper Mississippi Fan (Buffler and others, 1978; Shih and others, 1977; Walters, 1985; Weimer, 1989). Also in the Mississippi Fan area is another prominent foldbelt (the Mississippi Fan foldbelt), which is similar to the Perdido foldbelt (Weimer and Buffler, 1989a). The folds trend east-west and also affect the entire sedimentary section. They are characterized by southward- (basinward-) verging thrusts and asymmetrical folds cored by salt. Deformation occurred mainly in the middle Miocene through Pliocene and influenced early Mississippi Fan sedimentation. As with the Perdido foldbelt, the folding probably was the result of updip sedimentary loading. In this eastern Mississippi fan area a few seismic horizons can be followed across the deformed margin and tied to wells on the shelf (Addy and Buffler, 1984; Walters, 1985) (Fig. 5). This is the only location in the deep Gulf where this can be done with any success.

The southern and eastern margins of the deep Gulf are bounded by the Campeche and Florida Escarpments, which represent eroded and subsided Early Cretaceous carbonate platform margins (Bryant and others, 1969; Huerta, 1980; Locker and Buffler, 1983; Shaub, 1983a; Phair, 1984; Corso, 1987; Winker and Buffler, 1988) (Figs. 1, 2, and 7). These Early Cretaceous margins are interpreted to have been established along a regional tectonic hinge zone separating thick transitional crust from thin transitional crust (Buffler and Sawyer, 1985; Corso, 1987). Because of the steepness and eroded nature of the present escarpments, it is impossible to correlate reflectors across them. An exception to this is the MCSB of the deep Gulf, which merges with the escarpments and is correlated with a prominent mid-Cretaceous reflector/unconformity on the top of the banks (Figs. 8 and 18). As discussed earlier, this surface within the carbonate section in the banks also is mid-Cretaceous in age (Worzel and others, 1973; Mitchum, 1978; Huerta, 1980; Locker and Buffler, 1983; Shaub, 1983a; Corso, 1987) and, thus, also is designated MCSB.

Below the MCSB in both the West Florida and Yucatan Shelves is a thick, uniform, flat-lying sedimentary section inferred to be composed mainly of Lower Cretaceous shallow-water platform sediments. These rocks have been studied extensively in the northeastern Gulf and subdivided into four sequences (Fig. 5) (Corso, 1987). Commonly at the margin of these platforms there is a chaotic or reflector-free seismic-facies zone with diffractions, which has been interpreted as a "reef" facies or some kind of high-energy carbonate buildup at the platform margin (Locker and Buffler, 1983; Shaub, 1983a; Phair, 1984; Corso, 1987) (Fig. 18). This feature has paleo-relief and represents a geologic feature, although its exact origin is still not clear. In places, along the base of the escarpments, steeply dipping reflections probably represent talus or forereef material deposited along the base of the steep carbonate escarpment (Locker and Buffler, 1983; Corso, 1987; Corso and others, 1989). Along the Campeche platform margin, the MCSB is not as well defined as in the West Florida margin and is complicated by sets of shelfward-prograding reflections, which can be interpreted as shelfward-prograding carbonate banks, contour current buildups, or listric growth faults (Locker and Buffler, 1983). Along some of the escarpments, reflectors in the platform section as well as the talus deposits appear truncated, suggesting considerable erosion (Corso and others, 1989). This is supported by the fact that only shallow-water back-reef sediments have been recovered along the escarpment by dredging (Freeman-Lynde, 1983) and by the submersible *Alvin* (Paull and others, 1985). In the northeastern Gulf of Mexico the Lower Cretaceous section is underlain by a thick Jurassic section, which is equivalent to the Jurassic rocks in the West Florida basin and the deep southeastern Gulf (Corso, 1987) (Figs. 5 and 18). All of the subsided Early Cretaceous platform margins along the escarpments are capped by a thin Upper Cretaceous–Cenozoic deep-water pelagic carbonate section that thickens landward (Mitchum, 1978; Locker and Buffler, 1973; Worzel and others, 1973). These sediments were deposited following the mid-Cretaceous terminal drowning of the Early Cretaceous margins as sea level rose and the basin continued to subside.

REFERENCES CITED

Addy, S. K., and Buffler, R. T., 1984, Seismic stratigraphy of the shelf and slope, northeastern Gulf of Mexico: American Association of Petroleum Geologists Bulletin, v. 68, p. 1782–1789.

Amery, G. B., 1969, Structure of Sigsbee Scarp, Gulf of Mexico: American Association of Petroleum Geologists Bulletin, v. 53, p. 2480–2482.

Angstadt, D., 1983, Seismic stratigraphy and geologic history of the southeastern Gulf of Mexico/southwestern Straits of Florida [M.A. thesis]: University of Texas at Austin, 206 p.

Angstadt, D. M., Austin, J. A., Jr., and Buffler, R. T., 1983, Deep-sea erosional unconformity in the southeastern Gulf of Mexico: Geology, v. 11, p. 215–218.

—— , 1985, Seismic stratigraphy and geologic history of the southeastern Gulf of Mexico–southwestern Straits of Florida: American Association of Petroleum Geologists Bulletin, v. 69, p. 977–995.

Antoine, J. W., 1972, Structure of the Gulf of Mexico, *in* Rezak, R., and Henry, V. J., eds., Contributions on the geological and geophysical oceanography of the Gulf of Mexico: College Station, Texas A & M University Oceanography Studies, v. 3, p. 1–34.

Antoine, J. W., and Bryant, W. R., 1969, Distribution of salt and salt structures in Gulf of Mexico: American Association of Petroleum Geologists Bulletin, v. 53, p. 2543–2550.

Bertagne, A. J., 1980, Seismic stratigraphic investigation, western Gulf of Mexico [M.A. thesis]: University of Texas at Austin, 171 p.

—— , 1984, Seismic stratigraphy of Veracruz Tongue, deep southwestern Gulf of Mexico: American Association of Petroleum Geologists Bulletin, v. 68, p. 1894–1907.

Blickwede, J. F., and Queffelec, T. A., 1988, Perdido Foldbelt; A new deep-water frontier in western Gulf of Mexico [abs.]: American Association of Petroleum Geologists Bulletin, v. 72, p. 163.

Bouma, A. H., and others, 1984a, Drilling program on Mississippi Fan–DSDP Leg 96, *in* Characteristics of Gulf Basin deep-water sediments and their exploration potential: 5th Annual Gulf Coast Society of Economic Paleontologists and Mineralogists Research Conference Proceedings, p. 8–10.

Bouma, A. H., and others, 1984b, Seismic stratigraphy and sedimentology of Leg 96 Drilling on the Mississippi Fan, *in* Characteristics of Gulf Basin deep-water sediments and their exploration potential: 5th Annual Gulf Coast Society of Economic Paleontologists and Mineralogists Research Conference Proceedings, p. 11–22.

Bouma, A. H., and others, 1986, Initial reports of the Deep Sea Drilling Project: Washington, D.C., U.S. Government Printing Office, v. 96, 824 p.

Bryant, W., and others, 1968, Structure of Mexican continental shelf and slope, Gulf of Mexico: American Association of Petroleum Geologists Bulletin, v. 52, p. 1204–1228.

Bryant, W., Meyerhoff, A. A., Brown, N., Furrer, M., Pyle, T., and Antoine, J., 1969, Escarpments, reef trends, and diapiric structures, eastern Gulf of Mexico: American Association of Petroleum Geologists Bulletin, v. 53, p. 2506–2542.

Buffler, R. T., 1983a, Structure of the Sigsbee Scarp, Gulf of Mexico, *in* Bally, A. W., ed., Seismic expression of structural styles; A picture and work atlas: American Association of Petroleum Geologists Studies in Geology 15, v. 2, p. 2.3.2–50.

—— , 1983b, Structure of the Mexican Ridges Foldbelt, southwest Gulf of Mexico, *in* Bally, A. W., ed., Seismic exploration of structural styles; A picture and work atlas: American Association of Petroleum Geologists Studies in Geology 15, v. 2, p. 2.3.3–16.

—— , 1983c, Structure and stratigraphy of the Sigsbee Salt Dome Area, deep south-central Gulf of Mexico, *in* Bally, A. W., ed., Seismic expression of structural styles; A picture and work atlas: American Association of Petroleum Geologists Studies in Geology 15, v. 2, p. 2.3.2–56.

Buffler, R. T., and Sawyer, D. S., 1985, Distribution of crust and early history,

Gulf of Mexico basin: Gulf Coast Association of Geological Societies Transactions, v. 35, p. 333–344.

Buffler, R. T., Worzel, J. L., and Watkins, J. S., 1978, Deformation and origin of the Sigsbee Scarp–Lower Continental Slope, northern Gulf of Mexico: Houston, Texas, Offshore Technology Conference, Proceedings of 1978, v. 3, p. 1425–1433.

Buffler, R. T., Shaub, F. J., Watkins, J. S., and Worzel, J. L., 1979, Anatomy of the Mexican Ridges, southwestern Gulf of Mexico, *in* Watkins, J. S., and others, eds., Geological and geophysical investigations of continental margins: American Association of Petroleum Geologists Memoir No. 29, p. 319–327.

Buffler, R. T., Watkins, J. S., Worzel, J. L., and Shaub, F. J., 1980, Structure and early geologic history of the deep central Gulf of Mexico, *in* Pilger, R., ed., Proceedings of a Symposium on the Origin of the Gulf of Mexico and the Early Opening of the Central North Atlantic: Baton Rouge, Louisiana, Louisiana State University, p. 3–16.

Buffler, R. T., Shaub, F. J., Huerta, R., and Ibrahim, A. K., 1981, A model for the early evolution of the Gulf of Mexico basin: Oceanologica Acta, Proceedings 26th International Geological Congress, Geology of Continental Margins Symposium, Paris, July 1990, p. 129–136.

Buffler, R. T., and others, eds., 1984a, Gulf of Mexico, Ocean Margin Drilling Program Regional Atlas Series, Atlas No. 6: Marine Science International, Woods Hole, Massachusetts, 36 p.

Buffler, R. T., and others, 1984b, Initial reports of the Deep Sea Drilling Project: Washington, D.C., U.S. Government Printing Office, v. 77, 747 p.

Corso, W., 1987, Development of the Early Cretaceous northwest Florida carbonate platform [Ph.D. thesis]: University of Texas at Austin, 136 p.

Corso, W., Buffler, R. T., and Austin, J. A., Jr., 1989, Erosion of the southern Florida escarpment, *in* Bally, A. W., ed., Atlas of seismic stratigraphy: American Association of Petroleum Geologists Studies in Geology 27, v. 2, p. 149–157.

Ewing, J., Antoine, J., and Ewing, M., 1960, Geophysical measurements in the western Caribbean Sea and in the Gulf of Mexico: Journal of Geophysical Research, v. 65, p. 4087–4126.

Ewing, J. I., Worzel, J. L., and Ewing, M., 1962, Sediments and oceanic structural history of the Gulf of Mexico: Journal of Geophysical Research, v. 67, p. 2509–2527.

Ewing, M., and Antoine, J. W., 1966, New seismic data concerning sediments and diapiric structures in Sigsbee deep and continental slope, Gulf of Mexico: American Association of Petroleum Geologists Bulletin, v. 50, p. 479–504.

Ewing, M., Worzel, J. L., and others, 1969, Initial reports of the Deep Sea Drilling Project: Washington, D.C., U.S. Government Printing Office, v. 1, 672 p.

Faust, M., 1984, Seismic stratigraphy of the mid-Cretaceous unconformity (MCU) in the central Gulf of Mexico basin: Geophysics, v. 55, p. 868–884.

Feeley, M. H., Buffler, R. T., and Bryant, W. R., 1984, Seismic stratigraphic analysis of the Mississippi Fan [Ph.D. thesis]: Texas A & M University, Department of Oceanography Technical Report 84–2–T, 208 p.

Foote, R. Q., Martin, R. G., and Powers, R. B., 1983, Oil and gas potential of the maritime boundary region in the central Gulf of Mexico: American Association of Petroleum Geologists Bulletin, v. 67, p. 1047–1065.

Freeman-Lynde, R. P., 1983, Cretaceous and Tertiary samples dredged from the Florida Escarpment: Gulf Coast Association of Geological Societies Transactions, v. 33, p. 91–100.

Garrison, L. E., 1972, Acoustic-reflection profiles, western continental margin, Gulf of Mexico, 1970 cruise 70–02 of R/V Cadete Virgilio Uribe: U.S. Geological Survey and Universidad Nacional Autonoma de Mexico, USGS–GD–72–001, 19 p.

Garrison, L. E., and Martin, R. G., 1973, Geologic structure in the Gulf of Mexico basin: U.S. Geological Survey Professional Paper 773, 85 p.

Garrison, L. E., Kenyon, N. H., and Bouma, A. H., 1983, Channel systems and lobe construction in the Mississippi Fan: Geo-Marine Letters, v. 2, p. 31–39.

Haq, B. U., Hardenbol, J., and Vail, P. R., 1987, Chronology of fluctuating sea levels since the Triassic: Science, v. 235, p. 1156–1167.

Houtz, R., Ewing, J., and LePichon, X., 1968, Velocity of deep-sea sediments from sonobuoy data: Journal of Geophysical Research, v. 73, p. 2615–2641.

Huerta, R., 1980, Seismic stratigraphic and structural analysis of northeast Campeche Escarpment, Gulf of Mexico [M.A. thesis]: The University of Texas at Austin, 107 p.

Humphris, C. C., 1978, Salt movement on continental slope, northern Gulf of Mexico, in Bouma, A. H., Moore, G. T., and Coleman, J. M., eds., Framework facies and oil trapping characteristics of the upper continental margin: American Association of Petroleum Geologists Studies in Geology 7, p. 69–85.

Ibrahim, A. K., Carye, J., Latham, G., and Buffler, R. T., 1981, Crustal structure in the Gulf of Mexico from OBS refraction and multichannel reflection data: American Association of Petroleum Geologists Bulletin, v. 65, p. 1207–1229.

Jackson, M.P.A., and Seni, S. J., 1983, Geometry and evolution of salt structures in a marginal rift basin of the Gulf of Mexico, East Texas: Geology, v. 11, p. 131–135.

Jones, B. R., Antoine, J. W., and Bryant, W. R., 1967, A hypothesis concerning the origin and development of salt structures in the Gulf of Mexico sedimentary basin: Gulf Coast Association of Geological Societies Transactions, v. 17, p. 211–216.

Kastens, K. A., and Shor, A. N., 1985, Depositional processes of a meandering channel on Mississippi Fan: American Association of Petroleum Geologists Bulletin, v. 69, p. 190–202.

—— , 1986, Evolution of a channel meander on the Mississippi Fan: Marine Geology, v. 71, p. 165–175.

Klitgord, K. D., Hutchinson, D. R., and Schouten, H., 1988, U.S. Atlantic continental margin; Structural and tectonic framework, in Sheridan, R. E., and Grow, J. A., eds., The Atlantic Continental Margin, U.S.: Boulder, Colorado, Geological Society of America, The Geology of North America, v. I-2, p. 19–55.

Kirkland, D. W., and Gerhard, J. E., 1971, Jurassic salt, central Gulf of Mexico, and its temporal relation to circum-Gulf evaporites: American Association of Petroleum Geologists Bulletin, v. 55, p. 680–686.

Ladd, J. W., Buffler, R. T., Watkins, J. S., Worzel, J. L., and Carranza, A., 1976, Deep seismic reflection results from the Gulf of Mexico: Geology, v. 4, p. 365–368.

Lin, T., 1984, Seismic stratigraphy and structure of the Sigsbee Salt Basin, south-central Gulf of Mexico [M.A. thesis]: The University of Texas at Austin, 102 p.

Locker, S. D., and Buffler, R. T., 1983, Comparisons of Lower Cretaceous carbonate shelf margins, northern Campeche Escarpment and northern Florida Escarpment, Gulf of Mexico, in Bally, A. W., ed., Seismic expression of structural styles; A picture and work atlas: American Association of Petroleum Geologists Studies in Geology 15, v. 2, p. 2.2.3–123.

Long, J. M., 1978, Seismic stratigraphy of part of the Campeche Escarpment, southern Gulf of Mexico [M.A. thesis]: The University of Texas at Austin, 105 p.

Lord, J., 1986, Seismic stratigraphy and geologic history of the West Florida Basin, eastern Gulf of Mexico [M.A. thesis]: Houston, Texas, Rice University, 207 p.

—— , 1987, Seismic stratigraphy investigation of the West Florida Basin: Gulf Coast Association of Geological Societies Transactions, v. 37, p. 123–138.

Martin, R. G., 1980, Distribution of salt structures in the Gulf of Mexico; Map and descriptive text: U.S. Geological Survey Map MF-1213, 2 sheets.

—— , 1984, Diapiric trends in the deep-water Gulf basin, in Characteristics of Gulf Basin deep-water sediments and their exploration potential: Fifth Annual Gulf Coast Society of Economic Paleontologists and Mineralogists Research Conference Proceedings, p. 60–62.

Mitchum, R. M., 1978, Seismic stratigraphic investigation of west Florida slope, Gulf of Mexico, in Bouma, A. H., Moore, G. T., and Coleman, J. M., eds., Framework facies and oil-trapping characteristics of the upper continental margin: American Association of Petroleum Geologists Studies in Geology, v. 7, p. 193–224.

Moore, G. W., and del Castillo, L., 1974, Tectonic evolution of the southern Gulf of Mexico: Geological Society of America Bulletin, v. 85, p. 607–618.

Moore, G. T., and others, 1978, Mississippi Fan, Gulf of Mexico; Physiography, stratigraphy, and sedimentation patterns, in Bouma, A. H., Moore, G. T., and Coleman, J. M., eds., Framework, facies, and oil-trapping characteristics and upper continental margin: American Association of Petroleum Geologists Studies in Geology, v. 7, p. 155–191.

Moore, G. T., Woodbury, H. O., Worzel, J. L., Watkins, J. S., and Starke, G. W., 1979, Investigation of Mississippi Fan, Gulf of Mexico, in Watkins, J. S., and others, eds., Geological and geophysical investigations of continental margins: American Association of Petroleum Geologists Memoir 29, p. 383–402.

Paull, C. K., and others, 1985, Biological communities at the base of the Florida Escarpment resemble hydrothermal vent taxa: Science, v. 226, p. 965–967.

Pew, E., 1982, Seismic structural analysis of deformation in the southern Mexican Ridges [M.A. thesis]: The University of Texas at Austin, 102 p.

Phair, R. L., 1984, Seismic stratigraphy of the Lower Cretaceous rocks in the southwestern Florida Straits, southeastern Gulf of Mexico [M.A. thesis]: The University of Texas at Austin, 319 p.

Phair, R. L., and Buffler, R. T., 1983, Pre-Middle Cretaceous geologic history of the deep southeastern Gulf of Mexico, in Bally, A. W., ed., Seismic expression of structural styles; A picture and work atlas: American Association of Petroleum Geologists Studies in Geology 15, v. 2, p. 2.2.3–141.

Pindell, J., and Dewey, J. D., 1982, Permo-Triassic reconstructions of western Pangea and the evolution of the Gulf of Mexico/Caribbean region: Tectonics, v. 1, p. 179–211.

Pszczolkowski, A., 1978, Geosynclinal sequences of the Cordillera de Guaniquanico in western Cuba; Their lithostratigraphy, facies development, and paleogeography: Acta Geologica Polonica, v. 28, p. 1–96.

Rosenthal, D. B., 1987, Distribution of crust in the deep eastern Gulf of Mexico [M.A. thesis]: The University of Texas at Austin, 149 p.

Salvador, A., 1987, Late Triassic-Jurassic paleogeography and origin of Gulf of Mexico Basin: American Association of Petroleum Geologists Bulletin, v. 71, p. 419–451.

Schlager, W., 1989, Drowning unconformities on carbonate platforms, in Crevello, A., Wilson, J. L., Read, J. F., and Sarg, F. J., eds., Controls on carbonate platform and basin systems: Society of Economic Paleontologists and Mineralogists Special Publication 44, p. 15–25.

Schlager, W., Buffler, R. T., and Scientific Party of DSDP Leg 77, 1984a, DSDP Leg 77; Early history of the Gulf of Mexico: Geological Society of America Bulletin, v. 95, p. 226–236.

Schlager, W., Buffler, R. T., Angstadt, D., and Phair, R. L., 1984b, Geologic history of the southeastern Gulf of Mexico, in Initial reports of the Deep Sea Drilling Project: Washington, D.C., U.S. Government Printing Office, v. 77, p. 715–738.

Seekatz, J. G., 1977, Stratigraphic and structural features of part of the Sigsbee Escarpment, northwestern Gulf of Mexico [M.A. thesis]: The University of Texas at Austin, 107 p.

Shaub, F. J., 1983a, Origin of the Catoche Tongue, southeastern Gulf of Mexico, in Bally, A. W., ed., Seismic expression of structural styles; A picture and work atlas: American Association of Petroleum Geologists Studies in Geology 15, v. 2, p. 2.2.3–129.

—— , 1983b, Growth faults on the southwestern margin of the Gulf of Mexico, in Bally, A. W., ed., Seismic expression of structural styles; A picture and work atlas: American Association of Petroleum Geologists Studies in Geology 15, v. 2, p. 2.2.3–2.2.3-15.

—— , 1984, The internal framework of the southwestern Florida Bank: Gulf

Coast Association of Geological Societies Transactions, v. 34, p. 237–245.

Shaub, F. J., Buffler, R. T., and Parsons, J. G., 1984, Seismic stratigraphic framework of deep central Gulf of Mexico basin: American Association of Petroleum Geologists Bulletin, v. 68, p. 1790–1802.

Shih, T. C., Worzel, J. L., and Watkins, J. S., 1977, Northeastern extension of Sigsbee Scarp, Gulf of Mexico: American Association of Petroleum Geologists Bulletin, v. 61, p. 1962–1978.

Stuart, C. J., and Caughey, C. A., 1977, Seismic facies and sedimentology of terrigenous Pleistocene deposits in northwest and central Gulf of Mexico: American Association of Petroleum Geologists Memoir 26, p. 249–275.

Uchupi, E., and Emery, K. O., 1968, Structure of continental margin off Gulf Coast of the United States: American Association of Petroleum Geologists Bulletin, v. 52, p. 1162–1193.

Usmani, T. U., 1980, Seismic stratigraphic analysis of the southwestern abyssal Gulf of Mexico [M.S. thesis]: Houston, Texas, University of Houston, 91 p.

Vail, P. R., and others, 1977, Seismic stratigraphy and global changes in sea level, Parts 1–11, *in* Payton, C. E., ed., Seismic stratigraphy; Applications to hydrocarbon exploration: American Association of Petroleum Geologists Memoir 26, p. 51–212.

Walters, R. D., 1985, Seismic stratigraphy and salt tectonics of the Plio-Pleistocene deposits, continental slope and upper Mississippi Fan, northern Gulf of Mexico [M.A. thesis]: The University of Texas at Austin, 200 p.

Watkins, J. S., and others, 1975a, Multichannel seismic reflection investigations of the western Gulf of Mexico: Proceedings Offshore Technology Conference, OTC2729, p. 797–806.

Watkins, J. S., Worzel, J. L., Houston, M. H., Ewing, M., and Sinton, J. B., 1975b, Deep seismic reflection results from the Gulf of Mexico, Part 1: Science, v. 187, p. 834–836.

Watkins, J. S., Worzel, J. L., Shaub, F. J., Ladd, J. W., and Buffler, R. T., 1976, Seismic Section WG-3, Tamaulipas Shelf to Campeche Scarp, Gulf of Mexico: American Association of Petroleum Geologists Seismic Section 1, 1 sheet.

Watkins, J. S., Ladd, J. W., Buffler, R. T., Shaub, F. J., Houston, M. H., and Worzel, J. L., 1978, Occurrence and evolution of salt in the deep Gulf of Mexico, *in* Bouma, A. H., Moore, G. T., and Coleman, J. M., eds., Framework, facies, and oil-trapping characteristics of the upper continental margin: American Association of Petroleum Geologists Studies in Geology 7, p. 43–65.

Weimer, P., 1989, Sequence stratigraphy and depositional history of the Mississippi Fan, deep Gulf of Mexico [Ph.D. thesis]: The University of Texas at Austin, 300 p.

——, 1990a, Sequence stratigraphy, facies geometries, and depositional history of the Mississippi Fan, Gulf of Mexico: American Association of Petroleum Geologists Bulletin, v. 74, p. 425–453.

——, 1990b, Seismic facies, characteristics and variations in channel evolution, Mississippi Fan (Plio-Pleistocene), Gulf of Mexico, *in* Weimer, P., and Link, M. H., eds., Seismic facies and sedimentary processes of submarine fans and turbidite systems: New York, Springer-Verlag (in press).

Weimer, P., and Buffler, R. T., 1988, Distribution and seismic facies of the Mississippi Fan channels: Geology, v. 16, p. 900–903.

——, 1989a, Structural origin and evolution of Mississippi Fan foldbelt, Gulf of Mexico [abs.]: American Association of Petroleum Geologists Bulletin, v. 73, p. 425.

——, 1989b, Seismic definition of fan lobes, Mississippi Fan, Gulf of Mexico, *in* Bally, A. W., ed., Atlas of seismic stratigraphy: American Association of

Petroleum Geologists Studies in Geology 27, v. 3, p. 79–87.

Wilhelm, O., and Ewing, M., 1972, Geology and history of the Gulf of Mexico: Geological Society of America Bulletin, v. 83, p. 575–600.

Winker, C. D., 1982, Cenozoic shelf margins, northwestern Gulf of Mexico Basin: Gulf Coast Association of Geological Societies Transactions, v. 32, p. 427–448.

——, 1984, Clastic shelf margins of the post-Comanchean Gulf of Mexico; Implications for deep-water sedimentation, *in* Bebout, D. G., Mancini, E. A., and Perkins, B. F., eds., Characteristics of Gulf Basin deep-water sediments and their exploration potential: 5th Annual Gulf Coast Section Society of Economic Paleontologists and Mineralogists Research Conference Proceedings, p. 109–120.

Winker, C. D., and Buffler, R. T., 1988, Paleogeographic evolution of early deep-water Gulf of Mexico and margins, Jurassic to Middle Cretaceous (Comanchean): American Association of Petroleum Geologists Bulletin, v. 72, p. 318–346.

Winker, C. D., and Edwards, M. B., 1983, Unstable progradational clastic shelf margins, *in* Stanley, D. J., and Moore, G. T., eds., The shelfbreak; Critical interface on continental margins: Society of Economic Paleontologists and Mineralogists Special Publication 33, p. 139–157.

Worzel, J. L., and Burk, C. A., 1979, The margins of the Gulf of Mexico, *in* Watkins, J. S., Montadert, L., and Dickerson, P. W., eds., Geological and geophysical investigations of continental margins: American Association of Petroleum Geologists Memoir 29, p. 403–419.

Worzel, J. L., Leyden, R., and Ewing, M., 1968, Newly discovered diapirs in Gulf of Mexico: American Association of Petroleum Geologists Bulletin, v. 52, p. 1194–1203.

Worzel, J. L., and others, 1973, Initial reports of the Deep Sea Drilling Project: Washington, D.C., U.S. Government Printing Office, v. 10, 334 p.

MANUSCRIPT ACCEPTED BY THE SOCIETY DECEMBER 28, 1989

ACKNOWLEDGMENTS

The author wishes to dedicate this chapter to Dr. Joel Watkins, who was mainly responsible for initiating and then carrying out the early phases of deep Gulf of Mexico seismic investigations at the University of Texas. He has been my mentor over the years and continues to be a valued colleague, a good friend, and a source of inspiration. I also wish to thank Joe Worzel, from whom I learned a great deal, particularly about working at sea. I am particularly grateful to the many graduate students, who have been the main contributors to the synthesis presented here and have given me great satisfaction in my role as a faculty member. Special thanks also must go the many industry representatives who, over the years, have provided us with funds to collect, process, and interpret the seismic data and who have been extremely free about sharing their data, ideas, and enthusiasm with us whenever possible. The following people have provided critical review of the chapter, which has greatly improved its content and presentation: Curtis Humphris, Ray Martin, Robert Mitchum, Amos Salvador, and Charles Winker. Special appreciation also goes to Kathy Moser and Toni Lee Mitchell, who have typed too many drafts of this paper, and to Nancy Kelly who helped compile and draft many of the figures. Partial funding for work on this chapter came from the National Science Foundation Grant No. 8417771. This is University of Texas Institute for Geophysics contribution no. 786.

Chapter 14

Origin and development of the Gulf of Mexico basin

Amos Salvador
Department of Geological Sciences, The University of Texas at Austin, P.O. Box 7909, Austin, Texas 78713-7909

INTRODUCTION—A SHORT HISTORICAL REVIEW

For more than 100 years, geologists have speculated about the initial steps of the formation of the present Gulf of Mexico basin. A review of the literature, most likely incomplete, uncovered more than 70 publications on the subject by close to 80 authors.

Early workers (Schuchert, 1909; Willis, 1909) considered the Gulf of Mexico to be an ancient feature, a deep-water body in existence since the Precambrian. Willis (*in* Schuchert, 1935, p. 72) regarded the Gulf as representing "a mass of basalt which was erupted in Pre-Cambrian time. . .," a basin "of great antiquity."

With increased information on the area opinions progressively changed, and from the late 1910s to the early 1930s most contributors to the controversy came to favor a much later beginning for the Gulf of Mexico basin (Dumble, 1918; Miser, 1921; Schuchert, 1923, 1929, 1935; Sellards, 1932; and a number of others). They believed that during most of the Paleozoic a continental landmass or borderland, most commonly referred to as "Llanoria," occupied the northwestern part of the present Gulf of Mexico basin, from northeastern Mexico to Mississippi, and included a large but undetermined part of the present northwestern Gulf of Mexico (Fig. 1), an idea first implied by Edward Suess (1888) in his celebrated *Das Antlitz der Erde*. In the southeastern part of the present gulf, the borderland was thought to have been submerged and covered by a shallow sea; the entire area was regarded as a "neutral area, or better, a slightly negative one," an "ancient 'flat' plate," the "Gulf of Mexico Plate" (Schuchert, 1935). This "plate" or borderland was believed to have foundered during the late Paleozoic or the Mesozoic to form the present Gulf of Mexico basin. As late as 1954, Eardly accepted this interpretation.

The early workers were seriously handicapped in their attempts to interpret the early geologic history of the Gulf of Mexico basin by the fact that little or nothing was known at the time about the geologic make-up of the central part of the basin below the waters of the Gulf of Mexico.

This situation began to change in the late 1930s and early 1940s, when the oil industry started exploratory drilling and geophysical surveys in the shallow waters of the Gulf of Mexico, and particularly in the mid- to late 1950s and early 1960s, when refraction and reflection seismic surveys by academic institutions extended the knowledge of the basin to the deep, central gulf (see Chapter 13, this volume). The refraction surveys indicated con-

Figure 1. The context of Llanoria (from Schuchert, 1923).

Salvador, A., 1991, Origin and development of the Gulf of Mexico basin, *in* Salvador, A., ed., The Gulf of Mexico Basin: Boulder, Colorado, Geological Society of America, The Geology of North America, v. J.

vincingly that in the central part of the Gulf of Mexico basin, a thick sedimentary sequence was underlain by oceanic crust (M. Ewing and others, 1955; J. Ewing and others, 1960, 1962).

The recognition that oceanic crust underlies the central part of the Gulf of Mexico basin refuted the presence under the basin of a foundered continental-crust block or "plate." Uncertainty still remained, however, concerning the age of the oceanic crust. Was it the Precambrian basalt of Bailey Willis (*in* Schuchert, 1935), or a much younger crust of late Paleozoic or, more probably, of Mesozoic age? In other words, is the present Gulf of Mexico basin a "permanent" feature that has persisted in its present location since Precambrian time, or is it a much younger structure that did not come into being until the late Paleozoic or the Mesozoic?

There were contrary views. Meyerhoff (1967), for example, favored an "old" crust and contended that at the end of the Precambrian, the Gulf of Mexico was a basin "as it is today." Paine and Meyerhoff (1970, p. 10) considered the origin of the Gulf of Mexico basin as settled: "the Gulf basin has been an oceanic plate since the beginning of earth history." Shurbet and Cebull (1975, p. T26) also believed that "The Gulf of Mexico has been an oceanic and probably deep-water basin for most and perhaps all of its history, beginning in late Precambrian or Cambrian time."

Most workers, however, came to favor a younger age for the oceanic crust. Interpretation of marine multichannel seismic reflection profiles shot in the Gulf of Mexico during the mid-1970s appears to have answered the question of the age of the oceanic crust. Ladd and others (1976) established that the basal part of the thick sedimentary section that overlies the oceanic crust in the deep central part of the Gulf of Mexico could not be much older than Middle or Late Jurassic, and that the crust was probably of Early Jurassic age (see Chapter 13, this volume). These multichannel seismic reflection surveys also contributed strong evidence to establish that no salt seemed to be present in the stratigraphic section overlying the oceanic crust throughout the deep central Gulf of Mexico. This information would become most influential in the development of theories on the origin of the Gulf of Mexico basin.

The general agreement on a Mesozoic age for the oceanic crust underlying the central Gulf of Mexico basin, the apparent absence of salt in this central area, and the increasingly "mobilist" approach to regional tectonic interpretation resulting from the widespread acceptance of the principles of plate tectonics caused a shift in the theories about the origin of the Gulf of Mexico basin, from those favoring an old, "permanent" basin, or one ensuing from the foundering of a preexisting continental plate or landmass, to theories endorsing a Mesozoic "rift and drift" origin for the basin. The many publications on this subject during the past 15 years are almost unanimous in subscribing to such an interpretation.

However, while generally following the concepts of plate tectonics, the more recent views on the origin of the Gulf of Mexico basin often differ in the particulars of the timing and the

processes. And because they all accept that the first steps of the formation of the present basin—the initial rifting episodes—took place during the Mesozoic, they often do not concern themselves with the geologic history of the region during pre-Mesozoic time. Yet, stratigraphic and structural developments during the Mesozoic may well have been significantly influenced by pre-Mesozoic tectonic features and events, and may be better understood and explained if interpreted with some insight into the inherited structural fabric and prior geologic history.

Thus, before discussing the origin and development of the *present* Gulf of Mexico basin, I first speculate about the pre-Mesozoic geologic history of the area, specifically the pre–Late Triassic history.

PALEOZOIC—THE OPENING AND CLOSING OF THE PROTO-ATLANTIC OCEAN

Knowledge of pre–Upper Triassic rocks within the Gulf of Mexico basin is very limited (reviewed in Chapter 7, this volume). It is all but impossible to unravel the early geologic history of the area of the present basin on the basis of such limited knowledge. Only by making use of information peripheral to the margins of the basin can we attempt to reconstruct the sequence of pre–Late Triassic geologic events that affected the area that would become the present Gulf of Mexico basin.

It is particularly difficult to speculate about the events that took place before the late Mississippian or early Pennsylvanian (mid-Carboniferous). Outcrops of lower Paleozoic rocks in the highlands surrounding the basin are few, small in areal extent, and separated from each other by considerable distances. Except for northern Florida, southern Georgia, and southeastern Alabama, few wells within the basin have penetrated unmetamorphosed pre-Pennsylvanian rocks.

Those who have endeavored more recently to reconstruct the world's paleogeography during the Paleozoic in the context of plate tectonics concepts (Wilson, 1966; Smith and others, 1973; Kanasewich and others, 1978; Scotese and others, 1979; Ziegler and others, 1979; Scotese, 1984) have postulated that during the early Paleozoic the North American Plate (as part of the Northern Hemisphere continent of Laurasia) was bordered along what are today its eastern and southern margins by a large open ocean (the Proto-Atlantic or Iapetus Ocean) which separated the plate from what some have called the Southern Hemisphere continent of Gondwana—the aggregation of the South American, African, Indian, Antarctic, Australian, and several other smaller plates. The Proto-Atlantic is believed to have opened during the late Precambrian or early Cambrian as Laurasia and Gondwana drifted away from one another.

Thomas (1977, 1983, 1988, 1991) has suggested that the breakup of a Precambrian supercontinent and the opening of the Proto-Atlantic Ocean took place along a northeast-trending system of rifts that was offset by northwest-striking transform faults (Fig. 2). In the area of the present Gulf of Mexico basin two such major transforms, both with considerable right-lateral offset, have been recognized by Thomas: one extending from Alabama to Oklahoma, which he called the "Alabama-Oklahoma transform fault," and a second, the "Texas transform," roughly parallel to the lower course of the Rio Grande (Fig. 2). (Here, as throughout this chapter, geographic positions and directions within the various plates are those that the plates have at present.) King (1975) and Cebull and others (1976) had discussed the role of these two transform-fault zones in the late Proterozoic and early Paleozoic opening of the Proto-Atlantic, and had suggested a similar early tectonic history for the southern margin of the North American Plate. Walper and Rowett (1972) and Wood and Walper (1974) had also discussed these two transform faults or megashears, but in a somewhat different context. Many other authors also have discussed them in the past 10 to 15 years in connection with the tectonic history of southwestern United States and northern Mexico.

The Alabama-Oklahoma and Texas transform faults are believed to have shaped to a great extent the southern rifted margin of the North America Plate. They have remained fundamental, though perplexing, structural elements in the tectonic evolution of the Gulf of Mexico basin and surrounding regions. Their imprint can be recognized, in one way or another, when reconstructing the Mesozoic as well as the Paleozoic geologic history of the region.

Even though a late Precambrian–early Paleozoic period of rifting that resulted in the opening of the Proto-Atlantic has been generally accepted by most investigators on the basis of evidence from the southeastern United States, little or nothing is known about contemporaneous events that may have taken place in Mexico and northern Central America. The reason is that rocks of Precambrian and early Paleozoic age are very poorly represented in these regions (see Chapter 7, this volume). It is even possible that much of present-day Mexico and northern Central America underlain by continental crust were not part of the North American Plate during Precambrian and much of Paleozoic time.

During the early Paleozoic, after the opening of the Proto-Atlantic, sedimentation took place over most of the interior cratonic area of the detached North American continental plate and along the length of the submerged part of its irregularly shaped, rifted passive margins, as well as in the bordering deeper-water areas of the Proto-Atlantic.

The lower Paleozoic (Cambrian to lower Mississippian) sedimentary section deposited over the newly formed continental margins of the North American Plate in the southeastern United States, from the southern Appalachian Mountains to the Marathon uplift of southwest Texas, consists of shallow-marine carbonate-shelf deposits, generally thin and indicative of passive, stable conditions of deposition (Fig. 3). A basal sandstone interval is present in several areas, and chert is common in the section, particularly in the Devonian. Off-shelf, deep-water, lower Paleozoic sedimetnary rocks equivalent to the shallow-shelf deposits of the continental margin are known in the Ouachita Mountains, in the Marathon uplift, and in a number of wells along the Ouachita orogenic belt between them.

A - LATE PRECAMBRIAN

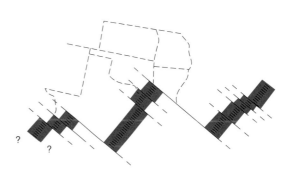

B - LATE CAMBRIAN

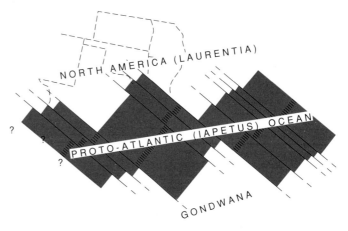

C - ORDOVICIAN or SILURIAN

Figure 2. Opening of the Proto-Atlantic Ocean in the late Precambrian and early Paleozoic (from Thomas, 1977, 1988). The Texas transform of Thomas is shown as a broader zone of transform faults that extends farther to the southwest in order to account for the apparent termination of the Ouachita orogenic belt in northeastern Mexico.

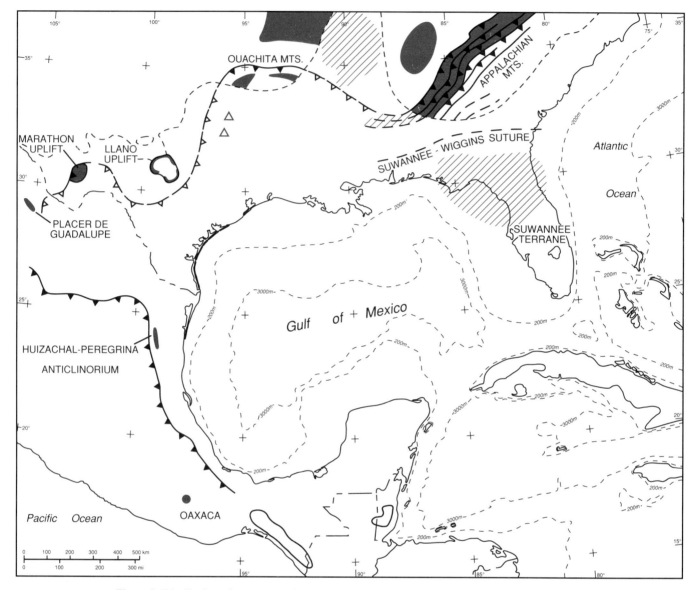

Figure 3. Distribution of unmetamorphosed lower Paleozoic rocks in the Gulf of Mexico basin and bordering areas. Solid areas are outcrops; diagonal rules show subsurface. Triangles represent wells.

In northern Central America and the Mexican part of the Gulf of Mexico basin and surrounding areas, rocks considered to be early Paleozoic in age are generally metamorphosed, except in three small and widely separated areas—Placer de Guadalupe in the state of Chihuahua, the Huizachal-Peregrina anticlinorium in central Tamaulipas, and the state of Oaxaca (Fig. 3). In the first two areas, the lower Paleozoic section has some similarity in lithologic composition and thickness to sections of the same age along the northern rim of the Gulf of Mexico basin in the United States (see Chapter 7, this volume). This similarity may indicate deposition under analogous and related sedimentary and tectonic conditions.

The upper Paleozoic (upper Mississippian–lower Permian)

section of the Gulf of Mexico basin records a marked and abrupt change in depositional character. Whereas the lower Paleozoic sequences are generally thin, indicate slow and stable deposition along the northern and northwestern rims of the basin, and are metamorphosed over most of its western and southern parts, the upper Paleozoic section is much thicker, unmetamorphosed over the greater part of the region, and reflects very fast deposition under a tectonically active regime. Distinctive of this upper Paleozoic section are turbidite sequences of thinly interbedded sandstones and shales (flysch), several thousand meters thick, which in the Ouachita Mountains have been dated as late Mississippian to mid-Pennsylvanian, in the Marathon uplift may range into the early Permian, and in northeastern Mexico seem to be entirely of

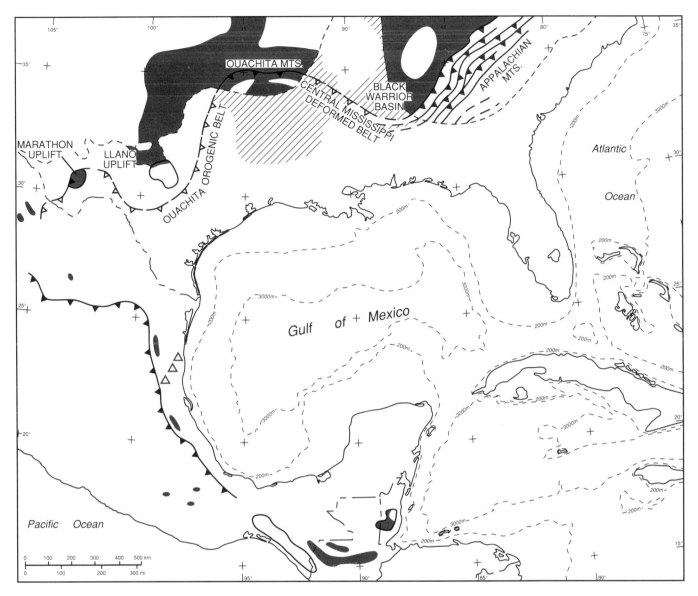

Figure 4. Distribution of upper Paleozoic rocks in the Gulf of Mexico basin and bordering areas. Solid areas are outcrops; diagonal rules show subsurface. Triangles represent wells.

early Permian age (Fig. 4). Sediment-dispersal markers in the flysch sequences of the Black Warrior basin, north of the Central Mississippi deformed belt, the Ouachita Mountains, and the Marathon uplift, indicate that these synorogenic clastic wedges were derived for the most part from an orogenically uplifted highland to the southwest, south, and southeast, respectively, of these three areas. The flysch clastic wedges prograded in a general northward and northwestward direction over the foreland basins bordering to the south the North American craton.

The abrupt change in the lithologic character and in the rate of deposition between the lower and upper Paleozoic sequences, from relatively thin and predominantly carbonate to thick and predominantly clastic sedimentary sections, has been interpreted to indicate that a major tectonic event took place along the southern margin of the North American Plate starting in the late Mississippian and lasting perhaps until the early Permian. This event is believed to be the closing of the Proto-Atlantic or Iapetus Ocean, and the resulting crustal collision and eventual welding of the North American Plate with the margin of the African and South American parts of Gondwana. This collision has been interpreted as a continent-continent collision, and, more recently, as an arc-continent collision between the North American continental margin and a volcanic island arc on the edge of another continental plate or microcontinent (Graham and others, 1975; Wickham and others, 1976; Mack and others, 1983). It appears, judging from recent literature, that the answer to the provocative question raised by Wilson (1966), Did the Atlantic close and re-open? is yes.

In addition to the striking change in lithologic character and rate of deposition between the lower and upper Paleozoic sedimentary sequences, several other lines of evidence have been interpreted to support a late Paleozoic plate collision, the closing of the Proto-Atlantic Ocean, and the conversion of the southeastern margin of the North American Plate of the time from a passive to a convergent margin.

1. Late Paleozoic thrusting directed toward the North American craton has been recognized from the southern Appalachian Mountains to the Ouachita Mountains and from there along the subsurface Ouachita orogenic belt to the Marathon uplift. The Ouachita-Marathon thrust belt as evidence of a late Paleozoic collision between the southern margin of the North American Plate and a southern continent or volcanic island arc has been questioned recently by Royden and others (1990). They believe that the Ouachita-Marathon thrust belt could have evolved "as an accretionary wedge developed during southward subduction of the passive margin of North America, and that continental collision was either lacking . . . or incomplete, involving only the distal edge of a continental region undergoing extension adjacent to the active plate boundary."

In Mexico, the late Paleozoic thrusting is not evident because it has been overprinted over extensive areas by the late Cretaceous–early Tertiary Laramide orogenic deformation. Nevertheless, in the Huizachal-Peregrina anticlinorium of northeastern Mexico, the Granjeno Schist and associated serpentinite have been interpreted as being an allochthonous block, metamorphosed during the Carboniferous or Permian and transported into the area, possibly from the east, in latest Paleozoic time (see Chapter 7, this volume).

2. Thick sequences of flysch of predominantly late Mississippian to mid-Pennsylvanian age in the Black Warrior basin and the Ouachita Mountains, late Mississippian to early Permian in the Marathon uplift, and of Permian age in Mexico, suggest the appearance of orogenic highlands along an uplift belt extending from the southern Appalachian Mountains to east-central Mexico (Fig. 4).

3. An unmetamorphosed lower Paleozoic section (Cambrian or Ordovician to Devonian) has been penetrated by numerous wells in the subsurface of northern Florida, southern Georgia, and southeastern Alabama (Fig. 3) (see Chapter 7, this volume). It contains a fauna and flora with distinct African affinities, unlike those occurring in beds of the same age farther to the north in the Appalachian Mountains. This has led a number of geologists to conclude that the area where the lower Paleozoic rocks have been identified represents a remnant of the African Plate (the so-called Suwannee terrane) that was welded to the North American Plate during their collision in late Paleozoic time. The Suwannee terrane was left behind when the two plates separated during the Mesozoic to form the present North Atlantic Ocean. A suture zone has been recognized that separates the North American Plate on the north from the welded terrane with African affinities to the south. This has been called the Suwannee-Wiggins suture (Fig. 3) (see Thomas and others, 1989; Plate 3 and Chapter 7, this volume).

4. From northeastern to east-central Mexico and south to Guatemala and Belize, the Paleozoic section is intruded by Permian-Triassic silicic plutonic rocks—granites, granodiorites, and tonalites (Fig. 5) (see Chapter 7, this volume). This belt of plutons could be interpreted as indicating a zone of collision and subduction, or as the magmatic core of a volcanic island arc.

The nature and timing of the collision of the North American Plate with a converging southern continent or magmatic arc and the resulting orogeny have been the subjects of considerable discussion.

The timing of the collision in the region from the southern Appalachian Mountains to the Marathon uplift can be established with some degree of confidence on the basis of the age of the upper Paleozoic flysch sequences, if it is agreed that they can be considered a testimony of the orogenic events brought about by the collision. This evidence places the orogeny as having started in the region during the late Mississippian (Meramecian), with tectonic activity probably continuing until mid-Pennsylvanian time (Atokan and Desmoinesian), except in the Marathon uplift, where it may have lasted until the early Permian (King, 1975, 1977).

An overall progressively younger age in a southwestward direction for the collision and for the suturing of the southern margin of the North American Plate has been favored by most authors (Graham and others, 1975; Walper and Miller, 1985, and many others). This somewhat diachronous nature of the collision and the initiation of the succeeding orogeny may be due, at least in part, to the irregular shape of the southern margin of the North American Plate; sharp bends were controlled by the configuration of the rifted continental crust, as shown diagramatically in Figure 2C. Collision took place earlier along protruding parts of the continental margin, and later in the embayments (Thomas, 1977, 1989).

Unmetamorphosed and little-deformed mid-Pennsylvanian-lower Permian sedimentary sections penetrated by several wells south of the Ouachita Mountains, in northeast Texas, north Louisiana, and south Arkansas (Fig. 4) may furnish additional evidence concerning the time of collision between the North American and the African and South American Plates. These upper Paleozoic rocks have been interpreted as post-orogenic sequences deposited in shallow-marine successor basins developed behind an orogenic front (see Chapter 7, this volume). Such an interpretation would imply that the orogeny resulting from the continental collision had ceased by mid-Pennsylvanian time in the Ouachita Mountains area, while it was still active in the Marathon uplift, and would support the progressively younger age of the collision in a southwestward direction.

Along the Central Mississippi and Ouachita deformed belts, the late Paleozoic plate collision has been interpreted to have resulted in the southward subduction of the North American continental plate under an accretionary prism and associated vol-

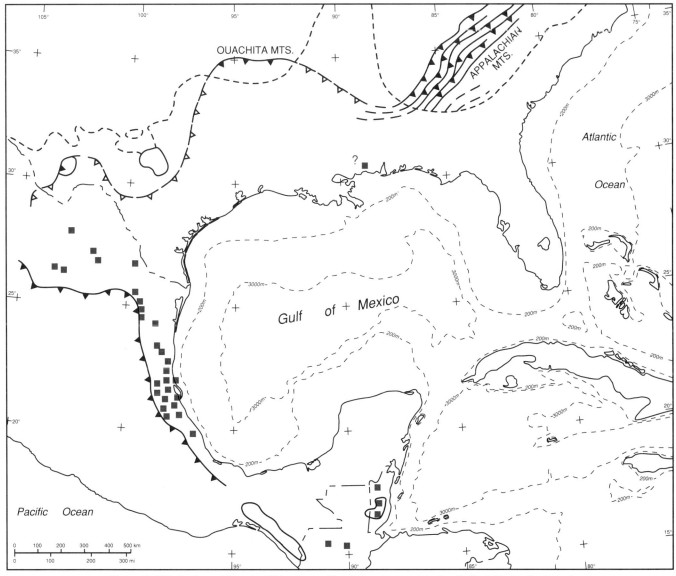

Figure 5. Distribution of Permian-Triassic granitic plutons (squares) in the Gulf of Mexico basin and bordering areas.

canic island arc that probably bordered the encroaching Gondwana continental plate. Orogenic events accompanying the collision culminated in large-scale thrusting toward the craton.

The evidence for the late Paleozoic crustal collision is much less conclusive in eastern and southern Mexico and northern Central America due to the limited information available about Precambrian and Paleozoic rocks in these regions. That the upper Paleozoic flysch sequences in northeastern Mexico have been dated as Permian may be interpreted to mean that the plate convergence, collision, and suturing took place later in northeastern Mexico. However, the uncertainty concerning how much of present-day Mexico was part of the North American Plate during the Paleozoic and what the relative position of continental Mex-

ico was with respect to the North American Plate during that time may make such an interpretation premature.

Goetz and Dickerson (1985) and Stewart (1988) have suggested that in late Proterozoic and Paleozoic time the southern margin of the North American plate was located in what is now northern Mexico and that it was not until the late Paleozoic or earliest Mesozoic that several continental blocks, of unknown history, were assembled and accreted to the North American Plate to form most of present Mexico.

Many aspects of the Paleozoic history of present Mexico are not known or are disputed. We do not known, for example, the timing, the magnitude, or the direction of the displacement along the Texas and related transforms in northeastern Mexico during the Paleozoic. It is not unreasonable, however, to think that the late Paleozoic collision may have reactivated the preexisting fractured and weakened late Proterozoic–early Paleozoic transform-fault zone zones. But we do not know that this was the case.

That the Texas transform-fault zone may have played a major role in the Paleozoic tectonic history of the northwestern part of the Gulf of Mexico basin is suggested by the important differences in the character of the Paleozoic sections on either side of this postulated fault trend, and by the fact that the problem of the continuation of the Ouachita orogenic belt into Mexico has not yet been resolved to anyone's complete satisfaction (for example, see Shurbet and Cebull, 1987; Handschy and others, 1987).

Unexplained, too, is the belt of Permian-Triassic silicic plutons that extends from northern Central America to northeastern Mexico, that is not known to extend across the Texas transform. Is this plutonic belt related to plate collision and possible subduction, or was the emplacement of the intrusive bodies governed by other, younger, tectonic events?

Finally, in the Huizachal-Peregrina anticlinorium in northeastern Mexico (Fig. 3), a fairly complete and unmetamorphosed lower Paleozoic section is 800 km from the nearest similar section to the northwest, from which it is separated by an area where the lower Paleozoic is believed to be represented by metamorphic rocks. This is another enigma that needs to be solved before a clearer picture can be obtained of the late Paleozoic collision events that marked the closure of the Proto-Atlantic Ocean along the southern margin of the North American plate.

However, in spite of the many uncertainties that still remain concerning the Paleozoic geologic history of the area that would become the Gulf of Mexico basin, there is almost general agreement that during the latest Paleozoic and earliest Mesozoic (early Triassic) the area was part of a very large emergent landmass, the supercontinent of Pangea, which, according to current thinking, included most of the continental plates of the Earth. The apparent absence of marine rocks of latest Permian, Early Triassic and Middle Triassic age in the Gulf of Mexico basin and surrounding areas supports this inference.

(A more complete treatment of the stratigraphic and structural history of the southern margin of the North American plate in Paleozoic time and extensive lists of references are in Hatcher and others [1989]; especially Chapters 10, 15, 16, 23, and 27.)

LATE PALEOZOIC–EARLIEST MESOZOIC— THE ASSEMBLY OF PANGEA

The fitting of the continental plates during the late Paleozoic and early Mesozoic (the continents of Laurasia and Gondwana) to form the supercontinent of Pangea has been accomplished with reasonable success, except along the adjoining margins of the North American and South American Plates of the time, the area that would become the Gulf of Mexico basin and the Caribbean region. Whereas the fit of the continental plates bordering the Atlantic has been generally accepted with only minor variations since the early part of this century, the fit of the North American

and South American Plates has been either ignored or been the subject of numerous and widely disparate interpretations, none of them entirely convincing. The well-known reconstruction around the Atlantic by Bullard and others (1965), for example, disregards the problem of fitting the North American and South American Plates; if their map were to be completed by adding the regions of southern Mexico and northern Central America known to be underlain by continental crust, there would be a disturbing amount of overlap between these regions and the northern part of the South American Plate. Other authors have attempted to solve this problem by bending or fragmenting Mexico and northern Central America into small plates or blocks and displacing them in various ways to eliminate the overlap.

The early attempts to reconstruct Pangea in the area of the present Gulf of Mexico basin used different configurations to place both the Yucatan block and the part of northern Central America underlain by continental or "transitional" crust (the so-called Chortís block or Honduras-Nicaragua block) into the open space left by Bullard and others (1965) between the North American and South American Plates (Carey, 1958, 1976; Dietz and Holden, 1970; Freeland and Dietz, 1971; Smith and Briden, 1977). Others (Walper and Rowett, 1972; Wood and Walper, 1974; Van der Voo and French, 1974; Van der Voo and others, 1976; Pilger, 1978, 1981; Walper and others, 1979; Walper, 1980; Van Siclen, 1984), preferred a close fit between the North American and South American Plates, which required placing central and southern Mexico as well as the Yucatan and Chortís blocks west of the South American Plate, and considerably west of their present positions. During or shortly after the opening of the Gulf of Mexico basin, these continental fragments would have moved eastward; accomplished by either rotating or bending them, or by translating them along a system of roughly northwest-trending, left-lateral, strike-slip faults or megashears of considerable displacement. The likelihood of large displacement along northwest-trending strike-slip faults active during the Paleozoic and early Mesozoic (Triassic and Early Jurassic) cannot, of course, be discarded, as discussed earlier in this chapter. Their activity during the rest of the Mesozoic, however, is not supported by any available evidence.

The disparity of the early attempts to reconstruct the fit of the North American and South American Plates in late Paleozoic or early Mesozoic time is due, in part, to the diverse perspectives brought to the solution of this controversial problem. The diversity of interpretations was also caused by the different types of geological and geophysical information on the Gulf of Mexico basin area with which the authors were most familiar and by personal emphases on particular aspects or elements of this information—the supposed composition of the crust, gravity data, or paleomagnetic data, for example—to the exclusion of other information. However, the diversity of the reconstructions also reflects the very limited knowledge of the Paleozoic rocks that underlie the present Gulf of Mexico basin area, and the lack of sufficient information at the time about the geologic make-up of its central part.

With more and better information available, particularly information on the nature of the crust and the overlying sedimentary section in the central part of the basin, more recent interpretations of the fit of Pangea in the area of the future Gulf of Mexico basin vary within considerably narrower limits. These more recent reconstructions of Pangea place the Yucatan block in a position between the North American and South American Plates, and place the Chortís block, when shown, somewhere in the Pacific region, west of Mexico (Moore and del Castillo, 1974; Owen, 1976, 1983; Humphris, 1978; Salvador and Green, 1980; Pindell and Dewey, 1982; Anderson and Schmidt, 1983; Pindell, 1985; Salvador, 1987; Dunbar and Sawyer, 1987; Ross and Scotese, 1988). Increasing emphasis has also been given to the non-rigid aspects of plate motions. It is now evident that continental crust extension played an important role in the opening of the Gulf of Mexico basin. In reconstructing Pangea, several attempts have been made, therefore, to restore the transitional crust to its original dimensions before being stretched during rifting and the opening of the Gulf of Mexico basin (Pindell, 1985; Buffler and Sawyer, 1985; Dunbar and Sawyer, 1987; Ross and Scotese, 1988).

A very generalized reconstruction of the western part of Pangea, which includes the adjacent regions of the North American, African, and South American Plates as they may have been welded together in late Permian or early Triassic time, is shown in Figure 6. The proposed reconstruction accounts for the stretching of the continental crust underlying the southern part of the U.S. Gulf Coastal Plain and the Yucatan block in a predominantly north-northwest–south-southeast direction and by an amount approximately equivalent to that favored by Dunbar and Sawyer (1987). It also takes into consideration the extension of the crust by brittle failure that marked the beginning of the breakup of Pangea and resulted, during the Late Triassic to Early or Middle Jurassic, in the formation of extensive graben systems in both the North American and South American Plates. The position of the Yucatan block shown in Figure 6 differs somewhat from that shown in the reconstructions of other authors in the degree of rotation from its present position.

How to fit the southern part of Mexico without somehow fragmenting and displacing it remains a nagging problem, a solution of which will not be attempted in this chapter. The possibility that major west-northwest–east-southeast–trending transform faults may have played an important role in the solution of this problem, as suggested by several authors (Pindell and Dewey, 1982; Anderson and Schmidt, 1983; Pindell, 1985; Ross and Scotese, 1988; and many others), cannot be disproved. However, evidence of major horizontal motion along such intracontinental transforms in the Gulf of Mexico basin region is lacking. Another possibility, discussed earlier, is that most of the present continental Mexico did not attain its present position with respect to the North American Plate until the late Paleozoic or early Mesozoic.

The position of the South American Plate shown in Figure 6

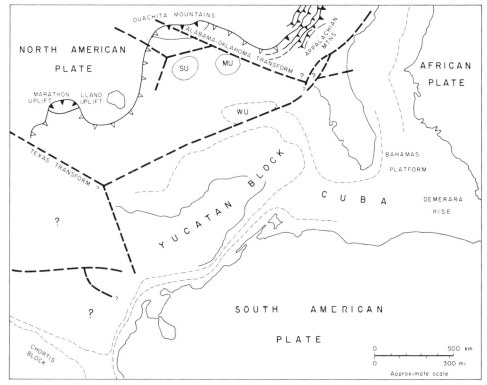

Figure 6. Reconstruction of western Pangea (future Gulf of Mexico basin region) in late Paleozoic or earliest Triassic time. SU, Sabine uplift; MU, Monroe uplift; WU, Wiggins uplift.

is only approximate; how to fit the margins of the North American and South American Plates during the late Paleozoic or early Mesozoic is controversial, and outside the scope of this volume.

Another unanswered question is the original affiliation of the Yucatan block: was it a part of the South American Plate that was left behind when this plate drifted southward away from the North American Plate, as favored by Buffler and Sawyer (1985) and Ross and Scotese (1988), or was it originally part of the North American Plate? Available information suggests a certain relation between the Yucatan block and the northeastern part of present Mexico: both areas include lower Paleozoic metamorphic sequences overlain by unmetamorphosed upper Paleozoic rocks (upper Mississippian, Pennsylvanian, and Permian), intruded by Permian-Triassic granitic plutons. On the other hand, the Paleozoic section of the Yucatan block has little if any affinity with rocks of the same age in northern South America. Still unresolved, of course, is the question of whether northeastern Mexico was originally part of the North American or of the South American Plate, or if the Yucatan block and parts of Mexico were components of a third separate plate.

New and better geological and geophysical information and more careful integration of all available data are necessary to solve this and other controversies and to arrive at increasingly better interpretations of the post-collision fit of the North American, African, and South American Plates.

LATE TRIASSIC TO LATE MIDDLE JURASSIC— THE INITIAL BREAKUP OF PANGEA[1]

Whereas the unraveling of the geological history of the southern margin of the North American Plate and of the region that would become the Gulf of Mexico basin during the Paleozoic and early Triassic is necessarily based on few facts and much speculation, the events that took place in the area starting in the Late Triassic, events which led to the formation of the present Gulf of Mexico basin, can be reconstructed with considerably more assurance based on much more ample and reliable information. These events began in Late Triassic time with the breakup of the supercontinent of Pangea in the area of the present southern margin of the North American Plate, and continued during the Early and Middle Jurassic. Evidence for such a long period of tensional deformation is recorded in the extensive and complex network of tensional fractures, linear grabens, and half grabens filled with Upper Triassic–Lower Jurassic red beds and associated volcanic rocks. The present location of these Late Triassic–Early Jurassic graben systems is shown in Figure 7. They have been mapped around the periphery of the Gulf of Mexico basin in

---

[1]The following sections of this chapter, and to a lesser extent some of the previous ones, rely significantly upon other chapters of this volume, particularly upon Chapters 6 to 13. The reader, therefore, is referred to them for more detailed discussions of specific intervals of the stratigraphic section or particular aspects of the geologic history of the Gulf of Mexico Basin, and for a complete list of references on the respective subjects. In this chapter, reference is made only to publications not referenced in other chapters, or indispensable to the discussions of this chapter.

eastern Mexico, and in the southeastern United States (see Chapter 8, this volume). The large and complex northeast–southwest–trending graben system in southern Georgia and northwestern Florida is known to extend along the east coast of the United States and Canada, both onshore and offshore, and to have played an important role in the early opening of the North Atlantic Ocean and in the separation of the North American Plate from the Africa–South America part of Pangea.

Seismic reflection records suggest the presence of a half graben northwest of the Campeche Escarpment and a small graben in the southeastern Gulf of Mexico, both interpreted to be filled with sedimentary rocks of probable Late Triassic–Early Jurassic age (Buffler and others, 1980; Schlager and others, 1984). A major east-northeast–west-southwest–trending graben system, active since the Late Triassic, has been postulated to be under the northern part of the Gulf of Mexico, offshore from Texas and Louisiana (Fig. 7) (Salvador and Green, 1980; Salvador, 1987), in which red beds were first deposited and which was the site of the formation of thick salt deposits during the later Middle Jurassic (Fig. 8). The great thickness and the complex deformation of the salt in the area has made it difficult, if not impossible, to detect by geophysical methods the presence of such a pre-salt graben.

Upper Triassic–Lower Jurassic red beds have been reported below the Middle Jurassic salt in the East Texas basin and it is likely that similar red beds are under the North Louisiana salt basin, under the deeper parts of the Mississippi salt basin, and under the thick salt deposits of the Bay of Campeche, though no red beds have been yet reached by wells or indicated by geophysical information.

In the southern part of the Florida Peninsula and the West Florida Shelf, numerous wells have penetrated volcanic rocks of Early Jurassic age below the Cretaceous section (Fig. 7) (see Chapter 6, this volume). The northern boundary of these volcanic rocks seems to correspond with a postulated northwest–southeast–trending lineament (fracture zone or fault?), perhaps the continuation of the Bahamas fracture zone in the Atlantic Ocean (Klitgord and others, 1984), which is also aligned with the northwest-trending Alabama-Arkansas fault system of Thomas (1988) and others; the southern limit of these rocks is not well known. Their origin (dikes, sills, and/or flows?) and association, if any, with the early stages of rifting in the Gulf of Mexico basin region are unknown.

May (1971) described swarms of diabase (dolerite) dikes of Late Triassic to Early Jurassic age from eastern North America, western Africa, and northeastern South America, that, if shown on a map of the assembled late Paleozoic–early Mesozoic Pangea, display a marked radial pattern centered roughly in the Bahamas Platform (Fig. 9). This system of radial diabase dikes has been interpreted to be the result of the rise of a mantle column or plume ("hot spot") that may have been instrumental in initiating the breakup of Pangea.

The location of the grabens and rift basins along which Pangea broke apart is believed to have been controlled to a great

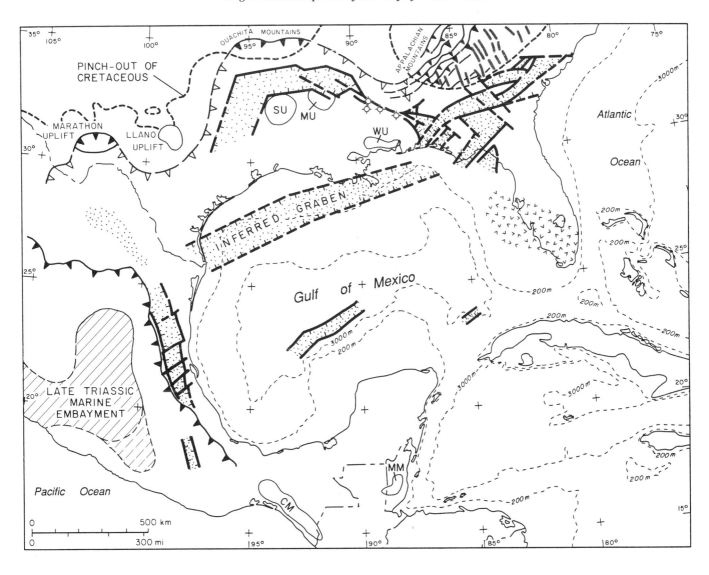

Figure 7. Distribution of Late Triassic–Early Jurassic rocks in the Gulf of Mexico basin. SU = Sabine uplift; MU = Monroe uplift; WU = Wiggins uplift; MM = Maya Mountains; CM = Chiapas massif.

extent by the preexisting structural fabric, trends of weakness determined by previous tectonic events. Some of these trends are very old—the rifts and transform faults involved in the late Proterozoic and early Cambrian rifting events leading to the opening of the Proto-Atlantic Ocean, which shaped the margins of the North American Plate (Thomas, 1988)—others are younger, related to the collision of Gondwana with the North American Plate during the late Paleozoic.

Late Triassic and Early Jurassic marine deposition in the area of the present Gulf of Mexico basin and neighboring regions

is known to have taken place only in an embayment of the Pacific Ocean that extended to central Mexico in the Late Triassic and reached east-central Mexico in the Early Jurassic (Fig. 7; Salvador, 1987; Chapter 8, this volume). In the early part of the Middle Jurassic, the Pacific embayment was restricted to southwestern Mexico, indicating either a lowering of the sea level at that time or a general uplift of central and east-central Mexico. In east-central Mexico, beds of early Middle Jurassic age are nonmarine red beds (the Cahuasas and La Joya Formations) deposited along a linear trend, possibly the same graben system active

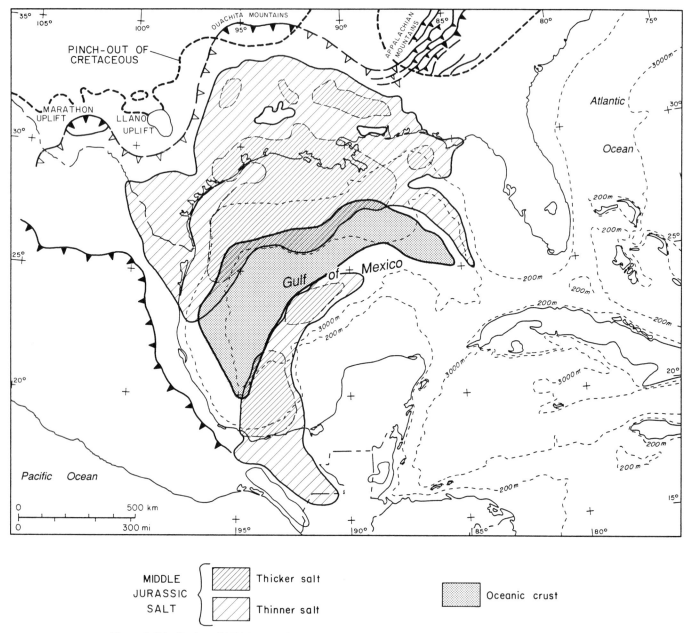

Figure 8. Distribution of Middle Jurassic salt deposits and of oceanic crust in the Gulf of Mexico basin.

during the Late Triassic and Early Jurassic. No rocks of early Middle Jurassic age have been identified anywhere else in the Gulf of Mexico basin area. The area presumably remained part of the emergent western part of Pangea where only red beds were being accumulated along those graben systems still active at the time.

Marine conditions extended to east-central Mexico again in late Middle Jurassic time, as indicated by the evaporites of the Huehuetepec Formation and the overlying marine calcarenites and shales of the Callovian Tepexic Formation. The irregular distribution of this formation indicates the presence of low islands separated by shallow tidal channels and arms of the sea in east-central Mexico during the Callovian. Alternating periods of higher and lower energy conditions resulted in the deposition of calcarenites and shales in these channels, around the islands, and along the mainland coast.

The first indication that seawater reached the present Gulf of Mexico basin through central Mexico is the occurrence of extensive and in places extremely thick salt deposits of probable late Middle Jurassic to very early Late Jurassic age (Salvador, 1987; Chapter 8, this volume). They were formed in very large shallow bodies of hypersaline water with limited communication with the ocean, wherein the evaporation exceeded the inflow of marine water. The climate was arid or semiarid, and the land around these bodies of water was probably low after a long period of emergence and erosion. No major rivers emptied into the hypersaline water bodies, and the supply of terrigenous clastic sediment was, therefore, very limited.

The salt deposits are in two separate regions of the Gulf of Mexico basin, northern and southern (Fig. 8; Salvador, 1987; Chapter 8, this volume). Between these two regions, salt seems to be absent in the western flank of the basin in east-central Mexico, in the western slope of the Gulf of Mexico, and in most of its deep central and southeastern parts.

In order to account for the large thickness of the Middle Jurassic salt in some areas and its much reduced development, or even absence, in others, it is necessary to assume that the thick sections of salt were formed in depressions (grabens) that subsided gradually and more or less continuously, while in other, less negative areas, thinner salt sections were formed. Some areas may have been covered by water during short and infrequent intervals, or remained above water level and, therefore, contain only thin layers of salt or no salt. The extremely incompetent and ductile nature of the salt and the extensive deformation it has undergone, particularly where it was very thick, makes it very difficult to estimate its original thickness and to determine the areas of its maximum accumulation. However, an approximate estimate of the relative thickness of the mother salt and the distribution of subsiding linear depressions or grabens during late Middle Jurassic and early Late Jurassic time, where the original salt deposits can be assumed to have been thickest, may be made on the basis of the present occurrence and structural configuration of the salt deposits (Fig. 8). Areas of maximum salt deposition were probably the interior salt basins of the U.S. Gulf Coast (East Texas, North Louisiana, and Mississippi), a belt along the Texas-Louisiana shelf and slope, and the Bay of Campeche (Fig. 8). The thickness of the mother salt in the interior salt basins may have been between 1,000 and 1,500 m, and may have reached 2,000 or 3,000 m in the Texas-Louisiana continental slope and the Bay of Campeche.

The evidence provided by the thickness distribution of the salt and by the Middle Jurassic sedimentary rocks of east-central Mexico, though indirect, suggests that at least some of the Late Triassic-Early Jurassic graben systems in the Gulf of Mexico basin area remained active during the Middle Jurassic, and perhaps as late as the earliest Late Jurassic. Other areas of thick salt deposition may represent new, younger trends of rifting and subsidence. They all indicate that the Gulf of Mexico basin area was still undergoing extension in the late Middle Jurassic and earliest Late Jurassic.

The initial steps of the breakup of Pangea and the early stage of the formation of the Gulf of Mexico basin, what can be called "the rift stage," lasted perhaps as much as 50 m.y., from the Late Triassic to the early Late Jurassic. This early stage was characterized by the prevalence of tensional deformation, resulting in the development of complex systems of linear grabens and half gra-

bens; these, in turn, effectively controlled the sedimentary processes active at the time—formation of predominantly nonmarine red bed sequences and associated volcanic rocks at first (Fig. 9), and the accumulation of extensive salt deposits toward the end, during the late Middle or early Late Jurassic (Fig. 10). All evidence now available indicates that the salt section was deposited over continental or extended continental (transitional) crust. It is also generally accepted that the salt deposits of the present northern and southern regions of the Gulf of Mexico basin area, though now separated by a belt in the central Gulf of Mexico where salt is absent, are of the same age and that they were originally formed in a single vast shallow basin or in closely located and related basins (Humphris, 1978; Salvador, 1987; Chapter 8, this volume).

The early, rift, stage of the formation of the Gulf of Mexico basin, from the Late Triassic to the late Middle Jurassic, also involved the beginning of the motion of the Yucatan block away from the North American Plate, and, as a result, a considerable stretching and necking of the continental crust in a predominantly north-northwest–south-southeast direction (Pindell, 1985; Buffler and Sawyer, 1985; Salvador, 1987; Dunbar and Sawyer, 1987). The amount of the north-south or north-northwest–south-southeast crustal extension has been estimated to have been 500–520 km (Pindell, 1985), 470 km (Buffler and Sawyer, 1985), or 480 km (Dunbar and Sawyer, 1987). This amount of extension represents 50% to 55% of the total displacement of the Yucatan block. Little extension seems to have taken place in an east-west direction.

Because the nonrigid aspects of plate motion and continental crust extension unquestionably played an important role in the formation of the Gulf of Mexico basin, not only the amount of crustal extension but also the time of such extension needs to be examined. For this, the evidence provided by the time and manner of formation of the extensive Middle Jurassic salt deposits of the basin can provide the best testimony: if it is accepted that the salt was formed in shallow hypersaline basins, the implication would be that during the time of the formation of the salt deposits, crustal extension must have been limited and probably restricted to sinking and broadening of the graben systems which previously had been the site of accumulation of Upper Triassic and Lower Jurassic red bed sequences. Rapid deposition of salt in these graben systems was able to keep up with subsidence and maintain the shallow-water conditions necessary for the formation of evaporitic deposits. More regional extension of the crust over most of the region that would become the Gulf of Mexico basin is unlikely to have taken place at the time of the deposition of the salt; the formation of a greatly extended and thinned crust would have resulted in substantial subsidence and in the development of deep-water conditions over most of the region. Such a deepening of the basin would have terminated the deposition of salt. It is logical to assume, therefore, that the major part of the crustal extension took place fast, toward the end or shortly after the formation of the salt, during the late Callovian and early Oxfordian. Previous to that time, extension must have taken

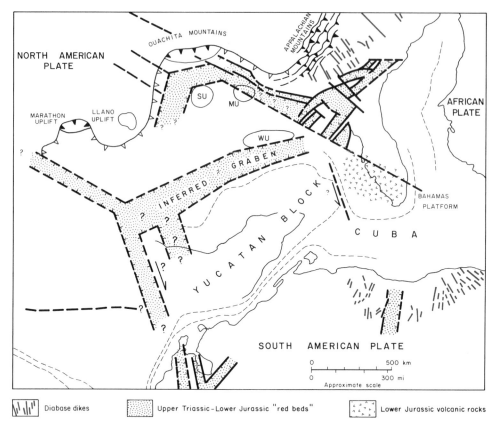

Figure 9. Reconstruction of western Pangea (future Gulf of Mexico basin region) for Late Triassic–Early Jurassic time.

place slowly and probably intermittently from Late Triassic to late Middle or early Late Jurassic time.

It has been suggested (Salvador and Green, 1980; Salvador, 1987) that an important part of the extensional activity along graben systems took place in one such system, trending east-northeast–west-southwest and located in the present northern Gulf of Mexico, below the thick and highly deformed Middle Jurassic salt deposits and roughly on trend with the graben system that extends all along the east coast of North America (Fig. 7).

LATE MIDDLE TO EARLY LATE JURASSIC— THE SOUTHWARD DRIFT OF THE YUCATAN BLOCK AND THE BIRTH OF THE GULF OF MEXICO BASIN

The broadening of the grabens and half grabens, the stretching of the continental crust, and the resulting displacement of the Yucatan block away from the North American Plate continued until the attenuated transitional crust was thinned to a critical thickness. Emplacement of oceanic crust and sea-floor spreading began at that time.

The time of this emplacement of oceanic crust can best be

established on the basis of the distribution and age of the widespread salt deposits of the Gulf of Mexico basin. It is necessary, however, to accept the following (see fig. 10).

1. The salt deposits in both the northern and southern parts of the basin were formed during the Callovian and the early Oxfordian.

2. The deposits were formed over continental or stretched (transitional) crust in a single shallow basin or in closely situated basins.

3. The salt deposits were bisected and separated to eventually reach their present location as a result of the emplacement of the oceanic crust and the concurrent sea-floor spreading.

This last assumption is supported by the present distribution of the oceanic crust in the Gulf of Mexico basin (see Chapter 4, this volume), which closely corresponds with the belt where salt is apparently absent between the two areas now containing the salt deposits (Fig. 8), and by the fact that the outline of the opposite boundaries of the salt deposits and their thickness distribution on either side of the salt-free area under the central part of the basin correspond well (Humphris, 1978; Salvador, 1987).

On this basis, it is possible to estimate that the time of the initiation of the emplacement of the oceanic crust took place toward the end or shortly after the formation of the salt deposits,

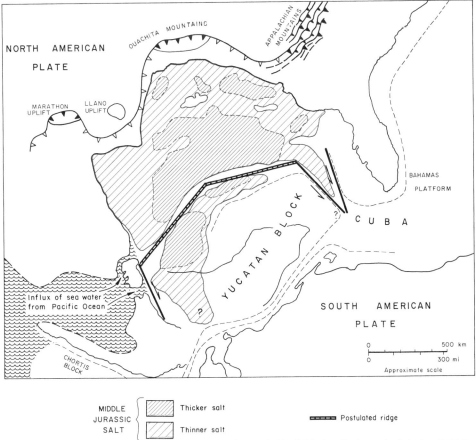

Figure 10. Reconstruction of western Pangea (future Gulf of Mexico basin region) in late-Middle Jurassic.

during the latest Callovian or early Oxfordian. The duration of crust emplacement seems to have been relatively brief: throughout the Gulf of Mexico basin, the Upper Jurassic sedimentary section (mid-Oxfordian and younger) shows considerable continuity, reflecting deposition during a period of marked tectonic stability (Salvador, 1987; Chapter 8, this volume), at least all along the northern, western, and southern flanks of the basin. This suggests that important tectonic disruption of the Upper Jurassic section had not taken place, and that sea-floor spreading had concluded and the Yucatan block had reached its present position. In addition, in the central, deep part of the Gulf of Mexico, the oceanic crust is overlain by a flat-lying sedimentary section, the base of which has been interpreted to be Late Jurassic in age. The time span of emplacement of the oceanic crust, like that of the final phase of stretching of the continental crust, may have been short, not much longer than 4 or 5 m.y.

To explain how the salt deposits were bisected and separated as a result of the emplacement of oceanic crust, the southward drift of the Yucatan block, and the resulting opening of the Gulf of Mexico basin, presents some problems, the solution of which may have a bearing on the time of emplacement of the oceanic crust. How was it possible to bisect the salt deposits,

probably across their thickest parts, and separate the two salt masses by about 300–350 km while the front of these detached salt masses was contained so the salt would not flow toward the newly opened space between them? It seems logical, even necessary, to postulate the formation of a roughly east-northeast–west-southwest–trending high (an oceanic spreading ridge?) across the central part of the present Gulf of Mexico basin sometime before, during, or shortly after the formation of the salt (Fig. 10). The formation of such a high may have announced the beginning of the emplacement of oceanic crust in the basin. If the high formed before or shortly after the beginning of the formation of the salt deposits, in late Bathonian or early Callovian (late Middle Jurassic), the salt would have been formed in two separate shallow basins or half grabens while oceanic crust was beginning to be emplaced between the two basins.

If, on the other hand, the high was not formed until after the formation of the salt (or was the event that put an end to its formation), oceanic crust material would have had to intrude the salt and set it apart into two separate bodies. No evidence is available to support the formation of the high or ridge, or to speculate about the process involved in the emplacement of the oceanic crust. Whereas the western boundary of the area of thick salt in the Bay of Campeche is sharp and shows no apparent lateral deformation, the salt along the rest of its boundaries on either side of the salt-free belt in the central Gulf of Mexico is highly distorted and squeezed into horizontally intruded tongues or nappes, making it impossible to ascertain the original nature of the boundaries.

How the Yucatan block moved from its former to its present position varies according to different interpretations of the origin of the Gulf of Mexico basin.

Evidence from magnetic-polarity lineations that could help to identify a spreading center responsible for the opening of the Gulf of Mexico basin, and to interpret the time and course of the Yucatan block as the oceanic crust was emplaced and the basin opened, is, unfortunately, ambiguous at best. The magnetic polarity lineations, if they exist, have been completely masked by the 10 to 15 km of sediments that overlie the oceanic crust in the central part of the basin. Magnetic anomalies that have been mapped in the basin are of low amplitude and lack a definite orientation pattern.

In the absence of reliable magnetic-polarity lineations that could serve as a guide, the course of the displacement of the Yucatan block favored by different authors depends on the position assigned to the block in reconstructions of Pangea in late Paleozoic or early Mesozoic time. This position, in turn, often depends on the particular evidence or type of information emphasized in the reconstruction.

The proposed amount of rotation undergone by the Yucatan block during its drift is one of the unresolved issues. Among the many attempts to interpret the origin of the Gulf of Mexico basin, Pindell and Dewey (1982) and Pindell (1985) proposed that the Yucatan block rotated 40° to 50° counterclockwise in its path from its late Paleozoic to its present position, in order to achieve an optimum fit of the continental plates during the Permian. Buffler and Sawyer (1985) and Dunbar and Sawyer (1987) favored similar counterclockwise rotation of 36° and 45°, respectively, based on the elimination of the wedged-shaped area of oceanic crust in the central part of the Gulf of Mexico basin, and on the restoration of the stretched transitional crust to its original thickness. Ross and Scotese (1988) also proposed a counterclockwise rotation of 50°. Salvador (1987) postulated a counterclockwise rotation of only 10° based on the fit of the boundaries of the northern and southern Middle Jurassic salt deposits and in order to attain a satisfactory alignment of the reassembled salt deposits and the probable channel of influx of seawater from the Pacific Ocean through east-central Mexico. His reconstruction of the position of the Yucatan block, however, did not take into consideration, as did the other above-mentioned interpretations, the stretching of the continental crust that took place prior to the emplacement of oceanic crust, sea-floor spreading, and the final stage of the displacement of the Yucatan block to its present position.

Gose and Sanchez-Barreda (1981), on the other hand, proposed a 22° ± 8° clockwise rotation of the Yucatan block during the early Mesozoic based on paleomagnetic data from Permian sedimentary rocks in southern Mexico, and Hall and others (1982), on the basis of their analysis of regional gravity data, favored a clockwise rotation of 25°.

Another aspect of the drift of the Yucatan block that has been treated somewhat differently, regardless of the amount and direction of rotation involved, is the nature of the boundaries between the Yucatan block and bordering crustal components along which the block was displaced to its present position. Salvador (1987) proposed that the southward drift of the Yucatan block occurred along two major, roughly parallel transform-fault zones trending north-northwest–south-southeast, one along the present east coast of Mexico and crossing the Isthmus of Tehuantepec, and the other along the present Florida Escarpment (Fig. 11). Evidence for these two transform-fault zones is scant. The straight trend of the southern part of the Florida Escarpment is the only indication of the possible presence of the eastern transform-fault zone. The postulated western fault zone corresponds with a steep and straight slope of the "basement" gulfward from the east coast of Mexico, and parallels the Late Triassic–Middle Jurassic graben system in east-central Mexico. Farther south, indirect evidence indicates that a major structural discontinuity may exist across the Isthmus of Tehuantepec—a radical change in tectonic style from eastward thrusting along the Sierra Madre Oriental of east-central Mexico northwest of the isthmus, to dominant strike-slip faulting parallel to the northeastern flank of the Chiapas massif to the southeast. Surface indications of a major fault zone across the isthmus are not available, but should not be expected if it is assumed that the transform was active only from the Late Triassic or Early Jurassic to the early Late Jurassic, and has been dormant since then (Salvador, 1987).

Pindell (1985) proposed a western transform, his "Tamaulipas–Golden Lane–Chiapas transform fault," along which the Yucatan block had drifted south in late Middle Jurassic time, but instead of having it cross the Isthmus of Tehuantepec, he depicted it as extending southeastward along the northeastern foothills of the Chiapas massif, an interpretation not supported by stratigraphic information in the area. Buffler and Sawyer (1985) also suggested that the drift of the Yucatan block took place along a western major transform and a lesser transform in the southeastern Gulf of Mexico, but did not designate the location of the western transform in present-day southern Mexico. Dunbar and Sawyer (1987) placed the southern limit of the Yucatan block at the "Chiapas fault of southern Mexico, after Pindell (1985)." Others have failed to specify the manner in which the Yucatan block was displaced to its present position.

The strongest disagreements concerning the breakup of Pangea, the drift of the Yucatan block to its present position, and the formation of the Gulf of Mexico basin have arisen, however, with respect to the southeastern part of the basin. The paleogeography and tectonic development of this area in the context of the origin of the basin presents problems not yet satisfactorily resolved; also in this case, the different interpretations proposed vary according to the position assigned to the Yucatan block before its separation from the North American Plate.

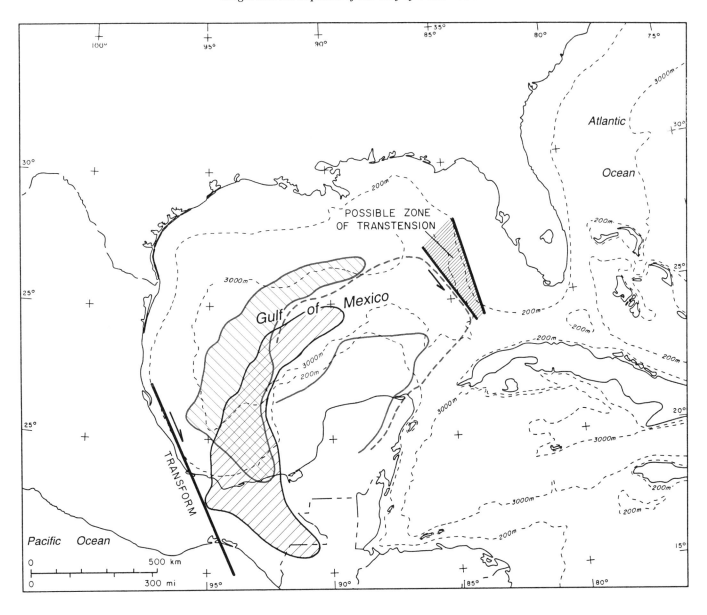

Displacement of the southern Middle Jurassic salt region and of the Yucatan Block

 Original position   Present position

Figure 11. Location of transform-fault zones along which the Yucatan block was displaced southward from its original to its present position.

Those who place the pre-drift Yucatan block in a more northeasterly position, fitting against the western margin of the Florida platform (Buffler and Sawyer, 1985; Salvador, 1987; Dunbar and Sawyer, 1987), favor a southward displacement of the eastern part of the block, with greater or lesser rotation, along a transform located approximately along the present Florida Escarpment. Transtensional displacement along this transform sometime during the Late Jurassic may account for the separation of the Yucatan block from the Florida platform and the establishment of a marine connection between the Gulf of Mexico and the Caribbean Sea.

Pindell (1985), on the other hand, placed the pre-rift Yucatan block farther to the west with respect to the North American plate (somewhere against present Texas and Louisiana), and filled the gap between the Yucatan block and the Florida platform with a continental crustal block (or blocks), his "Florida Straits block," which includes the southern part of the present Florida Shelf, south Florida, and the southwestern part of the Bahamas platform. During the Late Jurassic, the Florida Straits block is said to have moved some 300 km out of the eastern part of the Gulf of Mexico basin in an east-southeast direction along transform faults roughly parallel to fracture zones recognized in the North Atlantic Ocean, and to have reached its present position by Early Cretaceous time. Ross and Scotese (1988) accepted the existence of Pindell's Florida Straits block in their interpretation of the tectonic evolution of the Gulf of Mexico basin and the Caribbean region.

In spite of the many remaining uncertainties and differences of opinion concerning the timing and processes, most recent interpretations of the origin of the Gulf of Mexico basin agree that the initial stages of the formation of the basin correspond with the breakup of Pangea and the southward drift of the Yucatan block away from the southern margin of the North American Plate.

The initial rift and crustal extension phase, which started in the Late Triassic with the formation of complex graben systems, was followed by the sea-floor spreading phase and the beginning of emplacement of oceanic crust during which the Yucatan block was displaced to its present position. The separation of the Yucatan block from the southern margin of the North American Plate probably did not take place as a clean break along a single boundary zone: the northern part of the area that would become the Gulf of Mexico basin seems to have been characterized during the Late Triassic, Early Jurassic, and Middle Jurassic by the formation of a number of grabens and horsts, as clearly indicated by the location and thickness distribution of the Upper Triassic–Lower Jurassic red beds and the Middle Jurassic salt deposits (Salvador, 1987; Chapter 8, this volume). During the progressive southward displacement of the Yucatan block, crustal extension may have been greater in the graben areas and less intense in the horsts; as a result, some of the horsts were left behind by the Yucatan block and are represented today by positive structural elements such as the Sabine, Monroe, and Wiggins uplifts, in which the crust is apparently thicker and less extended than in surrounding rifted basinal areas floored by thinner and more extended crust; as, for example, the Mississippi salt basin (Worzel and Watkins, 1973; Nunn and others, 1984). The Angelina-Caldwell flexure may represent the boundary between thicker continental crust to the north and thinner stretched transitional crust to the south. It is possible that the structurally higher blocks of thicker continental crust have closer pre-drift geological and genetic affinities to the Yucatan block than to the North American Plate.

As a result of the drift of the Yucatan block, the Middle Jurassic salt deposits were split in two and the southern segment moved south. The process by which the salt deposits were separated is far from clear. As discussed previously, the existence of a high separating the two areas of salt occurrence seems most necessary, but no evidence is available at present about either the time or the process of its formation.

The southward motion of the Yucatan block must have

taken place along two roughly parallel transform faults. The position of the western transform fault, or fault zone, corresponds with the steep and straight slope of the "basement" offshore from the present coast of east-central Mexico (Plate 3) and crossed the Isthmus of Tehuantepec at a small angle (Fig. 11). The eastern zone of displacement is much more conjectural but probably was located along the present Florida Escarpment. Initially, movement along this eastern zone of displacement was probably of a transcurrent nature, but it may have been followed by transtensional displacement sometime during the Late Jurassic, perhaps shortly after the Yucatan block had reached its present position. This transtensional episode resulted in the separation of the Florida and Yucatan platforms, the formation of a complex system of horsts and grabens between the two platforms, and the opening of a channel between the Gulf of Mexico and the Caribbean. (The initial opening of the Gulf of Mexico to the Caribbean will be further discussed in the next section of this chapter).

The southward displacement of the Yucatan block, as interpreted in this chapter, involves a counterclockwise rotation of about 30° (Figs. 6 and 10). The original position of the Yucatan block was predicated on the basis of the preferred reassembly of the Middle Jurassic salt deposits, but also takes into consideration the present distribution of oceanic crust in the Gulf of Mexico basin, as well as the stretching and attenuation of the continental crust.

No displacement along the postulated late Proterozoic or early Paleozoic transform faults, the Alabama-Oklahoma and Texas transforms, can be documented in connection with the southward drift of the Yucatan block and the opening of the Gulf of Mexico basin in early Late Jurassic time. Neither can a clear tectonic relation be established between the events responsible for the formation of the basin and contemporaneous events in the North Atlantic region. Finally, even though the time of the separation of the Yucatan block from the southern margin of the North American Plate is believed to correspond with the time of the drift of the South American Plate away from the North American Plate, the relation and affinities of the Yucatan block to either one of these two major plates is far from clear. As discussed earlier, there is no evidence to indicate that the Yucatan block had been part of the South American Plate; its strongest affinities are with the eastern Mexico region. There is also no convincing evidence to link the Yucatan block with the North American Plate, or to presume that it was part of yet another continental plate, or a separate plate altogether.

LATE JURASSIC—A SUBSIDING BUT STABLE GULF OF MEXICO BASIN

Early in the Late Jurassic, probably not later than mid-Oxfordian, the emplacement of oceanic crust in the central part of the Gulf of Mexico basin ended, and the Yucatan block reached its present position to the east of southern Mexico. The newly emplaced oceanic crust cooled, contracted, and subsided. Subsidence also affected the stretched transitional crust; the thin transitional crust subsided faster than the adjacent thicker transitional crust, but less than the oceanic crust. Tensional deformation that had characterized the tectonic history of the Gulf of Mexico basin from the Late Triassic to early Late Jurassic ceased and was followed, starting in the Late Jurassic, by a second tectonic phase of intermittent subsidence that has continued until the present. Subsidence from mid-Oxfordian to mid-Cretaceous time was predominantly the result of cooling of the oceanic crust, the thermal subsidence phase; after the mid-Cretaceous, sedimentary loading is believed to have played an increasingly important role. After the mid-Oxfordian, major tectonic activity shifted southward to the Caribbean region.

The paleogeography of the area of the ancestral Gulf of Mexico basin changed markedly in the early Late Jurassic. The period of extensive salt formation in a shallow hypersaline basin or basins that covered large areas of the arid continental western Pangea was terminated by the emplacement of oceanic crust, the displacement of the Yucatan block, and the formation of a new, rapidly subsiding basin; it was followed by a widespread and prolonged marine transgression. The transgression was nearly continuous during the entire Late Jurassic and the earliest Early Cretaceous with only minor periods of regression or sea-level drop. Seawater entered the basin from the Pacific Ocean across central Mexico, and by mid-Oxfordian time, a large inland sea with unrestricted circulation and normal salinity, the ancestral Gulf of Mexico, was well developed. It was connected to the west with the Pacific Ocean but not opened yet to the Atlantic or the Caribbean, as the Florida and Yucatan platforms were low lands above sea level bordering the water-covered area to the east, southeast, and south. The Florida and Yucatan platforms remained emergent during the Late Jurassic; the Upper Jurassic section is known to pinch out against the base of the platforms along the Florida and Campeche escarpments.

During the tectonically stable period that characterized the Late Jurassic, the subsiding central part of the Gulf of Mexico basin was rimmed by broad stable ramps and shelves on which the predominantly marine Upper Jurassic section was deposited. The influx of coarse clastic sediments came mostly from the north and northwest, limited amounts at first, but copious during the latest Late Jurassic, indicating the presence in these areas of important rivers emptying into the Gulf of Mexico of the time. Toward the center of the basin, the Upper Jurassic section became progressively shalier, reflecting increasingly deeper water conditions of deposition.

The Late Jurassic transgression extended progressively over most of the area of the present Gulf of Mexico basin. The northern limit of Late Jurassic deposition in the United States part of the basin lies within the structural limits of the basin; in Mexico, on the other hand, Upper Jurassic sedimentary rocks are known from many areas in the central and western parts of the country, well beyond the western structural limits of the basin. Each Upper Jurassic stratigraphic unit oversteps the preceding one and pinches out progressively farther landward. The magnitude of this overstepping is nowhere considerable with the exception of north-

eastern and southern Mexico. In northeastern Mexico, the Oxfordian seems to be restricted to the Sabinas basin and surrounding areas, whereas Kimmeridgian and Tithonian sedimentary rocks have been recognized the length of the Chihuahua trough as far north as the El Paso area.

The initial beds deposited during the Late Jurassic transgression covered, with variable stratigraphic relations, an irregular surface composed of rocks of equally varied age and lithologic composition. In the central part of the basin the lower part of the sedimentary section overlying the oceanic crust is believed to be of Late Jurassic age; in east-central Mexico Upper Jurassic sedimentary rocks cover the Middle Jurassic marine and nonmarine section, for the most part conformably, and over large parts of the basin these sediments lie over the Middle Jurassic salt deposits. The contact between the salt and the overlying Upper Jurassic is probably conformable; the boundary is certainly not believed to represent an important depositional hiatus. Along the periphery of the basin, where it oversteps the salt, the Upper Jurassic lies unconformably over Paleozoic unmetamorphosed and metamorphic Paleozoic rocks, and over Upper Triassic–Lower Jurassic and Middle Jurassic red beds and associated volcanic rocks. Preexisting topographic features controlled to a considerable extent the early Late Jurassic sedimentation; these features were particularly influential in northeastern Mexico and in southern Alabama and northwest Florida. Also influential in the deposition of the initial Upper Jurassic section, as well as the rest of the Upper Jurassic, were contemporaneous movements along regional normal-fault zones, generally parallel to the periphery of the northern flank of the basin, and the flow of the underlying Middle Jurassic salt.

Terrigenous clastic sediments began to enter the new basin in early or mid-Oxfordian time along the northeastern regions of the basin, contributed by streams flowing south from the North American continental interior and the southern Appalachian Mountains (Fig. 13). The climate was still arid and the sediment load brought in by the rivers was accumulated in alluvial fans and deltaic complexes and subsequently redistributed by wind, which at the time blew predominantly toward the east and southeast, to form extensive dune fields. The lower or mid-Oxfordian fluvial, deltaic, and eolian deposits compose the Norphlet Formation. Accumulation of evaporite deposits may still have been taking place farther basinward. A precursor of the Mississippi River seems to have been entering the Gulf of Mexico in west-central Mississippi. The northwestern part of the Gulf of Mexico basin region, southwest of the ancestral Mississippi, remained a low land area or the site of the formation of evaporite deposits (see Chapter 8, this volume).

The widespread Late Jurassic marine transgression reached the northern Gulf of Mexico basin region after the deposition of the basal Oxfordian terrigenous clastic section, the Norphlet Formation of the northeastern part of the basin. In the northeastern region, the encroaching sea reworked the upper part of the previously deposited fluvial, deltaic, and eolian sediments and redeposited them as a thin, shallow-water, marine-sandstone unit.

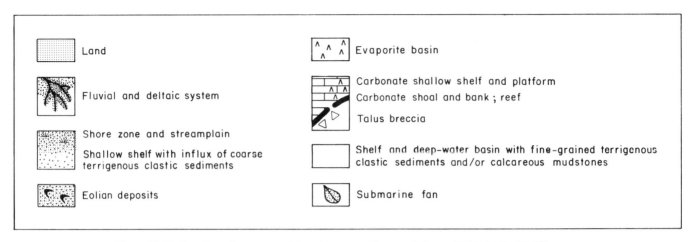

Figure 12. Explanation of patterns used in paleogeographic maps (Figures 13-18; 21-24; 27-30).

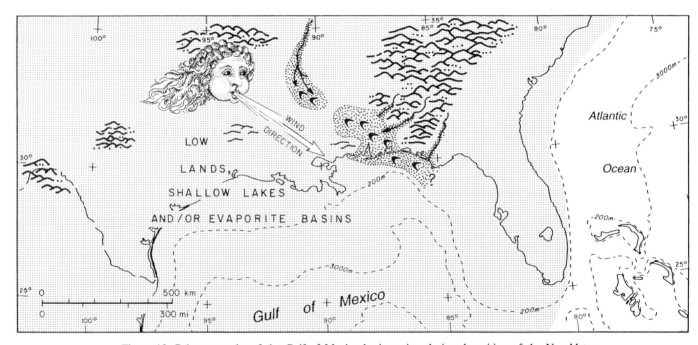

Figure 13. Paleogeography of the Gulf of Mexico basin region during deposition of the Norphlet Formation (early Late Jurassic). (For explanation of patterns, see Figure 12).

In the northern and northwestern parts of the basin, the transgressive sea deposited a generally thin section of lag materials over an irregularly eroded surface.

A similar situation probably prevailed along the southern part of the basin in southern Mexico, but very little information is available about the basal part of the Upper Jurassic section in the area. In east-central Mexico, in the eastern part of the corridor that connected the Pacific Ocean with the Gulf of Mexico basin region, marine deposition continued essentially uninterrupted from the Middle Jurassic to the Late Jurassic.

As the Late Jurassic transgression progressed over the shallow ramps or shelves surrounding the early Gulf of Mexico, the supply of coarse terrigenous clastic sediments seems to have diminished and the basal sandstones and lag deposits were covered by the mid-Oxfordian carbonate mudstones and micritic limestones of the lower part of the Smackover and Zuloaga Formations and equivalent stratigraphic sequences. They all reflect deposition under low-energy, severely restricted conditions—intertidal mud flats or coastal areas where the development of algal mats alternated with deposition of carbonate muds. The climate gradually

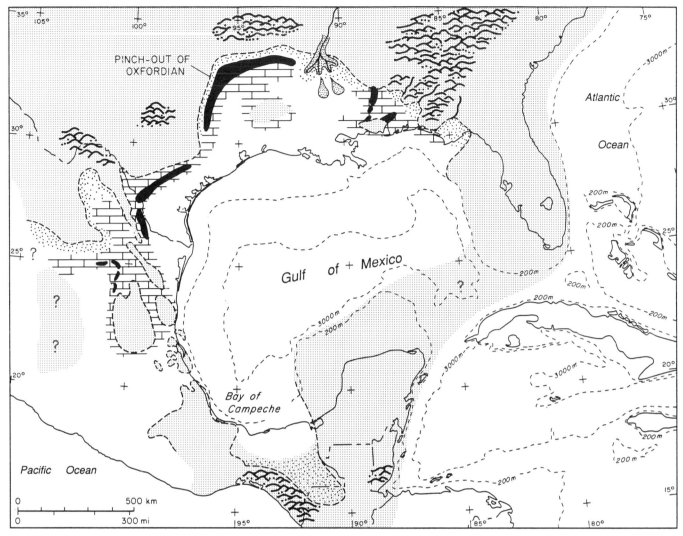

Figure 14. Paleogeography of the Gulf of Mexico basin region during the late Oxfordian. (For explanation of patterns, see Figure 12.)

became more humid, perhaps as a result of the presence of a large body of water in the region.

The Oxfordian transgression reached its maximum extent toward the end of the deposition of the carbonate mudstones and argillaceous limestones of the lower Smackover, Zuloaga, and equivalent sediments elsewhere in the basin. Deposition at this time was probably in low-energy open-marine subtidal environments.

The upper part of the Oxfordian section of the northern and northwestern parts of the Gulf of Mexico basin was deposited under high-energy conditions during a stillstand or moderate lowering of sea level (Fig. 14). For the most part, the margins of the basin were still broad, regional ramps sloping gently toward the deep, central part of the basin. Toward the end of the Oxfordian, the ramps were progressively modified by the formation of a widespread shoal system in which oolitic grainstones and packstones formed widespread linear offshore bars and beaches. These shoals separated a landward area of intertidal coastal plains and littoral lagoons from a deeper shelf or ramp and a basinal area where micritic limestones, carbonate muds, and shales were deposited. The margins of the basin evolved in this way from gentle basinward-sloping ramps to shelves. Patch reefs grew locally over sea-floor highs, and islands and embayments developed locally along the trend of the shoals. Local ductile deformation of the Middle Jurassic salt took place during the Oxfordian and is reflected in the lithofacies and thickness of the Oxfordian section.

During the Oxfordian, coarser-grained terrigenous clastic deposition was restricted to west-central Mississippi and adjoining areas in Louisiana and Arkansas, and to northeastern Mexico. In the west-central Mississippi area, sandstones are interbedded with the carbonates and calcareous shales of the Smackover Formation. Their areal distribution suggests the presence in the area of the mouth of a major stream entering the ancestral Gulf of Mexico from the north, probably the early Mississippi River. Some of the sandstones have been interpreted to have been deposited in deep water as turbidites or as submarine-fan complexes. In northeastern Mexico, the more basinward Oxfordian section is predominantly composed of shales and shaly limestones, but becomes increasingly sandy toward the bordering land areas of the time to the west and northwest.

In east-central Mexico, the conditions of deposition and paleogeography during the Oxfordian were considerably different from those in the northern and northwestern margins of the basin. The Oxfordian section was not underlain by a salt interval and does not, therefore, show the effect of early flowage and deformation of such an incompetent and ductile layer. No coarse-grained clastic intervals are present in the Oxfordian section of this area, indicating no appreciable influx of coarse sediments and the absence of important streams flowing into the western margin of the basin. The Oxfordian of the western Gulf of Mexico basin is predominantly composed of the marine shales of the Santiago Formation.

Much less is known of the paleogeography and environments of deposition in southern Mexico during the Oxfordian. The Oxfordian section of the Bay of Campeche and the Reforma area, predominantly composed of marine shales and shaly limestones, was deposited in a large embayment of the Oxfordian Gulf of Mexico which shallowed to the southeast. The lower part of the nonmarine red beds of the Todos Santos Formation of southern Mexico and northwestern Guatemala may be of Oxfordian age. The Oxfordian over most of southern Mexico is known to be underlain by Middle Jurassic salt and should, therefore, reflect the contemporaneous early deformation of the salt.

In contrast to the Upper Triassic, Lower Jurassic, and Middle Jurassic sedimentary sequences, which were characterized by deposition in tectonically active, geographically restricted, linear tensional depressions (grabens, half grabens), the Oxfordian section, as a whole, shows widespread distribution and no major evidence that tectonic activity influenced sedimentation. The Oxfordian sequences are of remarkably uniform lithology, do not show abrupt changes in thickness, and reflect sedimentation over extensive and stable shelves or ramps that plunged at low angles toward the deep center of the basin.

During the Oxfordian, the Florida and Yucatan platforms were emergent and probably connected to each other. The Gulf of Mexico was not yet in communication with the recently opened North Atlantic Ocean; its only communication with other marine areas was to the west toward the Pacific Ocean through central and perhaps northwestern Mexico, as indicated by the distribution and nature of the Oxfordian sedimentary sequences and the Pacific affinities of its ammonite faunas. In northeastern Mexico, a belt of islands, shoals, and shallow-water shelves extended from southwest Texas to east-central Mexico separating the Gulf of Mexico to the east from a smaller basin in central

Mexico. The belt of islands seems to correspond with, and may owe their presence to, the belt of granitic plutons that intruded the Paleozoic rocks during the late Permian and/or early Triassic (Figs. 5 and 14).

The trend toward shallower-water conditions initiated during the late Oxfordian continued during the early Kimmeridgian. Over the whole Gulf of Mexico basin, the lower Kimmeridgian section appears to reflect less marine or shallower marine environments of deposition than the underlying Oxfordian.

The linear grainstone shoals that developed during the late Oxfordian along the northern and northwestern flanks of the basin persisted during the early Kimmeridgian. They provided the restrictive barriers behind which hypersaline lagoons and sabkhas formed in an environment protected from waves and currents, landward from the shoals and parallel to the rim of the basin. Deposition of evaporites, low-energy carbonates, and some fine-grained terrigenous clastic sediments (the Buckner Member of the Haynesville Formation and lower part of the Olvido Formation) took place in these lagoons and sabkhas, which were bordered landward by nonmarine fluvial and deltaic environments in which a predominantly coarse clastic section was deposited. Basinward from the grainstone barriers, the Kimmeridgian section becomes shalier, and indicates deposition in increasingly deeper water marine environments (Fig. 15).

In east-central Mexico, the lower Kimmeridgian section, represented by the Tamán Formation, is also thought to have been deposited under somewhat shallower-water marine conditions than the underlying Santiago Formation.

The shallowing of the northern and western rims of the basin may have been due to a general uplift of these regions or, perhaps more likely, to a lowering of the sea level.

Widespread marine conditions returned to the northernmost Gulf of Mexico basin area during the late Kimmeridgian. The evaporitic pans were covered by clastic and carbonate sediments deposited in shallow-water open-marine shelves. Along the northeastern part of the basin, terrigenous clastic deposition predominated, particularly in northeastern Louisiana, southeastern Arkansas, and central Mississippi, where the presence of a deltaic complex—the delta of the ancestral Mississippi River—is presumed to have been present. To the southwest, in Texas and northeastern Mexico, the influx of terrigenous clastic sediments was not as dominant, and the upper Kimmeridgian section is predominantly composed of carbonates.

The western Gulf of Mexico basin area, in east-central Mexico, was also an open-marine shelf sloping eastward toward the deep central part of the early Gulf of Mexico, into which the influx of terrigenous clastics was minimal. The coast to the west of the shelf was irregular in shape and numerous islands still dotted the offshore. Interbedded limestones and shales were deposited in the deeper parts of the shelf, while oolitic and skeletal grainstones and calcarenites were formed in shallower and higher-energy environments at bars, banks, and shoals along the coast and around the islands.

In southern Mexico, the limited information available indi-

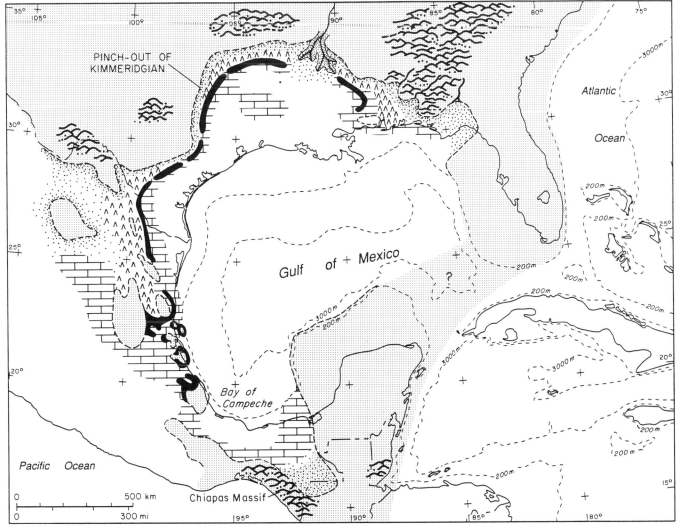

Figure 15. Paleogeography of the Gulf of Mexico basin region during the early Kimmeridgian. (For explanation of patterns, see Figure 12.)

cates that during the Kimmeridgian an open-marine embayed shelf persisted in the Bay of Campeche and Isthmus of Tehuantepec areas, and that the embayment became shallower toward a coastal area located to the southeast, along the northeastern front of the Chiapas massif. Influx of terrigenous clastics into this embayment was, as in east-central Mexico, very limited. In southernmost Mexico and northwestern Guatemala, the Kimmeridgian is probably represented within the continental red bed sequence of the Todos Santos Formation.

The Florida and Yucatan platforms were still exposed during the Kimmeridgian, but the possible opening of a connection between the ancestral Gulf of Mexico and the Atlantic Ocean during the Kimmeridgian (or earlier, during the Oxfordian) remains highly controversial. There is at present no evidence to support or disprove the presence of such a connection. On the basis of regional stratigraphic information, it has been assumed (in Chap-

ter 8, this volume) that a connection between the Gulf of Mexico and the Atlantic was not established until the late Kimmeridgian, after the period of restricted marine conditions that characterized the early Kimmeridgian. Not only early Kimmeridgian deposition, but Oxfordian deposition as well, seem to have been more restricted geographically than the subsequent late Kimmeridgian and Tithonian deposition. It is possible, however, that if the initial southward displacement of the Yucatan block during the early Oxfordian took place along a transform fault located approximately along the present Florida Escarpment, but was followed later in the Oxfordian by more transtensional displacement, the separation of the Florida and Yucatan peninsulas and the communication of the Gulf and the Atlantic could have been achieved earlier than the late Kimmeridgian.

The Late Jurassic transgression reached its peak during the Tithonian or earliest Cretaceous. The waters of the Gulf of Mex-

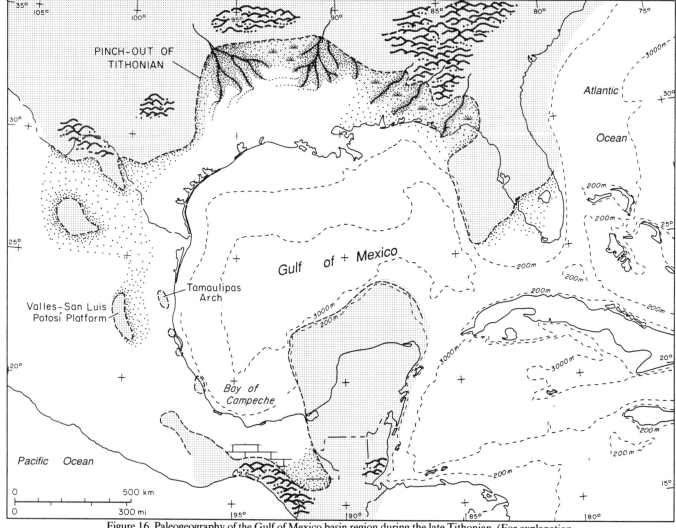

Figure 16. Paleogeography of the Gulf of Mexico basin region during the late Tithonian. (For explanation of patterns, see Figure 12.)

ico of the time advanced over the surrounding lands and covered most of the islands still remaining around the periphery of the gulf. During the latest Tithonian or earliest Cretaceous, marine waters also may have covered for the first time part of the southern Florida platform, and evidence exists that the Gulf of Mexico was at this time in communication with the Atlantic Ocean. The entire Yucatan platform, however, remained above sea level.

Locally, along the updip parts of the northern flank of the Gulf of Mexico basin, an unconformity has been recognized between the Kimmeridgian and the overlying Tithonian. Basinward, however, the unconformity dies out and the boundary becomes conformable and often transitional. This unconformity has been interpreted as having been the result of a eustatic lowering and subsequent rise of the sea level, but tectonic causes cannot be discarded as the unconformity has not been reported from other regions of the basin.

During the Tithonian, and continuing into the early part of the Early Cretaceous, the northern flank of the Gulf of Mexico basin witnessed a great influx of coarse terrigenous clastic sediments contributed by several major streams draining the Appalachian and Ouachita mountains and the continental interior of the United States. The delta of the ancestral Mississippi River can also be recognized in the Tithonian section of northeastern Louisiana and west-central Mississippi. Other important and complex deltas were present in northeast Texas (Fig. 16). The thick wedges of terrigenous clastic sediments deposited in these deltas were worked over by waves and redistributed along the coast by east-to-west longshore currents. This increase in the influx of terrigenous clastic sediments may reflect a regional uplift and/or a climatic change in the continental interior to the north that resulted in the increase of the load of the streams draining the region.

The nonmarine fluvial and deltaic environments of the northern periphery of the basin are replaced southward by an extensive shallow shelf bordered by lagoons, barrier beaches, bars, and barrier islands. East and west of the Sabine uplift, deeper marine depocenters were present, in which thick shale sections (Bossier Formation) were deposited.

The Tithonian transgression is also reflected in the sedimentary section of the northwestern flank of the Gulf of Mexico basin. The sea covered most of the islands which previously bordered the coast and advanced over the bordering land areas. The Tamaulipas arch was almost completely covered by the sea, as was the Valles–San Luis Potosí platform. Marine basinal shales characterize the more basinward Tithonian of northeastern Mexico, but landward the section becomes increasingly sandy, reflecting proximity to the sources of terrigenous clastic sediments.

In marked contrast to the thick Tithonian section of the northern and northwestern Gulf of Mexico basin, predominantly composed of coarse terrigenous clastics, the Tithonian is represented in the western and southern parts of the basin by a much thinner section of shales, calcareous shales, and argillaceous limestones. A broad marine embayment remained in existence in southern Mexico during the Tithonian. In the Bay of Campeche and Reforma areas, the Tithonian is composed of open-marine shales, which grade to shallow-water carbonates and then to nonmarine red beds toward the Chiapas massif and northwestern Guatemala.

The occurrence of numerous beds of bentonite, bentonitic shale, and chert in beds, lenses and nodules in the Tithonian section of east-central and southern Mexico indicates the presence of volcanic activity at the time, probably along the Pacific margin of Mexico. The occurrence near Ixtapan de la Sal, 90 km south-southwest of Mexico City, of Tithonian ammonites and tintinnids in slightly metamorphosed shales interbedded with lava flows confirms this interpretation. This late Jurassic volcanic activity may have restricted temporarily the connection between the Gulf of Mexico and the Pacific Ocean.

By the mid-Late Jurassic, and certainly by the end of the Jurassic, the basic stratigraphic, structural, and geographic elements of the Gulf of Mexico basin, as we know them today, were established. A deep central area was surrounded by shallow stable shelves, the Gulf of Mexico of the time was in communication with the Atlantic Ocean through a passage between the Florida and Yucatan platforms, the major parts of which were still above sea level. The connection with the Pacific Ocean persisted, though it may have been somewhat restricted by extensive volcanic activity along the Pacific margin of Mexico.

Since the end of the Jurassic, during the Cretaceous and Cenozoic, the stratigraphic and structural framework of the basin has undergone some modifications but has not been notably changed. The basin, particularly its central part, subsided persistently, and structural deformation was restricted to local flow of the Middle Jurassic salt and to displacement along listric normal (growth) faults around the periphery of sedimentary depocenters (see Chapter 5, this volume). Only the Laramide orogeny, active

during the latest Cretaceous and early Tertiary, involved the western flank of the basin in an episode of compressional deformation that produced the Sierra Madre Oriental of eastern Mexico.

The geography of the region has changed to a much greater extent. The sea has advanced at times, covering much of the Gulf of Mexico basin region, and has even extended beyond it. At other times, the sea has retreated and much of the stable shelves surrounding the central, deep part of the basin have been exposed. But the position and relation of the deep and the shelves, the main areas of influx of terrigenous clastic sediments, and the location and character of the Florida and Yucatan stable carbonate platforms have not changed appreciably.

EARLY CRETACEOUS—CONTINUED STABILITY

The Early Cretaceous was a time of remarkable tectonic stability in the Gulf of Mexico basin. Subsidence continued throughout the basin, but was most active in its central part and in the interior salt basins along the northern flank (East Texas basin, North Louisiana salt basin, and Mississippi salt basin). Winker and Buffler (1988) have estimated that by the end of the Early Cretaceous, the central part of the basin had reached a depth of 4.2 to 4.7 km below sea level. Apart from this general subsidence, tectonic activity was restricted to deformation of the Middle Jurassic salt, particularly in the areas of thicker salt in the northern interior salt basins, and to listric normal (growth) faulting around the rims of depositional centers and along progradational shelf margins, also in the northern flank of the basin. Best documented of these growth-fault trends are the Mexia, Talco, State Line, Pickens, Quitman, Gilbertown, and Pollard fault zones, some of which were initiated during the Late Jurassic (see Plate 2). The growth faults soled into detachment surfaces in shaly sequences or in the Middle Jurassic salt, and are often associated with downdip antithetic faults, which created distinctive symmetrical grabens (see Chapters 3 and 5, this volume).

The deep central part of the basin was bordered by stable shelves and ramps, broad to the north and northwest, but much narrower to the west.

Although the Early Cretaceous was a time of very limited tectonic deformation, a number of positive and negative topographic features, inherited from Jurassic and earlier tectonic episodes, were present in the basin, and influenced to some extent the Early Cretaceous sedimentary processes and paleogeography. Among the most prominent are the ridges projecting from the southern Appalachian Mountains and the Peninsular arch in the northeast, the Coahuila and Valles–San Luis Potosí platforms to the west, and the Chiapas massif and the Yucatan platform to the southeast.

During the Early Cretaceous, the area of the present Gulf of Mexico basin was part of a broad seaway that extended from the Atlantic Ocean on the east, over most of the Florida platform, to the Pacific Ocean on the west. Mexico became almost entirely flooded. Only the Coahuila and Yucatan platforms remained above sea level during the early part of the Early Cretaceous, but they also were covered by shallow water by mid-Early Cretaceous time—the Yucatan platform during the late Aptian or early Albian, and the Coahuila platform during the early Albian. By Albian time the seaway reached its maximum Early Cretaceous extent and advanced toward the northwest to establish communication with the Western Interior Seaway.

The main influx of terrigenous clastic sediments into the Gulf of Mexico basin during the Early Cretaceous, as during the Late Jurassic, was from the northeast, north, and northwest, brought in by streams draining the Ouachita and Appalachian mountains and the North American continental interior. Particularly during the early part of the Early Cretaceous, it resulted in a high rate of clastic sedimentation along the northeastern, northern, and northwestern flanks of the basin. The influx of terrigenous clastic sediments decreased progressively and by late Barremian time clastic deposition was much reduced and limited to the northernmost periphery of the basin. Away from the northern sources of terrigenous clastic sediments, carbonate deposition prevailed over the western and southern flanks of the basin during the early part of the Early Cretaceous.

During the late Early Cretaceous, as the influx of terrigenous clastic sediments decreased, the stable shelves, ramps, and platforms bordering the deep central part of the Gulf of Mexico basin became the site of widespread carbonate deposition, particularly during the Albian. Carbonate shelves and platforms extended at this time from the Bahamas region, through the Florida platform, and along the northern Gulf of Mexico basin to northeastern Mexico. Along the western flank of the basin, in east-central Mexico, carbonate platforms were restricted to more local developments—the Valles–San Luis Potosí, Tuxpan, Córdoba, and other smaller platforms that, unlike the northern platforms and shelves, remained active until the earliest Late Cretaceous (Cenomanian). A large area of carbonate and evaporite deposition was also present over the Yucatan platform.

The carbonate shelves and platforms were often fringed along their basinal margins by high-energy shoals, rudist-dominated reefal buildups, and barrier islands interrupted by tidal channels and passes. Behind them, banks, patch reefs, and occasional evaporites were often formed in intrashelf basins and back-reef lagoons. The reefal buildups and shoals were most distinctly developed where the shelf and platform margins faced the open sea and were exposed to high-energy conditions. In the sheltered embayments of the coast, as in the Sabinas basin and the Isthmus of Tehuantepec region, low-energy environments favored the development of gentle ramps devoid of barrier reefs and shoals.

The Early Cretaceous shelf-margin reefal buildups do not crop out along the northern Gulf of Mexico basin, and knowledge about them has been based on subsurface and seismic information. The reefal buildups and associated back-reef and basinal sections crop out in northeastern and east-central Mexico, where they have been studied in detail; extensive literature on the subject is available.

Several stratigrahic breaks have been recognized in the Lower Cretaceous section of the northern Gulf of Mexico basin. The earliest is represented by a marked unconformity at a level within the Valanginian. Scott and others (1988) summarized evidence for two other breaks: an "intra-Aptian" break at the contact between the Sligo Formation and the overlying shales of the Pearsall Formation, and an "intra-Albian" break at the contact between the Fredericksburg and Washita groups. Breaks contemporaneous with those in the northern region have not been reported from the rest of the basin, and it is not possible to conclude if the breaks are not present outside of the northern part of the basin, or if insufficient information has made it difficult to recognize them. The reduced thickness of the lower part of the Lower Creteaceous section (Berriasian to Aptian) in some carbonate platforms of east-central and southeastern Mexico may conceal unconformities within the section. If unconformities are indeed present in the platforms, they should die out toward the more basinal areas surrounding them. Several factors could have led to the development of the stratigraphic breaks: changes in the sea level, changes in the rate of subsidence of the basin, uplift of bordering land areas, fluctuations in the influx of terrigenous clastic sediments, or a combination of two or more of these factors. It is not always possible to establish which of these factors was most influential in the generation of any one break.

The "intra-Aptian" stratigraphic break, at the base of the Pearsall, and its equivalents in Mexico, the La Peña and Otates formations, has been extensively used in the northwestern Gulf of Mexico basin region to subdivide the Lower Cretaceous section into two units: the "Coahuilan Series" below the break, and the "Comanchean Series" above. While useful in the northwestern region where the break emphasizes an important event of the Early Cretaceous history of the region, this subdivision is not as readily applicable to other regions of the basin. For this reason, the subdivision of the Lower Cretaceous into Coahuilan and Comanchean "series" will not be utilized in the present regional synthesis of the development of the entire Gulf of Mexico basin.

During the Early Cretaceous, as during the Late Jurassic, and as during the remaining history of the Gulf of Mexico basin, three distinct stratigraphic and tectonic provinces can be recognized around the periphery of the basin, surrounding the deep central region.

1. To the northeast, north, and northwest, the early Early Cretaceous was characterized by the persistent influx of terrigenous coarse clastic sediments, and by the widespread development of broad carbonate shelves with low-relief margins during the late Early Cretaceous. At this later time, the influx of terrigenous clastic sediments was much reduced, particularly to the northwest. Several stratigraphic breaks have been recognized in this province.

2. To the west and southwest, the influx of coarse clastics was minimal and the region was occupied by several isolated platforms covered by shallow water and separated by deeper-water basins and straits. The platforms had high-relief margins along which great reef complexes were built. Shallow-water carbonates, and evaporites were deposited on the platforms. No

distinct stratigraphic breaks have been recognized in this province.

3. The third province, to the east and southeast, is occupied by the stable Florida and Yucatan platforms, emergent until the Early Cretaceous, and in which the Lower Cretaceous section is predominantly composed of shallow-water carbonates and evaporites.

The northern and northwestern province

Of the three stratigraphic-tectonic provinces, the northern and northwestern Gulf of Mexico basin province has the most eventful history. Conditions in this province during the earliest part of the Early Cretaceous (Berriasian and early Valanginian) did not change appreciably from those which had prevailed during the late Late Jurassic. Abundant influx of coarse terrigenous clastic sediments prograding from the north (upper part of the Cotton Valley Group) continued uninterrupted in nearshore areas. In northeastern Mexico, conglomerates and sandstones derived from the southern part of the Coahuila platform and the El Burro uplift were deposited around these emergent positive features. An embayment of the Gulf of Mexico of the time—the Sabinas basin—extended northwestward toward Chihuahua between the Coahuila platform and the North American mainland. Basinward, along shelfal areas, the clastic sections graded to shallow-water carbonates, and beyond to basinal carbonate mudstones and limestones. During the Berriasian and early Valanginian the shelves and platforms were narrower than those that would develop later in the Early Cretaceous.

A marked unconformity has been recognized along the northern Gulf of Mexico basin within the lower part of the Lower Cretaceous section. In the subsurface of the northern part of this region, the coarse-clastic Hosston section overlies unconformably the similar coarse-clastic sediments of the Cretaceous upper part of the Cotton Valley Group. The unconformity omits part or all of the Valanginian section in updip areas, but dies out in a downdip basinward direction, where seismic profiles show an apparently conformable and complete Lower Cretaceous section. This unconformity has not been reported from other areas of the basin, suggesting that it may be the result of gentle uplift of the North American cratonic region. Future detailed studies may uncover evidence for this unconformity in other areas of the Gulf of Mexico basin.

The supply of terrigenous clastic sediments from the northeast, north, and northwest continued during the Hauterivian, the Barremian, and the early part of the Aptian, but at a progressively reduced rate (Fig. 17). A clastic sequence (Hosston Formation) accumulated along the northern Gulf of Mexico basin over the eroded surface of the Berriasian section as the sea transgressed the coastal plain. The Hosston section was deposited in alluvial-valley, deltaic, interdeltaic, and coastal-plain environments. It graded basinward to, and was overlain by, a carbonate section (Sligo and Cupido formations), which developed a broad shelf fringed, particularly in early Aptian time, by well-developed rudist-dominated reefs (Sligo and Cupido reefs).

In northeastern Mexico, the Coahuila platform remained above sea level from Hauterivian to early Aptian time, and continued to provide terrigenous clastic sediments to the area surrounding it. Formation of shallow-water carbonates persisted in the Sabinas basin interrupted only by an episode of predominantly evaporite deposition (La Virgen Formation) during the Barremian (Fig. 17). Contribution of terrigenous clastic sediments to the northeastern Mexico region greatly diminished after the Hauterivian, but persisted farther northeast along the northern flank of the basin until the earliest Aptian.

During most of the early Aptian, carbonate deposition predominated over the northern and northwestern Gulf of Mexico basin. The shelf carbonates and the shelf-margin reefs, banks, and shoals of the Sligo and Cupido formations attained their maximum development at this time. The shelf margin has been shown to have prograded through time and to have maintained a gentle basinward slope. A well-defined shelf margin and related organic buildups have been recognized at the surface, in wells, and in seismic profiles from near Monterrey in northeastern Mexico to southern Louisiana; it probably extended the length of the present Florida Escarpment. Southwest of Monterrey, the location of the shelf and its margin is uncertain, and subject to several interpretations (see Smith, 1981, for a summary). Basinward from the organic banks and reefs that characterized the margins of the carbonate platforms, carbonate mudstones and limestones were deposited in open-marine, deeper-water environments.

The reasons for the localization of the platform margins and the development along them of the linear reefal buildups are not clearly understood. The location of the margins may have been determined by a change in the slope of the basement at a crustal hinge zone separating less-subsided, thicker transitional crust from more-subsided, thinner transitional crust (see Chapter 4, this volume).

In late Aptian time, large volumes of fine-grained terrigenous material (the Pearsall or Pine Island shales of the northern part of the basin, and the La Peña and Otates shale units of northeastern and east-central Mexico) entered the northern and northwestern Gulf of Mexico basin, probably as a result of a gentle uplift of the cratonic land areas to the north and west. The muddying of the water resulting from this sudden supply of fine-grained terrigenous material terminated the deposition of platform carbonates and may explain the demise of the Sligo-Cupido reefs. The Pearsall, La Peña and Otates shale section has been interpreted to represent a landward encroachment of deeper-water environments over the shallow-water Sligo-Cupido carbonate platform, which may have been due to an increase in the rate of subsidence of the basin and/or a widespread rise in sea-level during the late Aptian (Scott and others, 1988). Scott and others (1988) interpreted a hiatus of 0.5 m.y. at the Sligo-Pearsall contact.

After the deposition of the upper Aptian shales, the water cleared during the Albian and carbonate deposition resumed over the extensive shelf in the northern and northwestern Gulf of Mexico basin (Fig. 18). The margin of this shelf, as that of the

Aptian shelf, prograded basinward a considerable distance. It corresponded approximately with that of the underlying Aptian shelf except in south Texas and northeastern Mexico, where it was located farther landward. Rudist reefs and organic banks (Stuart City reef) developed along the margin of the shelf, and allochthonous breccias formed in front of the reefs. Seaward, toward the deeper parts of the basin, deposition of deep-water carbonate mudstones and limestones continued. In the early Albian, a thin but widespread evaporitic unit—the Ferry Lake Anhydrite—was deposited over much of the northern shelf, indicating the temporary development of a restricted-shelf environment.

Influx of terrigenous clastic sediments was most abundant along the northeastern part of the basin, generally restricted to its updip regions. Some coarse-clastic sediments reached considerable distances basinward along the northeastern and northwestern parts of the basin (for example, the Paluxy sandstones).

During the early Albian the Coahuila platform was covered by the sea and became the site of deposition of carbonates and evaporites in its center while rudist banks formed around its rim; in the late Albian, deposition of shallow-water limestones prevailed over the platform. Deeper-water limestones and carbonate mudstones were deposited in the Sabinas basin, between the Coahuila platform and the Albian carbonate platform to the northeast. The margin of this latter platform, fringed by the Stuart City reef, turned from a general southwestern trend in south Texas to a more westerly direction in northeastern Mexico (Fig. 18). Starting in mid-Albian time, a roughly circular evaporitic basin—the Maverick basin—developed behind the Stuart City reef; it lasted until the end of the Albian and accumulated a sequence of evaporites and limestones (McKnight Formation) and overlying carbonate mudstones (Salmon Peak Formation).

The termination of Albian carbonate deposition along the northern Gulf of Mexico basin and the death of the Stuart City reefs in much of this area was also due to the influx into the basin during the earliest Cenomanian of the terrigenous, fine-grained

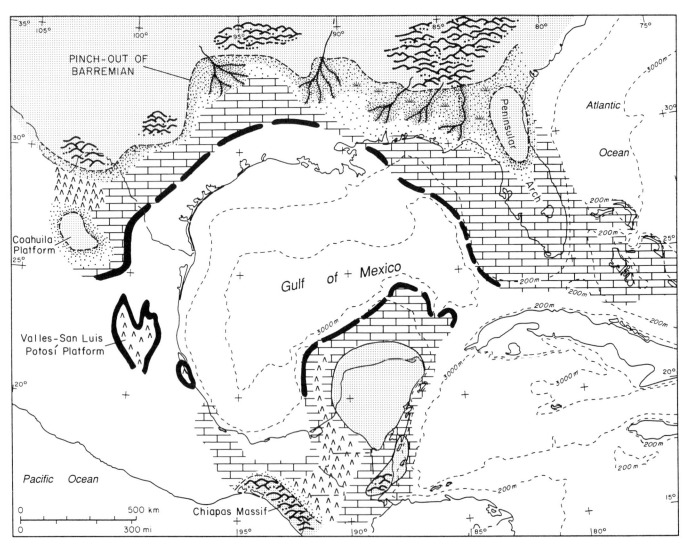

Figure 17. Paleogeography of the Gulf of Mexico basin region during the Barremian. (For explanation of patterns, see Figure 12.)

clastic sediments of the Del Rio and Grayson formations. This influx of mud was not as strong or as widespread as the late Aptian Pearsall–La Peña–Otates pulse, and as a result, the close of shelf-carbonate deposition and end of the building of the Stuart City reefs may not have been simultaneous everywhere. Cenomanian events will be discussed in more detail in the next section of this chapter.

The western and southwestern province

The western and southwestern province in east-central and southern Mexico had a much less eventful geologic history. Throughout the Early Cretaceous, the paleogeography of the area was characterized by the presence of isolated platforms covered by shallow water in which carbonates and occasional evaporites were deposited—the Valles–San Luis Potosí, Tuxpan, Córdoba, and some other smaller platforms (Figs. 17 and 18). Deeper-

water basins, straits, and channels separated the platforms. In these basins, the Lower Cretaceous section is composed predominantly of carbonate mudstones and argillaceous carbonates. Bentonites, tuffs, and occasional lavas in the lower part of the Lower Cretaceous of east-central and southern Mexico reflect continued volcanism probably along the present Pacific margin of Mexico.

Carbonate deposition in the platforms was slow during the early part of the Early Cretaceous (Berriasian to Aptian) but accelerated considerably during the Albian. In contrast to the low-relief margins of the shelves in the northern Gulf of Mexico basin, the platforms in the western and southwestern part of the basin were bordered during the Albian and early Cenomanian by high-relief margins along which great rudist reefs encircled the platform interior. The best understood and most extensively studied and described are the reefs fringing the Valles–San Luis Potosí and Tuxpan platforms. Those that may have developed on the Córdoba platform are poorly known, because the platform was strongly deformed during the Laramide orogeny, and its eastern

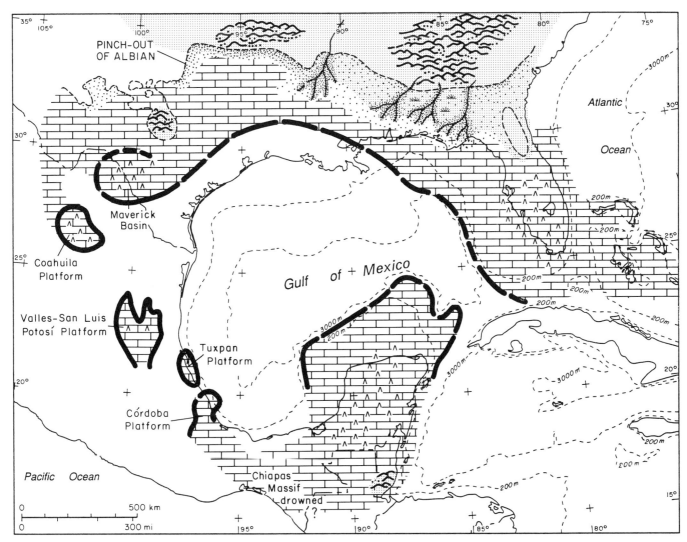

Figure 18. Paleogeography of the Gulf of Mexico basin region during the Albian. (For explanation of patterns, see Figure 12.)

margin was buried below the thick Cenozoic section of the Vera-cruz basin.

The reefal buildups around the margins of the Valles–San Luis Potosí and Tuxpan platforms—the El Abra Formation—are Albian and early Cenomanian in age; they grade toward the center of the platform to backreef and lagoonal sections composed predominantly of miliolid limestones, dolomites, and occasional evaporites. The carbonate mudstones and limestones deposited in the deeper-water basins surrounding the platforms are included in the Tamaulipas Superior (Upper Tamaulipas) Formation. As a result of the rapid vertical aggradation of their margins, the flanks of the platforms had slopes of as much as 20° to 45°, and a relief of up to 1000 m by the end of the Albian. Skeletal-fragment limestones, breccias, turbidite, and debris-flow deposits—the Tamabra Formation—were often deposited along the lower part of the foreslopes bordering the platforms.

The Albian–lower Cenomanian carbonate complexes of the Valles–San Luis Potosí and Tuxpan platforms are overlain unconformably by a variety of younger units ranging in age from Late Cretaceous to Oligocene. The El Abra reef was exposed to subaerial erosion and developed extensive caverns and karst features, particularly along the northern part of the platform-margin reef trend of the Tuxpan platform. Both the El Abra reefs and the forereef deposits of the Tamabra Formation have provided the reservoirs for important accumulations of oil—the Golden Lane (Faja de Oro) and Poza Rica fields (see Chapter 15, this volume).

In southern Mexico, a broad embayment was present between the Chiapas massif and the Yucatan platform during the Early Cretaceous. (Figs. 17 and 18). The lower part of the Lower Cretaceous section of this embayment is composed of basinal argillaceous carbonates to the north, which grade southward to shallower-water carbonates (Zacatera Group) and then to nonmarine red beds (Todos Santos Formation). The upper Lower Cretaceous (Albian) is represented by the lower part of the massive shelf carbonates of the Sierra Madre Group. Toward the Reforma area to the north, they grade to open-marine, deeper-water argillaceous limestones.

The Florida and Yucatan platforms

The Florida and Yucatan platforms remained stable during the Early Cretaceous, when they were progressively covered by the sea. The marine transgression over the Florida platform appears to have started in the south during the Late Jurassic and to have progressed northward with time. The Lower Cretaceous section thickens southward and gulfward and pinches out toward the Peninsular arch. At the end of the Early Cretaceous, only the northern, highest part of the arch, in northern Florida, was still above sea level (Figs. 17 and 18).

The transgression over the Yucatan platform did not start until later, during the late Aptian or early Albian, perhaps as a result of the increase in the rate of subsidence of the whole basin mentioned above. By the end of the Early Cretaceous the platform was completely covered by shallow-marine waters. The two submerged shallow platforms were separated by a deeper-water

channel, and the Yucatan platform was bordered to the southwest during mid-Early Cretaceous time by a shallow evaporitic embayment, the successor of a more marine Late Jurassic embayment.

Deposition of shallow-water carbonates and evaporites prevailed in the Florida and Yucatan platforms during the Early Cretaceous. The supply of terrigenous clastic sediments to the Florida platform was restricted to its northern part, in the areas surrounding the emergent part of the Peninsular arch; no terrigenous clastic rocks are present in the Lower Cretaceous section of the Yucatan platform.

No major breaks have been reported within the Lower Cretaceous sequence of the Florida and Yucatan platforms. However, if some of the stratigraphic hiatuses recognized in the Lower Cretaceous section along the northern and northwestern Gulf of Mexico basin, for example, the Valanginian unconformity, are indeed due to a lowering of the sea level, these breaks should be clearly reflected in the shallow-water stratigraphic section of the Florida and Yucatan carbonate platforms. More careful study may bring them to light.

During most of the Early Cretaceous, the margins of the Florida and Yucatan platforms facing the deep central part of the Gulf of Mexico basin were fringed by reefs predominantly built by rudists (Figs. 17 and 18). They are represented at present by steep submarine escarpments—the Florida and Campeche Escarpments—with relief from 1500 to 2000 m. Along the southern part of the Florida Escarpment, the reefal buildups seem to have been eroded by submarine currents, resulting in a retreat of the escarpment of perhaps as much as 5 to 10 km. Dredging along the escarpment did not encounter platform-margin reefal buildups, but recovered instead miliolid limestones and dolomites indicative of shallow-marine, low-energy, platform-interior environments (Freeman-Lynde, 1983). Rudist-reef limestones, however, have been reported from the breccia talus at the base of the escarpment.

LATE CRETACEOUS—THE GULF OF MEXICO BASIN STIRS

The tectonic stability that prevailed in the Gulf of Mexico basin during the Late Jurassic and Early Cretaceous was broken during the Late Cretaceous. Early in the Late Cretaceous an important event, or events, took place in the basin that produced a widespread stratigraphic break, an intra-Cenomanian unconformity, that has been recognized throughout most of the periphery of the basin. The unconformity is most evident in the northern Gulf of Mexico basin, along the gulfward margins of the Florida and Yucatan carbonate platforms, and in the Valles–San Luis Potosí and Tuxpan platforms. It represents a profound change in the depositional regime over most of the Gulf of Mexico basin. Because of its distinctness, the mid-Cenomanian stratigraphic break has been used in the northern Gulf of Mexico basin as the boundary between the "Lower Cretaceous" and "Upper Cretaceous," even though it does not correspond with the boundary between the two Cretaceous series that has been established internationally at the base of the Cenomanian. The break has also been used as the top of the provincial "Comanchean Series" and the base of the "Gulfian Series," which includes the rest of the Upper Cretaceous section.

The widespread distribution of the unconformity, from the north flank of the basin to the Florida and Yucatan platforms, and over provinces with contrasting stratigraphic and structural histories, suggests a major lowering of the sea level during the mid-Cenomanian in the region. This interpretation is reinforced by the recognition in many other regions of the world of a concurrent sea-level fall (Vail and others, 1977, and many subsequent publications).

The lowering of the sea level exposed the shallow shelves and platforms around the rims of the basin to subaerial erosion, virtually to their margins, but the central part of the basin as well as the intra-shelf basins and embayments around its periphery— the central part of the present East Texas basin, the Rio Grande embayment, the Sabinas, Tampico-Misantla, and Veracruz basins, and most of the basins in southern Mexico—remained covered by the sea. In these basinal areas, no unconformity is present, and deposition was continuous through the Cenomanian into the Turonian. In the East Texas basin, for example, the mid-Cenomanian unconformity is well developed along its flanks, but in the deep, central part of the basin the Cenomanian-Turonian section seems to be continuous and conformable—the Buda Formation is overlain conformably by the Maness Shale and this unit, in turn, is overlain conformably by the Woodbine Formation. Southern Mississippi may also have been below sea level at the time—the Dantzler Formation overlies the undifferentiated Fredericksburg-Washita section and underlies the Tuscaloosa, both conformably; the Cenomanian-Turonian sequence also in this area appears to lack any stratigraphic breaks (Figs. 19 and 20).

In the Burgos and Sabinas basins of northeastern Mexico, the boundary between the Cuesta del Cura Formation, of late Albian and early Cenomanian age, and the overlying Agua Nueva Formation, of mid-Cenomanian to Turonian age, is also clearly conformable and transitional.

The mid-Cenomanian unconformity also dies out toward the deep, central part of the Gulf of Mexico basin. Interpretation of seismic reflection profiles has made it possible to trace the unconformity basinward from the Campeche and Florida Escarpments, and from the northeastern part of the basin into a perceptibly conformable sequence (Faust, 1990; Chapter 13, this volume). The position within this conformable sequence that has been interpreted to correlate to the level at which the unconformity fades out, corresponds to a prominent high-amplitude reflector called the "mid-Cretaceous sequence boundary" or "MCSB" (Chapter 13, this volume). At this reflector there is a substantial downward increase in refraction velocity that is believed to reflect a sharp lithologic change, from a sequence composed of fine-grained terrigenous clastics above to a predominantly carbonate section below.

In the north-central Gulf of Mexico basin other factors must

have contributed to the development of the mid-Cenomanian unconformity. In this area (northeast Texas, northern Louisiana, Arkansas, and northwestern Mississippi), the Upper Cretaceous section above the unconformity onlaps the lowermost Upper Cretaceous (lower Cenomanian) and the Lower Cretaceous, and covers unconformably the eroded and tilted edges of Jurassic, Triassic, and Paleozoic rocks (Fig. 20). Regional uplift, tilting, and erosion, locally caused or enhanced by igneous intrusions and volcanism in the Mississippi embayment area, must have taken place before the deposition of the Upper Cretaceous sequence above the unconformity.

Extensive igneous activity is known to have taken place in the Mississippi embayment area during the early Late Cretaceous (100–80 Ma; Cenomanian to early Campanian). The Sabine and Monroe uplifts and the Jackson dome were formed or reactivated during this time, probably as the result of the emplacement of igneous intrusions. This igneous activity may have been responsible for at least part of the uplift of a broad area of the Mississippi embayment (which has been called the "Southern Arkansas uplift"; see Chapter 3, this volume), during mid-Cenomanian time, and for the enhancement of the mid-Cenomanian unconformity in that area.

This important intra-Cenomanian erosional episode effectively terminated the widespread shelf- and platform-carbonate deposition that had characterized the Early Cretaceous in the Gulf of Mexico basin. Platform-carbonate deposition continued during the Late Cretaceous only in parts of east-central and southern Mexico, principally in the Valles–San Luis Potosí and Córdoba platforms, as well as in the Florida and Yucatan platforms.

Early Late Cretaceous magmatic activity has been reported not only from the Mississippi embayment region, but also from the subsurface of the northern part of the Yucatan Peninsula. In addition, late Late Cretaceous volcanism (87–70 Ma; Santonian to mid-Maastrichtian) is known along the Balcones fault zone in central and south Texas, and from offshore Louisiana (see Chapter 6, this volume).

The most important events to affect the Gulf of Mexico basin during the Late Cretaceous were the initial episodes of the Laramide orogeny that are recorded in the upper part of the Upper Cretaceous section along the western and northwestern flanks of the basin in east-central and northeastern Mexico. The Laramide orogeny had a profound influence not only on the structural configuration, but also on the stratigraphy of the basin.

Subsidence of the central part of the Gulf of Mexico basin continued during the Late Cretaceous, as did local and intermittent deformation of the Middle Jurassic salt, and growth faulting. Shelves along the northern part of the basin were generally less well developed during the Late Cretaceous than they had been during the Early Cretaceous. Their margins roughly correspond with those of the Early Cretaceous, but they were devoid of rudist reefal buildups. The western platforms were no longer marked by steep, high margins, and the Florida and Yucatan carbonate platforms were also less distinct and somewhat reduced in size.

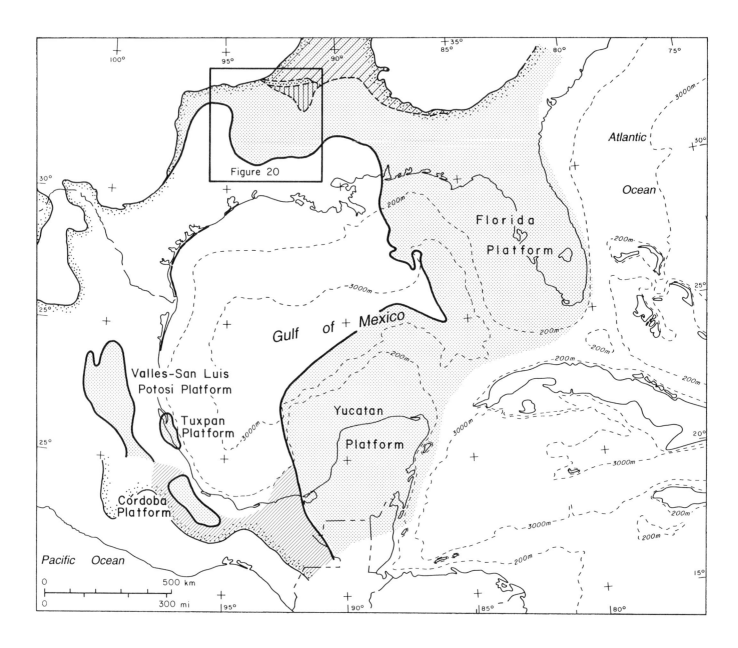

Distribution of the mid-Cenomanian unconformity

Subcrop below mid-Cenomanian unconformity

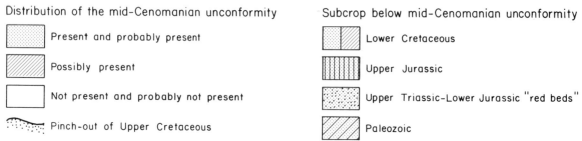

Figure 19. Distribution of the mid-Cenomanian unconformity in the Gulf of Mexico basin region, and subcrop below it.

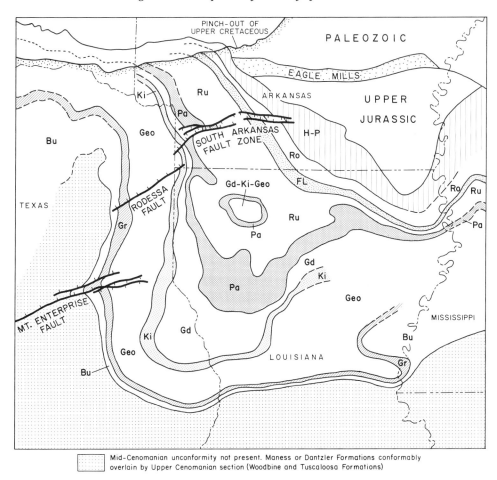

SUBCROP BELOW MID-CENOMANIAN UNCONFORMITY

WASHITA GROUP		FREDERICKSBURG GROUP			
Bu	Buda Formation	Ki	Kiamichi Formation	Ru	Rusk Formation
Gr	Grayson Formation	Gd	Goodland Formation	FL	Ferry Lake Anhydrite
Geo	Georgetown Formation	Pa	Paluxy Formation	Ro	Rodessa Formation
				H-P	Hosston and Pearsall Formations

Figure 20. Subcrop map below mid-Cenomanian unconformity, northeast Texas, southern Arkansas, and north Louisiana.

The lowstand of the sea during the Cenomanian considerably reduced the size of the extensive waterway that had covered the Gulf of Mexico basin region during the late Early Cretaceous and probably severed its connection with the epicontinental Western Interior Seaway. The sea-level fall was followed by a major marine transgression during Turonian to Campanian time that covered the areas exposed during the Cenomanian lowstand and reestablished communication of the Gulf of Mexico with the Western Interior Seaway. This communication finally vanished during the latest Late Cretaceous (Maastrichtian) after the onset of the Laramide orogeny.

The Upper Cretaceous section is strongly overprinted by the cyclic sea-level fluctuations that took place during the Late Cretaceous. Stratigraphic hiatuses and unconformities in the section are accentuated around the margins of the Gulf of Mexico basin, in the shallow, low-relief shelves, particularly along the northern part of the basin, and in the platforms bordering it; they diminish and disappear basinward. Some of the lesser and more local hiatuses and unconformities are undoubtedly not related to sea-level changes, but rather to local tectonic or igneous episodes.

During the latest Late Cretaceous, northwestern and western sources of terrigenous clastic sediments, also the product of the

Laramide orogeny, were added to the northern sources (the Appalachian and Ouachita Mountains and the continental interior) which had provided clastic sediments during the early part of the Late Cretaceous. Pelagic sedimentation continued uninterrupted during the entire Late Cretaceous in the deep, central part of the basin. Seismic reflection and refraction data have been interpreted to indicate that during the early-Late Cretaceous (mid-Cenomanian) the nature of the sedimentary section changed from carbonates to fine-grained terrigenous clastics.

Rocks of Late Cretaceous age cover the entire Gulf of Mexico basin and crop out around most of its rims, except in the Florida and Yucatan platforms. The three distinct stratigraphic and tectonic provinces that characterized the Gulf of Mexico basin during the Late Jurassic and Early Cretaceous can still be recognized in the Late Cretaceous record, although they are somewhat modified.

1. To the north and northwest, the Upper Cretaceous section is composed predominantly of coarse terrigenous clastics in its lower part, chalk and chalky marls in its middle part, and shales and calcareous shales in its upper part. Terrigenous material is an important component of the upper part of the Upper Cretaceous section in south Texas and northeastern Mexico.

In addition to the major mid-Cenomanian unconformity, other lesser breaks of only local significance have been recognized in the Upper Cretaceous section of the northern Gulf of Mexico basin. They are believed to indicate uplift of local tectonic features.

2. Along the western flank of the Gulf of Mexico basin, the lower part of the Upper Cretaceous section is composed predominantly of shales, calcareous shales, and argillaceous limestones. An influx of terrigenous clastic sediments started in the Campanian, and continued through the latest Late Cretaceous. The source of these terrigenous sediments was apparently a landmass in central Mexico, which during the Late Cretaceous progressively increased in size and expanded toward the east. The mid-Cenomanian unconformity has been recognized in the Valles–San Luis Potosí and Tuxpan platforms, but almost everywhere else the Upper Cretaceous section appears to be continuous and conformable.

3. Carbonate and evaporite deposition continued in the stable Florida and Yucatan platforms. The mid-Cenomanian unconformity has been clearly recognized along the margins of the platforms facing the Gulf of Mexico, but not within the platform interior, where, if the unconformity is due to a major lowering of the sea level, it is probably represented by a disconformity or very low angle unconformity difficult to identify in a nearly horizontal section, which is composed of alternating limestones and evaporites, both above and below the stratigraphic break.

Notwithstanding the marked continuity in the general nature of the stratigraphic and structural framework of the Gulf of Mexico basin from the Early Cretaceous to the Late Cretaceous, some definite changes were brought about by the mid-Cenomanian events: the broad carbonate shelves and platforms fringed by prominent rudist reefs and organic banks, that charac-

terized much of the Early Cretaceous, were greatly reduced, and the Late Cretaceous Gulf of Mexico advanced far inland toward the north along the Mississippi embayment after the Turonian. Toward the end of the Late Cretaceous, the initiation of the Laramide orogeny contributed to still more changes, principally by providing a western source of terrigenous clastic sediments and by reducing the extent of the connection between the Gulf of Mexico and the Pacific Ocean.

By the end of the Late Cretaceous, therefore, the geography of the Gulf of Mexico basin region, while still preserving the three distinctive stratigraphic-tectonic provinces recognized before, had undergone distinct modifications.

The northern and northwestern province

In the earliest Late Cretaceous (early Cenomanian), a strong influx of fine-grained terrigenous material entered the northern and northwestern Gulf of Mexico basin from the north and northwest. The clay muddied the sea, smothered platform-carbonate deposition, and killed the fringing Stuart City reefs. It deposited a thin but widespread shale unit, the Del Rio and Grayson formations, excellent lithologic markers that have been recognized from northeastern Mexico to north Louisiana. The lower Cenomanian section of the northeastern Gulf of Mexico basin (Mississippi, Alabama, and the western Florida panhandle) is composed predominantly of fine-grained clastics interbedded with some sandstones and limestones.

The deposition of the Del Rio and Grayson shale section may represent a deepening of the northern shelf of the basin, and, as a consequence, a marine transgression over the shelf. It was followed by the deposition of a thin and widespread marine limestone unit, the Buda Formation. The wide distribution and remarkably constant small thickness of the Del Rio, Grayson, and Buda formations attest to their deposition under very stable conditions over a broad shelf that bordered to the north and northwest the deep center of the Gulf of Mexico basin—characteristic conditions that prevailed during the Early Cretaceous and continued uninterrupted during the earliest Late Cretaceous. The boundary between the Lower Cretaceous and the Upper Cretaceous appears to be conformable over most of the Gulf of Mexico basin. The lower Cenomanian section represents the last episode of the long Early Cretaceous cycle of stable deposition.

In mid-Cenomanian time, these stable stratigraphic and tectonic conditions were disturbed by the important events that resulted in the uplift of at least the southern part of the Mississippi embayment and the development of the mid-Cenomanian unconformity.

The mid-Cenomanian lowering of the sea level was followed during the late Cenomanian and Turonian by a widespread marine transgression that extended inland over the northern shelf of the Gulf of Mexico basin to the foothills of the Ouachita and Appalachian mountains, covered the southern part of the Mississippi embayment, and reestablished a communication with the Western Interior Seaway. This transgression brought about new controls on the patterns of sediment supply and deposition, as well as important changes in the paleogeography of the region.

The initial deposits of the sequence overlying the mid-Cenomanian unconformity along the northern Gulf of Mexico basin are the fluvial and deltaic terrigenous clastics—sandstones, siltstones, and shales—of the Woodbine and Tuscaloosa Formations (Fig. 21). They covered the shelf from central Texas to Alabama, southern Georgia, and northern Florida. The terrigenous sediments of the Woodbine were principally derived from the Ouachita Mountains in southern Oklahoma and Arkansas; those of the Tuscaloosa were from the Appalachians and the North American continental interior. The grain size of these units decreased southward and southwestward away from the sources of sediments. The copious supply of clastic sediments to the Gulf of Mexico basin during the late Cenomanian to Turonian may reflect an uplift of the northern bordering lands. In deeper parts of the basin, some coarse-clastic sediments were deposited as turbidites in submarine fans.

In central and northeast Texas and neighboring parts of Louisiana, the Woodbine is overlain by the upper Cenomanian-Turonian marine shale section of the Eagle Ford Formation. The Woodbine pinches out at or near the shelf margin and at the San Marcos arch; the upper Cenomanian-Turonian section in south Texas, northeastern Mexico, and the region beyond the shelf margin is predominantly composed of shales and carbonates generally assigned to the Eagle Ford Formation.

The Woodbine and Tuscaloosa contain abundant volcaniclastic material, reflecting the magmatic activity in the southern Mississippi embayment region, from northern Louisiana to southeastern Mississippi. Volcanoes are believed to have been present in the area of the present Monroe uplift and Jackson dome, and to have supplied the surrounding areas with substantial amounts of volcanic material during the late Cenomanian. Igneous intrusion may have been the cause of the formation or reactivation of these structural features. The products of this volcanic activity are predominantly alkalic—peridotites, alkali basalts, phonolites, nephelinites, and carbonatites (see Chapter 6, this volume). The

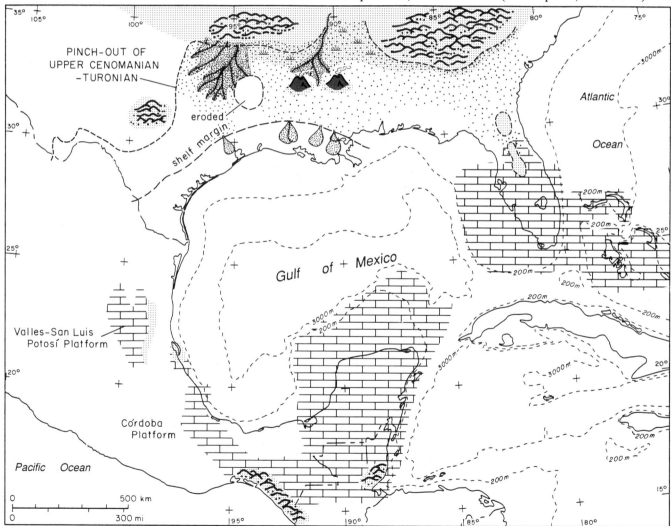

Figure 21. Paleogeography of the Gulf of Mexico basin region during the late Cenomanian–Turonian. (For explanation of patterns, see Figure 12.)

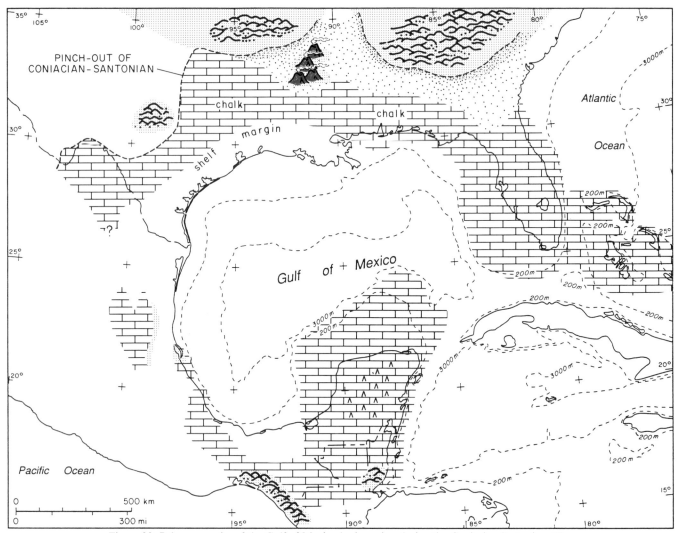

Figure 22. Paleogeography of the Gulf of Mexico basin region during the Coniacian-Santonian. (For explanation of patterns, see Figure 12.)

cause of this magmatic episode is not well understood. It has been speculated that it could have been the result of the reactivation during the Late Cretaceous of an old plate boundary or of the Paleozoic Mississippi Valley graben (or Reelfoot rift) due to an episode of rapid subsidence of the Gulf of Mexico basin.

The Sabine uplift area was gently uplifted in mid-Cenomanian time, as indicated by the thinning of the Woodbine Formation over it. The uplift was reactivated during the late Cenomanian or early Turonian, at which time the Woodbine and the part of the Eagle Ford Formation so far deposited were stripped from the uplifted area. In Coniacian time, the Austin chalk was deposited unconformably over the eroded edges of these older units.

In many parts of the Gulf of Mexico basin the late Turonian was a period of regression; but after that the Late Cretaceous sea continued its transgression during the Coniacian and Santonian, advancing northward along the Mississippi embayment as far as

southern Illinois, and maintaining communication with the Western Interior Seaway. The supply of terrigenous clastic sediments from the north decreased considerably, particularly from the sources in the Ouachita Mountains. Coniacian-Santonian clastic sequences deposited in nearshore, low-energy environments, therefore, are limited to the Mississippi embayment and the northeastern Gulf of Mexico basin (Fig. 22). In these updip areas, the Coniacian-Santonian section overlies unconformably Turonian and older rocks. To the south and southwest, from northeastern Mexico to Alabama, the Coniacian and Santonian were times of widespread deposition of chalk and chalky marls in a shallow-water, open-marine shelfal environment (the Austin Formation or Group and equivalent units). Farther south, beyond the margin of the shelf, the chalk and carbonate section graded basinward to dark calcareous shales.

The unconformity at the base of the Coniacian in the northern rim of the Gulf of Mexico basin dies out basinward except in

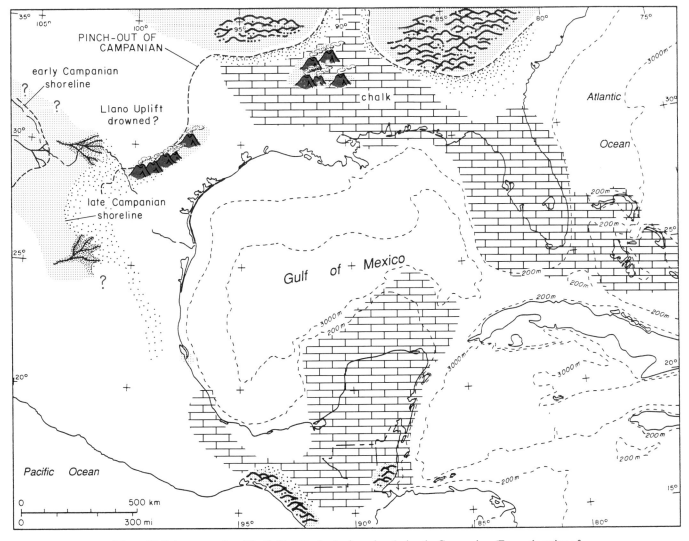

Figure 23. Paleogeography of the Gulf of Mexico basin region during the Campanian. (For explanation of patterns, see Figure 12.)

the Sabine uplift and the northern part of the East Texas basin, where an unconformity has been recognized at the base of the Austin chalk section. Minor, local stratigraphic breaks within the Austin may indicate recurrent tectonic (or magmatic) activity during the Coniacian and Santonian.

Volcanic activity continued in the Monroe uplift area of southern Arkansas, northern Louisiana, and Mississippi during the Coniacian and Santonian. Scattered bentonite beds in the Austin chalk section in northeast Texas indicate periodic ash falls derived from volcanoes to the northeast. More abundant volcaniclastic material occurs in the Coniacian-Santonian clastic sequences of the Mississippi embayment.

During the Campanian, deposition of chalk and chalky carbonates shifted eastward in the shelf of the northern Gulf of Mexico basin, and northward in the Mississippi embayment (the Selma Group of Louisiana, southern Mississippi, and southern

Alabama). Updip, toward the northern and northeastern sources of terrigenous clastic sediments, the chalky carbonates grade to inner-shelf terrigenous clastic sections (Fig. 23). In the upper Mississippi embayment in western Tennessee, the Campanian is represented by the sediments of a deltaic complex—the delta of the Mississippi River of that time. In southern Louisiana and northeast and central Texas, the chalky carbonates of the Selma Group grade to the marls and claystones of the Taylor Formation.

Along the northern Gulf of Mexico basin, deposition was generally continuous from Santonian to Campanian time. Several minor transgression-regression cycles and concomitant unconformities have been recognized in the Campanian section of the northern periphery of the basin. The unconformities, however, die out toward the center of the basin.

In south Texas and northeastern Mexico, the effects of the

early stages of the Laramide orogeny are clearly reflected in the Campanian section. The Austin chalk is overlain in this area by a sequence in which terrigenous clastics increase toward the top. The uppermost Campanian unit is the deltaic San Miguel Formation, composed predominantly of sandstones, sandy limestones, and siltstones derived from uplifted areas to the west and northwest. The early Campanian shore was well to the west of the Burgos and Sabinas basins, but prograded eastward throughout the Campanian (Fig. 23). The coarse-grained terrigenous clastic sections of south Texas and northeastern Mexico grade basinward to increasingly shaly sequences. In the deep central part of the basin deposition of pelagic fine-grained clastic sediments continued, but in its western part the increasing influx of clastics from the west is evident in the thickness patterns of the Upper Cretaceous section, shown in seismic reflection profiles of the area.

An important magmatic event took place in south and central Texas during the latest Cretaceous. More than 200 volcanic sites have been recognized in this area along the length of a northeast-trending arcuate belt, about 80 km wide and 400 km long—the so-called Balcones volcanic province. The age of the Balcones intrusive and extrusive rocks range from 87 to 70 Ma (Santonian to mid-Maastrichtian), but igneous activity of the province seems to have peaked during the Campanian. The rocks are predominantly alkalic in composition—alkali basalts, phonolites, and nephelinites—and occur as lava flows, laccoliths, scoria cones, volcanic plugs, sills, and dikes. Thin beds of bentonitic ash and tuffs have also been reported. The Balcones volcanic province is aligned along the strikes of regional faults of the Cretaceous(?)-Tertiary Balcones and Luling fault systems, and also follows roughly the trend of the Paleozoic Ouachita orogenic belt. Its origin, as that of the somewhat earlier magmatic episode in the southern Mississippi embayment, has not been well explained.

Over parts of south Texas, the skeletal-debris carbonate banks of the Anacacho Limestone were deposited over sea-floor elevations created by intrusive igneous bodies.

Volcanic activity in the southern Mississippi embayment may have persisted periodically until the Campanian, as indicated by the presence of volcaniclastic debris in the Campanian section of Mississippi and Alabama and by the renewed uplift of the Monroe uplift and the Jackson dome during or shortly after the Campanian. As a result, the Monroe uplift and the Jackson dome formed high grounds on which the so-called Monroe Gas Rock and Jackson Gas Rock sequences of carbonate shoals of latest Cretaceous age were built. In parts of the Monroe uplift, a pronounced angular unconformity separates Campanian from overlying Paleocene beds.

Deposition along the northern Gulf of Mexico basin continued uninterrupted from Campanian to Maastrichtian time. The geography and patterns of sediment deposition in this province during the Maastrichtian were essentially the same as those during the Campanian: chalk and chalky carbonates, deposited in a shallow, low-energy shelf, make up most of the section in the northeastern shelf of the basin; they grade to nearshore marine

and deltaic terrigenous clastic sequences—sandstones and siltstones—toward the northern sources of sediments, and to increasingly shaly sections toward the more basinal areas to the south and southwest. Stratigraphic breaks of perhaps only local significance or that are indicative of minor sea-level oscillations are also recorded as small disconformities in the Maastrichtian section of the northern rim of the basin, particularly in the upper Mississippi embayment.

In south Texas and northeastern Mexico, southwest of the San Marcos arch, the effects of the Laramide orogeny to the west and northwest became increasingly evident during the Maastrichtian; the Maastrichtian sequence is composed predominantly of terrigenous clastics that grade from fluvial and deltaic in the west to shallow-water marine to the east. Commercial coal deposits occur in the fluvial clastic section of the Maastrichtian Olmos Formation in northeastern Mexico (see Chapter 16, this volume). During the Maastrichtian, the shoreline of the western highlands advanced farther to the east (Fig. 24).

The Maastrichtian and the Cretaceous ended with a sea-level drop and a regression over the Gulf of Mexico basin. As a result, the Cretaceous-Tertiary boundary is represented by a stratigraphic hiatus and disconformity in the more positive areas of the northern part of the basin—shelves and platforms—but by essentially continuous sections in its deeper, more basinward parts. The connection between the Gulf of Mexico and the Western Interior Seaway, that had persisted until the Campanian, ceased to exist sometime during the Maastrichtian.

The western and southwestern province

During most of the Late Cretaceous the western and southwestern regions of the present Gulf of Mexico basin were part of a broad seaway that attained its maximum extent during the Turonian and connected the Gulf of Mexico with the Pacific Ocean. It was not until late in the Late Cretaceous (Campanian and Maastrichtian) that there was a progressive reduction of this marine connection. The upper Upper Cretaceous section of eastern and southeastern Mexico contains evidence of the appearance of uplifted land areas to the west which supplied terrigenous clastic sediments to the Gulf of Mexico basin region, and of the onset of the Laramide orogeny.

Except for the platforms bordering the Gulf of Mexico basin—the Valles-San Luis Potosí, Tuxpan, and Córdoba platforms—neither the mid-Cenomanian unconformity nor the other lesser stratigraphic breaks recognized in the northern part of the basin are reflected in the more basinal Cenomanian to Santonian sequences of the western part of the basin. The Cenomanian to Santonian section in this region is composed of dark calcareous shales, shaly limestones, and micritic limestone (upper part of the Cuesta del Cura and Tamaulipas Superior, and the Agua Nueva, Indidura, and San Felipe Formations) (Figs. 21 and 22). Bentonites, chert beds, and tuffaceous material are common in the section, most abundant in the late Cenomanian–Turonian Agua Nueva and Indidura Formations, but are also present in the Coniacian-Santonian San Felipe Formation. They reflect the

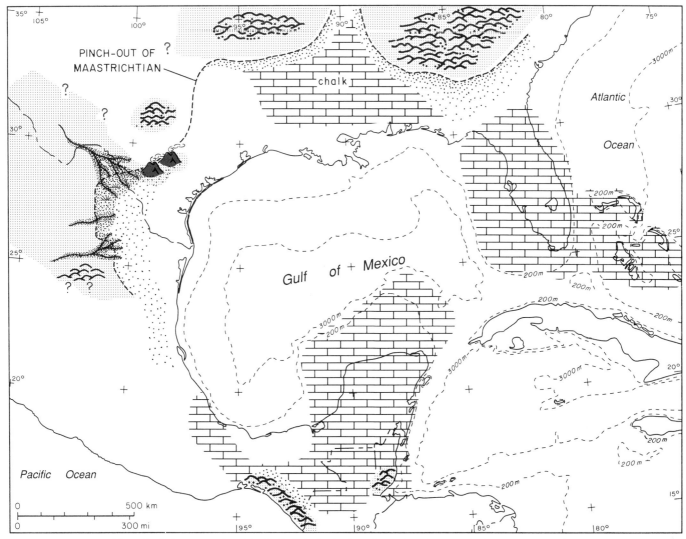

Figure 24. Paleogeography of the Gulf of Mexico basin region during the Maastrichtian. (For explanation of patterns, see Figure 12.)

presence of volcanic activity in nearby regions, probably along the Pacific margin of Mexico.

In the platforms, shallow-water carbonates and occasional reefal buildups formed during the early Cenomanian and persisted, though less extensively, from the late Cenomanian to the Santonian: the Tamasopo Formation in the Valles–San Luis Potosí platform, and the upper part of the Orizaba Formation and Guzmantla Formation in the Córdoba platform. Massive platform carbonates are also known to form the Cenomanian to Santonian section in the Sierra de Chiapas of southern Mexico (the upper part of the Sierra Madre Group).

The Campanian and Maastrichtian are represented in eastern and southeastern Mexico by the uniform and widespread shales and marls of the Mendez Formation, deposited in a low-energy shelf or shallow basin (Figs. 23 and 24). Greenish bentonites have been reported from the Mendez in several areas,

indicating the persistence of volcanic activity, probably in western Mexico. In the area of the Valles–San Luis Potosí platform, the Mendez grades westward to a coarse terrigenous clastic sequence (the Cárdenas Formation) indicative of the appearance of a western source of clastic sediments. The position of the eastern shore of this land area in central Mexico during the Campanian and Maastrichtian is not precisely known; it certainly must have been a considerable distance to the west of the western and southwestern Gulf of Mexico basin during the early Campanian, but it is thought to have migrated eastward during the rest of the Campanian and the Maastrichtian.

Along the eastern margin of the Valles–San Luis Potosí platform, the Mendez Formation overlies unconformably the eroded surface of the Tamabra Formation and the El Abra reef limestones, which often have well-developed karst features.

In southern Mexico, the Mendez Formation grades to platform carbonates in the Veracruz basin and Córdoba platform area (the Atoyac Formation), and toward the shelfal area in the Sierra de Chiapas (Angostura Formation). The Angostura, in turn, grades to the terrigenous clastic section of the Ocozocoautla Formation toward the Chiapas massif (Figs. 23 and 24).

In the central part of the Gulf of Mexico, deposition of fine-grained terrigenous clastics continued uninterrupted during the Campanian and Maastrichtian.

The Florida and Yucatan platforms

The deposition of shallow-water carbonates and evaporites in the broad Florida and Yucatan platforms, rimmed by reefal buildups and organic banks, continued from the Early Cretaceous into the earliest Late Cretaceous (early Cenomanian). As mentioned earlier, evidence of a major lowering of the sea level in mid-Cenomanian time is clear along the margins of the platforms facing the Gulf of Mexico, but it is not as clear in the interior of the platforms, which must also have been affected by the fall of sea level. Other sea-level oscillations during the Late Cretaceous should similarly be reflected in the Upper Cretaceous section of the Florida and Yucatan platforms, but they have not yet been reported.

During the rest of the Late Cretaceous, after the Florida and Yucatan platforms were again covered by the sea in the late Cenomanian, deposition of shallow-water shelf carbonates and evaporites resumed (Figs. 21–24). The platforms, however, were somewhat reduced in size and were no longer rimmed by rudist-dominated reefs and organic banks. Instead, they were bordered by distally steepened ramps in which open-marine shelf carbonates were deposited.

In the northern part of the Florida platform, an emergent part of the Peninsular arch persisted until the late Cenomanian or Turonian in the form of a few isolated islands, which provided a source for the calcareous sands that surrounded them (Fig. 21). Farther north, the Upper Cretaceous carbonate-evaporite section grades northward and northwestward to a section of terrigenous clastics—shales, siltstones, and fine-grained sandstones—shed from the Appalachian Mountains. The Maastrichtian section in the Suwannee saddle is very thin or absent; the area separates clastic deposits to the north from the carbonate-evaporite section predominant in the Florida platform (Fig. 24).

No terrigenous clastic sequences similar to those of the northern Florida platform are known from the Yucatan platform, where the Upper Cretaceous section is characterized by shallow-water carbonates and evaporites. Evaporites form most of the section in the central part of the platform, and carbonates predominate around its margins.

In the northern part of the Yucatan Peninsula, poorly known intrusive or volcanic rocks—andesites and andesitic tuffs—are apparently interbedded with Albian, Cenomanian, and Turonian limestones. Their total distribution and origin are not known.

CENOZOIC—VAST INFLUX OF PROGRADING TERRIGENOUS CLASTIC WEDGES FROM NORTH AND WEST

The culmination of the Laramide orogeny during the Paleocene and early Eocene brought important changes to the Gulf of Mexico basin region. The influx of terrigenous clastic sediments from the north and northwest, which had increased appreciably toward the end of the Late Cretaceous, reflecting the initial episodes of the Laramide orogeny, intensified significantly during the Paleocene and early Eocene as uplift and orogenic deformation progressed eastward in the Cordilleran region of the United States and in Mexico. The orogeny, presumably the result of nearly horizontal subduction of an oceanic plate or plates under the North American plate along its Pacific margin, resulted in the development of lengthy fold and thrust belts; in the Gulf of Mexico basin and immediate regions, they are represented by the Chihuahua tectonic belt (along which a left transpression component may have been significant) and the thrust belt that gave rise to the Sierra Madre Oriental of Mexico.

Influx of terrigenous clastic sediments into the Gulf of Mexico basin continued intermittently after the end of the Laramide orogeny in mid-Eocene time, and the locations of its greatest impulse shifted. The size and shape of Gulf of Mexico basin in the early Cenozoic was determined by the position of the Cretaceous carbonate platforms, but as large volumes of terrigenous clastic sediments continued to enter the basin, the shorelines of the Gulf of Mexico and its shelves and shelf margins migrated progressively basinward during the Cenozoic. Very thick sedimentary sections accumulated eventually over the continental slopes and filled increasingly deeper parts of the gulf. Subsidence of the basin continued during the Cenozoic, but it was predominantly the result of the loading of the crust by the prograding thick wedges of Cenozoic sediments.

The accumulation of thick Cenozoic sedimentary sections also contributed to the recurrent mobilization of the Middle Jurassic salt and of the lower Cenozoic overpressured plastic shales. As a result, numerous salt and shale diapirs and allochthonous, nearly horizontal salt tongues or nappes formed contemporaneously with deposition of the Cenozoic section. Around the perimeter of centers of rapid deposition, listric normal (growth) faults were common. Growth faults are progressively younger, and involve progressively younger sediments, in a basinward direction. In the Louisiana continental shelf and slope, the salt has been fundamental to the evolution of the growth faults. Offshore Texas, growth faulting does not appear to be as intimately associated with salt mobilization (Worral and Snelson, 1989).

During the Pleistocene, alternating glacial and interglacial periods were responsible for rapid changes in sea level and the consequent changing stratigraphic patterns and styles.

Carbonate and evaporite deposition in the Gulf of Mexico basin was restricted to the Florida and Yucatan platforms during the Cenozoic. Terrigenous clastic deposition dominated the rest of the basin, coarser around the periphery, and increasingly finer-grained toward its deep central part. The amount of the progradation of the Cenozoic clastic wedges varied; it was largest along the northern and northwestern flanks of the basin and smallest in the western flank. Clastic-wedge progradation and basinward migration of the shorelines and shelves was, however, a characteristic depositional pattern of the Cenozoic over most of the Gulf of Mexico basin. Modifications and deviations from this general pattern developed locally and will be discussed in the following pages.

As in the previous discussion of the Mesozoic history of the Gulf of Mexico basin, the description of the structural and stratigraphic development of the basin during the Cenozoic can best be treated by discussing separately each of the three distinct stratigraphic-structural provinces that have persisted with only minor modifications since shortly after the formation of the basin: the northern and northwestern province dominated by a prolific supply of terrigenous clastic sediments and characterized by a persistent basinward progradation of thick clastic wedges over a broad shelf; the western province, where the influx of clastic sediments was not as abundant, the shelf considerably narrower, and in which the Laramide orogeny produced during the early Cenozoic important modifications in the paleogeography of the area; and the very stable carbonate-evaporite province of the Florida and Yucatan platforms. In the deep, central part of the Gulf of Mexico basin, the Cenozoic is represented by as much as 5000 m of pelagic sediments about which very little is known.

The northern and northwestern progradational province

The Cenozoic stratigraphy and depositional history of the northern and northwestern province was distinguished by the abundant supply of terrigenous clastic sediments that entered the region from the north and northwest. This prolific supply of terrigenous clastic material to the northern and northwestern Gulf of Mexico basin varied in volume at different times during the Cenozoic, and it is believed to have been controlled by tectonic events beyond the limits of the basin—the culmination of the Laramide orogeny during the Paleocene and early Eocene, and later recurrent uplifts of the Rocky Mountains, the Colorado Plateau, or the Appalachian Mountains. The variable supply of terrigenous clastics resulted in an episodic pattern of sedimentation that records numerous transgressive and regressive episodes. Each depositional episode was characterized by initial progradation and the resulting basinward advance of the coast and shelf, followed by a transgression and landward retreat of the shore. The displacement of the shore zone, basinward or landward, during any one depositional episode was in the order of several tens of kilometers, but generally each successive progradation extended farther basinward than the previous one, and each transgression failed to reach as far landward as its predecessor. In this manner, each offlapping clastic wedge was deposited farther basinward in an imbricate pattern, extending the shelf margin toward the deep center of the Cenozoic Gulf of Mexico and accumulating a thick clastic section over the continental slope

(Fig. 25) (see Chapter 11, this volume). Growth faulting was common at or basinward of the building-outward shelf margins. Each depositional episode reflects a complex interplay of sediment supply, sea-level position, and degree of mobilization of the underlying salt deposits.

The supply of terrigenous clastic sediments not only fluctuated in volume during the Cenozoic, but the location of the main avenues of ingress and the thickest centers of deposition also changed. During the late Paleocene and early Eocene the main inlets and resulting thickest sediment accumulations were in northeast Texas, northeast Louisiana, southeast Arkansas, and western Mississippi; during the rest of the Eocene and the early Oligocene, the most active depocenters were in south Texas; during the late Oligocene and Miocene, the sites of thickest deposition shifted progressively from south Texas to southwest Louisiana, and to the present location of the Mississippi River delta. Finally, in Pliocene and Pleistocene time, the depocenters moved southwestward, to somewhat west of the Mississippi delta during the Pliocene and south of the coast of southeast Texas and southwest Louisiana during the Pleistocene (Fig. 26).

The maximum thickness attained by each of the various Cenozoic units are registered in the most active depocenters of the time. The total Cenozoic section has thicknesses of 10,000 m offshore the southern coast of Texas, and 11,000 m under the southern part of the delta of the Mississippi River. Of this total thickness, the combined Paleocene-Eocene section reaches a maximum thickness of about 6,000 m along the coast of south Texas, the Oligocene is about 5,000 m just offshore from south Texas, the Miocene is about 6,000 m under the present Mississippi Delta, the Pliocene is 3,600–4,000 m southwest of the delta, and the Pleistocene is 3,500–3,800 m offshore from southeast Texas and southwest Louisiana, and 3,000 m in the Mississippi fan (Fig. 26).

The Cretaceous ended with a drop in the level of the sea, and a considerable basinward retreat of the coast of the Gulf of Mexico. During the early Paleocene, the supply of terrigenous coarse clastics was much reduced; the lower Paleocene section is represented over most of the northern Gulf of Mexico basin by a transgressive sequence predominantly composed of shales—the Midway Group. Deposition was not accompanied by growth faulting. The lower Paleocene section becomes sandier in the Mississippi embayment and in southern Alabama; in the Florida panhandle, the lower Paleocene includes some carbonates at the base—the Clayton Formation.

During the late Paleocene and early Eocene, large volumes of terrigenous coarse-clastic sediments entered the northern Gulf of Mexico basin from the north through two main inlets: one in western Mississippi, southeast Arkansas, and northeast Louisiana, and another in northeast Texas and along the axis of the Houston embayment (Fig. 27). The existence of an eastern avenue of supply shows that the course of the Mississippi River in this region remained essentially the same in late Paleocene–early Eocene time as it is today. The rivers bringing clastic sediments into northeast Texas flowed into the area from the northwest. In both

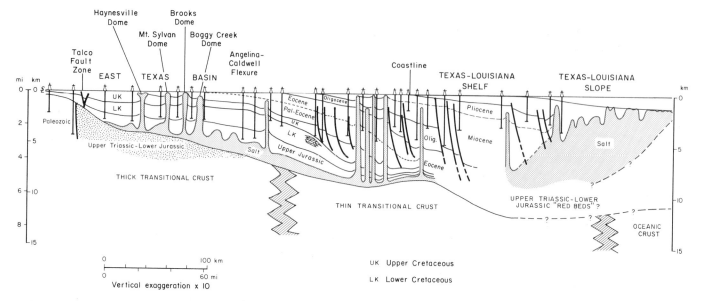

Figure 25. Cross section through the northern Gulf of Mexico basin showing prograding and imbricated depositional pattern of the Cenozoic section.

areas, large progradational deltaic complexes have been recognized. They are composed of sedimentary sequences deposited in a variety of environments—fluvial and delta plain landward, and delta front and prodelta in the downdip areas. Deep-water environments bordered the deltas farther basinward. Each of the delta systems comprises three phases of progradation and abandonment, and developed several major lobes (see Chapter 11, this volume).

Along depositional strike southwest of the delta complexes, barrier-bar and strand-plain systems with related landward lagoons have been mapped along the northwestern Gulf of Mexico basin (for example, see Fisher and McGowen, 1967; Galloway, 1968). The clastic sediments of these systems had sources in the delta lobes to the northeast and were transported by southwest-flowing longshore currents. The sediments of the delta complexes as well as those of the strand-plain and associated systems are generally included in the Wilcox Group.

The considerable sediment loading by the delta complexes was responsible for the first of many episodes of growth faulting and salt mobilization during the Cenozoic. The Wilcox fault zone is particularly well developed along the front of the deltas of the time (see Plate 2). In south Texas, it terminates downward in a detachment surface within the salt section.

At least one submarine canyon, the Lavaca canyon, is known to have developed during the late Paleocene. It formed on the progradational flank of a delta lobe as a result of submarine mass wasting caused by headward erosion at the shelf margin. The canyon was later filled with prodelta muds and delta-front sands. Several other canyons were formed in earliest Eocene time.

The Upper Paleocene–Lower Eocene clastic section of the Mississippi embayment and southern Alabama grades eastward

into a section composed predominantly of carbonates, the Cedar Keys Formation of the Florida platform.

During the Middle and Late Eocene the paleogeography and depositional framework of the northern Gulf of Mexico basin underwent only minor modifications: broad high-constructive fluvial-deltaic systems persisted in northeast Texas, and extended at times into northern Louisiana, southern Arkansas, and Mississippi. Southwest of the fluvial-deltaic systems in central and south Texas, the section is predominantly represented by a strand-plain–barrier-bar system composed of strike-trending sand bodies that had sources in the delta complexes to the northeast and that were transported by longshore currents. Basinward, both the delta system and the strand-plain–barrier-bay system grade to marine sequences.

The Middle and Upper Eocene section also reflects a number of cyclical transgressive and regressive-prograding events that resulted in the alternation of thick, sand-rich, prograding deltaic or strand-plain sequences and thin transgressive shaly sequences deposited over a marine shelf. The Middle and Upper Eocene is represented by the Claiborne and Jackson groups. The Claiborne includes four progradational tongues; from bottom to top, these are the Carrizo (considered by many as the uppermost unit of the underlying Wilcox Group), Queen City, Sparta, and Yegua formations, separated by three transgressive marine shaly units, from bottom to top, the Reklaw, Weches, and Cook Mountain formations (Guevara and Garcia, 1972; Ricoy and Brown, 1977) (see Plate 5). During the time of deposition of the Queen City and Sparta formations, a high-destructive, wave-dominated delta complex developed in south Texas and extended to northeastern Mexico.

The Jackson Group is, in terms of deposited sand volume,

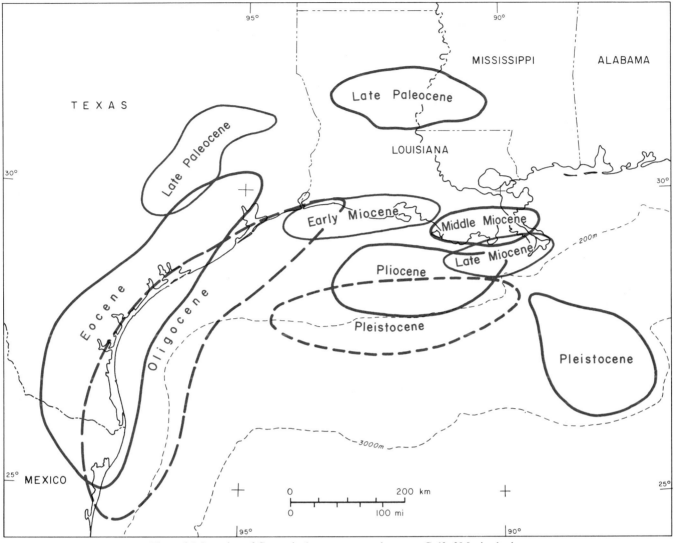

Figure 26. Location of Cenozoic depocenters, northwestern Gulf of Mexico basin.

the least important of the Middle and Late Eocene clastic wedges. It includes a northeastern fluvial-deltaic system characterized by the predominance of fine-grained clastic sediments, and a southwestern strand-plain–barrier-bar system (Fisher and others, 1970), but does not show appreciable basinward progradation. The sediments of the Jackson Group contain abundant volcaniclastic material, bentonites, and bentonitic claystones, reflecting late Eocene volcanic activity to the west and southwest.

In general, the volume of terrigenous clastic-sediment influx to the Gulf of Mexico basin during the Middle and Late Eocene was not as copious as that supplied to the basin during the late Paleocene–early Eocene (Wilcox Group). It also shifted geographically; while the prograding, lobate delta systems of the Sparta Formation generally extended from northeast Texas to adjoining areas of Louisiana, Arkansas, and even Mississippi,

those of the Queen City and Jackson were mostly restricted to south and northeast Texas.

In southeastern Mississippi and southern Alabama, the Middle and Upper Eocene is composed of shalier, marine-shelf sequences with limited coarse terrigenous clastic constituents. Farther east, the clastic sequences grade to the carbonate section of the Florida platform, the Oldsmar, Avon Park, and Ocala formations.

During the Oligocene, the main ingress of terrigenous clastic sediments shifted to the area of the Rio Grande embayment in south Texas and northeast Mexico, where the section attained its maximum thickness (Figs. 26 and 28). The Lower Oligocene Vicksburg Formation was deposited under conditions similar to those prevailing during the Middle and Late Eocene. The offlapping Vicksburg terrigenous clastic wedge was deposited farther ba-

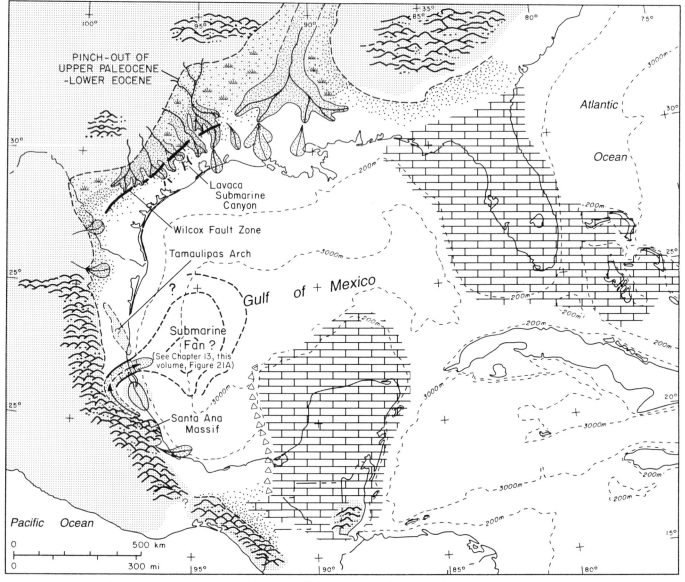

Figure 27. Paleogeography of the Gulf of Mexico basin region during late Paleocene–early Eocene (Wilcox) time. (For explanation of patterns, see Figure 12.)

sinward than previous Cenozoic wedges, and developed at the margin of the shelf, and downdip from the Wilcox fault zone, another marked growth-fault belt, the Vicksburg fault zone or "Vicksburg flexure." This fault zone is a very low angle listric-normal-fault zone with associated antithetic faults and pronounced thickening and rollover of the section in its downthrown side above the detachment surface, probably in the underlying Jackson shale section.

A major increase in the influx of terrigenous clastic sediments into the northern Gulf of Mexico Basin took place during the Late Oligocene (Fig. 28). This, the greatest progradational clastic wedge of the Cenozoic, was deposited at this time from north-

eastern Mexico to Mississippi. It is generally designated as the Frio Formation in the subsurface and believed to correspond in Texas to the surface Catahoula Formation. The relationship of the subsurface Frio with the type area of the surface Catahoula of Louisiana and Mississippi is not firmly established.

The main influx of terrigenous clastic sediments during the Late Oligocene continued to be into the Rio Grande embayment area of south Texas where the Frio Formation reaches a thickness of 4,500 m. During the deposition of this thick sedimentary wedge, basinward progradation over the Vicksburg shelf was as much as 80 km. The considerable sedimentary load caused reactivation of preexisting salt structures and the formation of new ones. Salt

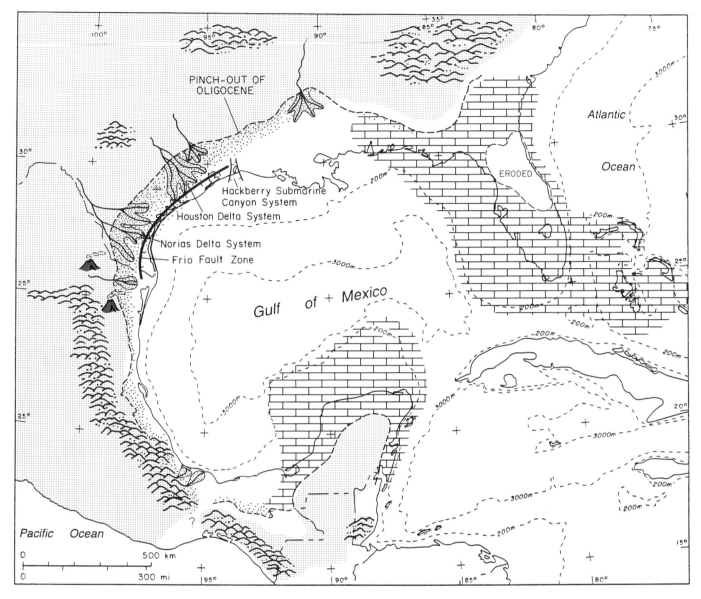

Figure 28. Paleogeography of the Gulf of Mexico basin region during Oligocene (Frio) time. (For explanation of patterns, see Figure 12.)

mobilization, in turn, affected sedimentary processes. Abundant volcaniclastic material in the Frio section gives evidence that the late Eocene volcanic activity to the west of the northern Gulf of Mexico continued with increased vigor during the Oligocene.

Two large fluvial-deltaic systems were active in late Oligocene time (Fig. 28). The largest was located in the Rio Grande embayment area, and the other in the Houston embayment. In between, a wave-dominated strand-plain–barrier-bar system developed in which long, strike-oriented linear sand bars separate downdip marine shales from updip lagoons (Boyd and Dyer, 1964). Growth faults developed on its basinward side.

The southern delta, the Norias delta system, unlike previous

deltas in the area, was fed by a single major river. The thick sedimentary sequence deposited in this delta expanded greatly at the margin of the underlying Vicksburg wedge and developed a broad growth-fault belt, the Frio fault zone or "flexure." Beyond the fault zone, the coarser-clastic deltaic sediments grade to a finer-grained offlapping sequence deposited over the continental slope. The Houston delta system is smaller, and prograded basinward a shorter distance, but developed a distinct growth-fault zone. It was fed by two or three moderate-size streams.

In southeast Texas and southwest Louisiana a major erosional feature, the so-called Hackberry embayment, formed during the late Oligocene (Fig. 28). It is believed to have been carved by submarine processes operating in the margin of the shelf. The "embayment" was later filled with a sequence of deep-water sands and shales up to 1,000 m thick that onlaps and pinches out updip against the flanks of the submarine-canyon–like feature. Uppermost Oligocene (Frio) and Miocene sandstones and shales cover the Hackberry embayment.

The thick east Texas progradational Oligocene Frio section grades eastward in southern Mississippi and Alabama into a much thinner aggradational sequence of shales, marls, and limestones deposited in a shallow shelf. Carbonate content increases eastward toward the Florida platform.

The latest Oligocene and earliest Miocene was marked in the northern province of the Gulf of Mexico basin by a widespread transgression during which the shaly section of the Anahuac Formation was deposited over the late Oligocene shelf.

During the Miocene, the main Cenozoic depocenters shifted again, this time toward the northeast, from the coastal Texas area during the Oligocene to southern Louisiana. By late Miocene the thickest section (more than 6,000 m) was deposited in the area of the present Mississippi Delta, supplied by the Mississippi River of the time (Figs. 26 and 29). The increased load of terrigenous clastic sediments brought into the basin by the Mississippi has been attributed to the reactivation of the uplift of the Rocky Mountains, the Colorado Plateau, and the Appalachian Mountains. The increase in the supply of terrigenous clastic sediments to the Gulf of Mexico during the Miocene extended as far east as the Florida platform, where carbonate deposition which had prevailed since the Early Cretaceous was interrupted by the prolific dispersal of clastic sediments from the north. A very thick Miocene section was also deposited in offshore Texas, where the

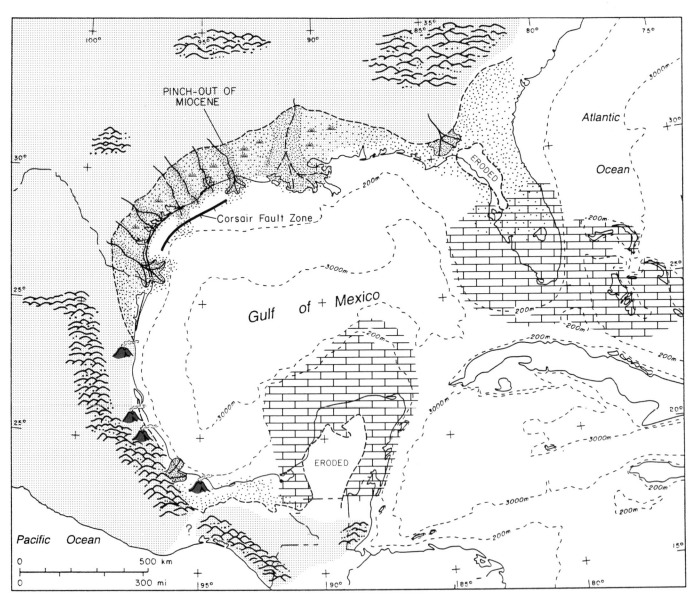

Figure 29. Paleogeography of the Gulf of Mexico basin region during the Miocene. (For explanation of patterns, see Figure 12.)

Middle and Upper Miocene attain 4,600 m in the downthrown block of the Corsair growth fault (Vogler and Robison, 1987).

The Miocene, as the earlier Cenozoic sequences, is represented by thick offlapping wedges of fluvial-deltaic terrigenous coarse-clastic sediments that prograded over the underlying late Oligocene (Frio) unstable shelf margin and graded basinward to thinner marine shalier sections. Delta systems and the margin of the shelf advanced southward by as much as 80 km during the Miocene. Large growth-fault zones with horizontal slip of as much as 30 to 40 km, rollover folding, and complex antithetic faults developed throughout the Miocene, particularly along the gulfward margin of the deltaic systems and around the periphery of the depocenters. Highly expanded sections were formed along the downthrown blocks of these growth faults, as well as in local salt-withdrawal basins, created by renewed mobilization of the salt as a result of the sedimentary load of the Miocene section.

The depositional and structural processes that had been active in the northern and northwestern Gulf of Mexico basin since the early Tertiary persisted during the Pliocene. Extensive and thick fluvial-deltaic systems remained active, the shelf margin prograded farther basinward, and important growth-fault zones developed along it. The main Pliocene depocenter was located under the continental shelf of the Gulf of Mexico, to the southwest of the present delta of the Mississippi River (Figs. 26 and 30). In it, as much as 3,600 to 4,000 m of terrigenous clastics were deposited in less than 4 m.y., at a rate of sedimentation much faster than any known in previous times during the Cenozoic. Elsewhere along the northern Gulf of Mexico basin, Pliocene sequences are thinner, particularly in the onshore coastal areas.

The continental glacial events that took place during the Pleistocene in the northern part of North America modified con-

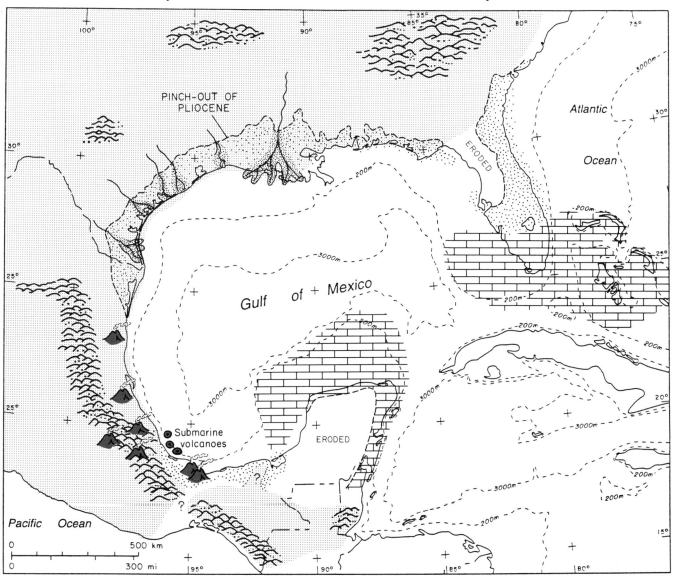

Figure 30. Paleogeography of the Gulf of Mexico basin region during the Pliocene. (For explanation of patterns, see Figure 12.)

siderably the sedimentary patterns that had previously characterized the Cenozoic of the northern Gulf of Mexico basin region.
Sea level dropped by as much as 150 m during Pleistocene glacial
periods, exposing large areas of the northern shelf of the Gulf of
Mexico, lowering the base level of the streams, and causing them
to entrench their valleys across the shelf and to deposit their
terrigenous-clastic load first on the outer part of the shelf and then
on the continental slope. The rise of the sea level during interglacial periods submerged again the shelf, raised the base level of the
steams, and shifted the areas of active deposition landward, causing the reworking and redistribution of some of the sediments
deposited during the previous glacial period (see Chapter 12, this
volume).

The main Pleistocene sedimentary depocenter was located
in the outer shelf of the Gulf of Mexico near the present shelf
margin, south of southeast Texas and southwest Louisiana, where
up to 3,800 m of sediments were deposited in less than 2 m.y.
(Fig. 26). This fast accumulation of a thick sedimentary section
was accompanied by growth faulting and triggered the mobilization of the underlying Middle Jurassic salt. The salt deformation,
in turn, affected the dispersal and deposition of the clastic sediments. Gravity slumping of sediment and cutting of submarine
canyons was also common in the more unstable areas, particularly on the basinward side of the shelf margin and around the
periphery of intrashelf depositional basins.

Very young syndepositional diapiric salt deformation,
growth faulting, and gravity slumping are still active today in the
outer shelf and particularly in the northern slope of the Gulf of
Mexico. Numerous small basins have formed between allochthonous salt massifs and ridges active throughout the Quaternary
(see Chapter 12, this volume). Subsidence rates in these small
basins are among the highest in the world (Worrall and Snelson,
1989). This irregular distribution of high salt structures and small
basins has given the northern slope of the Gulf of Mexico a
complex bathymetric configuration of small multiform highs and
lows (see Chapter 2, this volume; Plate 1).

Salt flow under the northern Gulf of Mexico in the form of
nearly horizontal, southward-moving nappes is believed to have
been active since at least the late Eocene, but to have been most
active during the Pliocene and Pleistocene. This southward advance of the salt nappes was probably driven by the sedimentary
load caused by the thick Cenozoic sedimentary wedges, particularly the Pliocene and Pleistocene deltaic systems, that progressively prograded southward and squeezed the salt ahead of them.
The horizontal displacement of some of the nappes may be more
than 150 km.

An important sedimentary and physiographic feature developed in the Gulf of Mexico basin during the latest Pliocene and
Pleistocene is the Mississippi submarine Fan (see Chapters 2 and
12, this volume). Located under the waters of the eastern Gulf of
Mexico, the fan is more than 4 km thick, covers an area of more
than 300,000 km^2, and contains a volume of sediments of close
to 300,000 km^3. It extends from the margin of the continental
shelf off southern Louisiana to the central abyssal plain (Fig. 31),

and is reflected in the floor of the gulf by a southward bowing of
the bathymetric contours (Plate 1). The Mississippi Fan is composed of a complex aggregation of numerous superposed and
overlapping elongate lobes. Accumulation of the fanlobes took
place in several stages, primarily during late Pleistocene glacial
periods, when sea level was low and the supply of terrigenous
sediments to the basin was most intense. During interglacial periods, the sea level rose, sedimentation rates were minimal in the
fan area, and only a thin layer of hemipelagic sediments was
deposited over the fan. The Mississippi Fan has been divided into
an upper, middle, and lower fan (Fig. 31); the upper fan starts at the
base of an incised submarine canyon, the Mississippi Canyon, cut
into the continental shelf and slope, and through which the clastic
sediments forming the fan have been funneled. Studies of reflection seismic data over the Mississippi fan have allowed Feeley
and others (1990) and Weimer (1990) to recognize 13 and 17
stacked seismic sequences, respectively, within the fan. The base
of each sequence is distinguished by an erosional unconformity,
which is often overlain by mass-transport deposits—turbidites,
slides, and debris-flow deposits. Within each sequence, a series of
channel, levee, and associated overbank deposits can be identified. Fluctuations of the sea level due to glacial episodes have
been interpreted to have been the major controlling factors in
determining the timing and geometry of the sedimentary sequences forming the Mississippi Fan.

The Quaternary deposits of the onshore region of the northern Gulf of Mexico basin, both the Pleistocene and the Holocene,
are principally related to the activity of the Mississippi and other
rivers draining into the Gulf of Mexico. Fluvial and eolian
deposits—fluvial terraces and loess deposits—have been recognized the length of the lower valley of the Mississippi River.
Lesser Quaternary deposits are also known from the alluvial
valleys of other rivers. Along the coast, the Quaternary is represented by the sediments forming lengthy barrier islands and associated lagoons, beaches, dunes, mudflats, marshes, swamps, and
deltas (see Chapter 2, this volume).

The western and southwestern province

The western Gulf of Mexico basin region was strongly influenced by the Laramide orogeny during the Cenozoic, particularly
during the early Cenozoic. Starting in the late Cretaceous, extensive areas of western Mexico began to be uplifted, restricting the
broad seaway that during the Earlier part of the Cretaceous had
extended from the Pacific Ocean, across Mexico, to the Gulf of
Mexico region. The orogeny progressed eastward during the Late
Cretaceous and early Tertiary and culminated during the Early or
Middle Eocene with the formation of the Sierra Madre Oriental
of eastern Mexico.

The Laramide orogeny modified drastically the paleogeography of the western Gulf of Mexico basin region, progressively
displacing the western coast of the Gulf of Mexico toward the
east from a location somewhere in western or central Mexico.
The uplift of the Sierra Madre Oriental, that extends from north-

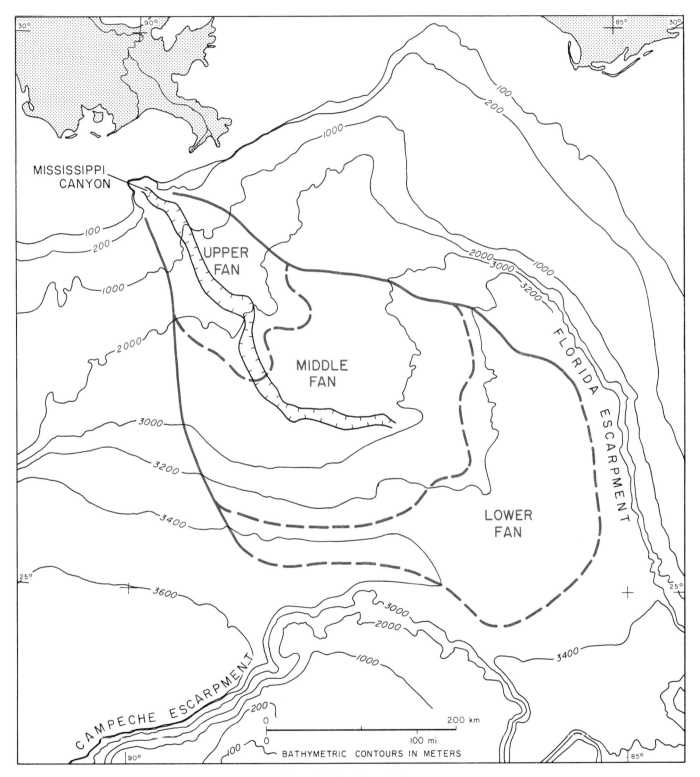

Figure 31. The Mississippi fan.

eastern Mexico to the Isthmus of Tehuantepec, was coupled with the development of a series of depressions or basins roughly parallel and to the east of the orogenic belt in which thick sequences of Cenozoic sediments were deposited. From north to south, these basins are known as the Burgos, Tampico-Misantla, and Veracruz basins. East of the Isthmus, three more basins were developed during the Cenozoic; the Isthmus Saline, Comalcalco, and Macuspana basins, known as the "southeastern basins" (see Plate 2). The structural characteristics of the southeastern basins makes it difficult to establish a clear connection between their development and the evolution of the Laramide orogeny and the uplift of the Sierra Madre Oriental.

The relation between the chronology of the Laramide uplifts and the initiation of the supply of coarse terrigenous clastic sediments to the Cenozoic basins of the western Gulf of Mexico basin region is not well understood. While it is generally believed that the Laramide orogeny started throughout most of Mexico during the late Cretaceous, and that it had its final, and perhaps most intense activity during the Middle Eocene, significant and widespread volumes of coarse clastics appear earlier in the northern basins and progressively later in those farther south: in the Burgos basin, Cenozoic coarse-clastic sections are first present during the latest Paleocene and early Eocene; in the Tampico-Misantla and Veracruz basins, during the late Eocene. (Submarine canyons cut during the Paleocene into the continental shelf and slope that border to the east of the Tampico-Misantla and Veracruz basins, were filled with upper Paleocene–lower Eocene turbiditic sequences predominantly composed of the shales and commonly argillaceous, fine-grained to very fine-grained sandstones of the Chicontepec Formation. In the shelves in which the canyons were cut, the Paleocene–lower Eocene section is entirely made up of marine shales.) In the southeastern basins, widespread deposition of coarse terrigenous clastics did not become prevalent until the early Miocene. Before the first influx of coarse terrigenous clastics, deposition of marine shales was dominant throughout eastern Mexico during the early Cenozoic.

This diachronous first supply of coarse clastics into the western Gulf of Mexico basin region, earlier in the north and progressively later toward the south, can be interpreted to have been governed by the distance from the uplifted provenance areas to the depositional basins, or to have been the result of a progressive southward advance of the Laramide orogenic activity. However, in attempting to explain the absence of coarse terrigenous sediments in the lower Cenozoic sequences of eastern Mexico, one must consider that the Sierra Madre Oriental, believed to be the source of at least part of the sediments in these sequences, is predominantly built of Cretaceous carbonates and shales, and is thus an improbable source of coarse terrigenous clastic sediments. It is conceivable that coarse clastics were not brought into the Cenozoic basins of eastern Mexico until the Sierra Madre Oriental had been deeply dissected, earlier in some areas and later in others, or until the rivers draining into these basins were able to cut back westward beyond the Sierra Madre Oriental. In any case, this is an interesting problem that needs further investigation.

The Cenozoic sedimentary basins of eastern Mexico are not all alike. They have different and distinctive stratigraphic and structural characteristics that make it almost necessary, certainly desirable and more convenient, to discuss them as three separate units: (1) the Burgos basin; (2) the Tampico-Misantla and Veracruz basins; and (3) the southeastern basins—the Isthmus Saline, Comalcalco, and Macuspana basins.

The Burgos basin. The name "Burgos basin" has been given to the part of the Rio Grande embayment that extends southward into northeastern Mexico (see Plate 2). The Cenozoic stratigraphy and depositional history of this area is very similar, therefore, to that of the rest of the Rio Grande embayment of south Texas (Figs. 27 to 30). The main differences result from the closer proximity of the Burgos basin to the Sierra Madre Oriental and, consequently, the stronger reflection in this basin of the tectonic activity of the Laramide orogeny.

As in south Texas, the lower Paleocene section is predominantly composed of a westward-transgressive shale unit (the Midway Formation) that covers the Upper Cretaceous section, unconformably in updip areas. Paleocene fluvial-deltaic complexes have been recognized along the western rim of the Burgos basin. The Midway shale section is overlain by the late Paleocene–early Eocene strand-plain–barrier-bar system of the Wilcox Formation. The Wilcox, in turn, is overlain by a succession of prograding clastic wedges of Eocene age that, like in south Texas, represent alternating thick fluvial-deltaic prograding systems, and thinner transgressive, generally finer-grained units deposited over the shelves built by the preceeding prograding wedges (the Reklaw, Queen City, Weches, Cook Mountain, Yegua and Jackson Formations). The deltaic coarse-clastic sections grade basinward to shalier marine sections. The same depositional patterns prevalent during the Eocene persisted, with limited modifications, during the Oligocene and Miocene (Vicksburg, Frio, Catahoula, and Oakville Formations).

Each coarse-clastic wedge prograded farther eastward into the basin, and, as a result, the shoreline and the shelf margin advanced in the same direction. The Pliocene-Quaternary section represents a short transgression followed by a regressive depositional episode that has continued to the present. The Cenozoic section in the Burgos basin reaches 10,000 m in thickness.

Growth-fault systems were developed at various times during the Cenozoic in front of the deltaic systems and along the margins of the shelf. For the most part they were the continuation into Mexico of the Cenozoic fault zones or "flexures" of south Texas—the Wilcox, Vicksburg, and Frio fault zones.

The Tampico-Misantla and Veracruz basins. The formation and development of the Tampico-Misantla and Veracruz basins during the Cenozoic were closely tied to the tectonic events of the Laramide orogeny. During the Late Cretaceous, the orogeny ended the period of carbonate deposition in east-central Mexico and caused the uplift of the Sierra Madre Oriental, and the formation of the two basins. The Sierra Madre Oriental is also believed to have been the source of the bulk of the sediments deposited in the basins during the Cenozoic. The maximum thickness of the Cenozoic section is 6,000 m in the Tampico-Misantla basin and 8,000 m in the Veracruz basin.

The western flank of both the Tampico-Misantla and Veracruz basins is characterized by a complex thrust-fold compressional belt, the front of the Sierra Madre Oriental. The east flank of the Tampico-Misantla basin, and its northern extension, the Magiscatzin basin, is a pre-Cenozoic high structural belt that includes the Tamaulipas arch, its offshore extension, and the Tuxpan platform (see Plate 2). A similar high structural trend did not develop in the Veracruz basin until Pliocene time.

The Paleocene section in the Tampico-Misantla and Veracruz basins is predominantly composed of marine shales, the lower and middle parts of the Velasco Formation. The area of the present two basins was then a continuous shelf bordering the deep central Gulf of Mexico. The uplift of the Santa Ana (Teziutlan) massif in late Paleocene time split the shelf in two, and resulted in the inception of the Tampico-Misantla basin to the north and the Veracruz basin to the south. It has been speculated that much of the Mesozoic section was eroded from the massif and that the instability resulting from its uplift contributed to the formation of submarine canyons at the shelf margin of the time, both north and south of the massif. The sediments derived from the erosion of the Santa Ana massif are believed to have contributed to the cutting of the submarine canyons (see Chapter 11, this volume).

During most of the Eocene, deposition of a marine section composed predominantly of shale continued in the shelfal areas of the Tampico-Misantla and Veracruz basins—the upper part of the Velasco, and the Aragon and Guayabal formations. The submarine canyons were filled with fine-grained clastic sediments, the Chicontepec Formation, in part of turbiditic character, derived from the west and from the Tuxpan platform, that had remained emerged during the Paleocene and Eocene (Fig. 27).

The late Middle or Late Eocene witnessed the first widespread supply of terrigenous coarse-clastic sediments into the Tampico-Misantla and Veracruz basins. This has been attributed to the last episodes of the Laramide orogeny and the continuing uplift of western uplands, this time in closer proximity to the present Cenozoic basins of east-central Mexico. A marked mid-Eocene unconformity in the Veracruz basin bears witness to these orogenic episodes (Mossman and Viniegra, 1976; Cruz Helú and others, 1977). The Middle and/or Upper Eocene clastics above the unconformity overlie Cretaceous to Lower and Middle Eocene units. The Middle to Late Eocene tectonic activity also caused the tilting of the Córdoba platform to the west of the Veracruz basin, and the renewed formation of compressional thrust and fold structures along the western flank of the basin.

The coarse-clastic Eocene sections, composing the Tantoyuca Formation, are best developed along the western flanks of the Tampico-Misantla and Veracruz basins; they grade eastward into finer-grained, marine shales, the Chapopote Formation. Two prograding deltaic complexes, the so-called Cazones and Tecolutla deltas (the first described in Chapter 11, this volume) have been recognized in the Tampico-Misantla basin; others undoubtedly were developed in the Veracruz basin.

Strong influx of terrigenous coarse clastic sediments continued in the Tampico-Misantla and Veracruz basins during the Oligocene, Miocene, and Pliocene (Figs. 28, 29, and 30).

In the Tampico-Misantla basin, the Oligocene is represented along the western flank of the basin by coarse-grained nonmarine and shallow-water marine units reflecting alternating transgressive and regressive episodes—the Mesón and Palma Real Formations—that grade eastward to marine shale sections, the Horcones and Alazán formations. A shallowing of the basin and a regressive episode took place at the end of the Oligocene, caused by either a further uplift of the region or by a lowering of the sea level. It was followed during the early Miocene by a trasngression that deposited the lower Miocene section—the Coatzintla and Tuxpan Formations—unconformably over the underlying Oligocene. The remaining Miocene section is represented by a regressive sequence of terrigenous clastic rocks. The coast of the Gulf of Mexico was progressively shifted eastward to its present position. Pliocene and Quaternary sediments are restricted to the present coast of east-central Mexico, and to the continental shelf and slope, under the waters of the Gulf of Mexico. In the Tampico-Misantla basin, as well as throughout the length of the eastern Mexico Gulf coastal plain, there is abundant evidence of volcanic activity during the Tertiary—small, local intrusions and considerable amounts of tuffaceous and volcanic ash in the Tertiary sedimentary section probably derived from volcanic centers in the Sierra Madre Occidental, along the Pacific margin of Mexico.

The Oligocene section of the Veracruz basin is also composed of coarse clastics to the west, the La Laja Formation, that grade to the marine shale sections of the Chapopote and Horcones formations to the east. In the early Miocene, thick prograding clastic wedges covered the basin, causing the development of growth-fault systems. Cruz Helú and others (1977) recognized a major unconformity near the base of the Miocene section, but the exact age of this unconformity and its cause is still debated. The "lower Miocene" onlaps westward over the eroded and complexly folded and thrust-faulted Cretaceous and lower Tertiary section. The Miocene and Pliocene are represented by a thick terrigenous clastic sequence difficult to subdivide into distinctive lithostratigraphic units due to the uniform lithologic character of the section. Quaternary clastic sediments are present along the foothills of the Sierra Madre Oriental, along the coast, and offshore. Volcanic activity in the Veracruz basin reached its peak during the Pliocene and the Quaternary, both to the north, along the Trans-Mexican Neovolcanic belt, and to the south, in the Los Tuxtlas volcanic center. In both areas, several volcanoes have been active in historical time. The Pliocene episodes of volcanic activity caused the uplift of the eastern side of the Veracruz basin.

Along the western continental slope of the Gulf of Mexico, offshore from the Tampico-Misantla and Veracruz basins, Miocene(?), Pliocene, and Pleistocene sediments are involved in a very large basinward gravity slump or decollement system that is bounded updip near the coast by a growth-fault zone, and below by a very low angle detachment surface, probably within the lower Tertiary section. The Miocene(?)-Pliocene-Pleistocene section is deformed into a series of subparallel folds, with wavelengths that average 10–12 km, and a relief that ranges from 300 m to as much as 1 km. These folds are reflected in the bathymetry of the east Mexico slope, at depth of 1500 to 3000 m, as prominent ridges—the so-called "Mexican Ridges" (see Chapter 2, this volume, and Plate 1).

The southeastern basins. The structural architecture of the Cenozoic basins of southeastern Mexico, east of the Isthmus of Tehuantepec, contrasts markedly with that of the Tampico-Misantla and Veracruz basins. While these two basins are bounded to the west by the compressive eastern front of the Sierra Madre Oriental, which reflects the effects of the Laramide orogeny, the structure of the southeastern basins is dominated by tensional and transcurrent deformation. The compressive front of the Sierra Madre Oriental cannot be prolonged across the isthmus, and the tectonic activity of the Laramide orogeny in southeastern Mexico is not clearly recognized. In addition, the Middle Jurassic salt underlies the southeastern basins, and has played an important role in the structural configuration and depositional patterns of their Cenozoic sedimentary section, particularly in the Isthmus Saline basin. In the Tampico-Misantla and Veracruz basins, on the other hand, the salt is not present, and salt deformation, therefore, plays no role in their structural development. The Comalcalco and Macuspana basins are bounded by large normal faults; and between them rises the Villahermosa uplift, a large horst (see Plate 2).

South of the southeastern basins is the Sierra de Chiapas, where the Jurassic, Cretaceous, and lower Tertiary section is folded and strike-slip faulted. South of the Sierra de Chiapas is the Chiapas massif, composed of pre-Mesozoic igneous and metamorphic rocks. The Tertiary section of the Sierra de Chiapas represents a transition between terrigenous clastic sequences to the south, probably derived from the Chiapas massif, and including turbidites deposited as submarine fans, and predominantly carbonate sections to the north, toward the Yucatan Peninsula, where the Tertiary grades into a section entirely composed of platform carbonates. The Quaternary consists mainly of pyroclastic rocks and andesitic lavas, the product of the eruption of several volcanoes in the Sierra de Chiapas.

In the southeastern basins—the Isthmus Saline, Comalcalco, and Macuspana basins—and in their extensions into the Bay of Campeche, the Paleocene, Eocene, and Oligocene are represented by a thick marine shale section, the Nanchital Shale and equivalents (Figs. 27 and 28). At the base of the Paleocene section in the areas closer to the Yucatan platform—the Bay of Campeche and the Macuspana basin—is a widespread breccia composed of fragments of Cretaceous carbonates derived from the platform. The breccia reflects a period of erosion of the Yucatan platform, probably during the lowering of the sea level that characterized the end of the Cretaceous in the Gulf of Mexico basin region. Breccia lenses are locally found higher in the Paleocene-Eocene section. Bentonitic shales and bentonites are common in the Paleogene section, particularly in its upper part (Eocene and Oligocene).

Terrigenous coarse-clastic sediments did not reach the southeastern basins until the Miocene. They are believed to have been derived from the south, from the Chiapas massif, and to have prograded northward over the present area of the coastal plain of the Gulf of Mexico, where they covered unconformably the older Tertiary section. This significant influx of coarse clastics has been interpreted to indicate an uplift of the Chiapas massif during the late Oligocene or early Miocene.

The Miocene, Pliocene, and Pleistocene sections of the southeastern basins are composed of a uniform and monotonous alternation of sandstones, siltstones, and shales containing abundant tuffaceous and bentonitic material indicative of volcanic activity in the region, probably along the Pacific margin of Mexico (Figs. 29 and 30). The northernmost Miocene sections in the Macuspana basin and in the Gulf of Campeche are composed predominantly of marine shales with subordinate coarser-clastic intervals, some of which have been interpreted as having been deposited in deep water as submarine fans. The Pleistocene section includes fluvial-deltaic systems that prograde toward the northwest and that were probably deposited by the ancestors of the Coatzacoalcos, Uzpanapa, Grijalva, and Usumacinta Rivers. Holocene barrier-island and lagoonal sediments are common along the coast. Farther inland the Holocene is represented by fluvial terraces along the course of the main rivers of the region.

The Cenozoic section attains a thickness of 10,000 m in the Comalcalco and Macuspana basins, and is somewhat less thick in the Isthmus Saline basin.

The Florida and Yucatan platforms

Since the Cretaceous, the Florida and Yucatan platforms have been extremely stable areas covered by very shallow water, where carbonates and occasional evaporites accumulated during the Cenozoic (Figs. 27 to 30).

In the Florida platform, the lower part of the Cenozoic (Paleocene, Eocene, and Oligocene) is composed predominantly of shallow-water limestones and dolomites. Evaporites are common in the Paleocene and Lower Eocene sections (the Cedar Keys Formation), but seem to be absent in the rest of the Eocene and Oligocene units (Oldsmar, Avon Park, Ocala, and Suwannee formations). In the western part of the Florida panhandle, terrigenous clastics increase as the lower Cenozoic section grades into the predominantly terrigenous-clastic Paleogene sequences of the northern province of the Gulf of Mexico basin.

A renewal of the uplift of the southern Appalachian Mountains during the early Miocene caused a strong southward flood of terrigenous coarse clastics to prograde over most of the Florida platform (Hawthorne Group and Pliocene units). Only in the southernmost part of the Florida Peninsula did carbonate deposition persist during the Miocene and Pliocene. A stratigraphic hiatus and disconformable relation have been reported at the base of the Miocene section in peninsular Florida. Clastic deposition predominated during the Miocene and Pliocene over most of the Florida platform; widespread carbonate deposition did not resume until the Quaternary.

The Holocene is represented along the coastal areas by alluvial and marine terraces, reefs, and dune, beach, and marsh deposits.

The Cenozoic section in the Florida platform ranges in thickness from a feather edge along the foothills of the Appalachian Mountains to the north, to almost 2,000 m in the South Florida basin.

It should be expected that eustatic sea-level changes would leave a clear record in the sedimentary record of a shallow submarine carbonate platform such as the Florida platform. Some stratigraphic hiatuses have been reported from this platform; others have been postulated (see Plate 5). The monotonous nature of the lithology of the Cenozoic carbonate section makes it difficult, however, to document these hiatuses in the absence of good biostratigraphic control. A stratigraphic hiatus at the base of the Tertiary seems to reflect the lowering of the sea level at the end of the Cretaceous. Another has been recognized at the base of the terrigenous-clastic Miocene section, as discussed above; others are less evident. Detailed geological and geophysical investigations may substantiate the presence and nature of these and other stratigraphic hiatuses and disconformities, and document the history of relative sea-level changes in the Florida platform region.

The Cenozoic section of the Yucatan platform is not as well known as that of the Florida platform. Only a few wells have been drilled through the section and information on them is scarce. The Cenozoic is composed predominantly of platform carbonates, generally light-colored and massive, with lesser amounts of interbedded evaporites in the lower part (Paleocene). Calcareous mudstones and argillaceous limestones become increasingly abundant toward the north. Terrigenous clastics are entirely absent from the section, because the platform was a long distance from possible terrigenous-clastic provenance areas, unlike the Florida platform. The upper Pliocene and Pleistocene section is characterized by bioclastic packstones.

The Cenozoic of the Yucatan platform thickens from less than 100 m in the southern Yucatan Peninsula in northern Guatemala and Belize, to 1,000 m in the northern part of the peninsula, and to about 2,000 m along the western margin of the Yucatan shelf, under the waters of the Gulf of Mexico.

The stratigraphic breaks recognized in the Florida platform that are believed to have resulted from sea-level oscillations during the Cenozoic have not been reported from the Yucatan platform, though they may be present. More stratigraphic information and more careful study may reveal them.

SEDIMENTARY CYCLES IN THE GULF OF MEXICO BASIN

In view of the recent strong interest in sedimentary sequences, depositional cyclicity, stratigraphic units bounded by major regional unconformities, and the possible relation between them and eustatic sea-level changes, it was originally planned that a separate chapter of this volume would summarize the evidence for such stratigraphic features and relations in the Gulf of Mexico basin. It was hoped that such a chapter would discuss the grounds for the recognition of regional sedimentary cyclicity in the stratigraphic section of the basin, and attempt to determine the geological processes that had controlled them, whether eustatic sea-level changes, tectonic events, or a combination of the two.

For numerous reasons, these objectives could not be attained: while many sedimentary cycles have been described from the stratigraphic section of the Gulf of Mexico basin, and many of them have been discussed in Chapters 8 to 13 of this volume, most have only been recognized in restricted regions of the basin, either because they have only a local development, or because the information necessary for their detection is not available or has not been properly interpreted.

In addition, the stratigraphy and structure of large areas of the basin is not propitious for the detection and establishment of sedimentary cycles: in the central part of the basin, sedimentation in deep water has been essentially continuous since the Late Jurassic, and only seismic information is available in the area; along the northern outer shelf and slope of the Gulf of Mexico, the sedimentary section has been affected, and to a great extent masked, by the recurrent and complex deformation of the Middle Jurassic salt; in the Florida and Yucatan carbonate platforms, the nearly uniform lithologic composition of major parts of the section does not lend itself to the ready recognition of stratigraphic breaks and sedimentary cycles, particularly in Yucatan, where only a few wells have been drilled.

Outcrop studies in the Gulf of Mexico basin have been limited to its rims, and even there the study of stratigraphic sections for the purpose of recognizing sedimentary cycles faces many restrictions, principally the onlap of the older sequences by upper Tertiary and Quaternary deposits. The Jurassic section, for example, crops out only along the western flank of the basin, in eastern Mexico; the Lower Cretaceous does not crop out east of the Mississippi River or in the Florida and Yucatan platforms; the Upper Cretaceous is not much better represented at the surface. In northeastern and east-central Mexico, where the best exposures of the Mesozoic section are present, much of this section represents essentially continuous deposition in deep-water environments.

Finally, the comparison of the sedimentary cycles and stratigraphic discontinuities recognized in the Gulf of Mexico basin with those reported from other regions of the world (Vail and others, 1977, and many other publications), is often made difficult, if not impossible, by the scarcity or absence of age-diagnostic fossils in many of the sedimentary sections of the basin. This is particularly true in the case of the extensive fluvial-deltaic systems common in the northern flank of the basin.

The northern Gulf of Mexico basin, the United States Gulf coastal plain region, and the northern shelf of the Gulf of Mexico, offer the best and most abundant sources of information for the study and recognition of sedimentary cyclicity in the basin. As a result, numerous publications are available that discuss the presence in the region of a considerable number of sedimentary sequences or cycles, with or without related unconformities or depositional breaks. Best documented are those in the Upper Jurassic and younger sections. Much less information is available about older rocks. Among the most evident and well-defined sedimentary cycles and their boundary unconformities or depositional breaks are, from older to younger, as follows.

1. A marked shallowing of the section in the upper part of the Oxfordian (Smackover and Zuloaga Formations) is apparent over most of the northern and northwestern Gulf of Mexico basin; it ended with the deposition of the Buckner and Olvido evaporites. A depositional break develops at this level in updip areas.

2. An unconformity has been reported between the Kimmeridgian and the Tithonian in the north-central part of the basin. The section above the unconformity (the Cotton Valley Group) onlaps and transgresses over the unconformity surface.

3. A regression during Valanginian time has been recorded by an unconformity separating the Cotton Valley below from the Hosston Formation above.

4. A "mid-Aptian break" has been recognized throughout the northwestern Gulf of Mexico basin. It represents an abrupt influx of fine-grained terrigenous clastic sediments into the basin and a marine transgression that terminated the widespread platform-carbonate deposition that had prevailed earlier in the area. This important break, at the base of the predominantly shale sections of the Pearsall, La Peña, and Otates Formations, has been chosen as the boundary between the provincial Coahuilan and Comanchean "Series."

5. Probably the most conspicuous stratigraphic break and unconformity of the Upper Jurassic and younger sedimentary section of the northern Gulf of Mexico basin is the mid-Cenomanian (or mid-Cretaceous) unconformity, discussed earlier in this chapter. It has been selected as the boundary between the Comanchean and Gulfian "Series" of local usage.

6. A stratigraphic break and disconformity is present at the Cretaceous/Tertiary boundary in the more positive areas around the periphery of the Gulf of Mexico basin.

7. The Tertiary stratigraphic section of the northern, as well as the western, flank of the Gulf of Mexico basin is composed of multiple sedimentary sequences, each characterized by initial progradation and the resulting basinward advance of the coast and shelf margin, followed by transgression and landward retreat of the shoreline (see Chapter 11, this volume). Each sequence represents a distinct depositional episode or cycle.

8. During the Pleistocene, fluctuations of the sea level, in response to continental glacial episodes, resulted in well-defined sedimentary cycles.

While distinct and generally accepted along the northern Gulf of Mexico basin, not all of these depositional breaks and unconformities, and the sedimentary sequences or cycles they bind, can be clearly detected in other parts of the basin. The most prominent exceptions are the mid-Cenomanian unconformity, the pre-Tertiary unconformity, and the Pleistocene sedimentary cycles controlled by glacial events. They are all recognizable throughout most of the basin.

It is conceivable that the stratigraphic breaks recognized in the northern flank of the basin are also present along its western flank and in the Florida and Yucatan platforms, but that insufficient available information, or the common acceptance of alternative interpretations of the stratigraphy of the areas away from the northern Gulf of Mexico basin, are the reasons for the failure to identify them. It is also possible that the sedimentary cycles and stratigraphic breaks of the northern Gulf of Mexico basin were caused not by basinwide events such as eustatic sea-level changes, but by local controlling factors, either tectonic or sedimentary— gentle uplift or downwarping of the northern flank of the basin, uplift resulting from igneous activity, variation in the supply of sediments to the basin, or shift of the fluvial-deltaic systems, among others. It is also very possible that some of the sedimentary cycles and breaks were the product of the interplay of sedimentary, tectonic, and eustatic causes, difficult to separate on the basis of present information (see Chapter 11, this volume).

The identification of regional sedimentary cycles recognizable over the entire basin, and their relation to worldwide events, proved not to be as easy as originally envisioned. To determine which of the sedimentary cycles are governed by regional causes, which are strictly local, which are due to tectonic or sedimentary events, and which are the response to eustatic sea-level changes, remains a difficult but inviting field for research.

REFERENCES CITED

Anderson, T. H., and Schmidt, V. A., 1983, The evolution of Middle America and the Gulf of Mexico-Caribbean Sea region during Mesozoic time: Geological Society of America Bulletin, v. 94, p. 941–966.

Boyd, D. B., and Dyer, B. F., 1964, Frio barrier bar system of south Texas: Gulf Coast Association of Geological Societies Transactions, v. 14, p. 309–322.

Buffler, R. T., and Sawyer, D. S., 1985, Distribution of crust and early history, Gulf of Mexico Basin: Gulf Coast Association of Geological Societies Transactions, v. 35, p. 333–344.

Buffler, R. T., Watkins, J. S., Shaub, F. J., and Worzel, J. L., 1980, Structure and early geologic history of the deep central Gulf of Mexico Basin, in Pilger, R. H., Jr., ed., The origin of the Gulf of Mexico and the early opening of the central North Atlantic Ocean: Baton Rouge, Louisiana State University, p. 3–16.

Bullard, E., Everett, J. E., and Smith, A. G., 1965, The fit of the continents around the Atlantic, in Blackett, P.M.S., Bullard, E., and Runcorn, S. K., eds., A symposium on continental drift: Royal Society of London, Philosophical Transactions No. 1088, p. 41–51.

Carey, S. W., 1958, The tectonic approach to continental drift, in Carey, S. W., ed., Continental drift—A symposium: Hobart, University of Tasmania, p. 177–355.

——, 1976, The expanding Earth (Developments in Geotectonics 10): New York, Elsevier Scientific Publishing Co., 488 p.

Cebull, S. E., Shurbet, D. H., Keller, G. R., and Russell, L. R., 1976, Possible role of transform faults in the development of apparent offset in the Ouachita-Southern Appalachian tectonic belt: Journal of Geology, v. 84, p. 107–114.

Cruz Helú, P., Berdugo V., R., and Bárcenas P., R., 1977, Origin and distribution of Tertiary conglomerates, Veracruz Basin, Mexico: American Association of Petroleum Geologists Bulletin, v. 61, p. 207–226.

Dietz, R. S., and Holden, J. C., 1970, Reconstruction of Pangaea: Breakup and dispersion of continents, Permian to Recent: Journal of Geophysical Research, v. 75, p. 4939–4956.

Dumble, E. T., 1918, The geology of east Texas: University of Texas Bulletin 1869, 388 p.

Dunbar, J. A., and Sawyer, D. S., 1987, Implications of continental crust extension for plate reconstruction: An example from the Gulf of Mexico: Tectonics, v. 6, p. 739–755.

Eardley, A. J., 1954, Tectonic relations on North and South America: American Association of Petroleum Geologists Bulletin, v. 38, p. 707–773.

Ewing, J., Antoine, J., and Ewing, M., 1960, Geophysical measurements in the western Caribbean and in the Gulf of Mexico: Journal of Geophysical Research, v. 65, p. 4087–4126.

Ewing, J. I., Worzel, J. L., and Ewing, M., 1962, Sediments and oceanic structural history of the Gulf of Mexico: Journal of Geophysical Research, v. 67, p. 2509–2527.

Ewing, M., Worzel, J. L., Ericson, D. B., and Heezen, B. C., 1955, Geophysical and geological investigations in the Gulf of Mexico, Part I: Geophysics, v. 20, p. 1–18.

Faust, M. J., 1990, Seismic stratigraphy of the mid-Cretaceous unconformity (MCU) in the central Gulf of Mexico Basin: Geophysics, v. 55, p. 868–884.

Feeley, M. H., Moore, T. C., Jr., Loutit, T. S., and Bryant, W. R., 1990, Sequence stratigraphy of Mississippi Fan related to oxygen isotope sea level index: American Association of Petroleum Geologists Bulletin, v. 74, p. 407–424.

Fisher, W. L., and McGowen, J. H., 1967, Depositional systems in the Wilcox Group of Texas and their relationship to occurrence of oil and gas: Gulf Coast Association of Geological Societies Transactions, v. 17, p. 105–125.

Fisher, W. L., Proctor, C. V., Jr., Galloway, W. E., and Nagle, J. S., 1970, Depositional systems in the Jackson Group of Texas, their relationship to oil, gas, and uranium: Gulf Coast Association of Geological Societies Transactions, v. 20, p. 234–261.

Freeland, G. L., and Dietz, R. S., 1971, Plate tectonic evolution of Caribbean–Gulf of Mexico region: Nature, v. 232, no. 5305, p. 20–23.

Freeman-Lynde, R. P., 1983, Cretaceous and Tertiary samples dredged from the Florida Escarpment: Gulf Coast Association of Geological Societies Transactions, v. 33, p. 91–100.

Galloway, W. E., 1968, Depositional systems of the lower Wilcox Group, north-central Gulf Coast Basin: Gulf Coast Association of Geological Societies Transactions, v. 18, p. 275–289.

Goetz, L. K., and Dickerson, P. W., 1985, A Paleozoic transform margin in Arizona, New Mexico, west Texas and northern Mexico, *in* Dickerson, P. W., and Muehlberger, W. R., eds., Structure and tectonics of Trans-Pecos Texas: West Texas Geological Society Field Conference, Publication 85-81, p. 173–184.

Gose, W. A., and Sanchez-Barreda, L. A., 1981, Paleomagnetic results from southern Mexico: Geofísica Internacional, v. 20, p. 163–175.

Graham, S. A., Dickinson, W. R., and Ingersoll, R. V., 1975, Himalayan-Bengal model for flysch dispersal in the Appalachian-Ouachita system: Geological Society of America Bulletin, v. 86, p. 273–286.

Guevara, E. H., and Garcia, R., 1972, Depositional systems and oil-gas reservoirs in the Queen City Formation (Eocene), Texas: Gulf Coast Association of Geological Societies Transactions, v. 22, p. 1–22.

Hall, D. J., Cavanaugh, T. D., Watkins, J. S., and McMillen, K. J., 1982, The rotational origin of the Gulf of Mexico based on regional gravity data, *in* Watkins, J. S., and Drake, C. L., eds., Studies in continental margin geology: American Association of Petroleum Geologists Memoir 34, p. 115–126.

Handschy, J. W., Keller, G. R., and Smith, K. J., 1987, The Ouachita System in northern Mexico: Tectonics, v. 6, p. 323–330.

Hatcher, R. D., Jr., Thomas, W. A., and Viele, G. W., eds., 1989, The Appalachian-Ouachita orogen in the United States: Boulder, Colorado, Geological Society of America, The Geology of North America, v. F-2, 767 p.

Humphris, C. C., Jr., 1978, Salt movement on continental slope, northern Gulf of Mexico: American Association of Petroleum Geologists Bulletin, v. 63, p. 782–798.

Kanasewich, E. G., Havskov, J., and Evans, M. E., 1978, Plate tectonics in the Phanerozoic: Canada Journal of Earth Science, v. 15, p. 919–955.

King, P. B., 1975, Ancient southern margin of North America: Geology, v. 3, p. 732–734.

———, 1977, Marathon revisited, *in* Stone, C. G., ed., Symposium on the Ouachita Mountains: Arkansas Geological Commission, p. 41–69.

Klitgord, K. D., Popenoe, P., and Schouten, H., 1984, Florida: A Jurassic transform plate boundary: Journal of Geophysical Research, v. 89, p. 7753–7772.

Ladd, J. W., Buffler, R. T., Watkins, J. S., Worzel, J. L., and Carranza, A., 1976, Deep seismic reflection results from the Gulf of Mexico: Geology, v. 4, p. 365–368.

Mack, G. H., Thomas, W. A., and Horsey, C. A., 1983, Composition of Carboniferous sandstones and tectonic framework of southern Appalachian-Ouachita orogen: Journal of Sedimentary Petrology, v. 53, p. 931–946.

May, P. R., 1971, Pattern of Triassic-Jurassic dikes around the North Atlantic in the context of pre-drift positions of the continents: Geological Society of America Bulletin, v. 82, p. 1285–1292.

Meyerhoff, A. A., 1967, Future hydrocarbon provinces of Gulf of Mexico–Caribbean region: Gulf Coast Association of Geological Societies Transactions, v. 17, p. 217–260.

Miser, H. D., 1921, Llanoria, the Paleozoic land area in Louisiana and eastern Texas: American Journal of Science, ser. 5, v. 2, p. 61–89.

Moore, G. W., and del Castillo, L., 1974, Tectonic evolution of the southern Gulf of Mexico: Geological Society of America Bulletin, v. 85, p. 607–618.

Mossman, R. W., and Viniegra, F., 1976, Complex fault structures in Veracruz Province of Mexico: American Association of Petroleum Geologists Bulletin, v. 60, p. 379–388.

Nunn, J. A., Scardina, A. D., and Pilger, R. H., Jr., 1984, Thermal evolution of the north-central Gulf Coast: Tectonics, v. 3, p. 723–740.

Owen, H. C., 1976, Continental displacement and expansion of the Earth during the Mesozoic and Cenozoic: Royal Society of London Philosophical Transactions, ser. A, v. 281, p. 223–291.

———, 1983, Atlas of continental displacement—200 million years to the present: Cambridge, Cambridge University Press, 159 p.

Paine, W. R., and Meyerhoff, A. A., 1970, Gulf of Mexico Basin: Interactions among tectonics, sedimentation, and hydrocarbon accumulation: Gulf Coast Association of Geological Societies Transactions, v. 20, p. 5–44.

Pilger, R. H., Jr., 1978, A close Gulf of Mexico, pre-Atlantic Ocean plate reconstruction and the early rift history of the Gulf and North Atlantic: Gulf Coast Association of Geological Societies Transactions, v. 28, p. 385–393.

———, 1981, The opening of the Gulf of Mexico: Implications for the tectonic evolution of the northern Gulf Coast: Gulf Coast Association of Geological Societies Transactions, v. 31, p. 377–381.

Pindell, J. L., 1985, Alleghenian reconstruction and subsequent evolution of the Gulf of Mexico, Bahamas and Proto-Caribbean: Tectonics, v. 4, p. 1–39.

Pindell, J. L., and Dewey, J. F., 1982, Permo-Triassic reconstruction of western Pangea and the evolution of the Gulf of Mexico/Caribbean region: Tectonics, v. 1, p. 179–211.

Ricoy, J. U., and Brown, L. F., Jr., 1977, Depositional systems in the Sparta Formation (Eocene), Gulf Coast Basin of Texas: Gulf Coast Association of Geological Societies Transactions, v. 27, p. 139–154.

Ross, M. I., and Scotese, C. R., 1988, A hierarchical tectonic model of the Gulf of Mexico and Caribbean region: Tectonophysics, v. 155, p. 139–168.

Royden, L. H., Burchfield, B. C., Ye, H., and Schuepbach, M. S., 1990, The Ouachita-Marathon thrust belt: Orogeny without collision: Geological Society of America Abstracts with Programs, v. 22, no. 7, p. A112.

Salvador, A., 1987, Late Triassic-Jurassic paleogeography and origin of Gulf of Mexico Basin: American Association of Petroleum Geologists Bulletin, v. 71, p. 419–451.

Salvador, A., and Green, A. R., 1980, Opening of the Caribbean Tethys (origin and development of the Caribbean and the Gulf of Mexico), *in* Aubouin, J., Debelmas, J., and Latreille, M., eds., Colloque C5—Geology of the Alpine Chains born of the Tethys (26th International Geological Congress, Paris, 1980): Bureau de Recherches Geologiques et Minieres Memoire 115, p. 224–229.

Schlager, W., Buffler, R. T., Angstadt, D., and Phair, R., 1984, Geologic history of the southeastern Gulf of Mexico, *in* Buffler, R. T., Schlager, W., and others, Initial reports of the Deep Sea Drilling Project, Volume 77: Washington, D.C., U.S. Government Printing Office, p. 715–738.

Schuchert, C., 1909, Paleogeography of North America: Geological Society of America Bulletin, v. 20, p. 427–606.

——, 1923, Sites and nature of the North American geosynclines: Geological Society of America Bulletin, v. 34, p. 151–230.

——, 1929, Geological history of the Antillean region: Geological Society of America Bulletin, v. 40, p. 337–360.

——, 1935, Historical geology of the Antillean-Caribbean region: New York, John Wiley & Sons, 811 p.

Scotese, C. R., 1984, An introduction to this volume: Paleozoic paleomagnetism and the assembly of Pangea, *in* Van der Voo, R., Scotese, C. R., and Bonhommet, N., eds., Plate reconstruction from Paleozoic paleomagnetism: American Geophysical Union Geodynamics Series, v. 12, p. 1–10.

Scotese, C. R., Bambach, R. K., Barton, C., Van der Voo, R., and Ziegler, A. M., 1979, Paleozoic base maps: Journal of Geology, v. 87, p. 217–277.

Scott, R. W., Frost, S. H., and Shaffer, B. L., 1988, Early Cretaceous sea-level curves, Gulf Coast and southeastern Arabia, *in* Wilgus, C. K., and five others, eds., Sea-level changes—An integrated approach: Society of Economic Paleontologists and Mineralogists Special Publication 42, p. 275–284.

Sellards, E. H., 1932, The pre-Paleozoic and Paleozoic Systems in Texas, *in* The geology of Texas, Volume I, Stratigraphy: University of Texas Bulletin 3232, p. 15–238.

Shurbet, D. H., and Cebull, S. E., 1975, The age of the crust beneath the Gulf of Mexico: Tectonophysics, v. 28, p. T25–30.

——, 1987, Tectonic interpretation of the westernmost part of the Ouachita-Marathon (Hercynian) orogenic belt, west Texas–Mexico: Geology, v. 15, p. 458–461.

Smith, A. G., and Briden, J. C., 1977, Mesozoic and Cenozoic paleocontinental maps: Cambridge, Cambridge University Press, 63 p.

Smith, A. G., Briden, J. C., and Drewry, G. E., 1973, Phanerozoic world maps, *in* Hughes, N. F., ed., Organisms and continents through time: Paleontological Association of London, Special Papers in Paleontology 12, p. 1–42.

Smith, C. I., 1981, Review of geologic setting, stratigraphy and facies distribution of the Lower Cretaceous in northern Mexico, *in* Lower Cretaceous stratigraphy and structure, northern Mexico Field trip guidebook: West Texas Geological Society Publication 81-74, p. 1–27.

Stewart, J. H., 1988, Late Proterozoic and Paleozoic southern margin of North America and the accretion of Mexico: Geology, v. 16, p. 186–189.

Thomas, W. A., 1977, Evolution of Appalachian-Ouachita salients and recesses from reentrants and promontories in the continental margin: American Journal of Science, v. 277, p. 1233–1278.

——, 1983, Continental margins, orogenic belts, and intracratonic structures: Geology, v. 11, p. 270–272.

——, 1988, Early Mesozoic faults of the northern Gulf Coastal Plain in the context of opening of the Atlantic Ocean, *in* Manspeizer, W., ed., Triassic-Jurassic rifting (Developments in Geotectonics 22): New York, Elsevier Scientific Publishing Co., p. 463–476.

——, 1989, The Appalachian-Ouachita orogen beneath the Gulf Coastal Plain between the outcrops in the Appalachian and Ouachita Mountains, *in* Hatcher, R. D., Jr., Thomas, W. A., and Viele, G. W., eds., The Appalachian-Ouachita orogen in the United States: Boulder, Colorado, Geological Society of America, The Geology of North America, v. F-2, p. 537–553.

——, 1991, The Appalachian-Ouachita rifted margin of southeastern North America: Geological Society of America Bulletin, v. 103, p. 415–431.

Vail, P. R., Mitchum, R. M., Jr., and Thompson, S., III, 1977, Seismic stratigraphy and global changes of sea level, Part 4: Global cycles of relative changes of sea level, *in* Payton, C. E., ed., Seismic stratigraphy—Applications to hydrocarbon exploration: American Association of Petroleum Geologists Memoir 26, p. 83–97.

Van der Voo, R., and French, R. B., 1974, Apparent polar wandering for the Atlantic-bordering continents: Late Carboniferous to Eocene: Earth-Science Reviews, v. 10, p. 99–119.

Van der Voo, R., Mauk, F. J., and French, R. B., 1976, Permian-Triassic continental configurations and the origin of the Gulf of Mexico: Geology, v. 4,

p. 177–180.

Van Siclen, D. C., 1984, Early opening of initially-closed Gulf of Mexico and central North Atlantic Ocean: Gulf Coast Association of Geological Societies Transactions, v. 34, p. 265–275.

Vogler, H. A., and Robison, B. A., 1987, Exploration for deep geopressured gas: Corsair trend, offshore Texas: American Association of Petroleum Geologists Bulletin, v. 71, p. 777–787.

Walper, J. L., 1980, Tectonic evolution of the Gulf of Mexico, *in* Pilger, R. H., Jr., ed., The origin of the Gulf of Mexico and the early opening of the central North Atlantic Ocean: Baton Rouge, Louisiana State University, p. 87–98.

Walper, J. L., and Miller, R. E., 1985, Tectonic evolution of Gulf Coast basins, *in* Perkins, B. F., ed., Habitat of oil and gas in the Gulf Coast: Gulf Coast Section, Society of Economic Paleontologists and Mineralogists, Fourth Annual Research Conference Proceedings, p. 25–42.

Walper, J. L., and Rowett, C. L., 1972, Plate tectonics and the origin of the Caribbean Sea and the Gulf of Mexico: Gulf Coast Association of Geological Societies Transactions, v. 22, p. 105–116.

Walper, J. L., Henk, F. H., Jr., Louden, E. J., and Raschilla, S. N., 1979, Sedimentation on a trailing plate margin: The northern Gulf of Mexico: Gulf Coast Association of Geological Societies Transactions, v. 29, p. 188–201.

Weimer, P., 1990, Sequence stratigraphy, facies geometries, and depositional history of the Mississippi Fan, Gulf of Mexico: American Association of Petroleum Geologists Bulletin, v. 74, p. 425–453.

Wickham, J., Roeder, D., and Briggs, G., 1976, Plate tectonics models for the Ouachita fold belt: Geology, v. 4, p. 173–176.

Willis, B., 1909, Paleogeographic maps of North America: Journal of Geology, v. 17, p. 203–208.

Wilson, J. T., 1966, Did the Atlantic close and then re-open?: Nature, v. 211, p. 676–681.

Winker, C. D., and Buffler, R. T., 1988, Paleogeographic evolution of early deep-water Gulf of Mexico and margins, Jurassic to middle Cretaceous (Comanchean): American Association of Petroleum Geologists Bulletin, v. 72, p. 318–346.

Wood, M. L., and Walper, J. L., 1974, The evolution of the interior Mesozoic Basin and the Gulf of Mexico: Gulf Coast Association of Geological Societies Transactions, v. 24, p. 31–41.

Worrall, D. M., and Snelson, S., 1989, Evolution of the northern Gulf of Mexico, with emphasis on Cenozoic growth faulting and the role of salt, *in* Bally, A. W., and Palmer, A. R., eds., The geology of North America—An overview: Boulder, Colorado, Geological Society of America, The Geology of North America, v. A, p. 97–138.

Worzel, J. L., and Watkins, J. S., 1973, Evolution of the northern Gulf Coast deduced from geophysical data: Gulf Coast Association of Geological Societies Transactions, v. 23, p. 84–91.

Ziegler, A. M., Scotese, C. R., McKerrow, W. S., Johnson, M. E., and Bambach, R. K., 1979, Paleozoic paleogeography: Annual Review of Earth and Planetary Sciences, v. 7, p. 473–402.

ACKNOWLEDGMENTS

This chapter has greatly benefited from discussions with other authors of this volume, particularly W. E. Galloway, E. McFarlan, Jr., D. S. Sawyer, and N. F. Sohl. They and J. K. Arbenz, J. C. Maxwell, C. I. Smith, W. A. Thomas, and G. W. Viele reviewed parts or the entire manuscript. Their thoughtful comments and suggestions for improving the text and figures of this chapter are deeply appreciated.

MANUSCRIPT ACCEPTED BY THE SOCIETY APRIL 18, 1991

The Geology of North America
Vol. J, The Gulf of Mexico Basin
The Geological Society of America, 1991

Chapter 15

Oil and gas resources

Richard Nehring
NRG Associates, P.O. Box 1655, Colorado Springs, Colorado 80901

INTRODUCTION

The oil and gas resources of the Gulf of Mexico basin are a major contribution to the geologically derived wealth of the basin. This chapter provides a concise, essentially quantitative description of those resources.

The description is divided into four parts. The first part indicates the overall amount of known petroleum resources within the basin and the distribution of these resources by product (crude oil, natural gas, and natural gas liquids), by geographic area (subprovince), and by gross field size category. The second part discusses the history of petroleum exploration, discovery, and production within the basin.

The third part, the bulk of the chapter, describes the petroleum resources of the basin by stratigraphic unit. This description is divided into five major units: the Upper Jurassic, Lower Cretaceous, Upper Cretaceous, Paleogene, and upper Cenozoic (Neogene/Pleistocene). Within each of these units the discussion is organized by series, stage, or groups of stages. The description of each grouping covers the known amounts of petroleum, the distribution of these resources by field size categories, the producing trends within the grouping, and the fundamental factors of petroleum accumulation—reservoir, trap, seal, and source—within these trends. The chapter concludes with a brief discussion of why the Gulf of Mexico basin is so productive.

The limits on the size of this chapter do not permit discussions of even a few individual fields. For those who are interested in such discussions, many excellent publications (other than the references cited here) on oil and gas fields in the Gulf of Mexico basin have been published by the geological societies within the area (especially the Corpus Christi, East Texas, Houston, Lafayette, Mississippi, New Orleans, Shreveport, and South Texas Geological Societies), the Gulf Coast Association of Geological Societies, the Gulf Coast Section of the Society of Economic Paleontologists and Mineralogists, the various state geological surveys (especially those of Alabama and Texas), and the American Association of Petroleum Geologists, particularly the relevant volumes in their *Atlas of Oil and Gas Fields*.

Many structural features—basins, uplifts, fault systems, etc.—are mentioned in the text of this chapter, but are not shown in Figures 5 through 9 (which show major producing trends for the various stratigraphic intervals). These features can be found on Plate 2. Figure 4, the stratigraphic chart within this chapter, is of necessity simplified. More detailed charts can be found in Chapters 8, 9, 10, and 11 and on Plate 5 of this volume.

Definitions

As indicated, the discussion has a strong quantitative focus. To simplify the discussion, several terms are used frequently. The key ones are defined as follows: *Petroleum resources* will be discussed here primarily in terms of *known recovery* (the sum of cumulative production and proved reserves). Both the known recovery of individual hydrocarbons (crude oil, natural gas, and of natural-gas liquids) and the combination of these hydrocarbons into *barrels of oil equivalent* (BOE) will be considered. Natural gas is converted into BOE at a rate of 6,000 ft^3 per barrel.

Province size. The Gulf of Mexico basin, its subprovinces, and its stratigraphic units will be compared by size to other petroleum provinces around the world. The province size categories used in these comparisons include *megaprovinces* (those with more than 100 billion BOE known recovery), *superprovinces* (25 to 100 billion BOE), and *other major provinces* (7.5 to 25 billion BOE).

Play and trend size. Petroleum resources will be discussed by plays and trends. Plays are groups of geologically similar reservoirs and prospects having basically the same source-reservoir-trap controls of oil and gas. Trends are groups of one or more plays within the same formation. The size of the various producing trends within the Gulf of Mexico basin is incorporated into the discussion using three broad size categories: *major*—greater than one billion BOE known recovery; *moderate*—between 100 and 1,000 million BOE known recovery; and *minor*—less than 100 million BOE known recovery.

Field size. Several field size classifications are used, including *significant*—greater than one million BOE known recovery; *large*—50 to 500 million BOE known recovery; *giant*—greater than 500 million BOE known recovery; and *supergiant*—greater than five billion BOE known recovery. All giant fields are shown and named on Plate 4. Plate 4 also shows all other fields larger

Nehring, R., 1991, Oil and gas resources, *in* Salvador, A., ed., The Gulf of Mexico Basin: Boulder, Colorado, Geological Society of America, The Geology of North America, v. J.

than five million BOE, with those bigger than 100 million BOE being shown in brighter colors.

THE PETROLEUM RESOURCES OF THE GULF OF MEXICO BASIN

The Gulf of Mexico basin is one of the foremost petroleum provinces of the world. As of the end of 1987, it has a demonstrated ultimate known recovery of 112.7 billion barrels of crude oil, 22.5 billion barrels of natural gas liquids (for a total of 136.6 billion barrels of petroleum liquids), and 523.8 trillion ft^3 of natural gas, for a total of 222.5 billion barrels oil equivalent (Table 1).

These current estimates of known recovery provide only a minimum estimate of the ultimate potential of the Gulf of Mexico basin. There are several more billion barrels of oil equivalent resources in fields within the basin that were discovered in the 1980s but have not yet been developed. Although the largest fields in the basin have most likely been discovered, the basin still has considerable exploration potential, particularly in deeper horizons below 3,000 m (10,000 ft) and in the offshore. Continued development efforts in known fields, including new pool discoveries, infill drilling programs, enhanced oil recovery operations, and the recent burst of horizontal drilling activity, are also likely to increase ultimate recovery in the basin substantially.

As a producing petroleum province, the Gulf of Mexico basin belongs in the same rank as the Arabian-Iranian province of the Middle East and the West Siberian province of the Soviet Union. The Gulf of Mexico basin contains approximately 9% of the world's known recovery of petroleum liquids (crude oil and natural gas liquids) and approximately 11% of the world's known recovery of natural gas. Only the Arabian-Iranian province (with nearly half of the world's total) contains more petroleum liquids. Only the Arabian-Iranian and West Siberian provinces contain more natural gas. No other province contains even 5% of the world's known conventional resources of petroleum liquids or natural gas. Only if unconventional heavy oil resources are included in the calculations, do any other provinces, namely the western Canada and the eastern Venezuela provinces, possibly belong in the same league.

The Gulf of Mexico basin is primarily an oil-producing petroleum province. Of the 222.5 billion barrels oil equivalent ultimate recovery as of the end of 1987, 50.7% were crude oil, 10.1% were natural gas liquids, and only 40.2% were natural gas. The relative importance of petroleum liquids and natural gas varies substantially across the basin. The southern half (in Mexico) is highly oil prone, with more than 81% of its oil equivalent known ultimate recovery consisting of crude oil and natural gas liquids. By comparison, the northern half (in the United States) is more gas prone, gas providing more than 52% of its oil equivalent ultimate known recovery. As will be discussed later, this difference is primarily the result of differences in the maturity and nature of the source rocks in the two halves of the basin.

The Gulf of Mexico basin has been an important center of

TABLE 1. THE KNOWN PETROLEUM RESOURCES OF THE GULF OF MEXICO BASIN AS OF DECEMBER 31, 1987*

	Cumulative Production	Proved Reserves	Ultimate Recovery
Crude Oil			
	Billions of barrels		
Mexico	14.66	47.18	61.84
United States	46.54	4.34	50.88
Total	61.20	51.52	112.72
Natural Gas			
	Trillions of cubic feet		
Mexico	22.36	75.21	97.57
United States	359.94	66.32	426.26
Total	382.30	141.53	523.83
Natural Gas Liquids			
	Billions of barrels		
Mexico	1.66	6.93	8.59
United States	11.57	2.35	13.92
Total	13.23	9.28	22.51
Oil Equivalent			
	Billions of barrels		
Mexico	20.05	66.64	86.69
United States	118.10	17.75	135.85
Total	138.15	84.39	222.54

*Sources: Publications of the American Gas Association, American Petroleum Institute, Petroleos Mexicanos, U.S. Energy Information Administration.
Note: The estimates of proved reserves for Mexico include 10.91 billion barrels of crude oil, 26.69 trillion cubic feet of natural gas, and 1.32 billion barrels of natural gas liquids for a total of 16.68 billion barrels oil equivalent reserves in the Chicontepec Formation of the Tampico-Misantla basin. These reserves are at best considered to be proven *undeveloped* reserves.

petroleum production for the world for more than the past 75 years. Its contributions to world petroleum supply are not, however, simply a matter of past history. The basin will be an important source of oil and gas for several decades to come. As of the end of 1987, 45.7% of the known ultimate recovery of crude oil, 37.6% of the natural gas liquids, and 27.0% of the natural gas still remain to be produced. These proportions vary substantially between the two halves of the basin, reflecting the quite different exploration, development, and production histories of Mexico and the United States. In Mexico, where the vast petroleum resources of the Reforma and Campeche areas were not discovered and developed until the 1970s and early 1980s, only 23% of the known ultimate recovery crude oil, natural gas, and natural gas liquids has been produced. In the United States, where most of the major discoveries were made from the 1920s to the 1950s, more than 91% of the known ultimate recovery of crude oil and more than 83% of the known ultimate recovery of natural gas and of natural gas liquids have already been produced.

As befits an area where the world's first offshore well out of sight of the adjacent coast was drilled, the Gulf of Mexico basin is one of the prime centers of offshore petroleum production in the world. Only the offshore fields in the Arabian-Iranian province have a greater known ultimate recovery. The North Sea is the only other major offshore producing area containing much the same amounts as the offshore Gulf of Mexico basin. Nearly a third of the known ultimate recovery of petroleum in the Gulf of Mexico basin is in offshore fields (Table 2). Because of its later period of discovery and development, the offshore portion of the Gulf of Mexico basin is relatively more important as a current source of proved reserves for basin production. More than half of proved crude oil reserves (55.7%), nearly a third of natural gas reserves (31.5%), and 43.1% of natural gas liquids reserves are in the offshore fields of the basin.

The Gulf of Mexico basin forms a single coherent geologic unit. However, for purposes of evaluating its petroleum resources, it is often divided into several subprovinces. These subprovinces, listed in a counterclockwise direction beginning in the northeast corner of the basin, are the South Florida, Mid-Gulf Coast, Arkla, East Texas, Gulf Coast Onshore, Gulf Coast Offshore, Burgos-Sabinas, Tampico-Misantla, Veracruz, Isthmus Saline, Reforma, Macuspana, and Campeche subprovinces (Fig. 1). (Often two or more of these subprovinces are combined into larger units, the most common and geologically coherent combinations being the Arkla–East Texas, the Gulf Coast Onshore and Offshore–Burgos, and the Reforma-Campeche groups.)

Many of these subprovinces are major sources of petroleum by themselves (Table 3). Three—the Gulf Coast Onshore, Gulf Coast Offshore, and Campeche—are each large enough to be classed as superprovinces (25 to 100 billion BOE known recovery), a distinction shared by only 12 other petroleum provinces in the world. Another four—Arkla, East Texas, Reforma, and Tampico-Misantla—are each large enough to be classed as major petroleum provinces (7.5 to 25 BOE known recovery), a distinction shared by only 18 other petroleum provinces worldwide. Thus, even the individual components of the Gulf of Mexico basin provide many of the major centers of petroleum production for the world.

The distribution of petroleum resources by product across these subprovinces illustrates the diversity in petroleum resources within the Gulf of Mexico basin: seven—South Florida, Mid-Gulf Coast, East Texas, Tampico-Misantla, Isthmus Saline, Reforma, and Campeche—are predominantly oil-prone areas; three—the Gulf Coast Offshore, Burgos-Sabinas, and Macuspana—are predominantly gas-prone areas; the remaining three—Arkla, Gulf Coast Onshore, and Veracruz—contain oil and gas resources that are approximately in balance, neither type strongly predominating.

In the major petroleum provinces of the world, petroleum resources are typically highly concentrated in a small number of world-class *giant* fields (fields with a known recovery of 500

TABLE 2. THE OFFSHORE AND ONSHORE PETROLEUM RESOURCES OF THE GULF OF MEXICO BASIN AS OF DECEMBER 31, 1987*

	Cumulative Production	Proved Reserves	Ultimate Recovery
Crude Oil			
		Billions of barrels	
Offshore	12.57	28.70	41.27
Onshore	48.63	22.82	71.45
Natural Gas			
		Trillions of cubic feet	
Offshore	97.70	44.65	142.35
Onshore	284.60	96.88	381.48
Natural Gas Liquids			
		Billions of barrels	
Offshore	2.67	4.00	6.67
Onshore	10.56	5.28	15.84
Oil Equivalent			
		Billions of barrels	
Offshore	31.52	40.14	71.66
Onshore	106.63	44.25	150.88

*Sources: Publications of the American Gas Association, American Petroleum Institute, Petroleos Mexicanos, U.S. Energy Information Administration.

million BOE or more). For example, more than 75% of the oil resources of the Arabian-Iranian province are in 25 *supergiant* fields (fields with a known recovery of five billion BOE or more). The immense natural gas resources of the West Siberian province are similarly concentrated in a few supergiant fields.

The distribution of petroleum resources by field size category in the Gulf of Mexico basin does not correspond to this pattern (Table 4). Although giant fields are important contributors to the petroleum resources of the basin, they do not predominate as they do in nearly every other major petroleum province of the world. The five supergiant (*A.J. Bermudez, Abkatun, Cantarell,* and *Ku* in Mexico; and *East Texas* in the United States) and 48 other giant oil and gas fields in the Gulf of Mexico basin contained 91.68 billion BOE as of the end of 1987, only 44.5% of the basin total. Almost as much of the petroleum resources of the basin, 76.80 billion BOE or 37.3%, has been found in 559 *large* fields, fields with a known recovery of 50 to 500 million BOE. Very substantial amounts (37.34 billion BOE, an amount exceeded by only a few other provinces worldwide) are in the thousands of smaller fields that have been discovered in the basin.

The distribution of petroleum resources by field size category varies substantially by product within the Gulf of Mexico basin. Crude oil resources are the most concentrated, with nearly 60% being found in giant fields. The distribution of natural gas liquids is roughly similar to the overall distribution, as 45% of the natural gas liquids are in giant fields. The natural gas resources of the basin are unusually dispersed. Large fields, with

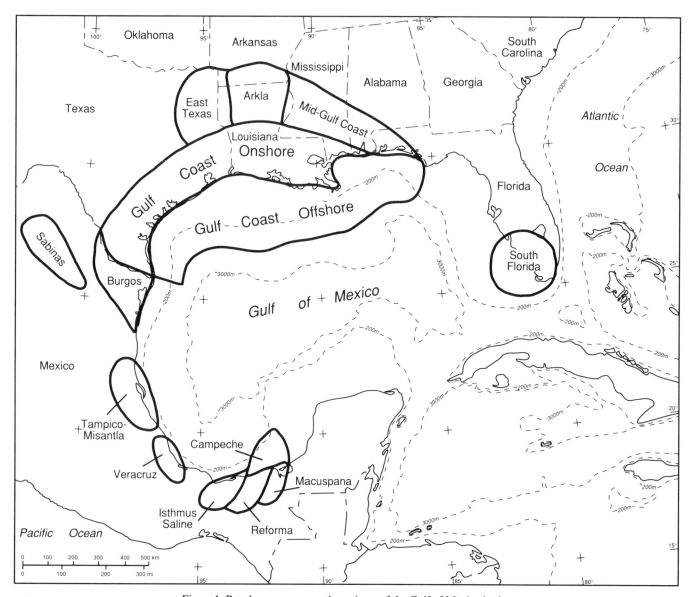

Figure 1. Petroleum resources subprovinces of the Gulf of Mexico basin.

48% of the total, predominate. Smaller fields actually contain slightly more natural gas than giant fields.

The geographic distribution of petroleum resources in the Gulf of Mexico basin varies even more dramatically. In the southern half of the basin (Mexico), petroleum resources are highly concentrated in giant fields (85% of the total), a proportion similar to the norm of most other major petroleum provinces. In the northern half of the basin (the United States), they are unusually dispersed. Less than 24% of the total amount is in giant fields, while 51% is in large fields and 25% is in smaller fields.

This geographic difference reflects a difference by gross stratigraphic interval. Mesozoic resources, which include all five supergiant fields in the Gulf of Mexico basin, are both highly concentrated in giant fields and are predominately in the southern half of the basin. Of the 95.34 billion BOE in Mesozoic formations, 70.16 billion BOE (73.6%) are in giant fields and 15.37 billion BOE (16.7%) are in large fields. In sharp contrast, Cenozoic resources, which are predominately in the northern half of the basin, are highly dispersed. Only 21.52 billion BOE (19.5%) of a 110.48 billion BOE Cenozoic total (less Chicontepec) are in giant fields, while 61.43 billion BOE (55.6% of the Cenozoic total) are in large fields. Nowhere else in the world has so much oil and gas been discovered in large and smaller fields as in the United States Gulf Coast Cenozoic.

TABLE 3. THE ULTIMATE RECOVERY OF PETROLEUM RESOURCES BY SUBPROVINCE AND PRODUCT IN THE GULF OF MEXICO BASIN AS OF DECEMBER 31, 1987

Subprovince	Crude Oil (Billion bbls)	Natural Gas (Trillion cf)	NGL (Billion bbls)	Oil Equivalent (Billion bbls)
South Florida	0.11	0.01	0	0.11
Mid-Gulf Coast	2.90	9.39	0.52	4.98
Arkla	4.16	29.44	0.75	9.82
East Texas	9.12	30.43	1.84	16.03
Gulf Coast				
Onshore	24.84	228.67	8.09	71.04
Offshore	9.77	128.32	2.71	33.87
Burgos-Sabinas	0.05	12.50	0.35	2.49
Tampico-Misantla	17.24	32.84	1.77	24.48
Veracruz	0.12	1.09	0.01	0.31
Isthmus Saline	2.12	2.48	0.16	2.69
Reforma	11.26	28.60	2.73	18.76
Macuspana	0	6.06	0.18	1.19
Campeche	31.05	14.01	3.40	36.78

*Sources: Publications of the American Gas Association, American Petroleum Institute, Petroleos Mexicanos, U.S. Energy Information Administration.
Note: The estimates for the Tampico-Misantla sub-province include the Chicontepec accumulation.

THE EXPLORATION AND PRODUCTION HISTORY OF THE GULF OF MEXICO BASIN

The most remarkable features of the exploration history of the Gulf of Mexico basin are the overall level of discoveries and their continuity over a lengthy period. Unlike most other major petroleum provinces, in which discoveries have been highly concentrated within a period of one to three decades, substantial numbers of significant new field discoveries have been made in the Gulf of Mexico basin for the past seven decades. Moreover, this unique record of success in significant discoveries is likely to continue into the 1990s.

Because of the relatively low number of obvious surface structures in the Gulf of Mexico basin, discoveries in the basin proceeded at a slow pace for 25 years after the first significant discovery of oil and gas was made at Corsicana in 1895 (Table 5). In no year during this period were more than seven significant new field discoveries made. Beginning with the first use of geophysical exploration methods during the 1920s, the rate of discovery in the Gulf of Mexico basin accelerated rapidly. Since then, it has been sustained by advances in geologic knowledge, continuous improvements in geophysical techniques, steady progress in drilling technology, the growth of the oil and gas market, and the initiation and further development of offshore exploration and production technology (see Chapter 1, this volume).

Few petroleum provinces have more than 500 significant fields. During the past half century, more than 500 significant new fields were discovered in the Gulf of Mexico basin every

TABLE 4. THE PETROLEUM RESOURCES IN THE GULF OF MEXICO BASIN BY FIELD SIZE CATEGORY AS OF DECEMBER 31, 1987

Size Category	Number of Fields	Crude Oil (Billion bbls)	Natural Gas (Trillion cf)	NGL (Billion bbls)	Oil Equivalent (Billion bbls)
Supergiant (5,000 million BOE+)					
Mexico	4	27.63	16.80	2.99	33.42
United States	1	5.46	1.50	0.62	6.33
Total	5	33.09	18.30	3.61	39.75
Giant (500 to 5,000 million BOE)					
Mexico	23	17.06	34.26	3.33	26.10
United States	25	10.81	74.52	2.60	25.83
Total	48	27.87	108.78	5.93	51.93
Large (50 to 500 million BOE)					
Mexico	45	4.89	11.76	0.64	7.49
United States	514	25.23	226.39	6.35	69.31
Total	559	30.12	238.15	6.99	76.80
Other (<50 million BOE)					
Mexico	...	1.33	7.99	0.30	2.96
United States	...	9.38	123.85	4.35	34.38
Total	...	10.71	131.84	4.65	37.34

Note: The amounts for Mexico exclude the Chicontepec accumulation because of the poor applicability of field designations and thus of field size categories to the petroleum resources of this area.

decade. In the 1950s and possibly in the 1980s, there were more than 1,000 significant new fields discovered in the basin.

The annual record of discoveries in the Gulf of Mexico basin indicates a unique continuity in exploration success. Since 1925, there have been at least ten significant new fields discovered every year in the basin. Since 1934, there have been at least 25 significant fields discovered every year. From 1937 through 1989, there were at least 50 significant fields discovered every year.

The peak periods of exploration success are just as remarkable. Any province with more than 100 significant fields is an important exploration area. In the Gulf of Mexico basin, more than 100 significant fields have been discovered in at least each of eleven years—1949, 1952, 1953, 1954, 1955, 1956, 1957, 1976, 1977, 1979, and 1984. When the size of recent discoveries is fully recognized, it is likely that more than 100 significant fields were also discovered in 1985 and 1988.

As Table 5 suggests, the number of significant discoveries in the Gulf of Mexico basin has undergone two distinct cycles. The number of significant fields being discovered increased at a relatively steady pace from the mid-1920s to a peak of 129 in 1955. After declining in the late 1950s and through the 1960s to a low of 51 in 1970, the number of significant fields being discovered

TABLE 5. SIGNIFICANT OIL AND GAS FIELD DISCOVERIES
BY DECADE IN THE GULF OF MEXICO BASIN*

Decade	Mexico	United States		Total	
Pre-1900	1			1	
1900-1909	6	16		22	
1910-1919	9	26		35	
1920-1929	13	111		124	
1930-1939	9	331		340	
1940-1949	14	674	(676)	688	(690)
1950-1959	74	975	(985)	1,049	(1,059)
1960-1969	69	663	(676)	732	(745)
1970-1979	42+	775	(797)	817	(839)
1980-1989	27+	587	(901)	614	(928)
Total	263+	4,159	(4,520)	4,422	(4,783)

*Primary sources: Petroleos Mexicanos (as partially reported in the *American Association of Petroleum Geologists Bulletin* and the *Oil and Gas Journal*), The Significant Oil and Gas Fields of the United States Data Base.
Note: From 1940 on, the sum of confirmed and probable significant discoveries in the United States portion of the Gulf of Mexico Basin is shown in parentheses. All significant discoveries prior to that time are confirmed. Because the total number of significant discoveries since 1970 in Mexico is likely to be understated, a plus sign appears following the identified number of discoveries.

TABLE 6. LARGE AND GIANT OIL AND GAS FIELD DISCOVERIES
BY DECADE IN THE GULF OF MEXICO BASIN*

Decade	Mexico	United States	Total	Percent of Significant
Pre-1900	0	1	1	100.0
1900-1909	2	13	15	68.2
1910-1919	0	14	14	40.0
1920-1929	1	37	38	30.6
1930-1939	3	120	123	36.2
1940-1949	5	125	130	18.9
1950-1959	14	92	106	10.1
1960-1969	10	60	70	9.6
1970-1979	23	75	98	12.0
1980-1989	14	26	40	6.5
Total	72	563	635	14.3

*Primary sources: Petroleos Mexicanos (as partially reported in the *American Association of Petroleum Geologists Bulletin* and the *Oil and Gas Journal*), The Significant Oil and Gas Fields of the United States Data Base.
Note: The total number of large and giant fields in the United States shown here is 23 more than the total shown in Table 4 because this table includes several discoveries made since 1975 which were not confirmed as large fields as of the end of 1987, but which have subsequently been demonstrated to exceed 50 million BOE. Including them in this table provides a more accurate indicator of the number of recent large and giant discoveries, although the estimates shown here for the 1980s are still likely to be understated.

has risen to a level of 75 to 125 per year from 1974 through 1989 (except for the slump of 1986).

As noted earlier, an unusually high number of large and giant oil and gas fields have been discovered in the Gulf of Mexico basin. The overall pattern of discovery for these fields is similar to that of all significant fields in the basin (Table 6). Discoveries of large and giant fields proceeded at a slow pace through the mid-1920s. As geophysical exploration methods began to be used extensively, the discovery rate soared, quickly reaching its peak in the late 1930s and 1940s. Discoveries of large and giant fields declined slowly during the 1950s, being sustained primarily by the first extensive exploration offshore and by a major exploration effort in the southern half of the basin in Mexico. After large and giant discoveries dropped sharply in the 1960s, exploration in deeper water offshore Louisiana and Texas and in the Reforma and Campeche areas of Mexico led to their marked resurgence in the 1970s. Although estimates for the 1980s are still preliminary, the early indications are that the rate of discovery has declined substantially, despite some major very deep and deepwater discoveries.

The large and giant discoveries in the Gulf of Mexico basin exhibit a unique record of continuity similar to that of all significant fields in the basin. Few petroleum provinces worldwide contain more than 50 large and giant fields. In the Gulf of Mexico basin, more than 50 large and giant fields were discovered in each of the five decades from the 1930s to the 1970s. This unique record of success may continue into the 1980s as well, once the full size of many recent discoveries becomes recognized. At least one large or giant field was discovered in the basin every year

from 1915 to 1986. (If any of the discoveries made in 1987 is eventually confirmed as a large field, this record of success extends through 1989.) In 52 of those years at least five large or giant fields were discovered each year, including 44 of the 50 years from 1931 to 1980. In 28 of the 46 years from 1934 to 1979, at least ten large or giant fields were discovered each year.

No other basin worldwide has even come close to producing so many major discoveries for such a long period. This unique record of continuous major exploration successes is attributable to (1) the substantial number of major structures in the Gulf of Mexico basin with good reservoirs and good source rocks; (2) the widespread distribution of these structures across the range of exploration environments encountered in the basin (e.g., onshore, deep, shallow-water offshore, deep-water offshore); and (3) the rate at which advances in exploration, drilling, and production technology made these large prospects identifiable and accessible.

The effects of developments in exploration and drilling technology on the rate of discoveries in the Gulf of Mexico basin show up most dramatically in the discovery patterns of world-class giant fields in the basin. Unlike the discovery patterns of significant fields or even of all large and giant fields, the dominant theme in the discovery patterns of giant fields in the basin is one of discontinuity. Giant discoveries in the basin are divided into four distinct phases, each with a distinct set of objectives and exploratory techniques (Table 7).

The first phase, from 1901 to 1922 with only four giant

discoveries, was essentially based on drilling over seeps and on surface structures. The second, the peak of onshore giant discoveries from 1928 to 1941 with 21 giant discoveries, resulted from the first extensive application of geophysical technology in United States Gulf Coast exploration. Most of the giant fields discovered during this period were associated with salt structures, salt-cored anticlines, or fault-controlled anticlines in the Louisiana and Texas Coastal Plain. The third phase, from 1948 to 1958 with ten giant discoveries, was restricted to shallow-water fields offshore Louisiana and to moderate depth fields in Mexico. This phase corresponds to the first major period of offshore exploration in the basin and the beginning of a major exploration effort by Petroleos Mexicanos. The fourth and most recent phase from 1970 to 1988 is, with 22 giant discoveries, surprisingly the most important, both in the number of giant discoveries and the

amount discovered in giant fields. It focused on deep- and deeper water offshore structures, primarily in the Reforma and Campeche areas. No giant discoveries were made in the years between any pair of these phases. Each group of giant discoveries was also made relatively early in the history of exploration of the larger group of fields to which they belong.

The amount of oil and gas discovered in the Gulf of Mexico basin by year and decade of discovery reflects the patterns of discovery of both all significant fields and giant fields within the basin (Table 8). Although the amounts discovered in both the 1900s and the 1910s were not paltry, they are relatively insignificant compared with the rapid increase in amounts discovered that began in the late 1920s and reached a peak in the 1930s when geophysics was first used extensively in the basin. Very large amounts were also discovered in the 1940s and 1950s when giant and large discoveries continued to be made at high levels. The amount discovered declined substantially in the 1960s, a decade during which no giant fields were discovered in the basin. However, it reached its historic peak in the 1970s because of the many giant and supergiant discoveries in Mexico. As these discoveries tapered off, the amount discovered in the 1980s declined sharply.

The decline shown for the 1980s is however overstated. Many of the reserved discovered during this period, particularly in the United States, have yet to be booked. Once all fields discovered from 1980 to 1989 are fully booked, the total discovered in the United States' portion of the Gulf of Mexico basin for this decade will most likely be between 8 to 12 billion BOE. This adjustment would place the amounts discovered in the 1980s in the entire basin not far behind the totals for the 1940s and 1950s.

TABLE 7. DISCOVERY PHASES OF GIANT FIELDS IN THE GULF OF MEXICO BASIN

First Phase (1901-1922)

1901 - Ebano-Panuco	1916 - Monroe
1909 - Naranjos-Cerro Azul (Golden Lane)	1922 - Smackover

Second Phase (1928-1941)

1928 - Agua Dulce-Stratton	1935 - Katy
1929 - Van	1936 - Carthage
1930 - Caillou Island	1937 - Bateman Lake
East Texas*	Bayou Sale
Poza Rica	Borregos-Seeligson-T.C.B.
1931 - Conroe	Webster
Thompson	1938 - Timbalier Bay
1933 - Greta-Tom O'Connor	1939 - La Gloria
1934 - Hastings	1940 - Hawkins
Old Ocean	1941 - Bastian Bay
West Ranch	

Third Phase (1948-1958)

1948 - Reynosa	1956 - Grand Isle 043
1949 - Bay Marchand 002	San Andres
West Delta 030	Vermillion 014
1950 - South Pass 024	1958 - Mound Point
1951 - Jose Colomo/Chilapilla	Tiger Shoal

Fourth Phase (1970-1986)

1970 - Jay-Little Escambia Creek	1980 - Cardenas
	Jujo
1971 - Eugene Island 330	Pol
1972 - Cactus-Nispero	1982 - Chuc
Sitio Grande	Muspac
1973 - A. J. Bermudez*	1984 - Batab
1976 - Agave	Caan
Cantarell*	Sen
1977 - Giraldas	1985 - Tecominoacan
Paredon	1986 - Uech
1979 - Abkatun*	
Ku*	
Maloob	

*Supergiant Field.

TABLE 8. KNOWN RECOVERY BY DECADE OF DISCOVERY IN THE GULF OF MEXICO BASIN*

Decade	Mexico (Billion BOE)	United States (Billion BOE)	Total (Billion BOE)
Pre-1900	0	0.05	0.05
1900-1909	2.49	1.88	4.37
1910-1919	0	3.44	3.44
1920-1929	0.13	9.30	9.43
1930-1939	3.16	41.70	44.86
1940-1949	1.39	25.95	27.34
1950-1959	4.73	21.47	26.20
1960-1969	1.82	13.40	15.22
1970-1979	44.05	14.68	58.73
1980-1987	12.20	3.98	16.18
Total	69.97	135.85	205.82

*Primary sources: Petroleos Mexicanos (as partially reported in the *American Association of Petroleum Geologists Bulletin* and the *Oil and Gas Journal*), The Significant Oil and Gas Fields of the United States Data Base.

Note: The time of discovery is defined here as the year in which each field was discovered. Because field designations and thus field discovery years are not clearly applicable to the petroleum resources of the Chicontepec Formation, the amounts from this accumulation are excluded from the totals for Mexico.

Like the trends in the number of significant discoveries and in the combined total of large and giant discoveries, there is a remarkable and unique continuity in the amounts of oil and gas discovered in the Gulf of Mexico basin. In the 60 years from 1928 to 1987, there were at most only seven years (1932, 1946, 1968, 1969, 1978, 1983, and 1987) in which less than one billion BOE was discovered. In only three of these (1932, 1983, and 1987) were less than 800 million BOE discovered; however, many of the reserves discovered in 1983 and 1987 have yet to be booked. No other province in the world has enjoyed such an uninterrupted record of exploratory success.

The peak years of discovery in the basin are also impressive. In at least seven years (1930, 1937, 1949, 1956, 1973, 1976, and 1979), at least five billion BOE were discovered each year. (With modest amounts of reserve growth, 1980 and 1984 may eventually qualify for this listing as well.) All of these years are associated with either supergiant discoveries or multiple giant discoveries.

Oil production (crude oil and natural gas liquids) in the Gulf of Mexico basin has essentially followed the path of discovery, but has also been affected significantly by the growth in demand for oil and extra-market forces (particularly prorationing

in the United States from the early 1930s to the early 1970s). Significant production began in 1901 with the discovery of *Spindletop* in the Houston embayment (Fig. 2). The first major increase in oil production resulted from the development of the Golden Lane fields in east-central Mexico from 1918 through 1922. (In 1918, Gulf of Mexico oil production averaged 288,000 barrels per day and accounted for 20.9% of world oil production. By 1922, it had nearly tripled to 844,000 barrels per day and accounted for 35.9% of world production.) However, during the late 1920s, oil production in both Mexico and the United States declined.

This decline was dramatically reversed in the United States by the discovery of the *East Texas* field in 1930. Oil production in the United States grew rapidly thereafter, more than quadrupling from 1929 to 1937 as many other major discoveries came on line. From the early 1930s to the late 1940s the Gulf of Mexico basin accounted for 25 to 30% of world oil production. With the discovery and development of the first large and giant field offshore and the continuous growth in world oil demand, Gulf of Mexico basin oil production increased rapidly again in the last 1950s to the early 1970s (from 2.67 million b/d in 1958 to 5.53 million b/d in 1971). However because world oil produc-

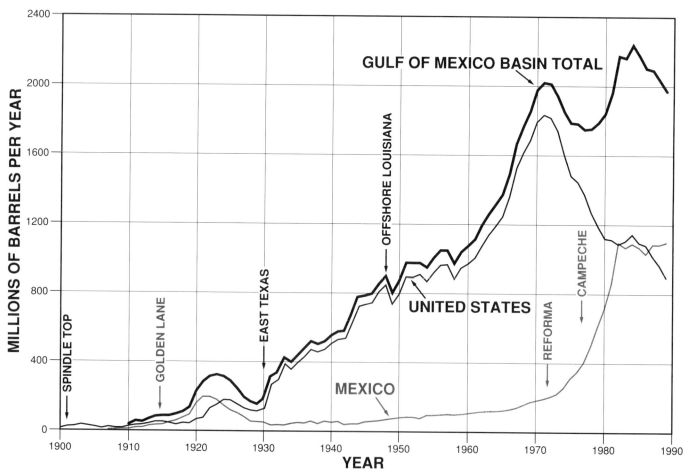

Figure 2. Oil production in the Gulf of Mexico basin, 1900 to 1989.

tion grew even more rapidly, the proportion of world oil production provided by the Gulf of Mexico basin was slowly decreasing, dropping below 10% in 1973 and reaching a low of 7.5% in 1979.

Following a peak production of 5.04 million b/d in 1971, oil production in the United States portion of the Gulf of Mexico basin has declined more or less steadily through the 1970s and 1980s to only 2.43 million b/d in 1989. However, production in the Gulf of Mexico basin managed to reach record levels in the early 1980s because of the rapid growth in oil production in the Reforma and Campeche areas of Mexico. From 1972 to 1982, oil production in Mexico grew from 0.53 million b/d to 2.99 million b/d, a level that has essentially been sustained through the remainder of the 1980s. Consequently Gulf of Mexico basin oil production reached its record level in 1984 (6.11 million b/d) and provided around 10% of world oil production throughout the 1980s.

Good historic data for natural gas production in the Gulf of Mexico basin are not available prior to 1960. The pattern since then has both similarities and differences to the pattern of oil production within the basin. Like oil, natural gas production in the Gulf of Mexico basin (predominantly in the United States) grew rapidly from 1960 to 1972, doubling from 6.75 to 13.60 trillion ft^3 per year (Fig. 3). After this historic peak, production has declined in the United States, albeit at a slower rate than oil production, most recent discoveries having been gas fields. Although natural gas production in Mexico increased during the 1970s and early 1980s, this increase (in what is essentially an oil-prone area) was insufficient to offset the decline in production in the United States. Nonetheless, gas production in the Gulf of Mexico basin was sustained at rates of 10 to 12 trillion ft^3 per year throughout the 1980s, providing 40 to 50% of total North American gas production.

THE PETROLEUM RESOURCES OF THE GULF OF MEXICO BASIN BY STRATIGRAPHIC UNIT

One of the more unusual features of the Gulf of Mexico basin is the distribution of petroleum resources throughout the stratigraphic section of the basin (Fig. 4). In almost all major petroleum provinces, the known petroleum resources are highly concentrated in one or two stratigraphic units. In the Gulf of Mexico basin, every one of the five producing units discussed in this chapter contains major amounts (on a worldwide scale) of petroleum resources (Table 9). Each unit has enough petroleum to be a major petroleum province by itself; each except the Upper Jurassic is large enough to be a superprovince by itself.

This section describes the petroleum resources of the Gulf of Mexico basin by stratigraphic unit. The description is divided into five gross intervals: Upper Jurassic, Lower Cretaceous, Upper Cretaceous, Paleogene, and upper Cenozoic (Neogene/Pleistocene). Each description begins with a summary of the known recovery of the unit, its distribution by stage or series, its broad geographic distribution, and its gross field size distribution. A simplified stratigraphic columnar section, illustrating the subdivi-

sions used in the discussion, is shown in Figure 4. More detailed stratigraphic charts are included in Chapters 8, 9, 10, and 11 and on Plate 5 of this volume.

Most of the discussion is devoted to the petroleum resources within each stage or group of stages. Each discussion summarizes the total amounts within each stage or grouping and the distribution of that amount by broad field size categories; identifies the producing trends within each grouping; indicates their location; reviews their discovery histories; describes the fundamental factors of petroleum accumulation—reservoir, trap, seal, and source—within them; and concludes with a brief discussion of their potential. The length of discussion of each grouping is determined both by the complexity of the grouping and by its relative importance. The sources used for each discussion, besides the literature cited, are, for the United States, the Significant Oil and Gas Fields of the United States data base issued by NehRinG Associates (which provided the information on discovery histories, field and reservoir sizes, and reservoir characteristics), and, for Mexico, various published sources, particularly the annual *Memoria de Labores,* originating from Pemex.

For some producing units, known recovery within them cannot be determined with certainty, either because formation boundaries do not coincide with stage boundaries or because vertical reservoir limits straddle formations and stages because of fracturing. In such cases, known recovery is allocated to the predominant series, stage, or formation. However, alternative allocations and their effects on the distribution of known recovery by stage and series are also discussed.

Upper Jurassic petroleum resources

Upper Jurassic rocks provide the geologically oldest commercially significant amounts of reservoired petroleum in the Gulf of Mexico basin (Fig. 4). Although Upper Jurassic petroleum resources in the Gulf of Mexico basin are sufficiently large to constitute a major petroleum province by themselves (7.5 billion barrels of crude oil, 24 to 36 trillion ft^3 of natural gas, and 2.0 billion barrels of natural gas liquids, or 13.5 to 15.5 billion barrels oil equivalent known recovery; Table 10), within the context of the huge petroleum resources of the Gulf of Mexico basin, they are only of modest importance. Only 6.7% of the crude oil, 4.5% of the natural gas, and 8.9% of the natural gas liquids, or 6.1% of the barrels of oil equivalent known recoverable resources of the Gulf of Mexico basin are reservoired in Upper Jurassic rocks.

Within the Upper Jurassic, the Kimmeridgian is clearly the most prolific unit. Carbonates of that age in the Campeche and Reforma areas of southeastern Mexico form part of one of the major oil producing centers of the world (Fig. 5). Rocks of Oxfordian age rank second. Both the Oxfordian upper Smackover carbonates (petroleum liquids and natural gas) and the lower Oxfordian Norphlet sandstones (natural gas) contain major amounts of petroleum. The youngest Upper Jurassic (Tithonian) is of least importance, although the sandstones of the Cotton Valley Group produce major amounts of natural gas.

If the discovered but undeveloped Norphlet gas reserves offshore Alabama are included in the calculations, the Upper Jurassic petroleum resources of the Gulf of Mexico basin are evenly divided between the northern and southern halves of the basin. Petroleum resources in the southern half are predominantly crude oil. Petroleum resources in the northern half are predominantly natural gas, although crude oil and natural gas liquids are not insignificant.

Giant and large fields contain most of Upper Jurassic known recovery in the Gulf of Mexico basin. Fourteen giant fields, including the supergiant *A. J. Bermudez* and *Cantarell* fields in southeastern Mexico, contain 7,830 million BOE (58.1%) of demonstrated known Upper Jurassic recovery in the basin. Another 39 large fields contain 3,640 million BOE, 27.0% of known recovery.

Lower Oxfordian (Norphlet Formation). Until the mid-1980s, the lower Oxfordian Norphlet Formation was the least important of all the oil and gas producing stratigraphic intervals from the Upper Jurassic through the Pleistocene in the Gulf of Mexico basin. As of the end of 1987, estimated known recovery

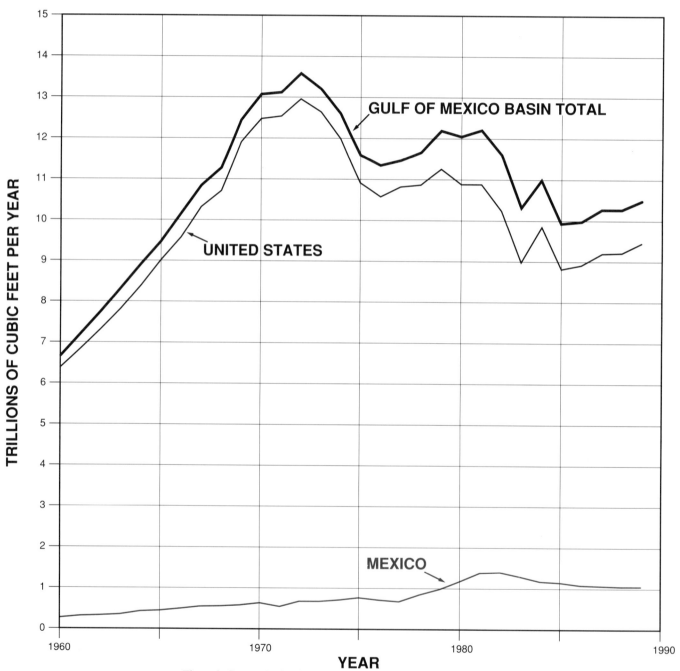

Figure 3. Gas production in the Gulf of Mexico basin, 1960 to 1989.

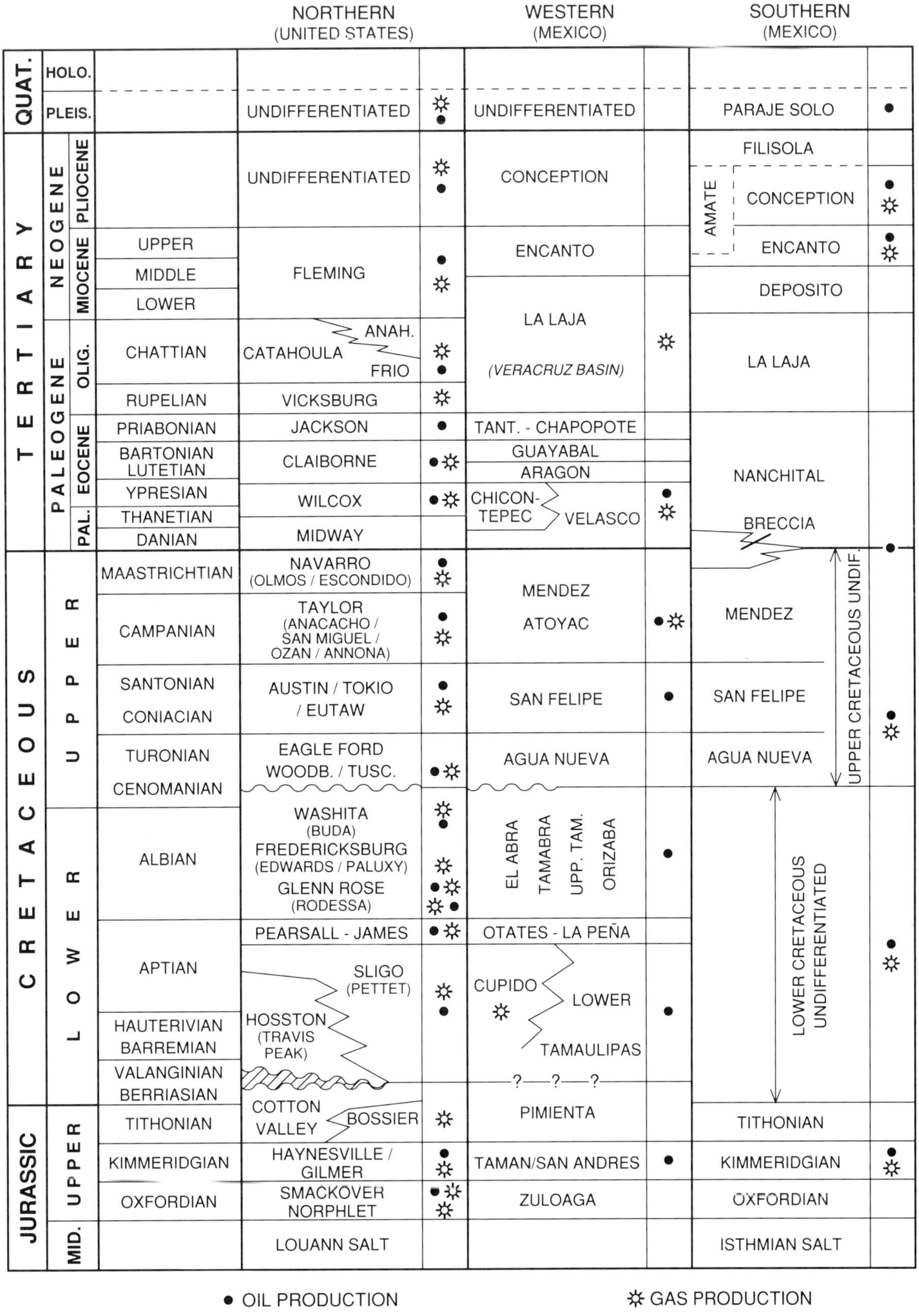

Period	Epoch	Stage / Age	NORTHERN (UNITED STATES)	N. prod.	WESTERN (MEXICO)	W. prod.	SOUTHERN (MEXICO)	S. prod.
QUAT.	HOLO.			☼				
QUAT.	PLEIS.		UNDIFFERENTIATED	☼ ●	UNDIFFERENTIATED		PARAJE SOLO	●
TERTIARY – NEOGENE	PLIOCENE		UNDIFFERENTIATED	☼ ●	CONCEPTION		FILISOLA / AMATE — CONCEPTION	● ☼
TERTIARY – NEOGENE	MIOCENE UPPER		FLEMING	●	ENCANTO		ENCANTO	● ☼
TERTIARY – NEOGENE	MIOCENE MIDDLE		FLEMING	☼	LA LAJA (VERACRUZ BASIN)		DEPOSITO	
TERTIARY – NEOGENE	MIOCENE LOWER		FLEMING		LA LAJA (VERACRUZ BASIN)		DEPOSITO	
TERTIARY – PALEOGENE	OLIG.	CHATTIAN	CATAHOULA (ANAH. / FRIO)	☼ ●	(VERACRUZ BASIN)	☼	LA LAJA	
TERTIARY – PALEOGENE	OLIG.	RUPELIAN	VICKSBURG	☼				
TERTIARY – PALEOGENE	EOCENE	PRIABONIAN	JACKSON	●	TANT. - CHAPOPOTE		NANCHITAL	
TERTIARY – PALEOGENE	EOCENE	BARTONIAN / LUTETIAN	CLAIBORNE	● ☼	GUAYABAL / ARAGON		NANCHITAL	
TERTIARY – PALEOGENE	EOCENE	YPRESIAN	WILCOX	● ☼	CHICONTEPEC / VELASCO	●	BRECCIA	☼
TERTIARY – PALEOGENE	PAL.	THANETIAN	MIDWAY		VELASCO		BRECCIA	
TERTIARY – PALEOGENE	PAL.	DANIAN	MIDWAY		VELASCO			
CRETACEOUS UPPER		MAASTRICHTIAN	NAVARRO (OLMOS / ESCONDIDO)	● ☼	MENDEZ		MENDEZ	●
CRETACEOUS UPPER		CAMPANIAN	TAYLOR (ANACACHO / SAN MIGUEL / OZAN / ANNONA)	● ☼	ATOYAC	● ☼	MENDEZ	
CRETACEOUS UPPER		SANTONIAN / CONIACIAN	AUSTIN / TOKIO / EUTAW	● ☼	SAN FELIPE	●	SAN FELIPE	☼
CRETACEOUS UPPER		TURONIAN / CENOMANIAN	EAGLE FORD WOODB. / TUSC.	● ☼	AGUA NUEVA		AGUA NUEVA	
CRETACEOUS LOWER		ALBIAN	WASHITA (BUDA) / FREDERICKSBURG (EDWARDS / PALUXY) / GLENN ROSE (RODESSA)	☼ ● ☼ ● ☼ ●	EL ABRA / TAMABRA / UPP. TAM. / ORIZABA	●	LOWER CRETACEOUS UNDIFFERENTIATED	● ☼
CRETACEOUS LOWER		APTIAN	PEARSALL - JAMES / SLIGO (PETTET)	● ☼	OTATES - LA PEÑA / CUPIDO / LOWER	☼	LOWER CRETACEOUS UNDIFFERENTIATED	
CRETACEOUS LOWER		HAUTERIVIAN / BARREMIAN	HOSSTON (TRAVIS PEAK)		TAMAULIPAS	●		
CRETACEOUS LOWER		VALANGINIAN / BERRIASIAN			? — ? — ?			
JURASSIC UPPER		TITHONIAN	COTTON VALLEY / BOSSIER	☼	PIMIENTA		TITHONIAN	
JURASSIC UPPER		KIMMERIDGIAN	HAYNESVILLE / GILMER	● ☼	TAMAN / SAN ANDRES	●	KIMMERIDGIAN	● ☼
JURASSIC UPPER		OXFORDIAN	SMACKOVER NORPHLET	● ☼ ☼	ZULOAGA		OXFORDIAN	
JURASSIC MID.			LOUANN SALT				ISTHMIAN SALT	

Note: A vertical bracket in the Southern column spans the Paleocene–Upper Cretaceous interval as "UPPER CRETACEOUS UNDIF." (with ● and ☼) and the Lower Cretaceous interval as "LOWER CRETACEOUS UNDIFFERENTIATED" (with ● and ☼).

● OIL PRODUCTION   ☼ GAS PRODUCTION

Figure 4. Principal oil and gas productive intervals of the Gulf of Mexico basin.

TABLE 9. THE DISTRIBUTION OF PETROLEUM RESOURCES IN THE GULF OF MEXICO BASIN BY MAJOR STRATIGRAPHIC UNIT

Unit	Crude Oil (Billion bbls)	Natural Gas (Trillion cf)	NGL (Billion bbls)	Oil Equivalent (Billion bbls)
Upper Cenozoic	22.66	204.64	6.58	63.36
Paleogene	25.62	192.84	6.08	63.84
Upper Cretaceous	30.75	37.41	3.67	40.65
Lower Cretaceous	26.16	65.24	4.18	41.21
Upper Jurassic	7.53	23.70	2.00	13.48
Total	112.72	523.83	22.51	222.54

TABLE 10. KNOWN RECOVERY OF CRUDE OIL, NATURAL GAS, AND NATURAL GAS LIQUIDS AS OF DECEMBER 31, 1987, IN THE GULF OF MEXICO BASIN BY STRATIGRAPHIC INTERVAL WITHIN THE UPPER JURASSIC

	Crude Oil (Million bbls)	Natural Gas (Bcf)	NGL (Million bbls)	BOE (Million bbls)
Tithonian				
Cotton Valley-N*	245	8,340	345	1,980
Kimmeridgian				
Haynesville-N*	140	2,010	40	515
San Andres-S*	780	840	55	975
Unnamed Campeche-S*[†]	3,000	1,590	385	3,650
Unnamed Reforma-S*[†]	2,040	3,720	355	3,015
Total	5,960	8,160	835	8,155
Oxfordian				
Smackover-N*	1,300	6,270	765	3,110
Norphlet-N*	25	930[§]	55	235[§]
Total	1,325	7,200	820	3,355
Upper Jurassic Total	7,530	23,700	2,000	13,480

*N = northern half of basin (United States); S = southern half of basin (Mexico).
[†]Includes some reservoirs that extend upwards into the Tithonian.
[§]Plus 6,000 to 12,000 Bcf (1,000 to 2,000 million BOE) discovered but not developed or booked.

in the Norphlet was only 930 billion ft^3 of natural gas, 55 million barrels of natural gas liquids, and 25 million barrels of oil, or 235 million barrels oil equivalent. However, during the 1980s substantial amounts of natural gas not yet included in proved reserves were discovered in the Norphlet offshore Alabama, Florida, and Mississippi (Fig. 5). Mink and others (1987) have estimated a known recovery of 4.9 to 8.1 trillion ft^3 of natural gas in Norphlet reservoirs in Alabama state offshore areas alone. If the discoveries in federal offshore areas are included, discovered Norphlet gas resources are at least 6 to 12 trillion ft^3, making Norphlet reservoirs one of the most important concentrations of natural gas in the entire Mesozoic sequence of the Gulf of Mexico basin.

Although several Norphlet fields have been discovered along and downdip of the Pickens-Gilbertown-Pollard fault system bordering the Mississippi salt basin to the northeast, Norphlet petroleum resources in the Gulf of Mexico basin are highly concentrated within and immediately south of Mobile Bay, offshore south Alabama (Fig. 5 and Plate 2). Within this area, the known and probable gas resources are highly concentrated in five large fields covering 2,000 to 6,000 ha (5,000 to 15,000 acres) each: *Bon Secour Bay, Fairway, Mary Ann, North Central Gulf/Mobile Block 827,* and *Northwest Gulf/Mobile Block 823.* Onshore, with the one exception of *Flomaton* in southwest Alabama, no Norphlet reservoir exceeds ten million barrels oil equivalent in size. *Mary Ann* field was discovered in 1979; the other four fields were discovered in the mid-1980s. Since that time Norphlet discoveries have continued to be made in the federal offshore waters at very high rates of success.

The Norphlet reservoirs are eolian sheet and dune sandstones (see Chapter 8, this volume). These eolian sands cover the northern two-thirds of the Mississippi salt basin and continue to the southeast for another 300 km almost to the Middle Ground arch. Net sandstone thicknesses within the Norphlet range from 30 to 150 m, with deposition in some localities exceeding 200 m (Marzano and others, 1988). In the main area of Norphlet production in and around Mobile Bay, the average net thicknesses of the productive sand intervals are 35 to 70 m. The reservoirs in this area are 6,000 to 7,000 m deep, while the smaller reservoirs in the updip areas of Norphlet production are 4,400 to 5,500 m deep (Bolin and others, 1989).

Despite these great reservoir depths, porosity in the Norphlet is excellent, averaging 10 to 15% in the largest fields with some values exceeding 20% (Bolin and others, 1989; Dixon and others, 1989). Some disagreement has existed in the literature about the mechanisms responsible for this porosity. A consensus appears to be forming that it is a result of a combination of preserved primary porosity and secondary enhanced porosity (Marzano and others, 1988; Vaughan and Benson, 1988; Dixon and others, 1989; Lock and Broussard, 1989).

Trapping of Norphlet reservoirs is entirely structural, including salt-cored anticlines, fractured salt-cored anticlines, and extensional fault traps associated with salt movement (Bolin and others, 1989; Mancini and others, 1985). The seals for the Norphlet reservoirs are provided primarily by a continuous tight zone at the top of the Norphlet and secondarily by the overlying basal Smackover calcareous shales. The Norphlet tight zone, which varies from less than 1 m to more than 50 m thick, consists of eolian sands that were reworked during the marine transgression that began Smackover time and were subsequently tightly cemented (Marzano and others, 1988).

The limits on Norphlet hydrocarbon production in the Gulf of Mexico basin result primarily from the extent of the eolian sandstones (see Chapter 8, this volume). Outside of the area of

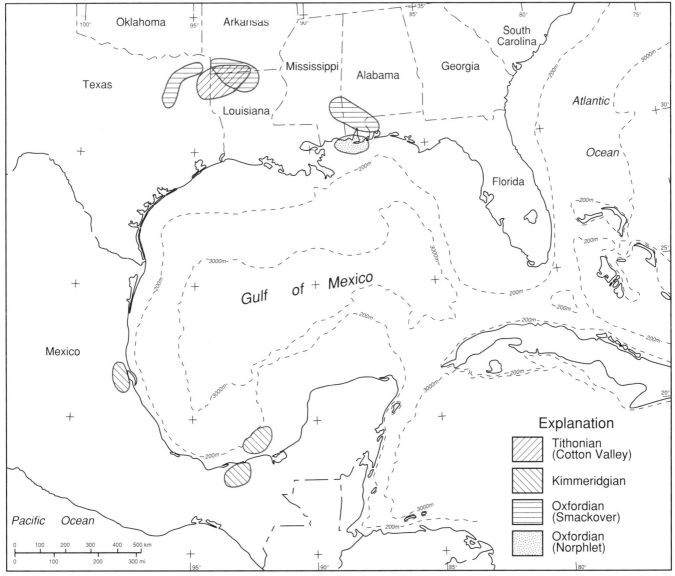

Figure 5. Major Upper Jurassic petroleum-producing trends of the Gulf of Mexico basin.

eolian sand deposition, reservoir quality rocks in the Norphlet are rare to nonexistent. Within this area, several factors can limit Norphlet production. These include limited petroleum-generating capacity where the overlying lower Smackover source rocks are thin, or barriers to migration, particularly during the later stages of hydrocarbon generation, created by the uppermost Norphlet tight zone. Preservation of Norphlet hydrocarbons is also problematic, particularly in the deepest and hottest parts of the favorable reservoir area. Outside of the greater Mobile Bay area, the size of future Norphlet discoveries may be limited by small areas of closure (less than 200 ha or 500 acres) in the available traps. However, given the low density of exploration drilling, there still appears to be substantial undiscovered Norphlet gas potential in the offshore area and in the immediately adjacent onshore area north of Mobile Bay at depths of 5,500 to 7,000 m.

Upper Oxfordian (Smackover Formation). The upper Oxfordian Smackover is the second most important oil and gas producing interval within the Upper Jurassic sequence in the Gulf of Mexico basin. As of the end of 1987, estimated known recovery in the Smackover was 1,300 million barrels of crude oil, 6,270 billion ft^3 of natural gas, and 765 million barrels of natural gas liquids, for a total of 3,110 million BOE.

Smackover oil and gas resources are centered in several areas along the north rim of the Gulf of Mexico basin (Fig. 5). One major center of crude oil, natural gas liquids, and natural gas production is found in the Conecuh (Covington) embayment and around the Mobile graben, both in southwest Alabama. The other major concentration is found in a series of essentially parallel east-west producing trends along the Arkansas-Louisiana boundary that contain predominantly natural gas and natural gas liquids downdip and crude oil updip. The largest moderate-size concentration, predominantly natural gas and natural gas liquids, is found within and immediately downdip of the Mexia-Talco fault zone on the north and west flank of the East Texas basin. A smaller moderate size concentration of crude oil, natural gas, and natural gas liquids is found along and downdip of the Gilbertown fault zone on the northeast flank of the Mississippi salt basin. Minor amounts of natural gas have also been discovered in the Smackover within the northwest corner of the Mississippi salt basin.

Although significant Smackover oil and gas have been found in more than 175 fields, most of the known recovery is concentrated in one giant (565 million BOE or 18.3%) and 17 large fields (1,485 million BOE or 48.0%). The one giant field (*Jay-Little Escambia Creek*) is located on the Florida-Alabama border (Plate 4). The large fields include *Big Escambia Creek* and *Blackjack Creek* in the Conecuh embayment; *Chunchula* and *Hatter's Pond* in the Mobile graben; *Pachuta Creek* in the Gilbertown fault zone; *Cotton Valley, Dorcheat-Macedonia, Haynesville, Magnolia, McKamie-Patton, Midway, Shongaloo, North-Red Rock, Shuler,* and *Walker Creek* along the Arkansas-Louisiana border; and *Bryans Mill, Eustace,* and *New Hope* along the north and west flanks of the East Texas basin.

Significant Smackover discoveries have occurred in several cycles, each cycle corresponding to the testing of various Smackover exploration concepts. The first Smackover discoveries were made in the late 1930s in south Arkansas and reached their peak in the various trends along the Arkansas-Louisiana border during the 1940s and 1950s. Smackover discoveries in the east Texas trends were concentrated in the 1960s and early 1970s, although small discoveries have continued in these trends into the 1980s. Smackover discoveries in the Mississippi salt basin were clustered in the late 1960s and early 1970s. The biggest Smackover discoveries in southwest Alabama occurred in the early 1970s, but smaller discoveries continued to be made there into the late 1980s.

The upper Smackover reservoirs were originally deposited on carbonate ramps, with high energy grainstones being deposited on shoals controlled by low relief salt anticlines, basement structures, or growth faults (see Chapter 8, this volume). In Arkansas, Louisiana, and Mississippi, limestones with both preserved primary and secondary porosity provide the Smackover reservoirs. In southern Arkansas, the upper Smackover reservoirs are in a well-developed oolitic grainstone, the Reynolds Oolite. The productive section in other areas is commonly in dolomitized carbonates. Dolomitization of the upper Smackover appears to be related to the presence of an evaporitic section directly above the productive interval (Collins, 1980; Moore and Druckman, 1981; Moore, 1984; Benson, 1988).

Smackover reservoir quality is generally good, but rarely excellent. In the southwest Alabama plays, the major Smackover reservoirs are 3,900 to 5,600 m deep, 5 to 30 m thick, and have 10 to 16% porosity with 20 to 100 md permeability. In the various Smackover trends along the Arkansas-Louisiana border, the reservoirs are typically 2,000 to 3,600 m deep and 5 to 40 m thick, with 8 to 20% porosity and 25 to 1,000 md permeability, the better quality reservoirs being found in the shallower, updip portion. In east Texas, Smackover reservoirs are typically 2,700 to 4,000 m deep and 5 to 60 m thick (mostly 15 to 20 m thick) with 8 to 20% porosity and 1 to 80 md permeability. Smackover reservoirs in the main trend in the Mississippi salt basin are 3,400 to 5,200 m deep, 5 to 30 m thick, and have 8 to 20% porosity with 5 to 150 md permeability, the better quality reservoirs being the shallower ones.

Trapping of Smackover hydrocarbons is predominantly structural, although in some plays stratigraphic factors are important as well. The most important type of trap is the low to intermediate relief salt-cored anticline. Fault closures, usually updip from the anticlinal structures, are another important trapping mechanism in Smackover fields. Both lithologic and diagenetic changes, typically influenced by paleostructure, can control the extent of the anticlinal accumulations (Collins, 1980; Baria and others, 1982; Presley and Reed, 1984; Mink and others, 1985; Bolin and others, 1989).

The primary seal for Smackover petroleum accumulations is the Buckner Anhydrite, the basal member of the overlying Haynesville Formation. The Buckner varies substantially in thickness, even being absent over some reservoir-quality Smackover, and thus is not always a competent seal, particularly when faulted (Moore, 1984). Tight Smackover carbonates also serve as a seal in some reservoirs.

Despite a substantial exploration history extending over a half-century, some potential for further Smackover discoveries still exists. Within existing plays, there are numerous small prospects remaining to be tested. Some deeper Smackover potential is likely in the Mississippi salt basin, although preservation is problematic. Some deep Smackover fields may also be found offshore of the Florida panhandle.

Kimmeridgian of southeast Mexico; San Andrés Member; and Haynesville Formation. The Kimmeridgian is easily the most important producing section within the Upper Jurassic in the Gulf of Mexico basin. As of the end of 1987, known recovery in the Kimmeridgian was approximately 5,960 million barrels of crude oil, 8,160 billion ft^3 of natural gas, and 835 million barrels of natural gas liquids, or 8,155 million BOE, 60% of the total known recovery in the Upper Jurassic.

The primary centers of Kimmeridgian crude oil, natural gas, and natural gas liquids production in the Gulf of Mexico basin are the southern half of the Bay of Campeche west of the Yucatan Peninsula and the northern half of the Reforma area onshore in Chiapas and Tabasco states immediately southwest of the Bay of Campeche (Fig. 5). Kimmeridgian oil in close to major quantities has also been discovered in the San Andrés Member of the Tamán Formation in a few fields both south and north of the

Tuxpan Platform in the Tampico-Misantla basin in east-central Mexico. Most of the remaining Kimmeridgian oil has been discovered in a moderate-size Haynesville sandstone trend along the Arkansas-Louisiana border. A moderate amount of Kimmeridgian natural gas has been discovered in two Gilmer Limestone trends in east Texas, the larger of these being found on the southwest flank of the East Texas basin and the smaller on the west flank of the Sabine uplift. Several minor Haynesville sandstone oil fields have also been discovered in southeast Mississippi and southwest Alabama.

Kimmeridgian petroleum resources are extremely concentrated in several giant and large fields. Two supergiant fields, *Cantarell* in the Bay of Campeche and *A. J. Bermudez* in Reforma, and eight other giant fields (*Batab, Pol,* and *Uech* in Campeche, *Cardenas, Jujo, Paredon,* and *Tecominoacan* in Reforma, and *San Andrés* in the Tampico-Misantla basin) provide 6,960 million BOE (85.7% of the total; Plate 4). The eight large fields (*Luna* in the Reforma area; *Arenque* and *Tamaulipas-Constituciones* in the Tampico-Misantla basin; *Haynesville, Shongaloo North-Red Rock,* and *Shuler* along the Arkansas-Louisiana border; and *Gladewater* and *Personville, North-Pokey* in the East Texas basin) contained another 910 million BOE (11.2%). Giant and large fields thus provide 97% of the Kimmeridgian petroleum resources in the Gulf of Mexico basin.

Despite their relatively small number, discoveries of Kimmeridgian reservoirs in the Gulf of Mexico basin span a period of more than 50 years. After the first Haynesville discovery at *Shuler* in 1937, there were only a few discoveries in the same trend into the early 1950s. All of the major San Andrés discoveries in the Tampico-Misantla basin in Mexico were made between the mid-1950s and mid-1960s, including the *San Andrés* field in 1956. Nearly all of the Gilmer Limestone discoveries were made from the late 1960s to the early 1980s. Since the early 1980s, there have been several small Haynesville sandstone discoveries along the northern and northeastern rim of the Gulf of Mexico basin. The discovery of Kimmeridgian petroleum in southeastern Mexico was initiated with the discovery of *A. J. Bermudez* in 1973. Subsequent discoveries have continued in Campeche and Reforma into the mid-1980s, including *Cantarell* in 1976, *Paredon* in 1977, *Cardenas, Jujo,* and *Pol* in 1980, *Batab* in 1984, *Tecominoacan* in 1985, and *Uech* in 1986.

Kimmeridgian reservoirs in the Gulf of Mexico basin exhibit a variety of lithologies, depositional environments, and reservoir characteristics. The Kimmeridgian limestones and dolomites in the Campeche and Reforma areas were deposited in various shallow-marine environments. Reservoir depths vary substantially offshore from 2,400 to 5,200 m and only moderately onshore from 5,100 to 6,400 m. As giant fields go, the Campeche and Reforma fields are relatively small in area, ranging from 1,200 to 10,000 hectares (3,000 to 25,000 acres). They obtain their immense volume through a very good to spectacular thickness of the reservoirs, ranging from 60 to 200 m offshore to 200 to 550 m onshore. Matrix porosity in the Kimmeridgian is low, usually around 4 to 6%, although where oolitic limestones and dolomites are present, porosity can be as high as 15%. Because of intense fracturing, permeability is excellent and initial per well production of 5,000 to 20,000 bbls per day is normal (Chavarria, 1980; Santiago Acevedo and Mejia Dautt, 1980; Viniegra-O., 1981; Santiago and Baro, 1991).

The limestones of the San Andrés Member were deposited in shoals within a high-energy, shallow-water environment (see Chapter 8, this volume). Porosity is of moderate quality, while permeability tends to be low. Reservoir depths in the San Andrés range from 1,800 to 3,400 m. The Gilmer Limestone is a shoal-water carbonate, generally deposited over salt structures. Porosity is low (usually 8 to 10%) and permeability is very low (less than 0.5 md), requiring that the reservoirs be acidized and fractured to produce commercially. Reservoir depths are typically between 3,400 and 4,000 m (Presley and Reed, 1984; Kosters and others, 1989). Haynesville sands are considered to be fluvial-deltaic sands that have been subject to some marine reworking (Mink and others, 1985; Mann and others, 1989). Both porosity and permeability can be good, although variable, ranging upwards to 20% and 400 md, respectively. Reservoir depths are typically between 2,200 and 3,200 m.

Kimmeridgian fields in the Campeche and Reforma areas are typically faulted anticlines, formed by movement of the underlying salt. Elsewhere, the Kimmeridgian reservoirs in the Gulf of Mexico basin are typically found in combination traps of various varieties. San Andrés traps combine the influence of basement structures with lithologic and diagenetic variations. Gilmer traps are predominantly stratigraphic; however, the presence of enhanced porosity is characteristically related to underlying salt structures (Kosters and others, 1989). Haynesville traps are generally anticlines and faulted anticlines with lithologic changes. Seals for the Campeche and Reforma Kimmeridgian reservoirs are usually provided by Tithonian shales. Seals for San Andrés reservoirs are provided by tighter carbonates within the San Andrés or the overlying Pimienta Formation shales. Gilmer reservoirs are sealed by the overlying Bossier Formation shales. The Haynesville sandstones are sealed by overlying anhydrites and nonpermeable sandstones.

Because few deep (>6,000 m) structures have yet to be drilled in the Campeche and Reforma areas, there are numerous possibilities for major Kimmeridgian discoveries there. Because of the intense reservoir fracturing, a lack of matrix permeability should not constrain gas and condensate accumulations to at least 7,500 m deep. Because of limited areas of reservoir quality rocks in the Kimmeridgian elsewhere, future discoveries are likely to be limited to small fields along the northern rim of the Gulf of Mexico basin.

Tithonian (Cotton Valley Group). The Tithonian Cotton Valley Group is a predominantly gas-producing unit of modest significance within the context of the total petroleum resources of the Gulf of Mexico Basin. As of the end of 1987, estimated known recovery from the Cotton Valley was 8,340 billion ft^3 of natural gas, 345 million barrels of natural gas liquids, and 245 million barrels of crude oil, or 1980 million barrels oil equivalent. (The upper part of the Cotton Valley Group is known to be of Early Cretaceous age [see Chapter 8, this volume]. Nevertheless, because most of the group is of Late Jurassic [Tithonian] age, it will be discussed here as belonging entirely to the Jurassic section.)

Cotton Valley petroleum resources are concentrated in several areas along the northern rim of the Gulf of Mexico basin (Fig. 5). The largest and only major concentration is a gas-condensate "blanket sandstone play" on the northwest rim of the North Louisiana salt basin. Immediately south and west of this area is the so-called "tight massive sandstone gas play." Immediately north of the "blanket sandstone play" in Arkansas and along the Pickens-Gilbertown fault zone in Mississippi are two moderate-size predominantly oil plays. Minor amounts of Cotton Valley gas have also been discovered on the southwest flank of the East Texas basin, the northwest flank of the Mississippi salt basin, and on the south flank of the Wiggins uplift in southern Mississippi.

There are more than 115 significant Cotton Valley reservoirs in the northern Gulf of Mexico basin. The concentration of Cotton Valley known recovery in large and giant fields is less than the norm for the Upper Jurassic sequence: 15.5% is in one giant field; another 54% is in 11 large fields. These giant and large fields are concentrated in the three largest producing trends within a relatively narrow band less than 250 km long, including *Athens-Sugar Creek, Benton, Calhoun, Cotton Valley, Hico-Knowles, Lisbon,* and *Ruston* in the "blanket sandstone play"; *Bethany, Carthage, Oak Hill,* and *Waskom* in the "tight massive sandstone play"; and *Dorcheat-Macedonia* in the updip oil and gas play in south Arkansas.

Cotton Valley exploration has gone through several distinct phases. Discoveries in the "blanket sandstone play" and the oil and gas play immediately updip to it were first made in the late 1930s and reached their peak in the 1940s and early 1950s. Discoveries in the Pickens-Gilbertown fault zone play were highly concentrated in the late 1960s and early 1970s. Discoveries in the tight Cotton Valley massive sandstones first occurred in the late 1960s and reached their peak in the 1970s as gas prices improved. A few minor Cotton Valley discoveries were made in the early 1980s. Although development of tight Cotton Valley reservoirs has continued at an active level, the Cotton Valley has not been a major center of Gulf Coast exploration recently.

Cotton Valley reservoirs consist of fluvial, deltaic, barrier bar, and strandplain sandstones (see Chapter 8, this volume). Reservoir depths in the Cotton Valley range from 1,000 m in the updip area to more than 6,000 m in southern Mississippi. Most Cotton Valley production, however, is concentrated in the 2,000 to 3,000 m depth interval. The various Cotton Valley sandstone plays vary widely in reservoir quality. The individual reservoirs in the "blanket sandstone play" tend to be thin (2 to 20 m), but have good porosity (12 to 18%) and good permeability. Reservoirs in the "tight massive sandstone play" are much thicker (10 to 120 m), but have poor porosity (6 to 10%) and such poor permeability (less than 0.5 md) that they require massive hydraulic fracturing to produce (Collins, 1980; Moore, 1983; Wescott, 1984; Eversull, 1985; Sydboten and Bowen, 1987; Kosters and others, 1989).

With the exception of Cotton Valley fields in Mississippi, where structural traps (anticlines and faulted anticlines) predominate, Cotton Valley fields are usually found in combination traps (anticlines, faulted anticlines, or salt-cored structures associated with lithologic and diagenetic variations). Seals are provided by shales, siltstones, and tightly cemented nonpermeable sandstones within the Cotton Valley Group itself.

The trends in recent Cotton Valley exploration(only a few relatively small new field discoveries were made during the 1980s) do not suggest that many more Cotton Valley discoveries remain to be made. However, with substantial increases in natural gas prices, some discoveries are certainly possible in the central East Texas basin and in deeper parts of the Mississippi salt basin (Presley and Reed, 1984). Moreover, there are still substantial development opportunities within the massive sandstone play, particularly if gas prices increase, fracturing technology improves, and horizontal drilling proves to be applicable within the play.

Upper Jurassic source rocks. The Upper Jurassic, the least important system in the Gulf of Mexico basin for reservoired petroleum, is by contrast the most important system within the basin in terms of source rocks. It contains the only identified "world-class" source rock within the basin, a basin in which other evidently significant sources have not been recognized. It also contains at least one other prominent source rock.

The "world-class" source rock is the Upper Jurassic, particularly the Tithonian black, calcareous, marine shales of southeastern and east-central Mexico. These shales are both very rich in total organic carbon (TOC) and thermally mature. They have long been thought to be the main source of the oil and gas produced from the giant fields of the Tampico-Misantla, Reforma, and Campeche petroleum subprovinces of Mexico, though little has been published on this subject.

Viniegra-O. (1981, p. 27) stated that chemical analyses of the oils from the Cretaceous Golden Lane, Poza Rica, and San Andrés fields in the state of Veracruz, east-central Mexico, as well as those from the Bermudez field in the Cretaceous Reforma area of southeastern Mexico, "indicate that all these oils belong to the same 'family' and that they are probably of Jurassic origin." Viniegra-O. does not provide the chemical analyses to support his interpretation but discusses geological evidence that indicates the Kimmeridgian-Tithonian section is the most likely oil source for much of the oil in east-central and southeastern Mexico. Guzman and Guzman (1981) also assert that the Upper Jurassic section contains the main source rocks for the oil and gas found in southeastern Mexico.

A well-documented case in favor of the Tithonian sediments being the most important source rocks in southeastern Mexico has been presented more recently by Holguin Quiñones

(1985[1988]). He summarizes the extensive geochemical work carried out by Petroleos Mexicanos in the last decade, which indicates that the Tithonian black calcareous shales and shaly limestones are the principal source rock of petroleum in southern Mexico. There is, moreover, a tight geographical fit between the presence of these shales and shaly limestones and the occurrence of major petroleum accumulations in Upper Jurassic (mainly Kimmeridgian) and Cretaceous reservoirs in southeastern and east-central Mexico. Where these source rocks are present in these areas, petroleum has been found; where they are absent, no sizeable Jurassic-Cretaceous accumulations have been discovered.

The Tithonian organic-rich shales, however, may not be the only source of oil and gas in east-central and southeastern Mexico; older Upper Jurassic dark shales and shaly limestones, of Kimmeridgian and perhaps Oxfordian age, should also be considered as possible petroleum sources in these areas. The entire Upper Jurassic section has all the physical characteristics of organic-rich source rocks. All together, the Jurassic source rocks in east-central and southeastern Mexico have provided more than 80 billion BOE known recovery and at least four times that amount in in-place petroleum resources.

The laminated calcareous mudstones of the lower part of the Oxfordian Smackover Formation of the northern flank of the Gulf of Mexico basin, from east Texas to Alabama and the Florida panhandle are the other prominent source rock in the Jurassic sequence. They are believed to have sourced the oil and gas discovered in Jurassic accumulations along the northern Gulf of Mexico basin (Erdman and Morris, 1974; Oehler, 1984; Mancini and others, 1986; Sassen and others, 1987a, b; Wade and others, 1987; Sassen and Moore, 1988; Driskill and others, 1988; Claypool and Mancini, 1989; Sassen, 1989,1990a; Sweeney, 1990). Dutton and others (1987), Sassen (1989), and Burgess (1990) have also suggested that oil accumulated in some Cretaceous reservoirs in northeast Texas was generated in the underlying Smackover Formation.

Current measurements of total organic carbon (TOC) within the lower Smackover do not indicate a particularly rich source rock—an average TOC content of only 0.5% (Oehler, 1984; Sassen and others, 1987a). However, because most of the lower Smackover has been thermally mature for some time and thus has generated and expelled substantial amounts of petroleum, current measurements of TOC probably understate original values considerably. Algal-derived amorphous and thus oil-prone kerogen is the dominant type of organic matter within the lower Smackover. Some kerogen of terrestrial origin is, however, present in southeastern Alabama (Oehler, 1984; Wade and others, 1987; Sassen and others, 1987a; Sassen and Moore, 1988; Sassen, 1990a).

The geological and geochemical evidence for the generation of a substantial amount of oil and gas by the calcareous mudstones of the lower Smackover is strong. On purely geological grounds, it has been argued (Sassen and others, 1987a, b) that the Norphlet and Smackover Formations are sandwiched between two impermeable evaporite intervals over much of the northern flank of the Gulf of Mexico basin—the Louann Salt below and the Buckner Member of the Haynesville Formation above—making it most likely that the oil and gas reservoired in the Norphlet sandstones and in the grain limestones of the upper Smackover were generated in the adjacent calcareous mudstones of the lower Smackover. The geochemical evidence, based on numerous isotope analyses for the characterization of the oils, and on oil-source rock correlations, strongly supports the geologic evidence.

Nearly ten billion BOE known recovery (or more, if all of the discovered Norphlet resources are included) in the Gulf of Mexico basin is believed to have been sourced by the lower Smackover.

The source of Norphlet petroleum is commonly considered to be the laminated carbonate mudstones of the immediately overlying lower Smackover. The composition of Norphlet reservoir fluids, geochemical source-rock evaluations, and oil-source rock correlations all support this conclusion (Sassen and Moore, 1988; Claypool and Mancini, 1989). Crude oil was generated from the Smackover during most of the Cretaceous and was subsequently cracked into natural gas during the Tertiary. Migration from the lower Smackover into the Norphlet is considered to be short-range lateral migration, in some cases being strongly influenced by local faulting (Sassen and others, 1987b; Claypool and Mancini, 1989).

The source of all upper Smackover production is also believed to be the lower Smackover calcareous mudstones. As indicated earlier, these source rocks contain oil-prone amorphous kerogen of algal origin. Migration from the source to the reservoir is considered to be short-range lateral and vertical migration (Oehler, 1984; Sassen and others, 1987b; Sassen and Moore, 1988; Claypool and Mancini, 1989).

In deep Smackover and Norphlet petroleum accumulations, the crude oil originally reservoired has been thermally cracked to gas-condensate and gas (mainly methane). With increased depth and thermal maturity, the methane and other hydrocarbons in sulfate-rich reservoirs have been further destroyed by thermochemical reactions to yield hydrogen sulfide, carbon dioxide, and nitrogen (Sassen and Moore, 1988; Sassen and others, 1987b; Claypool and Mancini, 1989). The hydrogen sulfide content of the gas in some Smackover fields in the Mississippi salt basin is as high as 75%.

The fluid composition of the upper Smackover fields, from east Texas to the western Florida panhandle, changes with depth: the shallower, updip fields principally produce crude oil; those further downdip, deeper in the basin, contain oil and condensate; the deepest accumulations contain mainly dry natural gas. This distribution in trends generally parallel to the rim of the basin has been explained (Mancini and others, 1985; Sassen and Moore, 1988; Driskill and others, 1988) in terms of depth of burial of the lower Smackover source rocks and the resulting degree of their thermal maturity, not to differences in the characteristics of the source rocks.

The source rocks for Kimmeridgian accumulations in the United States are not clearly established. No published references exist for sources for Gilmer Limestone gas reservoirs. Haynesville sandstone production is limited to a narrow band where the underlying Buckner anhydrite is thin or absent or where extensive faulting permits upward migration from the Smackover (Sassen, 1990a).

The sources of Cotton Valley hydrocarbons vary by play. Source rocks of the two oil plays are considered to be the underlying lower Smackover carbonate mudstones. Migration from the Smackover into Cotton Valley is considered to have occurred along faults or in areas where the intervening Buckner anhydrite is thin or absent (Collins, 1980; Sassen and others, 1987b; Sassen, 1990a). Sources for the oil, condensate, and some of the gas of the "blanket sandstone play" are most likely the lower Smackover mudstones, and for the rest of the gas, a combination of the downdip Bossier shales and shales and coals interbedded with the Cotton Valley sandstones (Collins, 1980; Eversull, 1985).

Lower Cretaceous petroleum resources

The Lower Cretaceous is the most important petroleum producing system within the Mesozoic section in the Gulf of Mexico basin. As of the end of 1987, it had a known recovery of 26.2 billion barrels of crude oil, 65.2 trillion ft^3 of natural gas, and 4.2 billion barrels of natural gas liquids, totalling 41.2 billion BOE (Table 11). Overall it provides 23.2% of the crude oil, 12.5% of the natural gas, 18.5% of the natural gas liquids, and 18.5% of the oil equivalent known recovery of the Gulf of Mexico basin. Treated as an individual petroleum province, the Lower Cretaceous of the Gulf of Mexico basin would be exceeded in size by no more than ten other petroleum provinces worldwide.

Because Lower Cretaceous carbonate reservoirs in southeast Mexico are extensively dolomitized and are therefore devoid of age diagnostic fossils, Lower Cretaceous production in that area cannot be differentiated by stages. Differentiation by stage is however possible within the Lower Cretaceous on the western and northern sides of the Gulf of Mexico basin. In these two areas, the upper Albian–lower Cenomanian is the most important, followed in order by the Hauterivian-Barremian–lower Aptian and the upper Aptian–lower Albian. The Berriasian and Valanginian intervals are not known to be productive of hydrocarbons.

Lower Cretaceous known recovery, as discussed here, includes small amounts of known recovery from the lower part of the Upper Cretaceous Cenomanian section. These amounts are assigned here to the Lower Cretaceous even though they are in formations that straddle the Lower Cretaceous–Upper Cretaceous boundary because their known recovery is predominantly Lower Cretaceous.

Lower Cretaceous petroleum in the Gulf of Mexico basin is highly concentrated in the southern half of the basin. Three-fourths of Lower Cretaceous known recovery is in Mexico. This concentration is particularly marked in liquids, Mexico having

87.6% of the Lower Cretaceous crude oil and 78.7% of the natural gas liquids. The northern half actually has the majority (55.3%) of Lower Cretaceous known recovery of natural gas.

Lower Cretaceous known recovery in the Gulf of Mexico basin is highly concentrated in giant and large fields. Nineteen giant fields, including the supergiant *Abkatun, A. J. Bermudez,* and *Cantarell* fields, contained 28,843 million BOE, 70.0% of Lower Cretaceous known recovery. Another 68 large fields contribute 6,733 million BOE, 16.3% of the Lower Cretaceous total.

Undifferentiated Lower Cretaceous (southeast Mexico). Lower Cretaceous petroleum production in southeast Mexico cannot be differentiated by stage, as indicated earlier. Nonetheless, the Lower Cretaceous in the Reforma area onshore and the adjacent Campeche area offshore is of world-class proportions, having an estimated known recovery of 17,845 million barrels of crude oil, 23,230 billion ft^3 of natural gas, and 2,865 million barrels of natural gas liquids or 24,222 million BOE as of December 31, 1987. These amounts constitute a majority of Lower Cretaceous known recovery in the Gulf of Mexico basin.

The overwhelmingly oil-prone Lower Cretaceous accumulations in the Bay of Campeche are by the slightest margin the

TABLE 11. KNOWN RECOVERY OF CRUDE OIL, NATURAL GAS, AND NATURAL GAS LIQUIDS IN THE GULF OF MEXICO BASIN BY STRATIGRAPHIC INTERVAL WITHIN THE LOWER CRETACEOUS

	Crude Oil (Million bbls)	Natural Gas (Bcf)	NGL (Million bbls)	BOE (Million bbls)
Upper Albian-lower Cenomanian				
Fredericksburg-Washita-N*	1,615	5,490	140	2,670
El Abra-Tamabra-Upper Tamaulipas-Orizaba-S*	4,830	5,440	405	6,142
Total	6,445	10,930	545	8,812
Upper Aptian-lower Albian				
Pearsall-Glen Rose-N*	1,050	9,390	270	2,885
Hauterivian-Barremian-lower Aptian				
Travis Peak-Hosston-Sligo-N*	580	21,180	480	4,590
Lower Tamaulipas-Cupido-S*	600	510	15	700
Total	1,180	21,690	495	5,290
Undifferentiated (SE Mexico)				
In Reforma-S*	7,085	18,940	1,825	12,067
In Campeche-S*	10,400	4,290	1,040	12,155
Total	17,485	23,230	2,865	24,222
Lower Cretaceous Total	26,160	65,240	4,175	41,209

*N = Northern (United States); S = Southern (Mexico).

more important of these two trends (Fig. 6). The Lower Cretaceous accumulations of Reforma, which are concentrated in the southern two-thirds of this area, are still oil-prone, albeit with substantially higher proportions of natural gas and natural gas liquids because of their greater depths. Each of these Lower Cretaeous producing areas has enough petroleum resources to be a major petroleum province by itself.

The Lower Cretaceous petroleum resources of Campeche and Reforma are highly concentrated in giant fields. Twelve giant fields—including the *A. J. Bermudez* supergiant in Reforma; the *Abkatun, Cantarell,* and (possibly) *Ku* supergiants in Campeche; *Chuc* and *Pol* in Campeche; and *Agave, Cactus-Nispero, Cardenas, Giraldas, Paredon,* and *Sitio Grande* in Reforma—contained an estimated 22,445 million BOE, 92.7% of the total in the undifferentiated Lower Cretaceous (Plate 4). Ten large fields (*Bellota,*

Cacho Lopez, Chiapas, Comoapa, Copano, Fenix, Iris, Mora, Mundo Nuevo, and *Rio Nuevo*), all in the Reforma area with another 1,640 million BOE (6.8% of the total), contained most of the remainder.

Exploration in these two trends was initiated with the discoveries of *Sitio Grande* and *Cactus-Nispero* in 1972. These play openers were followed by *A. J. Bermudez* (1973), *Agave* (1976), *Giraldas* and *Paredon* (both 1977), and *Cardenas* (1980) on-shore, along with other large discoveries in the late 1970s and early 1980s. Offshore, the initial Lower Cretaceous discovery occurred in *Cantarell* in 1976, with *Abkatun* and *Ku* (both 1979), *Pol* (1980), and *Chuc* (1982) coming next (Santiago Acevedo and Meijia Dautt, 1980; Santiago and Baro, 1991).

Lower Cretaceous reservoirs in the Reforma and Campeche areas are essentially found in a high-energy bank-edge sequence

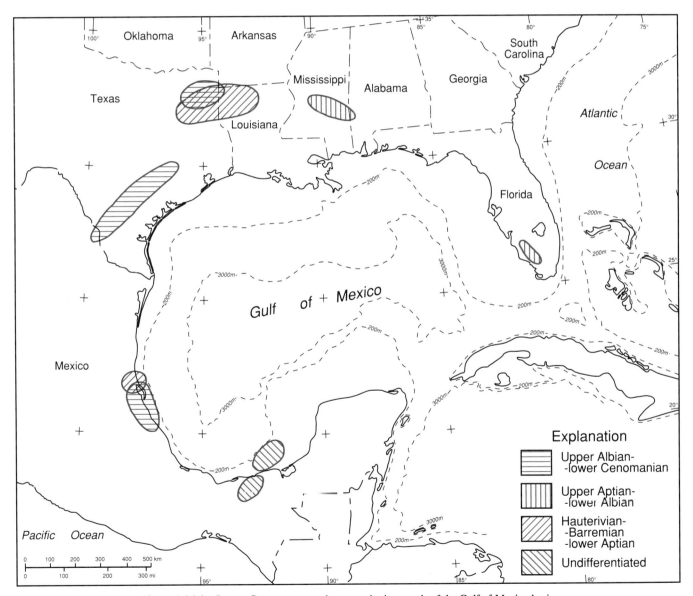

Figure 6. Major Lower Cretaceous petroleum-producing trends of the Gulf of Mexico basin.

west of the Yucatan carbonate platform (Viniegra O., 1981). The Lower Cretaceous accumulations in the Reforma area are 3,700 to 5,400 m deep. Together with the deeper Kimmeridgian reservoirs in the same fields, these constitute some of the deepest major oil fields worldwide. Producing areas of these fields range from 850 to 14,000 ha (2,100 to 34,500 acres). Net thicknesses are substantial, ranging from 165 to 400 m. Matrix porosity is poor (generally 3 to 7%), but because of the intensive microfracturing created by pillowing of the underlying salt, permeability is excellent.

The Lower Cretaceous accumulations in the Campeche area are noticeably shallower, being only 1,700 to 4,100 m deep. Like the Reforma area fields, producing areas are small relative to field size, being only 1,400 to 12,150 ha (3,500 to 30,000 acres). However, thicknesses here are also considerable, ranging from 45 to 230 m. At 5 to 10%, matrix porosity is slightly better than it is in the Reforma area. Here also permeability is excellent because of the intensive fracturing within the dolomite reservoirs (Santiago and Baro, 1991).

Trapping in both groups of fields is structural, consisting of anticlines, usually faulted, created by movement of the underlying salt. Seals for the Reforma fields are either Upper Cretaceous, Paleocene, or Eocene shales, depending on whether the Upper Cretaceous and Paleocene have been eroded in the immediate area. Seals for the Campeche fields are provided by Paleocene shales (Santiago Acevedo and Mejia Dautt, 1980; Viniegra O., 1981; Santiago and Baro, 1991).

Considerable exploratory potential, particularly for natural gas and condensate, exists for the Lower Cretaceous in Campeche and Reforma, primarily because many of the deeper (greater than 5,500 m) structures remain to be drilled in the area between the two trends straddling the coastline. As both source rocks and structures are known to be present, the main exploratory risk is whether the Lower Cretaceous contains favorable reservoir rocks to create a substantial accumulation.

Hauterivian-Barremian–lower Aptian (Hosston/Travis Peak, Sligo/Pettet, Cupido, and Lower Tamaulipas Formations). Hauterivian-Barremian–lower Aptian sandstones and limestones are an important producing interval within the Lower Cretaceous, particularly for natural gas. As of the end of 1987, known recovery from this sequence was 21,690 billion ft^3 of natural gas, 1,180 million barrels of crude oil, and 495 million barrels of natural gas liquids, for a total of 5,290 million barrels of oil equivalent. Although this is a significant amount on a world scale, it is only 12.8% of the total Lower Cretaceous known recovery (but 33.2% of Lower Cretaceous natural gas) and only 2.4% of known petroleum recovery in the Gulf of Mexico basin.

Hauterivian-Barremian–lower Aptian petroleum resources are spread across several formations in several areas of the northern and western Gulf of Mexico basin, but most are concentrated in just three trends (Fig. 6). The largest of these, a Sligo/Pettet limestone natural gas and natural gas liquids trend with nearly half of Hauterivian-Barremian–lower Aptian known recovery, straddles the Sabine uplift and the areas immediately east and

west of it in the North Louisiana salt basin and the East Texas basin, respectively. The only other major trend, a Travis Peak (Hosston) natural gas play, immediately underlies this Pettet trend. The largest moderate-size trend, in fractured Lower Tamaulipas limestones on the northern flank of the Tampico-Misantla basin, contains slightly more than half of Hauterivian-Barremian–lower Aptian oil. Besides these three trends, there is a moderate-size Sligo reef gas and liquids play southeast of the Sabine uplift (albeit in just the large *Black Lake* field), several minor Hosston/Travis Peak and Sligo/Pettet oil plays in the various fault systems along the northern rim of the Gulf of Mexico basin, a moderate-size deep Hosston gas play in the southern Mississippi salt basin and on the Wiggins uplift, a minor La Virgen Dolomite (Cupido) gas trend in the Sabinas basin, and a minor Sligo reef gas trend in south Texas.

Of the 230 fields in the Gulf of Mexico basin with significant Hauterivian-Barremian–lower Aptian production, less than 30 giant and large fields provide 4,150 million BOE known recovery (77.6%). Three giants—*Carthage* in east Texas, the *Ebano-Panuco* fields in east-central Mexico, and *Smackover* in Arkansas (with a small amount of Travis Peak production) provide 1,796 million BOE (33.4%; Plate 4). Another 26 large fields—most notably *Soso* in Mississippi; *Ada-Sibley, Athens-Sugar Creek, Bear Creek-Bryceland, Black Lake, Haynesville, Lisbon, Lucky,* and *Sligo* in north Louisiana; *Bethany-Longstreet, Greenwood-Waskom,* and *Joaquin-Logansport* on the Texas-Louisiana border; *Chapel Hill, New Hope, Opelika, Trawick, Tri-Cities, Whelan, Willow Springs,* and *Woodlawn* in east Texas; and *Tamaulipas-Constituciones* in east-central Mexico contain another 2,336 million BOE (44.2% of the total).

Although the first Hauterivian-Barremian–lower Aptian oil was discovered shortly after 1900 in the *Ebano-Panuco* fields, nearly all of the significant discoveries have occurred in the past 60 years. Discoveries in the major Pettet trend straddling the Sabine uplift span the period between 1930 and the mid-1980s, including *Carthage* in 1936. Discoveries in the underlying Travis Peak cover the period from 1935 to the mid-1980s including the Travis Peak reservoir in the *Carthage* field in 1942. Nearly all of the discoveries in the oil plays in the northern rim fault systems occurred between 1940 and 1970, including the Travis Peak reservoir in the *Smackover* field in 1951. *Black Lake* was discovered in 1964. The Hosston gas discoveries in Mississippi were concentrated in the 1970s.

The Hosston/Travis Peak reservoirs are predominantly terrigenous clastics deposited in fluvial and deltaic environments across a broad coastal plain. In the main Hosston/Travis Peak trend, the accumulations range from 1,400 to 3,400 m deep, have thicknesses of 2 to 70 m, porosities of 6 to 20%, and permeabilities of less than 0.1 to more than 100 md (Dutton, 1987; Kosters and others, 1989). In the deep Hosston of southern Mississippi, the accumulations are 4,300 to 5,000 m deep with thicknesses of 3 to 30 m. Reflecting the coarsening nature of Hosston clastics to the east, these reservoirs still have porosities of 8 to 13% and permeabilities of 2 to 25 md, despite being substantially deeper than the tighter Travis Peak reservoirs in east Texas. The predominantly clastic fault-line trends in Mississippi have accumulations 2,750 to 4,750 m deep that are 3 to 24 m thick, with 12 to 20% porosity and 30 to 250 md permeability. In Arkansas and Texas, the accumulations in these trends are shallower, being only 850 to 2,500 m deep with thicknesses of 2 to 30 m, porosities of 12 to 30%, and permeabilities of 20 to 2,000 md (porosity and permeability varying inversely with depth).

The Hosston/Travis Peak clastic section grades upward and gulfward (southward) into the Sligo/Pettet carbonates. The productive portions of the latter formation consists of oolitic limestones and rudist banks that formed in shoal areas on structural highs. In the main Sligo/Pettet trend straddling the Sabine uplift, the significant accumulations are 1,500 to 3,250 m deep and have thicknesses of 2 to 30 m, average porosities of 11 to 20%, and permeabilities of 5 to 300 md (Bebout and others, 1981; Kosters and others, 1989). The reefal *Black Lake* field to the southeast is 2,440 m deep with a reservoir averaging 18 m thick, with 17% porosity and 110 md permeability (Bailey, 1978). The Lower Tamaulipas limestones of the *Ebano-Panuco* fields are very low-porosity deep-water carbonates that produce through an extensive fracture system (Guzman and others, 1955).

Trapping for both Hosston/Travis Peak and Sligo/Pettet fields is a combination of structural and stratigraphic factors. A wide variety of structural traps are found including anticlines, anticlinal noses, fault traps, salt-cored anticlines, and salt domes. These are typically modified by lithologic changes, including reefing within the Sligo/Pettet carbonates (Galloway and others, 1983). Seals for these accumulations are usually provided by impermeable shales and mudstones within the overall interval.

All of the significant Hauterivian-Barremian–lower Aptian discoveries of the 1980s were small fields, suggesting that this objective is now in a mature exploration phase. Several possibilities still remain, including some deep Hosston-Sligo objectives in the Mississippi salt basin and deep Sligo reefs along the lightly drilled Sligo shelf margin in south Texas. Additional development of the tight Travis Peak sandstones in east Texas also has promise to add to Hauterivian-Barremian–lower Aptian reserves.

Upper Aptian–lower Albian of the northern Gulf of Mexico basin (Pearsall, James, Glen Rose, Rodessa, and Sunniland Formations). The immense size of the amounts of petroleum in both the Gulf of Mexico basin as a whole and the Lower Cretaceous within it are illustrated by the relative position of the upper Aptian–lower Albian sequence of the northern Gulf of Mexico basin and Florida within the overall context of the basin's petroleum resources. Although the petroleum resources of this interval are hardly insignificant, with known recovery being 1,050 million barrels of crude oil, 9,390 billion ft^3 of natural gas, and 270 million barrels of natural gas liquids for a total of 2,885 million BOE as of the end of 1987, they are only 7.0% of the petroleum resources of the Lower Cretaceous and 1.3% of the petroleum resources of the entire Gulf of Mexico basin.

Upper Aptian–lower Albian petroleum resources, as discussed here, include petroleum accumulations in the Pearsall,

James, Glen Rose, Rodessa, and (in only a minor degree) Pine Island Formations in the northern Gulf of Mexico basin and in the Sunniland Formation of southern Florida. Most (more than 75%) of the known petroleum resources in the upper Aptian–lower Albian of the northern Gulf of Mexico basin are concentrated in one Rodessa and James gas and liquids trend straddling the Sabine uplift and the adjacent basins to the east and west (Fig. 6). (Geographically this trend overlaps the main Sligo/Pettet and Hosston/Travis Peak producing trends in the Hauterivian-Barremian–lower Aptian.) Other upper Aptian–lower Albian producing trends include a moderate-size Rodessa oil and gas trend across the northern half of the Mississippi salt basin, a minimally moderate-size Sunniland oil trend in south Florida, a minor Mooringsport (upper Glen Rose) and James gas trend in southern Mississippi, and a minor Rodessa–Pine Island (Hogg Sand) fault-line oil trend along the northern rim of the basin in south Arkansas and east Texas.

The upper Aptian–lower Albian of the northern Gulf of Mexico basin is unusual within the Mesozoic sequence in that giant fields provide such a small proportion of a substantial known recovery from nearly 200 significant fields. The only contribution from giant fields is 102 million BOE from relatively small Rodessa reservoirs in the giant *Carthage* and *Hawkins* fields in the East Texas basin. Most of the known recovery (1,955 million BOE or 67.8%) in the upper Aptian–lower Albian comes from 25 large fields, including *Citronelle* in southwest Alabama; *Gwinville* and *Soso* in Mississippi; *Stephens-Wesson* in Arkansas, *Ada-Sibley*, *Athens-Sugar Creek*, and *Sligo* in north Louisiana; *Bethany-Longstreet*, *Greenwood-Waskom*, *Joaquin-Logansport*, and *Rodessa* straddling the Louisiana-Texas border; and *Cayuga*, *Chapel Hill*, *Fairway*, *Fort Trinidad*, *New Hope*, *Opelika*, *Quitman*, *Trawick*, *Whelan*, *Willow Springs*, and *Woodlawn* in east Texas. *West Felda* in the Sunniland trend in south Florida falls just short of large-field status.

The discovery history of the main Rodessa/James play exhibits an unusually long duration of healthy levels of discovery. Since the initial discoveries at *Sligo* in 1927 and *Rodessa* in 1930, there have been numerous significant fields found in each successive decade into the 1980s, including the Rodessa reservoirs in *Hawkins* in 1947 and in *Carthage* in 1950. In the other upper Aptian–lower Albian trends, discoveries have been more concentrated. Most of the Rodessa discoveries in the northern Mississippi salt basin occurred in the 1950s and 1960s. The Sunniland trend in south Florida peaked in the 1960s and 1970s. James and Mooringsport discoveries in south Mississippi occurred primarily in the late 1940s and 1950s (although the James is currently enjoying some activity). Most discoveries in the Arkansas–east Texas fault-line trend were made in the 1940s and 1950s.

Reservoirs within the main Rodessa/James trend include Rodessa limestones and sandstones and James limestones. Depositional environments range from deltaic and delta-front clastics to the north grading basinward into shallow to deep carbonate shelves (Bushaw, 1968; McFarlan, 1977; Chapter 9, this volume). Reservoirs on the Sabine uplift and in the North Louisiana salt basin to the east range from 1,050 to 2,750 m deep, are 1 to 40 m thick, and have average porosities of 12 to 25% and average permeabilities of 5 to 500 md. In the East Texas basin, the reservoirs are deeper (2,000 to 3,150 m), have thicknesses between 2 and 80 m, and have only 10 to 20% porosity and 2 to 400 md permeability.

Rodessa reservoirs in the northern Mississippi salt basin are sandstones deposited in fluvial-deltaic environments (Bolin and others, 1989). Depths range from 2,650 to 4,550 m; thicknesses are 3 to 30 m; porosities average 10 to 25% and permeabilities average 10 to 400 md.

Sunniland limestone reservoirs are predominantly back reef shoals (Richards, 1988). Depths in this trend are 3,500 to 3,600 m; average thicknesses are only 1 to 8 m. Porosities average 15 to 20% while permeabilities average 50 to 100 md.

Traps in the main Rodessa/James trend combine structural and stratigraphic elements, being predominantly anticlines or salt-cored anticlines, modified by lithologic changes. The Sunniland limestone reservoirs are stratigraphically trapped by lithologic changes. Other Rodessa and James plays have structural traps, mainly anticlines. The predominant seal is provided by the Ferry Lake Anhydrite at the top of the Rodessa (lower Glen Rose). Other seals are provided by various anhydrite and shale zones within the Rodessa (lower Glen Rose).

Given the continuing record of exploratory success in the main Rodessa/James trend, these objectives still have moderate exploratory potential, particularly in the East Texas basin. In southern Mississippi, the deeper James provides some possible targets for natural gas exploration. The Sunniland Formation in south Florida may have some minor potential as well.

Upper Albian–lower Cenomanian of the western and northern Gulf of Mexico basin (El Abra, Tamabra, Upper Tamaulipas, Orizaba, Fredericksburg, and Washita). The upper Albian–lower Cenomanian of the western and northern Gulf of Mexico basin is the most important differentiated petroleum producing interval to which known recovery can be assigned in the Lower Cretaceous sequence of the Gulf of Mexico basin. As of December 31, 1987, known recovery in the upper Albian–lower Cenomanian was 6,445 million barrels of crude oil, 10,930 billion ft^3 of natural gas, and 545 million barrels of natural gas, for a total of 8,812 million BOE, a size large enough to qualify the upper Albian–lower Cenomanian interval in the Gulf of Mexico basin as a major province by itself. This interval includes the two best known early oil discoveries in Mexico, the fields of the *Golden Lane* trend and the giant *Poza Rica* field (Plate 4).

The petroleum resources of the upper Albian–lower Cenomanian, as defined here, include in the northern Gulf of Mexico basin the accumulations in the formations of the Fredericksburg Group (primarily those in the Paluxy and Edwards Formations) and lesser amounts of oil and gas in the Buda Formation of the Washita Group. In the western part of the basin, in east-central Mexico, the upper Albian–lower Cenomanian includes accumulations in the El Abra, Tamabra, Upper Tamaulipas, and Orizaba

Formations, even though the lower part of these formations is known to be of lower Albian age (see Fig. 4 and Plate 5). In Mexico, the Albian and Cenomanian stages are grouped into what is called the "Middle Cretaceous" (see Chapters 9 and 10, this volume). Because the "Middle Cretaceous" is not a recognized formal subdivision of the Cretaceous and is not, therefore, used in this volume, and because the majority of petroleum production from this interval is in the Lower Cretaceous section as recognized in this volume, all of "Middle Cretaceous" known petroleum recovery in Mexico is assigned here to the Lower Cretaceous. If the known recovery from this interval could be divided between the Lower and the Upper Cretaceous, it would most likely make the latter series slightly more important than the Lower Cretaceous to the overall petroleum resources of the Gulf of Mexico basin.

Upper Albian–lower Cenomanian known recovery is spread across several major and moderate-size trends (Fig. 6). The largest is the Tamabra limestone oil trend immediately west of the Tuxpan platform in the center of the Tampico–Misantla basin. The other two major trends are the El Abra limestone oil fields on the perimeter of the Tuxpan platform and a set of parallel Edwards (Fredericksburg) gas and liquids plays in south and south-central Texas. The largest moderate-size trend is a Paluxy (Fredericksburg) oil and gas play in the northern half of the East Texas basin. Another trend of roughly similar size is a fractured upper Tamaulipas limestone oil play in the northern Tampico-Misantla basin. Other trends of moderate size include a Paluxy (Fredericksburg) oil and gas play on the east side of the Sabine uplift in northwest Louisiana, a Paluxy (Fredericksburg) oil and gas unconformity play in northeast Louisiana, a Paluxy (Fredericksburg)-Washita gas play on the western half of the Wiggins uplift, a Fredericksburg (Paluxy)-Washita oil and gas play on the northeast flank of the Mississippi salt basin, and an Orizaba limestone oil play in the Veracruz basin. Of the several minor upper Albian–lower Cenomanian producing trends, a Buda (Washita) oil play in south-central Texas and a Paluxy (Fredericksburg) oil play in southwest Arkansas are the most important.

Like most of the Mesozoic producing intervals, known recovery in the more than 200 significant upper Albian–lower Cenomanian oil and gas fields is concentrated in a few giant and large fields. Four giant fields in the Tampico-Misantla basin—the *Ebano-Panuco* fields, the *Naranjos-Cerro Azul* fields of the Old Golden Lane trend, *Poza Rica*, and *San Andrés*—contained 4,500 million BOE, 51.1% of the upper Albian–lower Cenomanian total (Plate 4). Another 31 large fields, including *Dexter* and *Hub* in Mississippi; *Caddo-Pine Island, Delhi–Big Creek, Red River–Bull Bayou,* and *Sligo* in north Louisiana; *Chapel Hill, Coke, Quitman, Sand Flat–Shamburger Lake,* and *Talco* in east Texas; *Darst Creek, Fashing, Jourdantown, Luling-Branyon, Person–Panna Maria,* and *Salt Flat–Tenney Creek* in south Texas; *Atun, Bagre, Hallazgo, Ordonez, Santa Agueda, Tamaulipas-Constituciones,* and *Tres Hermanos* in the Tampico-Misantla basin; and *Matapionche* in the Veracruz basin, provide another

2,896 million BOE, 32.9% of upper Albian–lower Cenomanian known recovery.

Significant discoveries of upper Albian–lower Cenomanian reservoirs span almost the entire discovery history of the Gulf of Mexico basin. The first upper Albian–lower Cenomanian oil was discovered in the *Ebano-Panuco* fields shortly after 1900. This discovery was quickly followed by most of the discoveries in the *Old Golden Lane* El Abra fields during the next 15 years. The first discoveries in the Paluxy occurred in the 1910s on the Sabine uplift, a play in which discoveries have continued sporadically into the 1980s. The first Edwards discoveries in south Texas occurred in the shallow updip portion of that trend in the 1920s, peaked as exploration moved downdip from the 1940s through the 1970s, and have persisted into the 1980s with continued deep discoveries. The first and largest Tamabra discovery, *Poza Rica,* was made in 1930; most of the subsequent Tamabra discoveries occurred in the 1940s and 1950s. *New Golden Lane* discoveries along the southwest edge of the Tuxpan platform were concentrated in the 1950s, while the *Marine Golden Lane* fields along the eastern edge of that platform were discovered in the 1960s. Paluxy discoveries in east Texas span the period from 1930 into the 1980s, while Paluxy and Fredericksburg-Washita discoveries in northeast Louisiana and Mississippi were concentrated in the 1940s, 1950s, and 1960s.

The Tamabra reservoirs are shallow-water limestones, essentially debris from reefs located on paleohighs. Average porosities throughout the trend vary from 8 to 20 percent, while permeabilities exhibit great variability (5 to 1,000 md). In *Poza Rica,* the dominant field in the Tamabra trend, the reservoir is 2,000 m deep and 80 m thick, and has an average porosity of 17 percent and an average permeability of 63 md (Ayala Nieto and Perez Matus, 1974). Trapping is predominantly stratigraphic, the result of a loss of porosity because of lithologic changes within the limestone. However, the location of porosity is highly influenced by paleostructure. Seals are provided by Upper Cretaceous Agua Nueva shales (Guzman, 1967; Coogan and others, 1972).

The El Abra is a shallow shelf limestone in which tremendous secondary porosity and permeability were created by subaerial erosion during the Eocene and Oligocene along the margins of the Tuxpan platform. Because this platform currently tilts to the southeast, reservoir depths vary from 500 to 800 m in the *Old Golden Lane* fields on the northwest edge to 2,000 to 2,200 m in the *Marine Golden Lane* fields on the southeast edge. Traps are limited to erosional highs confined to the narrow margin of the platform edge. Seals are provided by Upper Cretaceous, Eocene, and Oligocene shales (Guzman, 1967; Viniegra-O., and Castillo-Tejero, 1970; Coogan and others, 1972).

The upper Tamaulipas limestones are low porosity limestones deposited in a deep, open-marine environment. They are productive only where fractured. Accumulations in the upper Tamaulipas are concentrated on a regional anticlinal nose, the southern end of the Sierra de Tamaulipas (Guzman and others, 1955). Oil production in the extensively overthrust Orizaba limestones in the Veracruz basin is found in zones of secondary porosity at depths of 2,500 to 2,800 m, well below the major unconformity between the thrust plates and the overlying Tertiary sediments. Prospective traps directly underlying this unconformity typically lack seals or have been flushed by meteoric water (Mossman and Viniegra, 1976).

The Edwards reservoirs in south and south-central Texas consist of tidal-flat, shallow-marine, and shelf-margin carbonates. Depths to Edwards production vary widely, from as shallow as 300 m updip in the Luling fault zone to as deep as 4,925 m downdip in the Stuart City Reef trend. Reservoir thicknesses within the Edwards vary from 4 to 75 m. Porosity is highly variable, ranging from 5 to 30 percent, depending both upon depth and depositional environment. Permeability varies from less than 1 to more than 75 md. Trapping in the Edwards is predominantly by faulting, with the accumulations often being limited by lithologic and diagenetic changes within the Edwards. Downdip, in the Stuart City Reef trend, reefs provide the predominant trap. Seals are provided by the younger Georgetown (Washita) limestones and Taylor shales and by tight carbonates within the Edwards itself (Rose, 1972; Bebout and others, 1977; McFarlan, 1977; Cook, 1979; Galloway and others, 1983; Kosters and others, 1989).

Paluxy reservoirs are predominantly fluvial and deltaic sandstones. In east Texas, Paluxy reservoirs are 1,200 to 2,400 m deep and 1 to 45 m thick, with good porosities of 16 to 28 percent and good to excellent permeabilities of 100 to 3,000 md. Paluxy reservoirs on the east side of the Sabine uplift are 750 to 1,,675 m deep and 1 to 10 m thick, with porosities of 25 to 30% and permeabilities of only 100 to 250 md. In the Paluxy oil and gas unconformity play in northeast Louisiana, reservoir depths range from 950 to 2,775 m; thicknesses are 3 to 27 m; while porosities average 25 to 30% and permeabilities average 400 to 1,400 md. Paluxy traps range from faults and salt-related structures, often modified by lithologic changes, in east Texas, south Arkansas, and Mississippi to unconformities in northeast Louisiana. Seals are provided by the overlying Fredericksburg section and by shales within the Paluxy itself (Bushaw, 1968; Caughey, 1977; Galloway and others, 1983).

Because the upper Albian–Lower Cenomanian has been subject to extensive exploration for many decades, few exploratory possibilities still remain. The best chances for further discoveries are for a few downdip Paluxy reservoirs in the East Texas basin and for a few deeper Edwards reservoirs in south and south-central Texas.

Lower Cretaceous source rocks. Despite being a prolific producing interval, the Lower Cretaceous of the Gulf of Mexico basin does not contain any major recognized source rocks. Onshore, there are only a few source units within the Lower Cretaceous that may have made modest contributions to Lower Cretaceous known recovery. Offshore, however, the deep-water Flex trend oils occurring in Pleistocene reservoirs have been suggested to be of Lower Cretaceous origin on the basis of comparisons between their carbon and sulfur isotope ratios and those of Type II kerogens encountered at Deep Sea Drilling Project

(DSDP) Sites 535 and 540 in the Straits of Florida. The latter may be representative of contemporaneous basinal sediments throughout the Gulf of Mexico basin (Thompson and others, 1990). As the following discussion indicates, most petroleum currently reservoired in the Lower Cretaceous is believed to have originated in the Upper Jurassic.

The primary source rocks for Lower Cretaceous reservoirs in the Campeche and Reforma areas of southern Mexico are the prolific Upper Jurassic (Tithonian) black shales and argillaceous limestones (Santiago Acevedo and Mejia Dautt, 1980; Viniegra O., 1981; Holguin Quiñones, 1985 [1988]; Santiago and Baro, 1991), although a possible contribution by Lower Cretaceous source rocks cannot be discarded (Viniegra-O., 1981). In the Tampico-Misantla basin, the shales of the Pimienta Formation, of the same age, are often mentioned as the source of the lower and upper Tamaulipas, Tamabra, and El Abra accumulations (Guzman and others, 1955; Guzman, 1967; Viniegra O. and Castillo-Tejero, 1970). In both areas, migration from source to reservoir was short-distance vertical migration.

Source rocks for the Hauterivian-Barremian–lower Aptian accumulations in the northern Gulf of Mexico basin vary primarily by hydrocarbon type. The fault-system oil plays along the northern rim, were probably sourced from the Jurassic Smackover (Sassen, 1990a; Burgess, 1990). The Hosston/Travis Peak gas plays straddling the Sabine uplift were most likely sourced from the underlying Upper Jurassic Bossier shales of the Cotton Valley Group (Dutton, 1987). Shales within the Hosston/Travis Peak may also be a gas source, although the thermally mature Type III kerogens of the Travis Peak have a generally low TOC (Dutton and others, 1987).

Organic shales within the Hosston Formation have also been mentioned as possible source rocks in southern Mississippi (Evans, 1987), but no conclusive geochemical evidence has been presented to support this claim.

The Jurassic Bossier Formation may have also been a partial source for the Sligo/Pettet gas and liquids trends, but the higher liquids content of these accumulations suggests another source as well, possibly within the Sligo/Pettet carbonates themselves (although none has yet been identified).

In the upper Aptian–lower Albian interval of the northern Gulf of Mexico basin, a major source of Glen Rose reservoirs, particularly the liquids, are the underlying Jurassic lower Smackover calcareous mudstones (Sassen, 1990a). Shales, such as the Pine Island, and carbonates within the lower Glen Rose may also be a source of oil and gas for other Glen Rose reservoirs. In the upper Albian–lower Cenomanian trends of the northern Gulf of Mexico basin, oil and gas in the Edwards is most likely sourced from within the Edwards itself (Bebout and others, 1977; Cook, 1979; Galloway and others, 1983). Sources for Paluxy oil and gas range from the Upper Jurassic lower Smackover for the fault-line traps to the Upper Cretaceous Tuscaloosa for the unconformity traps (Evans, 1987; Sassen, 1990a).

The source of the oil accumulations in the Lower Cretaceous Sunniland Formation of south Florida is believed to be the part of the section closely associated with the reservoirs. Geological reasoning and geochemical studies indicate that the Sunniland and associated carbonates, particularly the organic rich, argillaceous limestones in the lower part of the formation, are the probable source of the oil (Palacas, 1978, 1984; Palacas and others, 1984). Lower Cretaceous carbonates both above and below the Sunniland Formation may also be considered as potential oil source rocks (Palacas, 1978; Palacas and others, 1981).

Upper Cretaceous petroleum resources

Upper Cretaceous reservoirs are a major contributor to the petroleum resources of the Gulf of Mexico basin. At 30.7 billion barrels of crude oil, 37.4 trillion ft^3 of natural gas, and 3.7 billion barrels of natural gas liquids, or 40.6 billion barrels of oil equivalent known recovery (Table 12), the Upper Cretaceous alone in the Gulf of Mexico basin is exceeded in size by no more than ten other petroleum provinces worldwide. Overall it provides 27.3% of the crude oil, 16.3% of the natural gas liquids, 7.1% of the

TABLE 12. KNOWN RECOVERY OF CRUDE OIL, NATURAL GAS, AND NATURAL GAS LIQUIDS AS OF DECEMBER 31, 1987, IN THE GULF OF MEXICO BASIN BY STRATIGRAPHIC INTERVAL WITHIN THE UPPER CRETACEOUS

	Crude Oil (Million bbls)	Natural Gas (Bcf)	NGL (Million bbls)	BOE (Million bbls)
Maastrichtian				
Navarro-N*	1,140	8,970	30	2,665
Campanian				
Taylor-N*	760	930	20	935
Atoyac-S*	20	120	0	40
Total	780	1,050	20	975
Coniacian-Santonian				
Austin/Tokio/Eutaw-N*	995	3,420	165	1,730
San Felipe-S*	200	240	5	245
Total	1,195	3,660	170	1,975
Mid-Cenomanian-Turonian				
Woodbine/Tuscaloosa-N*	8,525	9,990	940	11,130
Undifferentiated (SE Mexico)				
Agua Nueva-San Felipe-S*	1,455	5,610	535	2,925
Unnamed-S*†	17,650	8,130	1,972	20,977
Total	19,105	13,750	2,507	23,902
Upper Cretaceous Total	30,745	37,410	3,667	40,647

*N = Northern (United States); S = Southern (Mexico).
†Includes some amounts within the Lower Paleocene.

natural gas, and 18.3% of the oil equivalent known recovery in the Gulf of Mexico basin.

Like Lower Cretaceous resources, Upper Cretaceous petroleum resources in southeastern Mexico cannot be differentiated by stage. These resources are however the most important concentration of Upper Cretaceous petroleum in the basin. In the northern and western parts of the basin, on the other hand, four distinct stratigraphic intervals have been recognized: mid-Cenomanian–Turonian, Coniacian-Santonian, Campanian, and Maastrichtian (Fig. 4). Of these four units, the most productive interval is the oldest, which contains the predominantly mid-Cenomanian Woodbine and Tuscaloosa sandstones. None of the other three intervals has more than 1.5% of the total known recovery of the Gulf of Mexico basin. Of these three, the Maastrichtian (Navarro) sandstones and limestones are the most important, followed closely behind by the Coniacian-Santonian Eutaw and Tokio sandstones and Austin and San Felipe carbonates. The Campanian sandstones are the least important productive stage within the Gulf of Mexico basin.

The majority of Upper Cretaceous known recovery is in the southern half of the Gulf of Mexico basin in southeastern Mexico. Fields in the southern half contained 59.5% of known recovery in the Upper Cretaceous, compared to 40.5% in the northern half. The predominant share of Upper Cretaceous crude oil (62.9%) and natural gas liquids (68.5%) is also in the southern half. Most (61.7%) of the known recovery of natural gas in the Upper Cretaceous, however, is in fields in the northern half of the basin.

Even more than the Upper Jurassic and the Lower Cretaceous, Upper Cretaceous oil and gas production in the Gulf of Mexico basin is dominated by a few giant fields. Twenty giant fields have 33,485 million BOE known recovery (or 82.4% of the total in the Upper Cretaceous). These giants include all the supergiant fields in the Gulf of Mexico Basin—*Abkatun, A. J. Bermudez, Cantarell,* and *Ku* in southeastern Mexico and *East Texas* in the United States. Another 46 large fields contribute 4,992 million BOE, an additional 12.3% of Upper Cretaceous known recovery.

Undifferentiated Upper Cretaceous (southeast Mexico). The most important Upper Cretaceous producing intervals in the Gulf of Mexico basin are also undifferentiated by individual stage. As of December 31, 1987, known recovery in the undifferentiated Upper Cretaceous of the Campeche area and in the Agua Nueva–San Felipe section of the Reforma area was approximately 19,105 million barrels of crude oil, 13,740 billion ft^3 of natural gas, and 2,507 million barrels of natural gas liquids, or 23,902 million BOE, nearly 60% of the Upper Cretaceous total in the Gulf of Mexico basin.

The two areas discussed here are undifferentiated for slightly different reasons. The main producing Upper Cretaceous reservoir of the Campeche area is a fractured dolomitized breccia that is in upward continuity with the lower Paleocene (Santiago and Baro, 1991). Because these breccias are continuously productive, are predominantly of Late Cretaceous age, and are derived from the detritus of Cretaceous carbonates, they are considered in this

chapter as solely Upper Cretaceous. In the Reforma area, the productive Agua Nueva and San Felipe limestones are also one continuous reservoir, connected by an extensive fracture system. Because they span the entire upper Cenomanian-Turonian-Coniacian-Santonian interval, they are considered in this chapter as an undifferentiated Upper Cretaceous reservoir, although they clearly belong to just the lower half of the Upper Cretaceous.

As indicated previously, undifferentiated Upper Cretaceous production is divided into two trends, both of major proportions (Fig. 7). The most important is the Upper Cretaceous covering much of the Bay of Campeche. This trend is essentially a heavy to medium gravity (20 to 36° API) oil trend characterized by low gas-oil ratios (150 to 1,000 ft^3 per barrel). On the other hand, the Agua Nueva–San Felipe trend in the southeast half of the Reforma area is a light (37 to 51° API) oil trend with very high gas-oil ratios (2,500 to 10,000 ft^3 per barrel).

Giant fields contain nearly all of the petroleum resources in these two areas. Fourteen giant fields—the *Abkatun, Cantarell,* and (possibly) *Ku* supergiants in Campeche; the *A. J. Bermudez* supergiant in Reforma; the *Batab, Caan, Chuc, Ek, Maloob,* and *Pol* giants in Campeche; and the *Agave, Cactus-Nispero, Muspac,* and *Sen* giants in Reforma—provide an estimated 23,450 million BOE (Plate 4). The remainder is in one large field, *Bacab* in Campeche, which may prove to be a giant when it is fully developed.

Discoveries in both areas are relatively recent. Upper Cretaceous oil in the Reforma area was first found in *Cactus-Nispero* in 1972 and followed by subsequent discoveries in *A. J. Bermudez* (1973), *Agave* (1976), *Muspac* (1982), and *Sen* (1984). In the Campeche area oil was first found in *Chac* (now part of *Cantarell*) in 1976 with subsequent discoveries in *Abkatun, Ku,* and *Maloob* in 1979, *Ek* and *Pol* in 1980, *Chuc* in 1982, and *Batab* and *Caan* in 1984 (Santiago Acevedo and Meija Dautt, 1980; Santiago and Baro, 1991).

The dolomitized Upper Cretaceous breccias in Campeche are derived from calcareous detritus eroded off the Yucatan Platform to the east and deposited in a marine slope and basin environment. These Upper Cretaceous accumulations are found at depths of 1,400 to 4,000 m. The areal extent of individual accumulations is relatively small for giant fields, ranging from 1,500 to 12,000 ha (3,750 to 30,000 acres). But like the Jurassic and Lower Cretaceous accumulations in Campeche, reservoir thicknesses are considerable (75 to 275 m). Matrix porosity is relatively good, ranging from 9 to 13% before being enhanced by both fracturing and some dissolution cavities. The intensive fracturing provides excellent permeability (Chavarria G., 1980; Santiago and Baro, 1991).

The San Felipe and Agua Nueva limestones in the Reforma area are deep-water deposits. They exhibit some dolomitization and in the *Sen* field, the Upper Cretaceous accumulation in Reforma closest to the Campeche fields, include breccias as well. Depths for these Reforma accumulations range from 2,600 to 4,600 m. The productive accumulations cover only 1,800 to 6,500 hectares (4,500 to 16,000 acres), but have substantial

470   *R. Nehring*

thicknesses (130 to 250 m). Matrix porosity is only 5 to 12%, but because of the intensive microfracturing, permeability is excellent (Viniegra-O., 1981; Santiago and Baro, 1991).

Trapping for both the Campeche and Reforma fields is structural, all fields being anticlines, typically with interior and/or bounding faults. Seals for the Campeche accumulations are provided by Paleocene shales. Seals for the Reforma accumulations are the Upper Cretaceous Mendez shales and Paleocene shales (Santiago Acevedo and Mejia Dautt, 1980; Viniegra-O., 1981; Santiago and Baro, 1991).

Because little exploratory drilling has occurred in the Campeche and Reforma areas since the early 1980s, both areas, particularly Campeche, hold substantial promise for future giant and large discoveries. The most promising area appears to be that between *Abkatun* on the northeast and *A. J. Bermudez* on the southwest (see Plate 4). Although the potential in the overall Mesozoic section is excellent here, the specific potential for the Upper Cretaceous is uncertain, primarily because of the possibility that the Upper Cretaceous may be thin or absent on many prospects.

Mid-Cenomanian - Turonian (Woodbine and Tuscaloosa Formations). The mid-Cenomanian - Turonian Woodbine and Tuscaloosa sandstones (with minor contributions from Eagle Ford sandstones) are the dominant petroleum producing interval into which known recovery can be assigned in the Upper Cretaceous sequence of the northern and western Gulf of Mexico basin. As of December 31, 1987, estimated known recovery in the Woodbine/Tuscaloosa was 8,525 million barrels of crude oil, 9,990 billion ft^3 of natural gas, and 940 million barrels of natural gas liquids, for a total of 11,130 million BOE, approximately

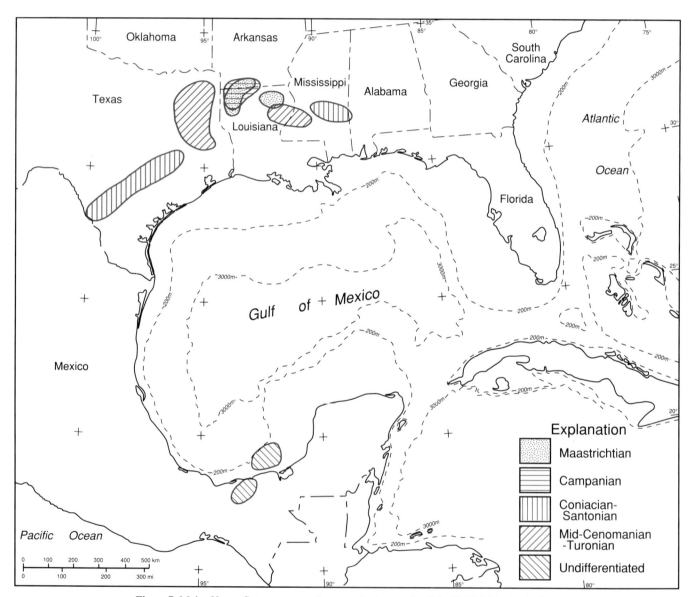

Figure 7. Major Upper Cretaceous petroleum-producing trends of the Gulf of Mexico basin.

two-thirds of the known recovery within the Upper Cretaceous sequence of the northern and western Gulf of Mexico basin and 27% of the known recovery within the entire Upper Cretaceous. Nearly all of these resources are in reservoirs of Cenomanian age, the Woodbine being entirely Cenomanian and the predominant producing section in the Tuscaloosa being the Cenomanian lower Tuscaloosa. The Woodbine is the reservoir of one of the most famous of Gulf of Mexico basin fields, the supergiant *East Texas* field.

Although Woodbine/Tuscaloosa accumulations are common across the northern margin of the Gulf of Mexico basin, they are heavily concentrated in the East Texas basin (Fig. 7). The largest concentration here, predominantly oil and natural gas liquids within the Woodbine, is located on the west flank of the Sabine uplift. The other major trend, oil and natural gas in the Woodbine, is found in the center of the East Texas basin. A third moderate-size Woodbine oil trend is found within the Mexia-Talco fault zone on the western and northern margins of that basin. A moderate-size Woodbine gas, oil, and natural gas liquids trend occurs south and downdip of the East Texas basin. The largest Tuscaloosa concentration, a moderate-size oil and gas trend, straddles the area between the north Louisiana and Mississippi salt basins. A second moderate-size concentration, the deep gas and natural gas liquids producing "deep Tuscaloosa trend," is immediately downdip of the Late Cretaceous shelf edge in east-central Louisiana. An oil and gas play of similar size occurs on the western half of the Wiggins uplift in southern Mississippi.

Petroleum resources are highly concentrated in a few large and giant fields of the 200 fields with significant Woodbine/Tuscaloosa production. Three Woodbine giants—the 53,500-ha (132,000-acre) supergiant *East Texas* and the giant *Hawkins* and *Van* fields contain 7,965 million BOE (71.6% of the total; Plate 4). Twenty-one other large fields (*Bryan, Cayuga, Grapeland, Long Lake, Mexia, Neches, Powell,* and *Quitman* in Texas; *Lake St. John, Moore-Sams, Morganza, Port Hudson,* and *Richland* in Louisiana; and *Baxterville, Brookhaven, Cranfield, Heidelberg, Hub, Little Creek, Maxie,* and *Pistol Ridge* in Mississippi) provide another 2,060 million BOE (18.5% of the total).

Significant discoveries in the Woodbine/Tuscaloosa span seven decades. The first major discoveries were made in the Mexia-Talco fault zone in the early 1920s. Exploration quickly moved eastward into the East Texas basin, with most of the Woodbine pools there being discovered in the 1930s, 1940s, and 1950s (including *Van* in 1929, *East Texas* in 1930, and *Hawkins* in 1940). Most of the Tuscaloosa discoveries along the Wiggins uplift and in the updip trend in northeastern Louisiana and southwest Mississippi were also made in the 1940s and 1950s. Discoveries in the downdip trend in Texas occurred primarily from the early 1970s to the early 1980s. Discoveries in the deep Tuscaloosa trend were concentrated in the late 1970s.

The sandstones of both the Woodbine and Tuscaloosa Formations were deposited in similar depositional environments (see Chapter 10, this volume). Updip in east Texas, northeast Louisiana, and southwest Mississippi, the reservoirs are predominantly

fluvial-deltaic sandstones. Downdip in southeast Texas and east central Louisiana, the reservoirs are shelf and slope sandstones, including submarine fans, offshore bars, and turbidites. Sediment sources of the Woodbine/Tuscaloosa are the Ouachita Mountains in Oklahoma and Arkansas and the Appalachians to the northeast (Karges, 1962; Oliver, 1971; Foss, 1979; Harrison, 1980; Galloway and others, 1983; Smith, 1985; Kosters and others, 1989).

Woodbine/Tuscaloosa reservoir quality is generally excellent. In the East Texas basin, Woodbine reservoirs are 850 to 1,800 m deep, 2 to 40 m thick, and have 20 to 30% porosity with 100 to 3,000 md permeability. In the updip area of Louisiana and Mississippi, depths to the Tuscaloosa reservoirs range from 600 to 3,600 m, thicknesses are 2 to 15 m, yet porosities still average 18 to 30% with permeabilities of 50 to 1,500 md. In the downdip trend of southeast Texas, Woodbine reservoirs are from 2,400 to 4,600 m deep, 2 to 15 m thick, with porosities between 12 to 20% and permeabilities of 5 to 200 md. In the deep (4,600 to 5,800 m) downdip Tuscaloosa of east-central Louisiana, reservoir thickness are typically 10 to 50 m, and porosities are 20 to 25% with permeabilities of 10 to 500 md. Porosity preservation in the deeper Tuscaloosa is attributed to chlorite coatings on the sand grains, which inhibited both compaction and cementation (Thomson, 1979; Smith, 1985).

Trapping in the Woodbine/Tuscaloosa exhibits nearly every type of trap possible for clastic reservoirs. The *East Texas* field is the classic erosional pinchout trap, with the seals provided by the overlying unconformable Austin Chalk and the underlying Washita limestones. Elsewhere in the East Texas basin, Woodbine traps are predominantly structural (salt-cored anticlines, anticlines, and faults) with some being modified by lithologic changes. Trapping in the downdip Woodbine trend in southeast Texas is primarily stratigraphic, the traps being formed by both lithologic and diagenetic changes. Trapping in the deep Tuscaloosas Trend is primarily structural, mostly in anticlines on the upthrown side of growth faults. Trapping in the Tuscaloosa fields of northeast Louisiana and southwest Mississippi is both structural and stratigraphic, including anticlines, lithologic changes, and unconformities. Other than the *East Texas* field and the Mexia-Talco fault zone fields, seals for Woodbine/Tuscaloosa accumulations are provided by Eagle Ford, Tuscaloosa, or Woodbine shales (Halbouty and Halbouty, 1982; Galloway and others, 1983; Smith, 1985; Kosters and others, 1989).

Because successful Woodbine/Tuscaloosa exploration has occurred for seven decades and has covered nearly all of the geologically favorable areas, major Woodbine/Tuscaloosa discoveries in the future are unlikely. The best possibilities for further discoveries are for small stratigraphic traps at intermediate depths (2,500 to 4,500 m) along the Louisiana/Mississippi border and in the southeast quarter of the East Texas basin.

Coniacian-Santonian (Austin, Eutaw, Tokio, and San Felipe Formations). Until the recent application of horizontal drilling within the Austin Chalk, the Coniacian-Santonian interval provided only relatively modest amounts of oil and gas to the

Gulf of Mexico basin total. Known recovery as of the end of 1987 was 1,195 million barrels of crude oil, 3,660 billion ft^3 of natural gas, and 170 million barrels of natural gas liquids, a total of 1,975 million BOE. As horizontal drilling continues to be successfully applied in the Austin Chalk, the Coniacian-Santonian could soon surpass the Maastrichtian in importance among the various intervals of the Upper Cretaceous.

Coniacian-Santonian oil and gas production occurs in several areas along the northern and western margins of the Gulf of Mexico basin (Fig. 5). Until the recent growth in Austin Chalk reserves, the most important area was a moderate-size oil and gas trend in Eutaw sandstones across the northern half of the Mississippi salt basin. The moderate-size oil, natural gas, and natural gas liquids Austin Chalk play on the northwest margin of the basin in south and southeast Texas was the only other important concentration of petroleum. Four other Coniacian-Santonian trends deserve mention: a moderate-size San Felipe limestone oil trend in the northern Tampico-Misantla basin in east-central Mexico, a moderate-size Tokio sandstone oil and gas play in southwest Arkansas and northwest Louisiana, a minor Eutaw gas play on the Wiggins uplift, and a minor San Felipe gas play in the Veracruz basin.

There are no giant fields as yet among the 70 significant producing Coniacian-Santonian fields. As horizontal drilling activity proceeds, the areally immense (263,000 ha or 650,000 acres) *Giddings* field in the Austin Chalk trend will soon become a giant. The San Felipe provides a large reservoir within the giant *Ebano-Panuco* group of fields in east central Mexico. These two fields, together with fourteen other large fields—*Baxterville, Eucutta, Gwinville, Heidelberg, Maxie, Soso,* and *Tinsley* in Mississippi; *Cotton Valley, Greenwood-Waskom, Shongaloo, North-Red Rock* and *Sligo* in Louisiana; and *Luling-Branyon, Pearsall,* and *Salt Flat-Tenney Creek* in Texas—contained 1,637 million BOE, 82.9% of Coniacian-Santonian known recovery.

Exploration and development in the Coniacian-Santonian spans almost the entire period of petroleum exploration in the Gulf of Mexico basin. The first discoveries were made shortly after 1900 in both the *Ebano-Panuco* area and in some of the Tokio sandstone fields. Tokio discoveries continued on a sporadic basis into the early 1960s. Updip Austin Chalk discoveries were concentrated in the 1920s and 1930s. Eutaw discoveries occurred primarily in the 1940s and 1950s. Extensive downdip Austin Chalk discoveries and development only began in 1975 after the increase in oil prices. Although few Austin discoveries have occurred since the early 1980s, development drilling has intensified since the late 1980s with the widespread application of horizontal drilling in the larger Austin Chalk fields, particularly in the *Pearsall* field in south Texas.

Coniacian-Santonian reservoirs vary widely in quality, differing primarily by lithology. The sandstone reservoirs are of very good to excellent quality while the carbonate reservoirs generally require natural fracturing to be commercial. The transgressive marine sandstones of the Eutaw Formation in the Mississippi salt basin are found in reservoirs 800 to 2,200 m deep and 3 to 20 m

thick, with porosities of 25 to 30% and permeabilities of 200 to 1,500 md. The transgressive, reworked, marine Tokio sandstones are in reservoirs 500 to 1,000 m deep and only 1 to 10 m thick, with even better porosities (27 to 35%) and permeabilities (100 to 2,500 md). Austin Chalk reservoirs range from 500 to 3,500 m deep and have net thicknesses of 5 to 70 m. Their porosities range from 5 to 25%, the best porosities being found in the updip reservoirs, while matrix permeabilities are less than 1 md (Stapp, 1977; Galloway and others, 1983). San Felipe limestone reservoirs in the *Ebano-Panuco* fields are 300 to 600 m deep and are very tight, being productive only because of extensive natural fracturing (which often continues into the underlying Turonian Agua Nueva Formation) (Guzman and others, 1955).

Trapping in the Coniacian-Santonian also varies by lithology. Both Eutaw and Tokio accumulations are predominantly structurally trapped. Eutaw traps are predominantly faults and faulted anticlines related to underlying salt movement. These traps are sealed by marly chalk in the upper Eutaw. Tokio traps are anticlines, with accumulations often modified by lithologic changes. The Tokio accumulations are sealed by the overlying Brownstown Marl. Trapping in both the Austin and the San Felipe depends heavily on the presence of natural fractures. The *Ebano-Panuco* fields lie above a basement arch with the fractured San Felipe limestone being sealed by overlying Mendez marls (Guzman and others, 1955). The downdip Austin Chalk reservoirs lie along a hingeline that has created sufficient stress to form the micro-fractures which make them productive. Updip the reservoirs are predominantly fault traps in the Luling fault zone (Galloway and others, 1983).

The Coniacian-Santonian interval still has substantial potential for future oil and gas reserve additions, primarily from horizontal development drilling within the Austin Chalk. (Horizontal drilling may also have potential within the San Felipe in the *Ebano-Panuco* fields.) Its exploration potential is, however, modest, being limited to a few remaining opportunities within the Austin Chalk and the Eutaw Formation.

Campanian (Taylor Group and Atoyac Formation). Rocks of Campanian age provide the least important petroleum-producing unit in the Gulf of Mexico basin. As of the end of 1987, estimated known recovery in the Campanian Taylor Group and Atoyac Formation was only 780 million barrels of crude oil, 1,050 billion ft^3 of natural gas, and 20 million barrels of natural gas liquids, or 975 million BOE.

Campanian production is concentrated in three areas (Fig. 7): a moderate-size Ozan sandstones and Annona Chalk (Taylor Group) oil and gas trend in southwest Arkansas and northwest Louisiana, a moderate size group of San Miguel sandstones and Anacacho Limestone (Taylor Group) oil and gas plays in south Texas, and a minor oil and gas play in the Atoyac Limestone in the Veracruz basin in east-central Mexico. Within these three areas are only 60 fields with significant Campanian production. Campanian known recovery is nearly too small to even provide an opportunity for a giant field. But production within it is still concentrated in a handful of fields. Six large fields—*Caddo-Pine*

Island, Elm Grove, Haynesville, and *Homer* in Louisiana; *Stephens-Wesson* in Arkansas; and *Big Wells* in south Texas—contained 648 million BOE known recovery, 69.7% of the Campanian total.

Because Campanian production is concentrated on shallow structures, the first Campanian discovery was made early in the exploration history of the Gulf of Mexico basin (*Caddo-Pine Island* in 1905). Nearly all of the Ozan and Annona significant discoveries were made between 1915 and 1960. Most Anacacho discoveries were made from 1915 to 1940. The deeper downdip San Miguel fields were discovered primarily from 1955 to 1980.

Campanian reservoirs vary widely in quality. The shallow (600 to 850 m deep) Ozan sandstones (locally known as the Baker, Buckrange, Graves, and Meakin sands) are thin (1 to 12 m thick), but have excellent porosity (25 to 35%) and permeability (100 to 5,000 md). The shallow (400 to 1,000 m) Annona Chalk reservoirs are thicker (10 to 40 m) with very good porosity (25%) and poor permeability (less than 5 md). The Anacacho Limestone reservoirs are also shallow (200 to 850 m), of modest thickness (5 to 45 m) and possess adequate porosity (12 to 20%) and permeability (1 to 100 md). The deeper (350 to 1,750 m) deltaic San Miguel sandstones vary from 1 to 50 m in thickness, have good porosity (18 to 26%), but only adequate permeability (1 to 100 md).

Campanian traps are predominantly combination structural and stratigraphic traps. Typically Ozan fields are anticlines or faulted anticlines modified by lithologic changes and sealed by shales within the Ozan. Annona traps combine structure with fracturing covered by an overlying shale-marl seal. Anacacho and San Miguel traps include faults, lithologic changes, and drapes over volcanic plugs. Seals are provided primarily by San Miguel shales (Simmons, 1967; Lewis, 1977; Weise, 1980; Galloway and others, 1983; Kosters and others, 1989). Traps in the overthrust Atoyac Limestone are fault traps sealed by overlying Eocene shales (Mossman and Viniegra, 1976).

As most Campanian prospects in the Gulf of Mexico basin are shallow, they have been subjected to intensive exploration for more than 75 years. Small traps in San Miguel sandstones provide the only remaining significant Campanian exploration objectives.

Maastrichtian (Navarro Group: Nacatoch, Olmos, and Escondido sandstones; Monroe Gas Rock). Because it contains two giant fields, the Maastrichtian constitutes the second most important producing interval to which petroleum resources can be assigned within the Upper Cretaceous sequence in the northern and western Gulf of Mexico basin. As of the end of 1987, estimated known recovery from the Maastrichtian was 8,970 billion ft^3 of natural gas, 1,140 million barrels of crude oil, and 30 million barrels of natural gas liquids, or 2,665 million barrels BOE.

Maastrichtian production is concentrated in a few areas on the north and northwest margins of the Gulf of Mexico basin (Fig. 7). Nearly half of the known recovery is found in a major Monroe Gas Rock natural gas trend on the Monroe uplift in northeastern Louisiana. Most of the oil has been found in a moderate-size Nacatoch Sandstone (Navarro Group) trend in south Arkansas and on the east flank of the Sabine uplift in northwestern Louisiana. The only other moderate-size concentration has been found in a number of Olmos and Escondido oil and gas plays in south Texas. There is also a minor oil and gas play in sandstones of the Navarro Group along the Mexia fault zone.

Like the other Upper Cretaceous intervals, Maastrichtian production is highly concentrated in a few giant and large fields. Of the 80 significant producing Maastrichtian fields, two giants—the 100,000-ha (250,000-acre) *Monroe* field in northeast Louisiana and the 17,800-ha (44,000-acre) *Smackover* field in Arkansas—dominate, with 1,825 million BOE known recovery (68.5% of the Maastrichtian total; Plate 4). Another nine large fields—*El Dorado, South* in Arkansas; *Homer* and *Red River-Bull Bayou* in northwest Louisiana; *Bethany, Corsicana,* and *Waskom* in east Texas; and *A.W.P., Big Foot,* and *Charlotte* in south Texas—contained another 440 million BOE (16.5% of the total).

Because of the shallow depths of many of the Maastrichtian reservoirs, many were discovered early in the exploration history of the Gulf of Mexico basin. The first commercially significant discovery in the northern Gulf of Mexico basin, *Corsicana* in 1895, was a Nacatoch Sandstone field in the Mexia fault zone. (The first discovery, of *Nacogdoches* in 1867, was in a small, low-productivity Mount Selma [Eocene Claiborne] reservoir.) Most of the other Nacatoch and Monroe Gas Rock reservoirs were discovered between 1900 and 1940, including *Monroe* in 1916 and *Smackover* in 1922. The first Olmos and Escondido discoveries in south Texas occurred in the 1920s, but most of the significant discoveries in these trends occurred in the 1950s, 1960s, and 1970s, with a few continuing into the 1980s.

Reservoir conditions vary considerably across the various Maastrichtian trends. The reservoir in the *Monroe* field is a sandy limestone around 650 m deep, and 10 to 15 m thick, with an average porosity of 25% and widely variable permeability. The Nacatoch reservoirs are shelf sands ranging from 100 to 1,100 m in depth and 2 to 30 m in thickness, with porosities of 20 to 35% and permeabilities of 20 to 2,000 md (McGowen and Lopez, 1983). The Olmos and Escondido sandstones were deposited over a variety of depositional environments (coastal-plain, deltaic, barrier-strandplain, and marine shelf). Because of variations in depositional conditions and in depths (100 to 3,200 m), they vary widely in reservoir quality, being 1 to 20 m thick and having porosities of 15 to 30% and permeabilities of less than one to more than 100 md (Tyler and Ambrose, 1986; Dennis, 1987).

Trapping mechanisms also vary among the various Maastrichtian fields. Gas in *Monroe* field is trapped over a large basement uplift with the irregular downdip limits of the accumulation being determined by lithologic and diagenetic changes. Nacatoch traps are predominantly combination traps: anticlinal and fault-related accumulations modified by lithologic variations. Olmos and Escondido traps range from the purely structural (anticlines, faults, and sands draped over volcanic mounds) to the purely stratigraphic (lithologic changes and unconformities) (Tyler and

Ambrose, 1986). Seals for the Maastrichtian fields are provided by Navarro shales or, in the case of the *Monroe* field, by overlying Paleocene shales.

Future exploration opportunities within the Maastrichtian section appear to be limited to Olmos and Escondido sandstones in south Texas. Various opportunities exist both in the updip and downdip trends (Tyler and Ambrose, 1986). Because Nacatoch and Monroe Gas Rock reservoirs are so shallow and because the favorable areas have been subject to extensive drilling, further significant discoveries in these formations are highly unlikely.

Upper Cretaceous source rocks. Because the undifferentiated Upper Cretaceous carbonates of the Campeche and Reforma areas of southeast Mexico are the most important concentration of Upper Cretaceous petroleum in the Gulf of Mexico basin, their probable source rocks—the Upper Jurassic Tithonian black calcareous shales—are the most important source for Upper Cretaceous petroleum in the basin. Migration from these sources into the Upper Cretaceous reservoirs was primarily short-distance vertical migration.

Surprisingly little has been published on possible source rocks for the Upper Cretaceous oil and gas accumulations in the United States part of the Gulf of Mexico basin, even though reservoirs of this age contain 11.4 billion barrels of oil, 1.15 billion barrels of natural gas liquids, and 23.3 Tcf of gas, and include the *East Texas* field, the largest oil field in the contiguous 48 states.

The Eagle Ford dark shales have often been mentioned as the source for the oil and gas in the Upper Cretaceous accumulations of the East Texas basin, the most prolific area of petroleum production from reservoirs of this age in the northern Gulf of Mexico basin. No substantiating geochemical evidence has been published, however, to support this view. Surles (1985, p. 309) states, for instance, that "geochemical analysis of the Eagle Ford Group throughout the [East Texas] basin suggests that the Eagle Ford shales may be a major source-rock for petroleum in Austin, Eagle Ford, Woodbine and possible Buda Formations," but the geochemical data has not yet been made available. The Upper Jurassic Smackover has also been proposed as the source of the oil and gas in some of the Upper Cretaceous fields of the East Texas basin (Dutton and others, 1987; Burgess, 1990).

In Louisiana, Mississippi, and Alabama, both the shales within the Tuscaloosa Formation (Mancini and Payton, 1981; Sassen, 1990b), and the Upper Jurassic lower Smackover calcareous mudstones (Sassen, 1990a), have been proposed as likely petroleum source rocks for the petroleum accumulations in Tuscaloosa reservoirs, the likelihood of each source varying by location of the reservoirs.

Koons and others (1974) concluded that the oils in lower Tuscaloosa reservoirs in southern Mississippi and southwestern Alabama belong to two different families on the basis of their chemical composition. One of them appeared to have been derived from a deep source, which they did not identify, but the other was believed to be indigenous to the Tuscaloosa.

The Eagle Ford and Tuscaloosa organic-rich shales may account for more than ten billion BOE known recovery. They contain kerogen of marine origin and may have reached sufficient maturity to start generating oil in the late Late Cretaceous.

The gas content of Woodbine/Tuscaloosa reservoirs increases with depth. As the Woodbine and Tuscaloosa sandstones can be excellent conduits, substantial lateral migration is believed to have occurred from downdip source to the updip traps (Waples, 1985; Evans, 1987; Sassen and others, 1987b; Walters and Dusang, 1988; Claypool and Mancini, 1989; Sassen, 1990b).

Source rocks for the Coniacian-Santonian reservoirs are reasonably well established. The Eutaw in the northern Mississippi salt basin and the Tokio in Arkansas and Louisiana were sourced by the Upper Jurassic lower Smackover, the oil migrating upward through an extensive fault network (Sassen, 1990a). The chalks of the Austin Formation, which are organic-rich and oil-prone, were the most likely source rock for Austin accumulations, with most of the migration occurring through the extensive system of micro-fractures within the Austin Chalk (Grabowski, 1984; Hunt and McNichol, 1984). The source rocks for the San Felipe oil in the *Ebano-Panuco* fields are postulated to be either the immediately underlying Turonian Agua Nueva calcareous shales or the Upper Jurassic Pimienta shales (Guzman and others, 1955).

Because of the relatively small amounts of petroleum discovered in the Campanian, little research has been conducted in possible source rocks for it. The possibilities vary widely by formation and area. The lower Smackover is the most likely source for most of the Ozan and possibly Annona oil near the northern basin margin where faults provide conduits for vertical migration (Sassen, 1990a). Anacacho and San Miguel oil was most likely generated in the Upper Cretaceous Austin Chalk that migrated upwards through faults (Grabowski, 1984). San Miguel gas could have originated within San Miguel shales. Atoyac oil is considered to have an Upper Jurassic source (Mossman and Viniegra, 1976).

Knowledge of Maastrichtian source rocks is slight, partly because the stage is gas-prone. Crude-oil source rock correlations and the location of most of the Nacatoch Sandstone oil fields near basin-margin fault systems, strongly support the Jurassic lower Smackover as the source for Nacatoch oil (Sassen, 1990a). The source for the dry *Monroe* gas is not clear. The Austin Chalk is the most likely source for Olmos and Escondido oil (Grabowski, 1984). The source of Olmos and Escondido gas is not clear.

Paleogene petroleum resources

Paleogene reservoirs are by a narrow margin the most important petroleum-producing interval in the Gulf of Mexico basin. As of the end of 1987, Paleogene known recovery in the basin was 25.6 billion barrels of crude oil, 192.8 trillion ft^3 of natural gas, and 6.1 billion barrels of natural gas liquids, for a total of 63.8 billion barrels of oil-equivalent resources (Table 13).

This total probably understates the actual amount in the Paleogene, as some of the petroleum resources in the Campeche area assigned in this discussion to the undifferentiated Upper Cretaceous are actually in reservoirs of lower Paleocene age.

The Paleogene of the Gulf of Mexico basin is thus by itself one of the most important concentrations of petroleum in the world. No more than six other provinces worldwide exceed it in size. Within the Gulf of Mexico basin, it provides 22.7% of the total crude oil, 36.8% of the natural gas, 27.0% of the natural gas liquids, and 28.7% of the oil equivalent resources. Unlike the other four major producing intervals in the basin in which either petroleum liquids or natural gas is clearly the dominant hydrocarbon, the petroleum resources of the Paleogene are almost equally divided between liquids and gases.

Each of the Paleogene stages within the Gulf of Mexico basin contains major amounts of petroleum. The gas-prone Oligocene predominates, with slightly more than half of the total resources. The oil-prone Paleocene with slightly more than a quarter of the total, is next in importance. The Eocene, while the least important of the three series, is still a major contributor, as measured on a world scale, to the overall total.

In contrast to the underlying Cretaceous petroleum resources, which are heavily concentrated in the southern half of

TABLE 13. KNOWN RECOVERY OF CRUDE OIL, NATURAL GAS, AND NATURAL GAS LIQUIDS AS OF DECEMBER 31, 1987, IN THE GULF OF MEXICO BASIN BY STRATIGRAPHIC INTERVAL WITHIN THE PALEOGENE

Unit	Crude Oil (Million bbls)	Natural Gas (Bcf)	NGL (Million bbls)	BOE (Million bbls)
Oligocene				
Catahoula (Frio/ Anahuac)-N and S*	9,950	107,510	3,030	30,898
Vicksburg-N and S*	325	9,900	325	2,300
La Laja-S*	0	700	2	119
Total	10,275	118,110	3,357	33,317
Eocene				
Jackson-N and S*	880	1,800	30	1,210
Claiborne-N and S*	1,966	20,460	715	6,091
Wilcox-N and S*	1,550	25,710	655	6,490
Total	4,396	47,970	1,400	13,791
Paleocene				
Chicontepec-S*	10,930	26,760	1,325	16,715
Midway-N*	15	0	0	15
Total	10,945	26,760	1,325	16,730
Paleogene				
Total	25,616	192,840	6,082	63,838

*N = Northern (United States); S = Southern (Mexico).

the Gulf of Mexico basin, the petroleum resources of the Paleogene are highly concentrated in the northern half of the basin (69.8% of the total). This concentration is particularly marked in natural gas, the United States having 79.4% of the Paleogene total. It also contains 57.1% of the Paleogene crude oil and 72.4% of the natural gas liquids. This is due primarily to the fact that the Paleogene section in east-central and southern Mexico is composed predominantly of marine shales and is therefore devoid of good reservoir rocks (see Plate 5).

Also, unlike the Mesozoic intervals, Paleogene known recovery is not concentrated in giant fields. The 12 Paleogene giant fields contain 12.20 billion BOE known recovery, 25.9% of the Paleogene total (excluding the Chicontepec Formation). Paleogene petroleum is mostly concentrated in 165 large fields, with 17.94 billion BOE, 38.1% of the Paleogene total (excluding Chicontepec).

Paleocene. Paleocene sandstones are an important producing interval within the Paleogene sequence, particularly for crude oil. As of the end of 1987, Paleocene known recovery was 10,945 million barrels of crude oil, 26,760 billion ft^3 of natural gas, and 1,325 million barrels of natural gas liquids, for a total of 16,730 million barrels oil equivalent resources.

The relative importance of the Paleocene in the Paleogene sequence and in the Gulf of Mexico basin is more uncertain than these estimates indicate. On the one hand, the Paleocene contribution could be substantially larger than stated above. As discussed earlier, the producing zones in the Upper Cretaceous in the Bay of Campeche extend into the lower Paleocene. Possibly 5 to 20% of the undifferentiated Upper Cretaceous (or 1 to 4 billion BOE) should properly be assigned to the Paleocene instead. The Wilcox Group (assigned in this discussion entirely to the Eocene) extends into the upper Paleocene. Thus some unknown proportion of its petroleum resources properly belongs to the Paleocene.

On the other hand, the Paleocene could be substantially less important than indicated here. Practically all of the petroleum assigned to the Paleocene is from the Chicontepec Formation. Yet the Chicontepec Formation extends upward into the lower Eocene, suggesting that some unknown amount of its petroleum should be assigned to the Eocene instead. More importantly, nearly all of the known recovery from the Chicontepec Formation is at best classified as proved undeveloped reserves. If most of these reserves are never developed, if recovery rates are not as good as anticipated, or if oil-in-place is less than originally calculated, the estimates for Chicontepec known recovery could be sharply reduced. Paleocene known recovery could thus be substantially more or substantially less than indicated above.

Assigned Paleocene petroleum resources are almost entirely in the Chicontepec Formation. The Chicontepec Formation consists of turbidites derived from the Sierra Madre Oriental to the west and the Tuxpan platform to the east. It was deposited in a deep erosional channel west and southwest of the Tuxpan platform and the Poza Rica trend in the Tampico-Misantla basin of east-central Mexico (Fig. 8). It covers an area of 3,300 km^2.

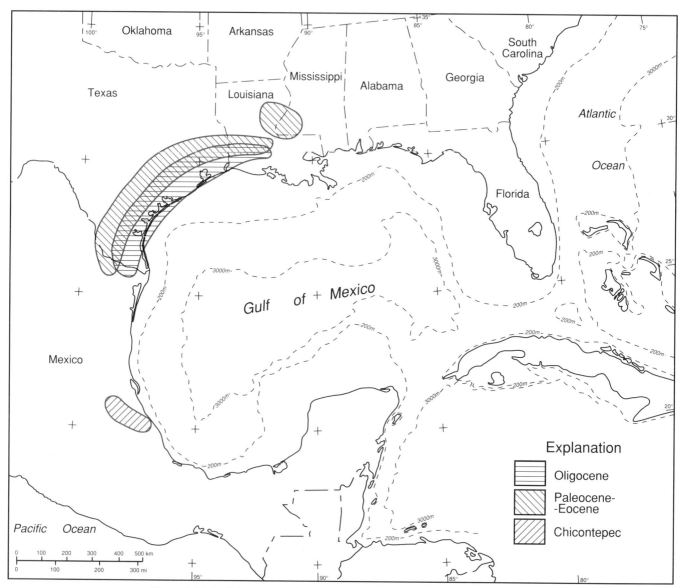

Figure 8. Major Paleogene petroleum-producing trends of the Gulf of Mexico basin.

Similar channels exist to the southeast (Nautla-Ayotoxco) and north (Bejuco–La Laja) of the Tuxpan platform, but their petroleum resources, if any, have yet to be evaluated. The only other assigned Paleocene petroleum resources are the very minor oil resources of the Poth (Midway) sandstone in south Texas (Hopf, 1967).

Because of the unusual characteristics of the Chicontepec Formation and its general lack of development, no field size distribution within it is possible. Given the asserted continuity of possible production, it could be considered as one giant or supergiant field. Given the nature of deposition, trapping however is essentially stratigraphic. The Chicontepec Formation was first penetrated in the 1940s but was not recognized as having major productive potential until the 1970s.

Chicontepec reservoirs are generally of poor quality. Reservoir sediments consist of fine grained sandstones that grade into siltstones. Average porosity is 7%; average permeability is less than one millidarcy; and average water saturation is 44%. Moreover, these averages mask a high degree of reservoir heterogeneity. Depths to the top of the Chicontepec Formation are 1,750 to 2,800 m. Net sand thicknesses within the Chicontepec area vary from 0 to 80 m. (The available literature does not indicate the proportion of this net sand that is oil-saturated.) Because of the low permeability and high clay content of the formation, Chicontepec wells must be fractured, using crude oil as the fracturing medium, to be economic completions (Renteria Curiel, 1977; Busch and Govela S., 1978; Metheny, 1979).

The Chicontepec Formation contains a substantial proportion of the known oil-in-place in the Gulf of Mexico basin. However, because of its poor producing characteristics, the petroleum resources it contains are essentially petroleum resources for the twenty-first century when oil prices are more likely to be sufficiently high to justify a major development effort.

Eocene (Wilcox, Claiborne, and Jackson Groups). Although they are the least important interval in the Paleogene sequence, Eocene sandstones in the Gulf of Mexico basin are an important producing unit in world terms. As of the end of 1987, Eocene known recovery (counting the entire Wilcox Group as Eocene), was 4,396 million barrels of crude oil, 47,970 billion ft^3 of natural gas, and 1,400 million barrels of natural gas liquids for a total of 13,791 million barrels oil equivalent. Of the three Eocene main units, the gas-prone Wilcox is the most important with 6,490 million BOE, 47.1% of the Eocene total. The overlying liquids-prone Claiborne Group (producing from the Reklaw, Queen City, Sparta, Cook Mountain, and Yegua Formations) is next with 6,091 million BOE, 44.2% of the total. The predominantly oil-prone Jackson Group is the smallest unit with only 1,210 million BOE, 8.8% of the total.

Eocene production is found in several major and moderate-size trends in the onshore northwestern Gulf of Mexico basin (Fig. 8). The location of each of these trends corresponds approximately to the location of the major sandstone depocenters for each stratigraphic unit. The proportion of hydrocarbons by type within each trend is determined primarily by the kerogen type of the predominant source rock for each trend. There are three major Wilcox trends. The largest, a highly gas-prone trend, is found in south Texas and the Burgos basin of Mexico. The second largest, a gas and oil trend, straddles the Houston embayment from the San Marcos arch to across the Louisiana-Texas border. The third, an oil trend, straddles the Mississippi River in east-central Louisiana and southwestern Mississippi. A moderate-size Wilcox oil and gas trend is found downdip of this latter trend in south central Louisiana.

The formations of the Claiborne Group are productive in several major and moderate-size trends. The only Queen City Formation trend, a moderate-size gas trend, is found in south Texas and the adjoining Burgos basin of Mexico. The Sparta Formation produces in a moderate-size oil and gas trend in east central Louisiana. A major gas and oil Yegua trend extends from the San Marcos arch to the Mississippi River. There is also a moderate-size oil and gas Yegua trend in south Texas and the Burgos basin. The Carrizo, Reklaw, Cane River, and Cook Mountain Formations are only minor producing units. The only major trend in the Jackson Group is an oil-producing trend in south Texas and the Burgos basin. A moderate-sized Jackson oil trend also occurs in the Houston embayment.

In contrast to most of the Cenozoic and Mesozoic producing intervals, Eocene production is not highly concentrated in giant and large fields. There are only two giant fields that produce from the Eocene, *Conroe* and *Katy* in the Houston embayment, which together contained 2,540 million BOE, 18.4% of the Eocene total (Plate 4). Of the more than 1,000 significant fields producing from the Eocene, 48 are large fields, providing 4,386 million BOE, 31.8% of the Eocene total. The large Eocene fields produc-

ing from the Wilcox include *Lake St. John, Nebo-Hemphill, Olla,* and *Tullos-Urania* in east-central Louisiana; *Reddell* and *Ville Platte* in south central Louisiana; *Columbus, Helen Gohlke, Provident City,* and *Sheridan* in the Houston embayment; and *Burnell, Clayton Northeast, Fandango, Gato Creek–Lundell, Hagist Ranch, J. C. Martin, Karon, Laredo, Seven Sisters East, Thompsonsville Northeast,* and *Tulsita* in south Texas. Large Claiborne fields include *Bammell, Fairbanks, Houston North, Hull-Merchant, Liberty-Dayton, Moores Dome, Raccoon Bend, Silsbee, Tomball,* and *Village Mills East* in the Houston embayment and *Krotz Springs* in south-central Louisiana. Large Jackson fields include *Conoco-Driscoll, Government Wells, Hoffman, Loma Novia,* and *Seven Sisters* in south Texas and *Humble* in the Houston embayment.

The distribution of petroleum resources by field size categories varies substantially by grouping within the Eocene of the Gulf of Mexico basin. Because the two Eocene giant fields produce primarily from the Claiborne, 63% of Claiborne known recovery is in giant and large fields. By comparison, only 40% of Wilcox known recovery and 38% of Jackson known recovery has come from giant and large fields.

Eocene discoveries in the Gulf of Mexico basin span the past 75 years. The first discoveries occurred in the shallow Jackson in 1914. Most Jackson discoveries were made from 1920 to 1960, although a few have continued into the 1960s, 1970s, and 1980s. Discoveries in the Yegua were concentrated between 1930 and 1970, including *Conroe* in 1931 and *Katy* in 1935. However, Yegua discoveries have continued into the 1970s and 1980s, particularly for downdip objectives. The Queen City has been explored more recently, discoveries within it being concentrated from the 1950s into the 1980s. Sparta discoveries occurred sporadically from 1930 to 1980. The two largest Wilcox trends have yielded substantial numbers of significant discoveries every decade from the 1940s to the 1980s. Discoveries in the Wilcox oil trend straddling the Mississippi River were concentrated between 1950 and 1970.

Wilcox reservoirs in south Texas and the Burgos basin are predominantly deltaic and delta-front sandstones. Because Wilcox reservoirs occur across a relatively broad band in this area, reservoir depths vary enormously, from as shallow as 700 m updip to as deep as 5,100 m downdip. Thickness of the productive sands vary substantially as well, from as little as 2 m to as much as 200 m combined net thicknesses in a single field. Wilcox porosity is generally good, even at depth, ranging from 16 to 27%, while permeabilities vary from poor (less than 1 md) to good (more than 500 md). Trapping is predominantly structural, being determined primarily by the pervasive growth faulting within the trend. Lithologic variations and erosional unconformities also affect many traps within the trend. Seals are provided by Wilcox shales (Fisher and McGowen, 1967; Bebout and others, 1982; Galloway and others, 1983; Ewing, 1986; Long, 1986; Loucks and others, 1986; Galloway, 1989; Kosters and others, 1989; Chapter 11 of this volume).

The Wilcox within the Houston embayment is a fluvial-deltaic sequence. Like the south Texas Wilcox, reservoir depths within this trend vary enormously, from less than 500 m to as much as 4,650 m. Thicknesses vary by two orders of magnitude (1 to 100 m). Because of differences in reservoir depth, grain composition, and diagenesis, reservoir quality is highly variable, porosities ranging from 13 to 38% and permeabilities varying from 0.1 to 2,500 md. As the productive trend corresponds to the Wilcox growth fault system, trapping is predominantly structural, with fault traps, faulted rollover anticlines, and occasional salt structures providing the trapping mechanisms. Many of these traps are, however, modified by lithologic variations. Seals are provided by Wilcox shales (Fisher and McGowen, 1967; Bebout and others, 1982; Galloway and others, 1983; Ewing, 1986; Loucks and others, 1986; Galloway, 1989; Kosters and others, 1989).

Wilcox production in east-central Louisiana and southwest Mississippi occurs in a fluvial-deltaic sequence at a termination of the ancestral Mississippi River drainage system. Compared to Wilcox production to the southwest, this trend is relatively shallow, with most reservoirs being between 1,000 and 2,000 m deep. Reservoirs within this trend are typically thin (1 to 15 m) though a few are as much as 50 m thick. Because the reservoirs are shallow and have not undergone substantial compaction or diagenesis, both porosities (averaging 25 to 35%) and permeabilities (150 to 1,500 md) are very good. Combination traps predominate in this trend, the typical trap being formed by lithologic variations across a structural nose. Shales within the Wilcox provide the seals.

Queen City (Claiborne Group) production in south Texas and the Burgos basin is concentrated along the downdip margin of a highly destructive, wave-dominated delta system. Depths of Queen City reservoirs range from 600 to 2,900 m, while thicknesses range from 1 to 50 m. Reservoir quality is generally fair to good, with porosities from 16 to 33% and permeabilities from 40 to 60 md. Traps are predominantly structural, being fault traps or rollover anticlines along normal faults. In many cases, the extent of the trap is shaped by lithologic variations. Seals are provided by shales within the Queen City (Guevara and Garcia, 1972; Galloway, 1989; Kosters and others, 1989).

The few scattered reservoirs in the deltaic Sparta Formation (Claiborne Group) in east-central Louisiana range in depth from 575 to 3,500 m. The sediment source for Sparta sandstones was the ancestral Mississippi River drainage system to the north. Thicknesses of Sparta reservoirs vary from 1 to 70 m. Reservoir quality is generally good, as porosities range from 21 to 34% and permeabilities vary from 60 to more than 1,000 md. Trapping is essentially structural, the typical trap being a faulted anticline associated with the commonplace Gulf Coast Tertiary growth faults. Seals are provided by shales within the Sparta itself.

Because of the breadth of the major Yegua (Claiborne Group) trend in southeast Texas and south Louisiana, it has a tremendous variation in reservoir depths, ranging from as shallow as 350 m updip to as deep as 4,350 m downdip. Depositional environments vary substantially as well, encompassing fluvial,

deltaic, shelf, and slope (turbidite) systems. Yegua sediments in this trend originated both in the ancestral Mississippi River drainage and in the southern Rocky Mountains. Reservoir thicknesses range from 1 to 100 m. The prolific production of this trend results from good to excellent reservoir quality, as evidenced by porosities of 10 to 38% and permeabilities of 20 to 3,000 md. Yegua trapping is predominantly structural, being created by both normal faults and salt structures. Many traps are also modified by lithologic variations. Seals are provided by interbedded Yegua shales and (occasionally) overlying Jackson shales (Galloway and others, 1983; Ewing and Fergeson, 1989; Galloway, 1989; Kosters and others, 1989).

The south Texas–Burgos basin Yegua trend is found in a strandplain/barrier bar depositional system. As a result, traps within it are both stratigraphic and structural, ranging from sand pinchouts to the usual Gulf Coast Tertiary growth fault traps. Reservoirs within this trend are from 325 to 2,575 m deep and 1 to 120 m thick. Reservoir quality is very good, with porosities of 23 to 33% and permeabilities of 40 to 800 md (Galloway and others, 1983; Kosters and others, 1989).

The Jackson trend of south Texas and the Burgos basin is also a strandplain/barrier bar play. The play is characterized by many shallow reservoirs, the range of reservoir depths being 115 to 1,750 m. Reservoir thicknesses vary from 1 to 100 m, with many exceeding 30 m. Because the reservoirs are shallow, reservoir quality is very good to excellent. Typical porosities are 25 to 40% while typical permeabilities range from 200 to 2,000 md. Trapping ranges from purely stratigraphic pinchouts and other lithologic variations to purely structural (faults and rollover anticlines), with no single trap type predominating. Seals are provided by Jackson shales (West, 1963; Fisher and others, 1970; Galloway and others, 1983; Kosters and others, 1989).

The Gulf Coast Eocene is in general a mature exploration trend. Some good exploration potential remains, particularly in downdip sections of the Wilcox and Yegua in both south Texas and the Houston embayment. Numerous small discoveries are still possible in the updip Wilcox, Yegua, Queen City, and Cook Mountain.

Oligocene (La Laja, Vicksburg, Frio, and Anahuac Formations). Oligocene sandstones are not only the most important petroleum producing interval in the Paleogene sequence in the Gulf of Mexico basin; they are also highly significant on a world scale. With known recoveries as of the end of 1987 of 118,110 billion ft^3 of natural gas, 10,275 million barrels of crude oil, and 3,357 million barrels of natural gas liquids (33,317 million barrels oil equivalent), the Oligocene contained 15% of the total petroleum resources of the Gulf of Mexico basin. No more than 12 other provinces worldwide have more petroleum than the Gulf of Mexico basin Oligocene.

The Oligocene is a gas-prone interval, particularly in the Vicksburg and La Laja Formations. Of the total oil-equivalent resources in the Oligocene, 59.1% is natural gas (73.0% in the Vicksburg and La Laja). Though relatively inferior, on an absolute scale the volumes of Oligocene petroleum liquids are substantial. Nearly all of Oligocene crude oil is concentrated in reservoirs within the dominant Catahoula Group (Frio and Anahuac Formations).

The large amounts of Oligocene petroleum are distributed across several major producing trends (Fig. 8). Like the Eocene, the locations of these major trends correspond approximately to the locations of the major sandstone depocenters for each stratigraphic unit. The largest Frio trend, a gas and liquids play, is located in south Texas and the Burgos basin of Mexico. An oil and gas Frio trend that is nearly as large spreads across much of the Houston embayment. Each of these two trends is large enough to qualify as a major petroleum producing province. Other major Frio trends include a gas and oil play straddling the San Marcos arch and another gas and liquids trend spread across south Louisiana. The only major Anahuac trend is a gas and liquids play in south Louisiana. Most of the remaining Anahuac production is concentrated in a gas and liquids trend of near major proportions in south Texas and the Burgos basin.

There is only one major Vicksburg trend, a gas and liquids trend in south Texas and the Burgos basin. A Vicksburg oil and gas trend of just barely moderate proportions is found in the Houston embayment. The only Oligocene trend of economic significance not located in the northwest quadrant of the Gulf of Mexico basin is a La Laja gas trend of moderate proportions in the Veracruz basin.

Unlike the Eocene, Oligocene known recovery is substantially concentrated in giant and large fields. But like the Eocene, large fields predominate. The ten giant fields producing from the Oligocene—*Reynosa* in the Burgos basin; *Agua Dulce-Stratton, Borregos-Seeligson-T.C.B.*, and *La Gloria* in south Texas; *Greta–Tom O'Connor* and *West Ranch* on the San Marcos arch; and *Hastings, Old Ocean, Thompson,* and *Webster* in the Houston embayment—account for 9,660 million barrels known recovery, 29.0% of the Oligocene total (Plate 4). All produce from the Frio; most also produce from the Vicksburg or Anahuac. Out of the more than 1,000 significant fields producing from the Oligocene, 125 are large fields, containing 13,550 million BOE, 40.6% of the Oligocene total. Large Vicksburg fields include *Kelsey Area, Jeffress, McAllen Ranch,* and *Sun* in south Texas. Large Frio fields include *Brasil, Francisco Cano,* and *Trevino* in the Burgos basin; *Alazan North, Arnold-David-Luby, Kelsey Area, McAllen-Pharr, San Salvador, Sun,* and *Viboras* in south Texas; *Flour Bluff Area, Heyser, Laguna Larga, Lake Pasture, Palacios, Placedo, Plymouth, Portilla,* and *White Point* straddling the San Marcos arch; *Anahuac, Barbers Hill, Bay City East, Chocolate Bayou, Dickinson-Gillock, Fannett Dome, Fig Ridge-Seabreeze, Lovells Lake, Magnet Withers, Markham North-Bay City North, Oyster Bayou, Pledger, Port Neches North, Red Fish Reef, Stowell, Trinity Bay,* and *West Columbia* in the Houston embayment; and *Rayne* and *Shuteston* in south Louisiana. Large Anahuac fields include *Mustang Island–Red Fish Bay* and *Willamar Area* in south Texas; *Saxet* on the San Marcos arch; and *Hackberry East, Hackberry West, Iowa, Jennings, Lake Arthur, Mud Lake East,* and *Thornwell South* in south Louisiana.

In contrast to the Eocene, in which the distribution of known recovery in large and giant fields varies substantially by stage, the proportion of Oligocene known recovery in large and giant fields is quite even by interval, the Anahuac, Frio, and Vicksburg each having around 65 to 70% of their known recovery in giant and large fields. Among the various major trends the distribution of known recovery by field size class varies somewhat, being around 85% in the Houston embayment Frio and the south Texas–Burgos basin Anahuac, and between 60 to 70% in all the other major trends.

Significant discoveries in the Gulf of Mexico basin Oligocene span the past 70 years. The initial discovery in the Vicksburg occurred in 1927; substantial levels of discovery did not begin until 1940 and have continued into the 1980s. The first discovery in the south Texas–Burgos basin Frio trend occurred in 1925 with most of the rest occurring during the next 45 years, including *Agua Dulce–Stratton* in 1928, *Borregos-Seeligson-T.C.B.* in 1933, *La Gloria* in 1939, and *Reynosa* in 1948. Significant discoveries in the Frio have continued in this trend into the 1970s and 1980s but at substantially lower levels. Discoveries in the San Marcos arch Frio trend began in 1930 and peaked in the following ten years, including *Greta-Tom O'Connor* in 1933 and *West Ranch* in 1934. After continuing at substantial levels from 1940 to 1970, they have been declining in the 1970s and 1980s. Discoveries in the Houston embayment Frio trend first occurred in the 1920s, peaked in the 1930s with the finding of *Thompson* (1931), *Hastings* (1934), *Old Ocean* (1934), and *Webster* (1937), and continued at substantial levels into the 1960s before declining in the 1970s and 1980s. Discoveries in the south Louisiana Frio were concentrated from 1930 to 1970, but have continued at a declining level into the 1980s. Discoveries in both of the sizeable Anahuac trends were spread out between 1930 and 1980.

The Vicksburg reservoirs of south Texas and the Burgos basin are found in a predominantly deltaic and delta-front depositional environment (see Chapter 11, this volume). Reservoir depths in the Vicksburg range from 1,250 to 4,200 m. Thicknesses vary from 1 to 150 m, the thicker reservoirs being found in the expanded downdip section of the Vicksburg. Reservoir quality in the Vicksburg varies tremendously. Porosity, which is mostly secondary, ranges from 13 to 28%, while permeabilities range from 0.1 to 1,500 md. Because most Vicksburg reservoirs are located in and around the Vicksburg fault zone, trapping is primarily structural (faults, faulted rollover anticlines, and the like). Lithologic variations bound many reservoirs, however. Seals are provided by Vicksburg shales (Loucks and others, 1986; Galloway, 1989; Kosters and others, 1989; Coleman and Galloway, 1990). No published information is available about the reservoir characteristics of the La Laja Formation clastics in the Veracruz basin.

The four major Frio trends in the Gulf of Mexico basin not surprisingly exhibit many similarities in depositional environment, reservoir characteristics, and trap types. The south Texas–Burgos basin Frio encompasses a fluvial depositional environment updip and a deltaic one downdip. Because of the breadth of the trend and the total thickness of the Frio section, reservoir depths vary from less than 100 m to more than 4,150 m. Reservoir thicknesses are generally modest (1 to 50 m), though in a few cases they exceed 200 m. Because of variations in reservoir depth and subsequent diagenesis, reservoir quality varies enormously, with porosities anywhere from 12 to 36% and permeabilities of 1 to 2,000 md. Trapping is predominantly structural, the result of growth faulting and shale ridges. Stratigraphic traps are common in the fluvial system updip from the major growth faults. Frio shales provide the seals (Galloway and others, 1982; Galloway and others, 1983; Loucks and others, 1986; Galloway, 1989; Kosters and others, 1989).

The Frio straddling the San Marcos arch separates the two major Frio deltaic depocenters in south Texas and the Houston embayment. This trend consists of an updip streamplain environment and a downdip strandplain/barrier bar environment, the sediments for the latter originating in the two Frio delta systems that flank it. Reservoir depths in this trend range from 650 to 3,950 m deep; the reservoirs are typically thin (1 to 50 m). Reservoir quality is moderate to good, with porosities of 20 to 26% and permeabilities of 25 to 2,500 md. Trapping is largely structural, being determined by growth faults and shale ridges. Many of these traps are however modified by lithologic variations. Seals are the Frio shales (Galloway and others, 1982; Galloway and others, 1983; Loucks and others, 1986; Kosters and others, 1989).

The Houston embayment Frio encompasses the second major Frio depocenter, a fluvial-deltaic system whose sediments originated to the northwest in the southern Rocky Mountains. Like the other Frio trends, this one exhibits substantial variations in reservoir depths (625 to 4,450 m) and has generally thin reservoirs (2 to 40 m). Reservoir quality is generally good to excellent in this Frio depocenter, with porosities of 16 to 36% and permeabilities of 50 to 3,000 md. Trapping is determined by the ubiquitous growth faults and salt structures, though lithologic controls on reservoir extent are common. Seals are provided by Frio and Anahuac shales (Galloway and others, 1982; Galloway and others, 1983; Loucks and others, 1986; Galloway, 1989; Kosters and others, 1989).

The south Louisiana Frio encompasses a variety of depositional environments, including a strandplain/barrier bar one to the west, Hackberry submarine channel sands in southwestern Louisiana, and a fluvial-deltaic one to the east, the sediment sources for the latter being the ancestral Mississippi River. It is the deepest of all Frio trends, with reservoir depths ranging from 1,600 to 5,125 m. Reservoirs within this trend are thicker than average for the Frio, thicknesses varying from 2 to 150 m. The quality of these reservoirs is good to excellent, as porosities range from 20 to 35% and permeabilities range from 50 to 2,500 md. Trapping is primarily structural, the typical trap being either a faulted rollover anticline or a salt structure. Seals are provided by Frio and Anahuac shales (Galloway and others, 1982; Galloway and others, 1983; Kosters and others, 1989).

The Anahuac of south Louisiana is similar to the underlying Frio. The reservoirs are concentrated in a deltaic depositional environment. Reservoir depths range from 1,200 to 4,875 m. Net sand thicknesses are on average slightly better than the Frio, varying from 3 to 200 m. Recorded porosities are very good (25 to 35%) while permeabilities vary widely (10 to 2,000 md). Trapping is structural, the predominant trap types being faulted anticlines associated with normal faults and salt structures. Anahuac shales provide the seals.

The south Texas–Burgos basin Anahuac is located in a deltaic and delta-front depositional environment continuous with the underlying Frio. The reservoirs are relatively shallow, being only 650 to 2,450 m deep. Reservoir thicknesses are modest (3 to 45 m). Because of the shallow depths, reservoir quality is very good (with 23 to 33% porosity and 40 to 1,500 md permeability). Reservoirs are typically combination traps (faulted anticlines with lithologic variations).

The Oligocene is in most cases a mature to supermature exploration area. Drilling densities for Oligocene objectives are quite high, particularly for shallower objectives. However, some excellent exploration opportunities remain, particularly in the deep downdip Frio of south Louisiana and the deep downdip Vicksburg of south Texas.

Paleogene source rocks. The source rocks both within the Paleogene and for Paleogene reservoirs are not well understood, largely because of some noticeable biases in geochemical research in the Gulf of Mexico basin, biases which have only recently been addressed (Jones, 1990). First, despite the clear importance of the Paleogene to the petroleum resources of the Gulf of Mexico basin, sources in and for the Paleogene have been little studied relative to those in and for the Mesozoic and Upper Cenozoic. Secondly, to the extent that Paleogene sources have been examined, the emphasis has been on the oil-generating potential within the Paleogene, even though most of the producing groups within the Paleogene are gas-prone (the exceptions being the Chicontepec and Sparta Formations, and the Jackson Group). The potential source rocks and their characteristics are thus poorly correlated with the distribution of hydrocarbons by type in the various Paleogene trends. Given the state of the literature, a review of this nature can only summarize what is known and discuss the more likely possibilities for filling the gaps in our knowledge.

The oil in the Chicontepec Formation of the Tampico-Misantla basin is the only Paleogene petroleum with a Mesozoic source. The predominant, if not the only source, of Chicontepec petroleum are the organic rich Upper Jurassic (Tithonian) shales and calcareous shales of the underlying Pimienta Formation. Because the southeastern half of the Chicontepec erosional channel cuts into the Upper Jurassic, direct pathways were available for oil to migrate from this source both vertically and laterally into the Chicontepec sandstones. Moreover, the Upper Jurassic is also thermally mature where it was eroded during the early Paleocene (Busch and Govela S., 1978).

Crude oils in the Wilcox and younger Paleogene formations of the northwestern and northern Gulf of Mexico basin have distinctly different chemical composition than Cretaceous and Jurassic-sourced oils, providing strong evidence that they were generated in the Paleogcnc (Sassen and others, 1988; Sassen, 1990b; Wenger and others, 1990).

Shales within the Wilcox Formation, from northeastern Mexico to Mississippi, have been reported to fit the requirements of potential source rocks—adequate organic-matter content (0.8 to 1.5% TOC) and thermal maturity (Evans, 1987; Dow and others, 1988; Sassen and others, 1988; Walters and Dusang, 1988). Oil and gas reservoired in the Wilcox Formation could, therefore, have been generated in the Wilcox.

The increase in the proportion of gas to oil in the various Wilcox trends going from south Louisiana to south Texas has been attributed to increasing maturity attendant upon greater depths of burial (Thompson and others, 1990). Migration into the Wilcox reservoirs was mostly short-distance lateral and vertical migration, except into the Wilcox oil trend in east-central Louisiana and southwestern Mississippi, where lateral migration of up to 150 km from deeper downdip sources is believed to have occurred (Sassen and others, 1988; Sassen, 1990b; Wenger and others, 1990).

Sources for the oil and gas accumulations in the Eocene Claiborne and Jackson Groups are more obscure. The underlying Wilcox shales and the organic-rich shales of the Sparta Formation (Claiborne Group) have been most commonly mentioned as the possible sources of at least some of the oil and gas in the Claiborne and Jackson reservoirs (Sassen, 1990b).

The source of the oil and gas reservoired in the Oligocene Vicksburg and Catahoula/Frio section is controversial. La Plante (1974) reported that the Oligocene section in south Louisiana, onshore and near-offshore, contains disseminated, terrestrially derived kerogen capable of generating hydrocarbons if subject to sufficiently high temperatures. Bissada and others (1990) concluded, on the other hand, that the content of organic carbon, mostly Type III, in Oligocene and younger sediments in offshore Texas and Louisiana indicates that these sediments should not be considered adequate petroleum source rocks.

Tanner and Feux (1990) reported that geochemical data and oil-source rock correlations suggest that in south Texas the shales of the Upper Eocene Jackson Group have generated the oil and gas reservoired in the Oligocene Vicksburg and Frio sandstones and that shales in the lower part of the Vicksburg Group appear to have generated minor amounts of gas.

According to Galloway and others (1982a), the source rocks for the oil and gas accumulated in the Frio Formation of Texas are either the lower Frio shales underlying the accumulations or the Frio downdip section laterally equivalent to the productive intervals. In either case, source-to-reservoir migration distance, vertical or lateral, would be small. The Frio is lean in organic matter but judged an adequate source by Galloway and others (1982a), particularly in view of the considerable total volume of organic matter in the thick Frio section.

Given the distribution of oil and gas by type in the various Oligocene trends, one would expect differences in the characteris-

tics and mix of sources, ranging from more oil-prone (marine) sources in south Louisiana and the Houston embayment to more gas-prone (terrestrial) sources in south Texas and the Burgos basin.

In general, source-reservoir relationships in the Eocene and Oligocene appear to follow the model of late migration from a thermally mature source rock into recently formed traps in younger overlying reservoirs (Curtis, 1989). A small proportion of Eocene and Oligocene gas may also be biogenic. Mixtures of sources for the same reservoir are also a possibility. It is very likely that as more geochemical studies are conducted and published, more evidence will become available indicating that the Paleogene section contains substantial volumes of potential oil and gas source rocks.

Upper Cenozoic petroleum resources

Upper Cenozoic (Neogene and Pleistocene) reservoirs are only slightly less important than the dominant Paleogene as a petroleum-producing system in the Gulf of Mexico basin. As of the end of 1987, upper Cenozoic known recovery in the basin was 22.7 billion barrels of crude oil, 204.6 trillion ft^3 of natural gas, and 6.6 billion barrels of natural gas liquids, for a total of 63.4 billion barrels of oil equivalent resources (Table 14). Given the extensive number of recent offshore discoveries in the Plio-

TABLE 14. KNOWN RECOVERY OF CRUDE OIL, NATURAL GAS, AND NATURAL GAS LIQUIDS AS OF DECEMBER 31, 1987, IN THE GULF OF MEXICO BASIN BY STRATIGRAPHIC INTERVAL WITHIN THE UPPER CENOZOIC

Interval	Crude Oil (Million bbls)	Natural Gas (Bcf)	NGL (Million bbls)	BOE (Million bbls)
Pleistocene				
Undifferentiated-N*	1,200	28,410	438	6,373
Paraje Solo-S*	105	109	7	130
Total	1,305	28,519	445	6,503
Pliocene				
Undifferentiated-N*	2,440	22,922	474	6,734
Upper Amate/ Concepcion-S*	465	4,756	111	1,369
Total	2,905	27,678	585	8,103
Miocene				
Fleming-N*	16,490	144,435	5,312	45,888
Lower Amate/ Encanto-S*	1,960	4,008	238	2,866
Total	18,450	148,443	5,550	48,754
Upper Cenozoic Total	22,660	204,640	6,580	63,360

*N = Northern (United States); S = Southern (Mexico).

cene and Pleistocene, the confirmed upper Cenozoic total is likely to be in the range of 65 to 70 billion BOE by the early 1990s.

The Gulf of Mexico basin upper Cenozoic is thus by itself one of the largest concentrations of petroleum in the world. No more than six other provinces elsewhere exceed its contributions. Among the five petroleum producing systems of the Gulf of Mexico basin, it has the largest shares of natural gas (39.1% of the basin total), and of natural gas liquids (29.2% of the total). It also has 20.1% of the crude oil and 28.5% of the oil-equivalent resources. The upper Cenozoic is the only one of the five petroleum producing systems of the Gulf of Mexico basin in which natural gas is the predominant hydrocarbon.

Each of the upper Cenozoic units contains major amounts of petroleum. The Miocene clearly predominates, having more than three-fourths of the total resources. (The Miocene is the only unit in the upper Cenozoic in which liquids and natural gas are of roughly equal importance.) The slightly gas-prone Pliocene is moderately larger than the highly gassy Pleistocene interval.

Upper Cenozoic petroleum resources are highly concentrated in the northern half of the Gulf of Mexico basin. More than 93% of Upper Cenozoic known recovery is in the United States. This concentration is particularly marked in natural gas (95.6% of the total) and natural gas liquids (94.6%).

Eighteen of the giant fields in the Gulf of Mexico basin produce from upper Cenozoic reservoirs. Yet these giant fields contain only 9.32 billion BOE known recovery, 14.7% of the Upper Cenozoic total. The upper Cenozoic of the Gulf of Mexico basin has the greatest amount of petroleum in large fields of any major petroleum producing province worldwide. The 325 large fields producing from upper Cenozoic reservoirs contain 43.50 BOE known recovery, 68.7% of the Upper Cenozoic total.

Miocene. Miocene sandstones are the single most important producing unit in the Gulf of Mexico basin. As of the end of 1987, known recovery in the Miocene was 148,443 billion ft^3 of natural gas, 18,450 million barrels of crude oil, and 5,550 million barrels of natural gas liquids, for 48,754 million BOE. The Miocene alone thus contains 21.9% of the oil equivalent petroleum resources of the Gulf of Mexico basin (28.3% of the natural gas, 24.7% of the natural gas liquids, and 16.4% of the crude oil). Only seven other provinces worldwide contain more petroleum than the Gulf of Mexico basin Miocene.

The Miocene is a gas-prone interval, but only by the slightest of margins. Just 50.8% of its known recovery is natural gas, 37.8% is crude oil, and 11.4% is natural gas liquids. This even distribution between gas and liquids contrasts with the considerably more gas-prone intervals of the underlying Oligocene and the overlying Pliocene and Pleistocene.

During the Oligocene and Miocene the major depocenters of sedimentation in the northern Gulf of Mexico basin shift gradually to the northeast from the Rio Grande embayment (Vicksburg) to the Mississippi embayment (upper Miocene) (Limes and Stipe, 1959; Galloway, 1989; Chapter 14, this volume). These shifting depocenters define the major trends of Miocene petro-

leum production in the northern Gulf of Mexico basin (Fig. 9). Substantial lower Miocene production has been found along the entire length of the Texas and Louisiana coasts, including a moderate-size fluvial-deltaic gas trend in south Texas and the adjacent offshore, a major coastal plain/barrier bar/strandplain gas trend straddling the upper gulf coast of Texas, a major fluvial-deltaic gas and oil trend in southwest Louisiana and the adjacent offshore (the largest in the Lower Miocene), and a major deltaic gas and oil trend in southeast Louisiana. In the Middle Miocene section, the largest concentrations of production are evenly divided between two major deltaic gas and oil trends in southeast Louisiana and straddling the southwest Louisiana coastline. Together all of the Middle Miocene reservoirs straddling the Texas gulf coast form a major trend. In the Upper Miocene section, there is no production of consequence along the Texas coast.

Upper Miocene production is dominated by a major deltaic oil and gas trend straddling the southeast Louisiana coast, with nearly 40% of all Miocene known recovery. The Upper Miocene fluvial-deltaic gas and oil trend in southwest Louisiana is the second largest of all the Miocene producing trends, but is only approximately one-third the size of the Upper Miocene trend in southeast Louisiana. Known recovery from the Miocene in the northern Gulf of Mexico basin thus increases moving upward and shifts in relative concentration eastward from the Lower Miocene to the Upper Miocene. In the southern Gulf of Mexico basin there are two Miocene trends, a major Encanto oil trend in the Isthmus Saline basin and a moderate-size gas-condensate lower Amate trend in the Macuspana basin.

There are 17 giant fields that produce from Miocene reservoirs: *Bastian Bay, Bateman Lake, Bayou Sale,* and *Caillou Island* in south Louisiana; *Bay Marchand 002* and *Timbalier Bay* straddling the Louisiana coastline; *Grand Isle 043, Mound Point, South Pass 024, Tiger Shoal, Vermilion 014,* and *West Delta 030* offshore Louisiana; *Agua Dulce-Stratton, Greta-Tom O'Connor, Thompson,* and *West Ranch* in Texas; and *Jose Colomo/Chilapilla* in the Macuspana basin (Plate 4). Yet because these are only minimal size giants (500 to 1,000 million BOE) or because Miocene reservoirs contain only a relatively small proportion of their known recovery, these giant fields provided no more than 8,040 million BOE, only 16.5% of the Miocene total.

Miocene known recovery is concentrated in the 254 large fields with Miocene reservoirs. Together these fields contained 34,045 million BOE, 69.8% of Miocene known recovery. This amount is the greatest amount of oil and gas in large fields found anywhere in the world. These large fields, limiting the list to only those fields between 200 and 500 million BOE known recovery from the Miocene, include *Bay Ste. Elaine, Bayou Penchant, Belle Isle, Deep Lake, Erath, Garden City, Garden Island Bay, Gibson, Grand Bay, Hollywood-Houma, Lafitte, Lake Barre, Lake Pagie, Lake Pelto, Lake Raccourci, Lake Sand, Lake Washington, Lirette, Paradis, Patterson, Pecan Island, Quarantine Bay, Valentine, Venice, Weeks Island,* and *West Bay* in south Louisiana; *East Cameron 064, Grand Isle 016, Grand Isle 047, Main Pass 041, Main Pass 069, Rabbit Island, Ship Shoal 176, South Pass*

027, South Pass 061, South Timbalier 135, Vermilion 039, Vermilion 180, West Cameron 180, West Cameron 192, West Delta 027, West Delta 073, and *West Delta 079* in offshore Louisiana; and *Bacal, Cinco Presidentes, Cuichapa, Ogarrio,* and *Sanchez Magallanes* in the Isthmus Saline basin.

Discoveries of the more than 1,200 fields producing from the Miocene span nearly the entire exploration history of the Gulf of Mexico basin, beginning in 1901 and continuing into the late 1980s. Most however have occurred since 1930. The first Miocene discoveries were in shallow Lower Miocene objectives associated with salt domes in the Houston embayment, including *Thompson* (1931). The bulk of Lower Miocene discoveries onshore in Louisiana and Texas was concentrated between 1930 and 1970, including *West Ranch* (1953). Lower Miocene exploration was extended downdip offshore in the 1960s, 1970s, and 1980s.

Middle Miocene discoveries onshore first occurred in force in the 1930s, including *Greta–Tom O'Connor* (1933), *Agua Dulce-Stratton* (1937), and *Bayou Sale* (1937), and continued at high levels in the 1940s, 1950s, 1960s, and 1970s. Middle Miocene discoveries first occurred offshore Louisiana in the 1950s, including *Vermillion 014* (1956) and have continued off both Louisiana and Texas throughout the 1960s, 1970s, and 1980s.

Upper Miocene discoveries in south Louisiana were concentrated between 1920 and 1960, including *Bateman Lake* (1937), *Timbalier Bay* (1938), *Bastian Bay* (1942), *Caillou Island* (1941), and *Bay Marchand 002* (1949). They have continued at diminishing rates into the 1960s, 1970s, and 1980s. The initiation of substantial offshore exploration in the Gulf of Mexico in the late 1940s was the beginning of Upper Miocene discoveries offshore as well, including *West Delta 030* (1949), *South Pass 024* (1950), *Grande Isle 043* (1956), *Mound Point* (1958), and *Tiger Shoal* (1958). Upper Miocene discoveries have persisted offshore at generally healthy rates from the 1950s into the 1980s.

Most of the discoveries in the Miocene Encanto occurred in the Isthmus Saline basin in the 1950s and 1960s, although the first major ones were in the 1920s. Lower Amate production was discovered in 1956 at *Jose Colomo/Chilapilla* in the Macuspana basin.

Because the Miocene in the northern Gulf of Mexico basin is a more maturely explored series than the Pliocene or Pleistocene, it has not been as thoroughly characterized using the recent insights of sequence stratigraphy (e.g., as outlined by Sangree and others, 1990). But those insights can be readily applied to older characterizations of Miocene reservoirs. Most Miocene reservoirs originated as deltaic sandstones. These deltas not only provided future reservoirs, they also were the sediment source for various shelf sand deposits both downdip from and between the major deltas. Some Miocene reservoirs are also found in updip fluvial and strandplain/streamplain environments (Limes and Stipe, 1959; Rainwater, 1964; Galloway and others, 1986; Morton and others, 1988; Reise and others, 1990).

Because of the characteristics of Miocene depositional environments and the high rate of Miocene sedimentation, Miocene

reservoir quality is generally high. In the Lower Miocene trends of south Louisiana, reservoir depths are mostly between 1,000 and 4,650 m, though some are as shallow as 450 m and as deep as 5,400 m. Most reservoirs are 5 to 80 m thick, though in a few fields in this main Lower Miocene depocenter combined net thickness exceed 150 m. Reservoir quality is good to excellent, varying both with depth and depositional environment from 20 to 35% porosity and 80 to 5,000 md permeability. Along the Texas gulf coast, depth ranges in the Lower Miocene are both narrower and on average shallower, most reservoirs occurring between 200 and 3,650 m, though some are as shallow as 100 m and some as deep as 4,680 m. These reservoirs are on average thinner (mostly 2 to 50 m) than the Lower Miocene in south Louisiana, reflecting both a greater predominance of shelf/strandplain reservoirs and deltas that were not as thick. Reservoir quality is still good to excellent, with porosities of 22 to 35% and permeabilities of 30 to 2,000 md.

The Middle Miocene reservoirs of south Louisiana are found across an extensive depth range (1,000 to 5,200 m). Most are 5 to 100 m thick, though some have combined net thicknesses up to 200 m. Reflecting the range in depths, porosities are 20 to 35% while permeabilities are 100 to 2,000 md. Along and offshore the Texas coast, Middle Miocene reservoirs are mostly 300 to 3,500 m deep, though a few are found as deep as 5,100 m. Like their Lower Miocene counterparts, they are thinner than the Middle Miocene reservoirs in Louisiana, ranging between 1 to 75 m thick. Porosities are generally very good to excellent (25 to 35%), reflecting the relatively shallow reservoir depths. Permeabilities vary widely from 25 to 2,500 md.

Upper Miocene reservoirs in south Louisiana, the major center of Miocene deposition and production, occur across a very extensive depth range (300 to 5,400 m). To date relatively few have been found however below 4,000 m. Thicknesses are generally 5 to 100 m, though in a few fields in southeast Louisiana combined net thicknesses in the Upper Miocene exceed 500 m. Reservoir quality in these sands is good to excellent, with porosities generally ranging from 20 to 35% and permeabilities from 50 to 2,500 md.

Trapping in the Miocene is predominantly structural. Throughout the northern Gulf of Mexico basin Miocene trends, growth faults provide numerous fault traps and rollover anticlines. Salt structures, usually piercement salt domes, have created Miocene traps in south Louisiana, the adjacent offshore, the Houston embayment, and the Isthmus Saline basin. Depositional variations are however frequently important in determining reservoir limits. Because of the substantial thicknesses and extensive faulting, field areas tend to be small, most being between 100 and 4,000 ha (250 to 10,000 acres). In a few Miocene fields, all of which are giants, the producing area approaches 8,000 ha (20,000 acres). Transgressive shales provide areal seals, but many local seals are present as well within the Miocene section.

Because Miocene objectives in the Gulf of Mexico basin have been pursued since 1900, remaining exploration potential within the Miocene is a small fraction of the petroleum discovered within it to date. Some potential still exists, however, concentrated primarily in deeper, subtler traps in the Upper and Middle Miocene offshore Louisiana and Texas.

Pliocene. Although it is large enough to qualify as a major province in world terms, the Pliocene of the Gulf of Mexico basin is essentially a middle-range producing interval within the context of the overall resources of the basin. As of the end of 1987, Pliocene known recovery within the basin was 27,678 billion ft^3 of natural gas, 2,905 million barrels of crude oil, and 585 million barrels of natural gas liquids for a total of 8,103 million barrels oil equivalent.

Pliocene production is essentially divided into four trends, two major-size trends offshore Louisiana in the major Pliocene depocenter in the northern Gulf of Mexico and two moderate size trends in southeast Mexico (Fig. 9). The largest of these, a slope trend seaward of the late Pliocene shelf edge, is almost evenly divided between natural gas and petroleum liquids, gas being slightly predominant. The other large Pliocene trend, a trend covering the ancestral Pliocene continental shelf, and the recent Mississippi Delta, is slightly more gas-prone. The largest of the Pliocene trends in southeastern Mexico is an upper Amate gas trend of near-major proportions in the Macuspana basin. The Isthmus Saline basin hosts a sizeable Concepcion oil trend. There is also a minor Pliocene oil trend onshore in south-central Louisiana.

Pliocene petroleum is unusually concentrated in large fields, even by comparison to the other intervals of the Gulf Coast Cenozoic where such a concentration is commonplace. There are only two giant fields with Pliocene production, *Jose Colomo/- Chilapilla* and *Bay Marchand 002* (Plate 4). Together, the Plio-

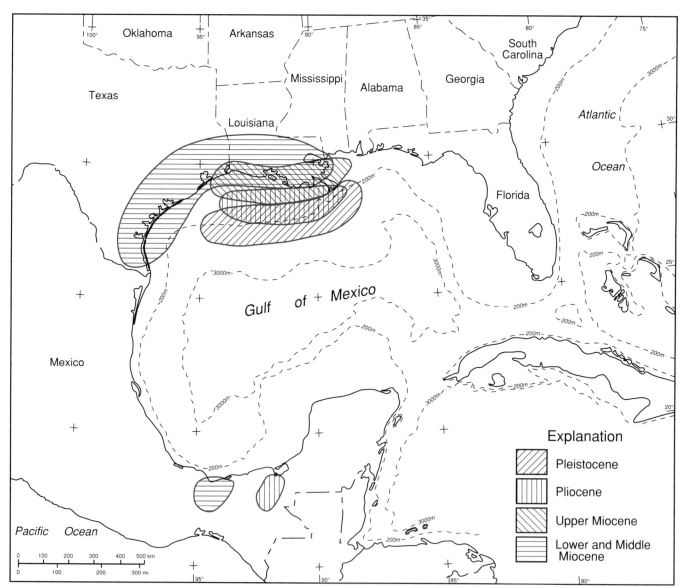

Figure 9. Major upper Cenozoic petroleum-producing trends of the Gulf of Mexico basin.

cene reservoirs in these two giants contained only 653 million BOE, only 8.1% of the Pliocene total. By comparison, of a total of more than 170 significant fields, there are 54 large fields with Pliocene reservoirs, containing altogether 5,836 million BOE, a staggering 72.1% of all Pliocene known recovery. These large fields include *Lake Barre* in south Louisiana; *East Cameron 231, Eugene Island 175, Eugene Island 238, Eugene Island 258, Eugene Island 266, Eugene Island 276, Eugene Island 292, Eugene Island 306, Main Pass 306, Main Pass 311, Mississippi Canyon 194, Ship Shoal 207, Ship Shoal 208, Ship Shoal 222, South Marsh Island 048, South Marsh Island 066, South Marsh Island 073, South Marsh Island 130, South Pass 062, South Pass 078, South Pass 089, South Pelto 20, South Timbalier 172,* and *Vermilion 255* in offshore Louisiana; *Hormiguero* and *Usumacinta* in the Macuspana basin; and *El Golpe, Ogarrio, Santuario,* and *Tupilco* in the Isthmus Saline basin.

Although the first Pliocene discovery was made as early as 1901 at *Jennings* in south Louisiana, Pliocene exploration and discovery in the Gulf of Mexico basin is concentrated in the past 40 years. Extensive Pliocene shelf discoveries were initiated with the finding of *Bay Marchand 002* in 1949 and have gradually accelerated in number from the 1950s into the 1980s. Discoveries in the Pliocene slope trend began in the 1960s and have continued at a steady pace into the 1980s. Pliocene discoveries in the Macuspana basin were concentrated in the 1950s, being initiated by the discovery of *Jose Colomo/Chilapilla* in 1951. Most of the Pliocene reservoirs in the Isthmus Saline basin were discovered in the 1950s and 1960s.

In the northern Gulf of Mexico basin, the Pliocene reservoirs are generally of good to excellent quality. The most productive reservoirs are concentrated in deltaic, delta-front, and basin floor environments deposited during regressive (lowstand) periods within the area of the main Pliocene depocenter. Pliocene production has been found in the interdeltaic areas east and west of this depocenter in turbidite-channel and slope-fan sandstones (Reed and others, 1987; Geitgey, 1988, Pacht and others, 1990b). Reservoir depths in the Pliocene shelf trend vary from 300 to 5,750 m, though most occur between 900 and 3,950 m. Reservoir thicknesses range from 2 to 135 m, with nearly all between 5 and 60 m. Reservoir depths in the Pliocene slope trend range from 750 to 5,150 m, with nearly all concentrated between 1,500 and 3,950 m. Producing sand thicknesses in this trend vary from 2 to 100 m, nearly all between 5 and 50 m the thicker measurements usually being associated with more than one pool.

Trapping in Pliocene reservoirs in the northern Gulf of Mexico basin is determined primarily by growth faults and piercement salt domes, although the producing limits of many reservoirs are shaped by depositional variations. Seals are provided by shales in the Pliocene sequence, typically at the base of transgressive units (Woodbury and others, 1973; Geitgey, 1988).

The quality of the Concepcion reservoirs in the Isthmus Saline basin of southeast Mexico varies widely, the sands being highly lenticular. Trapping is determined by salt domes and salt anticlines, as modified by depositional variations in individual reservoirs. The quality of the upper Amate reservoirs in the Macuspana basin also is widely variable. Combined sand thicknesses can be substantial, in the best cases exceeding 100 m. Traps are predominantly faulted anticlines with depositional variations (Guzman and others, 1955; Beebe, 1968).

Although the exploration history of the Pliocene is relatively recent within the overall exploration history of the Gulf of Mexico basin, the Pliocene is by now a mature trend. The possibilities that remain, probably still a considerable number, are mostly deep (4,000 m or more), subtle, below salt overhangs or tongues, or small, but still concentrated geographically within the main area of Pliocene deposition on the northern Gulf of Mexico shelf (Pratsch, 1989; Pacht and others, 1990b).

Pleistocene. Although the Pleistocene is the least important producing interval within the entire Cenozoic section in the Gulf of Mexico basin, in world terms it is not an insignificant unit. Known recovery in the Pleistocene as of the end of 1987 was 28,519 billion ft^3 of natural gas, 1,305 million barrels of crude oil, and 445 million barrels of natural gas liquids, for a total of 6,503 million barrels oil equivalent. These amounts clearly understate the Pleistocene total. In the middle and late 1980s, more than 80 significant Pleistocene fields were discovered that were undeveloped and thus unbooked as of the end of 1987. The subsequent development of these recent discoveries is likely to increase the Pleistocene total to more than eight billion BOE, the largest concentration of Quaternary petroleum in the world.

Pleistocene production is concentrated in two major trends; both of which are offshore in the northern Gulf of Mexico basin in the one major depocenter of Pleistocene sedimentation (Fig. 9). The largest, a highly gas-prone trend with nearly 75% of Pleistocene known recovery, is located on the southern half of the continental shelf off Louisiana and Texas. The other, also gas-prone but with a greater proportion of oil, is located downdip of the first on the northern Gulf of Mexico continental slope. Moderate amounts of Pleistocene oil have also been found in the Paraje Solo Formation in several fields in the Isthmus Saline basin of southeast Mexico.

There is only one predominantly Pleistocene giant oil and gas field, *Eugene Island 330* offshore Louisiana (Plate 4). This field, when combined with minor amounts of Pleistocene production from *Caillou Island* and *Timbalier Bay* fields, provides a known recovery of 628 million barrels BOE, only 9.7% of the Pleistocene total. Of the more than 175 fields producing from the Pleistocene, 45 are large fields, containing 3,648 million BOE, 56.1% of the Pleistocene total. These large fields include *East Cameron 271, East Cameron 299, East Cameron 334, Eugene Island 296, Eugene Island 306, Eugene Island 342, Grande Isle 095, Mississippi Canyon 311, Ship Shoal 246, Ship Shoal 274, South Marsh Island 107, South Marsh Island 128, South Marsh Island 137, Vermilion 245, Vermilion 320, West Cameron 533, West Cameron 565, West Cameron 587, West Cameron 617,* and *West Cameron 639* in offshore Louisiana; *High Island 370, High Island 563, High Island 571,* and *High Island 573* in offshore Texas; and *El Plan* and *Sanchez Magallanes* in Mexico. Several more Pleistocene fields on the Gulf of Mexico slope will qualify as large fields when developed. Reflecting the small average size of Pleistocene large fields (even smaller than that of the Pliocene), *Eugene Island 330* is the only field with Pleistocene known recovery exceeding 225 million BOE.

Pleistocene discoveries are relatively recent, because most Pleistocene fields are offshore in water depths greter than 50 m. A few scattered shallow Pleistocene discoveries were made onshore in south Louisiana and southeast Mexico beginning in the 1930s. Discoveries began offshore on the shelf trend in the 1960s, but did not reach their peak until the 1970s and 1980s (including the discovery of *Eugene Island 330* in 1970). Exploration of the slope trend began in the 1970s, with discoveries not reaching their peak to date until the 1980s.

Pleistocene reservoirs are generally of good to excellent quality. The best reservoirs were deposited in outer shelf and upper slope environments during regressive episodes. The most common productive depositional environments include alluvial channel fills, delta-front sheets, and submarine basin-floor fans (Morton and Jirik, 1989; Pacht and others, 1990a, b). Reservoir depths in the shelf trend range from 600 to 4,650 m. Net thicknesses vary from 2 to 80 m, though most are from 5 to 50 m. The greater thicknesses are generally combined net thicknesses from several stacked pools. In the slope trend reservoir depths range from 1,000 to 3,500 m, while thicknesses vary from 5 to 90 m.

Traps in the shelf trend are structural or combination traps. The structural element is provided by both salt and shale structures and growth fault systems. The limits of many reservoirs are determined by depositional variations. Most of the traps in the slope trend are related to salt structure and movement, some of which verges on the chaotic. Productive areas of individual fields are small to moderate in size, ranging from 200 to 4,000 ha (500 to 10,000 acres). Seals are provided by interbedded shales, the best being in transgressive units (Pearcy and Ray, 1986; Reed and others, 1987; Ray, 1988, Pacht and others, 1990a, b). Because of the chaotic structural conditions on the slope, these seals are not always competent, as evidenced by the numerous seeps found across much of the northern Gulf of Mexico slope (Kennicutt and Brooks, 1990).

Although some Pleistocene objectives (e.g., those associated with salt diapirs on the shelf) have been extensively explored, overall the interval still has substantial exploration potential. Numerous slope objectives still remain to be drilled; exploration and development of these prospects will however be constrained by the high costs of deep-water operations. On the shelf, there are many smaller fields left to be discovered, particularly deeper objectives in the growth fault systems or in stratigraphic traps. As the knowledge of Pleistocene sequence stratigraphy is refined and becomes more site specific, many of these traps should be identified (Pacht and others, 1990b).

Upper Cenozoic source rocks. The source rocks within the upper Cenozoic and for upper Cenozoic reservoirs are a subject of considerable debate. This debate has succeeded in eliminating some formerly postulated sources. It has not narrowed the field sufficiently to identify specific sources for particular productive trends. Critical information is lacking, forcing many of the decisive arguments to be inferential. A discussion of upper Cenozoic source rocks can thus only state what is known and lay out the various justifiable arguments where the debate depends upon inferences.

There is general agreement that the shales surrounding most of the producing Miocene, Pliocene, and Pleistocene reservoirs are not the sources for oil and thermogenic gas in each of these intervals. Because of the extremely high rates of sedimentation in the major depocenters (averaging more than 1,000 m/m.y. in the upper Miocene, Pliocene, and Pleistocene), these shales are usually too lean (<0.5 TOC) to be an effective source. Moreover, when they are sufficiently rich, they contain terrestrial organic matter (Type III kerogen) most likely to have generated only gas (La Plante, 1974; Young and others, 1977; Dow, 1978, 1984; Rice, 1980; Holland and others, 1980; Nunn and Sassen, 1986; Bissada and others, 1990). In addition, because these sediments are of recent origin, they are thermally immature, not having been buried deep enough long enough to generate thermogenic gas or what have typically been analyzed as mature oils (Dow, 1978; Morton and others, 1988; Pasley and others, 1988). At best, the shales are generally believed to be a source of only biogenic gas, particularly for the younger Pliocene and Pleistocene reservoirs (Rice, 1980; Rice and Claypool, 1981; Walters, 1990). Huc and Hunt (1980), however, have reported source rocks of Miocene age from two wells offshore south Texas, and Walters and Cassa (1985), based on biomarker analysis, discuss petroleum source rocks of early to middle Miocene age in the eastern part of offshore Louisiana and of Pliocene age in the western part.

The evidence that the oil and most of the gas in upper Cenozoic reservoirs has originated in and migrated upwards from deeper sources is compelling (Hanor and Sassen, 1990). Chemical analyses of oils from stacked reservoirs covering an extensive depth range within the same field have consistently indicated that in each field all reservoirs have the same source, whether the fields being studied are in the Miocene (Price, 1990), Pliocene (Requejo and Halpern, 1990), or Pleistocene (Holland and others, 1980). However, because these potential sources are deep (in many cases 6,000 m or more), they have not been penetrated by the drill. Hence, information is lacking to identify and evaluate source rocks directly. Thus several potential sources have typically been proposed for each producing interval. In general, these proposals follow the same conceptual model, namely generation occurs long after deposition of the source rock and migration occurs shortly after traps are created in shallower and younger reservoirs (Curtis, 1989; Sassen, 1990b).

Having to resort to deeper sources has, in turn, made it necessary to rely on some kind of migration mechanism or mechanisms to move the oil and gas from the source beds to the reservoir. Deep-seated faults have most commonly been proposed as conduits for migration (Dow, 1978, 1984; Jones, 1981;

Dinkelman and Curry, 1988; Bissada and others, 1990). Oil is believed to have migrated up the faults as a continuous liquid phase or as a solute in a gas phase.

Piercement salt domes with their surrounding complex fault systems have also been suggested as pathways for oil and gas migration from deep sources to shallower reservoirs. In addition, the high thermal conductivity of the salt has been proposed to explain the preferred occurrence of oil trapped in upper Cenozoic reservoirs around salt domes in contrast to the predominant occurrence of gas in reservoirs of this age in other types of traps. The salt domes are believed to have acted as geological "radiators" that increased the temperature around the diapiric structures and raised, as a result, the thermal maturity of the organic matter in the surrounding upper Cenozoic sediments (Williams and Lerche, 1987, 1988; O'Brien and Lerche, 1984, 1987, 1988).

For the Miocene of south Louisiana, the leading candidates as the primary source rock are the shales of the Paleocene Midway, and Eocene Wilcox and Sparta Formations. These shales have been analyzed directly just updip of the main Miocene producing trends in south Louisiana. They have fair to good levels of total organic carbon (TOC); the organic matter within them is a mixture of Type II and III kerogens and is thus capable of generating both oil and gas; underneath the Miocene these shales are thermally mature; and they generate oils similar in type to those found in Miocene reservoirs (Sassen, 1990b; Sassen and Chinn, 1990). Oligocene shales, particularly those from the lower Anahuac Formation (Galloway, 1990), are a second possibility; however, no geochemical evaluations have been published of this interval. For the Upper Miocene, particularly in southeast Louisiana and the adjacent offshore, Lower Miocene shales deposited during transgressive episodes in intraslope basins are another plausible candidate. Such shales are likely to contain at least fair concentrations of TOC; and the Lower Miocene is thermally mature in this area (Dow, 1984). The objection that the at-best fair-quality (0.5 to 2.0% TOC) source rocks likely to be encountered in the Lower Miocene are insufficient to generate the substantial quantities of oil and gas reservoired in the Upper Miocene is countered by the observation that source rocks of similar quality, volume, and age have generated substantial quantities of oil and gas in the geologically analogous Niger Delta (Bustin, 1988), though no geochemical analytical data was presented to substantiate this argument.

The Cretaceous has been postulated as a source for the oil in the Miocene (Bissada and others, 1990). However (1) the geochemical dissimilarity between the oils of the Upper Cretaceous Tuscaloosa, which is a prolific source updip, and Tertiary oils (Walters and Dusang, 1988; Wenger, Sassen, and Schumacher, 1990); (2) the current supermaturity of the Cretaceous under the main Miocene trend; and (3) the complicated sequence of earlier generation and migration, temporary trapping, and subsequent remigration that proponents of this hypothesis must employ all argue against a Cretaceous source.

For the gas-prone Miocene trends straddling the Texas coast, the Paleocene and Eocene are less likely sources, simply because there are approximately 3,000 to 4,500 m of intervening Oligocene sediments (Williamson, 1959). The Oligocene, particularly the Frio-Anahuac, is the most likely source of thermogenic hydrocarbons, especially for the more prolific lower Miocene trends. It contains terrestrial (gas-prone) organic matter, though mostly in marginal concentrations, and it is thermally mature under all of the highly gas-prone Miocene producing trends of Texas (Galloway and others, 1982b; Galloway and others, 1986). For the less prolific Middle and Upper Miocene trends, both the Oligocene and the Lower Miocene are the most likely source rocks (Morton and others, 1988). Biogenic gas from surrounding shales may also be a source for some reservoirs.

Sediments ranging in age from Mesozoic to Lower Pliocene have been proposed as the sources for the crude oils in Pliocene and Pleistocene reservoirs along the eastern Texas shelf and the Louisiana shelf and slope. Dow (1984) and Dow and others (1990) favor the Middle Miocene and Lower Pliocene transgressive sequence as the preferred source for the oil in uppermost Cenozoic reservoirs. The shales in these sequences are believed to have been buried deep enough to have reached thermal maturity. Based on biomarker analysis, Walters and Cassa (1985) suggest that the oils reservoired in the Pliocene and Pleistocene of nearshore eastern offshore Louisiana were probably generated in the underlying Lower and Middle Miocene, while oils produced from reservoirs of the same age in the western Louisiana shelf were probably generated in Lower Pliocene sources. Thompson and others (1990) distinguished two groups of oils and condensates produced from Pliocene and Pleistocene reservoirs in offshore Texas and Louisiana. They are believed to be derived from different lithofacies of the same source-rock section, probably the Mesozoic (Jurassic or Cretaceous), accounting for their identical carbon isotope ratios. The oils of the first group, distributed across the Louisiana and eastern Texas shelf, were probably derived from clastic rocks, while the oils of the second group, reservoired in fields along the Louisiana shelf edge and continental slope, were most likely generated in carbonate or carbonate-rich sources.

The geographic localization of crude oil within the Pliocene and Pleistocene trends suggests several controlling factors, all of which could be operative in varying degrees: (1) geographic differences in kerogen type (predominantly terrestrial versus predominantly marine, the latter only occurring in localized intraslope basins); (2) substantial oil migration only in the immediate vicinity of piercement salt domes and growth faults; (3) local barriers to oil migration by salt tongues and sills, particularly in the Pleistocene trend, a possibility only recognized recently (Brooks, 1989; Pratsch, 1989). The calculated productivity of areally small intraslope basins is sufficient to account for known Pliocene and Pleistocene emplaced oil (Williams and Lerche, 1988).

Analysis of Pliocene and Pleistocene gases indicate likely multiple sources: (1) biogenic gas for adjacent shales, particularly in the Pleistocene where reservoirs with 98 to 99% methane are common, and (2) thermogenic gas from the same deep sources from which oil was generated (Rice, 1980; Rice and Claypool, 1981; Walters, 1990). Paleogene sources for the Pliocene and Pleistocene outer shelf hydrocarbons are less likely than they were for the Miocene, simply because the Paleogene intervals are currently supermature in this area and thus face many of the same difficulties as do the Cretaceous sources for the Miocene, particularly given indications of relatively recent migration into Pliocene and Pleistocene reservoirs (Holland and others, 1980).

Sources for most Pleistocene slope hydrocarbons may differ considerably from those on the shelf. Oils from most of the slope fields form a distinctly different group from shelf oils, the properties of the former suggesting both a highly anoxic and a carbonate-rich source (Thompson and others, 1990). The sediments deposited in Middle Miocene intraslope basins are also a possible source for at least some of the oil and gas in the slope fields (Dow and others, 1990). Any determination of sources for northern Gulf of Mexico slope hydrocarbons must, however, be considered highly tentative at this time given the paucity of information about potential source rock characteristics in the 4,500 to 7,500-m depth range across the slope and an accompanying lack of comprehensive knowledge about how salt sills and tongues have locally controlled oil and gas migration.

The source or sources for the oil and gas in the upper Cenozoic reservoirs of the Isthmus Saline, Comalcalco, and Macuspana basins of southern Mexico is uncertain. Miocene sediments in these basins are reported to have a high organic content, but their thermal maturity, except when deeply buried, ranges only from moderately immature to moderately mature (Holguin Quiñones, 1985 [1988]). Oil and gas accumulated in the upper Cenozoic reservoirs in southern Mexico is believed, therefore, to have migrated from deeper sources—most likely the prolific Upper Jurassic dark shales that are known to be the source of the oil and gas of the giant fields of the Reforma and Gulf of Campeche areas. A combination of short-distance horizontal and vertical migration routes between it and the producing reservoirs is possible. However, the "radiator effect" of the numerous intrusive salt structures in the Isthmus Saline basin may have resulted in increased maturation of the upper Cenozoic shales and thus of their enhanced potential as oil source rocks. In the Macuspana basin, where salt structures play a small role and gas is the main fluid produced, the source could be the thick Upper Cretaceous or Paleogene shales within the basin.

THE PRODUCTIVITY OF THE GULF OF MEXICO BASIN

The Gulf of Mexico basin is one of three petroleum megaprovinces in the world. This chapter has concentrated on describing in some detail the high productivity of this basin. It is appropriate to conclude the chapter with a brief explanation of this high productivity.

The basic reasons for the very large petroleum resources of the Gulf of Mexico basin become clearer when we examine the approximate distribution of these petroleum resources by both the age of their reservoir rocks and the age of their inferred source rocks (Fig. 10). The distributions shown should be considered *approximations*. As the previous discussion indicated, the distribution of known recovery by reservoir age is uncertain in the Cretaceous, Paleocene, and Eocene. The distribution of the amount of oil and gas supplied by individual source rocks is even more uncertain, particularly for petroleum reservoired in Cenozoic reservoirs. However, even allowing for a range of error in these distributions of ±25%, the gross outlines are still clear.

Reservoired petroleum in the Gulf of Mexico basin, though present in substantial quantities throughout the entire stratigraphic column from the Upper Jurassic Oxfordian to the Quaternary Pleistocene, is concentrated in two broad intervals: the Oligocene-Miocene and, to a lesser degree, the Albian-Cenomanian. The distribution of the probable oil and gas supplied by individual source rocks is even more concentrated, the

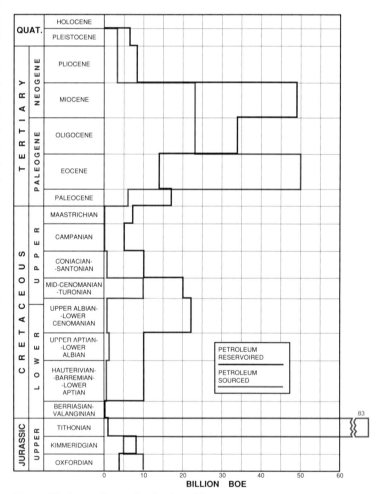

Figure 10. Approximate distribution of known petroleum recovery by reservoir age and by source-rock age in the Gulf of Mexico basin.

Upper Jurassic Tithonian being the dominant source and the Eocene most likely being the other major source. Unlike the distribution of reservoired known recovery, productive source rocks appear to be missing from many intervals in the stratigraphic column.

These two distributions suggest a division of Gulf of Mexico basin petroleum resources into two broad petroleum megasystems. (A petroleum system includes all those elements that are essential for an oil and gas accumulation to exist in nature—a source rock, migration path, reservoir rock, seal, and trap [Magoon, 1988]. Petroleum systems are generally considered to be closed systems.) These two megasystems are (1) the Gulf of Mexico basin Mesozoic, plus Chicontepec and the Isthmus Saline basin (hereinafter the Gulf of Mexico basin Mesozoic); and (2) the Gulf of Mexico basin Cenozoic, less Chicontepec and the Isthmus Saline basin (hereinafter the Gulf of Mexico basin Cenozoic). The division is essentially by age of the source rocks, the sources for Chicontepec and the Isthmus Saline basin being the Upper Jurassic Tithonian black shales. These two megasystems are approximately equal in size, the Gulf of Mexico basin Mesozoic having a known recovery of 117 billion BOE (including discovered but unbooked resources in the Norphlet) and the Gulf of Mexico basin Cenozoic having a known recovery of 107 billion BOE. These two megasystems have several common elements; however, they differ substantially in two areas, source rocks and reservoirs.

The primary reason for the great productivity of the Gulf of Mexico basin Mesozoic is the presence of excellent oil-prone source rocks, particularly the Upper Jurassic Tithonian black shales of Mexico and secondarily the Upper Jurassic Oxfordian lower Smackover and the Upper Cretaceous Cenomanian-Turonian Eagle Ford/Tuscaloosa in the United States. The Tithonian shales are the sources for more than 80 billion BOE known recovery in the Gulf of Mexico basin and probably at least one-third of a trillion BOE of in-place petroleum (not even considering the substantial amounts likely to have escaped into the atmosphere). The lower Smackover and Eagle Ford/Tuscaloosa each are the sources for around 10 to 21 billion BOE known recovery.

In contrast to the excellence of the source rocks, reservoirs in the Gulf of Mexico basin Mesozoic are generally of poor quality. More than two-thirds of known recovery in the Mesozoic megasystem are in reservoirs with low primary porosity and/or permeability, including carbonates requiring natural fracturing to be productive (e.g., the reservoirs of southeast Mexico), carbonates made productive only by subaerial erosion (e.g., the El Abra of the Tampico-Misantla basin), or low permeability sandstones requiring artificial fracturing to be commercial (e.g., Chicontepec and the Cotton Valley and Travis Peak sandstones of east Texas). In only a few intervals is reservoir quality very good to excellent, most notably in the Woodbine/Tuscaloosa and Norphlet sandstones. Because the source rocks are so rich, they have generally expelled enough petroleum to fill any available reservoir, irre-

spective of its quality. Moreover, enough reservoir volume exists to accommodate very large amounts of petroleum.

Emplacement of petroleum in reservoirs of the Mesozoic megasystem has been facilitated by the fact that most secondary migration in the megasystem has been short-distance vertical and lateral migration (usually less than 5 km). Even medium distance lateral migration (more than 50 km) is rare, the most notable examples of this being migration within the Chicontepec and Woodbine/Tuscaloosa Formations. Traps are plentiful, the result of movement of the underlying salt, faulting, or the presence of basement structures. Seals, whether of shales, anhydrites, or tight sandstones and carbonates, are plentiful, and generally good, such as the very thick shale seals in southeastern Mexico, except where fracturing or faulting has made them less than wholly competent (e.g., in both southeast Mexico and the Tampico-Misantla basin, where many of the oil seeps in the Gulf of Mexico basin exist). Thus, of the five elements generally considered to be necessary for the presence of oil and gas, four—source, migration, trap, and seal—are satisfied to a high degree. The fifth—reservoir—can only be considered as adequate.

In sharp contrast to the Mesozoic megasystem, the great productivity of the Gulf of Mexico basin Cenozoic megasystem primarily results from the presence of extremely large volumes of very good to excellent reservoirs. Throughout the Eocene to Pleistocene section, high-quality sandstone reservoirs exist. On the other hand, identified Cenozoic source rocks are only marginally adequate to good. There are large volumes of potential source rocks within the Cenozoic megasystem, but because of high rates of deposition, the proportion of total organic matter within them is generally low, limiting the amount of hydrocarbons they are capable of generating. Because they have only been recently deposited, substantial volumes of potential Cenozoic source rocks are also thermally immature and thus incapable of generating anything but biogenic gas.

However, migration within the Cenozoic has overwhelmingly been short-distance vertical migration, leaving relatively little generated petroleum to be dispersed along migration paths. Early migration into Cenozoic reservoirs also helped preserve the excellent qualities of these reservoirs. The rapid large-volume sedimentation that produced Cenozoic reservoirs also resulted in subsequent movement of the underlying salt and the creation of numerous growth-faults, the two major mechanisms of trap formation in the Cenozoic section. Local and areal seals, formed by transgressive shales, are also common throughout the Cenozoic section. The general absence of major seeps (other than on the structurally chaotic slope where seeps are plentiful) in a faulted, gas-prone area is solid evidence for the adequacy of these seals. Good to excellent reservoir, migration, trap, and seal elements are thus present in the Gulf of Mexico basin Cenozoic megasystem. The favorable combination of these four elements led to the highly efficient trapping of the petroleum that was generated by fair to good Cenozoic source rocks.

REFERENCES CITED

Ayala Nieto, S. R., and Perez Matus, A., 1974, Proyecto para incrementar la produccion de aceite, Yacimiento Tamabra—Campo Poza Rica: Ingenieria Petrolera, v. 14, p. 195–202.

Bailey, J., 1978, Black Lake Field, Natchitoches Parish, Louisiana; A review: Gulf Coast Association of Geological Societies Transactions, v. 28, p. 11–24.

Baria, L. R., Stoudt, D. L., Harris, P. M., and Crevello, P. D., 1982, Upper Jurassic reefs of Smackover Formation, United States Gulf Coast: American Association of Petroleum Geologists Bulletin, v. 66, p. 1449–1482.

Bebout, D. G., Schatzinger, R. A., and Loucks, R. G., 1977, Porosity distribution in the Stuart City trend, Lower Cretaceous, south Texas, *in* Bebout, D. G,. and Loucks, R. G., eds., Cretaceous carbonates of Texas and Mexico; Applications to subsurface exploration: University of Texas at Austin, Bureau of Economic Geology Report of Investigations 89, p. 234–256.

Bebout, D. G., Budd, D. A., and Schatzinger, R. A., 1981, Depositional and diagenetic history of the Sligo and Hosston Formations (Lower Cretaceous) in south Texas: University of Texas at Austin, Bureau of Economic Geology Report of Investigations 109, 70 p.

Bebout, D. G., Weise, B. R., Gregory, A. R., and Edwards, M. B., 1982, Wilcox Sandstone reservoirs in the deep subsurface along the Texas Gulf Coast: University of Texas at Austin, Bureau of Economic Geology Report of Investigations 117, 125 p.

Beebe, B. W., 1968, Occurrence of natural gas in Mexico, *in* Beebe, B. W., ed., Natural gases of North America: Tulsa, American Association of Petroleum Geologists Memoir 9, v. 1, p. 209–232.

Benson, D. J., 1988, Depositional history of the Smackover Formation in southwest Alabama: Gulf Coast Association of Geological Societies Transactions, v. 38, p. 197–205.

Bissada, K. K., Katz, B. J., Barnicle, S. C., and Schunk, D. J., 1990, On the origin of hydrocarbons in the Gulf of Mexico basin; A reappraisal, *in* Schumacher, D., and Perkins, B. F., eds., Gulf Coast oils and gases; Their characteristics, origin, distribution, and exploration and production significance: Society of Economic Paleontologists and Mineralogists Foundation, Gulf Coast Section, 9th Annual Research Conference Proceedings, Austin, p. 163–171.

Bolin, D. E., Mann, S. D., Burroughs, D., Moore, H. E., Jr., and Powers, T. J., 1989, Petroleum atlas of southwestern Alabama: Tuscaloosa, Alabama, Geological Survey of Alabama, Atlas 23, 218 p.

Brooks, R. O., 1989, Horizontal component of Gulf of Mexico salt tectonics, *in* Gulf of Mexico salt tectonics, associated processes and exploration potential: Society of Economic Paleontologists and Mineralogists Foundation, Gulf Coast Section, 10th Annual Research Conference Proceedings, Houston, p. 22–24.

Burgess, J. D., 1990, Correlation of Cretaceous and Jurassic oils from Hunt and Kaufman Counties, northeast Texas, in Schumacher, D., and Perkins, B. F., eds., Gulf Coast oils and gases; Their characteristics, origin, distribution, and exploration and production significance: Society of Economic Paleontologists and Mineralogists Foundation, Gulf Coast Section, 9th Annual Research Conference Proceedings, Austin, p. 69–75.

Busch, D. A., and Govela S., A., 1978, Stratigraphy and structure of Chicontepec turbidites, southeastern Tampico-Misantla basin, Mexico: American Association of Petroleum Geologists Bulletin, v. 62, p. 235–246.

Bushaw, D. J., 1968, Environmental synthesis of the east Texas Lower Cretaceous: Gulf Coast Coast Association of Geological Societies Transactions, v. 18, p. 416–438.

Bustin, R. M., 1988, Sedimentology and characteristics of dispersed organic matter in Tertiary Niger Delta; Origin of source rocks in a deltaic environment: American Association of Petroleum Geologists Bulletin, v. 72, p. 277–298.

Caughey, C. A., 1977, Depositional systems in the Paluxy Formation (Lower Cretaceous), northeast Texas; Oil, gas, and groundwater resources: University of Texas at Austin, Bureau of Economic Geology Geological Circular 77-8, 59 p.

Chavarria, G., J., 1980, Geology of the Campeche Sound: 11th World Energy Conference, Proceedings, v. 1A, Munich, p. 362–371.

Claypool, G. E., and Mancini, E. A., 1989, Geochemical relationships of petroleum in Mesozoic reservoirs to carbonate source rocks of Jurassic Smackover Formation, southwest Alabama: American Association of Petroleum Geologists Bulletin, v. 73, p. 904–924.

Coleman, Janet, and Galloway, W. E., 1990, Petroleum geology of the Vicksburg Formation, Texas: Gulf Coast Association of Geological Societies Transactions, v. 40, p. 119–130.

Collins, S. E., 1980, Jurassic Cotton Valley and Smackover Reservoir trends, east Texas, north Louisiana, and south Arkansas: American Association of Petroleum Geologists Bulletin, v. 64, p. 1004–1013.

Coogan, A. H., Bebout, D. G., and Maggio, C., 1972, Depositional environments and geologic history of Golden Lane and Poza Rica trend, Mexico, An alternative view: American Association of Petroleum Geologists Bulletin, v. 6, p. 1419–1447.

Cook, T. D., 1979, Exploration history of South Texas Lower Cretaceous carbonate platform: American Association of Petroleum Geologists Bulletin, v. 63, p. 32–49.

Curtis, D. M., 1989, A conceptual model for sources of oils in Gulf Coast Cenozoic reservoirs: Gulf Coast Association of Geological Societies Transactions, v. 39, p. 37–56.

Dennis, J. G., 1987, Depositional environments of A.W.P. Olmos Field, McMullen County, Texas: Gulf Coast Association of Geological Societies Transactions, v. 37, p. 55–63.

Dinkelman, M., and Curry, D. J., 1988, Seismic-stratigraphic geochemical model of occurrence of oils in Gulf of Mexico flexure trend [abs.]: American Association of Petroleum Geologists Bulletin, v. 72, p. 179.

Dixon, S. A., Summers, D. M., and Surdam, R. C., 1989, Diagenesis and preservation of porosity in Norphlet Formation (Upper Jurassic), southern Alabama: American Association of Petroleum Geologists Bulletin, v. 73, p. 707–728.

Dow, W. G., 1978, Petroleum source beds on continental slopes and rises: American Association of Petroleum Geologists Bulletin, v. 62, p. 1584–1606.

—— , 1984, Oil source beds and oil prospect definition in the upper Tertiary of the Gulf Coast: Gulf Coast Association of Geological Societies Transactions, v. 34, p. 329–339.

Dow, W. G., Mukhopadhyay, P. K., and Jackson, T., 1988, Source rock potential and maturation of deep Wilcox from south-central Texas [abs.]: American Association of Petroleum Geologists Bulletin, v. 72, p. 179.

Dow, W. G., Yukler, M. A., Senftle, J. T., Kennicutt, M. C., II, and Armentrout, J. M., 1990, Miocene oil source beds in the East Breaks basin, flex-trend, offshore Texas, *in* Schumacher, D., and Perkins, B. F., eds., Gulf Coast oils and gases: Their characteristics, origin, distribution, and exploration and production significance: Society of Economic Paleontologists and Mineralogists Foundation, Gulf Coast Section, 9th Annual Research Conference Proceedings, Austin, p. 139–150.

Driskill, B. W., Nunn, J. A., Sassen, R., and Pilger, R. H., Jr., 1988, Tectonic subsidence, crustal thinning and petroleum generation in the Jurassic trend of Mississippi, Alabama, and Florida: Gulf Coast Association of Geological Societies Transactions, v. 38, p. 257–265.

Dutton, S. P., 1987, Diagenesis and burial history of the Lower Cretaceous Travis Peak Formation, east Texas: University of Texas at Austin, Bureau of Economic Geology Report of Investigations 164, 58 p.

Dutton, S. P., Finley, R. J., and Herrington, K. L., 1987, Organic geochemistry of the Lower Cretaceous Travis Peak Formation, East Texas basin: Gulf Coast Association of Geological Societies Transactions, v. 37, p. 65–74.

Erdman, J. G., and Morris, D. A., 1974, Geochemical correlation of petroleum: American Association of Petroleum Geologists Bulletin, v. 58, p. 2326–2337.

Evans, R., 1987, Pathways of migration of oil and gas in the South Mississippi salt basin: Gulf Coast Association of Geological Societies Transactions, v. 37, p. 75–86.

Eversull, L. G., 1985, Depositional systems and distributions of Cotton Valley blanket sandstones in north Louisiana: Gulf Coast Association of Geological Societies Transactions, v. 35, p. 49–57.

Ewing, T. E., 1986, Structural styles of the Wilcox and Frio growth-fault trends in Texas: University of Texas at Austin, Bureau of Economic Geology Report of Investigations 154, 86 p.

Ewing, T. E., and Fergeson, W. G., 1989, The downdip Yegua trend; An overview: Gulf Coast Association of Geological Societies Transactions, v. 39, p. 75–83.

Fisher, W. L., and McGowen, J. H., 1967, Depositional systems in the Wilcox Group of Texas and their relationship to occurrence of oil and gas: Gulf Coast Association of Geological Societies Transactions, v. 17, p. 105–125.

Fisher, W. L., Proctor, C. V., Jr., Galloway, W. E., and Nagle, J. S., 1970, Depositional systems in the Jackson Group of Texas; Their relationship to oil, gas, and uranium: Gulf Coast Association of Geological Societies Transactions, v. 20, p. 234–261.

Foss, D. C., 1979, Depositional environment of Woodbine sandstones, Polk County, Texas: Gulf Coast Association of Geological Societies Transactions, v. 29, p. 83–94.

Galloway, W. E., 1989, Genetic stratigraphic sequences in basin analysis II: Application to northwest Gulf of Mexico Cenozoic basin: American Association of Petroleum Geologists Bulletin, v. 73, p. 143–154.

—— , 1990, Paleogene depositional episodes, genetic stratigraphic sequences, and sediment accumulation rates northwest Gulf of Mexico basin, *in* Sequence stratigraphy as an exploration tool; Concepts and practice in the Gulf Coast: Society of Economic Paleontologists and Mineralogists Foundation, Gulf Coast Section, 11th Annual Research Conference Proceedings, Houston, p. 165–176.

Galloway, W. E., Hobday, D. K., and Magara, K., 1982, Frio Formation of the Texas Gulf Coast basin; Depositional systems, structural framework, and hydrocarbon origin, migration, distribution, and exploration potential: University of Texas at Austin, Bureau of Economic Geology Report of Investigations 122, 78 p.

Galloway, W. E., Ewing, T. E., Garrett, C. M., Tyler, N., and Bebout, D. G., 1983, Atlas of major Texas oil reservoirs: University of Texas at Austin, Bureau of Economic Geology, 139 p.

Galloway, W. E., Jirik, L. A., Morton, R. A., and DuBar, J. R., 1986, Lower Miocene (Fleming) depositional episode of the Texas Coastal Plain and continental shelf: University of Texas at Austin, Bureau of Economic Geology, 50 p., 7 pl.

Geitgey, J. E., 1988, Plio-Pleistocene evolution of central offshore Louisiana: Gulf Coast Association of Geological Societies Transactions, v. 38, p. 151–156.

Grabowski, G. J., Jr., 1984, Generation and migration of hydrocarbons in Upper Cretaceous Austin Chalk, south-central Texas, *in* Palacas, J. G., ed., Petroleum geochemistry and source rock potential of carbonate rocks: American Association of Petroleum Geologists Studies in Geology 18, Tulsa, p. 97–115.

Guevara, E. H., and Garcia, R., 1972, Depositional systems and oil-gas reservoirs in the Queen City Formation (Eocene), Texas: Gulf Coast Association of Geological Societies Transactions, v. 22, p. 1–22.

Guzman, E. J., 1967, Reef type stratigraphic traps in Mexico, *in* Origin of oil, geology, geophysics; Proceedings, 7th World Petroleum Congress, v. 2: London, p. 461–470.

Guzman, E. J., and Guzman, A. E., 1981, Petroleum geology of Reforma area, southeast Mexico, and exploratory effort in Baja California, northwest Mexico, *in* Halbouty, M. T., ed., Energy resources of the Pacific region: American Association of Petroleum Geologists Studies in Geology 12, p. 1–11.

Guzman, E. J., Saurez, R., and Lopez-Ramos, E., 1955, Outline of the petroleum geology of Mexico, *in* Bullard, F. M., ed., Proceedings, Conference on Latin American Geology: Austin, University of Texas, p. 1–30.

Halbouty, M. T., and Halbouty, J. J., 1982, Relationships between East Texas Field region and Sabine uplift in Texas: American Association of Petroleum Geologists Bulletin, v. 66, 1042–1054.

Hanor, J. S., and Sassen, R., 1990, Evidence for large-scale vertical and lateral migration of formation waters, dissolved salt, and crude oil in the Louisiana

Gulf Coast, *in* Schumacher, D., and Perkins, B. F., eds., Gulf Coast oils and gases: Their characteristics, origin, distribution, and exploration and production significance: Society of Economic Paleontologists and Mineralogists Foundation, Gulf Coast Section, 9th Annual Research Conference Proceedings, Austin, p. 283–296.

Harrison, F. W., Jr., 1980, Louisiana Tuscaloosa versus southeast Texas Woodbine: Gulf Coast Association of Geological Societies Transactions, v. 30, p. 105–111.

Holguin Quinones, N., 1985 (1988), Evaluacion geoquimica del sureste de Mexico: Associacion Mexicana de Geologos Petroleros Boletin, v. 37, no. 1, p. 3–48.

Holland, D. S., Nunan, W. E., Lammkin, D. R., and Woodhams, R. L., 1980, Eugene Island Block 330 Field, offshore Louisiana, *in* Halbouty, M. T., ed., Giant oil and gas fields of the decade 1968–1978: Tulsa, American Association of Petroleum Geologists Memoir 30, p. 253–280.

Hopf, R. W., 1967, The Poth Sand trend of southwest Texas, *in* Ellis, W. G., ed., Contributions to the geology of south Texas: San Antonio, South Texas Geological Society, p. 46–55.

Huc, A. Y., and Hunt, J. M., 1980, Generation and migration of hydrocarbons in offshore south Texas Gulf Coast sediments: Geochimica et Cosmochimica Acta, v. 44, p. 1081–1089.

Hunt, J. M., and McNichol, A. P., 1984, The Cretaceous Austin Chalk of south Texas; A petroleum source rock, *in* Palacas, J. G., ed., Petroleum geochemistry and source rock potential of carbonate rocks: American Association of Petroleum Geologists Studies in Geology 18, Tulsa, p. 117–125.

Jones, R. W., 1981, Some mass balance and geological constraints on migration mechanisms: American Association of Petroleum Geologists Bulletin, v. 65, p. 103–122.

—— , 1990, Conference highlights, *in* Schumacher, D., and Perkins, B. F., eds., Gulf Coast oils and gases; Their characteristics, origin, distribution, and exploration and production significance: Society of Economic Paleontologists and Mineralogists Foundation, Gulf Coast Section, 9th Annual Research Conference Proceedings, Austin, p. 1–8.

Karges, H. E., 1962, Significance of Lower Tuscaloosa Sand patterns in southwest Mississippi: Gulf Coast Association of Geological Societies Transactions, v. 12, p. 171–185.

Kennicutt, M. C., II, and Brooks, J. M., 1990, Seepage of gaseous and liquid petroleum in the northern Gulf of Mexico, *in* Schumacher, D., and Perkins, B. F., eds., Gulf Coast oils and gases; Their characteristics, origin, distribution, and exploration and production significance: Society of Economic Paleontologists and Mineralogists Foundation, Gulf Coast Section, 9th Annual Research Conference Proceedings, Austin, p. 309–310.

Koons, C. B., Bond, J. G., and Pierce, F. L., 1974, Effects of depositional environment and post depositional history on chemical composition of lower Tuscaloosa oils: American Association of Petroleum Geologists Bulletin, v. 58, p. 1271–1280.

Kosters, E. C., and 9 others, 1989, Atlas of major Texas gas reservoirs: University of Texas at Austin, Bureau of Economic Geology, 161 p.

La Plante, R. E., 1974, Hydrocarbon generation in Gulf Coast Tertiary sediments: American Association of Petroleum Geologists Bulletin, v. 58, p. 1281–1289.

Lewis, J. O., 1977, Stratigraphy and entrapment of hydrocarbons in the San Miguel Sands of southwest Texas: Gulf Coast Association of Geological Societies Transactions, v. 27, p. 90–98.

Limes, L. L., and Stipe, J. C., 1959, Occurrence of Miocene oil in south Louisiana: Gulf Coast Association of Geological Societies Transactions, v. 9, p. 77–90.

Lock, B. E., and Broussard, S. W., 1989, The Norphlet reservoir in Mobile Bay; Origin of deep porosity: Gulf Coast Association of Geological Societies Transactions, v. 39, p. 187–194.

Long, J., 1986, The Eocene Lobo gravity slide, Webb and Zapata Counties, Texas, *in* Stapp, W. L., ed., Contributions to the geology of south Texas: San Antonio, South Texas Geological Society, p. 270–293.

Loucks, R. G., Dodge, M. M., and Galloway, W. E., 1986, Controls on porosity and permeability of hydrocarbon reservoirs in lower Tertiary sandstones along the Texas Gulf Coast: University of Texas at Austin, Bureau of Economic Geology Report of Investigations 149, 78 p.

Magoon, L. B., ed., 1988, Petroleum systems of the United States: Washington, D.C., U.S. Geologic Survey Bulletin 1870, 68 p.

Mancini, E. A., and Payton, J. W., 1981, Petroleum geology of South Carlton Field, lower Tuscaloosa "Pilot Sand," Clarke and Baldwin Counties, Alabama: Gulf Coast Association of Geological Societies Transactions, v. 31, p. 139–147.

Mancini, E. A., Mink, R. M., Bearden, B. L., and Wilkerson, R. P., 1985, Norphlet Formation (Upper Jurassic of southwestern and offshore Alabama), environments of deposition and petroleum geology: American Association of Petroleum Geologists Bulletin, v. 69, p. 881–898.

Mancini, E. A., Mink, R. M., and Bearden, B. L., 1986, Integrated geological, geophysical and geochemical interpretation of Upper Jurassic petroleum trends in the eastern Gulf of Mexico: Gulf Coast Association of Geological Societies Transactions, v. 36, p. 219–226.

Mann, S. D., Mink, R. M., Bearden, B. L., and Schneeflock, R. D., Jr., 1989, The "Frisco City Sand": A new Jurassic reservoir in southwest Alabama: Gulf Coast Association of Geological Societies Transactions, v. 39, p. 195–205.

Marzano, M. S., Pense, G. M., and Andronaco, P., 1988, A comparison of the Jurassic Norphlet Formation in Mary Ann Field, Mobile Bay, Alabama to onshore regional Norphlet trends: Gulf Coast Association of Geological Societies Transactions, v. 38, p. 85–100.

McFarlan, E., Jr., 1977, Lower Cretaceous sedimentary facies and sea level changes, U.S. Gulf Coast, in Bebout, D. G., and Loucks, R. G., eds., Cretaceous carbonates of Texas and Mexico; Applications to subsurface exploration: University of Texas at Austin, Bureau of Economic Geology Report of Investigations 89, p. 5–11.

McGowen, M. K., and Lopez, C. M., 1983, Depositional systems in the Nacatoch Formation (Upper Cretaceous), northwest Texas and southwest Arkansas: University of Texas at Austin, Bureau of Economic Geology Report of Investigations 137, 59 p.

Metheny, S. L., Jr., 1979, Giant Chicontepec given 42% of Mexican oil reserves: Oil and Gas Journal, August 20, 1979, p. 82–85.

Mink, R. M., Bearden, B. L., and Mancini, E. A., 1985, Regional Jurassic geologic framework of Alabama coastal waters area and adjacent federal waters area: Tuscaloosa, Geological Survey of Alabama and State Oil and Gas Board, Oil and Gas Report 12, 58 p.

Mink, R. M., Hamilton, R. P., Bearden, B. L., and Mancini, E. A., 1987, Determination of recoverable natural gas reserves for the Alabama coastal waters area: Tuscaloosa, Geological Survey of Alabama and State Oil and Gas Board, Oil and Gas Report 13, 74 p.

Moore, C. H., 1984, Regional patterns of diagenesis, porosity evolution, and hydrocarbon production, upper Smackover of the Gulf rim, in Presley, M. W., ed., The Jurassic of east Texas: Tyler, East Texas Geological Society, p. 55.

Moore, C. H., and Druckman, U., 1981, Burial diagenesis and porosity evolution, Upper Jurassic Smackover, Arkansas and Louisiana: American Association of Petroleum Geologists Bulletin, v. 65, p. 597–628.

Moore, T., 1983, Cotton Valley depositional systems of Mississippi. Gulf Coast Association of Geological Societies Transactions, v. 33, p. 163–167.

Morton, R. A., and Jirik, L. A., 1989, Structural cross sections, Plio-Pleistocene series, southeastern Texas continental shelf: University of Texas at Austin, Bureau of Economic Geology, 7 p., 13 pl.

Morton, R. A., Jirik, L. A., and Galloway, W. E., 1988, Middle-upper Miocene depositional sequences of the Texas Coastal Plain and continental shelf: Geological framework, sedimentary facies, and hydrocarbon plays: University of Texas at Austin, Bureau of Economic Geology, 40 p., 7 pl.

Mossman, R. W., and Viniegra, F., 1976, Complex fault structures in Veracruz Province of Mexico: American Association of Petroleum Geologists Bulletin, v. 60, p. 379–388.

Nunn, J. A., and Sassen, R., 1986, The framework of hydrocarbon generation and migration, Gulf of Mexico continental slope: Gulf Coast Association of Geological Societies Transactions, v. 36, p. 257–262.

O'Brien, J. J., and Lerche, I., 1984, The influence of salt domes on paleotemperature distributions: Geophysics, v. 49, p. 2032–2043.

——, 1987, Heat flow and thermal maturation near salt domes, in Lerche, I., and O'Briene, J. J., eds., Dynamic geology of salt and related structures: Academic Press, p. 711–750.

——, 1988, Impact of heat flow anomalies around salt diapirs and salt sheets in the Gulf Coast on hydrocarbon maturity; Models and observations: Gulf Coast Association of Geological Societies Transactions, v. 38, p. 231–243.

Oehler, J. H., 1984, Carbonate source rocks in the Jurassic Smackover trend of Mississippi, Alabama, and Florida, in Palacas, J. G., ed., Petroleum geochemistry and source rock potential of carbonate rocks: Tulsa, American Association of Petroleum Geologists Studies in Geology 18, p. 63–69.

Oliver, W. B., 1971, Depositional systems in the Woodbine Formation (Upper Cretaceous), northeast Texas: University of Texas at Austin, Bureau of Economic Geology Report of Investigations 73, 28 p.

Pacht, J. A., Bowen, B. E., Beard, J. H., and Shaffer, B. L., 1990a, Sequence stratigraphy of Plio-Pleistocene depositional facies in the offshore Louisiana south additions: Gulf Coast Association of Geological Societies Transactions, v. 40, p. 643–659.

Pacht, J. A., Bowen, B. E., Shaffer, B. L., and Pottorf, B. R., 1990b, Sequence stratigraphy of Plio-Pleistocene strata in the offshore Louisiana Gulf Coast; Applications to hydrocarbon exploration, in Sequence stratigraphy as an exploration tool; Concepts and practice in the Gulf Coast: Society of Economic Paleontologists and Mineralogists Foundation, Houston Gulf Coast Section, 11th Annual Research Conference Proceedings, p. 269–285.

Palacas, J. G., 1978, Preliminary assessment of organic carbon content and petroleum source rock potential of Cretaceous and Tertiary carbonates, South Florida basin: Gulf Coast Association of Geological Societies Transactions, v. 28, p. 357–381.

——, 1984, Carbonate rocks as sources of petroleum; Geological and chemical characteristics and oil-source correlations: London, 11th World Petroleum Congress Proceedings, v. 2, p. 31–43.

Palacas, J. G., Daws, T. A., and Applegate, A. V., 1981, Preliminary petroleum source rock assessment of pre-Punta Gorda rocks (lowermost Cretaceous - Jurassic?) in south Florida: Gulf Coast Association of Geological Societies Transactions, v. 31, p. 369–376.

Palacas, J. G., Anders, D. E., and King, J. D., 1984, South Florida basin; A prime example of carbonate source rocks of petroleum, in Palacas, J. G., ed., Petroleum geochemistry and source rock potential of carbonate rocks: Tulsa, American Association of Petroleum Geologists Studies in Geology 18, p. 71–96.

Pasley, M. A., Ferrell, R. E., and Sassen, R., 1988, Thermal maturity and kerogen type of the Robulus "L" Sands, offshore Vermillion Parish, Louisiana: Gulf Coast Association of Geological Societies Transactions, v. 38, p. 139–144.

Pearcy, J. R., and Ray, P. K., 1986, The production trends of the Gulf of Mexico: Exploration and development: Gulf Coast Association of Geological Societies Transactions, v. 36, p. 263–273.

Pratsch, J. C., 1989, Salt in oil and gas exploration offshore Gulf Coast region, U.S.A., in Gulf of Mexico salt tectonics, associated processes and exploration potential: Society of Economic Paleontologists and Mineralogists Foundation, Gulf Coast Section, 10th Annual Research Conference Proceedings, Houston, p. 111–114.

Presley, M. W., and Reed, C. H., 1984, Jurassic exploration trends of east Texas, in Presley, M. W., ed., The Jurassic of east Texas: Tyler, East Texas Geological Society, p. 11–22.

Price, L. C., 1990, Crude oil characteristization at Caillou Island, Louisiana by "generic" hydrocarbons, in Schumacher, D., and Perkins, B. F., eds., Gulf Coast oils and gases: Their characteristics, origin, distribution, and exploration and production significance: Society of Economic Paleontologists and Mineralogists Foundation, Gulf Coast Section, 10th Annual Research Conference Proceedings, Austin, p. 237–261.

Rainwater, E. H., 1964, Regional stratigraphy of the Gulf Coast Miocene: Gulf Coast Association of Geological Societies Transactions, v. 14, p. 81–124.

Ray, P. K., 1988, Lateral salt movement and associated traps on the continental slope of the Gulf of Mexico: Gulf Coast Association of Geological Societies Transactions, v. 38, p. 217–223.

Reed, J. C., Leyendecker, C. L., Khan, A. S., Kinler, C. J., Harrison, P. F., and Pickens, G. P., 1987, Correlation of Cenozoic sediments, Gulf of Mexico outer continental shelf, part I; Galveston area offshore Texas through Vermillion area offshore Louisiana: New Orleans, Minerals Management Service, 1987.

Reise, W. C., Hill, W. A., Rosen, R. N., Olsen, R. S., and Sudduth, D. N., 1990, Sequence stratigraphy of the Miocene system, offshore Texas; Alternative models and their global implications, *in* Sequence stratigraphy as an exploration tool; Concepts and practice in the Gulf Coast: Society of Economic Paleontologists and Mineralogists Foundation, Gulf Coast Section, 11th Annual Research Conference Proceedings, Houston, p. 299–306.

Renteria Curiel, S., 1977, "El Chicontepec," Un reto a La Ingerieria: Revista Mexicana del Petroleo, v. 15, n. 257, p. 47–56.

Requejo, A. G., and Halpern, H. I., 1990, A geochemical study of oils from the South Pass 61 Field, offshore Louisiana, *in* Schumacher, D., and Perkins, B. F., eds., Gulf Coast oils and gases; Their characteristics, origin, distribution, and exploration and production significance: Society of Economic Paleontologists and Mineralogists Foundation, Gulf Coast Section, 9th Annual Research Conference Proceedings, Austin, p. 219–235.

Rice, D. D., 1980, Chemical and isotopic evidence of the origins of natural gases in offshore Gulf of Mexico: Gulf Coast Association of Geological Societies Transactions, v. 30, p. 203–213.

Rice, D. D., and Claypool, G. E., 1981, Generation, accumulation, and resource potential of biogenic gas: American Association of Petroleum Geologists Bulletin, v. 65, p. 5–25.

Richards, J. A., 1988, Depositional history of the Sunniland Limestone (Lower Cretaceous), Raccoon Point Field, Collier County, Florida: Gulf Coast Association of Geological Societies Transactions, v. 38, p. 473–483.

Rose, P. R., 1972, Edwards Group, surface and subsurface, central Texas: University of Texas at Austin, Bureau of Economic Geology Report of Investigations 74, 198 p.

Sangree, J. B., Vail, P. R., and Mitchum, R. M., Jr., 1990, A summary of exploration applications of sequence stratigraphy as an exploration tool; Concepts and practice in the Gulf Coast: Society of Economic Paleontologists and Mineralogists Foundation, Gulf Coast Section, 11th Annual Research Conference Proceedings, Houston, p. 321–327.

Santiago Acevedo, J., and Mejia Dautt, O., 1980, Giant fields in the southeast of Mexico: Gulf Coast Association of Geological Societies Transactions, v. 30, p. 1–31.

Santiago, J., and Baro, A., 1991, Mexico's giant fields, 1978–1988 decade, *in* Halbouty, M. T., ed., Giant oil and gas fields of the decade 1978–1988: American Association of Petroleum Geologists Memoir (in press).

Sassen, R., 1989, Migration of crude oil from the Smackover source rock to Jurassic and Cretaceous reservoirs of the northern Gulf rim: Organic Geochemistry, v. 14, p. 51–60.

——, 1990a, Geochemistry of Carbonate Source Rocks and Crude Oils in Jurassic Salt Basins of the Gulf Coast, in Brooks, J., ed., Classic Petroleum Provinces: Geological Society Special Publication No. 50, London, p. 265–277.

——, 1990b, Lower Tertiary and Upper Cretaceous source rocks in Louisiana and Mississippi; Implications to Gulf of Mexico crude oil: American Association of Petroleum Geologists Bulletin, v. 74, p. 857–878.

Sassen, R., and Chinn, E. W., 1990, Implications of Lower Tertiary source rocks in south Louisiana to the origin of crude oil, Offshore Louisiana, *in* Schumacher, D., and Perkins, B. F., eds., Gulf Coast oils and gases; Their characteristics, origin, distribution, and exploration and production significance: Society of Economic Paleontologists and Mineralogists Foundation, Gulf Coast Seciton, 9th Annual Research Conference Proceedings, Austin, p. 175–179.

Sassen, R., and Moore, C. H., 1988, Framework of hydrocarbon generation and destruction in eastern Smackover trend: American Association of Petroleum Geologists Bulletin, v. 72, p. 649–663.

Sassen, R., Moore, C. H., and Meendsen, F. C., 1987a, Distribution of hydrocarbon source potential in the Jurassic Smackover Formation: Organic Geochemistry, v 11, p. 379 383.

Sassen, R., Moore, C. H., Nunn, J. A., Meendsen, F. C., and Heydari, E., 1987b, Geochemical studies of crude oil generation, migration, and destruction in the Mississippi salt basin: Gulf Coast Association of Geological Societies Transactions, v. 37, p. 217–224.

Sassen, R., Tye, R. S., Chinn, E. W., and Lemoine, R. C., 1988, Origin of crude oil in the Wilcox trend of Louisiana and Mississippi; Evidence for long-range migration: Gulf Coast Association of Geological Associations Transactions, v. 38, p. 27–34.

Simmons, K. A., 1967, A primer on 'Serpentine Plugs' in south Texas, *in* Ellis, W. G., ed., Contributions to the geology of south Texas: San Antonio, South Texas Geological Society, p. 125–132.

Smith, G. M., 1985, Geology of the deep Tuscaloosa (Upper Cretaceous) gas trend in Louisiana, *in* Perkins, B. F., and Martin, G. B., eds., Habitat of oil and gas in the Gulf Coast: Society of Economic Paleontologists and Mineralogists Foundation, Gulf Coast Section, 4th Annual Research Conference Proceedings, p. 153–190.

Stapp, W. L., 1977, Geology of the fractured Austin and Buda Formations in the subsurface of south Texas: Gulf Coast Association of Geological Societies Transactions, v. 27, p. 208–229.

Surles, M. A., Jr., 1985, Petroleum and source rock potential of Eagle Ford Group (Upper Cretaceous), East Texas basin [abs.]: American Association of Petroleum Geologists Bulletin, v. 69, p. 309.

Sweeney, R. E., 1990, Chemical alteration of Smackover oils as a function of maturation, *in* Schumacher, D., and Perkins, B. F., eds., Gulf Coast oils and gases; Their characteristics, origin, distribution, and exploration and production significance: Society of Economic Geologists and Mineralogists Foundation, Gulf Coast Section, 9th Annual Research Conference Proceedings, Austin, p. 23–30.

Sydboten, B. D., Jr., and Bowen, R. L., 1987, Depositional environments and sedimentary tectonics of the subsurface Cotton Valley Group (Upper Jurassic), west-central Mississippi: Gulf Coast Association of Geological Societies Transactions, v. 37, p. 239–245.

Tanner, J. A., and Fuex, A. N., 1990, Chemical and isotopic evidence of the origin of hydrocarbons and source potential of rocks from the Vicksburg and Jackson Formations of Slick Ranch area, Starr County, Texas, *in* Schumacher, D., and Perkins, B. F., eds., Gulf Coast oils and gases; Their characteristics, origin, distribution, and exploration and production significance: Society of Economic Paleontologists and Mineralogists Foundation, Gulf Coast Section, 10th Annual Research Conference Proceedings, Austin, p. 79–97.

Thompson, K.F.M., Kennicutt, M. C., II, Brooks, J. M., 1990, Classification of offshore Gulf of Mexico oils and gas condensates: American Association of Petroleum Geologists Bulletin, v. 74, p. 187–198.

Thomson, A., 1979, Preservation of porosity in the deep Woodbine/Tuscaloosa trend, Louisiana: Gulf Coast Association of Geological Societies Transactions, v. 29, p. 396–403.

Tyler, N., and Ambrose, W. A., 1986, Depositional systems and oil and gas plays in the Cretaceous Olmos Formation, south Texas: University of Texas at Austin, Bureau of Economic Geology Report of Investigations 152, 42 p.

Vaughan, R. L., Jr., and Benson, D. J., 1988, Diagenesis of the Upper Jurassic Norphlet Formation, Mobile and Baldwin Counties and offshore Alabama: Gulf Coast Association of Geological Societies Transactions, v. 38, p. 543–551.

Viniegra-O., F., 1981, Great Carbonate Bank of Yucatan, southern Mexico: Journal of Petroleum Geology, v. 3, p. 247–278.

Viniegra O., F., and Castillo-Tejero, C., 1970, Golden Lane fields, Veracruz, Mexico, *in* Halbouty, M. T., ed., Geology of giant petroleum fields: American Association of Petroleum Geologists Memoir 14, p. 309–325.

Wade, W. J., Sassen, R., and Chinn, E. W., 1987, Stratigraphy and source potential of the Smackover Formation in the northern Manila Embayment, southeastern Alabama: Gulf Coast Association of Geological Societies

Transactions, v. 37, p. 277–286.

Walters, C. C., 1990, Organic geochemistry of gases and condensates from Block 511A High Island South Addition offshore Texas, Gulf of Mexico, *in* Schumacher, D., and Perkins, B. F., eds., Gulf Coast oils and gases; Their characteristics, origin, distribution, and exploration and production significance: Society of Economic Paleontologists and Mineralogists Foundation, Gulf Coast Section, 9th Annual Research Conference Proceedings, Austin, p. 185–198.

Walters, C. C., and Cassa, M. R., 1985, Regional organic geochemistry of offshore Louisiana: Gulf Coast Association of Geological Societies Transactions, v. 35, p. 277–286.

Walters, C. C., and Dusang, D. D., 1988, Source and thermal history of oils from Lockhart Crossing, Livingston, Parish, Louisiana: Gulf Coast Association of Geological Societies Transactions, v. 38, p. 37–44.

Waples, D. W., 1985, Geochemistry in petroleum exploration: International Human Resources Development Corporation, Boston, 232 p.

Weise, B. R., 1980, Wave-dominated delta systems of the Upper Cretaceous San Miguel Formation, Maverick basin, south Texas: University of Texas at Austin, Bureau of Economic Geology Report of Investigations 107, 39 p.

Wenger, L. M., Sassen, R. ,and Schumacher, D., 1990, Molecular characteristics of Smackover, Tuscaloosa, and Wilcox-reservoired oils in the eastern Gulf Coast, *in* Schumacher, D., and Perkins, B. F., eds., Gulf Coast oils and gases; Their characteristics, origin, distribution, and exploration and production significance: Society of Economic Paleontologists and Mineralogists Foundation, Gulf Coast Section, 9th Annual Research Conference Proceedings, Austin, p. 37–57.

Wescott, W. A., 1984, Diagenesis of a tight gas formation; The Jurassic Cotton Valley Sandstone, East Texas basin, *in* Presley, M. W., ed., The Jurassic of east Texas: Tyler, East Texas Geological Society, p. 118–126.

West, T. S., 1963, Typical stratigraphic traps, Jackson trend of south Texas: Gulf Coast Association of Geological Societies Transactions, v. 13, p. 67–78.

Williams, D. F., and Lerche, I., 1987, Salt domes, organic-rich source beds and reservoirs in intraslope basins of the Gulf Coast Region, *in* Lerche, I., and O'Brien, J. J., eds., Dynamic geology of salt and related structures: Academic Press, p. 751–786.

—— , 1988, Organic-rich source beds and hydrocarbon production in the Gulf Coast region: Gulf Coast Association of Geological Societies Transactions, v. 38, p. 145–150.

Williamson, J.D.M., 1959, Gulf Coast Cenozoic history: Gulf Coast Association of Geological Societies Transactions, v. 9, p. 14–29.

Woodbury, H. O., Murray, I. B., Jr., Pickford, P. J., and Akers, W. H., 1973, Pliocene and Pleistocene depocenters, Outer Continental Shelf, Louisiana and Texas: American Association of Petroleum Geologists Bulletin, v. 57, p. 2428–2439.

Young, A., Monaghan, P. H., and Schweisberger, R. T., 1977, Calculation of ages of hydrocarbons in oils-physical chemistry applied to petroleum geochemistry I: American Association of Petroleum Geologists Bulletin, v. 61, p. 573–600.

ACKNOWLEDGMENTS

Research for this chapter was aided substantially by the contributions of Tammy Skufca and Sheila Steele. Jeanette Reese did her usual excellent job in preparing the manuscript.

I appreciate the many helpful comments from the reviewers of this chapter: W. F. Fisher, W. E. Galloway, R. Gonzalez, J. Martinez, D. D. Rice, R. Sassen, and K.F.M. Thompson, and particularly the editor of this volume, A. Salvador.

MANUSCRIPT ACCEPTED BY THE SOCIETY AUGUST 19, 1991

The Geology of North America
Vol. J, The Gulf of Mexico Basin
The Geological Society of America, 1991

Chapter 16

Mineral resources and geopressured-geothermal energy

Stanley R. Riggs
Department of Geology, East Carolina University, Greenville, North Carolina 27858
Samuel P. Ellison, Jr., William L. Fisher, and William E. Galloway
Department of Geological Sciences, The University of Texas at Austin, P.O. Box 7909, Austin, Texas 78713-7909
Mary L. W. Jackson and Robert A. Morton
Bureau of Economic Geology, The University of Texas at Austin, University Station, Box X, Austin, Texas 78713

INTRODUCTION

The Gulf of Mexico basin is best known for its vast and widespread oil and gas resources. They have been described in the preceding chapter of this volume. The basin, however, also contains important deposits of phosphate, lignite, and sulfur and small deposits of uranium. In addition, salt from several salt domes is produced by underground and solution mining and is used principally as a chemical feedstock for the manufacture of many industrial products. Large volumes of geopressured-geothermal water are also known from the Tertiary sediments of the Gulf of Mexico basin, particularly around its northern margin. It often contains natural gas in solution. This overpressured, gas-bearing hot water may someday be an important source of thermal and kinetic energy; it is now just a gleam in the eye of imaginative energy tacticians.

The phosphate deposits of Florida and southeastern Georgia, the Florida Phosphogenic Province, represent about 75 percent of the total domestic phosphate production, and ranged between 34 and 28 percent of the total world production between 1983 and 1987.

Important lignite deposits, for the most part of Eocene age, are known from the Gulf of Mexico basin. Two-thirds of the lignite is found in Texas, but it occurs also in parts of northeastern Mexico, Louisiana, Arkansas, Tennessee, Mississippi, and Alabama. Modest resources of Upper Cretaceous bituminous coal are found in northeastern Mexico.

Once an important industry, the production of sulfur from the caprocks of some of the many salt domes in the U.S. Gulf Coastal Plain and shallow waters of the Gulf of Mexico in Texas and Louisiana, as well as in the Isthmus of Tehuantepec region of southern Mexico, has declined since its peak years during the late 1960s and early 1970s. Sulfur derived from the processing of high-sulfur "sour" oil and gas—referred to as "recovered sulfur"—has to a great extent replaced mined caprock sulfur in the market.

Of interest is that metallic sulfide concentrations have also been reported recently from salt-dome caprocks. They are the result of the interaction of metalliferous formation waters with reduced sulfur produced or trapped in the caprock.

PHOSPHATE

Stanley R. Riggs

INTRODUCTION

Phosphate, a critical world resource

Agricultural resources, followed closely and now intimately by energy resources, are the most basic ingredients of modern society. As world population continues its exponential explosion, we become increasingly more dependent on a high fertilizer- and energy-intensive agriculture. The United States produced 51, 40, and 41 million metric tons of phosphate in 1985, 1986, and 1987, respectively (Stowasser, 1989). Nearly 90 percent of this was utilized for fertilizer and came from Atlantic Coastal Plain deposits in Florida and North Carolina; more than 75 percent was derived from numerous mines in the central and north Florida phosphate districts (Fig. 1 and Plate 4 in pocket). These coastal plain deposits represent one of the most important phosphate resources in the world, accounting for between 34 and 28 percent of total world production between 1983 and 1987 (Stowasser, 1989).

The phosphate resource and reserve potential for the southeastern United States is extremely large. However, mounting land use and environmental pressures cause much speculation about the future availability of land-based phosphate reserves, particularly in Florida. This mounting pressure on an important world deposit is generating increased interest in the deeper and lower grade deposits of other portions of the southeast system, including

Riggs, S. R., Ellison, S. P., Jr., Fisher, W. L., Galloway, W. E., Jackson, M.L.W., and Morton, R. A., 1991, Mineral resources and geopressured-geothermal energy, *in* Salvador, A., ed., The Gulf of Mexico Basin: Boulder, Colorado, Geological Society of America, The Geology of North America, v. J.

the extensive phosphate resources on the Atlantic continental shelf.

Origin of sedimentary marine phosphorites

Most sediments and sedimentary rocks contain less than 0.3 percent P_2O_5. However, periodically through geologic time, phosphorites (5 percent P_2O_5 or greater as defined by Riggs, 1979a) have formed on the sea floor in response to specialized oceanic conditions and have accumulated in sufficient concentrations to form major stratigraphic units of regional extent. Most workers have interpreted phosphate deposition to represent periods of low rates of sedimentation in combination with large supplies of nutrient phosphorus derived through upwelling on a broad, shallow, marine shelfal environment in the low to mid-latitudes. Cold, nutrient-enriched upwelling currents are considered to be prerequisite to supply nutrient elements for production of large volumes of organic matter; low oxygen concentrations at the sediment-water interface are necessary to accumulate organic matter in the sediments (Sheldon, 1981). Phosphorus is then concentrated by bacterial processes at either the sediment-water interface or within interstitial pore waters (O'Brien and others, 1981; Riggs, 1979a, 1980, 1982). This process leads to the primary formation and growth of phosphate grains, which may remain where they formed or be transported as clastic particles within the environment of formation. During subsequent periods of time, some primary phosphate grains may be physically reworked into another sediment unit in response to different environmental processes (Baturin, 1982; Bentor, 1980; Riggs, 1979a, b, 1980).

Such anomalous events happened during a portion of the Miocene, and to a lesser extent during the Pliocene, throughout the southeastern U.S. continental margin to produce one of the most extensive and important phosphate sequences on the Earth's surface. The resulting deposits occur on the emerged coastal plain and are fairly well known geologically. Emerged deposits usually extend seaward under the continental shelf, occasionally cropping out on the sea floor where phosphate grains may be reworked into Holocene sediments.

Phosphates within the Gulf of Mexico basin

Major economic phosphate sediments within the Gulf of Mexico basin are restricted to the Miocene to early Pliocene Hawthorn Group (Riggs, 1979a, b; Scott, 1988) of Florida and southeast Georgia. These extensive phosphorites were deposited on the Florida platform in response to environmental controls and sedimentation processes associated with the southeastern U.S. Atlantic continental margin system; they constitute the Florida Phosphogenic Province (Fig. 1). Extensive drilling and stratigraphic work throughout this province by phosphate companies and other workers (including Altschuler and others, 1964; Cathcart, 1968a, b; Clarke, 1972; Olson, 1966; Riggs, 1979a, b; Scott, 1982, 1983, 1988; Sever and others, 1967; etc.) have established

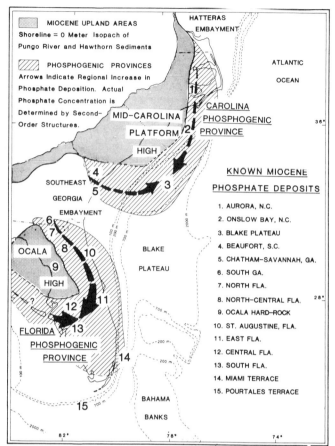

Figure 1. Map of southeastern United States showing the regional distribution of cumulative phosphate formed during the Neogene and location of major phosphate provinces and known deposits (from Riggs, 1984).

that the major and most extensive phosphorites and phosphatic sediments are of Neogene age and occur in the early Miocene to Pliocene Hawthorn Group, with significantly lesser amounts occurring locally in Pleistocene sediments.

The Hawthorn sediments underlie much of the southeastern U.S. Coastal Plain (Fig. 2), extending from the southern tip of Florida, through southeast Georgia, and into southern South Carolina. The Hawthorn Group also extends eastward under the entire Atlantic continental shelf from the offshore Pourtales Terrace (Fig. 1; Gorsline and Milligan, 1963), northward along the Miami Terrace (Mullins and Neumann, 1979) and the Florida, Georgia, and South Carolina continental shelf (Charm and others, 1969; Emery and Uchupi, 1972; Hathaway and others, 1970, 1976; Poag, 1978; Popenoe, 1983; Popenoe and Meyer, 1983). The very broad West Florida Shelf is also probably underlain by phosphatic sediments of the Hawthorn Group (Riggs, 1984); however, little is known about the Tertiary sediments in this shelf area.

Phosphatic sediments do occur locally within a few other stratigraphic units within the Gulf of Mexico basin; however, phosphate concentrations are generally minor with negligible economic potential. Cathcart (1968a, b) summarized the units that contain phosphatic sediments in the Gulf of Mexico basin, other than the Miocene-Pliocene deposits of Florida and Georgia, as follows:

1. Eocene Tallahatta Formation, Georgia (phosphate grains in a quartz sand);

2. Eocene Clairborne Group, Alabama (phosphatic shell marl);

3. Paleocene Midway Group, Texas (phosphate pellets in greensands and marls);

4. Cretaceous Prairie Bluff Formation of Selma Group and equivalents, Georgia, Alabama, and Mississippi (phosphate pellets in chalk); and

5. Cretaceous Eagle Ford Formation, Texas (phosphate conglomerate at base of a shale).

PHOSPHORITES OF THE FLORIDA PHOSPHOGENIC PROVINCE

Regional structural framework

The general patterns of Neogene sedimentation throughout the southeastern U.S. continental margin were controlled and defined by the pre-Neogene structural framework and erosional paleotopography (Fig. 2). Two scales or orders of structural and topographic features controlled the primary formation and subsequent deposition of phosphorites (Riggs, 1981, 1984). First-order, Mesozoic features define the regional setting and determine location, size, and geometry of the phosphogenic provinces. These large-scale features include the Peninsular arch under the Ocala high and Mid-Carolina platform high and adjacent depositional centers, including the South Florida basin, Southeast Georgia embayment, and Baltimore canyon trough (Fig. 2). Maximum concentrations of Neogene phosphate formed and accumulated on the shelf environments around the nose and flanks of these first-order structural platform highs (Riggs, 1984), producing the Florida and Carolina Phosphogenic Provinces, respectively (Fig. 1). Neogene phosphate formation decreased away from the first-order structural noses to minimums in adjacent first-order embayments.

Superimposed upon this regional structural framework are a series of second-order, Cenozoic structural and paleotopographic features that define the various phosphate districts (Figs. 2 and 3). The geographic location of each second-order feature, within the first-order structural framework, determines the amount and type of phosphate formed and type of associated sediment components. The specific size and geometry of each district is dictated by genesis of the basin where the phosphates accumulated (i.e., structural deformation, primary depositional processes, subaerial or submarine erosion, etc.) and post-depositional erosional history of the basin.

The Ocala high is a NW-SE–trending, second-order feature (Fig. 2) with a core of Eocene limestone of the Avon Park

Formation, which controlled the deposition of all Neogene sediments (Riggs, 1979b). Two major second-order sediment basins occur in association with the Ocala high. The southern portion of the Florida Peninsula is occupied by the large Okeechobee basin, which accumulated over 225 m of Hawthorn Group phosphatic carbonate sediments during the Neogene (Freas, 1968). To the north, the Ocala high is terminated in Georgia by the Gulf trough or Suwanee straits (Patterson and Herrick, 1971; Riggs, 1984). Prior to the Neogene, the Gulf trough or Suwanee straits was open between the Gulf of Mexico and the Atlantic Ocean, preventing southward transport of terrigenous sediments to peninsular Florida. However, as the basinal trough filled by the end of the Oligocene, a flood of terrigenous sediments began to move southward, diluting the inner shelf phosphorites of south Georgia and Florida (Riggs, 1979b).

Miocene phosphorites were deposited in association with the topographically or structurally produced second-order features around the perimeter of the Ocala high (Figs. 1 and 3). The southern end of the Ocala high forms the broad and extensive Central Florida platform, which plunges gently into the subsurface (Fig. 3); the crest and adjacent flanks of this structure contain the world's largest producing phosphate districts, the central and

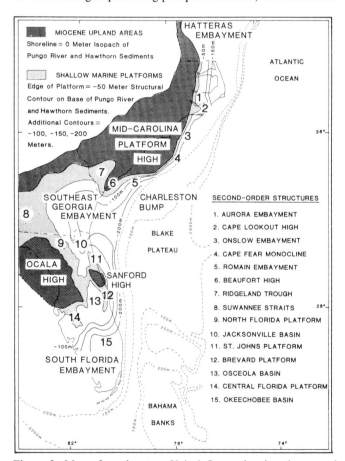

Figure 2. Map of southeastern United States showing the general Miocene shoreline and major first- and second-order structural and paleotopographic features that controlled Neogene phosphate sedimentation (from Riggs, 1984).

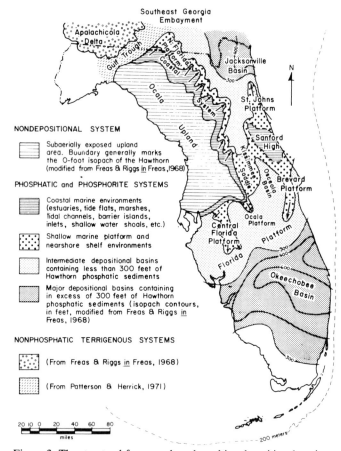

Figure 3. The structural framework and resulting depositional environments for the phosphogenic portion of the Miocene in Florida (from Riggs, 1979b).

south Florida deposits (Fig. 1; Altschuler and others, 1964; Cathcart, 1968a, b). Along the eastern side of the Ocala high is a broad, irregular, and shallow continental shelf system referred to as the North Florida platform (Fig. 3). The southern half of the North Florida platform contains only a thin sequence of Neogene sediments with minor amounts of phosphate due to the shadow effect of the Sanford high on major oceanographic current systems. The northern half of the North Florida platform contains an irregular to moderately thick section of Neogene sediments, which are as much as 25 m thick and contain high concentrations of phosphate within the second-order embayments. Seaward, off the flank of the platform, Neogene phosphatic sands and clays increase to more than 150 m in thickness in the Jacksonville basin (Fig. 3; Freas, 1968; Scott, 1983, 1988).

Another positive element of extreme importance to Neogene sedimentation is the Sanford high (Figs. 2 and 3). This narrow NW-SE–trending, second-order feature is separated from the Ocala high by the southern end of the Jacksonville basin on the northwest, the Kissimmee saddle on the west, and the Osceola basin on the southwest (Fig. 3). Long and narrow platforms plunge off the north and south ends of the Sanford high. These

platforms contain thick, extensive sequences of phosphate-rich sediments that extend into the adjacent Jacksonville, Osceola, and Okeechobee basins. Phosphorites also extend east off the Sanford high and associated platforms onto the Florida continental shelf. Core holes off Jacksonville, Florida, show thick and relatively phosphate-rich downdip sections of the Miocene stratigraphic sequence (Charm and others, 1969; Hathaway and others, 1970).

The western margin of the Ocala high occurs on the West Florida Shelf in the Gulf of Mexico. Little is known about the Tertiary geology in this area; however, the southwestern limit to the Ocala high is probably coincident with the northwest-trending Bahama fault zone, as delineated by Klitgord and Popenoe (1984). Riggs (1984) stated that if Neogene phosphorites exist on the West Florida Shelf, they should have a similar distribution (Fig. 1) and relation to first- and second-order structural and paleotopographic features as elsewhere within the Florida Phosphogenic Province. Birdsall (1978) found minor concentrations of phosphate in the surface sediments throughout much of the West Florida Shelf, suggesting that major Neogene phosphorites might exist in the shallow subsurface.

The occurrence of extensive phosphate pavements and slabs on the upper Miami Terrace with abundant phosphate nodules on the lower Miami Terrace has been described by Mullins and Neumann (1979). They present evidence suggesting multiple episodes of phosphatization associated with deposition of the Hawthorn sediments during the lower to middle Miocene. Similar phosphorites have been described on the Pourtales Terrace (Gorsline and Milligan, 1963). All of these deposits are situated off the continental shelf on the southeast and south nose of the Florida Peninsula (Figs. 1 and 2).

Depositional sequences

The Hawthorn sediments form a wedge-shaped unit that ranges from a feather edge on the flanks of the Ocala and Sanford highs to greater than 225 m in some of the downslope basins. Riggs (1979b) and Scott (1988) subdivided the Hawthorn Group in peninsular Florida into the Arcadia and Peace River Formations on the basis of lithology, stratigraphic distribution, phosphate petrology, and sedimentary structures. These two formations are partially contemporaneous Miocene units and represent the time of major phosphate formation. The Peace River Formation is dominated by terrigenous sediments and constitutes the major phosphorite unit in Florida. It was deposited in complex coastal-marine and inner-shelf environments around the Ocala High and into the northern basins (Fig. 3). The Arcadia Formation is dominantly a dolomite with varying amounts of phosphate and interbedded fine terrigenous sediment deposited as a middle- and outer-shelf facies (Fig. 3). It occurs below and down the depositional slope to the south and east of the Peace River Formation. The Miocene-Pliocene Bone Valley Member of the Peace River Formation contains highly variable concentrations of phosphate. It occurs as complex fluvial facies in the upslope area around the Ocala high and grades downslope into estuarine, open bay, and shallow-marine facies containing decreasing amounts of phosphate and often bearing other stratigraphic names.

The Miocene Arcadia and Peace River Formations of Florida are dominated by three major sediment components (phosphate, carbonate, and terrigenous sediment) that display several significant regional patterns (Riggs, 1979b). Minor concentrations of phosphate formed everywhere within the marine environment, but areas of optimum formation were associated with shallow-water structural platforms that projected onto the continental shelf. The phosphate grains, which were transported as clastic particles, were subsequently deposited and accumulated on the platforms or in adjacent entrapment basins. Some phosphate formed within the major depositional basins where it was significantly diluted by high depositional rates of the terrigenous and carbonate components.

The distribution pattern of terrigenous sediments suggests a major source from the north across the Gulf trough (Fig. 3) and into the shallow-coastal and inner-shelf environments around the Ocala high (Riggs, 1979b). The sands and clays were then transported east into the Jacksonville Basin and south along the Ocala and Sandford highs, with only minor amounts moving into southern Florida. The regional distribution pattern of the carbonate component, which is predominantly dolosilt, is opposite that of the terrigenous sediments. The carbonate is an authigenic/diagenetic sediment formed in the middle- to outer-shelf marine environments where it is the dominant component. Carbonate rapidly decreases in abundance into the inner-shelf environments around the Ocala and Sanford highs where it becomes a subordinate component.

In addition to the regional distribution patterns within the Hawthorn Group, Riggs (1979b) described a highly interbedded lithologic character to these sediments. This suggests a strong cyclical nature resulting from alternating sequences of chemical and terrigenous sedimentation through time. Major pulses of terrigenous sedimentation from the north and carbonate sedimentation from the south and east interact with the processes of phosphate sedimentation within shallow-shelf environments to produce the complex Hawthorn stratigraphy. Thus, the quantitative importance of phosphate, terrigenous, and carbonate components at any point in time is (1) dependent on the regional location with respect to the major structural elements and sediment sources, and (2) a function of the distinctive pulses of both chemical and terrigenous sedimentation through time. Details of the cyclic patterns have not yet been worked out, but they do exist and appear to be similar to those described in the Pungo River Formation in North Carolina (Riggs, 1984; Riggs and others 1982, 1985).

Some phosphate formed in most portions of the continental margin around the Ocala and Sanford highs; however, there are definite regional patterns of the different phosphate grain types within the Hawthorn sediments (Riggs, 1979a, b, 1980). Phosphate grains on the inner-shelf are dominantly medium to fine

sand-sized intraclasts (rounded but irregularly shaped grains). Locally, laminae and beds of microcrystalline phosphate mud (microsphorite) and phosphate intraclast gravels occur on the shallow inner-shelf platforms associated with hiatus surfaces characterized by sediment bypass around the nose of structural platforms (i.e., the Central Florida platform in Fig. 3). In outer-shelf and deeper basinal environments, phosphate is dominated by well-sorted, fine to very fine sand-size peloids (very regularly shaped geometric grains). The dominant phosphate component on the Blake Plateau and Miami Terrace (Fig. 2) consists of thick and extensive pavements of indurated microcrystalline phosphate mud and coarse intraclastic plates and pebbles of pavement material (Manheim and others, 1980; Mullins and Neumann, 1979).

DEPOSITIONAL MODEL FOR NEOGENE PHOSPHORITE SEDIMENTATION

The Neogene stratigraphic sequence of the southeastern United States is characterized by complex depositional patterns with significant variations in (1) net volume and concentration of phosphate, (2) phosphate grain types, (3) associated carbonate and terrigenous components, and (4) stratigraphic cycles, including deposition and erosion and changing lithologic patterns (Riggs, 1979b). Riggs (1984) related these sedimentologic patterns to the interaction of at least four environmental parameters operating through the last 20 m.y.:

1. size, shape, and structural framework of the continental shelf depositional system;

2. intensity, stability, and duration of the oceanographic conditions and associated climates that control the physics, chemistry, and biology of the shelf environment (i.e., upwellings, boundary currents, and eustatic sea-level oscillations associated with glaciation-deglaciation);

3. amount and rate of diluent terrigenous and carbonate sedimentation resulting from changing climatic and oceanographic conditions; and

4. additional secondary factors (i.e., subsequent history of deposition, structural deformation, erosion, burial, diagenetic history, and weathering of the sediments), which severely modify the stratigraphic unit and included phosphate sediment.

RESOURCE POTENTIAL

Phosphate resources

Since initial phosphate mining began in the Central Florida Phosphate District in the 1880s, there have been innumerable estimates of the phosphate resources within the Florida Phosphogenic Province. Cathcart and others (1984) have summarized the latest estimates that include those of Zellers and Williams (1978), Fountain and Hayes (1979), and Mayberry (1981), as well as their own estimates. Table 1 summarizes the resource information developed by Cathcart and others by economic resource category (as defined by the U.S. Geological Survey, 1980) and by geographic region within Florida.

TABLE 1. SUMMARY OF PHOSPHATE RESOURCES OF FLORIDA, ESTIMATED BY VARIOUS AUTHORS*

| | Zellers and Williams, 1978 | | Fountain and Hayes, 1979 | | Mayberry, 1981 | | Cathcart and others, 1984 | |
	Phosphate Conc. (million tons)	P$_2$O$_5$ (million tons)	Phosphate Conc. (million tons)	P$_2$O$_5$ (million tons)	Phosphate Conc. (million tons)	P$_2$O$_5$ (million tons)	Phosphate Conc. (million tons)	P$_2$O$_5$ (million tons)
Identified Economic Resources[†]								
North Florida	845	260	1,056§	320			900	280
Central Florida	780	250	884	280	1,270	400	800	260
South and East Florida	1,000	300	3,052	920			2,000	600
Identified Marginally Economic Resources[†]								
North Florida	404	120	512§	160			450	140
Central Florida	25	10	31	10	1,820	560	20	tr.
South and East Florida	527	160	1,600	490			1,000	300
Identified Subeconomic Resources[†]								
North Florida	20	tr.	33§	tr.§			20	tr.
Central Florida	32	tr.	10	tr.	1,270	390	10	tr.
South and East Florida	166	50	500	150			400	120
Total Identified Resources	3,979	1,210	7,650§	2,240	9,060	2,770	5,600	1,700
Average Percent P$_2$O$_5$	30.4		30.6		30.6		30.3	
Total Hypothetical Resources**								
North Florida							900	250
Northeast Florida							650	200
South Florida							1,425	400
East Florida							2,200	650
Total Florida	180	60	No Data		4,700	1,420	5,175	1,500
Grand Total Resource	4,159	1,270	7,650	2,240	13,760	4,190	10,775	3,200

*Modified from Cathcart and others, 1984.

[†]Identified Resources = Resources whose tonnage and grade have been determined by drilling and chemical analysis and in which the phosphate product can be recovered using existing technology (Cathcart and others, 1984). The basis for subdividing identified resources into economic, marginally economic, and subeconomic categories is discussed in Cathcart and others (1984).

§Resource numbers include South Georgia.

**Hypothetical Resources = Resources for which drilling information is sparse (less than one drill hole per section), only lithologic logs of drill holes are available, or where geologic inference indicates that deposits are present (Cathcart and others, 1984).

tr. = trace.

Identified phosphate resource estimates (Table 1) range from a low of about 4 billion metric tons of phosphate concentrate to a high of more than 9 billion metric tons, which are either currently economic, marginally economic, or subeconomic (Cathcart and others, 1984). Mayberry (1981) and Cathcart and others (1984) are the only authors whose estimates realistically consider the hypothetical resource potential within Florida; both papers estimate about 5 billion additional metric tons of phosphate concentrate.

It should be noted, however, that even these hypothetical estimates do not include most of the vast amounts of "deep" phosphate that occurs in discrete beds within deeper sediments of the Okeechobee, Osceola, and Jacksonville basins (Fig. 3; Marvasti and Riggs, 1989; Riggs, 1986b), nor any potential resources that occur on the continental shelf (Marvasti and Riggs, 1987; Riggs and Manheim, 1988). Extrapolating on the basis of the minimal data that do exist for these areas, it is safe to conclude that "speculative resource" estimates would probably increase the potential phosphate resources by at least an order of magnitude. Of course, none of this hypothetical or speculative resource is available with conventional mining. However, it does represent a future potential resource that will become available as shallow reserves are depleted and, more important, as competing land-use and environmental pressures increasingly conflict with present surface-mining techniques (Marvasti and Riggs, 1989; Riggs, 1986b). It will require new mining technology, such as slurry

TABLE 2. ESTIMATED URANIUM RESOURCES OCCURRING IN THE
PHOSPHATE DEPOSITS IN FLORIDA AND SOUTHERN GEORGIA*

Phosphate District	Total Uranium Resource (m tons)	Recoverable Uranium (m tons)	Average U in Phosphate Concentrate (ppm)
Central Florida†§	634,000	124,000	110
South Florida†§	931,000	336,000	110
North Florida-South Georgia†	1,223,000	123,000	70
East Florida	1,305,000	272,000	102
Totals	4,093,000	855,000	

*From Fountain and Hayes, 1979.
†Phosphate-producing districts.
§Uranium-producing districts.

mining procedures currently under development, to ever realize this vast potential resource.

Byproduct resources

Sedimentary phosphates of the Hawthorn Group are uncommonly rich in numerous elements (Altschuler, 1980; Altschuler and others, 1958, 1967; Cathcart, 1956, 1978) that may be of economic significance in two different ways. First, some of the elements may themselves have important economic byproduct potential. Second, abundances of other elements may have significant impacts on the environment and human health through the wastes of processing or through the use of fertilizer products. These latter effects are complex and poorly known; some of them may be good and some bad. Thus, in some cases it may be beneficial to recover certain elements to alleviate the environmental and human health risks or to add to the resource value.

Altschuler (1980) listed 11 elements occurring in world phosphates that are significantly enriched compared to average marine shales (Cd = 60X; U = 30X; Ag = 30X; Y = 10X; Se = 8X; Yb = 5X; Mo = 4X; La = 4X; Sr = 2X; Pb = 2X; and Zn = 2X). Many of these elements, including uranium and most of the rare earth elements, occur in anomalous concentrations in most phosphate deposits in the southeastern United States and warrant further investigations as potential byproduct resources.

The Central Florida Phosphate District is presently a major uranium producer as byproduct recovery from phosphoric acid production. In 1980, the district had a total annual recovery capacity of over 3 million pounds of U_3O_8 (Reaves, 1984). The substantial but diminishing uraniferous phosphate resource of central Florida is expected to be depleted gradually over the next 15 to 25 years, according to Fountain and Hayes (1979). However, they established an extensive uranium potential within the phosphate resources of Florida and southern Georgia (Table 2). The 4.09 million metric tons of uranium resource represents many billions of tons of potential phosphate concentrate without any recovery factors. The actual amount of this resource that is potentially recoverable is totally dependent on many rapidly changing economic and technological factors.

LIGNITE AND COAL

William L. Fisher and Mary L. W. Jackson

INTRODUCTION

Lignite and coal deposits are extensive in the northwestern flank of the Gulf of Mexico basin. The lignite province extends some 1,800 km from northeast Mexico to Georgia (Fig. 4 and Plate 4, in pocket). In northeastern Mexico, lignite and coal deposits are present in the states of Chihuahua, Coahuila, Nuevo Leon, and Tamaulipas and are Tertiary and Late Cretaceous in age (Flores Galicia, 1988). In the United States, the province contains predominantly lignite and includes portions of nine states: Texas, Louisiana, Arkansas, Missouri, Kentucky, Tennessee, Mississippi, Alabama, and Georgia. Lignite in these states occurs principally in the Paleocene-Eocene Wilcox Group and the Eocene Claiborne and Jackson Groups; less extensive deposits occur in the Paleocene Midway Group (Table 3). Gulf Coast lignites and coals formed in swamps and marshes principally associated with fluvially dominated delta systems, their associated updip fluvial systems, and their laterally associated strandplain lagoonal systems (Kaiser, 1978; Ayers and Lewis, 1985; Finkelman and Casagrande, 1986; Kaiser and others, 1986; Finkelman and others, 1987).

The quality and character of Gulf of Mexico basin lignites commonly vary significantly even within individual mining areas and seams. However, meaningful regional trends and variations have been correlated with depositional environment, and to a lesser extent, with age and geography. Both Cretaceous and Tertiary coals in Mexico are bituminous C in rank. The best quality Gulf Coast lignites occur in fluvial and deltaic systems of the Wilcox Group, classed as Lignite A using the ASTM coal classification. The generally poorer quality lignites, or Lignite B, occur in Wilcox strandplain-lagoonal systems, as well as in fluvial, deltaic, and lagoonal systems of the Claiborne and Jackson Groups (Luppens, 1979; Tewalt, 1986). In general, lignite quality decreases across the northern part of the Gulf of Mexico basin from west to east (Table 4).

Near-surface resources of Gulf of Mexico basin coals and lignites are estimated at 31,450 million t (Table 5). Two-thirds of the estimated resource exists in Texas, and the balance is more or less evenly distributed in the remaining part of the southeastern United States and northwestern Mexico. Historically, lignite resource estimates have tended to increase as data bases expanded and understanding of the regional geology improved.

NORTHEASTERN MEXICO

Tertiary lignite and coal deposits of potential economic value are known in the Burgos basin in the northeastern Mexico

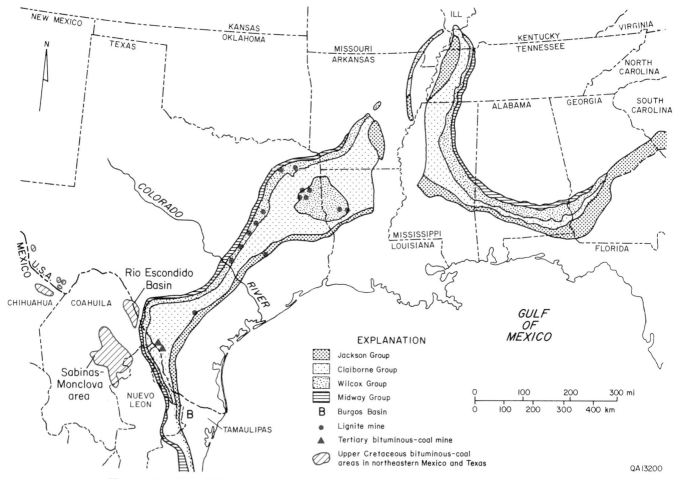

Figure 4. Location of Tertiary lignite-bearing outcrops in the northwestern flank of the Gulf Coast. Coal-bearing regions in Mexico (Flores Galacia, 1988), and U.S. bituminous-coal and lignite mines (Kaiser, 1985; Friedman and others, 1988) are also shown.

states of Nuevo Leon and Tamaulipas (Fig. 4). The lignite and coal occur principally in the Eocene Bigford Formation and El Pico Clay of the Claiborne Group (Flores Galicia, 1988). Bituminous-C coal beds as much as 1.5 m thick are present in the Wilcox Group. In the Claiborne, proved reserves of 76 million t and a total "resource" of 252 million t have been estimated in beds thicker than 80 cm and buried at a depth of less than 40 m (Flores Galicia, 1988). No mining operations have yet been undertaken in these deposits.

In Upper Cretaceous rocks, commercial coal deposits occur in Coahuila; coal of potential economic value is known in beds of the same age in Chihuahua, Nuevo Leon, and Tamaulipas. Most of the coal deposits are in fluvial beds of the Olmos Formation and have been mined since the turn of the century in both underground and open pit mines. The coal is bituminous C in rank. The coal-bearing section is 13 to 30 m thick, and contains up to 15 coal seams that range in thickness from 2 cm to 2 m.

Coal resources in northeastern Mexico have been estimated for the Upper Cretaceous seams of the Sabinas-Monclova area and Rio Escondido basin (Fig. 4; Ojeda-Rivera, 1978; Flores

Galicia, 1988). Proved reserves in the Sabinas-Monclova area were estimated in 1985 at 1,256 million t, probable and possible reserves at 120 million t, and "resources" at 1,180 million t, for a total of 2,556 million t (Flores Galicia, 1988). The coal in the Sabinas-Monclova area is used principally to manufacture metallurgical coke. Production from this area in the mid-1970s was about 4.25 million t, and plans called for production of 10 million t by 1980 (Ojeda-Rivera, 1978). Through 1973, the region had produced about 110 million t. More recent production statistics are not available. In the Rio Escondido basin area (Fig. 4), proved coal reserves have been estimated at 600 million t, probable plus possible reserves at 116 million t, and "resources" at 500 million t (Flores Galicia, 1988). The coal from the Rio Escondido basin area is used locally as fuel for electric-power generation.

TEXAS

Near-surface lignite resources in Texas total 21,200 million t and occur in the Wilcox and Jackson Groups and in the Yegua

TABLE 3. STRATIGRAPHIC OCCURRENCE OF TERTIARY GULF COAST LIGNITES
(adapted from Luppens, 1979)

			South Texas and Northeast Mexico	Central-East Texas		Louisiana	Arkansas	Tennessee	North-Central Mississippi	East-Central Mississippi	West Alabama
Tertiary	Eocene	Jackson Group	upper	Whitsett		Yazoo	Undivided	Undivided	Undivided	Yazoo	Yazoo
			middle	Manning*							
				Wellborn							
			lower*	Caddell		Moodys Branch				Moodys Branch	Moodys Branch
		Claiborne Group	Yegua Laredo	Yegua* Cook Mountain		Cockfield Cook Mountain	Cockfield* Cook Mountain	Undivided*	Cockfield* Cook Mountain	Cockfield* Cook Mountain	Gosport
			El Pico Clay	Sparta Sand Weches		Sparta Sand	Sparta Sand*		Zilpha	Zilpha	Lisbon
			Bigford	Queen City Reklaw		Cane River	Cane River		Winona	Winona	
			Carrizo Sand	Carrizo Sand		Carrizo Sand	Carrizo Sand		Tallahatta	Tallahatta	Tallahatta
	Paleocene	Wilcox Group	upper	Calvert* Bluff	(East) upper*	Sabinetown Pendleton* Marthaville Hall Summit* Lime Hill Converse	Undivided*	Undivided*	Undivided*	Hatchetigbee	Hatchetigbee
			middle	Simsboro						Tuscahoma*	Tuscahoma*
			lower*	Hooper*	lower*	Cow Bayou Dolet Hills* Naborton*				Nanafalia	Nanafalia
		Midway Group	Wills Point	Wills Point		Porters Creek	Porters Creek	Porters Creek	Naheola	Naheola*	Naheola*
			Kincaid	Kincaid		Kincaid/ Clayton	Kincaid	Clayton	Porters Creek Clayton	Porters Creek Clayton	Porters Creek Clayton

* Principal occurrences of lignite

QA 13201

Formation of the Claiborne Group (Kaiser and others, 1980). The lignite-bearing strata crop out as three bands parallel to the present coastline and in a circular area associated with the Sabine uplift along the Texas-Louisiana border. For delineation of resources, these geologic strata have been divided geographically into seven units (Table 6). Ninety-two percent of the total resource tonnage occurs northeast of the Colorado River in east Texas. Geologically, 71 percent of the total tonnage is in the Wilcox Group, and the remainder is mostly in the Jackson Group. On a Btu basis, the Wilcox Group accounts for 77 percent of the state's total (Kaiser and others, 1980).

Coal rank in Texas varies from Lignite B in the Claiborne and Jackson Groups to Lignite A in the Wilcox Group in east Texas. Analyses of Wilcox "lignite" in the Sabine uplift area have shown that it has a rank of subbituminous C (Mukhopadhyay, 1989; Tewalt, 1986).

The thickest and most laterally continuous near-surface lignite seams in Texas are located in Wilcox fluvial-deltaic deposits of central Texas and the southern half of the Sabine uplift area and in deltaic deposits of the Jackson in east Texas (Fig. 4). In these two regions, individual seams have been identified that are continuous for up to 32 and 48 km, respectively. Typical seam thicknesses are 0.6 to 3 m, but may locally exceed 4.6 m. Re-

source blocks of up to 900 million t are present in the Jackson and Wilcox of east Texas. The next most important resource areas are the fluvial deposits of the northeast Wilcox and northern Sabine Uplift areas. Although seams are generally thinner than those of the Wilcox and Jackson of east Texas (commonly less than 2.4 m), resource blocks of up to 450 t are present. Jackson Group deposits southwest of the Colorado River are smaller than those to the northeast; individual deposits of up to 270 million t are indicated. Lignites also occur in the Yegua Formation northeast of the Colorado River but are thinner (commonly less than 1.8 m) and more discontinuous; resource block sizes in this region are estimted at 135 million t (Kaiser and others, 1980).

In addition to the near-surface resources, there are large resources in seams greater than 1.5 m thick deeper in the basin, down to a depth of 610 m. An estimated 25 billion t of resource exists in the deep basin and may be partially recoverable using in-situ gasification, deep surface mining, or conventional underground mining. Of this total tonnage, 60 percent is in the Wilcox, and 40 percent is in the Jackson (Kaiser, 1989).

Lignite production in Texas began in the 1880s, increased to 1.3 million t in 1914, and then declined again through the 1930s and 1940s with the advent of oil as an energy source

TABLE 4. TYPICAL AS-RECEIVED ANALYSES OF
GULF COAST LIGNITES*

State and Region	Moisture (%)	Ash (%)	Volatile Matter (%)	Fixed Carbon (%)	Sulfur (%)	Btu/lb
Texas						
Central Wilcox	32	15	29	24	1.0	6,593
Northeast Wilcox[†]	33	15	27	25	0.9	6,465
East Jackson	41	21	23	15	1.3	4,729
South Jackson	23	42	22	13	1.9	3,972
East Claiborne	37	18	26	19	1.0	5,761
Louisiana						
Wilcox	30	12	29	29	0.7	7,073
Arkansas						
Southwest Wilcox	34	15	26	25	0.5	6,171
Northeast Wilcox	42	13	25	20	0.6	5,465
Southwest Claiborne	37	11	31	21	1.0	6,733
Tennessee						
Claiborne	45	12	27	16	0.6	5,379
Mississippi						
North Wilcox	44	12	25	19	0.5	5,396
South Wilcox	43	12	24	21	1.2	5,509
Claiborne	42	14	30	14	0.5	5,855
Alabama						
West Midway	46	8	24	22	2.7	5,738
Central Wilcox	50	9	22	19	2.1	5,156

*Texas data from Kaiser and others (1980), all other data from Luppens (1979).
[†]Includes Sabine Uplift.

TABLE 5. NORTHWESTERN GULF OF MEXICO BASIN
COAL AND LIGNITE RESOURCES

	Resources* (million t)
Mexico (coal)	
Burgos Basin	250
Sabinas–Monclova area	1,180
Rio Escondido Basin	500
United States (lignite)	
Texas	21,200
Louisiana	900
Arkansas	2,250
Tennessee	900
Mississipi	4,500
Alabama	1,800
Total	31,450

*U.S. data from White and others (1982). Mexico data from Flores Galicia (1988). In the U.S., estimated in-place tonnage is for seams ≥9.0 m thick to 61 m deep.

(Fisher, 1963). Production has more recently increased from about 2 million t in 1970 to 47.2 million t in 1988 (Fig. 5). Currently, 13 mines, all open-pit operations, are in production (Fig. 4). Nearly all the lignite produced in these mines is used for electric-power generation in plants near the mines. Lignite-burning power plants generate approximately 18 percent of the electricity consumed in Texas. One mine, however, has produced lignite for the manufacture of activated carbon since the early 1930s.

Cannel coal has been produced for many years from two mines in south Texas (Evans, 1974; Fig. 4). Production is from the Bigford and El Pico Formations of the middle Eocene Claiborne Group. Production reached a maximum of 341,000 t in 1984 but has decreased since then to 274,000 t in 1988 (Railroad Commission of Texas, 1989).

In 1989, Texas ranked sixth in the United States as a coal-producing state. Approximately 99 percent of the production in Texas is Wilcox and Jackson lignite. The remainder is south Texas cannel coal.

LOUISIANA

Lignite resources of commercial significance in Louisiana are restricted to lower Wilcox Group strata on the Sabine uplift (Table 3). As in Texas, these sediments are fluvial-deltaic in origin (Arthur, 1985). The Louisiana Geological Survey (Johnston and Mulcahy, 1982) has estimated lignite resources at 900 million t; however, because of limited availability of data in some areas of known lignite occurrence, the actual resource number may be higher by at least 450 million t. Most seams are less than 2.4 m thick. The Chemard Lake Lentil in De Soto Parish covers about 207 km², is continuous for over 24 km, and contains in excess of 450 million t of resources (Roland and others, 1976). Other potentially commercial deposits occur in Red River, Natchitoches, Bienville, Caddo, and Bossier Parishes.

The Dolet Hills surface mine in De Soto Parish (Fig. 4) started operation in 1985; annual production reached 2.6 million t in 1988. In November 1989, the Oxbow mine opened in nearby Red River Parish. Both mines provide fuel for an electric power plant in De Soto Parish.

ARKANSAS

Lignites occur in southwest, south-central, and northeast Arkansas in both the Wilcox and Claiborne Groups (Woodward and Gilreath, 1976; Prior and others, 1985; Table 3). Wilcox lignites were deposited in fluvial environments, whereas Claiborne lignites were deposited in deltaic environments; ASTM rank of both is Lignite A (Williamson, 1986). On the basis of extensive drilling conducted by the Arkansas Geological Commission, total

TABLE 6. TEXAS NEAR-SURFACE LIGNITE RESOURCES
BY REGION

Region	Resources* (million t)
Central Wilcox	5,878
Northeast Wilcox	4,625
Sabine Uplift Wilcox	3,548
South Wilcox	978
East Jackson	4,080
South Jackson	687
East Claiborne	1,406
Total	21,202

*Data from Kaiser and others (1980).

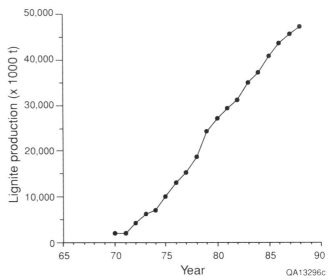

Figure 5. Lignite production in Texas since 1970. Data from the Railroad Commission of Texas (1989).

lignite in place (irrespective of seam thickness) to a depth of 46 m is estimated at 12 billion t (Holbrook, 1980). Phillips Coal Company has estimated resources at 2,250 million t in seams greater than 0.9 m thick at depths up to 61 m (Luppens, 1979). Most seams are less than 2.1 m thick. Resource blocks of up to 495 million t are indicated (Hamner, 1978). No commercial mining of lignite has yet been undertaken in Arkansas.

TENNESSEE, MISSISSIPPI, AND ALABAMA

Lignites located in western Tennessee total an estimated 900 million t and are found in the Wilcox and Claiborne Groups. Seam thicknesses are generally less than 2.7 m (Luppens, 1979). No information on the size of individual deposits is available.

Mississippi lignite resources occur principally in the Wilcox and Claiborne Groups (Williamson, 1976; Cleaves, 1980; Table 3). Total resources are estimated at 4,500 million t and occur mainly in the fluvial and delta-plain sequences of the Wilcox Group (Luppens, 1979; Williamson, 1986). Of principal significance in the counties of east-central Mississippi, from Calhoun southeast through Lauderdale, are Wilcox lignites typically up to 2.7 m thick and in resource blocks of up to 450 million t. Lignite resources in the Claiborne are less extensive and thinner (typically less than 2.1 m) than in the Wilcox; however, resource blocks of up to 450 million t are indicated in Quitman County and to the south. In the area of Lauderdale County, several lignite seams are also present in the Midway Group as an extension of lignites in southwestern Alabama.

Total lignite resources in Alabama are estimated at 1,800 million t (Daniel, 1973). More than one-half of these resources are located in western Alabama in the Naheola Formation (Oak Hill Member) of the Midway Group. Deep-basin resources of more than 4 billion t may also be present in this member (Williamson, 1986). The Oak Hill Member consists of a single seam of 0.3 to 4.3 m in thickness that can be followed from the Mississippi-Alabama state line southeasterly into Wilcox County,

a distance of approximately 120 km. Lesser lignites occur in Pike, Coffee, and Crenshaw Counties of southeastern Alabama in the Nanafalia and Tuscahoma Formations of the Wilcox Group (Tolson, 1985). Unique occurrences of lignite in this area are highly lenticular, thick (as much as 15 m) deposits formed in depressions produced by the collapse of caverns in the Clayton Formation limestone of the Midway Group. These deposits rarely exceed 1.6 km in width and are of limited commercial significance. Thin, lenticular seams also occur in the middle Wilcox Group.

The lignite deposites of western Tennessee, Mississippi, and Alabama, as well as resources in Kentucky, Missouri, and Georgia, although identified, have not been mined.

SULFUR

Samuel P. Ellison, Jr.

INTRODUCTION

Native sulfur has been produced in the Gulf of Mexico region since 1894 when sulfur deposits were discovered in the cap rock of the Sulfur Mines Dome, Calcasieu Parish, southwestern Louisiana. The first underwater mine, in Lake Peigneur (Jefferson Island Dome), also in Louisiana, started production in 1932. Herman Frasch employed superheated steam and compressed air to melt the sulfur in place and flow it through wells to the surface as a hot amber to brown liquid. Upon reaching the surface of the ground the liquid cooled into solid yellow sulfur and was ready for shipment. In recent years, most sulfur has been shipped as liquid, with only a minor amount shipped as solid. The Frasch method of extraction through wells was also used at Bryan Mound, Brazoria County, Texas, to initiate Frasch mining in Texas in 1912. Similarly, the same mining method was used in

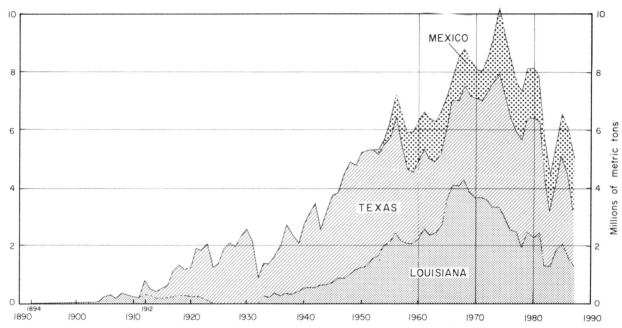

Figure 6. History of Frasch sulfur production in the Gulf of Mexico basin region. From U.S. Bureau of Mines (1931 to 1988), Consejo de Recursos Minerales, Mexico (1964 to 1988), Haynes (1959), Myers (1968b), and Bodenlos (1973).

1953 at San Cristobal, state of Veracruz, Mexico. Chronology of the production of sulfur in the Gulf of Mexico basin region through 1987 is shown in Figure 6.

Thirty-six salt-dome cap rocks have produced native sulfur since 1894; 29 in the United States and eight in Mexico. Of these, 25 were active in 1969, but by 1989 only four of them were still in operation in the United States (Caminada, Garden Island Bay, and Grande Isle in Louisiana, and Boling in Texas), and four in Mexico (Jaltipan and Texistepec, in production for a number of years, and Otapan and Amexquite [Mezquital] from which production started in 1989) (Fig. 7, and Plate 4). The remaining salt-dome cap rocks have been abandoned or shut down because of depletion or not being economical to operate. Additional salt plugs and salt intrusions are known to have cap rocks that contain sulfur but have not been mined. In 1988, sulfur in the cap rock of a salt dome was found east of the Mississippi River Delta, the first commercial discovery in 30 years and apparently one of the region's largest. No doubt, future exploration will find more cap rocks with native sulfur both onshore and offshore.

Discovery of sulfur in the cap rock of the Challenger Knoll in 3,572 m of water in the Sigsbee Plain of the Gulf of Mexico (Fig. 7; Ewing and others, 1969; Davis and Kirkland, 1979) suggests that cap rocks can be made by the dissolving action of sea water as well as by meteoric water or connate water. At the Challenger Knoll the drill ship penetrated 143.9 m into the ocean floor and found characteristic cap rock material composed of anhydrite, gypsum, calcite, and sulfur. A portion of the calcitic core had 19 percent native sulfur. This also suggests that many of the offshore deeply buried salt masses could carry sulfur if cap rocks occur.

Important volumes of sulfur have also been produced in both the United States and Mexican parts of the Gulf of Mexico basin region from "sour" crude oils and natural gas—commonly referred to as "recovered sulfur." Production of recovered sulfur in the United States, which in the mid-1960s amounted to only about 1 to 1.3 million metric tons per year (half of it in Texas) and represented 10 to 13 percent of the total domestic production of sulfur in all forms, has increased steadily to more than 6 million metric tons per year in 1987, two-thirds of the total domestic sulfur production. In Mexico, similarly, production of recovered sulfur has increased from less than 5 percent of the total sulfur production in the mid-1970s, to 25 percent in the late 1980s. Shelton (1980) expected that the volume of recovered sulfur will increase to 75 percent of the total U.S. sulfur production in the not too distant future.

This rapid incease in the recovery of sulfur as a nondiscretionary byproduct of the production and refining of "sour" oil and natural gas is mainly responsible for the marked decrease in the last few years of the exploration for and production of sulfur from salt-dome cap rocks (Fig. 6). The future level of these exploration and production activities will depend on the volumes of "sour" oil and natural gas produced by the petroleum industry (to an extent dependent on the price of these products) and on the demand for sulfur, an important feedstock in the manufacture of fertilizers, sulfuric acid and other chemicals, medicines, paper, and many other products.

GEOLOGY

Salt domes with sulfur in their cap rocks are known in an east-west belt on land near the coast and immediately offshore in the northern part of the Gulf of Mexico basin (Fig. 7) (Halbouty, 1979). These occurrences extend from western Alabama (the McIntosh Dome) westward across southern Louisiana and into Texas, with the most westerly site at Palangana Dome, Duval County, Texas. No general structural pattern can be deduced from the locations of these sulfur-bearing cap rocks. The concentration of sulfur occurrences in the Isthmus of Tehuantepec, Mexico, is in the most westerly part of the Isthmus Saline basin (Fig. 7; Salas, 1988). Salt domes in the three interior salt basins (East Texas, North Louisiana, and Mississippi salt basins) are known to have cap rocks, but no occurrence of sulfur has been reported from any of them.

The geologic setting of native sulfur in the cap rocks of salt domes and salt intrusions is ideally shown in a cross section of the Sulfur Mines Dome, Calcasieu Parish, Louisiana (Fig. 8). Here

the sulfur mass occurs across the top of the entire salt dome immediately above the anhydrite cap. The native sulfur is in both the very porous calcitic limestone and the borderline anhydrite. However, sulfur deposits may not be continuous across the cap rock of a dome, but may be patchy and irregular as in the cap rocks of Hoskins Mound, Brazoria County (Fig. 8) and Boling Dome, Wharton County, both in Texas, and Jefferson Island Dome, Iberia Parish, Louisiana (Fig. 9).

Greater irregularity of the sulfur deposits occurs in the Mexican domes at Amexquite and Jaltipan (Fig. 10). These domes show more than one zone of concentration of sulfur, and apparently local faulting of the cap rock creates an irregular distribution of the sulfur.

Thicknesses of sulfur in the cap rock have a wide range, from a few millimeters to more than 220 m. Similarly, the depths

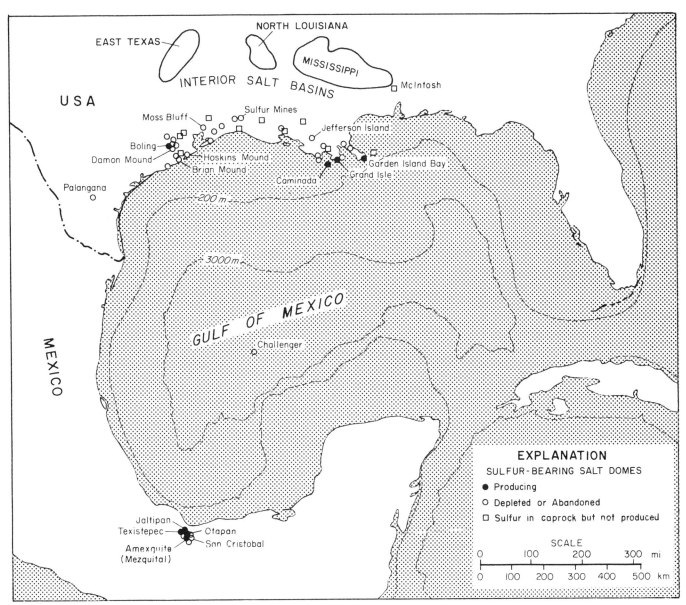

Figure 7. Sulfur-bearing salt domes in the Gulf of Mexico basin region.

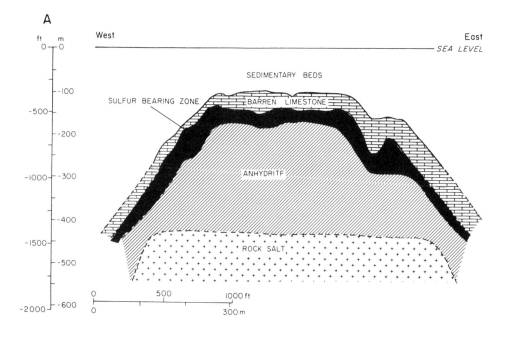

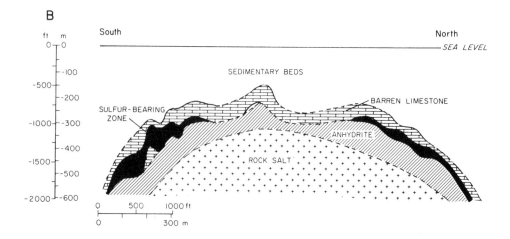

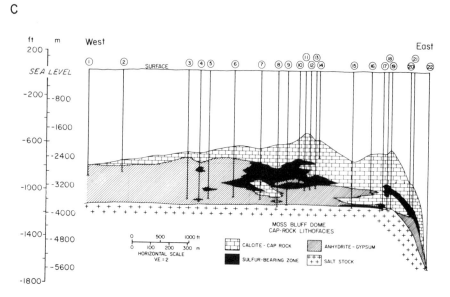

Figure 8. A, Sulfur Mines dome, Calcasieu Parish, Louisiana (from Myers, 1968a, after Kelley, 1925); B, Hoskins dome, Brazoria County, Texas (from Myers, 1968a, after Marks, 1936); C, Moss Bluff dome, Liberty County, Texas (from Seni, 1987).

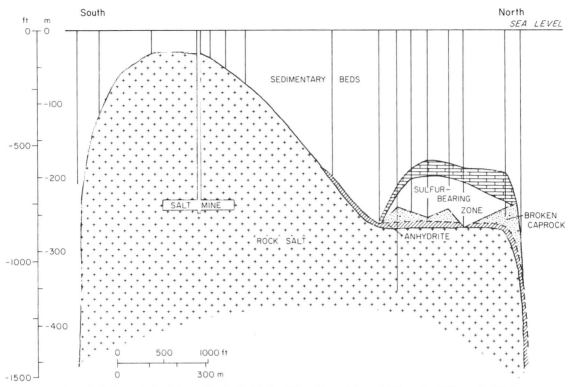

Figure 9. Jefferson Island dome, Iberia Parish, Louisiana (from Myers, 1968a, after O'Donnell, 1935).

to the sulfur-bearing portions of the cap rocks in producing domes range from near the surface to more than 732 m. The average depth to the sulfur deposits in producing domes onshore is 297 m. The depth below sea level to sulfur production may increase if the sulfur occurrence on Challenger Knoll ever becomes commercial.

Normally, a cap rock covers a salt dome that reached upward into sediments where waters dissolved the halite and initiated the development of an insoluble-residue cap rock consisting principally of anhydrite. The size of the cap rock depends on the size of the salt intrusion. The largest salt dome currently being mined for sulfur is Boling Dome, Wharton County, Texas (Wolf, 1933); it is approximately 11 km across in an east-west direction and covers more than 2,098 hectares (5,182 acres). Only a portion of this dome has a sulfur-bearing caprock, and the sulfur deposits extend a small distance down the flanks of the dome. The Jaltipan Dome in Mexico has a comparable size. The dimensions of other producing or abandoned sulfur-bearing domes range from 9 km to less than a kilometer in diameter and 1,200 to 30 hectares in area. The Sulfur Mines Dome, Calcasieu Parish, Louisiana, (Fig. 8A) is among the smallest, with a diameter of 1.4 km and an area of 34 hectares.

The salt masses in Mexico have much gentler flanks than those in the northern Gulf of Mexico basin region and appear to be salt intrusions of an irregular shape rather than the ideal pipelike pillars in the U.S. Gulf Coast area (Contreras V. and Castillon B., 1968; Salas, 1988; Fig. 10).

The mineral composition of salt-dome cap rocks is variable, but most of the producing cap rocks exhibit three general mineralogical zones (Taylor, 1938): (1) an upper calcite zone, (2) a middle transitional zone with gypsum and sulfur, and (3) a lower anhydrite zone.

These zones are not layered in the typical stratigraphic sense but generally do exist in this order and may grade from one to another. The anhydrite cap rocks commonly are well banded as a result of cycles of dissolution and cap-rock accretion. This underplating mechanism results in an inverted anhydrite cap-rock stratigraphy (Kyle and others, 1987). The basal contact between the anhydrite and the salt is sharp. Any of the cap rocks may be granulated, brecciated, and sheared. They may contain fragments of adjacent formations, and all zones may extend down the flanks of the salt plug.

The calcite zone, commonly the upper zone, may be thicker toward the edges of the salt plug. In addition to calcite, it may contain dolomite, celestite, strontianite, barite, quartz, native sulfur, pyrite, marcasite, pyrrhotite, sphalerite, and galena. The upper part of the calcite zone is commonly known to the sulfur operators as the "barren cap" and is typically fine-grained except where cut by coarse-grained calcite veins. Porosity, with vugs and caverns, is variable, and petroleum or petroleum residues are normally ubiquitous.

A transitional zone commonly marks the change from mostly calcite above to mostly anhydrite below, and this zone, too, is typically thickest toward the margin of the cap rock. Most

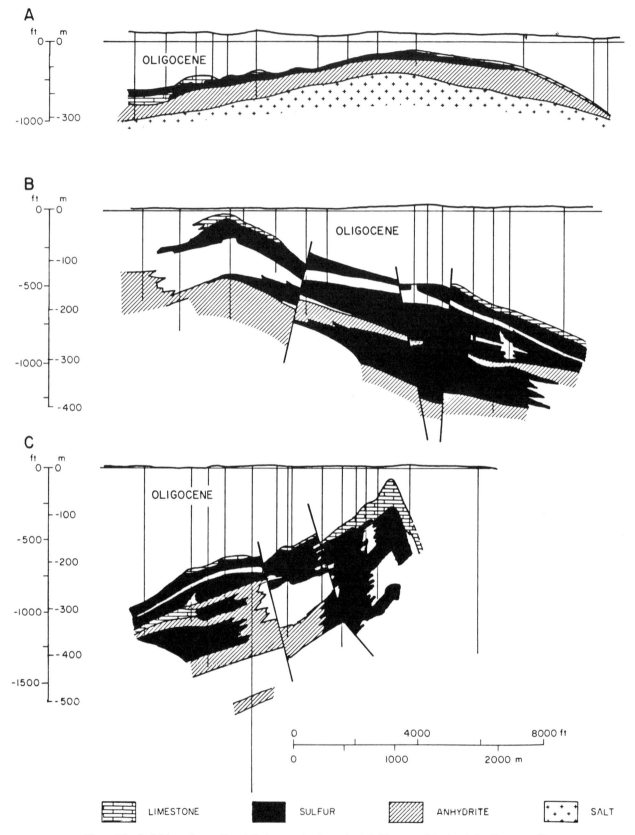

Figure 10. A, Jaltipan dome; B and C, Amexquite dome, both in Veracruz, Mexico (after Contreras V. and Castillon B., 1968).

workers consider this to be the most active zone of alteration of anhydrite to calcite and that commonly the alteration is still in progress. Petrologic studies show that the outlines of the replaced anhydrite grains are retained by the gypsum and calcite. Further, it appears that sulfur is most common in this zone, with sulfur replacing anhydrite, calcite, and gypsum. The porosity and permeability are normally highest in this zone. Petroleum or petroleum residues are also abundant.

Arguments exist about the age relations of the various minerals, but calcite and sulfur seem to be essentially contemporaneous. The sulfur in the transitional zone is commonly of the "disseminated type" but vein calcite and vein sulfur, a still younger phase, may cut the "disseminated" sulfur. Caverns, vugs, banding, minor amounts of celestite, strontianite, barite, dolomite, marcasite, pyrite, pyrrhotite, sphalerite, galena, and acanthite (Ag_2S) may exist.

The anhydrite zone occurs immediately above the salt, normally forms a mantle over the entire salt plug, and may extend down the flanks. The anhydrite is massive, saccharoidal near the salt contact, typically banded, and brecciated. Small quantities of quartz, celestite, barite, sulfur, calcite, dolomite, marcasite, pyrite, pyrrhotite, sphalerite, galena, and acanthite also occur, with the concentrations of the heavier metal minerals nearest the rim of the cap rock. Price and Kyle (1983) showed schematic concentrations of the heavier metal minerals in the anhydrite zone of the Hockley Dome, Harris County, Texas. This dome, however, does not contain sulfur.

The sulfur-bearing cap rocks occur within strata ranging in age from Oligocene to Pliocene. The salt below is probably of Middle Jurassic (Callovian) age (see Chapter 8, this volume). The area of sulfur-bearing domes along the U.S. Gulf Coast coincides with a belt of oil and gas occurrences in upper Tertiary beds. More than 90 percent of the salt domes in this belt have associated petroleum accumulations, while only 30 percent of the salt domes outside this belt are productive. If bacterial activity still persists in the Holocene, and if the ground-water dissolving action of the salt is still in operation, the ages of the cap rocks would range from Oligocene to Holocene.

ORIGIN OF THE SULFUR

One of the clearest statements on the origin of cap rock sulfur is presented by Davis and Kirkland (1979) in their summary of the origin of bioepigenetic sulfur deposits. They summarize by stating, "Sulfur deposits in caprocks over salt domes. . . .are clearly bioepigenetic. Anhydrite, introduced by salt diapirs into geologically younger sediments, is converted biogenetically to sulfur and calcite in the presence of petroleum in caprock. . . ."

Thode and others (1954) outlined the basic biogenetic process in the formation of sulfur in the cap rock of salt domes. They proposed that colonies of sulfate-reducing bacteria such as *Desulfovibrio desulfuricans* feed on petroleum and anhydrite resulting in the formation of hydrogen sulfide and calcite. The

hydrogen sulfide was then oxidized by bacterial action or meteoric ground water to elemental sulfur. Sassen (1980) described similar biodegradation of crude oil and sulfate reduction to explain the origin of calcite and elemental sulfur in the cap rock of the Damon Mount Dome, Brazoria County, Texas (Fig. 7).

Feely and Kulp (1957) performed experiments and made isotopic analyses showing that the $^{32}S/^{34}S$ ratios of the sulfate from the cap rocks are like those from the Clarke County, Alabama, Jurassic salt beds. However, the $^{32}S/^{34}S$ ratios in the hydrogen sulfide of the cap rocks are consistently higher (enriched from 2.7 to 4.8 percent in ^{32}S) than those from the nearby anhydrites. The $^{13}C/^{12}C$ ratios of the calcite cap rock are lower by 4.3 percent than ordinary sedimentary carbonates. Sassen (1980) confirmed these results in his study of the carbon isotope values of the secondary calcite in the cap rock of the Damon Mound Dome: he determined that the secondary calcite was isotopically light and most likely to have been derived from the oxidation of a crude oil of similar isotopic composition, not as the result of the recrystallization of a marine limestone. These isotopic studies provide strong evidence for the biochemical origin of the cap-rock sulfur. Feely and Kulp (1957) thought that oxidation of hydrogen sulfide to native sulfur could implicate another bacterium, *Theobacillus theiooxidus*, but no such bacterium has been found in the Gulf of Mexico cap rocks. Oxidation by meteoric ground water is still an accepted means of converting hydrogen sulfide to native sulfur by bacteria.

PRODUCTION AND RESERVES

Through 1987, accumulative Frasch sulfur production from salt-dome cap rocks reached 112 million metric tons in Louisiana, 186 million metric tons in Texas, and about 45 million metric tons in Mexico, for an accumulated grand total of about 343 million metric tons for the entire Gulf of Mexico basin region. The history of sulfur production from salt-dome cap rocks is shown in Figure 6.

Various estimates of future native sulfur reserves in salt-dome cap rocks in Texas and Louisiana range from 90 to 115 million metric tons, and those of Mexico amount to approximately 5 million metric tons (Shelton, 1980). Since about 343 million metric tons have been produced in the Gulf of Mexico basin region by 1987 a small reserve of only 120 million metric tons seems to be a low estimate. Many of the salt domes and salt-intrusion cap rocks have not yet been explored for sulfur by drilling, as neither have the many salt masses in the northern Gulf of Mexico and the Bay of Campeche area.

SUMMARY

Thirty-six salt-dome cap rocks have produced native sulfur in the Gulf of Mexico basin region at one time or another, but only eight are still producing sulfur, four in the United States, and four in Mexico. About 343 million metric tons of sulfur had been produced in the region by 1987, but only 120 million metric tons of reserves are estimated. Much of the production comes from the transitional zone of the cap rocks near the base of the calcite zone and at the top of the anhydrite zone. The age of the cap rocks ranges from Oligocene to Holocene. The native sulfur is the result of bacterial conversion of petroleum and anhydrite to hydrogen sulfide, which is subsequently oxidized to native sulfur.

URANIUM

William E. Galloway

INTRODUCTION

The South Texas Uranium Province lies along the low-relief inner coastal plain of central and southern Texas and extends into the state of Tamaulipas, Mexico (Plate 4). Total area of the active province exceeds 40,000 km^2.

Uranium was discovered in the Texas Coastal Plain in 1954 when a radiometric anomaly was detected by an airborne survey flown for petroleum exploration. A flurry of exploration followed, and several shallow oxidized deposits were discovered and mined. Deeper, unoxidized, roll-type deposits were discovered by drilling in 1963. During a decade of intense exploration and development in the 1970s, numerous deposits were located across the entire province. Depths of potentially commercial deposits exceed 300 m. Open-pit mining dominated early development and is still utilized for many of the shallow deposits. In 1975, the first pilot in-situ leach facility began production. The process has proved particularly appropriate for much of the south Texas ore, which occurs in unconsolidated, highly permeable sand, and has been extensively used for extraction of both shallow and deep deposits. More than 25 separate leach sites have been mined. The level of activity has remained moderate throughout the late 1980s with two to four leach sites and one open-pit mine in operation throughout the period.

Exploration and mining have focused on the Upper Eocene Whitsett Formation (Jackson Group) and the Oligocene–Lower Miocene Catahoula and Oakville Formations (Fig. 11). However, more recent exploration has delineated deposits within the Carrizo Sandstone (Lower Eocene) and Goliad Formation (Middle to Upper Miocene), significantly expanding the known stratigraphic distribution of the uranium mineralization.

With the exception of the Whitsett Formation, host sandstone units were deposited in coastal plain fluvial systems (Galloway, 1977, 1982; Hoel, 1982; Hamlin, 1983). The Carrizo Sandstone constitutes a bed-load fluvial system (Hamlin, 1983) that fed a major offlapping upper Wilcox delta system of the deep subsurface. Deposition of the Whitsett Formation as shore-zone facies upon a submerged platform of older Eocene sediments presaged the middle and late Tertiary episodes of clastic influx and basin margin offlap. Outcrop and shallow subsurface deposits of these later episodes consist of the remnant inner coastal plain fluvial systems that fed basinward into deltaic and interdeltaic

coastal systems. A shift from the uniform, humid paleoclimate that dominated the early Tertiary to a more arid and geographically zoned paleoclimate in late Tertiary time accompanied the onset of Catahoula deposition.

The structure of the shallow uranium-bearing sandstones is simple. Large-scale growth faults segment the deeply buried stratigraphic foundation of the basin fill, but the few faults that penetrate the shallow section have displacements of a few tens of meters or less. Spatial association of many uranium deposits with both growth-fault and Mesozoic-rooted fault trends led to early proposals for a genetic relation (Eargle and Weeks, 1961).

DEPOSIT GENESIS

The uranium-bearing fluvial systems of the South Texas Uranium Province also include several major aquifers of the coastal plain. Formation of epigenetic uranium deposits records the evolution of the hydrodynamics and hydrochemistry of an aquifer system that either contained or established hydraulic continuity with a uranium-rich source (Galloway and others, 1979; Walton and others, 1981). The most prominent and well-documented source of uranium in the Tertiary section is the abundant vitric volcanic debris that was reworked and deposited with Upper Eocene through Lower Miocene strata (Galloway and Kaiser, 1980; Walton and others, 1981). Paleosols of the Catahoula are highly leached, oxidized, and uranium-depleted. Pedogenic alteration and leaching of volcanic ash–rich Oligocene and Miocene flood-plain muds could have provided a charge of uranium to older, confined aquifers, such as the Carrizo and Whitsett, that subcropped beneath a thin, and perhaps ephemeral, cover of such younger ash (Galloway and others, 1979). Sourcing within the Catahoula and Oakville fluvial systems likely occurred very early, beginning in the early burial phase, and decreasing as deeper burial restricted ground-water circulation. The development of widespread alteration tongues and associated mineralization fronts within these aquifers suggests that final concentration of uranium occurred within semiconfined to confined aquifers, likely while uranium release continued at the depositional surface. The source of uranium in the younger Miocene Goliad Formation remains uncertain. Uranium was possibly recycled from underlying metals-rich aquifers (Adams and Smith, 1981) continued to be released from stratigraphically older volcanics, or was released in sufficient quantity from the sparsely interbedded volcanic debris. Modern ground waters contain very low concentrations of uranium (typically < 5 ppb) and appear incapable of forming new deposits (Henry and others, 1982; Galloway, 1982).

Once uranium had been leached, it moved into the most transmissive hydrostratigraphic units and then basinward in response to the regional hydraulic gradient. As the oxidizing, uranium-charged ground water moved down-gradient, oxidant was consumed by reactions with reductants within the aquifer matrix. Uranium and commonly associated trace metals such as selenium and molybdenum, were concentrated at a sharply defined Eh fence. The typical occurrence of marcasite as an ore-

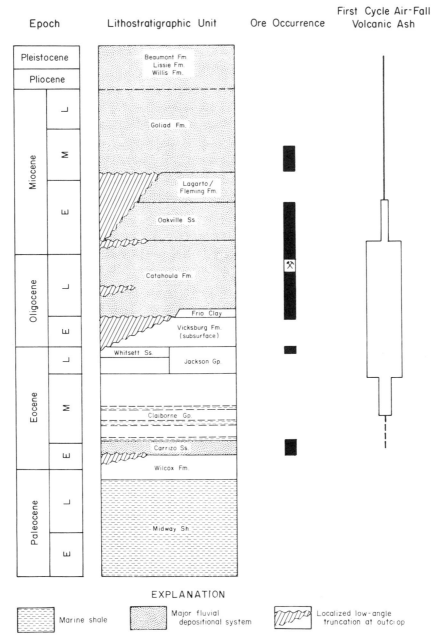

Figure 11. Stratigraphic section and distribution of uranium deposits, South Texas Uranium Province.

stage sulfide mineral phase indicates a concurrent drop in pH at the oxidation-alteration boundary (Goldhaber and Reynolds, 1979; Galloway and Kaiser, 1980).

Uranium deposits are irregular roll-like or tabular bodies that commonly lie along linear to highly sinuous trends, which may extend for several kilometers (Fig. 12). The uranium occurs as discrete coffinite and less-common uraninite grains, concentrations of titanium-oxide grains, and amorphous phases intimately mixed with dispersed alumino-silicate blebs and grain coats and less commonly with carbonaceous organic debris. Consistently associated metals include molybdenum and selenium. Deposits

are zoned, with successive Se-U-Mo concentration peaks defining the polarity of the Eh gradient during ore genesis (Harshman, 1974). Zonation of mineralogic and isotopic composition of iron disulfide minerals have been well documented by Goldhaber and others (1978, 1979) and Reynolds and Goldhaber (1983).

The alteration zonation pattern typical of the fluvially hosted south Texas uranium deposits is shown in Figure 13. Geometric, mineralogic, and isotopic patterns combine to reveal a complex history of alteration of shallow aquifers that includes partial to total syngenetic oxidation, epigenetic reduction and sulfidization, epigenetic oxidation and metallogenesis, post-ore resulfidizing alteration, and young, epigenetic oxidation (Gold-haber and others, 1979; Galloway and Kaiser, 1980; Galloway, 1982). This complex history of successive oxidizing and reducing events records alternating flushing of portions of the aquifer by shallow meteoric ground water (oxidizing) and thermobaric waters (reducing) derived from the deep basin.

The only known ground-water reservoir containing abundant isotopically heavy sulfide lies within the buried Mesozoic carbonate section, which is more than 5 km below the shallow aquifers (Galloway, 1982; Goldhaber and others, 1979). Spatial association of epigenetic sulfide enrichment with deep-seated growth faults is strong circumstantial evidence that such structures provide the vertical conduits for expulsion of thermobaric waters.

The constructional phase of the generalized south Texas uranium cycle (Fig. 12) must therefore include options for varia-

ble mechanisms of host reduction. Units such as the Whitsett Formation contain abundant detrital organic material and synge-netic sulfide minerals (Reynolds and others, 1982). Other units, such as many of the sand bodies in the Catahoula and Oakville, are dependent on an early introduction of epigenetic sulfide to establish reduced "islands" within an otherwise oxidized aquifer. In such units, the areas of mineralization are likely to be spatially associated with structural features.

The dynamic interaction of the deep and shallow ground-water regimes is also responsible for wide-scale modification of simple alteration tongue geometry (Fig. 12). Upon stagnation of meteoric flow, large portions of rereduced aquifer may remain, obscuring the critical mineralization boundary. Similarity of pre- and post-ore sulfides makes their recognition extremely difficult. Primary reduced ground preserves sulfide replacement textures, visible with reflection microscopy, that are not present in

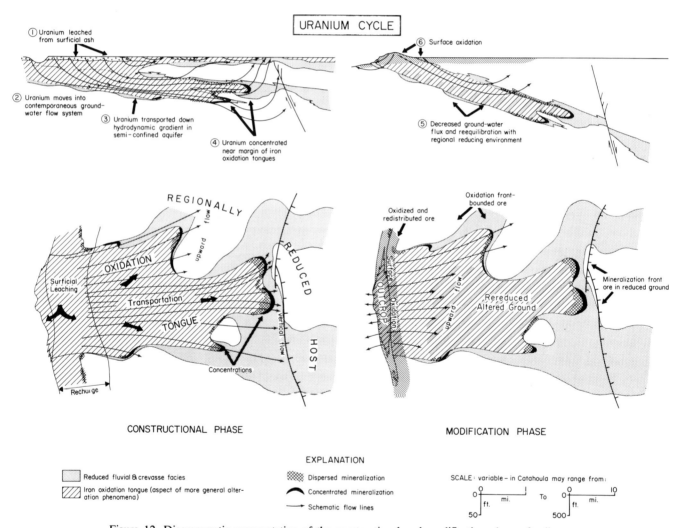

Figure 12. Diagrammatic representation of the constructional and modification phase of roll-type uranium deposit genesis. Constructional events include primary mobilization, migration, and concentration of uranium within a semiconfined to confined aquifer matrix. Modification of mineralization fronts includes resulfidization of oxidized ground or remobilization and possible destruction of shallow deposits. From Galloway and others (1979).

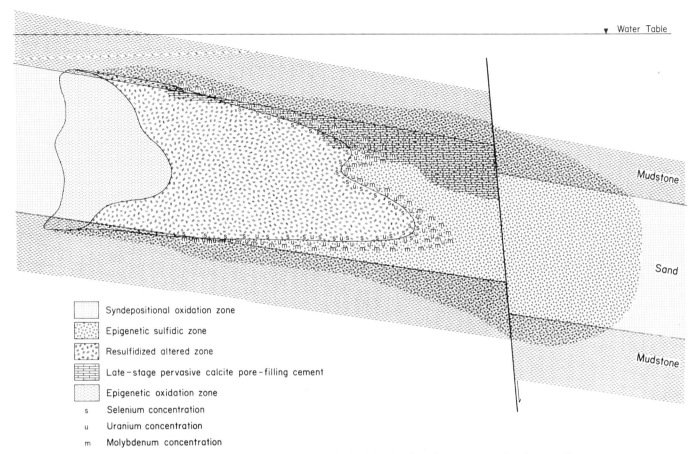

Water Table

Mudstone

Sand

Mudstone

☐ Syndepositional oxidation zone

☐ Epigenetic sulfidic zone

☐ Resulfidized altered zone

☐ Late-stage pervasive calcite pore-filling cement

☐ Epigenetic oxidation zone

s Selenium concentration

u Uranium concentration

m Molybdenum concentration

Figure 13. Generalized alteration zonation typical of Catahoula and younger uranium host aquifers. Multiple epigenetic alteration events include pre- and post-mineralization sulfidization as well as ore-stage epigenetic oxidation and modern oxidation. Metals are zoned across the mineralization front. Alteration and mineralization are typically associated with fault zones. From Galloway (1982).

rereduced zones (Reynolds and Goldhaber, 1983). Resulfidization stabilized the uranium deposit, and may be an isotopically datable event (Ludwig and others, 1982). Multiple, nested mineralization fronts show that successive episodes of epigenetic oxidation/mineralization and sulfidization may recur (Galloway, 1982).

Other parts of the mineralization front may remain actively flushed by meteoric waters, with resultant remobilization and redistribution of uranium and daughter products. Radiometric disequilibrium is a common problem in ore-grade determination in the South Texas Uranium Province. Near the outcrop, weathering and active oxidation and remobilization by shallow meteoric circulation can completely redistribute orebodies. Small oxidized ore pods consisting of autunite and various unusual minerals such as umohoite have been mined.

REPRESENTATIVE DEPOSITIONAL/ MINERALIZATION SYSTEMS

Sandstones of the Whitsett and Oakville Formations illustrate the diversity of depositional and mineralization styles of South Texas uranium hosts and their contained deposits.

Whitsett Formation

The Whitsett Formation is the topmost unit of the Eocene Jackson Group (Fig. 1). Uranium occurs within coastal barrier-bar and associated sand facies where the northern, updip margin of the regional, strike-oriented barrier sand belt intersects the outcrop (Fig. 14). More than 30 open-pit uranium mines have exposed deposits in coastal-barrier core and back-barrier, tidal inlet, cuspate delta, and suspended-load distributary channel sand facies. In plan view, the geometry of mineralization fronts ranges from linear and strike-parallel to highly sinuous and dip-oriented, reflecting the dominant geometry of the host sand body (Galloway and others, 1979).

Volcanic ash is abundant within the Whitsett Formation. However, much of the ash is vitric and retains several parts per million of uranium. Ash deposited in areas of shallow water table or in subaqueous barrier/strandplain environments shows little evidence of effective leaching of metals. Stratigraphic relations further confirm the dependence of Whitsett mineralization on external sources. As shown in Figure 14, commercial uranium deposits occur only in a very limited segment of the Jackson

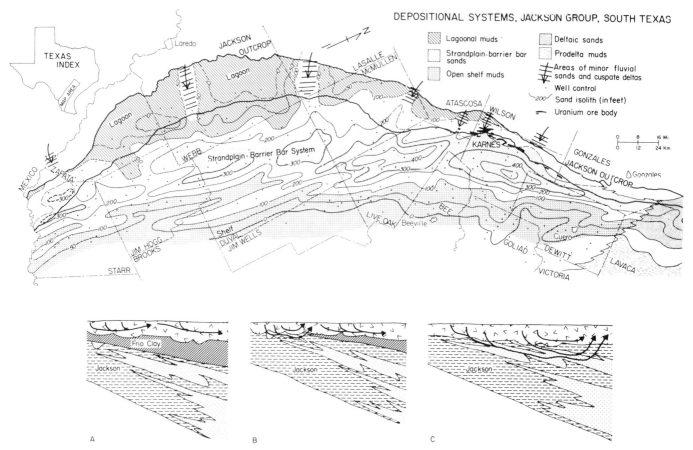

Figure 14. Net-sand isolith and facies map of the Jackson Group (Eocene), South Texas Coastal Plain. The uranium district is localized in Karnes County, where uppermost sands of the strandplain-barrier bar system subcrop directly beneath overlying Catahoula tuffs. Modified from Galloway and others (1979).

barrier/strandplain system. The schematic stratigraphic cross sections show that Jackson sandstones are largely hydraulically isolated from overlying Catahoula tuffs by the overlying Frio clay (Fig. 14A) or by the updip Jackson lagoonal mudstone (Fig. 14B). Only where the upper Whitsett sands lie directly beneath the Catahoula Formation in slight angular discordance (Fig. 14C), as in the Karnes County uranium district, do roll-type deposits occur within the Jackson sand (Galloway and others, 1979).

Typical of shore-zone deposits, Whitsett sands are intrinsically reduced, containing dispersed, syngenetic pyrite and carbonaceous organic matter (Galloway and others, 1979; Reynolds and others, 1982). It is ironic that the Whitsett orebodies, which inpart lie along shallow faults and overlie sour gas fields, inspired the theory of extrinsic reduction (Eargle and Weeks, 1961) that applies so well to many of the younger South Texas districts but not to the district of its origin.

Oakville Formation sandstones

The sandstones of the Oakville Formation (Oligocene–Lower Miocene) were deposited by several large-to-small coastal-plain rivers, which collectively form the Oakville bed-load fluvial system (Galloway, 1982). Major depositional elements include three principal fluvial axes produced by large, extrabasinal rivers and numerous small fluvial complexes and relatively mud-rich bounding facies (Figs. 15a and b).

Permeable, highly transmissive framework elements of the Oakville sandstones consist of bed-load and mixed-load channel fills and associated sheetflood splays. Flood-plain muds and silts are the prinicpal confining facies. Oakville uranium deposits lie within and along the margins of the transmissive elements in the aquifer (Fig. 15c). What is more important is that the size of the deposits can be positively correlated with the relative transmissivity of the associated host. The largest Oakville uranium districts (reserves exceeding 5,000 t) lie within the George West and Hebbronville bed-load axes. Smaller commercial deposits occur in the intermediate-transmissivity New Davy axis, and subcommercial orebodies have been reported to occur in the least transmissive Moulton and Burton/Penn trends.

Oakville uranium districts are closely associated with fault zones. Iron sulfide is typically abundant, and studies in the two largest districts show this sulfide to be mainly the isotopically

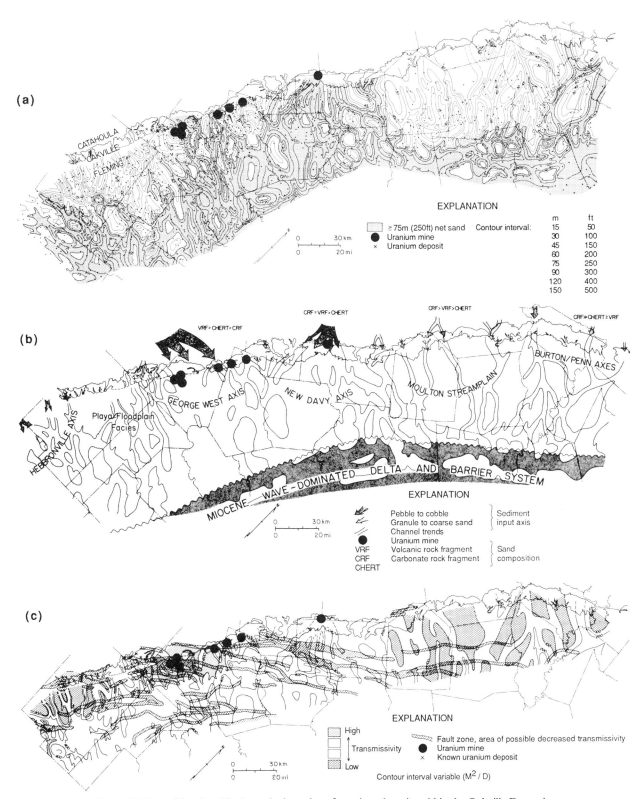

Figure 15. Depositional and hydrogeologic setting of uranium deposits within the Oakville Formation, which is representative of post-Eocene uranium hosts and mineralization style. a, net-sand map; b, depositional elements of the Oakville fluvial system; c, Regional semiquantitative transmissivity map showing the distribution of uranium districts and satellite deposits around and within highly transmissive fluvial axes. From Galloway and others (1982).

heavy, epigenetic pre- and post-mineralization type (Goldhaber and others, 1979; Galloway, 1982). Resulfidized roll-type deposits are common.

Comparison of Whitsett and Oakville mineralization styles illustrates a general trend that appears to characterize the South Texas Uranium Province. Orebodies in older, Eocene host aquifers (Whitsett) are typically simple roll-type sandstone deposits. Successively younger hosts (Oligocene–Lower Miocene Catahoula, Oakville) were increasingly oxidized and leached as climatic aridity became more extreme later in the Tertiary. Consequently, host preparation, in the form of epigenetic sulfidic reduction, defines areas of potential uranium entrapment, and uranium deposits are increasingly tied locally to leak points along faults. Further, repeated influx of reducing formation waters produces complex alteration patterns that cannot be distinguished by simple criteria.

STATUS OF URANIUM MINING IN SOUTH TEXAS

Although the domestic uranium market is depressed, and is likely to remain so for a number of years, modest levels of exploration and mining activity continue in the South Texas Uranium Province. Chevron Resources (60 percent) and Total Minerals (40 percent) have initiated open pit mining at the Rhode Ranch

deposit (Oakville Formation) and are processing the ore at Chevron's mill in Panna Maria, 130 km from the mine. Solution mining continues at a few sites. Restoration of several earlier solution mining projects has been successfully completed.

GEOPRESSURED-GEOTHERMAL ENERGY

Robert A. Morton

INTRODUCTION

Geopressured-geothermal energy is a vast but dilute hydrologic resource contained in deep aquifers beneath the Gulf Coastal Plain and adjacent continental shelf (Fig. 16). For decades, wells drilled along the northwestern margin of the Gulf of Mexico basin encountered abnormally high temperatures and pressures that presented a challenge to drillers, but their energy potential was not recognized until the mid 1960s (Stuart, 1970); at that time the possibility of converting the thermal and kinetic energy to power seemed remote. Interest in this unconventional source of energy rose to prominence a decade later when rising fuel prices, increasing import dependence, and changing federal policies spurred a nationwide search for alternative energy supplies.

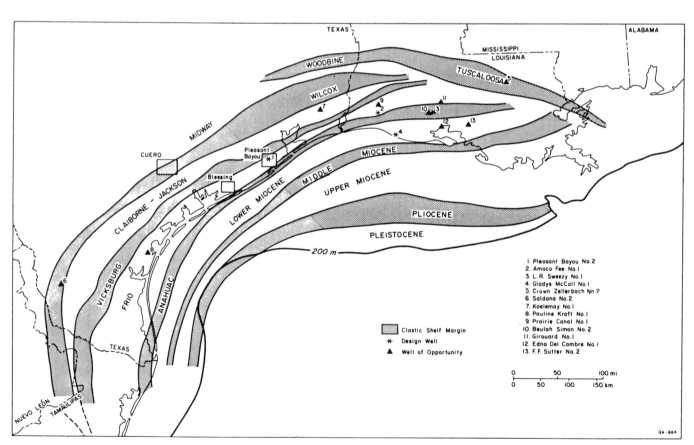

Figure 16. Locations of design wells and wells of opportunity in relation to ancient shelf margins. From Winker and others (1983).

GEOLOGIC SETTING

The term geopressure refers to pore fluid pressure greater than that exerted by a column of seawater at a particular depth (Dickinson, 1953; Stuart, 1970; Jones, 1975); it is manifested as pressure gradients that exceed 10.5 kPa/m and approach 22.6 kPa/m (Fig. 17). The top of geopressure, which commonly coincides with an increase in thermal gradient (Lewis and Rose, 1970; Jones, 1975), is largely controlled by growth faults and sedimentary facies. It typically occurs beneath the massive sandstones and near the top or within the interval of interbedded sandstones and shales (Fig. 17); the underlying massive shales are also geopressured (Dickinson, 1953; Stuart, 1970). Geopressured reservoirs are commonly bounded by faults and sealed on the updip (downthrown) side by shelf-slope shales and on the downdip (upthrown) side by transgressive shales; reservoirs also are overlain and underlain by transgressive shales or interbedded sandstones and shale (Dickinson, 1953; Winker and others, 1983). Because of facies dependency, the regional top of geopressure is expressed as a subparallel series of highs and lows. The upper boundary is depressed beneath sand-rich depocenters and is elevated basinward above massive shales. The top of geopres-

sure represents an important geologic and hydrologic boundary because it controls fluid migration, hydrocarbon accumulation, and diagenetic reactions. In the geopressured zone, sediments are gas prone, fault density increases, fault angles decrease, clay minerals are transformed from smectite to illite, and organic matter matures to form hydrocarbons.

There is general agreement that rapid sedimentation and hydraulic discontinuity of the sandstones retard expulsion of interstitial fluids during compaction and burial. As a result, pore fluids partially support the overburden, causing abnormal formation pressure (Dickinson, 1953; Jones, 1975). There is less agreement regarding sources of heat. Extreme temperatures associated with the underlying thin transitional crust, igneous intrusions, salt domes, and radiation are all cited as possible heat sources. Regional variations in thermal gradients and heat flow are usually attributed to the insulating capacity of water and to mass transfer of fluids (Lewis and Rose, 1970; Jones, 1975).

Prospective geopressured-geothermal reservoirs occur within thick regressive sequences deposited during rapid progradation of the continental margin. Notable among the major clastic wedges in the Gulf of Mexico basin are the Wilcox Group and Frio Formation in Texas, and the Tuscaloosa, Frio, and Fleming

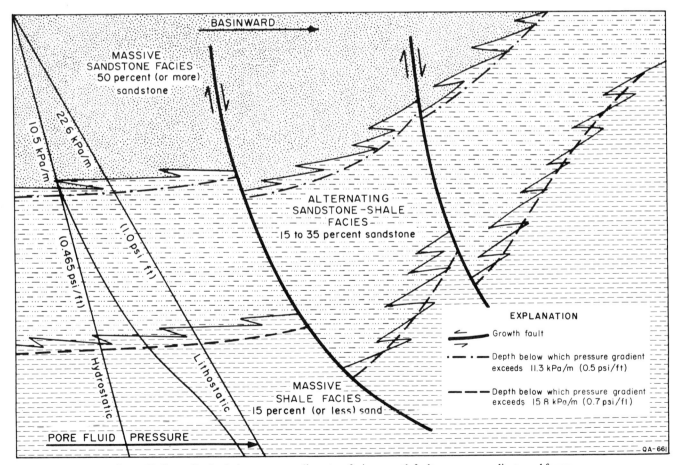

Figure 17. Generalized relations among sedimentary facies, growth faults, pressure gradients, and formation temperatures, northern Gulf Coast Basin. Modified from Stuart (1970) and Norwood and Holland (1974).

Formations in Louisiana (Fig. 16). These stratigraphic units exhibit physical, chemical, and sedimentological characteristics that are favorable for exploitation of geopressured-geothermal energy. Oligocene and younger sediments beneath the continental shelf also have suitable characteristics (Wallace and others, 1978), but are less attractive than onshore deposits because offshore operating costs are high.

RESOURCE ASSESSMENT

The hot, overpressured brines contain three forms of energy: heat, hydraulic energy, and dissolved gases rich in methane. Initial interest in the resource focused on heat extraction using binary converters and mechanical turbines to generate electric power. Other direct applications considered the low-process heat available for refining sugar, milling paper, and producing gasohol; during the late 1970s, the potential for marketing gas was a primary reason for continued research and field tests.

Estimates of the in-place resource are widely divergent because assumptions vary greatly and because different methods and geographic areas are used to define the resource. The enormous sediment volume of the Gulf of Mexico basin and optimistic assumptions (including contributions from fluids contained in shales) led to extremely high estimates that range from 1,680 to 4,480 trillion m^3 (60,000 to 160,000 Tcf) of methane alone (Brown, 1976). Resource estimates have subsequently undergone negative revisions because detailed geological studies indicate thinner, fewer, and less porous sandstones; lower methane concentrations; and lower permeabilities than those used in preliminary calculations. But even conservative resource estimates are substantial compared to the annual consumption of 0.56 trillion m^3 (20 Tcf) in the United States in the late 1980s. Recovery efficiencies, estimated at 2 to 5 percent of the in-place resource, represent a critical but unknown factor related to long-term production.

SUMMARY OF TEST RESULTS

Industrial applications of geopressured-geothermal energy will require large-diameter wells capable of producing hot (> 150°C), overpressured (> 15.8 kPa/m) brines at high rates (6,300 m^3/day) for extended periods (20 years or more). To determine the feasibility of those requirements the U.S. Department of Energy (DOE) conducted field tests that provided both regional and site-specific information regarding characteristics of the reservoirs and fluids.

Design wells

Four wells (Fig. 16, Table 7) were drilled at sites specifically selected to test geopressured aquifers under optimum conditions. The well sites were selected primarily on the basis of expected net sandstone thicknesses, formation temperatures and pressure gradients, reservoir quality, methane concentrations, possible environmental damage, and leasing arrangements. Geologic conditions encountered at each of the sites were essentially

as predicted, with the exception of higher salinities and consequently lower methane solubilities.

Of the four wells, three were planned for multiyear tests; however, the L. R. Sweezy well was expected to deplete a limited reservoir in nine months. This short-term test investigated the effects of pressure drawdown (possible reservoir compaction, shale dewatering, and changes in fluid composition and concentration) that might occur late in the production history of large reservoirs. The field tests indicate that reservoirs may experience high yields; however, they will require prolonged production periods or increased fluid extraction to reach critical gas saturation—theoretically the pressure at which dispersed gas bubbles coalesce within the reservoir and form free gas produced in association with water. Each of the design wells also produced some liquid hydrocarbons in association with the thermal brines and solution gas. These unexpected concentrations of aromatic hydrocarbons (benzene and toluene) may be dissolved in the brine or may occur in the reservoir in a free phase as a residual product of primary hydrocarbon migration.

Wells of opportunity

The DOE also acquired 13 exploratory oil and gas wells that penetrated geopressured sandstones but were nonproductive and subsequently abandoned by industry operators. Successful short-term (two-week) tests of eight wells (Table 7) provided a spectrum of physical and chemical properties for the formations of interest (Fig. 16, Table 7). The other reentries were unsuccessful due to mechanical problems.

The Edna Del Cambre No. 1 and Saldana No. 1 wells produced gas in quantities greater than saturation, suggesting exsolution of free gas or drawdown of a nearby gas cap. Another well, the Koelemay No. 1, produced commercial quantities of oil and was returned to the operator. Methane concentrations in the other reservoirs were at or near saturation for in-situ conditions, except for the Crown Zellerbach No. 1, which was undersaturated with respect to methane.

POSSIBLE CONSTRAINTS TO LARGE-SCALE EXPLOITATION

Reservoir properties

The producing life of a geothermal well is determined principally by the physicochemical properties of the formation and the interstitial fluids. These critical properties vary greatly from area to area, and no single formation has all the characteristics of an ideal reservoir: one that is thick, areally extensive, highly permeable, hot, under extreme pressure, and contains low-salinity water saturated with methane and liquid hydrocarbons. Drainage area and attendant well life depend on reservoir size and reservoir quality. Reservoir size is mainly a function of sand-body geometry and fault-block area. Two different concepts of reservoir interval have been used to assess the resource and identify prospective areas. One type of interval consists of multiple sandstones several hundred to a thousand meters thick that are

TABLE 7. RESERVOIR AND FLUID CHARACTERISTICS, DEPARTMENT OF ENERGY DESIGN WELLS AND WELLS OF OPPORTUNITY*

	Stratigraphic Unit	Perforations (m)	Temperature (°C)	Pressure (kPa)	Porosity (%)	Permeability (md)	Salinity (mg/L)	Gas/Water (m³/m³)	Carbon Dioxide (mole %)
Design Wells									
General Crude Oil - DOE Pleasant Bayou No. 2	Lower Frio *Anomalina bilateralis*	4,466–4,485	149	76,245	19	150	131,000	4.0	10
Magma Gulf-Technadril - DOE Amoco Fee No. 1	Middle Frio *Miogypsinoides*	4,693–4,701	148	83,305	19	400	165,000	4.5	8.6
Dow Chemical - DOE L. R. Sweezy No. 1	Upper Frio *Cibicides jeffersonensis*	4,072–4,089	113	79,292	22	300	100,000	3.0	2.4
Technadril - Fenix and Scisson - DOE Gladys McCall No. 1	Lower Miocene *Siphonina davisi*	4,758–4,767	156	86,187	26	64	97,800	5.0	9.5
Wells of Opportunity									
Martin Exploration Co. Crown Zellerbach No. 2	Upper Cretaceous Tuscaloosa	5,100–5,109	164	69,517	17	17	32,000	5.9	23
Riddle Oil Co. Saldana No. 2	Upper Wilcox	2,972–2,995	149	45,726	16	17	12,800	8.5–9.7§	26
Lear Petroleum Inc. Koelemay No. 1	Yegua	3,550–3,593	127	65,205	20	100	15,000	5.4–5.7§**	7
Ross-Pope Drilling Co. Pauline Kraft No. 1†	Lower Frio *Anomalina bilateralis*	3,889–3,922	128	76,535	23	39	......	......	......
Houston Oil and Minerals Corp. Prairie Canal No. 1	Middle Frio Hackberry	4,508–4,520	146	89,300	28	95	42,600	7.4–9	9
Southport Exploration Inc. Beulah Simon No. 2	Upper Frio *Camerina A*	4,476–4,505	140	89,803	19	30	106,000	3.8	8
Wainoco Oil and Gas Co. Girouard No. 1	Upper Frio *Marginulina texana*	4,497–4,520	134	91,101	26	200	23,500	7.2	6
Coastal States Gas Co. Edna Del Cambre No. 1	Lower Miocene *Planulina* sp.	3,804–3,844	112	74,851	29	103	133,300	4.0§	2
Nuehoff Oil Co. Fairfax Foster Sutter No. 2	Lower Miocene *Marginulina ascensionensis*	4,813–4,856	132	83,628	19	20	190,900	4.1	8

*Data compiled from unpublished U.S. Department of Energy files, Eaton Operating Co. (1982), and Dorfman and Morton (1985).
†Low flow rate.
§Exceeds saturation.
**Commercial producer.

genetically related but vertically separated. These extensive aquifer systems contain 10^6 to 10^8 km^3 of sandstone (Morton and others, 1983a) that could be produced only by multiple sequential completions. Another type of reservoir interval involves individual sandstones a few decimeters to perhaps 200 m thick. These laterally continuous aquifers contain 10^4 to 10^6 km^3 of sandstone and are typical of the production reservoirs in the design wells and wells of opportunity.

Geopressured sediments are displaced by numerous faults and fault-block sizes vary greatly. Within the Wilcox and Frio trends of Texas, fault-block areas range from less than 1 to 135 km^2. The distribution is strongly skewed toward areas of 26 km^2 or less (Morton and others, 1983a). Because fault compartments tend to be small, the orientation of optimum sandstone development with respect to fault-block shape is critical to reservoir continuity (Fig. 18). It follows that barrier and deltaic sandstones having uniform thicknesses or having depositional axes parallel to the fault-block axes would likely have greater continuity than fluvial channel or submarine fans having sandstone axes normal to the fault-block trend (Fig. 18). Faults in the geopressured zone are not necessarily barriers to fluid flow, especially when displacement is less than the thickness of correlative sand bodies.

Porosity and permeability of Gulf Coastal Plain sandstones systematically decrease with depth; however, regional differences in thermal gradient, sandstone composition, and burial history allow preservation of primary pore properties in some areas, whereas they are occluded in others. As a result, reservoir quality varies greatly between south Texas and south Louisiana. The low permeabilities (<10 md) of deep reservoirs in south Texas are generally attributed to high thermal gradients and an assemblage of unstable volcanic rock fragments that reacted with pore waters to form authigenic cements including chlorite, albite, and late-stage ferroan calcite (Loucks and others, 1981). Higher sandstone permeabilities in southeast Texas are partly explained by lower thermal gradients and more stable detrital minerals, but more important was the leaching of calcite cement and feldspar grains at depth to form secondary porosity (Loucks and others, 1981). The trend of generally higher porosities and permeabilities and lower thermal gradients extends into Louisiana (Bebout and Gutierrez, 1981) where deeply buried reservoirs are commonly composed of highly permeable sandstone. High permeability of poorly consolidated sand is a negative quality when it results in excessive sand production.

Fluid chemistry

The chemical composition and concentration of formation waters and solution gas also are important to geopressured-geothermal resource development. Methane solubility above 100°C is inversely related to salinity; furthermore, potential scaling and corrosion of production equipment also are related to hydrochemistry.

Formation waters in the Gulf of Mexico basin are generally NaCl types (Kharaka and others, 1977) except for the CaNaCl waters found in south Texas (Morton and others, 1983b; Morton

and Land, 1987) and CaCl waters in the Mississippi Salt basin (Stuart, 1970). Total dissolved solids of these waters range from 10,000 to more than 300,000 mg/L. Regional hydrochemical data indicate that many geopressured formations contain waters having dissolved solids two to six times more concentrated than seawater. These highly saline waters are enriched by dissolution of nearby salt domes in the Houston and Mississippi embayments and by vertical migration of deep-basin fluids formerly in contact with Mesozoic sediments.

A common misconception is that salinities decrease with depth below the top of geopressure, and therefore, salinities are lower in the geopressure zone than in the hydropressure zone (Jones, 1975). This appears to be true where the zone of intermediate pressure gradients (10.4 to 15.8 kPa/m) is less than 500 m thick (Fig. 19A), but these conditions are atypical. Commonly, intermediate pressure gradients persist over a depth interval greater than 500 m. A typical salinity profile (Fig. 19B) shows a decrease near the top of geopressure, an increase at intermediate pressure gradients, and a decrease at extremely high pressure gradients. The maximum and minimum salinity concentrations in both the hydropressure and geopressure zones are comparable (Morton and others, 1983b). The coincidence of high salinities with intermediate pressure gradients and intermediate to high sandstone percents partly explains the higher than expected brine concentrations in the four design wells (Table 7).

Methane accounts for 71 to 91 percent of the total volume of gas dissolved in the hot brines. Carbon dioxide is also abundant in some aquifers, accounting for up to 26 percent of the total volume of produced gas (Table 7); nitrogen and heavier hydrocarbons make up the remaining few percent. Like methane, CO_2 concentrations depend on temperature, pressure, and salinity of the formation water; but unlike methans, CO_2 concentrations are also related to sediment age (Table 7); waters in older sediments generally are enriched in CO_2 and H_2S. Despite the differences in thermobaric properties and geologic settings of the test wells, methane and heavier hydrocarbons in solution have not exceeded 8 m^3 of gas per m^3 of water (Table 7; Fig. 20). The upper limit of these hydrocarbon concentrations is substantially less than originally expected, primarily because salinities of formation waters are high.

Brine disposal

Large-scale commercial production of geothermal aquifers will require sustained, high-volume injection of brine; a condition that may not be met easily because of potential formation damage within the disposal zone. Field tests suggest that plugging of disposal zones by carbonate precipitates or produced solids could limit the period of trouble-free injection and could require occasional workover of the disposal wells. Pore plugging also could result if injected brines are incompatible with formation waters, causing clay dehydration in the host rocks. Although adding to the cost of operation, these problems of injectivity are routinely handled in thousands of disposal wells throughout the Gulf Coastal Plain.

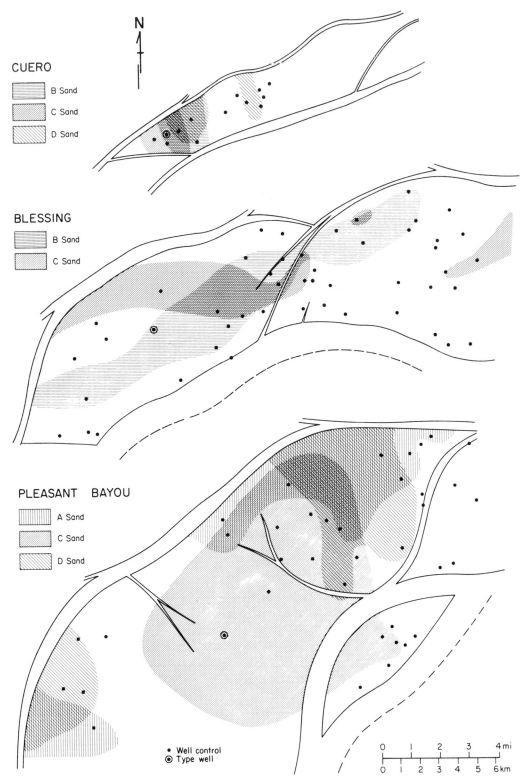

Figure 18. Extent of optimum sandstone development, mapped from electric-log patterns (thickness, absence of shale breaks, lateral continuity), for three geopressured-geothermal study areas. Locations of areas shown in Figure 16. From Winker and others (1983).

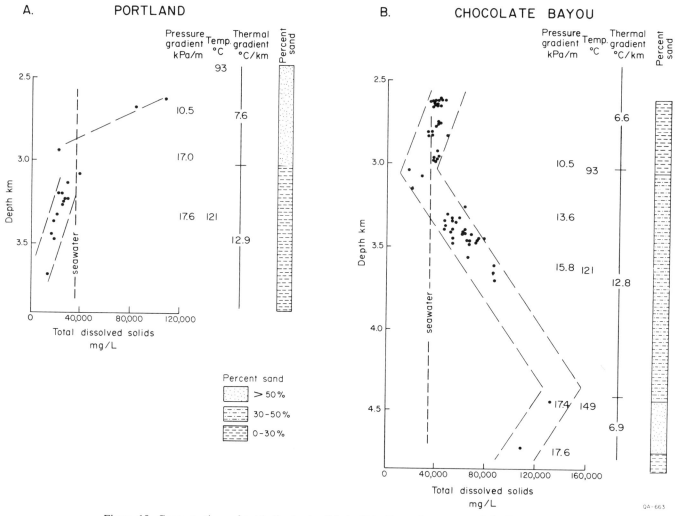

Figure 19. Concentrations of total dissolved solids in Frio Formation waters of the Portland and Chocolate Bayou fields, Texas. Chemical data from unpublished files, Phillips Petroleum Company, Kharaka and others (1977), and Morton and others (1983b). Modified from Morton and Land (1987).

Environmental concerns

A major environmental concern is that long-term high-volume production and injection might cause land-surface subsidence and microseismic events. Estimates of potential subsidence vary widely because appropriate rock compressibilities and compaction coefficients are not precisely known. The nonlinear deformation of reservoir rocks and overburden strata attendant with pressure reduction (Jogi and others, 1981) makes subsidence prediction difficult. Microseismic monitoring of design wells revealed several types of seismological phenomena, including microearthquakes and harmonic tremors (Mauk, 1982). Some of the microearthquakes occur along growth faults, suggesting possible shear behavior. Possible dilatant behavior associated with episodic fluid movement along the fault planes is suggested by the harmonic tremors (Mauk, 1982). It is uncertain whether or not fluid transfer, rupture, and attendant microseismic events are related to subsurface production and injection activities.

Economic analyses

Although the geopressured-geothermal resource contains 65 percent more thermal energy than methane, the methane has greater economic value; therefore economic analyses are related to gas prices. Estimates of gas prices necessary to break even or to be slightly profitable have ranged from $5 to $15/Mcf. In the late 1980s the resource remained undeveloped primarily because the risks associated with conventional gas production were substantially less than those for solution gas. Economically attractive geopressured-geothermal plant designs call for integrated systems that (1) generate electric power with the hydraulic and thermal energy, (2) utilize waste gas for on-site power, and (3) market the pipeline-quality gas. Other economic strategies are directed

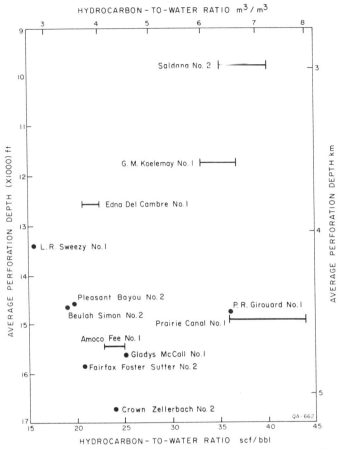

Figure 20. Hydrocarbon-to-water ratios for reservoirs tested by the Department of Energy design wells and wells of opportunity, based on production averages. Does not include minor volumes (<1 m³) of gas that remain in the disposal water at separator conditions.

toward increasing methane concentrations; for example, by using the design-well technology to produce watered-out gas reservoirs. Undoubtedly, solution gas and dispersed free gas will eventually be produced from geopressured aquifers just as water and gas are presently being coproduced to enhance gas recovery from abandoned hydropressured reservoirs.

SUMMARY

Future development of geopressured-geothermal energy will require not only solving technical problems related to reservoir behavior and production equipment, but it also will depend on mitigating environmental impacts, formulating regulatory policies, and achieving favorable economic returns on investments. The latter requirements will be difficult to achieve in the foreseeable future because demand for unconventional energy supplies will continue to depend on the availability of conventional petroleum supplies and their attendant prices. Integrated power systems run by geopressured-geothermal energy are technically feasible for site-specific applications that can utilize all three forms of energy. However, economic constraints will probably limit the scope of exploitation.

REFERENCES CITED

Adams, S. S., and Smith, R. B., 1981, Geology and recognition criteria for sandstone uranium deposits in mixed fluvial-shallow marine sedimentary sequences, South Texas, final report: U.S. Department of Energy Open-File Report GJBX-4, 145 p.

Altschuler, Z. S., 1980, The geochemistry of trace elements in marine phosphorites; Part 1, characteristic abundances and enrichments, *in* Bentor, Y. K., ed., Marine phosphorites: Society of Economic Paleontologists and Mineralogists Special Publication 29, p. 19–30.

Altschuler, Z. S., Clarke, R. S., and Young, E. J., 1958, Geochemistry of uranium in apatite and phosphorites: U.S. Geological Survey Professional Paper 314-D, p. 45–90.

Altschuler, Z. S., Cathcart, J. B., and Young, E. J., 1964, The geology and geochemistry of the Bone Valley Formation and its phosphate deposits, west central Florida, *in* Geological Society of America Annual Meeting Guidebook, Field Trip 6: U.S. Geological Survey, 68 p.

Altschuler, Z. S., Berman, S., and Cuttitta, F., 1967, Rare earths in phosphorites; Geochemistry and potential recovery: U.S. Geological Survey Professional Paper 575B, p. B1–B9.

Arthur, H. L., 1985, Petrography and depositional environment of the Chemard Lake lignite, De Soto, Louisiana [M.S. thesis]: Monroe, Northeast Louisiana University, 173 p.

Ayers, W. B., Jr., and Lewis, A. H., 1985, The Wilcox Group and Carrizo Sand (Paleogene) in east-central Texas; Depositional systems and deep-basin lignite: Austin, University of Texas Bureau of Economic Geology Special Publication, 19 p.

Baturin, G. N., 1982, Phosphorites on the sea floor: Amsterdam, Elsevier, 343 p.

Bebout, D. G., and Gutierrez, D. R., 1981, Geopressured geothermal resources in Texas and Louisiana; Geological constraints: Baton Rouge, Louisiana, Proceedings, Fifth Gulf Coast Geopressured-Geothermal Energy Conference, p. 13–24.

Bentor, Y. K., 1980, Phosphorites; The unsolved problems, *in* Bentor, Y. K., ed., Marine phosphorites: Society of Economic Paleontologists and Mineralogists Special Publication 29, p. 3–18.

Birdsall, B. C., 1978, Eastern Gulf of Mexico; Continental shelf phosphorite deposits [M.S. thesis]: St. Petersburg, University of South Florida, 87 p.

Bodenlos, A. J., 1973, Sulfur; U.S. mineral resources: U.S. Geological Survey Professional Paper 820, p. 605–618.

Brown, W. M., 1976, 100,000 quads of natural gas?: Hudson Institute Research Memorandum 31, HI-2451/2P, July 1976.

Cathcart, J. B., 1956, Distribution and occurrence of uranium in the calcium phosphate zone of the land-pebble phosphate district of Florida: U.S. Geological Survey Professional Paper 300, p. 489–494.

—— , 1968a, Florida-type phosphate deposits of the United States; Origin and techniques for prospecting, *in* Seminar on Sources of Mineral Raw Materials for the Fertilizer Industry in Asia and the Far East: Proceedings, United Nations ECAFE, Mineral Resource Development Series 32, p. 178–186.

—— , 1968b, Phosphate in the Atlantic and Gulf Coastal Plains, *in* Brown, L. F., ed., Proceedings 4th Forum of Geology of Industrial Minerals: Austin, University of Texas Press, p. 23–24.

—— , 1978, Uranium in phosphate rock: U.S. Geological Survey Professional Paper 988-A, 6 p.

Cathcart, J. B., Sheldon, R. P., and Gulbrandsen, R. A., 1984, Phosphate-rock resources of the United States: U.S. Geological Survey Circular 888, 48 p.

Charm, W. B., Nesteroff, W. D., and Valdes, S., 1969, Detailed stratigraphic description of the JOIDES cores on the continental margin off Florida: U.S. Geological Survey Professional Paper 581-D, 13 p.

Clarke, D. S., 1972, Stratigraphy, genesis, and economic potential of the southern part of the Florida land-pebble phosphate field [Ph.D. thesis]: Rolla, University of Missouri, 182 p.

Cleaves, A. W., 1980, Depositional systems and lignite prospecting models; Wilcox Group and Meridian Sandstone of northern Mississippi: Gulf Coast

Association of Geological Societies Transactions, v. 30, p. 283–307.

Consejo de Recursos Minerales, 1964 to 1988: Mexico City, Anuario Estadistico de la Mineria Mexicana.

Contreras V., H., and Castillon B., M., 1968, Morphology and origin of salt domes of Isthmus of Tehuantepec, *in* Braunstein, J., and O'Brien, G. D., eds., Diapirism and diapirs: American Association of Petroleum Geologists Memoir 8, p. 244–260.

Daniel, T. W., Jr., 1973, A strippable lignite bed in south Alabama: Geological Survey of Alabama Bulletin 101, 101 p.

Davis, J. B., and Kirkland, D. W., 1979, Bioepigenetic sulfur deposits: Economic Geology, v. 74, p. 462–468.

Dickinson, G., 1953, Geological aspects of abnormal reservoir pressures in Gulf Coast Louisiana: American Association of Petroleum Geologists Bulletin, v. 37, p. 410–432.

Dorfman, M. H., and Morton, R. A., eds., 1985, Proceedings of the Sixth U.S. Gulf Coast Geopressured-Geothermal Energy Conference: New York, Pergamon Press, 344 p.

Eargle, D. H., and Weeks, A.M.D., 1961, Possible relation between hydrogen sulfide-bearing hydrocarbons in fault-line oil fields and uranium deposits in the southeast Texas Coastal Plain: U.S. Geological Survey Professional Paper 424-D, p. 7–9.

Eaton Operating Company, 1982, Wells of Opportunity Program: Houston, Eaton Operating Company Final Contract Report DOE-ET-27081-9, 90 p.

Emery, K. O., and Uchupi, E., 1972, Western North Atlantic Ocean; Topography, rocks, structure, water, life, and sediments: American Association of Petroleum Geologists Memoir 17, 532 p.

Evans, T. J., 1974, Bituminous coal in Texas: Austin, University of Texas Bureau of Economic Geology Guidebook 14, 122 p.

Ewing, M., and 7 others, 1969, Initial reports of the Deep Sea Drilling Project: Washington, D.C., U.S. Government Printing Office, v. 1, p. 84–111.

Feeley, H. W., and Kulp, J. L., 1957, Origin of Gulf Coast salt-dome sulphur deposits: American Association of Petroleum Geologists Bulletin, v. 41, p. 1802–1853.

Finkleman, R. B., and Casagrande, D. J., eds., 1986, Geology of Gulf Coast lignite geology; Biennial Lignite Symposium, Dallas, Texas: Reston, Virginia, Environmental and Coal Associates, 283 p.

Finkleman, R. B., Casagrande, D. J., and Benson, S. A., eds., 1987, Gulf Coast lignite geology; 14th Biennial Lignite Symposium, Dallas, Texas; Reston, Virginia, Environmental and Coal Associates, 283 p.

Fisher, W. L., 1963, Lignites of the Texas Gulf Coastal Plain: Austin, University of Texas Bureau of Economic Geology Report of Investigations 50, 164 p.

Flores Galacia, E., 1988, Geologia y reservas de los yacimientos de carbon en la Republica Mexicana, *in* Salas, G. P., ed., Geologia Economica de Mexico: Mexico, D.F., Fondo de Cultura Economica, p. 175–217.

Fountain, R. C., and Hayes, A. W., 1979, Uraniferous phosphate resources of the southeastern United States, *in* DeVoto, R. H., and Stevens, D. N., eds., Uraniferous phosphate resources, United States and Free World: U.S. Department of Energy Publication GJBX-110 (79), p. 65–122.

Freas, D. H., 1968, Exploration for Florida phosphate deposits, *in* Seminar on Sources of Mineral Raw Material for the Fertilizer Industry in Asia and the Far East: Proceedings, United States ECAFE, Mineral Resource Development Series 32, p. 187–200.

Friedman, S. A., Treworgy, C. G., Smith, C. J., and Aylsworth, J. A., 1988, Developments in coal in 1987; American Association of Petroleum Geologists Bulletin, v. 72, no. 10B, p. 368–377.

Galloway, W. E., 1977, Catahoula Formation of the Texas Coastal Plain; Depositional systems, composition, structural development, ground-water flow history, and uranium distribution: Austin, University of Texas Bureau of Economic Geology Report of Investigations 87, 59 p.

—— , 1982, Epigenenic zonation and fluid flow history of uranium-bearing fluvial aquifer systems, South Texas Uranium Province: Austin, University of Texas Bureau of Economic Geology Report of Investigations 119, 31 p.

Galloway, W. E., and Kaiser, W. R., 1980, Catahoula Formation of the Texas Coastal Plain; Origin, geochemical evolution, and characteristics of uranium deposits: Austin, University of Texas Bureau of Economic Geology Report of Investigations 100, 81 p.

Galloway, W. E., Kreitler, C. W., and McGowen, J. H., 1979, Depositional and ground-water flow systems in the exploration for uranium: Austin, University of Texas Bureau of Economic Geology, 267 p.

Goldhaber, M. B., and Reynolds, R. L., 1979, Origin of marcasite and its implications regarding the genesis of roll-type uranium deposits: U.S. Geological Survey Open-File Report 79–1696, 35 p.

Goldhaber, M. B., Reynolds, R. L., and Rye, R. O., 1978, Origin of a South Texas roll-type uranium deposit; 2, Petrology and sulfur isotope studies: Economic Geology, v. 73, p. 1690–1705.

—— , 1979, Formation and resulfidization of a South Texas roll-type uranium deposit: U.S. Geological Survey Open-File Report 79–1651, 41 p.

Gorsline, D. S., and Milligan, D. B., 1963, Phosphatic deposits along the margin of the Pourtales Terrace, Florida: Deep-Sea Research, v. 10, p. 259–262.

Halbouty, M. T., 1979, Salt domes, Gulf region, United States and Mexico, 2nd ed.: Houston, Texas, Gulf Publishing Company, 561 p.

Hamlin, H. S., 1983, Fluvial depositional systems of the Carrizo–upper Wilcox in south Texas: Gulf Coast Association of Geological Societies Transactions, v. 33, p. 281–287.

Hamner, J. M., 1978, How much lignite? Where?, *in* Proceedings, Ozarks Regional Commission Lignite Conference, Little Rock, Arkansas: Ozarks Regional Commission, p. 14–16.

Harshman, E. N., 1974, Distribution of elements in some roll-type uranium deposits, *in* Formation of uranium ore deposits: International Atomic Energy Agency Panel Proceedings No. STI/PUB 374, p. 169–183.

Hathaway, J. C., McFarlan, P. F., and Ross, D. A., 1970, Mineralogy and origin of sediments from drill holes on the continental margin off Florida: U.S. Geological Survey Professional Paper 581-E, 26 p.

Hathaway, J. C., Schlee, J. S., and Poag, C. W., 1976, Preliminary summary of the 1976 Atlantic margin coring project of the USGS: U.S. Geological Survey Open-File Report 76–844, 218 p.

Haynes, W., 1959, Brimstone, the stone that burns: New York, D. van Nostrand Company, 308 p.

Henry, C. D., and 5 others, 1982, Geochemistry of ground water in the Oakville Sandstone, a major aquifer and uranium host of the Texas Coastal Plain: Austin, University of Texas Bureau of Economic Geology Report of Investigations 118, 63 p.

Hoel, H. D., 1982, Goliad Formation of the South Texas Gulf Coastal Plain; Regional genetic stratigraphy and uranium mineralization [M.A. thesis]: Austin, University of Texas, 173 p.

Holbrook, D. F., 1980, Arkansas lignite investigations: Arkansas Geological Commission, Preliminary Report, 157 p.

Jogi, P. N., Gray, K. E., Ashman, T. R., Thompson, T. W., 1981, Compaction measurements on cores from the Pleasant Bayou wells: Baton Rouge, Proceedings Fifth Gulf Coast Geopressured-Geothermal Energy Conference, p. 75–81.

Johnston, J. E., and Mulcahy, S. A., 1982, Louisiana description of seams, *in* Keystone Coal Industry Manual: New York, McGraw Hill, p. 574.

Jones, P. H., 1975, Geothermal and hydrocarbon regimes, northern Gulf of Mexico Basin: Austin, Proceedings First Geopressured-Geothermal Energy Conference, p. 15–89.

Kaiser, W. R., 1985, Texas lignite; Status and outlook to 2000: Austin, University of Texas Bureau of Economic Geology Mineral Resource Circular 76, 17 p.

—— , ed., 1978, Proceedings 1976 Gulf Coast Lignite Conference: Austin, University of Texas Bureau of Economic Geology Report of Investigations 90, 276 p.

—— , 1989, Texas; Description of seams, *in* Keystone coal industry manual: Stamford, Connecticut, Maclean Hunter, p. 571–578.

Kaiser, W. R., Ayers, W. B., Jr., and LaBrie, L. W., 1980, Lignite resources in Texas: Austin, University of Texas Bureau of Economic Geology Report of Investigations 104, 52 p.

Kaiser, W. R., and 17 others, 1986, Geology and ground-water hydrology of deep-basin lignite in the Wilcox Group of east Texas: Austin, University of

Texas Bureau of Economic Geology Special Report, 182 p.

Kelley, P. K., 1925, The Sulphur salt dome, Louisiana: American Association of Petroleum Geologists Bulletin, v. 9, p. 479–496.

Kharaka, Y. K., Callender, E., and Carothers, W. W., 1977, Geochemistry of geopressured geothermal waters from the Texas Gulf Coast: Lafayette, Louisiana, Proceedings, Third Gulf Coast Geopressured-Geothermal Energy Conference, v. 2, p. 121–164.

Klitgord, K. D., and Popenoe, P., 1984, Florida; A Jurassic transform plate boundary: Journal Geophysical Research, v. 89, p. 7753–7772.

Kyle, J. R., Ulrich, M. R., and Gose, W. A., 1987, Textural and paleomagnetic evidence for the mechanism and timing of anhydrite cap rock formation, Winnfield salt dome, Louisiana, in Lerche, I., and O'Brien, J. J., eds., Dynamical geology of salt and related structures: Orlando, Florida, Academic Press, p. 497–542.

Lewis, C. R., and Rose, S. C., 1970, A theory relating high temperatures and overpressures: Journal of Petroleum Technology, v. 22, p. 11–16.

Loucks, R. G., Richmann, D. L., and Milliken, K. L., 1981, Factors controlling reservoir quality in Tertiary sandstones and their significance to geopressured geothermal production: Austin, University of Texas Bureau of Economic Geology Report of Investigations 111, 41 p.

Ludwig, K. R., Goldhaber, M. B., Reynolds, R. L., and Simmons, K. R., 1982, Uranium-lead isochron age and preliminary sulfur isotope systematics of the Felder uranium deposit, South Texas: Economic Geology, v. 77, p. 557–563.

Luppens, J. A., 1979, Exploration for Gulf Coast United States lignite deposits; Their distribution, quality, and reserves, in Argall, G. O., Jr., ed., Proceedings, Second International Coal Exploration Symposium: San Francisco, Miller Freeman Publishers, Inc., p. 1195–1210.

Manheim, R. M., Pratt, R. M., and McFarlin, P. F., 1980, Composition and origin of phosphorite deposits of the Blake Plateau, in Bentor, Y. K., ed., Marine phosphorites: Society of Economic Paleontologists and Mineralogists Special Publication 29, p. 117–137.

Marks, A. H., 1936, Hoskins Mound salt dome, Brazoria County, Texas: American Association of Petroleum Geologists Bulletin, v. 20, p. 155–178.

Marvasti, A., and Riggs, S. R., 1987, Potential for marine mining of phosphate within the U.S. Exclusive Economic Zone (EEZ): Marine Mining, v. 6, p. 291–300.

—— , 1989, The U.S. phosphate industry; Pressures for evolutionary change: Marine Technology Society Journal, v. 23, p. 27–36.

Mauk, F. J., 1982, Microseismic monitoring of Chocolate Bayou, Texas, the Pleasant Bayou No. 2 geopressured-geothermal energy test well program: Teledyne Geotech Technical Report 82-2, 74 p.

Mayberry, R. C., 1981, Phosphate reserves, supply and demand, southeastern Atlantic coastal states, 1980–2000 A.D.: Denver, Society of Mining Engineers, American Institute of Mining Engineering, 21 p. (preprint).

Morton, R. A., and Land, L. S., 1987, Regional variations in formation water chemistry, Frio Formation (Oligocene), Texas Gulf Coast: American Association of Petroleum Geologists Bulletin, v. 71, p. 191–206.

Morton, R. A., Ewing, T. E., and Tyler, N., 1983a, Continuity and internal properties of Gulf Coast sandstones and their implications for geopressured energy development: Austin, University of Texas Bureau of Economic Geology, Report of Investigations 132, 70 p.

Morton, R. A., Han, J. H., and Posey, J. S., 1983b, Variations in chemical compositions of Tertiary formation waters, Texas Gulf Coast: Austin, University of Texas Bureau of Economic Geology Report to the Department of Energy, Contract DE-AC08-79ET27111, p. 63–135.

Mukhopadhyay, P. K., 1989, Organic petrography and organic geochemistry of Texas Tertiary coals in relation to depositional environment and hydrocarbon generation: Austin, University of Texas Bureau of Economic Geology Report of Investigations 188, 118 p.

Mullins, H. T., and Neumann, A. C., 1979, Geology of the Miami Terrace and its paleo-oceanographic implications: Marine Geology, v. 30, p. 205–231.

Myers, J. C., 1968a, Gulf Coast sulfur resources, in Brown, L. F., Jr., ed., Proceedings of the Fourth Forum on Geology of Industrial Minerals: Austin, University of Texas Bureau of Economic Geology Special Publication, p. 57–65.

—— , 1968b, Sulfur; Its occurrence, production, and economics, in Beebe, B. W., and Curtis, B. F., eds., Natural gases of North America: American Association of Petroleum Geologists Memoir 9, v. 2, p. 1948–1956.

Norwood, E. M., Jr., and Holland, D. S., 1974, Lithofacies mapping; A descriptive tool for ancient delta systems of the Louisiana outer continental shelf: Gulf Coast Association of Geological Societies Transactions, v. 24, p. 175–188.

O'Brien, G. W., Harris, J. R., Milnes, A. R., and Veeh, H. H., 1981, Bacterial origin of east Australian continental margin phosphorites: Nature, v. 294, p. 442–444.

O'Donnell, L., 1935, Jefferson Island salt dome, Iberia Parish, Louisiana: American Association of Petroleum Geologists Memoir 9, v. 2, p. 1948–1956.

Ojeda-Rivera, J., 1978, Main coal regions of Mexico, in Kottlowski, F. E., Cross, A. T., and Meyerhoff, A. A., eds., Coal resources of the Americas: Geological Society of America Special Paper 179, p. 73–84.

Olson, N. K., 1966, Geology of the Miocene and Pliocene series in the north Florida–south Georgia area: Southeastern Geological Society Guidebook, 12th Annual Field Conference, 94 p.

Patterson, S. H., and Herrick, S. M., 1971, Chattahoochee Anticline, Appalachicola Embayment, Gulf Trough and related structural features, southwestern Georgia, fact or fiction: Georgia Geological Survey Information Circular 41, 16 p.

Poag, C. W., 1978, Stratigraphy of the Atlantic continental shelf and slope of the United States: Annual Review of Earth and Planetary Science, v. 6, p. 251–280.

Popenoe, P., 1983, High-resolution seismic reflection profiles collected August 4–28, 1979, between Cape Hatteras and Cape Fear and off Georgia and north Florida (Cruise GS-7903-6): U.S. Geological Survey Open-File Report 83–512, 3 p.

Popenoe, P., and Meyer, F. W., 1983, Description of single channel high-resolution seismic reflection data collected from continental shelf/slope and upper rise between Cape Hatteras, North Carolina, and Norfolk, Virginia, and Vero Beach to Miami, Florida (Cruise 80-G-9): U.S. Geological Survey Open-File Report 83–515, 4 p.

Price, P. E., and Kyle, J. R., 1983, Metallic sulfide deposits in Gulf Coast salt dome caprocks: Gulf Coast Association of Geological Societies Transactions, v. 33, p. 189–193.

Prior, W. L., Clardy, B. F., Baker, Q. M., 1985, Arkansas lignite investigations: Arkansas Geologic Commission Information Circular 28-C, 214 p.

Railroad Commission of Texas, 1989, Report on activities January through December, 1988: Austin, Railroad Commission of Texas, 1 p.

Reaves, R. J., 1984, The importance of by-product uranium to phosphate rock producers, in Harben, P. W., and Dickson, E. M., eds., Phosphates; What prospect for growth?: New York, Metal Bulletin, Inc., p. 247–252.

Reynolds, R. L., and Goldhaber, M. B., 1983, Iron disulfide minerals and the genesis of roll-type uranium deposits: Economic Geology, v. 78, p. 105–120.

Reynolds, R. L., Goldhaber, M. B., and Carpenter, D. J., 1982, Biogenic and nonbiogenic ore-forming processes in the South Texas Uranium District; Evidence from the Panna Maria deposit: Economic Geology, v. 77, p. 541–556.

Riggs, S. R., 1979a, Petrology of the Tertiary phosphorite system of Florida: Economic Geology, v. 74, p. 195–220.

—— , 1979b, Phosphorite sedimentation in Florida; A model phosphogenic system: Economic Geology, v. 74, p. 285–314.

—— , 1980, Intraclast and pellet phosphorite sedimentation in the Miocene of Florida: Journal of Geological Society of London, v. 137, p. 741–748.

—— , 1981, Relation of Miocene phosphorite sedimentation to structure in the Atlantic continental margin, southeastern United States: American Association of Petroleum Geologists Bulletin, v. 65, p. 1669.

—— , 1982, Phosphatic bacteria in the Neogene phosphorites of the Atlantic Coastal Plain–continental shelf system: Geological Society of America Abstracts with Programs, v. 14, p. 77.

—— , 1984, Paleoceanographic model of Neogene phosphorite deposition, U.S. Atlantic continental margin: Science, v. 223, p. 123–131.

——, 1986a, Future frontier for phosphate in the Exclusive Economic Zone; Continental shelf of southeastern United States, *in* Lockwood, M., and Hill, G., eds., Exploring the New Ocean Frontier; Proceedings of the Exclusive Economic Zone Symposium: Rockville, Maryland, National Oceanic and Atmospheric Administration, p. 97–101.

——, 1986b, Future U.S. phosphate resources; A new perspective, *in* Bush, W. R., eds., Economics of internationally traded minerals: Littleton, Colorado, Society of Mining Engineers, p. 153–159.

——, 1987, Model of Tertiary phosphorites on the world's continental margins, *in* Teleki, P. G., Dobson, M. R., Moore, J. R., and von Stackelberg, U., eds., Marine minerals: Dordricht, Holland, D. Reidel Publishing Company, p. 99–118.

Riggs, S. R., and Manheim, F. T., 1988, Mineral resources of the U.S. Atlantic continental margin, *in* Sheridan, R. E., and Grow, J. A., eds., The Atlantic Continental Margin, U.S.: Boulder, Colorado, Geological Society of America, The Geology of North America, v. I-2, p. 501–520.

Riggs, S. R., Lewis, D. W., Scarborough, A. K., and Snyder, S. W., 1982, Cyclic deposition of Neogene phosphorites in the Aurora area, North Carolina, and their possible relationship to global sea-level fluctuations: Southeastern Geology, v. 23, p. 189–204.

Riggs, S. R., and 5 others, 1985, Geologic framework of phosphate resources in Onslow Bay, North Carolina continental shelf: Economic Geology, v. 80, p. 716–738.

Roland, H. L., Jr., Jenkins, G. M., and Pope, D. E., 1976, Lignite; Evaluation of near-surface deposits in northwest Louisiana: Louisiana Geological Survey Mineral Resources Bulletin 2, 39 p.

Salas, G. P., 1988, Historia de la exploracion en busca del azufre en el sureste de Mexico, *in* Salas, G. P., ed., Geologia Economica de Mexico: Mexico, D.F., Fondo de Cultura Economica, 544 p.

Sassen, R., 1980, Biodegradation of crude oil and mineral deposition in a shallow Gulf Coast salt dome: Organic Geochemistry, v. 2, p. 153–166.

Scott, T. M., 1982, A comparison of the cotype localities and cores of the Miocene Hawthorn Formation in Florida, *in* Scott, T. M., and Upchurch, S. B., eds., Miocene of the southeastern United States: Southeastern Geological Society and Florida Bureau of Geology Special Publication 25, p. 237–246.

——, 1983, The Hawthorn Formation of northeastern Florida: Florida Bureau of Geology Report of Investigation 94.

——, 1988, The lithostratigraphy of the Hawthorn Group (Miocene) of Florida: Florida Geological Survey Bulletin 59, 148 p.

Seni, S. J., 1987, Evolution of Boling Dome cap rock with emphasis on included terrigenous clastics, Ford Bend and Wharton Counties, Texas, *in* Lerche, I., and O'Brien, J. J., eds., Dynamical geology of salt and related structures: Orlando, Florida, Academic Press, p. 543–591.

Sever, C. W., Cathcart, J. B., and Patterson, S. H., 1967, Phosphate deposits of south-central Georgia and north-central peninsular Florida: Georgia Division of Conservation South Georgia Minerals Project Report 7, 62 p.

Sheldon, R. P., 1981, Ancient marine phosphorites: Annual Review of Earth and Planetary Science, v. 9, p. 251–284.

Shelton, J. E., 1980, Sulfur, *in* Mineral facts and problems: U.S. Bureau of Mines Bulletin 671, p. 877–898.

Stowasser, W. F., 1989, Phosphate rock, *in* Minerals yearbook; [Metals and minerals]: Washington, D.C., U.S. Government Printing Office, U.S. Bureau of Mines, v. 1, p. 673–688.

Stuart, C. A., 1970, Geopressures; Supplement to the Proceedings of the Second Symposium on Abnormal Subsurface Pressure: Baton Rouge, Louisiana State University, 121 p.

Taylor, R. E., 1938, Origin of the caprock of Louisiana salt domes: Department of Conservation, Louisiana Geological Survey Bulletin 11, 189 p.

Tewalt, S. J., 1986, Chemical characterization of Texas lignites: Austin, University of Texas Bureau of Economic Geology Geological Circular 86-1, 52 p.

Thode, H. G., Wanless, R. K., and Wallouch, R., 1954, The origin of native sulphur deposits from isotope fractionation studies: Geochimica et Cosmochimica Acta, v. 5, p. 286–298.

Tolson, J. S., 1985, Alabama lignite: Geological Survey of Alabama Bulletin 123, 216 p.

U.S. Bureau of Mines, 1931 to 1988, Minerals Yearbook: Washington, D.C., U.S. Department of the Interior (sulfur chapters of each year).

U.S. Geological Survey, 1980, Principles of a resource/reserve classification for minerals: U.S. Geological Survey Circular 831, 5 p.

Wallace, R. H., Jr., Kraemer, T. F., Taylor, R. E., and Wesselman, J. B., 1978, Assessment of geopressured resources in the northern Gulf of Mexico Basin, *in* Muffler, L.J.P., ed., Assessment of geopressured resources of the United States, 1978: U.S. Geological Survey Circular 790, p. 132–155.

Walton, A. W., Galloway, W. E., and Henry, C. D., 1981, Release of uranium from volcanic glass in sedimentary sequences; An analysis of two systems: Economic Geology, v. 76, p. 69–88.

White, D. M., Kaiser, W. R., and Groat, C. G., 1982, Status of Gulf Coast lignite activity, *in* Kube, W. R., Sondreal, E. A., and White, D. M., compilers, Technology and use of lignite: Grand Forks, North Dakota, Grand Forks Energy Technology Center, U.S. Department of Energy, GFETC/IC—83/1 (DE82015926), p. 107–141.

Williamson, D. R., 1976, An investigation of the Tertiary lignites of Mississippi: Mississippi Geologic Economic and Topographic Survey Information Series MGS-74-1, 147 p.

——, 1986, Lignites of Alabama, Arkansas, and Mississippi, *in* Finkelman, R. B., and Casagrande, D. J., eds., Geology of Gulf Coast lignites: Houston, Texas, Environmental and Coal Associates, p. 115–125.

Winker, C. D., Morton, R. A., Ewing, T. E., and Garcia, D. D., 1983, Depositional setting, structural style, and sandstone distribution in three geopressured geothermal areas, Texas Gulf Coast: Austin, University of Texas Bureau of Economic Geology Report of Investigations 134, 60 p.

Wolf, G. A., 1933, The Boling Dome, Texas, *in* Oklahoma and Texas: Washington, D.C., 16th International Geological Congress Guidebook to Excursion A-6, p. 86–91.

Woodward, M. B., and Gilreath, L., 1976, Arkansas lignite characteristics: Arkansas Geological Commission Report prepared for the U.S. Bureau of Mines.

Zellers, M. E., and Williams, J. M., 1978, Evaluation of the phosphate deposits of Florida using minerals availability system: U.S. Bureau of Mines Open-File Report 112–78, 106 p.

Manuscript Accepted by the Society July 31, 1990

ACKNOWLEDGMENTS

By S. R. Riggs

I acknowledge the following persons for reviewing this manuscript and offering many helpful suggestions: Frank Brown of the Texas Bureau of Economic Geology, Thomas Scott of the Florida Bureau of Geology, and Amos Salvador of the University of Texas at Austin.

By W. L. Fisher and M.L.W. Jackson

Earle F. McBride contributed information about the lignite and coal deposits in northeast Mexico, and W. R. Kaiser, J. A. Luppens, and D. R. Williamson supplied information on lignite deposits and production in the United States. L. F. Brown and R. A. Morton reviewed the manuscript.

By S. P. Ellison, Jr.

Frank Brown, Alan F. Edward, Riley Davenport, and Ray Young read the manuscript and offered many comments and suggestions for its improvement. Their contribution is gratefully acknowledged.

By R. A. Morton

Critical reviews of the section on geopressured-geothermal energy were provided by Myron Dorfman, Earl Ingerson, and Amos Salvador and are gratefully acknowledged.

Printed in U.S.A.

Chapter 17

Ground water

John M. Sharp, Jr.
Department of Geological Sciences, The University of Texas at Austin, Austin, Texas 78713-7909
Charles W. Kreitler
Texas Bureau of Economic Geology, The University of Texas at Austin, Austin, Texas 78713
Juan Lesser
Lesser y Asociados, S.A., Seminario 119, Queretaro 76050, Qro., Mexico

INTRODUCTION

Ground water is an important natural resource in the Gulf of Mexico region. Although the region as a whole is relatively rich in water resources, water availability and quality vary dramatically throughout the area. Water resources include rivers—the Chattahoochee, Alabama, Mississippi, Sabine, Trinity, Brazos, Colorado, Rio Grande, and Grijalva—and the immense ground-water resources that are the focus of this chapter. Many major cities in the Gulf of Mexico basin (e.g., Miami, in Florida; Memphis, in Tennessee; Gulfport and Baton Rouge, in Louisiana; and Houston and San Antonio, in Texas) rely chiefly on ground water. Climate varies from moist temperature or subtropical to semiarid conditions, and while population densities (and water-resource demands) vary from minimal to intense, it can be stated that fresh, potable water is the mineral resource in greatest demand. Its continued availability in the region will require special attention in the near future. As elsewhere, the foremost challenge is the provision of adequate quantities of good-quality water, but several special problems exist in the region, including salt-water intrusion and subsidence. An understanding of the hydrogeologic setting of aquifers in the Gulf of Mexico basin is required to preserve and fully utilize this valuable resource.

Aquifers in the Gulf of Mexico basin area (Fig. 1) may be grouped into the following categories: clastic sediments dipping toward the center of the basin; the major carbonate systems of Florida, Texas, and Yucatan; and less importantly, major alluvial aquifers, island aquifers, and volcanic aquifers.

CLASTIC AQUIFERS

The thick section of predominantly Cenozoic clastic sediments is perhaps the chief characteristic of the Gulf of Mexico basin. The sandy units, where part of the meteoric hydrodynamic regime, serve as productive aquifers. Figure 2 is a cross section (B-B' in Fig. 1) indicating the major features of this section and is adapted from the studies of Fogg and Kreitler (1982), Galloway

Figure 1. Major aquifers in the Gulf of Mexico basin. Circled letters refer to hydrostratigraphic columns in Table 1.

and others (1982), Wesselman (1983), Donnelly (1988), Grubb and Carrillo R. (1988), Hosman (1988), Pettijohn and others (1988), and Sharp and others (1988). The aquifers are bounded by their outcrop, by lower and, for confined systems, upper clay-rich units, and on occasion, by either geopressures or brackish waters. The major sources of salinity are dissolution of disseminated evaporite minerals within the aquifer, "connate" marine waters, brines migrating upward from the geopressured section, or brines derived from salt-dome dissolution. In most cases, the 10,000 mg/l TDS (total dissolved solids) isocon is shallower than the top of geopressures, but the existence of geopressures is a key

Sharp, J. M., Jr., Kreitler, C. W., and Lesser, J., 1991, Ground water, *in* Salvador, A., ed., The Gulf of Mexico Basin: Boulder, Colorado, Geological Society of America, The Geology of North America, v. J.

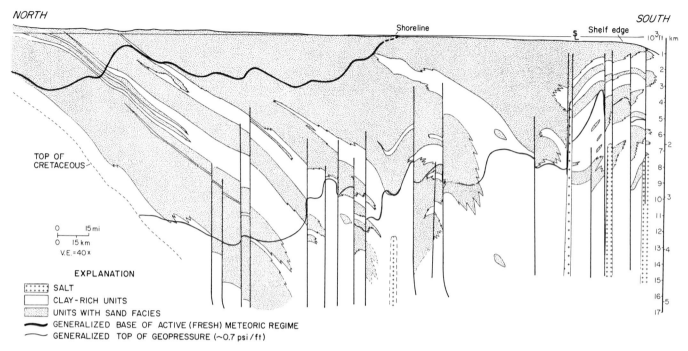

Figure 2. Generalized cross section B-B′ through the northern Gulf Coast clastic section.

element in the clastic units. As demonstrated in a comparison of Figure 2 with Figures 3 and 4, fluid pressures in the meteoric zone are nearly hydrostatic, but at depths of 2 to 4 km the sediments commonly become highly over-pressured.

The geologic evolution of the sediments and sedimentary rocks that make up the aquifers has been treated in detail elsewhere (e.g., Blanchard, 1987; Sharp and others, 1988; Kreitler, 1989; Sharp and McBride, 1989). The sedimentary rocks are exposed to a variety of hydrodynamic systems that leave their imprint on the aquifers. Figure 5 is a flow chart of the possible hydrodynamic systems encountered. The aquifer systems discussed below have been exposed primarily to meteoric water diagenesis. Some have been buried deeply enough to have been altered by deep-burial processes, possibly augmented by free convection.

Generalized hydrostratigraphic columns are given in Table 1. The oldest clastic units of interest are of Cretaceous age. Between Linares and Ciudad Victoria, northwestern Mexico, is a large outcrop of Upper Cretaceous fractured shale. Because of the lack of alternative water resources in the area, this thin section (<50 m) of fractured shale is an important local aquifer. In the upper Mississippi embayment, the Cretaceous McNairy or Ripley Formation is the lowermost important clastic aquifer (Groshkopf, 1955; Brahana and Mesko, 1988), and in central Alabama and adjacent states, the lowermost clastic aquifer is in the sediments of the Cretaceous Tuscaloosa Group (Renken, 1984; Miller and others, 1987).

In most areas of the coastal plain, Cenozoic sandy units are the major aquifers; the shales are not fractured sufficiently to serve as aquifers and serve solely as aquitards. The oldest

Cenozoic clastic units of interest are of Paleocene age. During the Paleocene a major transgression created the widespread blanket of clay-rich rocks of the Midway Group. These serve as a confining unit for underlying aquifers throughout the Gulf of Mexico basin. The overlying Paleocene-Eocene Wilcox Group is mostly of nonmarine, fluvial/deltaic origin. Numerous sand lenses of the Wilcox form the lower portions of the Texas Coastal Uplands

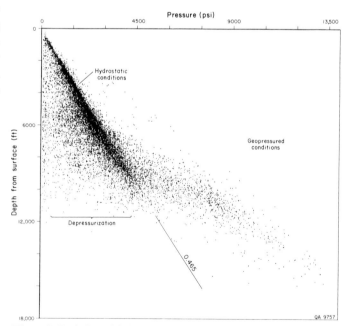

Figure 3. Variation of fluid pressures with depth in the Gulf of Mexico basin in the Texas Frio (Oligocene) Formation (after Kreitler, 1989).

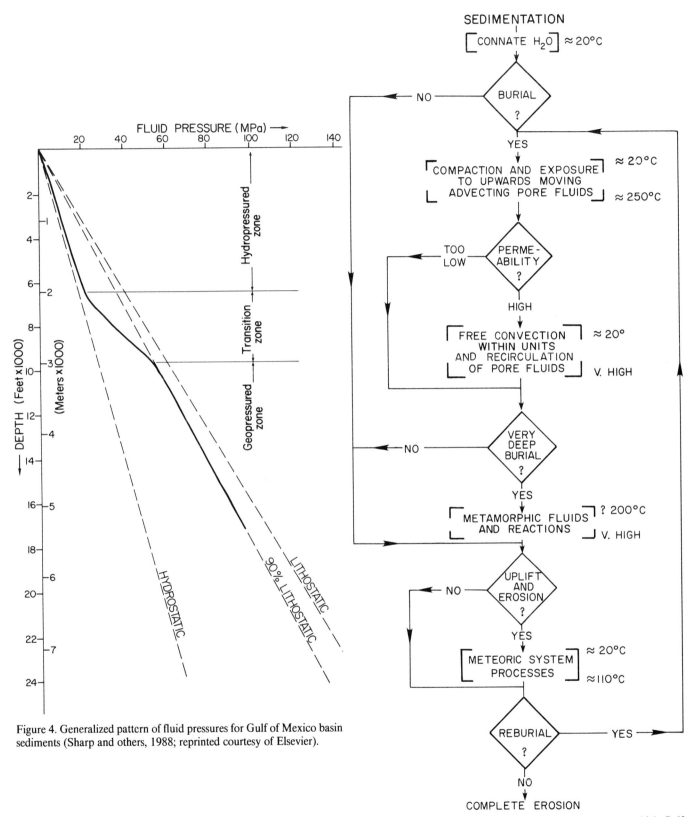

Figure 4. Generalized pattern of fluid pressures for Gulf of Mexico basin sediments (Sharp and others, 1988; reprinted courtesy of Elsevier).

Figure 5. Flow chart of possible hydrodynamic systems to which Gulf Coast aquifers are potentially exposed during their evolution (Sharp and McBride, 1989; reprinted courtesy of Elsevier).

TABLE 1. GENERAL HYDROSTRATIGRAPHIC COLUMNS FOR GULF OF MEXICO CLASTIC UNITS*

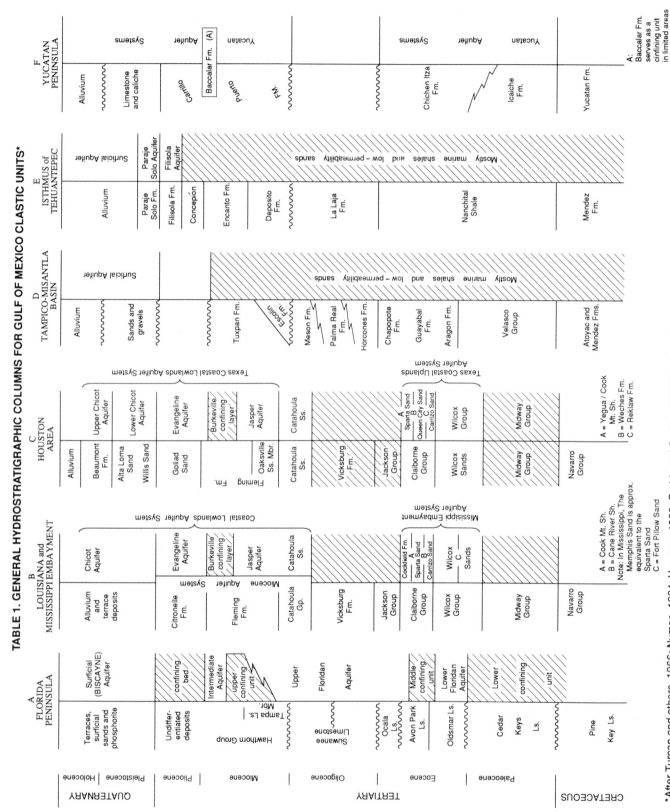

*After Turcan and others, 1966; Nyman, 1984; Hosman, 1988; Grubb and Carillo, 1988; Martin and others, 1988; and Scott, 1989.

and Mississippi embayment aquifer systems of Grubb (1984). In south Texas, the Upper Wilcox is an important regional aquifer (Hamlin, 1988). The Wilcox contains many such important aquifers, but the sands, except for the the Fort Pillow sand in the Mississippi embayment, tend to be regionally discontinuous. The overlying Claiborne Group Carrizo Sandstone, which is also nonmarine, contains more regionally extensive sands, as well as more regionally extensive clay layers. The Claiborne forms the upper portion of the Texas Coastal Uplands and Mississippi embayment aquifer systems (Grubb and Carrillo, 1988). The Claiborne's areally extensive sand/sandstone layers, the Cockfield Formation, the Sparta Sand, and the Memphis Sand, are regional aquifers. Farther south, along the coastal plain in northeast Mexico, the Claiborne Group forms a multilayered brackish-water aquifer.

During the late Eocene and Oligocene, a regional transgression deposited the massive marine clays of the Jackson and Vicksburg Groups, which serve as confining layers. Overlying these clays is a thick sequence of Miocene and Pliocene fluvial, deltaic, and shallow-marine deposits—a sequence of alternating, lensoidal, high- and low-permeability units. These are part of the Coastal Lowlands aquifer system in the United States (Grubb, 1984), but do not form significant aquifers along the Mexican portion of the Gulf of Mexico Coastal Plain. The Tertiary section in east-central Mexico is composed mainly of shales; the sands are few and discontinuous. In Texas, locally important units in the Coastal Lowlands aquifer system include the Oakville and Catahoula Sandstones, which also host significant uranium deposits, and the Goliad Sand. Also included in this sequence are the Evangeline and Jasper aquifers of Texas and the Miocene (actually Miocene-Pliocene) aquifer of Louisiana and Mississippi (Hosman, 1988). The Veracruz area is supplied by an alluvial aquifer composed of 100 to 200 m of Tertiary sand and gravel. In the Isthmus of Tehuantepec area of southern Mexico, the Filisola Sand forms an important aquifer near Villahermosa. The Paraje Solo Sand near Veracruz has good potential for future development. Finally, Quaternary alluvial, terrace, and coastal deposits cover the shallowest, coastward portions of the Gulf of Mexico basin. These form important local aquifers onshore and on the barrier islands, and the regionally important Chicot aquifer of the Texas/Louisiana coast. The geologic age of the aquifers generally becomes younger toward the coast, although in some areas, two or more aquifers of different ages are stacked.

Recharge and discharge to these clastic systems correspond to topography. Recharge occurs on the topographic divides, and discharge is generally to the major river systems (Fogg and Kreitler, 1982; Smith and others, 1982; Grubb and Carrillo R., 1988). In the very shallow-dipping, very low-relief regions near the coastline, there is discharge directly to the Gulf of Mexico in bays and lagoons from the shallow unconfined units, and possibly by cross-formational flow from confined aquifers. In general, however, flow is to the river systems (Grubb and Carrillo R., 1988, their Figs. 4, 5, and 6). This limits the amount of flow down the structural dip of these large homoclinal sedimentary

packages and also cross-formational flow. Pleistocene sea-level changes may also have had a significant impact on the depth of penetration of fresh ground water. A sea-level drop of approximately 100 m would have caused the shoreline to migrate approximately 320 km gulfward in the northern part of the basin. The continental shelf, which is beneath the Gulf of Mexico, would have been subaerially exposed. The extent of meteoric systems in the Gulf Coastal Plain would have been significantly greater during low sea-level stands, but today the coast generally represents the limit of fresh waters (Kreitler and others, 1977). This indicates rapid flushing of most of the shallow systems, but the degree of flushing of deeper units by meteoric waters remains a controversy (Blanchard, 1987; Bethke, 1989).

The salinity distribution in the clastic aquifers is described in detail by Pettijohn and others (1988), among others. Figure 6 depicts these variations for some of the clastic aquifer units of the northern Gulf of Mexico Coastal Plain, including data for portions of the continental shelf. Low TDS waters occur significantly deeper in the highly transmissive sandstones. The general downflow increase in salinity is evident, as are zones of brines (TDS $\sim 10^5$ mg/l) in many cases, but not in all, which originate by dissolution of salt diapirs (Hanor and others, 1986; Hanor, 1987; Sharp and others, 1988; Kreitler and others, 1989). Major growth faults, however, may limit the coastward extent of fresh ground water (Rollo, 1969; Kreitler, 1977a, b; Kreitler and others, 1977).

Three outcrop maps are shown: the middle Claiborne aquifer (Fig. 7), the mid-Miocene aquifer system (Fig. 8), and upper Pleistocene-Holocene deposits, including the Mississippi River alluvium (Fig. 9).

The middle Claiborne aquifer crops out in a narrow belt near the margin of Tertiary sediments and around the Sabine uplift. Freshwaters (salinities < 1,000 mg/l) are generally present in the outcrop area and for some distance downdip. In the Mississippi embayment, the freshwater section is areally extensive. The extent of the freshwater section in the middle Claiborne aquifer diminishes southward; south of San Antonio, brackish waters exist even in the outcrop. This is caused by an increase in the amount of shale present and greater aridity. Several local zones of brackish water, one west of the Sabine uplift and one in the Mississippi embayment, represent cross-formational flow, mixing with saline waters leaking from underlying units. The more saline or briny (> 70,000 mg/l) zones in southern Louisiana represent salt-dome dissolution.

The mid-Miocene aquifer system (Fig. 8) shows similar patterns. The freshwater section is most extensive in Mississippi, Alabama, and Louisiana. Note also the greater extent of the briny facies. The shifting of outcrop and freshwater zones gulfward is the general rule in these clastic units.

Figure 9 depicts the outcrop and salinity zones of upper Pleistocene-Holocene sediments. The freshwater–brackish-water line lies roughly at the coast. It extends slightly gulfward east of New Orleans and inland in south Texas. Again, greater aridity and greater proportions of clay-rich sediment are the probable causes. Salinities of less than mean seawater ($\sim$ 35,000 mg/l) are

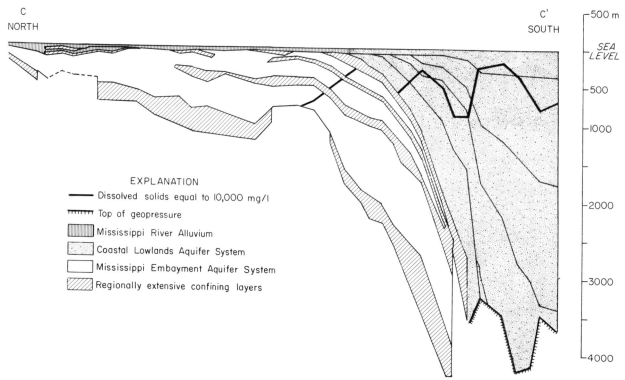

Figure 6. Generalized cross section (C-C') through the Mississippi embayment (after Grubb and Carillo R., 1988).

not found far offshore from Texas, which indicates the difficulty in flushing saline waters from the system even under greatly lowered sea levels. On the other hand, Pettijohn and others (1988) indicate dissolved solids of less than 35,000 mg/l to the limit of the Louisiana continental shelf, suggesting different conditions perhaps related to the greater rate of Pleistocene sediment deposition in offshore Louisiana. Local areas of brackish waters in the Mississippi embayment represent mixing with ground water from underlying units.

The Coastal Lowland aquifer system (Grubb, 1984) contains geologically young sediments. Grubb and Carrillo R. (1988) extend this unit to the edge of the continental shelf and include all sediments from the land surface of sea floor to the shallower of the top of the Vicksburg-Jackson confining unit or the top of the geopressured zone. The presence of brackish waters in the Beaumont Clay and the thin freshwater section in the Pleistocene of southern Louisiana, and the absence of fresh waters in shallow, confined offshore units, strongly suggest the presence of original ("connate") waters of deposition and inadequate geologic time to flush the "connate" waters.

MAJOR CARBONATE SYSTEMS

Flanking the clastic aquifers of the Cenozoic discussed above are three important carbonate aquifer systems: The Floridan aquifer system, the Edwards (Balcones fault zone) aquifer of Texas, and the Yucatan carbonate aquifer (see Fig. 1). These aquifers are extremely productive and the sole sources of

water in their respective areas. The Floridan aquifer supplies the needs of most of Florida (except for the Biscayne aquifer in the greater Miami area); the Yucatan carbonates supply the peninsula and the island of Cozumel. The Edwards aquifer supplies over 2 million people, including the greater San Antonio area, with water.

The *Floridan aquifer system* is a thick (up to 1 km), complex system of Paleocene to Miocene carbonates (Stringfield, 1936; Hanshaw and others, 1965; LeGrand and Stringfield, 1966; Miller, 1986; Johnston and Miller, 1988; Bush and Johnston, 1988; Johnston and Bush, 1988). Four major hydrostratigraphic units are delineated: (1) Lower Eocene and older limestones and clastics that typically contain saline waters; (2) Middle Eocene to Middle Miocene limestones; (3) the confining Hawthorn Group; and (4) the unconfined Quaternary limestones that form the Biscayne aquifer of the Miami area. The Eocene units are the most important hydrogeologically. The aquifer is unconfined along the Florida panhandle, northern Florida, and southern Georgia (Fig. 10). Recharge is abundant in the unconfined and semiconfined areas because of high rainfall (as much as 1,400 mm/yr) and the region's flat topography. Flow is generally from the outcrop and central Florida areas toward the coast. Discharge (mostly via springs) is to several of the large rivers, such as the Suwannee, and by cross-formational flow from confined portions near and even offshore in submerged springs in the Gulf known as "blue holes." The freshwater "lens" of the Floridan aquifer is approximately 600 m (2,000 ft) in north-central Florida, but is very thin in southern Florida.

MIDDLE CLAIBORNE AQUIFER

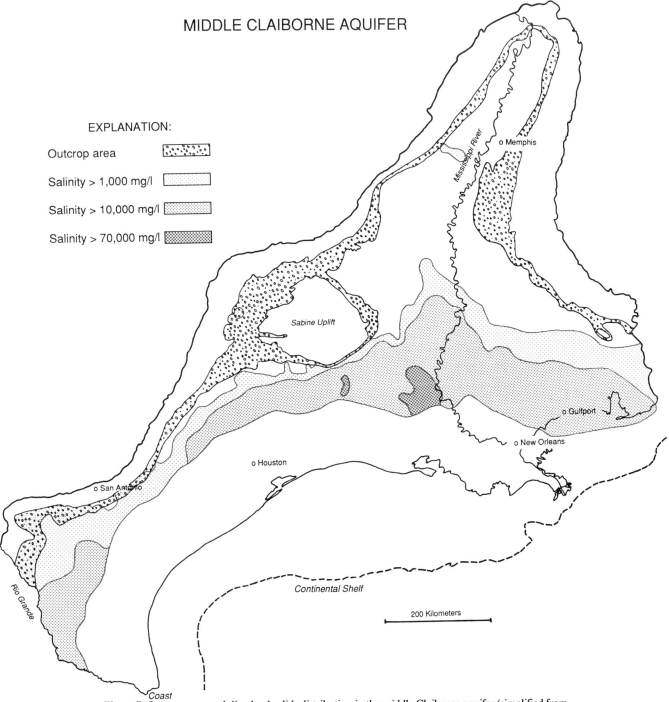

Figure 7. Outcrop area and dissolved solids distribution in the middle Claiborne aquifer (simplified from Pettijohn and others, 1988).

Permeability in the Floridan aquifer is controlled by present and paleokarstification episodes. Dissolution during Miocene and Pleistocene lowstands of sea level created a deep (to 152 m; 500 ft) zone of karstification. In addition, dolomitization, mostly in the freshwater/saltwater mixing area, has both increased and decreased permeability (Thayer and Miller, 1984). Finally, the hydrogeologic effects of the Hawthorn Group must be con-

sidered. The Hawthorn limits discharge rates of the underlying units, and thus, depth to the freshwater/saltwater interface in the Floridan aquifer is much deeper than in the Yucatan platform, discussed below. In addition, the greater thickness of the Hawthorn in southern Florida has limited the development of the regional flow system and inhibited karstification of the aquifer. Consequently, the Floridan aquifer is not a significant water resource in south Florida, and the unconfined Biscayne aquifer must be relied upon.

The *Edwards aquifer* (Livingston and others, 1936; Sayre and Bennett, 1942; Woodruff and Abbott, 1979; Maclay and Land, 1988; Sharp, 1990) is found in Cretaceous rocks, deposited on a broad carbonate platform. Miocene-age faulting created a series of down-dropped blocks that control the position of the aquifer (Fig. 11). About 85 percent of the recharge is from

ephemeral, losing streams flowing across the outcrop. Flow is generally subparallel to strike, and discharge is via majo springs where the main gulfward-flowing rivers have cut into or through the aquifer (Fig. 12). The southern and eastern boundary of the aquifer is the "bad-water line," the 1,000 mg/l TDS isocon. Unlike the Floridan and Yucatan aquifer systems, where this isocon represents the mixing of fresh water with marine waters, the bad-water line of the Edwards aquifer is formed by three

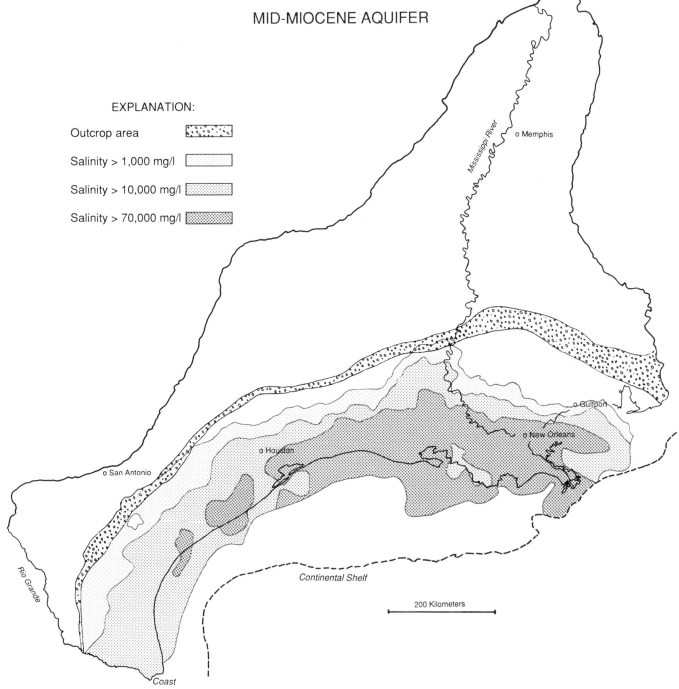

Figure 8. Outcrop area and dissolved solids distribution in the mid-Miocene aquifer system (simplified from Pettijohn and others, 1988).

processes: dissolution of disseminated gypsum and anhydrite (this predominates in the western portions near the Rio Grande); mixing of oil-field–like brines from down-dip portions of the Edwards or from underlying units (predominant in the San Antonio–Austin area); and possible cross-formational flow of sulfate-facies water from underlying Cretaceous units, which becomes more important in areas north of Austin, Texas (Sharp and Clement, 1988).

The great permeability of the Edwards aquifer is derived from karstification episodes—one in the Cretaceous and one in the late Cenozoic (late Miocene through Holocene). These episodes effectively leached all halide minerals, selectively dissolved gypsum and anhydrite in evaporite-rich beds and rudistid reef deposits; formed numerous dolines; and enhanced permeability parallel to fault lines.

The *Yucatan Peninsula* possesses an extremely productive

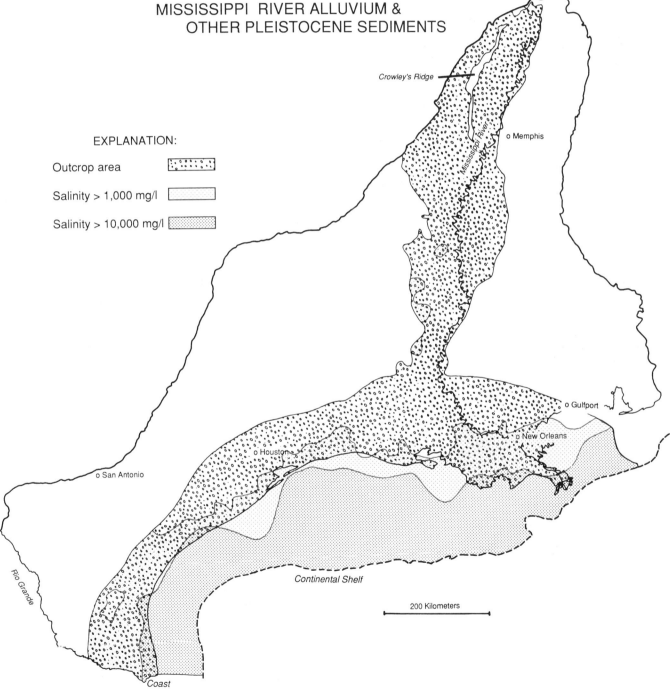

Figure 9. Outcrop area and dissolved solids distribution in the Mississippi River alluvium and other Pleistocene sediments (simplified from Pettijohn and others, 1988).

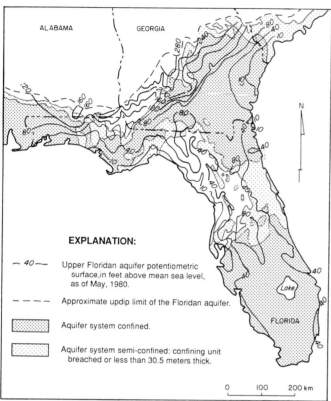

Figure 10. The Floridan aquifer system, showing areas of confinement and semiconfinement. Generalized equipotentials are also depicted (after Johnston and Miller, 1988).

aquifer system (Lesser, 1976; Lesser and others, 1978; Back and Lesser, 1981; Marin and others, 1990). The peninsula is generally situated at less than 50 m above sea level in a humid tropical climate (800 to 1,700 mm/yr precipitation), but there is both a wet and a dry season. The Tertiary carbonates possess a maximum thickness of about 1,000 m (Lesser and Weidie, 1988) and overlie Cretaceous carbonates and evaporites. The major aquifers are developed in Eocene or Miocene/Pliocene units. In many respects, the Yucatan carbonates are similar to those in Florida, but the low relief and the absence of significant confining layers have led to a vastly different hydrogeologic setting.

The exposed carbonates are subject to intense chemical weathering, which leads to the formation of the cenotes (vertical limestone shafts, open to the surface, which contain standing water), collapsed bays (caletas) along the coasts, elongated depressions along fault trends, and the formation of tropical soils. Rainwater enters the aquifer still unsaturated with respect to calcite. In the eastern Yucatan, coastal geomorphology is controlled by fracture trends and calcite dissolution, which results from the mixing of meteoric and marine waters (Back and others, 1979). No confining layers are present, unlike in Florida. The aquifer is therefore extremely permeable. Heads in the aquifer vary from only a few meters in the central part of the Peninsula to a few centimeters above sea level near the coast (Fig. 13), where the freshwater aquifer is relatively thin because of both the very high transmissivity and sea-water intrusion. At Chichén Itzá, for example, 80 km from the coast, the land surface and the water table are at elevations of 30 and 1.2 m above sea level, respectively. The lens of fresh water, up to several tens of meters thick, thins toward the coast (Fig. 14). Salinity is controlled by sea-water intrusion on a massive scale and dissolution of evaporite minerals. The intrusion occurs cyclically. During the dry season, the freshwater/saltwater interface moves inland from 100 m to as much as 12 km in Dzidzantun (north-central part of Yucatan; Lesser and Weidie, 1988) because of high pumping, lack of re-

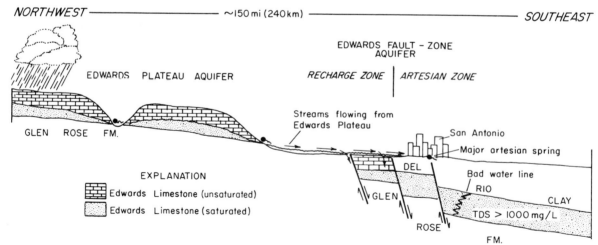

Figure 11. Cross section through the Edwards aquifer, depicting fault controls of aquifer geometry (from Clement and Sharp, 1988; reprinted courtesy of the National Water Well Association, Dublin, Ohio).

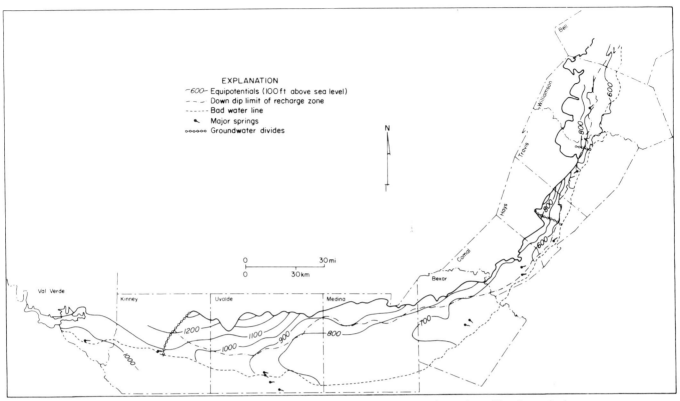

Figure 12. Edwards aquifer's confined and unconfined areas and major springs. Also shown is the bad-water line (Clement and Sharp, 1988; reprinted courtesy of the National Water Well Association, Dublin, Ohio).

charge, high permeability, and lack of effective storage in the carbonates. During the rainy season, the interface moves coastward. The drops of sea level during Pleistocene and earlier low sea-level stands, and therefore drops in the regional base level, have probably been a major control on developing the very high permeabilities observed in both the Floridan and Yucatan carbonate aquifers.

QUATERNARY ALLUVIAL AQUIFERS

The large rivers flowing through the Gulf Coastal Plain are commonly associated with thick bands of alluvium (Rosenshein, 1988; Sharp, 1988). The flood-plain alluvial systems are very important water resources (Ackerman, 1989). The most important alluvial system in terms of potential production is associated with the Mississippi River. The Mississippi River alluvium consists of several hundred meters of clastic sediment deposited during the late Pleistocene (Fisk, 1944; Boswell and others, 1968; Sharp, 1988). Thick top-stratum clays create a significant confined aquifer in western Mississippi; in other sites, the aquifer is unconfined. The Brazos, Red, and Colorado Rivers also form important but local aquifers.

Water quality in alluvial systems, although hard, is generally suitable for most uses. Iron, above drinking-water standards, is occasionally present, and in some areas, salinities become high

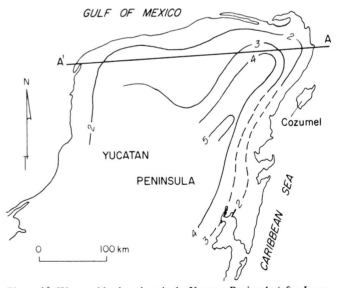

Figure 13. Water-table elevations in the Yucatan Peninsula (after Lesser and Weidie, 1988).

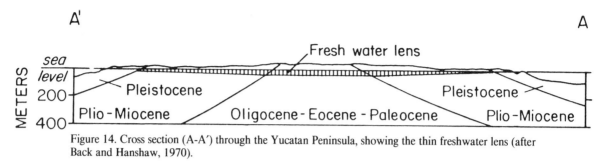

Figure 14. Cross section (A-A') through the Yucatan Peninsula, showing the thin freshwater lens (after Back and Hanshaw, 1970).

because of significant flow from adjoining clastic units. The sediments in these alluvial aquifers reflect the paleodepositional controls of varying sea level and stream discharge during the Pleistocene. The rivers tend to be underfit, and the most productive, coarsest sediment is often found in the substratum near the base of the alluvial fill.

OTHER AQUIFERS

In east-central Mexico, Pliocene-age basalt flows and alluvial-valley sediments are the major ground-water providers (Grubb and Carrillo R., 1988). The basalt flows cover several thousand square kilometers and are several tens of meters thick. The alluvial systems cut across these basalt flows. The alluvium may be several tens of meters thick.

On the barrier islands that parallel much of the northwestern and northern coast of the Gulf of Mexico, freshwater lenses are important local aquifers. Well-sorted beach sands provide aquifers with saturated thicknesses of up to tens of meters overlying saline water. Highly permeable limestones, such as the Island of Cozumel, have much thinner lenses. Consequently, moderate (10^{-2} to 10^{-4} cm/sec) hydraulic conductivity sediments provide the best aquifers. Water on the islands is limited because of the potential for salt-water intrusion and the thin freshwater lenses, but the water is typically of good quality.

CRITICAL PROBLEMS

While the ground-water resources of the Gulf of Mexico basin are vast, there are, as alluded to above, significant problems, both existing and future. These include water availability, natural water quality, pollution, subsidence, and potential rising sea levels.

Water availability

Where surface-water resources are not available, ground water can be relied upon in the Gulf of Mexico basin, except where thick, clayey units are present at the surface. These clayey units include some Upper Cretaceous units (such as the Navarro and Taylor Formations of central Texas), the Midway Group, the Jackson Group, the Vicksburg Group, Eocene to Miocene units in the northeastern Mexican coastal plain, and some Pleistocene

clays. Even in the thick confining units, however, thin, discontinuous lenses of permeable sands are generally present so that water availability often is dependent upon quality considerations. Growth faults serve to truncate some units and limit the extent of some aquifers. High permeabilities are formed by dissolution along fault surfaces in carbonate units. Nevertheless, with the exception of the Edwards aquifer, structural controls are generally less important than stratigraphic controls.

Natural water quality

Because of the vicinity of the Gulf of Mexico (aerosols in the form of salt crystals), the dissolution of salt domes and disseminated evaporite minerals, the abundance of young marine and young coastal sediments with saline/brackish pore fluids, and perhaps, the mixing with migrated oil-field waters (including brines), water salinity is a major quality problem. In organic-rich sediments, reducing conditions generate abundant dissolved iron and organic constituents that can be a problem.

Heavy pumping of coastal aquifers has caused salt-water intrusion in clastic and, more significantly, in the highly permeable carbonate systems. This, coupled with subsidence risks, has caused many coastal municipalities to switch to surface-water resources where possible.

Pollution

Aquifer systems of the Gulf of Mexico basin are highly susceptible to contamination. This is especially demonstrated in the Yucatan Peninsula, where sewage treatment plants are rare. Consequently, municipal, industrial, agricultural, and domestic wastes rapidly enter the highly permeable limestones because soils are thin and numerous karst features extend to the surface. There is little sorption or retention of contaminants, which are readily transported in fractures. Fecal coliform bacteria can survive for several years in warm, Yucatan Peninsula ground waters; enteritis and other intestinal diseases were a factor in over 50 percent of the deaths in the area (Doehring and Butler, 1974), and the Mexican government is engaged in a massive campaign to provide safe municipal and rural water supplies. The geologic situation is similar in the Edwards and Floridan aquifers, but pollution is less today because significant portions of these aquifers are confined and because there are more and better waste treatment facilities. In the Edwards aquifer system, considerable efforts are being made to control the quality in the streams that naturally recharge the aquifer. Such control of recharge is less feasible in the Florida and Yucatan Peninsulas because of the large areal extent of their recharge zones.

The clastic sedimentary aquifers have a high potential for contamination because of industrial development along the Gulf of Mexico Coastal Plain. Numerous industrial-waste disposal sites are present at the land surface, which provide potential point sources of pollution. The greater Houston area has a large number of such sites from previous improper chemical-waste disposal; there are numerous other sites throughout the Gulf of Mexico basin. An alternative method of disposal of hazardous chemicals is by deep injection into saline, clastic formations in Louisiana and Texas (Kreitler and others, 1989). Texas alone has over 100 industrial waste-injection wells, which inject more than 6 billion gallons of liquid hazardous wastes each year. Leaks in poorly constructed or failed injection or petroleum-production wells and underground storage tanks have caused aquifer pollution in both Mexico and the United States. Finally, Texas and Louisiana have potentially significant air pollution. Aerosols may be causing low levels of widespread but significant aquifer contamination (Brown and Sharp, 1989).

Subsidence

The extraction of subsurface fluids has created severe subsidence along the Texas and Louisiana coasts (Gabrysch, 1982, 1984). More than 2 m of subsidence has occurred in the greater Houston area since the mid-1940s, primarily because of ground-water pumping. Other areas of notable subsidence include northeast of Corpus Christi, Texas. In these areas, surface-water resources have been sought. Unfortunately, reservoirs trap sediment and reduce sediment input to the Gulf. This is causing still-unknown changes in coastal geomorphology and may accelerate the general coastline retreat. Subsidence has caused the submergence of roads, houses, and port facilities, and differential subsidence has caused or accelerated fault movement and, thus, disrupted roads and utilities systems. In the Yucatan Peninsula, but more importantly in Florida, pumping has led to catastrophic sinkhole collapses (Beck, 1984). Identification of potential areas of sinkhole collapse is extremely difficult and thus exacerbates a very dangerous problem.

Finally, high rates of local subsidence have been traced to petroleum production (Pratt and Johnson, 1926; Kreitler, 1977b; and Holzer and Bluntzer, 1984). The more regional subsidence observed in the upper Texas coast has been also attributed to depressurization of petroleum reservoirs (Ewing, 1985; Germiat and Sharp, 1990; Sharp and Germiat, 1990).

Rising sea levels

The general historical retreat of the Gulf of Mexico shoreline is accelerated by subsidence and may be further accelerated by future eustatic sea-level rise. Gornitz and Lebedeff (1987) suggest a worldwide rate of sea-level rise of approximately 1.5 mm/yr. Tidal gauges along the Florida Gulf Coast show a higher rate (2.2 to 2.4 mm/yr) since 1908. Sharp and Germiat (1990) attribute this to downwarping of the crust by the thick pile of Cenozoic sediment. Eustatic sea-level rises of as much as 5 m by 2100 have been predicted. Sharp and Germiat's delphic analysis suggests a 50 percent chance for a rise of more than 1 m by 2100. The projections indicate that the Gulf of Mexico coastal areas will experience exacerbated shoreline retreat and inundation of wetlands in the next century. These will have drastic hydrogeologic, economic, and societal impacts.

CONCLUSIONS

The availability of ground water in the Gulf of Mexico basin has been and will continue to be critical in the social and economic development of this region by providing water for agriculture, industry, and small and large municipalities. Abundant supplies have been found predominantly in the Mesozoic to Cenozoic limestones and Cenozoic sandstones.

The best limestone aquifer is the Floridan aquifer, because of its high transmissivities and rapid, abundant recharge. It has a potential for additional production, but increased development over the aquifer may result in its contamination. Future ground-water production of the Edwards aquifer in Texas is more limited because of the drier climate, lower rates of recharge, its less extensive recharge area, potential overdevelopment, and environmental concerns. Farmers, municipalities, river authorities, and environmentalists are already fighting over the water rights for the aquifer. Future ground-water production from limestone aquifers of the Yucatan limestones will be limited by the thin freshwater lens.

Sand and sandstone aquifers in the Gulf of Mexico basin region are areally extensive and provide water supplies for major cities, small communities, industry, and agriculture. Ground water from Tertiary sandstones will continue to provide a major percentage of water for these users. Much of this production is mining water thousands of years old from confined aquifers, but the resource is vast. The potential for surface contamination of confined aquifers is minor, but the shallow unconfined portions of the systems are susceptible. Differential land subsidence, inundation, and saltwater intrusion require careful utilization of these water supplies. Inundation and intrusion, however, are coastal problems; future development may require new sites that are further inland.

542   *J. M. Sharp, Jr., and Others*

REFERENCES CITED

Ackerman, J. D., 1989, Hydrology of the Mississippi River valley alluvial aquifer, south-central United States: U.S. Geological Survey Water-Resources Investigations Report 88–4028, 74 p.

Back, W., and Hanshaw, B. B., 1970, Comparison of chemical hydrogeology of the carbonate peninsulas of Florida and Yucatan: Journal of Hydrology, v. 10, p. 330–368.

Back, W., and Lesser, J. M., 1981, Chemical constraints on groundwater management in the Yucatan Peninsula, Mexico, in Beard, L. R., ed., Water for survival: Journal of Hydrology, v. 51, p. 119–130.

Back, W., Hanshaw, B. B., Pyle, T. E., Plummer, N., and Weidie, A. E., 1979, Geochemical significance of groundwater discharge and carbonate solution to the formation of Caleta Xel Ha, Quintana Roo, Mexico: Water Resources Research, v. 15, no. 6, p. 1521–1535.

Beck, B. F., ed., 1984, Sinkholes; Their geology, engineering, and environmental impact: Boston, Massachusetts, A. A. Balkema, 429 p.

Bethke, C. M., 1989, Modeling subsurface flow in sedimentary basins: Geologische Rundschau, v. 78, p. 129–154.

Blanchard, P. E., 1987, Fluid flow in compacting sedimentary basins [Ph.D. thesis]: Austin, University of Texas, 190 p.

Boswell, E. H., Cushing, E. M., and Hosman, R. L., 1968, Quaternary aquifers of the Mississippi Embayment: U.S. Geological Survey Professional Paper 448-E, 29 p.

Brahana, J. V., and Mesko, T. O., 1988, Hydrogeology and preliminary assessment of regional ground-water flow in the Upper Cretaceous and adjacent aquifers, northern Mississippi Embayment: U.S. Geological Survey Water-Resources Investigations Report 87–4000, 65 p.

Brown, T. J., and Sharp, J. M., Jr., 1989, Modeling the effect of aerosol dispersal on ground-water systems: Geological Society of America Abstracts with Programs, v. 21, p. 496.

Bush, P. W., and Johnson, R. H., 1988, Ground-water hydraulics, regional flow, and ground-water development of the Floridan aquifer system in Florida, and in parts of Georgia, South Carolina, and Alabama: U.S. Geological Survey Professional Paepr 1403-C, 80 p.

Clement, T. J., and Sharp, J. M., Jr., 1988, Hydrochemical facies in the bad-water zone of the Edwards aquifer, central Texas, in Proceedings of the Ground Water Geochemistry Conference: Dublin, Ohio, National Water Wall Association, p. 127–149.

Doehring, D. O., and Butler, J. H., 1974, Hydrogeologic constraints on Yucatan's development: Science, v. 186, p. 591–595.

Donnelly, A.C.A., 1988, Meteoric water penetration in the Frio Formation, Texas Gulf Coast [M.A. thesis]: Austin, University of Texas, 137 p.

Ewing, T. E., 1985, Subsidence and surface faulting in the Houston-Galveston area, Texas–related to deep fluid withdrawal?, in Dorfman, M. H., and Morton, R. A., eds., Geopressured-geothermal energy: Austin, Texas, University of Texas, Proceedings of the 6th U.S. Gulf Coast Geopressured-Geothermal Energy Conference, p. 289–298.

Fisk, H. N., 1944, Geological investigation of the alluvial valley of the lower Mississippi River: U.S. Army Corps of Engineers Mississippi River Commission, 78 p.

Fogg, G. E., and Kreitler, C. W., 1982, Ground-water hydraulics and hydrochemical facies in Eocene aquifers of the East Texas Basin: Texas Bureau of Economic Geology Report of Investigation 127, 75 p.

Gabrysch, R. K., 1982, Groundwater withdrawals and land surface subsidence in the Houston-Galveston region, Texas, 1906–1980: U.S. Geological Survey Open-File Report 72–571, 68 p.

—— , 1984, Case history 9.12; The Houston-Galveston region, Texas, U.S.A., in Poland, J. F., ed., Guidebook to studies of land subsidence due to groundwater withdrawal: Paris, UNESCO, p. 253–262.

Galloway, W. E., Hobday, D. K., and Magara, K., 1982, Frio Formation of the Texas Gulf Coast basin; Depositional systems, structural framework, and hydrocarbon origin, migration, distribution, and exploration potential: Texas Bureau of Economic Geology Report of Investigation 122, 78 p.

Germiat, S. J., and Sharp, J. M., Jr., 1990, Assessment of future coastal land loss along the upper Texas Gulf coast: Bulletin of the Association of Engineering Geologists, v. 19 (in press).

Gornitz, V., and Lebedeff, S., 1987, Global sea-level changes during the past century, in Nummedal, D., and others, eds., Sea-level fluctuations and coastal evolution: Society of Economic Paleontologists and Mineralogists Special Publication 41, p. 3–16.

Groshkopf, J. G., 1955, Subsurface geology of the Mississippi Embayment of southeast Missouri: Missouri Division of the Geological Survey and Water Resources, v. 37, 2nd series, 133 p.

Grubb, H. F., 1984, Planning report for the Gulf Coast regional aquifer; System analysis in the Gulf of Mexico Coastal Plain, United States: U.S. Geological Survey Water-Resources Investigations Report 84–4219, 30 p.

Grubb, H. F., and Carrillo R., J. J., 1988, Region 23, Gulf of Mexico Coastal Plain, in Back, W., Rosenshein, J. S., and Seaber, P. R., eds., Hydrogeology: Boulder, Colorado, Geological Society of America, The Geology of North America, v. O-2, p. 219–228.

Hamlin, H. S., 1988, Depositional and ground-water flow systems of the Carrizo–upper Wilcox, southern Texas: Texas Bureau of Economic Geology Report of Investigation 175, 61 p.

Hanor, J. S., 1987, Kilometer-scale thermohaline overturn of pore waters in the Louisiana Gulf Coast: Nature, v. 327, p. 501–503.

Hanor, J. S., Bailey, J. E., Rogers, M. C., and Milner, L. R., 1986, Regional variations in physical and chemical properties of south Louisiana oil field brines: Transactions of the Gulf Coast Association of Geological Societies, v. 36, p. 143–149.

Hanshaw, B. B., Back, W., and Rubin, M., 1965, Carbonate equilibrium and radiocarbon distribution related to groundwater flow in the Floridan limestone aquifer, U.S.A., in Hydrology of fractured rocks, vol. 1: UNESCO, International Association of Scientific Hydrology, p. 601–614.

Holzer, T. L., and Bluntzer, R. L., 1984, Land subsidence near oil and gas fields, Houston, Texas: Ground Water, v. 22, p. 450–459.

Hosman, R. L., 1988, Geohydrologic framework of the Gulf Coastal Plain: U.S. Geological Survey Hydrological Investigation Atlas HA-695, 2 sheets, scale 1:2,500,000.

Johnson, R. H., and Bush, P. W., 1988, Summary of the hydrology of the Floridan aquifer system in Florida and in parts of Georgia, South Carolina, and Alabama: U.S. Geological Survey Professional Paper 1403-A, 24 p.

Johnston, R. H., and Miller, J. A., 1988, Region 24, Southeastern United States, in Back, W., Rosenshein, J. S., and Seaber, P. R., eds., Hydrogeology: Boulder, Colorado, Geological Society of America, The Geology of North America, v. O-2, p. 229–236.

Kreitler, C. W., 1977a, Fault control of subsidence, Houston-Galveston, Texas: Ground Water, v. 15, no. 3, p. 203–214.

—— , 1977b, Faulting and land subsidence from groundwater and hydrocarbon production, Houston-Galveston, Texas, in 2nd International Land Subsidence Symposium Proceedings, Anaheim, California, 1976: International Association of Hydrological Sciences Publication 121, p. 435–445.

—— , 1989, Hydrogeology of sedimentary basins: Journal of Hydrology, v. 106, p. 29–53.

Kreitler, C. W., Guevera, E., Granata, G., and McKalips, D., 1977, Hydrogeology of Gulf Coast aquifers, Houston-Galveston area, Texas: Transactions of the Gulf Coast Association of Geological Societies, v. 27, p. 72–89.

Kreitler, C. W., Akhter, M. S., and Donnelly, A.C.A., 1989, Hydrologic-hydrochemical characterization of Texas Gulf Coast saline formations used for deep-well injection of chemical wastes, in Proceedings of the International Symposium on Class I and II Injection Wells: Underground Injection Practice Research Foundation, p. 177–197.

LeGrand, H. E., and Stringfield, V. T., 1966, Development of permeability and storage in the Tertiary limestones of the southeastern states, USA: International Association of Scientific Hydrology Bulletin, v. 11, p. 61–73.

Lesser, H., Azpeitia, J., and Lesser, J. M., 1978, Geohidrología de la Isla de

Cozumel, Quintana Roo: Recursos hidráulicos, v. VII, no. 1, 18 p.

Lesser, J. M., 1976, Bosquejo geológico e hidrogeoquímico de la Península de Yucatán: Secretaría de Recursos Hidráulicos Boletín, no. 10, June, 11 p.

Lesser, J. M., and Weidie, A. E., 1988, Region 25, Yucatan Peninsula, *in* Back, W., Rosenshein, J. S., and Seaber, P. R., eds., Hydrogeology: Boulder, Colorado, Geological Society of America, The Geology of North America, v. O-2, p. 237–241.

Livingston, P., Sayre, A. N., and White, W. N., 1936, Water resources of the Edwards Limestone in the San Antonio area, Texas: U.S. Geological Survey Water-Supply Paper 773-B, 55 p.

Maclay, R. W., and Land, L. F., 1988, Simulation of flow in the Edwards aquifer, San Antonio region, Texas, and refinement of storage and flow concepts: U.S. Geological Survey Water-Supply Paper 2336, 48 p.

Marin, L. E., and 5 others, 1990, Hurricane Gilbert; Its effects on the aquifer in northern Yucatan, Mexico, *in* Simpson, E. S., and Sharp, J. M., Jr., eds., Selected Papers on Hydrogeology: Heise, Hannover International Association of Hydrogeologists, v. 1, p. 111–127.

Martin, A., Jr., Whiteman, C. D., Jr., and Becnel, M. J., 1988, Generalized potentiometric surface of the upper Jasper and equivalent aquifers in Louisiana, 1984: U.S. Geological Survey Water-Resources Investigations Report 87–4139, 2 sheets, scale 1:500,000.

Miller, J. A., 1986, Hydrogeologic framework of the Floridan aquifer system in Florida and in parts of Georgia, Alabama, and South Carolina: U.S. Geological Survey Professional Paper 1403-B, 91 p.

Miller, J. A., Barker, R. A., and Renken, R. A., 1987, Hydrology of the southeastern Coastal Plain aquifer system; An overview, *in* Vecchioli, J., and Johnson, A. I., eds., Aquifers of the Atlantic and Gulf Coastal Plain: American Water Resources Association Monograph 9, p. 53–77.

Nyman, D. J., 1984, The occurrence of high concentrations of chloride in the Chicot aquifer system of southwestern Louisiana: Louisiana Department of Transportation and Development Office of Public Works, Water Resources Technical Report 33, 75 p.

Pettijohn, R. A., Weiss, J. S., and Williamson, A. K., 1988, Distribution of dissolved-solids concentrations and temperature in ground water of the Gulf Coast aquifer systems, south-central United States: U.S. Geological Survey Water-Resources Investigations Report 88-4082, 5 sheets.

Pratt, W. E., and Johnston, D. W., 1926, Local subsidence of the Goose Creek oil field: Journal of Geology, v. 34, p. 577–590.

Renken, R. A., 1984, The hydrogeologic framework of the southeastern Coastal Plain aquifer system of the United States: U.S. Geological Survey Water-Resources Investigations Report 84–4243, 26 p.

Rollo, J. R., 1969, Salt-water encroachment in aquifers of the Baton Rouge area, Louisiana: Louisiana Department of Conservation, Geological Survey, and Department of Public Works Water Resources Bulletin 13, 45 p.

Rosenshein, J. S., 1988, Hydrogeology of North America, region 18; Alluvial valleys, *in* Back, W., Rosenshein, J. S., and Seaber, P. R., eds., Hydrogeology: Boulder, Colorado, Geological Society of America, The Geology of North America, v. O-2, p. 165–176.

Sayre, A. N., and Bennett, R. P., 1942, Recharge, movement, and discharge in the Edwards Limestone reservoir, Texas: American Geophysical Union Transaction, pt. 1, p. 19–27.

Scott, T. M., 1989, The lithostratigraphy and hydrostratigraphy of the Floridan aquifer system in Florida, *in* Scott, T. M., Arthur, J., Ruport, F., Upchurch, S., and Randazzo, A., eds., The lithostratigraphy and hydrostratigraphy of the Floridan aquifer system in Florida; 28th International Geological Congress Guidebook T185, American Geophysical Union, p. 2–9.

Sharp, J. M., Jr., 1988, Alluvial aquifers along major rivers, *in* Back W., Rosenshein, J. S., and Seaber, P. R., eds., Hydrogeology: Boulder, Colorado, Geological Society of America, The Geology of North America, v. O-2, p. 273–282.

——, 1990, Stratigraphic, geomorphic, and structural controls of the Edwards aquifer, Texas, U.S.A., *in* Simpson, E. S., and Sharp, J. M., Jr., eds., Selected Papers on Hydrogeology: Heise, Hannover International Association of Hydrogeologists, v. 1, p. 67–82.

Sharp, J. M., Jr., and Clement, T. J., 1988, Hydrochemical facies as hydraulic boundaries in karstic aquifers; The Edwards aquifer, U.S.A., *in* Karst hydrogeology and karst environment protection, Proceedings of the 21st International Association of Hydrogeologists Congress: Guilin, Peoples Republic of China, v. 2, p. 841–845.

Sharp, J. M., Jr., and Germiat, S. J., 1990, Risk assessment and causes of subsidence along the Texas Gulf Coast, *in* Fairbridge, R. W., and Paepe, R., eds., Greenhouse effect, sea level, and drought: Dordrecht, Kluwer Academic Publishers (in press).

Sharp, J. M., Jr., and McBride, E. F., 1989, Sedimentary petrology; A guide to paleohydrogeologic analyses; Example of sandstones from the northwest Gulf of Mexico: Journal of Hydrology, v. 108, p. 367–386.

Sharp, J. M., Jr., and 12 others, 1988, Diagenetic processes in northwest Gulf of Mexico sediments, *in* Chilingarian, G. V., and Wolf, K. H., eds., Diagenesis II: Elsevier Science Publishers, p. 43–113.

Smith, G. E., Galloway, W. E., and Henry, C. D., 1982, Regional hydrodynamics and hydrochemistry of the uranium-bearing Oakville aquifer (Miocene) of south Texas: Texas Bureau of Economic Geology Report of Investigation 124, 31 p.

Stringfield, V. T., 1936, Water in the Florida Peninsula: U.S. Geological Survey Water-Supply Paper 773-C, p. 116–195.

Thayer, P. A., and Miller, J. A., 1984, Petrology of lower and middle Eocene carbonate rocks: Transactions of the Gulf Coast Association of Geological Societies, v. 34, p. 421–434.

Turcan, A. N., Jr., Wesselman, J. B., and Kilburn, C., 1966, Interstate correlation of aquifers, southwestern Louisiana and southeastern Texas: U.S. Geological Survey Professional Paper 550-D, p. D231–D236.

Wesselman, J. B., 1983, Structure, temperature, pressure, and salinity of Cenozoic aquifers of south Texas: U.S. Geological Survey Hydrological Investigations Atlas HA-654, 1 sheet.

Woodruff, C. M., Jr., and Abbott, P. L., 1979, Drainage-basin evolution and aquifer development in a karstic limestone terrain, south-central Texas, U.S.A.: Earth Surface Processes, v. 4, p. 319–334.

MANUSCRIPT ACCEPTED BY THE SOCIETY MAY 7, 1990

ACKNOWLEDGMENTS

We thank Bill Back, Hayes Grubb, Tony Randazzo, and, of course, Amos Salvador for their reviews and encouragement. Acknowledgment is also made to the donors of the Petroleum Research Fund of the American Chemical Society for partial support of the background research. Manuscript preparation was funded by the Owens-Coates Fund of the Geology Foundation, the University of Texas at Austin. Rosemary Brant assisted with the editing; Jeff Horowitz and Karen Bergeron drafted the figures.

The Geology of North America
Vol. J, The Gulf of Mexico Basin
The Geological Society of America, 1991

Chapter 18

Summary; Current knowledge and unanswered questions

Amos Salvador
Department of Geological Sciences, The University of Texas at Austin, P.O. Box 7909, Austin, Texas 78713-7909

INTRODUCTION

The information contained in the preceding chapters of this volume is clear evidence of the significant progress that has been made in understanding the geologic and geophysical composition and geologic history of the Gulf of Mexico basin. Many uncertainties remain, however; many questions are still unanswered, and many fundamental problems are yet to be solved.

Our current advanced knowledge of the basin is the result of the persistent collection during the last 100 years of a great volume of geological and geophysical information, most of it by the petroleum industry in its search for oil and gas accumulations, but also by geologists and geophysicsts from national and state geological surveys and academic institutions.

It is safe to say that, for many years now, the Gulf of Mexico basin region has been the home of the largest concentrations of geologists and geophysicists anywhere in the world. The region probably can claim to have the densest seismic coverage—unfortunately not all of it in the public domain—and close to 700,000 wells have been drilled throughout the basin, many of them to considerable depths. The amount of geological and geophysical information on the basin is, therefore, voluminous, as are the number of publications on its geologic composition and history.

As mentioned in Chapter 1 of this volume, the collection of geological and geophysical information on the Gulf of Mexico basin progressed from the periphery to the center, from the investigation of the rock outcrops around its margins by the geologists of the oil companies and the geological surveys, to the interpretation of the increasing volumes of subsurface geological and geophysical information on the basin. The early subsurface information was restricted to shallow depths and to onshore areas. With time, drilling reached greater depths, and more sophisticated geophysical methods recorded increasingly deeper and more accurate information. In time, drilling and geophysical surveys extended into the waters of the Gulf of Mexico. At present, most of even the deepest parts of the Gulf, little known a few decades ago, are covered by regional seismic reflection and refraction surveys conducted by government agencies, academic institutions, and the oil industry. Invaluable information also has been contributed by the drilling results of legs 1, 10, 77, and 96 of the Deep Sea Drilling Project (DSDP).

As should be expected, interpretations and ideas concerning the geology of the Gulf of Mexico basin, its structure, origin, and geologic history, evolved progressively as new and better information became available. Controversies and disparate speculations concerning the origin and development of the basin, common when little or nothing was known of the nature of its central part, progressively were resolved. While some theories and interpretations are still debatable—and debated!—and still need additional documentation, the disagreements are more closely confined and less vehement.

DEVELOPMENT OF CURRENT KNOWLEDGE AND MAJOR FINDINGS

Detailed understanding of the geology and geophysical framework of the Gulf of Mexico basin has developed gradually during the last 100 years. Additions to the knowledge of the basin have generally been incremental, though many of them were decisive in the conception, development, and testing of significant geological ideas: sedimentary models (particularly of fluvial and deltaic systems), listric-normal-fault mechanics, salt and shale tectonics, basin-analysis techniques, organic geochemistry theories, and many others.

The introduction and general acceptance of certain techniques of data acquisition and interpretation played a decisive role in improving the understanding of the geology of the basin: micropaleontological studies in the early 1920s, the reflection seismograph in the late 1920s and early 1930s, and the wire-line logging of wells in the early 1930s. In addition, major "break-throughs" took place periodically.

In the late 1930s and early 1940s, petroleum exploration wells in southern Arkansas and northern Louisiana discovered that the Lower Cretaceous sediments, known from outcrops and previous shallower drilling, were underlain in the subsurface of the region by a section that nowhere cropped out along the northern flank of the Gulf of Mexico basin. This previously un-

Salvador, A., 1991, Summary; Current knowledge and unanswered questions, *in* Salvador, A., ed., The Gulf of Mexico Basin: Boulder, Colorado, Geological Society of America, The Geology of North America, v. J.

known section was eventually dated as of Late Jurassic age and correlated with sections of similar age known in Mexico. In addition, as these Upper Jurassic rocks contained sizable accumulations of oil and gas, the discovery opened an important new petroleum play in the United States Gulf Coastal Plain region.

Also in the late 1930s and early 1940s, exploratory wells in northern Louisiana and northern Florida encountered a considerable section of unmetamorphosed Paleozoic rocks not expected before to be present in the subsurface of the northern Gulf of Mexico basin. Furthermore, those in northern Florida were found to contain a fauna and flora with definite Afro-European affinities, which led many geologists to speculate that the Paleozoic section of this region is a remnant of west Africa welded to North America in the late Paleozoic and left behind when the North American and African Plates separated during the opening of the present Atlantic Ocean (see Chapter 7 of this volume).

A major contribution to the knowledge of the geologic composition of the Gulf of Mexico basin was provided by reflection and refraction seismic surveys conducted in the deeper, central part of the Gulf of Mexico from the mid-1950s to the early 1960s by academic institutions. These surveys indicated that in this central part of the Gulf, a thick sedimentary section was underlain by oceanic-type crust, and that the extensive Middle Jurassic salt deposits known from the northern and southern parts of the Gulf of Mexico basin appeared to be absent in its central, deep part, as they are absent in east-central Mexico (see Chapters 8, 13 and 14, this volume).

In 1961, a "red bed" sequence (the Eagle Mills Formation) known since the late 1930s to underlie in southern Arkansas the Upper Jurassic section and to overlie Paleozoic beds, was dated as of Late Triassic age and interpreted to represent deposition in a graben or half-graben active at that time. Moreover, the Eagle Mills "red beds" were recognized to have a marked similarity to the "red beds" of the Newark Supergroup of the eastern United States. Correlaton with "red bed" sequences of the same age in east-central Mexico soon followed. This information, in addition to the new knowledge about the central part of the basin, provided invaluable evidence for the interpretation of the early tectonic development of the Gulf of Mexico basin.

In the late 1960s and early 1970s, more and better reflection and refraction seismic data in the Gulf of Mexico made possible the establishment of a stratigraphic framework, and the delineation and characterization of structural provinces in the Gulf, including a more precise picture of the distribution of the salt deposits and the recognition of lateral salt tectonics (see Chapter 13, this volume).

This new and more precise information on the distribution and tectonic behavior of the salt deposits, prompted the notion that the salt may have been formed in a single evaporitic basin, which was later spread apart by the formation of new oceanic crust in the central part of the Gulf of Mexico, and by the southward movement of the Yucatan block to its present position. This concept has been incorporated since then, in one way or another, into all modern interpretations of the origin of the Gulf of Mexico basin.

The geological and geophysical information now available provides a reasonably good picture of the stratigraphy and structure of the Gulf of Mexico basin—the types of crust that underlie it and their general distribution, the overall outline of the basin and the location and form of its many second-order structural features, and the nature of the thick sedimentary section filling the basin. Three stratigraphic provinces are now clearly recognized: a carbonate-evaporite province to the east and southeast, a province to the southwest and west predominantly composed of carbonates and fine-grained clastic sediments, and a province to the northwest and north where coarse-grained terrigenous clastic rocks form an important part of the section. Major regional unconformities or hiatuses have been identified, and attempts have been made to relate them to worldwide sea-level changes. From the tectonic point of view, three distinct types of margins can be distinguished in the basin: a stable margin to the east and southeast, corresponding to the carbonate-evaporite province of the Florida and Yucatan platforms; a compressional margin to the west following the length of the Sierra Madre Oriental of Mexico; and a northwestern and northern margin characterized by sustained subsidence and sedimentary progradation. Reasonable interpretations of the origin and development of the basin have been developed.

There are, however, many still unanswered questions and unsolved or poorly understood problems.

UNANSWERED QUESTIONS AND REMAINING PROBLEMS

The least understood aspects of the geologic history of the Gulf of Mexico basin region are undoubtedly those concerning the pre-Triassic events. There are good reasons why such is the case. As discussed in Chapter 7 of this volume, pre-Triassic rocks are known from only a few and widely separated outcrop areas around the periphery of the basin and from less than 250 wells within the basin, none of them in its central, deeper part, and few of them having penetrated more than a few tens of meters of pre-Triassic rocks.

Many unanswered questions remain, therefore, concerning the pre-Triassic geologic history of the region. Particularly important to the interpretation of the origin and early development of the present Gulf of Mexico basin during the Late Triassic and Early Jurassic is the understanding of the late Paleozoic and earliest Triassic events that took place during and following the postulated closing of the Proto-Atlantic Ocean and the collision of Gondwana with the North American Plate. It is most likely that these geologic events may have had considerable influence on the stratigraphic and structural developments that followed during the remaining part of the Mesozoic. It is, therefore, important to be able to answer the following questions.

1. What was the timing of the collision of the plates?
2. How did the North American and African/South American Plates fit as they came together in late Paleozoic time to form the western part of the supercontinent of Pangea? Where were

the Yucatan and Chortis blocks located in late Paleozoic time with respect to the North American and African/South American Plates?

3. Where is the continuation of the Ouachita orogenic belt in Mexico?

4. Do the Mojave-Sonora megashear and other similar transform faults in Mexico extend into the Gulf of Mexico basin area? If so, when was there displacement along them? And, in more general terms, how important have transcurrent faults been in the tectonic development of the Gulf of Mexico basin area, during the early Mesozoic as well as during the Paleozoic?

Unanswered questions and remaining problems are not limited, however, to the understanding of pre–late Triassic events. Also critical are solutions to many outstanding problems concerning the age, origin, composition, type, structure, and tectonic history of the crust underlying the Gulf of Mexico basin, the distribution of the Late Triassic–Early Jurassic graben systems filled with nonmarine "red beds", and the time and manner of formation and subsequent deformation of the Middle Jurassic salt deposits.

More specifically:

1. What is the age, composition, and depositional history of the sediments directly overlying the (oceanic) crust in the central, deep part of the Gulf of Mexico? An answer to this question would supply valuable information concerning the age of the crust.

2. Is there a major Late Triassic–Jurassic graben system trending east-northeast to west-southwest, and filled with "red beds" in the northern part of the Gulf of Mexico under the thick and highly deformed Louann Salt section that could have played a major role in the southward displacement of the Yucatan block and in the accumulation of the thick salt section? If not, what underlies the salt in the northern part of the Gulf of Mexico?

3. When, why, and how did the subsidence of the Gulf of Mexico basin area start? What has been the cause or causes of the persistent subsidence prevalent in the basin from sometime in the Mesozoic to the present?

4. Can direct geological, geophysical, or geochemical evidence be developed to determine more accurately the age of the salt deposits of both the northern and southern parts of the Gulf of Mexico basin? The age of the salt and the manner of its accumulation is unquestionably a fundamental clue to establishing the timing of the origin and early development of the basin.

5. If the salt deposits were indeed deposited in an essentially continuous large, shallow basin, as currently postulated, how was it possible to bisect them across the area of their thickest development and to separate the two parts about 300 km while the front of the two detached salt masses was contained so the salt would not flow toward the newly opened space between them? Was a ridge present in the central part of the Gulf of Mexico basin area before or during the accumulation of the salt? If so, what caused the ridge to form? Were the salt deposits, then, formed in two separate but contemporaneous basins (half grabens) with their deeper parts near the ridge?

6. Were the salt deposits everywhere in the basin accumulated over continental or stretched transitional crust? What is the chronologic relation between the accumulation of the salt deposits and the formation of the oceanic crust underlying the central part of the basin?

7. What is the structural configuration of the western flank of the Gulf of Mexico basin, offshore of the eastern coast of Mexico between the Mexico–U.S. border and the Isthmus of Tehuantepec? No deep-penetration reflection seismic surveys have been conducted in this critical area where the "basement" dips steeply eastward. Does this steep "basement" slope reflect the presence of a major NNW-SSE fault zone in this area?

8. What stratigraphic intervals in the Gulf of Mexico basin have generated the vast volumes of oil and gas known to be present and yet to be found in the basin? In particular, what are the source rocks for the oil accumulations in upper Tertiary and Pleistocene sediments in the northern part of the Gulf of Mexico, offshore of Louisiana? The organic matter contained in these young sediments is, according to conventional critiera, thermally immature and predominantly "gas-prone", and, consequently, an unlikely source for the oil, and perhaps even the gas, identified in this northern Gulf of Mexico region.

9. What have been the paths of migration of the oil and gas from their source rocks to their current accumulations? What role have faults (normal, transcurrent), salt domes, or other structural features played in the migration of oil and gas and in the circulation of water up and out of the basin?

Not until reasonable answers to these questions can be given will it be possible to interpret with any confidence the more fundamental aspects of the stratigraphic and tectonic development—the geologic history—of the Gulf of Mexico basin, its origin, and subsequent development.

NEEDED ADDITIONAL INFORMATION AND NEW STUDIES

Many of the unanswered questions and unresolved problems discussed above will not be settled easily or soon. They will remain a challenge to geologists and geophysicists studying the Gulf of Mexico basin for many years to come. The following programs of study and research projects, some requiring new data but others based on existing information, will make considerable contributions to answering the questions and resolving the problems.

1. Additional deep-penetration, high-resolution, regional reflection seismic surveys, both onshore and in the Gulf of Mexico. Efforts should be made to connect regional onshore and offshore seismic reflection profiles in order to obtain a complete stratigraphic and structural representation of the basin, from rim to rim. Particularly necessary are seismic surveys of the western flank of the basin in order to be able to properly interpret the deep structure of the area. It would also be most desirable to attempt to obtain good reflection seismic data below the thick and highly deformed salt deposits in the northern part of the Gulf

of Mexico. Additional high-resolution reflection surveys in the southeastern part of the Gulf of Mexico, between the Florida and Yucatan platforms could conceivably help determine more precisely the time during the Late Jurassic when a connection was established between the Gulf of Mexico and the Atlantic Ocean.

2. Additional refraction seismic surveys, both onshore and in the Gulf of Mexico to better characterize the different types of crust (oceanic, transitional, and continental) underlying the basin, and to more precisely establish the boundaries between these different types of crust.

3. Deep drilling down to, and a substantial distance into the "basement" at strategic locations within the Gulf of Mexico basin. A deep well or wells in the central, deep part of the basin that would reach the (oceanic) crust could contribute invaluable information concerning the nature of the crust and the age of the sediments directly overlying it, and, consequently, the time of emplacement of the (oceanic) crust. Such information would help greatly in interpreting the origin and early history of the Gulf of Mexico basin. Similarly, a deep well through the pre-Cretaceous sediments overlying the "basement" in the southeastern part of the Gulf of Mexico, between the Florida and the Yucatan platforms, would define the deep stratigraphy of the area and help determine the time when the Gulf of Mexico became in communication with the Atlantic Ocean. Site 535 of Leg 77 of the DSDP was chosen with these objectives in mind but was unable to reach the desired penetration. Other significant locations could be selected in areas where no deep wells have been drilled down to the "basement" by the petroleum industry, and where the results would be particularly pertinent to the solution of important geological or physical problems. All deep wells, naturally, should be closely controlled by, and tied to deep-penetration, high-resolution seismic reflection surveys.

4. Regional geological studies, structural as well as stratigraphic, using the latest concepts—plate tectonics, sequence stratigraphy, any others—and the best, most recent age determinations obtained from either paleontologic studies or isotopic methods. (Many problems still remain concerning the precise dating of both sedimentary and igneous rocks of the Gulf of Mexico basin region.) Special consideration should be given to regional unconformities and their possible meaning in terms of the tectonic events or the eustatic sea-level changes that may have caused them. These studies should result in the preparation of detailed eustatic and tectonic subsidence charts and of paleogeographic maps of time intervals as short as the available stratigraphic information would permit. Such maps and charts would be invaluable in the reconstruction, step by step, of the geologic history of the area.

Studies of the pre-Triassic rocks, as mentioned earlier, would be critical in understanding not only the early history of the region, but also the influence that the early stratigraphic and structural framework of the region has on subsequent, Late Triassic and Jurassic, geologic developments. Unfortunately, only limited information is now available on these older rocks, and not

much new information is likely to be forthcoming. Valuable interpretations are possible, however, on the basis of what sources are now accessible if studies are undertaken with a regional, basinwide point of view.

Considerable information about Cretaceous and Cenozoic regional stratigraphic hiatuses or unconformities may be obtained from detailed lithostratigraphic and biostratigraphic studies (and perhaps seismic-stratigraphic interpretation) of the Florida and Yucatan carbonate platforms. Their Cretaceous to Holocene stratigraphic section, composed predominantly of shallow-water carbonates and evaporites, and deposited under extremely stable tectonic conditions, should reflect admirably important eustatic changes in sea-level and the corresponding stratigraphic hiatuses and sedimentary cycles.

5. Geochemical studies to identify the source rocks for the important oil and gas accumulations of the Gulf of Mexico basin. A great amount of work has been done already, some of it published and much unpublished, but much more remains to be done. The geochemical studies should be integrated with burial and maturation history research, and with the interpretation of the time and means of the generation, migration, and accumulation of the oil and gas.

6. Studies aimed at urgent environmental problems will also be needed. Foremost, perhaps, will be those necessary in response to the growing concern about the availability of fresh water. The persistent increase of the population along the Gulf of Mexico Coastal Plains, has resulted in a corresponding increase in the consumption of water. As discussed in the previous chapter, excessive withdrawal of ground water in some areas has resulted in land subsidence and salt-water encroachment; in the Florida and Yucatan peninsulas, the lowering of the water table has resulted in harmful sinkhole collapses. Also of great concern is the potential contamination of the ground water by herbicides, pesticides, fertilizers, and toxic industrial wastes. Shortages and pollution of ground water may become major long-term societal concerns in the communities surrounding the Gulf of Mexico. Geologists will need to play a key role in guaranteeing the proper supply of this precise resource.

Many other regional geological, geophysical, and geochemical studies could be mentioned that would contribute substantially to a better understanding of the geology of the Gulf of Mexico basin: the mechanics of salt and shale deformation and of listric normal ("growth") faults, the influence of the Pleistocene glacial episodes on the depositional events of the time, the causes of depositional cyclicity, and the processes that take place in the transition zone between the shelf and the continental slope in actively prograding continental margins.

The Gulf of Mexico basin offers such a variety of stratigraphic and structural features and conditions—clastic and carbonate sedimentary sequences, evaporites, large reef trends, a major delta, thrust belts, strike-slip and listric normal fault systems, complex salt-deformation features, one of the foremost petroleum provinces of the world—that, as it has in the past, it will offer geologists, geophysicists, and geochemists for many decades a unique opportunity for study, and an excellent natural laboratory in which to pursue their research.

The youth of the basin fill, for example, makes possible the study of many modern sedimentary processes, and permits the examination of active hydrologic and thermal phenomena, as well as their products, including the diagenesis of the sands, muds, and organic matter that takes place during the early consolidation history of sedimentary basins. The results of these studies of modern sediments can advantageously be extended back in time and used in the interpretation of older segments of the stratigraphic sequence not only in the Gulf of Mexico basin, but in many other basins of the world.

Many of these regional geological, geophysical, and geochemical studies need not wait for new and better information to be provided by coming surface and subsurface mapping projects, as well as by new drilling, and new geophysical surveys, or by the publication of now unavailable proprietary subsurface geological and geophysical information, desirable as this new information may be. A great volume of reliable information has been already gathered, and much of it has been published. Much can be accomplished by studying, interpreting, or reinterpreting this currently available information. The proper understanding of the fundamental aspects of the geology of the Gulf of Mexico basin, however, will be accomplished only by comprehensive studies that embrace the entire or large sections of the basin. Such regional syntheses should include and be consistent with the total geological and geophysical, stratigraphic and structural information from both the United States and Mexico.

ACKNOWLEDGMENTS

William E. Galloway, Ray G. Martin, E. McFarlan, Jr., Robert A. Morton, and Raymond D. Woods contributed excellent ideas for the preparation of this chapter. Ray G. Martin, Robert A. Morton, Grover E. Murray, and Raymond D. Woods reviewed the manuscript. Their thoughtful comments and suggestions greatly improved the contents of the chapter.

Index

[Italic page numbers indicate major references]

Typeset by WESType Publishing Services, Inc., Boulder, Colorado
Printed in U.S.A. by Malloy Lithographing, Inc., Ann Arbor, Michigan

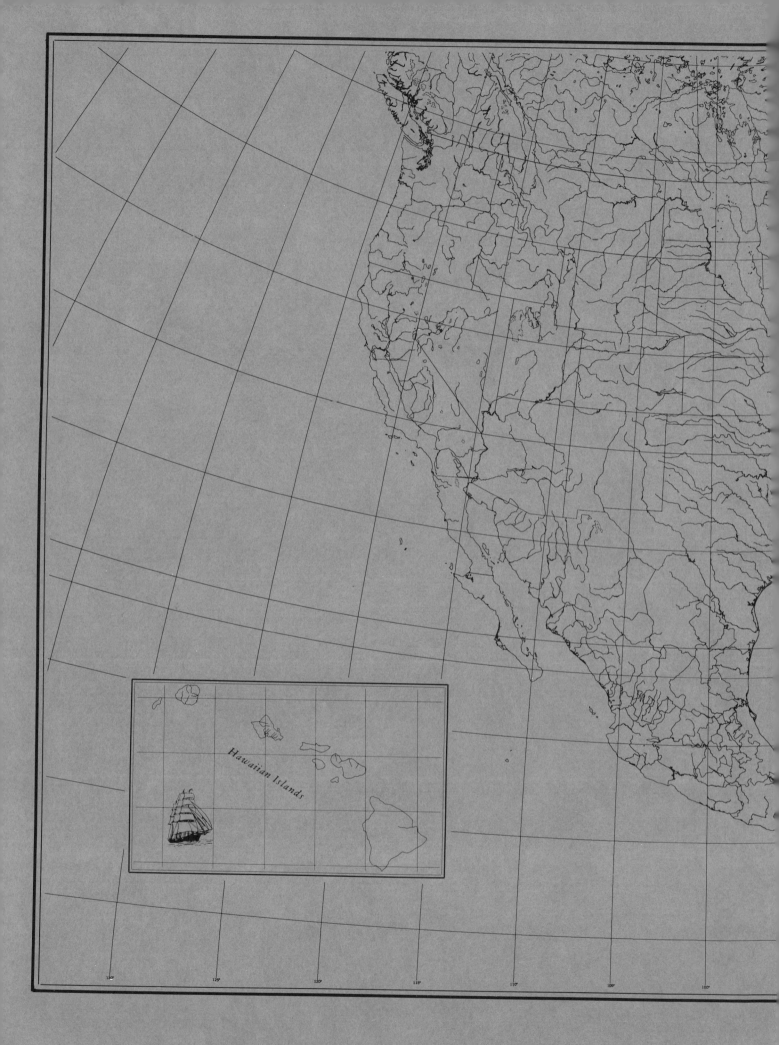

Hawaiian Islands